The yeasts

The yeasts

a taxonomic study

third revised and enlarged edition

edited by

N.J.W. Kreger-van Rij
Groningen, The Netherlands

1984

Elsevier Science Publishers B.V. – Amsterdam

ISBN 0-444-80421-8

Publishers:
ELSEVIER SCIENCE PUBLISHERS B.V.
P.O. BOX 211, 1000 AE AMSTERDAM
THE NETHERLANDS

Sole distributors for the USA and Canada:
ELSEVIER SCIENCE PUBLISHING COMPANY, INC.
52 VANDERBILT AVENUE
NEW YORK, N.Y. 10017

Sole distributors for Japan:
MARUZEN CO.LTD
3-10, NIHONBASHI 2-CHOME
CHUO-KU
TOKYO 103
JAPAN

Library of Congress Cataloging in Publication Data
Main entry under title:

The Yeasts.

Bibliography: p.
Includes index.
1. Yeast fungi--Classification. 2. Fungi--Classification. I. Kreger-Van Rij, N. J. W.
QK617.5.Y48 1984 589.2'33 84-4150
ISBN 0-444-80421-8 (U.S.)

PRINTED IN THE NETHERLANDS

Contributors

D.G. AHEARN, Department of Biology, Georgia State University, Atlanta, Georgia 30303, U.S.A.

R.J. BANDONI, Department of Botany, University of British Columbia, Vancouver, B.C., Canada V6T 1W5

J.W. FELL, School of Marine and Atmospheric Science, University of Miami, Miami, Florida 33149, U.S.A.

E. JOHANNSEN, Microbiology Research Group, Council of Scientific and Industrial Research, Pretoria, South Africa

N.J.W. KREGER-VAN RIJ, Laboratory for Medical Microbiology, State University, 9713 EZ Groningen, The Netherlands

C.P. KURTZMAN, Northern Regional Research Center, United States Department of Agriculture, Peoria, Illinois 61604, U.S.A.

K.J. KWON-CHUNG, Clinical Mycology Section, National Institute of Allergy and Infectious Diseases, National Institutes of Health, Bethesda, Maryland 20014, U.S.A.

Sally A. MEYER, Department of Biology, Georgia State University, Atlanta, Georgia 30303, U.S.A.

M.W. MILLER, Department of Food Science and Technology, University of California, Davis, California 95616, U.S.A.

H.J. PHAFF, Department of Food Science and Technology, University of California, Davis, California 95616, U.S.A.

L. RODRIGUES DE MIRANDA, Yeast Division, Centraalbureau voor Schimmelcultures, 2628 BC Delft, The Netherlands

R.L. SCHLITZER, Department of Biology, Georgia State University, Atlanta, Georgia 30303, U.S.A.

M.TH. SMITH, Yeast Division, Centraalbureau voor Schimmelcultures, 2628 BC Delft, The Netherlands

A. STATZELL TALLMAN, School of Marine and Atmospheric Science, University of Miami, Miami, Florida 33149, U.S.A.

J.P. VAN DER WALT, Microbiology Research Group, Council of Scientific and Industrial Research, Pretoria, South Africa

D. YARROW, Yeast Division, Centraalbureau voor Schimmelcultures, 2628 BC Delft, The Netherlands

Preface

The present, third edition of **'The Yeasts – a taxonomic study'** follows the first (J. Lodder & N.J.W. Kreger-van Rij, 1952) and the second edition (J. Lodder ed., 1970) which were preceded by monographs on the ascosporogenous yeasts (Stelling-Dekker, 1931), and the anascosporogenous yeasts, part one (J. Lodder, 1934) and part two (H.A. Diddens & J. Lodder, 1942).

Since the 1970 edition there have been many new developments in yeast taxonomy: new species have been described from new isolates, mating experiments have yielded perfect states of ascomycetous and basidiomycetous yeasts, and new characters as well as changed evaluation of old characters have led to reclassifications. The 1970 edition included 39 genera and 349 species while the present edition includes 60 genera and 500 species.

Knowledge of the basidiosporogenous yeasts has increased considerably in recent years. These yeasts are now classified in two taxonomically different groups, the teliospore-forming yeasts and the Filobasidiaceae. There are also other basidiomycetous fungi, such as the Tremellales, with a yeast phase in their life cycle. The descriptions of the yeast states of several of these species have been included in this edition.

Of the new characters, DNA base composition and DNA base sequence homology are of great importance for yeast taxonomy, specifically for the delimitation of species. Promising new characters for generic differentiation are, for instance, the ultrastructure of the cell wall and the chemical composition of coenzyme Q.

The aims of the third edition are identical to those of former editions, first, in presenting and discussing a classification of yeasts including diagnoses of genera and descriptions of species, and second, in providing methods for the identification of yeast strains.

The present taxonomic system has developed gradually and, as pointed out by the authors, is not yet entirely satisfactory. Unfortunately, reclassifications are often confusing for the users. Therefore, some authors believe in reticence and wait for a more definite scheme of classification before changing names of species.

The rapid development in yeast taxonomy requires continuous critical reviews. It is, therefore, very difficult to present an up-to-date treatise. The taxonomic system proposed in this edition may be considered as a stage in the evolution towards a satisfactory classification.

Identification of yeasts has become more complicated with the increase in the number of species and the increase in characters used for the standard descriptions. Traditionally, in this treatise there is a key to the genera and keys to the species within each genus. To facilitate the identification of yeast strains, extra keys are supplied directly to the species.

The description of genera and species in this edition has been slightly abridged from the 1970 edition. Historical surveys of the genera and original descriptions of the species have been omitted, but references to them have been provided. The data on the origin of the strains studied are condensed, but references to ecological details, if available, are given. Condensation of the text was necessary to obtain a monograph of a reasonable size.

The present monograph is in contents and shape a direct descendant of the former editions. This made it easier to undertake the editorship which Dr. J. Lodder delegated to me. I am extremely grateful to her for help and advice which she gave from a long experience in yeast taxonomy both as an author and editor.

The pleasant co-operation of the authors is thankfully acknowledged.

N.J.W. Kreger-van Rij

Acknowledgements

Research and manuscript preparations by J.W. Fell and A. Statzell Tallman were funded by the National Science Foundation, Program in Systematic Biology (Grant No. 8040752).

N.J.W. Kreger-van Rij is much indebted to Dr. J.W. Fell, Dr. H.J. Phaff and Dr. C.P. Kurtzman for critically reading her contributions to the monograph. Dr. Fell suggested to include yeast forms of Tremellales in this monograph. The advice on nomenclature problems given by Dr. J. van Brummelen and by Dr. J.A. Leussink is gratefully acknowledged. Many thanks are due to Dr. R.T. Moore for corrections of the Glossary. Part of the work for this monograph was supported by The Netherlands Organization for the Advancement of Pure Research (Z.W.O.).

C.P. Kurtzman gratefully acknowledges the technical assistance of Mrs. M.J. Smiley. The new figures for the genera *Hansenula, Issatchenkia, Lipomyces, Pachysolen* and *Pichia* were prepared by Mr. C.E. Needham.

Sally A. Meyer gratefully acknowledges the assistance of Stephanie Condron Prevost and Lindsay R. Layton.

H.J. Phaff gratefully acknowledges the technical assistance of Mary Miranda and Joanne Tredick. Part of this work was supported by Public Health Service Grant GM-16370 from the Institute of General Medical Sciences.

M. Th. Smith, L. Rodrigues de Miranda and D. Yarrow thank Mr. G.A. Poot and Mrs. I. Paanakker for their scrupulous and patient technical and clerical assistance, and also Mr. K.J.L. Bosman and his assistants of the library of the Laboratorium voor Microbiologie, Technische Hogeschool, Delft, for their help.

The sending of many yeast cultures from the Collection of the Yeast Division of the Centraalbureau voor Schimmelcultures, Delft, to the authors is gratefully acknowledged.

The Glossary has been prepared by N.J.W. Kreger-van Rij with the assistance of J.W. Fell and C.P. Kurtzman.

The figures in this monograph were partly taken from "**The Yeasts**", edition I and II. The figures of edition I and the figures belonging to the genera *Endomycopsis, Debaryomyces* and *Pichia* which were new in edition II were made by Mrs. P.J. Klinkhamer-Hellendoorn. The figures belonging to the genera *Nematospora, Candida, Oosporidium, Torulopsis* and *Trichosporon* of edition II were prepared by Mrs. L. Assis-Lopes and photographed by Mrs. C. Barroso-Lopes. The figures belonging to the genera *Dekkera, Kluyveromyces, Lodderomyces, Wingea* and *Brettanomyces* of edition II were made by Dr. J.P. van der Walt. The other figures of edition II, as far as their origin is not mentioned, were prepared by Mr. J.A. Schuur either from drawings made by Miss W.Ch. Slooff or from photographs. Figures 1–5 and 7–14 of edition III were prepared by Dr. N.J.W. Kreger-van Rij and redrawn by Mr. C.E. Needham.

Directions for the use of the book

This monograph provides the criteria and methods for the classification and identification of yeasts.

In Chapter I, a system of classification in families, subfamilies and genera is presented. Each genus is briefly characterized. The criteria for the distinction of genera and species are discussed and special attention is given to newer characteristics.

Chapter II gives (1) directions for isolation and maintenance of yeasts, (2) descriptions in detail of the tests of the standard description with an explanation of morphological terms, and (3) suggestions for the procedure of identification. A list of the tests of the standard description with a page reference to the methods is given on p. 000, and an alphabetical list of media with a page reference for their preparation can be found on p. 000.

Chapters III, IV and V contain a systematic discussion of the genera of the ascosporogenous, basidiosporogenous (in three groups) and imperfect yeasts, respectively. The genera and species of each group are presented in alphabetical sequence.

The description of a genus comprises: (a) diagnosis, (b) type species, (c) list of accepted species with the date of first publication, (d) a key to the species of the genus with a page reference to the standard description, a table of species with key characters if more than five species are involved, (e) systematic discussion of the species, and, if required, (f) comments on the genus in which the authors give their views of the composition and taxonomic position of the genus, especially the latest developments. Under g. species are listed which have not been accepted by the authors, either because they do not fit in the genus, or because no strains were available at the time of the conclusion of the manuscript.

The description of each species includes: synonyms, standard description, comments if necessary, origin of the strains studied and the type strain. Where possible, a connection between perfect and imperfect species is indicated at the beginning of the description of each species.

The standard description consists of the basic features, similar for all species and additional features of specific importance for individual genera or species, both established with standardized tests.

Explanation of symbols used in the standard description and in the tables:

+	reaction positive
+ w	reaction weakly positive
+ s	reaction slow (latent)
−	reaction negative
+ (−)	reaction positive, seldom negative
− (+)	reaction negative, seldom positive
v	some strains give a positive reaction, others a negative one

The magnification of the drawings of vegetative cells and asci or ascospores is 1000 ×, unless otherwise indicated. The magnification of the drawings of pseudomycelium and true mycelium is 500 ×, unless otherwise indicated. The magnification of the new *Candida* drawings is arbitrary.

Orthographic errors against the International Rules of Botanical Nomenclature in the names of the species have been corrected by the authors.

In Chapter VI keys are given, one to the genera, others to species or groups of species. A list of the keys is at the beginning of this chapter on p. 967.

A glossary to the terms used (p. 1007), a list of references (p. 1011) and an index of the names of the taxa (p. 1052) complete this book.

Abbreviations

The following abbreviations are used for names of genera:

Aciculoconidium	*Ac.*	*Pachysolen*	*Pa.*
Ambrosiozyma	*A.*	*Pachytichospora*	*P'spora*
Arthroascus	*Ar.*	*Phaffia*	*Ph.*
Brettanomyces	*Br.*	*Pichia*	*P.*
Bullera	*B.*	*Pityrosporum*	*Pit.*
Candida	*C.*	*Rhodosporidium*	*Rhodosp.*
Citeromyces	*Cit.*	*Rhodotorula*	*Rh.*
Clavispora	*Cl.*	*Saccharomyces*	*Sacch.*
Cryptococcus	*Cr.*	*Saccharomycodes*	*S'codes*
Cyniclomyces	*Cyn.*	*Saccharomycopsis*	*S.*
Debaryomyces	*Deb.*	*Sarcinosporon*	*Sar.*
Dekkera	*D.*	*Schizoblastosporion*	*Schizobl.*
Fibulobasidium	*Fib.*	*Schizosaccharomyces*	*Schiz.*
Filobasidiella	*Fil.*	*Schwanniomyces*	*Schw.*
Filobasidium	*F.*	*Sirobasidium*	*Sir.*
Guilliermondella	*G.*	*Sporidiobolus*	*Sporid.*
Hanseniaspora	*H'spora*	*Sporobolomyces*	*Sp.*
Hansenula	*H.*	*Sporopachydermia*	*Sporop.*
Holtermannia	*Holt.*	*Stephanoascus*	*Steph.*
Issatchenkia	*I.*	*Sterigmatomyces*	*St.*
Kloeckera	*Kl.*	*Sympodiomyces*	*Symp.*
Kluyveromyces	*K.*	*Torulaspora*	*T'spora*
Leucosporidium	*Leu.*	*Torulopsis*	*T.*
Lipomyces	*L.*	*Tremella*	*Trem.*
Lodderomyces	*Lod.*	*Trichosporon*	*Tr.*
Malassezia	*Mal.*	*Trigonopsis*	*Trig.*
Metschnikowia	*M.*	*Wickerhamia*	*W.*
Nadsonia	*N.*	*Wickerhamiella*	*Wick.*
Nematospora	*Nem.*	*Wingea*	*Wi.*
Oosporidium	*O.*	*Zygosaccharomyces*	*Zygosacch.*

The following abbreviations are used for the names of type culture collections:

ATCC	The American Type Culture Collection
CBS	Centraalbureau voor Schimmelcultures; Yeast Division, Delft
LKB	Laboratory of the Kodama Brewing Company
NRRL	The Northern Regional Research Laboratory

Contents

Chapter I

General classification of the yeasts

A. Delimitation of the yeasts

by N.J.W. Kreger-van Rij

A yeast may be defined as a unicellular fungus reproducing by budding or fission. Flegel (1977) has pointed to the advantages of this morphological definition which leaves out other, historical considerations discussed by Lodder and Kreger-van Rij (1952). Budding yeast cells may be a stage in the life cycle of multicellular fungi.

The yeast species in the present monograph are all characterized by a similar set of features, both morphological and physiological. This type of description, in which physiological characters are important, distinguishes yeast taxonomy from other fungal taxonomy.

The choice of the genera included in the monograph was made for practical reasons and has developed from the history of yeast taxonomy. Included are only organisms with hyaline cells or which produce red, orange or yellow pigments. Excluded are fungi such as *Ajellomyces (Blastomyces) dermatitidis, Histoplasma capsulatum, Aureobasidium pullulans* and many of the smut-fungi that may have yeast phases under specific conditions. These yeast phases are usually included in the systematic description of the organisms.

The yeasts are taxonomically diverse and include ascomycetes and basidiomycetes. A third group, the imperfect yeasts, have both ascomycetous and basidiomycetous affinities.

B. Systems of classification of the yeasts

by N.J.W. Kreger-van Rij

1. History

Lodder et al. (1958) and Lodder (1970) have reviewed previous systems of classification of the yeasts. The reader is referred to these publications.

2. Present system of classification

The system of classification used in this edition is given in Tables 1,2,3 and 4, followed by diagnoses of families and subfamilies and short characteristics of each genus. A full diagnosis of the genera is part of the systematic discussion in chapters III, IV and V.

Table 1

Classification of the yeasts in the Eumycota

Ascomycotina	
Hemiascomycetes	
Endomycetales	Spermophthoraceae
	Saccharomycetaceae
Basidiomycotina	Filobasidiaceae
Ustilaginales	Teliospore-forming yeasts
Tremellales	Sirobasidiaceae
	Tremellaceae
Deuteromycotina	
Blastomycetes	Cryptococcaceae
	Sporobolomycetaceae

Table 2

Classification of the ascosporogenous yeasts

Spermophthoraceae	*Coccidiascus*	
	Metschnikowia	
	Nematospora	
Saccharomycetaceae		
Schizosaccharomycetoideae	*Schizosaccharomyces*	
Nadsonioideae	*Hanseniaspora*	
	Nadsonia	
	Saccharomycodes	
	Wickerhamia	
Lipomycetoideae	*Lipomyces*	
Saccharomycetoideae	*Ambrosiozyma*	*Pachysolen*
	Arthroascus	*Pachytichospora*
	Citeromyces	*Pichia*
	Clavispora	*Saccharomyces*
	Cyniclomyces	*Saccharomycopsis*
	Debaryomyces	*Schwanniomyces*
	Dekkera	*Sporopachydermia*
	Guilliermondella	*Stephanoascus*
	Hansenula	*Torulaspora*
	Issatchenkia	*Wickerhamiella*
	Kluyveromyces	*Wingea*
	Lodderomyces	*Zygosaccharomyces*

a. Families and genera of the ascosporogenous yeasts

Spermophthoraceae (cited from Gäumann, 1964)

Haploid, non-septate hyphae; sporangiospores may be formed which can copulate and form diploid mycelium. This mycelium is usually uninucleate and can form eight to twelve spindle-shaped ascospores.

Genus *Coccidiascus* Chatton (1913)

Budding yeast cells.

Asci large, banana- or crescent-shaped; ascospores spindle-shaped, intertwined in a helix, liberated.

Genus *Metschnikowia* Kamienski (1899)

Multilateral budding; pseudomycelium rudimentary, rarely lacking.

Asci much larger than vegetative cells, clavate, sphaeropedunculate or ellipsoidopedunculate; ascospores needle-shaped, attenuate at one or both ends, without whip-like appendage, liberated or not.

Fermentation present or absent; nitrate not assimilated.

Genus *Nematospora* Peglion (1897)

Multilateral budding; true mycelium.

Asci larger than vegetative cells; ascospores spindle-shaped with a whip-like appendage, eight per ascus, liberated.

Fermentation; nitrate not assimilated.

Saccharomycetaceae

Mycelium, pseudomycelium, arthrospores and budding cells side by side or alone. Vegetative reproduction by fission or by budding. Asci may arise after isogamous or heterogamous conjugation; often a stage of vegetative reproduction between diploidization and ascus formation. Ascospores of various shapes, but not needle-shaped. Besides an oxidative dissimilation a fermentative dissimilation is often present.

I. Subfamily Schizosaccharomycetoideae

Mycelium and (or) arthrospores; vegetative reproduction by fission; ascus formation after conjugation; no gametangiogamy; ascospores spherical, oval or reniform; dissimilation oxidative and fermentative.

Genus *Schizosaccharomyces* Lindner (1893)

Fission cells; true hyphae and arthrospores may occur.

Asci conjugated; ascospores spherical, ovoidal or reniform, smooth or warty, liberated.

Fermentation; nitrate not assimilated.

II. Subfamily Nadsonioideae

Budding cells and occasionally pseudomycelium, no mycelium; vegetative reproduction by bipolar budding on a more or less broad base; dissimilation oxidative and (or) fermentative.

Genus *Hanseniaspora* Zikes (1911)

Bipolar budding, cells usually lemon-shaped.

Asci unconjugated; ascospores hat-shaped and liberated, or spherical with or without ledge and smooth or warty and not liberated.

Fermentation; nitrate not assimilated.

Genus *Nadsonia* Sydow (1912)

Bipolar budding on a broad base, cells usually lemon-shaped; no pseudomycelium.

Asci formed after conjugation between mother cell and bud, either the mother cell or a bud at the other pole becomes the ascus; ascospores spherical, warty and brown, not liberated.

Fermentation present or absent; nitrate not assimilated. No growth above 30°C.

Genus *Saccharomycodes* Hansen (1904)

Bipolar budding on a broad base, cells usually lemon-shaped; pseudomycelium absent or poorly developed.

Asci unconjugated; ascospores spherical with a narrow ledge, not liberated.

Fermentation; nitrate not assimilated.

Genus *Wickerhamia* Soneda (1960)
Bipolar budding, cells usually lemon-shaped; no pseudomycelium.
Asci unconjugated; ascospores cap-shaped, liberated.
Fermentation; nitrate not assimilated.

III. Subfamily Lipomycetoideae
Budding cells only, multilateral budding. Asci sac-like, developed from mother cell as a protuberance or after conjugation of two cells; ascospores spherical to oval, amber-colored. Dissimilation strictly oxidative.

Genus *Lipomyces* Lodder et Kreger-van Rij (1952)
Multilateral budding.
Asci developed from active buds, from mother cell or from conjugation of two cells; ascospores smooth, warty or with ridges, amber to brown, usually liberated from the ascus.
No fermentation; nitrate not assimilated.

IV. Subfamily Saccharomycetoideae
Mycelium and budding cells, pseudomycelium and (or) single budding cells; vegetative reproduction by fission and budding, or budding only; conjugation may or may not immediately precede ascus formation, ascospores of various shapes; dissimilation from merely oxidative to predominantly fermentative.

Genus *Ambrosiozyma* van der Walt (1972)
Budding yeast cells, pseudomycelium, abundant true mycelium, hyphal septa with a dolipore.
Asci on the hyphae; ascospores hat-shaped.
Fermentation; nitrate assimilated or not.

Genus *Arthroascus* von Arx (1972)
Budding on a broad base, often at the two poles of the cells; true mycelium and arthrospores, cross walls with a closure line. Cells mostly elongate.
Asci conjugated, generally spindle-shaped by a swelling in the middle of the cell; ascospores hat- or Saturn-shaped, smooth or warty, liberated.
No fermentation; nitrate not assimilated.

Genus *Citeromyces* Santa María (1957)
Multilateral budding; no pseudomycelium.
Ascospores spherical and warty, not liberated.
Fermentation; nitrate assimilated.

Genus *Clavispora* Rodrigues de Miranda (1979)
Multilateral budding.
Asci conjugated; ascospores club-shaped, warty, liberated.
Fermentation; nitrate not assimilated.

Genus *Cyniclomyces* van der Walt et Scott (1971b)
Budding yeast cells; pseudomycelium.

Asci unconjugated; ascospores oval to cylindrical, generally not liberated.

Fermentation. Growth between 30 and 40°C; an increased level of CO_2 in the gas atmosphere is required. In stomach of rabbits.

Genus *Debaryomyces* Lodder et Kreger-van Rij nom. cons. (Lodder & Kreger-van Rij, 1978)

Multilateral budding.

Asci conjugated, usually mother–daughter cell conjugation. Ascospores spherical or oval, warty or with ridges, not liberated.

Fermentation absent or present, not vigorous; nitrate not assimilated.

Genus *Dekkera* van der Walt (1964)

Multilateral budding, pseudomycelium.

Asci conjugated; ascospores hat-shaped, liberated.

On malt agar slow growth and cells short-lived. Under aerobic conditions usually strong acetic acid production from glucose.

Fermentation; nitrate assimilated or not.

Genus *Guilliermondella* Nadson et Krassilnikov (1928)

Budding yeast cells and abundant true mycelium. Hyphal septa with plasmodesmata.

Conjugation precedes ascus formation on the hyphae; ascospores lunate, liberated.

Fermentation; nitrate not assimilated.

Genus *Hansenula* H. et P. Sydow (1919)

Multilateral budding; pseudo- or true mycelium may be present.

Ascospores hat-, Saturn-shaped or hemispherical, generally liberated.

Fermentation present or absent; nitrate assimilated.

Genus *Issatchenkia* Kudriavzev (1960)

Multilateral budding; pseudomycelium.

Asci unconjugated; ascospores spherical and warty, not liberated.

Pellicle formation on liquid media.

Fermentation; nitrate not assimilated.

Genus *Kluyveromyces* van der Walt (1956a)

Multilateral budding.

Asci one- to multi-spored; ascospores spherical, ellipsoidal, reniform or crescentiform, liberated.

Fermentation; nitrate not assimilated.

Genus *Lodderomyces* van der Walt (1966a)

Multilateral budding, pseudomycelium.

Asci unconjugated; ascospores oblong or fusiform, not liberated.

Fermentation; nitrate not assimilated.

Genus *Pachysolen* Boidin et Adzet (1957)

Multilateral budding; pseudomycelium poorly developed or absent.

Asci conjugated or unconjugated, thick-walled with a tube; the end of the tube is thin-walled and contains the spores; ascospores hat-shaped, liberated.

Fermentation; nitrate assimilated.

Genus *Pachytichospora* van der Walt (1978)

Multilateral budding, pseudomycelium rudimentary or absent.

Asci unconjugated; ascospores spheroidal to ellipsoidal with an eccentrically thickened wall, not liberated.

Fermentation; nitrate not assimilated.

Genus *Pichia* Hansen (1904)

Multilateral budding, often pseudomycelium, true hyphae if present, usually scarce, exceptionally abundant; arthrospores may be formed.

Asci conjugated or unconjugated, they are single yeast cells or pseudomycelial cells; ascospores spherical, hemispherical, hat- or Saturn-shaped, smooth or warty, generally liberated.

Fermentation present or absent; nitrate not assimilated.

Genus *Saccharomyces* Meyen ex Reess (Meyen, 1838; Reess, 1870)

Multilateral budding, pseudomycelium may be formed.

Asci unconjugated; ascospores spherical or oval, smooth, seldom warty, not liberated.

No pellicle formation on liquid media.

Fermentation vigorous; nitrate not assimilated.

Genus *Saccharomycopsis* Schiönning (1903)

Budding yeast cells, abundant true mycelium, pseudomycelium and arthrospores may occur. Septa in the hyphae with plasmodesmata, *S. lipolytica* excepted.

Asci on the mycelial hyphae; ascospores spherical, hat- or Saturn-shaped, smooth or warty, the ledge may be irregular; ascospores liberated or not.

Fermentation present or absent; nitrate not assimilated.

Genus *Schwanniomyces* Klöcker (1909a)

Multilateral budding; pseudomycelium rudimentary or absent.

Conjugated asci, generally conjugation between mother cell and bud; ascospores spherical with an equatorial ledge, warty, not liberated.

Fermentation; nitrate not assimilated.

Genus *Sporopachydermia* Rodrigues de Miranda (1978)

Multilateral budding, no pseudomycelium.

Conjugated or unconjugated asci; ascospores spherical or oval with a thick wall, liberated.

No fermentation; nitrate not assimilated; inositol assimilated.

Genus *Stephanoascus* Smith, van der Walt et Johannsen (1976)

Budding yeast cells, abundant true mycelium, pseudomycelium, no arthrospores.

Hyphal septa with plasmodesmata.
Asci are part of a hypha formed after fusion of two cells; they are spherical with a small top cell; ascospores hemispherical with a thick wall at the convex side, not liberated.
No fermentation; nitrate not assimilated.

Genus *Torulaspora* Lindner (1904)
Multilateral budding; pseudomycelium may be formed.
Asci formed after conjugation, usually between mother cell and bud; ascospores spherical or ellipsoidal, warty or smooth, not liberated.
Fermentation vigorous; nitrate not assimilated.

Genus *Wickerhamiella* van der Walt (van der Walt & Liebenberg, 1973a)
Multilateral budding, no pseudomycelium.
Asci conjugated; ascospores oblong and rugose, liberated.
No fermentation; nitrate is assimilated.

Genus *Wingea* van der Walt (1967)
Multilateral budding, no pseudomycelium.
Asci conjugated or unconjugated; ascospores lentiform, light brown, not liberated.
Fermentation; nitrate not assimilated.

Genus *Zygosaccharomyces* Barker (1901)
Multilateral budding; pseudomycelium may be formed.
Conjugated asci; ascospores spherical or ellipsoidal, not liberated.
Fermentation vigorous; nitrate not assimilated.

Table 3

Classification of the basidiosporogenous yeasts

Filobasidiaceae	*Chionosphaera* *Filobasidiella* *Filobasidium*
Teliospore-forming yeasts	*Leucosporidium* *Rhodosporidium* *Sporidiobolus*
Sirobasidiaceae	*Fibulobasidium* *Sirobasidium*
Tremellaceae	*Holtermannia* *Tremella*

b. Families and genera of the basidiosporogenous yeasts

Filobasidiaceae (Olive, 1968), type species: *Filobasidium floriforme*

Basidia arising in loose or dense groups from a mycelium with clamp connections, slender, non-septate, bearing sessile thin-walled basidiospores terminally; thick-walled probasidia lacking, blastospores present.

Genus *Chionosphaera* Cox (1976)

Budding yeast cells, pseudomycelium, true mycelium with clamp connections. No ballistospores.

Basidia in synnemata (fruiting bodies), sessile basidiospores on the apex of the basidium.

No living or dead material is available of the single species *C. apobasidiales*.

Genus *Filobasidiella* Kwon-Chung (1975)

Budding yeast cells, no pseudomycelium, true mycelium with clamp connections. Hyaline cultures.

Basidiospores in four chains on the apex of the basidium.

Genus *Filobasidium* Olive (1968)

Budding yeast cells; pseudomycelium may occur, true mycelium with clamp connections. Hyaline cultures.

Sessile basidiospores in a whorl on the apex of the basidium.

Teliospore-forming yeasts

Genus *Leucosporidium* Fell, Statzell, Hunter et Phaff (1969)

Budding yeast cells, pseudomycelium may occur, true mycelium with and without clamp connections. Hyaline cultures. No ballistospores.

Teliospores. Metabasidia with basidiospores.

Fermentation may occur; nitrate is assimilated.

Genus *Rhodosporidium* Banno (1967)

Budding yeast cells; pseudomycelium may occur, true mycelium with and without clamp connections. Yellow, orange or red cultures due to carotenoid pigments. No ballistospores.

Teliospores. Metabasidia with basidiospores.

No fermentation; nitrate is assimilated or not.

Genus *Sporidiobolus* Nyland (1949)

Budding yeast cells, pseudomycelium, true mycelium with clamp connections. Pink or red cultures due to carotenoid pigments. Ballistospores.

Teliospores. Metabasidia with basidiospores.

No fermentation; nitrate is assimilated or not.

Tremellales with a yeast state (Sirobasidiaceae and Tremellaceae) see Introduction p. 541.

Table 4

Classification of the imperfect yeasts

Cryptococcaceae	*Aciculoconidium*	*Rhodotorula*
	Brettanomyces	*Sarcinosporon*
	Candida	*Schizoblastosporion*
	Cryptococcus	*Sterigmatomyces*
	Kloeckera	*Sympodiomyces*
	Malassezia	*Trichosporon*
	Oosporidium	*Trigonopsis*
	Phaffia	
Sporobolomycetaceae	*Bullera*	*Sporobolomyces*

c. Families and genera of the imperfect yeasts

Cryptococcaceae

Budding yeast cells always present; moreover pseudomycelium, true mycelium and arthrospores may be formed. Cells hyaline, or red, orange or yellow due to carotenoid pigments, very seldom brown or black.

Dissimilation strictly oxidative or oxidative and fermentative.

Genus *Aciculoconidium* King et Jong (1976a)

Budding yeast cells, abundant true mycelium with short-oval blastospores and terminal needle-shaped blastospores, rounded at the base and pointed at the other end.

Fermentation; nitrate not assimilated.

Genus *Brettanomyces* Kufferath et van Laer (1921)

Multilateral budding; pseudomycelium and non-septate true mycelium may be formed.

On malt agar slow growth and cells short-lived. Under aerobic conditions usually strong acetic acid production from glucose.

Fermentation; nitrate assimilated or not.

Genus *Candida* Berkhout (1923)

Multilateral budding, polar budding if present on a narrow base; pseudo- and true mycelium may be formed.

Budding cells not apiculate, flask-shaped or triangular; no formation of sympodia or sterigmata. No strong acid formation from glucose. Inositol positive strains form pseudomycelium. Visible pigmentation due to carotenoid pigments absent.

Fermentation or not; nitrate assimilated or not; inositol assimilated or not.

Genus *Cryptococcus* Kützing (1833)

Multilateral budding; pseudomycelium absent or rudimentary. Most strains have capsulated cells; the cultures on solid media are generally mucous; they are hyaline, or red or orange due to carotenoid pigments; in one species brown or black cultures occur under special conditions.

No fermentation; nitrate assimilated or not; inositol assimilated.

Genus *Kloeckera* Janke (1928)

Bipolar budding on a broad base, cells lemon-shaped (apiculate). Pseudomycelium may be formed.

Fermentation; nitrate not assimilated.

Genus *Malassezia* Baillon (1889)

Monopolar budding on a broad base, cells spherical or bottle-shaped; true hyphae may occur.

No fermentation; natural oils may stimulate growth; optimum temperature of growth 35–37°C. Isolated from human and warm-blooded animal skin.

Genus *Oosporidium* Stautz (1931)

Multilateral budding on a broad base, true mycelium, no arthrospores. Cultures have a pink or orange-yellow color of a non-carotenoid nature.

Asexual endospores.

No fermentation; nitrate assimilated.

Genus *Phaffia* Miller, Yoneyama et Soneda (1976b)

Budding cells; rudimentary pseudomycelium may be present; chlamydospores may be formed. Cultures orange or red due to carotenoid pigments.

Fermentation; nitrate is not assimilated; inositol is not assimilated.

Genus *Rhodotorula* Harrison (1928)

Multilateral budding, pseudo- and true mycelium may occur. The cultures are red or yellow due to carotenoid pigments, the cultures are often mucous.

No fermentation, nitrate is assimilated or not, inositol is not assimilated, starch-like compounds are not produced, urease is positive.

Genus *Sarcinosporon* King et Jong (1975)

Budding yeast cells and septate hyphae. Septation in yeast cells and hyphae in different planes which gives sarcina-like agglomerates of cells; arthrospores.

Asexual endospores.

No fermentation, nitrate is not assimilated, inositol is assimilated.

Genus *Schizoblastosporion* Ciferri (1930)

Bipolar budding on a broad base, pseudomycelium absent or primitive.

No fermentation, nitrate not assimilated.

Genus *Sterigmatomyces* Fell (1966)

Yeast cells form as buds on sterigmata; separation of the cells in mid-region of the sterigma or adjacent to the daughter cell, no pseudomycelium, true mycelium may be formed. No ballistospores.

The cultures are hyaline or orange due to the presence of carotenoid pigments.

No fermentation, nitrate is assimilated or not, inositol is assimilated or not, urease is positive.

Genus *Sympodiomyces* Fell et Statzell (1971)

The yeast cell produces a conidiophore with a terminal conidium; the conidiophore elongates to develop a succession of terminal conidia. Some true mycelium is formed. No ballistospores.

No fermentation, nitrate is not assimilated, urease is negative.

Genus *Trichosporon* Behrend (1890)

Budding yeast cells, pseudomycelium and abundant true mycelium with arthrospores.

Asexual endospores may be formed.

Fermentation or not, nitrate assimilated or not, inositol assimilated or not.

Genus *Trigonopsis* Schachner (1929)

Cells ellipsoidal, triangular, tetrahedral or rhombohedral, buds are formed from the angles of the triangular or tetrahedral cells, no pseudomycelium.

No fermentation, nitrate is not assimilated.

Sporobolomycetaceae (cited from Lodder & Kreger-van Rij, 1952)

Mycelium, pseudomycelium and budding yeast cells. Vegetative reproduction by fission or budding. The vegetative cells may form aerial sterigmata that are single or bifurcated. Smooth hyaline spores are formed in an oblique position to the sterigmata. When mature, the spores (ballistospores) are ejected by a drop-excretion mechanism. Dissimilation strictly oxidative.

Genus *Bullera* Derx (1930)

Budding yeast cells; pseudo- and true mycelium may be formed. The cultures are hyaline.

Ballistospores are rotationally symmetrical, spherical, obovoid, turbinate, apiculate or ampulliform.

No fermentation, nitrate is assimilated or not, inositol is assimilated or not, starch-like compounds are produced or not.

Table 5

Physiological and biochemical characteristics of *Endomyces* species

	Fermentation			Assimilation					
	G	Ga	Su	Ga	Su	Ce	Ra	X	Man
'E. decipiens' (*G. armillariae*)	−			+	−	+	−	+	+
E. magnusii	+	+ w	+	+	+	−	+	−	+
E. ovetensis	−			+	−	−	−	−	−
E. tetrasperma	+	−	−	+	−	−	−	−	−
E. geotrichum	+ vw/−	+ vw/−	−	+	−	−	−	+	+
E. reessii	−			+	−	−	−	+	−

The following compounds are not assimilated: maltose, trehalose, lactose, soluble starch, L-arabinose, D-ribose, L-rhamnose, erythritol, ribitol, inositol and nitrate.

G = glucose, Ga = galactose, Su = sucrose, Ce = cellobiose, Ra = raffinose, X = D-xylose, Man =

Genus *Sporobolomyces* Kluyver et van Niel (1925)

Budding yeast cells; pseudo- and true mycelium may be formed. The cultures are red or pink due to carotenoid pigments.

Ballistospores are bilaterally symmetrical, usually obovoid, pyriform or reniform. No fermentation, nitrate is assimilated or not, inositol is assimilated or not, starch-like compounds are produced or not, urease is positive.

C. The genera *Endomyces* and *Geotrichum*

by N.J.W. Kreger-van Rij

The ascomycetous yeasts are classified in the Hemiascomycetes which is in several respects a heterogeneous group. The genera vary in the method of diploidization (somatogamy and gametangiogamy), the shape of the ascus (e.g. proliferation of the ascus), the shape of the ascospores, the number of spores per ascus, the number of nuclei per cell, and the ultrastructure of the septum. One of the genera of the Hemiascomycetes not included in this monograph is *Endomyces*. Since some species of *Endomyces* and the similar imperfect genus *Geotrichum* resemble yeast species, the two genera will be briefly discussed here.

Both genera are characterized by the formation of branched, septate hyphae breaking up into arthrospores. A fundamental problem of the taxonomy of the genus *Endomyces* concerns the identity of the type species, *E. decipiens*. Moreover, various reclassifications of the other *Endomyces* species have been proposed.

Kreger-van Rij and Veenhuis (1972), in an ultrastructural study of hyphae and ascospores distinguished three groups among the *Endomyces* species:

(1) *E. decipiens:* no gametangiogamy; ascospores hat-shaped, four per ascus, dehisced; plasmodesmata in a random arrangement,

(2) *E. magnusii, E. ovetensis* and *E. tetrasperma:* gametangiogamy; ascospores oval with a wide wall, mostly four per ascus, dehisced; plasmodesmata arranged in a circle.

				vit	37	25	ur	G + C	Co-Q
Suc	Cit	Sor	Glu						
+	−	v	−	+	−	v	−	37	
v	−	+	+	−	−	+	−	34	9[1]
+	−	+	−	−	−	+	+	48.5	9[2]
+	+	+	+	−	+	+	−	38	
+	+	+	+	v	v	+	−	37.5	
+	+	v	−	v	−	+	−	34.5	9[1]

D-mannitol, Suc = succinic acid, Cit = citric acid, Sor = L-sorbose, Glu = D-glucitol, vit = growth without the addition of vitamins, 37 = growth at 37°C, 25 = growth at 25°C, ur = urease, G + C = G + C content from Guého (1979), Co-Q = coenzyme Q from [1] Yamada et al. (1973a) and [2] Yamada et al. (1976b).

(3) *E. geotrichum* and *E. reessii:* gametangiogamy; ascospores oval with a lobed wall, mostly one per ascus, not dehisced; plasmodesmata in a random arrangement.

Constriction at the septa was observed in species of group 2. This gave the appearance of buds formed on a broad base (Kreger-van Rij & Veenhuis, 1972, Fig. 11). Ascospores have not been observed in the strains of *E. decipiens* studied, including the CBS strain described by Stelling-Dekker (1931) as well as recent isolates from *Armillaria mellea.* However, asci with hat-shaped ascospores were described by de Bary (1859) and Brefeld (1891) in the mold growing on *Armillaria mellea* and to which the name *Hypomyces decipiens* was originally given by Tulasne and Tulasne (1865).

Redhead and Malloch (1977) studied *Endomyces decipiens* occurring on *Armillaria mellea* in an herbarium specimen in which asci with hat-shaped spores were present. These authors came to the conclusion that this fungus, by the absence of arthrospores, differed from the pure cultures labelled *Endomyces decipiens* and described, e.g., by Stelling-Dekker (1931). Thus two different species occurring on *Armillaria mellea* had been described as *Endomyces decipiens,* one from the herbarium specimen with hat-shaped ascospores and without arthrospores, and one isolated in pure culture with arthrospores which Redhead and Malloch indicated as *Geotrichum* species, later named *Geotrichum armillariae* by von Arx (1977). In this connection, it is remarkable that in the descriptions of *E. decipiens* by de Bary (1859) and Brefeld (1891) the formation of hat-shaped ascospores and of arthrospores in the same organism is mentioned and shown in a drawing.

In 1979 I observed asci in a fungus growing on the gills of *Armillaria mellea* (unpublished). The 6–8 spores within the asci were angular with a narrow, irregular wall; when dehisced, they were oval with a wide wall. The asci were conical, long-oval and cylindrical, attached at the bottom to two hyphal cells, typical of asci formed after gametangiogamy. The ascospores resembled those of *Endomyces* group 2, but the arrangement of the plasmodesmata in the septa was, in contrast, random. The isolated fungus had arthrospores and agreed also in other respects with the culture labelled *E. decipiens* described by Stelling-Dekker. It failed to sporulate in culture. These observations corroborate Redhead and Malloch's conclusion that two different species have been described as *Endomyces decipiens.* EM observations of an herbarium specimen of *Armillaria mellea* with a fungus with hat-shaped ascospores showed that plasmodesmata in the hyphal septa had a random arrangement (unpublished). Isolation in pure culture of this fungus seems a requirement before a decision can be made about its classification.

Redhead and Malloch (1977) brought the other *Endomyces* species into the following genera: *E. magnusii* Ludwig (1886) into *Magnusiomyces, E. tetrasperma* Macy et Miller (1971) and *E. ovetensis* Peláez et Ramirez (1956b) into *Zendera,* and *E. geotrichum* Butler et Petersen (1972) and *E. reessii* van der Walt (1959c) into *Galactomyces.* Von Arx (1977) brought all these species into the genus *Dipodascus.*

In Table 5 physiological and biochemical properties of *Endomyces* species are given.

Whereas the difference between the *Endomyces* species, with the exception of *E. decipiens* with hat-shaped ascospores, and the perfect yeasts of this yeast monograph lies in the presence or absence of gametangiogamy, the difference between the imperfect genus *Geotrichum* and the imperfect yeast genera is based on vegetative criteria and is not always distinct. *Geotrichum* is distinguished by the formation of hyphae breaking up into arthrospores and by the absence of loose budding cells and blastospores. The type species of *Geotrichum* is *Geotrichum candidum.* Butler and Petersen (1970, 1972) discovered the perfect state of this species which they described as *Endomyces geotrichum.* This species belongs to group 3 of the *Endomyces* species mentioned previously. The imperfect state of *Endomyces* species of group 2 may be classified as *Geotrichum* species, but bud formation on a broad base, which is occasionally observed, may also bring these species into the imperfect genus *Trichosporon.* This is, for instance, the case with *E. ovetensis* and the imperfect state *Tr. sericeum.* Additionally, the imperfect state of *Dipodascus* species may be identified as *Geotrichum* species. Von Arx (in von Arx et al., 1977) and Weijman (1979) have transferred several *Trichosporon* species to the genus *Geotrichum* (see Comments on the genus *Trichosporon*), including species with blastospores on a narrow base, which has made the genus *Geotrichum* more heterogeneous. In this monograph, these species have been provisionally retained in the genus *Trichosporon.*

For a survey of physiological properties and DNA base composition of *Geotrichum* species, see Guého (1979).

D. Principles of classification

In this section the use of characters by which genera and species are differentiated will be discussed, and the criteria by which families and subfamilies are recognized will be mentioned.

1. The standard description

by N.J.W. Kreger-van Rij

The standard description of a species gives the characteristics by which it is recognized. These characteristics are in general similar for all species. They are morphological, physiological, biochemical, and sometimes ecological; for sporogenous species they include the features of the sexual cycle. The taxonomic value of the descriptive characters depends (1) on their ability to differentiate and (2) on their being constant. Most of the characters are defined as the result of a particular test, e.g. fermentation and assimilation. Since the results of the tests may be influenced by the test conditions, the latter have to be strictly standardized. This holds not only for physiological and biochemical properties, but also for morphological features which may vary with the medium of growth. The mechanism leading to the result of a test, or the genotype responsible for the phenotype, may be simple or complex and is often unknown. Since the taxonomic value of a character lies primarily in the differences it shows between yeasts, the mode of the test applied is usually of secondary importance, provided it is standardized. Initially, in the Delft yeast monographs, simple tests were used for the standard description and identification of the yeasts.

Over the years, there have been changes in the methods of the tests and the number of characters increased considerably. For example, the use of a liquid medium for assimilation tests took the place of the auxanographic method (Wickerham & Burton,

1948). Wickerham (1951) extended the number of carbon compounds from 5 to 31 which contributed to a better species delimitation. He selected the carbon compounds of the assimilation tests for their differentiating value. Experience has shown that several of the carbon compounds chosen by Wickerham may be dispensed with; those which are often weakly or slowly assimilated may lead to inconsistent results and others, such as ethanol and glycerol, are utilized by most yeasts and thus have a low differentiating value. In the 1970 edition of 'The Yeasts' the number of carbon compounds routinely tested was 30; in the present edition the number is reduced to 18. However, in some genera extra carbon compounds are required for species differentiation, e.g. melibiose and melezitose in *Hansenula* and *Rhodotorula,* and in *Candida* the number of carbon compounds used for species differentiation is 27.

Research in the chemical composition, the ultrastructure and nuclear DNA base composition of yeasts has added characters to the standard description which are considered to be of fundamental importance. Generally, these characters are not routinely determined, but are the work of specialists. As further discussed under D.3., this work has had a great impact on taxonomy, not only for separating species, but also for genera.

A difference in one of the easily determinable features of the standard description may or may not correlate with a notable difference in DNA homology leading to a distinction of species. If there is a positive correlation, the feature in question acts as an indicator that is useful for the characterization of a species and indispensable as a key character.

2. *The species*

by N.J.W. Kreger-van Rij

A yeast species may be defined as an assemblage of clonal populations, i.e. a collection of strains (Lodder & Kreger-van Rij, 1952; Stanier et al., 1971). This means that the standard description of a species is preferably based on several strains.

A solid base for species differentiation is interfertility among the strains of one species and absence of interfertility with strains of another species. In heterothallic yeasts, the formation of viable spores after pairing of compatible mating types that results in a zygote which develops into an ascus or (meta)basidium, is a useful criterion for the delimitation of a species. DNA–DNA reassociation tests have made possible the establishment of conspecificity among homothallic and asporogenous yeasts. Moreover, DNA base composition (G + C content), when at variance, indicates that the yeasts belong to different species.

Since the use of the characters for differentiation often varies among the taxa, it will be treated in relation to the delineation of families, genera and species (D.4., D.5. and D.6.). A few characters of general importance to be discussed in the following section (D.3.) include: DNA composition and relatedness, similarity of enzymes and cell wall structure and composition.

3. *DNA, enzymes and cell wall*

by H.J. Phaff

Introduction

Because fossil records for microorganisms are largely unavailable and/or difficult to interpret, microbiologists and microbial evolutionists have taken recourse to constructing phenetic trees and classifying microorganisms on the basis of the similarity or dissimilarity of their genetic macromolecules, namely DNA and/or the products of DNA, i.e., ribosomal RNA (rRNA) and proteins. Prokaryotic microorganisms have been studied extensively by such methods and the techniques developed for bacteria were later extended to the classification of yeasts. These include nuclear DNA (nDNA) homologies (see Price et al., 1978, for a review), rRNA–DNA homology (Bicknell & Douglas, 1970), allozyme patterns (Baptist & Kurtzman, 1976; Holzschu, 1981), immunological comparisons of proteins (Lachance & Phaff, 1979), serological analyses of cell walls (Tsuchiya et al., 1974; Fukazawa et al., 1980), and coenzyme Q comparisons (Yamada & Kondo, 1972, 1973; Yamada et al., 1973a,b,c; Yamada et al., 1976a,b; Yamada et al., 1977; Yamada et al., 1980; Yarrow & Nakase, 1975). These studies have provided important data for classification but each is limited to yielding data pertinent at only certain levels of taxonomic organization.

a. DNA composition and relatedness

Nucleic acid comparisons among yeasts were originally limited to DNA base composition determinations (Martini et al., 1972; Meyer & Phaff, 1970; Nakase & Komagata, 1970a,b, 1971a,b,c,d,e,f,g). Most investigators used thermal denaturation as a means to determine the base composition of the nuclear DNA, reported as mol% G + C (Marmur & Doty, 1962). Bak (1973) showed that results obtained by this method can be greatly affected by sample impurities and/or minor DNA species, and must therefore be interpreted with caution. Schildkraut et al. (1962) utilized a method based on the buoyant density properties of nucleic acid molecules in cesium salt gradients established by ultracentrifugation to determine base composition. This method is less prone to error than the melting procedure, but, in many cases, careful use of thermal procedures has produced comparable data to those determined by buoyant density. The primary utility of data gathered by either method is exclusionary. Numerous studies on DNA–DNA homology have shown that two organisms with base composition values differing by more than 1.5–2.0 mol% G + C do not have large sequences of DNA in common and therefore are not closely related (Price et al., 1978; Holzschu et al., 1979a). However, determination of G + C content cannot distinguish the relationships among yeasts with similar DNA base compositions. Subsequently, DNA–DNA complementarity analysis was developed to quantitate the similarity of convergent yeasts.

DNA base composition values continue to be useful for description of new yeast species (Phaff et al., 1978; Starmer et al., 1979), but another primary usage of these data is screening of organisms for use in DNA–DNA complementarity experiments (Price et al., 1978). Several methods have been developed to examine DNA sequence relatedness among yeasts. Bak and Stenderup (1969) utilized the nitrocellulose filter binding assay for separation of single- and double-stranded DNA formed during reannealing experiments. This method has been widely used by other investigators to examine a wide variety of yeasts, including species of *Saccharomyces, Kloeckera, Hanseniaspora, Candida, Torulopsis,* and *Metschnikowia* (Meyer & Phaff, 1972; Meyer et al., 1978) (see also Price et al., 1978, for a review) but it has been largely replaced by other techniques in recent years. The difficulties encountered with the filter binding assay include: the DNA fixed to the filter papers leeches away during incubation, making quantification of double-stranded DNA imprecise; in addition, single strands that are bound to the filter paper cannot fully reassociate due to steric hindrance (Price et al., 1978). Conditions that affect the renaturation of sheared DNA strands can be controlled best when the molecules are free to react in solution (Wetmur & Davidson, 1968). A free solution system allows the products of the reaction to be quantified by several alternative methods, i.e., the S_1 nuclease method (Crosa et al., 1973), hydroxylapatite (HA) batch method developed by Brenner et al. (1969), passing the completed reaction mixture over a HA column as described by Lachance (1980), or by optically monitoring the renaturation reaction (Martini & Phaff, 1973; Kurtzman et al., 1980a).

DNA–DNA homology experiments have proved to be most valuable for species delimitation (Price et al., 1978). Results published by many authors show that high complementarity, i.e. greater than ca. 80%, indicates conspecificity; low complementarity values, i.e., less than ca. 20%, indicate that the strains are only distantly related (Phaff & Holzschu, 1981). Complementarity values ranging from 65–80% are rather commonly observed (Phaff & Holzschu, 1981; Holzschu et al., 1981) and are generally regarded to represent more differentiated strains within a species. Taxonomic interpretation of values ranging from 20–35% is difficult. Kurtzman et al. (1980a) consider *Pichia amylophila* and *P. mississippiensis* as separate species although their genomes show about 25% DNA sequence complementarity. Ascospores from interspecific matings (sensu Kurtzman) of these species were not viable. The low DNA sequence homology between the above species supported the sexual hybridization data. This study was the first to show that low DNA–DNA homology correlated with lack of interfertility among yeast strains and therefore provided a basis for evaluating DNA homology with the definition of a biological species, i.e., species "can be defined as Mendelian populations, or arrays of populations, which are reproductively isolated from other population arrays" (Dobzhansky, 1976). However, Kurtzman et al. (1980b) went on to describe a different relationship between biochemical characterization and cellular hybridization among varietal strains of *Pichia scutulata.* A low percentage of viable spores is produced from matings of compatible strains of *P. scutulata* var. *scutulata* and *P. scutulata* var. *exigua,* whose DNAs show only about 25% homology. Clearly then, the lower limit of DNA–DNA homology values suggesting species delimitation is not well defined. Further emphasis on this point is the report by Johannsen and van der Walt (1978) of interfertility among *Kluyveromyces* species. These authors mated auxotrophic yeast strains, attained by *N*-methyl-*N'*-nitro-*N*-nitrosoguanidine mutagenesis, on media which supported growth of prototrophic recombinant clones. Utilizing this method they observed from various combinations prototrophic colonies, albeit at a low frequency (10^{-3} to 10^{-8}). From mixes of *K. thermotolerans* and *K. marxianus,* which have significantly different nuclear DNA base composition values, 46.2 mol% G + C and 41.2 mol% G + C, respectively, and essentially no DNA homology

(Presley & Phaff, unpublished data; Martini & Phaff, 1973), they also recovered recombinant clones. The F_1 generation from a cross of *K. marxianus* and *K. thermotolerans* appears to be most similar to *K. marxianus* on the basis of DNA base composition, i.e., F_1 40.6 mol% G + C; *K. marxianus* 41.2 mol% G + C; *K. thermotolerans* 46.2 mol% G + C (Martini et al., 1972; H. Presley, personal communication). These data would seem to indicate that the F_1 generation isolated from mixes of *K. marxianus* and *K. thermotolerans* by Johannsen and van der Walt (1978) represents an aneuploid yeast strain. Additionally, other laboratories have not been able to confirm the fertility of the F_1 generation isolated from this cross (S.R. Snow, personal communication) or in a similar cross between *K. fragilis* and *K. dobzhanskii* (Douglas et al., 1969). Apparently, recombination at the interspecific level can occur as reported. However, the event probably does not take place by a random meiotic process, but by a process resembling protoplast fusion with subsequent degeneration of some genetic material resulting in an aneuploid nucleus. This phenomenon has been noted by many investigators using various yeast protoplast fusion systems (Ferenczy, 1981; Gunge & Tamaru, 1978; Russel & Stewart, 1979). Johannsen and van der Walt did not specifically consider *K. marxianus* and *K. thermotolerans* to represent a single taxon but did suggest that interfertility be used to demarcate species. Generally, however, strains showing a very low degree of DNA homology have been considered to represent separate species. Reassociation values of 35 to 65%, which would indicate an intermediate divergence in genetic material but relatively close affinity, have not been reported, except in the genus *Hanseniaspora* (Meyer et al., 1978). This is not the case in prokaryotic DNA–DNA homology studies which exhibit a continuum of complementarity values (Brenner, 1973). The reasons for the gap in reassociation values in yeasts are not clear but may reflect the basic dynamics of yeast speciation, conditions under which reannealing experiments are done, or, perhaps, the selection of strains used in experiments.

Reports comparing yeasts by DNA–DNA homology appearing prior to 1978 have been reviewed in detail by Price et al. (1978) and will not again be reviewed here. In 1979, Fuson et al. provided evidence that the presence or absence of hyphal development is not necessarily a good trait for species differentiation. This conclusion was based on the observation that the type strain of *Hansenula canadensis* which is differentiated from *H. wingei* by its homothallism, ascospore number and lack of true hyphae has a higher degree of DNA homology (77.6%) with the type strain of *H. wingei* than do some strains identified as *H. wingei* (62.2%). More recently Fuson et al. (1980) examined the use of ascospore morphology as a trait for species differentiation. Their data show greater than 98% sequence complementarity between 4 strains of *Pichia ohmeri,* including an asexual form, 2 strains designated as having h^+ mating systems which produce either round or hat-shaped spores when mated to the same h^- strain, and the designated h^- strain itself. This result certainly was expected in view of the mating compatibility among the sexually active strains. These observations suggest that the morphology of ascospores may not always be a sound basis for species separation. Possibly a genetic analysis of resulting progeny could clarify the genetic basis of ascospore morphology among strains of *P. ohmeri.*

Holzschu et al. (1979a) used DNA–DNA complementarity to unequivocally identify strains of *Candida lusitaniae* (an opportunistic invader of humans) which showed variable physiological profiles, presumably due to exposure to antibiotics. This type of analysis may prove valuable in the future since yeasts which are exposed to antimycotic agents often display altered physiologies which make their identification difficult (Ahearn, 1978; Pappagianis et al., 1979).

Price and Phaff (1979) placed *Debaryomyces cantarellii* and *Deb. phaffii* into synonymy with *Pichia polymorpha* while retaining species status for *P. pseudopolymorpha* based on DNA complementarity values among these yeasts. The authors transferred *Pichia polymorpha* and *P. pseudopolymorpha* to the genus *Debaryomyces* as *Deb. polymorphus* and *Deb. pseudopolymorphus* on the basis of spore topography and presence of CoQ-9.

Phaff and Holzschu (1981) and Starmer et al. (1980) presented preliminary data showing strain relatedness within the species *Pichia heedii, P. amethionina,* and *P. opuntiae.* The results of these investigations indicated that strains of *P. heedii* are very closely related, the varieties of *P. amethionina* are more distantly related, and the varieties of *P. opuntiae* are the most differentiated among these species.

Recently, the DNA–DNA homology technique has been extended to taxonomic studies of basidiomycetous yeasts. Aulakh et al. (1981) demonstrated that *Filobasidiella neoformans* and *Fil. bacillispora* have reduced interspecific DNA–DNA sequence relatedness, 55 to 63%, in agreement with earlier data suggesting that these species represent closely related taxa. Holzschu et al. (1981) compared strains of *Sporidiobolus* species by DNA–DNA homology and proposed a new species, *Sporid. ruinenii.* This study also presented renaturation kinetics of nuclear DNA isolated from *Sporid. johnsonii* which indicated that basic differences in the organization of nuclear DNA may exist between ascomycetous and basidiomycetous yeasts, i.e., nuclear DNA from basidiomycetous yeasts contains a greater amount of repetitive DNA sequences which appear to be distributed among many families of repeated sequences.

Ribosomal RNA (rRNA)–DNA homology studies have thus far been made with species belonging to only a few genera, mainly *Saccharomyces* and *Kluyveromyces* (Bicknell & Douglas, 1970). The data indicate that intrageneric relationships obtained by this method are not meaningful because of the high degree of conservation of the DNA sequences coding for rRNA. However, such homology studies appear useful and very promising in revealing intergeneric distances and thus more distant relationships between yeast species.

b. Similarity of enzymes

Coenzyme Q analysis (Yamada & Kondo, 1972; Yamada et al., 1973a,b,c; Yamada & Kondo, 1973; Yamada et al., 1976a,b; Yamada et al., 1977; Yamada et al., 1980) has revealed that the coenzymes in yeast vary from Q-6 to Q-10. In many genera, all species have the same Q system, although a few exceptions occur in some genera, such as in *Pichia* and *Hansenula.* This may indicate that the atypical species have a close affinity to species in other genera. Studies on the coenzyme Q system of asporogenous yeasts are useful in correlating the taxonomic relationship of imperfect yeasts with certain groups of perfect genera.

Taxonomy by allozyme analysis is based on the premise that genetic diversification is a function of time. This means that organisms that are recently separated should show a high degree of similarity among macromolecules and those more distantly separated should exhibit an increasing number of differences in their DNA and proteins. Allozyme analysis as a technique is much more expressive of intraspecific variation than is DNA–DNA sequence complementarity (Holzschu, 1981). The proportion of point mutations that are electrophoretically detectable is estimated at approximately 0.27 owing to the redundancy in the genetic code and the large proportion of amino acids that are electrically neutral (Baptist et al., 1971; Selander, 1976). The nearly universal occurrence of extensive protein polymorphism in natural populations has led some workers to believe that variation would prove physiologically important and therefore to be under selective control; others regard it as without phenotypic effect and thus selectively neutral (Selander, 1976). Although this controversy still rages, data derived by this technique do provide insight into evolutionary processes and taxonomy.

Electrophoresis of proteins or enzymes using the zymogram method has been found very useful in examining variation and genetic diversity in many kinds of natural populations, but relatively few of such studies have been applied to yeasts. Shechter (1973) attempted to identify *Candida* species by general protein electrophorograms but found that results were highly dependent on culture condition, the age of the culture, and the morphological stage of the yeasts in question. Baptist and Kurtzman (1976) utilized comparative enzyme patterns to separate sexually active strains of *Cryptococcus laurentii* var. *laurentii* from nonreactive strains and from the varieties *magnus* and *flavescens*. Two new species, *Cr. magnus* and *Cr. heveanensis,* were proposed from the strains comprising the variety *magnus*. This study suffered from a paucity of enzymes studied making interpretation of the results quite subjective and sensitive to the effects of population variation. Holzschu (1981) studied the evolutionary relationships among some 400 strains of various cactophilic species of *Pichia*. He studied the banding patterns in starch gels of 14 metabolic enzymes and determined genetic distances among the various yeast populations. He discovered specific diagnostic alleles that could be used for identification of all *Pichia* species and their varieties included in the study.

c. *The cell wall*

1. ULTRASTRUCTURE OF THE CELL WALL

by N.J.W. Kreger-van Rij

In transmission electron microscopy (TEM) of sections of suitably fixed cells, the wall shows electron-light and electron-dense layers. Structural details of taxonomic importance may be observed in the lateral wall of vegetative cells, the hyphal septum and the ascospore wall. The taxonomic value attached to the ultrastructure of the septum and the ascospore wall is discussed under D.4. and D.5., that of the lateral wall follows here.

A distinct difference has been observed between the wall of ascomycetous and basidiomycetous yeasts (Kreger-van Rij & Veenhuis, 1971a): the wall of ascomycetous yeasts in sections of fixed material has a broad light inner layer and a thinner dark outer layer; basidiomycetous yeasts have a wall composed of a variable number of thin dark and light layers giving it a lamellar appearance. The imperfect basidiomycetous and ascomycetous yeasts showed the same difference in structure and, as a consequence, the ultrastructure of the wall may be used to recognize a non-sporulating yeast as ascomycetous or basidiomycetous.

Ascomycetous and basidiomycetous yeasts also differ in the wall structure of bud formation (see p. 62). Of the types of vegetative reproduction found among the ascomycetous yeasts, bipolar budding, a characteristic of the subfamily Nadsonioideae, needs special mention. In contrast to multilateral budding, where each bud is formed at a separate site, successive budding by bipolar species is found to occur at the same cell sites. As a result, the cell produces multiple scars, first described by Streiblová et al. (1964) from primuline stained preparations that were visualized by fluorescence microscopy. These multiple scars are a progression of concentric ridges of residual cell wall material and have been studied in TEM by Kreger-van Rij and Veenhuis (1971c). Multiple scars also occur in ascomycetous yeasts not belonging to the Nadsonioideae, e.g. in *Schizosaccharomyces pombe* (Streiblová & Wolf, 1972), in hyphal species such as *Arthroascus javanensis,* in some *Saccharomycopsis* species (Kreger-van Rij & Veenhuis, 1973a), and in *Ambrosiozyma cicatricosa* where the dactiloid cells produce a series of buds from their top (Scott & van der Walt, 1971b). With the exception of *Schiz. pombe,* the cells in all these species also produce buds from other sites than the poles.

In some basidiomycetous yeasts, buds are formed at the two poles or at one pole of the cells and more or less distinct scars occur. For instance, multiple scars are formed on cells of *Phaffia rhodozyma.* This type of bud formation, however, is not considered to have special taxonomic value.

2. CHEMICAL COMPOSITION OF THE CELL WALL

by H.J. Phaff

Among the newer approaches to taxonomy and phylogeny of yeasts belongs the composition of the polysaccharides that make up the cell wall and the capsule of capsulated yeasts. Such information is of value mainly at the generic level. Of the numerous species of yeast known only a small number has been investigated in depth with regard to structural details of the component polysaccharides. The cell wall composition of most species is usually inferred from information obtained with related species.

A historical review of the development of our knowledge of the yeast cell envelope was given by Phaff (1971, 1977). Most information based on chemical analyses has been derived from studies with cell walls from baker's yeast, *Saccharomyces cerevisiae,* and a few closely related species of *Saccharomyces.* It is generally assumed that other budding ascomycetous yeasts have a somewhat similar composition, although there is evidence from enzymatic studies that there are quantitative differences (Tanaka et al., 1966) and structural differences in the various polysaccharides.

Ascomycetous budding yeasts contain an alkali-insoluble, acid-insoluble β-(1→3)-glucan as the major structural component. Associated with this is an alkali-insoluble, acid-soluble β-(1→6)-glucan and an alkali-soluble β-(1→3)-glucan with a significant number of β-(1→6)-bonds in its branched structure. The remaining major component is a mannan-protein complex. Very significant differences have been demonstrated in the carbohydrate moiety, particularly in the side chains of the branched mannan molecule (Ballou, 1976).

Kocourek and Ballou (1969), on the basis of earlier chemical studies of yeast mannan by Ballou and his coworkers, adapted the method of acetolysis of yeast mannans for the purpose of 'fingerprinting' these polysaccharides. With controlled acetolysis, the α-(1→6) linkages of the mannan backbone are selectively cleaved, and the oligosaccharides thus formed (representing the side-chains including one mannose residue originally part of the backbone) can be separated according to size by gel filtration. The elution pattern thus obtained appears to be reproducibly characteristic of a particular species. The molar ratios of the fragments can be determined from the areas under the elution peaks. A comparison of the elution patterns of a significant number of species indicated that this approach will be useful in yeast systematics. Although this method provides information on the proportion and size of side-chains in various mannans, it does not reveal the types and distribution of the linkages between the mannose units in the side-chains.

Another approach to the fingerprinting of mannans was developed by Gorin and Spencer (see Gorin & Spencer, 1970, for a comprehensive review). They determined and made a comparison of the proton magnetic resonance spectra of the isolated mannans of a large number of yeast species. These spectra furnished very sensitive criteria for distinguishing between the mannans of different species or for showing their similarity. The H-1 protons give complex signals in the τ 4.0–5.0 region. As the H-1 and DOH signals are often very close or superimposed at room temperature, the measurements are made at 70°C since the temperature-dependent DOH signal is then shifted upfield. With this technique, a large number of water-soluble mannose-containing polysaccharides can be distinguished. The chemical shifts of the H-1 proton signals depend largely on: (i) the structure of the parent anhydrohexose unit; (ii) the position(s) substituted in the unit; (iii) the structure of the substituent unit; (iv) the structure of the aglycone unit if one is present; and (v) the position(s) substituted in such an aglycone unit. Gorin et al. (1969b) concluded that polysaccharides with similar proton magnetic resonance spectra have related chemical structures, and thus mannans from different species with similar spectra may indicate that these species have a close phylogenetic relationship.

This technique has been very useful in confirming or rejecting postulated imperfect and perfect forms of the same species of yeast (Manachini, 1979). The technique has also been helpful in forming tentative groups of seemingly related species in heterogeneous asporogenous genera such as *Candida* (Spencer & Gorin, 1969a). Proton magnetic resonance spectra have also been used to support or reject various kinds of evidence for proposed phylogenetic lines of *Hansenula* and *Pichia* species (Spencer & Gorin, 1969b); to obtain presumptive evidence for the occurrence of β-linkages in the side-chains of the mannans as their presence produces proton magnetic resonance spectra with distinctive H-1 signals at higher field than τ 4.55 (Gorin et al., 1969a); and to detect sugars other than mannose in the mannans (Gorin & Spencer, 1968).

Budding ascomycetous yeasts contain only a small amount of chitin (1–2% of the cell wall, dry weight basis), which occurs mainly in the budscars. The chitin content of the walls of filamentous ascomycetous yeasts appears to be significantly higher, e.g., in species of *Saccharomycopsis* and *Eremascus*. Also in *Nadsonia* the chitin content is higher and in species of these three genera chitin appears to be a structural component of the cell wall. The higher chitin content in *Nadsonia* than in the other bipolarly budding genera (*Hanseniaspora* and *Saccharomycodes*) suggests that *Nadsonia* is not closely related to the last two genera. The cell wall composition of the fission yeasts (*Schizosaccharomyces*) differs in several important respects from that of the budding ascomycetous yeasts. Species of this genus lack chitin and contain in addition to the β-glucans of the budding yeasts another major structural polysaccharide, i.e., α-(1→3)-glucan (Bacon et al., 1968; Meyer & Phaff, 1980). The mannan of *Schizosaccharomyces* species has single galactose units attached to the backbone rather than chains of mannose residues as in other ascomycetous yeasts (Ballou, 1976). The fission yeasts thus have a unique cell wall composition and they do not appear closely related to any other group of ascomycetous yeasts. Weijman (1977) studied the cell wall composition of *Dipodascus* and *Dipodascopsis* and found that the monotypic genus *Dipodascopsis* contains glucuronic acid in the cell wall, whereas this compound is lacking in *Dipodascus*. His data therefore gave further support for the separation of *Dipodascopsis* from *Dipodascus*.

The cell walls of yeasts related to basidiomycetous species generally contain β-glucan and chitin as major structural components (Bartnicki-Garcia, 1968). Species of *Cryptococcus* and *Phaffia rhodozyma* contain in addition major proportions of α-(1→3)-glucan (Bacon et al., 1968; Meyer & Phaff, 1977) but this polysaccharide appears lacking in *Rhodotorula* species (Meyer & Phaff, 1977). Von Arx and Weijman (1979) placed yeast species with heterobasidiomycetous relationships in two families depending on the presence or absence of D-xylose in their cell walls. The Sporobolomycetaceae (lacking xylose) included species of *Rhodotorula, Rhodosporidium, Sporobolomyces* and some *Candida* species. The Filobasidiaceae (containing xylose) included species of *Bullera, Cryptococcus, Filobasidium, Filobasidiella, Phaffia rhodozyma,* and some species of *Candida* and *Trichosporon.*

The composition of capsular polysaccharides has also been found useful as a systematic aid. For example, the slimy phosphomannans found in certain species of *Hansenula* and *Pichia* point to their common ancestry (Wickerham & Burton, 1962). The presence of linear mannans with alternating β-(1→3) and β-(1→4) bonds in *Rhodotorula* species, and an acidic heteropolysaccharide containing xylose, mannose and glucuronic acid in *Cryptococcus,* has supported the transfer of several species from *Rhodotorula* to *Cryptococcus* (Phaff & Spencer, 1969) and has established a relationship between *Cryptococcus* and the heterobasidiomycetous genus *Tremella* (Slodki et al., 1966). Slodki and Wickerham (1966) showed that *Lipomyces starkeyi* and *L. lipofer,* which were difficult to separate by classical criteria, could be separated by the presence of D-galactose in the capsule of *L. starkeyi* and its absence in *L. lipofer*. However, later studies by Gorin et al. (1978) have shown that this difference may not hold true in all cases due to the occurrence of mutants in which the capsule thickness is reduced (see Comments under the genus *Lipomyces*).

Immunological comparisons of cell wall antigens have been considered reliable criteria for specific and rapid characterization, but mainly of medically important yeasts (Tsuchiya et al., 1974; Fukazawa et al., 1980). However, Ballou (1974) working with mutant strains in studies on cell wall synthesis has found that one genetic change can greatly affect carbohydrate surface antigens and lead to improper identification.

Immunological studies of protein similarity are very sparse for yeasts and the resultant information is highly dependent on the kinds of protein studied. Lachance and Phaff (1979) used exo-β-glucanases from species of *Kluyveromyces* but found that the enzyme was poorly conserved and immunological distances between most species were too great to be reliable for determining evolutionary relationships in that genus.

4. *Classification of the ascosporogenous yeasts*

by N.J.W. Kreger-van Rij and C.P. Kurtzman

The Hemiascomycetes, which include the ascosporogenous yeasts, are differentiated from the other Ascomycotina by their lack of ascocarps and ascogenous hyphae (Ainsworth, 1973).

a. Families In the Endomycetales (Table 1), one of the orders of the Hemiascomycetes, two of the four families comprise yeast genera: the Spermophthoraceae and the Saccharomycetaceae. The Spermophthoraceae are distinguished from the other families by their needle-shaped ascospores. The family Saccharomycetaceae is distinguished from the Ascoideaceae and Endomycetaceae by the presence of single cells reproducing by budding and fission; species with the latter feature do not form gametangia. Cells forming buds also occur in the families of the Ascoideaceae and Endomycetaceae, but the buds occur only on the hyphae as blastospores (blastoconidia). There is probably a close relationship between the three families.

b. Subfamilies Subdivision of the Saccharomycetaceae (Table 2) into four subfamilies is based on features of sexual and vegetative reproduction.

The Lipomycetoideae are unique in their method of ascus formation. Asci form from special 'active' buds which may first conjugate with either the mother cell or another bud before ascosporogenesis. Since vegetative reproduction is by multilateral budding, it is the unusual method of ascus formation that serves to distinguish this subfamily.

The other three subfamilies are distinguished from each other by their method of vegetative reproduction: the Schizosaccharomycetoideae by fission, the Nadsonioideae by bipolar budding on a broad base, and the Saccharomycetoideae by multilateral budding. Some species of the Saccharomycetoideae may also form true mycelium. Vegetative reproduction in the Schizosaccharomycetoideae occurs through elongation at one end of a cell followed by cross wall formation, without constriction, between the original half and the new growth. This wall is split. Subsequent outgrowths may occur at the same site of the original cell leading to multiple scars on the cell wall. Bipolar budding in the Nadsonioideae resembles fission and may also produce multiple scars, but the outgrowth of the bud on the mother cell is on a somewhat narrower base than in the Schizosaccharomycetoideae. In multilateral budding, which is found in the Saccharomycetoideae, buds may be formed at various sites on the cells. In elongate cells, budding is often confined to the poles and shoulders of the cells, but multiple scars are not formed since each site gives rise to only one bud.

c. Genera The Spermophthoraceae include three yeast genera (Tables 2 and 6) which may be distinguished on the basis of slight differences in the morphology of their long, slender ascospores. The single species of the genus *Coccidiascus, C. legeri,* has been observed as a parasite in the intestines of *Drosophila* spp., but has not yet been obtained in pure culture. The genera *Metschnikowia* and *Nematospora* differ not only in the shape

Table 6

Characters of the genera of the ascosporogenous yeasts

	No. spec	TH	sept	Shape of ascospores	sp free	No. of spores	h/d	NO_3	F	Co-Q
Ambrosiozyma	4	+	dol	hat-shaped	v	1–4	h/d?	v	+	7^5
Arthroascus	1	+		hat-, Saturn-shaped, smooth or warty	+	1–4	h	–	–	8^5
Citeromyces	1	–		spherical, warty	–	1–2	h/d	+	+	8^3
Clavispora	1	–		clavate, warty	+	1–4	h	–	+	8^1
Coccidiascus	1	–		spindle-shaped	+	2–8	d?			
Cyniclomyces	1	–		oval, cylindrical	v	1–4	d		+	
Debaryomyces	9	–		spherical, oval, warty or with ridges	–	1–4	h	–	v	9^5
Dekkera	2	–		hat-shaped	+	1–4	d	v	+	9^7
Guilliermondella	1	+	pl	lunate	+	2–4	h	–	+	8^5
Hanseniaspora	6	–		hat-shaped, spherical with and without ledge, smooth or warty	v	1–4	d	–	+	6^4
Hansenula	30	v		hat-, Saturn-shaped, hemi-spherical	v	1–4	h/d	+	v	$7,8^3$
Issatchenkia	4	–		spherical, warty	–	1–4	d	–	+	7^3
Kluyveromyces	11	–		spherical, ellipsoidal, reniform, crescentiform	+	1-many	h/d	–	+	6^5
Lipomyces	5	–		oval, oblong, smooth, warty or with ridges	+(–)	1–30	h	–	–	9^6
Lodderomyces	1	–		oblong, fusiform	–	1–2	d	–	+	9^6
Metschnikowia	6	–		needle-shaped	v	1–2	h/d	–	v	9^6

Nadsonia	3	–		spherical, warty	–	1–2	h	–	v	6[4]
Nematospora	1	+		spindle-shaped	+	8	h?	–	+	6[6]
Pachysolen	1	–		hat-shaped	+	4	h/d	+	+	8[3]
Pachytichospora	1	–		ellipsoidal, thick-walled	–	1–2	d	–	+	6
Pichia	56	v		spherical, hemi-spherical, hat-, Saturn-	+	1–4	h/d	–	v	7,8[3],9[5]
Saccharomyces	7	–		spherical, short-oval, smooth or warty	–	1–4	d	–	+	6[5]
Saccharomycodes	1	–		spherical with narrow ledge	–	2–4	d	–	+	6[4]
Saccharomycopsis	7	+	pl(–)	spherical, hat-, Saturn-shaped, smooth or warty	v	1–4	h(d)	–	v	8[1],9[5]
Schizosaccharomyces	4	v		spherical, ellipsoidal, reniform, smooth or warty	+	2–8	h	–	+	9,10[2]
Schwanniomyces	1	–		spherical with equatorial ledge, warty	–	1–2	h	–	+	9[6]
Sporopachydermia	2	–		spherical, ellipsoidal, thick-walled	+	1–4	h/d	–	–	
Stephanoascus	1	+	pl	hemi-spherical, thick-walled	–	1–4	h	–	–	
Torulaspora	3	–		spherical, ellipsoidal, smooth or warty	–	1–4	h	–	+	6[5]
Wickerhamia	1	–		cap-shaped	+	1–2	d	–	+	9[4]
Wickerhamiella	1	–		oblong, rugose	+	1	h	+	–	9
Wingea	1	–		lentiform	–	1–4	h	–	+	9[3]
Zygosaccharomyces	8	–		spherical, ellipsoidal	–	1–4	h	–	+	6[5]

No. spec = number of species in the genus, TH = true mycelium, sept = ultrastructure of the septum, dol = dolipore, pl = plasmodesmata, sp free = spores liberated from the ascus, No. of spores = number of spores per ascus, h/d = vegetative cells haploid or diploid, NO_3 = assimilation of nitrate, F = fermentative ability, Co-Q = coenzyme Q from [1]Yamada & Kondo, 1972; [2]Yamada et al., 1973a; [3]Yamada et al., 1973c; [4]Yamada et al., 1976a; [5]Yamada et al., 1976b; [6]Yamada et al., 1977; [7]Yamada et al., 1980.

of the ascospores, but also in the number of spores per ascus and the ability to form true mycelium.

Of the subfamilies of the Saccharomycetaceae (Tables 2 and 6), the Lipomycetoideae and Schizosaccharomycetoideae each comprise a single genus, *Lipomyces* and *Schizosaccharomyces* respectivelly. In addition to special methods of ascus formation, the genus *Lipomyces* is typified by the shape, structure and color of the ascospores, and the number of spores per ascus.

The four genera of the Nadsonioideae are diverse, but some of them may be more closely related to the Saccharomycetoideae than assignment to another subfamily would suggest. The four genera are primarily distinguished by the shape and structure of the ascospores, but there are other differences, such as the method of diploidization. For instance, in *Saccharomycodes* the spores germinate and fuse while still within the ascus, and in *Nadsonia* ascus formation is more complicated and involves conjugation between mother cell and bud.

The subfamily of the Saccharomycetoideae is a large group with 24 genera, 12 of which are monotypic. The main characteristics for differentiating the genera are mentioned in Table 6. Those considered to be most important are: production of true mycelium and structure of the septum, shape and structure of the ascospores, liberation of the spores from the ascus, assimilation of nitrate, and coenzyme Q. Of less importance are: the number of spores per ascus, ploidy of the vegetative cells and fermentative ability. The genera mostly differ by a combination of characters but a few by only a single characteristic.

The characters of Table 6, which includes all ascosporogenous genera, are discussed below.

Production of abundant true mycelium is an important generic criterion that characterizes five yeast genera of the Saccharomycetoideae: *Ambrosiozyma, Arthroascus, Guilliermondella, Saccharomycopsis* and *Stephanoascus*. In two other genera, *Hansenula* and *Pichia,* some species produce a limited amount of true mycelium, with one exception: *Pichia burtonii* forms true hyphae abundantly. Kurtzman (see p. 315) retains this species in the genus *Pichia,* but von Arx and van der Walt (1976) have classified it in the monospecific genus *Hyphopichia*. Another feature of taxonomic importance is the *ultrastructure of the septum.* The different types allow a division of the hyphal species. For instance, the dolipore characterizes the genus *Ambrosiozyma;* plasmodesmata are found in *Guilliermondella, Stephanoascus* and nearly all species of *Saccharomycopsis,* but, as far as known, not in *Hansenula* and *Pichia,* including *P. burtonii.* Species in the latter two genera have a closure line, the remnants of plasmalemma where the centripetally formed septum was closed. The single species of the genus *Arthroascus, Ar. javanensis,* and one of *Saccharomycopsis, S. lipolytica,* recently described as *Yarrowia lipolytica* by van der Walt and von Arx (1980), also have a closure line although in *S. lipolytica* a narrow pore with endoplasmic reticulum through the pore has also been observed (Kreger-van Rij, 1980). Gow et al. (1980) observed an open connection with a diameter of 25 nm in hyphal septa of *Candida albicans* and indicated it as a micropore. In contrast with the typical ascomycete pore of higher fungi with a tapered edge, the micropore has blunt edges. Closure lines also occur in septa with plasmodesmata.

The *shape of the ascospores,* including ornamentation with ledges and warts, varies considerably among the yeasts but is fairly constant in one species. Scanning EM has made a closer view of the surface of the spore possible and the *internal structure* of the wall may be examined in sections with transmission EM. Development and germination of the spores may be included in the examination. In addition to the ornamentation of the spore wall, there are variations in the number and thickness of the layers observed in sections of material fixed with $KMnO_4$ or OsO_4. For instance, in some types of spores the layer, which becomes the wall of the vegetative cell after germination, is already present, but in others the wall is formed during germination. The ultrastructural features of the spore wall are still under investigation, and studies may produce several models for the spore wall that will be useful for taxonomic differentiation. Although knowledge of ultrastructure is still incomplete, shape and structure of the ascospores are generally considered among the most important characters for differentiating genera. This is evident from the number of genera, several monospecific, which are mainly or exclusively differentiated by spore morphology, e.g. *Guilliermondella, Citeromyces* and *Sporopachydermia.* Most genera have one type of ascospore, occasionally with small variations such as the size and site of the ledge in hat- and Saturn-shaped spores, both of which occur in the genera *Pichia, Hansenula* and *Saccharomycopsis.* In others, such as *Issatchenkia,* there are differences in the layering among the species, but Kurtzman (see p. 223) does not consider the genus phylogenetically heterogeneous because other characteristics seem to unite it. In contrast, the three types of spore wall layering found among the species of the genus *Zygosaccharomyces* induced Yarrow (see p. 394) to consider the proposed classification provisional. Although differences in shape and structure of the ascospores are used for the distinction of genera, similarity does not always indicate a close relationship, e.g. hat-shaped spores occur in several different genera. In these cases, other features are co-decisive for the distinction of the genus.

Liberation of the spores from the ascus at maturity is another generic character. The spores are set free by a partial or complete dissolution of the ascus wall. The alternative is germination of the spores within the ascus followed by penetration of the ascus wall. Spores may germinate and conjugate within the ascus as found in *Saccharomycodes ludwigii* and *Saccharomyces cerevisiae.* In heterothallic strains this reestablishes heterothallic diploids. Ascospore liberation is typical of all *Kluyveromyces* species and this is one of the features distinguishing *Kluyveromyces* from *Saccharomyces* and *Zygosaccharomyces.* Table 6 shows that nearly all hat- and Saturn-shaped spores are liberated. In a few species such as *Pichia carsonii* and *Cyniclomyces guttulatus* liberation of the ascospores varies among the strains.

The *assimilation of nitrate* has long been used for separating ascosporogenous genera. Four genera: *Citeromyces, Hansenula, Pachysolen* and *Wickerhamiella,* are nitrate-positive. However, variations of this character are accepted by some authors. For instance, in the genus *Dekkera* the ability to assimilate nitrate varies among the strains of certain species and in *Ambrosiozyma* only one of the four species is nitrate-positive. The latter is not accepted by all authors: von Arx (1972) classified the nitrate-positive species *Ambrosiozyma platypodis* in the separate genus *Hormoascus.* In a few nitrate-negative genera, such as *Debaryomyces* and *Pichia,* nitrite-positive species or strains occur. The difference between nitrate and nitrite assimilation is very small, namely one enzymatic step. For this reason, Kurtzman (see p. 376) proposed to disregard nitrate assimilation as a criterion for the separation of either species or genera. As a result, all species in *Hansenula* and some in *Pichia* will be reassigned.

Fermentative ability, particularly a vigorous fermentation of sugars, is a main characteristic of the genera *Saccharomyces, Zygosaccharomyces* and *Torulaspora.* A few ascosporogenous genera are non-fermentative, in others fermentation varies from absent through weak to strong. These transitions, e.g. in the genus *Hansenula,* demonstrate that fermentative ability is generally not decisive for generic differentiation.

The *coenzyme Q system,* indicated as the number of isoprene units per ubiquinone molecule, has been established for many yeast species by Yamada and coworkers (see p. 20). The coenzyme Q structure shows a very good correlation with existing genera, i.e. the species in many genera have the same number. Exceptions may be found among the species in *Hansenula* and *Pichia.* Some authors consider the Co-Q number as a valuable generic characteristic. For instance, the number is included in the diagnoses of the genera *Kluyveromyces* and *Issatchenkia.* The Co-Q number among the ascosporogenous genera ranges between 6 and 10; genera with the same number may differ markedly in other properties.

In addition to the features for generic differentiation mentioned above, there are a few others, used incidentally or as a supplementary character. Among these is the *method of ascus formation.* In hyphal yeast species the asci may be formed exclusively on the hyphae, as e.g. in *Saccharomycopsis* and *Stephanoascus* species, or the asci are loose cells, often resulting from the conjugation of two yeast cells. The latter is the case in *Pichia burtonii.* This character, along with the structure of the septum, seems of consequence in the distinction of the genera *Saccharomycopsis* and *Pichia.* Ascus formation in several *Saccharomycopsis* species and in *Guilliermondella* is special in that two cells fuse and are separated again by a wall. The two cells or cells formed on them as buds turn into asci. In some genera knowledge of the life cycle, especially concerning nuclear events during diploidization is incomplete or lacking. For example, ascus formation in some *Zygosaccharomyces* and *Torulaspora* species occurs without apparent conjugation, and in several diploid species it is not yet known whether they are homo- or heterothallic.

A second character that is used to define genera is *the shape of the ascus.* Two genera are noteworthy: *Pachysolen,* according to Kurtzman (see p. 290), has asci borne on an elongate, refractile, tube-like ascophore, while *Stephanoascus* produces asci with an apical cell.

The *number of spores per ascus* is generally one to four or exceptionally eight. In a few species, such as *Kluyveromyces polysporus* and *Lipomyces* spp. multi-spored asci occur with more than eight spores. Some species have only one spore per ascus, such as *Debaryomyces hansenii* and *Pachytichospora transvaalensis.* Although perhaps not an important character, the number of spores per ascus may be typical of a genus.

A fourth generic feature is the *ploidy of vegetative cells.* Although many genera are characterized as either haploid or diploid, others include both ploidy levels, occasionally in one species. Ploidy is generally a non-decisive character for generic differentiation, with the exception of the genera *Saccharomyces, Zygosaccharomyces* and *Torulaspora* where ploidy is used as one of the main features for their generic distinction.

The description of new genera is usually based on unique features felt to be of fundamental importance by their authors. However, the lack of consensus among taxonomists as to which characteristics represent fundamental differences is reflected in the generic descriptions in this book. Many features initially felt to be of consequence are later shown to be trivial.

The phylogenetic theories of Guilliermond (1909) and Wickerham (1951) have had great influence on yeast taxonomy. Guilliermond proposed that yeast species with true hyphae were more primitive than those without hyphae. In this respect *Saccharomycopsis* species are more primitive than *Saccharomyces* species. Wickerham considered haploid species more primitive than diploid ones. The change from haploid to diploid was associated with the ability to ferment an increased number of carbon compounds, decreased dependence on external vitamins, and formation of more hyphae. The latter seems to be in striking contrast with Guilliermond's ideas.

Wickerham stressed the importance of ascospore shape for generic differentiation and this principle is followed by many taxonomists. Ascospore shape is supposed to be a character that remained stable during evolutionary change of other properties and thus a binding element. The Co-Q number, in the opinion of several authors, may be even of more evolutionary significance for generic differentiation than ascospore shape. Other generic characters have in the course of time decreased in importance, such as ploidy of the vegetative cells, formation of pseudomycelium, pellicle formation and fermentative ability.

Since the 1970 edition of 'The Yeasts', the number of ascosporogenous genera has increased from 22 to 33, partly as genera separated from other ones, partly from new species. Undoubtedly, many new perfect states will be found in the large reservoir of imperfect species, either by crossing experiments or by the use of new conditions for sporulation.

d. Species Species are generally differentiated by a combination of physiological and morphological properties. Both types of characters are elaborated in the description of the species as the result of tests under standard conditions. Other conditions may give different results which could lead to erroneous identifications when compared with the standard descriptions.

Of the morphological properties, the shape of the ascospores is of primary importance and, occasionally, the number of spores per ascus. The size of the vegetative cells is very seldom useful for differentiation, but the presence of true hyphae is currently of consequence.

Among the physiological properties, the assimilation of carbon compounds is most important. The taxonomic value of fermentation reactions is generally considered to be lower than that of assimilation reactions. For example, some strains of a species may ferment a certain sugar weakly while other strains are unable to ferment this sugar. Of course, the assimilation of some compounds may also be variable.

The carbon compounds used for the fermentation and assimilation tests have been chosen for their reliability under standard conditions. However, Scheda and Yarrow (1966) have shown that mutants occur and may be easily selected when a strain is grown in shaken solution with a sugar that originally was not assimilated. This finding has led to the merging of many species in the genera *Saccharomyces, Zygosaccharomyces* and *Torulaspora* which were previously separated by the fermentation or assimilation of a single sugar. DNA complementarity studies have confirmed the wisdom of many of these decisions.

Some nitrogen compounds are useful for the distinction of species in certain genera:

ethylamine and cadaverine in *Saccharomyces, Zygosaccharomyces* and *Torulaspora,* and glucosamine in *Issatchenkia.*

Apart from the assimilation and fermentation reactions, other physiological properties in the standard description may be decisive for definition of a species. Vidal-Leiria et al. (1979) have shown that the maximum temperature of growth is a valuable species characteristic. Additionally, vitamin requirements may be useful for taxonomy. Generally, this is described as the ability to grow in a vitamin-free medium; occasionally, special vitamins are used as key characters, such as inositol in *Zygosaccharomyces.* A third type of criterion used for the differentiation of species in a few genera, such as *Kluyveromyces* and *Saccharomyces,* is the resistance to the antibiotic cycloheximide (actidione). Finally, osmotolerance, tested in two media, 50% glucose and 10% NaCl + 5% glucose, is typical for several species.

Discovery of life cycles and mating types has brought fundamental new elements to yeast taxonomy. In heterothallic species, conspecificity is indicated when the crossing of compatible strains results in fertile post-meiotic progeny.

In 1970, van der Walt (1970c,e) accepted interfertility among certain species in *Saccharomyces* and in *Kluyveromyces.* In 1978, Johannsen and van der Walt hybridized mutant strains of *Kluyveromyces* species. They obtained hybrids between *K. marxianus* and six other *Kluyveromyces* species and have, consequently, reduced the latter to varieties of *K. marxianus.*

Studies on DNA relatedness have made an extremely important contribution to the distinction of species since they have allowed comparisons among species that either have no known sexual stage or that fail to hydridize for other reasons such as homothallism or differences in ploidy. In some genera, DNA studies have corroborated existing species, such as in *Hanseniaspora* (Meyer et al., 1978), reduced the number of species as in *Schwanniomyces* (Price et al., 1978), or detected new species as in *Pichia* and *Issatchenkia* (Kurtzman et al., 1980a,b).

Both genetic and DNA relatedness studies have allowed some assessment of the importance of the various characters used in the standard description. These characters need not be the same for all genera, since species differentiation varies among the genera. In some genera, such as *Saccharomycopsis,* species are very dissimilar, while in others, such as *Pichia,* there are groups of similar species. Small genera are generally homogeneous in morphological properties.

5. *Classification of the basidiosporogenous yeasts*

by J.W. Fell and N.J.W. Kreger-van Rij

The basidiosporogenous genera are classified among the yeasts due the presence of a unicellular budding growth phase similar to that found in the ascomycetes. The basidiomycetous yeasts are distinguished by the presence of basidia with basidiospores, a criterion that traditionally separates the filamentous basidiomycetes from the ascomycetes which have an ascus within which ascospores are formed. The yeast phase of the basidiomycetes can be distinguished by several characteristics that are usually not found in the ascomycetes. These typical characters which are found in many, but not all basidiomycetes, are morphological, chemical and physiological.

The morphological characters include capsulated cells which result in mucous cultures, the presence of carotenoid pigments, ballistospores, distinct ultrastructures of cell walls including bud formation and hyphal septa (Kreger-van Rij & Veenhuis, 1971a), and nuclear behavior during mitosis (McCully & Robinow, 1972a,b). Chemical characters include the composition of the capsule and cell wall (Slodki et al., 1966; Bastide et al., 1979; von Arx & Weijman, 1979), a G + C mol% higher than 50 (Meyer & Phaff, 1970) and a color reaction with diazonium blue B (van der Walt & Hopsu-Havu, 1976). Physiological characters include the production of starch-like compounds and the presence of the enzymes urease, extracellular DNAse (Cazin et al., 1969) and tryptophan synthetase (Hütter & DeMoss, 1967). Details of most of these characteristics are presented in Chapter II.

The life cycles of many of the basidiomycetous yeasts are similar to the filamentous basidiomycetes such as rusts, smuts and mushrooms. What, if any, distinction exists between many of the yeasts and some of these basidiomycetes is not clear. The species presented in this third edition of 'The Yeasts' range from yeasts with a basidiomycetous life cycle to basidiomycetes with a yeast-like growth phase. Included in the latter group are the Tremellales with the genera *Tremella, Holtermannia, Sirobasidium* and *Fibulobasidium.* While this inclusion of filamentous fungi with yeast phases is far from complete, the intent is to foster an interest in bridging the gap of knowledge between the yeasts and other basidiomycetes. The reason for the gap is partially due to the different research approaches in the two disciplines. Yeast biologists tend to concentrate on biochemical and physiological responses of the organisms while basidiomycete research has evolved through a variety of other avenues such as morphology and phytopathology.

The relationship of the yeasts to the basidiomycetes was first suggested by Kluyver and van Niel (1925) when they noted the similarity of the sporobolomycete ballistospores to basidiospores. The basidiomycetous sexual life cycles were reported from two observations: (1) isolation of organisms in the mycelial dikaryotic state from which basidia and basidiospores were produced; (2) mating of yeast cells which resulted in the sexual state. The first reports of strains in the dikaryotic state included *Sporidiobolus johnsonii* Nyland (1949) and *Filobasidium floriforme* Olive (1968). Successful mating studies were first accomplished by Banno (1967) with strains of *Rhodotorula glutinis* resulting in the original description of *Rhodosporidium.* Subsequently *Leucosporidium* was described (Fell et al., 1969) using heterothallic strains of *Candida scottii.* This was followed by the discovery of the perfect states of *Cryptococcus,* including *Cr. neoformans,* and the establishment of the genera *Filobasidium* and *Filobasidiella* (Kwon-Chung, 1975, 1976, 1977).

The basidiomycetous yeasts include three groups: the teliospore-forming yeasts, and the yeasts of the Filobasidiaceae and the Tremellales. An introduction to each group is given by the authors: Fell (p. 491), Kwon-Chung (p. 468) and Bandoni (p. 541). The first mentioned group produce teliospores which are thick-walled cells where karyogamy takes place. Germination of the teliospores results in meiosis and the formation of metabasidia with basidiospores. The Filobasidiaceae lack teliospores, but have long thread-like basidia with basidiospores on the apex. The Tremellales also lack teliospores, they are distinguished by fruit-bodies with a hymenium in which the basidia are formed.

The three genera of the teliospore-forming yeasts, *Leucosporidium, Rhodosporidium* and *Sporidiobolus* (Table 7), are differentiated by two characters: the formation of ballistospores and the presence of carotenoid pigments. *Leucosporidium* species lack carotenoid pigments and ballistospores, are nitrate-positive and most species have a low (19°C) maximum temperature of growth. The species are physiologically differentiated by the assimilation of carbon compounds such as maltose, melibiose, cellobiose and rhamnose; some species have fermentative ability. *Rhodosporidium* species have carotenoid pigments and lack ballistospores. The species are differentiated by the methods of sexual reproduction, such as different types of homothallism and heterothallism, the shape of the teliospores, and by such physiological features as the assimilation of nitrate, melezitose, melibiose and inositol. *Sporidiobolus* species have carotenoid pigments and ballistospores. The species are differentiated by homo- or heterothallism and the assimilation of raffinose and nitrate.

Table 7

Characteristics of the genera of the basidiosporogenous yeasts

	No. of species	Basidiospores		I	ur	NO_3	F	St pr	Co-Q
Filobasidiaceae									
Chionosphaera	1	sessile in a whorl							
Filobasidiella	1	in four chains		+	+	−	−	+	10^2
Filobasidium	3	sessile in a whorl		+	+	v	v	+	9,$10^{1,2}$
		Ballistosp	carot						
Teliospore-forming yeasts									
Leucosporidium	6	−	−	v	+	+	v	v	8,9,10^1
Rhodosporidium	9	−	+	v	+	v	−	v	8,9,10^2
Sporidiobolus	4	+	+	−	+	v	−	−	10^2
Tremellales									
Fibulobasidium	1	+	−	+		−	−	+	
Holtermannia	1	−	−	+		−	−	+	
Sirobasidium	1	+	−	+		−	−	+	
Tremella	9	−	−	v	+	−	−	v	

I = assimilation of inositol, ur = urease, NO_3 = assimilation of nitrate, F = fermentative ability, St pr = starch formation, Co-Q = coenzyme Q from: [1]Yamada & Kondo, 1972; [2]Yamada & Kondo, 1973; Ballistosp = ballistospores, carot = visible carotenoid pigments.

Of the three genera of the Filobasidiaceae, *Chionosphaera* is distinguished by the arrangement of basidia in synnemata. The genera *Filobasidiella* and *Filobasidium* differ in the method of basidiospore formation; in *Filobasidiella* four chains of basidiospores are formed on the apex of the basidium, in *Filobasidium* the four to ten sessile basidiospores are arranged in a whorl on the apex of the basidium. *Filobasidiella* is monospecific with two varieties which differ serologically and may be recognized by a few physiological characters; the three *Filobasidium* species are differentiated by the shape of the basidiospores, fermentative ability and the assimilation of nitrate.

The four genera of the Tremellales have their imperfect states in three genera, although not all of these Tremellales may be identified with described species of imperfect yeasts. The two genera with ballistospores, *Sirobasidium* and *Fibulobasidium,* fit the diagnosis of *Bullera.* Those without ballistospores, *Tremella* and *Holtermannia,* can be identified as species of *Cryptococcus* and *Candida.* The inositol-positive strains are similar to species of *Cryptococcus* while the inositol-negative species resemble *Candida* species. The yeast state of the various species of Tremellales is described in this monograph and references are given to the perfect states.

6. *Classification of the imperfect yeasts*

by N.J.W. Kreger-van Rij

This group of the fungi imperfecti includes yeasts for which the perfect state has not been described. This state may include either asci or basidia. The yeasts may be heterothallic haploids for which no mating types have been found, or they may be haploid or diploid yeasts with unknown conditions for sporulation. Theoretically, in the future all imperfect species will disappear when the perfect states have been found; in practice, there will always be yeasts which fail to sporulate and for these, it is convenient to have a taxonomic system of imperfect species.

Several imperfect yeasts closely resemble the perfect species in morphological, physiological and biochemical properties and, for that reason, are considered to be the imperfect state of the perfect species (Table 8). Moreover, DNA-homology tests have quantitatively established the relationship between perfect and imperfect strains and thus corroborated a presumed identity. For most imperfect species the relationship to a perfect state is unknown; they may belong to one of the described perfect genera or to undescribed genera.

The imperfect yeasts include species classified in seven ascomycetous and seven basidiomycetous genera and three genera of mixed affinity (Table 9). Since it has become possible to determine the ascomycetous or basidiomycetous nature of a yeast with ultrastructural or chemical methods, some authors have proposed to retain ascomycetous yeasts in the genus *Candida* and basidiomycetous yeasts in *Trichosporon,* and to reclassify the remaining species. For instance, Moore (1980) combined the basidiomycetous *Candida* species in a new genus *Vanrija,* and von Arx (von Arx et al., 1977) and Weijman (1979) placed most of the ascomycetous *Trichosporon* species in the genus *Geotrichum.* A disadvantage of these reclassifications is that both *Vanrija* and *Geotrichum,* although purely basidiomycetous and ascomycetous, are in other respects heterogeneous: *Vanrija* may include imperfect states of teliospore-forming species and species of the Filobasidiaceae and perhaps still others. The authors of the mixed genera *Candida, Cryptococcus* and *Trichosporon* in this edition have refrained from reclassifications. They have, however, indicated which species are ascomycetous and basidiomycetous.

Classification of the imperfect yeasts is, in the absence of the taxonomically important characters of sexual reproduction, defective in that a genus may include the imperfect state of species of various genera. Differentiation of imperfect species and genera that are analogous to perfect species and genera is often impossible because suitable distinguishing characters are lacking. For instance, strains of *Rhodotorula glutinis* may be the imperfect state of three different *Rhodosporidium* species. Because of this problem, the maintenance of a practical rather than a natural generic classification of the imperfect yeasts is preferable, provided adequate keys to the species are made available.

In the present classification of imperfect yeasts several small groups are characterized by a single morphological property and by a combination of a few morphological and physiological features, and there is a large remainder included in the genus *Candida*.

The division into two families follows this principle. The first family, Sporobolomycetaceae, is distinguished by the formation of ballistospores. All other genera are included in a second family, Cryptococcaceae, although the heterogeneity of this large family makes this classification rather unrealistic.

The Sporobolomycetaceae include two genera: *Sporobolomyces* and *Bullera*, both basidiomycetous. They are distinguished by the color of the streak cultures: red to orange in *Sporobolomyces* and cream-colored to yellowish in *Bullera*. This distinction differs from that used in edition II of 'The Yeasts', where the shape of the ballistospores was decisive: 'symmetrical' in *Bullera* and 'asymmetrical' in *Sporobolomyces*. As a consequence, at that time *Sporobolomyces* included hyaline species, at present reclassified in *Bullera*. The ballistospores in *Bullera* are now described as 'usually rotationally symmetrical', those of *Sporobolomyces* as 'bilaterally symmetrical'. *Sporobolomyces* species may be the perfect state of *Sporidiobolus* species. The perfect state of two *Sporobolomyces* species, *Sp. salmonicolor* and *Sp. shibatanus*, was found after successful mating experiments. The type strain of *Aessosporon salmonicolor* has been shown by Fell and Tallman (1980b) to belong to *Sporidiobolus salmonicolor*.

The perfect states of the six *Bullera* species are unknown. However, the yeast states of *Fibulobasidium inconspicuum* and *Sirobasidium magnum*, both Sirobasidiaceae (Tremellales), fit the diagnosis of *Bullera* but they differ slightly in the standard descriptions from the accepted *Bullera* species. This possible relationship between *Bullera* and Tremellales stresses the distinction between *Bullera* and *Sporobolomyces*, at least some species of the latter being related to the teliospore-forming yeasts. *Bullera* species are differentiated by the assimilation of nitrate and several carbon compounds and by the temperature of growth.

The Cryptococcaceae include yeast species classified in 15 genera seven of which are monospecific. Most genera are distinguished by criteria of vegetative reproduction, a few by physiological features. The differentiation of the genera will be discussed in view of these criteria (see Table 9).

1. *Monopolar budding.* This feature is typical of *Malassezia* species. In addition to budding cells, hyphae which break up may form in one of the two species of the genus. Slooff (1970c) described the two species in the genus *Pityrosporum*, and considered *Pityrosporum* as a form genus of *Malassezia*. *Malassezia* is considered to be basidiomycetous (Moore, 1980). The species have a high mol% G + C (see Yarrow & Ahearn in this edition) and a DBB reaction that is positive for one species (Hagler & Ahearn, 1981). However, bud formation which is phialidic (Kreger-van Rij & Veenhuis, 1970) is unusual for basidiomycetous yeasts.

2. *Bipolar budding.* Two ascomycetous genera are characterized by this feature, one *fermenting: Kloeckera*, and one *non-fermenting: Schizoblastosporion*. The six *Kloeckera* species

have their perfect state in six *Hanseniaspora* species which was shown by DNA homology tests (Meyer et al., 1978). As far as known, *Kloeckera* species are homothallic and suitable conditions for sporulation must be found in order to reclassify them as *Hanseniaspora* species. The perfect state of the single *Schizoblastosporion* species is unknown.

3. *Formation of conidia on sympodia* is typical of the genus *Sympodiomyces,* with one species. The genus is ascomycetous and the septa have a closure line. No relationship to a perfect species is known.

4. *Formation of buds on sterigmata* characterizes the genus *Sterigmatomyces;* the buds are not ejected like ballistospores. The six species are basidiomycetous, one of them with red cultures, the others hyaline. The species vary in the assimilation of nitrate, even among the strains of one species, of inositol, and in starch production. The perfect state of a new *Sterigmatomyces* species was recently described by Kraepelin & Schulze (1982) as *Sterigmatosporidium polymorphum.*

Table 8

Perfect species and their imperfect states

Perfect	Imperfect
Citeromyces matritensis	*Candida globosa*
Clavispora lusitaniae	*Candida lusitaniae*
Debaryomyces hansenii	*Candida famata*
Dekkera bruxellensis	*Brettanomyces bruxellensis*
Dekkera intermedia	*Brettanomyces intermedius*
Hanseniaspora guilliermondii	*Kloeckera apis*
Hanseniaspora occidentalis	*Kloeckera javanica*
Hanseniaspora osmophila	*Kloeckera corticis*
Hanseniaspora uvarum	*Kloeckera apiculata*
Hanseniaspora valbyensis	*Kloeckera japonica*
Hanseniaspora vineae	*Kloeckera africana*
Hansenula americana	*Candida bimundalis* var. *americana*
Hansenula anomala	*Candida pelliculosa*
Hansenula bimundalis	*Candida bimundalis* var. *bimundalis*
Hansenula canadensis	*Candida melinii*
Hansenula capsulata	*Candida molischiana*
Hansenula fabianii	*Candida fabianii*
Hansenula holstii	*Candida silvicola*
Hansenula jadinii	*Candida utilis*
Hansenula sydowiorum	*Candida nitrativorans*
Issatchenkia occidentalis	*Candida sorbosa*
Issatchenkia orientalis	*Candida krusei*
Kluyveromyces marxianus var. *marxianus*	*Candida kefyr*
Kluyveromyces marxianus var. *lactis*	*Candida sphaerica*
Kluyveromyces thermotolerans	*Candida dattila*
Metschnikowia pulcherrima	*Candida pulcherrima*
Metschnikowia reukaufii	*Candida reukaufii*
Pichia burtonii	*Candida variabilis, Trichosporon variabile*
Pichia fermentans	*Candida lambica*
Pichia guilliermondii	*Candida guilliermondii*

Perfect	Imperfect
Pichia humboldtii	*Candida ingens*
Pichia membranaefaciens	*Candida valida*
Pichia norvegensis	*Candida norvegensis*
Saccharomyces exiguus	*Candida holmii*
Saccharomyces telluris	*Candida pintolopesii*
Saccharomycopsis (Yarrowia) lipolytica	*Candida lipolytica*
Stephanoascus ciferrii	*Candida ciferrii*
Torulaspora delbrueckii	*Candida colliculosa*
Wickerhamiella domercqii	*Candida domercqii*
Filobasidiella neoformans var. *neoformans*	*Cryptococcus neoformans* var. *neoformans*
Filobasidiella neoformans var. *bacillispora*	*Cryptococcus neoformans* var. *gattii*
Filobasidium capsuligenum	*Candida japonica*
Filobasidium floriforme	*Cryptococcus albidus* var. *albidus*
Filobasidium uniguttulatum	*Cryptococcus uniguttulatus*
Leucosporidium frigidum	*Candida frigida*
Leucosporidium gelidum	*Candida gelida*
Leucosporidium nivalis	*Candida nivalis*
Leucosporidium scottii	*Candida scottii*
Rhodosporidium capitatum	*Cryptococcus infirmo-miniatus*
Rhodosporidium dacryoidum	*Rhodotorula minuta*
Rhodosporidium diobovatum	*Rhodotorula glutinis*
Rhodosporidium infirmominiatum	*Cryptococcus infirmo-miniatus*
Rhodosporidium malvinellum	*Rhodotorula graminis*
Rhodosporidium paludigenum	*Rhodotorula graminis*
Rhodosporidium sphaerocarpum	*Rhodotorula glutinis*
Rhodosporidium toruloides	*Rhodotorula glutinis*
Sporidiobolus pararoseus	*Sporobolomyces shibatanus*
Sporidiobolus salmonicolor	*Sporobolomyces salmonicolor*

In several cases, the connection between perfect and imperfect species is uncertain (see systematic discussion of the relevant species).

The production of triangular cells is a criterion of the ascomycetous genus *Trigonopsis* with a single species.

6. *Buds formed multilaterally on a broad base and true mycelium without arthrospores* are present in the ascomycetous genus *Oosporidium* with one species. The cultures have a light orange color which is not due to carotenoid pigments. The hyphal septa have an ascomycete pore. No relationship with a perfect state is known.

7. *Formation of true mycelium with arthrospores in addition to single budding cells or pseudomycelium* is characteristic of the genus *Trichosporon* which includes six basidiomycetous and nine ascomycetous species. The basidiomycetous species are rather similar and, as mentioned before, some authors restrict the genus to these species. No perfect state is known. Kreger-van Rij & Veenhuis (1971b) observed a dolipore in the septa of *Tr. cutaneum* which probably indicates that this species does not belong to the teliospore forming yeasts which have a simple pore.

The ascomycetous species are a heterogeneous group with the perfect states belonging to various genera, e.g. *Pichia* and those *Endomyces* species with septal plasmodesmata in a ring (Kreger-van Rij & Veenhuis, 1972).

7a. One of the *Trichosporon* species, *Tr. inkin,* forms cross walls in different planes of the cell which gives *sarcina-like agglomerates of arthrospores.* Smith, in this edition, following King and Jong (1975) considers this species to belong to a separate genus *Sarcinosporon* and indicates the agglomerates as sporangia. This concept is not generally accepted. Therefore, *Tr. (Sarcinosporon) inkin* is also mentioned among the *Trichosporon* species and included in the *Trichosporon* key.

7b. Another *Trichosporon* species, *Tr. aculeatum,* is accepted by Smith in this edition as the single species of the new genus *Aciculoconidium* by King and Jong (1976a). These authors did not consider the disarticulation of the mycelium in *Tr. aculeatum* as the formation of arthrospores and, therefore, removed the species from the genus *Trichosporon.* Since this view is questionable, *Tr. aculeatum* is also mentioned among the species in the genus *Trichosporon* and in the *Trichosporon* key. The species has typical *needle-shaped blastospores. Tr. aculeatum* is ascomycetous and the structure of the septa with a closure line resembles that of *Tr. melibiosaceum* and *Tr. fennicum.*

8. A combination of *red or yellow cultures due to the presence of carotenoid pigments,* the *inability to assimilate inositol* and the *absence of fermentation* is typical of the basidiomycetous genus *Rhodotorula.* The cells are usually capsulated and a few strains may produce pseudo- or true mycelium. For some of the species the perfect state in the genus *Rhodosporidium* is known. For a discussion of the differentiation of the eight species see Comments on the genus *Rhodotorula.*

9. The recently described genus *Phaffia* is *similar to Rhodotorula* but with *fermentative ability.* In the description of the single species, monopolar budding and the occurrence of chlamydospores is mentioned.

10. A genus *similar to Rhodotorula* is the genus *Cryptococcus.* The distinctive difference between the two is the *assimilation of inositol,* which is *positive* in *Cryptococcus.* Most *Cryptococcus* species have capsulated cells and they are non-fermentative. The genus is basidiomycetous with the exception of a single species, *Cr. melibiosum.* The perfect states of *Cr. lactativorus* and *Cr. cereanus* have been found and they are now classified in the ascomycetous genus *Sporopachydermia. Cr. melibiosum* may follow in due time.
At present, this species has been reclassified in the genus *Myxozyma* as *M. melibiosi* by van der Walt et al. (1981), and in the genus *Candida* as *C. melibiosophila* by Golubev et al. (1981b). Golubev et al. (1974) considered the assimilation of glucuronic acid a better character for the distinction of the genus *Cryptococcus* than the assimilation of inositol. All basidiomycetous *Cryptococcus* species are glucuronic acid positive, and the ascomycetous species are negative. By this test the authors excluded the latter from the genus *Cryptococcus.* Rodrigues de Miranda, in this monograph, pointed out that several *Rhodotorula* species are also glucuronic acid positive, which would make a reclassification of these species necessary.

The criterion of inositol assimilation brought three species with red cultures into the genus, although the other species have hyaline or slightly yellow cultures. For one of the red cultures, *Cr. infirmo-miniatus,* the perfect state is a *Rhodosporidium* species. Two hyaline *Cryptococcus* species, *Cr. uniguttulatus* and *Cr. neoformans,* were found to have their perfect state in the genera *Filobasidium* and *Filobasidiella* respectively. Possibly, other *Cryptococcus* species are related to *Tremella* and *Holtermannia* species. *Cryptococcus* presently consists of a heterogeneous group of species.

In the 1970 edition of 'The Yeasts', the genera *Cryptococcus* and *Torulopsis* were distinguished from the genus *Candida* by the absence of pseudo- and true mycelium, and *Cryptococcus* differed from *Torulopsis* in the assimilation of inositol, positive in the former. At present, *Torulopsis* has been merged in *Candida* and the distinction between *Cryptococcus* and *Candida* is more complicated. The distinction is formulated as follows:

Table 9

Characteristics of the genera of the imperfect yeasts

	No. of species		carot	TH	arthr	PH	ballistosp	I	ur	NO_3	F	St pr	Co-Q
Aciculoconidium	1	asc	−	+	+		−	−	−	−	+	−	
Brettanomyces	9	asc	−	v	−	v	−	−		v	+		9[4]
Bullera	6	bas	−	v	−	v	+	v		v	−	v	
Candida	196	asc/bas	−	v	−	v	−	v	v	v	v	v	6,7,8,9,10[1]
Cryptococcus	19	bas/asc	v	−	−	−	−	+		v	−	v	8,9,10[2]
Kloeckera	6	asc	−	−	−	v	−	−		−	+	−	6[3]
Malassezia	2	bas	−	v			−				−		
Oosporidium	1	asc	−	+	−	−	−	−	−	+	−	+	
Phaffia	1	bas	+	−	−	v	−	−	+	−	+	+	
Rhodotorula	8	bas	+	v	−	v	−	−	+	v	−	−	9,10[2]
Sarcinosporon	1	bas	−	+	+	+	−	+		−	−		
Schizoblastosporion	1	asc	−	−	−	v	−	−		−	−		
Sporobolomyces	7	bas	+	v	−	v	+	v	+	v	−	v	10[2]
Sterigmatomyces	6	bas	v	v	−	−	−	v	+	v	−	v	
Sympodiomyces	1	asc	−	+	−	−	−	+	−	−	−	−	
Trichosporon	15	bas/asc	−	+	+	+	−	v	v	v	v	v	
Trigonopsis	1	asc	−	−	−	−	−	−		−	−		

carot = visible carotenoid pigments, TH = true mycelium, arthr = arthrospores, PH = pseudomycelium, I = assimilation of inositol, ur = urease, NO_3 = assimilation of nitrate, F = fermentative ability, St pr = production of starch-like compounds, Co-Q = coenzyme Q number from: [1] Yamada & Kondo, 1972; [2] Yamada & Kondo, 1973; [3] Yamada et al., 1976a; [4] Yamada et al., 1980; asc = ascomycetous, bas = basidiomycetous.

species which combine the characters assimilation of inositol and absence of pseudo- and true mycelium are *not* included in *Candida* but in *Cryptococcus*.

11. *Brettanomyces* is chiefly characterized by the physiological property *vigorous production of acetic acid* from glucose. On malt agar growth is slow and the cultures are short-lived. An additional morphological feature is the frequent occurrence of *ogival cells*. The genus is ascomycetous and two of the nine species have their perfect state in the genus *Dekkera* which is also characterized by acid formation in addition to the formation of hat-shaped ascospores. Yamada et al. (1980) found that all *Brettanomyces* species have a Co-Q-9 system. The differentiation of the *Brettanomyces* species is by fermentation and assimilation reactions and by the formation of non-septate mycelium. The assimilation of nitrate varies within some of the species.

12. All imperfect species which do not fit in the 14 previously mentioned genera are classified in the genus *Candida*. This genus includes ascomycetous and basidiomycetous species. Table 8 shows that *Candida* comprises the imperfect state of at least 14 ascomycetous and two basidiomycetous genera, the latter, *Filobasidium* and *Leucosporidium* respectively from the Filobasidiaceae and the teliospore forming yeasts. In edition II of 'The Yeasts', *Candida* included only yeasts which produced pseudomycelium; in the present edition yeasts without pseudomycelium of the former genus *Torulopsis* are also classified in *Candida*.

Many ascosporogenous genera are distinguished by characteristics of ascospore formation. In the absence of these criteria, a distinction of corresponding imperfect ascomycetous genera is very difficult or impossible. Although several characters such as Co-Q number, production of hyphae and septal ultrastructure, fermentative ability and nitrate assimilation may indicate in which perfect genus a certain *Candida* species should be classified if spores were formed, these features may not be sufficient for generic differentiation and it seems probable that the genus *Candida* will remain heterogeneous with regard to ascomycetous species.

The 27 basidiomycetous *Candida* species accepted in this monograph may be the imperfect states of two groups of basidiosporogenous yeasts, those forming teliospores and those not forming teliospores. Although data are yet scarce, ultrastructural and chemical investigations suggest that two types of characters correlate with this distinction: (1) the ultrastructure of the hyphal septum, namely a simple pore in the teliospore forming yeasts and a dolipore in the others, (2) the chemical composition of cells with xylose absent in the teliospore formers and present in the others. Based on the chemical data, von Arx and Weijman (1979) distinguished two basidiomycetous families: Sporobolomycetaceae (xylose-absent) and Filobasidiaceae (xylose-present), both including perfect and imperfect species. These authors reclassified *Candida* species of the first group in the genus *Rhodotorula* (*Rh. ingeniosa, Rh. muscorum, Rh. bogoriensis, Rh. diffluens* and *Rh. javanica*) and of the second group in the genus *Apiotrichum* (*A. humicola* and *A. curvatum*).

Golubev and Blagodatskaya (1978) studied composition of extracellular compounds in glucuronic acid positive *Candida* species. The ascomycetous species could be distinguished from the others by a negative urease test and the absence of xylose. Golubev (1981b) considered the use of formation of hyphae for distinguishing between *Candida* and *Cryptococcus* incorrect. Therefore, he (1981b) reclassified a number of basidiomycetous *Candida* species in the genus *Cryptococcus,* namely *Cr. amylolentus, Cr. curvatus, Cr. humicolus, Cr. marinus* and *Cr. podzolicus*. These species produce pseudomycelium and some of them true mycelium. According to Golubev and Blagodatskaya (1978), these species are distinguished from the other basidiomycetous *Candida* species by the composition of the extracellular polysaccharides and by the absence of fermentative ability. Barnett et al. (1983) followed Golubev's reclassifications.

From the above it appears that there is no consensus of opinion on the reclassification of basidiomycetous *Candida* species. With the exception of Moore (1980), all authors leave in *Candida* both ascomycetous and basidiomycetous species. However, the new delimitations of the genera *Candida, Cryptococcus* and *Rhodotorula* are not yet clearly defined.

For the identification of *Candida* species many features must be used and several keys are possible. The authors of *Candida* in this edition have split the genus into eight groups. One group includes the basidiomycetous species, distinguished by a positive DBB reaction. Characteristics for distinguishing the other groups are: assimilation of nitrate, inositol, raffinose, sucrose, maltose, mannitol and erythritol. Key characters for the species of each group vary. Most of them are physiological: fermentation and assimilation reactions, temperature of growth, vitamin requirements and starch production. The total number of carbon compounds tested for assimilation is 27. In four of the eight groups the formation of pseudomycelium is used as a key character.

Included in the *Candida* key are several perfect species which may be isolated in the haploid state.

E. Nomenclature and type strain

by N.J.W. Kreger-van Rij

Nomenclature and typification of yeasts are settled by the International Code of Botanical Nomenclature in Rules and Recommendations (Stafleu et al., 1978).

Some articles of primary importance are:

Art. 36. In order to be validly published, a name of a new taxon (..) published on or after 1 Jan. 1935 must be accompanied by a Latin description or diagnosis (...).

Art. 37. Publications on or after 1 Jan. 1958 of the name of a new taxon of the rank of family or below is valid only when the nomenclatural type is indicated.

Art. 33, paragraph 2. A new combination or a new name for a previously recognized taxon published on or after 1 Jan. 1953 is not validly published unless its basionym (name-bringing or epithet-bringing synonym) or the replaced synonym (when a new name is proposed) is clearly indicated and a full and direct reference is given to its author and original publication with page or plate reference and date.

Art. 9 states that "type specimens cannot be living cultures", and that "the type of a name of a species is a single specimen".

Two serious objections have been raised against art. 9 (Lodder, 1970). First, for yeasts preservation of the type as a living culture is required since yeast species are partly characterized by physiological and biochemical properties. In this respect, art. 18a of the International Code of Nomenclature of Bacteria (Lapage et al., 1975) would be suitable for yeasts. It states: "A type strain is made up of living cultures of an organism which are descended from the strain designated as the nomenclatural type. The strain should agree closely in its characters with those of the original description".

Yeasts are not the only organisms which require living type material and of which the nomenclature is ruled by the Botanical Code. Lewin and Fogg (1978) made the following proposal (9) for microscopic algae and fungi: Recommendation 7.C. "For microorganisms, type material should be of a pure clone, preserved so as to conserve as many taxonomically important features as possible. In cases where taxonomically important microscopic or physiological features cannot be retained in dead and preserved specimens, every effort should be made to isolate, designate and conserve in an established culture collection viable material of a single pure type culture, if technically possible in a state (e.g., freeze-dried or under liquid nitrogen) which can be expected to minimize mutation or other changes with time". This Recommendation is suitable for yeasts and yeast-like organisms, although a more binding instruction in a Rule would be preferable.

Lack of living type material makes it difficult or impossible to be certain about the identity of older described species. In some cases this may induce authors to use this doubt as a motive for changing well-established names. Art. 14 of the Code makes it possible to conserve names to "serve stability in nomenclature". A proposal for conservation must be submitted to the General Committee (see Botanical Code) for a decision by an International Botanical Congress.

The second objection to art. 9 of the Botanical Code concerns the designation of a single type specimen in the case of haploid heterothallic yeasts. Lodder (1970) supported Wickerham's suggestion (1965c) to indicate two strains of opposite sex as type specimens. Many yeast taxonomists follow this suggestion. Fell (personal communication) has proposed to designate the two types of a heterothallic species as type and allotype, the latter word derived from zoological nomenclature (Mayr, 1969). The allotype is a paratype of the opposite sex of the holotype.

F. The future of yeast taxonomy

by N.J.W. Kreger-van Rij

In this monograph various new developments in yeast taxonomy are mentioned and discussed. Starting from there, the following speculations and desirabilities for future development may be considered.

In the sporogenous yeasts, DNA homology together with crossing experiments may solve many of the species problems, although considerable work will be required. The imperfect yeasts provide a rich source for the discovery of new sexual species. In this respect, species delimitation of the imperfects guided by G + C content is useful.

Generic differentiation is still problematic, although a number of criteria are considered promising for a less subjective distinction of the genera. For example, the use of ribosomal RNA–DNA homology, coenzyme Q numbers, and complex features such as the structure and chemical composition of the cell wall. Practical questions in this respect are: where to classify a new species, in an existing genus or a new one; what are the links between the many monospecific genera and is it possible to combine some of them; and what should be done with an apparently unrelated species provisionally classified in a special genus?

The authors of the basidiosporogenous yeasts in this monograph have pointed out that several life cycles still require explanation. In the relatively new field of these yeasts many new developments are to be expected, such as the elucidation of connections between 'yeasts' and other basidiomycetes. Insight into the relationship between ascosporogenous yeasts and the other Hemiascomycetes is still a desideratum. Comparative studies of ultrastructure and sexual reproduction have already provided criteria for distinction and similarities. Supplementary studies of the other hemiascomycetous species with the methods used in yeast taxonomy may be useful.

Since the appearance of the first Delft monographs, the number of characters of the standard description has increased, especially the number of carbon compounds of the assimilation tests. More compounds give more additional means for differentiation; also correlation of features may have higher value for the distinction of species than the assimilation of a single compound. Moreover, many of these features may serve as key characters. On the other hand, some of the characters are useful in very restricted groups; in others they may have no differentiating value, because they are either positive or negative in all species, or vary within one species. Therefore, it may be desirable to restrict the basic standard description to a number of generally applicable criteria and use extra features for special groups. Within a few species, strains occur which differ notably in their G + C content and therefore belong to different species. However, they cannot be distinguished by the characteristics of the standard description and for these species new characters will have to be found.

Where possible, simplification of methods seems to be a desirable goal. For instance, it may be feasible to shorten the duration of a number of tests without losing important means for differentiation.

Finally, a critical aspect of taxonomy is the identification of yeasts through the use of keys. As a result of the increased number of species, the keys have become more elaborate. In the foregoing editions of 'The Yeasts', the procedure for the identification was to first find the genus via the key to the genera and then locate the species via keys in each genus. The key to the genera might offer difficulties since some of the characters used, e.g. those of sexual reproduction, are not always easily established. Therefore, other keys have been prepared, e.g. by Barnett and coworkers (Barnett & Pankhurst, 1974; Barnett et al., 1979), directly to the species and restricted to the use of physiological properties. Keys to a limited number of species have also been made, for instance to medical and industrial yeasts. For these selected species, a reduced number of key characters is required. The features in some keys have been established with methods adapted in order to obtain rapid results. These modified characters may differ from those of the standard description.

Apparently, there is a need for various types of keys and, although there is a tendency for yeast taxonomists to specialize in certain groups of yeasts, the authors familiar with species and criteria may be required to give more time to the production of these keys.

Chapter II

Methods for the isolation, maintenance, classification and identification of yeasts

by J.P. van der Walt and D. Yarrow

I. The isolation and maintenance of yeast cultures

Yeasts occur widely in nature and have been recovered from widely differing terrestrial as well as marine sources. Certain yeasts are more or less ubiquitous while others appear to be restricted to very specific habitats. Yeasts seldom occur in the absence of either molds or bacteria. For their recovery, recourse is usually had to enrichment techniques using media which, while permitting the development of yeasts, suppress mold and bacterial growth. The composition of such enrichment media is determined by the fact that yeasts are, as a rule, capable of development at hydrogen-ion concentrations and water activities which either suppress or inhibit bacterial growth. Antibiotics may also be used to suppress bacteria, provided that these compounds are inactive against yeasts. Fungistatic agents, for suppression of molds, should be used with caution, since such compounds may also inhibit some yeasts.

As the majority of yeasts are mesophilic, elective or enrichment cultures are usually incubated at 20–28°C. For the isolation of psychrophilic taxa, lower temperatures (4–15°C) are a prerequisite. In the case of yeasts which are strictly associated with warm-blooded sources, higher isolation temperatures (30–37°C) are required.

Certain yeasts, e.g. species of the genera *Cyniclomyces* and *Malassezia,* have exceptional nutritional requirements. For details of isolating and maintaining these species, the reader should consult the relevant chapters dealing with these taxa.

Anyone interested in the recovery of yeasts from natural substrates is also referred to the informative publications by Beech and Davenport (1969, 1971), and Davenport (1980) dealing with isolation of non-pathogenic yeasts. For the isolation of pathogenic yeasts reference is made to the publications by Buckley et al. (1969) and Buckley (1971).

A. Isolation based on the use of acidic media (pH 3.5–5.0)

For acidifying media, preference is given to either hydrochloric or phosphoric acid.

The use of organic acids such as acetic acid is not recommended for general isolation purposes. At pH 3.5–4.0 such acids are only slightly dissociated and the high concentrations of undissociated acids have an inhibitory effect on most yeasts. Notable exceptions are *Zygosaccharomyces bisporus* and *Zygosacch. bailii.*

1. Solid media for direct isolation

When yeasts are present in high numbers, they may be isolated by direct plating of the material, or suspensions of the material, usually onto either acidified YM agar or malt extract agar. As agar itself undergoes rapid hydrolysis when autoclaved at pH 3.5–4.0, a predetermined volume of 1 N hydrochloric acid is added to the sterilized, molten agar, only after it has been cooled to 45°C. As a rule, the addition of circa 0.7% (v/v) 1 N hydrochloric acid to YM agar gives the desired pH 3.7–3.8. After rapid mixing, plates are immediately poured and allowed to set.

For quantitative studies, use may be made of the dilution-pour plate technique.

2. Liquid media for enrichment purposes

When yeasts are present in rather low numbers, their isolation from natural substrates may require prior enrichment. In such cases the material is introduced into either YM broth or malt extract which have been adjusted to pH 3.7–3.8 by the addition of either hydrochloric or phosphoric acid.

The development of molds may be restricted by the exclusion of air which may be accomplished by carefully pouring a layer of sterile pharmaceutical paraffin about 1 cm deep over the surface of the media and incubating until adequate yeast growth has occurred. This procedure favors the development of fermentative species.

Alternatively, the inoculated flasks may be incubated on a rotary shaker for one or two days (Wickerham, 1951). The molds which are prevented from sporulation, aggregate in pellets and are rapidly outnumbered by the multiplying yeast cells. Separation of the yeast cells is effected either by direct plating or by first removing the fungal pellets by means of aseptic filtration through a loose plug of sterile, absorbent cotton or glass wool. If bacterial contamination persists in the enrichment as a result of an upward drift of the pH during incubation, plating on acidified YM agar or malt extract is recommended. The use of the shake-culture technique leads to the recovery of both fermentative as well as non-fermentative yeasts.

Whereas many yeast species may be recovered at pH 3.7–3.8, some species, notably those of the genus *Schizosaccharomyces,* are nonetheless inhibited by very acidic media and are best isolated from moderately acid media (pH 4.5–5.0) in which bacterial development is further suppressed by increasing the sugar content of the medium, i.e., to 30–40% (w/v).

An extremely useful medium for the isolation of yeasts from soil and insect frass was developed by Wickerham (1969b). This medium, which he refers to as IM, contains yeast nitrogen base (p. 100) with the addition of glucose and six other carbon sources. The pH is not adjusted, but drops after inoculation as a result of the utilization of the ammonium sulphate provided as source of nitrogen. For further details the publication by Wickerham (1969b) must be consulted.

B. Isolation based on the use of media with elevated sugar concentrations

Most yeasts are capable of development on media with moderately high sugar concentrations, which tend to suppress bacterial development.

1. Solid media for direct isolation

For the recovery of osmotolerant and osmophilic yeasts from foodstuffs of low water activity, YM agar containing 30–50% glucose and 2% agar is recommended. The elective action of such media may be enhanced by adjusting the pH to 4.5–5.0. In most cases the osmophilic species recovered in this way may be successfully subcultured on 30, 10 and 1% glucose media.

2. Liquid media for enrichment purposes

As not all yeasts grow equally well at pH 3.5–4.0, isolations may also be made using YM broth or malt extract of which the glucose concentrations have been raised to 30 or 40%. The use of high sugar concentrations nevertheless tends to select both osmophilic and osmotolerant species. As osmotolerant molds are not inhibited at these sugar concentrations, incubation on a rotary shaker is again recommended.

Another extremely useful medium for the isolation of yeasts from soil and insect frass, also described by Wickerham (1969b), and which he refers to as D-20, contains yeast nitrogen base to which 20% glucose and 0.1% each of yeast extract and malt extract have been added. For further details Wickerham's (1969b) publication must be consulted.

C. Isolation by the use of antibiotics and other inhibitory compounds

A number of antibiotic-containing media for the isolation of yeasts have been published (Davenport, 1980). Frequently, such media have been designed for the isolation of particular species or genera. These special-purpose techniques usually rely on the use of antibiotics and inhibitors which permit the development of the desired yeast species, e.g. the isolation of *Brettanomyces* and *Dekkera* species by employing actidione and sorbic acid at pH 4.8 (van der Walt & van Kerken, 1961a). Selected nitrogen sources, favoring the development of particular species may also be employed, e.g. the isolation of *Cryptococcus neoformans* using creatinine with the addition of diphenyl to suppress mold growth (Kwon-Chung et al., 1978). Kwon-Chung has also given another medium containing niger seed (*Guizotia abyssinica*), chloramphenicol and diphenyl for the recovery of this species (see p. 52). Likewise, use may also be made of selected carbon sources, e.g. the recovery of methanol-utilizing yeasts employing methanol as sole source of carbon with the addition of cycloserine and penicillin G to suppress bacterial growth (van Dijken & Harder, 1974).

The use of media containing antibiotics such as actidione, actinomycin, aureomycin, chlorotetracycline, chloramphenicol, penicillin, α-hydroxyquinoline and streptomycin has also been reviewed by Beech and Davenport (1969, 1971) and Beech et al. (1980).

D. *Isolation by membrane filtration techniques*

Yeasts may also be recovered and cultured from liquid substrates by the membrane filtration technique (Mulvany, 1969), using agar media which permit the selective development of yeasts (e.g., acidified malt extract agar or YM agar). For further details, the technical bulletins of firms supplying such membranes for bacteriological purposes should be consulted.

E. *Purification and maintenance of yeast cultures*

Axenic or pure cultures of isolates deriving from the enrichment cultures are routinely obtained by replating on either YM agar or malt extract agar. Should bacterial contamination persist, recourse may be had to the use of acidulated agar or an agar containing a suitable antibiotic, e.g. penicillin. Colonies developing on these primary plates are inspected for their macromorphology, preferably under low magnification (16 ×) which facilitates the selection of different colonial forms. Selected, single, well isolated colonies are then brought into culture by replating. As a rule two platings are adequate. It must, however, be pointed out that where more than one colonial form appear during the replating of a single, well isolated colony, these may represent morphological or sexual variants. In such cases it is necessary to examine these different colonial forms in detail, including the possibility that they might represent mating types.

For the maintenance and storage of yeast cultures, the recent publication by von Arx and Schipper (1978) must be consulted. When storing yeast cultures for any length of time, such strains are best maintained on GPY agar which contains glucose as sole carbohydrate source, and reduces the risk of possible mutational changes in their fermentative and assimilatory patterns due to the selection of mutants (Scheda, 1966b). The majority of yeasts may be stored at 5°C and subcultured at 5–6 monthly intervals. Psychrophilic species are best stored at 3–10°C. Some yeasts, e.g. *Malassezia* species and *Saccharomyces telluris,* need to be transferred monthly, while in the case of *Dekkera* and *Brettanomyces,* cultures are transferred every two months. In the case of strains of the latter two genera which produce excessive amounts of acetic acid, the incorporation of 2% calcium carbonate and 1% (v/v) of a sterile, 100-fold concentrated vitamin stock-solution (see p. 79) to maintenance media has been found to be beneficial (van der Walt, 1970a). Some basidiomycetous yeasts are, as a rule, maintained on 2% malt extract agar.

Yeast cultures may also be stored at 5°C or at room temperature under mineral oil, sterilized by dry heat (170–180°C) (Henry, 1947; Beech, 1957) using either screw-capped bottles or self-sealing plastic covers. With this technique it is imperative to ensure that the entire agar surface is adequately covered by the oil.

For the lyophilization of yeast cultures reference is made to the publications of Haynes et al. (1955) and von Arx and Schipper (1978). Although lyophilization has been reported to induce respiratory deficient cells and changes in vitamin requirements (Wynants, 1962), there is no published evidence that lyophilization effects changes in the diagnostically important characteristics such as the fermentative and assimilatory patterns or sporulation of strains (Wickerham & Flickinger, 1946; Kirsop, 1955, 1974; Haynes et al., 1955; Bréchot et al., 1958; Brady, 1960; Martin, 1964). Lyophilization is, in fact, a convenient procedure for maintaining most ascogenous yeasts, some of which, on prolonged maintenance on agar slopes, show impairment of their ability to form ascospores.

Media

1. Yeast extract-malt extract broth referred to as YM broth (Wickerham, 1951)

Dissolve 3 g yeast extract, 3 g malt extract, 5 g peptone and 10 g glucose in 1 liter water. The reaction of the medium ranges between pH 5 and 6 depending on the batch of ingredients.

After dispensing, the medium is sterilized at 15 lbs overpressure for 15 minutes.

2. Yeast extract-malt extract agar referred to as YM agar (Wickerham, 1951)

This medium is prepared by adding 2% agar to yeast extract-malt extract broth before sterilization. The reaction of the medium ranges between pH 5 and 6. The medium is sterilized at 15 lbs overpressure for 15 minutes.

3. Malt extract (Lodder & Kreger-van Rij, 1952)

To make malt extract, 1 kg of malt mixed with 2.6 liter tap water is heated in a water bath at 45°C for 3 hours and is stirred continuously. The temperature is then raised to 63°C and maintained for 1 hour, after which the mixture is pressed through a hair sieve or cheesecloth. The filtrate, after being autoclaved for 15 minutes at 15 lbs overpressure is filtered through paper and diluted to a density of 15° Balling*, and adjusted to pH 5.4 if this is necessary, after which it is dispensed and sterilized for 15 minutes at 10 lbs overpressure.

Sometimes it is possible to obtain unhopped malt extract from breweries. The wort is adjusted to pH 5.4 and then diluted to 15° Balling. After dispensing, the malt extract is sterilized for 15 minutes at 10 lbs overpressure. If brewery wort is not available powdered malt extract may be used. Dissolve 15% (w/w) powdered malt extract in demineralized water by heating in a boiling water bath. After cooling the extract is adjusted to pH 5.4.

4. Malt agar or malt extract agar

Malt extract is diluted at 10° Balling to which 2% agar is added. After dispensing, the medium is sterilized at 10 lbs overpressure for 15 minutes. If brewery wort is not available a 10% w/w solution of powdered malt extract in demineralized water may be used.

5. GPY agar (von Arx & Schipper, 1978)

To 500 ml demineralized water add 40 g glucose, 5 g peptone, 500 ml yeast autolysate and 20 g agar. Dissolve the agar by steaming and distribute in tubes. Autoclave at 10 lbs overpressure for 15 minutes. For the preparation of yeast autolysate, p. 79.

6. Sabouraud's 4% glucose agar

In 1000 ml distilled water dissolve 10 g peptone and 40 g glucose and after adjustment to pH 7 add 20 g agar. Dissolve the agar by steaming and distribute in tubes. Sterilize by autoclaving at 10 lbs overpressure for 15 minutes.

---

*The specific gravity of a malt extract expressed in degrees Balling is determined by the use of a Balling saccharometer.

7. Niger seed agar (*Guizotia abyssinica* medium)

The Niger seed agar formulated by Shields and Ajello in 1966 has been modified by Kwon-Chung as follows:

Distilled water	800 ml
Guizotia abyssinica extract	200 ml
Agar	20 g
Glucose	1 g

Autoclave at 15 lbs pressure for 15 minutes. The extract of *Guizotia abyssinica* seed (known as Thistle or Niger seed) is prepared by pulverizing the seeds in a Waring blender. 70 g of seed powder is added to 350 ml distilled water and autoclaved for 10 minutes at 10 lbs pressure. Filter through gauze. Some pressure may be required to force sufficient liquid through the gauze.

8. Niger seed agar with chloramphenicol and diphenyl

Chloramphenicol, 40 mg, should be added to 1000 ml Niger seed agar prior to sterilization. When the agar is cooled to about 55°C, add 10 ml of diphenyl solution (0.1 g of diphenyl in 10 ml of 95% ethanol) aseptically, mix well, and then pour into sterile Petri dishes.

II. Procedures for the classification and identification of yeasts

Workers, when setting about identifying yeast strains, should at all times convince themselves that they are dealing with pure cultures before proceeding with the determination of the morphological, sexual and physiological properties of the strains. Experience has taught that no yeast can be identified or classified with confidence until a systematic study of its morphological, cultural, sexual and physiological characteristics has been made, under standardized conditions. The characteristics employed in the standard description of each species are listed on p. 102.

A. Characteristics of vegetative reproduction

1. Modes of vegetative reproduction

Vegetative or asexual reproduction occurs in yeasts by budding, by fission or a combination of these two processes. Careful observation of conidiogenesis* or the process of conidium formation as a rule provides considerable information which greatly facilitates the identification of strains.

Buds may arise either on yeast cells or on hyphae. Reproduction by budding is initiated by the formation of a small evagination or bud at a point on the surface of the cell. During subsequent development, the cell remains more or less constant in size, while the bud or blastospore (blastoconidium) increases in size to form a new cell which, after some time, separates from the mother cell. Depending on how the bud is formed in terms of the ultrastructure of the cell wall, budding may be either

*Workers wishing to familiarize themselves with modern mycological terminology, which is also applicable to the yeasts, are referred to the publications by von Arx (1979), von Arx & Weijman (1979), Cole & Samson (1979) and von Arx (1981).

holoblastic or enteroblastic. In holoblastic budding all layers of the mother cell or conidiogenous cell, are involved in the formation of the bud. After separation of the bud, usually on a rather narrow base, a scar remains through which no further budding occurs. Von Arx and Weijman (1979) consider holoblastic budding to be characteristic of the Saccharomycetales and related anamorphic states. In enteroblastic budding, the first bud arises from a rupture in the wall of the mother cell, through which the innermost layer of the mother cell evaginates and ultimately grows out to form the outermost layer of the bud which is ultimately abstricted. After a number of buds have arisen from the same site, this site on the mother cell is surrounded by a collarette due to the recurrent formation and abstriction of the successive buds arising from the inner layer of the mother cell. Enteroblastic budding is characteristic of basidiomycetous yeasts and related anamorphic states.

Budding may also be referred to in terms of the position of the budding sites. If budding is restricted to one pole of the mother cell it is referred to as monopolar budding (Fig. 1) and may occur on a rather broad base, e.g. in *Malassezia*. If buds are formed exclusively at the distal poles of the mother cell, it is referred to as bipolar budding (Fig. 2). The buds are abstricted on a rather broad base by the formation of a cross wall, and is also referred to as 'budding on broad base' or 'bud-fission'. Recurrent budding leads to the formation of multiple, transverse scars or annellations at the poles of the mother cell (Streiblová, 1971). Bipolar budding is characteristic of the apiculate yeasts. Multilateral or multipolar budding (Fig. 3) implies budding at different sites on the mother cell.

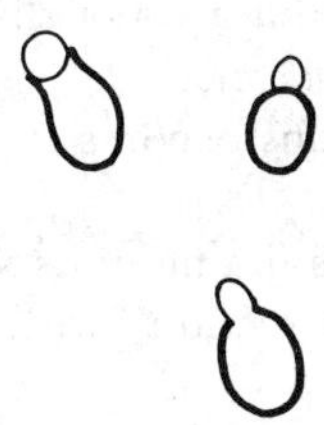

Fig. 1 – Monopolar budding

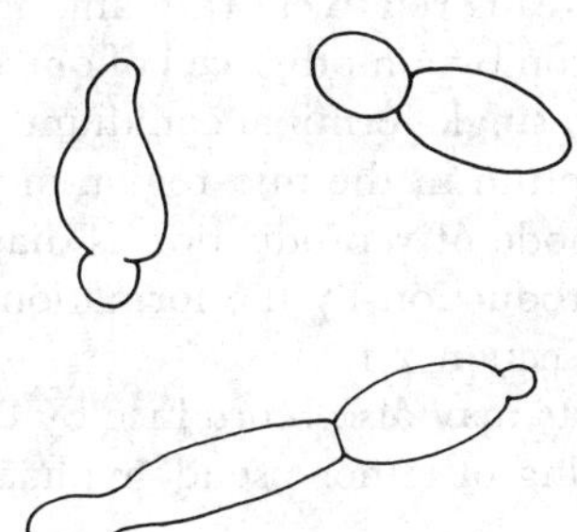

Fig. 2 – Bipolar budding

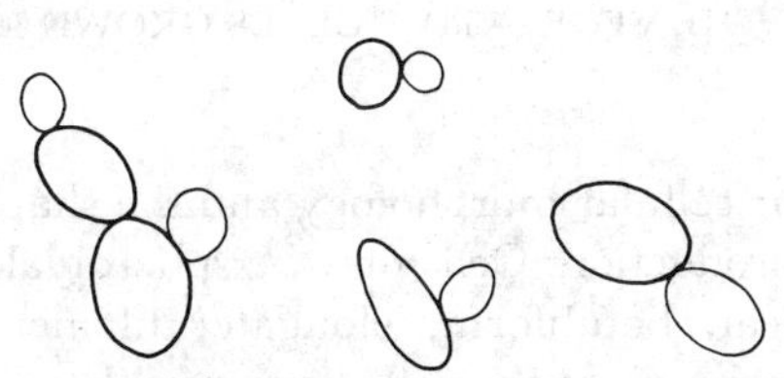

Fig. 3 – Multilateral budding

Budding may also be described in terms of the formation of the succession of buds. Sympodial budding describes the process by which new buds appear just behind and adjacent to the previous bud site. Acropetal budding describes the formation of a succession of buds which arise in chains with the youngest at the apex of the chain.

Basipetal budding refers to the formation of a succession of buds in which the oldest cell is at the apex of the chain.

Reproduction by fission (Fig. 4) implies the duplication of a vegetative cell by means of the ingrowth from cell wall of a transverse septum which bisects the long axis of the cell. The newly formed fission cells, which are arthroconidia (arthrospores), elongate and the process repeats itself. Recurrent fission by cells may give rise to transverse, multiple scars or annellations (Streiblová, 1971). Reproduction by fission is characteristic of the genus *Schizosaccharomyces.*

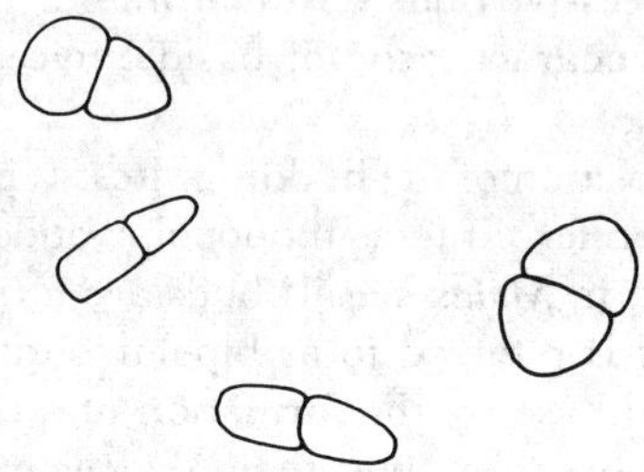

Fig. 4 – Fission

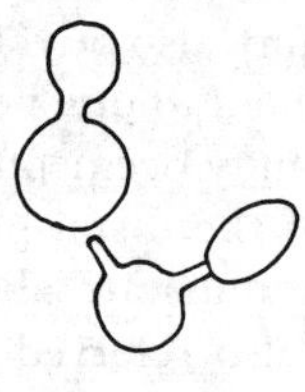

Fig. 5 – Conidia on stalks

Vegetative reproduction by the formation of conidia borne on stalk-like, tubular structures is relatively rare among the yeasts. This mode of reproduction entails the formation by a mother cell of one or more tubular protuberances, each of which gives rise to a single terminal conidium (Fig. 5). After maturation the conidium is disjointed at a septum in the mid-region of the protuberance. The conidia are not discharged. This mode of reproduction is diagnostic of the genus *Sterigmatomyces.*

Reproduction by the formation of discharged spores or ballistospores is discussed under section 2.f.

Yeasts may also reproduce by the formation of more or less filamentous structures consisting of either pseudohyphae or true hyphae (see under section 2.b.).

2. *Characteristics of vegetative cells*

a. THE MORPHOLOGY OF THE VEGETATIVE CELLS GROWN IN LIQUID AND ON SOLID MEDIA

Yeasts may vary in their cellular morphology and the shape of vegetative cells may relate to their mode of reproduction. Cells may be spheroidal, subglobose, ellipsoidal, ovoid, obovoid, cylindrical, botuliform, elongate, filamentous, apiculate, ogival, lunate or triangular. In many cases the cell shape may be characteristic of particular genera or species. The lemon-shaped cells of the apiculate yeasts, the bottle-shaped cells of *Malassezia,* the lunate cells of *Metschnikowia lunata* and *Candida peltata,* and the triangular cells of *Trigonopsis* are typical examples. It should, however, be borne in mind that both the shape and the size of cells of different strains of a species are liable to variation, e.g., in *Saccharomyces cerevisiae.*

Methods

1. Growth in liquid media

As a rule the cellular morphology and mode of reproduction of strains are examined in liquid culture. For this purpose malt extract or 2–4% glucose-yeast extract-peptone water is used. Some genera e.g. *Malassezia* and *Cyniclomyces* require special media.

Material from a young actively growing slant is inoculated in 30 ml of either 15° Balling malt extract or 2–4% glucose-yeast extract-peptone water in 100-ml, cotton-plugged Erlenmeyer flasks. After 2–3 days incubation at 25 or 28°C the culture is examined.

The shape and mode of reproduction of the vegetative cells are noted; whether the cells occur singly, in pairs or larger aggregates. It is further noted whether or not cells are encapsulated. The length and width of the cells are measured and the extreme values obtained from the measurement of at least 20 cells are recorded.

Cultural characteristics are noted after 2–3 days incubation at 25 or 28°C and again after 4 weeks at either 20°C or at room temperature.

Yeast growth in liquid media may result in the formation of a compact, coherent, flocculent or mucoid sediment, a ring, islets or a pellicle. The pellicle which may be dull, is frequently wrinkled and creeping and formed within one or two days. Alternatively, it may be glistening and moist, forming only on prolonged cultivation. Should true hyphae be formed, a thick, tenacious surface growth may be produced which may eventually result in the formation of a mucoid mass. The development of any characteristic smell (e.g., esters) is also noted.

Pellicle formation is also noted in the carbon assimilation tests (see p. 81).

2. Growth on solid media

Some laboratories prefer the use of morphology agar for the study of morphological and cultural characteristics of strains. When this medium is used, both slants and Dalmau plates are prepared which are inoculated from actively growing slant cultures. (For the preparation of Dalmau plates see p. 59.) The tubes and plates are incubated at 25°C for 7–10 days. The shape, mode of reproduction and cellular dimensions are recorded as described above. The cultural characteristics of the growth are noted as outlined below.

If morphology agar is not used, strains are examined for their cultural characteristics on either malt extract agar or 2% glucose-yeast extract-peptone agar. When dealing with strains of *Dekkera* and *Brettanomyces* species which produce excessive amounts of acetic acid, malt extract-2% calcium carbonate agar is used. Actively growing material is inoculated as a streak culture on slants of these media in plugged tubes and incubated at 20°C or room temperature for 4 weeks after which the cultural characteristics of the strain are noted. In the case of *Dekkera* and *Brettanomyces* strains where cultures are slow growing, the slants are examined after 6 weeks.

When cultures are examined on solid media, attention is given to the following characteristics:

Texture, i.e. whether the growth is mucoid or viscous, butyrous, friable, coherent or tenacious. Mucoid growth is frequently associated with encapsulation of cells as a result of the production of extracellular polysaccharide material, while matted, coherent growth is generally associated with the formation of an abundance of pseudohyphae or true hyphae.

Color, noting distinctive colors, such as yellow, orange or red. Growth may be pigmented by the production of pigments of certain types. The presence of yellow, orange, and red carotenoid pigments is, for instance, characteristic of the genera *Rhodotorula, Sporobolomyces, Phaffia, Rhodosporidium* and *Sporidiobolus,* while the production of a non-carotenoid, Bordeaux-red pigment, pulcherrimin, is typical of only certain yeasts, e.g. *Metschnikowia pulcherrima* and certain *Kluyveromyces* species. The majority of yeasts, however, produce growth ranging from white to cream or light shades of brown.

Surface, i.e. whether the growth is glistening or dull, smooth, sectored, striated, pulvinate, verrucose, plicate, crispulate, rugose, rough or hirsute. Some yeasts are, however, characterized by both rough and smooth forms of growth. Strains may be smooth when isolated and become rough during maintenance. Some strains when plated, may yield a mixture of rough and smooth colonies or colonies with rough and smooth sectors. A change from the smooth to the rough form may also involve a change in the shape of the cells from spherical to oval and elongate; it may also lead to the formation of pseudohyphae and a pellicle where none was formed before. For example, strains of *Candida albicans* are nearly always smooth when first isolated but may become rough and tenacious when maintained on laboratory media for a long period. Both the smooth and rough forms produce pseudohyphae, although there may be a difference in structure. Freshly isolated strains of *Trichosporon cutaneum (beigelii)* may be mucoid and soft whereas rough forms are tenacious and hairy.

Elevation, i.e. whether the growth is flat and spreading, raised and restricted.

Margin, i.e. whether the edge of the streak culture is smooth or entire, undulating, lobate, rhizoid, erose or fringed with pseudohyphae or true hyphae.

Media

1. Malt extract, see p. 51.

2. 2% (w/v) glucose-yeast extract-peptone water

Dissolve 20 g glucose, 10 g peptone and 5 g yeast extract in 1 liter of distilled water. The pH is not adjusted.

After dispensing, the medium is autoclaved for 15 minutes at 15 lbs overpressure.

3. 2% or 4% (w/v) glucose-yeast extract-peptone agar with olive oil (for the cultivation of *Malassezia furfur*). A few drops of olive oil are added to 2 or 4% glucose-yeast extract-peptone agar prior to sterilization.

4. 1% (v/v) yeast autolysate-4% (w/v) glucose-1% (w/v) proteose peptone water (for the cultivation of *Cyniclomyces guttulatus*). The ingredients are dissolved in distilled water. The reaction of the medium is adjusted to pH 4.5 prior to sterilization.

5. Malt extract agar or malt agar, see p. 51.

6. Malt extract agar-2% (w/v) calcium carbonate is prepared by adding 20 g calcium carbonate, which had previously been sterilized in a cotton-plugged flask by dry heat (160–180°C) for 2 hours, to 10° Balling malt extract containing 20 g agar. The medium is sterilized by autoclaving for 15 minutes at 10 lbs overpressure.

7. 2% (w/v) glucose-yeast extract-peptone agar. This medium is prepared by adding 2% (w/v) agar to 2% (w/v) glucose-yeast extract-peptone water. After dispensing, the medium is autoclaved for 15 minutes at 15 lbs overpressure.

8. Morphology agar

The composition of this medium is given in Table 10 on p. 100. To rehydrate the medium, suspend 35 g of Bacto yeast morphology agar in 1 liter cold distilled water. Heat to dissolve the ingredients completely. Dispense in tubes or flasks and sterilize by autoclaving for 15 minutes at 15 lbs overpressure. The final reaction of the medium is about pH 5.6.

b. FORMATION OF PSEUDOMYCELIUM AND TRUE MYCELIUM

If a yeast reproduces exclusively by budding, the mature bud may either detach itself immediately or remain attached to the mother cell and eventually give rise to either clusters of cells or chains of cells. The tendency of some yeasts to form chains of cells, results in the formation of pseudohyphae or pseudomycelium. Pseudohyphae or pseudomycelium consequently refers to the formation of elongated cells which arise by budding.

Pseudohyphae may be rudimentary in which case they consist of elongated cells of more or less equal dimensions, or they may be differentiated into elongated cells, each of which may produce blastospores in a regular and more or less characteristic arrangement. In earlier taxonomic systems (see van der Walt, 1970a) the arrangement of the blastospores was used for the differentiation of certain genera which are no longer accepted. These generic names have nevertheless been adopted to designate the different types of pseudohyphae. Langeron and Talice (1932) described in detail and illustrated the development of the various types of pseudohyphae. They are as follows:

'*Mycotorula*-type': pseudohyphae with compact, more or less spherical clusters of blastospores around the junction of adjacent, elongated pseudohyphal cells (Fig. 6a).

'*Mycotoruloides*-type': pseudohyphae with blastospores in loose branched verticils (Fig. 6b).

'*Candida*-type': pseudohyphae with chains of blastospores arising at the junction of adjacent pseudohyphal cells (Fig. 6c).

'*Mycocandida*-type': strongly branching pseudohyphae with often only one or two symmetrically arranged blastospores or with blastospores arranged in small whorls or verticils at the junction of adjacent pseudohyphal cells (Fig. 6d).

'*Blastodendrion*-type': stalagmoid blastospores arranged in a more or less penicillium-like structure (Fig. 6e).

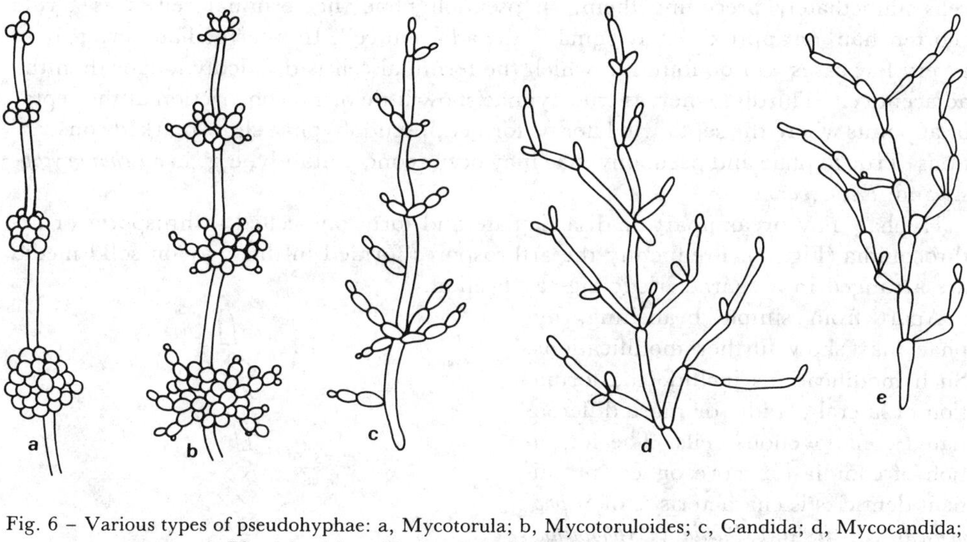

Fig. 6 – Various types of pseudohyphae: a, Mycotorula; b, Mycotoruloides; c, Candida; d, Mycocandida; e, Blastodendrion. (Redrawn from Diddens & Lodder, 1942)

It must, however, be accepted that transitional types do occur and that the formation of pseudohyphae may be markedly affected by cultural conditions (see van der Walt, 1970a).

An unusual expression of pseudohyphae not frequently encountered is the formation of blastese. This term describes the so-called pseudogermination of blastospores in which the germ-tube results in the formation of filamentous, non-septate, elongations (Langeron & Guerra, 1939/40b). This term has been broadened to include the formation of pseudohyphae consisting of a single filamentous cell which may branch but in which septum formation does not occur. Blastese formation is characteristic of *Brettanomyces anomalus*.

Although yeasts do not reproduce by the exclusive formation of hyphae, numerous taxa under suitable conditions produce true septate branching hyphae more or less copiously. True hyphae proliferate by continuous growth of the hyphal tip, followed by the formation of septa. Septation lags behind the growth of the hyphal tip to such a degree that the terminal cell measured from the tip to the first septum is often longer than the preceding cell which is measured from the first to the second septum (Wickerham, 1951).

In practice, since budding and fission may occur concurrently, it is frequently difficult to differentiate between true hyphae, pseudohyphae and intermediate forms. To distinguish between these three forms Wickerham (1951) applies three criteria, basing observations on the terminal cells of the hyphal structures. Firstly, true hyphae usually have refractive, straight septa which can generally be differentiated by their greater thickness and refractivity from the edges of the vacuoles. Pseudohyphae have no discernible septa, and the ends of intercalary cells are curved and not refractive. Intermediate forms usually show only a small percentage of cells separated by septa. Secondly, in true hyphae many of the terminal cells are considerably longer than the cells immediately preceding them. In pseudohyphae the terminal cell is as a rule shorter than, or approximately equal to the adjacent cell. In intermediate forms, only a very few cases will be found in which the terminal cell is decidedly longer than the adjacent cell. Thirdly, whereas true hyphae show little or no constriction at the septa, or at points where the septa will later be formed, pseudohyphae show marked constrictions. True hyphae and pseudohyphae may occur concomitantly, e.g. in *Saccharomycopsis* and *Trichosporon*.

Hyphae may break apart or disarticulate and form one-celled arthrospores or arthroconidia (Fig. 7). Frequently the arthrospores formed in this way on solid media are arranged in a characteristic zig-zag fashion.

Apart from simple branching, hyphae may show further modifications. Such modifications include the formation of lateral conidia or more differentiated conidiogenous cells. The formation of conidia on more or less prominent denticles is characteristic of *Stephanoascus ciferrii* and *Pichia (Hyphopichia) burtonii* and the formation of conidiogenous cells showing repetitive monopolar budding, of *Ambrosiozyma cicatricosa*. Hyphae may also form clamp connections. These arise from outgrowths which at cell division make a connection between the resulting two cells by fusion with the lower. Such clamp connections are characteristic of the dikaryotic phase of basidiomycetous taxa. Hyphae may also form anastomoses by the fusion of branches of the same or different hyphae. Anastomosing hyphae are characteristic of *Ambrosiozyma (Hormoascus) platypodis*.

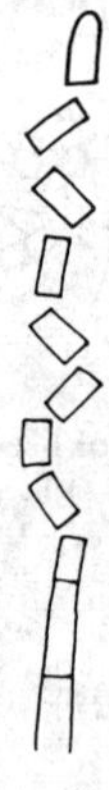

Fig. 7 – Arthrospores

The ultrastructure of hyphal septa shows variation. Except for the presence of visible septal pore bodies, direct light microscopy does not reveal much detail of the hyphal septa of filamentous yeasts. For details of the ultrastructural features of such septa, section 2.g. (p. 62) must be consulted.

When examining cultures for the formation of pseudohyphae or true hyphae, it should also be noted whether the structures are recumbent, intramatrical or erect. Aerially-borne chains of cells are characteristic of several *Candida* species.

For the routine study of pseudohyphae and true hyphae formation corn meal agar, potato-glucose agar (also referred to as potato agar or potato-dextrose agar), rice agar or morphology agar are normally used. Either the slide culture method of Rivalier and Sydel (1932a,b) or the plate technique developed by Dalmau (1929) may be employed.

Methods

1. Slide culture procedure

A Petri dish, containing a U-shaped glass-rod supporting two glass slides, is sterilized by dry heat at 160–180°C for 2 hours.

Corn meal agar, potato agar, morphology agar or rice agar is melted and poured into a second Petri dish. The glass slides are quickly removed from the glass rod with a flame-sterilized pair of tweezers, and are dipped into the agar after which they are replaced on the glass rod support. The agar may also be poured directly onto the slides.

After solidification of the agar on the slides, the yeast is inoculated very lightly in one or two lines along each slide and a sterile coverslip is placed over part of the lines. Some sterile water is poured into the Petri dish to prevent the agar from drying out. The culture is then incubated at 25°C for 4–5 days. For observation the slides are taken out of the Petri dish and the agar is wiped off the back of the slide. The edges of the streak under and around the coverslip are examined microscopically.

2. The Dalmau plate technique (Wickerham, 1951)

Corn meal agar, potato agar, rice agar or morphology agar may be used. Freshly poured plates are set aside for 1–2 days in order to allow the surface to dry. A single streak inoculation is made near one side of the plate (as from the relative positions 10 o'clock to 2 o'clock). The inoculum should be light and taken from a fresh slant culture. In addition to the single streak inoculation, two point inoculations are made near the other side of the plate (as at positions 4 o'clock and 8 o'clock). A central section of the streak and one of the point inoculations are covered with sterile coverslips. The preparations are studied microscopically using a 3-mm dry objective and 10 × oculars after incubation at 25°C for 7 or 10 days.

Media

1. Corn meal agar (Bernhardt, 1946)

12.5 g yellow corn meal is stirred in 300 ml water and heated in a water bath at 60°C for 1 hour and then filtered through paper. The volume of the filtrate is made up to 300 ml and 3.8 g agar is added. The mixture is autoclaved for 15 minutes at 15 lbs overpressure. The hot liquid is refiltered through absorbent cotton wool, dispensed and sterilized for 15 minutes at 15 lbs overpressure.

Suitable commercial brands of corn meal agar are also available.

2. Potato agar (potato-glucose agar)

100 g washed, peeled and finely grated potatoes are soaked in 300 ml tap water for several hours in a cold room or refrigerator. The mass is filtered through cloth and autoclaved for 1 hour at 15 lbs overpressure. For the preparation of the agar, 230 ml of this potato extract is added to 770 ml tap water containing 20 g glucose and 20 g agar. After the agar is dissolved by heating, the medium is dispensed and sterilized for 15 minutes at 15 lbs overpressure.

Potato-glucose agar is also commercially available.

3. Rice agar (Taschdjian, 1953)

Cover 20 g of white rice with 1 liter of tap water and bring to boil. Simmer over a low flame for 45 minutes, keeping the vessel covered. Filter the liquid through gauze and discard the rice. After cooling add tap water to the filtrate to make up 1 liter. Add 20 g agar, dissolve by steaming. Distribute in tubes or bottles and sterilize by autoclaving for 15 minutes at 15 lbs overpressure.

c. FORMATION OF ASEXUAL ENDOSPORES

The formation of asexual endospores is not a common phenomenon but has been observed in the genera *Trichosporon, Candida, Cryptococcus* and *Oosporidium.* These endospores are vegetative cells which are delimited within cells or hyphae. These cells, unlike chlamydospores or ascospores cannot be stained selectively. These structures are usually observed in old cultures on YM agar, malt extract agar, potato agar and corn meal agar, which have been maintained at room temperature.

No special media have been devised which favor the specific development of endospores. For a more detailed discussion of the formation of these spores the publication of do Carmo-Sousa (1969) should be consulted.

d. FORMATION OF CHLAMYDOSPORES

The chlamydospore has been defined as a thick-walled, non-deciduous, intercalary or terminal, asexual spore formed by the rounding off of a cell or cells (Ainsworth, 1971). The asexual character of the chlamydospore distinguishes it from the teliospore of the Uredinales and Ustilaginales, from which the basidium is produced.

As chlamydospores are generally rich in lipid material, they are well adapted for maintaining vitality through periods of dormancy. Mature chlamydospores have particular affinities for certain dyes. This is clearly evidenced by the fact that these cells, in contrast to normal vegetative cells, show a marked degree of acid-fastness on staining – a characteristic which is shared by ascospores (see van der Walt, 1970a). Another characteristic of these cells when observed in older cultures is that they shed their outer membranes just prior to or during germination.

Chlamydospore formation which is characteristic of *Candida albicans* and the ascogenous genus *Metschnikowia* and its imperfect or anamorphic states, is occasionally also observed in old slant cultures of other taxa, including *Trichosporon cutaneum (beigelii)* and *Cryptococcus* species. In the genus *Metschnikowia,* however, the chlamydospores or dormant cells fulfil a dual function insofar as they may either germinate giving rise to the budding diplophase, or may, by the formation of a peduncle give rise to clavate or ellipsoidopedunculate asci. In this case the modified chlamydospores are the site of reduction division in certain seemingly imperfect yeasts, e.g. *Candida (Entelexis) magnoliae,* which, although lacking ascigerous states, are nevertheless characterized by different ploidy levels (see van der Walt & Johannsen, 1973).

e. FORMATION OF GERM TUBES BY *CANDIDA ALBICANS*

The formation of germ tubes is at present accepted as a reliable property for identifying *Candida albicans* and is used in medical laboratories because of its rapidity. The germ tube of *Candida albicans* is a thin filamentous outgrowth from the cell without a constriction at its point of origin. Presumptive identifications based on the formation of germ tubes are 95–100% accurate when compared with conventional identification (Stenderup & Thomsen, 1964; Ahearn et al., 1966; Joshi & Gavin, 1974). The formation of germ tubes is, however, influenced by temperature, amount of inoculum, composition of the medium, and strain. The publication of Ogletree et al. (1978) should be consulted for an evaluation of the various techniques.

Method

A simple test for germ tubes may be performed by suspending cells from a 24-hour-old culture (10^5–10^6 cells per ml) in bovine serum. The cells are examined microscopically after 1–3 hours incubation at 37°C.

f. FORMATION OF BALLISTOSPORES

The formation of discharged spores or ballistospores (ballistoconidia) is a rather specialized mode of reproduction which is characteristic of certain basidiomycetous genera e.g. *Sporobolomyces.* Ballistospores are produced on sterigmata that protrude from vegetative cells (Fig. 8) and are discharged into the air by the so-called drop mechanism. Although this mechanism is also operative in the discharge of basidiospores of certain Basidiomycetes the ballistoconidium is not accepted to be homologous with the discharged basidiospore. Ballistospores may show two types of symmetry: rotational and bilateral. The first type may be: globose, ovoidal, obovoid, apiculate, turbinate or ampulliform; the second type: reniform or allantoid.

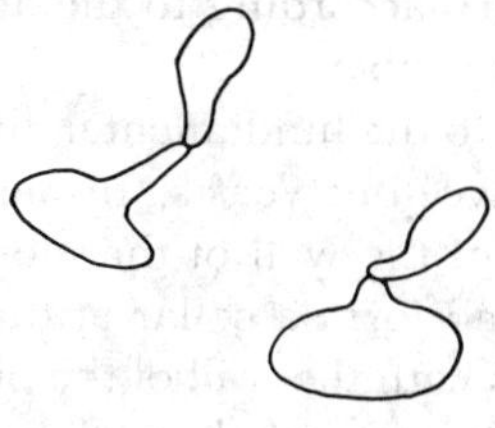

Fig. 8 – Ballistospores

Methods

Ballistospores are generally detected as mirror images on the lower side of an inverted Petri dish in which the culture is growing. The image is caused by entrapment of discharged spores. More refined methods for the detection of ballistospores have, however, been described by Lodder and Kreger-van Rij (1952) and more recently by do Carmo-Sousa and Phaff (1962).

Lodder and Kreger-van Rij (1952) recommend slide cultures for the detection of mirror images. For this purpose only one of the slides used in the preparation of a slide culture as described on p. 59 is covered with malt extract agar or potato agar. It is inoculated and then placed surface down over the second slide with a thin U-shaped glass support between them. After incubation, preferably at 18–20°C, both slides can easily be examined under the microscope. The lower of the two slides will show the images of the inoculation lines.

Do Carmo-Sousa and Phaff (1962) recommend the following procedure:

10 ml molten corn meal agar is poured into a Petri dish and after setting is inoculated with the test organism along two diameters at right angles. The dish is inverted over another Petri dish bottom containing malt agar on which is placed a sterile slide. One of the lines is positioned over the slide and the two dishes are taped together along the entire circumference. A piece of moistened sterile cotton may be placed inside to raise the humidity. Incubation is best at 18–20°C. The discharged spores germinate to form colonies on the bottom dish. The sterile glass collects discharged spores for direct microscopy.

Media

1. Malt extract agar, see p. 51.

2. Potato agar, see p. 59.

3. Corn meal agar, see p. 59.

g. ULTRASTRUCTURE OF THE CELL WALL AND HYPHAL SEPTA

As has been shown by Kreger-van Rij and Veenhuis (1971a) ascomycetous and basidiomycetous yeasts differ fundamentally in the ultrastructure of their cell walls when examined in ultrathin sections by transmission electron microscopy (TEM). In material fixed in either aqueous potassium permanganate or glutaraldehyde, post-fixed in osmium tetroxide, and stained with uranyl acetate and Reynold's lead citrate, the cell walls of the ascomycetous yeasts consist of a single, thin, dark outer and a broad light inner layer. The cell wall of basidiomycetous yeasts appears laminate and consists of thin, light or greyish and dark layers and the outer edge is rather diffuse. The thickness of the wall and the number of layers vary according to the age of the cell, in very young buds only a single greyish layer is visible.

Kreger-van Rij and Veenhuis (1971a) drew attention to the fundamental differences in the mode of budding of basidiomycetous and ascomycetous yeasts. In the former, the wall of the bud is continuous with the inner layer of the wall of the mother cell. The outer layers of the wall of the mother cell break and form a collar at the base of the bud. In ascomycetous yeasts, a new wall continuous with the wall of the bud arises from under the wall of the mother cell. In sections, the part of the wall within the mother cell is often very light and a thin dark line may be visible separating it from the wall of the mother cell. Part of the new wall is separated from the bud and remains in the mother cell at the perimeter of the scar left when the bud separates (Kreger-van Rij & Veenhuis, 1969b, 1975a; Kreger-van Rij, 1978b).

The transmission electron microscopical studies of Kreger-van Rij and Veenhuis (1969a, 1971a, 1973a) showed that hyphal septa may differ markedly in their ultrastructure. The following types have been observed in ascomycetous yeasts:

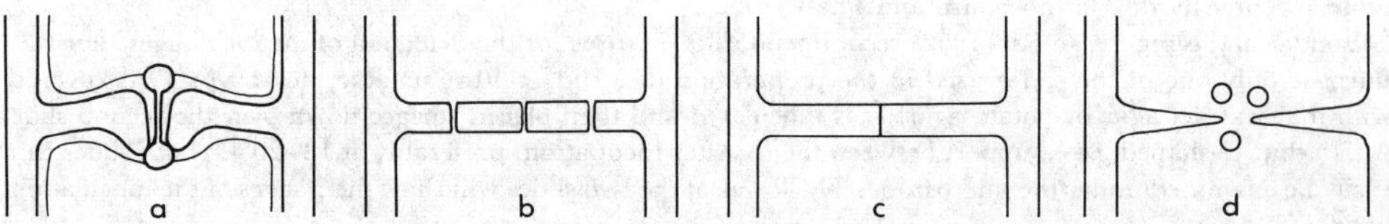

Fig. 9 – Diagrammatic drawings of longitudinal sections through the septa of ascomycetous fungi: a, plugged dolipore; b, plasmodesmata; c, closure line; d, simple pore with Woronin bodies

(1) Septa with a plugged dolipore (Fig. 9a), i.e. a pore with a thickened edge and endoplasmic reticulum through the pore with dilations at both ends to form plugs. These are found in *Ambrosiozyma* species.

(2) Septa with plasmodesmata (Fig. 9b), narrow connections between the protoplasts of adjacent cells. They may contain strands of endoplasmic reticulum. The

plasmodesmata may be arranged in a ring or may be random. The latter is found, e.g., in *Saccharomycopsis capsularis*.

(3) Septa with a single central connection indicated by a closure line (Fig. 9c) formed by the two plasma membranes being retained in the middle of the septum when it is closed by centripetal growth. This type of connection is found in *Arthroascus javanensis*.

Septa of the following types have been observed in basidiomycetous yeasts:

(1) Septa with a dolipore. Endoplasmic reticulum may form a cap or parenthesome at both ends of the pore. The cap may be primitive (Fig. 10a) or vesicular (Fig. 10b). Dolipores with or without caps are found in the Filobasidiaceae.

(2) Septa with a simple pore with a tapered edge (Fig. 10c); these are found in the teliospore-forming yeasts.

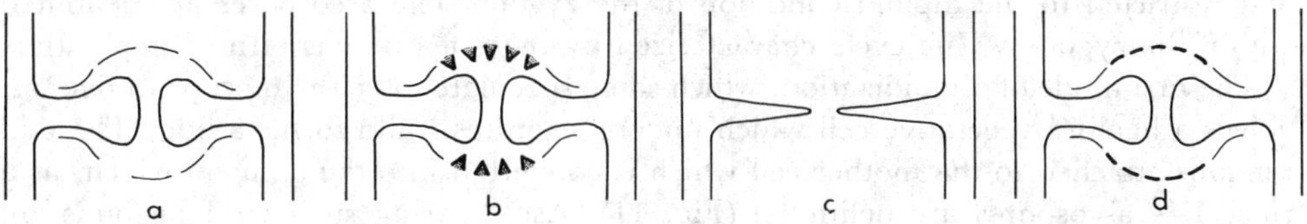

Fig. 10 – Diagrammatic drawings of longitudinal sections through the septa of basidiomycetous fungi: a, dolipore with a primitive cap; b, dolipore with a vesicular cap; c, simple pore; d, dolipore with a well-developed cap

The kinds of septa found in yeasts differ from the typical ascomycete and basidiomycete pore structures of other fungi. The Ascomycetes have a simple pore which often has Woronin bodies around it (Fig. 9d), and the Basidiomycetes have a dolipore with well developed caps (Fig. 10d) (Moore & McAlear, 1962).

B. Sexual characteristics

Many yeasts, although not all, are characterized by sexual reproduction involving an alternation of generations with the formation of characteristic cells in which reduction division occurs. In the ascogenous yeasts this site of meiosis is the ascus in which the haploid generation or ascospores are delimited internally by so-called 'free-cell' formation, i.e., the process by which the cytoplasm surrounding the meiotic nuclei becomes enveloped by a wall. In the basidiomycetous yeasts reduction division is restricted to the basidium on which the haplophase is externally delimited as basidiospores. Since the ascus and the basidium are homologous structures, they are collectively referred to as dangeardiens (Moreau, 1949; van der Walt & Johannsen, 1974a).

If a yeast is characterized by the formation of either asci or basidia, it is referred to as a perfect yeast or a perfect state. Yeasts which lack sexual stages, are termed imperfect yeasts or imperfect states. In modern mycological parlance (Hennebert & Weresub, 1977) a perfect state is referred to as a teleomorph or teleomorphic state and an imperfect state, as an anamorph or anamorphic state. The combined states are referred to as the holomorph.

1. Characteristics of ascospore formation

The ascogenous yeasts may be either homothallic or heterothallic and may be vegetatively stabilized in either the haplophase or diplophase or may exist as a mixture of the two. The existence of higher degrees of ploidy have also been reported. Depending on the ploidy level of the vegetative phase, asci may be conjugated or unconjugated.

a. CHARACTERISTICS OF ASCI AND ASCOSPORES

In the case of homothallic yeasts which are vegetatively stabilized exclusively in the haplophase, plasmogamy, karyogamy and meiosis occur within the zygote which is as a rule constituted by the fusion of two vegetative cells. The diplophase is short-lived and restricted to the diploid condition of the zygote. The ascospores are delimited within the zygote. A life cycle characterized by these features is termed haplontic.

The first mode of diploidization, which some speculate to be the most primitive, involves a haploid vegetative cell which undergoes mitosis and forms a bud. The bud remains attached to the mother cell which is converted into the ascus in which, as a rule, 1–4 ascospores are delimited (Fig. 11). Asci bearing such vestigial buds are characteristic of the genera *Schwanniomyces, Torulaspora, Debaryomyces, Wingea,* and certain species of the genera *Pichia* and *Hansenula.* Diploidization is effected by the fusion of two daughter nuclei within the confirmed structure of the mother cell and its attached bud. Kreger-van Rij and Veenhuis (1975a, 1976b) and Kreger-van Rij (1977a) have submitted evidence that the bud, attached to the mother cell, is first abstricted with the subsequent dissolution of the cross wall which separates the two cells, prior to the fusion of the two daughter nuclei. Other workers (van der Walt et al., 1977) maintain that the actual abstriction of the bud is not a prerequisite of the process. This mode of ascus formation is referred to as mother–daughter cell conjugation or bud-meiosis. As the process involves the fusion of two sister nuclei, it does not constitute heterogamy. As mother–daughter cell conjugation does not involve the fusion of independent cells, strains of species in which diploidization is effected exclusively by this process would be predominantly inbreeding, as appears to be the case in the genus *Schwanniomyces* (Johannsen & van der Walt, 1980). A process comparable to mother–daughter cell conjugation appears to be operative in the apiculate yeast genus *Nadsonia,* where karyogamy is effected by the fusion of the nuclei of a bud and mother cell. The contents of the zygote move into a second bud at the opposite pole. The second bud which is abstricted by a septum, becomes the ascus.

Fig. 11 – Ascus formed by 'mother-daughter cell conjugation'

Diploidization may also be effected by the fusion of two independent, free-occuring, haploid cells. The cells may fuse directly giving rise to more or less amoeboid, conjugated asci, e.g. in *Schizosaccharomyces.* Alternatively, the cells may form elongated copulatory processes or conjugation tubes which fuse, resulting in conjugated

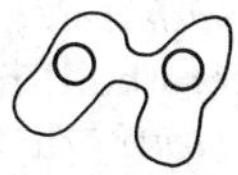

Fig. 12 – Conjugated ascus

Fig. 13 – Ascus with abortive conjugation tube

dumbbell-shaped asci (Fig. 12), e.g. in *Debaryomyces, Torulaspora, Wingea* and *Zygosaccharomyces.* Occasionally the conjugation tubes fail to fuse. Cells bearing such abortive conjugation tubes may nevertheless convert into 1–2 spored asci (Fig. 13). In this event it is presumed that the haploid nuclei of such cells with abortive conjugation tubes, undergo mitosis and the sister nuclei fuse in a manner comparable to the process operative in mother-daughter cell conjugation. As the protuberance is not abstricted from the cell, the process constitutes somatogamous autogamy. Asci bearing either vestigial buds or abortive conjugation tubes may be formed concomitantly, e.g. in strains of *Debaryomyces, Torulaspora* and *Wingea.*

Fig. 14 – Unconjugated ascus

In the case of homothallic yeast strains which are vegetatively stabilized in the diplophase, a single diploid vegetative cell may undergo reduction division and be converted direct into an unconjugated ascus (Fig. 14), e.g. as in the genus *Saccharomyces.* The diploid condition is rapidly restored either by the conjugation of germinating ascospores within the ascus or by autogamous fusion of daughter nuclei at the conclusion of the first mitotic division within a germinating ascospore (Winge & Laustsen, 1937). The latter process has been referred to as direct diploidization or autodiploidization. In such a cycle the haploid condition is short-lived and restricted to the ascosporal stage. A life cycle characterized by these features is referred to as diplontic.

Should some zygotes in the case of the haplontic cycle proceed to reproduce mitotically, or in the case of the diplontic cycle diploidization of the ascospores be delayed, these modifications would result in a vegetative phase consisting of both diploid and haploid cells. Such a mixed vegetative phase would then give rise to both conjugated as well as unconjugated asci. A life cycle characterized by such features is referred to as diplohaplontic.

If heterothallism is operative the diplophase is normally heterozygous for the mating type genes and individual cells are bisexual. The existence of unisexual diploid material has, nevertheless, been reported (Wickerham, 1958; Oshima & Takano, 1972). If the diplophase is stabilized, the asci remain unconjugated and unisexual haploid ascospores of both mating types are formed. Ascospores of opposite mating type may conjugate directly within the ascus giving rise to the diplophase, e.g. in *Saccharomycodes,* or the ascospores may germinate and establish haploid vegetative phases which may be of opposite mating types. Normally the diplophase can then only be initiated if conjugation of haploid cultures of opposite mating types has occurred, e.g. in *Hansenula* and *Pichia.* In this event conjugated asci are frequently produced. Slant cultures of mating types are not invariably stable and may revert to sporulating cultures as a result of mutation of the mating type alleles (Hawthorne, 1963; Takano & Oshima, 1967, 1970). In certain species, e.g. *Pichia membranaefaciens* and *Pichia spartinae,* strains may be either homothallic or heterothallic. Heterothallism may also be associated with sexual agglutination, e.g. in *Hansenula canadensis (wingei)* and *Saccharomyces kluyveri* (Wickerham, 1956, 1958). When opposite mating types of sexually agglutinative species are mixed, the cells agglutinate in a manner analogous to an antigen–antibody reaction. In *Hansenula wingei* this agglutination has been shown to be mediated by complementary glycoproteins present on the outer surface of the opposite mating types (Brock, 1959; Taylor, 1965; Crandall & Brock, 1968).

It must, however, be borne in mind that deviations from the normal alternation of the haplophase and diplophase may occur. Winge and Roberts (1954), for instance, demonstrated the formation of binucleate ascospores in material which normally remains haploid in single-ascospore cultures. In dealing with heterothallic strains of *Saccharomyces kluyveri,* Wickerham (1958) reported the formation of triploids and tetraploids by the conjugation of unisexual diploid cells with either unisexual haploid or diploid cells of opposite mating types. Ascospores were reported as having been produced at each ploidy level above haploid, and unisexual strains with more than one set of chromosomes as having sporulated in unconjugated asci. Takahashi et al. (1958) and Takahashi and Ikeda (1959) found diploid homothallic strains of *Saccharomyces cerevisiae* with a unisexual or bisexual mating reaction.

In the filamentous yeasts, the mycelial phase may be either haploid or diploid, which in turn determines the mode of ascus formation. In the case of the haploid heterothallic species *Stephanoascus ciferrii,* for example, the fusion of two hyphae of opposite sex or anastomoses of lateral branches of two such hyphae precedes ascus formation. If, on the other hand, the hyphae are diploid, the asci may be borne in characteristic clusters (e.g. *Ambrosiozyma monospora*) or in terminal chains (e.g. *Ambrosiozyma (Hormoascus) platypodis*) or hyphal units may be directly converted into intercalar asci, as in *Saccharomycopsis capsularis.* In the latter species ascus formation may, however, also be initiated by the fusion of presumed haploid cells (Kreger-van Rij & Veenhuis, 1975b). Only part of the cell wall between the cell-pairs appears to dissolve. After the nuclear migrations, the opening in the cross wall closes and both cells are converted into asci. When examining filamentous species it should be noted whether the asci are borne on recumbent or erect hyphal structures. In most filamentous species of yeast asci are formed exclusively on the mycelial hyphae. In a few species, such as *Pichia (Hyphopichia) burtonii*, the asci are loose cells.

The form of the ascus can be characteristic of a genus. In *Lipomyces,* ascospores are delimited within sac-like appendages or structures, resulting from enlarged buds or from the conjugation of buds or from the conjugation of a bud with a cell that is not the mother cell. Likewise the formation of large clavate, ellipsoidopedunculate asci is typical of *Metschnikowia*, while among the filamentous yeasts the characteristically crowned asci of *Stephanoascus* readily differentiate this genus from others.

Asci may be persistent in which case the ascus is only ruptured when the spores germinate, or evanescent when the walls of the asci lyse with the ensuing liberation of the ascospores. Evanescent asci are, for instance, characteristic of the genera *Clavispora, Dekkera, Kluyveromyces, Sporopachydermia, Wickerhamia* and certain representatives of the genera *Pichia, Hanseniaspora* and *Hansenula.* Dehiscing asci which rupture at a specific site of the ascus, are rare and have only been observed in the genus *Wickerhamiella*.

Ascus formation as a rule is induced by conditions which restrict vegetative reproduction and is moreover affected by such factors as the nutritional status of the cells, the composition and pH of the sporulation medium to which cultures are transferred, temperature and degree of aeration.

As considerable difficulty is sometimes encountered in inducing poorly nourished cells to form asci it may be necessary to culture some strains on a rich medium before these are transferred to the sporulation media. On the other hand, many yeasts sporulate without any special preparation on commonly employed stock culture media such as YM agar and malt extract agar.

A variety of special sporulation media have been described in the literature; these include Gorodkowa agar, acetate agar, diluted V8 agar, Herman's growth-restriction agar, 2% malt extract agar, Kleyn's agar, V8 vegetable juice agar, corn meal agar, wedges of carrot and potato and blocks of gypsum or cement. In most of these media the concentration of available carbohydrate is limited so that the transference of actively growing, well nourished cells to these media restricts further vegetative reproduction.

Some genera appear to sporulate best on one particular medium. For *Saccharomyces*, acetate agar has been recommended (Adams, 1949; Fowell, 1952; Kleyn, 1954; McClary et al., 1959). On the other hand, species of the genus *Metschnikowia* are observed to sporulate only on diluted V8 agar, while for *Dekkera* the use of vitamin-enriched media is recommended (van der Walt & van Kerken, 1960a). Osmophilic species, like *Zygosaccharomyces rouxii* are reported to sporulate best on media containing 2% sodium chloride (Wickerham & Burton, 1960). For the genera *Sporopachydermia*, *Pichia* and *Hansenula* 5% malt agar (Difco) has proven to be particularly useful. Some workers prefer 2% malt extract agar. In special cases, e.g. the genus *Lipomyces*, sporulation is favored by transferring cells to media deficient in nitrogen.

Temperature may have a profound effect on ascus formation. For the majority of species a temperature of 20–25°C is usually optimal. *Debaryomyces* species nevertheless sporulate best at 20°C or lower. The genus *Metschnikowia*, however, requires temperatures of 12–15°C. In general ambient room temperatures are suitable for ascus formation by most other yeasts.

Some yeasts are, however, slow in forming ascospores, and cultures may require 3–6 weeks before ascospores are delimited. In view of the importance of ascospore formation as taxonomic criterion, a strain may only be regarded as anascogenous or imperfect when (a) it has failed to yield ascospores on a wide variety of media and (b) it has been established that it does not represent a haploid mating type of some or other heterothallic species. It must also be borne in mind that sometimes yeast strains deteriorate in their ability to sporulate and eventually lose the ability.

Ascospores delimited within the ascus may vary in shape, ornamentation, color, dimensions and the number formed per ascus.

The number of ascospores delimited per ascus may vary from one to many. Single- or bispored asci are common in *Lodderomyces*, *Schwanniomyces* and certain representatives of the genus *Debaryomyces*. Multispored asci (i.e., more than eight per ascus) are usually characteristic of the genus *Lipomyces* and certain species of the genus *Kluyveromyces*.

The shape of the ascospores observed in yeasts varies considerably; they may be spheroidal, ellipsoidal, lentiform (oblate-ellipsoidal), cylindrical with obtuse ends, clavate, reniform to crescentiform, hat-shaped (galeate), cap-shaped, saturniform or walnut-shaped, to spheroidal with a tangential brim, falcate, fusiform or acerose to spindle-shaped with whip-like appendages. The surface of the ascospores may be smooth (glabrous) or warty (verrucose). However, the brims and ledges of hat-shaped or saturniform ascospores and the surface ornamentation, may occasionally be reduced to the extent that these structures cannot be detected by direct light microscopy. Recourse must then be had to observations either by scanning electron microscopy (SEM) or transmission electron microscopy (TEM). The ascospores of various genera have been studied comparatively and references are given in a succeeding paragraph to some of these studies.

Scanning electron microscopy allows the surface of ascospores to be studied and for this the spores have to be free of the ascus. This may be achieved by dissolving the ascus wall with an enzyme such as Glusulase; the technique of preparing specimens has been described by Kurtzman et al. (1974a). With this technique the surface of the ascospore may be seen either to be smooth, warty, ridged or to have a ledge.

Transmission electron microscopy of suitable fixed sections of ascospores reveals the layering of their walls. Moens (1971) observed that in the very early stages of the development of the ascospore, the wall consists of two membranes lying close together (the prospore wall) between which the wall of the ascospore itself develops. A ledge can sometimes be seen early on (Kreger-van Rij & Veenhuis, 1975b). The mature ascospore wall has an electron-dense outer layer which may be either thick or thin, and which can have dark warts. The inner part of the wall is usually less electron-dense and may consist of more than one layer and of varying electron density. In some species the outer part or the inner layer is narrow at first and later broadens, e.g. *Sporopachydermia* and *Stephanoascus* (Kreger-van Rij, 1980, 1978a). When warts are present, they are formed by either the single light inner layer or the outermost of two inner layers.

Kawakami (1960) was the first to publish electronmicrographs of ascospores. Besson (1966) and Kreger-van Rij (1969c) examined ascospores from various genera. Ascospores of *Pichia* and *Hansenula* were studied by Kodama (1969), of *Hansenula* by Black and Gorman (1971), of *Hanseniaspora* by Kreger-van Rij and Ahearn (1968) and by Kreger-van Rij (1977b), of *Debaryomyces* by Kreger-van Rij and Veenhuis (1975a), of *Schwanniomyces* by Kreger-van Rij (1977a), of *Torulaspora* by Kreger-van Rij and Veenhuis (1976b), of *Saccharomyces* and *Kluyveromyces* by Kreger-van Rij (1978b, 1979), of *Issatchenkia* by Kurtzman et al. (1980b). Kurtzman et al. (1972, 1975) and Kurtzman and Smiley (1974) observed the ascospores of *Schwanniomyces, Debaryomyces, Saccharomyces* and *Pichia* with the scanning electron microscope.

As will become evident from the keys to the genera and species, ascosporal morphology may be of particular diagnostic value. Thus, cap-shaped ascospores are characteristic of the genus *Wickerhamia,* spindle-shaped ascospores of *Nematospora,* verrucose, saturniform to walnut-shaped ascospores of *Schwanniomyces* and hat-shaped ascospores of the genera *Ambrosiozyma, Dekkera,* and of some species of the genera *Hansenula* and *Pichia* – to cite but a few examples. It should, however, also be borne in mind that some variation in ascosporal shape may be observed within a species, e.g. *Pichia ohmeri,* where both hat-shaped as well as spheroidal ascospores are encountered.

In certain genera, e.g. *Lipomyces, Nadsonia, Saccharomycopsis, Pichia, Hansenula* and *Wingea* the ascospores may be pigmented with the result that actively sporulating material assumes an amber, brown or reddish-brown color.

Ascospores liberated from evanescent asci may have the tendency to remain in agglutinated masses. This is common in the genera *Kluyveromyces, Dekkera* and *Hanseniaspora* and among certain *Pichia* and *Hansenula* species.

The mature ascospore wall has particular affinities for certain basic and acid dyes. This is clearly evidenced by the fact that the mature ascospores, in contrast to the vegetative cells, show a marked degree of acid-fastness – a property which, remarkably, is shared by chlamydospores. Ascospores like chlamydospores are also well adapted to retain vitality under conditions which do not permit vegetative reproduction. Characteristic of the genus *Schizosaccharomyces* is the fact that owing to the presence of amyloid material, the ascospore wall in some species stains blue with diluted Lugol's iodine solution. As has been demonstrated by Beijerinck (1897, 1898a), and Wickerham and Burton (1954a) ascospores are to some extent more heat resistant than normal vegetative cells and generally they are able to survive temperatures of 52–65°C for longer periods than vegetative cells. Beijerinck (1898a) and Wickerham and Burton (1954a) pointed out that actively sporulating cultures can be recovered by the heat treatment of homothallic strains in which the ability to sporulate has deteriorated. Wickerham and Burton (1954a) also showed that the heat treatment of ascosporic material provides a means of determining whether species are homothallic or heterothallic (see p. 73).

When studying isolates, the dimensions of the asci and ascospores should be recorded.

Methods and media

The culture to be studied is first brought to a state of active growth and optimal nutrition either by subculturing on a special presporulation medium, YM agar or malt agar for 1–2 days at 25–28°C. The sporulation media are then lightly inoculated with the culture.

If gypsum blocks are employed, cells from a 2- to 3-day-old vigorous culture in liquid medium, such as malt extract or YM broth, may also be used.

The inoculated sporulation media are then incubated at 25°C for 3 days before being examined microscopically for the first time. Material that has not sporulated is then maintained at room temperature and examined at weekly intervals for at least 4–6 weeks.

Ascospore formation can be verified by staining the heat-fixed preparations. Several procedures have been recommended for the differential staining of ascospores.

Staining procedures

Schaeffer-Fulton's modification of the Wirtz method (Schaeffer & Fulton, 1933)

Heat-fixed preparations are flooded with 5% aqueous malachite green for 30–60 sec, and heated to steaming 3–4 times. The excess stain is rinsed off under running water for about half a minute. The preparation is then counterstained with 0.5% safranine for about 30 sec. The mature ascospores stain blue-green and the vegetative cells red. Wickerham (1951) recommends staining in hot (80°C) 5% aqueous malachite for 3–5 minutes and reported that in *Hansenula* mature Saturn-shaped ascospores with thick rings retain both stains.

Modified Kufferath carbol-fuchsine stain

Heat-fixed preparations are flooded with Ziehl–Neelsen carbol-fuchsine and steamed gently for about 2–5 minutes. Slides are then decolorized with either 2% lactic acid or 95% ethanol containing 1% concentrated hydrochloric acid. The time required for the decoloration varies. The slides are rinsed in water and counter-stained with either 1% methylene blue, thionin or Nile blue hydrochloride. The mature ascospores stain red and the vegetative cells blue. Wickerham (1951) reports that the ascospores of *Schizosaccharomyces* species are not acid fast.

Wickerham (personal communication) states that ascospores of *Hansenula* and *Pichia* are acid fast while still in the ascus but cease to be acid fast once they have swollen after liberation retaining, however, their characteristic shape.

Presporulation media

1. Medium of Lindegren (1949)

This medium contains 10 ml beet leaves extract (100 g per 100 ml boiling water), 10 ml beetroot extract (100 g per 100 ml boiling water), 35 ml canned apricot juice, 16.5 ml bottled grape juice, 2 g dried yeast, 2.5 ml glycerol, 1 g calcium carbonate and 3 g agar. Water is added to give a final volume of 100 ml. After solution of the agar by steaming for 10 minutes, the medium is dispensed in tubes and autoclaved for 15 minutes at 15 lbs overpressure.

2. Grape juice

Freshly expressed grape juice of any grape variety is diluted to 8–10° Balling and sterilized in flowing steam for 10 minutes.

Sporulation media

1. Vegetable juice agar (Mrak et al., 1942a), V8 agar

Equal weights of washed, unpeeled carrots, beetroots, cucumbers and potatoes are minced or grated finely and mixed with a quantity of water equal to the weight of vegetables used. The mixture is autoclaved for 10 minutes at 10 lbs overpressure and expressed through cheesecloth. The extract has a pH of circa 5.7. To the extract are added 2% (w/v) dried yeast and 2% (w/v) agar. After solution of the agar by steaming, the medium is dispensed in tubes and autoclaved for 15 minutes and 15 lbs overpressure. The tubes are slanted.

Wickerham et al. (1946) recommend the following vegetable juice agar: 14 g of agar is melted in 340 ml demineralized water; in another container is placed 350 ml of V8 juice from a can containing according to the label 12 fl. oz., and 5 g of moist compressed yeast dispersed in 10 ml distilled water. The suspension is stirred. The pH is adjusted to 6.8. The suspension is then heated for 10 minutes in flowing steam. The pH is again adjusted until a cooled sample has a pH of 6.8. The hot medium is then poured into the hot agar solution, stirred and dispensed into tubes. Sterilization is at 15 lbs overpressure for 15 minutes. Slants are prepared as needed, avoiding unnecessary heating when melting the medium. (The canned vegetable juice containing a blend of tomatoes, carrots, celery, beets, parsley, lettuce, spinach and watercress is marketed by Campbell Soup Company, Camden, NJ, U.S.A. under the proprietary name 'V8' vegetable juice.)

2. Dilute V8 agar (Pitt & Miller, 1968)

A can of V8 juice is diluted with an equal volume of demineralized water and adjusted to pH 5.5 with sodium hydroxide. The diluted juice is filtered through Whatman No. 1 paper and diluted further from 1:2, 1:9, 1:19 to 1:29 to which dilutions 2% agar is added. After steaming to dissolve the agar the media are tubed and autoclaved.

3. Gorodkowa agar (modified)

To tap water add 0.1% (w/v) glucose, 1% peptone, 0.5% sodium chloride and 2% agar. Steam until the agar dissolves and dispense in tubes. Autoclave for 15 minutes at 15 lbs overpressure and slant. Some workers advocate 0.25% glucose and the addition of 1% meat extract.

4. Fowell's acetate agar (Fowell, 1952)

To demineralized water add 0.5% sodium acetate trihydrate and adjust the pH to 6.5–7.0. Add 2% agar and steam to dissolve the agar. Dispense in tubes, autoclave for 15 minutes at 15 lbs overpressure and slant.

5. Kleyn's acetate agar (Kleyn, 1954)
To demineralized water add 0.25% (w/v) Bacto tryptose, 0.062% glucose, 0.062% sodium chloride, 0.5% sodium acetate trihydrate and 1.5% agar. Dissolve the constituents by heating and dispense in tubes. Sterilize by autoclaving for 15 minutes at 15 lbs overpressure and slant. Initial pH after sterilization 6.9–7.1.

6. McClary's acetate agar (McClary et al., 1959)
To demineralized water add 0.1% (w/v) glucose, 0.18% potassium chloride, 0.25% yeast extract, 0.82% sodium acetate trihydrate and 1.5% agar. Dissolve the constituents by heating and dispense in tubes. Sterilize by autoclaving for 15 minutes at 15 lbs overpressure and slant.

7. Malt extract agar or ME agar (Wickerham, 1951)
Dissolve 12 g agar in 400 ml demineralized water by steaming. Then dissolve 20 g powdered malt extract in the hot solution. Dispense in tubes or bottles and sterilize by autoclaving for 15 minutes and 15 lbs overpressure and slant. Slants are kept in the refrigerator until used. Minimal heating should be used in melting the sterilized medium.

8. Potato agar, see p. 59.

9. Corn meal agar, see p. 59.

10. YM agar, see p. 51.

11. Yeast extract-2% glucose agar
To demineralized water add 0.5% (w/v) of a commercial powdered yeast extract, 2% glucose and 2% agar. Dissolve by steaming and dispense in tubes. Sterilize by autoclaving for 15 minutes at 15 lbs overpressure and slant.

12. Vegetable wedges
Carrot, potato, beet, cucumber and turnip wedges may be used. The vegetables are thoroughly cleaned by washing. Long cylinders about 1 cm in diameter are cut from the vegetables with the aid of a cork borer. The wedges are prepared by cutting the cylinders obliquely. The wedges are rinsed in cold water and placed in suitable tubes containing some water to prevent dessication. The tubes are sterilized by autoclaving for 15 minutes at 10 lbs overpressure.

13. Gypsum blocks and wedges
Mix 8 parts of gypsum (calcium sulphate hemihydrate) with 3 parts of water. Cast the paste into cylindrical or truncate conical forms about 3–4 cm high. After setting the blocks are placed in suitable, sterile glass dishes with covers and maintained at 110–120°C for at least 2 hours. Before use, sterile water is added to the dishes to give a depth of about 1 cm. The use of a solution of a mannitol-dipotassium hydrogen phosphate mixture (18 ml 0.5–2% mannitol + 2 ml 5% K_2HPO_4 has also been advocated). A thin layer of well nourished yeast cells is spread on the top of the block. The gypsum may also be placed in tubes. Gypsum and water are mixed to a creamy paste which is poured into sterile, cotton-plugged tubes by means of a funnel. The tubes are then slanted and allowed to harden at 50°C for 24–48 hours. The tubes are then autoclaved.

14. Aqueous agar
Dissolve 20 g agar in 1 liter demineralized water. Distribute and autoclave at 10–15 lbs overpressure.

15. Yeast infusion (yeast water) agar
To yeast infusion add 1.5% agar. Sterilize by autoclaving for 15 minutes at 15 lbs overpressure. For the preparation of yeast infusion see p. 79.

16. Starkey's ethanol medium (modified)
This medium contains 0.5% ethanol, 0.10% dipotassium hydrogen phosphate, 0.025% potassium dihydrogen phosphate, 0.025% magnesium sulphate heptahydrate, 0.005% calcium chloride and 2% agar in tap water. Autoclave for 15 minutes at 10 lbs overpressure. Ethanol is added after sterilization.

17. Rice agar, see p. 59.

18. Restricted growth (RG) medium (Herman, 1971)
To 250 ml demineralized water is added 0.05 g yeast extract, 0.05 g peptone, 0.25 g glucose and 5 g agar. Dissolve by steaming and dispense in tubes. Sterilize by autoclaving for 15 minutes at 15 lbs overpressure.

19. YM agar + 2% sodium chloride
Medium is prepared by adding 2% sodium chloride to YM agar.

20. Oatmeal agar
To 100 ml of tap water add 4 g oatmeal, boil for 1 hour and filter through cheesecloth so as to permit the passage of starch. The filtrate is made to 100 ml and 1.5 g agar is added. Dissolve by steaming and dispense in tubes. Sterilize by autoclaving for 15 minutes at 15 lbs overpressure.

21. 2% Glucose-yeast autolysate agar (Phaff et al., 1960)
To 1000 ml yeast autolysate add 20 g glucose and 20 g agar. Dissolve the agar by steaming and distribute in tubes. Autoclave at 10 lbs overpressure for 15 minutes.

22. Adams' acetate agar (Adams, 1949)
To demineralized water add 0.04% glucose, 1.4% anhydrous sodium acetate and 2% agar. Dissolve by heating, dispense in tubes, autoclave for 15 minutes at 15 lbs overpressure and slant.

b. METHODS FOR THE ISOLATION OF MATING TYPES

When dealing with strains of which the perfect or imperfect states are not known and for the purpose of testing for interfertility, it is necessary to be able to isolate mating types from haploid and diploid heterothallic species. Occasionally it may also be desirable to isolate diploid vegetative cultures from haploid heterothallic species.

The following paragraphs have been prepared in collaboration with Dr. L.J. Wickerham:

Methods

Isolation of mating types of heterothallic haploid yeast species

A species that produces few or no ascospores when isolates are tested individually on sporulation media, should be tested in mixtures of four strains. The larger the total number of strains thus tested, the greater the chances of success in obtaining mating types among the more primitive species of yeasts. To make mixtures, approximately equal amounts of actively growing cells are mixed in the center of a slant of sporulation medium and the mixture is spread out. Observations should be made for the presence of zygotes and ascospores after a few days and then weekly over a period of three weeks. If a sexual reaction is observed which is in excess of the reaction that is seen to occur in the individual strains, the four strains of the reacting groups should be mixed in six pairs to determine which of these strains are cross mating. Details of these procedures are given by Wickerham (1952, 1960, 1965a,b and 1969a), Wickerham and Burton (1952, 1954a,b and 1962), Phaff and Yoneyama (1961) and Slooff (1964).

Occasionally a natural substrate which is rich in both the mating types of a heterothallic yeast may be streaked directly on plates of a medium on which the species is known to react sexually. While zygotes and ascospores may later be observed in the confluent areas of growth, very few or no zygotes or ascospores will be observed in individual colonies. Such colonies should be numbered, and equal portions of each colony mixed together to include a total of some 12–16 colonies in three or four mixtures. If a sexual reaction is obtained, cells from the remnants of the pertinent colonies may be subcultured on YM agar slants, and then transferred separately and mixed in pairs on a sporulation medium for the recognition of mating types. This procedure may be followed whenever microscopic observations of colonies on isolation plates suggest that a heterothallic species is involved. Wickerham and Burton (1960) first became aware of heterothallism in *Zygosaccharomyces rouxii* through examination of colonies streaked from miso, a fermented Japanese food.

Isolation of mating types of heterothallic diploid species

It is only rarely that the diplophase and the two sexes of a species which normally exists in the diploid state, are recovered simultaneously either from a natural habitat or from an elective culture. When plated on a suitable sporulation medium, colonies of the haplophase and the presence of zygotes in confluent areas of adjacent haploid colonies may indicate the possible presence of mating types. Portions of the presumed haploid colonies, as indicated by their smaller cell size, may be mixed and observed regularly during 6 days for the presence of zygotes.

More commonly, however, mating types of diploid species are obtained by heat treatment of a suspension of a sporulating culture (Wickerham & Burton, 1954a). The optimal temperature varies with the species from 52 to 65°C. The temperature selected should permit vegetative cells to remain alive for about 6–10 minutes, after which only ascospores should remain viable, so that plates streaked with the heated suspension should at intervals after 10 minutes only yield colonies derived from ascospores. The heating should be continued until all ascospores are inactivated and there is no further development of colonies. Colonies on the terminal plates, i.e. plates that yielded colonies at the longest time interval are examined for the ability to produce ascospores. If all the colonies form ascospores the strain is homothallic. If only a few or no ascospores are produced, the strain is heterothallic. Nonsporulating isolates are then vitalized by subculturing and mixed by fours and then in pairs to obtain the most active mating types. Heat treatment may be used for isolating the mating types of any genus of yeasts that contains heterothallic ascosporogenous species. The procedure described above may also be applied to the recovery of actively sporulating cultures from homothallic strains in which the ability to form ascospores has deteriorated.

Apart from the heat treatment method, mating types may be obtained by isolating single ascospores by micromanipulation. This method is quick and particularly useful when dealing with material that yields a high percentage of viable ascospores.

Isolation of diploid vegetative cultures from haploid heterothallic species

It is presumed by Wickerham and Burton (1962) that all species of the genus *Hansenula,* no matter how primitive, form both diploid and haploid vegetative cells, though the most primitive species may form so few diploid cells that the latter escape detection. As the species of *Hansenula* presumably evolved consecutively they are assumed to have produced an ever increasing number of diploid cells until a point was reached above which all species were predominantly stabilized in the diplophase. A species placed in the category just below this transition point, e.g. *Hansenula fabianii,* produces many zygotes when mating types of opposite sexuality are mixed on ME slants. The zygotes often yield diploid progeny as indicated by the presence of unconjugated asci. The diplophase may be isolated from the longest hyphal tips that grow out from a mixture of mating types on a ME agar slant inoculated in a truncate manner (Wickerham, 1965a,b). It is possible to maintain pure diploid cultures as long as sporulation is prevented.

2. *Characteristics of basidiospore formation*

The basidiomycetous yeasts occur either as the budding haplophase, the dikaryotic mycelial phase or the self-sporulating diplophase. Although septate dikaryotic mycelium with clamp connections is as a rule characteristic of the teleomorphic states of basidiomycetous taxa, it is not always expressed.

Sexual reproduction in the basidiomycetous yeasts may be either heterothallic or homothallic. The sexual systems of these yeasts are as yet incompletely understood. In the heterothallic species, which may have either bipolar or tetrapolar incompatibility systems, the dikaryotic mycelium is produced by one of the conjugants after mating of a pair of compatible cells. The dikaryotic mycelium ultimately forms large, inflated, frequently lipid-rich, clamped cells in which karyogamy occurs. In view of this function these cells have been interpreted as probasidia. They may be intercalary, lateral or terminal, and may be either thick-walled or not. In the homothallic or self-fertile strains, two kinds of homothallism are found and they are known as primary and secondary homothallism (Fell, 1974). In strains with primary homothallism the mycelium is uninucleate and lacks clamp connections, in those with secondary homothallism it is dikaryotic and has clamp connections.

However, the subsequent development of the large lipid-rich cells into the basidial structures may be highly diversified in the different taxa, involving both functional as well as structural modifications. The terminology of the basidium has not been standardized (see Donk, 1973) and for a comparative illustrated use of the descriptive terms, Ainsworth (1971) may be consulted.

In the ustilaginaceous genera *Rhodosporidium* and *Leucosporidium,* the thick-walled cells have been referred to as teliospores, teleutospores and ustilospores. These spores may vary in shape, ranging from spheroidal to ovoidal and angular. They may be markedly pigmented in the genus *Rhodosporidium.* After maturation and a period of dormancy the teliospore forms a germ tube, or promycelium. In *Leucosporidium scottii* the diploid nucleus migrates into the promycelium where it undergoes reduction division and the four haploid nuclei distribute themselves within the promycelium. In *Rhodosporidium toruloides,* however, meiosis occurs within the teliospore itself and the four haploid nuclei migrate into the promycelium. After reduction division the promycelium forms transverse septa which separate the haploid nuclei. Each nucleus of the septated promycelium then divides mitotically and one of them migrates into a bud which usually develops laterally on each of the promycelial cells. These haploid, sessile buds are referred to as sporidia. In the heterothallic species the genetic factors controlling compatibility segregate during meiosis so that the sporidia give rise to yeast phases of different mating types. In other species, e.g. *Leucosporidium gelidum,* the promycelium does not become septate and the sporidia develop terminally. Some interpret the budding septate promycelium as a phragmometabasidium.

In the heterothallic genus *Filobasidium* the basidia are either terminal or lateral on hyphae with clamp connections. Thin-walled cells slightly broader than the hyphae bearing them – referred to by Olive (1968) as probasidia, but this term is misapplied according to Fell (personal communication) – elongate to form long slender non-septate basidia which taper apically. The tip is expanded and bears, as a rule, 6–8 sessile basidiospores arranged in a characteristic petal-like whorl. No information on the karyology is available. The individual basidiospores give rise to yeast phases of opposite mating types. In the related genus *Filobasidiella,* the basidia usually arise in groups either terminally or laterally on septate dikaryotic hyphae. The basidia are slender non-septate elongations with swollen apices. Up to 20 haploid basidiospores are budded off in a basipetal chain from each of four sites on an apex. Karyogamy and meiosis occur in the swollen apices (Kwon-Chung, 1980). The sessile basidiospores give rise to yeast phases of opposite mating types.

Methods

The heterothallic basidiomycetous yeasts are rarely isolated from their source as teleomorphic or sexual states but usually as budding haploid yeast phases. Compatible strains need to be mated to obtain the dikaryotic clamp-bearing mycelial phase. This involves thoroughly mixing loopsful of cells of potential opposite mating types on agar plates. The mixing should be confined to as small an area as possible. The plates are incubated and frequently examined under low magnification (60–100 ×) for clamped hyphae growing out into the agar around the mated colonies. When the clamped hyphae have extended well beyond the colonies, their tips are carefully dug out of the agar and transferred to suitable media for further observation. Cultural conditions, such as temperature and illumination, may affect results of mating tests appreciably.

A variety of media such as malt agar, corn meal agar, hay infusion agar and more complex media have been recommended for the mating tests. For further details the publications of Banno (1967), Fell et al. (1969, 1973), Fell and Statzell Tallman (1980b), Kwon-Chung (1975, 1976a, 1977) and Rodrigues de Miranda (1972) as well as the descriptive chapters must be consulted.

Whereas the maturation and subsequent development of the basidia of *Filobasidium* and *Filobasidiella* as well as the teliospores of some ustilaginaceous species, e.g. *Rhodosporidium toruloides,* may be observed directly on the media on which the dikaryotic, clamped mycelium is produced, the teliospores of other ustilaginaceous species, e.g. *Leucosporidium antarticum* and *Leucosporidium frigidum,* fail to germinate under these conditions. As the germination of the teliospores and the subsequent development of the basidial structure are decisive for the classification of these yeasts, special techniques have been developed to accelerate the germination of these spores. These techniques usually involve soaking bits of agar on which the spores are present, in demineralized water at temperatures ranging from 5°C to room temperature, sometimes for as long as 10 weeks. Conditions for germination vary from species to species. For further details, the descriptive chapters must be consulted.

Media

1. Hay infusion agar

Mix 50 g of decomposing hay with 1 liter of demineralized water, autoclave at 15 lbs overpressure for 30 minutes and then filter. Dissolve 2 g potassium monohydrogen phosphate and 15 g agar in 1 liter of the filtrate and adjust the pH to 6.2. Sterilize by autoclaving at 15 lbs overpressure for 20 minutes.

2. Sucrose-yeast extract agar (Kwon-Chung, 1976a)

To 1 liter of demineralized water add 1 g potassium dihydrogen phosphate, 0.5 g magnesium sulphate heptahydrate, 0.1 g calcium chloride, 0.1 g sodium chloride, 5.0 μg biotin, 20 g sucrose, 0.5 g yeast extract and 40 g agar and dissolve by heating. Sterilize by autoclaving at 15 lbs overpressure for 20 minutes.

3. Hybridization techniques for taxonomic purposes

The use of hybridization studies as guide-lines for the delimitation of perfect yeast species was first suggested by Winge and Laustsen (1939). The implementation of this approach to the delimitation of species, however, stems from a series of investigations by Wickerham and Burton (1952, 1954a,b) and Wickerham (1958), who were the first to show that several imperfect species (e.g. *Candida guilliermondii, Candida melibiosi, Candida pelliculosa, Candida chodati, Torulopsis globosa* and *Torulopsis sphaerica*) constituted mating types of heterothallic ascogenous species. In the case of the basidiomycetous yeasts, Banno (1967), using the prototrophic selection technique, was the first to establish that certain strains of the genus *Rhodotorula* constituted haploid mating types of a previously unknown perfect state which he assigned to the genus *Rhodosporidium.*

The discovery of heterothallism in the yeast domain has had far-reaching taxonomic implications. Not only did the discovery of mating type systems rationalize the classification and nomenclature of both the imperfect as well as perfect species by confirming the synonomy of numerous taxa, but the systematic search for mating types among the anamorphic taxa also led to the recovery of several new heterothallic genera (e.g., *Stephanoascus, Sporopachydermia, Leucosporidium* and *Filobasidiella*).

The delimitation of species on the basis of interfertility accepts that if two yeast strains are able to cross or hybridize with the formation of fertile hybrid progeny, such strains are representative of the same species. The fertility of the hybrid progeny is assessed on the basis of the viability of the ascospores (in the case of the basidiomycetous taxa, the sporidia or basidiospores) of the initial crosses and further by checking the mating ability of one further generation. Back-crossing with parental strains may be used to detect the possible amphiploidy of strains (Kurtzman et al., 1980a,b).

Several techniques for the hybridization of yeasts have been developed.

When dealing with heterothallic taxa hybridization studies hinge on the availability of mating types. The mating types of the taxa under study are mass-mated in all possible combinations on suitable media and interfertility is assessed by the formation of fertile perfect states expressed as either asci or in the case of basidiomycetous yeasts, teliospores or basidia, in compatible crosses. The experimental details of such crosses have been dealt with under sections B.1. and 2. For further information the recent publications by Kurtzman et al. (1980a,b) may be consulted.

When dealing with homothallic strains, techniques are more laborious and involve the procedures developed for the hybridization of *Saccharomyces cerevisiae* and its biotypes. These techniques have been fully reviewed by Fowell (1969). The first method involves micromanipulation and relies either on the mating of individual ascospores or haploid, vegetative cells of the strains under study. The pairing of individual ascospores finds particular application when dealing with strains characterized by a diplontic life cycle and high ascosporal viability. The second method involves the prototrophic selection technique of Pomper and Burkholder (1949). This method relies on the mass-mating of either two complementary auxotrophic parental mutant strains or a prototroph and an auxotrophic strain, where the auxotroph utilizes a carbon source not utilized by the prototroph. After mating and allowing for hybrid formation, the mated cultures are plated on a minimal medium which permits the development of recombinants only. Interfertility is assessed by the recovery of fertile hybrid progeny with combined parental phenotypes and by comparison of recombination frequencies observed in intrastrain and intraspecific crosses. This method has the advantage in that it obviates problems arising from low ascosporal viability of parental strains and permits the detection of recombination at very low frequencies. For experimental details, the publications of Banno (1967), Johannsen and van der Walt (1978, 1980), van der Walt and Johannsen (1979) and Johannsen (1981) must be consulted.

Interfertility between strains as criterion for their conspecificity can only be applied to species in which the alternation of generations involves the fusion of free-occurring, independent cells. This criterion cannot be applied to completely inbreeding strains as for example those of the genus *Schwanniomyces* (Johannsen & van der Walt, 1980).

As already outlined in Chapter I, available evidence suggests that the species delimited, either on the basis of interfertility of strains or on the reassociation of their fragmented, single-stranded nDNA, does not invariably result in the same taxon. For an evaluation of these procedures the relevant sections of Chapter I should be consulted.

C. *Physiological and biochemical characteristics*

Physiological properties primarily serve to describe, differentiate and identify yeast strains. They also serve to describe, characterize, and differentiate species and to a lesser extent, genera. The characteristics which have proved most useful for routine identification purposes are those associated with the utilization of carbon and nitrogen sources, growth factor requirements, growth at elevated temperatures and on media of high sugar or sodium chloride content, the formation of typical characteristic metabolites and the susceptibility of the yeast to antibiotics. The diagnostic value these various characteristics carry, is discussed in Chapter I and will become evident from the descriptive chapters.

As the results of tests to establish the various physiological characteristics are frequently dependent on the techniques employed, the tests must be standardized and there must be rigorous adherence to the prescribed procedures. The testing for the various physiological characteristics should only be undertaken on fresh yeast material grown under optimum nutritional conditions.

1. Utilization of carbon compounds

A yeast developing on a carbohydrate source must either be able to ferment it or to utilize it by respiration. With very few exceptions, it has been found that when a yeast strain utilizes a carbohydrate fermentatively, it is also capable of utilizing it oxidatively. The reverse does, however, not hold.

a. FERMENTATION OF CARBOHYDRATES

Yeasts vary in their ability to ferment sugars. For example, the genera *Kluyveromyces, Saccharomyces, Torulaspora* and *Zygosaccharomyces* are all characterized by the vigorous fermentation of at least glucose, while other genera, e.g. *Lipomyces* and *Sterigmatomyces* are strictly nonfermentative, while in other genera, e.g. *Hansenula,* an entire range from nonfermentative to strongly fermentative species may be found.

A number of tests have been devised to determine the presence or absence of fermentation as indicated by the production of carbon dioxide from carbohydrates under conditions which permit the fermentative dissimilation of the sugar (see van der Walt, 1970a). Heavy inocula must be employed in the closed systems as the availability of oxygen is limited, and in consequence cell proliferation is restricted. The restriction of cell proliferation, however, is undesirable since it hampers the detection of the fermentation that takes place as a result of the formation of adaptive enzyme systems. For routine purposes Durham tubes containing 2% (w/v) sugar solutions have been found to be the most useful.

The basal medium in which the carbohydrates are dissolved must contain an adequate amount of available nitrogen and other nutrients, including trace elements and vitamins essential for growth so that optimal conditions obtain for the synthesis of the enzyme systems involved in the transport and hydrolysis of the carbohydrates. The basal medium employed must be free of other fermentable sugars, e.g. trehalose which may give spurious results as sometimes happens when using yeast infusion as a basal medium. In practice both dilute yeast autolysate and the peptone-yeast extract base proposed by Wickerham (1951) have given good results.

For identification purposes the ability to ferment glucose, galactose, sucrose, maltose, lactose and raffinose is routinely examined. The fermentation of other carbohydrates, including trehalose, melibiose, and inulin is occasionally reported.

If fermentation occurs, glucose will always be fermented. Kluyver (1914) first pointed out that the ability to ferment glucose was invariably associated with the fermentation of both fructose and mannose. This may be accounted for on the basis that yeast hexokinase not only mediates the phosphorylation of glucose, but that of fructose and mannose as well. Certain osmophilic yeasts, however, have been reported to ferment fructose preferentially from a mixture of glucose and fructose. This characteristic, however, is not made use of in routine identifications (see van der Walt, 1970a). Yeasts which ferment sucrose with few exceptions also ferment raffinose. Similarly, the ability to ferment lactose appears to be invariably associated with the fermentation of galactose. Yeasts which ferment lactose with few exceptions do not ferment maltose.

As earlier but limited experience indicated that the ability to ferment a specific sugar was, by and large, constant for a given yeast strain, the fermentation pattern of a species had come to be regarded as stable and immutable. However, since the appearance of the earlier monographs on yeast taxonomy, evidence has accumulated which indicates that the stability of the fermentation pattern is by no means a hard and fast rule. The so-called instability of the fermentation pattern can be accounted for by mutational changes as evidenced in certain yeast strains as a result of their prolonged maintenance on malt agar or as a result of exposing cells to specific carbohydrates for prolonged periods (see van der Walt, 1970a).

Methods and media

Only high grade sugars should be used.

The fermentation of all sugars used is tested in 2% (w/v) solutions in Durham tubes except in the case of raffinose where a 4% (w/v) solution is normally employed. As melibiose is relatively expensive its fermentation need only be tested when raffinose is also fermented or when a fermentative yeast is known to assimilate melibiose.

The sugars are dissolved either in dilute (1:10) yeast autolysate or in a 0.5% solution of commercial powdered yeast extract. A yeast infusion is less satisfactory. 5–6-ml aliquots of the solution are dispensed into plugged, sterile 150 × 12 mm tubes carrying insert tubes of approximately 50 × 6 mm which are sterilized by autoclaving for 15 minutes at 10 lbs overpressure. Wickerham (1951) advocates the use of a basal medium consisting of peptone and a yeast extract containing bromothymol blue. 2-ml aliquots of the basal medium are pipetted into sterile, plugged Durham tubes and autoclaved for 15 minutes at 15 lbs overpressure. After sterilization 1-ml quantities of the various sterile sugar solutions are added aseptically to the tubes. 6% aqueous solutions of the sugars are prepared except in the case of raffinose which is made up in a 12% solution. The sugar solutions are sterilized by filtration and stored in the refrigerator.

The fermentation media are inoculated with about 0.1 ml of a suspension of yeast cells made by suspending the growth of a 24–48 hour YM or malt extract agar culture in 4.5 ml sterile water. Tubes may also be more laboriously inoculated directly from actively growing slant cultures by means of a stout platinum loop. For the slow growing and nutritionally fastidious genera *Brettanomyces* and *Dekkera*, it has been found that by preparing inocula in stock vitamin solutions more rapid results are obtained.

The tubes are incubated at 25 or 28°C. The tubes are regularly shaken and observed for the accumulation of gas in the inserts and for change in color of the indicator (if used) over a period of 14 days. Depending on the time required for the formation of visible amounts of gas, the fermentation is rated as:

+	strong, with gas filling the insert tube within 1–3 days
+ w	weak, with the insert tube only partially filled
+ vw	very weak, with only a bubble formed in the insert tube
+ s	slow or delayed, but with gas filling the insert tube
–	absent
v	some strains of a species ferment a sugar and others not.

1. Yeast autolysate

Mix equal weights of compressed yeast and water in a flask. A thin layer of toluene may be added. Incubate at 50–55°C for 3 days stirring occasionally. Boil for a few minutes and allow to cool. The pH is adjusted to 6 by the addition of N KOH. The supernatant is cleared either by the addition of egg white or by filtration with use of a filter aid, e.g. Hyflo Supercel. The clear filtrate represents full strength yeast autolysate.

2. Yeast infusion or yeast water (Lodder & Kreger-van Rij, 1952)

Mix 200 g compressed yeast with 1 liter tap water, add some egg white and autoclave at 15 lbs overpressure for 15 minutes. The mixture is filtered, while hot, through thick filter paper, twice if necessary.

3. Fermentation basal medium (Wickerham, 1951)

Dissolve 4.5 g powdered yeast extract (a batch should be used which will support the growth of markedly vitamin-deficient yeasts, e.g. *Kloeckera apiculata*) and 7.5 g peptone in 1 liter demineralized water. Add bromothymol blue to give a sufficiently dense green color. 2-ml aliquots are placed in 150 × 12 mm tubes carrying insert tubes. The tubes are sterilized by autoclaving for 15 minutes at 15 lbs overpressure. When cool 1-ml concentrated, filter-sterilized sugar solutions are added aseptically.

4. Stock vitamin solution (van der Walt & van Kerken, 1961a)

Dissolve 0.2 mg biotin, 40 mg calcium pantothenate, 0.2 mg folic acid, 200 mg inositol, 40 mg niacin, 20 mg *p*-aminobenzoic acid, 40 mg pyridoxine hydrochloride, 20 mg riboflavin and 100 mg thiamine in 1 liter demineralized water and sterilize by filtration.

The sterile solution is dispensed in 10-ml aliquots stored at − 10 to − 15°C.

b. ASSIMILATION OF CARBON COMPOUNDS

A number of growth tests have been devised to determine the ability or inability of yeasts to utilize carbon compounds.

In the first type of test, the compound is added to a complete liquid basal medium in which the compound serves as the sole source of carbon. This type of test has been standardized by Wickerham and Burton (1948) and finds wide application. As the test runs over several weeks, it permits the adaptation of the yeast to certain compounds. Erroneous results are avoided by the use of a blank consisting of the inoculated basal medium devoid of any carbon source. Aeration or agitation gives more rapid results, as has been shown by Barnett and Ingram (1955) and Ahearn et al. (1960). Currently, use is frequently made of the Rollordrum equipment used for tissue culture and marketed by the New Brunswick Scientific Co. Inc.

The second type of test embodies the use of a medium as employed in the first test, but with the addition of purified agar. The inoculum is applied either as a streak or dot. This type of test also demands the use of a blank since agar not infrequently contains traces of assimilable compounds which may support growth. This technique has been adapted by Shifrine et al. (1954) and Beech et al. (1955) for screening large numbers of yeasts on a single plate. This method also lends itself to the Lederberg replica-plating technique.

In the third type of test – the familiar auxanographic method of Beijerinck (1889a) – a cooled, molten basal agar medium devoid of any carbon source, is seeded with the yeast inoculum. A pour-plate is then prepared on which small amounts of the compounds to be tested are placed after solidification of the agar. This procedure has the advantage that it is possible to record results within 2–4 days. As the agar tends to dry out on prolonged incubation, it has the disadvantage that compounds that require an adaptation period or which are only slowly utilized, cannot be tested by this method.

While most yeasts can grow in the routinely employed mineral media, there are exceptions, e.g. the genus *Cyniclomyces* which requires a very complex nitrogen source as well as carbon dioxide for growth.

While the Dutch monographers Lodder (1934) and Diddens and Lodder (1942) introduced the use of glucose, fructose, mannose, galactose, sucrose, maltose, lactose and ethanol in assimilation tests, it was mainly as a result of Wickerham's monograph on the genus *Hansenula,* that the use of a greater number of carbon compounds gained general acceptance (Wickerham, 1951).

Wickerham and Burton (1948), after screening 70 compounds using 100 yeast strains representative of 22 genera, selected some 30 compounds which held most promise for taxonomic purposes. The compounds they selected were extensively used in the second edition of 'The Yeasts'.

In the present edition, the utilization of 18 specific carbon compounds is given in the standard description of each species. In special cases, for instance when these compounds do not suffice for differentiation of the species, the assimilation of additional compounds is mentioned. The 18 compounds and the additional compounds are the following:

Hexoses: galactose, L-sorbose

Disaccharides: sucrose, maltose, cellobiose, trehalose, lactose, melibiose

Trisaccharides: raffinose, melezitose

Polysaccharides: soluble starch, inulin

Pentoses: D-xylose, L-arabinose, D-ribose, L-rhamnose, D-arabinose

Alcohols: erythritol, ribitol (adonitol), D-mannitol, inositol, methanol, ethanol, glycerol, galactitol (dulcitol), D-glucitol (sorbitol)

Organic acids: succinic acid, citric acid, DL-lactic acid, (−)-malic acid, glucuronic acid, 2-ketogluconate, 5-ketogluconate

Glycosides: α-methyl-D-glucoside, arbutin, salicin

Other compounds: glucono-δ-lactone, D-glucosamine hydrochloride, decane, hexadecane.

As glucose is utilized by all yeasts, its utilization is not reported in the descriptive chapters. It is nevertheless always included as a standard for comparing the rate at which other carbon sources are utilized and also to assess whether or not strains are capable of growth in a mineral medium.

In the descriptive chapters the utilization of carbon sources is listed as:

+	rapid
+ w	weak
+ s	slow or latent
+ w or s	weak or slow
+ or w	strong or weak
+ or s	strong or slow
+ (−)	+, seldom −, i.e. most strains studied are +, a few −
− (+)	−, seldom +
+ s (−)	slow, seldom −
+ s (+)	slow, seldom strong

It must be pointed out that as in the case of fermentation, strains or certain species may acquire the ability to utilize certain sugars after prolonged maintenance on media such as malt agar.

Methods

For the carbon assimilation tests it is imperative to use only high-grade, standardized products devoid of any contamination that might give rise to spurious results. For example certain batches of marketed brands of maltose have on occasion been found to contain traces of glucose, which can be detected auxanographically by testing with any yeast which does not utilize maltose. When employing partially degraded polysaccharides, e.g. soluble starch, care should be taken not to use material which has been degraded beyond a certain limit and which is consequently more readily attacked.

1. Liquid medium assimilation tests (Wickerham & Burton, 1948; Wickerham, 1951; Lodder & Kreger-van Rij, 1952)

Preparation of the inoculum

Before proceeding with the tests, the strain to be tested must first be brought to a state of active growth. This is effected by transferring the strain once or twice on e.g. YM agar at 25–28°C at 2–3 day intervals depending on its growth rate.

Cell suspensions are usually used to inoculate the media used in the assimilation tests.

Some laboratories prefer to use cell suspensions made up with sterile tap water or sterile yeast nitrogen base. Material is removed directly and aseptically from an actively growing slant culture by means of a stout platinum loop, and is thoroughly dispersed in about 3 ml sterile tap water or demineralized water in a 16-mm tube. The suspension is aseptically diluted with sterile water until the black lines approximately ¾ mm wide drawn on white cardboard become visible through the tube as dark bands. Each of the tubes containing the different carbon sources is then inoculated with one drop of such a suspension delivered from a Pasteur pipette. A similarly inoculated blank tube containing the basal medium without any carbon source serves as control. When dealing with the fastidious genera *Dekkera* and *Brettanomyces* it is advisable to prepare cell suspensions for inoculation purposes in sterile, concentrated stock vitamin solution. In making direct suspensions of this kind, great care must be exercised not to transfer agar particles or syneresis exudate formed in the tubes which may introduce assimilable material into tests and so give rise to spurious results. The use of a blank tube is therefore imperative.

Wickerham (1951) described the use of a special inoculation medium. This inoculation medium which has the same composition as the medium used in the liquid carbon assimilation test, contains 0.1% glucose and is dispensed in 10-ml aliquots in tubes. One ml of this medium is withdrawn aseptically and is used to suspend the growth from a 24–48 hour slant culture of the strain being studied. Depending on the abundance of growth on the slant, 0.2–0.4 ml of the suspension is returned to the inoculation medium which is then incubated at 25°C for 48 hours. The inoculation is then diluted with the base medium which contains no glucose. With the aid of a photometer, the density is made equal to that imparted by approximately 10×10^6 cells per ml of *Saccharomyces cerevisiae* (NRRL Y-567) or 15×10^6 cells per ml of *Candida utilis* (NRRL Y-900). These concentrations of cells when contained in 18-mm (outside diameter) test tubes transmit about 55% of the light that has passed through a blue filter (420 nm) in a Lumitron photometer. If a photometer or other means for determining density is not available, 5 ml of the inoculum in 16-mm tubes should be diluted until black lines approximately ¾ mm wide drawn on white cardboard, become visible through the tubes as dark bands – generally about two volumes of diluent are required per volume of culture. Each tube in the assimilation tests receives 0.1 ml of the diluted inoculum.

After inoculation the tubes containing the various carbon sources, are incubated for 3 weeks at either 25 or 28°C. Psychrophilic species are tested at lower temperatures, often 12°C. The tubes may be placed upright or slanted. To provide better aeration and more rapid utilization of the carbon sources, some laboratories prefer incubating the assimilation tubes either on a rotary shaker or a Rollordrum. Stationary cultures are examined after 3 weeks. Cultures on rotary shakers or on Rollordrums are examined after 1 week and again after 3 weeks. Some laboratories using stationary cultures examine the tubes at weekly intervals when the tubes are also shaken manually. The tubes are examined for the presence of pellicles and the possible secretion of visible amounts of riboflavin which are produced by certain yeasts, e.g. species of the genus *Debaryomyces*.

The degree to which the various carbon compounds are utilized is determined by placing the tubes which have been vigorously shaken to disperse all yeast growth, against a white card bearing lines approximately ¾ mm wide, drawn with India ink. If growth in the tubes completely obliterates the lines, it is recorded as 3 + ; if the lines appear as diffuse bands, the growth is rated as 2 + ; if the bands are distinguishable as such but have indistinct edges it is recorded as 1 + ; while the absence of growth is indicated as − . A 3 + or 2 + reaction within 3 weeks is considered to be positive, and a 1 + reaction as very weak or negative if the carbon source is known to carry assimilable impurities. In cases where the utilization of a carbon source is doubtful it may be rechecked by inoculating a loopful of the culture into a fresh tube of the relevant carbon source. At the final reading, several tubes originally containing sugars are tested for the presence of amyloid compounds, as will be described later in this chapter. Tubes may not be incubated in incubators in which fermentation tests are being run concomitantly since traces of accumulating ethyl alcohol formed during fermentation could lead to growth in the assimilation tubes and thus give rise to spurious results.

Media

1. Inoculation medium (Wickerham, 1951)

Dissolve 6.7 g of Bacto yeast nitrogen base* and 1 g glucose in 100 ml demineralized water. This 10-fold concentrated solution is filter-sterilized and is stored in the refrigerator. 1-ml aliquots of this basal medium are then pipetted in 9-ml amounts of sterile demineralized water in plugged 16-mm tubes.

2. Stock vitamin solution (van der Walt & van Kerken, 1961a), see p. 79.

3. Nitrogen basal medium for carbon assimilation tests in liquid medium

The composition of the basal medium is given in Table 10, p. 100.

A 10-fold concentrated medium is prepared by dissolving 6.7 g of Bacto yeast nitrogen base and the appropriate amount of the carbon compound equivalent to glucose (i.e., containing the same amount of carbon as 5 g glucose) in 100 ml demineralized water. When raffinose is the carbon source it is used at twice this concentration. It may be necessary to warm the demineralized water somewhat to effect complete solution. Where organic acids or their acid salts are used as the carbon source, the solution is adjusted to pH 5.6 by the addition of sodium hydroxide pellets. The solution is then sterilized by filtration and the sterile solution is kept in a refrigerator or deep-freeze cabinet until required. Aliquots of 0.5 ml of the sterile solution are pipetted aseptically into clear, plugged, 16-mm tubes, containing 4.5 ml demineralized water which has been sterilized by autoclaving for 15 minutes at 15 lbs overpressure. After shaking the tubes are ready for use.

When *soluble starch* or *inulin* is used as the carbon source, it is dissolved in distilled water by gentle heating over an open flame as otherwise it may be lost to a considerable extent if filters were used for sterilization. The solutions containing 0.5% of the carbohydrate are then pipetted in 4.5-ml amounts into sterile plugged 16-mm tubes. The tubes are sterilized by autoclaving for 15 minutes at 10 lbs overpressure. After cooling each sterilized tube receives 0.5 ml of the sterile basal medium in 10-fold concentration. After shaking the tubes are ready for use.

When testing the ability to utilize *hexadecane,* the following procedures described by Scheda (1966a) and Scheda and Bos (1966) may be adopted.

To a 100-ml cotton-plugged flask containing 18 ml sterile demineralized water, 2 ml of 10-fold concentrated filter-sterilized yeast nitrogen base and 2% (v/v) of sterile alkane are added aseptically. (The hexadecane can be sterilized by autoclaving.) The flasks are inoculated and incubated at 25–28°C and regularly shaken. A flask devoid of any carbon source serves as control. Utilization of the alkane is assessed by the degree of proliferation and growth of the yeast. The test may also be conducted on the Rollordrum in 16-mm assimilation tubes using 5 ml of sterile nitrogen base and 1 or 2 drops of sterile alkane.

The ability to utilize *hydrocarbons* may also be tested by using an agar-based mineral medium (Scheda & Bos, 1966; Markovetz & Kallio, 1964). The basal medium used is prepared by adding 2 g of high grade, purified agar, 1 ml hexadecane and 0.25 g taurocholate (as emulsifying agent) to 90 ml demineralized water, after which the medium is autoclaved for 15 minutes at 10 lbs overpressure. After cooling to circa 50°C,

*For the composition of this medium, see Table 10, p. 100.

10 ml of filter-sterilized yeast nitrogen base is added aseptically. The medium is then carefully mixed to disperse the hexadecane (avoiding excessive frothing) after which plates are poured. When testing a large number of strains, replica plates are made from spot colonies of the strains grown on malt extract agar for 1–2 days at 25°C. For control purposes, a replica is also made on the mineral agar-based medium from which the alkane has been omitted. The plates are incubated at 25°C for 1 week. Slants of the basal medium may also be used, in which case the sterilized medium is aseptically distributed in 16-mm sterile plugged tubes. Markovetz and Kallio (1964) also using slants, omit both taurocholate and the alkane from the medium but add 1 ml of the sterile alkane to the slants after inoculation. For further details the publication by Markovetz and Kallio (1964) must be consulted.

2. Assimilation tests on agar medium

The utilization of carbon sources with an agar medium may be tested either in tubes or by employing the replica-plating technique with the use of a multiple inoculation device.

A set of tubes containing the various carbon sources in a basal medium solidified with 2% agar is inoculated with an actively growing culture of the yeast strain under study. A light inoculum is applied as a streak over the central surface of the agar. A tube containing the nitrogen base without any carbon source serves as control. The tubes are incubated at the desired temperature and the growths on the various carbon sources are regularly compared with the growth in the control tube over a period of 3 weeks.

The tubes are prepared by adding aseptically 0.5-ml aliquots of the filter-sterilized, 10-fold concentrated solutions of the various carbon compounds, prepared as for the liquid assimilation tests, to 4.5-ml amounts of sterile, molten, 2.25% washed or high-grade, purified agar in 16-mm plugged tubes. Mix carefully to avoid frothing. Slant the tubes to provide a maximum agar surface and allow the medium to set.

For the assimilation of hydrocarbons on agar medium see p. 82.

For details regarding the use of the replica-plating method, the publications by Shifrine et al. (1954) and Beech et al. (1955) must be consulted.

Medium

Washed agar

Dissolve 20 g agar in 500 ml demineralized water by autoclaving. Pour into a shallow container. After solidification the 4% agar is cut up into small blocks and placed in 1 liter of demineralized water in the refrigerator. Change the water daily for 7 days. The agar blocks are then rinsed several times and allowed to drain completely. Demineralized water is then added to make up 1 kg after which the agar is autoclaved for 15 minutes at 10 lbs overpressure. Commercially available, high-grade, specially purified agar may, as a rule, be used instead of washed agar.

3. Auxanographic method

Aliquots of 18–20 ml of sterile, synthetic basal medium devoid of a carbon source are melted and allowed to cool to about 40°C and poured into a Petri dish containing about 2 ml of a suspension of the yeast under test in sterile tap water. If the basal medium I (see below) is not supplemented with growth factors, one drop of yeast autolysate or 20% (w/v) sterile aqueous solution of commercial yeast extract or one drop of a 100-fold concentrated stock vitamin solution must also be added. The liquid is thoroughly mixed and the plate allowed to set. After solidification the plate is kept, lid-side up, at 25°C for a few hours to obtain a dry agar surface. Small amounts of the various carbon compounds are then deposited at different, evenly spaced sites on the agar. These sites are marked on the outside of the smaller dish. The plates are then incubated at 25°C lid-side down. Results are observed after 2 or 3 days.

Media

1. Basal medium I (Lodder & Kreger-van Rij, 1952)

Dissolve 0.5% ammonium sulphate, 0.1% potassium dihydrogen phosphate and 0.05% magnesium sulphate heptahydrate in 2% washed or purified high-grade agar. The medium is dispensed in 18–20-ml aliquots and sterilized by autoclaving for 15 minutes at 10 lbs overpressure.

2. Basal medium II
Dissolve 6.7 g Bacto yeast nitrogen base and 20 g washed agar in 1 liter demineralized water, after which dispense in 18–20-ml aliquots and then sterilize by autoclaving for 15 minutes at 10 lbs overpressure.

c. SPLITTING OF ARBUTIN

The splitting of arbutin is a test for confirming β-glucosidase activity in yeast strains. The test depends upon the fact that if a yeast strain hydrolyzes arbutin, the aglucone moiety, hydroxyquinone, gives a brown color with any soluble ferric salts incorporated in the medium. While the ability to split arbutin is frequently associated with the ability to utilize cellobiose, this is not invariably the case.

Method and medium

A slant of arbutin agar is inoculated with an actively growing culture of the yeast under test and incubated at 25 or 28°C. An uninoculated tube serves as control. If arbutin is split, a dark brown color develops in the agar, usually within 2–7 days.

Arbutin agar
Dissolve 0.5% (w/v) arbutin and 2% agar in yeast infusion or in dilute (1:10) yeast autolysate by heating. After the ingredients have dissolved, the medium is dispensed in 5–7-ml aliquots in 16-mm cotton-plugged tubes and sterilized by autoclaving for 15 minutes at 15 lbs overpressure. Immediately after sterilization 2–3 drops of a sterile 1% ferric ammonium citrate solution is added aseptically to the molten agar in each tube. The tubes are carefully shaken to avoid frothing and then slanted.

2. *Assimilation of nitrogen compounds*

Yeasts are capable of utilizing a diversity of nitrogen sources. For identification purposes, however, the utilization of nitrate, nitrite, ethylamine hydrochloride, cadaverine dihydrochloride, L-lysine, creatinine and creatine as sole source of nitrogen have been found to be most useful. The utilization of imidazol may, however, find wider application as diagnostic test than has been anticipated (LaRue & Spencer, 1967).

The ability or inability to utilize nitrate-nitrogen, as diagnostic criterion, is particularly valuable for determinative purposes. Many genera are characterized by their inability to utilize nitrates, e.g. *Saccharomyces, Kluyveromyces, Pichia* and *Debaryomyces,* while in other genera, e.g. *Hansenula, Pachysolen, Citeromyces* and *Wickerhamiella* all species utilize nitrate. Among the imperfect genera both nitrate-positive and nitrate-negative species occur, e.g. in *Candida* and *Trichosporon.* With few exceptions, the utilization of nitrate is stable at least within the species.

Species which utilize nitrate are also able to utilize nitrite, but the reverse does not invariably apply; for instance, certain species of the genera *Debaryomyces* (e.g. *Deb. hansenii*) and *Pichia* (e.g. *P. pinus*) utilize nitrite but not nitrate. In testing the assimilation of nitrite, toxic effects due to nitrous acid formation may occur in media with a pH below 6. Recourse must therefore be had to testing nitrite in low concentrations, a procedure which unfortunately does not always give clearcut results. In view of this, the assimilation of nitrite may best be tested by the auxanographic technique. Auxanograms may nevertheless still show a zone of inhibition immediately around the site where the sodium nitrite has been applied. For testing in liquid medium Wickerham (1957) advocates the use of 0.26 g sodium nitrite per 100 ml 10-fold concentrated medium, but even at this concentration inhibition due to the toxicity may be manifested.

Whereas the utilization of nitrate may serve to differentiate both species as well as genera, the utilization of ethylamine hydrochloride, cadaverine dihydrochloride and L-lysine only serve to discriminate species.

The ability to utilize a nitrogen source is assessed by determining growth response. For this purpose either the auxanographic technique of Beijerinck (1889a) or the liquid assimilation test devised by Wickerham (1946, 1951) may be used. The auxanographic technique has the disadvantage that the agar medium tends to dry out, in which case the slow utilization of a nitrogen source may escape detection. The utilization of creatinine and creatine is nevertheless preferably tested by the auxanographic technique (see van der Walt, 1970a).

Since, as a result of the excretion of assimilable nitrogen, yeast cells which are initially well nourished may proliferate considerably even when an external nitrogen source is absent, mere multiplication in the liquid test medium does not provide conclusive evidence of the utilization of the nitrogen source. In using the liquid test medium it is therefore necessary to verify the assimilation of a specific compound by transferring a small inoculum from the first tube to a second. The use of a blank tube is therefore obligatory.

Methods and media

Only high grade, preferably analytical grade compounds should be used as nitrogen sources in the nitrogen assimilation tests. When using amine salts it is advisable to ascertain that these are free from traces of ammonium compounds.

1. Assimilation in liquid medium (Wickerham, 1946, 1951)

The liquid medium employed for testing the ability of yeasts to utilize the various nitrogen sources is similar in composition to that used in the tests on carbon assimilation. Its composition is given in Table 10, p. 100.

The medium is prepared in a 10-fold concentration by dissolving 11.7 g Bacto yeast carbon base together with the requisite amount of the nitrogen source in 100 ml demineralized water. Where the nitrogen source is nitrate, 0.78 g potassium nitrate is used; where it is nitrite, 0.26 g sodium nitrite is used; and where it is primary amino nitrogen, 0.64 g ethylamine hydrochloride, 0.68 g cadaverine dihydrochloride and 0.56 g L-lysine are used. It may be necessary to warm the water somewhat to effect complete solution of the constituents. The solution is then filter-sterilized and stored in the refrigerator until required.

The final medium is prepared by transferring aseptically 0.5-ml aliquots of the sterile 10-fold concentrated medium into 4.5-ml amounts of sterile demineralized water in 16-mm plugged tubes, which have been autoclaved for 15 minutes at 15 lbs overpressure. Blank tubes are prepared by adding 0.5 ml of a filter-sterilized basal solution. After the tubes have been shaken to mix the water and the denser, 10-fold concentrated solutions, the media are ready for use.

The nitrogen assimilation test tubes receive the same inoculum as is used in the carbon assimilation tests and they are incubated at the same temperature.

After the tubes have been incubated for a week, even though the nitrogen sources may not have been utilized, signs of growth are detected, probably owing to the soluble nitrogenous compounds excreted by the cells and to the small amount of ammonium sulphate in the inoculation medium. Because growth due to this source of nitrogen may mask the results, a second tube is inoculated with one loopful from the first.

If the second tube shows a 2 + to 3 + reaction (as scored for the carbon assimilation tubes, see p. 82) the yeast strain is capable of utilizing the nitrogen source; if a 1 + reaction is scored the yeast is considered to be incapable of utilizing the nitrogen source.

2. Assimilation in liquid medium (Nickerson, 1944a)

Doubtful nitrate reactions in the test according to Wickerham are seldom encountered, but if they do occur they may be confirmed by the chemical test for nitrite.

Methods and reagents

Four tubes of nitrate assimilation medium are inoculated by needle from a slant and tested after approximately 3, 5, 10 and 15 days. If the yeast under test utilizes nitrate, one or more of the tubes will give a positive reaction for nitrite. The nitrite test of which the description follows, is described in the 'Manual of Methods for Pure Culture Study of Bacteria'.

Reagent 1
Dissolve 8 g sulphanilic acid in 1 liter of 5 N acetic acid (1 part glacial acetic acid to 2.5 parts of water) or in 1 liter of dilute sulphuric acid (1 part concentrated acid to 20 parts of water).

Reagent 2
Dissolve 5 g α-naphthylamine in 1 liter of 5 N acetic acid or in dilute sulphuric acid (1 part concentrated acid to 20 parts of water). Or dissolve 6 ml of dimethyl-α-naphthylamine in 1 liter of 5 N acetic acid. The latter reagent gives a permanent red color in the presence of high concentrations of nitrite.

Put a few drops of each of these reagents in nitrate cultures. The development of a distinct pink or red color indicates the presence of nitrite.

Absence of nitrite in the presence of good growth may, however, indicate complete consumption of the nitrate. Test, therefore, for nitrate by adding a pinch of zinc dust to the tube to which the reagents 1 and 2 have been introduced and allow it to stand for a few minutes. If nitrate is still present it will be reduced to nitrite with the development of the characteristic pink color.

3. Auxanographic method (Lodder & Kreger-van Rij, 1952)

The procedure described on p. 83 is employed. In this case, however, the basic agar medium is devoid of a nitrogen source and glucose serves as the source of carbon. If medium I is used, one drop of the 100-fold concentrated vitamin solution is added but yeast autolysate is omitted. Ammonium sulphate is included as a source of nitrogen for comparative purposes. As very little sodium nitrite and ethylamine hydrochloride are required for auxanograms, overdosing of the plates can be avoided by dipping only the tip of an inoculating needle into saturated solutions of these salts and then touching it to the surface of the seeded plate.

After application of the various nitrogen sources the Petri dishes are incubated at 25°C and the results can be read after 2–3 days.

1. Basal Medium I (Lodder & Kreger-van Rij, 1952)
The basal medium for the nitrogen auxanographic test contains 2% glucose, 0.1% potassium dihydrogen phosphate, 0.05% magnesium sulphate heptahydrate and 2% washed or high grade, purified agar. The medium is distributed in 18–20-ml aliquots and sterilized by autoclaving for 15 minutes at 10 lbs overpressure.

2. Basal Medium II
Dissolve 11.7 g Bacto yeast carbon base and 20 g washed or high grade, purified agar in 1 liter demineralized water. After the constituents have dissolved, the medium is dispensed in 18–20-ml aliquots and sterilized by autoclaving for 15 minutes at 10 lbs overpressure.

3. Growth in vitamin-free medium, vitamin requirements

The use of the ability or inability to grow in a mineral medium devoid of vitamins as a diagnostic property, was introduced by Wickerham (1951). The complete list of vitamins employed by him and those that have been generally adopted include biotin, calcium pantothenate, folic acid, *m*-inositol, niacin, *p*-aminobenzoic acid, pyridoxine hydrochloride, riboflavin and thiamine hydrochloride.

Yeasts vary in their ability to grow in a vitamin-free medium. Strains of the genera *Hanseniaspora* and *Kloeckera* for example, have been found to have an absolute requirement for *m*-inositol and pantothenic acid, while *Dekkera, Metschnikowia* and *Brettanomyces* species require biotin and thiamine. Other genera comprise both auxo-autotrophic and auxo-heterotrophic species. Some species may also vary in their vitamin requirements.

The detection of vitamin requirements is based upon cell proliferation in standardized, complete, mineral media devoid of vitamins. Yeast cells that are initially well nourished, may proliferate considerably in the absence of an external vitamin source because of a carry-over of previously acquired growth factors. Proliferation in the first tube of liquid test medium does not provide conclusive evidence that this proliferation is independent of an external vitamin source. Therefore, before the ability of a yeast strain to grow in the absence of an extraneous vitamin source can be assessed, it is necessary to continue subculturing in the vitamin-free medium until this internal source of growth factors is depleted.

Apart from absolute vitamin requirements, the addition of certain vitamins sometimes stimulates growth to a marked degree.

Method and medium

The composition of the medium is given in Table 10, p. 100. The 10-fold concentrated medium is prepared by dissolving 16.7 g Bacto vitamin-free yeast base in 100 ml demineralized water which may have to be warmed slightly to effect complete solution of the ingredients. The final solution is prepared by pipetting aseptically 0.5 ml of the concentrated solution into 4.5 ml sterile water in 16-mm plugged tubes. When the solutions have been mixed by shaking they are ready for inoculation.

If the requirements of a yeast for a particular vitamin or combination of vitamins is to be determined, the media are prepared either by adding the requisite amount of vitamins to the concentrated basal medium before filter-sterilization or by filter-sterilizing the concentrated basal medium and the desired vitamins separately. This is accomplished by first distributing 0.5-ml aliquots of the 10-fold concentrated vitamin-free medium in sterile 16-mm plugged tubes. Sufficient quantities of the desired vitamins to make up 100 ml of the final medium are dissolved in 90 ml demineralized water and the solution is filter-sterilized. The final medium is prepared by adding 4.5 ml of the sterile vitamin solution to each of 0.5 ml of the 10-fold concentrated basal medium under aseptic conditions.

The media are inoculated in the same way as the carbon assimilation tubes using either Wickerham's inoculation medium or cell suspensions in sterile demineralized water or a light inoculum may be used by transferring the cells with an inoculating needle directly from a slant culture.

The tubes are normally incubated at 25°C. The tubes usually show growth after 7 days. As growth may be due to vitamins carried over from the inoculum a second vitamin-free tube is always inoculated with the standard loop from the first. After a similar period of incubation growth in the second tube is scored in the same way as followed for the carbon and nitrogen assimilation tests. Yeast strains which give a 3 + reaction in the second tube after 7 days are capable of synthesizing all the vitamins required for their development. Tubes which score 2 + or less are considered as incapable of synthesizing all the vitamins for growth.

4. Growth on 50% glucose-yeast extract agar and in 10% sodium chloride plus 5% glucose in yeast nitrogen base

Yeasts recovered from substrates with high sugar or salt content are generally resistant to high osmotic pressures. While a large variety of yeast species grow well in glucose concentrations of up to 40% by weight, few species are capable of development with sugar concentrations of between 50–70%. The ability to grow on high concentrations of sugar is currently tested by growth on agar media containing 50% (w/w) glucose.

Wickerham (1951) introduced the use of a liquid medium containing 5% glucose and 10% sodium chloride for testing the ability to grow at moderately high osmotic pressures. Some laboratories prefer to use this medium.

In dry climates, culture media in tubes or in Petri dishes are inclined to lose water rather rapidly as a result of evaporation. In media of high osmotic pressure such water losses can lead to critical changes in the concentration of the solutes. Water losses can be reduced by incubating tubes either in inflated plastic bags or with self-sealing plastic caps, which, while permitting gaseous exchange, nevertheless reduce evaporation.

Methods and media

a. Growth on 50% glucose-yeast extract agar

Actively growing cultures are lightly inoculated as a streak on 50% (w/w) glucose agar in tubes. The tubes are incubated at 25°C and are then examined for growth after 5 days.

The medium is made up by dissolving 50 g of glucose in 50 ml of either yeast infusion or 1% (v/v) yeast autolysate. Note the volume. Add 3% (w/v) agar and dissolve by steaming or in a water bath. After dissolving the agar, the medium is dispensed in 5–6-ml aliquots in 16-mm plugged tubes. The tubes are autoclaved for 10 minutes at 10 lbs overpressure and then slanted. Over-autoclaving leads to a noticeable browning of the medium. Any tubes showing browning after autoclaving have to be discarded.

Certain laboratories test the ability to grow on 50% (w/w) glucose agar by inoculating the poured plates with a multipoint inoculating device.

b. Growth in 10% sodium chloride plus 5% glucose medium (Wickerham, 1951)

The medium used in this test is prepared by dissolving 5 g glucose and 10 g sodium chloride in 100 ml demineralized water. This solution is dispensed in 4.5-ml aliquots in 16-mm plugged tubes and sterilized by autoclaving for 15 minutes at 15 lbs overpressure. To each tube is added 0.5 ml of a filter-sterilized solution of 6.7 g Bacto yeast nitrogen base dissolved in 100 ml distilled water. The solutions are mixed by shaking and are then ready for use.

The same inoculum (0.1 ml) is used as in the assimilation tests. Growth amounting to 1 + after 21 days incubation at 25 or 28°C is recorded as weak, a 2 + cell density is regarded as moderate and 3 + as good. (See p. 82 for particulars on this system of scoring.) A negative or 1 + reaction after 7 days incubation followed by a 2 + or 3 + reaction at 21 days is regarded as slow.

5. Growth at 37°C and at other temperatures; maximum temperature of growth

By and large, temperatures between 20 and 28°C are optimal for the growth of most yeast species. There are, however, exceptions, particularly those species which have specific and restricted habitats. Yeasts from the polar regions proliferate very poorly at 20°C, but grow well at 15°C or 4°C. On the other hand, some yeasts associated with warm-blooded sources fail to grow at 20°C, e.g. *Cyniclomyces guttulatus*. Certain species, particularly of the genera *Kluyveromyces* and *Hansenula* are capable of growth at 45°C or even 48°C. As such temperatures are critical, the accuracy of water baths and incubators should be regularly checked when determining these temperatures.

Growth at 37°C and at other temperatures may be determined either in liquid or on agar media.

Methods and media

a. Growth at 37°C and at other temperatures

1. Growth in liquid medium (Wickerham, 1951)

A tube of the usual glucose assimilation medium is inoculated in the usual way and incubated at 37°C or at other temperatures. Growth is scored in the usual way after 1 and 3 weeks. The results are reported in the same manner as in the test for the ability to grow in the osmotic pressure medium described by Wickerham (see p. 88).

2. Growth on solid medium (Kreger-van Rij, 1964b)

The yeast strain under test is grown either on malt extract agar or YM agar at 37°C or at other temperatures for 2–4 days. In case of weak growth a subculture is made which is incubated for 2–4 days at the same temperature. The results of the latter test are taken as decisive.

b. Maximum temperature of growth (van Uden & Farinha, 1958)

Tubes containing glucose-yeast extract-peptone water are inoculated with the strain under test as in the carbon assimilation tests. The inoculated tubes are incubated in water baths or incubators kept at a constant temperature each operating at a different temperature at intervals of 1°C. The maximum temperature for growth is considered to be that located between the lowest temperature at which no growth occurred after 7 days and the next lowest temperature where growth had occurred.

For the preparation of glucose-yeast extract-peptone water see p. 56.

6. *Acid production from glucose*

Most, if not all, yeast cultures produce traces of volatile and non-volatile acids. However, it is only when excessive amounts of acetic acid are produced that acid formation can be of diagnostic use and then only when the test for this property is performed under standard conditions (Custers, 1940; van der Walt, 1970a). For this purpose an agar medium containing 5% glucose and 0.5% calcium carbonate is employed. Acid production is determined by the amount of calcium carbonate that goes into solution. Production of acid on the standard medium is of rather restricted use. It is fairly characteristic of the genera *Dekkera* and *Brettanomyces* and is most marked in the strongly fermentative species which give a vinegary smell as a result of the excessive production of acetic acid. Acid production may be weak or even absent in weakly fermentative species, e.g. *Br. custersianus.*

The limited value of this test is evident from the fact that certain strains of some *Hansenula* and *Kloeckera* species also produce detectable amounts of acid on the standard medium. Certain *Candida* species which produce citric acid, give a weakly positive reaction.

Method and medium

The yeast strain under test is inoculated onto slants of glucose-chalk agar and incubated at 25 or 28°C. Acid production on this medium is reported as positive if sufficient acid is formed to clarify the opaque medium. If incomplete solution of the calcium carbonate is effected, the reaction is reported as weakly positive.

5% Glucose-0.5% calcium carbonate agar

To 1 liter yeast infusion or dilute (1:10) yeast autolysate or 0.5% powdered yeast extract in demineralized water, add 50 g glucose, 5 g chalk (calcium carbonate, precipitated, light) and 20 g agar. Sterilize by autoclaving for 15 minutes at 15 lbs overpressure. Cool to 50°C, swirl gently to suspend the chalk evenly throughout the medium. Distribute aseptically in sterile, plugged tubes and slant on a cold surface to avoid settling of the chalk. Plates may be prepared if inoculation is done with a multipoint inoculating device.

7. *Formation of extracellular, amyloid compounds ('starch' test)*

Aschner et al. (1945) and Mager and Aschner (1947) studied the production of extracellular, amyloid or starch-like material by encapsulated yeasts. They found that under suitable cultural conditions several yeast strains formed extracellular polysaccharides which give a blue or greenish blue color with iodine solution. One of the conditions was that the reaction of the medium is maintained below pH 5.0. The formation of starch-like polysaccharides may, however, in some species also be detected at higher pH values.

Methods and media

a. Testing in liquid culture at pH 5.6 (Wickerham, 1951, 1952)
After 21 days the carbon assimilation test cultures which initially contained sugars or polyhydric alcohols are examined for the presence of amyloid compounds. Some laboratories prefer using 1% instead of the customary 0.5% glucose. One drop of dilute Lugol's iodine solution is added per tube and the tube shaken. Strains producing amyloid material will show color reactions varying in intensity from blue to purple or green; the color may reside in the medium, in the cells or in both. In doubtful cases, strains may be retested by cultivation on a rotary shaker or Rollordrum at 25°C and by using 3% glucose. In some genera the relatively large amounts of glycogen synthesized may interfere with the test, especially if iodine solutions stronger than 0.02 N are added. The brown color typical of glycogen may mask a weak 'starch' reaction. In doubtful cases the tubes should be allowed to stand at room temperature for a few hours or overnight. The brown color indicating glycogen disappears while the blue color indicating amyloid material remains.

b. Testing on solid media
Strains are either streaked singly on suitable agar plates or cultures may be applied to the agar medium by the multiple inoculation technique. Some laboratories prefer to test the formation of amyloid material on agar slants in tubes.
After incubation for 1–2 weeks the plates or tubes are flooded with dilute iodine solution.

1. Medium I (Lodder & Kreger-van Rij, 1952)
Dissolve 0.2% ammonium sulphate, 0.2% potassium dihydrogen phosphate, 0.1% magnesium sulphate heptahydrate and 2% glucose in demineralized water. Adjust the solution to pH 4.5. Prepare an equal volume of 4% agar in demineralized water. The two separate portions of the media are autoclaved for 15 minutes at 10 lbs overpressure. Add the two portions together aseptically while the agar is still molten. Gently mix the solutions to avoid frothing. Prepare plates in Petri dishes containing 1 drop of yeast autolysate or of a 20% (w/v) solution of commercial yeast extract and 1 drop of a 100-fold concentrated vitamin solution.
If slants are required the growth factor supplement is added to the liquefied agar which is then dispensed aseptically in 5–6-ml aliquots in sterile plugged tubes.

2. Medium II
Dissolve 10 g glucose and 6.7 g Bacto yeast carbon base in 100 ml demineralized water and filter-sterilize. Add aseptically ½ volume of the 10-fold concentrated medium to 4½ volumes of sterile, liquefied 2¼% agar. The solutions are mixed and plates or tubes poured aseptically.

3. Lugol's iodine solution (Gram's modification)
Dissolve 1 g iodine and 2 g potassium iodide in 300 ml demineralized water.

8. Production of ammonia from urea (urease test)

Practically all yeasts can utilize urea in low concentrations as sole source of nitrogen provided that adequate amounts of vitamins are supplied. Nevertheless, yeasts differ in their ability to hydrolyze high concentrations of urea to ammonia in complete media containing an organic nitrogen source such as peptone. Seeliger (1956) investigated the hydrolysis of urea by a number of ascogenous and anascogenous species on Christensen's urea agar (1946). The results of this investigation showed that urease activity is generally lacking in the ascogenous species, whereas it is particularly marked in the basidiomycetous genera *Cryptococcus* and *Rhodotorula* (Abadie, 1967; Hagler & Ahearn, 1981). Although the majority of the ascogenous yeasts fail to give a positive urease test, *Schizosaccharomyces pombe, Saccharomycopsis (Yarrowia) lipolytica* and species of the genus *Lipomyces* are notable exceptions.

Although urease activity is frequently linked with 'starch' formation, this is not an empiric rule since both urease-positive/starch-negative and urease-negative/starch-positive yeasts are known. The substrate specificity of the urease reaction makes it a useful characteristic to employ as a criterion in yeast taxonomy. Its greatest application is as a confirmatory test for imperfect taxa.

The urease test may be performed on solid medium (Seeliger, 1956), or in liquid medium (Roberts et al., 1978). A prescription for the test on solid medium is given here.

Method and medium (Seeliger, 1956)

The yeast strain under test is inoculated onto Christensen's urea agar and incubated at 25°C. Cultures are observed daily up to 5 days. The reactions are recorded as positive after the appearance of a deep pink color.

Christensen's urea agar (Christensen, 1946)

Dissolve 1 g peptone, 1 g glucose, 5 g sodium chloride and 2 g potassium dihydrogen phosphate and 0.012 g phenol red in 1 liter demineralized water and adjust the reaction to pH 6.8. Add 20 g agar and dissolve after which 4.5-ml aliquots of the medium are dispensed in 16-mm plugged tubes. Autoclave for 15 minutes at 15 lbs overpressure. Immediately after autoclaving 0.5 ml of a 20%, filter-sterilized urea solution is added. After mixing the tubes are slanted and allowed to set.

Christensen's urease test medium is also commercially available.

9. Splitting of fat

Although lipase activity has been demonstrated in several yeast species, e.g. *Candida lipolytica, C. rugosa, C. steatolytica* and *Trichosporon pullulans,* this property is usually only employed as a confirmatory test.

Numerous methods have been proposed for the detection of lipolytic activity in microorganisms using substrates such as tributyrin, olive oil, tallow, Tween 40, 60, 80, as well as partial glyceryl esters of higher fatty acids (see Bulder, 1955; Sierra, 1957; Tuynenburg-Muys & Willemse, 1965). Tuynenburg-Muys and Willemse compared the various methods and concluded that for routine purposes none was, in fact, superior to the test developed by Eijkman (1901). They did, however, recommend the use of partial glycerol esters of higher fatty acids which have a higher melting point than beef tallow.

Method and medium

Fresh beef suet is melted out, filtered and the tallow sterilized by autoclaving. An 0.5-ml aliquot of the sterile fat is pipetted into a sterile warm Petri dish which is tilted so that the fat forms a thin, uniform layer on the bottom. The surplus material is pipetted off. The dish is then placed in the refrigerator for 2 hours. Thereafter, 18–20 ml sterile Gorodkowa agar containing 0.1% chalk is melted, cooled to 40°C and carefully layered over the fat after which the medium is left to set. The plates are inoculated and incubated at 25°C. A positive result may be detected within a few days by the appearance of an opaque deposit of the calcium salts formed by the liberated fatty acids under the inoculation streak. If partial glyceryl esters of higher fatty acids are available, these may be used instead of beef tallow.

10. Ester production

Although fermentative yeasts may form a variety of esters in varying amounts, ester production has for general taxonomic purposes been found to have very limited application as a diagnostic criterion. It is only when these compounds are produced in sufficient amounts to be easily detected by smell, that this property can be used for the characterization of species. Ethyl acetate appears to be the commonest and most readily detectable ester formed by yeasts.

For routine testing purposes, ester formation is usually detected by the odor produced by plate cultures on either YM agar or malt extract agar.

Method and medium

A plate culture of the yeast strain under test is prepared on YM agar and incubated at 25–28°C. The culture is judged by smell after 24–48 hours.

11. Cycloheximide (actidione) resistance

Whiffen (1948) was the first to report that yeasts varied in their sensitivity towards the antibiotic actidione (cycloheximide). Her results, based on the study of 22 species, indicated that yeasts could be divided into three categories on the basis of their sensitivity towards cycloheximide: species that are markedly sensitive (inhibited by 1 μg/ml) such as *Saccharomyces cerevisiae,* those that are moderately sensitive (inhibited by 25 μg/ml) such as *Schizosaccharomyces pombe,* and those species which are tolerant (not inhibited by concentrations as high as 1000 μg/ml) such as *Kluyveromyces lactis.*

Because of the possibility that strains may become adapted to low concentrations of cycloheximide (see van der Walt, 1970a) the resistance to this compound finds limited application in the characterization of species when tested in liquid medium at concentrations of 100 and 1000 ppm.

Method and medium

The basal medium is prepared at 10-fold strength. Dissolve either 0.1 g (for 100 ppm) or 1 g (for 1000 ppm) of cycloheximide in 2.5 ml of acetone. The acetone solution is added to a solution of 6.7 g of Difco yeast nitrogen base and 10 g glucose in 100 ml demineralized water. After thorough mixing the combined solutions are filter-sterilized. To prepare for use, 0.5-ml aliquots of the 10-fold concentrated basal medium are pipetted aseptically into 16-mm plugged tubes containing 4.5 ml demineralized water which has been sterilized by autoclaving for 15 minutes at 15 lbs overpressure. After mixing the solutions, the tubes are inoculated in the same manner as for the carbon and nitrogen assimilation tests. The tubes are incubated at 25–28°C (with agitation) for 3 weeks.

A 2+ or 3+ reading on the 7th day of incubation is scored as +. If a 2+ or 3+ reading is delayed until the 14th or 21st day, the reading is scored as +s. The notation +(−) indicates that most strains are +, a few −.

12. Tolerance of 1% acetic acid

Although Sand (1973) and Pitt (1974) showed that yeasts could be distinguished on the basis of their resistance to benzoic acid and acetic acid, tolerance to these acids finds very limited application for routine identification purposes. Resistance to 1% acetic acid has nevertheless proven to be a useful criterion for the separation of *Zygosacch. rouxii* from *Zygosacch. bisporus* and *Zygosacch. bailii.*

Method and medium

Dissolve 10 g glucose, 1 g tryptone, 1 g yeast extract and 2 g agar in 100 ml demineralized water by autoclaving for 15 minutes at 15 lbs overpressure. Cool the sterilized medium to 45–50°C and add 1 ml of glacial acetic acid. After rapid mixing, plates are immediately poured and allowed to set.

The plates are inoculated with 1 ml of a suspension of cells diluted to a 2+ density as rated for the carbon assimilation tests.

The plates are incubated at 25–28°C and examined after 3 and 6 days for the development of visible growth.

13. Gelatin liquefaction

The ability to liquefy gelatin is of very limited diagnostic value as very few yeasts (e.g., *Cryptococcus macerans*) are strongly proteolytic. It is only in cases where rapid and complete liquefaction is effected in about one week that this property is applied as confirmatory test.

Stelling-Dekker (1931), Lodder (1934) and Diddens and Lodder (1942) employed a 10° Balling malt extract-20% gelatin medium, while Wickerham (1951) advocated the use of a synthetic medium containing 10% gelatin.

Methods and media

1. Synthetic medium (Wickerham, 1951)

Dissolve 10 g gelatin in 90 ml hot, demineralized water. The solution is dispensed in 4.5-ml amounts in 15-mm plugged tubes and autoclaved for 15 minutes at 15 lbs overpressure. After cooling to about 37°C, 0.5-ml quantities of the 10-fold concentrated Bacto yeast carbon base solution containing 5% glucose (i.e., as used in the assimilation test) are added aseptically. The final medium contains 0.5% glucose, 0.5% ammonium sulphate in addition to trace elements and vitamins. The tubes are allowed to gel in a vertical position. Each tube receives the same amount of inoculum as employed in the assimilation tests (see p. 81). After spreading the inoculum over the solid surface of the medium, the tube is incubated at 25°C at which temperature the medium remains solid unless acted upon by the yeasts under study. At approximately 7 and 21 days the depth of the liquid layer, if any, is measured in millimeters.

2. Malt extract gelatin

Dissolve 100 g powdered malt extract and 120 g gelatin in 1 liter demineralized water by moderate heating and dispense either in tubes or bottles. Sterilize by autoclaving for 15 minutes at 10 lbs overpressure. Gelatin is very susceptible to acid hydrolysis and excessive heating should be avoided. The melting and solidification temperature of the medium is 25–27°C. The liquefaction of the medium is tested either in tubes or in Petri dishes using a multipoint inoculating device. The medium is incubated at 18°C.

14. Diazonium Blue B (DBB) color test

On the basis of the study of 70 yeast strains representative of 26 ascogenous and 4 basidiomycetous genera, van der Walt and Hopsu-Havu (1976) reported that all strains of the basidiomycetous yeasts as well as those anamorphs which, on the basis of the ultrastructure of their cell walls as observed by TEM related them to the basidiomycetous taxa, gave an intense dark red color with a buffered solution of the dye Diazonium Blue B (DBB). This color reaction was, however, not observed with any cultures of either the ascogenous yeasts or the anamorphs which, on the basis of the ultrastructure of their cell walls as observed by TEM related them to the ascomycetous yeasts.

Hagler and Ahearn (1981) introduced modifications of the DBB test which they applied to some 100 strains of yeasts, filamentous fungi and algae. The results obtained with the modified test reconfirmed that the color reaction was restricted to basidiomycetous taxa. Hagler and Ahearn (1981) stressed that while the alkali-ethanol-DBB procedure is a simple, apparently reliable test for determining the basidiomycetous affinity of a yeast, the validity of the modified test should be further evaluated. For further details the publication by Hagler and Ahearn (1981) must be consulted.

Although the mechanism of the reaction has not been elucidated, this color reaction is nonetheless applied as diagnostic feature for differentiating species of some imperfect genera. The quality of the dye also affects the intensity of the color reaction.

Methods

a. DBB test on solid media

The strains under investigation are grown as spot cultures on plates of Sabouraud's 4% glucose agar supplemented with 0.5% yeast extract. The plates are incubated at 25°C or at 15°C in the case of psychrophilic strains. As originally published (van der Walt & Hopsu-Havu, 1976) the method requires that the cultures be cultivated for 3 weeks. Hagler and Ahearn (1981), however, proposed reducing the time of cultivation to 5–7 days and then incubating the plates at 55–60°C for 16 hours. The freshly prepared, chilled DBB reagent (1–2 drops) is then applied directly to the surface of colonies, ensuring thorough contact between the colonies and the reagent. A positive reaction is recorded when the colonies develop a dark red or violet-red color, usually within 1–2 minutes at room temperature.

b. DBB test in liquid medium (alkali-ethanol-DBB test; Hagler & Ahearn, 1981)

The strains are grown in 5.0 ml of yeast nitrogen base with 0.5% glucose (pH 7.0) for 3 days at 25°C on a Rollordrum rotating at 30 rpm. Psychrophilic strains are cultured at 15°C. Slow-growing mesophilic strains may have to be cultured for 6 days to obtain sufficient growth. The cultures are spun down in a clinical centrifuge and the supernatant is decanted. The pelleted cells are suspended in 0.5 ml of 0.05 N KOH and placed in a boiling water bath for 10 minutes. The cell suspension is cooled to ambient temperature and mixed well with 2.5 ml of 95% ethanol. This suspension is centrifuged briefly at low speed after which the supernatant is discarded. The pellet is suspended in 0.3 ml of the DBB reagent. A violet color which develops within a few seconds is regarded as a positive DBB reaction. Immediately after the color is formed, 0.1 ml of 95% ethanol is added. By this treatment the violet color is retained by the cells for several hours.

Media

1. Sabouraud's 4% glucose-0.5% yeast extract agar
Dissolve by steaming 40 g glucose, 20 g peptone, 5 g yeast extract and 20 g agar in 1 liter demineralized water. Distribute in containers and sterilize by autoclaving for 15 minutes at 15 lbs overpressure.

2. Yeast nitrogen base + 0.5% glucose (see Table 10, p. 100).

3. DBB reagent
The DBB reagent is prepared by dissolving 15 mg of Diazonium Blue B salt (marketed by Sigma as *o*-dianisidine tetrazotized, Practical Grade, 20% pure, or by Farbwerke Hoechst AG as Fast Blue B salt, or by Imperial Chemical Industries as Brentamine Blue B) in 15 ml chilled 0.25 M Tris buffer, pH 7. The Diazonium Blue B salt is unstable in warm moist conditions, but 15 mg portions of the dye may be stored in sealed test tubes at 4°C for several weeks. The freshly prepared reagent is kept in an ice bath and used before it turns dark yellow, or within about 30 minutes.

15. Canavanine-glycine-bromothymol blue (CGB) agar medium for the differentiation of Cryptococcus neoformans var. neoformans and Cryptococcus neoformans var. gattii

Method

Cryptococcus neoformans var. *neoformans* can be differentiated from *Cr. neoformans* var. *gattii* within 2–5 days when cultures on CGB agar are incubated at 25°C. Isolates of *Cr. neoformans* var. *gattii* hydrolyze glycine and are resistant to L-canavanine. As a result of glycine hydrolysis, the pH of the medium is elevated and bromothymol blue, the color indicator, turns the medium blue. Most isolates of *Cr. neoformans* var. *neoformans*, on the other hand, do not hydrolyze glycine, and a few which hydrolyze glycine are sensitive to L-canavanine. The culture medium stays greenish-yellow in color.

Medium

1. Stock solution (A) (10 × strength)
Dissolve 30 mg L-canavanine sulphate (Sigma, St. Louis), 10 g glycine, 1 g potassium dihydrogen phosphate, 1 g magnesium sulphate heptahydrate and one drop of Bejectal with vitamin C solution (Abbott, Chicago) in 100 ml demineralized water. Adjust the pH to 5.6 and filter-sterilize. The Bejectal solution can be replaced by 1 mg thiamine hydrochloride.

2. Bromothymol blue (0.4 %) solution
Dissolve 0.4 g bromothymol blue in 64 ml 0.01 N NaOH. Add 36 ml demineralized water. When sodium bromothymol blue is used, dissolve 0.4 g in 100 ml demineralized water.

3. Bromothymol blue agar base (B)
To 880 ml demineralized water add 20-ml 0.4% bromothymol blue solution. Add 20 g agar and autoclave for 15 minutes at 15 lbs overpressure and cool to about 55°C.

Add 100 ml of stock solution A to 900 ml molten bromothymol blue agar base (B), mix well and pour into Petri dishes or tubes to make slants.

The positive test with CGB medium indicates a change in pH from 5.8 ± 0.1 (greenish-yellow) to at least 7 (cobalt blue).

16. The mean molar percentage of guanine plus cytosine of nDNA (mol% G + C)

As outlined in Chapter I (p. 17), the determination of the mean molar % of guanine + cytosine of nDNA finds particular application in comparative studies of yeast strains which cannot, on the basis of their phenotypic properties be unambiguously assigned to definite species. As the so-called G + C content of yeast strains does not provide information on the sequence of the four bases in the DNA strands, additional criteria are required to establish the possible identity of strains with similar nDNA base compositions.

DNA base compositions of strains are determined either by thermal denaturation procedures or by buoyant density determinations which fall beyond the scope of most routine laboratories. For details of the various methods, the recent publications of Meyer et al. (1975, 1977), Price et al. (1978) and Kurtzman et al. (1980a,b) must be consulted.

17. Deoxyribonucleic acid reassociation

As outlined in Chapter I (p. 17), nDNA polynucleotide sequence studies aim at determining the relatedness between yeast strains on the basis of the degree to which their sheared, fragmented, single-stranded nDNA reanneals. Several methods, including both radio-isotopic as well as spectrophotometric techniques, have been employed for determining such reannealment. Although nDNA base sequence complementarity studies have greatly contributed to the rationalization of the classification of both perfect and imperfect yeast taxa, these studies fall beyond the scope of most routine laboratories. For details of the methods used, the recent publications by Meyer et al. (1975, 1977), Price et al. (1978), Kurtzman et al. (1980a,b), and Holzschu et al. (1979a) must be consulted.

18. Coenzyme Q structure

Yamada and Kondo (1972, 1973) and Yamada et al. (1973a,b; 1976a,b; 1977, 1980) have shown that yeasts vary in the structure of the ubiquinone system or coenzyme Q in their electron transport system, namely in the number of isoprene units per molecule. In the ascomycetous yeasts the Co-Q number ranges from 6 to 10, in the basidiomycetous yeasts from 8 to 10. For further details in connection with the application of the coenzyme Q system for taxonomic purposes, reference is made to Chapter I, p. 20 and p. 30.

19. Staining of yeast nuclei

Although several methods have been recommended for staining of yeast nuclei, none has been found to be superior to the Feulgen–Giemsa technique of Ganesan and Swaminathan (1958) as modified by Robinow (1961). This method, if followed correctly, gives extremely good preparations and has, in fact, been applied for chromosomal counts.

The method involves the hydrolysis of albumin-mounted cells in 1 N HCl at 60°C followed by washing in a buffer and the application of the Giemsa stain. When following this method it should be borne in mind that the optimal time of hydrolysis is not only critical but also varies for different yeasts. The optimal hydrolysis time needs therefore to be determined. For further details the publications by Robinow (1961) and McCully and Robinow (1972a,b) should be consulted.

20. *Melanin synthesis (phenoloxidase activity); DOPA medium*

The formation of melanin-like pigments from various *ortho-* and *para-*diphenols is used for rapid presumptive identification of *Cryptococcus neoformans*. A variety of substrates may be used: extracts of the seeds of *Guizotia abyssinica,* catechol, chlorogenic acid, protocatechuic acid, norepinephrine, dopamine, hydroxymetanilamine, dihydroxymetanilamide, and dihydroxyphenylalanine (DOPA) (Staib, 1962; Chaskes & Tyndall, 1975, 1978; Nurudeen & Ahearn, 1979). The medium of Chaskes & Tyndall (1975) that contains DOPA as a specific substrate and modified to contain a low glucose concentration is given below.

Method and medium

Young cells of the strain to be tested are streaked onto the agar and incubated at a temperature between 20 and 30°C for 24–72 hours. The colonies of *Cryptococcus neoformans* are black.

The medium is prepared by mixing 200 ml of a solution containing the substrate from which the pigment is formed with 800 ml of a basal agar at 55°C just before pouring into Petri dishes or tubes. To avoid oxidation of the pigment-producing substrate when preparing the complete medium the solution should not be held at 55°C for longer than is strictly necessary.

To prepare the solution first dissolve 0.04 g dihydroxyphenylalanine (DOPA), then 1.0 g asparagine, 1.0 g L-glutamine, and 1.0 g glycine in 200 ml of demineralized water. The pH is adjusted to 5.5 with 1 M K_2HPO_4 (about 2.5 ml is needed) and the solution is filter-sterilized.

The basal agar comprises 4.0 g KH_2PO_4, 2.5 g $MgSO_4 \cdot 7H_2O$, 10 mg thiamine-HCl, 20 μg biotin, 0.5 g glucose and 25 g agar in 800 ml of demineralized water. The pH is adjusted to 5.5 with approximately 7 ml of 1 M K_2HPO_4. The basal agar is sterilized by autoclaving at 15 lbs overpressure for 15 minutes. The complete medium should be stored in a refrigerator and used within one week of preparation.

21. *Tetrazolium indicator medium (TTC medium)*

This test is used in some medical laboratories as an aid to identification of some species. *Candida tropicalis* characteristically produces dark red to maroon colonies owing to the reduction of the tetrazolium salt whereas the yeasts which are incapable of reducing the tetrazolium salt, such as *Candida albicans,* produce pale pink to whitish colonies. The range of coloration from pink to orange and red which develops with various species may be useful with additional tests for screening and presumptive identification.

Method and medium

A heavy inoculum of cells from a 24–48-hour-old culture is spotted onto the agar. The plate is incubated for 24 hours and then inspected for color formation.

TTC stock solution is prepared by dissolving 1 g of 2,3,5-triphenyltetrazolium chloride in 10 ml of demineralized water. The solution is filter-sterilized and stored at 4°C in a dark-glass container.

The basal agar is prepared by adding to 1 liter of demineralized water 10 g of soybean peptone, 10 g of glucose, and 15 g of agar and dissolving them by heating. The medium is sterilized by autoclaving at 15 lbs overpressure for 15 minutes. After cooling the agar to 50°C 0.1 ml of the TTC stock solution is added giving a final concentration of 100 μg per ml and the complete medium is poured into Petri dishes. The final pH is 7.0.

III. Some practical suggestions on the routine identification of yeasts

Before commencing with the determination of any yeast strain its freedom from contamination must be established, preferably by plating. However, when dealing with ascogenous cultures, particularly with strains freshly recovered from natural substrates, it should be borne in mind that such strains may be heterothallic. In such a case, the injudicious selection of single colonies after repeated plating of an actively sporulating culture might lead to recovery of a single mating type and the loss of the sporulating diploid state. In this event, the isolation of both mating types must be undertaken.

When examining large numbers of undetermined strains, the worker is advised to adhere to a scheduled programme. Since many of the determination tests run over several weeks, the use of a diary in which the dates are noted when the results are to be recorded, is recommended.

Because the different criteria employed in classification of yeasts are not equally important in the different taxa, the tests may be divided into two categories, i.e. those used for primary differentiation and those which are used more or less for confirmation purposes.

Tests used for primary differentiation which permit the classification to genus level and frequently to the species as well, include the following:

1. A detailed examination of the mode of vegetative reproduction, cellular morphology, the formation of pseudohyphae (pseudomycelium), true hyphae (mycelium) and ballistospores as well as cultural aspects in liquid medium and on solid medium.
2. The determination of the presence or absence of ascospore formation, using several sporulation media (including Gorodkowa agar, McClary's acetate agar, YM agar, V8 agar, 2% malt extract agar or ME agar, growth restriction agar and potato-dextrose agar). Sight should not be lost of the fact that some genera, e.g. *Lipomyces,* require special sporulation media. Careful attention should be given to the characteristics of both the asci and ascospores, if detected.
3. Testing for the fermentation of glucose, galactose, sucrose, maltose, lactose and raffinose.
4. Testing in liquid medium for the assimilation of the 18 carbon compounds of the standard description.
5. The assimilation of nitrate in liquid medium or by auxanogram.
6. Growth at 37°C.
7. Growth in vitamin-free medium.
8. Testing for the production of extracellular, amyloid material by the addition of dilute iodine solution to the carbon assimilation tests in liquid medium, i.e. after the final results of the latter have been recorded.

On completion of these tests and after having consulted the keys to the genera and species, it will become apparent which of the confirmatory tests are still required for the identification of the strains. These confirmatory tests may include the following:

1. The fermentation of trehalose, melibiose, inulin and starch.
2. The assimilation of 2-ketogluconate, 5-ketogluconate, glucono-δ-lactone, glucosamine, gluconate, melibiose, melezitose, α-methyl-D-glucoside, D-glucitol, methanol, glycerol, hexadecane, and malic acid.

3. The utilization of ethylamine hydrochloride, cadaverine dihydrochloride, L-lysine (if not included in the primary tests) and nitrite and creatinine by the auxanographic technique.

4. The production of extracellular amyloid compounds (if not included in the primary tests).

5. Growth on 50% glucose-yeast extract agar or in 10% sodium chloride plus 5% glucose in yeast nitrogen base.

6. Maximum growth temperature.

7. Urease test.

8. The splitting of fat.

9. Resistance to cycloheximide (if not included in the primary tests).

10. Tolerance of 1% acetic acid.

As it is generally desirable to complete as many tests as possible on a strain at one time, some workers prefer to include some of the confirmatory tests (e.g., cycloheximide resistance, utilization of ethylamine hydrochloride) in the primary series.

The following programme applicable to a 5½-day week was adopted for the study of the genus *Kluyveromyces*. While it may serve as a guide, it should not be regarded as a prescribed procedure, since other genera, e.g. *Dekkera* and *Brettanomyces,* require other schedules.

On Fridays and Saturdays fresh slant cultures of the strains to be classified are prepared from actively growing material. They are to be used for the preparation of inocula to be used on Mondays and Tuesdays. It is desirable to prepare two slants of each strain.

On Mondays and Tuesdays, the following inoculations are made with each of the strains under test:

1. Malt extract flasks and malt extract agar slants for the examination of morphological and cultural properties. The flask and slant cultures are examined on the following Thursday and Friday, respectively, and again 4 weeks after inoculation.

2. Dalmau plates with corn meal agar for the examination of the presence of pseudomycelium and true mycelium. The plates are examined 10 days after inoculation.

3. Sporulation media which are examined after 3 days and again after 1, 2, 3 and 4 weeks.

4. Slants of arbutin agar which are examined after 5 days.

5. Slants of 50% glucose-yeast extract-agar which are examined after 5 days.

6. Slants of YM agar which are incubated at 37°C and examined after 3 days.

7. Full sets of Durham tubes for the fermentation of glucose, galactose, sucrose, maltose, lactose and raffinose. The fermentation sets are examined after 1, 2, 3, 4, 7, 10 and 14 days.

8. Full sets of carbon assimilation tubes of the 18 compounds tested which are examined after 1 week and again after 3 weeks.

9. Sets of nitrogen assimilation media for testing the utilization of nitrate and ethylamine hydrochloride. The sets are examined at weekly intervals up to 3 weeks after inoculation.

10. Tubes of vitamin-free medium which are examined after 1 week and 3 weeks.

11. Tubes of yeast nitrogen base containing 0.5% glucose and 100 ppm cyclohex-

Table 10

Composition of chemically defined media[1] for taxonomic study of yeasts

Formula ingredients per liter	Yeast morphology agar	Yeast nitrogen base for carbon assimilation tests	Yeast carbon base for nitrogen assimilation tests	Vitamin-free yeast base for vitamin requirement test
Nitrogen sources:				
Ammonium sulphate	3.5 g	5 g	none	5 g
Asparagine	1.5 g	none	none	none
Carbon source:				
Dextrose	10 g	none	10 g	10 g
Amino acids:				
L-Histidine monohydrochloride	10 mg	10 mg	1 mg	10 mg
DL-Methionine	20 mg	20 mg	2 mg	20 mg
DL-Tryptophan	20 mg	20 mg	2 mg	20 mg
Vitamins:				
Biotin	20 μg	20 μg	20 μg	none
Calcium pantothenate	2,000 μg	2,000 μg	2,000 μg	none
Folic acid	2 μg	2 μg	2 μg	none
Inositol	10,000 μg	10,000 μg	10,000 μg	none
Niacin	400 μg	400 μg	400 μg	none
p-Aminobenzoic acid	200 μg	200 μg	200 μg	none
Pyridoxine hydrochloride	400 μg	400 μg	400 μg	none
Riboflavin	200 μg	200 μg	200 μg	none
Thiamine hydrochloride	400 μg	400 μg	400 μg	none
Compounds supplying trace elements:				
Boric acid	500 μg	500 μg	500 μg	500 μg
Copper sulphate	40 μg	40 μg	40 μg	40 μg
Potassium iodide	100 μg	100 μg	100 μg	100 μg
Ferric chloride	200 μg	200 μg	200 μg	200 μg
Manganese sulphate	400 μg	400 μg	400 μg	400 μg
Sodium molybdate	200 μg	200 μg	200 μg	200 μg
Zinc sulphate	400 μg	400 μg	400 μg	400 μg
Salts:				
Potassium phosphate monobasic	0.85 g	0.85 g	0.85 g	0.85 g
Potassium phosphate dibasic	0.15 g	0.15 g	0.15 g	0.15 g
Magnesium sulphate	0.5 g	0.5 g	0.5 g	0.5 g
Sodium chloride	0.1 g	0.1 g	0.1 g	0.1 g
Calcium chloride	0.1 g	0.1 g	0.1 g	0.1 g

Continued on page 101

Formula ingredients per liter	Yeast morphology agar	Yeast nitrogen base for carbon assimilation tests	Yeast carbon base for nitrogen assimilation tests	Vitamin-free yeast base for vitamin requirement test
Agar	18 g	none	none	none
Amount of final medium from 100 g dehydrated medium	2.8 l	14.9 l	8.5 l	5.9 l
Amount of dehydrated medium per liter of finished medium	35.0 g	6.7 g	11.7 g	16.7 g

[1] These media are marketed by the Difco Laboratories of Detroit (Michigan) in dehydrated form. They are based on the formulae published by Wickerham (1951) but since then have been fortified with greater amounts of biotin, inositol and calcium pantothenate. The pH of the media has been raised from 5.2 to 5.6 by the phosphates listed in the above table.

imide which are examined at weekly intervals up to 3 weeks after inoculation.

With adequate assistance and depending on the number of confirmatory tests required, a worker can handle 10–20 yeast strains per week. The time required for adequate microscopic examination of the cultures is, however, the limiting factor.

For the rapid recording and analysis of results use may be made of stencilled or printed descriptive charts which provide for the tabulation of the morphological, sexual and physiological characteristics of the strains studied. If collections of any size are to be maintained, a punched card system may be used in conjunction with such descriptive charts.

If a strain is recovered which, after repeated testing, cannot be identified with any described species and appears to be representative of an unpublished species, the worker is strongly urged before describing a new species to either obtain specialist opinion from a competent taxonomist who has experience with the genus in question, or to submit the culture to a recognized taxonomical center for an opinion. Failure to do so frequently results in the introduction into the literature of yet another synonym of an already described species.

In preparing the description of new species it is advisable to adhere as closely as possible to the scheme of the standard descriptions adopted in the second edition of 'The Yeasts'. For the preparation of a Latin diagnosis the use of the book by Stearn (1973) is recommended. Unless the worker is competent in Latin, it is advisable to submit the Latin diagnosis to a classical scholar for editing.

Strains of accepted new species must be deposited with at least one internationally recognized culture collection.

IV. List of observations and tests included in the standard description with page reference to the methods

V. List of media, reagents and stains

Media

Reagents

Stains

Chapter III

Discussion of the genera belonging to the ascosporogenous yeasts

Genus 1. *Ambrosiozyma* van der Walt

by N.J.W. Kreger-van Rij

a. DIAGNOSIS OF THE GENUS

Abundant formation of true mycelium with blastospores; in addition, budding yeast cells and pseudomycelium. The septa have a central pore surrounded by a thickening of the wall (dolipore); the pore is plugged. In light microscopy, the plugged pore shows as a small dark dot in the middle of the septum.

Asci are generally formed on the hyphae. Ascospores are hat-shaped.

b. TYPE SPECIES OF THE GENUS

Ambrosiozyma monospora (Saito) van der Walt

c. SPECIES ACCEPTED IN THE GENUS

1. *Ambrosiozyma cicatricosa* (Scott et van der Walt) van der Walt (Scott & van der Walt, 1971b)
2. *Ambrosiozyma monospora* (Saito) van der Walt (Saito, 1932)
3. *Ambrosiozyma philentoma* van der Walt, Scott et van der Klift (in van der Walt, 1972)
4. *Ambrosiozyma platypodis* (Baker et Kreger-van Rij) van der Walt (Baker & Kreger-van Rij, 1964)

d. KEY TO THE SPECIES OF THE GENUS

1a	Nitrate assimilated	
	A. platypodis . . . . p. 111	
b	Nitrate not assimilated	2
2a	Mostly one to two spores per ascus	3
b	Mostly four spores per ascus	
	A. philentoma . . . . p. 110	

e. SYSTEMATIC DISCUSSION OF THE SPECIES

1. AMBROSIOZYMA CICATRICOSA (SCOTT ET VAN DER WALT) VAN DER WALT (1972)

Synonym

Pichia cicatricosa Scott et van der Walt (1971b)

Standard description of *Ambrosiozyma cicatricosa*

Growth in malt extract: After 3 days at 25°C cells are short-oval to elongate, pear- and lemon-shaped, (3–8) × (6–14) μm; single or in pairs. Buds are formed on a broad base; more than one bud may be formed at the same site leading to multiple scars on the mother cell. Abundant branched mycelium with a diameter of 2.5–5 μm is present. The culture has a fragrant odor. A flocculent sediment is formed. After one month at 17°C a thick, lumpy mass is present.

Growth on malt agar: After one month at 17°C the streak culture is cream colored to yellow, tough, dull and raised in the middle. The edge is fringed with mycelium.

Slide cultures on potato- and corn meal agar: Branched mycelium is formed with groups of spherical to elongate blastospores at the end of the hyphae or alongside them, often in loose groups. Some blastospores produce a series of buds at the top and show polar elongation with multiple scars, the dactyloid cells (Scott & van der Walt, 1971b). Pseudomycelium may be present. The septa have a dolipore which is visible as a small dark dot in the light microscope.

Formation of ascospores: The asci are spherical or short-oval and situated singly or in groups, occasionally among blastospores, at the end of the hyphae or alongside them at the septa. They contain one to four, often two hat-shaped spores which are generally not liberated. The spores measure (5–6) × (10–12) μm, including the brim (Scott & van der Walt, 1971b).

Spores were observed on V8-, potato- and corn meal agar.

Fermentation:

Glucose	+	Maltose	+ s
Galactose	–	Lactose	–
Sucrose	+ s	Raffinose	–

Assimilation of carbon compounds:

Galactose	–	Raffinose	–	Erythritol	+
Sucrose	+	Soluble starch	–	Ribitol	+
Maltose	+	D-Xylose	+	D-Mannitol	+
Cellobiose	+	L-Arabinose	+	Succinic acid	+
Trehalose	+	D-Ribose	v	Citric acid	+
Lactose	–	L-Rhamnose	–	Inositol	–

Splitting of arbutin: +

Assimilation of nitrate: –

Growth in vitamin-free medium: –

Growth on 50% (w/w) glucose-yeast extract agar: +w

Growth at 37°C: +w or –

Origin of the strains studied

Material from tunnels of *Xyleborus torquatus* and *Platypus externedentatus* in *Cussonia umbellifera* and *Macaranga capensis* (2) (Scott & van der Walt, 1971b).

Type strain: CBS 6157, isolated by Scott and van der Walt.

2. *AMBROSIOZYMA MONOSPORA* (SAITO) VAN DER WALT (1972)

Synonyms

Endomycopsis monospora Saito (1932)
Endomycopsis fibuligera (Lindner) Dekker var. *monospora* (Saito) Lodder et Kreger-van Rij (1952)
Endomyces bispora Verrall (1943) [nec *Endomyces bisporus* Beck (1922)]
Endomycopsis fasciculata Batra (1963)
Pichia fasciculata (Batra) Boidin, Abadie et Lehodey (1965a)

Standard description of *Ambrosiozyma monospora*

Growth in malt extract: After 2 days at 25°C the culture consists of mycelial hyphae with side branches, with a diameter of 1.5–5.5 μm, and spherical to short-oval, budding yeast cells measuring (7–9) × (7–13) μm; single, in pairs or in chains. Pseudomycelium may also be present. The culture has a fragrant odor. A loose, flocky sediment is formed. After one month at 17°C a thick, lumpy sediment, filling up the liquid and a broad ring are present.

Growth on malt agar: After one month at 17°C the streak culture is yellow to yellowish-brown, tough, raised, dull, delicately wrinkled over the whole surface. The margin is fringed with mycelium.

Slide cultures on potato- and corn meal agar: True, branched mycelium is formed. The septa have a dolipore, visible as a small dark dot with the light microscope. Spherical to short-oval blastospores occur in chains or in clusters at the end of the mycelial hyphae or alongside them (Fig. 15). Small protuberances may occur on the hyphae. Pseudomycelium may be present.

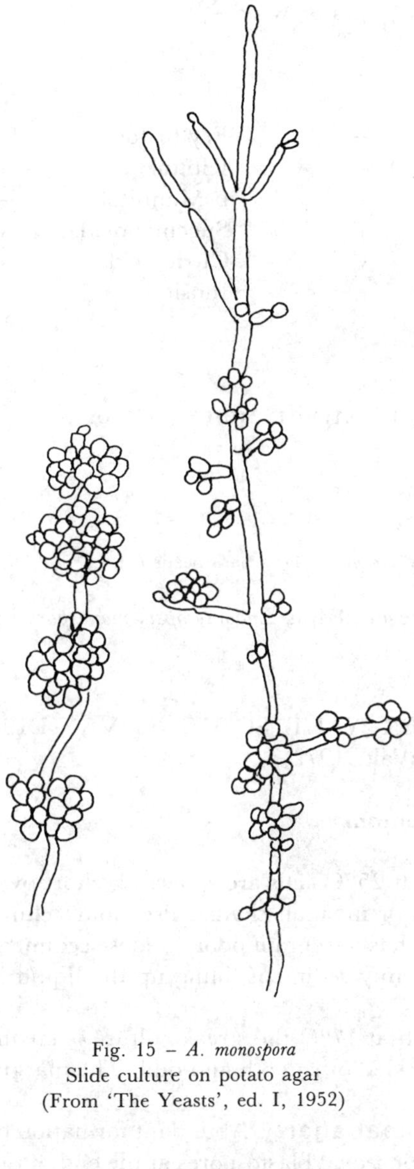

Fig. 15 – *A. monospora*
Slide culture on potato agar
(From 'The Yeasts', ed. I, 1952)

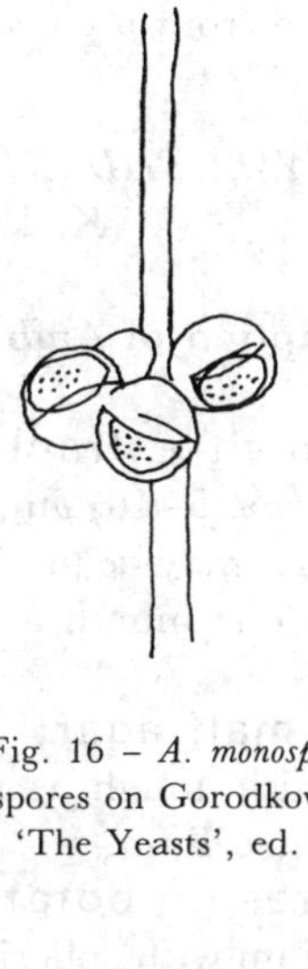

Fig. 16 – *A. monospora*
Ascospores on Gorodkowa agar
(From 'The Yeasts', ed. I, 1952)

Formation of ascospores: The asci are spherical or oval. They are situated in groups at the end of the mycelial hyphae or alongside them at the septa (Fig. 16). Single spores give cultures which sporulate again (Batra, 1963). The spores are hat-shaped; one to two are formed per ascus. They measure, including the brim, (3–4) × (6–8) μm. They are liberated from the ascus. The presence of many spores gives the culture a brown appearance.

Spores were observed on Gorodkowa-, YM-, potato- and corn meal agar.

Fermentation:

Glucose	+ s or w	Maltose	+ s or w or –
Galactose	–	Lactose	–
Sucrose	+ s or w or –	Raffinose	–

Assimilation of carbon compounds:

Galactose	–	Raffinose	–	Erythritol	+
Sucrose	+	Soluble starch	–	Ribitol	+
Maltose	+	D-Xylose	+	D-Mannitol	+
Cellobiose	+	L-Arabinose	+	Succinic acid	+
Trehalose	+	D-Ribose	v	Citric acid	+
Lactose	–	L-Rhamnose	–	Inositol	–

Splitting of arbutin: +

Assimilation of nitrate: –

Growth in vitamin-free medium: –

Growth on 50% (w/w) glucose-yeast extract agar: + w or –

Growth at 37°C: + w

Origin of the strains studied

Exudate of a coconut palm (1), *Pinus strobus* (1), *Platypus compositus* (1), a bark beetle (1).

Type strain: CBS 2554, the original strain described as *Endomycopsis monospora* by Saito and isolated from a coconut palm.

3. *AMBROSIOZYMA PHILENTOMA* VAN DER WALT, SCOTT ET VAN DER KLIFT (in van der Walt, 1972)

Standard description of *Ambrosiozyma philentoma*

Growth in malt extract: After 2 days at 25°C cells are spherical, short-oval and oval, (3.5–7) × (5–10) μm; single, in pairs or in small chains. Pseudomycelium and true mycelium may be present. The culture has a fragrant odor. A loose sediment is present. After one month at 17°C a thick lumpy sediment filling up the liquid is present.

Growth on malt agar: After one month at 17°C the streak culture is cream colored to brownish, tough, raised, slightly wrinkled or smooth and dull. The margin is fringed with mycelium.

Slide cultures on potato- and corn meal agar: Abundant formation of branched mycelium with spherical to oval and long-oval blastospores at the end of the hyphae or along them, at the septa and in between, often in small branched chains. The septa have a dolipore, visible as a small dark dot with the light microscope. Pseudomycelium may be present.

Formation of ascospores: Fusion between loose yeast cells occurs. The asci are spherical or oval, situated terminally or laterally on the hyphae, often in small chains. They contain one to four, usually four, hat-shaped spores which are easily liberated. According to van der Walt (1972), the species is homothallic.

Spores were observed on V8-, malt- and corn meal agar, and, after growth in malt extract, in water.

Fermentation:

Glucose	+ w or s	Maltose	–
Galactose	–	Lactose	–
Sucrose	+ w or s	Raffinose	–

Assimilation of carbon compounds:

Galactose	–	Raffinose	–	Erythritol	+
Sucrose	+	Soluble starch	–	Ribitol	+
Maltose	+	D-Xylose	+	D-Mannitol	+
Cellobiose	+	L-Arabinose	–	Succinic acid	+
Trehalose	+	D-Ribose	+	Citric acid	+
Lactose	–	L-Rhamnose	+	Inositol	–

Splitting of arbutin: +

Assimilation of nitrate: –

Growth in vitamin-free medium: –

Growth on 50% (w/w) glucose-yeast extract agar: + w

Growth at 37°C: –

Origin of the strains studied

Material from tunnels of *Xyleborus aemulus* in *Rapanea melanophoeos* (2).

Type strain: CBS 6276, one of the above strains.

4. AMBROSIOZYMA PLATYPODIS (BAKER ET KREGER-VAN RIJ) VAN DER WALT (1972)

Synonyms

Endomycopsis platypodis Baker et Kreger-van Rij (1964)
Hansenula platypodis (Baker et Kreger-van Rij) Fiol (1967)
Hormoascus platypodis (Baker et Kreger-van Rij) von Arx (1972)

Standard description of *Ambrosiozyma platypodis*

Growth in malt extract: After 3 days at 25°C the cells are spherical 5–9 μm; single or in pairs. Mycelial hyphae and pseudomycelium are also present. The culture has a sour fruity odor. A sediment and a ring are formed. After one month at 17°C a sediment and a ring are present.

Growth on malt agar: After one month at 17°C the streak culture is yellow with an orange tinge, tough, raised in the middle, dull and strongly plaited; the edge is fringed with mycelium.

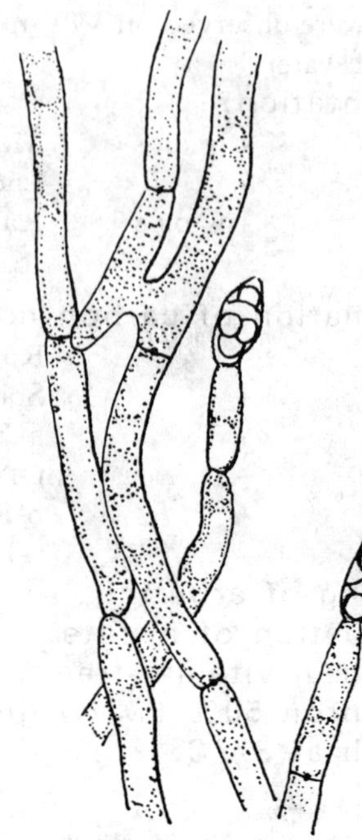

Fig. 17 – *A. platypodis*
Slide culture on corn meal agar, 1000 ×
(From 'The Yeasts', ed. II, 1970)

Slide cultures on potato- and corn meal agar: True branched mycelium with spherical, oval to long-oval blastospores is abundantly formed. The blastospores occur singly, in clusters or in short chains near the septa or in between. The septa have a dolipore, visible as a small dark dot with the light microscope. Anastomoses between hyphae may occur (Fig. 17). Pseudomycelium may be present.

Formation of ascospores: Conjugation between yeast cells may occur. The zygote forms buds or hyphae; the former may turn directly into asci, the latter may bear spherical or oval asci terminally or laterally, often in small chains (Fig. 17). Anastomoses between hyphae may also occur. The asci contain two to four, usually four, hat-shaped spores which are easily liberated. The presence of many spores gives the culture a brown color. Single spores give cultures which sporulate again.

Spores were observed on potato-, corn meal-, malt-, Gorodkowa- and YM agar.

Fermentation:

Glucose	+ w or s	Maltose	–
Galactose	–	Lactose	–
Sucrose	–		

Assimilation of carbon compounds:

Galactose	–	Raffinose	–	Erythritol	+
Sucrose	+	Soluble starch	v	Ribitol	+
Maltose	+	D-Xylose	+	D-Mannitol	+
Cellobiose	+	L-Arabinose	–	Succinic acid	+
Trehalose	+	D-Ribose	+	Citric acid	+
Lactose	–	L-Rhamnose	+	Inositol	–

Splitting of arbutin: +

Assimilation of nitrate: +

Growth in vitamin-free medium: –

Growth on 50% (w/w) glucose-yeast extract agar: –

Growth at 37°C: –

G + C: 28.5–33.4 mol%, 4 strains (Nakase & Komagata, 1969)

Comments

Ambrosiozyma platypodis is the only one of the four *Ambrosiozyma* species assimilating nitrate. Since it is very similar to the other species, especially to *A. philentoma,* I have retained it in the genus *Ambrosiozyma* and not reclassified it in a new genus *Hormoascus* as proposed by von Arx (1972).

Origin of the strains studied

Frass tunnels of *Platypus cylindrus* (1) (Baker, 1963); frass of *Tsuga canadensis* (1); South Santiam River, Oregon, U.S.A. (1).

Type strain: CBS 4111, isolated by Baker.

Genus 2. *Arthroascus* von Arx

by N.J.W. Kreger-van Rij

a. DIAGNOSIS OF THE GENUS

Reproduction by budding on a broad base and by the formation of true mycelium breaking up into arthrospores. Both budding cells and arthrospores are mostly elongate. Cross walls with a closure line, i.e. the remnants of plasmalemma after centripetal closure during development of the wall.

Asci formed directly after conjugation between two cells. They are generally spindle-shaped. The ascospores have a ledge.

b. TYPE SPECIES OF THE GENUS

Arthroascus javanensis (Klöcker) von Arx

c. SPECIES ACCEPTED IN THE GENUS

1. *Arthroascus javanensis* (Klöcker) von Arx (Klöcker, 1909b)

e. SYSTEMATIC DISCUSSION OF THE SPECIES

1. *ARTHROASCUS JAVANENSIS* (KLÖCKER) VON ARX (1972)

Synonyms

Endomyces javanensis Klöcker (1909b)
Schwanniomyces javanensis (Klöcker) Zender (1925)
Endomycopsis javanensis (Klöcker) Dekker (Stelling-Dekker, 1931)
Schizosaccharomyces javanensis (Klöcker) Streiblová (1963)
Pichia nonfermentans Nakase (1971b)

Standard description of *Arthroascus javanensis*

Growth in malt extract: After 2 days at 25°C the cells are long-oval and lemon-shaped to elongate, (2.5–5) × (8–15) μm (Fig. 18). Buds are formed on a more or less broad base at the extremities of the cells, seldom in the middle. More than one bud may be formed at the same site of the cell. Short mycelial hyphae with cross walls may be present. Cross walls between yeast cells and in hyphae have a closure line (Kreger-van Rij & Veenhuis, 1973a). A sediment is formed. After one month at 17°C a sediment and a ring which easily drops are present.

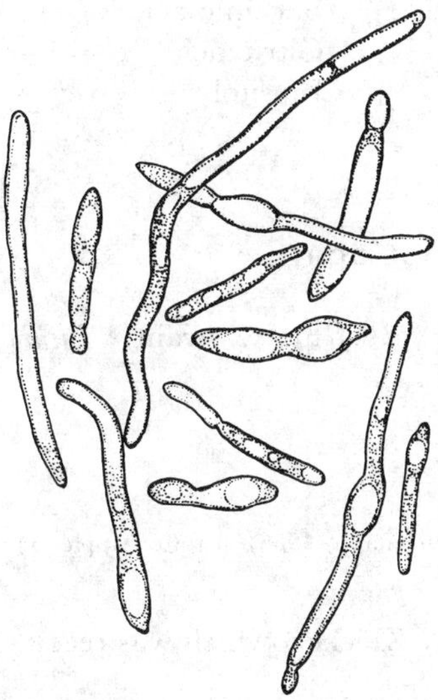

Fig. 18 – *Ar. javanensis*
After 3 days in malt extract
(From 'The Yeasts', ed. I, 1952)

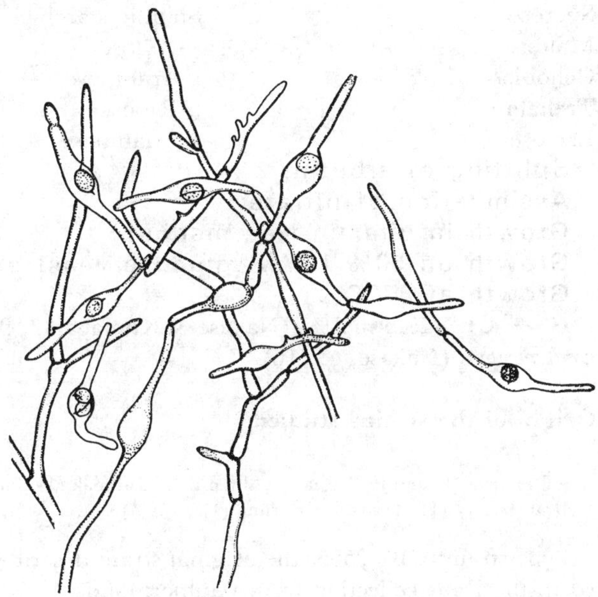

Fig. 19 – *Ar. javanensis*
Slide culture on potato agar, 1000 ×
(From 'The Yeasts', ed. I, 1952)

Growth on malt agar: After one month at 17°C the streak culture is yellowish-white, soft, dull, flat and very delicately wrinkled.

Slide cultures on potato- and corn meal agar: Pseudomycelium consisting of elongate cells in chains is formed. Mycelial hyphae with side branches and cross walls also occur. Short protuberances may give the cells a dentate aspect. Contact between cells via the denticles leads to changes in the wall of the contacted cell, but not to fusion. Dead cells may be penetrated by denticles grown out to stalks (Kreger-van Rij & Veenhuis, 1973b). Both pseudomycelium and true mycelium break up easily.

Formation of ascospores: Conjugation between separate cells and between connected cells, the latter by dissolving of the cross wall between them (Kreger-van Rij & Veenhuis, 1976a), precedes ascus formation. One to four spores are formed in one of the fused cells which is often swollen in the middle (Fig. 19). The spores are spherical or somewhat flattened with a ledge in the middle or on one side. The ledge is not always clearly visible with the light microscope. The type strain (CBS 2555) produced mostly one spore per ascus with a warty appearance, the other strains studied often had four smooth spores per ascus. All spores were easily liberated from the ascus. Single spores gave cultures which sporulated again.

Spores were observed on potato-, YM-, V8- and malt agar.

Fermentation: –

Assimilation of carbon compounds:

Galactose	–	Raffinose	–	Erythritol	–
Sucrose	–	Soluble starch	v	Ribitol	–
Maltose	+w or –	D-Xylose	–	D-Mannitol	–
Cellobiose	–	L-Arabinose	–	Succinic acid	+
Trehalose	v	D-Ribose	–	Citric acid	–
Lactose	–	L-Rhamnose	–	Inositol	–

Splitting of arbutin: –

Assimilation of nitrate: –

Growth in vitamin-free medium: –

Growth on 50% (w/w) glucose-yeast extract agar: –

Growth at 37°C: –

G + C: 29.5 mol%, (Nakase & Komagata, 1968b); 34.1 mol% strain *Pichia nonfermentans* (Nakase, 1971b)

Origin of the strains studied

Soil in Java (1) and in Peoria (1), slime flux of an oak (1) (Phaff & Knapp, 1956), a rotten apple (1), soybean factory (1), decayed tree trunk (1).

Type strain: CBS 2555, the original strain described by Klöcker, which was received in the Delft collection from Guilliermond.

Genus 3. *Citeromyces* Santa María

by C.P. Kurtzman

a. DIAGNOSIS OF THE GENUS

Asexual reproduction is by multilateral budding on a narrow base. Cells are spheroidal to ellipsoidal. Neither pseudohyphae nor true hyphae are produced.

Asci are persistent and contain one or infrequently two roughened spheroidal ascospores. Asci arise from diploid cells or from conjugation between complementary mating types.

Sugars are fermented.

Nitrate is assimilated.

b. TYPE SPECIES OF THE GENUS

Citeromyces matritensis (Santa María) Santa María

c. SPECIES ACCEPTED IN THE GENUS

1. *Citeromyces matritensis* (Santa María) Santa María (Santa María, 1956a)

e. SYSTEMATIC DISCUSSION OF THE SPECIES

1. CITEROMYCES MATRITENSIS (SANTA MARÍA) SANTA MARÍA (1957)

Imperfect state: *Candida globosa* Yarrow et Meyer

Synonyms

Hansenula matritensis Santa María (1956a)
Torula globosa Olson et Hammer (1935) nom. nud.
Torulopsis globosa (Olson et Hammer) Lodder et Kreger-van Rij (1952) nom. nud.
Candida globosa Yarrow et Meyer (1978)

Standard description of *Citeromyces matritensis*

Growth on 5% malt extract agar: After 3 days at 25°C, the cells are spheroidal to ellipsoidal (3.0–9.5) × (4.0–10.5) μm, occur as singles, pairs, or in threes, and may have one or two buds. Growth is butyrous and white to faintly tannish-white.

Growth on the surface of assimilation media: Pellicles are not formed.

Dalmau plate culture on morphology agar: After 7 days at 25°C, growth under the coverglass lacks pseudohyphae or true hyphae. Aerobic growth is cream colored with a faintly tan center, slightly glistening with fine striations, butyrous, and low convex with a depressed center. The margin is entire or finely lobed.

Formation of ascospores: Many natural isolates are diploid and the vegetative cells convert directly into asci. The asci have thick persistent walls and contain one or, infrequently, two spheroidal ascospores with a roughened wall and a prominent lipid droplet (Fig. 20). Sporulation is usually abundant on YM agar, and well sporulated cultures turn red to brown. Wickerham (1958) demonstrated *Cit. matritensis* to be heterothallic by heat treatment of a sporogenous culture of the type strain. In addition, he showed that the type strain of *Torulopsis globosa* (NRRL Y-1506, CBS 162) conjugated with one of the mating types (NRRL Y-2407-9), but apparently represents a unisexual diploid form of this species because it also sporulated alone on media with high sugar concentrations. Complementary mating types show a strong sexual agglutination reaction if grown in liquid medium rather than on an agar medium (Wickerham, 1958). Complementary mating types derived from the type strain are CBS 7103 (NRRL Y-2407-9) and CBS 7104 (NRRL Y-2407-10).

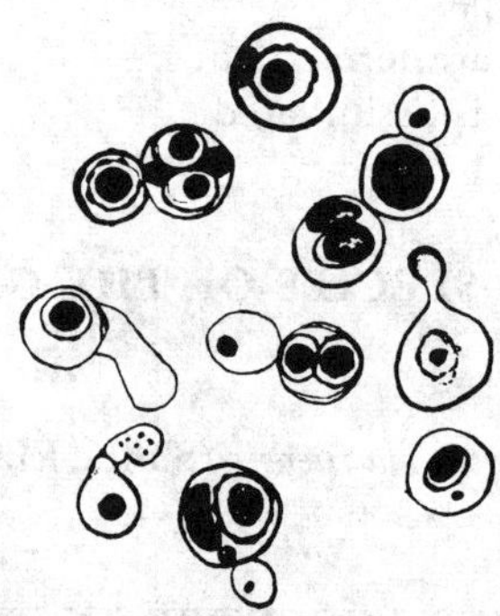

Fig. 20 – *Cit. matritensis*
CBS 2835 × CBS 2836. After 7 days on YM agar
(From 'The Yeasts', ed. II, 1970)

Spores were observed on YM-, malt extract-, and V8 agar.

Fermentation:

Glucose	+	Lactose	–
Galactose	–	Raffinose	v
Sucrose	+ or w	Trehalose	–
Maltose	v		

Assimilation of carbon compounds:

Galactose	–	Raffinose	+	Erythritol	–
Sucrose	+	Soluble starch	–	Ribitol	v
Maltose	+	D-Xylose	–	D-Mannitol	+
Cellobiose	–	L-Arabinose	–	Succinic acid	–
Trehalose	+	D-Ribose	–	Citric acid	v
Lactose	–	L-Rhamnose	–	Inositol	–

Assimilation of nitrate: +
Growth in vitamin-free medium: + or w
Growth in 10% sodium chloride plus 5% glucose in yeast nitrogen base: +
Growth at 37°C: –
G + C: 45.5 mol%, one strain (Nakase & Komagata, 1968a)

Origin of the strains studied

Fruit preserved in syrup (1), sugar (1), condensed milk (3), flux from *Myoporum sandwicense,* Hawaii, U.S.A. (5).

Type strain: CBS 2764 (NRRL Y-2407) received from Santa María as the type strain of *Hansenula matritensis.* It was isolated from fruit preserved in syrup.

f. COMMENTS ON THE GENUS

Citeromyces matritensis is unique among the nitrate-positive ascomycetous yeasts in that it is the only species to form rugose spheroidal ascospores. Possible relationships of this species with nitrate-negative yeasts having a similar spore morphology were discussed earlier by Wickerham (1970a).

Genus 4. *Clavispora* Rodrigues de Miranda

by L. Rodrigues de Miranda

a. DIAGNOSIS OF THE GENUS

Cells of various shape which reproduce by multilateral budding. Pseudomycelium may be formed.

Conjugation of cells of opposite sex precedes ascus formation. The ascospores are clavate, sometimes containing a small oil droplet; one to four ascospores are formed per ascus. The spore wall has small warts which may not be visible under the light microscope. The spores are easily liberated from the ascus.

Metabolism oxidative and fermentative.

External source of vitamins not required.

b. TYPE SPECIES OF THE GENUS

Clavispora lusitaniae Rodrigues de Miranda

c. SPECIES ACCEPTED IN THE GENUS

1. *Clavispora lusitaniae* Rodrigues de Miranda (1979)

e. SYSTEMATIC DISCUSSION OF THE SPECIES

1. *CLAVISPORA LUSITANIAE* RODRIGUES DE MIRANDA (1979)

Imperfect state: *Candida lusitaniae* van Uden et do Carmo-Sousa

Synonyms

Candida lusitaniae van Uden et do Carmo-Sousa (1959)
Candida parapsilosis (Ashf.) Langeron et Talice var. *obtusa* nom. nud. (Dietrichson, 1954)
Candida obtusa (Dietrichson) van Uden et do Carmo-Sousa ex van Uden et Buckley (1970)
Saccharomyces carmosousae Montrocher (1967)

Standard description of *Clavispora lusitaniae*

Growth on malt extract: After 3 days at 25°C, the cells are subglobose to ovoidal, (1.5–6.0) × (2.5–10.0) μm, single, in pairs or in short chains. After one month a ring and sediment are formed. Some islets may be present.

Growth on malt agar: After one month at 20°C the streak culture is cream colored, glistening, soft and smooth.

Slide culture on potato agar: A well developed pseudomycelium is formed consisting of branched chains of slender pseudohyphae bearing short chains of blastospores in a verticillated position.

Formation of ascospores: After conjugation of cells of opposite sex, asci are formed. The ascospores are clavate, the wider end being 2–3 times wider than the narrow end, between one and four occurring per ascus (Fig. 21). The wall has small warts, visible in sections in the electron microscope (N.J.W. Kreger-van Rij, personal communication); under the light microscope the spore wall may appear to be smooth. The spores are easily liberated from the ascus.

5% Difco malt extract agar is a suitable medium for conjugation and spore formation.

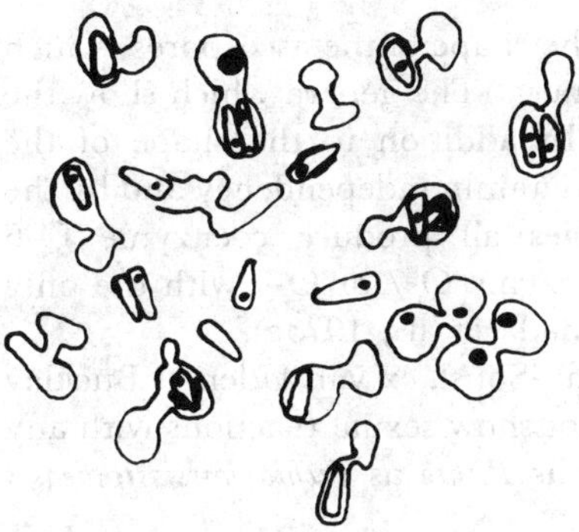

Fig. 21 – *Cl. lusitaniae*
CBS 4413 × CBS 6936
Ascospores on malt extract agar

Fermentation:

Glucose	+	Maltose	– or + vw
Galactose	v	Lactose	–
Sucrose	+ or w	Raffinose	–

Assimilation of carbon compounds:

Galactose	v	Raffinose	–	Erythritol	–
Sucrose	+	Soluble starch	–	Ribitol	+
Maltose	+	D-Xylose	+	D-Mannitol	+
Cellobiose	+	L-Arabinose	v	Succinic acid	+ or w
Trehalose	+	D-Ribose	+ or w	Citric acid	+ w or –
Lactose	–	L-Rhamnose	+	Inositol	–

Assimilation of nitrate: –

Assimilation of nitrite: –

Growth in vitamin-free medium: + w

Growth on 50% (w/w) glucose-yeast extract agar: –

Growth at 37°C: +

G + C: 46.1 mol%, type strain *Candida lusitaniae*, CBS 4413; 46.3 mol%, CBS 5901; 45.9 mol%, type strain *C. obtusa*, CBS 1944 (all by Meyer & Phaff, 1972); 43.4 mol%, type strain *C. lusitaniae*, CBS 4413 (Nakase & Komagata, 1971f)

Origin of the strains studied

Pig (3), sputum (2), human skin (1), corn meal (1), urine (1), mastitis milk (1), excrements of *Corvus albus* (1), citrus peel juice (1).

Type strain: CBS 6936, isolated by Mateles from citrus peel juice; strain CBS 4413, isolated by van Uden from a pig is the isotype.

f. COMMENTS ON THE GENUS

The genus *Clavispora* is mainly characterized by the shape of the ascospores, which is clavate. The shape is not found in other yeast genera. The genera which show the closest resemblance are *Kluyveromyces* and *Pichia.* In addition to the shape of the ascospores *Clavispora* differs from *Kluyveromyces* by its vitamin independency and by the production of coenzyme Q-8. *Kluyveromyces* species all produce coenzyme Q-6 (Yamada et al., 1976b). *Pichia* species produce coenzyme Q-7 or Q-9 with the only exception of *Pichia pastoris,* that produces Q-8 (Yamada et al., 1973c).

Candida obtusa (Dietrichson) van Uden et do Carmo-Sousa ex van Uden et Buckley var. *arabinosa* Montrocher (Montrocher, 1967) did not show sexual reactions with any of the mating types. It is now classified in the genus *Pichia* as *Pichia mississippiensis* Kurtzman et al.

Genus 5. *Coccidiascus* Chatton

by H.J. Phaff

a. DIAGNOSIS OF THE GENUS

Vegetative cells spherical to ovoidal, 5–15 μm in length, with a large vacuole and nucleus. Vegetative reproduction by budding. Mycelium, pseudomycelium, or ballistospores are lacking. Vegetative cells which are transformed into asci elongate, forming banana-shaped to crescent-shaped asci arranged within parasitophorous vacuoles of *Drosophila* intestinal epithelium cells in the manner of coccidian schizonts. Thick-walled asci observed in host tissue range from 13–18 μm tip to tip. Mature asci contain two relatively thin-walled ascospores, closely intertwined in a helix. Asci from dead flies have been observed to contain up to eight ascospores. Ascospores extruded from asci are spindle-shaped, bilaterally flattened, without apical cytoplasmic appendages, each with an apical nucleus. Fermentation and assimilation reactions are unknown.

b. TYPE SPECIES OF THE GENUS

Coccidiascus legeri Chatton

c. SPECIES ACCEPTED IN THE GENUS

1. *Coccidiascus legeri* Chatton (1913)

e. SYSTEMATIC DISCUSSION OF THE SPECIES

1. *COCCIDIASCUS LEGERI* CHATTON (1913)

This organism has not been cultured and the description of this species therefore is the same as the generic description.

f. COMMENTS ON THE GENUS

Chatton (1913) observed and named this species, which parasitized the intestinal cells of *Drosophila funebris* Fabr. in France. He was unable to cultivate it and as far as is known no type material was deposited in a herbarium. This species was rediscovered by Lushbaugh et al. (1976) as a parasite within vacuoles of the intestinal epithelium of *Drosophila melanogaster* Meigen in the state of Georgia, U.S.A. They provided an expanded description of *Coccidiascus legeri,* illustrated by electron micrographs and light microscope photographs of material in situ. They were also unable to culture the organism and, in the absence of a known holotype or lectotype, they designated a neotype which was deposited in the New York Botanical Garden Cryptogamic Herbarium (New York, U.S.A.).

Genus 6. *Cyniclomyces* van der Walt et Scott

by H.J. Phaff and M.W. Miller

a. DIAGNOSIS OF THE GENUS

Cells are long-oval to cylindrical, single, in pairs or in chains. Vegetative reproduction is by budding at the poles or on a broad shoulder. Pseudomycelium is formed in liquid media.

Ascospores are oval to cylindrical, one to four per ascus. No conjugation of cells prior to sporulation. Upon germination an exosporium is evident.

In liquid media a sediment is formed.

Fermentation of sugars is weak.

Growth occurs only between 30 and 40°C. Amino acids and an increased level of CO_2 in the gas atmosphere are required for growth. The cells are short-lived.

b. TYPE SPECIES OF THE GENUS

Cyniclomyces guttulatus (Robin) van der Walt et Scott

c. SPECIES ACCEPTED IN THE GENUS

1. *Cyniclomyces guttulatus* (Robin) van der Walt et Scott (Robin, 1853)

e. SYSTEMATIC DISCUSSION OF THE SPECIES

1. CYNICLOMYCES GUTTULATUS (ROBIN) VAN DER WALT ET SCOTT (1971b)

Synonyms

Cryptococcus guttulatus Robin (1853)
Saccharomyces guttulatus (Robin) Winter (1884)
Saccharomycopsis guttulata (Robin) Schiönning (1903)
Atelosaccharomyces guttulatus (Robin) de Beurmann, Gougerot et Vaucher (1909)

Standard description of *Cyniclomyces guttulatus*

Growth in liquid medium (yeast autolysate (1% w/v)-proteose peptone (1%)-glucose (2%), pH 4.5): After one day at 37°C cells are long-ovoidal to cylindrical (Fig. 22). Budding occurs at the shoulders of the cells resulting in branched chains of pseudomycelium. In many cases the original mother cell appears dead by the time the chain is formed, as judged by complete granulation

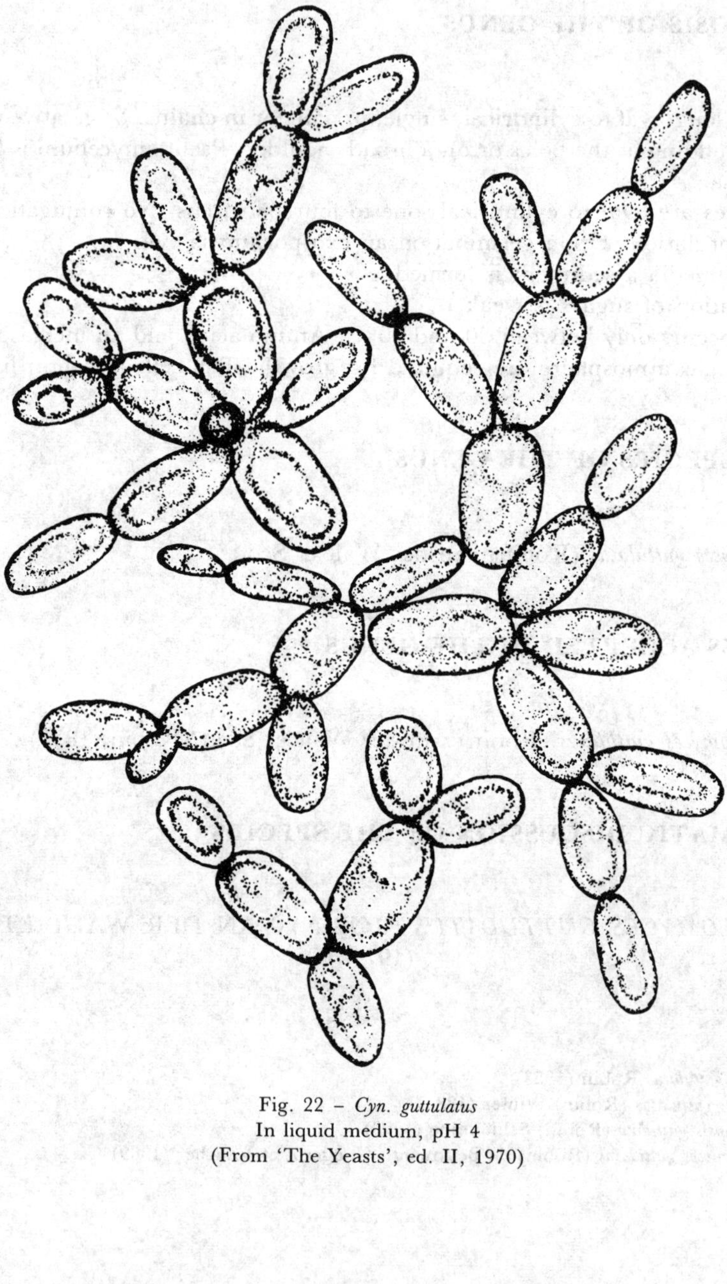

Fig. 22 – *Cyn. guttulatus*
In liquid medium, pH 4
(From 'The Yeasts', ed. II, 1970)

of its protoplasm. After 48 hours the range of cell dimensions is (4.4–6.1) × (14–21) μm. No ring or pellicle is formed, even after an extended period of incubation. The cells have a tendency to grow on the glass wall of a tube, about 1 cm below the surface of the liquid. In older cultures the cells form a loose, often granular sediment.

Growth on solid medium: When a plate of yeast autolysate-proteose peptone-glucose agar (pH 4.5) is inoculated with cells of *Cyn. guttulatus,* smooth, semiglossy colonies appear after 24 hours provided the plates are incubated in an atmosphere high in CO_2 content (optimal concentration 15%). Cells are oval to cylindrical, (4.7–7.9) × (8.8–19.5) μm, mainly single or in pairs (Fig. 23).

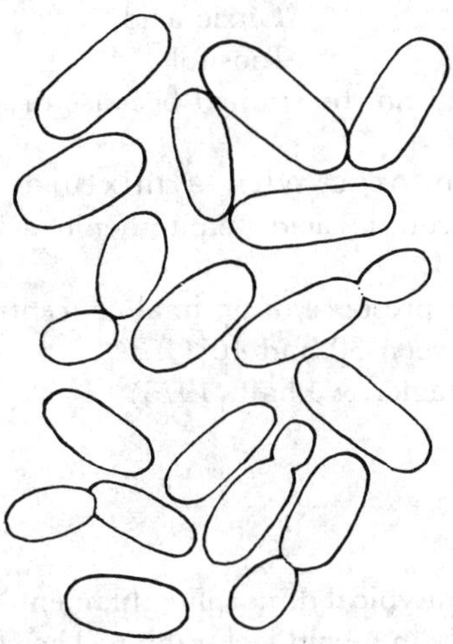

Fig. 23 – *Cyn. guttulatus*
On agar medium, pH 5
(From 'The Yeasts', ed. II, 1970)

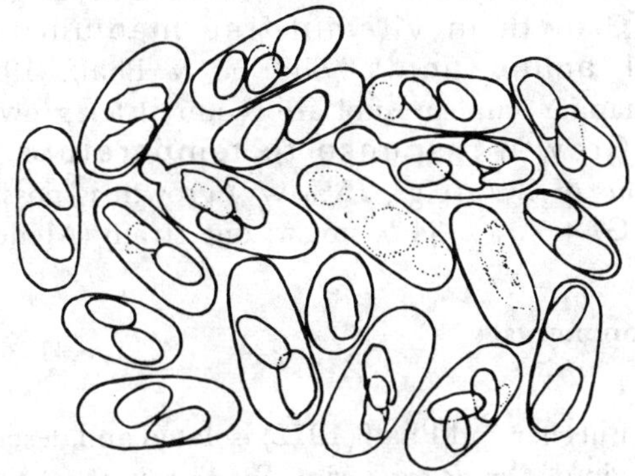

Fig. 24 – *Cyn. guttulatus*
Asci with spores
(From 'The Yeasts', ed. II, 1970)

Formation of ascospores: No conjugation immediately preceding ascospore formation; the vegetative cells are presumed to be diploid. Asci contain one to four elongate to spheroidal ascospores, (0.9–1.6) × (1.0–2.8) μm (Fig. 24), which are lying free in the ascus. This arrangement is quite different from the tightly fitting spores in asci of species of *Saccharomyces*. Asci rarely rupture at maturity. Ascospores germinate directly or after conjugation with another spore. Germination tubes are formed with simultaneous shedding of an outer spore coat (exosporium). Germination is followed by formation of buds.

Ascospores can be obtained only at lower temperatures, best at 18°C, where vegetative growth is not possible. Slants are prepared of a suitable sporulation medium of the following composition: 0.3% yeast autolysate, 0.3% malt extract, 0.15% Bacto-peptone, 1% glucose and 2% agar, and inoculated with about 1 ml sedimented cell material grown in a liquid medium with a pipet. The slants are incubated in a horizontal position for 5–7 days at 18°C.

Fermentation:

Glucose	+ w	Lactose	–
Galactose	–	Melibiose	–
Sucrose	+ w	Raffinose	–
Maltose	–		

Assimilation of carbon compounds tested in 1% (w/v) yeast autolysate at 37°C:

Galactose	–	Raffinose	+	Erythritol	–
Sucrose	+	Soluble starch	–	Ribitol	–
Maltose	–	D-Xylose	–	D-Mannitol	–
Cellobiose	–	L-Arabinose	–	Succinic acid	–
Trehalose	– (+s)	D-Ribose	–	Citric acid	–
Lactose	–	L-Rhamnose	–	Inositol	–

Assimilation of nitrate: This property could not be studied because of the complex nature of the media required for growth.

Growth in vitamin-free medium supplemented with a mixture of 21 amino acids (Shifrine & Phaff, 1959): Nicotinic acid, pantothenic acid, thiamine, and inositol are required for growth.

Growth response to temperature: In the presence of optimal concentrations of CO_2 (e.g., 15% v/v) growth is possible between 30 and 40°C.

G + C: 34.5 mol%, one strain (Mendonça-Hagler & Phaff, 1975)

Comments

Buecher and Phaff (1972) isolated and described an atypical dimorphic, filamentous strain of *Cyniclomyces* (syn. *Saccharomycopsis) guttulatus* from a wild jack rabbit. The colonies of this strain were irregular, rough, and dull and the vegetative growth consisted of septate, branched hyphae, radiating from the central origin of growth. New filaments originated by fragmentation of older aggregates. The atypical strain was not CO_2-dependent at 37°C but did require oxygen. At 30°C, the lower limits for growth, a morphogenetic change took place from the filamentous form to the yeast form, which was stimulated by anaerobic conditions. The budding form of the filamentous strain was relatively stable in culture, and at either 30 or 37°C it was CO_2-dependent. Spontaneous reversions from the budding to the filamentous form took place sporadically.

One of the problems of working with *Cyniclomyces guttulatus* is the high death rate of the cells. The low pH required for the isolation of this yeast causes early granulation and death. If sufficient CO_2 is not present and the inoculum contains low cell numbers, the cells will die before they can grow and generate their own CO_2 in liquid media. This species is therefore difficult to maintain in culture. A lyophilized preparation of a young sporulating culture is recommended for extended storage. Sporulating slants may be stored in a refrigerator for 6–12 months. Re-inoculation of sporulating cells in liquid growth medium results in good growth after 5–10 days.

Because *Cyn. guttulatus* is short-lived and thus easily lost in collections, a brief description is given of a reliable isolation procedure.

Isolation principle: The yeast is most easily isolated from stomach contents of a rabbit, but because fecal material is more readily available, the procedure described below is based on fecal pellets. The yeast is exacting in its growth requirements, needing a balanced mixture of amino acids in the medium (as is found in yeast autolysate), grows best between 35 and 40°C, and, for growth on solid media, it requires a high concentration of CO_2 in the atmosphere. Additionally, it is very tolerant to low pH values of the medium. Use is made of these properties to isolate pure cultures.

Method of isolation: Samples of fecal pellets of rabbits, preferably samples of soft feces (because the latter normally have a higher yeast content) are inspected microscopically for the presence of the large *Cyn. guttulatus* cells by suspending small samples in a drop of water on a slide. Material in which the yeast is present in significant numbers gives the best results. The isolation medium contains 1% powdered yeast autolysate, 1% proteose peptone (Difco), 2% glucose and the pH is adjusted to 3.0 with 1 N HCl. 125-ml Erlenmeyer flasks with 50 ml of this medium are sterilized by exposing them to flowing steam for 10 min.

1. Add one or two rabbit pellets to a flask and incubate for 2–3 days at 37°C. A whitish, granular sediment of yeast forms at the bottom, which should consist of clusters and chains of the characteristic, large cells of *Cyn. guttulatus*.
2. If the sediment appears satisfactory, transfer 1 ml of the sediment with a sterile pipet to a second flask of the same medium and incubate 2–3 days at 37°C.
3. After growth has developed, transfer 0.1 ml (or 2 drops) of the sediment to an agar plate (1% yeast autolysate, 1% proteose peptone, 2% glucose, pH 5.0) and spread with a glass spreader. The plates are placed in a desiccator with water in the bottom to raise the humidity. Flush the desiccator with CO_2 from a cylinder or place a short piece of lighted candle in the desiccator next to the plates, after which the lid is put in place. After a few minutes, when the CO_2 level has increased, the candle goes out and the desiccator is incubated at 37°C for 3 days.
4. An isolated colony is then inspected for purity. If bacteria had not been eliminated during the two previous steps, they would develop on the rich medium at pH 5.0. If the plate is pure, the colony is transferred to a slant and to a liquid medium of the same composition as used in step 3 and incubated as described above. Note: Growth in liquid medium is always in the form of branched chains of cells, whereas on solid media the cells occur more typically in short chains, pairs and singly.

Origin of the strains studied

Feces or stomach content of various domesticated rabbits (15), wild rabbits (3), and wild jack rabbit (1).

Type strain: CBS 5913, isolated by Phaff.

f. COMMENTS ON THE GENUS

The long history of the genus *Cyniclomyces* has been reviewed by Phaff (1970b) and van der Walt and Scott (1971b).

Genus 7. *Debaryomyces* Lodder et Kreger-van Rij nom. cons.

by N.J.W. Kreger-van Rij

a. DIAGNOSIS OF THE GENUS

Vegetative reproduction by multilateral budding. A primitive, or occasionally a well developed, pseudomycelium may be formed.

Heterogamous conjugation, between mother cell and bud, generally precedes ascus formation. Isogamous conjugation also occurs. The spores are spherical or oval. They have warty walls or, in one species, ridges on the wall. One to two spores are usually formed per ascus; in some species up to four spores are present in the ascus.

Fermentation rather slow, weak or absent; never vigorous.

Nitrate not assimilated, nitrite may be assimilated.

b. TYPE SPECIES OF THE GENUS

Debaryomyces hansenii (Zopf) Lodder et Kreger-van Rij

c. SPECIES ACCEPTED IN THE GENUS

1. *Debaryomyces castellii* Capriotti (1958e)
2. *Debaryomyces coudertii* Saëz (1960)
3. *Debaryomyces hansenii* (Zopf) Lodder et Kreger-van Rij (Zopf, 1889)
4. *Debaryomyces marama* di Menna (1954a)
5. *Debaryomyces melissophilus* (van der Walt et van der Klift) Kurtzman et Kreger-van Rij (van der Walt & van der Klift, 1972)
6. *Debaryomyces polymorphus* (Klöcker) Price et Phaff (Klöcker, 1912a)
7. *Debaryomyces pseudopolymorphus* (Ramirez et Boidin) Price et Phaff (Ramirez & Boidin, 1953a)
8. *Debaryomyces tamarii* Ohara et Nonomura (1954c)
9. *Debaryomyces vanriji* (van der Walt et Tscheuschner) Abadie, Pignal et Jacob (van der Walt & Tscheuschner, 1956a)

   Debaryomyces vanriji (van der Walt et Tscheuschner) Abadie, Pignal et Jacob var. *vanriji*

   Debaryomyces vanriji (van der Walt et Tscheuschner) Abadie, Pignal et Jacob var.

yarrowii (Santa María et García Aser) Kreger-van Rij var. nov. (Santa María & García Aser, 1971)

d. KEY TO THE SPECIES OF THE GENUS

1a Sucrose assimilated 2
 b Sucrose not assimilated
   Deb. coudertii .... p. 133

2a Maltose assimilated 3
 b Maltose not assimilated
   Deb. tamarii .... p. 142

3a Xylose assimilated 4
 b Xylose not assimilated
   Deb. melissophilus .... p. 138

4a Erythritol assimilated 5
 b Erythritol not assimilated 9

5a Added vitamins required 6
 b Added vitamins not required 8

6a Fermentation of glucose and sucrose
   Deb. pseudopolymorphus .... p. 141
 b Glucose and sucrose are not or very weakly fermented 7

7a Spores spherical; one to two per ascus
   Deb. hansenii .... p. 135
 b Spores oval; one or four per ascus
   Deb. marama .... p. 137

8a Fermentation of glucose and sucrose
   Deb. polymorphus .... p. 139
 b Glucose and sucrose not or very weakly fermented
   Deb. vanriji var. *vanriji* .... p. 143

9a Added vitamins required 10
 b Added vitamins not required
   Deb. vanriji var. *yarrowii* .... p. 144

10a Fermentation of glucose and sucrose
   Deb. castellii .... p. 132
 b Glucose and sucrose are not or very weakly fermented
   Deb. hansenii .... p. 135

Table 11

Key characters of the species in the genus *Debaryomyces*

	Fermentation		Assimilation				Vit	Shape of the spores	Number of spores per ascus
	G	Su	Su	Ma	X	Er			
Deb. castellii	+	+	+	+	+	–	–	spherical	1–3(1)
Deb. coudertii	–		–	+	+	+	–	spherical	1
Deb. hansenii	+ vw/ –	+ vw/ –	+	+	+	v	–	spherical	1–2(1)
Deb. marama	+ vw/ –	–	+	+	+	+	–	oval	1–4(2)
Deb. melissophilus	–		+	+	–	+	–	spherical	1–4
Deb. polymorphus	+	+	+	+	+	+	+	spherical	1–2(1)
Deb. pseudopolymorphus	+	+	+	+	+	+	–	spherical	1–4
Deb. tamarii	+ vs	+ vw	+	–	–	–	–	spherical	1–2(1)
Deb. vanriji var. *vanriji*	+ vw/ –	+ vw/ –	+	+	+	+	+	spherical	1–4
Deb. vanriji var. *yarrowii*	+ vw/ –	+ vw/ –	+	+	+	–	+	spherical	1–3

Numbers in parentheses refer to the number of spores most frequently observed. G = glucose, Su = sucrose, Ma = maltose, X = D-xylose, Er = erythritol, Vit = growth without the addition of vitamins.

e. SYSTEMATIC DISCUSSION OF THE SPECIES

1. *DEBARYOMYCES CASTELLII* CAPRIOTTI (1958e)

Synonyms

Zymodebaryomyces castellii (Capriotti) Novák et Zsolt (1961)
Torulaspora castellii (Capriotti) van der Walt et Johannsen (1975a)

Standard description of *Debaryomyces castellii*

Growth in malt extract: After 3 days at 25°C, cells are spherical to short-oval, (3.8–7.5) × (5–8.5) μm; single, in pairs or in groups (Fig. 25). A sediment and a thin, dull, creeping pellicle are formed. After one month at 17°C a sediment and a pellicle are present.

Growth on malt agar: After one month at 17°C the streak culture is yellowish-white, dull-glistening, smooth, with an occasionally sinuous margin.

Slide cultures on potato- and corn meal agar: No formation of pseudomycelium.

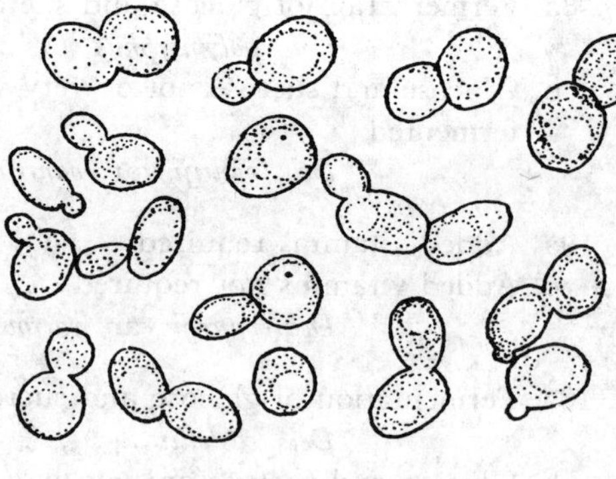

Fig. 25 – *Deb. castellii*
After 2 days in malt extract
(From 'The Yeasts', ed. II, 1970)

Formation of ascospores: Conjugation between mother cell and bud precedes ascus formation. The spores are spherical with a warty wall. The warts are not always distinctly visible with the light microscope. One to three, generally one spore per ascus.

Spores were observed on Gorodkowa- and acetate agar.

Fermentation:

Glucose	+	Lactose	–
Galactose	–	Melibiose	+w
Sucrose	+	Raffinose	+
Maltose	+w or s		

Assimilation of carbon compounds:

Galactose	+	Raffinose	+	Erythritol	–
Sucrose	+	Soluble starch	+	Ribitol	+
Maltose	+	D-Xylose	+	D-Mannitol	+
Cellobiose	+	L-Arabinose	+	Succinic acid	+
Trehalose	+	D-Ribose	–	Citric acid	+
Lactose	+	L-Rhamnose	+	Inositol	–

Splitting of arbutin: +

Assimilation of nitrate: –

Growth in vitamin-free medium: –

Growth on 50% (w/w) glucose-yeast extract agar: +

Growth at 37°C: –

G + C: 37.1 mol%, one strain (Price et al., 1978)

Origin of the strains studied

Soil in Sweden (2).

Type strain: CBS 2923, isolated by Capriotti.

2. *DEBARYOMYCES COUDERTII* SAËZ (1960)

Synonym

Torulaspora coudertii (Saëz) van der Walt et Johannsen (1975a)

Standard description of *Debaryomyces coudertii*

Growth in malt extract: After 2 days at 25°C, the cells are spherical to short-oval, (3.5–5) × (4–5.8) μm; single or in pairs. A very thin sediment is formed. After one month at 17°C a sediment and a ring are present.

Growth on malt agar: After one month at 17°C the streak culture is yellowish-white, dull-shiny, soft, slightly raised in the middle and smooth.

Slide cultures on potato- and corn meal agar: No formation of pseudomycelium.

Formation of ascospores: Conjugation between mother cell and bud precedes ascus formation. The spores are spherical with a warty wall; one is formed per ascus. The presence of many spores gives the culture a brown appearance.

Spores were observed on V8 agar at 20°C.

Fermentation: –

Assimilation of carbon compounds:

Galactose	+	Raffinose	–	Erythritol	+
Sucrose	–	Soluble starch	–	Ribitol	+
Maltose	+	D-Xylose	+	D-Mannitol	+
Cellobiose	+	L-Arabinose	+	Succinic acid	+
Trehalose	+	D-Ribose	+	Citric acid	–
Lactose	–	L-Rhamnose	–	Inositol	–

Splitting of arbutin: +

Assimilation of nitrate: –

Growth in vitamin-free medium: –

Growth on 50% (w/w) glucose-yeast extract agar: +

Growth at 37°C: –

G + C: 37.4 mol%, one strain (Price et al., 1978)

Origin of the strain studied

Feces of *Aptenodytes patagonica,* the emperor penguin (1).

Type strain: CBS 5167, isolated by Saëz.

3. *DEBARYOMYCES HANSENII* (ZOPF) LODDER ET KREGER-VAN RIJ (1952)

Imperfect state: *Candida famata* (Harrison) Meyer et Yarrow

Synonyms

Saccharomyces hansenii Zopf (1889)
Torulaspora hansenii (Zopf) van der Walt et Johannsen (1975a)
Debaryomyces tyrocola Konokotina (1913)
Debaryomyces tyrocola Konokotina var. *hansenii* (Zopf) Dekker (Stelling-Dekker, 1931)
Debaryomyces membranaefaciens Naganishi (1917)
Debaryomyces kloeckeri Guilliermond et Péju (1919)
Debaryomyces matruchoti Grigoraki et Péju (1921)
Debaryomyces hudeloi da Fonseca (1922)
Debaryomyces kloeckeri Guilliermond et Péju var. *hudeloi* (da Fonseca) Dekker (Stelling-Dekker, 1931)
Debaryomyces hildegaardi Ota (1923)
? *Debaryomyces laedegaardi* Ota (1923)
? *Debaryomyces leopoldi* Ota (1923)
? *Debaryomyces lundsgaardi* Ota (1923)
Debaryomyces fabryi Ota (1924b)
Debaryomyces gruetzii Ota (1924b)
? *Debaryomyces tremoniensis* Ota (1924b)
? *Debaryomyces fabryi* Ota var. *tremoniensis* (Ota) Dodge (1935)
? *Debaryomyces emphysematosus* Ota (1924b)

Debaryomyces matruchoti Grigoraki et Péju var. *cesarii* Dekker (Stelling-Dekker, 1931)
Debaryomyces guilliermondii Dekker (Stelling-Dekker, 1931)
Debaryomyces guilliermondii Dekker var. *nova-zeelandicus* Lodder (1932)
Debaryomyces membranaefaciens Naganishi var. *hollandicus* Lodder (1932)
Debaryomyces kloeckeri Guilliermond et Péju var. *major* Lodder (1932)
Debaryomyces sake Saito et Ota (1932)
Eutorulopsis subglobosa Zach (in Wolfram & Zach, 1934)
Debaryomyces matruchoti Grigoraki et Péju var. *subglobosus* Zach
Debaryomyces subglobosus (Zach) Lodder et Kreger-van Rij (1952)
Debaryomyces miso Mogi (1938)
Debaryomyces nicotianae Giovannozzi (1939)
Debaryomyces nicotianae Giovannozzi var. *minor* Giovannozzi (1939)
Debaryomyces marylandii Giovannozzi (1941)
Debaryomyces cavensis Giovannozzi (1941)
Debaryomyces japonicus Naganishi (1941a)
Debaryomyces orientalis Naganishi (1941a)
Debaryomyces fukuyamaensis Naganishi (1941b)
Paratorulopsis banheggii Galgóczy et Novák (1962)
Torulopsis westerdijkii Novák et Vitéz (1964)
Debaryomyces nepalensis Goto et Sugiyama (1968)
Pichia adzetii Jacob (1969b)
Torulopsis armenti Kocková-Kratochvílová, Sláviková et Beránek (1977a)

Standard description of *Debaryomyces hansenii*

Growth in malt extract: After 2 days at 25°C, the cells are spherical to short-oval, (2–7) × (2.4–8.5) μm; single, in pairs or in short chains (Fig. 26). A sediment, a ring and, in some strains, a smooth or wrinkled, dry pellicle or islets are formed. After one month at 17°C a sediment, a ring, and, occasionally, a smooth or wrinkled pellicle are present.

Growth on malt agar: After one month at 17°C the streak culture is greyish-white to yellowish, soft, shiny or dull, smooth or partly or entirely striped or wrinkled.

Slide cultures on potato- and corn meal agar: Pseudomycelium is absent, or it is very primitive. Exceptionally, it is well developed.

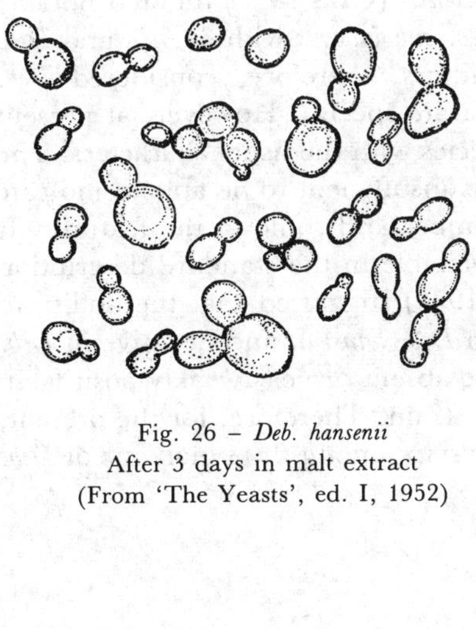

Fig. 26 – *Deb. hansenii*
After 3 days in malt extract
(From 'The Yeasts', ed. I, 1952)

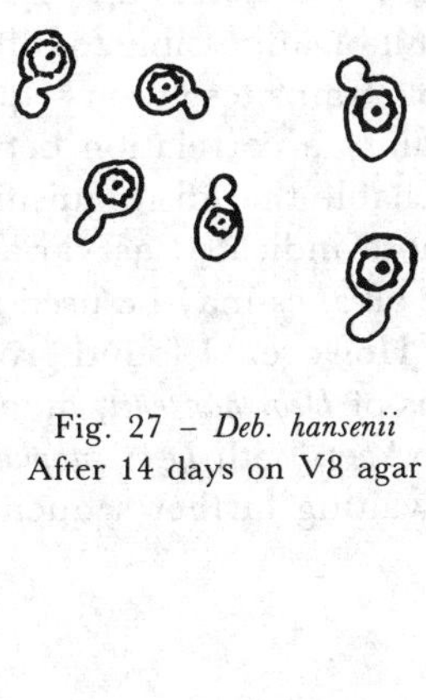

Fig. 27 – *Deb. hansenii*
After 14 days on V8 agar

Formation of ascospores: Usually conjugation between mother cell and bud precedes ascus formation; conjugation between separate cells may also occur. The spores are spherical with a warty wall (Fig. 27). The wartiness is not always distinct under the light microscope. In scanning micrographs the warts appear as small blunt protuberances or as small ridges (Kurtzman et al., 1975). Usually one, seldom two spores are formed per ascus. The presence of many spores gives the culture a brown color.

Spores were observed on V8-, Gorodkowa-, acetate- and potato agar at 20°C.

Fermentation:

Glucose	+ vw or –	Maltose	+ vw or –
Galactose	+ vw or –	Lactose	–
Sucrose	+ vw or –		

Assimilation of carbon compounds:

Galactose	+	Raffinose	+	Erythritol	v
Sucrose	+	Soluble starch	+	Ribitol	+
Maltose	+	D-Xylose	+	D-Mannitol	+
Cellobiose	+	L-Arabinose	+	Succinic acid	+
Trehalose	+	D-Ribose	v	Citric acid	v
Lactose	v	L-Rhamnose	v	Inositol	–

Splitting of arbutin: +

Assimilation of nitrate: –

Assimilation of nitrite: v

Assimilation of creatine: + (–)

Growth in vitamin-free medium: –

Growth on 50% (w/w) glucose-yeast extract agar: +

Growth at 37°C: v

G + C: 38.2–38.6 mol%, 6 strains; 39.1 mol%, *Deb. nepalensis,* CBS 5921; 37.3 mol%, *Deb. subglobosus,* CBS 792 (Price et al., 1978)

Comments

Price et al. (1978), in a study of DNA sequence relatedness, found that the original strains of *Deb. subglobosus* (CBS 792) and *Deb. nepalensis* (CBS 5921) differed notably from the type strain of *Deb. hansenii* (CBS 767): *Deb. subglobosus* with 39.7% and *Deb. nepalensis* with 15.4% relative binding. The authors, therefore, considered *Deb. hansenii, Deb. subglobosus* and *Deb. nepalensis* as separate species. However, at present it is difficult or impossible to differentiate these species with the usual characters. The number of strains tested on sequence relatedness is insufficient to be able to indicate with certainty a correlation between sequence number and some of the features. It seems probable that distinguishing features are available in the standard description among those indicated as variable. Price et al. (1978) suggested that the ability to assimilate alkanes may be used as such: positive in *Deb. hansenii* and negative in *Deb. nepalensis.* However, I found growth on hexadecane absent or very weakly positive in five strains of *Deb. hansenii,* among which the type strain. Therefore, for the present, I prefer to keep both *Deb. subglobosus* and *Deb. nepalensis* among the synonyms of *Deb. hansenii* awaiting further sequence studies.

Origin of the strains studied

Cheese (2), sausages (3) (Wickerham, 1957); salted beans (1), pickled cucumbers (1), sake moto (4), edo-miso (1), rennet (2), mushrooms (1), air (1), tobacco (5), soil (1), tomato puree (1), white wine (1), human skin (3), diseased nails (3), interdigital mycosis (1), patient with angina (1), sputum (1), clinical material (1), tan liquid (1), unknown (8).

Type strain: CBS 767, a strain named *Saccharomyces hansenii* received in the Delft collection from the Carlsberg Laboratory.

4. *DEBARYOMYCES MARAMA* DI MENNA (1954a)

Synonyms

Pichia marama (di Menna) Phaff (1956)
Zymodebaryomyces marama (di Menna) Novák et Zsolt (1961)

Standard description of *Debaryomyces marama*

Growth in malt extract: After 2 days at 25°C, the cells are spherical to short-oval to oval, (2–4) × (3.5–9) μm; single, in pairs or in chains. A sediment, a ring and, occasionally, a thin dull pellicle are formed. After one month at 17°C a sediment, a ring, and, occasionally, a more or less complete pellicle are present.

Growth on malt agar: After one month at 17°C the streak culture is yellowish-white, soft, shiny to dull, and smooth.

Slide cultures on potato- and corn meal agar: No pseudomycelium is formed. Short chains of cells may be present.

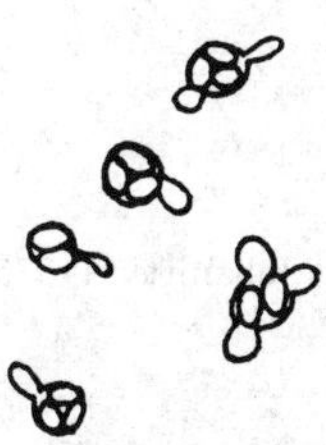

Fig. 28 – *Deb. marama*
After 8 days on V8 agar
(From 'The Yeasts', ed. II, 1970)

Formation of ascospores: Conjugation between mother cell and bud precedes ascus formation. The spores are oval and in scanning micrographs have spiral ridges (Kurtzman et al., 1975). Under the light microscope they may have a smooth appearance. One to four, usually two, spores are formed per ascus (Fig. 28). The presence of many spores gives the culture a brown color.

Spores were observed on V8 agar at 20°C and in old malt agar cultures.

Fermentation:

Glucose	+ vw or –	Maltose	–
Galactose	–	Lactose	–
Sucrose	–		

Assimilation of carbon compounds:

Galactose	+	Raffinose	+	Erythritol	+
Sucrose	+	Soluble starch	+	Ribitol	+
Maltose	+	D-Xylose	+	D-Mannitol	+
Cellobiose	+	L-Arabinose	+	Succinic acid	+
Trehalose	+	D-Ribose	v	Citric acid	+
Lactose	v	L-Rhamnose	–	Inositol	–

Splitting of arbutin: +

Assimilation of nitrate: –

Growth in vitamin-free medium: –

Growth on 50% (w/w) glucose-yeast extract agar: +w

Growth at 37°C: –

G + C: 39.1 mol%, one strain (Price et al., 1978)

Comments

Deb. marama is the only *Debaryomyces* species with spiral ridges on the spore wall; the spores in all other species have warts.

Origin of the strains studied

Air in New Zealand (1), cider (1), probably from a case of human disease (1).

Type strain: CBS 1958, isolated by di Menna in New Zealand.

5. *DEBARYOMYCES MELISSOPHILUS* (VAN DER WALT ET VAN DER KLIFT) KURTZMAN ET KREGER-VAN RIJ (1976)

Synonyms

Pichia melissophila van der Walt et van der Klift (1972)
Torulaspora melissophila (van der Walt et van der Klift) van der Walt et Johannsen (1975a)

Standard description of *Debaryomyces melissophilus*

Growth in malt extract: After 3 days at 25°C, the cells are short-oval to oval, (1.5–4) × (2.5–5.5) μm; single, in pairs or in short chains. A sediment is formed. After one month at 17°C a sediment and a ring are present.

Growth on malt agar: After one month at 17°C the streak culture is cream colored, soft, smooth and dull-shiny.

Slide cultures on potato- and corn meal agar: Pseudomycelium is not formed.

Formation of ascospores: Conjugation between mother cell and bud or between separate cells precedes ascus formation. The spores are spherical with a warty wall which is distinct in scanning electron microscopy and in ultrathin sections (Kurtzman & Kreger-van Rij, 1976), but not under the light microscope. One to four spores are formed in the zygote, either in one of the fused cells or distributed between both. The cells may form protuberances by which conjugation takes place. Single spores give cultures which sporulate again (van der Walt & van der Klift, 1972; Kurtzman & Kreger-van Rij, 1976).

Spores were observed on V8-, YM- and malt agar.

Fermentation: –

Assimilation of carbon compounds:

Galactose	+	Raffinose	–	Erythritol	+
Sucrose	+	Soluble starch	–	Ribitol	+
Maltose	+	D-Xylose	–	D-Mannitol	+
Cellobiose	v	L-Arabinose	–	Succinic acid	–
Trehalose	–	D-Ribose	–	Citric acid	+
Lactose	–	L-Rhamnose	–	Inositol	–

Splitting of arbutin: v

Assimilation of nitrate: –

Growth in vitamin-free medium: –

Growth on 50% (w/w) glucose-yeast extract agar: +

Growth at 37°C: –

G + C: 39.8 mol%, one strain (Price et al., 1978)

Origin of the strains studied

Digestive tract of the African honey bee (*Apis mellifera* var. *adansonii*) (1), soil in South Africa (1).

Type strain: CBS 6344, isolated from an African bee.

6. *DEBARYOMYCES POLYMORPHUS* (KLÖCKER) PRICE ET PHAFF (1979)

Synonyms

Pichia polymorpha Klöcker (1912a)
Saccharomyces polymorphus (Klöcker) Novák et Zsolt (1961)
Torulaspora polymorpha (Klöcker) van der Walt et Johannsen (1975a)
Debaryomyces cantarellii Capriotti (1961a)
Debaryomyces phaffii Capriotti (1961b)
Torulaspora phaffii (Capriotti) van der Walt et Johannsen (1975a)

Standard description of *Debaryomyces polymorphus*

Growth in malt extract: After 3 days at 25°C, the cells are oval, long-oval and cylindrical, (1.5–7.5) × (3.5–13.5) μm and longer; single or in pairs. A sediment and a thin, dull, creeping pellicle are formed. After one month at 17°C a sediment, a ring and a thin pellicle are present.

Growth on malt agar: After one month at 17°C the streak culture is cream-colored to yellowish, soft, flat, smooth or partly wrinkled, dull to shiny.

Slide cultures on potato- and corn meal agar: Pseudomycelium is lacking or it consists of elongate cells in a tree-like formation.

Formation of ascospores: Conjugation between mother cell and bud precedes ascus formation. The spores are spherical with a warty wall (Fig. 29). The warts are not always distinctly visible with the light microscope but they are clearly perceptible in scanning micrographs (Kurtzman & Smiley, 1974; Kurtzman et al., 1975). One to two, generally one spore per ascus.

Spores were observed on V8-, Gorodkowa- and corn meal agar.

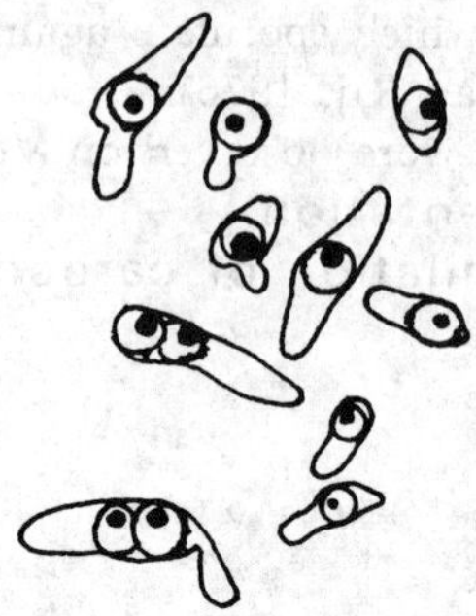

Fig. 29 – *Deb. polymorphus* After 8 days on V8 agar (From 'The Yeasts', ed. II, 1970)

Fermentation:

Glucose	+, s or w	Lactose	–
Galactose	+ w, s or –	Melibiose	+ w or –
Sucrose	+ or w	Raffinose	+ or w
Maltose	+ w, s or –		

Assimilation of carbon compounds:

Galactose	+	Raffinose	+	Erythritol	+
Sucrose	+	Soluble starch	+	Ribitol	+
Maltose	+	D-Xylose	+	D-Mannitol	+
Cellobiose	+	L-Arabinose	v	Succinic acid	+
Trehalose	+	D-Ribose	+	Citric acid	+
Lactose	v	L-Rhamnose	–	Inositol	–

Splitting of arbutin: +

Assimilation of nitrate: –

Growth in vitamin-free medium: +

Growth on 50% (w/w) glucose-yeast extract agar: +

Growth at 37°C: v

G + C: 35.7–35.9 mol%, 5 strains (Price et al., 1978)

Origin of the strains studied

Soil in Denmark (3), in Finland (4), in South Africa (1), mushroom in Germany (1), ant-hill (1) (Golubev & Babjeva, 1972).

Type strain: CBS 186, from Klöcker's collection.

7. *DEBARYOMYCES PSEUDOPOLYMORPHUS* (RAMIREZ ET BOIDIN) PRICE ET PHAFF (1979)

Synonyms

Pichia pseudopolymorpha Ramirez et Boidin (1953a)
Saccharomyces pseudopolymorphus (Ramirez et Boidin) Novák et Zsolt (1961)
Torulaspora pseudopolymorpha (Ramirez et Boidin) van der Walt et Johannsen (1975a)

Standard description of *Debaryomyces pseudopolymorphus*

Growth in malt extract: After 3 days at 25°C, the cells are long-oval to cylindrical, (3–6.5) × (4.5–16) μm; single or in pairs. A sediment and a dull creeping pellicle are formed. After one month at 17°C a sediment and pellicle are present.

Growth on malt agar: After one month at 17°C the streak culture is white to yellowish-white, soft, dull, delicately wrinkled, and somewhat raised in the middle. The edge is fringed with pseudomycelium.

Slide cultures on potato- and corn meal agar: Pseudomycelium in tree-like formations of elongate cells or of elongate cells with short-oval cells arranged along them.

Formation of ascospores: Conjugation between mother cell and bud precedes ascus formation. The spores are spherical; they have a warty wall (Kurtzman & Smiley, 1974) which is distinct in the scanning electron microscope, but not under the light microscope. One to four spores are formed per ascus, one to two on Gorodkowa agar, two to four on acetate agar.

Spores were observed on acetate-, Gorodkowa- and V8 agar.

Fermentation:

Glucose	+ s	Lactose	–
Galactose	+ vw or –	Melibiose	+ w or –
Sucrose	+ s	Raffinose	+ w
Maltose	+ vw or –		

Assimilation of carbon compounds:

Galactose	+	Raffinose	+	Erythritol	+
Sucrose	+	Soluble starch	+	Ribitol	+
Maltose	+	D-Xylose	+	D-Mannitol	+
Cellobiose	+	L-Arabinose	+	Succinic acid	+
Trehalose	+	D-Ribose	+	Citric acid	+
Lactose	+	L-Rhamnose	+	Inositol	–

Splitting of arbutin: +

Assimilation of nitrate: –

Assimilation of nitrite: +

Growth in vitamin-free medium: –

Growth on 50% (w/w) glucose-yeast extract agar: +

Growth at 37°C: –

G + C: 35.7 mol%, 2 strains (Price et al., 1978)

Origin of the strains studied

Tanning liquor (2).

Type strain: CBS 2009.

8. *DEBARYOMYCES TAMARII* OHARA ET NONOMURA (1954c)

Synonym

Torulaspora tamarii (Ohara et Nonomura) van der Walt et Johannsen (1975a)

Standard description of *Debaryomyces tamarii*

Growth in malt extract: After 2 days at 25°C the cells are spherical to short-oval, (2.5–4.2) × (2.7–5) μm; single, in pairs or in small groups. A sediment is formed. After one month at 17°C a sediment and a ring are present.

Growth on malt agar: After one month at 17°C the streak culture is cream colored, soft, smooth and dull-shiny.

Slide cultures on potato- and corn meal agar: No pseudomycelium is formed.

Formation of ascospores: According to Ohara & Nonomura (1954c) heterogamous conjugation precedes ascus formation. The spores are spherical with a warty wall. The warts are not always distinctly visible. One to two spores are formed per ascus.

No spores were observed in the strain studied.

Fermentation:

Glucose	+ vs	Maltose	–
Galactose	+ vs	Lactose	+ vw
Sucrose	+ vw	Raffinose	–

Assimilation of carbon compounds:

Galactose	+	Raffinose	+	Erythritol	–
Sucrose	+	Soluble starch	–	Ribitol	–
Maltose	–	D-Xylose	–	D-Mannitol	+
Cellobiose	+	L-Arabinose	–	Succinic acid	+
Trehalose	+	D-Ribose	–	Citric acid	–
Lactose	+	L-Rhamnose	–	Inositol	–

Splitting of arbutin: +

Assimilation of nitrate: –

Growth in vitamin-free medium: –

Growth on 50% (w/w) glucose-yeast extract agar: + w

Growth at 37°C: –

G + C: 46.4 mol%, one strain (Price et al., 1978)

Comments

The mol% G + C of *Deb. tamarii* (46.4) is much higher than that of the other *Debaryomyces* species which lie between 33.0 and 39.8.

Origin of the strain studied

Tamari soya (1).

Type strain: CBS 4333.

9. *DEBARYOMYCES VANRIJI* (VAN DER WALT ET TSCHEUSCHNER) ABADIE, PIGNAL ET JACOB (1963)

This species has two varieties:

Debaryomyces vanriji (van der Walt et Tscheuschner) Abadie et al. var. *vanriji*
Debaryomyces vanriji (van der Walt et Tscheuschner) Abadie et al. var. *yarrowii* (Santa María et García Aser) Kreger-van Rij var. nov.

Synonyms of the variety *vanriji*

Pichia vanriji van der Walt et Tscheuschner (1956a)
Azymomyces vanriji (van der Walt et Tscheuschner) Novák et Zsolt (1961)
Torulaspora vanriji (van der Walt et Tscheuschner) van der Walt et Johannsen (1975a)
Debaryomyces konokotinae Kudriavzev (1960)
Debaryomyces formicarius Golubev et Babjeva (1972)
Torulaspora formicaria (Golubev et Babjeva) van der Walt et Johannsen (1975a)

Synonyms of the variety *yarrowii*

Debaryomyces yarrowii Santa María et García Aser (1971) An. Inst. Nac. Invest. Agron. (Madrid) Series No. 1, p. 89 (basionym)
Torulaspora yarrowii (Santa María et García Aser) van der Walt et Johannsen (1975a)

Standard description of *Debaryomyces vanriji* var. *vanriji*

Growth in malt extract: After 2 days at 25°C the cells are spherical to short-oval, (3–8.5) × (3–20) μm; single or in groups. A sediment and a white, wrinkled creeping pellicle are formed. After one month at 17°C a sediment and a dull pellicle are present.

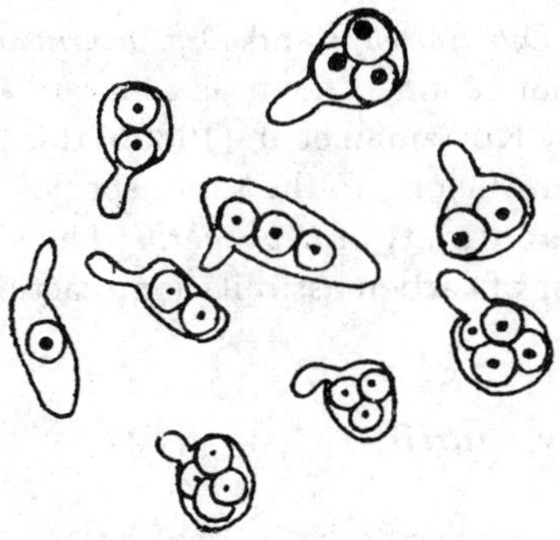

Fig. 30 – *Deb. vanriji* var. *vanriji*
After 14 days on V8 agar
(From 'The Yeasts', ed. II, 1970)

Growth on malt agar: After one month at 17°C the streak culture is yellow to yellowish-brown, dull or dull-glistening, smooth with a sinuous margin.

Slide cultures on potato- and corn meal agar: A primitive pseudomycelium is abundantly formed.

Formation of ascospores: Conjugation between mother cell and bud precedes ascus formation. The spores are spherical; one to four are formed per ascus (Fig. 30). The spore wall is warty which is not always visible under the light microscope, but distinct in scanning micrographs (Kurtzman et al., 1975), and in ultrathin sections (Santa María & García Aser, 1971).

Spores were observed on V8-, acetate-, arbutin-, YM-, malt-, potato- and corn meal agar and in the pellicle on malt extract.

Fermentation:

Glucose	+ vw or –	Maltose	–
Galactose	–	Lactose	–
Sucrose	+ vw or –		

Assimilation of carbon compounds:

Galactose	+	Raffinose	+	Erythritol	+
Sucrose	+	Soluble starch	+	Ribitol	+
Maltose	+	D-Xylose	+	D-Mannitol	+
Cellobiose	+	L-Arabinose	+ (vw)	Succinic acid	+
Trehalose	+	D-Ribose	+	Citric acid	+
Lactose	–	L-Rhamnose	v	Inositol	–

Splitting of arbutin: +

Assimilation of nitrate: –

Growth in vitamin-free medium: +

Growth on 50% (w/w) glucose-yeast extract agar: + (w)

Growth at 37°C: +

G + C: 33.2–33.3 mol%, 4 strains (Price et al., 1978)

Supplementary description of *Debaryomyces vanriji* var. *yarrowii*

Assimilation of carbon compounds:

Trehalose	–	Erythritol	–
Soluble starch	–	Citric acid	–

Growth at 37°C: –

G + C: 33.0 mol%, one strain (Price et al., 1978)

Comments

Price et al. (1978) found the type strains of *Deb. yarrowii* and *Deb. vanriji* to show 68.4% DNA base sequence complementarity but retained them as separate species. However, in view of a more recent publication by Kurtzman et al. (1980b) this percentage relatedness may indicate that the two strains belong to the same species. In the present description, *Deb. yarrowii* is reclassified as a variety of *Deb. vanriji.* The varieties *vanriji* and *yarrowii* are distinguished by a number of carbon assimilation reactions and by a difference in growth at 37°C.

Origin of the strains belonging to the variety *vanriji*

Soil in South Africa (2), slime flux of an oak (1), exudate of slippery elm (1), ant-hill (1).

Type strain: CBS 3024, isolated by van der Walt from soil in South Africa.

Origin of the strain belonging to the variety *yarrowii*

Excrements of insects on *Paulownia imperialis* in Madrid (1) (Santa María & García Aser, 1971).

Type strain: CBS 6264, isolated by the authors of *Deb. yarrowii*.

f. COMMENTS ON THE GENUS

Lodder and Kreger-van Rij (1978) made an official proposal for conservation of the name *Debaryomyces* Lodder et Kreger-van Rij against *Debaryomyces* Klöcker. The type of the latter, *Debaryomyces globosus* Klöcker, had been classified in the genus *Saccharomyces* (van der Walt, 1970e). After publication of this proposal, van der Walt and Johannsen (1978) gave a new name *Debaryozyma* for the *Debaryomyces* species. However, in accordance with Recommendation 15A of the International Code of Botanical Nomenclature the name *Debaryomyces* is applicable pending a decision about the proposal by the General Committee (Lodder & Kreger-van Rij, 1979).

The genus *Debaryomyces* Lodder et Kreger-van Rij (1952; Kreger-van Rij, 1970a) is distinguished from other ascomycetous yeast genera by the special structure of the spore wall, multilateral budding, inability to assimilate nitrate, and fermentation absent or not vigorous. All *Debaryomyces* species are haploid and conjugation between mother cell and bud generally precedes ascus formation.

Kreger-van Rij and Veenhuis (1975a) have described the internal ultrastructure of the spore wall of *Deb. hansenii* which is, as far as known, also typical of the other *Debaryomyces* species. Spherical warty spores in other nitrate-negative genera such as *Torulaspora, Saccharomyces* and *Zygosaccharomyces,* have a different structure of the wall with the exception of *Zygosaccharomyces rouxii* (Kreger-van Rij, 1979). In contrast with *Debaryomyces,* fermentation in *Saccharomyces, Zygosaccharomyces* and *Torulaspora* species is vigorous. *Issatchenkia* species also have warty spores, but the asci are generally unconjugated. *Saccharomyces, Zygosaccharomyces, Torulaspora* and *Issatchenkia* differ from *Debaryomyces* in the coenzyme Q number: *Saccharomyces, Zygosaccharomyces, Torulaspora,* 6; *Issatchenkia,* 7; and *Debaryomyces,* 9 (Yamada et al., 1973c, 1976b).

A proposal by van der Walt and Johannsen (1975a) to combine *Debaryomyces* and *Torulaspora* species in the genus *Torulaspora* was rejected by Kreger-van Rij and Veenhuis (1976b). A distinct difference in the structure of the spore wall between *Debaryomyces* and *Torulaspora* species correlates with other differences between them. Price et al. (1978) gave additional arguments for keeping *Debaryomyces* separate from the *Torulaspora* species, among others the Q-coenzyme. Kocková-Kratochvílová et al. (1978) applied numerical taxonomy to *Debaryomyces* species and species of the *Torulaspora* group. The latter were clearly distinguishable from the *Debaryomyces* species.

Price et al. (1978) made an important study of DNA sequence relatedness among *Debaryomyces* and some similar *Pichia* species which clarified taxonomic relations and resulted in merging of some species, e.g., *Deb. cantarellii* and *Deb. phaffii* with *Pichia polymorpha* to *Deb. polymorphus*.

Genus 8. *Dekkera* van der Walt

by J.P. van der Walt

a. DIAGNOSIS OF THE GENUS

Vegetative cells spheroidal, subglobose to ellipsoidal, frequently ogival, cylindrical or elongate, reproducing by budding. Pseudomycelium.

Asci unconjugated, evanescent, one- to four-spored. Ascospores hat-shaped or somewhat spheroidal with tangential brims, tending to agglutinate when released.

In malt extract and on malt agar slow growth with a characteristic aroma, cultures as a rule short-lived. Under aerobic conditions usually a vigorous production of acetic acid from glucose.

Fermentation usually stimulated by molecular oxygen.

Coenzyme Q-9 system present.

Extraneous vitamin source required.

b. TYPE SPECIES OF THE GENUS

Dekkera bruxellensis van der Walt

c. SPECIES ACCEPTED IN THE GENUS

1. *Dekkera bruxellensis* van der Walt (1964)
2. *Dekkera intermedia* van der Walt (1964)

d. KEY TO THE SPECIES OF THE GENUS

1a Galactose and cellobiose assimilated
D. intermedia . . . . p. 148

b Galactose and cellobiose not assimilated
D. bruxellensis . . . . p. 147

e. SYSTEMATIC DISCUSSION OF THE SPECIES

1. *DEKKERA BRUXELLENSIS* VAN DER WALT (1964)

Imperfect state: *Brettanomyces bruxellensis* Kufferath et van Laer

Standard description of *Dekkera bruxellensis* (van der Walt, 1970b)

Growth in malt extract: After 5 days at 28°C the cells are occasionally spheroidal, usually ellipsoidal, frequently ogival, or cylindrical to elongate, $(2.0-6.5) \times (4.0-22.0)$ μm, reproducing by budding, and occur singly, in pairs, short chains or small clusters (Fig. 31). Pseudomycelium is usually produced. A sediment is formed. Occasionally a thin ring or islets or a a thin pellicle may be present. After 10 days at room temperature the sediment is usually floccose to somewhat mucoid. A thin ring, islets or a thin pellicle may be formed.

A characteristic aroma usually develops.

Growth on malt agar: After 6 weeks at room temperature the streak culture is usually restricted, cream-colored to light brown, occasionally somewhat salmon colored, usually shiny, smooth to crispulate and somewhat raised. The margin may be entire, undulating or lobiform.

Growth on malt agar 2% calcium carbonate: After 6 weeks at room temperature the streak culture is usually restricted, cream-colored, usually shiny, smooth or pitted to crispulate and somewhat raised. The margin may be entire, undulating to lobiform or occasionally fringed with pseudomycelium. Acid is produced.

Dalmau plate culture on corn meal agar: Pseudomycelium is usually abundantly produced under both aerobic and anaerobic conditions. It consists of filamentous cells with or without blastospores. The blastospores may be arranged in chains or small, slightly branched verticils. Under anaerobic conditions, filamentous cells not infrequently terminate in a vesicle.

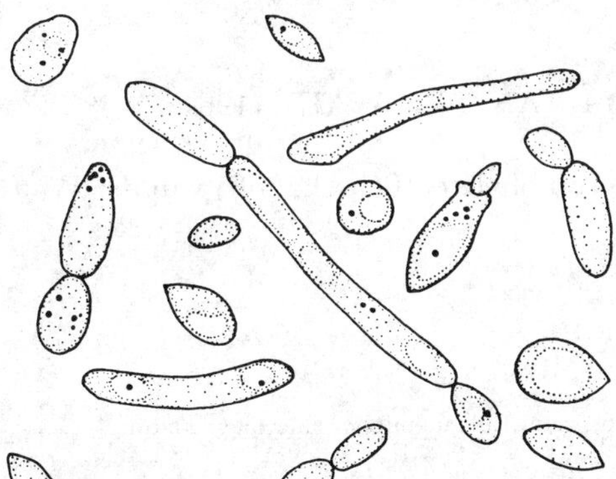

Fig. 31 – *D. bruxellensis*
After 5 days in malt extract
(From 'The Yeasts', ed. II, 1970)

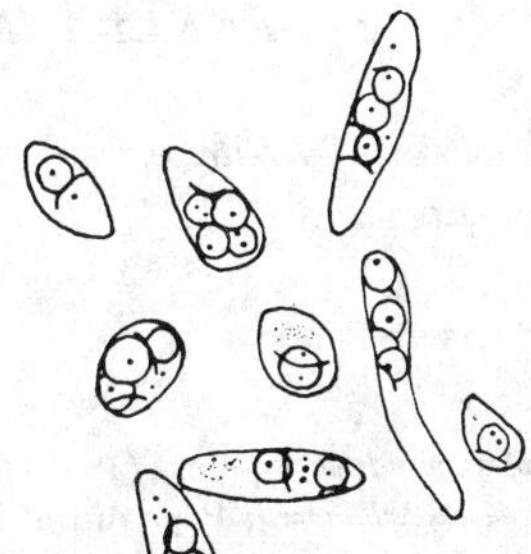

Fig. 32 – *D. bruxellensis*
After 14 days on YM agar + vitamins
(From 'The Yeasts', ed. II, 1970)

Formation of ascospores: Diploid, vegetative cells are directly transformed into asci, producing one to four ascospores. The ascospores are usually more or less spheroidal with tangential brims (Fig. 32). The ascospores are liberated soon after maturation and tend to agglutinate. Sporulation may be sparse. As mating types could not be recovered from heat-treated ascosporic material, the strains studied are presumed to be homothallic.

Sporulation is observed in cultures at room temperature after 7–15 days on YM agar supplemented with vitamins.

Fermentation:

Glucose	+	Maltose	+
Galactose	–	Lactose	–
Sucrose	+	Raffinose	–

Assimilation of carbon compounds:

Galactose	–	Raffinose	+w	Erythritol	–
Sucrose	+	Soluble starch	–	Ribitol	v
Maltose	+	D-Xylose	–	D-Mannitol	–
Cellobiose	–	L-Arabinose	–	Succinic acid	v
Trehalose	+	D-Ribose	v	Citric acid	–
Lactose	–	L-Rhamnose	–	Inositol	–

Assimilation of nitrate: v

Growth in vitamin-free medium: –

Growth on 50% (w/w) glucose-yeast extract agar: –

Growth at 37°C: +

Acid production from glucose (according to the standard method): +

Origin of the strains studied

Stout beer (3) (Custers, 1940).

Type strain: CBS 74, isolated by Custers from Belgian stout beer.

2. *DEKKERA INTERMEDIA* VAN DER WALT (1964)

Imperfect state: *Brettanomyces intermedius* (Krumbholz et Tauschanoff) van der Walt et van Kerken

Synonyms

Torulopsis cylindrica Walters (1943)
Brettanomyces schanderlii Peynaud et Domercq (1956), nom. illegit. public. sine diagn. latine

Standard description of *Dekkera intermedia*

Growth in malt extract: After 5 days at 28°C the cells are occasionally spherical, usually ellipsoidal, frequently ogival, or cylindrical to elongate,

(2.0–7.0) × (3.5–18.0) μm, reproducing by budding, and occur singly, in pairs, short chains or small clusters. A sediment is formed. A slight ring, islets or a thin, dull pellicle may be formed. After 10 days at room temperature a sediment is present which may be dusty, coherent, floccose or occasionally rather mucoid. A ring, islets or a dull, creeping pellicle may be present.

A characteristic aroma is usually produced.

Growth on malt agar: After 6 weeks at room temperature the streak culture is rather restricted, raised, cream colored to light brown, shiny to somewhat dull, smooth, pitted or crispulate. The margin may be entire, undulating or lobiform, occasionally with some structure.

Growth on malt agar 2% calcium carbonate: After 6 weeks at room temperature the streak culture is usually raised, cream colored to light brown, shiny to dull, smooth, finely pitted, crispulate or occasionally somewhat rugose. The margin may be entire, undulating, lobiform with some structure or occasionally somewhat fimbriate. Acid is produced.

Dalmau plate culture on corn meal agar: Pseudomycelium is usually abundantly produced and often better developed under anaerobic conditions. It consists of filamentous cells with or without blastospores. The blastospores may be arranged in chains or in somewhat branched verticils.

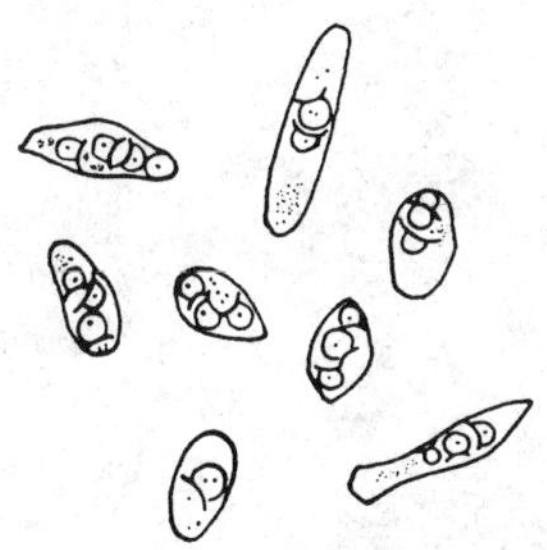

Fig. 33 – *D. intermedia*
After 21 days on YM agar + vitamins
(From 'The Yeasts', ed. II, 1970)

Formation of ascospores: Diploid, vegetative cells are directly transformed into asci, producing one to four ascospores. The spores are hat-shaped, liberated soon after maturation and tend to agglutinate (Fig. 33). Sporulation may be sparse. As mating types could not be recovered from heat-treated ascosporic material, the strains studied are presumed to be homothallic.

Sporulation is best observed in cultures at room temperature after 7–15 days on YM agar supplemented with vitamins.

Fermentation:

Glucose	+	Maltose	+ or s
Galactose	+ or s	Lactose	–
Sucrose	+ or s	Raffinose	v

Assimilation of carbon compounds:

Galactose	+	Raffinose	v	Erythritol	–
Sucrose	+ or s	Soluble starch	–	Ribitol	v
Maltose	+ or s	D-Xylose	–	D-Mannitol	–
Cellobiose	+	L-Arabinose	–	Succinic acid	v
Trehalose	+	D-Ribose	v	Citric acid	–
Lactose	–	L-Rhamnose	–	Inositol	–

Assimilation of nitrate: v

Growth in vitamin-free medium: –

Growth on 50% (w/w) glucose-yeast extract agar: –
Growth at 37°C: +
Acid production from glucose (according to the standard method): +

Origin of the strains studied

Beer (1) (Walters, 1943); wine (2) (Peynaud & Domercq, 1956); wine (2) (van der Walt & van Kerken, 1959); winery equipment (2) (van der Walt & van Kerken, 1960a); 'tea beer fungus' (1).

Type strain: CBS 4914, isolated by van der Walt from 'tea beer fungus' deriving from Mozambique.

Genus 9. *Guilliermondella* Nadson et Krassilnikov

by N.J.W. Kreger-van Rij

a. DIAGNOSIS OF THE GENUS

Formation of true mycelium with blastospores and of budding yeast cells. The septa in the hyphae have plasmodesmata.

Ascus formation is preceded by cell fusion and, after that, separation of the fused cells by a wall. The ascospores are spherical, oval or kidney-shaped; the wall at the two poles of the spores is thickened which gives the appearance of two appendages.

b. TYPE SPECIES OF THE GENUS

Guilliermondella selenospora Nadson et Krassilnikov

c. SPECIES ACCEPTED IN THE GENUS

1. *Guilliermondella selenospora* Nadson et Krassilnikov (1928)

e. SYSTEMATIC DISCUSSION OF THE SPECIES

1. GUILLIERMONDELLA SELENOSPORA NADSON ET KRASSILNIKOV (1928)

Synonym

Endomycopsis selenospora (Nadson et Krassilnikov) Dekker (Stelling-Dekker, 1931)

Standard description of *Guilliermondella selenospora*

Growth in malt extract: After 3 days at 25°C branched mycelial hyphae of 2–3 μm diameter with cross walls are formed. Oval and long-oval budding cells, (2–4.5) × (7–12) μm, occur singly or in pairs, or on the hyphae at the ends or alongside them. A sediment is formed. After one month at 17°C a flocky sediment and, occasionally, a ring are present.

Growth on malt agar: After one month at 17°C the streak culture is yellow-brown, dull, tough, raised, smooth or with slight folds; the margin is fringed with mycelium.

Slide cultures on potato- and corn meal agar: True mycelium with side branches is formed. The cross walls contain plasmodesmata. Blastospores may occur on the hyphae. Short protuberances on the hyphae give them a dentate aspect (Fig. 34). Contact between cells via denticles leads to submicroscopic changes in the wall of the contacted cell. Dead cells may be penetrated by denticles grown out to stalks (Kreger-van Rij & Veenhuis, 1973b).

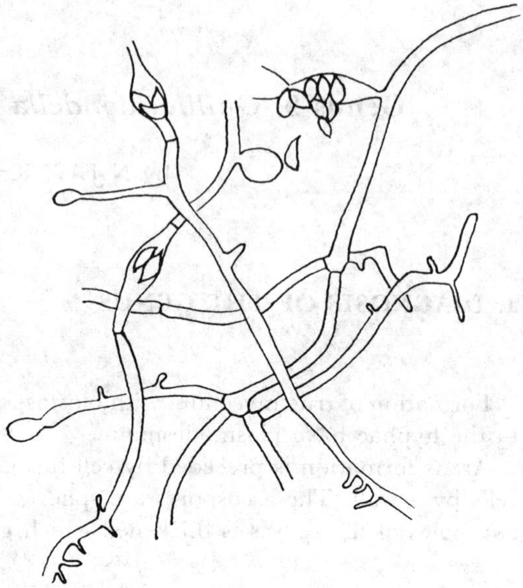

Fig. 34 – *G. selenospora*
Slide culture on corn meal agar, 1000 ×
(From 'The Yeasts', ed. II, 1970)

Formation of ascospores: Asci are formed after conjugation of cells in different hyphae or of adjacent hyphal cells, and subsequent separation of the fused cells by a cross wall. Both cells may turn into asci or buds are formed on them which become asci (Kreger-van Rij & Veenhuis, 1976a). The asci are situated in intercalary fashion, as buds on the hyphal cells or between two hyphae. The spores are spherical, oval, kidney-, or sickle-shaped; two to four are formed per ascus (Fig. 34). The various shapes were observed in one strain. Two appendages, which are outgrowths of the spore wall, may be visible at the extremities of the spores. The spores are easily liberated from the ascus. Single spores give cultures which sporulate again.

Spores were observed on potato- and corn meal agar.

Fermentation:

Glucose	+	Maltose	–
Galactose	–	Lactose	–
Sucrose	–		

Assimilation of carbon compounds:

Galactose	+	Raffinose	–	Erythritol	–
Sucrose	–	Soluble starch	v	Ribitol	+
Maltose	–	D-Xylose	+	D-Mannitol	–
Cellobiose	–	L-Arabinose	+	Succinic acid	+
Trehalose	–	D-Ribose	+ w	Citric acid	–
Lactose	–	L-Rhamnose	–	Inositol	–

Splitting of arbutin: –
Assimilation of nitrate: –
Growth in vitamin-free medium: –
Growth on 50% (w/w) glucose-yeast extract agar: –
Growth at 37°C: +w or –

Origin of the strains studied

Tan liquid (1), unknown (1).

Type strain: CBS 2562, of unknown origin, received in the Delft collection from Guilliermond.

Genus 10. *Hanseniaspora* Zikes

by M.Th. Smith

a. DIAGNOSIS OF THE GENUS

Cells are apiculate (lemon-shaped), oval to long-oval or elongate. Vegetative reproduction by bipolar budding in basipetal succession. Usually a well developed pseudomycelium is formed, sometimes a rudimentary or no pseudomycelium. Vegetative cells are diploid.

Three types of ascospores are found in the genus *Hanseniaspora*. Ascospores are either hat- or helmet-shaped, two to four per ascus and released at maturity; or spherical with an equatorial ledge, smooth or warty; or spherical with warts. Spherical ascospores are not released from the ascus.

The ascospores germinate by budding; the nucleus divides, followed by karyogamy so that after mitosis the first bud re-establishes the diplophase (Phaff, 1970a).

Fermentation is present.

Nitrate is not assimilated.

All strains have an absolute requirement for inositol and pantothenate (Phaff, 1970a).

b. TYPE SPECIES OF THE GENUS

Hanseniaspora valbyensis Klöcker

c. SPECIES ACCEPTED IN THE GENUS

1. *Hanseniaspora guilliermondii* Pijper (1928)
2. *Hanseniaspora occidentalis* Smith (1974)
3. *Hanseniaspora osmophila* (Niehaus) Phaff, Miller et Shifrine (Niehaus, 1932)
4. *Hanseniaspora uvarum* (Niehaus) Shehata, Mrak et Phaff (Niehaus, 1932)
5. *Hanseniaspora valbyensis* Klöcker (1912b)
6. *Hanseniaspora vineae* van der Walt et Tscheuschner (1957b)

d. KEY TO THE SPECIES IN THE GENUS

1a Ascospores hat-shaped ... 2
b Ascospores spherical ... 3

2a Growth at 37°C
H'spora guilliermondii .... p. 156
b No growth at 37°C
H'spora valbyensis .... p. 161

3a Maltose assimilated ... 4
b Maltose not assimilated ... 5

4a Growth at 34°C
H'spora vineae .... p. 162
b No growth at 34°C
H'spora osmophila .... p. 158

5a Sucrose assimilated
H'spora occidentalis p.157
b Sucrose not assimilated
H'spora uvarum p.159

Table 12

Key characters of the species in the genus *Hanseniaspora* and their imperfect states in the genus *Kloeckera*

Perfect	Imperfect	Assimilation				34°C	37°C	Spore shape	No. of spores/ ascus
		Su	Ma	Gly	2-k Gl				
H'spora valbyensis		–	–	–	–		–	hat	1–4 (2)
	Kl. japonica	–	–	–	–		–		
H'spora guilliermondii		–	–	–	+		+	hat	1–4 (4)
	Kl. apis	–	–	–	+		+		
H'spora uvarum		–	–	–	+		–	spherical, warty with equatorial ledge	1–2
	Kl. apiculata	–	–	–	+		–		
H'spora occidentalis		+	–	+	–		–	spherical, smooth with equatorial ledge	1–2
	Kl. javanica	+	–	+	–		–		
H'spora osmophila		v	+	–	–	–	–	spherical, warty	1–2
	Kl. corticis	v	+	–	–	–	–		
H'spora vineae		v	+	–	–	+	–	spherical, warty	1–2
	Kl. africana	v	+	–	–	+	–		

Su = sucrose, Ma = maltose, Gly = glycerol, 2-k Gl = 2-ketogluconate; the numbers in parentheses refer to the number of spores most frequently observed.

e. SYSTEMATIC DISCUSSION OF THE SPECIES

1. *HANSENIASPORA GUILLIERMONDII* PIJPER (1928)

Imperfect state: *Kloeckera apis* Lavie ex Smith, Simione et Meyer

Synonyms

Hanseniaspora melligeri Lodder (1932)
Hanseniaspora apuliensis Castelli (1948)

Standard description of *Hanseniaspora guilliermondii*

Growth in glucose-yeast extract-peptone water: After 2 days at 25°C the cells are apiculate, oval or elongate, (2.2–5.8) × (4.5–10.2) μm, occasionally longer, single or in pairs. Reproduction by bipolar budding. A sediment is present. After one month a thin ring has formed.

Growth on glucose-yeast extract-peptone agar: After one month the streak culture is white to cream-colored, smooth, glossy; in cross-section a raised center and a flat periphery.

Slide culture on potato agar: Pseudomycelium is rudimentary or absent.

Formation of ascospores: One to four, mostly four, hat- to helmet-shaped ascospores are formed, normally released at maturity from the ascus (Fig. 35). The liberated spores often aggregate.

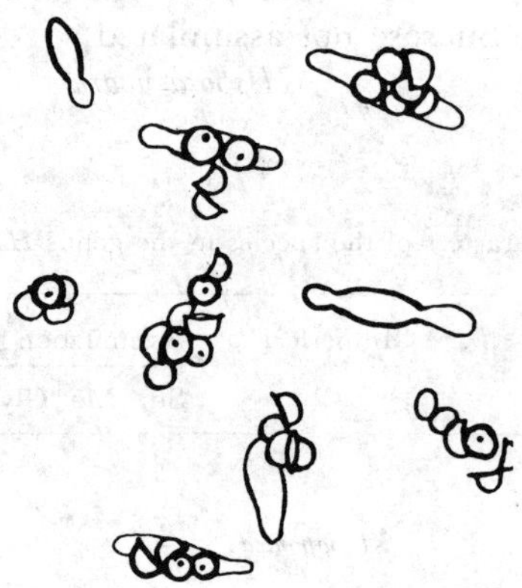

Fig. 35 – *H'spora guilliermondii*
On malt agar
(From 'The Yeasts', ed. II, 1970)

Abundant sporulation was observed usually on 5% Difco malt extract agar and potato agar at low temperature (15°C).

Fermentation:

Glucose	+	Maltose	–
Galactose	–	Lactose	–
Sucrose	–		

Assimilation of carbon compounds:

Galactose	–	Raffinose	–	Erythritol	–
Sucrose	–	Soluble starch	–	Ribitol	–
Maltose	–	D-Xylose	–	D-Mannitol	–
Cellobiose	+	L-Arabinose	–	Succinic acid	–
Trehalose	–	D-Ribose	–	Citric acid	–
Lactose	–	L-Rhamnose	–	Inositol	–

Additional carbon compounds tested: Glycerol –, 2-ketogluconate +
Assimilation of nitrate: –
Growth in vitamin-free medium: –
Growth on 50% (w/w) glucose-yeast extract agar: +
Growth at 37°C: +
G + C: 33.2–34.2 mol%, 10 strains (Meyer et al., 1978)

Comments

The species *H'spora apuliensis*, *H'spora guilliermondii* and *H'spora melligeri* placed in synonymy with *H'spora valbyensis* by Lodder and Kreger-van Rij (1952) and maintained as synonyms by Phaff (1970a) were separated from *H'spora valbyensis* by Meyer et al. (1978). The latter authors demonstrated in DNA–DNA reassociation experiments a low degree of reassociation of *H'spora valbyensis* with *H'spora apuliensis*, *H'spora guilliermondii* and *H'spora melligeri*. A high degree of reassociation was revealed between *H'spora apuliensis*, *H'spora guilliermondii* and *H'spora melligeri*.

Origin of the strains studied

Diseased nail (1), tomatoes (1), dates (1), grapes (3), cecum of baboon (1), culture contaminant (1), unknown (2).

Type strain: CBS 465, isolated by Pijper from diseased nail.

2. *HANSENIASPORA OCCIDENTALIS* SMITH (1974)

Imperfect state: *Kloeckera javanica* (Klöcker) Janke

Synonyms

Pseudosaccharomyces occidentalis Klöcker (1912b)
Kloeckera occidentalis (Klöcker) Janke (1928)

Standard description of *Hanseniaspora occidentalis*

Growth in glucose-yeast extract-peptone water: After 2 days at 25°C the cells are lemon-shaped, oval, sometimes round, (1.8–6.2) × (3.0–11.0) μm, single or in pairs. Reproduction by bipolar budding. A sediment has formed. After one month a very thin ring is present.

Growth on glucose-yeast extract-peptone agar: After one month the streak culture is white to cream-colored, smooth, glossy, with a raised center and a flat periphery.

Slide culture on potato agar: Pseudomycelium is generally lacking, but some strains produce a rudimentary or well developed pseudomycelium.

Formation of ascospores: Per ascus one or two ascospores are formed which are spherical and smooth with an equatorial ledge. Ascospores not released from ascus.

Sporulation occurred on 5% Difco malt extract agar.

Fermentation:

Glucose	+	Maltose	–
Galactose	–	Lactose	–
Sucrose	+		

Assimilation of carbon compounds:

Galactose	–	Raffinose	–	Erythritol	–
Sucrose	+	Soluble starch	–	Ribitol	–
Maltose	–	D-Xylose	–	D-Mannitol	–
Cellobiose	+	L-Arabinose	–	Succinic acid	–
Trehalose	–	D-Ribose	–	Citric acid	–
Lactose	–	L-Rhamnose	–	Inositol	–

Additional carbon compounds tested: Glycerol +, 2-ketogluconate –

Assimilation of nitrate: –

Growth in vitamin-free medium: –

Growth on 50% (w/w) glucose-yeast extract agar: –

Growth at 37°C: –

G + C: 34.9–35.9 mol%, 3 strains (Meyer et al., 1978)

Origin of the strains studied

Soil (1), *Drosophila* (1), orange juice (2).

Type strain: CBS 2592, isolated by Klöcker from soil in St. Croix, France.

3. *HANSENIASPORA OSMOPHILA* (NIEHAUS) PHAFF, MILLER ET SHIFRINE COMB. NOV.

Imperfect state: *Kloeckera corticis* (Klöcker) Janke

Synonyms

Kloeckeraspora osmophila Niehaus (1932) Zentralbl. Bakteriol. Parasitenk. Abt. II *86*, p. 145 (basionym)
Hanseniaspora osmophila (Niehaus) Phaff, Miller et Shifrine (1956) nom. inval.

Standard description of *Hanseniaspora osmophila*

Growth in glucose-yeast extract-peptone water: After 2 days at 25°C the cells are lemon-shaped, oval or long-oval, (3.5–6.0) × (7.2–18.2) μm, single or in pairs. Reproduction by bipolar budding. A sediment is present. After one month a thin ring has formed.

Growth on glucose-yeast extract-peptone agar: The streak culture is white to cream-colored, smooth, glossy; center is raised, periphery is flat.

Slide culture on potato agar: Pseudomycelium in a branched, tree-like, formation is formed.

Formation of ascospores: Per ascus one to two ascospores are formed which are spherical and warty and are not released from the ascus.

Sporulation occurred on 5% Difco malt extract agar.

Fermentation:

Glucose	+	Maltose	–
Galactose	–	Lactose	–
Sucrose	–		

Assimilation of carbon compounds:

Galactose	–	Raffinose	–	Erythritol	–
Sucrose	v	Soluble starch	–	Ribitol	–
Maltose	+	D-Xylose	–	D-Mannitol	–
Cellobiose	+	L-Arabinose	–	Succinic acid	–
Trehalose	–	D-Ribose	–	Citric acid	–
Lactose	–	L-Rhamnose	–	Inositol	–

Additional carbon compounds tested: Glycerol –, 2-ketogluconate –

Assimilation of nitrate: –

Growth in vitamin-free medium: –

Growth on 50% (w/w) glucose-yeast extract agar: +

Growth at 37°C: –; **at 34°C:** –

G + C: 40.5 mol%, one strain (Meyer et al., 1978)

Origin of the strain studied

Berries of Riesling grapes in Germany (1).

Type strain: CBS 313, isolated by Niehaus from grapes.

4. *HANSENIASPORA UVARUM* (NIEHAUS) SHEHATA, MRAK ET PHAFF COMB. NOV.

Imperfect state: *Kloeckera apiculata* (Reess emend. Klöcker) Janke

Synonyms

Kloeckeraspora uvarum Niehaus (1932) Zentralbl. Bakteriol. Parasitenk., Abt. II, *86,* p. 146 (basionym)
Hanseniaspora uvarum (Niehaus) Shehata, Mrak et Phaff (1955) nom. inval.
Pseudosaccharomyces austriacus Klöcker (1912b)
Kloeckera austriaca (Klöcker) Janke (1928)
Kloeckera brevis Lodder (1934)
Kloeckera brevis Lodder var. *rohrbachense* nom. inval. (Szilvinyi & Kaulich, 1948)
Kloeckera lodderi van Uden et Assis-Lopes (1953a)

Standard description of *Hanseniaspora uvarum*

Growth in glucose-yeast extract-peptone water: After 2 days at 25°C the cells are apiculate, round to oval, elongate, (1.5–5.0) × (2.5–11.5) μm, single or in pairs. Reproduction by bipolar budding. A sediment is present. After one month a very thin ring has formed.

Growth on glucose-yeast extract-peptone agar: After one month the streak culture is white to creamy, smooth, glossy, with a raised center and a flat periphery.

Slide culture on potato agar: Pseudomycelium is present (in a tree-like formation), rudimentary or absent.

Formation of ascospores: Per ascus one to two spherical ascospores are formed which are warty with an equatorial or subequatorial ledge (Fig. 36). Warts and ledge may not be observed distinctly by light microscopy. Ascospores not released from the ascus.

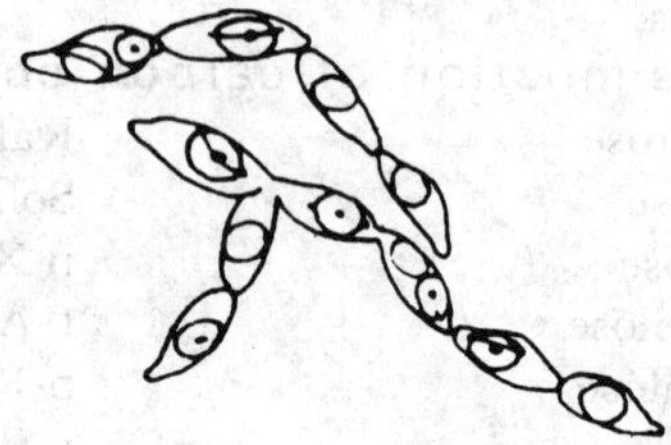

Fig. 36 – *H'spora uvarum* On potato agar under coverslip (From 'The Yeasts', ed. II, 1970)

Sporulation occurred on 5% Difco malt extract agar.

Fermentation:

Glucose	+	Maltose	–
Galactose	–	Lactose	–
Sucrose	–		

Assimilation of carbon compounds:

Galactose	–	Raffinose	–	Erythritol	–
Sucrose	–	Soluble starch	–	Ribitol	–
Maltose	–	D-Xylose	–	D-Mannitol	–
Cellobiose	+	L-Arabinose	–	Succinic acid	–
Trehalose	–	D-Ribose	–	Citric acid	–
Lactose	–	L-Rhamnose	–	Inositol	–

Additional carbon compounds tested: Glycerol –, 2-ketogluconate +

Assimilation of nitrate: –

Growth in vitamin-free medium: –

Growth on 50% (w/w) glucose-yeast extract agar: v

Growth at 37°C: –

G + C: 32.9–35.1 mol%, 16 strains (Meyer et al., 1978)

Origin of the strains studied

Grapes (3), soil (2), *Drosophila* (2), seawater (2), fruit must (2), cacao (1), cucumber brine (1), caterpillar (1), cider (1), throat (1), banana (1), baker's yeast (1).

Type strain: CBS 314, isolated by Niehaus from grapes.

5. *HANSENIASPORA VALBYENSIS* KLÖCKER (1912b)

Imperfect state: *Kloeckera japonica* Saito et Ohtani

Synonym

Endomyces valbyensis (Klöcker) Zender (1925)

Standard description of *Hanseniaspora valbyensis*

Growth in glucose-yeast extract-peptone water: After 2 days at 25°C the cells are apiculate, round to oval or elongate, (2.0–5.5) × (3.0–10.2) μm, single or in pairs (Fig. 37). Reproduction by bipolar budding. A sediment is present. After one month a thin ring may be formed.

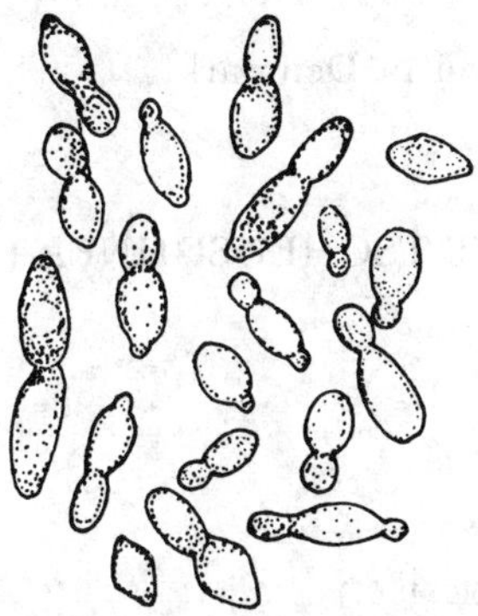

Fig. 37 – *H'spora valbyensis*
After 3 days in malt extract
(From 'The Yeasts', ed. I, 1952)

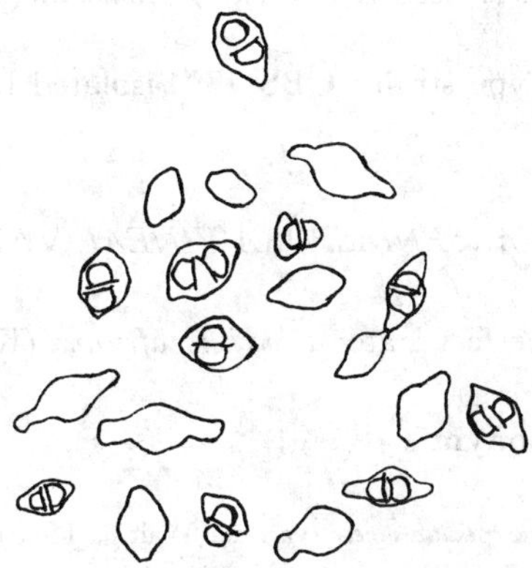

Fig. 38 – *H'spora valbyensis*, CBS 6618
On 5% malt extract agar

Growth on glucose-yeast extract-peptone agar: After one month the streak culture is white to cream-colored, smooth, glossy; center is somewhat raised and periphery is flat.

Slide culture on potato agar: Pseudomycelium is present (tree-like formation), rudimentary or absent.

Formation of ascospores: Two to four, mostly two, hat- to helmet-shaped ascospores are formed per ascus and usually released at maturity (Fig. 38). Liberated ascospores often aggregate.

Ascospores were observed on 5% Difco malt extract agar and potato agar.

Fermentation:

Glucose	+	Maltose	–
Galactose	–	Lactose	–
Sucrose	–		

Assimilation of carbon compounds:

Galactose	–	Raffinose	–	Erythritol	–
Sucrose	–	Soluble starch	–	Ribitol	–
Maltose	–	D-Xylose	–	D-Mannitol	–
Cellobiose	+	L-Arabinose	–	Succinic acid	–
Trehalose	–	D-Ribose	–	Citric acid	–
Lactose	–	L-Rhamnose	–	Inositol	–

Additional carbon compounds tested: Glycerol –, 2-ketogluconate –
Assimilation of nitrate: –
Growth in vitamin-free medium: –
Growth on 50% (w/w) glucose-yeast extract agar: – (+)
Growth at 37°C: –
G + C: 28.8–30.0 mol%, 5 strains (Meyer et al., 1978)

Origin of the strains studied

Soil (1), beer (1), tomato (1), unknown (2).

Type strain: CBS 479, isolated by Klöcker from soil in Denmark.

6. *HANSENIASPORA VINEAE* VAN DER WALT ET TSCHEUSCHNER (1957b)

Imperfect state: *Kloeckera africana* (Klöcker) Janke

Synonym

Vanderwaltia vineae (van der Walt et Tscheuschner) Novák et Zsolt (1961)

Standard description of *Hanseniaspora vineae*

Growth in glucose-yeast extract-peptone water: After 2 days at 25°C the cells are apiculate or round to oval or elongate, (2.0–7.5) × (4.0–14.5) μm, mostly single, sometimes in pairs. Reproduction by bipolar budding. A sediment has formed. After one month a very thin ring may be present.

Growth on glucose-yeast extract-peptone agar: After one month the streak culture is white to cream-colored, smooth, glossy; center is raised and periphery is flat.

Slide culture on potato agar: Rudimentary or well developed pseudomycelium with or without clusters of blastospores is formed.

Formation of ascospores: Per ascus one to two ascospores are produced, which are spherical and warty (Fig. 39). Ascospores not released from the ascus.

Sporulation was observed on 5% Difco malt extract agar.

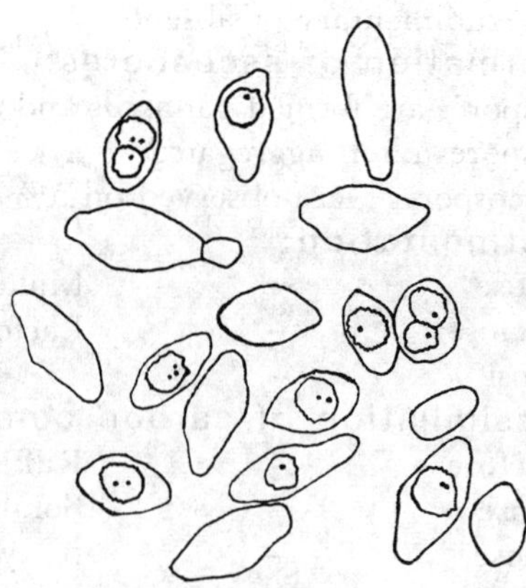

Fig. 39 – *H'spora vineae,* CBS 2171
On YM agar

Fermentation:

Glucose	+	Maltose	–
Galactose	–	Lactose	–
Sucrose	–		

Assimilation of carbon compounds:

Galactose	–	Raffinose	–	Erythritol	–
Sucrose	v	Soluble starch	–	Ribitol	–
Maltose	+	D-Xylose	–	D-Mannitol	–
Cellobiose	+	L-Arabinose	–	Succinic acid	–
Trehalose	–	D-Ribose	–	Citric acid	–
Lactose	–	L-Rhamnose	–	Inositol	–

Additional carbon compounds tested: Glycerol –, 2-ketogluconate –

Assimilation of nitrate: –

Growth in vitamin-free medium: –

Growth on 50% (w/w) glucose-yeast extract agar: +

Growth at 37°C: –; at 34°C: +

G + C: 39.0–40.7 mol%, 5 strains (Meyer et al., 1978)

Comments

The species *H'spora vineae* was placed in synonymy with *H'spora osmophila* by Miller and Phaff (1958) and maintained as a synonym by Phaff (1970a). However, Meyer et al. (1978) separated *H'spora vineae* from *H'spora osmophila,* since they demonstrated an average reassociation value of 48% in DNA–DNA reassociation experiments between *H'spora vineae* and *H'spora osmophila.*

Origin of the strains studied

Soil (2), *Drosophila* (1), unknown (2).

Type strain: CBS 2171, isolated by van der Walt and Tscheuschner from soil in South Africa.

f. COMMENTS ON THE GENUS

Miller and Phaff (1958) made a detailed study of the history, nomenclature, physiology, morphology and life cycle of all *Hanseniaspora* species.

Tsuchiya et al. (1966) and Tsuchiya and Imai (1968) carried out antigenic analyses of some species of the genera *Kloeckera* and *Hanseniaspora* by the slide agglutination method. Obtained results were discussed from the point of view of yeast classification.

Kreger-van Rij and Ahearn (1968) and Kreger-van Rij (1977b) performed a comparative study on the ultrastructure of the different types of ascospores produced in *Hanseniaspora.*

Spencer and Gorin (1968) carried out PMR spectrometry studies of mannans isolated from species of the genera *Hanseniaspora* and *Kloeckera*, and discussed the relationship between both genera.

Yamada et al. (1976a) studied the coenzyme Q system in the classification of apiculate yeast genera, namely *Hanseniaspora, Kloeckera, Nadsonia, Saccharomycodes* and *Wickerhamia*. The Q-6 system has been found in all the species of these genera except for the only species in the genus *Wickerhamia,* which demonstrated the Q-9 system. The authors discussed the results obtained in connection with other criteria like PMR spectra of cell wall polysaccharides, DNA base composition and serological characteristics.

Fiol and Billon-Grand (1978) examined the production of intracellular osidases, nitrite and nitrate reductases in some species of *Hanseniaspora* and *Kloeckera*. Taxonomical relationships between both genera are discussed.

Nakase and Komagata (1970a) and Meyer et al. (1978) studied the significance of DNA base composition in the classification of *Hanseniaspora*. The last mentioned authors performed DNA–DNA reassociation experiments to evaluate the status of the different *Hanseniaspora* species.

Genus 11. *Hansenula* H. et P. Sydow

by C.P. Kurtzman

a. DIAGNOSIS OF THE GENUS

Asexual reproduction is by multilateral budding on a narrow base. Cells are spheroidal, ellipsoidal, or elongate and occasionally may be tapered, but they are not ogival in shape. Pseudohyphae or true hyphae may be present.

Asci produce one to four ascospores that may be hat-shaped, hemispheroidal, or Saturn-shaped. Generally asci are dehiscent, but occasionally they are persistent. Asci are unconjugated, or if conjugated, they may show conjugation between bud and parent or between independent cells. Asci are not borne on ascophores. Species are homothallic or heterothallic.

Sugars may or may not be fermented.

Nitrate is assimilated.

b. TYPE SPECIES OF THE GENUS

Hansenula anomala (Hansen) H. et P. Sydow

c. SPECIES ACCEPTED IN THE GENUS

1. *Hansenula alni* Phaff, Miller et Miranda (1979)
2. *Hansenula americana* (Wickerham) Kurtzman (Wickerham, 1965b)
3. *Hansenula anomala* (Hansen) H. et P. Sydow (Hansen, 1891)
4. *Hansenula beckii* Wickerham (1951)
5. *Hansenula beijerinckii* van der Walt (1957)
6. *Hansenula bimundalis* Wickerham et Santa María (in Wickerham, 1965b)
7. *Hansenula californica* (Lodder) Wickerham (Lodder, 1932)
8. *Hansenula canadensis* Wickerham (1951)
9. *Hansenula capsulata* Wickerham (1951)
10. *Hansenula ciferrii* Lodder (1932)
11. *Hansenula dimennae* Wickerham (1969b)
12. *Hansenula dryadoides* Scott et van der Walt (1971a)
13. *Hansenula fabianii* Wickerham (1965a)
14. *Hansenula glucozyma* Wickerham (1969b)

15. *Hansenula henricii* Wickerham (1969b)
16. *Hansenula holstii* Wickerham (1960)
17. *Hansenula jadinii* (A. et R. Sartory, Weill et Meyer) Wickerham (Sartory et al., 1932)
18. *Hansenula lynferdii* van der Walt et Johannsen (1975b)
19. *Hansenula minuta* Wickerham (1951)
20. *Hansenula mrakii* Wickerham (1951)
21. *Hansenula muscicola* (Nakase et Komagata) Yarrow (Nakase & Komagata, 1966)
22. *Hansenula nonfermentans* Wickerham (1969b)
23. *Hansenula petersonii* Wickerham (1964b)
24. *Hansenula philodendra* van der Walt et Scott (in Scott & van der Walt, 1970b)
25. *Hansenula polymorpha* de Morais et Maia (1959)
26. *Hansenula saturnus* (Klöcker) H. et P. Sydow (Klöcker, 1902)
    Hansenula saturnus (Klöcker) H. et P. Sydow var. *saturnus*
    Hansenula saturnus (Klöcker) H. et P. Sydow var. *subsufficiens* Wickerham (1969b)
27. *Hansenula silvicola* Wickerham (1951)
28. *Hansenula subpelliculosa* Bedford (1942)
29. *Hansenula sydowiorum* Scott et van der Walt (1970a)
30. *Hansenula wickerhamii* Capriotti (1961c)

d. KEY TO THE SPECIES

1a Glucose not fermented ... 2
b Glucose fermented; other sugars not fermented ... 8
c Glucose and galactose fermented; other sugars not fermented ... 18
d Glucose and either, or both, sucrose and maltose fermented ... 20

2a Sucrose assimilated ... 3
b Sucrose not assimilated ... 4

3a Strong growth at 37°C; true hyphae present or absent
H. canadensis .... p. 181
b Little or no growth at 37°C; true hyphae present
H. beckii .... p. 175
c No growth at 37°C; true hyphae absent
H. alni .... p. 170

4a Cellobiose assimilated ... 5
b Cellobiose not assimilated ... 7

5a Ribitol assimilated ... 6
b Ribitol not assimilated
H. dryadoides .... p. 187

6a Rhamnose assimilated
H. henricii .... p. 191
b Rhamnose not assimilated
H. nonfermentans .... p. 201

7a Rhamnose assimilated
H. wickerhamii .... p. 211
b Rhamnose not assimilated
H. philodendra .... p. 203

8a Sucrose assimilated 9
b Sucrose not assimilated 14

9a Starch assimilated
H. holstii .... p. 192
b Starch not assimilated 10

10a Melezitose assimilated 11
b Melezitose not assimilated
H. californica .... p. 180

11a Erythritol assimilated
H. polymorpha .... p. 204
b Erythritol not assimilated 12

12a Strong fermentation of glucose 13
b Weak fermentation of glucose
H. beckii .... p. 175

13a Growth at 37°C
H. bimundalis .... p. 178
b No growth at 37°C
H. americana .... p. 171

14a Maltose assimilated
H. capsulata .... p. 183
b Maltose not assimilated 15

15a Trehalose assimilated 16
b Trehalose not assimilated 17

16a Rhamnose assimilated
H. glucozyma .... p. 190
b Rhamnose not assimilated
H. minuta .... p. 197

17a Growth in vitamin-free medium
H. mrakii .... p. 198
b No growth in vitamin-free medium
H. dimennae .... p. 186

18a Sucrose assimilated 19
b Sucrose not assimilated
H. muscicola .... p. 199

19a Starch assimilated
H. holstii .... p. 192
b Starch not assimilated
H. silvicola .... p. 207

20a Maltose assimilated 22
b Maltose not assimilated 21

21a Growth in vitamin-free medium
H. saturnus var. *saturnus* .... p. 206
b No growth in vitamin-free medium
H. saturnus var. *subsufficiens* .... p. 207

22a Erythritol assimilated 23
b Erythritol not assimilated 27

23a Growth in vitamin-free medium 24
b No growth in vitamin-free medium
H. subpelliculosa .... p. 208

24a Rhamnose assimilated 25
b Rhamnose not assimilated 26

25a Melibiose assimilated
H. sydowiorum .... p. 210
b Melibiose not assimilated
H. ciferrii .... p. 185

26a Starch assimilated
H. anomala .... p. 174
b Starch not assimilated
H. lynferdii .... p. 195

27a Growth in vitamin-free medium 28
b No growth in vitamin-free medium 29

Salient characteristics of the species in the genus *Hansenula*

	Fermentation					Assimilation											37°C	Vit	SS	mol%	Co–Q
	G	Ga	Su	Ma	Ra	Su	Ma	Ce	Tr	Me	Mz	St	Rh	Er	Rib	Cit				G + C	
H. alni	–	–	–	–	–	+	+	+	s	–	+	–	+	–	v	+	–	–	hat	41.9	
H. americana	+	–	–	–	–	+	+	+	+	–	+	–	+	–	–	+	–	–	hat	43.6	7
H. anomala	+	v	+	v	v	+	+	+	+	–	+	+	–	+	v	+	v	+	hat	36.7	7
H. beckii	v	–	–	–	–	+	+	+	+	–	+	–	+	–	v	+	v	–	hat	37.1	7
H. beijerinckii	+	–	+	–	w	+	+	+	v	–	+	–	v	–	–	–	v	+	sat	44.5	7
H. bimundalis	+	–	–	–	–	+	+	+	+	–	+	–	+	–	–	+	+	–	hat	41.1	7
H. californica	+	–	–	–	–	+	v	+	v	–	–	–	v	–	–	v	–	–	sat	43.0	7
H. canadensis	–	–	–	–	–	+	+	+	+,w	–	+	–	+	–	v	+	+	–	hat	40.3	7
H. capsulata	+	–	–	–	–	–	+	+	+	–	v	+	v	+	+	–	v	–	hat	47.0	8
H. ciferrii	+	w	+	v	w	+	+	+	+	–	+	+	+	+	+	+	v	+	hat	32.2	7
H. dimennae	+	–	–	–	–	–	–	+	–	–	–	–	v	–	–	v	–	–	sat	43.1	7
H. dryadoides	–	–	–	–	–	–	–	+	–	–	–	–	–	–	–	+	v	+	hat	31.8	
H. fabianii	+	–	+	w	w	+	+	+	+	–	+	+	–	–	–	+	+	–	hat	45.0	7
H. glucozyma	+	–	–	–	–	–	–	+	+	–	–	–	+	+	+	v	+	–	hat	45.1	7
H. henricii	–	–	–	–	–	–	–	+	v	–	–	–	+	+	+	–	+	–	hat	49.6	7
H. holstii	+	v	–	–	–	+	+	+	+	–	+	+	+	v	+	+	v	–	hat	37.0	8
H. jadinii	+	–	+	–	w	+	+	+	+	–	+	–	–	–	–	+	+	+	hat	45.5	7
H. lynferdii	+	–	+	–	w	+	+	s	s	–	+	–	–	+	s	+	–	+	hat	39.3	
H. minuta	+	–	–	–	–	–	–	+	+	–	–	–	–	–	+	v	–	–	hat	46.7	7
H. mrakii	+	–	–	–	–	–	–	+	–	–	–	–	v	–	–	–	v	+	sat	44.3	7
H. muscicola	+	s	–	–	–	–	v	+	s	–	–	–	+	s	+	+	+	–	hat	37.3	7
H. nonfermentans	–	–	–	–	–	–	–	+	+	–	–	–	–	–	+	v	+	–	hat	45.3	7
H. petersonii	+	–	+	–	w	+	+	+	+	–	+	–	+	–	–	+	+	–	hat	44.9	7
H. philodendra	–	–	–	–	–	–	–	–	+	–	–	–	–	+	s	s	+	–	hat	44.0	
H. polymorpha	+	–	–	–	–	+	+	v	+	–	+	–	v	+	+	v	+	–	hat	48.1	7
H. saturnus var. *saturnus*	+	–	+	–	w	+	–	+	–	–	–	–	v	–	–	v	v	+	sat	44.5	7
H. saturnus var. *subsufficiens*	+	–	+	–	w	+	–	+	–	–	–	–	+	–	–	v	v	–	sat	43.6	7
H. silvicola	+	+	–	–	–	+	+	+	v	–	v	–	+	–	+	v	v	–	hat	34.6	7
H. subpelliculosa	+	–	+	v	v	+	+	v	+	–	v	v	–	+	v	+	v	–	hat	33.7	7
H. sydowiorum	+	+	+	v	s	+	+	s	s	+	+	v	s	+	s	s	–	+	hat	36.2	
H. wickerhamii	–	–	–	–	–	–	–	–	+	–	–	–	+	+	+	–	–	–	hat	45.4	7

G = glucose, Ga = galactose, Su = sucrose, Ma = maltose, Ra = raffinose, Ce = cellobiose, Tr = trehalose, Me = melibiose, Mz = melezitose, St = soluble starch, Rh = L-rhamnose, Er = erythritol, Rib = ribitol, Cit = citric acid; Vit = growth in vitamin-free medium, SS = spore shape, mol% G + C = mol% G + C guanine + cytosine in the nuclear DNA (average value; range given with each species description), Co–Q = type of coenzyme Q (ubiquinone) in the electron-transport system (data from Yamada et al., 1973c); hat = hat-shaped, sat = Saturn-shaped.

28a Citric acid assimilated
H. jadinii .... p. 194
b Citric acid not assimilated
H. beijerinckii .... p. 177

29a Rhamnose assimilated
H. petersonii .... p. 202
b Rhamnose not assimilated
H. fabianii .... p. 189

e. SYSTEMATIC DISCUSSION OF THE SPECIES

1. *HANSENULA ALNI* PHAFF, MILLER ET MIRANDA (1979)

Standard description of *Hansenula alni*

Growth on 5% malt extract agar: After 3 days at 25°C the cells are ovoidal and measure (2.3–4.8) × (2.9–5.7) μm. They occur singly, in pairs, or in small clusters. Growth is butyrous, glistening, and cream-colored to white.

Growth on the surface of assimilation media: A thin, climbing pellicle is formed.

Dalmau plate culture on morphology agar: After 7 days at 25°C growth under the coverglass is sparse and pseudohyphae are poorly developed. True hyphae are not produced. Aerobic growth is tannish-white, smooth to faintly striated, glistening and with a slightly depressed center. Colony margins are entire to broadly lobed.

Formation of ascospores: This species is heterothallic and has been isolated from nature only in the haploid state. Ascospores develop in diploid cells resulting from conjugation between complementary mating types. Infrequently, the zygote itself will serve as the ascus. There are two to four hat-shaped spores per ascus, and asci are dehiscent. Complementary mating types are NRRL Y-11625 (CBS 6986, type strain) and NRRL Y-11626 (CBS 6987).

Spores were observed on 5% malt extract agar.

Fermentation: –

Assimilation of carbon compounds:

Galactose	–	Raffinose	–	Erythritol	–
Sucrose	+	Soluble starch	–	Ribitol	– or +s
Maltose	+	D-Xylose	+	D-Mannitol	+
Cellobiose	+	L-Arabinose	–	Succinic acid	+
Trehalose	+s	D-Ribose	–	Citric acid	+
Lactose	–	L-Rhamnose	+	Inositol	–

Additional carbon compounds tested: L-Sorbose –, melibiose –, melezitose +, inulin –, D-arabinose –, α-methyl-D-glucoside +, salicin +, DL-lactic acid +

Assimilation of nitrate: +
Growth in vitamin-free medium: – ; thiamine and pyridoxine are required
Growth in 10% sodium chloride plus 5% glucose in yeast nitrogen base: –
Growth at 37°C: – ; maximum temperature for growth: 33°C
G + C: 41.8–42.0 mol%, 3 strains (Phaff et al., 1979)

Comments

Phaff et al. (1979) reported limited mating between the mating types of *H. alni* and *H. wingei*, but ascospores were not formed. In an accompanying study, Fuson et al. (1979) showed the two species to exhibit only 6% DNA base sequence complementarity, thus confirming them as separate taxa. All isolates of *H. alni* have been from deciduous trees while *H. wingei* (= *H. canadensis*) occurs exclusively on coniferous trees. It would appear that the two species have recently evolved from a common ancestor.

Minor differences in carbon assimilation were noted between the original description and the present study. Phaff et al. (1979) reported weak growth on soluble starch and no growth on ribitol. In the present study, none of the strains grew on soluble starch, and one of the strains (NRRL Y-11626) gave a strong but latent reaction on ribitol.

Origin of the strains studied

Flux from *Alnus rubra* Bongard, British Columbia, Canada (2); flux from *A. rubra,* Washington, U.S.A. (1).

Type strain: CBS 6986 (UCD 68–928A, NRRL Y-11625), isolated by Phaff et al. from flux of *A. rubra,* B.C., Canada.

2. *HANSENULA AMERICANA* (WICKERHAM) KURTZMAN (1984)

Imperfect state: *Candida bimundalis* Wickerham et Santa María var. *americana* Wickerham

Synonyms

Hansenula bimundalis Wickerham et Santa María var. *americana* Wickerham (1965b)
Candida bimundalis Wickerham et Santa María var. *americana* Wickerham (1965b)

Standard description of *Hansenula americana*

Growth on 5% malt extract agar: After 3 days at 25°C, the cells are spheroidal to ellipsoidal (1.8–4.0) × (1.9–4.2) μm and single or in pairs. Growth is butyrous and light cream colored.

Growth on the surface of assimilation media: Only occasional strains form pellicles.

Dalmau plate culture on morphology agar: After 7 days at 25°C, growth under the coverglass shows abundant and moderately well differentiated pseudohyphae. In addition, true hyphae are present. Aerobic growth is white to tannish-white, butyrous, and glistening or dull. Hyphae or pseudohyphae may be found along the edges of colonies.

Formation of ascospores: This species is heterothallic and only haploid strains have been isolated from nature. Following conjugation between complementary mating types, zygotes usually serve as asci, although occasional diploid cells bud from the zygote and become asci. The ascospores are hat-shaped and there are usually four per ascus. Asci are dehiscent. Complementary mating types are NRRL Y-2156 (CBS 5644) and NRRL Y-2157 (CBS 5645).

Spores were observed on 5% malt extract agar.

Fermentation:

Glucose	+	Maltose	–
Galactose	–	Lactose	–
Sucrose	–	Raffinose	–

Assimilation of carbon compounds:

Galactose	–	Raffinose	–	Erythritol	–
Sucrose	+	Soluble starch	–	Ribitol	–
Maltose	+	D-Xylose	+	D-Mannitol	+
Cellobiose	+	L-Arabinose	v	Succinic acid	+
Trehalose	+	D-Ribose	–	Citric acid	+
Lactose	–	L-Rhamnose	+	Inositol	–

Additional carbon compounds tested: L-Sorbose –, melibiose –, melezitose +, inulin –, D-arabinose –, α-methyl-D-glucoside v, salicin +, DL-lactic acid +

Assimilation of nitrate: +

Growth in vitamin-free medium: –

Growth in 10% sodium chloride plus 5% glucose in yeast nitrogen base: –

Growth at 37°C: –

G + C: 43.6 mol%, type strain (Fuson et al., 1979)

Comments

In 1965, Wickerham (1965b) described the species *Hansenula bimundalis* and its variety *americana*. Strains of *H. bimundalis* var. *bimundalis* are associated with coniferous trees of Europe and Asia while strains of *H. bimundalis* var. *americana* are found with coniferous trees of southwestern United States. The variety *bimundalis* assimilates D-arabinose and grows at 37°C, but the variety *americana* does not. Wickerham also noted that hyphae of the variety *bimundalis* are about twice as long as hyphae of the variety *americana*. In mating studies, zygotes formed from intervarietal crosses, but ascospores were never produced.

Fuson et al. (1979) examined the nuclear DNA of certain tree inhabiting species of *Hansenula* and noted that the G + C content of *H. bimundalis* var. *americana* was 2.5% higher than that of *H. bimundalis* var. *bimundalis*. On the basis of previous experience, this was highly suggestive that the varieties represented distinct species. Kurtzman (1984) demonstrated by reassociation studies that the two varieties showed only 21% base sequence complementarity. This relatively low DNA relatedness coupled with the infertility of intervarietal crosses provided sufficient evidence for elevation of *H. bimundalis* var. *americana* to species status.

Origin of the strains studied

Frass from a Jeffrey pine, *Pinus jeffreyi* Grev. et Balf., California, U.S.A. (3); frass from a ponderosa pine, *Pinus ponderosa* Laws., New Mexico, U.S.A. (2).

Type strain: CBS 5644 (NRRL Y-2156), one of the strains from frass from a Jeffrey pine, Cal., U.S.A.

3. *HANSENULA ANOMALA* (HANSEN) H. ET P. SYDOW (1919)

Imperfect state: *Candida pelliculosa* Redaelli

Synonyms

Saccharomyces anomalus Hansen (1891)
Willia anomala Hansen (1904)
Endomyces anomalus (Hansen) Zender (1925)
Saccharomyces sphaericus v. Nägeli (1879)
Hansenula sphaerica (v. Nägeli) H. et P. Sydow (1919)
Hansenula anomala (Hansen) H. et P. Sydow var. *sphaerica* (v. Nägeli) Dekker (Stelling-Dekker, 1931)
Saccharomyces acetaethylicus Beijerinck (1892)
Monilia javanica Went et Prinsen Geerligs (1894)
Willia javanica Groenewege (1921a)
Hansenula javanica (Groenewege) Dekker (Stelling-Dekker, 1931)
Endoblastoderma pulverulentum Fischer et Brebeck (1894)
Mycoderma cerevisiae Desm. var. *pulverulentum* nom. nud.
Willia odessa Weber (1922)
Willia schneggii Weber (1922)
Hansenula schneggii (Weber) Dekker (Stelling-Dekker, 1931)
Hansenula anomala (Hansen) H. et P. Sydow var. *schneggii* (Weber) Wickerham (1970b)
Monilia productiva nom. nud. (evidently named by Beijerinck but not published by him)
Willia productiva Berkhout (1923)
Hansenula anomala (Hansen) H. et P. Sydow var. *productiva* Dekker (Stelling-Dekker, 1931)
Endomyces margaritae Zender (1925)
Willia margaritae (Zender) Guilliermond (1928)
Candida pelliculosa Redaelli (1925)
Mycocandida pelliculosa (Redaelli) Guerra (1935)
Willia bispora Mattlet (1926)
Hansenula bispora (Mattlet) Nannizzi (1934)
Hansenula anomala (Hansen) H. et P. Sydow var. *longa* Dekker (Stelling-Dekker, 1931)
Hansenula anomala (Hansen) H. et P. Sydow var. *robusta* Dekker (Stelling-Dekker, 1931)
Hansenula nivea Castelli (1933)
Hansenula panis Castelli (1933)
Saccharomyces aceris-sacchari Fabian et Hall (1933)
Hansenula anomala (Hansen) H. et P. Sydow var. *heteromorpha* Bedford (1942)

Candida pelliculosa Red. var. *cylindrica* Diddens et Lodder (1942)
Candida beverwijkii Novák et Vitéz (1964)

Standard description of *Hansenula anomala*

Growth on 5% malt extract agar: After 3 days at 25°C, the cells are spheroidal to elongate (1.9–4.1) × (2.1–6.1) μm and occur singly, in pairs, or in small clusters. Growth is butyrous and faintly tan in color.

Growth on the surface of assimilation media: Depending upon the strain, pellicles may or may not be present. When present, they vary from thin and smooth to thick and folded.

Dalmau plate culture on morphology agar: After 7 days at 25°C, growth under the coverglass may show abundant, branched pseudohyphae, although some strains form no pseudohyphae. None of the isolates forms true hyphae. Aerobic growth is white to tannish-white and generally butyrous. Some strains are smooth and glistening while others are dull and somewhat chalky. Colony margins range from entire to lobed and occasionally show a fringe of pseudohyphae. The cultures usually produce a faintly pleasant odor.

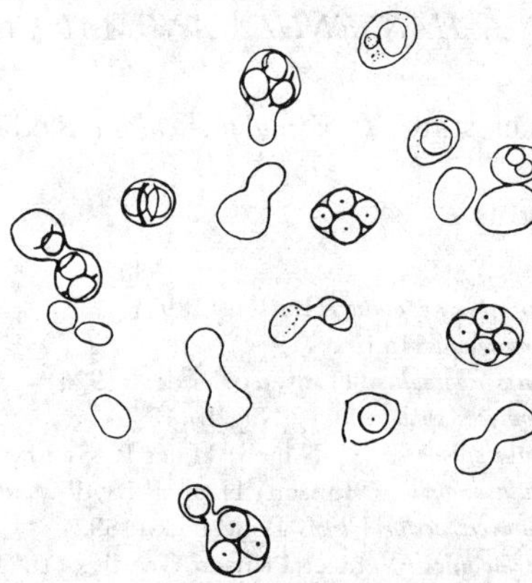

Fig. 40 – *H. anomala*
CBS 1984 × CBS 1982
After 1 week on malt extract agar
(From 'The Yeasts', ed. II, 1970)

Formation of ascospores: *Hansenula anomala* is heterothallic, but the sporogenous diploid form is usually isolated from nature. Diploid cells directly convert to asci and form one to four hat-shaped spores (Fig. 40). Asci are dehiscent. Complementary mating types are NRRL Y-366-8 (CBS 1984) and NRRL Y-2153-4 (CBS 1982).

Spores were observed on 5% malt extract agar, V8 agar, and sterilized carrot slices.

Fermentation:

Glucose	+	Maltose	v
Galactose	v	Lactose	–
Sucrose	+ or w	Raffinose	– or + w

Assimilation of carbon compounds:

Galactose	v	Raffinose	v	Erythritol	+
Sucrose	+	Soluble starch	+	Ribitol	v
Maltose	+	D-Xylose	v	D-Mannitol	+
Cellobiose	+	L-Arabinose	v	Succinic acid	+
Trehalose	+	D-Ribose	v	Citric acid	+
Lactose	–	L-Rhamnose	–	Inositol	–

Additional carbon compounds tested: L-Sorbose –, melibiose –, melezitose +, inulin –, D-arabinose –, α-methyl-D-glucoside +, salicin +, DL-lactic acid +

Assimilation of nitrate: +

Growth in vitamin-free medium: +

Growth in 10% sodium chloride plus 5% glucose in yeast nitrogen base: v

Growth at 37°C: v

G + C: 35.8–36.9 mol%, 3 strains of the former variety *anomala*; 36.7–37.4 mol%, 3 strains of the former variety *schneggii* (Kurtzman, 1984)

Comments

Wickerham (1970b) separated *H. anomala* var. *schneggii* from *H. anomala* var. *anomala* because the former is less capable of fermenting sucrose than the latter, and it does not grow in osmotic medium. Additionally, the variety *schneggii* does not assimilate raffinose. Perhaps even more striking, the variety *schneggii* produces cylindrical and often markedly elongated thread-like cells while the variety *anomala* does not. However, sexual reactions tend to obscure the other differences between the varieties. Reactions between strains of the variety *schneggii* are weak, but mixtures that contain one strain of each variety are stronger.

Kurtzman (1984) compared *H. anomala* var. *anomala* and *H. anomala* var. *schneggii* using DNA reannealing experiments. The two varieties showed 94% base sequence complementarity. Consequently, in view of the great similarity of these strains at the molecular level, the varietal designations have not been maintained.

Origin of the strains studied

Contaminants from industrial or pilot plant fermentation (12); ragi, Java (10); soil (9); gums of trees, principally *Prunus* (9); stored grain (6); ensilage (4); lakes or streams (3); warm-blooded animals (3); tree exudates (2); frass (2); fermenting mushrooms (2); sewage (2); fruit (1); from various collections and investigators (58).

Type strain: CBS 5759 (NRRL Y-366), received from Fabian, U.S.A., as *Willia anomala* Saito.

4. HANSENULA BECKII WICKERHAM (1951)

Synonyms

Endomyces bisporus Beck (1922)
Endomycopsis bispora (Beck) Dekker (Stelling-Dekker, 1931)

Standard description of *Hansenula beckii*

Growth on 5% malt extract agar: After 3 days at 25°C, the cells are ovoidal to elongate and measure (1.7–5.8) × (2.8–11.3) μm. They occur singly, in pairs, or in small clusters. Growth is light tan in color, and, depending upon the isolate, is either butyrous or mycelial.

Growth on the surface of assimilation media: Pellicles or rings are not formed.

Dalmau plate culture on morphology agar: After 7 days at 25°C, growth under the coverglass shows abundant and well branched pseudohyphae and true hyphae. Aerobic growth is light tan in color and may be butyrous or mycelial. Colony margins are entire or finely lobed.

Formation of ascospores: Beck (1922) reported the asci to be free cells in which two hat-shaped spores were formed. Subsequent investigators found Beck's strain (CBS 1890) to be asporogenous (Kreger-van Rij, 1970b), but her original description leaves no doubt about the presence of ascospores. Recently, an additional strain (CBS 6468, NRRL Y-11610) of *H. beckii* became available, and this culture formed two to four hat-shaped spores per ascus. Asci are single, unconjugated cells which become dehiscent upon maturation of the spores. Colonies of single-spore isolates from four-spored asci proved sporogenous and the species is presumed to be homothallic.

Spores were observed on YM agar.

Fermentation:

Glucose	– or +w	Maltose	–
Galactose	–	Lactose	–
Sucrose	–	Raffinose	–

Assimilation of carbon compounds:

Galactose	–	Raffinose	–	Erythritol	–
Sucrose	+	Soluble starch	–	Ribitol	v
Maltose	+	D-Xylose	+	D-Mannitol	v
Cellobiose	+	L-Arabinose	–	Succinic acid	+
Trehalose	+	D-Ribose	v	Citric acid	+
Lactose	–	L-Rhamnose	+	Inositol	–

Additional carbon compounds tested: L-Sorbose –, melibiose –, melezitose +, inulin –, D-arabinose –, α-methyl-D-glucoside +, salicin +, DL-lactic acid v

Assimilation of nitrate: +

Growth in vitamin-free medium: –

Growth in 10% sodium chloride plus 5% glucose in yeast nitrogen base: –

Growth at 37°C: – or +w

G + C: 37.1 mol%, type strain (Fuson et al., 1979)

Comments

In their electron microscopy study of septa in ascomycetous yeasts, Kreger-van Rij and Veenhuis (1973a) showed *H. beckii* (*Endomycopsis bispora*) to have hyphal septa with a closure line typical of other mycelial *Hansenula* species. In view of this and its ability to assimilate nitrate, *H. beckii* has been returned to *Hansenula* as proposed by Wickerham (1951).

The recently isolated sporogenous strain of *H. beckii* (NRRL Y-11610) was found to differ from the type strain by giving a weak and late fermentation of glucose and by its inability to assimilate ribitol, mannitol, and lactic acid. In a comparison of their nuclear DNAs, the two strains exhibited 97% base sequence complementarity and clearly represent the same species (Kurtzman et al., unpublished).

Separation of *H. beckii, H. canadensis,* and *H. alni* using standard tests is problematical; however, DNA reassociation studies clearly show the three to be distinct species (Fuson et al., 1979) despite their many phenotypic similarities. Although recourse to other tests might be desirable, the good growth shown by *H. canadensis* at 37°C serves to separate it from *H. alni*, which fails to grow at this temperature, and from *H. beckii* which grows poorly or not at all at 37°C. The presence of true hyphae in cultures of *H. beckii* distinguishes it from *H. alni*.

Origin of the strains studied

Frass from tunnels in the bark of a spruce, Beck, Austria (1); human scalp, Laaksonen, Finland (1).

Type strain: CBS 1890 (NRRL Y-1482).

5. *HANSENULA BEIJERINCKII* VAN DER WALT (1957)

Synonym

Hansenula coprophila Soneda (1959a)

Standard description of *Hansenula beijerinckii*

Growth on 5% malt extract agar: After 3 days at 25°C, the cells are spheroidal to elongate (2.8–5.5) × (3.5–10.1) μm and occur singly, in pairs, or in small clusters. Colonies are white or cream colored, butyrous, and have a dull surface.

Growth on the surface of assimilation media: Climbing mat pellicles are formed and they usually are folded.

Dalmau plate culture on morphology agar: After 7 days at 25°C, growth under the coverglass reveals infrequent and poorly developed pseudohyphae. True hyphae are not formed. Aerobic growth is white or cream colored, butyrous, and the surface is dull. Colony margins are entire to finely lobed.

Formation of ascospores: All recognized strains of this species are sporogenous. Asci form from single, unconjugated vegetative cells, and occasionally become dehiscent. Each ascus contains two to four Saturn-shaped spores. Rings are usually thin and difficult to see. Single-spore isolates from four-spored asci produce sporogenous colonies which suggests the species to be homothallic.

Spores were observed on YM- and 5% malt extract agar.

Fermentation:

Glucose	+	Maltose	–
Galactose	–	Lactose	–
Sucrose	+	Raffinose	+ w

Assimilation of carbon compounds:

Galactose	−	Raffinose	+	Erythritol	−
Sucrose	+	Soluble starch	−	Ribitol	−
Maltose	+	D-Xylose	+	D-Mannitol	v
Cellobiose	+	L-Arabinose	−	Succinic acid	+
Trehalose	v	D-Ribose	−	Citric acid	−
Lactose	−	L-Rhamnose	v	Inositol	−

Additional carbon compounds tested: L-Sorbose −, melibiose −, melezitose +, inulin +, D-arabinose −, α-methyl-D-glucoside v, salicin v, DL-lactic acid +

Assimilation of nitrate: +

Growth in vitamin-free medium: +

Growth in 10% sodium chloride plus 5% glucose in yeast nitrogen base: −

Growth at 37°C: v

G + C: 44.4–44.6 mol%, 3 strains (Kurtzman et al., unpublished)

Comments

The apparent relationship of this species with other Saturn-spored species of *Hansenula* is discussed in the section on *H. saturnus*.

Origin of the strains studied

Brook water, Maryland, U.S.A (1); swamp water, Michigan, U.S.A. (1); soil, van der Walt, South Africa (2); soil, Kodama, Japan (2); elephant dung, Soneda, Japan (1).

Type strain: CBS 2564 (NRRL YB-4312), received from van der Walt, South Africa.

6. *HANSENULA BIMUNDALIS* WICKERHAM ET SANTA MARÍA (in Wickerham, 1965b)

Imperfect state: *Candida bimundalis* Wickerham et Santa María var. *bimundalis*

Synonym

Candida bimundalis Wickerham et Santa María 1965 var. *bimundalis* (in Wickerham, 1965b)

Standard description of *Hansenula bimundalis*

Growth on 5% malt extract agar: After 3 days at 25°C, the cells are spheroidal to ellipsoidal (1.8–4.3) × (2.0–5.2) μm and single or in pairs. Growth is butyrous and light cream colored.

Growth on the surface of assimilation media: Usually pellicles are not formed, but occasional strains give waxy climbing pellicles.

Dalmau plate culture on morphology agar: After 7 days at 25°C, growth under the coverglass consists of abundant and moderately well differentiated pseudohyphae. True hyphae are also present. Aerobic growth is white to tannish-white, butyrous, and glistening or mat rugose. Colonies may be fringed with hyphae or pseudohyphae.

Formation of ascospores: This species is heterothallic and only haploid strains have been isolated from nature. Following conjugation between complementary mating types, zygotes may serve as asci or diploid cells arising from the zygotes become asci. The ascospores are hat-shaped with prominent brims and there are usually four per ascus. Asci are dehiscent. Complementary mating types are NRRL Y-5343 (CBS 5642) and NRRL Y-5354 (CBS 5643).

Spores were observed on 5% malt extract agar.

Fermentation:

Glucose	+	Maltose	–
Galactose	–	Lactose	–
Sucrose	–	Raffinose	–

Assimilation of carbon compounds:

Galactose	–	Raffinose	–	Erythritol	–
Sucrose	+	Soluble starch	–	Ribitol	–
Maltose	+	D-Xylose	+	D-Mannitol	+
Cellobiose	+	L-Arabinose	v	Succinic acid	+
Trehalose	+	D-Ribose	–	Citric acid	+
Lactose	–	L-Rhamnose	+	Inositol	–

Additional carbon compounds tested: L-Sorbose –, melibiose –, melezitose +, inulin –, D-arabinose +, α-methyl-D-glucoside v, salicin +, DL-lactic acid +

Assimilation of nitrate: +

Growth in vitamin-free medium: –

Growth in 10% sodium chloride plus 5% glucose in yeast nitrogen base: –

Growth at 37°C: +

G + C: 41.1 mol%, type strain (Fuson et al., 1979)

Comments

The close relationship between *H. bimundalis* and *H. americana* is discussed in the treatment of the latter species.

Origin of the strains studied

Larvae of the beetle *Ergates faber* and larvae of three other genera living principally in stumps of Scotch pine, *Pinus sylvestris* Linn., Spain (8); frass of a fir tree, *Abies firma* Sieb. et Zucc., Japan (1).

Type strain: CBS 5642 (NRRL Y-5343), isolated from a larva of *Ergates faber* (Wickerham, 1965b).

7. *HANSENULA CALIFORNICA* (LODDER) WICKERHAM (1951)

Synonyms

Zygohansenula californica Lodder (1932)
Zygowilliopsis californicus (Lodder) Kudriavzev (1960)
Williopsis californica (Lodder) von Arx (in von Arx et al., 1977)
Endomycopsis fukushimae Soneda (1962)
Pichia saturnospora Soneda (1962)
Hansenula californica (Lodder) Wickerham var. *maltosa* nom. nud. (Capriotti, 1959)

Standard description of *Hansenula californica*

Growth on 5% malt extract agar: After 3 days at 25°C, the cells are spheroidal to ellipsoidal (2.4–6.2) × (3.0–8.1) μm and single or in pairs. Growth is butyrous and grey-white in color.

Growth on the surface of assimilation media: Very thin pellicles, or rarely, no pellicles.

Dalmau plate culture on morphology agar: After 7 days at 25°C, growth under the coverglass shows neither pseudohyphae nor true hyphae. Aerobic growth is white to greyish-white, moderately glistening, and butyrous. Colony margins are finely lobed.

Formation of ascospores: Conjugation occurs between cells and their buds, between independent cells, or rarely asci are unconjugated. Asci contain one to four elongated Saturn-shaped ascospores which become moderately refractile upon release from their dehiscent asci. Single-ascospore isolates from four-spored asci form sporogenous colonies, thus suggesting the species to be homothallic.

Spores were observed on 5% malt extract agar.

Fermentation:

Glucose	+	Maltose	–
Galactose	–	Lactose	–
Sucrose	–	Raffinose	–

Assimilation of carbon compounds:

Galactose	–	Raffinose	–	Erythritol	–
Sucrose	+	Soluble starch	–	Ribitol	–
Maltose	v	D-Xylose	+	D-Mannitol	+
Cellobiose	+	L-Arabinose	–	Succinic acid	+
Trehalose	v	D-Ribose	–	Citric acid	v
Lactose	–	L-Rhamnose	v	Inositol	–

Additional carbon compounds tested: L-Sorbose +, melibiose –, melezitose –, inulin –, D-arabinose –, α-methyl-D-glucoside +, salicin v, DL-lactic acid +

Assimilation of nitrate: +

Growth in vitamin-free medium: –; biotin required

Growth in 10% sodium chloride plus 5% glucose in yeast nitrogen base: –

Growth at 37°C: –

G + C: 42.9–43.1 mol%, 3 strains (Kurtzman et al., unpublished)

Comments

Hansenula californica is quite similar to *H. dimennae* and can be distinguished from the latter in standard tests only by its ability to assimilate sucrose. Through use of nutritional markers, Wickerham (1969a) was able to detect hybrids in crosses between these two homothallic species, but nearly all ascospores produced by the hybrids were poorly formed. [**Note in proof:** The type strains of *H. californica* and *H. dimennae* showed 100% DNA complementarity, thus demonstrating these two taxa to be conspecific (Kurtzman et al., unpublished).]

Origin of the strains studied

Soil from North America, Europe, New Zealand, and Australia (27), a lichen growing on a rock (1), sludge from a sewage disposal plant (1), bodies of fresh water (3), slime flux of a deciduous tree (1), frass of deciduous trees (3), frass of a coniferous tree (1), sawdust (1), dung of a Japanese brown bear (2), ox dung (1), shrimp (1).

Type strain: CBS 5760 (NRRL Y-1680), isolated from soil, Murray Bridge, South Australia.

8. *HANSENULA CANADENSIS* WICKERHAM (1951)

Imperfect state: *Candida melinii* Diddens et Lodder

Synonyms

Azymohansenula canadensis (Wickerham) Novák et Zsolt (1961)
Hansenula wingei Wickerham (1956)
Endomycopsis wingei (Wickerham) Novák et Zsolt (1961)

Standard description of *Hansenula canadensis*

Growth on 5% malt extract agar: After 3 days at 25°C, the cells are spheroidal to elongate (1.1–4.6) × (2.8–8.0) μm and single or in pairs. Growth is butyrous and tannish-white.

Growth on the surface of assimilation media: Pellicles are absent or barely visible.

Dalmau plate culture on morphology agar: After 7 days at 25°C, growth under the coverglass shows abundant and highly branched pseudohyphae. True hyphae are also present. Aerobic growth is light tan in color, glistening, and butyrous. Margins of colonies are entire to lobed and may be fringed with pseudohyphae.

Formation of ascospores: Asci of the type strain are unconjugated single cells that contain two or rarely three hat-shaped spores. Asci dehisce to liberate the spores. Single-spore isolates from two-spored asci gave sporogenous colonies. The low frequency of three-spored asci has prevented their analysis for mating types, and it is not certain whether the type strain is homothallic as originally described (Wickerham, 1951).

Strains of *H. wingei*, now considered conspecific with *H. canadensis*, may be haploid or diploid when isolated from nature. Asci usually contain four hat-shaped spores and become dehiscent upon maturity. *H. wingei* is heterothallic and is the first fungus in which sexual agglutination was shown (Wickerham, 1956). Complementary mating types are NRRL Y-2340-5 (CBS 2432) and NRRL Y-2340-21 (CBS 2433).

Spores were observed on 5% malt extract agar.

Fermentation: –

Assimilation of carbon compounds:

Galactose	–	Raffinose	–	Erythritol	–
Sucrose	+	Soluble starch	–	Ribitol	v
Maltose	+	D-Xylose	+	D-Mannitol	+ or w
Cellobiose	+	L-Arabinose	–	Succinic acid	+
Trehalose	+ or w	D-Ribose	–	Citric acid	+
Lactose	–	L-Rhamnose	+	Inositol	–

Additional carbon compounds tested: L-Sorbose –, melibiose –, melezitose +, inulin –, D-arabinose –, α-methyl-D-glucoside v, salicin +, DL-lactic acid +

Assimilation of nitrate: +

Growth in vitamin-free medium: –

Growth in 10% sodium chloride plus 5% glucose in yeast nitrogen base: –

Growth at 37°C: +

G + C: 39.7–40.8 mol%, 7 strains (Fuson et al., 1979)

Comments

Wickerham (1956, 1970b) noted the phenotypic similarity of *H. canadensis* and *H. wingei* but distinguished between the two observing that *H. canadensis* produced no true hyphae and appeared homothallic while *H. wingei* formed true hyphae and was heterothallic. Indeed, *H. wingei* was quite remarkable in that mixtures of opposite mating types showed strong sexual agglutination (Wickerham, 1956). Subsequently, this process was studied in detail by various investigators (Brock, 1965; Conti & Brock, 1965; Crandall & Brock, 1968; Herman et al., 1966; Hunt & Carpenter, 1963; Taylor & Orton, 1968).

More recently, Fuson and coworkers (1979) compared the DNA relatedness of several *Hansenula* species and showed *H. canadensis* and *H. wingei* to share 78% of their nuclear DNA base sequences. On the basis of previously established guidelines (Price et al., 1978), these workers suggested the two taxa to be conspecific. Kurtzman et al. (unpublished), using somewhat different techniques, found the two species to have 76% DNA complementarity, thus confirming this unexpected relatedness. Because *H. canadensis* was described 5 years earlier than *H. wingei,* the former species epithet has priority of usage.

Extensive single-spore isolations failed to detect heterothallism in *H. canadensis,* and, therefore, precluded demonstration of conspecificity with *H. wingei* through mating studies. However, all single-spore isolates were obtained from two-spored asci, and it is entirely possible that these spores were diploid. By comparison, because of the predominance of diploid spores, we nearly failed to demonstrate heterothallism in *Pichia amylophila* (Kurtzman et al., 1980a). Another possibility would be that the type strain of *H. canadensis* is homothallic. Regardless of which possibility is correct, considering the mating system as a key character for taxonomic separation seems unsound. It has been shown that the mutation of a single locus in *Saccharomyces cerevisiae* (Hicks & Herskowitz, 1976) and in *Schizosaccharomyces pombe* (Leupold, 1950) will change the mating behavior from heterothallic to homothallic.

Origin of the strains studied

Frass from a red pine, *Pinus resinosa* Ait., Ontario, Canada (1); frass from an eastern white pine, *Pinus strobus* L., New Hampshire (1); dead conifer, Wyoming (1); from fungi cultured from a live bark beetle obtained from an Engelmann spruce (*Picea engelmannii* Parry), Colorado (2); Engelmann spruce, Wyoming (3); frass from a dead coniferous tree, Quebec, Canada (1).

Type strain: CBS 1992 (NRRL Y-1888), isolated by Wickerham from frass of a red pine.

9. *HANSENULA CAPSULATA* WICKERHAM (1951)

Imperfect state: On the basis of phenotypic similarity, *Candida molischiana* (Zikes) Meyer et Yarrow appears to represent the imperfect state of *H. capsulata*.

Standard description of *Hansenula capsulata*

Growth on 5% malt extract agar: After 3 days at 25°C, the cells are spheroidal to ellipsoidal (1.2–5.0) × (2.4–5.8) μm and single, in pairs, or occasionally in small clusters. Growth is butyrous and tannish-white in color.

Growth on the surface of assimilation media: Infrequently, thin pellicles may be produced and mucoid rings are occasionally formed.

Dalmau plate culture on morphology agar: After 7 days at 25°C, growth under the coverglass is devoid of pseudohyphae or true hyphae. Aerobic growth is white to tannish-white, glistening, and mucoid. Cultures usually produce a faint pleasant odor.

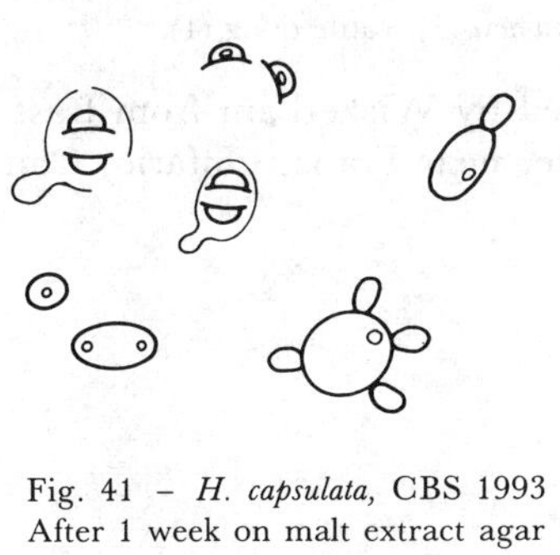

Fig. 41 – *H. capsulata*, CBS 1993
After 1 week on malt extract agar

Formation of ascospores: There are one or two hat-shaped ascospores per ascus (Fig. 41). Conjugation usually precedes sporulation and may be between a cell and its bud or infrequently between independent cells. In the latter case, conjugants commonly are a cell and the bud of another cell or a bud and the conjugation tube of a different cell. Asci are dehiscent and spores are liberated soon after formation. Single-spore isolates form sporogenous cultures, but because no more than two spores are formed per ascus, this is not conclusive evidence that the species is homothallic. However, the presence of parent-bud conjugation is suggestive of homothallism.

Spores were observed on YM- and 5% malt extract agar.

Fermentation:

Glucose	+	Maltose	–
Galactose	–	Lactose	–
Sucrose	–	Raffinose	–

Assimilation of carbon compounds:

Galactose	–	Raffinose	–	Erythritol	+
Sucrose	–	Soluble starch	+	Ribitol	+
Maltose	+	D-Xylose	+	D-Mannitol	+
Cellobiose	+	L-Arabinose	v	Succinic acid	v
Trehalose	+	D-Ribose	+	Citric acid	–
Lactose	v	L-Rhamnose	v	Inositol	–

Additional carbon compounds tested: L-Sorbose –, melibiose –, melezitose v, inulin –, D-arabinose v, α-methyl-D-glucoside v, salicin +, DL-lactic acid –

Assimilation of nitrate: + or w

Growth in vitamin-free medium: –

Growth in 10% sodium chloride plus 5% glucose in yeast nitrogen base: – or +w

Growth at 37°C: v

G + C: 46.8–47.1 mol%, 2 strains (Nakase & Komagata, 1971a)

Comments

Hansenula capsulata produces copious amounts of extracellular polysaccharides. The structures of the mannans and phosphomannans synthesized by this species have been studied in detail by Slodki et al. (1972), Gorin (1973), and Seymour et al. (1976).

Origin of the strains studied

Frass from mainly coniferous trees (53), soil on humus (3), lichen (1), cattle dung (1).

Type strain: CBS 1993 (NRRL Y-1842), isolated by Wickerham from frass of a coniferous tree on the shore of Wabatongushi Lake, near Franz, Ontario, Canada.

10. HANSENULA CIFERRII LODDER (1932)

Synonyms

Hansenula anomala (Hansen) H. et P. Sydow var. *ciferrii* Lodder et Kreger-van Rij (1952)
Endomycopsis ciferrii (Lodder) Novák et Zsolt (1961)

Standard description of *Hansenula ciferrii*

Growth on 5% malt extract agar: After 3 days at 25°C, the cells are spheroidal to ellipsoidal (3.3–5.0) × (3.9–9.5) μm and single, in pairs, or in small clusters. Growth is butyrous and white in color.

Growth on the surface of assimilation media: Pellicles are thin, smooth, and waxy.

Dalmau plate culture on morphology agar: After 7 days at 25°C, growth under the coverglass shows abundant and highly branched pseudohyphae. Mycelial variants derived from the type strain, such as NRRL Y-1322, form abundant true hyphae. Aerobic growth is white, faintly glistening to dull, and butyrous. Colony margins are usually lobed and may be fringed with pseudohyphae.

Formation of ascospores: Diploid cells convert to asci and form one to four hat-shaped ascospores. Asci are dehiscent. This species is heterothallic and complementary mating types are NRRL Y-1031-11 (CBS 1990) and Y-1031-27 (CBS 1991). Wickerham and Burton (1954a) obtained these mating types by heat-treating the sporogenous diploid type culture.

Spores were observed on 5% malt extract agar and sterilized carrot slices.

Fermentation:

Glucose	+	Maltose	v
Galactose	+w	Lactose	–
Sucrose	+	Raffinose	+w

Assimilation of carbon compounds:

Galactose	+	Raffinose	+	Erythritol	+
Sucrose	+	Soluble starch	+	Ribitol	+
Maltose	+	D-Xylose	+ or w	D-Mannitol	+
Cellobiose	+ or w	L-Arabinose	+ or w	Succinic acid	+
Trehalose	+	D-Ribose	+	Citric acid	+
Lactose	–	L-Rhamnose	+ or w	Inositol	–

Additional carbon compounds tested: L-Sorbose –, melibiose –, melezitose +, inulin –, D-arabinose –, α-methyl-D-glucoside +, salicin +, DL-lactic acid +

Assimilation of nitrate: +

Growth in vitamin-free medium: +

Growth in 10% sodium chloride plus 5% glucose in yeast nitrogen base: +w

Growth at 37°C: – or +w

G + C: 32.2 mol%, one strain (Nakase & Komagata, 1971a)

Comments

On standard tests, *H. ciferrii* differs from *H. anomala* only in its ability to assimilate rhamnose. Wickerham (1970b) considered the species distinct, however, because they showed no mating response following mixture of their mating types, and *H. ciferrii* produces crystals of tetraacetylphytosphingosine (Wickerham & Stodola, 1960; Stodola & Wickerham, 1960; Wickerham & Burton, 1962), but *H. anomala* has never been observed to produce this compound. Additionally, Nakase and Komagata (1971a) reported *H. ciferrii* to have a G + C content nearly 4% lower than that of *H. anomala*, and this suggests an almost complete lack of DNA base sequence complementarity.

Origin of the strain studied

Pod from the Tonka-bean tree, *Dipteryx odorata,* Dominican Republic (1).

Type strain: CBS 111 (NRRL Y-1031), isolated by Ciferri from Tonka-bean.

11. *HANSENULA DIMENNAE* WICKERHAM (1969b)

Standard description of *Hansenula dimennae*

Growth on 5% malt extract agar: After 3 days at 25°C, the cells are spheroidal or ellipsoidal (2.2–5.0) × (3.4–6.6) μm and single or in pairs. Growth is butyrous and white in color.

Growth on the surface of assimilation media: Thin pellicles are occasionally formed on certain carbon sources.

Dalmau plate culture on morphology agar: After 7 days at 25°C, growth under the coverglass lacks hyphae or pseudohyphae. Aerobic growth is white, glistening, and butyrous. Colony margins may be entire or lobed.

Formation of ascospores: Preceding sporulation, conjugation occurs between cells and their buds or between independent cells. Asci produce one to four elongated Saturn-shaped ascospores. Asci become dehiscent at maturity.

Spores were observed on 5% malt extract agar.

Fermentation:

Glucose	+ or s	Maltose	–
Galactose	–	Lactose	–
Sucrose	–	Raffinose	–

Assimilation of carbon compounds:

Galactose	–	Raffinose	–	Erythritol	–
Sucrose	–	Soluble starch	–	Ribitol	–
Maltose	–	D-Xylose	+	D-Mannitol	+
Cellobiose	+	L-Arabinose	–	Succinic acid	+
Trehalose	–	D-Ribose	–	Citric acid	– or + w
Lactose	–	L-Rhamnose	v	Inositol	–

Additional carbon compounds tested: L-Sorbose +, melibiose –, melezitose –, inulin –, D-arabinose –, α-methyl-D-glucoside v, salicin +, DL-lactic acid +

Assimilation of nitrate: +

Growth in vitamin-free medium: –; biotin required

Growth in 10% sodium chloride plus 5% glucose in yeast nitrogen base: –

Growth at 37°C: –

G + C: 42.8–43.3 mol%, 2 strains (Kurtzman et al., unpublished)

Comments

The relationship of *H. dimennae* with *H. californica* is discussed in the section on the latter species.

Origin of the strains studied

Soil from India (1), soil from New Zealand (2).

Type strain: CBS 5762 (NRRL YB-3239), isolated from soil from India.

12. *HANSENULA DRYADOIDES* SCOTT ET VAN DER WALT (1971a)

Standard description of *Hansenula dryadoides*

Growth on 5% malt extract agar: After 3 days at 25°C, the cells are spheroidal or ellipsoidal (2.5–5.3) × (2.6–7.7) μm and single, in pairs, and in small clusters. Growth is butyrous and tannish-white in color.

Growth on the surface of assimilation media: Pellicles are not formed but occasional incomplete rings may be observed.

Dalmau plate culture on morphology agar: After 7 days at 25°C, growth under the coverglass shows only rudimentary pseudohyphae. True hyphae are not produced. Aerobic growth is tannish-white, glistening, and butyrous. Margins vary from entire to lobed.

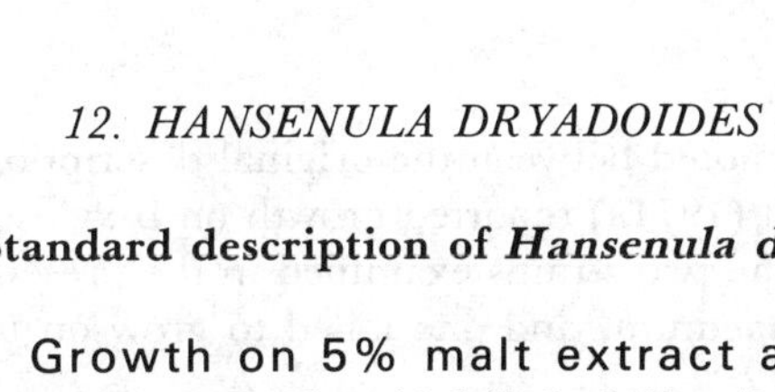

Fig. 42 – *H. dryadoides*, CBS 6154
After 1 week on malt extract agar

Formation of ascospores: Diploid cells convert directly to asci and each ascus contains two, or less frequently one, hat-shaped ascospores with prominent brims (Fig. 42). Asci are persistent. Scott and van der Walt (1971a) reported that heat treatment of sporogenous cultures gave colonies which sporulated, thus suggesting this species to be homothallic. However, spores from two-spored asci may be diploid.

Spores were observed on 5% malt extract-, morphology-, and V8 agar.

Fermentation: –

Assimilation of carbon compounds:

Galactose	–	Raffinose	–	Erythritol	–
Sucrose	–	Soluble starch	–	Ribitol	–
Maltose	–	D-Xylose	– or + s	D-Mannitol	+
Cellobiose	+	L-Arabinose	–	Succinic acid	+
Trehalose	–	D-Ribose	–	Citric acid	+
Lactose	–	L-Rhamnose	–	Inositol	–

Additional carbon compounds tested: L-Sorbose –, melibiose –, melezitose –, inulin –, D-arabinose –, α-methyl-D-glucoside –, salicin –, DL-lactic acid +

Assimilation of nitrate: +

Growth in vitamin-free medium: +

Growth in 10% sodium chloride plus 5% glucose in yeast nitrogen base: –

Growth at 37°C: v

G + C: 31.8 mol%, type strain (Miller & Baharaeen, 1979)

Comments

Minor differences in growth reactions were noted between the original description and the present study. Scott and van der Walt (1971a) reported growth on D-xylose, salicin, in osmotic medium and at 37°C. Of the two strains examined in the present study, neither grew on salicin or in osmotic medium, and one failed to grow on D-xylose and at 37°C.

Scott and van der Walt noted that *Candida berthetii* is rather similar to *H. dryadoides*, but the 5% difference in G + C that Poncet (1973) detected between the two taxa would preclude conspecificity.

Origin of the strains studied

From tunnels of *Platypus externedentatus* Fairm. infesting *Ficus* spp., Natal, South Africa (2).

Type strain: CBS 6154 (NRRL Y-10990), isolated from *Ficus sycomorus* L.

13. *HANSENULA FABIANII* WICKERHAM (1965a)

Imperfect state: *Candida fabianii* Kodama, Kyono, Iida et Onoyama

Synonym

Candida fabianii Kodama, Kyono, Iida et Onoyama (1964a)

Standard description of *Hansenula fabianii*

Growth on 5% malt extract agar: After 3 days at 25°C, the cells are spheroidal to ellipsoidal (1.0–4.2) × (2.0–6.5) μm and single, in pairs, or in small clusters. Growth is butyrous and tannish-white in color.

Growth on the surface of assimilation media: Thin pellicles and rings are produced.

Dalmau plate culture on morphology agar: After 7 days at 25°C, growth under the coverglass shows occasional pseudohyphae. True hyphae are not produced. Aerobic growth is tannish-white, glistening, and butyrous. Margins are entire or lobed; pseudohyphae are not formed around the periphery of aerobic colonies.

Formation of ascospores: This species is heterothallic and only asporogenous haploid isolates have been obtained from nature. Following the mixture of complementary mating types, zygotes may directly convert to asci or they may bud diploid cells which can become asci. The asci contain one to four hat-shaped ascospores and become dehiscent upon maturity. Complementary mating types are NRRL Y-1871 (CBS 5640) and NRRL Y-1873 (CBS 5641).

Spores were observed on 5% malt extract agar.

Fermentation:

Glucose	+	Maltose	+ w
Galactose	–	Lactose	–
Sucrose	+	Raffinose	+ w

Assimilation of carbon compounds:

Galactose	–	Raffinose	+	Erythritol	–
Sucrose	+	Soluble starch	+	Ribitol	–
Maltose	+	D-Xylose	+	D-Mannitol	+
Cellobiose	+	L-Arabinose	–	Succinic acid	+
Trehalose	+	D-Ribose	–	Citric acid	+
Lactose	–	L-Rhamnose	–	Inositol	–

Additional carbon compounds tested: L-Sorbose – , melibiose – , melezitose + , inulin – , D-arabinose – , α-methyl-D-glucoside + , salicin + , DL-lactic acid +

Assimilation of nitrate: +

Growth in vitamin-free medium: –

Growth in 10% sodium chloride plus 5% glucose in yeast nitrogen base: +

Growth at 37°C: +

G + C: 44.4–45.6 mol%, 6 strains (Nakase & Komagata, 1971a)

Comments

Of the heterothallic species of *Hansenula, H. fabianii* is most like *H. subpelliculosa.* The two are dissimilar only in their assimilation of erythritol, negative for *H. fabianii,* but Nakase and Komagata (1971a) reported their G + C content to differ by nearly 11% and this would preclude conspecificity.

Origin of the strains studied

A contaminant of a fermentor inoculated with *Aerobacter aerogenes* for production of butylene glycol, U.S.A. (1), alcohol fermentation plant, Cheju Island, near Korea (3), rice koji, Japan (4).

Type strain: CBS 5640 (NRRL Y-1871), isolated from a fermentor used for butylene glycol production.

14. HANSENULA GLUCOZYMA WICKERHAM (1969b)

Standard description of *Hansenula glucozyma*

Growth on 5% malt extract agar: After 3 days at 25°C, the cells are spheroidal or ellipsoidal (2.5–7.1) × (2.5–7.6) μm and are single, in pairs, or in small clusters. Growth is generally mucoid and tannish-white in color.

Growth on the surface of assimilation media: Pellicles are not produced.

Dalmau plate culture on morphology agar: After 7 days at 25°C, growth under the coverglass shows no evidence of hyphae or pseudohyphae. Aerobic growth is pink from the presence of ascospores, smooth, moderately glistening, and butyrous. Colony margins are entire or lobed. A faint sweet odor may be present.

Formation of ascospores: Conjugation usually, but not always, precedes ascus formation. Generally, conjugation is between a parent cell and a bud, but may also occur between independent cells. Asci are dehiscent and may produce from one to four hat-shaped ascospores. Single-spore isolates from four-spored asci form sporogenous cultures; consequently, the species is regarded to be homothallic.

Spores were observed on 5% malt extract agar.

Fermentation:

Glucose	+	Maltose	–
Galactose	–	Lactose	–
Sucrose	–	Raffinose	–

Assimilation of carbon compounds:

Galactose	–	Raffinose	–	Erythritol	+
Sucrose	–	Soluble starch	–	Ribitol	+
Maltose	–	D-Xylose	+ (if shaken)	D-Mannitol	+
Cellobiose	+	L-Arabinose	–	Succinic acid	–
Trehalose	+	D-Ribose	+	Citric acid	v
Lactose	–	L-Rhamnose	+	Inositol	–

Additional carbon compounds tested: L-Sorbose –, melibiose –, melezitose –, inulin –, D-arabinose –, α-methyl-D-glucoside –, salicin + (if shaken), DL-lactic acid –

Assimilation of nitrate: + (if shaken)

Growth in vitamin-free medium: –

Growth in 10% sodium chloride plus 5% glucose in yeast nitrogen base: +w

Growth at 37°C: +

G + C: 45.1 mol%, type strain (Nakase & Komagata, 1971a)

Comments

Hansenula glucozyma is a particularly interesting species because of its similarity to *Pichia pinus*. In standard tests, the species differ only in that *H. glucozyma* has the ability to assimilate nitrate. Kreger-van Rij (1964b) reported that although none of the strains of *P. pinus* that she studied could assimilate nitrate, some could assimilate nitrite. Despite their great phenotypic likeness, Kurtzman et al. (unpublished) found the type cultures of *H. glucozyma* and *P. pinus* to share only 3% common DNA base sequences. Nonetheless, these two species may be of importance to an understanding of the relationship between *Hansenula* and *Pichia*.

Origin of the strain studied

The only known strain of *H. glucozyma* was isolated from frass from an Engelmann spruce (*Picea engelmannii* Parry) growing in the Medicine Bow Mountains, Wyoming, U.S.A.

Type strain: CBS 5766 (NRRL YB-2185).

15. HANSENULA HENRICII WICKERHAM (1969b)

Standard description of *Hansenula henricii*

Growth on 5% malt extract agar: After 3 days at 25°C, the cells are spheroidal or ellipsoidal (3.1–5.4) × (3.4–7.5) μm and occur singly, in pairs, and in small clusters. Growth is butyrous and tannish-white in color.

Growth on the surface of assimilation media: Neither pellicles nor rings are produced.

Dalmau plate culture on morphology agar: After 7 days at 25°C, growth under the coverglass shows neither pseudohyphae nor true hyphae. Occasional short chains of cells are observed. Aerobic growth is tannish-white, smooth, faintly glistening, and butyrous. Colony margins are entire.

Formation of ascospores: Asci may be unconjugated or arise from conjugation between independent cells. Ascospores are hat-shaped and the asci are dehiscent. Usually only two spores are formed per ascus on YM- or 5% malt extract agar, but on morphology agar, three- or four-spored asci are common. The species apparently is homothallic as indicated by the fact that single-spore isolates from four-spored asci give sporogenous colonies.

Spores were observed on YM-, 5% malt extract-, and morphology agar.

Fermentation: –

Assimilation of carbon compounds:

Galactose	–	Raffinose	–	Erythritol	+
Sucrose	–	Soluble starch	–	Ribitol	+
Maltose	–	D-Xylose	+	D-Mannitol	+
Cellobiose	+	L-Arabinose	–	Succinic acid	v
Trehalose	v	D-Ribose	+	Citric acid	–
Lactose	–	L-Rhamnose	+	Inositol	–

Additional carbon compounds tested: L-Sorbose –, melibiose –, melezitose –, inulin –, D-arabinose –, α-methyl-D-glucoside –, salicin +, DL-lactic acid –

Assimilation of nitrate: +

Growth in vitamin-free medium: –

Growth in 10% sodium chloride plus 5% glucose in yeast nitrogen base: – or +w

Growth at 37°C: +

G + C: 49.6 mol%, type strain (Fuson et al., 1979)

Origin of the strains studied

Bird (?) feces collected in the Medicine Bow Mountains, Wyoming, U.S.A. (1); frass from a lodgepole pine (*Pinus contorta* Dougl.) collected near Markleeville, California, U.S.A. (1); soil, Medicine Bow Mountains, Wyoming (1).

Type strain: CBS 5765 (NRRL YB-2194), from bird (?) feces, Medicine Bow Mountains, Wyoming.

16. HANSENULA HOLSTII WICKERHAM (1960)

Imperfect state: *Candida silvicola* Shifrine et Phaff

Synonym

Candida silvicola Shifrine et Phaff (1956)

Standard description of *Hansenula holstii*

Growth on 5% malt extract agar: After 3 days at 25°C, the cells are ellipsoidal to elongate (1.7–4.0) × (2.1–6.9) μm and single, in pairs, or in small clusters. Growth is mucoid to butyrous and white in color.

Growth on the surface of assimilation media: Infrequently, thin pellicles or rings may be formed.

Dalmau plate culture on morphology agar: After 7 days at 25°C, growth under the coverglass shows abundant and well-branched pseudohyphae as well as true hyphae. Aerobic growth is white to tannish-white, dull to glistening, and mucoid to butyrous. Colony margins are entire or lobed and generally bordered by pseudohyphae and true hyphae. Freshly isolated strains sometimes have cells that produce tapered outgrowths at one or both ends. The outgrowths may attain a length of several times that of the main portion of the cell.

Formation of ascospores: *H. holstii* is heterothallic, and only asporogenous haploid strains have been isolated from nature. Asci are formed following conjugation of complementary mating types, and they contain two to four hat-shaped ascospores. The asci are dehiscent. Many natural isolates lack mating competence or show a very weak reaction (Wickerham, 1960; Kurtzman et al., 1973a). Following mixture of complementary mating types, 2 or 3 weeks may pass before conjugants are observed. Few isolates from frass of *Pinus* show mating competence, but a much higher percentage of isolates from gums of *Prunus* spp. are sexually reactive. Herman (1971) reported a minimal medium that allowed increased mating frequency, and increased sporulation has been noted in shaken cultures employing a medium developed to enhance extracellular polymer production (Slodki et al., 1972). Complementary mating types are NRRL Y-2154 (CBS 4141) and NRRL Y-2155 (CBS 4140).

Spores were observed on 5% malt extract agar and in shaken cultures of polymer production medium.

Fermentation:

Glucose	+	Maltose	–
Galactose	– or +w	Lactose	–
Sucrose	–	Raffinose	–

Assimilation of carbon compounds:

Galactose	+	Raffinose	–	Erythritol	v
Sucrose	+	Soluble starch	+	Ribitol	+
Maltose	+	D-Xylose	+	D-Mannitol	+
Cellobiose	+	L-Arabinose	+	Succinic acid	+
Trehalose	+	D-Ribose	+	Citric acid	+
Lactose	–	L-Rhamnose	+	Inositol	–

Additional carbon compounds tested: L-Sorbose +, melibiose –, melezitose +, inulin –, D-arabinose +, α-methyl-D-glucoside +, salicin +, DL-lactic acid –

Assimilation of nitrate: +

Growth in vitamin-free medium: –

Growth in 10% sodium chloride plus 5% glucose in yeast nitrogen base: + or w

Growth at 37°C: v

G + C: 36.8–37.1 mol%, 2 strains (Nakase & Komagata, 1971a)

Comments

As with *H. capsulata, H. holstii* produces large quantities of extracellular polysaccharides. Slodki and coworkers (Slodki et al., 1972; Seymour et al., 1976) have shown the mannans and phosphomannans of these two species to be basically different.

Origin of the strains studied

Several genera of coniferous and broadleaf trees from throughout the U.S. and Canada as well as from lichens, soil, lakes, and streams (154); cadaver (1).

Type strain: CBS 4140 (NRRL Y-2155), isolated by Wickerham (1960) from frass of an apparent yellow spruce (*Picea rubens?*) from the Gaspé Peninsula, Canada.

17. HANSENULA JADINII (A. ET R. SARTORY, WEILL ET MEYER) WICKERHAM (1951)

Imperfect state: *Candida utilis* (Henneberg) Lodder et Kreger-van Rij

Synonyms

Saccharomyces jadinii A. et R. Sartory, Weill et Meyer (1932)
Torula utilis Henneberg (1926)
Torulopsis utilis (Henneberg) Lodder (1934)
Cryptococcus utilis (Henneberg) Anderson et Skinner (1947)
Candida utilis (Henneberg) Lodder et Kreger-van Rij (1952)
Candida guilliermondii (Cast.) Langeron et Guerra var. *nitratophila* Diddens et Lodder (1942)
Torulopsis utilis (Henneberg) Lodder var. *major* Thaysen et Morris (1943).

Standard description of *Hansenula jadinii*

Growth on 5% malt extract agar: After 3 days at 25°C, the cells are ellipsoidal to elongate (2.5–8.0) × (4.1–11.2) μm and occur singly or in pairs. Growth is butyrous and tannish-white in color.

Growth on the surface of assimilation media: Thin pellicles may form infrequently on some media.

Dalmau plate culture on morphology agar: After 7 days at 25°C, growth under the coverglass shows occasional pseudohyphae and these may be highly branched. True hyphae are not produced. Aerobic growth is white to tannish-white, smooth, glistening, and butyrous. Colony margins are entire or infrequently lobed.

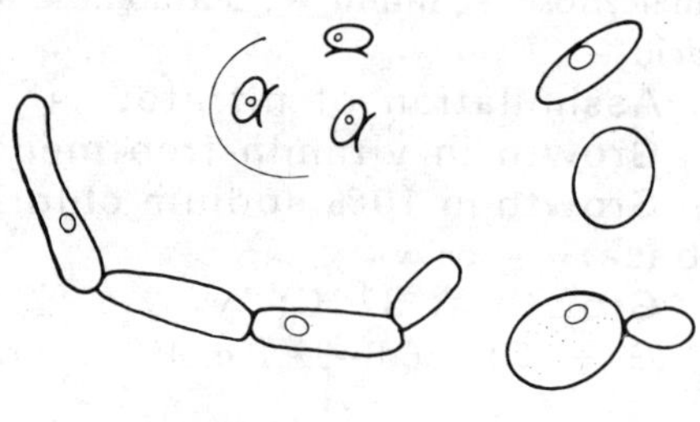

Fig. 43 – *H. jadinii*, CBS 1600
After 1 month on malt extract agar

Formation of ascospores: *H. jadinii* forms one to four hat-shaped ascospores in unconjugated dehiscent asci (Fig. 43). Single-spore isolates gave sporogenous colonies and the species is presumed to be homothallic. Sporulation is quite poor in the type strain and single-spore isolates from this culture showed no improvement in this respect (Kurtzman et al., 1979).

Spores were observed on 5% malt extract agar and on sterilized carrot slices.

Fermentation:

Glucose	+ or w	Maltose	–
Galactose	–	Lactose	–
Sucrose	+	Raffinose	+ w

Assimilation of carbon compounds:

Galactose	–	Raffinose	+	Erythritol	–
Sucrose	+	Soluble starch	–	Ribitol	–
Maltose	+	D-Xylose	+	D-Mannitol	+

Cellobiose	+	L-Arabinose	−	Succinic acid	+
Trehalose	+	D-Ribose	−	Citric acid	+
Lactose	−	L-Rhamnose	−	Inositol	−

Additional carbon compounds tested: L-Sorbose −, melibiose −, melezitose +, inulin +, D-arabinose −, α-methyl-D-glucoside +, salicin +, DL-lactic acid +

Assimilation of nitrate: +

Growth in vitamin-free medium: +

Growth in 10% sodium chloride plus 5% glucose in yeast nitrogen base: −

Growth at 37°C: +

G + C: 45.1–45.8 mol%, 3 strains (Kurtzman et al., 1979)

Comments

Candida utilis has been proposed as the imperfect state of *H. jadinii* because of phenotypic similarities (Wickerham, 1970b). Kurtzman et al. (1979) showed that single-spore isolates from the type strain of *H. jadinii* either sporulated no better than the parent or not at all. Because the parent culture sometimes failed to sporulate and because mixtures of asporogenous single-spore isolates did not sporulate, it was concluded that *H. jadinii* is homothallic. Further comparisons were made through DNA reassociation studies, and it was found that *C. utilis* and *H. jadinii* share 85% common base sequences. These data, confirmed by Manachini (1979), show *C. utilis* to be the imperfect form of *H. jadinii*. By contrast, *Hansenula petersonii* shared only 9% common base sequences with *C. utilis* and *H. jadinii* and, therefore, must be considered a separate species.

Origin of the strains studied

Pus from a human abscess (1), the udder of a cow with mastitis (1), industrial fermentations (4).

Type strain: CBS 1600 (NRRL Y-1542), isolated by Sartory et al. (1932) from the pus of a human abscess.

18. HANSENULA LYNFERDII VAN DER WALT ET JOHANNSEN (1975b)

Synonym

Candida lynferdii van der Walt et Johannsen (1975b)

Standard description of *Hansenula lynferdii*

Growth on 5% malt extract agar: After 3 days at 25°C, the cells are spheroidal to ellipsoidal (2.1–5.8) × (2.7–7.5) μm and single or in pairs. Growth is butyrous and tannish-white in color.

Growth on the surface of assimilation media: Thin creeping pellicles are formed.

Dalmau plate culture on morphology agar: After 7 days at 25°C, growth under the coverglass shows occasional moderately well developed pseudohyphae. True hyphae are not formed. Aerobic growth is tannish-white, slightly glistening, and butyrous. Colony margins are entire to finely lobed.

Formation of ascospores: Asci are unconjugated and arise directly from vegetative cells. There are one to four hat-shaped ascospores per ascus, and asci are dehiscent. Van der Walt and Johannsen (1975b) reported that heat-treatment of sporogenous cultures gave only sporogenous colonies suggesting the species to be homothallic.

Spores were observed on YM-, 5% malt extract- and V8 agar.

Fermentation:

Glucose	+	Maltose	–
Galactose	–	Lactose	–
Sucrose	+	Raffinose	+ w

Assimilation of carbon compounds:

Galactose	+	Raffinose	+	Erythritol	+
Sucrose	+	Soluble starch	–	Ribitol	+ s
Maltose	+	D-Xylose	–	D-Mannitol	+
Cellobiose	+ s	L-Arabinose	–	Succinic acid	+
Trehalose	+ s	D-Ribose	+ s	Citric acid	+
Lactose	–	L-Rhamnose	–	Inositol	–

Additional carbon compounds tested: L-Sorbose –, melibiose –, melezitose +, inulin +, D-arabinose –, α-methyl-D-glucoside +, salicin +, DL-lactic acid +

Assimilation of nitrate: +

Growth in vitamin-free medium: +

Growth in 10% sodium chloride plus 5% glucose in yeast nitrogen base: +

Growth at 37°C: –

G + C: 39.3 mol%, type strain (Miller & Baharaeen, 1979)

Comments

Hansenula lynferdii shares many characteristics with *H. jadinii* and *H. petersonii* including a tolerance to embalming fluid (van der Walt & Johannsen, 1975b). The three species can be separated by their assimilation reactions on L-rhamnose, galactose, and erythritol.

Van der Walt and Johannsen (1975b) reported *H. lynferdii* to give slow, weak growth on L-arabinose, but in the present study, no growth was detected on this compound.

Origin of the strain studied

The only known strain of this species was recovered from uncultivated grassland soil collected near Pretoria, South Africa.

Type strain: CBS 6695 (NRRL Y-7723), isolated from soil in South Africa and received from van der Walt.

19. *HANSENULA MINUTA* WICKERHAM (1951)

Standard description of *Hansenula minuta*

Growth on 5% malt extract agar: After 3 days at 25°C, the cells are spheroidal to ellipsoidal (1.0–2.6) × (1.0–3.2) μm and occur singly or in pairs. Growth is mucoid and light tan in color.

Growth on the surface of assimilation media: Pellicles are not produced and generally there are no rings.

Dalmau plate culture on morphology agar: After 7 days at 25°C, neither hyphae nor pseudohyphae are detected under the coverglass. Aerobic growth is light tan, glistening, and butyrous to mucoid. Colony margins are entire.

Formation of ascospores: Ascospore formation is preceded by conjugation between a parent cell and a bud. Asci produce one to four hat-shaped ascospores and become dehiscent at maturity. The spores are quite small but increase in size following their release. Because of the mode of ascus formation and the observation that single-spore isolates from four-spored asci are sporogenous, the species is presumed to be homothallic.

Spores were observed on 5% malt extract agar and on YM agar.

Fermentation:

Glucose	+	Maltose	–
Galactose	–	Lactose	–
Sucrose	–	Raffinose	–

Assimilation of carbon compounds:

Galactose	–	Raffinose	–	Erythritol	–
Sucrose	–	Soluble starch	–	Ribitol	+
Maltose	–	D-Xylose	+	D-Mannitol	+
Cellobiose	+	L-Arabinose	–	Succinic acid	v
Trehalose	+	D-Ribose	+	Citric acid	v
Lactose	–	L-Rhamnose	–	Inositol	–

Additional carbon compounds tested: L-Sorbose –, melibiose –, melezitose –, inulin –, D-arabinose –, α-methyl-D-glucoside –, salicin +, DL-lactic acid –

Assimilation of nitrate: +

Growth in vitamin-free medium: –

Growth in 10% sodium chloride plus 5% glucose in yeast nitrogen base: –

Growth at 37°C: –

G + C: 46.7 mol%, type strain (Kurtzman, unpublished)

Comments

Because of the small size of vegetative cells and the presence of parent-bud conjugations, *H. minuta* and *H. nonfermentans* are morphologically quite similar. In standard tests, the two species are differentiated by the ability of *H. minuta* to ferment glucose. Additionally, *H. minuta* does not grow at 37°C, but *H. nonfermentans* does.

Origin of the strain studied

The only known strain of this species was isolated from fermenting mushrooms (*Mycena pura* Fries) by Henrici.

Type strain: CBS 1708 (NRRL Y-411).

20. *HANSENULA MRAKII* WICKERHAM (1951)

Standard description of *Hansenula mrakii*

Growth on 5% malt extract agar: After 3 days at 25°C, the cells are spheroidal to ellipsoidal (2.2–7.2) × (2.4–9.5) μm and may be single or in pairs. Growth is butyrous and tannish-white in color.

Growth on the surface of assimilation media: Pellicles vary from thin and smooth to folded. Rings may reach 20 mm in height.

Dalmau plate culture on morphology agar: After 7 days at 25°C, growth under the coverglass shows abundant and highly branched pseudohyphae. True hyphae are not formed. Aerobic colonies are tannish-white, dull to faintly glistening, and butyrous. Margins are generally lobed.

Formation of ascospores: Generally, asci are formed directly from vegetative cells without prior conjugation although conjugation may occur between independent cells. Ascospores are Saturn-shaped and one to four are formed per ascus. The asci are dehiscent. Single ascospores from four-spored asci from the type strain formed sporogenous colonies, thus indicating the species to be homothallic.

Spores were observed on 5% malt extract agar.

Fermentation:

Glucose	+	Maltose	–
Galactose	–	Lactose	–
Sucrose	–	Raffinose	–

Assimilation of carbon compounds:

Galactose	–	Raffinose	–	Erythritol	–
Sucrose	–	Soluble starch	–	Ribitol	–
Maltose	–	D-Xylose	+	D-Mannitol	v
Cellobiose	+	L-Arabinose	–	Succinic acid	+
Trehalose	–	D-Ribose	–	Citric acid	–
Lactose	–	L-Rhamnose	v	Inositol	–

Additional carbon compounds tested: L-Sorbose −, melibiose −, melezitose −, inulin −, D-arabinose −, α-methyl-D-glucoside −, salicin +, DL-lactic acid +

Assimilation of nitrate: +

Growth in vitamin-free medium: +

Growth in 10% sodium chloride plus 5% glucose in yeast nitrogen base: −

Growth at 37°C: v

G + C: 44.3 mol%, 3 strains (Kurtzman et al., unpublished)

Comments

Recently, Kurtzman et al. (unpublished) discovered *H. mrakii* and *P. sargentensis* to show 68% DNA base sequence complementarity. These data suggest the two taxa to be conspecific and the significance of this finding is discussed under *Pichia*, Comments on the genus.

Origin of the strains studied

Soil from the Dobodura area of New Guinea (1); water from a swamp, Pell Lake, Wisconsin, U.S.A. (1); bog water, Pole Mountains, Wyoming, U.S.A. (1); soil, near Dehra Dun, Uttar Pradesh, India (1); cool effluent, Fales Hot Spring, California, U.S.A. (1).

Type strain: CBS 1707 (NRRL Y-1364), isolated from soil in New Guinea.

21. *HANSENULA MUSCICOLA* (NAKASE ET KOMAGATA) YARROW (1972a)

Synonym

Endomycopsis muscicola Nakase et Komagata (1966)

Standard description of *Hansenula muscicola*

Growth on 5% malt extract agar: After 3 days at 25°C, the cells are spheroidal, ellipsoidal, or elongate (1.0–4.0) × (1.2–9.5) μm and may occur singly or in pairs. Growth is somewhat butyrous to mycelial and light tan in color.

Growth on the surface of assimilation media: Pellicles are not formed but incomplete rings may be present.

Dalmau plate culture on morphology agar: After 7 days at 25°C, growth under the coverglass exhibits abundant highly branched pseudohyphae as well as true hyphae. Aerobic growth is tannish-white, dull, moist, and mycelial. Single colonies are nearly flat with entire margins. Cultures have a faint, pleasant odor.

Formation of ascospores: Asci are usually unconjugated, although conjugation can occur between independent cells. The asci may be free, attached to hyphae, or occur as swollen hyphal cells. Ascospores are hat-shaped and number from one to four per ascus. Asci are generally dehiscent, but swollen hyphal cells that convert to asci are persistent. Nakase and Komagata (1966) reported colonies from single-spore isolates to be sporogenous and the species is presumed to be homothallic.

Spores were observed on 5% malt extract agar.

Fermentation:

Glucose	+	Maltose	–
Galactose	+ s	Lactose	–
Sucrose	–	Raffinose	–

Assimilation of carbon compounds:

Galactose	+	Raffinose	–	Erythritol	+ s
Sucrose	–	Soluble starch	–	Ribitol	+
Maltose	v	D-Xylose	+ s	D-Mannitol	+
Cellobiose	+	L-Arabinose	+ s	Succinic acid	+
Trehalose	+ s	D-Ribose	+ s	Citric acid	+
Lactose	–	L-Rhamnose	+	Inositol	–

Additional carbon compounds tested: L-Sorbose –, melibiose –, melezitose –, inulin –, D-arabinose –, α-methyl-D-glucoside +s, salicin +, DL-lactic acid +

Assimilation of nitrate: +

Growth in vitamin-free medium: –

Growth in 10% sodium chloride plus 5% glucose in yeast nitrogen base: –

Growth at 37°C: +

G + C: 37.3 mol%, one strain (Nakase & Komagata, 1971a)

Comments

Hansenula muscicola was originally placed in the genus *Endomycopsis* because it produced true hyphae. Once *Endomycopsis* was shown to be invalid, Yarrow (1972a) transferred the species to *Hansenula* rather than *Saccharomycopsis* in keeping with presently accepted criteria for demarcation of these two genera.

In the original description, Nakase and Komagata (1966) reported *H. muscicola* to be unable to assimilate citrate, but in the present study, growth was noted on this compound.

Origin of the strains studied

Moss, Tottori Prefecture, Japan (3).

Type strain: CBS 5800 (NRRL Y-7005).

22. *HANSENULA NONFERMENTANS* WICKERHAM (1969b)

Standard description of *Hansenula nonfermentans*

Growth on 5% malt extract agar: After 3 days at 25°C, the cells are spheroïdal or ellipsoidal (1.3–2.5) × (2.2–4.4) μm and may be single, in pairs, or in small clusters. Growth is mucoid and light tan in color.

Growth on the surface of assimilation media: Pellicles are absent and rings are rarely observed.

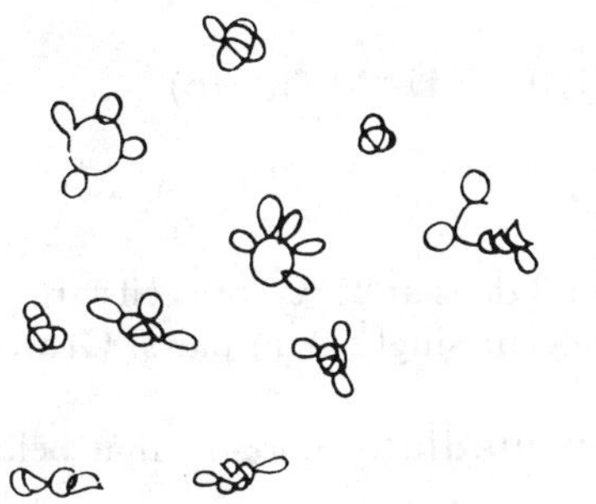

Fig. 44 – *H. nonfermentans,* CBS 5764
After 10 days on malt extract agar
(From 'The Yeasts', ed. II, 1970)

Dalmau plate culture on morphology agar: After 7 days at 25°C, neither hyphae nor pseudohyphae are produced under the coverglass. Aerobic growth is tannish-white, smooth, glistening, and slightly mucoid to butyrous. Colony margins are entire or occasionally lobed.

Formation of ascospores: Conjugation generally precedes sporulation and usually is between a parent cell and a bud but may occur between independent cells. Ascospores are hat-shaped and one to four are formed per ascus (Fig. 44). Asci are dehiscent and the rather small spores enlarge following liberation. Single-spore isolates from four-spored asci from the type strain formed sporogenous colonies indicating the species to be homothallic.

Spores were observed on 5% malt extract agar and on YM agar.

Fermentation: –

Assimilation of carbon compounds:

Galactose	–	Raffinose	–	Erythritol	–
Sucrose	–	Soluble starch	–	Ribitol	+
Maltose	–	D-Xylose	+	D-Mannitol	+
Cellobiose	+	L-Arabinose	–	Succinic acid	v
Trehalose	+	D-Ribose	+	Citric acid	v
Lactose	–	L-Rhamnose	–	Inositol	–

Additional carbon compounds tested: L-Sorbose –, melibiose –, melezitose –, inulin –, D-arabinose –, α-methyl-D-glucoside –, salicin +, DL-lactic acid –

Assimilation of nitrate: +

Growth in vitamin-free medium: –

Growth in 10% sodium chloride plus 5% glucose in yeast nitrogen base: –

Growth at 37°C: +

G + C: 45.3 mol%, type strain (Fuson et al., 1979)

Comments

The relationship of *H. nonfermentans* with *H. minuta* is discussed under the latter species.

Origin of the strain studied

The only known strain of this species was isolated from Libby Creek near Laramie, Wyoming, U.S.A.

Type strain: CBS 5764 (NRRL YB-2203).

23. *HANSENULA PETERSONII* WICKERHAM (1964b)

Standard description of *Hansenula petersonii*

Growth on 5% malt extract agar: After 3 days at 25°C, the cells are ellipsoidal to elongate (1.2–4.1) × (2.8–7.1) μm and occur singly or in pairs. Growth is butyrous and light tan in color.

Growth on the surface of assimilation media: Smooth, mat pellicles and rings develop on certain media.

Dalmau plate culture on morphology agar: After 7 days at 25°C, abundant, highly branched pseudohyphae are formed under the coverglass. True hyphae are also produced by the mat oxidative forms of this species, but not by the smooth glistening types, such as NRRL Y-3807. Aerobic growth is light tan, smooth, dull to glistening, and butyrous. Margins of single colonies may be entire or lobed and surrounded by hyphal outgrowth.

Formation of ascospores: Vegetative cells directly convert to asci without prior conjugation. Asci form one to four hat-shaped ascospores which are freed through dehiscence of the ascus. Both heat-treated sporogenous cultures and single-spore isolates give rise to sporogenous colonies and the species is presumed to be homothallic. Wickerham (1964b) reported the presence of glistening colony forms, such as NRRL Y-3807, that were asporogenous.

Spores were observed on 5% malt extract agar and on YM agar.

Fermentation:

Glucose	+	Maltose	–
Galactose	–	Lactose	–
Sucrose	+	Raffinose	+ w

Assimilation of carbon compounds:

Galactose	–	Raffinose	+	Erythritol	–
Sucrose	+	Soluble starch	–	Ribitol	–
Maltose	+	D-Xylose	+	D-Mannitol	+
Cellobiose	+	L-Arabinose	–	Succinic acid	+
Trehalose	+	D-Ribose	–	Citric acid	+
Lactose	–	L-Rhamnose	+	Inositol	–

Additional carbon compounds tested: L-Sorbose –, melibiose –, melezitose +, inulin v, D-arabinose −, α-methyl-D-glucoside +, salicin +, DL-lactic acid +

Assimilation of nitrate: +

Growth in vitamin-free medium: −

Growth in 10% sodium chloride plus 5% glucose in yeast nitrogen base: −

Growth at 37°C: +

G + C: 44.9 mol%, type strain (Kurtzman et al., 1979)

Comments

Wickerham (1964b, 1970b) discussed the similarities of *H. petersonii, H. jadinii,* and *Candida utilis* and suggested that their primary habitat may be warm-blooded animals. All three species are resistant to toxic chemicals, especially embalming fluid, and van der Walt and Johannsen (1975b) observed a similar chemotolerance for *H. lynferdii.* As noted earlier, Kurtzman et al. (1979) demonstrated through DNA reassociation studies that *H. jadinii* and *C. utilis* are conspecific while *H. petersonii* is a separate species.

Origin of the strains studied

Human cadavers preserved with embalming fluid (2).

Type strain: CBS 5555 (NRRL Y-3808).

24. *HANSENULA PHILODENDRA* VAN DER WALT ET SCOTT (in Scott & van der Walt, 1970b)

Standard description of *Hansenula philodendra*

Growth on 5% malt extract agar: After 3 days at 25°C, the cells are spheroidal to ellipsoidal (1.8–4.8) × (1.8–6.9) μm and single, in pairs, or less frequently, in small clusters. Growth is butyrous and tannish-white.

Growth on the surface of assimilation media: Pellicles are not formed.

Dalmau plate culture on morphology agar: After 7 days at 25°C, neither true hyphae nor pseudohyphae are detected under the coverglass. Aerobic growth is white, smooth to finely striated, glistening, and butyrous. Colony margins are usually lobed.

Formation of ascospores: Vegetative cells may convert directly to asci, or more frequently, ascus formation is preceded by conjugation between independent cells. Asci form one to four hat-shaped ascospores, and the spores are released through ascus dehiscence. Single-spore isolates from four-spored asci gave sporogenous colonies and the species is presumed to be homothallic.

Spores were observed on 5% malt extract-, YM-, and V8 agar.

Fermentation: −

Assimilation of carbon compounds:

Galactose	–	Raffinose	–	Erythritol	+
Sucrose	–	Soluble starch	–	Ribitol	+ s
Maltose	–	D-Xylose	+	D-Mannitol	+
Cellobiose	–	L-Arabinose	–	Succinic acid	+ s
Trehalose	+	D-Ribose	+ s	Citric acid	+ s
Lactose	–	L-Rhamnose	–	Inositol	–

Additional carbon compounds tested: L-Sorbose + s, melibiose –, melezitose –, inulin –, D-arabinose –, α-methyl-D-glucoside –, salicin –, DL-lactic acid –

Assimilation of nitrate: +

Growth in vitamin-free medium: –

Growth in 10% sodium chloride plus 5% glucose in yeast nitrogen base: +

Growth at 37°C: +

G + C: 44.0 mol%, type strain (Miller & Baharaeen, 1979)

Origin of the strains studied

Frass from tunnels of the Bostrichid, *Xylion adustus* Fahr., which was infesting a moribund specimen of *Ficus sycomorus* L., near St. Lucia, Natal, South Africa (2).

Type strain: CBS 6075 (NRRL Y-7210).

25. *HANSENULA POLYMORPHA* DE MORAIS ET MAIA (1959)

Synonym

Hansenula angusta Teunisson, Hall et Wickerham (1960)

Standard description of *Hansenula polymorpha*

Growth on 5% malt extract agar: After 3 days at 25°C, the cells are spheroidal to elongate (1.5–4.6) × (1.9–5.3) μm and single, in pairs, or in small clusters. Growth is butyrous and tannish-white in color.

Growth on the surface of assimilation media: Very thin pellicles are occasionally produced.

Dalmau plate culture on morphology agar: After 7 days at 25°C, neither hyphae nor pseudohyphae are detected under the coverglass. Aerobic growth is white to slightly tannish-white, smooth to finely striated, glistening, and butyrous. Colony margins are entire or finely lobed.

Formation of ascospores: Asci are unconjugated, or exhibit conjugation between parent and bud, or less frequently between independent cells. One to four hemispheroidal or narrow brimmed hat-shaped ascospores are formed per ascus. The asci are dehiscent. Single-spore isolates from four-spored asci from the type strain gave sporogenous colonies, and the species is presumed homothallic.

Spores were observed on 5% malt extract agar and on YM agar.

Fermentation:

Glucose	+	Maltose	–
Galactose	–	Lactose	–
Sucrose	–	Raffinose	–

Assimilation of carbon compounds:

Galactose	– or +w	Raffinose	–	Erythritol	+
Sucrose	+	Soluble starch	–	Ribitol	+
Maltose	+	D-Xylose	v	D-Mannitol	+
Cellobiose	v	L-Arabinose	–	Succinic acid	v
Trehalose	+	D-Ribose	+	Citric acid	v
Lactose	–	L-Rhamnose	v	Inositol	–

Additional carbon compounds tested: L-Sorbose v, melibiose –, melezitose +, inulin –, D-arabinose v, α-methyl-D-glucoside v, salicin v, DL-lactic acid –

Assimilation of nitrate: +

Growth in vitamin-free medium: –

Growth in 10% sodium chloride plus 5% glucose in yeast nitrogen base: + or w

Growth at 37°C: +

G + C: 47.8–48.3 mol%, 4 strains (Nakase & Komagata, 1971a)

Origin of the strains studied

Soil (1), concentrated orange juice (1), *Drosophila* spp. (5), intestinal tract of swine (1), 'alpechin' (1), insect frass (19).

Type strain: CBS 4732 (NRRL Y-5445), strain TB-3 isolated by Morais and Maia (1959) from soil irrigated with waste water from distilleries.

26. *HANSENULA SATURNUS* (KLÖCKER) H. ET P. SYDOW (1919)

This species has two varieties:

Hansenula saturnus (Klöcker) H. et P. Sydow var. *saturnus*
Hansenula saturnus (Klöcker) H. et P. Sydow var. *subsufficiens* Wickerham (1969b)

Synonyms of the variety *saturnus*

Saccharomyces saturnus Klöcker (1902)
Willia saturnus (Klöcker) Hansen (1904)
Williopsis saturnus (Klöcker) Zender (1925)
Pichia suaveolens Klöcker (1912a)
Hansenula suaveolens (Klöcker) Dekker (Stelling-Dekker, 1931)

Standard description of *Hansenula saturnus* var. *saturnus*

Growth on 5% malt extract agar: After 3 days at 25°C, the cells are spheroidal, ellipsoidal, or occasionally elongate (2.5–5.2) × (3.5–8.5) μm and occur singly, in pairs, or small clusters. Growth is butyrous and tannish-white in color.

Growth on the surface of assimilation media: Pellicles vary from thin and smooth to thick, folded, and mat. Rings may reach a height of 20 mm.

Dalmau plate culture on morphology agar: After 7 days at 25°C, growth under the coverglass shows only occasional primitive pseudohyphae. True hyphae are not produced. Aerobic growth is tannish-white, smooth, dull, and butyrous. Colony margins may be entire or finely lobed.

Formation of ascospores: Asci are generally unconjugated but occasionally conjugation between independent cells is observed. One to four Saturn-shaped ascospores are formed per ascus (Fig. 45). The thickness of the rings on the spores varies, and spores from one strain may even lack rings (Wickerham, 1970b). Asci are persistent. Single-spore isolates formed sporogenous colonies indicating the species to be homothallic.

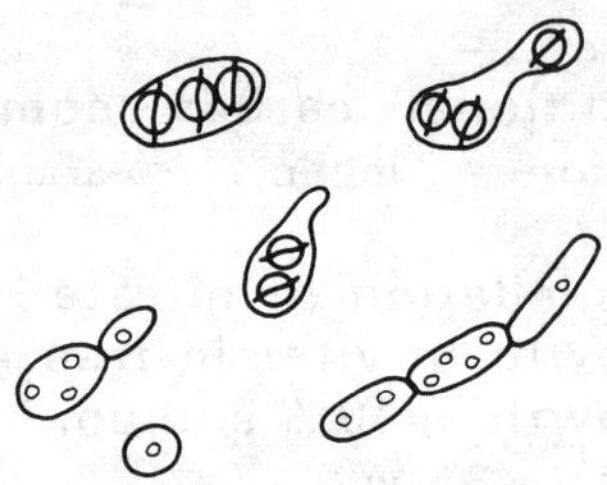

Fig. 45 – *H. saturnus* var. *saturnus*, CBS 5761
After 2 weeks on malt extract agar

Spores were observed on 5% malt extract agar and on YM agar.

Fermentation:

Glucose	+	Maltose	–
Galactose	–	Lactose	–
Sucrose	+	Raffinose	+ w

Assimilation of carbon compounds:

Galactose	–	Raffinose	+	Erythritol	–
Sucrose	+	Soluble starch	–	Ribitol	–
Maltose	–	D-Xylose	+	D-Mannitol	v
Cellobiose	+	L-Arabinose	–	Succinic acid	+
Trehalose	–	D-Ribose	–	Citric acid	v
Lactose	–	L-Rhamnose	v	Inositol	–

Additional carbon compounds tested: L-Sorbose –, melibiose –, melezitose –, inulin v, D-arabinose –, α-methyl-D-glucoside –, salicin +, DL-lactic acid +

Assimilation of nitrate: +

Growth in vitamin-free medium: +

Growth in 10% sodium chloride plus 5% glucose in yeast nitrogen base: –

Growth at 37°C: v

G + C: 44.4–44.5 mol%, 2 strains of the variety *saturnus*; 43.5–43.7 mol%, 2 strains of the variety *subsufficiens* (Kurtzman et al., unpublished)

Supplementary description of *Hansenula saturnus* var. *subsufficiens*

In standard tests, the variety *subsufficiens* differs from the variety *saturnus* only by its failure to grow in vitamin-free medium. As with the variety *saturnus*, single-spore isolates gave sporogenous colonies, and the variety is presumed to be homothallic.

Comments

Reactions of the Saturn-spored species of *Hansenula* on standard fermentation and assimilation tests are quite similar. As noted earlier, *H. californica* and *H. dimennae* can be distinguished from each other only by the sucrose assimilation reaction. Wickerham (1969a) reported that these two species hybridize, but nearly all spores produced by the hybrids were defective. Similarly, Wickerham noted that *H. beijerinckii* hybridized with *H. mrakii, H. saturnus* var. *saturnus,* and *H. saturnus* var. *subsufficiens,* but again, spores appeared poorly formed. It is anticipated that DNA reassociation studies should effectively clarify speciation.

Origin of the strains belonging to the variety *saturnus*

Soil from North America, New Zealand, and Africa (36), fresh water (8), frass (3), cow dung (1), diverse and uncertain origins (12).

Type strain: CBS 5761 (NRRL Y-1304), designated by Wickerham (1951) as the neotype.

Origin of the strains belonging to the variety *subsufficiens*

Soil from Liberia (2).

Type strain: CBS 5763 (NRRL YB-1657).

27. *HANSENULA SILVICOLA* WICKERHAM (1951)

Standard description of *Hansenula silvicola*

Growth on 5% malt extract agar: After 3 days at 25°C, the cells are ellipsoidal or elongate (1.7–4.0) × (2.4–9.0) μm and occur singly, in pairs, or in small clusters. Growth is butyrous and tannish-white in color.

Growth on the surface of assimilation media: Exceptionally thin pellicles are formed and thin rings about 2 mm high may be produced.

Dalmau plate culture on morphology agar: After 7 days at 25°C, growth under the coverglass shows abundant and highly branched pseudohyphae. True hyphae are occasionally detected. Aerobic growth is white to tannish-white, smooth, glistening, and butyrous. Colony margins are entire or occasionally lobed and devoid of hyphal outgrowths.

Formation of ascospores: Conjugation usually precedes sporulation. Generally, conjugation occurs between parent and bud, occasionally between independent cells, or rarely the asci are unconjugated. Asci form one to four hat-shaped ascospores that are freed when the ascus dehisces. Single-spore isolates obtained from four-spored asci of the type strain formed sporogenous colonies, and the species is presumed to be homothallic.

Spores were observed on 5% malt extract agar.

Fermentation:

Glucose	+	Maltose	–
Galactose	+	Lactose	–
Sucrose	–	Raffinose	–

Assimilation of carbon compounds:

Galactose	+	Raffinose	–	Erythritol	–
Sucrose	+	Soluble starch	–	Ribitol	+
Maltose	+	D-Xylose	+	D-Mannitol	v
Cellobiose	+	L-Arabinose	+	Succinic acid	+
Trehalose	v	D-Ribose	v	Citric acid	– or +w
Lactose	–	L-Rhamnose	+	Inositol	–

Additional carbon compounds tested: L-Sorbose +, melibiose –, melezitose v, inulin –, D-arabinose –, α-methyl-D-glucoside +, salicin +, DL-lactic acid v

Assimilation of nitrate: +

Growth in vitamin-free medium: –

Growth in 10% sodium chloride plus 5% glucose in yeast nitrogen base: –

Growth at 37°C: v

G + C: 34.1–35.1 mol%, 4 strains (Nakase & Komagata, 1971a)

Origin of the strains studied

Gums from wild black cherry trees, *Prunus serotina* Ehrh., Peoria, Illinois, U.S.A. (3); gums from wild pin cherry trees, *Prunus pensylvanica* Linn., Wisconsin and Vermont, U.S.A. (4).

Type strain: CBS 1705 (NRRL Y-1678), isolated from gum of *P. serotina,* Peoria, Illinois.

28. *HANSENULA SUBPELLICULOSA* BEDFORD (1942)

Synonym

Endomycopsis subpelliculosa (Bedford) Novák et Zsolt (1961)

Standard description of *Hansenula subpelliculosa*

Growth on 5% malt extract agar: After 3 days at 25°C, the cells are ellipsoidal (1.9–5.1) × (3.4–9.2) μm and occur singly or in pairs. Growth is white to tannish-white and butyrous.

Growth on the surface of assimilation media: Thin pellicles are occasionally formed.

Dalmau plate culture on morphology agar: After 7 days at 25°C, growth under the coverglass usually shows abundant and highly branched pseudohyphae, although some strains fail to produce any pseudohyphae. True hyphae are formed, but this is not common. Aerobic growth is white to tannish-white, smooth, glistening, and butyrous. Colony margins are usually lobed and devoid of pseudohyphal outgrowths. A faint, pleasant odor may be detected.

Formation of ascospores: Asci are unconjugated and form one to four hat-shaped spores. The asci become dehiscent at maturity. Wickerham and Burton (1954a) demonstrated this species to be heterothallic by heat-treatment of sporogenous cultures. Complementary mating types are NRRL Y-1683-11 (CBS 1997) and NRRL Y-1822-12 (CBS 1996).

Spores were observed on 5% malt extract agar.

Fermentation:

Glucose	+	Maltose	– or + w
Galactose	–	Lactose	–
Sucrose	+	Raffinose	– or + w

Assimilation of carbon compounds:

Galactose	v	Raffinose	+	Erythritol	+
Sucrose	+	Soluble starch	v	Ribitol	v
Maltose	+	D-Xylose	+ or w	D-Mannitol	+
Cellobiose	v	L-Arabinose	v	Succinic acid	+
Trehalose	+	D-Ribose	v	Citric acid	+
Lactose	–	L-Rhamnose	–	Inositol	–

Additional carbon compounds tested: L-Sorbose –, melibiose –, melezitose v, inulin –, D-arabinose – or + w, α-methyl-D-glucoside +, salicin +, DL-lactic acid +

Assimilation of nitrate: +

Growth in vitamin-free medium: –

Growth in 10% sodium chloride plus 5% glucose in yeast nitrogen base: +

Growth at 37°C: v

G + C: 33.4–33.9 mol%, 4 strains (Nakase & Komagata, 1971a)

Comments

Hansenula subpelliculosa, H. anomala, and *H. fabianii* are quite similar when compared by the standard tests. *H. anomala* is separated from the other two species by its growth in vitamin-free medium, and *H. subpelliculosa* is separated from *H. fabianii* by its ability to assimilate erythritol. Wickerham and Burton (1954a) indicated that these three heterothallic taxa do not interbreed.

Origin of the strains studied

Bedford's strains from sugar-egg concentrate (1), syrup (1), grape juice (1), and dried prunes (1); received from Etchells and isolated from fermenting cucumbers and from brined unshelled peas and green beans (27); soy sauce manufacturing, China (1); miso (1).

Type strain: CBS 5767 (NRRL Y-1683), from fermenting cucumbers.

29. *HANSENULA SYDOWIORUM* SCOTT ET VAN DER WALT (1970a)

Imperfect state: *Candida nitrativorans* van der Walt, Scott et van der Klift

Synonym

Candida nitrativorans van der Walt, Scott et van der Klift (1972)

Standard description of *Hansenula sydowiorum*

Growth on 5% malt extract agar: After 3 days at 25°C, the cells are spheroidal to ellipsoidal (1.8–5.3) × (2.7–7.8) μm and occur singly or in pairs. Growth is butyrous and light tannish-white in color.

Growth on the surface of assimilation media: Pellicles are smooth and waxy.

Dalmau plate culture on morphology agar: After 7 days at 25°C, growth under the coverglass shows abundant and well branched pseudohyphae. True hyphae are not produced. Aerobic growth is white, smooth, dull to moderately glistening, and butyrous. Colony margins are generally lobed.

Formation of ascospores: Asci are unconjugated. Scott and van der Walt (1970a) reported vegetative cells to first become lipid-rich and thick-walled, and following a dormant period, an outer membrane is shed to form the ascus. One to four hat-shaped ascospores are formed per ascus, and asci are dehiscent. Heat-treatment of sporogenous cultures gave only sporogenous colonies, and the species is presumed to be homothallic (Scott & van der Walt, 1970a).

Spores were observed on 5% malt extract agar.

Fermentation:

Glucose	+	Maltose	v
Galactose	+ or w	Lactose	–
Sucrose	+	Raffinose	+ s

Assimilation of carbon compounds:

Galactose	+	Raffinose	+	Erythritol	+
Sucrose	+	Soluble starch	– or + s	Ribitol	+ s
Maltose	+	D-Xylose	v	D-Mannitol	+
Cellobiose	+ s	L-Arabinose	+ s	Succinic acid	+
Trehalose	+ s	D-Ribose	+ s	Citric acid	+ s
Lactose	–	L-Rhamnose	+ s	Inositol	–

Additional carbon compounds tested: L-Sorbose −, melibiose +, melezitose +, inulin −, D-arabinose −, α-methyl-D-glucoside +, salicin +, DL-lactic acid +s

Assimilation of nitrate: +

Growth in vitamin-free medium: +

Growth in 10% sodium chloride plus 5% glucose in yeast nitrogen base: +

Growth at 37°C: −

G + C: 36.2 mol%, type strain (Miller & Baharaeen, 1979)

Comments

Hansenula sydowiorum is rather similar to *H. ciferrii* and *H. anomala,* but the three species can be separated by their assimilation reactions on L-rhamnose and melibiose. As noted earlier, *H. ciferrii* forms tetraacetylphytosphingosine while *H. sydowiorum* synthesizes 2-D-hydroxyhexadecanoic acid (Vesonder et al., 1970). Crystalline compounds have not been observed in cultures of *H. anomala.*

In contrast to the original description which reported the assimilation of soluble starch and D-xylose to be positive, two of the four strains examined in the the present study gave negative reactions on these compounds. In addition, two of the strains failed to ferment maltose.

Origin of the strains studied

Frass from tunnels of the Bostrichid beetle, *Sinonylon ruficorne* Fahr., infesting *Combretum apiculatum* Sond. (3); frass from tunnels in *Mimusops caffra* (1), all from South Africa.

Type strain: CBS 5995 (NRRL Y-7130), from tunnels in *S. ruficorne.*

30. *HANSENULA WICKERHAMII* CAPRIOTTI (1961c)

Standard description of *Hansenula wickerhamii*

Growth on 5% malt extract agar: After 3 days at 25°C, the cells are spheroidal or ellipsoidal (1.4–3.3) × (1.4–3.5) μm and single or in pairs. Growth is butyrous and tannish-white in color.

Growth on the surface of assimilation media: Pellicles are not produced.

Dalmau plate culture on morphology agar: After 7 days at 25°C, neither hyphae nor pseudohyphae are detected under the coverglass. Aerobic growth is tannish-white, smooth, glistening, and butyrous. Colony margins are entire to finely lobed.

Formation of ascospores: Asci generally form after conjugation between independent cells, but occasionally they are unconjugated or a parent cell may conjugate with a bud. The asci usually produce three or four rather small hat-shaped spores that are soon released by dehiscence (Fig. 46). Single-spore isolates give sporogenous colonies and the species appears to be homothallic.

Spores were observed on 5% malt extract agar.

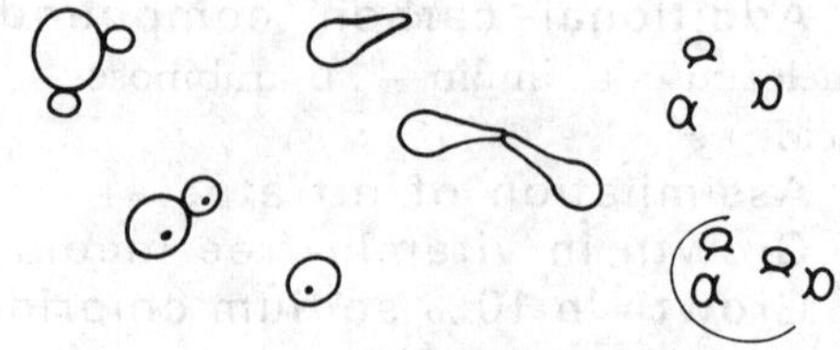

Fig. 46 – *H. wickerhamii,* CBS 4307
After 2 weeks on malt extract agar

Fermentation: –

Assimilation of carbon compounds:

Galactose	–	Raffinose	–	Erythritol	+
Sucrose	–	Soluble starch	–	Ribitol	+
Maltose	–	D-Xylose	+	D-Mannitol	+
Cellobiose	–	L-Arabinose	–	Succinic acid	–
Trehalose	+	D-Ribose	+	Citric acid	–
Lactose	–	L-Rhamnose	+	Inositol	–

Additional carbon compounds tested: L-Sorbose +, melibiose –, melezitose –, inulin –, D-arabinose +, α-methyl-D-glucoside –, salicin –, DL-lactic acid –

Assimilation of nitrate: +

Growth in vitamin-free medium: –

Growth in 10% sodium chloride plus 5% glucose in yeast nitrogen base: –

Growth at 37°C: –

G + C: 45.4 mol%, one strain (Nakase & Komagata, 1971a)

Origin of the strains studied

Soil from a swamp in which spruce and broad-leafed trees were growing, Hÿyttiala Province of Tavastia Australis, Finland (2).

Type strain: CBS 4307 (NRRL YB-4943), Capriotti's strain 241-A which he had designated as the type.

f. COMMENTS ON THE GENUS

Wickerham's (1970b) comprehensive treatment of the genus *Hansenula* has provided the foundation for the present study, and his phylogenetic scheme for the classification of species, which has served as a model for the evaluation of some of the newer taxonomic techniques, may be found in that treatise.

Since the 1970 treatment, species of *Hansenula* have been examined by DNA reassociation, determination of coenzyme Q, numerical analysis, proton magnetic resonance, and serology of cell wall mannans. The extensive studies by Yamada and coworkers (1972, 1973c, 1976b, 1977) on the type of coenzyme Q or ubiquinone present in the electron transport system of various yeast taxa showed all species of *Hansenula*, with the exception of *H. holstii* and *H. capsulata*, to have the Q-7 form of these quinones. The significance of the Q-8 form found in the latter two species is unknown, but the possibility that they may have a different phylogenetic origin than the remainder of the *Hansenula* species should be considered.

The limited DNA studies carried out to date generally support Wickerham's (1969c, 1970b) proposal for speciation in *Hansenula*. One exception has been the demonstration of the conspecificity of *H. canadensis* and *H. wingei* (Fuson et al., 1979). Since *H. wingei* is heterothallic and *H. canadensis* may be homothallic, consideration of the type of mating system as a taxonomic criterion, as noted earlier, could prove unwise.

The DNA studies have not proved entirely supportive of conclusions drawn from numerical analysis, from proton magnetic resonance, or from serology of cell wall mannans. From numerical analysis, Campbell (1973) proposed reduction of *H. wingei, H. canadensis,* and both varieties of *H. bimundalis* to a single species. DNA studies by Fuson and coworkers (1979) showed *H. canadensis* and *H. wingei* to be conspecific, but *H. wingei* and *H. bimundalis* showed little relatedness. Kurtzman (1984) demonstrated only limited DNA complementarity between the two varieties of *H. bimundalis*. Spencer and Gorin (1969b) found the proton magnetic resonance spectra of *H. beckii* and *H. canadensis* to be quite similar, but different from the pattern shared by *H. wingei* and the varieties of *H. bimundalis*. Tsuchiya et al. (1974) reported the cell surface antigens of *H. wingei, H. canadensis,* and *H. beckii* to be indistinguishable. Thus, the preceding two studies are somewhat contradictory and neither is in keeping with DNA complementarity studies which show little relatedness between *H. beckii* and *H. canadensis*. The reasons for the discrepancies are not at all clear, but Ballou (1974) has shown for *Saccharomyces* mannans that single gene changes can impart differences in immunological reaction.

The single test separating *Hansenula* from *Pichia*, the nitrate reaction, was recently shown to be unsuitable for delimiting either genera or species when it was found that *H. mrakii* and *P. sargentensis* show 68% DNA relatedness (Kurtzman et al., unpublished). The anticipated reassignment of the species in *Hansenula* is discussed under *Pichia*, Comments on the genus.

g. SPECIES NOT ACCEPTED IN THE GENUS

Species for which no strains were available:
Hansenula amylofaciens Dietrichson 1954
Hansenula ellipsoidospora Luteraan 1954
Hansenula ofunaensis Makiguchi et Asai 1976 (in Asai et al., 1976)

Species for which published descriptions were not located:
Hansenula arabitolgenes Fang nom. nud.?
Hansenula misumaiensis Sasaki et Yoshida nom. nud.?

Genus 12. *Issatchenkia* Kudriavzev

by C.P. Kurtzman

a. DIAGNOSIS OF THE GENUS

Asexual reproduction is by multilateral budding on a narrow base. Cells are spheroidal, ellipsoidal or elongate, and pseudohyphae are present. Pellicles form early on liquid media.

Asci are unconjugated or may be conjugated if formed by the pairing of complementary mating types. Asci are persistent and form one to four roughened spheroidal ascospores. The protuberances causing ascospore roughness may not be visible under the light microscope.

Sugars may be fermented.

Nitrate is not assimilated.

b. TYPE SPECIES OF THE GENUS

Issatchenkia orientalis Kudriavzev

c. SPECIES ACCEPTED IN THE GENUS

1. *Issatchenkia occidentalis* Kurtzman, Smiley et Johnson (1980b)
2. *Issatchenkia orientalis* Kudriavzev (1960)
3. *Issatchenkia scutulata* (Phaff, Miller et Miranda) Kurtzman, Smiley et Johnson (Phaff et al., 1976)
   Issatchenkia scutulata (Phaff, Miller et Miranda) Kurtzman, Smiley et Johnson var. *scutulata*
   Issatchenkia scutulata (Phaff, Miller et Miranda) Kurtzman, Smiley et Johnson var. *exigua* (Phaff, Miller et Miranda) Kurtzman, Smiley et Johnson (Phaff et al., 1976)
4. *Issatchenkia terricola* (van der Walt) Kurtzman, Smiley et Johnson (van der Walt, 1957)

d. KEY TO THE SPECIES OF THE GENUS

1a Growth in vitamin-free medium 2
 b No growth in vitamin-free medium
 I. terricola .... p. 221

2a D-Glucosamine assimilated 3
 b D-Glucosamine not assimilated 4

3a Growth at 40°C
 I. orientalis .... p. 218
 b No growth at 40°C
 I. occidentalis .... p. 216

4a Growth in osmotic medium (10% sodium chloride plus 5% glucose in yeast nitrogen base)
 I. scutulata var. *scutulata* .... p. 219
 b No growth in osmotic medium
 I. scutulata var. *exigua* .... p. 220

Table 14

Key characters of the species in the genus *Issatchenkia*

	Ferm G	Assimilation			Vit	Osm med	37°C	40°C
		DGM	Gly	Cit				
I. scutulata var. *exigua*	w/s	–	–	–	+	–	–	–
I. scutulata var. *scutulata*	+	–	+ /s	–	+	+	–	–
I. terricola	+ /w	–	+	w/ –	–	–	v	–
I. occidentalis	+	+	+	–	+	+	+	–
I. orientalis	+	+	+	+ /w	+	+	+	+

Ferm G = fermentation of glucose, DGM = D-glucosamine, Gly = glycerol, Cit = citric acid, Vit = growth in vitamin-free medium, Osm med = 10% sodium chloride plus 5% glucose in yeast nitrogen base.

e. SYSTEMATIC DISCUSSION OF THE SPECIES

1. *ISSATCHENKIA OCCIDENTALIS* KURTZMAN, SMILEY ET JOHNSON (1980b)

Imperfect state: *Candida sorbosa* Hedrick et Burke ex van Uden et Buckley

Synonym

Candida sorbosa Hedrick et Burke ex van Uden et Buckley (1970)

Standard description of *Issatchenkia occidentalis*

Growth on 5% malt extract agar: After 3 days at 25°C, the cells are ovoidal to elongate (2.2–4.3) × (3.2–9.3) μm and single or in pairs. Growth is butyrous and light cream colored.

Growth on the surface of assimilation media: Dry climbing pellicles are formed.

Dalmau plate culture on morphology agar: After 7 days at 25°C, growth under the coverglass shows abundant and moderately well branched pseudohyphae. Aerobic growth is white to tannish-white, dull, butyrous, striated, and slightly raised with a flattened center. The margin is smooth, finely serrate, or with small lobes. Colonies are fringed with pseudohyphae. A faintly acidic odor is present.

Formation of ascospores: Asci are unconjugated, persistent, and form one or two roughened ascospores (Fig. 47). Electron microscopy may be required to discern the protuberances on spore walls (Kurtzman & Smiley, 1974; Kurtzman et al., 1980b). Isolation of single spores demonstrated this species to be heterothallic (Kurtzman & Smiley, 1976). Following conjugation of complementary mating types, zygotes usually give rise to a succession of diploid cells that sporulate; consequently, conjugated asci are rarely, if ever, observed. Complementary mating types of *I. occidentalis* are NRRL Y-7552-61 (CBS 6887) and NRRL Y-7552-62 (CBS 6888).

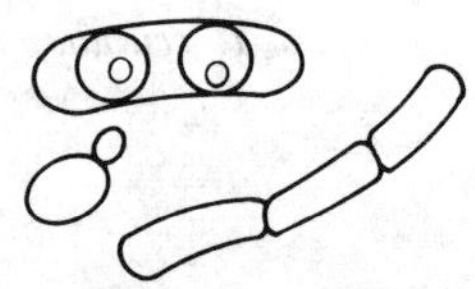

Fig. 47 – *I. occidentalis,* CBS 5459
After 1 month on a cucumber wedge

Spores were observed on cucumber and carrot wedges. Sporulation is usually poor.

Fermentation:

Glucose	+	Lactose	–
Galactose	–	Raffinose	–
Sucrose	–	Trehalose	–
Maltose	–		

Assimilation of carbon compounds:

Galactose	–	Raffinose	–	Erythritol	–
Sucrose	–	Soluble starch	–	Ribitol	–
Maltose	–	D-Xylose	–	D-Mannitol	–
Cellobiose	–	L-Arabinose	–	Succinic acid	+
Trehalose	–	D-Ribose	–	Citric acid	–
Lactose	–	L-Rhamnose	–	Inositol	–

Additional carbon compounds tested: L-Sorbose v, D-glucosamine·HCl +, ethanol +, glycerol +

Assimilation of nitrate: –

Growth in vitamin-free medium: +

Growth in 10% sodium chloride plus 5% glucose in yeast nitrogen base: +
Growth at 37°C: +; at 40°C: –
Starch formation: –
Gelatin liquefaction: –
G + C: 40.9–41.1 mol%, 2 strains (Kurtzman et al., 1980b)

Comments

Issatchenkia occidentalis was initially classified as a strain of *Pichia kudriavzevii (I. orientalis)*. Kreger-van Rij (1970c) noted that this strain differed from the others in being sorbose-positive and citrate-negative, but left it in *P. kudriavzevii* because it was otherwise similar. Kurtzman and Smiley (1976) demonstrated this strain to be heterothallic. One of the mating types conjugated with the type strain of *Candida sorbosa* and formed spores. The other strains of *P. kudriavzevii* and several strains of *Candida krusei* also conjugated with one of the mating types, but the frequency was extremely low and no spores formed.

DNA reassociation studies by Kurtzman et al. (1980b) showed *I. occidentalis* and *C. sorbosa* to have 98% base sequence complementarity, thus confirming the perfect-imperfect relationship demonstrated by mating tests. The extent of DNA reassociation among strains of *I. orientalis* and *Candida krusei* was 93–100%, but when DNA from these strains was compared with that from *I. occidentalis*, only 8% common sequences were detected. Consequently, evidence from both mating tests and DNA studies indicates the presence of two taxa, but a close relationship is suggested.

The heterothallic diploid type strain of *I. occidentalis* was shown to be heterozygous for L-sorbose assimilation; therefore, this compound is of questionable utility for separating *I. occidentalis* from *I. orientalis*.

Origin of the strains studied

Unknown substrate, Phaff, U.S.A. (1): feces of *Drosophila crucigera*, Hedrick, Hawaii, U.S.A. (1): 'tea fungus,' Boedijn, Indonesia (1). The first strain was originally identified as *P. kudriavzevii*, whereas the latter two strains were described as *C. sorbosa*.

Type strain: CBS 5459 (NRRL Y-7552), received from Phaff, U.S.A.

2. *ISSATCHENKIA ORIENTALIS* KUDRIAVZEV (1960)

Imperfect state: *Candida krusei* (Cast.) Berkhout

Synonyms

Pichia orientalis (Kudriavzev) Kreger-van Rij (1964b) [nec *Saccharomyces orientalis* Beijerinck (1898a); nec *Pichia orientalis* (Beijerinck) Guilliermond (1912)]
Pichia kudriavzevii Boidin, Pignal et Besson (1965b)

Standard description of *Issatchenkia orientalis*

Growth on 5% malt extract agar: After 3 days at 25°C, the cells are ovoidal to elongate (1.3–6.0) × (3.3–14.0) μm and single or in pairs. Growth is butyrous and light cream colored.

Growth on the surface of assimilation media: Heavy, dry climbing pellicles are formed.

Dalmau plate culture on morphology agar: After 7 days at 25°C, growth under the coverglass shows abundant and moderately branched pseudohyphae. Aerobic growth is tannish-white, dull to occasionally almost powdery, butyrous, low convex with a flattened center, margin varying from smooth to lobed, and fringed with pseudohyphae. Some cultures have a faintly acidic odor.

Formation of ascospores: Ascospores were not observed in the present study. Kudriavzev (1960) reported the spores to be spherical, smooth under the light microscope, and one per ascus. Asci were persistent. Two of the three strains observed by Kreger-van Rij (1970c) had occasional asci with one to two spherical ascospores when grown on V8 agar.

Fermentation:

Glucose	+	Lactose	–
Galactose	–	Raffinose	–
Sucrose	–	Trehalose	–
Maltose	–		

Assimilation of carbon compounds:

Galactose	–	Raffinose	–	Erythritol	–
Sucrose	–	Soluble starch	–	Ribitol	–
Maltose	–	D-Xylose	–	D-Mannitol	–
Cellobiose	–	L-Arabinose	–	Succinic acid	+
Trehalose	–	D-Ribose	–	Citric acid	+ or w
Lactose	–	L-Rhamnose	–	Inositol	–

Additional carbon compounds tested: L-Sorbose –, D-glucosamine·HCl +, ethanol +, glycerol +

Assimilation of nitrate: –

Growth in vitamin-free medium: +

Growth in 10% sodium chloride plus 5% glucose in yeast nitrogen base: +

Growth at 37°C: +; **at 40°C:** +

Starch formation: –

Gelatin liquefaction: – or +w

G + C: 40.1–40.3 mol%, 3 strains (Kurtzman et al., 1980b)

Comments

Candida krusei is considered to represent the imperfect form of *I. orientalis* because the type strains, as well as other isolates of both species, showed significant (93–100%) DNA base sequence complementarity (Kurtzman et al., 1980b). *Candida requinyii* also showed significant base sequence relatedness with *C. krusei* and *I. orientalis* but not with

I. occidentalis. Lack of mating competence has prevented demonstration of the perfect–imperfect relationship by genetic studies. The relationship between *I. orientalis* and *I. occidentalis* is discussed in the section on the latter species.

Origin of the strains studied

Fruit juices or berries, Kudriavzev, U.S.S.R. (1); tea beer, Netherlands (1); unknown, Suomalainen, Finland (1).

Type strain: CBS 5147 (NRRL Y-5396), received from Kudriavzev, U.S.S.R.

3. *ISSATCHENKIA SCUTULATA* (PHAFF, MILLER ET MIRANDA) KURTZMAN, SMILEY ET JOHNSON (1980b)

This species has two varieties:

Issatchenkia scutulata (Phaff, Miller et Miranda) Kurtzman, Smiley et Johnson var. *scutulata*

Issatchenkia scutulata (Phaff, Miller et Miranda) Kurtzman, Smiley et Johnson var. *exigua* (Phaff, Miller et Miranda) Kurtzman, Smiley et Johnson (1980b)

Synonym of the variety *scutulata*

Pichia scutulata Phaff, Miller et Miranda var. *scutulata* (1976)

Synonym of the variety *exigua*

Pichia scutulata Phaff, Miller et Miranda var. *exigua* Phaff, Miller et Miranda (1976)

Standard description of *Issatchenkia scutulata* var. *scutulata*

Growth on 5% malt extract agar: After 3 days at 25°C, the cells are ovoidal to elongate (2.9–5.5) × (3.5–10.0) μm and single or in short chains. Growth is butyrous and tannish-white in color.

Growth on the surface of assimilation media: Moderate to heavy, dry climbing pellicles are formed.

Dalmau plate culture on morphology agar: After 7 days at 25°C, growth under the coverglass is composed of moderately branched pseudohyphae. Aerobic growth is tannish-white, butyrous, low convex with a depressed center, and with a finely lobed margin. There is a faintly acidic odor.

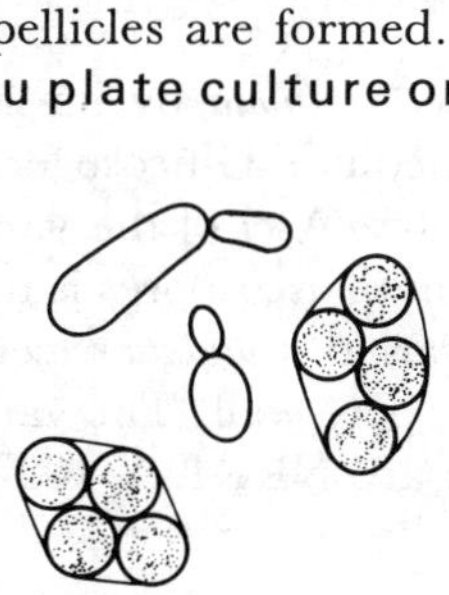

Fig. 48 – *I. scutulata* var. *scutulata*, CBS 6644
After 2 weeks on YM agar

Formation of ascospores: All natural isolates have been diploid with unconjugated asci. There are usually four (one to four) spheroidal spores per ascus and they frequently group into diamond-shaped tetrads (Fig. 48). Asci are persistent. The species is heterothallic and four-spored asci give two spores of each mating type. Asci arising from a mixture of complementary mating types are generally unconjugated because of proliferation of the diplophase before sporulation. Ascospore roughness is visible only under the electron microscope and results from small protuberances originating in the inner spore wall (Kurtzman et al., 1980b). Complementary mating types of *I. scutulata* var. *scutulata* are NRRL Y-7663-2 (CBS 6672) and NRRL Y-7663-3 (CBS 6673).

Spores were observed on YM- and 5% Difco malt extract agar.

Fermentation:

Glucose	+	Lactose	–
Galactose	–	Raffinose	–
Sucrose	–	Trehalose	–
Maltose	–		

Assimilation of carbon compounds:

Galactose	–	Raffinose	–	Erythritol	–
Sucrose	–	Soluble starch	–	Ribitol	–
Maltose	–	D-Xylose	–	D-Mannitol	–
Cellobiose	–	L-Arabinose	–	Succinic acid	+
Trehalose	–	D-Ribose	–	Citric acid	–
Lactose	–	L-Rhamnose	–	Inositol	–

Additional carbon compounds tested: L-Sorbose –, D-glucosamine · HCl –, ethanol +, glycerol + or s

Assimilation of nitrate: –

Growth in vitamin-free medium: +

Growth in 10% sodium chloride plus 5% glucose in yeast nitrogen base: +

Growth at 37°C: –

Starch formation: –

Gelatin liquefaction: –

G + C: 32.6–33.0 mol%, 3 strains of the variety *scutulata*; 32.4–32.6 mol%, 3 strains of the variety *exigua* (Phaff et al., 1976)

Supplementary description of *Issatchenkia scutulata* var. *exigua*

The variety *exigua* can be separated from the variety *scutulata* by its inability to assimilate glycerol and by its failure to grow in osmotic medium. Glucose fermentation is slow or weak. Morphologically, the varieties are similar. Asci of the variety *exigua* are frequently diamond-shaped, but a linear arrangement of ascospores is not uncommon. Ascospore ultrastructure is similar to that of the variety *scutulata* except that the protuberances seem to arise from the outer layer of the spore wall. The variety *exigua* is also heterothallic, and complementary mating types are NRRL Y-10920-3 (CBS 8067) and NRRL Y-10920-4 (CBS 8068).

Comments

Phaff et al. (1976) separated the two varieties because of the physiological differences enumerated above and on the basis of geographically separate habitats.

Kurtzman et al. (1980b) determined the extent of nuclear DNA base sequence complementarity between the two varieties to be only 25%, and this would suggest that they be regarded as separate species (Price et al., 1978, and references cited therein). However, intervarietal matings gave asci with spores that appeared normal and a small number of the spores proved viable. Sib-matings between these progeny gave viable spores. On the basis of low DNA complementarity and the reduced fertility of intervarietal crosses, the variety designations have been maintained.

Origin of the strains belonging to the variety *scutulata*

Slime flux of *Myoporum sandwicense,* Hawaii, U.S.A. (3).

Type strain: CBS 6644 (UCD-FST 71-102, NRRL Y-7663) received from Phaff.

Origin of the strains belonging to the variety *exigua*

Insect boring in *Picea sitchensis,* Washington, U.S.A. (1); slime flux of *Populus trichocarpa,* British Columbia, Canada (1); slime flux of *Salix* sp., British Columbia, Canada (1).

Type strain: CBS 6836 (UCD-FST 68-979B1, NRRL Y-10920), isolated from an insect boring in *Picea sitchensis,* Washington, U.S.A.; received from Phaff.

4. *ISSATCHENKIA TERRICOLA* (VAN DER WALT) KURTZMAN, SMILEY ET JOHNSON (1980b)

Synonyms

Pichia terricola van der Walt (1957)
Saccharomyces terricolus (van der Walt) Novák et Zsolt (1961)

Standard description of *Issatchenkia terricola*

Growth on 5% malt extract agar: After 3 days at 25°C, the cells are ovoidal to elongate (2.4–5.8) × (3.0–14.7) μm and single or in pairs. Growth is butyrous and tannish-cream in color.

Growth on the surface of assimilation media: Moderate to heavy, dry climbing pellicles are formed.

Dalmau plate culture on morphology agar: After 7 days at 25°C, growth under the coverglass shows poorly to moderately branched pseudohyphae. Aerobic growth is white to tannish-white, dull to faintly glistening, butyrous, low convex with a flattened center, and a lobed margin. The odor is faintly acidic.

Formation of ascospores: Natural isolates have been diploid, and asci are unconjugated. There are one to four, but usually two, spheroidal spores per ascus. Asci are persistent. Ascospores are roughened as a result of protuberances which originate in the outer spore wall, and this roughness can usually be perceived in the light microscope (Fig. 49). Kurtzman and Smiley (1976) demonstrated this species to be heterothallic. During laboratory cultivation, diploid cultures frequently become haploid to the exclusion of one mating type, and the type strain as well as several other cultures are now asporogenous haploids. Asci arising from a mixture of complementary mating types are usually unconjugated because of the proliferation of diploid cells before sporulation. Complementary mating types of *I. terricola* are NRRL YB-4310 (CBS 2617) and NRRL Y-7549-25 (CBS 6893).

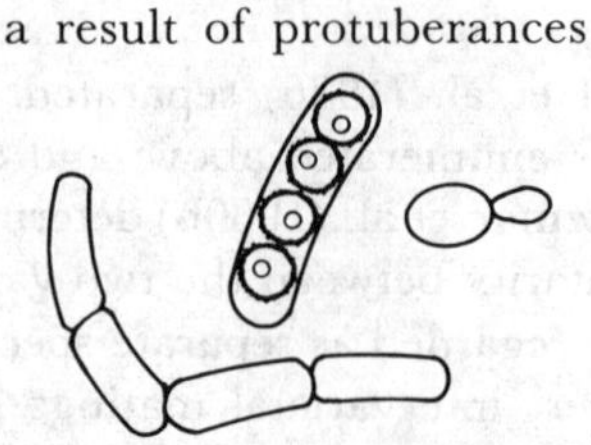

Fig. 49 – *I. terricola*
CBS 2617 × CBS 6893
After 10 days on YM agar

Spores were observed on YM-, 5% Difco malt extract-, and V8 agar.

Fermentation:

Glucose	+ or w	Lactose	–
Galactose	+	Raffinose	–
Sucrose	–	Trehalose	–
Maltose	–		

Assimilation of carbon compounds:

Galactose	–	Raffinose	–	Erythritol	–
Sucrose	–	Soluble starch	–	Ribitol	–
Maltose	–	D-Xylose	–	D-Mannitol	–
Cellobiose	–	L-Arabinose	–	Succinic acid	+
Trehalose	–	D-Ribose	–	Citric acid	+ w or –
Lactose	–	L-Rhamnose	–	Inositol	–

Additional carbon compounds tested: L-Sorbose –, D-glucosamine · HCl –, ethanol +, glycerol +

Assimilation of nitrate: –

Growth in vitamin-free medium: –

Growth in 10% sodium chloride plus 5% glucose in yeast nitrogen base: –

Growth at 37°C: v; **at 40°C:** –

Starch formation: –

Gelatin liquefaction: –

G + C: 37.9–38.2 mol%, 2 strains (Kurtzman et al., 1980b)

Origin of the strains studied

Soil, South Africa (1); soil, Italy (1); soil, Louisiana, U.S.A. (2); cherry juice (1); dregs of pressed grapes, Italy (1); sea water (1); spoiled figs, U.S.A. (1); old wine (1).

Type strain: CBS 2617 (NRRL YB-4310), soil isolate received from van der Walt, South Africa.

f. COMMENTS ON THE GENUS

Kudriavzev (1960) described the monotypic genus *Issatchenkia* on the basis of certain strains isolated from berries and fruit juices. The cultures formed spores that were spherical and smooth when observed with the light microscope. Asci were persistent and contained one spore. Kreger-van Rij (1964b) transferred the single species to the genus *Pichia* as *P. orientalis,* but Boidin et al. (1965b) pointed out that the name had been used previously and proposed the new name *Pichia kudriavzevii.*

Pichia terricola and *P. scutulata* were two other species phenotypically similar to *P. kudriavzevii.* Scanning electron microscopy showed all three to have warty ascospore walls, and in this respect, to differ from other round-spored species of *Pichia* which have essentially smooth spore walls (Kurtzman & Smiley, 1974; Phaff et al., 1976). Despite their similarity in ascospore surface structure to species of *Debaryomyces* and *Torulaspora* (Kurtzman et al., 1975), these three species do not exhibit the mother–daughter conjugation typical of *Debaryomyces* or the protuberances on asci that are associated with sporulation in *Torulaspora.* Furthermore, all species now assigned to *Issatchenkia* have Q-7 ubiquinone in the electron transport system while all *Saccharomyces* and *Torulaspora* species have Q-6 ubiquinone. Species of *Debaryomyces* and the remainder of the round-spored *Pichia* species have Q-9 ubiquinone (Yamada et al., 1973c, 1976b, 1977; Yamada, personal communication). Asci of the round-spored species now remaining in *Pichia* are either dehiscent or show mother–daughter conjugation or have both of these characteristics. Thus, the four species now classified in *Issatchenkia* appear to represent a natural grouping on the basis of morphological and molecular criteria (Kurtzman et al., 1980b). Spores of *I. orientalis* are not well characterized, but emendation of the genus was possible through comparison of the other sporogenous species included in the genus. Transmission electron microscopy revealed some differences in spore wall layering among the species, but considering their other similarities, ascospore wall layering was not considered to be a cardinal point for generic assignment (Kurtzman et al., 1980b).

Ascospores of *Saccharomyces telluris* van der Walt have surface topography and internal wall layering much like that shown by the majority of *Issatchenkia* species (Kurtzman, unpublished). However, little is known about the sequence of ascus formation by this species, and because it has a Q-6 ubiquinone system, it was not transferred to *Issatchenkia. Sacch. telluris* may be separated from known species of *Issatchenkia* on the basis of its extremely limited ability to assimilate carbon compounds.

Genus 13. *Kluyveromyces* van der Walt emend. van der Walt

by J.P. van der Walt and E. Johannsen

a. DIAGNOSIS OF THE GENUS

Vegetative cells spheroidal, ovoid, ellipsoidal, cylindrical to elongate, reproducing by budding.

Asci conjugated or unconjugated, evanescent, uni- to multispored. Ascospores crescentiform, reniform, oblong with obtuse ends, ellipsoidal or spheroidal, tending to agglutinate after liberation.

Fermentation.

Nitrate not utilized as sole source of nitrogen.

Coenzyme Q-6 system present.

External vitamin source required.

b. TYPE SPECIES OF THE GENUS

Kluyveromyces polysporus van der Walt

c. SPECIES ACCEPTED IN THE GENUS

1. *Kluyveromyces aestuarii* (Fell) van der Walt (Fell, 1961)
2. *Kluyveromyces africanus* van der Walt (1956b)
3. *Kluyveromyces blattae* Henninger et Windisch (1976a)
4. *Kluyveromyces delphensis* (van der Walt et Tscheuschner) van der Walt (van der Walt & Tscheuschner, 1956b)
5. *Kluyveromyces lodderi* (van der Walt et Tscheuschner) van der Walt (van der Walt & Tscheuschner, 1957a)
6. *Kluyveromyces marxianus* (Hansen) van der Walt (Hansen, 1888b)

6a. *Kluyveromyces marxianus* (Hansen) van der Walt var. *marxianus*

6b. *Kluyveromyces marxianus* (Hansen) van der Walt var. *bulgaricus* (Santa María) Johannsen et van der Walt (Santa María, 1956b)

6c. *Kluyveromyces marxianus* (Hansen) van der Walt var. *dobzhanskii* (Shehata, Mrak et Phaff) Johannsen et van der Walt (Shehata et al., 1955)

6d. *Kluyveromyces marxianus* (Hansen) van der Walt var. *drosophilarum* (Shehata, Mrak et Phaff) Johannsen et van der Walt (Shehata et al., 1955)

6e. *Kluyveromyces marxianus* (Hansen) van der Walt var. *lactis* (Dombrowski) Johannsen et van der Walt (Dombrowski, 1910)
6f. *Kluyveromyces marxianus* (Hansen) van der Walt var. *vanudenii* (van der Walt et Nel) Johannsen et van der Walt (van der Walt & Nel, 1963)
6g. *Kluyveromyces marxianus* (Hansen) van der Walt var. *wikenii* (van der Walt, Nel et van Kerken) Johannsen et van der Walt (van der Walt et al., 1966)
7. *Kluyveromyces phaffii* (van der Walt) van der Walt (van der Walt, 1963)
8. *Kluyveromyces polysporus* van der Walt (1956a)
9. *Kluyveromyces thermotolerans* (Filippov) Yarrow (Filippov, 1932)
10. *Kluyveromyces waltii* Kodama (1974)
11. *Kluyveromyces wickerhamii* (Phaff, Miller et Shifrine) van der Walt (Phaff et al., 1956)

d. KEY TO THE SPECIES AND VARIETIES OF THE GENUS

1a Lactose assimilated, occasionally slowly ... 2
b Lactose not assimilated ... 6

2a Ascospores crescentiform to reniform ... 3
b Ascospores spheroidal to ellipsoidal ... 4

3a Raffinose assimilated
K. marxianus var. *marxianus* .... p. 234
b Raffinose not assimilated
K. wickerhamii .... p. 250

4a Inulin fermented, occasionally slowly
K. marxianus var. *bulgaricus* .... p. 236
b Inulin not fermented ... 5

5a Growth in the presence of 100 ppm cycloheximide; lactose fermented
K. marxianus var. *lactis* .... p. 241
b No growth in the presence of 100 ppm cycloheximide; lactose not fermented
K. aestuarii .... p. 228

6a Ascospores crescentiform to reniform ... 7
b Ascospores spheroidal to ellipsoidal ... 13

7a Sucrose fermented ... 8
b Sucrose not fermented ... 11

8a Mannitol assimilated ... 9
b Mannitol not assimilated ... 10

9a Maltose fermented
K. marxianus var. *dobzhanskii* .... p. 238
b Maltose not fermented
K. marxianus var. *drosophilarum* .... p. 239

10a Asci multispored, large
K. polysporus .... p. 246
b Asci one- to four-spored
K. lodderi .... p. 232

11a Galactose fermented 12
b Galactose not fermented
K. delphensis .... p. 231

12a Asci multispored (one to 16 or more ascospores)
K. africanus .... p. 229
b Asci one- to four-spored
K. phaffii .... p. 245

13a Sucrose fermented 14
b Sucrose not fermented
K. blattae .... p. 230

14a Galactose assimilated 15
b Galactose not assimilated
K. waltii .... p. 249

15a Cellobiose assimilated
K. marxianus var. *vanudenii* .... p. 242
b Cellobiose not assimilated 16

16a Maltose assimilated; growth on 50% glucose agar
K. thermotolerans .... p. 248
b Maltose not assimilated; no growth on 50% glucose agar
K. marxianus var. *wikenii* .... p. 244

Table 15

Salient characteristics of the species and varieties in the genus *Kluyveromyces*

	Ascospores	Fermentation						Assimilation								mol% G + C*
		G	Ga	Su	Ma	La	Ra	Ga	Su	Ma	Ce	Tr	La	Ra	Ri	
K. delphensis	1–4, R	+	–	–	–	–	–	–	–	–	–	+ s	–	–	–	40.2
K. phaffii	1–4, R	+	+	–	–	–	–	+	–/+w	–	–	+ w	–	–	–	35.3
K. africanus	1–16+, R	+	+	–	–	–	–	+	–/+w	–	–	–/+w	–	–	–	38.5
K. blattae	1–8+ , S	+	+	–	–	–	–	+	–	–	–	–	–	–	+	33.1
K. waltii	1–4, S	+	–	+	–	–	+	–	+	–	–	–	–	+	–	45.6**
K. lodderi	1–4, R	+	+	+	–	–	+	+	+	–	–	+	–	+	–	35.4 – 35.7
K. polysporus	numerous, R	+	+	+	–	–	+	+	+	–	–	+ s	–	+	–	35.3
K. thermotolerans	1–4, S	+	v	+	v	–	+	+	+	+	–	+	–	+	–	46.2
K. wickerhamii	1–4, R	+	+ s	+ s	–	–	–	+	+	–	+	v	+	–	–	41.9 – 42.4
K. aestuarii	1–4, S	+	v	+	–	–	+	+	+	v	+	v	+	+	–	39.8 – 39.9
K. marxianus																
var. *wikenii*	1–4, S	+	+	+	–	–	+	+	+	–	–	–	–	+	–	41.1 – 41.2
var. *drosophilarum*	1–4, R	+	+	+	–	–	+	+	+	v	v	v	–	+	–	39.9 – 40.9
var. *vanudenii*	1–4, S	+	+	+	v	–	+	+	+	v	+	v	–	+	–	40.1 – 40.3
var. *dobzhanskii*	1–4, R	+	+ s	+	+	–	+	+	+	+	+	+	–	+	–	42.5 – 42.6
var. *marxianus*	1–4, R	+	+/s	+	–	v	+	+	+	–	+	–	+/s	+	v	41.2 – 41.3
var. *bulgaricus*	1–4, S	+	+	+	–	v	+	+	+	–	+ (s)	–/+w	+	+	–/+w	41.3 – 41.5
var. *lactis*	1–4, S (H)	+	+	v	v	+	v	+	+	v	+	+	+	v	–	40.3

G = glucose, Ga = galactose, Su = sucrose, Ma = maltose, Ce = cellobiose, Tr = trehalose, La = lactose, Ra = raffinose, Ri = D-ribose, S = spheroidal to ellipsoidal, R = reniform to crescentiform, (H) = heterothallic. * Data from Phaff (1978, personal communication). ** Data from Kodama (1974).

e. SYSTEMATIC DISCUSSION OF THE SPECIES

1. *KLUYVEROMYCES AESTUARII* (FELL) VAN DER WALT (1971)

Synonyms

Saccharomyces aestuarii Fell (1961)
Kluyveromyces aestuarii (Fell) van der Walt (1965c) nom. inval.

Standard description of *Kluyveromyces aestuarii* (van der Walt, 1970c)

Growth in malt extract: After 3 days at 28°C the cells are spheroidal to ellipsoidal, (2.0–5.5)×(2.5–6.5) μm, reproducing by budding, and occur singly or in pairs. A slight ring and sediment and occasionally islets or a thin pellicle are formed. After one month at room temperature a ring and sediment are present.

Growth on malt agar: After one month at room temperature the streak culture is butyrous, greyish-white, occasionally tinged with brown, shiny to dull, flat, spreading, smooth or verruculose towards the center; the margin is smooth, undulating to lobiform.

Dalmau plate culture on corn meal agar: Tree-like structures or a rudimentary pseudomycelium may be formed. It is often better developed under anaerobic conditions.

Formation of ascospores: Conjugation usually immediately precedes ascus formation. The asci are evanescent. One to four spheroidal to ellipsoidal ascospores are formed per ascus. Released ascospores tend to agglutinate.

Sporulation is observed on 2% malt extract agar, YM agar and the commonly employed sporulation media.

Fermentation:

Glucose	+	Lactose	–
Galactose	v	Raffinose	+
Sucrose	+	Inulin	–
Maltose	–		

Assimilation of carbon compounds:

Galactose	+	Raffinose	+	Erythritol	–
Sucrose	+	Soluble starch	–	Ribitol	–
Maltose	v	D-Xylose	– or +w	D-Mannitol	+
Cellobiose	+	L-Arabinose	–	Succinic acid	+
Trehalose	v	D-Ribose	–	Citric acid	–
Lactose	+	L-Rhamnose	–	Inositol	–

Assimilation of nitrate: –

Growth in vitamin-free medium: –

Growth on 50% (w/w) glucose-yeast extract agar: –

Growth at 37°C: –

Growth in the presence of 100 ppm cycloheximide: –

G + C: 39.5 mol%, one strain (Martini et al., 1972); 35.4–35.6 mol%, one strain (Poncet & Fiol, 1972); 39.8–39.9 mol%, 2 strains (Phaff, 1978; personal communication)

Origin of the strains studied

Estuarine mud from Biscayne Bay (1) (Fell, 1961); sea water from the Torres Strait area (1) (van Uden & Zobell, 1962).

Type strain: CBS 4438, isolated by Fell from estuarine mud.

2. *KLUYVEROMYCES AFRICANUS* VAN DER WALT (1956b)

Standard description of *Kluyveromyces africanus* (van der Walt, 1970c)

Growth in malt extract: After 3 days at 28°C the cells are ellipsoidal to almost cylindrical, (1.5–5.0) × (2.5–8.5) μm, reproducing by budding, and occur singly, in pairs or occasionally in small clusters or short chains. Conjugating cells may be present. A sediment is formed. After one month at room temperature only a sediment is present.

Growth on malt agar: After one month at room temperature the streak culture is butyrous, brownish-cream, smooth, flat, glistening. The margin is entire to somewhat lobiform.

Dalmau plate culture on corn meal agar: No pseudomycelium is formed. A few chains of cells may be formed.

Formation of ascospores: Diploid vegetative cells are apparently transformed into evanescent asci; one to 16 or more ascospores are formed per ascus. The ascospores are stoutly reniform to oblong. Released ascospores tend to agglutinate (Fig. 50).

Sporulation is observed on 2% malt extract agar, YM agar and the commonly employed sporulation media.

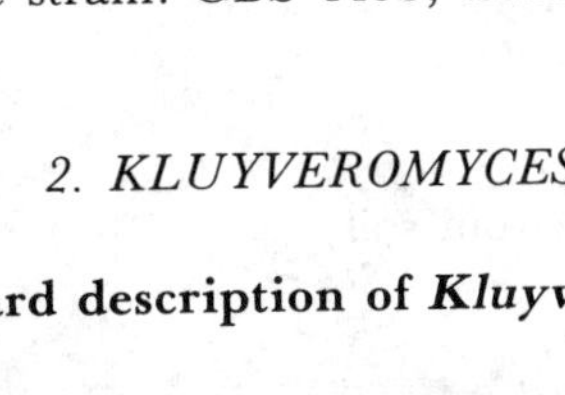

Fig. 50 – *K. africanus*
After 3 days on YM agar
(From 'The Yeasts', ed. II, 1970)

Fermentation:

Glucose	+	Maltose	–
Galactose	+	Lactose	–
Sucrose	–		

Assimilation of carbon compounds:

Galactose	+	Raffinose	–	Erythritol	–
Sucrose	– or +w	Soluble starch	–	Ribitol	–
Maltose	–	D-Xylose	–	D-Mannitol	–
Cellobiose	–	L-Arabinose	–	Succinic acid	–
Trehalose	– or +w	D-Ribose	–	Citric acid	–
Lactose	–	L-Rhamnose	–	Inositol	–

Assimilation of nitrate: –

Growth in vitamin-free medium: –

Growth on 50% (w/w) glucose-yeast extract agar: –

Growth at 37°C: –

G + C: 38.0 mol%, one strain (Martini et al., 1972); 35.4–35.6 mol%, one strain (Poncet & Fiol, 1972); 38.5 mol%, one strain (Phaff, 1978; personal communication)

Origin of the strains studied

Soil from Southern Africa (4) (van der Walt, 1956b).

Type strain: CBS 2517, isolated by van der Walt from soil.

3. *KLUYVEROMYCES BLATTAE* HENNINGER ET WINDISCH (1976a)

Standard description of *Kluyveromyces blattae*

Growth in malt extract: After 3 days at 28°C the cells are subglobose to ellipsoidal, (3.0–9.0) × (4.5–10.5) μm, reproducing by budding, and occur singly or in pairs. A sediment is formed. After one month only a sediment is present.

Growth on malt agar: After one month at room temperature the streak culture is butyrous, creamish-brown, shiny, flat, slightly raised along the center, smooth with some structure along the margin. The margin is entire, undulating to lobiform.

Dalmau plate culture on corn meal agar: No pseudomycelium is formed.

Formation of ascospores: Vegetative diploid cells are directly transformed into asci. The asci are evanescent. One to eight or more spheroidal to ellipsoidal ascospores are formed per ascus. Released ascospores tend to agglutinate.

Sporulation is observed on 2% malt extract agar and McClary's acetate agar.

Fermentation:

Glucose	+	Maltose	–
Galactose	+	Lactose	–
Sucrose	–		

Assimilation of carbon compounds:

Galactose	+	Raffinose	–	Erythritol	–
Sucrose	–	Soluble starch	–	Ribitol	–
Maltose	–	D-Xylose	–	D-Mannitol	–
Cellobiose	–	L-Arabinose	–	Succinic acid	–
Trehalose	–	D-Ribose	+	Citric acid	–
Lactose	–	L-Rhamnose	–	Inositol	–

Assimilation of nitrate: –

Growth in vitamin-free medium: –

Growth on 50% (w/w) glucose-yeast extract agar: –

Growth at 37°C: +w

G + C: 33.1 mol%, one strain (Phaff, 1978; personal communication)

Origin of the strains studied

Intestinal tract of *Blatta orientalis* (2) (Henninger & Windisch, 1976a).

Type strain: CBS 6284, isolated by Henninger and Windisch from *Blatta orientalis*.

4. *KLUYVEROMYCES DELPHENSIS* (VAN DER WALT ET TSCHEUSCHNER) VAN DER WALT (1971)

Synonyms

Saccharomyces delphensis van der Walt et Tscheuschner (1956b)
Dekkeromyces delphensis (van der Walt et Tscheuschner) Novák et Zsolt (1961) nom. illegit., comb. non valid.
Guilliermondella delphensis (van der Walt et Tscheuschner) Boidin, Abadie, Jacob et Pignal (1962)
Kluyveromyces delphensis (van der Walt et Tscheuschner) van der Walt (1965c) nom. inval.

Standard description of *Kluyveromyces delphensis* (van der Walt, 1970c)

Growth in malt extract: After 3 days at 28°C the cells are ellipsoidal, $(1.0–4.5) \times (1.5–5.5)$ μm, reproducing by budding, and occur singly, in pairs or short chains or small clusters (Fig. 51). A sediment and very slight ring are formed. After one month at room temperature a sediment and ring are present.

Growth on malt agar: After one month at room temperature the streak culture is butyrous, yellowish-grey, shiny, flat, rather spreading, somewhat raised in the center with some structure in the margin. The margin is undulating to lobiform.

Dalmau plate culture on corn meal agar: No pseudomycelium is formed.

Formation of ascospores: Conjugation as a rule immediately precedes ascus formation. The asci are evanescent. One to four reniform ascospores are formed per ascus. Released ascospores tend to agglutinate (Fig. 52).

Ascospore formation is observed on 2% malt extract agar, YM agar and the commonly employed sporulation media.

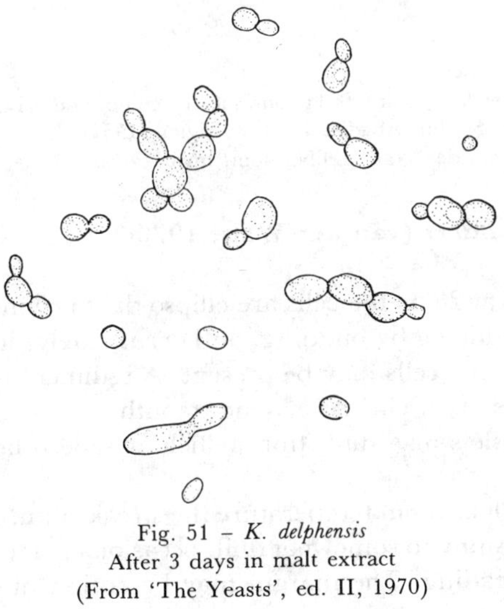

Fig. 51 – *K. delphensis*
After 3 days in malt extract
(From 'The Yeasts', ed. II, 1970)

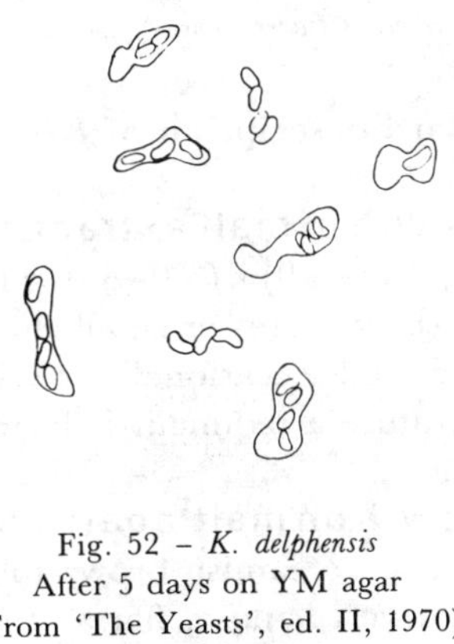

Fig. 52 – *K. delphensis*
After 5 days on YM agar
(From 'The Yeasts', ed. II, 1970)

Fermentation:

Glucose	+	Maltose	–
Galactose	–	Lactose	–
Sucrose	–		

Assimilation of carbon compounds:

Galactose	–	Raffinose	–	Erythritol	–
Sucrose	–	Soluble starch	–	Ribitol	–
Maltose	–	D-Xylose	–	D-Mannitol	–
Cellobiose	–	L-Arabinose	–	Succinic acid	–
Trehalose	+ s	D-Ribose	–	Citric acid	–
Lactose	–	L-Rhamnose	–	Inositol	–

Assimilation of nitrate: –

Growth in vitamin-free medium: –

Growth on 50% (w/w) glucose-yeast extract agar: + w

Growth at 37°C: +

G + C: 39.8 mol%, one strain (Martini et al., 1972); 36.3 mol%, one strain (Poncet & Fiol, 1972); 40.2 mol%, one strain (Phaff, 1978; personal communication)

Origin of the strain studied

Dried figs (1) (van der Walt & Tscheuschner, 1956b).

Type strain: CBS 2170, isolated by van der Walt and Tscheuschner from dried figs.

5. *KLUYVEROMYCES LODDERI* (VAN DER WALT ET TSCHEUSCHNER) VAN DER WALT (1971)

Synonyms

Saccharomyces lodderi van der Walt et Tscheuschner (1957a)
Dekkeromyces lodderi (van der Walt et Tscheuschner) Novák et Zsolt (1961) nom. illegit., comb. non valid.
Guilliermondella lodderi (van der Walt et Tscheuschner) Boidin, Abadie, Jacob et Pignal (1962)
Kluyveromyces lodderi (van der Walt et Tscheuschner) van der Walt (1965c) nom. inval.

Standard description of *Kluyveromyces lodderi* (van der Walt, 1970c)

Growth in malt extract: After 3 days at 28°C the cells are ellipsoidal to cylindrical, (2.5–6.0) × (3.0–8.0–11.0) μm, reproducing by budding, and occur singly, in pairs, short chains or small clusters. Conjugating cells may be present. A sediment is formed and occasionally a ring and islets as well. After one month at room temperature a sediment is formed. A ring, islets or a dull, thin pellicle may also be present.

Growth on malt agar: After one month at room temperature the streak culture is butyrous, creamish-brown to buff-colored, shiny to somewhat dull, occasionally sectored or verruculose, flat and somewhat spreading. The margin may be entire, undulating or lobiform.

Dalmau plate culture on corn meal agar: A rudimentary pseudomycelium is usually formed which is frequently better developed under anaerobic conditions.

Formation of ascospores: Conjugation immediately precedes ascus formation or diploid, vegetative cells may be directly transformed into asci. The asci are evanescent. One to four ascospores are formed per ascus. The ascospores are stoutly reniform or oblong with obtuse ends. Released ascospores tend to agglutinate.

Sporulation is observed on 2% malt extract agar, YM agar and the commonly employed sporulation media.

Fermentation:

Glucose	+	Maltose	–
Galactose	+	Lactose	–
Sucrose	+	Raffinose	+

Assimilation of carbon compounds:

Galactose	+	Raffinose	+	Erythritol	–
Sucrose	+	Soluble starch	–	Ribitol	–
Maltose	–	D-Xylose	–	D-Mannitol	–
Cellobiose	–	L-Arabinose	–	Succinic acid	– or +w
Trehalose	+	D-Ribose	–	Citric acid	–
Lactose	–	L-Rhamnose	–	Inositol	–

Assimilation of nitrate: –

Growth in vitamin-free medium: –

Growth on 50% (w/w) glucose-yeast extract agar: –

Growth at 37°C: –

G + C: 37.7 mol%, one strain (Martini et al., 1972); 34.4–34.6 mol%, one strain (Poncet & Fiol, 1972); 35.4–35.7 mol%, 2 strains (Phaff, 1978; personal communication)

Origin of the strains studied

South African soils (8) (van der Walt & Tscheuschner, 1957a).

Type strain: CBS 2757, isolated by van der Walt and Tscheuschner from soil.

6. *KLUYVEROMYCES MARXIANUS* (HANSEN) VAN DER WALT (1971)

K. marxianus, delimited on the basis of interfertility, comprises seven syngamous varieties (Johannsen, 1980). They will be described in the same way as species.

6a. *KLUYVEROMYCES MARXIANUS* (HANSEN) VAN DER WALT VAR. *MARXIANUS*

Imperfect state: *Candida kefyr* (Beijerinck) van Uden et Buckley

Synonyms

Saccharomyces marxianus Hansen (1888b)
Zygosaccharomyces marxianus (Hansen) Guilliermond et Negroni (1929)
Fabospora marxiana (Hansen) Kudriavzev (1960) nom. illegit. non rite public.
Dekkeromyces marxianus (Hansen) Novák et Zsolt (1961) nom. illegit., comb. non valid.
Guilliermondella marxiana (Hansen) Boidin, Abadie, Jacob et Pignal (1962)
Kluyveromyces marxianus (Hansen) van der Walt (1965c) nom. inval.
Kluyveromyces marxianus (Hansen) van der Walt (1971)
Saccharomyces kefir Beijerinck (1889b)
Saccharomyces fragrans Beijerinck (1895)
Saccharomyces muciparus Beijerinck (1908)
Saccharomyces fragilis Jörgensen (1909)
Fabospora fragilis (Jörgensen) Kudriavzev (1960) nom. illegit. non rite public.
Dekkeromyces fragilis (Jörgensen) Novák et Zsolt (1961) nom. illegit., comb. non valid.
Guilliermondella fragilis (Jörgensen) Boidin, Abadie, Jacob et Pignal (1962)
Dekkeromyces fragilis (Jörgensen) Santa María et Sanchez (1970) nom. illegit., comb. non valid.
Kluyveromyces fragilis (Jörgensen) van der Walt (1965c) nom. inval.
Kluyveromyces fragilis (Jörgensen) van der Walt (1971)
Zygosaccharomyces ashbyi Cordro'ch (1937) nom illegit. public. sine diagn. latine
Saccharomyces macedoniensis Diddens et Lodder (1939)
Fabospora macedoniensis (Diddens et Lodder) Kudriavzev (1960) nom. illegit. non rite public.
Dekkeromyces macedoniensis (Diddens et Lodder) Novák et Zsolt (1961) nom. illegit., comb. non valid.
Saccharomyces chevalieri (Guilliermond) var. *atypicus* Dietrichson (1954) nom. illegit. public. sine diagn. latine
Hansenula pozolis Herrera, Ulloa et Fuentes (1973)

Standard description of *Kluyveromyces marxianus* var. *marxianus* (van der Walt, 1970c)

Growth in malt extract: After 3 days at 28°C the cells are subglobose, ellipsoidal to cylindrical, (2.0–6.0) × (3.0–10.0) μm, reproducing by budding, and occur singly, in pairs or occasionally in short chains (Fig. 53). Pseudomycelium is usually formed to a greater or lesser extent. A sediment is formed and occasionally a slight ring as well. After one month a sediment is present. A ring and occasionally islets or a slight pellicle may be present.

Growth on malt agar: After one month at room temperature the streak culture is butyrous, cream-colored to brownish-cream, occasionally somewhat pinkish, flat and rather spreading, occasionally somewhat raised in the center, smooth or somewhat striated or sectored, shiny or occasionally somewhat dull. The margin may be undulating or lobiform, occasionally fringed with pseudomycelium.

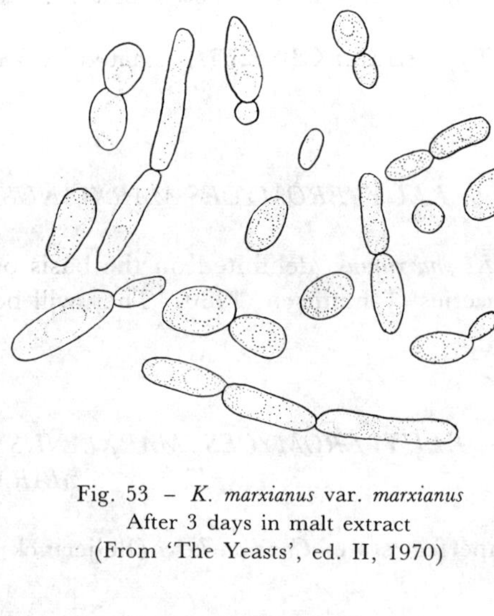

Fig. 53 – *K. marxianus* var. *marxianus*
After 3 days in malt extract
(From 'The Yeasts', ed. II, 1970)

Dalmau plate culture on corn meal agar: Pseudomycelium is usually abundantly formed and is frequently better developed under anaerobic conditions. The pseudomycelium is strongly branched (Fig. 54). Occasionally blastospores may be arranged in small clusters along the filamentous cells.

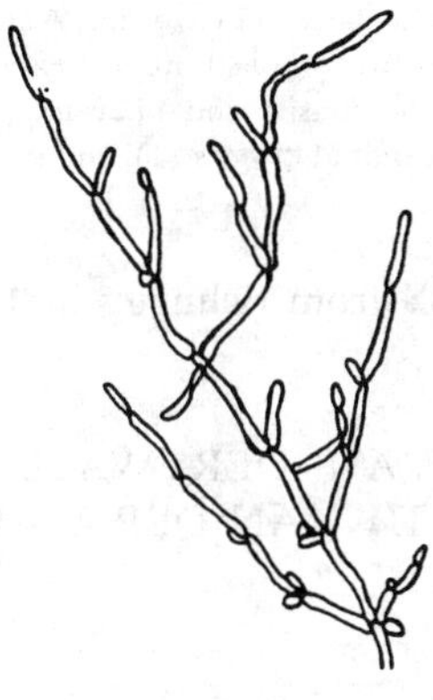

Fig. 54 – *K. marxianus* var. *marxianus*
Dalmau plate culture on corn meal agar
(From 'The Yeasts', ed. II, 1970)

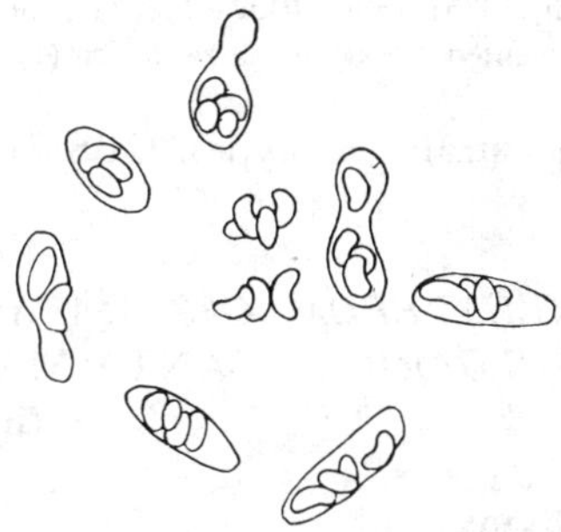

Fig. 55 – *K. marxianus* var. *marxianus*
After 4 days on YM agar
(From 'The Yeasts', ed. II, 1970)

Formation of ascospores: Conjugation may precede ascus formation or diploid vegetative cells may be directly transformed into asci. The asci are evanescent. One to four crescentiform to reniform ascospores are formed per ascus. Released ascospores tend to agglutinate (Fig. 55).

Ascospore formation is observed on 2% malt extract agar, YM agar and the commonly employed sporulation media.

Fermentation:

Glucose	+	Lactose	v
Galactose	+ or s	Raffinose	+
Sucrose	+	Inulin	+
Maltose	–		

Assimilation of carbon compounds:

Galactose	+	Raffinose	+	Erythritol	–
Sucrose	+	Soluble starch	–	Ribitol	+
Maltose	–	D-Xylose	+	D-Mannitol	+
Cellobiose	+	L-Arabinose	+	Succinic acid	+
Trehalose	–	D-Ribose	v	Citric acid	v
Lactose	+ or s	L-Rhamnose	–	Inositol	–

Assimilation of nitrate: –

Growth in vitamin-free medium: –

Growth on 50% (w/w) glucose-yeast extract agar: –
Growth at 37°C: +
G + C: 40.0–41.3 mol%, one strain (Martini et al., 1972); 38.8–39.3 mol%, one strain (Poncet & Fiol, 1972); 40.0–41.0 mol%, 4 strains (Nakase & Komagata, 1971d); 41.2–41.3 mol%, 6 strains (Phaff, 1978; personal communication)

Origin of the strains studied

South African soil (2); effluent of a sugar factory (1) (Sacchetti, 1939a); compressed yeast (1) (Beijerinck, 1908); decaying sisal leaf (1) (Cordro'ch, 1937); South African sorghum beer (1) (van der Walt, 1956c); maize meal (1); yoghurt (4); koumiss (1); mastitic milk (1); bovine mastitis (1); asthmatical expectoration (1) (Dietrichson, 1954); sputum (4); tubercular lung (1); lesion of the tonsils and pharynx (Ciferri & Redaelli, 1935); lung of an infant (1); thoracic cavity (1); section material (1); feces (1); unknown origin (4); fermented Mexican maize dough (1) (Herrera et al., 1973).

Type strain (neotype): CBS 712, received by the CBS from Schnegg in 1922.

6b. *KLUYVEROMYCES MARXIANUS* (HANSEN) VAN DER WALT VAR. *BULGARICUS* (SANTA MARÍA) JOHANNSEN ET VAN DER WALT (in Johannsen, 1980)

Synonyms

Saccharomyces fragilis (Jörgensen) var. *bulgaricus* Santa María (1956b)
Kluyveromyces cicerisporus van der Walt, Nel et van Kerken (1966)
Kluyveromyces bulgaricus (Santa María) van der Walt (1970c) nom. inval.
Kluyveromyces bulgaricus (Santa María) van der Walt (1971)

Standard description of *Kluyveromyces marxianus* var. *bulgaricus* (van der Walt, 1970c)

Growth in malt extract: After 3 days at 28°C the cells are subglobose, spheroidal or ellipsoidal to cylindrical, (2.5–6.0) × (3.5–11.0) μm, reproducing by budding, and occur singly or in pairs. Pseudomycelium may be present and cells may measure up to 15 μm in length. An incomplete ring and sediment are formed. After one month at room temperature an incomplete ring and sediment are present.

Growth on malt agar: After one month at room temperature the streak culture is creamish-brown, flat, smooth or faintly verruculose in the center or with structures along the margin, rather dull. The margin may be undulating, lobiform or erose.

Dalmau plate culture on corn meal agar: Pseudomycelium is produced. It may be rudimentary to rather well ramified and is usually better developed in anaerobic areas.

Formation of ascospores: Conjugation usually immediately precedes ascus formation or diploid, vegetative cells may be directly transformed into asci. The asci are evanescent. One to four spheroidal to ellipsoidal ascospores are formed per ascus. Mature ascospores usually contain lipid globules and are rapidly released from the asci, tending to agglutinate.

Sporulation is observed on 2% malt extract agar, YM agar and Kleyn's acetate agar.

Fermentation:

Glucose	+	Lactose	v
Galactose	+	Raffinose	+
Sucrose	+	Inulin	+ or s
Maltose	−		

Assimilation of carbon compounds:

Galactose	+	Raffinose	+	Erythritol	−
Sucrose	+	Soluble starch	−	Ribitol	+ s
Maltose	−	D-Xylose	+ s	D-Mannitol	v
Cellobiose	+ (s)	L-Arabinose	v	Succinic acid	+
Trehalose	− or + w	D-Ribose	− or + w	Citric acid	−
Lactose	+	L-Rhamnose	−	Inositol	−

Assimilation of nitrate: −

Growth in vitamin-free medium: −

Growth on 50% (w/w) glucose-yeast extract agar: −

Growth at 37°C: +

G + C: 42.4–43.4 mol%, 2 strains (Martini et al., 1972); 33.9, 39.2–39.5 mol%, 2 strains (Poncet & Fiol, 1972); 41.3–41.5 mol%, 2 strains (Phaff, 1978; personal communication)

Origin of the strains studied

Yoghurt (1) (Santa María, 1956b); sorghum beer (1); stracchino cheese (1); back of a cow (Klokke & Kamp, 1962); unknown origin (3).

Type strain: CBS 2762, isolated by Santa María from yoghurt.

6c. *KLUYVEROMYCES MARXIANUS* (HANSEN) VAN DER WALT VAR. *DOBZHANSKII* (SHEHATA, MRAK ET PHAFF) JOHANNSEN ET VAN DER WALT (in Johannsen, 1980)

Synonyms

Saccharomyces dobzhanskii Shehata, Mrak et Phaff (1955)
Dekkeromyces dobzhanskii (Shehata, Mrak et Phaff) Novák et Zsolt (1961) nom. illegit., comb. non valid.
Dekkeromyces dobzhanskii (Shehata, Mrak et Phaff) Santa María et Sanchez (1970) nom. illegit., comb. non valid.
Guilliermondella dobzhanskii (Shehata, Mrak et Phaff) Boidin, Abadie, Jacob et Pignal (1962)
Kluyveromyces dobzhanskii (Shehata, Mrak et Phaff) van der Walt (1965c) nom. inval.
Kluyveromyces dobzhanskii (Shehata, Mrak et Phaff) van der Walt (1971)

Standard description of *Kluyveromyces marxianus* var. *dobzhanskii* (van der Walt, 1970c)

Growth in malt extract: After 3 days at 28°C the cells are spheroidal, subglobose, ellipsoidal to cylindrical, (2.5–6.5) × (3.0–11.0) μm, reproducing by budding, and occur singly, in pairs, occasionally in small clusters. Pseudomycelium may be formed and elongate cells measuring up to 18 μm in length may be present. A sediment is formed. After one month at room temperature a sediment is present and usually a ring as well.

Growth on malt agar: After one month at room temperature the streak culture is brownish-cream to brown, sectored, smooth to faintly verruculose, with structures along the margin, flat, somewhat dull. The margin is erose.

Dalmau plate culture on corn meal agar: Pseudomycelium is abundantly formed and is usually well ramified.

Formation of ascospores: Conjugation immediately precedes ascus formation or diploid, vegetative cells may be directly transformed into asci. The asci are evanescent. One to four spheroidal to ellipsoidal ascospores are formed per ascus. The mature ascospores usually contain lipid globules and are rapidly released from the asci and tend to agglutinate.

Sporulation is observed on 2% malt extract agar and YM agar.

Fermentation:

Glucose	+	Maltose	–
Galactose	+	Lactose	–
Sucrose	+	Raffinose	+

Assimilation of carbon compounds:

Galactose	+	Raffinose	+	Erythritol	–
Sucrose	+	Soluble starch	–	Ribitol	+ s
Maltose	–	D-Xylose	+ s	D-Mannitol	+ s
Cellobiose	–	L-Arabinose	v	Succinic acid	+
Trehalose	–	D-Ribose	–	Citric acid	–
Lactose	–	L-Rhamnose	–	Inositol	–

Assimilation of nitrate: –

Growth in vitamin-free medium: –

Growth on 50% (w/w) glucose-yeast extract agar: –

Growth at 37°C: +

G + C: 43.1 mol%, one strain (Martini et al., 1972); 34.6–35.6 mol%, one strain (Poncet & Fiol, 1972); 41.1–41.2 mol%, 2 strains (Phaff, 1978; personal communication)

Origin of the strains studied

Drosophila pseudoobscura (1) (Shehata et al., 1955); fleshy fungi (2).

Type strain: CBS 2104, isolated by Shehata et al. from *Drosophila pseudoobscura.*

6d. *KLUYVEROMYCES MARXIANUS* (HANSEN) VAN DER WALT VAR. *DROSOPHILARUM* (SHEHATA, MRAK ET PHAFF) JOHANNSEN ET VAN DER WALT (in Johannsen, 1980)

Synonyms

Zygosaccharomyces mrakii Shehata et Mrak (1952) nom. illegit. public. sine diagn. latine
Saccharomyces drosophilarum Shehata, Mrak et Phaff (1955)
Dekkeromyces drosophilarum (Shehata, Mrak et Phaff) Novák et Zsolt (1961) nom. illegit., comb. non valid.
Guilliermondella drosophilarum (Shehata, Mrak et Phaff) Boidin, Abadie, Jacob et Pignal (1962)
Dekkeromyces drosophilarum (Shehata, Mrak et Phaff) Santa María et Sanchez (1970) nom. illegit., comb. non valid.
Kluyveromyces drosophilarum (Shehata, Mrak et Phaff) van der Walt (1965c) nom. inval.
Kluyveromyces drosophilarum (Shehata, Mrak et Phaff) van der Walt (1971)
Saccharomyces drosophilarum (Shehata, Mrak et Phaff) var. *acellobiosus* Phaff, Miller et Shifrine (1956) nom. illegit. public. sine diagn. latine
Saccharomyces phaseolosporus Shehata, Mrak et Phaff (1955)
Dekkeromyces phaseolosporus (Shehata, Mrak et Phaff) Novák et Zsolt (1961) nom. illegit., comb. non valid.
Guilliermondella phaseolospora (Shehata, Mrak et Phaff) Boidin, Abadie, Jacob et Pignal (1962)
Dekkeromyces phaseolosporus (Shehata, Mrak et Phaff) Santa María et Sanchez (1970) nom. illegit., comb. non valid.
Kluyveromyces phaseolosporus (Shehata, Mrak et Phaff) van der Walt (1965c) nom. inval.
Kluyveromyces phaseolosporus (Shehata, Mrak et Phaff) van der Walt (1971)
Zygofabospora krassilnikovii Kudriavzev (1960)
Dekkeromyces krassilnikovii (Kudriavzev) Novák et Zsolt (1961) nom. illegit., comb. non valid.

Standard description of *Kluyveromyces marxianus* var. *drosophilarum* (van der Walt, 1970c)

Growth in malt extract: After 3 days at 28°C the cells are spheroidal to ellipsoidal, seldom somewhat cylindrical, (2.0–5.0) × (2.5–6.5)μm, reproducing by budding, and occur singly, in pairs or short chains or clusters. Conjugating cells and asci may be present. A ring and sediment and occasionally a thin pellicle are formed. After one month at room temperature a ring, a coherent to somewhat floccose sediment and islets or a thin pellicle are present.

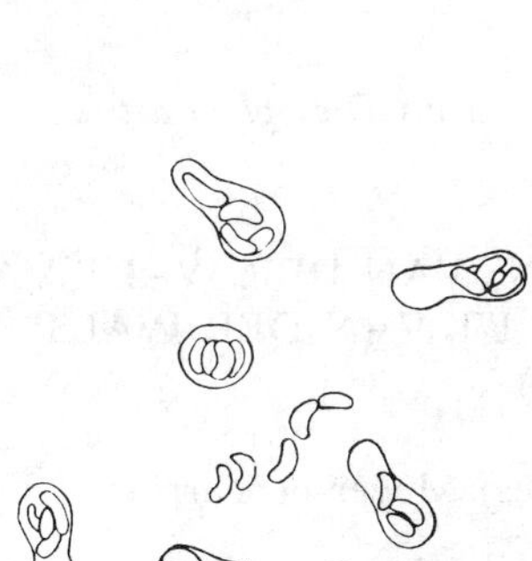

Fig. 56 – *K. marxianus* var. *drosophilarum*
After 5 days on YM agar
(From 'The Yeasts', ed. II, 1970)

Growth on malt agar: After one month at room temperature the streak culture is cream-colored to greyish-buff, occasionally with a pinkish tinge, flat, rather spreading, somewhat dull, smooth or verruculose, occasionally sectored or striated near the margin. The margin may be smooth, undulating or lobiform.

Dalmau plate culture on corn meal agar: Tree-like structures or a rudimentary pseudomycelium are formed.

Formation of ascospores: Conjugation immediately precedes ascus formation or diploid, vegetative cells may be directly transformed into asci. The asci are evanescent. One to four reniform to crescentiform ascospores are formed per ascus. Released ascospores tend to agglutinate (Fig. 56).

Sporulation is observed on 2% malt extract agar, YM agar and the commonly employed sporulation media.

Fermentation:

Glucose	+	Maltose	–
Galactose	+	Lactose	–
Sucrose	+	Raffinose	+

Assimilation of carbon compounds:

Galactose	+	Raffinose	+	Erythritol	–
Sucrose	+	Soluble starch	–	Ribitol	v
Maltose	v	D-Xylose	+ s	D-Mannitol	+
Cellobiose	v	L-Arabinose	–	Succinic acid	+
Trehalose	v	D-Ribose	–	Citric acid	v
Lactose	–	L-Rhamnose	–	Inositol	–

Assimilation of nitrate: –

Growth in vitamin-free medium: –

Growth on 50% (w/w) glucose-yeast extract agar: –

Growth at 37°C: +

G + C: 40.7–40.8 mol%, 4 strains (Martini et al., 1972); 38.5–39.0 mol%, 2 strains (Poncet & Fiol, 1972); 40.0 mol%, one strain (Nakase & Komagata, 1971d); 39.9–40.9 mol%, 8 strains (Phaff, 1978; personal communication)

Origin of the strains studied

Drosophila species (3) (Shehata et al., 1955); bovine cecum (1) (van Uden & do Carmo Sousa, 1957); slime flux of an oak (1) (Kudriavzev, 1960); South African soil (1).

Type strain: CBS 2105, isolated by Shehata et al. from *Drosophila azteca.*

6e. *KLUYVEROMYCES MARXIANUS* (HANSEN) VAN DER WALT VAR. *LACTIS* (DOMBROWSKI) JOHANNSEN ET VAN DER WALT (in Johannsen, 1980)

Imperfect state: *Candida sphaerica* (Hammer et Cordes) Meyer et Yarrow

Synonyms

Saccharomyces lactis α Dombrowski (1910)
Zygosaccharomyces lactis α Dombrowski (1910)
Guilliermondella lactis (Dombrowski) Boidin, Abadie, Jacob et Pignal (1962)
Dekkeromyces lactis (Dombrowski) Santa María et Sanchez (1970) nom. illegit., comb. non valid.

Kluyveromyces lactis (Dombrowski) van der Walt (1965c) nom. inval.
Kluyveromyces lactis (Dombrowski) van der Walt (1971)
Zygosaccharomyces casei Sacchetti (1932a, b)
Zygosaccharomyces versicolor Sacchetti (1933)
Saccharomyces sociasii Ramirez (1954)

Standard description of *Kluyveromyces marxianus* var. *lactis* (van der Walt, 1970c)

Growth in malt extract: After 3 days at 28°C the cells are usually spheroidal, ellipsoidal, or occasionally cylindrical, (2.0–6.5) × (3.0–8.0) μm, reproducing by budding, and occur singly, in pairs or occasionally in small clusters or agglutinated masses. Amoeboid cells may be formed. A sediment is formed and occasionally a ring, islets or a thin pellicle. After one month at room temperature, a sediment and ring are formed. A pellicle is usually present.

Growth on malt agar: After one month at room temperature the streak culture is brownish-cream to creamish-grey, frequently tinged with pink, flat and rather spreading, smooth, occasionally sectored or somewhat verruculose, somewhat shiny or occasionally rather dull. The margin is entire, undulating, or lobiform.

Dalmau plate culture on corn meal agar: A rudimentary pseudomycelium is usually formed. It is often better developed under anaerobic conditions. Under aerobic conditions it may be absent.

Formation of ascospores: Conjugation immediately precedes ascus formation or diploid, vegetative cells are directly transformed into asci. The asci are evanescent. One to four spheroidal to ellipsoidal ascospores are formed per ascus (Fig. 57). The mature ascospores usually contain lipid globules and are rapidly released from the asci, tending to agglutinate. Sporulating strains are as a rule heterothallic.

Sporulation is observed on 2% malt extract agar and YM agar.

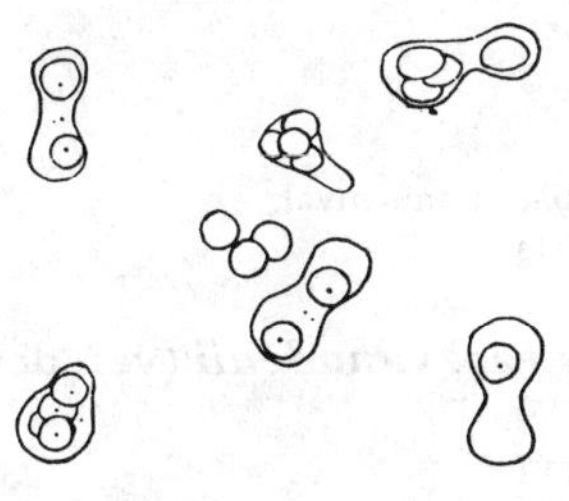

Fig. 57 – *K. marxianus* var. *lactis*
After 5 days on YM agar
(From 'The Yeasts', ed. II, 1970)

Fermentation:

Glucose	+	Lactose	+
Galactose	+	Raffinose	v
Sucrose	v	Inulin	–
Maltose	v		

Assimilation of carbon compounds:

Galactose	+	Raffinose	v	Erythritol	–
Sucrose	+	Soluble starch	–	Ribitol	v
Maltose	v	D-Xylose	v	D-Mannitol	+
Cellobiose	+	L-Arabinose	–	Succinic acid	+
Trehalose	+	D-Ribose	–	Citric acid	–
Lactose	+	L-Rhamnose	–	Inositol	–

Assimilation of nitrate: –
Growth in vitamin-free medium: –
Growth on 50% (w/w) glucose-yeast extract agar: –
Growth at 37°C: – or +s
Growth in the presence of 100 ppm cycloheximide: +
G + C: 41.4 mol%, 4 strains (Martini et al., 1972); 38.0–38.8 mol%, 2 strains (Poncet & Fiol, 1972); 39.3–40 mol%, 2 strains (Nakase & Komagata, 1971d); 40.3 mol%, 2 strains (Phaff, 1978; personal communication)

Origin of the strains studied

Gassy cheese (1) (Allen & Thornley, 1929); milk (1) (Trüper, 1928); Italian cheese (2) (Sacchetti 1932a, b; 1933); buttermilk (1); cream (4) (Laffer, 1936); fermented milk from Manchuria (1); unknown origin (1) (NRRL Y-1140; Wickerham & Burton, 1956); stools (1) (Ramirez, 1954).

Type strain (neotype): CBS 683, isolated by Allen and Thornley from gassy cheese.

6f. *KLUYVEROMYCES MARXIANUS* (HANSEN) VAN DER WALT VAR. *VANUDENII* (VAN DER WALT ET NEL) JOHANNSEN ET VAN DER WALT (in Johannsen, 1980)

Synonyms

Saccharomyces vanudenii van der Walt et Nel (1963)
Kluyveromyces vanudenii (van der Walt et Nel) van der Walt (1965c) nom. inval.
Kluyveromyces vanudenii (van der Walt et Nel) van der Walt (1971)

Standard description of *Kluyveromyces marxianus* var. *vanudenii* (van der Walt, 1970c)

Growth in malt extract: After 3 days at 28°C the cells are spheroidal, subglobose, ellipsoidal, occasionally cylindrical, (2.5–6.0) × (3.0–8.5) μm, reproducing by budding, and occur singly, in pairs, short chains or small clusters. A sediment is formed and occasionally an incomplete ring as well. After one month at room temperature a sediment and ring are formed. A pellicle may also be present.

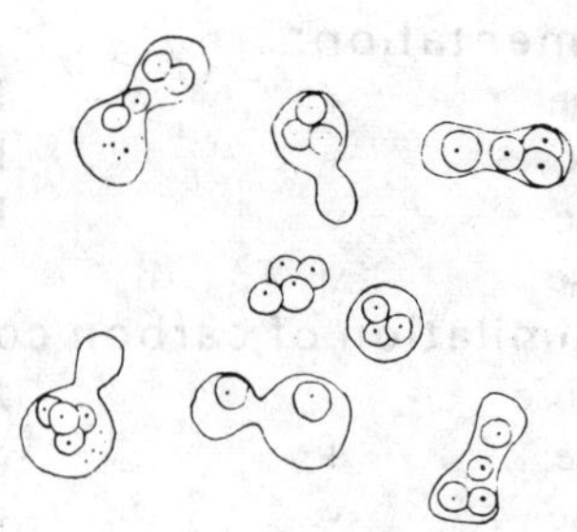

Fig. 58 – *K. marxianus* var. *vanudenii*
After 5 days on YM agar
(From 'The Yeasts', ed. II, 1970)

Growth on malt agar: After one month at room temperature the streak culture is creamish-brown, flat, spreading, somewhat dull, sectored, striated or slightly verruculose and somewhat raised towards the center. The margin may be undulating, lobiform or erose.

Dalmau plate culture on corn meal agar: Pseudomycelium is produced. It may be rudimentary or well ramified.

Formation of ascospores: Conjugation may immediately precede ascus formation or diploid, vegetative cells may be directly transformed into asci. The asci are evanescent. One to four spheroidal to ellipsoidal ascospores are formed per ascus (Fig. 58). Mature ascospores usually contain lipid globules and are rapidly released from the asci, tending to agglutinate.

Sporulation is observed on 2% malt extract agar, YM agar and commonly employed sporulation media.

Fermentation:

Glucose	+	Maltose	v
Galactose	+	Lactose	–
Sucrose	+	Raffinose	+

Assimilation of carbon compounds:

Galactose	+	Raffinose	+	Erythritol	–
Sucrose	+	Soluble starch	–	Ribitol	– or + w
Maltose	v	D-Xylose	+ or s	D-Mannitol	+
Cellobiose	+	L-Arabinose	– or + s	Succinic acid	+
Trehalose	v	D-Ribose	–	Citric acid	–
Lactose	–	L-Rhamnose	–	Inositol	–

Assimilation of nitrate: –

Growth in vitamin-free medium: –

Growth on 50% (w/w) glucose-yeast extract agar: –

Growth at 37°C: +

G + C: 41.7 mol%, one strain (Martini et al., 1972); 38.0–38.3 mol%, one strain (Poncet & Fiol, 1972); 40.1–40.3 mol%, 2 strains (Phaff, 1978; personal communication)

Origin of the strains studied

Winery equipment (3) (van der Walt & Nel, 1963); sorghum beer (2).

Type strain: CBS 4372, isolated by van der Walt and Nel from winery equipment.

6g. KLUYVEROMYCES MARXIANUS (HANSEN) VAN DER WALT VAR. *WIKENII* (VAN DER WALT, NEL ET VAN KERKEN) JOHANNSEN ET VAN DER WALT (in Johannsen, 1980)

Synonym

Kluyveromyces wikenii van der Walt, Nel et van Kerken (1966)

Standard description of *Kluyveromyces marxianus* var. *wikenii* (van der Walt, 1970c)

Growth in malt extract: After 3 days at 28°C the cells are spheroidal, subglobose, ellipsoidal to cylindrical, (2.5–6.5) × (3.0–11.0) μm, reproducing by budding, and occur singly, in pairs, occasionally in small clusters. Pseudomycelium may be formed and elongate cells measuring up to 18 μm in length may be present. A sediment is formed. After one month at room temperature a sediment is present and usually a ring as well.

Growth on malt agar: After one month at room temperature the streak culture is brownish-cream to brown, sectored, smooth to faintly verruculose, with structures along the margin, flat, somewhat dull. The margin is erose.

Dalmau plate culture on corn meal agar: Pseudomycelium is abundantly formed and is usually well ramified.

Formation of ascospores: Conjugation immediately precedes ascus formation or diploid, vegetative cells may be directly transformed into asci. The asci are evanescent. One to four spheroidal to ellipsoidal ascospores are formed per ascus. The mature ascospores usually contain lipid globules and are rapidly released from the asci and tend to agglutinate.

Sporulation is observed on 2% malt extract agar and YM agar.

Fermentation:

Glucose	+	Maltose	–
Galactose	+	Lactose	–
Sucrose	+	Raffinose	+

Assimilation of carbon compounds:

Galactose	+	Raffinose	+	Erythritol	–
Sucrose	+	Soluble starch	–	Ribitol	+ s
Maltose	–	D-Xylose	+ s	D-Mannitol	+ s
Cellobiose	–	L-Arabinose	v	Succinic acid	+
Trehalose	–	D-Ribose	–	Citric acid	–
Lactose	–	L-Rhamnose	–	Inositol	–

Assimilation of nitrate: –

Growth in vitamin-free medium: –

Growth on 50% (w/w) glucose-yeast extract agar: –

Growth at 37°C: +

G + C: 43.1 mol%, one strain (Martini et al., 1972); 34.6–35.6 mol%, one strain (Poncet & Fiol, 1972); 41.1–41.2 mol%, 2 strains (Phaff, 1978; personal communication)

Origin of the strains studied

Sorghum beer (3) (van der Walt et al., 1966).

Type strain: CBS 5671, isolated by van der Walt et al. from sorghum beer.

7. *KLUYVEROMYCES PHAFFII* (VAN DER WALT) VAN DER WALT (1971)

Synonyms

Fabospora phaffii van der Walt (1963)
Kluyveromyces phaffii (van der Walt) van der Walt (1965c) nom. inval.

Standard description of *Kluyveromyces phaffii* (van der Walt, 1970c)

Growth in malt extract: After 3 days at 28°C the cells are ellipsoidal to almost cylindrical, (2.0–4.5) × (3.0–8.0) μm, reproducing by budding, and occur singly or in pairs. A sediment is formed. After one month at room temperature only a sediment is present.

Growth on malt agar: After one month at room temperature the streak culture is butyrous, greyish-brown, flat, rather spreading, somewhat shiny and smooth to verruculose. The margin is undulating to lobiform.

Dalmau plate culture on corn meal agar: No pseudomycelium is formed. Chains of cells of equal length may be produced.

Formation of ascospores: Diploid, vegetative cells are directly transformed into asci. The asci are as a rule evanescent. One to four ascospores are formed per ascus. The ascospores are stoutly reniform to oblong with obtuse ends. Released ascospores tend to agglutinate.

Ascospore formation is observed on 2% malt extract agar, YM agar and the commonly employed sporulation media.

Fermentation:

Glucose	+	Maltose	–
Galactose	+	Lactose	–
Sucrose	–		

Assimilation of carbon compounds:

Galactose	+	Raffinose	–	Erythritol	–
Sucrose	– or +w	Soluble starch	–	Ribitol	–
Maltose	–	D-Xylose	–	D-Mannitol	–
Cellobiose	–	L-Arabinose	–	Succinic acid	–
Trehalose	+w	D-Ribose	–	Citric acid	–
Lactose	–	L-Rhamnose	–	Inositol	–

Assimilation of nitrate: –

Growth in vitamin-free medium: –

Growth on 50% (w/w) glucose-yeast extract agar: –

Growth at 37°C: –

G + C: 37.4 mol%, one strain (Martini et al., 1972); 33.6 mol%, one strain (Poncet & Fiol, 1972); 35.3 mol%, one strain (Phaff, 1978; personal communication)

Origin of the strain studied

Soil (1) (van der Walt, 1963).

Type strain: CBS 4417, isolated by van der Walt from South African soil.

8. *KLUYVEROMYCES POLYSPORUS* VAN DER WALT (1956a)

Synonym

Saccharomyces polysporus (van der Walt) Campbell (1972)

Standard description of *Kluyveromyces polysporus* (van der Walt, 1970c)

Growth in malt extract: After 3 days at 28°C the cells are subglobose, ellipsoidal to almost cylindrical, (2.0–5.0) × (3.0–7.0) μm, reproducing by budding, and occur singly, in pairs, short chains or small clusters. Conjugating cells and zygotes are usually present. Such cells vary considerably in size and may measure up to 25 × 28 μm or more (Fig. 59). A sediment and slight ring are formed and islets may be present. After one month at room temperature a sediment and occasionally a thin ring, islets or a thin pellicle are present.

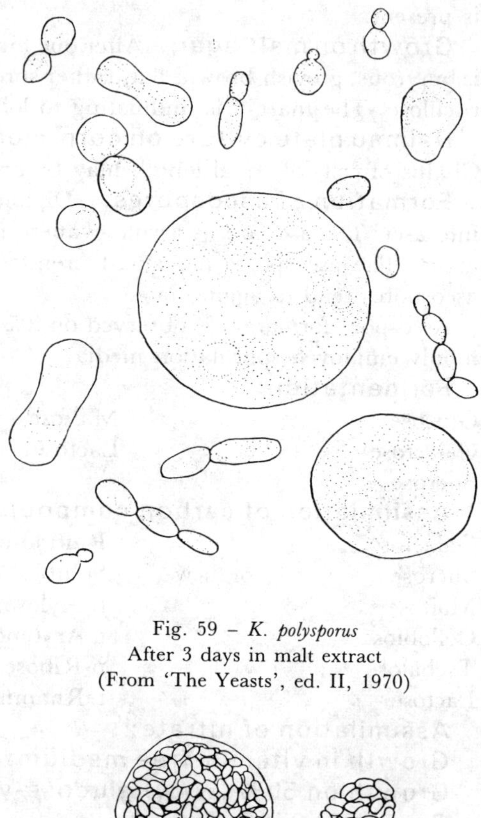

Fig. 59 – *K. polysporus*
After 3 days in malt extract
(From 'The Yeasts', ed. II, 1970)

Growth on malt agar: After one month at room temperature the streak culture is butyrous, brownish-cream, flat, somewhat spreading, shiny, smooth, occasionally sectored. The margin is entire, undulating or lobiform.

Dalmau plate culture on corn meal agar: Pseudomycelium is usually absent but tree-like structures or a rudimentary pseudomycelium are occasionally formed in stab-cultures.

Formation of ascospores: Conjugation immediately precedes ascus formation. Occasionally the zygotes form a bud. The zygotes enlarge appreciably. The evanescent asci are large and multispored. The ascospores are reniform to oblong with obtuse ends. Released ascospores tend to agglutinate (Fig. 60).

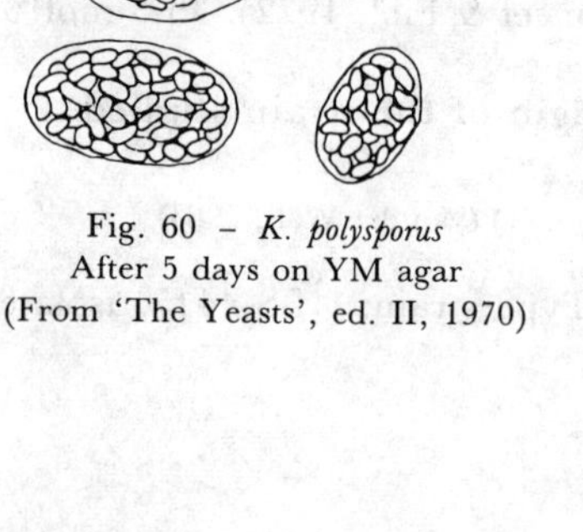

Fig. 60 – *K. polysporus*
After 5 days on YM agar
(From 'The Yeasts', ed. II, 1970)

Ascospore formation is observed on 2% malt extract agar, YM agar and the customary sporulation media.

Fermentation:

Glucose	+	Maltose	–
Galactose	+	Lactose	–
Sucrose	+	Raffinose	+

Assimilation of carbon compounds:

Galactose	+	Raffinose	+	Erythritol	–
Sucrose	+	Soluble starch	–	Ribitol	–
Maltose	–	D-Xylose	–	D-Mannitol	–
Cellobiose	–	L-Arabinose	–	Succinic acid	+ s
Trehalose	+ s	D-Ribose	–	Citric acid	+ s
Lactose	–	L-Rhamnose	–	Inositol	–

Assimilation of nitrate: –

Growth in vitamin-free medium: –

Growth on 50% (w/w) glucose-yeast extract agar: –

Growth at 37°C: –

G + C: 35.3 mol%, one strain (Martini et al., 1972); 34.4–34.6 mol%, one strain (Poncet & Fiol, 1972); 33.4 mol%, one strain (Nakase & Komagata, 1971d); 35.3 mol%, one strain (Phaff, 1978; personal communication)

Origin of the strains studied

South African soils (5) (van der Walt, 1956a).

Type strain: CBS 2163, isolated by van der Walt from soil.

9. *KLUYVEROMYCES THERMOTOLERANS* (FILIPPOV) YARROW (1972a)

Imperfect state: *Candida dattila* (Kluyver) Meyer et Yarrow

Synonyms

Zygosaccharomyces thermotolerans Filippov (1932)
Saccharomyces thermotolerans (Filippov) Campbell (1972)
Torula dattila Kluyver (1914) (see van der Walt & Johannsen, 1974b)
Mycotorula dattila (Kluyver) Harrison (1928)
Torulopsis dattila (Kluyver) Lodder (1934)
Cryptococcus dattilus (Kluyver) Skinner (1947a)
Saccharomyces veronae Lodder et Kreger-van Rij (1952)
Kluyveromyces veronae (Lodder et Kreger-van Rij) van der Walt (1970c) nom. inval.
Kluyveromyces veronae (Lodder et Kreger-van Rij) van der Walt (1971)
Zygosaccharomyces drosophilae Shehata et Mrak (1952) nom. illegit. public. sine diagn. latine

Standard description of *Kluyveromyces thermotolerans* (van der Walt, 1970c)

Growth in malt extract: After 3 days at 28°C the cells are spheroidal to ellipsoidal, (3.0–6.5) × (5.5–8.0) μm, reproducing by budding, and occur singly, in pairs, occasionally in short chains or small clusters. Cylindrical or even elongate cells which may measure up to 16 μm in length may be formed. A sediment is formed and occasionally a slight ring. After one month at room temperature a sediment and ring are present.

Growth on malt agar: After one month at room temperature the streak culture is butyrous, cream-colored to brownish-cream, shiny to somewhat dull, and smooth to crispulate. The margin is entire to undulating, occasionally lobate, frequently fringed with pseudomycelium.

Dalmau plate culture on corn meal agar: Pseudomycelium is formed more or less abundantly. The pseudomycelium may be very primitive or strongly branched. Blastospores may be arranged in verticils or clusters along the pseudomycelium. The pseudomycelium is usually better developed under anaerobic conditions.

Formation of ascospores: Conjugation by individual cells of equal or unequal size usually immediately precedes ascus formation or diploid, vegetative cells are directly converted to asci. The asci are evanescent. One to four spheroidal ascospores containing lipid globules are formed per ascus. Released ascospores tend to agglutinate.

Sporulation is observed on 2% malt extract agar and YM agar.

Fermentation:

Glucose	+	Maltose	v
Galactose	v	Lactose	–
Sucrose	+	Raffinose	+

Assimilation of carbon compounds:

Galactose	+	Raffinose	+	Erythritol	–
Sucrose	+	Soluble starch	–	Ribitol	v
Maltose	+	D-Xylose	–	D-Mannitol	+
Cellobiose	–	L-Arabinose	–	Succinic acid	v
Trehalose	+	D-Ribose	–	Citric acid	–
Lactose	–	L-Rhamnose	–	Inositol	–

Assimilation of nitrate: –

Growth in vitamin-free medium: –

Growth on 50% (w/w) glucose-yeast extract agar: +

Growth at 37°C: – or + vw

G + C: 47.4 mol%, 3 strains (Martini et al., 1972); 44.0–44.5 mol%, one strain (Poncet & Fiol, 1972); 44.4–45.9 mol%, 2 strains (Nakase & Komagata, 1971d); 46.2 mol%, one strain (Phaff, 1978; personal communication)

Origin of the strains studied

Dates (1) (CBS 137, type of *Torulopsis dattila;* Kluyver, 1914); plum jam (1) (Filippov, 1932); grapes (2) (CBS 2803, type of *Saccharomyces veronae;* Lodder & Kreger-van Rij, 1952); *Drosophila* (2); soil (2).

Type strain: CBS 6340, isolated by Filippov from fermenting plum jam.

10. KLUYVEROMYCES WALTII KODAMA (1974)

Standard description of *Kluyveromyces waltii*

Growth in malt extract: After 3 days at 28°C the cells are spheroidal, ovoid to ellipsoidal, (3.0–6.0) × (3.5–10.0) μm, reproducing by budding, and occur singly, in pairs, occasionally in small clusters. Short chains of cylindrical to botuliform cells, (2.0–3.0) × (5.0–10.0–16) μm also occur. A sediment is formed. After one month at room temperature a sediment is present.

Growth on malt agar: After one month at room temperature the streak culture is butyrous, brownish-cream, shiny, flat, slightly raised along the center, smooth with some structure along the margin. The margin is entire to undulating.

Dalmau plate culture on corn meal agar: Pseudomycelium is produced but is more abundantly formed under aerobic conditions. Clusters of blastospores which may branch are formed at the nodes of the filamentous cells.

Formation of ascospores: Conjugation by cells of more or less equal or unequal size precedes ascus formation. The asci are evanescent. One to four spheroidal ascospores are formed per ascus. The mature ascospores containing lipid globules are rapidly released from the asci and tend to agglutinate.

Ascospore formation is observed on 2% malt extract agar and YM agar.

Fermentation:

Glucose	+	Maltose	–
Galactose	–	Lactose	–
Sucrose	+	Raffinose	+

Assimilation of carbon compounds:

Galactose	–	Raffinose	+	Erythritol	–
Sucrose	+	Soluble starch	–	Ribitol	+ s
Maltose	–	D-Xylose	–	D-Mannitol	+
Cellobiose	–	L-Arabinose	–	Succinic acid	–
Trehalose	–	D-Ribose	–	Citric acid	–
Lactose	–	L-Rhamnose	–	Inositol	–

Assimilation of nitrate: –

Growth in vitamin-free medium: –

Growth on 50% (w/w) glucose-yeast extract agar: +

Growth at 37°C: –

G + C: 45.6 mol% (Kodama, 1974)

Origin of the strain studied

Exudate of *Ilex integra* in Japan (1) (Kodama, 1974).

Type strain: CBS 6430, isolated by Kodama.

11. KLUYVEROMYCES WICKERHAMII (PHAFF, MILLER ET SHIFRINE) VAN DER WALT (1971)

Synonyms

Saccharomyces wickerhamii Phaff, Miller et Shifrine (1956)
Dekkeromyces wickerhamii (Phaff, Miller et Shifrine) Novák et Zsolt (1961) nom. illegit., comb. non valid.
Guilliermondella wickerhamii (Phaff, Miller et Shifrine) Boidin, Abadie, Jacob et Pignal (1962)
Dekkeromyces wickerhamii (Phaff, Miller et Shifrine) Santa María et Sanchez (1970) nom. illegit., comb. non valid.
Kluyveromyces wickerhamii (Phaff, Miller et Shifrine) van der Walt (1965c) nom. inval.

Standard description of *Kluyveromyces wickerhamii* (van der Walt, 1970c)

Growth in malt extract: After 3 days at 28°C the cells are spheroidal, ellipsoidal to cylindrical, (1.5–4.5) × (2.5–6.5) μm, reproducing by budding, and occur singly, in pairs, or occasionally in short chains. Conjugating cells may be present. A sediment is formed. A dull, creeping pellicle and ring are present. After one month at room temperature a coherent sediment, a ring and a thin, dull pellicle are present.

Growth on malt agar: After one month at room temperature the streak culture is butyrous, creamish-brown, shiny, flat, spreading, smooth or with some fine structure. The margin is lobiform.

Dalmau plate culture on corn meal agar: A rudimentary pseudomycelium is formed and is somewhat better developed and more ramified under anaerobic conditions.

Formation of ascospores: Conjugation immediately precedes ascus formation or diploid, vegetative cells are directly transformed into asci. The asci are evanescent. One to four reniform or crescentiform ascospores are formed per ascus. Released ascospores tend to agglutinate.

Ascospore formation is observed on 2% malt extract agar, YM agar and the commonly employed sporulation media.

Fermentation:

Glucose	+	Lactose	–
Galactose	+ s	Raffinose	–
Sucrose	+ s	Inulin	–
Maltose	–		

Assimilation of carbon compounds:

Galactose	+	Raffinose	–	Erythritol	–
Sucrose	+	Soluble starch	–	Ribitol	–
Maltose	–	D-Xylose	+	D-Mannitol	–
Cellobiose	+	L-Arabinose	–	Succinic acid	+
Trehalose	v	D-Ribose	–	Citric acid	–
Lactose	+	L-Rhamnose	–	Inositol	–

Assimilation of nitrate: –

Growth in vitamin-free medium: –

Growth on 50% (w/w) glucose-yeast extract agar: –

Growth at 37°C: – or + vw

G + C: 41.5 mol% 4 strains (Martini et al., 1972); 40.2–40.9 mol%, one strain (Poncet & Fiol, 1972); 41.9–42.4 mol%, (Phaff, 1978; personal communication)

Origin of the strains studied

Drosophila montana (1) (Phaff et al., 1956); exudate from *Salix* (1); exudate from *Acer* (1); exudate from *Betula* (1).

Type strain: CBS 2745, isolated by Phaff et al. from *Drosophila montana.*

f. COMMENTS ON THE GENUS

The history of the genus *Kluyveromyces* and the synonymy of the genera *Dekkeromyces, Fabospora* and *Zygofabospora* have been reviewed by Johannsen (1978). For an earlier account, the second edition of 'The Yeasts' may be consulted (van der Walt, 1970c).

Species delimitation in *Kluyveromyces* is currently based on hybridization studies (Johannsen, 1980). Consequently, a number of taxa accepted in the second edition of 'The Yeasts' as distinct species (*K. bulgaricus, K. cicerisporus, K. dobzhanskii, K. drosophilarum, K. fragilis, K. lactis, K. phaseolosporus, K. vanudenii* and *K. wikenii*) have been relegated to seven syngamous varieties of *K. marxianus.* As these varieties readily hybridize, the recovery of hybrids with combined parental phenotypes is possible.

Authors intending to publish new species in *Kluyveromyces* should, before introducing new species names, examine such putative new taxa for their ability to hybridize with already described species. For this purpose the prototrophic selection technique of Pomper and Burkholder (1949) may be used. For experimental details the publications of Johannsen (1980) and van der Walt and Johannsen (1979) may be consulted.

As already pointed out by van der Walt (1970c), the genus *Kluyveromyces* has a very wide distribution and its representatives have been recovered from substrates as divergent as sea water, soil, insects, plant material, fresh fruit, jams and other fruit preserves, dairy and panary products, as well as from breweries, wineries and mammalian sources. With the exception of *K. aestuarii* which appears to be restricted to marine environments, there is at present insufficient evidence to accept that other elements of the genus are characterized by any stringent habitat specificity.

Genus 14. *Lipomyces* Lodder et Kreger-van Rij

by H.J. Phaff and C.P. Kurtzman

a. DIAGNOSIS OF THE GENUS

Cells ellipsoidal or globose, reproducing by multilateral budding or by bud-fission. Infrequently cells on a broad base develop in short chains; individual cells are then separated by septa. Oval to cylindrical giant cells may occur. Pseudomycelium absent or rudimentary. Cells of most species are surrounded by a thick mucous capsule, but in others it is reduced or lacking. Most strains form starch-like polysaccharides in this capsule if mucous. Cells in older cultures usually include a large fat globule.

Asci develop in various ways; an 'active bud' is most frequently involved in the formation of the ascus. This may or may not be preceded by a discernable conjugation.

(1) When the asci develop from active buds only, they often enlarge to an elongate pear shape (Fig. 61a, c, d). The wall of this type of ascus is thin. Two asci or more may arise from one mother cell (Fig. 61b). The entire ascus may liberate itself from the cell when mature (Fig. 61f). The number of spores may be high, up to 16 or more per ascus; in some species the number of spores is eight or four or less per ascus.

(2) When the ascus is the result of conjugation of an active bud and a protuberance of a cell, the ascus wall is thicker (Fig. 61g, j). Complicated groups of these cell-bud conjugations are frequently present on sporulation media.

(3) A mother cell, after having produced an ascus through an active bud,

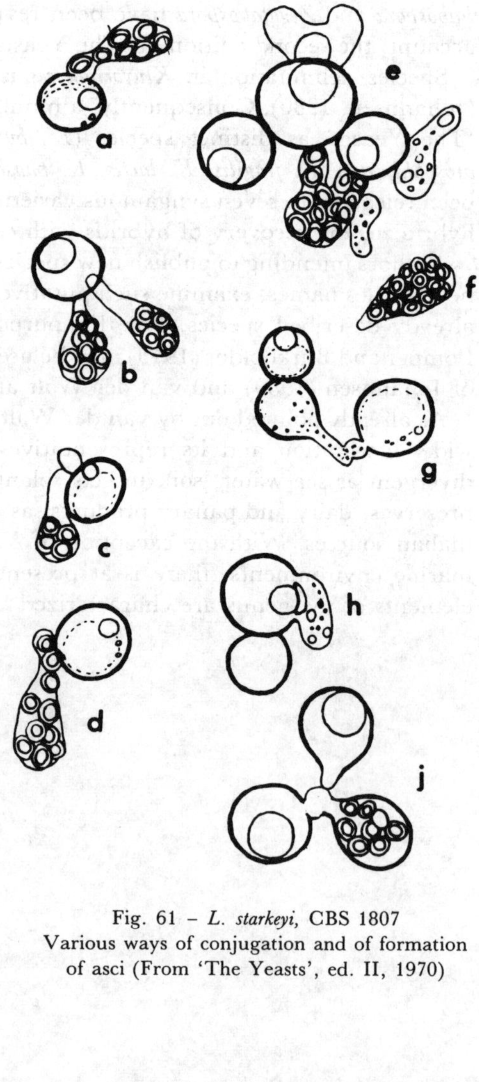

Fig. 61 – *L. starkeyi*, CBS 1807
Various ways of conjugation and of formation of asci (From 'The Yeasts', ed. II, 1970)

may itself turn into an ascus. Fully grown cells containing ascospores have been observed. These asci possess thick walls.

Ascospores are ellipsoidal to oblong-ellipsoidal, generally liberated at maturity. The pigmented walls are of a light amber to brown color. The ascospore wall may be smooth, or it bears denticles or ridges running longitudinally on the surface; two of these ridges circle over the poles of the spore. These ridges may be visible with the light microscope in the larger and strongly pigmented spores.

Germination of ascospores is by swelling to the size of a small cell followed by budding. Conjugation of swollen spores has been observed.

There is no fermentative ability.

b. TYPE SPECIES OF THE GENUS

Lipomyces starkeyi Lodder et Kreger-van Rij

c. SPECIES ACCEPTED IN THE GENUS

1. *Lipomyces anomalus* Babjeva et Gorin (1975)
2. *Lipomyces kononenkoae* Nieuwdorp, Bos et Slooff (1974)
3. *Lipomyces lipofer* Lodder et Kreger-van Rij ex Slooff (1970a)
4. *Lipomyces starkeyi* Lodder et Kreger-van Rij (1952)
5. *Lipomyces tetrasporus* Nieuwdorp, Bos et Slooff (1974)

d. KEY TO THE SPECIES OF THE GENUS

1a Erythritol not assimilated ... 2
b Erythritol assimilated, often latently ... 3

2a Strong growth at 33°C and on sucrose
L. kononenkoae ... p. 256
b No growth on sucrose; maximum growth temperature 28°C
L. anomalus ... p. 254

3a Ridges on the ascospore under the light microscope or in the electron microscope. Generally not more than four spores per ascus
L. tetrasporus ... p. 261
b Ridges on spore surface lacking ... 4

4a Ascospore wall smooth; assimilation of galactitol negative or weak; galactose absent in the extracellular polysaccharides or galactose/mannose ratio < 0.05:1
L. lipofer . . . . p. 257

b Prominent warts on the ascospore surface, visible clearly in the electron microscope; galactitol assimilated; galactose present in the extracellular polysaccharides with a galactose/mannose ratio > 0.4:1
L. starkeyi . . . . p. 259

Table 16

Key characters of the species in the genus *Lipomyces*

	Assimilation			Galactose in extracellular polysaccharide	Ascospore wall	Number of spores per ascus
	Er	Gal	Su			
L. anomalus	–	–	–	+	reticulate pattern of ridges	4–30 (numerous)
L. kononenkoae	–	v	+	+	smooth	2–16 (8)
L. lipofer	+s	v	+	– or trace	smooth	3–20 (varies but often 8)
L. starkeyi	+	+	+	+	warts	4–20 (4 or 8, strain dependent)
L. tetrasporus	+	+	+	+	ridges	1–4 (4)

The number of spores in parentheses is the most common number per ascus. Er = erythritol; Gal = galactitol; Su = sucrose.

e. SYSTEMATIC DISCUSSION OF THE SPECIES

1. *LIPOMYCES ANOMALUS* BABJEVA ET GORIN (1975)

Standard description of *Lipomyces anomalus*

Growth in malt extract: Very slow. After 10 days at 20°C a rudimentary pseudomycelium is present, the cells of which are spheroidal or ovoidal, (4.3–7.0) × (5.4–9) μm. After one month at 20°C only a small amount of sediment is present.

Growth on malt agar: Growth is slow and becomes visible only after one week at 20°C. After 10 days at 20°C vegetative reproduction is mainly by budding on a broad base and the cells occur mostly as pseudohyphae. Cells are spherical or oval, (4.5–7.0) × (5.3–9.0) μm and occasionally elongated, (4–10) × (8–18) μm. No capsules are formed. Young cells usually contain one large vacuole and several small dark granules. After one month at 20°C the streak culture is cream colored, mat, corrugated. Big spherical or oval cells with many buds are common.

Slide culture on potato agar: A rudimentary pseudomycelium consisting of rows of cells formed by bud-fission is abundantly formed (Fig. 62).

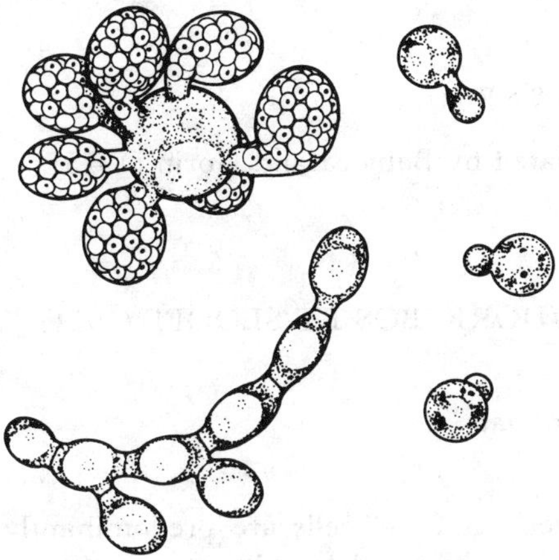

Fig. 62 – *L. anomalus*
After Babjeva & Gorin (1975)

Formation of ascospores: Spores were no longer produced by the strain investigated. According to the original description this species is homothallic. The asci are usually formed by conjugation of two buds on the same cell or on different cells. The conjugating buds are always separated from the mother cells by septa. This process closely resembles gametangiogamy. After conjugation the larger bud may separate from the mother cell. Several sac-like asci may be formed on the same cell (Fig. 62). Sometimes ascospores are formed within the pseudomycelial cells or in the cells with a great number of buds. The number of spores per ascus varies from four to 30 and even more, but multi-spored asci are most frequent. The ascospores are spherical, rarely oval, with one to two small oil droplets. Their surface has a reticulate pattern of ridges which is clearly seen in electron microscopic preparations and even on some spores under the light microscope.

Spores were originally observed on YNB maltose, lactose and sucrose agar media after 3–4 weeks at 18°C. Maintenance of stock cultures on malt agar results in loss of the ability to sporulate.

Fermentation: –

Assimilation of carbon compounds:

Galactose	–	Raffinose	–	Erythritol	–
Sucrose	–	Soluble starch	+ w	Ribitol	+ s
Maltose	+ s or w	D-Xylose	+ w	D-Mannitol	–
Cellobiose	+ s or w	L-Arabinose	–	Succinic acid	–
Trehalose	–	D-Ribose	–	Citric acid	–
Lactose	+ s or w	L-Rhamnose	–	Inositol	–

Additional carbon compounds tested: Galactitol –, glucosamine · HCl –, methanol –, *n*-hexadecane –

Assimilation of nitrate and of nitrite: –

Assimilation of ethylamine · HCl: –

Growth in vitamin-free medium: –

Growth on 50% (w/w) glucose-yeast extract agar: –

Growth on 5% sodium chloride-glucose-yeast extract agar: –

Growth at 37°C: –; **maximum temperature for growth:** 27–28°C; **optimum:** 16–18°C

Starch formation: +

Urease: –

Growth in the presence of 100 ppm of cycloheximide: +

G + C: 43.80 ± 0.18 mol%, type strain (Phaff & Holzschu, unpublished)

Origin of the strain studied

Podzolic soil in Komi ASSR (North Taiga subzone, U.S.S.R.) (1).

Type strain: CBS 6740 (BKM-Y-1969) isolated by Babjeva and Gorin (1975).

2. *LIPOMYCES KONONENKOAE* NIEUWDORP, BOS ET SLOOFF (1974)

Standard description of *Lipomyces kononenkoae*

Growth in malt extract: After one week at 25°C cells are predominantly globose 5–8.5 μm in diameter. Only few ellipsoidal cells are present, (4–5.8) × (5.5–8.5) μm. The cell walls and contents are faintly refractile, except the walls of the larger ones. Only a few small oil drops are present. After one month a sediment, a ring, and occasionally a thin, moist pellicle are present.

Growth on malt agar: After one week at 25°C cells are mostly globose, the diameter varying from 4–10 μm. Some ellipsoidal cells may be present, (4–6) × (6.2–8) μm. Several cells have produced one bud. After one month the streak culture is creamy white, glistening and mucous.

Slide cultures on potato- and corn meal agar: No mycelium or pseudomycelium is present.

Formation of ascospores: On corn meal agar or on V8 medium conjugation of two 'active buds' (Slooff, 1970a), initiated by one mother cell, frequently precedes the formation of very thin-walled asci. Globose spores are formed often lying in a cluster in close contact with the mother cell (Fig. 63). Pear-shaped asci also occur, possibly arising from diploidized single buds. Two or more asci on one mother cell may occur. YM agar stimulates the development of protuberances; fully grown cells with protuberances may be transformed directly into asci. The usual number of spores per ascus is eight but varies from two to 16. The ascospore wall is smooth.

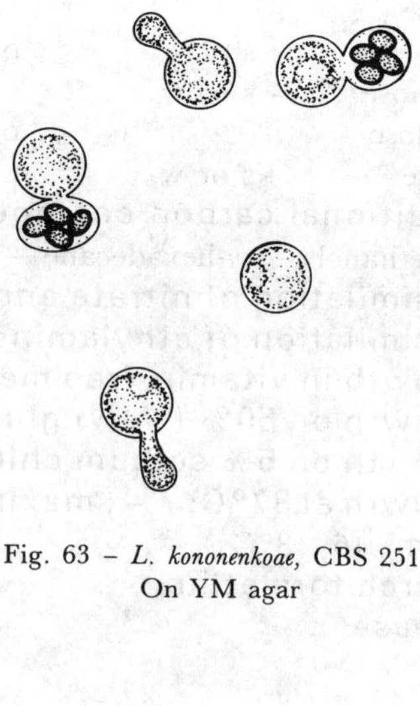

Fig. 63 – *L. kononenkoae*, CBS 2514
On YM agar

Fermentation: –

Assimilation of carbon compounds:

Galactose	+	Raffinose	+	Erythritol	–
Sucrose	+	Soluble starch	+	Ribitol	–
Maltose	+	D-Xylose	v	D-Mannitol	+
Cellobiose	+ s	L-Arabinose	–	Succinic acid	–
Trehalose	+	D-Ribose	–	Citric acid	+ w
Lactose	–	L-Rhamnose	–	Inositol	–

Additional carbon compounds tested: Galactitol v, glucosamine · HCl –, methanol –, *n*-hexadecane –

Assimilation of nitrate and of nitrite: –

Assimilation of ethylamine · HCl: –

Growth in vitamin-free medium: +

Growth on 50% (w/w) glucose-yeast extract agar: –

Growth on 5% sodium chloride-glucose-yeast extract agar: + w

Growth at 37°C: + w or –; **at 39°C:** –; **at 33°C:** +

Starch formation: + or w

Urease: –

Growth in the presence of 100 ppm of cycloheximide: +

G + C: 48.3 ± 0.4 mol%, type strain (Phaff & Holzschu, unpublished)

Origin of the strains studied

Soil from citrus grove in Trinidad (1), soil in Nigeria (1).

Type strain: CBS 2514, isolated by Becking from soil in Trinidad.

3. *LIPOMYCES LIPOFER* LODDER ET KREGER-VAN RIJ EX SLOOFF (1970a)

Synonyms

Torula lipofera den Dooren de Jong (1927)
Torulopsis lipofera (den Dooren de Jong) Lodder (1934)
Cryptococcus lipoferus (den Dooren de Jong) Skinner (1947b)

Standard description of *Lipomyces lipofer*

Growth in malt extract: After one week at 25°C growth is scant. Cells are granular, sometimes vacuolated, globose to ellipsoidal, reproducing by multilateral budding, (3.2–8.0) × (5–12) μm. Walls and cell contents are faintly refractile, sometimes containing a few small lipid globules. A thin ring may be present. After one month some sediment has formed and a few slimy islets may be present.

Growth on malt agar: After one week at 25°C cells are globose to ellipsoidal, (4–8) × (5–14) μm, containing vacuoles and small lipid granules. Large spheroidal cells containing one large lipid globule occur occasionally. After one month the streak culture is white to cream colored, glistening and pasty or mucous.

Slide culture on corn meal agar: Pseudomycelium absent.

Formation of ascospores: Active buds formed on lipid-containing cells become transformed into asci on V8- or YM agar in one to two weeks as follows (Fig. 64):

(1) An active bud enlarges, the contents turn granular and divide into spores. The number of spores is four to 20, sometimes less than four.

(2) Two active buds on one cell fuse; the fusion product enlarges and develops into an ascus. The number of spores is three to 16.

(3) Two active buds on separate cells or on neighboring cells copulate; this case is infrequently observed in *L. lipofer*.

(4) An active bud growing lengthwise fuses with the conjugation tube of another cell; the two cells joined by a conjugation tube turn into one ascus. In this case the recipient cell frequently contains the greater part of the ascospores: eight to 12 per ascus. Ascospores are ellipsoidal and contain an oil drop. The color is olive green to light amber. Malachite green stain is readily taken. The spore wall in the electron microscope is smooth.

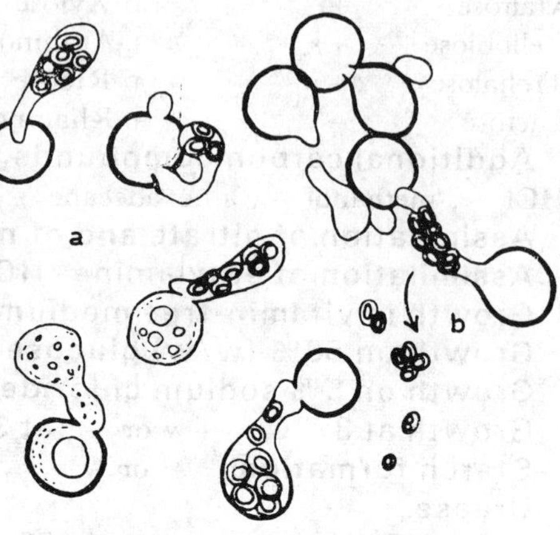

Fig. 64 – *L. lipofer*, CBS 944
On Starkey's ethanol agar
(From 'The Yeasts', ed. II, 1970)

Fermentation: –

Assimilation of carbon compounds:

Galactose	+	Raffinose	+	Erythritol	+ s
Sucrose	+	Soluble starch	+	Ribitol	–
Maltose	+	D-Xylose	+ s	D-Mannitol	+
Cellobiose	+ s (w)	L-Arabinose	+ (–)	Succinic acid	–
Trehalose	– (+ w)	D-Ribose	–	Citric acid	+ w
Lactose	+ s (–)	L-Rhamnose	–	Inositol	–

Additional carbon compounds tested: Galactitol – (+ w), glucosamine · HCl –, methanol –, *n*-hexadecane –

Assimilation of nitrate and of nitrite: –

Assimilation of ethylamine · HCl: –

Growth in vitamin-free medium: v; biotin is required by the negative strains

Growth on 50% (w/w) glucose-yeast extract agar: –

Growth on 5% sodium chloride-glucose-yeast extract agar: + s, w; **on 7.5% sodium chloride:** –

Growth at 37°C: –; **at 30°C:** –

Starch formation: A greenish color develops upon addition of iodine

Urease: –

Growth in the presence of 100 ppm of cycloheximide: +

G + C: 49.3 ± 0.1 mol%, type strain (Phaff & Holzschu, unpublished)

Comments

Although *Lipomyces lipofer* was first described by Lodder and Kreger-van Rij (1952), these authors failed to provide a description in Latin. According to Article 59 of the International Code of Botanical Nomenclature their binominal *Lipomyces lipofer* was not validly published as a new species. Slooff (1970a) provided a Latin description, thus explaining the authorities for the perfect state.

Origin of the strains studied

Soil in The Netherlands (2), soil in Wales (1), and a strain isolated by A.C. Thayson, presumably from soil (1).

Type strain: CBS 944, isolated by den Dooren de Jong from garden soil in The Netherlands.

4. *LIPOMYCES STARKEYI* LODDER ET KREGER-VAN RIJ (1952)

Standard description of *Lipomyces starkeyi*

Growth in malt extract: After one week at 25°C growth is scanty. Cells are granular, some have vacuoles; lipid globules are scarce and small; cells are globose to ellipsoidal, (4–7) × (5–11) μm and reproduce by multilateral budding (Fig. 65). Cell walls and cell contents are faintly refractile, except in some large globose cells and cells that contain a single large lipid globule. A thin ring may be present. After one month there are some small moist islets on the surface.

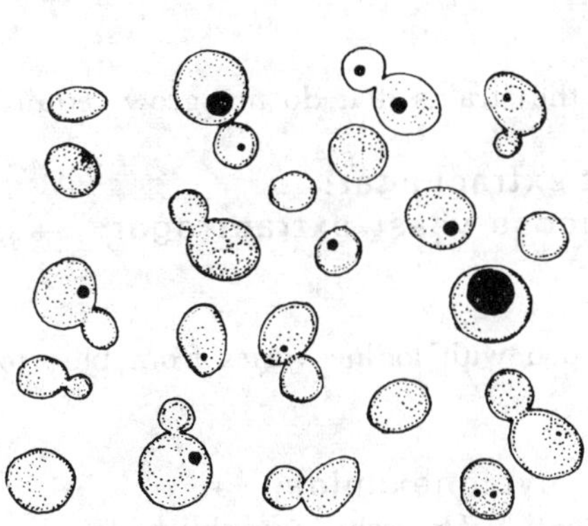

Fig. 65 – *L. starkeyi*
After 3 days in malt extract
(From 'The Yeasts', ed. I, 1952)

Growth on malt agar: After one week the cells are similar as in malt extract, except that occasional large round cells with a single large lipid globule occur and the cells are only moderately granulated. After one month the streak culture is white to light cream colored, glistening, smooth, and very mucous.

Slide culture on corn meal agar: Pseudomycelium absent.

Formation of ascospores: Active buds developed by lipid-containing cells are transformed into asci on V8 or YM agar in one to two weeks by the same four processes as described for *L. lipofer*. The highest numbers of spores were found in asci derived from one active bud (Fig. 61f) or from the fusion of two active buds (Fig. 61e): 16 or more per ascus. The usual number in the type strain was eight spores, but it varied from four to 20. Other strains developed two to six spores per ascus, the usual number being four. Starkey's strain number 72, formerly sporulating with one to four spores per ascus could no longer be induced to sporulate. In those cases where conjugation between bud and cell was involved the number of spores did not exceed eight. Ascospores are ellipsoidal and contain an oil drop. The color is light amber to brown. The malachite green stain is readily taken. The spore wall in an electron micrograph of an ultrathin section is double layered. The inner layer shows a light grey part, surrounded by a dark grey zone. This inner wall is enclosed by a white or light grey outer wall; locally it may be thin or lacking, in other places it develops blunt or pointed denticles on a broader or smaller base. In the cross section the denticles may reach a length equal to one fifth of the spore's diameter. A black line covers the outside of the cross section. In the type strain the wartiness of the spore wall is faintly discernable with the light microscope when the spore is viewed looking down at the pole.

Fermentation: –

Assimilation of carbon compounds:

Galactose	+	Raffinose	+	Erythritol	+
Sucrose	+	Soluble starch	+	Ribitol	v
Maltose	+	D-Xylose	+	D-Mannitol	+
Cellobiose	+ (–)	L-Arabinose	v	Succinic acid	v
Trehalose	+	D-Ribose	v	Citric acid	v
Lactose	– (+s)	L-Rhamnose	v	Inositol	v

Additional carbon compounds tested: Galactitol +, glucosamine · HCl –, methanol –, *n*-hexadecane –

Assimilation of nitrate and of nitrite: –

Assimilation of ethylamine · HCl: –

Growth in vitamin-free medium: v; the strains that do not grow require biotin

Growth on 50% (w/w) glucose-yeast extract agar: –

Growth on 5% sodium chloride-glucose-yeast extract agar: +s, w; **on 7.5% sodium chloride** : –

Growth at 37°C: –; **at 33°C:** +w

Starch formation: +; the color developed with iodine varies from blue to greenish

Urease: –

Growth in the presence of 100 ppm of cycloheximide: +

G + C: 48.1 ± 0.2 mol%, type strain (Phaff & Holzschu, unpublished)

Origin of the strains studied

Soil in the U.S.A. (3); soil, South Africa (1); soil, The Netherlands (1).

Type strain: CBS 1807 (Starkey's strain 74) isolated by Starkey from soil in the U.S.A.

5. *LIPOMYCES TETRASPORUS* NIEUWDORP, BOS ET SLOOFF (1974)

Synonyms

Zygolipomyces tetrasporus nom. nud. (Krassilnikov et al., 1967)
Zygolipomyces lactosus nom. nud. (Krassilnikov et al., 1967)

Standard description of *Lipomyces tetrasporus*

Growth in malt extract: After one week at 25°C the cells are globose to short-ellipsoidal, (4–7) × (4–9) μm, occurring singly, in pairs or in small clusters. A thin ring may be present. After one month, there is usually a ring and sometimes a mucous pellicle as well as a small amount of sediment.

Growth on malt agar: The cells are similar to those in malt extract. After one month the streak culture is white to light cream colored, glistening, smooth, and very mucous.

Slide culture on corn meal agar: Pseudomycelium absent.

Formation of ascospores: The ascus develops on YM agar in one to two weeks from conjugating 'active buds' or, more frequently, from a young cell conjugating by a tube with a bud from another cell. Often fully grown cells develop ascospores without evident previous conjugation. There are one to four spores per ascus (Fig. 66). Upon maturity the spores are released from the ascus. They are amber to brown and have lengthwise ridges, which may be visible in the light microscope.

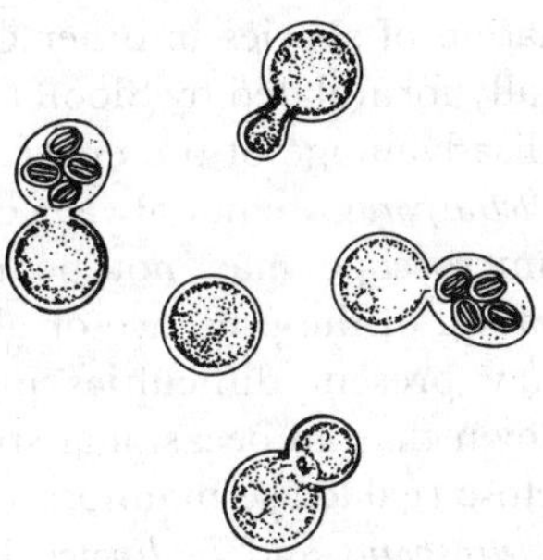

Fig. 66 – *L. tetrasporus*, CBS 5910
On YM agar

Fermentation: –

Assimilation of carbon compounds:

Galactose	+	Raffinose	+	Erythritol	+
Sucrose	+	Soluble starch	+	Ribitol	+ (s)
Maltose	+	D-Xylose	+	D-Mannitol	+
Cellobiose	– (+)	L-Arabinose	– (+)	Succinic acid	+ (s)
Trehalose	– (+w)	D-Ribose	– (+s)	Citric acid	+ s
Lactose	– (+)	L-Rhamnose	– (+)	Inositol	– (+)

Additional carbon compounds tested: Galactitol + (s), glucosamine · HCl –, methanol –, *n*-hexadecane –

Assimilation of nitrate and of nitrite: –

Assimilation of ethylamine · HCl: –

Growth in vitamin-free medium: + or w

Growth on 50% (w/w) glucose-yeast extract agar: –

Growth on 5% sodium chloride-glucose-yeast extract agar: +, w or s; on 7.5% sodium chloride: –

Growth at 37°C: –; at 33°C: + or w

Starch formation: + (the color with iodine varies from deep blue to greenish)

Urease: –

Growth in the presence of 100 ppm of cycloheximide: +

G + C: 49.1 ± 0.2 mol%, type strain (Phaff & Holzschu, unpublished)

Origin of the strains studied

Soil in the U.S.S.R. (4); soil, U.S.A. (1); soil, The Netherlands (1); soil, Africa (1).

Type strain: CBS 5910, isolated by Babjeva from soil in the U.S.S R.

f. COMMENTS ON THE GENUS

Because of the considerable variability among strains to assimilate carbon compounds that have been found useful in the differentiation of species in other genera, we found it necessary to continue using the key originally formulated by Slooff (1970a) and later expanded by Nieuwdorp et al. (1974). A disadvantage of the criteria used by these authors is that ridges on the ascospores of *L. tetrasporus* are not always distinct in the light microscope and a scanning electron microscope may not be readily available in all laboratories. Similarly, the demonstration of the presence or absence of galactose in the extracellular polysaccharides may present difficulties in some laboratories. In addition Gorin et al. (1978) have shown that an occasional strain of *L. tetrasporus* contains an unusually low level of galactose (galactose/mannose ratio of 0.07:1), thus making the differentiation between *L. tetrasporus* and *L. lipofer* difficult on this basis. A further complication pointed out by these same authors is the observation that highly capsulated strains of *Lipomyces* sometimes give rise to strains containing only a microcapsule with different galactose content. For example, a capsulated strain of *L. tetrasporus* with a galactose/mannose ratio of 0.29 showed a ratio of only 0.06 in a microcapsulated variant. Conversely, a capsulated strain of *L. lipofer* with a ratio of < 0.05 changed to a microcapsulated variant with a ratio of 0.28. Exploration of simpler and more reliable phenotypic criteria for the separation of *L. lipofer* and *L. tetrasporus* are, therefore, highly desirable.

It is noteworthy that the base compositions of the nuclear DNA of the *Lipomyces* species (with the exception of *L. anomalus*) are rather close. This suggests that the mol% G + C be determined of more strains and DNA/DNA reassociation experiments as described by Price et al. (1978) be conducted to determine the validity and relationship among the presently accepted species.

Genus 15. *Lodderomyces* van der Walt

by J.P. van der Walt

a. DIAGNOSIS OF THE GENUS

Vegetative cells spheroidal, ellipsoidal to cylindrical, reproducing by budding.

Asci unconjugated, persistent, one- to two-spored. Ascospores oblong with obtuse ends, fusiform to somewhat allantoid, glabrous.

Fermentation.

Higher alkanes utilized as sole source of carbon.

Coenzyme Q-9 system present.

Nitrate not utilized.

b. TYPE SPECIES OF THE GENUS

Lodderomyces elongisporus (Recca et Mrak) van der Walt

c. SPECIES ACCEPTED IN THE GENUS

1. *Lodderomyces elongisporus* (Recca et Mrak) van der Walt (Recca & Mrak, 1952)

e. SYSTEMATIC DISCUSSION OF THE SPECIES

1. LODDEROMYCES ELONGISPORUS (RECCA ET MRAK) VAN DER WALT (1971)

Synonyms

Saccharomyces elongisporus Recca et Mrak (1952)
Lodderomyces elongisporus (Recca et Mrak) van der Walt (1966a) nom. inval.

Standard description of *Lodderomyces elongisporus* (van der Walt, 1970d)

Growth in malt extract: After 3 days at 28°C the cells are ellipsoidal to cylindrical, occasionally elongate, (2.5–6.0) × (4.0–9.5) μm, reproducing by budding, and occur singly, in pairs, occasionally in short chains or small clusters. Pseudomycelium may be present (Fig. 67). A ring and sediment are formed. After one month at room temperature a thick ring and sediment are present.

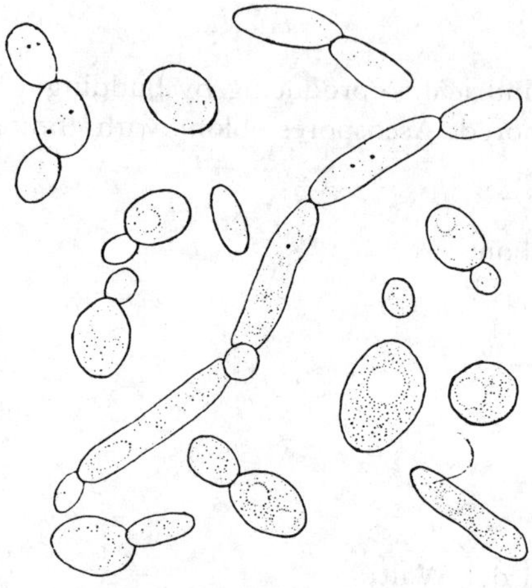

Fig. 67 – *Lod. elongisporus*
After 3 days in malt extract
(From 'The Yeasts', ed. II, 1970)

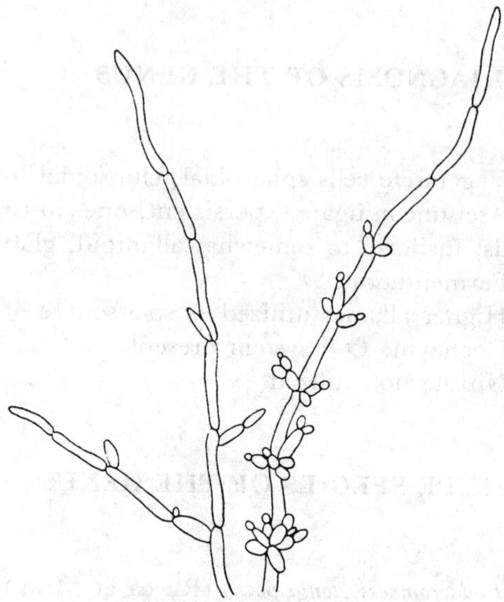

Fig. 68 – *Lod. elongisporus*
Dalmau plate culture on corn meal agar
(From 'The Yeasts', ed. II, 1970)

Growth on malt agar: After one month at room temperature the streak culture is butyrous, creamish-yellow to cream-colored, somewhat shiny to dull, smooth or crispulate to rugose in the center, rather spreading. The margin may be entire or fringed with pseudomycelium.

Dalmau plate culture on corn meal agar: Pseudomycelium is abundantly produced. The pseudomycelium may be branched with the formation of blastospores either in clusters or branching clusters (Fig. 68).

Formation of ascospores: Diploid, vegetative cells are directly transformed into persistent asci. One or occasionally two ascospores are formed per ascus. The ascospores are oblong with obtuse ends, occasionally somewhat fusiform or botuliform, almost filling the ascus (Fig. 69). As mating types could not be recovered from heat-treated ascosporic material, the strains studied are presumed to be homothallic.

Sporulation is best observed on Gorodkowa agar.

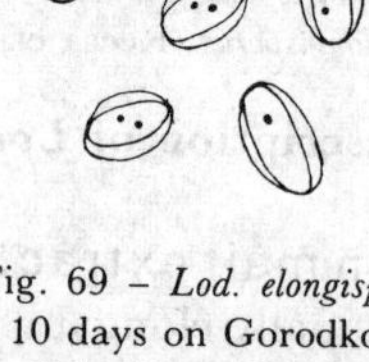

Fig. 69 – *Lod. elongisporus*
After 10 days on Gorodkowa agar
(From 'The Yeasts', ed. II, 1970)

Fermentation:

Glucose	+ s	Maltose	v
Galactose	v	Lactose	–
Sucrose	v	Raffinose	–

Assimilation of carbon compounds:

Galactose	+	Raffinose	–	Erythritol	–
Sucrose	+	Soluble starch	–	Ribitol	+
Maltose	+	D-Xylose	+ s	D-Mannitol	+
Cellobiose	–	L-Arabinose	–	Succinic acid	+
Trehalose	+	D-Ribose	–	Citric acid	+ s
Lactose	–	L-Rhamnose	–	Inositol	–

Additional carbon compounds tested: *n*-Decane + s, *n*-hexadecane +
Assimilation of nitrate: –
Growth in vitamin-free medium: v
Growth on 50% (w/w) glucose-yeast extract agar: + w or –
Growth at 37°C: +

Origin of the strains studied

Orange concentrate (1) (Recca & Mrak, 1952); soil (1).

Type strain: CBS 2605, isolated by Recca and Mrak from orange concentrate.

Genus 16. *Metschnikowia* Kamienski

by M.W. Miller and H.J. Phaff

a. DIAGNOSIS OF THE GENUS

Vegetative reproduction by multilateral budding. Cells spheroidal to ellipsoidal, also pyriform, cylindrical or lunate. Pseudomycelium rudimentary, rarely lacking.

Asci elongate: clavate, sphaeropedunculate or ellipsoidopedunculate. Ascospores needle-shaped, attenuate at one or both ends, without whip-like appendage. One or two spores per ascus.

Some species are parasitic in invertebrates, also free-living in aquatic habitats; others terrestrial, free-living.

b. TYPE SPECIES OF THE GENUS

Metschnikowia bicuspidata (Metschnikoff) Kamienski

c. SPECIES ACCEPTED IN THE GENUS

1. *Metschnikowia bicuspidata* (Metschnikoff) Kamienski (Metschnikoff, 1884)
   Metschnikowia bicuspidata (Metschnikoff) Kamienski var. *bicuspidata*
   Metschnikowia bicuspidata (Metschnikoff) Kamienski var. *australis* Fell et Hunter (1968)
   Metschnikowia bicuspidata (Metschnikoff) Kamienski var. *californica* Pitt et Miller (1970a)
   Metschnikowia bicuspidata (Metschnikoff) Kamienski var. *chathamia* Fell et Pitt (1969)
2. *Metschnikowia krissii* (van Uden et Castelo-Branco) van Uden (van Uden & Castelo-Branco, 1961)
3. *Metschnikowia lunata* Golubev (1977)
4. *Metschnikowia pulcherrima* Pitt et Miller (1968)
5. *Metschnikowia reukaufii* Pitt et Miller (1968)
6. *Metschnikowia zobellii* (van Uden et Castelo-Branco) van Uden (van Uden & Castelo-Branco, 1961)

d. KEY TO THE SPECIES OF THE GENUS

1a Asci formed by differentiation of large, thick-walled refractile chlamydospores; isolate of terrestrial origin ... 2

b Asci formed from vegetative cells; isolate of aquatic origin ... 4

2a Asci formed from spheroidal cells ('pulcherrima' cells), asci sphaeropedunculate ... 3

b Asci formed from ellipsoidal to cylindroidal cells ('reukaufii' cells), asci ellipsoidopedunculate to clavate. Pulcherrimin pigment not produced
M. reukaufii ... p. 275

3a Vegetative cells often lunate. Red pulcherrimin pigment not produced
M. lunata ... p. 272

b Vegetative cells not lunate. Red pulcherrimin pigment usually produced
M. pulcherrima ... p. 273

4a Asci clavate, one ascospore, pointed at one end only ... 5

b Asci clavate, two acicular ascospores, visible when ascus is crushed
M. bicuspidata ... p. 269

5a Glucose fermented, galactose and glucitol assimilated
M. zobellii ... p. 276

b Glucose not fermented, galactose and glucitol not assimilated
M. krissii ... p. 271

Table 17

Key characters of the species in the genus *Metschnikowia*

	Fermentation and assimilation					Number of spores per ascus	Hetero-thallic	Habitat
	G	Ga	α-G	Gδl	Glu			
M. bicuspidata var. *bicuspidata*	vF[1]	+	−	+	+	2	−[2]	aquatic
M. bicuspidata var. *australis*	+	+	−	−	+	2	+	aquatic
M. bicuspidata var. *californica*	F	+	+	+	+	2	+	aquatic
M. bicuspidata var. *chathamia*	F	+	+	−	+	2	+	aquatic
M. krissii	+	−	+	−	−	1	?[3]	aquatic
M. lunata[4]	F	+	+ s	+	+	1–2	−	terrestrial
M. pulcherrima	F	vF	+	+	+	2	+	terrestrial
M. reukaufii	F	vF	v	+	+	2	+	terrestrial
M. zobellii	F	+	+	+	+	1	?[3]	aquatic

G = glucose, Ga = galactose, α-G = α-methyl-D-glucoside, Gδl = glucono-δ-lactone, Glu = D-glucitol.
[1]F = fermentation, vF = variable fermentation, + = assimilation, − = no assimilation.
[2]Heterothallic haploid strains may arise by parasexuality.
[3]Sexual responses not known, as active haploids were not obtained.
[4]Vegetative cells often lunate.

e. SYSTEMATIC DISCUSSION OF THE SPECIES

1. *METSCHNIKOWIA BICUSPIDATA* (METSCHNIKOFF) KAMIENSKI (1899)

This species has four varieties:

Metschnikowia bicuspidata (Metschnikoff) Kamienski var. *bicuspidata*

Metschnikowia bicuspidata (Metschnikoff) Kamienski var. *australis* Fell et Hunter (1968)

Metschnikowia bicuspidata (Metschnikoff) Kamienski var. *californica* Pitt et Miller (1970a)

Metschnikowia bicuspidata (Metschnikoff) Kamienski var. *chathamia* Fell et Pitt (1969)

Synonyms of the variety *bicuspidata*

Monospora biscuspidata Metschnikoff (1884)
Monosporella bicuspidata (Metschnikoff) Keilin (1920)
Metschnikowiella bicuspidata (Metschnikoff) Kudriavzev (1960)
Metschnikowia kamienskii Spencer, Phaff et Gardner (1964)

Standard description of *Metschnikowia bicuspidata* var. *bicuspidata*

Growth in malt extract: After 3 days at 25°C vegetative cells are mostly ellipsoidal to cylindroidal, (3–8) × (3–13) μm, in some strains globose, 5–8 μm, or elongate, 23–35 μm in length. Reproduction is by multilateral budding, cells occurring singly or catenulate. After one month large refractile globose cells, 7–10 μm, containing a lipid globule, may be present. A ring and abundant sediment are formed, a pellicle is absent.

Growth on malt agar: After one month at 25°C the growth is cream-colored, the surface is smooth, or in occasional strains rugose, the edge entire or crenulate.

Dalmau plate culture on corn meal agar: Rudimentary pseudomycelium is formed, usually profusely.

Formation of ascospores: Asci arise from vegetative cells and are generally clavate, although less regular shapes are also formed, measuring (3–7) × (15–45) μm, sometimes to 60 μm, containing two, often appearing one, acicular ascospores, (0.7–1.1) × (15–50) μm. Asci have been observed to dehisce.

Sporulates on V8 agar or on diluted V8 agar.

Fermentation:

Glucose	+ (–)	Maltose	–
Galactose	–	Lactose	–
Sucrose	–	Raffinose	–

Assimilation of carbon compounds:

Galactose	+	Raffinose	–	Erythritol	–
Sucrose	+	Soluble starch	–	Ribitol	–
Maltose	+	D-Xylose	+ or w	D-Mannitol	+
Cellobiose	+	L-Arabinose	–	Succinic acid	v
Trehalose	+	D-Ribose	+ w or –	Citric acid	–
Lactose	–	L-Rhamnose	–	Inositol	–

Additional carbon compounds tested: α-Methyl-D-glucoside –, D-glucitol +, glucono-δ-lactone +

Assimilation of nitrate: –

Growth in vitamin-free medium: –

Growth on 50% (w/w) glucose-yeast extract agar: –

Growth at 37°C: –

G + C: 48.0 ± 0.2 mol%, type strain

Supplementary description of *Metschnikowia bicuspidata* var. *australis*

Fermentation: Glucose –

Assimilation of carbon compounds (in yeast nitrogen base containing 2% NaCl):

Ribitol	+ or s
α-Methyl-D-glucoside	–
Glucono-δ-lactone	–

G + C: 44.8 ± 0.2 mol%, type strain (Meyer & Phaff, 1969)

Supplementary description of *Metschnikowia bicuspidata* var. *californica*

Assimilation of carbon compounds:

α-Methyl-D-glucoside +
Glucono-δ-lactone +

G + C: 47.6 ± 0.3 mol%, type strain

Supplementary description of *Metschnikowia bicuspidata* var. *chathamia*

Assimilation of carbon compounds:

α-Methyl-D-glucoside +
Glucono-δ-glucoside –

G + C: 47.5 ± 0.2 mol%, type strain

Comments

Strains of the four varieties comprising the species *M. bicuspidata* are morphologically and physiologically quite similar, and related by their aquatic habitat, ascus shape, shape and number of ascospores and interfertility between mating types. The variety *bicuspidata* is generally homothallic although heterothallic haploids can be obtained (Wickerham, 1964a), apparently by a parasexual mechanism as described by Pitt and Miller (1970b). These haploids are capable of hybridizing with the other heterothallic varieties and producing sparsely sporulating hybrids.

Some strains of the variety *bicuspidata,* i.e. those designated as *M. kamienskii* in general had slightly lower maximum temperatures for growth (27–28°C) than the strains obtained from sporocysts which generally grew at 30–34°C.

The nine strains of the variety *bicuspidata* which we isolated from diseased *Artemia salina* taken from salt ponds (ca. 10% NaCl) in southern California did not grow in the usual assimilation media unless these were supplemented with 2% NaCl.

Ascospores have been observed to dehisce from the tip of the peduncle, occasionally quite forcefully. The ecological implication and the possibility of active mechanical predation for the variety *bicuspidata* have been discussed by Lachance et al. (1976), who misidentified these strains as the variety *australis*.

Origin of the strains belonging to the variety *bicuspidata*

Sporocysts of *Diplostomum flexicaudum* (5), *Artemia salina* (13), *Castolia odorata* (1).

Type strain: The neotype, as selected by Wickerham (1964a) CBS 5575 (NRRL Y-4993) is retained as the type of the variety *bicuspidata*.

Origin of the strains belonging to the variety *australis*

Waters of Antarctic Ocean (6).

Type strain: Strain CBS 5847, mating type *a*, has been maintained as the holotype of this variety. CBS 5848, mating type α is the allotype.

Origin of the strains belonging to the variety *californica*

Waters of Pacific Ocean, California, U.S.A. (6).

Type strain: CBS 6010 is the type of this variety.

Origin of the strains belonging to the variety *chathamia*

Ocean water at Chatham Island, New Zealand (3).

Type strain: CBS 6011 is the type of this variety.

2. *METSCHNIKOWIA KRISSII* (VAN UDEN ET CASTELO-BRANCO) VAN UDEN (1962a)

Synonym

Metschnikowiella krissii van Uden et Castelo-Branco (1961)

Standard description of *Metschnikowia krissii*

Growth in malt extract: After 3 days at 25°C the cells are spheroidal to ellipsoidal, (4.5–6) × (6–11) μm; elongate cells, (4.5–6) × (11–13) μm also occur. A thin ring may be formed.

Growth on malt agar: After one month at 25°C the streak culture is yellowish, semidull, soft, delicately punctilate to smooth.

Dalmau plate culture on corn meal agar: A rudimentary pseudomycelium is formed.

Formation of ascospores: The asci arise from vegetative cells and are generally clavate, 18–26 μm long and contain a single, needle-shaped ascospore, pointed at one end and nearly as long as the respective ascus.

Asci were observed on V8 agar.

Fermentation: –

Assimilation of carbon compounds:

Galactose	–	Raffinose	–	Erythritol	–
Sucrose	+	Soluble starch	–	Ribitol	–
Maltose	+	D-Xylose	–	D-Mannitol	+
Cellobiose	+	L-Arabinose	–	Succinic acid	+ or w
Trehalose	+	D-Ribose	–	Citric acid	–
Lactose	–	L-Rhamnose	–	Inositol	–

Additional carbon compounds tested: α-Methyl-D-glucoside +, D-glucitol –, glucono-δ-lactone –

Assimilation of nitrate: –

Growth in vitamin-free medium: –

Growth on 50% (w/w) glucose-yeast extract agar: –

Growth at 37°C: –

G + C: 46.2 mol% (Meyer & Phaff, 1970)

Origin of the strains studied

Seawater off California (4).

Type strain: CBS 4823, isolated by van Uden and Castelo-Branco.

3. *METSCHNIKOWIA LUNATA* GOLUBEV (1977)

Synonym

Schizoblastosporion kobayashii Soneda et Uchida (1971)

Standard description of *Metschnikowia lunata*

Growth in malt extract: After 3 days at 25°C the cells are lunate, (2–4) × (5–8) μm, or rarely ovoidal, (2.5–5) × (3.5–8) μm, single or in pairs. A slight sediment is formed. After one month at 20°C a sediment and ring are present. Pulcherrima cells (8–10.5) × (8–12) μm present which may have two small, pointed protrusions.

Growth on malt agar: After one month the streak culture is cream colored, mat, smooth surface, pasty, slightly convex with an entire border. Pulcherrimin is not produced.

Dalmau plate culture on corn meal agar: A rudimentary pseudomycelium is formed.

Formation of ascospores: Asci arise from spheroidal chlamydospores, which generally contain a single, prominent lipid globule. The lipid globule is no longer present in the ascus. Asci are sphaeropedunculate, spherical part (7–10) × (7–12) μm, peduncles cylindrical (2–4) × (13–40) μm, containing one to two acicular ascospores.

Asci are observed on dilute V8 agar at 15°C.

Fermentation:

Glucose	+	Maltose	–
Galactose	–	Lactose	–
Sucrose	–	Raffinose	–

Assimilation of carbon compounds:

Galactose	+	Raffinose	–	Erythritol	–
Sucrose	+	Soluble starch	–	Ribitol	+ s
Maltose	+	D-Xylose	+ s	D-Mannitol	+
Cellobiose	+	L-Arabinose	–	Succinic acid	+ s
Trehalose	+	D-Ribose	–	Citric acid	–
Lactose	–	L-Rhamnose	–	Inositol	–

Additional carbon compounds tested: α-Methyl-D-glucoside + s, D-glucitol +, glucono-δ-lactone +

Assimilation of nitrate: –

Growth in vitamin-free medium: –

Growth on 50% (w/w) glucose-yeast extract agar: + w or s

Growth at 37°C: –

G + C: 49.2 mol%, one strain (Golubev, 1977)

Comments

We consider *Schizoblastosporion kobayashii* a synonym of *M. lunata,* even though ascospores were not observed in the sphaeropedunculate asci produced in the former. All other phenotypic properties were identical.

Origin of the strains studied

Flower of *Vicia cracca* L. (1), flower of Casuarina tree (1).

Type strain: CBS 5946, isolated by Egorova.

4. *METSCHNIKOWIA PULCHERRIMA* PITT ET MILLER (1968)

Imperfect state: *Candida pulcherrima* (Lindner) Windisch

Synonyms

Torula pulcherrima Lindner (1901)
Torulopsis pulcherrima (Lindner) Sacc. (1906)
Saccharomyces pulcherrimus (Lindner) Beijerinck (1912)
Rhodotorula pulcherrima (Lindner) Harrison (1928)
Candida pulcherrima (Lindner) Windisch (1940)
Chlamydozyma pulcherrima Wickerham (1964a)
Torula rubefaciens Grosbüsch (1915)
Monilia (Cryptococcus) castellanii Re (1925)
Cryptococcus castellanii (Re) Castellani (1928)
Torulopsis castellanii (Re) Castellani et Jacono (1933)
Castellania castellanii (Re) Dodge (1935)
Torulopsis pulcherrima (Lindner) Sacc. var. *variabilis* Lodder (1934)
Torulopsis pulcherrima (Lindner) Sacc. var. *rubra* Castelli (1940)
Asporomyces uvae Mrak et McClung (1940)
? *Torulopsis rosea* Berlese (1895)
Torulopsis burgeffiana Benda (1962)
Chlamydozyma reukaufii Wickerham (1964a)

Standard description of *Metschnikowia pulcherrima*

Growth in malt extract: After 3 days at 25°C vegetative cells are globose to ellipsoidal, (2.5–6) × (4–10) μm, usually (2.5–5) × (4–7) μm, single, reproducing by multilateral budding. Chlamydospores (pulcherrima cells) are not present. After one month at 25°C pulcherrima cells are usually present, highly refractile, globose, (4) × (7–11) μm or subglobose, containing a single lipid globule, occasionally more than one; a thin ring and abundant sediment are present, a pellicle is absent.

Growth on malt agar: After one month at 25°C growth is cream colored, or if pulcherrimin pigment is produced, red-brown. The surface is smooth and glistening, in some strains sparsely papillate, edge entire. Pulcherrimin pigment is usually present in the medium.

Dalmau plate culture on corn meal agar: Aerobically, pseudomycelium is not produced; anaerobically, rudimentary pseudomycelium is formed, often profusely.

Formation of ascospores: Asci arise from spheroidal chlamydospores, which generally contain a single, prominent lipid globule. The lipid globule is no longer present in the ascus. Asci are sphaeropedunculate, (4–11) × (15–55) μm, usually (6–8) × (20–42) μm, peduncles are cylindrical, (1.0–2.5) × (10–45) μm, usually (1.5–2) × (15–30) μm, containing two, sometimes one, acicular to filiform ascospores, (0.4–1.5) × (9–27) μm (Fig. 70). In some strains, asci were observed to dehisce at the peduncle extremity.

Asci are observed on dilute, filtered V8 juice agar, and other dilute media, after incubation at 12–21°C (Pitt & Miller, 1968).

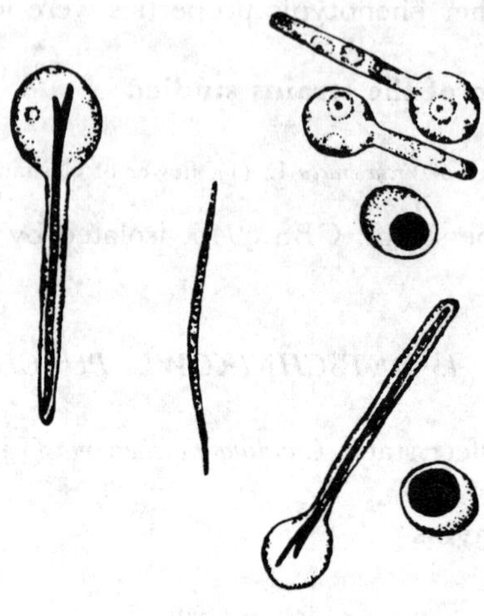

Fig. 70 – *M. pulcherrima*
Asci, ascospores and pulcherrima cells
After Pitt & Miller (1968)
(From 'The Yeasts', ed. II, 1970)

Fermentation:

Glucose	+	Maltose	–
Galactose	+ vw or –	Lactose	–
Sucrose	–	Raffinose	–

Assimilation of carbon compounds:

Galactose	+	Raffinose	–	Erythritol	–
Sucrose	+	Soluble starch	–	Ribitol	+
Maltose	+	D-Xylose	+	D-Mannitol	+
Cellobiose	+	L-Arabinose	–	Succinic acid	+
Trehalose	+	D-Ribose	v	Citric acid	v
Lactose	–	L-Rhamnose	–	Inositol	–

Additional carbon compounds tested: α-Methyl-D-glucoside +, D-glucitol +, glucono-δ-lactone +

Assimilation of nitrate: –

Growth in vitamin-free medium: –

Growth on 50% (w/w) glucose-yeast extract agar: + w or –

Growth at 37°C: v

G + C: 48.3 mol% (Meyer & Phaff, 1969)

Origin of the strains studied

Flowers (8), grapes (7), *Aesculus turbinata* (1), creek water (1), *Drosophila* spp. (4), *Salix* sp. (2), *Betula* sp. (1).

Type strain: CBS 5833, isolated by Mrak and McClung from grapes.

5. *METSCHNIKOWIA REUKAUFII* PITT ET MILLER (1968)

Imperfect state: *Candida reukaufii* (Grüss) Diddens et Lodder

Synonyms

Anthomyces reukaufii Grüss (1918)
Nectaromyces reukaufii (Grüss) H. et P. Sydow (1918)
Candida reukaufii (Grüss) Diddens et Lodder (1942)
Nectaromyces cruciatus Schoellhorn (1919)
Chlamydozyma zygota Wickerham (1964a)
Metschnikowia zygota Fell et Hunter (1968)

Standard description of *Metschnikowia reukaufii*

Growth in malt extract: After 3 days at 25°C vegetative cells are ellipsoidal to cylindroidal, (2–6) × (4–18) μm, usually (2–4) × (6–12) μm, single or catenulate, reproducing by multilateral budding. In some strains cells may be much longer, up to 35 μm. Chlamydospores are not present. After one month at 25°C somewhat elongate chlamydospores are usually present, (4–10) × (8–30) μm, usually (8–10) × (10–20) μm. They are highly refractile, containing several to numerous lipid droplets; a thin ring and abundant sediment are present, a pellicle is absent.

Growth on malt agar: After one month at 25°C growth is cream-colored, the surface is smooth and glistening, in some strains sparsely papillate, the edge is entire to crenulate.

Dalmau plate culture on corn meal agar: Aerobically, pseudomycelium is not produced; anaerobically, rudimentary pseudomycelium is profusely formed.

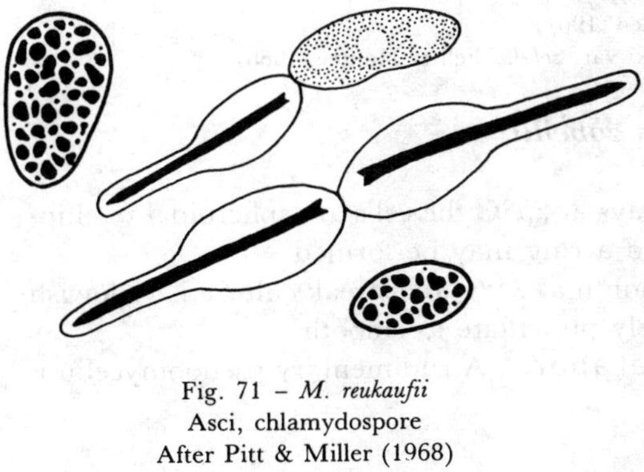

Fig. 71 – *M. reukaufii*
Asci, chlamydospore
After Pitt & Miller (1968)

Formation of ascospores: Asci arise from somewhat elongated chlamydospores that contain several to numerous lipid globules. Asci are ellipsoidopedunculate to clavate, (5–9) × (20–43) μm, usually (5–8) × (30–37) μm, containing two, rarely one, acicular to acerose ascospores (0.7–1.1) × (7–30) μm (Fig. 71). Asci do not dehisce.

Asci are observed on dilute, filtered V8 juice agar, and other dilute media at 12–21°C (Pitt & Miller, 1968).

Fermentation:

Glucose	+	Maltose	–
Galactose	+vw or –	Lactose	–
Sucrose	–	Raffinose	–

Assimilation of carbon compounds:

Galactose	+(–)	Raffinose	–	Erythritol	–
Sucrose	+	Soluble starch	–	Ribitol	+
Maltose	+	D-Xylose	+(–)	D-Mannitol	+
Cellobiose	+	L-Arabinose	–	Succinic acid	+
Trehalose	+	D-Ribose	+(–)	Citric acid	–
Lactose	–	L-Rhamnose	–	Inositol	–

Additional carbon compounds tested: α-Methyl-D-glucoside v, D-glucitol +, glucono-δ-lactone +

Assimilation of nitrate: –

Growth in vitamin-free medium: –

Growth on 50% (w/w) glucose-yeast extract agar: +w or –

Growth at 37°C: –

G + C: 42.2 mol% (Meyer & Phaff, 1969)

Origin of the strains studied

Flowers (12), unknown (1).

Type strain: CBS 5834, isolated by Spencer from *Epilobium angustifolium,* NWT, Canada.

6. *METSCHNIKOWIA ZOBELLII* (VAN UDEN ET CASTELO-BRANCO) VAN UDEN (1962a)

Synonyms

Metschnikowiella zobellii van Uden et Castelo-Branco (1961)
Metschnikowia bicuspidata (Metschnikoff) Kamienski var. *zobellii* Fell et Hunter (1968)

Standard description of *Metschnikowia zobellii*

Growth in malt extract: After 3 days at 25°C the cells are spheroidal to ellipsoidal, (6–11)×(7–11) μm. A pellicle and a ring may be formed.

Growth on malt agar: After one month at 25°C the streak culture is yellowish or greyish, semidull to dull, soft, delicately punctilate to smooth.

Dalmau plate culture on corn meal agar: A rudimentary pseudomycelium is formed.

Formation of ascospores: Asci arise from vegetative cells and are generally clavate, 18–24 μm long and contain a single, needle-shaped ascospore, pointed at one end and nearly as long as the respective ascus (Fig. 72).

Asci were observed on V8 agar.

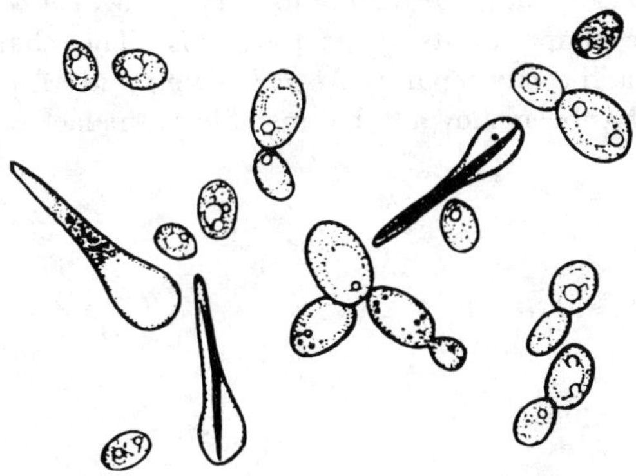

Fig. 72 – *M. zobellii.* Asci and vegetative cells on V8 medium
After van Uden & Castelo-Branco (1961) (From 'The Yeasts', ed. II, 1970)

Fermentation:

Glucose	+ or w	Maltose	–
Galactose	+	Lactose	–
Sucrose	–	Raffinose	–

Assimilation of carbon compounds:

Galactose	+	Raffinose	–	Erythritol	–
Sucrose	+	Soluble starch	–	Ribitol	+
Maltose	+	D-Xylose	+ or w	D-Mannitol	+
Cellobiose	+	L-Arabinose	–	Succinic acid	+
Trehalose	+	D-Ribose	–	Citric acid	–
Lactose	–	L-Rhamnose	–	Inositol	–

Additional carbon compounds tested: α-Methyl-D-glucoside +, D-glucitol +, glucono-δ-lactone +

Assimilation of nitrate: –

Growth in vitamin-free medium: –

Growth on 50% (w/w) glucose-yeast extract agar: +w or –

Growth at 37°C: –

G + C: 48 mol%, one strain (Meyer & Phaff, 1970)

Origin of the strains studied

Sea water (6), *Macrocystis pyrifera* (1), *Atherinopis affinis littoralis* (1).

Type strain: CBS 4821, isolated by van Uden and Castelo-Branco from sea water.

f. COMMENTS ON THE GENUS

The differences between *M. krissii* and *M. zobellii* consist of a number of one-way deficiences: inability of *M. krissii* to ferment glucose and to assimilate galactose, ribitol and D-glucitol.

Metschnikowia lunata can readily be distinguished from *M. pulcherrima* and *M. reukaufii* by the characteristic shape of its vegetative cells. The characteristic shapes of chlamydospores and asci easily separate *M. pulcherrima* and *M. reukaufii*. The production of pulcherrimin has been shown to be unstable so the lack of pigment production is not diagnostic.

Genus 17. *Nadsonia* Sydow

by M.W. Miller and H.J. Phaff

a. DIAGNOSIS OF THE GENUS

Cells lemon-shaped, oval, or elongate. Vegetative reproduction by bud-fission at both poles; this process involves the formation of a bud-like structure on a very wide neck. The bud is separated by the formation of a cross wall, followed by fission. Pseudomycelium is not formed; a few chains of elongate cells may be present.

Two mechanisms of ascospore formation are known. (1) After a heterogamic conjugation between the mother cell and a bud, the contents of the zygote move into another bud formed at the opposite end of the mother cell. This second bud is then delimited by a septum and becomes the ascus. (2) After a heterogamic conjugation between the mother cell and a bud, the bud is normally delimited by a septum and the mother cell becomes the ascus. One, or more rarely two, spherical, brownish, spiny- to warty-walled spores are formed. Spores contain a prominent lipid globule.

On malt extract a pellicle is usually formed.

Fermentation of glucose present or absent.

Nitrate is not utilized.

No growth above 30°C.

b. TYPE SPECIES OF THE GENUS

Nadsonia fulvescens (Nadson et Konokotina) Sydow

c. SPECIES ACCEPTED IN THE GENUS

1. *Nadsonia commutata* Golubev (1973)
2. *Nadsonia elongata* Konokotina (1913)
3. *Nadsonia fulvescens* (Nadson et Konokotina) Sydow (Nadson & Konokotina, 1911)

d. KEY TO THE SPECIES OF THE GENUS

1a Fermentation lacking, ascospores formed in the mother cell
N. commutata .... p. 280
b Fermentation present, ascospores formed in a bud, delimited by a septum 2

2a Maltose assimilated
N. fulvescens .... p. 283
b Maltose not assimilated
N. elongata .... p. 281

e. SYSTEMATIC DISCUSSION OF THE SPECIES

1. *NADSONIA COMMUTATA* GOLUBEV (1973)

Standard description of *Nadsonia commutata*

Growth in malt extract: After 3 days at 15°C cells are ovoidal to apiculate, (4.2–7.7) × (5.6–11.2) μm, single, in pairs or in short chains; budding is bipolar on a broad base. After one month a ring and sediment are formed, sometimes a pellicle.

Growth on malt agar: Cell shapes and dimensions similar to those observed in malt extract. After one month at 18°C the streak culture is cream-colored (brownish in sporulating cultures), smooth, semi-glossy; texture is pasty and the border is entire to lobate.

Dalmau plate culture on potato agar: Pseudomycelium is not formed.

Formation of ascospores: Asci are formed following conjugation between a mother cell and its bud. One to two spheroidal ascospores with a large lipid droplet are formed in the mother cell or less commonly in the bud (Fig. 73). Triad formation characteristic of the other species is not observed.

Sporulation takes place on YM or malt agar at 18°C.

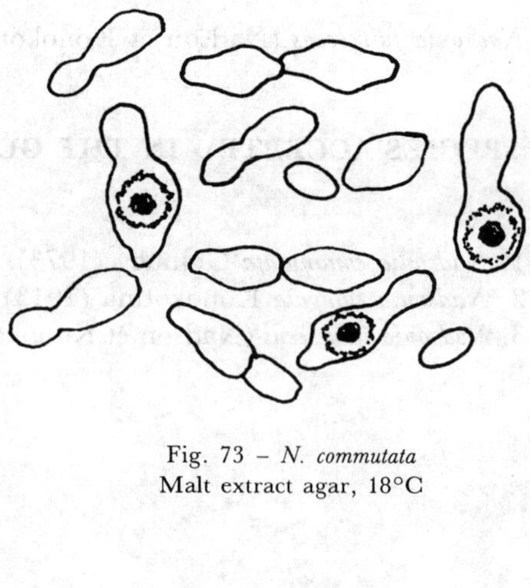

Fig. 73 – *N. commutata*
Malt extract agar, 18°C

Fermentation: –

Assimilation of carbon compounds:

Galactose	–	Raffinose	–	Erythritol	–
Sucrose	+s or –	Soluble starch	–	Ribitol	–
Maltose	– or +s	D-Xylose	–	D-Mannitol	–
Cellobiose	–	L-Arabinose	–	Succinic acid	+
Trehalose	–	D-Ribose	–	Citric acid	–
Lactose	–	L-Rhamnose	–	Inositol	–

Additional carbon compounds tested: Glucono-δ-lactone –, 2-ketogluconate –, 5-ketogluconate –

Assimilation of nitrate: –

Growth in vitamin-free medium: –

Growth on 50% (w/w) glucose-yeast extract agar: –

Growth at 37°C: –; maximum temperature for growth 24°C

G + C: 40.0 ± 0.3 mol%, type strain

Origin of the strain studied

Field soil from East Falkland Island, U.K. (South Atlantic Ocean) (1).

Type strain: CBS 6640 (BKM Y-1573) isolated from soil in 1972.

2. *NADSONIA ELONGATA* KONOKOTINA (1913)

Synonyms

Guilliermondia elongata Konokotina (1913)
Nadsonia richteri Kostka (1927)

Standard description of *Nadsonia elongata*

Growth in malt extract: After 3 days at room temperature cells are ovoidal to elongate or lemon-shaped, (4–8) × (7–18) μm, single, in pairs, or in short chains. Budding is bipolar on a very broad base. A septum forms across the constriction separating bud and mother cell and the two are separated by fission. A thin, cream-colored, slightly wrinkled, dull, creeping pellicle is present. After one month there is a thin pellicle; a moderate sediment is also present.

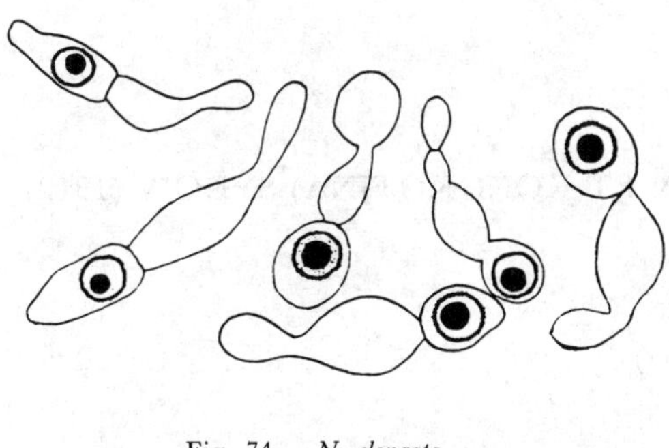

Fig. 74 – *N. elongata*
After 9 days on gypsum block
(From 'The Yeasts', ed. I, 1952)

Growth on malt agar: After 3 days at room temperature the cells are similar to those in malt extract. After one month at 18°C the streak culture is amber to brownish if spore formation is abundant. In poorly sporulating strains the color is greyish-cream. The surface is dull, usually very finely wrinkled, raised. The texture is pasty and the border entire to lobate.

Dalmau plate culture on potato agar: Pseudomycelium is absent or rudimentary; in the last case short chains of undifferentiated, elongate cells are formed.

Formation of ascospores: Sporulation takes place in the second bud as described under the diagnosis of the genus (Fig. 74). When an ascus is mature it normally separates by fission from the two empty cells of a triad.

Sporulation occurs abundantly on malt agar at 18°C often giving the growth an amber or brownish color due to the dark color of the spores.

Fermentation:

Glucose	+	Maltose	–
Galactose	–	Lactose	–
Sucrose	–		

Assimilation of carbon compounds:

Galactose	–	Raffinose	–	Erythritol	–
Sucrose	–	Soluble starch	–	Ribitol	–
Maltose	–	D-Xylose	–	D-Mannitol	–
Cellobiose	–	L-Arabinose	–	Succinic acid	+ or w
Trehalose	–	D-Ribose	–	Citric acid	–
Lactose	–	L-Rhamnose	–	Inositol	–

Additional carbon compounds tested: Glucono-δ-lactone + or w, 2-ketogluconate –, 5-ketogluconate –

Assimilation of nitrate: –

Growth in vitamin-free medium: +

Growth on 50% (w/w) glucose-yeast extract agar: –

Growth at 37°C: –; maximum temperature approx. 26°C

G + C: 41.8 ± 0.3 mol%, type strain

Origin of the strains studied

From slime fluxes of the following trees in Japan: *Aesculus* (1), *Alnus* (3), *Betula* (18), *Carpinus* (2), *Cornus* (1), *Eurya* (1), *Fagus* (1), *Prunus* (5) and *Sinoarundinaria* (1); in Russia: *Betula* (1), received by CBS from Nadson in 1927 labelled *N. fulvescens* (1), unknown (2).

Type strain: CBS 2594; it had been isolated by Konokotina in 1913 and was received by CBS in 1925.

3. *NADSONIA FULVESCENS* (NADSON ET KONOKOTINA) SYDOW (1912)

Synonym

Guilliermondia fulvescens Nadson et Konokotina (1911)

Standard description of *Nadsonia fulvescens*

Growth in malt extract: After 3 days at room temperature the cells are ovoidal to elongate or lemon-shaped, (5–8) × (8–16) μm and occasionally longer; single, in pairs, or in short chains (Fig. 75). Budding is bipolar on a very broad base. A septum forms across the constriction connecting bud and mother cell and the two are separated by fission. A thin, smooth, dry creeping pellicle is present. After one month a thin pellicle and a sediment are present.

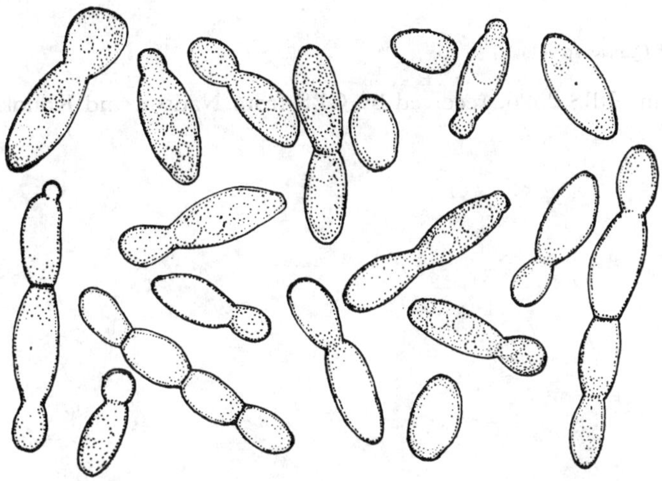

Fig. 75 – *N. fulvescens*
After 3 days in malt extract (From 'The Yeasts', ed. I, 1952)

Growth on malt agar: After 3 days at room temperature the cells are similar to those in malt extract. After one month at 18°C the streak culture is greyish to cream colored, nearly smooth, pasty, semidull, raised, border entire or lobulate.

Dalmau plate culture on potato agar: Pseudomycelium absent or rudimentary; in the last case short chains of undifferentiated elongate cells are formed.

Formation of ascospores: Sporulation, although not observed, takes place according to the original description as described for *N. elongata*. When an ascus is mature it normally separates from the two empty cells of a triad by fission.

Fermentation:

Glucose	+	Maltose	+ (w)
Galactose	+	Lactose	–
Sucrose	+	Raffinose	–

Assimilation of carbon compounds:

Galactose	+	Raffinose	–	Erythritol	–
Sucrose	+	Soluble starch	–	Ribitol	+
Maltose	+	D-Xylose	–	D-Mannitol	+
Cellobiose	–	L-Arabinose	–	Succinic acid	+
Trehalose	–	D-Ribose	–	Citric acid	+
Lactose	–	L-Rhamnose	–	Inositol	–

Additional carbon compounds tested: Glucono-δ-lactone +, 2-ketogluconate −, 5-ketogluconate −

Assimilation of nitrate: −

Growth in vitamin-free medium: −

Growth on 50% (w/w) glucose-yeast extract agar: −

Growth at 37°C: −

G + C: 41.5 ± 0.2 mol%, type strain

Origin of the strain studied

Slime flux of *Quercus* sp. (1)

Type strain: CBS 2596, received by CBS from Nadson and Konokotina in 1913.

Genus 18. *Nematospora* Peglion

by M.W. Miller and H.J. Phaff

a. DIAGNOSIS OF THE GENUS

Cells of various shapes reproducing vegetatively by multipolar budding. Well developed pseudomycelium usually present. True mycelium composed of sparsely septate, branched hyphae. No arthrospores.

Sporulating cells elongate, larger than the vegetative cells, dehiscing early after attaining maturity.

Spores spindle-shaped provided with a whip-like appendage; usually eight per ascus arranged in two bundles of four. Occasionally less or more than eight spores per ascus. Many spores give the impression of being septate near the equator of the spindle.

Fermentation present.

Nitrate not assimilated.

b. TYPE SPECIES OF THE GENUS

Nematospora coryli Peglion

c. SPECIES ACCEPTED IN THE GENUS

1. *Nematospora coryli* Peglion (1897)

e. SYSTEMATIC DISCUSSION OF THE SPECIES

1. *NEMATOSPORA CORYLI* PEGLION (1897)

Synonyms

? *Nematospora lycopersici* Schneider (1917)
? *Nematospora phaseoli* Wingard (1922)
Nematospora nagpuri Dastur (in Dastur & Singh, 1930)

Standard description of *Nematospora coryli*

Growth in malt extract: After 3 days at 25°C, the cells are ovoid, elongate, globose, or of bizarre shapes (Fig. 76). They occur singly, in pairs or in small chains. The relative proportions of cells of different shape vary from strain to strain. The size of the ovoid and globose cells varies between (1.7–13.9) × (4.2–14.0) μm. A butyrous sediment is present; some granular growth may occur.

Growth on malt agar: After one month at 25°C the streak culture is light cream colored; usually abundantly wrinkled, sometimes coarsely folded; glossy; pasty or membranous; slightly raised; border fringed with mycelium.

Dalmau plate culture on corn meal agar: True hyphae with few septa are present. Arthrospores are not formed. Pseudomycelium is usually abundant (Fig. 77). The blastospores are ovoid or cylindrical, forming short chains in verticillated positions.

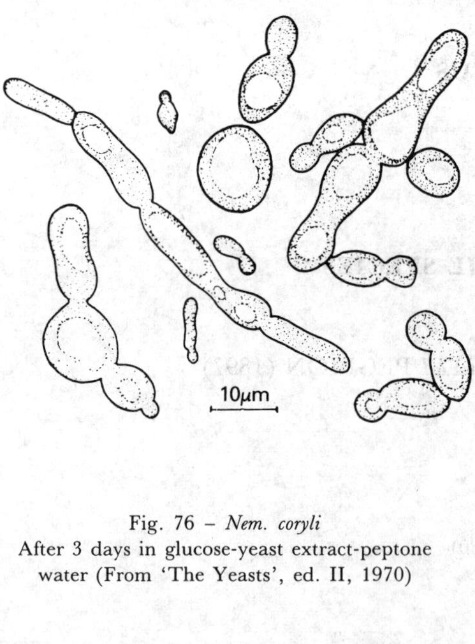

Fig. 76 – *Nem. coryli*
After 3 days in glucose-yeast extract-peptone water (From 'The Yeasts', ed. II, 1970)

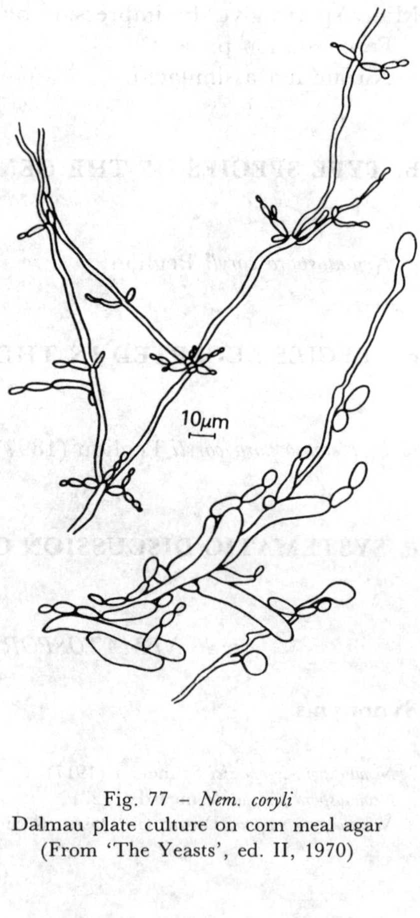

Fig. 77 – *Nem. coryli*
Dalmau plate culture on corn meal agar (From 'The Yeasts', ed. II, 1970)

Formation of ascospores: Conjugation preceding spore formation is not apparent. Very long, cylindrical sporulating cells with rounded ends, sometimes of irregular shape and slightly curved, occur. Usually eight spores are formed per sporulating cell. Cells with two and four spores occasionally occur. Rarely, cells containing 12 or 16 spores are present. The sporulating cells dehisce very soon after attaining maturity. The spores are spindle-shaped, smooth, provided with a very long, thin, whip-like appendage at one end (Fig. 78). Direct examination under the light microscope reveals the presence of vacuoles and a difference of refringence between the two halves on each side of the equator of the spindle. Staining with the periodic-Schiff technique shows that the wall of the half to which the whip is attached takes a deep pink shade whereas the other half remains almost unstained. When stained with the Ziehl-Neelsen or the Schaeffer-Fulton malachite green technique the half supporting the whip-like appendage is not acid fast and does not retain the malachite green; the other half is acid fast and retains the malachite green (do Carmo-Sousa, 1970a). Electron micrographs of longitudinal sections have shown that the spores lack a septum (Kreger-van Rij, 1969a). The spores are usually grouped in two bundles of four within the ascus. The whip-free ends of the spore bundles are directed towards the poles of the mother cell. Germination of the spores usually takes place near the equator of the spindle on the side of the whip-supporting half and gives rise either to buds or hyphae.

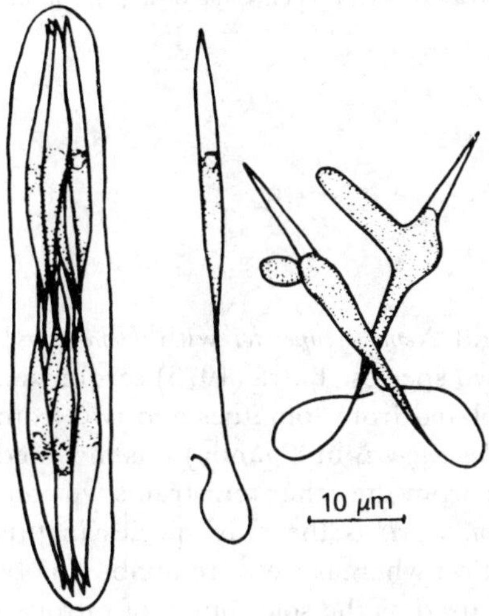

Fig. 78 – *Nem. coryli*
After 3 days on V8 agar
(From 'The Yeasts', ed. II, 1970)

Sporulation was observed after 2–3 days on V8 agar in five out of nine strains. Some strains also sporulated on malt agar and glucose-yeast extract-peptone agar.

Fermentation:

Glucose	+ or w	Lactose	–
Galactose	–	Melibiose	–
Sucrose	+ or w	Raffinose	+ w
Maltose	+ or w	Trehalose	v

Assimilation of carbon compounds:

Galactose	–	Raffinose	+ or w	Erythritol	–
Sucrose	+ or w	Soluble starch	+ or w	Ribitol	–
Maltose	+ or w	D-Xylose	–	D-Mannitol	–
Cellobiose	–	L-Arabinose	–	Succinic acid	+
Trehalose	+	D-Ribose	–	Citric acid	–
Lactose	–	L-Rhamnose	–	Inositol	–

Assimilation of nitrate: –

Growth in vitamin-free medium: –
Growth on 50% (w/w) glucose-yeast extract agar: –
Growth at 37°C: +
G + C: 43.4 ± 0.2 mol%, type strain

Origin of the strains studied

Diseased hazelnut (1), diseased cotton bolls (4), tomato (1), *Phaseolus vulgaris* (1), Lima bean (1), sputum (1), soybean (1), unknown (1).

Type strain: CBS 2608 isolated by Peglion.

f. COMMENTS ON THE GENUS

The question of synonymy for *Nem. phaseoli* and *Nem. lycopersici* with *Nem. coryli* results from the lack of authentic strains of these two species. Batra (1973) recognized *Nem. lycopersici* as a species based on one strain isolated from tomatoes and two from beans. One of these (CMI No. 77-422 from *Phaseolus vulgaris* in Uganda) was included in our study. Since its characteristics did not differ from the other ten strains we concluded, as did do Carmo-Sousa (1970a) that *Nem. coryli* is the only species in this genus. Batra also reported a positive nitrate reduction whereas we were unable to obtain growth of any of the strains when nitrate was used as the sole source of nitrogen in the medium.

Species of *Nematospora* are parasitic on plants, causing diseases known as stigmatomycosis. They have been found in many species of plants throughout the world and are always associated with insect punctures of the plant tissue made by insects with piercing-sucking mouth parts. Frazer (1944) has made a very careful study of the infection of cotton bolls by the cotton stainer, *Dysdercus*. The spores of *Nematospora* appear to be harbored in the stylet pouches of *Dysdercus suturellus*.

Genus 19. *Pachysolen* Boidin et Adzet

by C.P. Kurtzman

a. DIAGNOSIS OF THE GENUS

Asexual reproduction by multilateral budding on a narrow base. Cells are spheroidal to ellipsoidal and may have one or two small apiculi. Pseudohyphae may be poorly developed or absent.

Asci are quite unusual and their presence allows immediate recognition of this genus. Generally, a tube grows from a vegetative cell and the end of the tube forms an ascus usually containing four ascospores. The walls of the cell and tube become quite thickened and refractile except for the end bearing the ascospores. Here the wall is thin and dehisces to allow liberation of the ascospores. The ascospores are hemispheroidal with a narrow ledge at the base.

Glucose is fermented.

Nitrate is assimilated.

b. TYPE SPECIES OF THE GENUS

Pachysolen tannophilus Boidin et Adzet

c. SPECIES ACCEPTED IN THE GENUS

1. *Pachysolen tannophilus* Boidin et Adzet (1957)

e. SYSTEMATIC DISCUSSION OF THE SPECIES

1. PACHYSOLEN TANNOPHILUS BOIDIN ET ADZET (1957)

Synonyms

Hansenula tannophilus (Boidin et Adzet) Campbell (1973)
Pachysolen pelliculatus Boidin et Adzet (1957)

Standard description of *Pachysolen tannophilus*

Growth on 5% malt extract agar: After 3 days at 25°C, the cells are spheroidal to ellipsoidal (1.5–5.0) × (2.0–7.0) μm and usually have one or two buds. Growth is mucoid to butyrous and tannish-white.

Growth on the surface of assimilation media: Pellicles are formed by some strains.

Dalmau plate culture on morphology agar: After 7 days at 25°C, growth under the coverglass exhibits simple undifferentiated pseudohyphae as well as occasional highly branched strands. Aerobic growth is tannish-white, glistening, mucoid, low convex with margin entire and with a faint smell of esters.

Formation of ascospores: The ascus forms when a vegetative cell produces a stout tube, generally at one end. Depending upon the culture medium, and perhaps other factors, tubes may be quite short or up to 60 μm in length and are straight or curved. The tip of the tube enlarges to form the ascus, and consequently, the tube may be regarded as an ascophore. Asci contain up to four hemispheroidal ascospores that have a narrow ledge at the base. The ascus wall dehisces releasing the spores. Once this happens, it is seen that the ascus formed within a V-shaped notch at the end of the ascophore. Ascophore walls appear greatly thickened and become quite refractile. This characteristic of highly refractile ascophores bearing terminal asci makes *Pachysolen* readily identifiable under the light microscope even at moderate magnifications (Fig. 79). Asci may be conjugated or unconjugated depending upon the strain examined. The type strain NRRL Y-2460 forms unconjugated asci while NRRL Y-2461, Y-2462, and Y-2463 have conjugated asci. All cultures sporulated on YM agar at 25°C after about 1 week. Single spore isolates from NRRL Y-2460 gave sporogenous colonies, therefore, the species appears homothallic (Wickerham, 1970c).

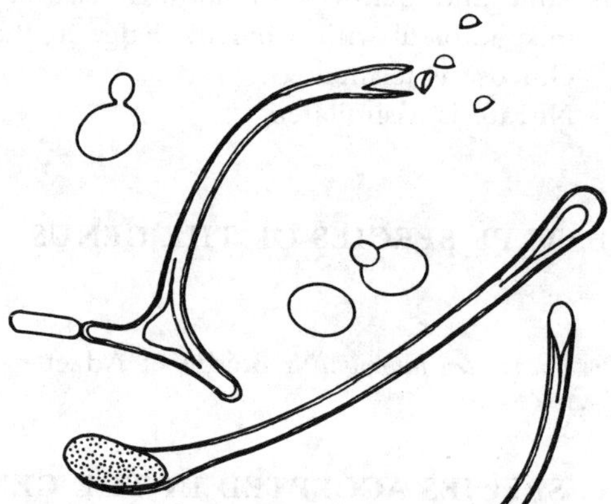

Fig. 79 – *Pa. tannophilus*
After 2 weeks on YM agar

Spores were observed on YM-, Gorodkowa-, and malt extract agar.

Fermentation:

Glucose	+	Maltose	–
Galactose	–	Lactose	–
Sucrose	–	Raffinose	–

Assimilation of carbon compounds:

Galactose	+	Raffinose	–	Erythritol	–
Sucrose	–	Soluble starch	–	Ribitol	+

Maltose	–	D-Xylose	+	D-Mannitol	+
Cellobiose	+ or s	L-Arabinose	+	Succinic acid	+
Trehalose	–	D-Ribose	– or +s	Citric acid	–
Lactose	–	L-Rhamnose	–	Inositol	–

Assimilation of nitrate: +

Growth in vitamin-free medium: –

Growth in 10% sodium chloride plus 5% glucose in yeast nitrogen base: +

Growth at 37°C: +

G + C: 43.0 mol%, one strain (Nakase & Komagata, 1968a)

Comments

Pachysolen tannophilus was originally separated from *Pa. pelliculatus* because the latter species formed a mat growth on solid media, produced a thin, dry pellicle, had greater development of pseudohyphae, and gave a weaker fermentation of glucose. These differences seem insufficient to differentiate the two species and *Pa. pelliculatus* is regarded to be a synonym. Wickerham (1970c) showed that a mat form could be selected from *Pa. tannophilus* by allowing an amoeba to feed on the cultures. The amoeba preferentially ingested mucoid colonies and allowed proliferation of the mat form.

Origin of the strains studied

Concentrated tanning liquors (4).

Type strain: CBS 4044 (NRRL Y-2460), designated by Boidin and Adzet (1957) as the type.

Genus 20. *Pachytichospora* van der Walt

by J.P. van der Walt

a. DIAGNOSIS OF THE GENUS

Heterothallic.

Diploid, vegetative cells subglobose, ellipsoidal to cylindrical, reproducing by budding. Haploid cells of opposite mating types sexually agglutinative.

Asci unconjugated, persistent, one- to two-spored. Ascospores spheroidal to ellipsoidal, glabrous, with eccentrically thickened, refringent walls, occasionally laterally indented.

Fermentation.

Coenzyme Q-6 system present.

Nitrate not utilized.

b. TYPE SPECIES OF THE GENUS

Pachytichospora transvaalensis (van der Walt) van der Walt

c. SPECIES ACCEPTED IN THE GENUS

1. *Pachytichospora transvaalensis* (van der Walt) van der Walt (van der Walt, 1956d)

e. SYSTEMATIC DISCUSSION OF THE SPECIES

1. PACHYTICHOSPORA TRANSVAALENSIS (VAN DER WALT) VAN DER WALT (1978)

Synonym

Saccharomyces transvaalensis van der Walt (1956d)

Standard description of *Pachytichospora transvaalensis*

Growth in malt extract: After 3 days the cells are subglobose, ellipsoidal, cylindrical to elongate, (2.5–6.0) × (4.5–9.0–22) μm, reproducing by budding, and occur singly or in pairs (Fig. 80). A sediment is formed. After one month at room temperature, a sediment is present and occasionally a ring as well.

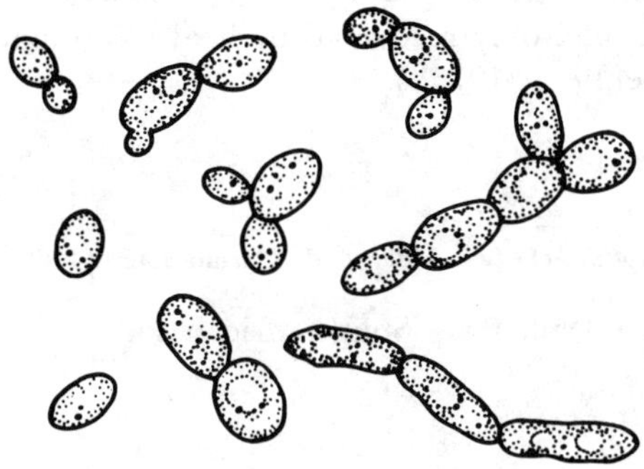

Fig. 80 – *P'spora transvaalensis*
After 3 days in malt extract

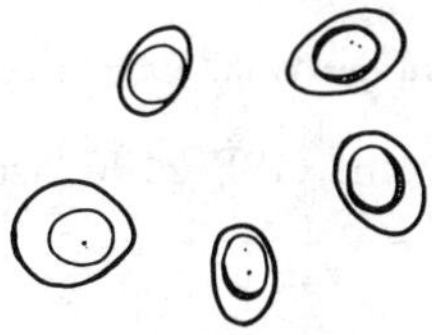

Fig. 81 – *P'spora transvaalensis*
After 7 days on YM agar

Growth on malt agar: After one month at room temperature the streak culture is butyrous, light-brown to brown, shiny or somewhat shiny, flat, smooth to finely crispulate. The margin is entire, undulating or lobiform.

Dalmau plate culture on corn meal agar: Under anaerobic conditions a rudimentary pseudomycelium is formed.

Formation of ascospores: Diploid, vegetative cells are directly transformed into persistent asci. Usually one or infrequently two ascospores are formed per ascus. The ascospores are spheroidal to ellipsoidal, with stout, refringent walls (Fig. 81). When examined in ultrathin sections by TEM the spores have eccentrically thickened walls, and occasionally appear to be laterally indented. Heat treatment of ascosporic material leads to the recovery of sexually agglutinative mating types. Ascospore formation is best observed on YM- and acetate agar.

Fermentation:

Glucose	+	Maltose	–
Galactose	+	Lactose	–
Sucrose	–		

Assimilation of carbon compounds:

Galactose	+	Raffinose	–	Erythritol	–
Sucrose	–	Soluble starch	–	Ribitol	–
Maltose	–	D-Xylose	–	D-Mannitol	–
Cellobiose	–	L-Arabinose	–	Succinic acid	–
Trehalose	+	D-Ribose	–	Citric acid	–
Lactose	–	L-Rhamnose	–	Inositol	–

Assimilation of nitrate: –
Growth in vitamin-free medium: –
Growth on 50% (w/w) glucose-yeast extract agar: –
Growth at 37°C: v

Comments

Pachytichospora transvaalensis is recovered from nature as the diploid, bisexual state. Sexually agglutinative mating types are, as a rule, only recovered by heat treatment of ascosporic material (van der Walt & Liebenberg, 1973b).

Origin of the strains studied

Soil (2) (van der Walt, 1956d); cecum of *Papio cyanocephalus* (1) (van Uden & do Carmo-Sousa, 1962).

Type strain: CBS 2186, isolated by van der Walt from South African soil.

Genus 21. *Pichia* Hansen

by C.P. Kurtzman

a. DIAGNOSIS OF THE GENUS

Asexual reproduction is by multilateral budding on a narrow base. Some species may also form arthrospores. Cells are spheroidal, ellipsoidal, or elongate and occasionally may be tapered, but they are not ogival in shape. Pseudohyphae or, to a limited extent, true hyphae may be present.

Asci produce one to four (infrequently more) ascospores that may be hat-shaped, hemispheroidal, Saturn-shaped, or spheroidal with a smooth surface. Generally asci are dehiscent, but occasionally they are persistent. Asci are unconjugated, or if conjugated, they may show conjugation between bud and parent or between independent cells. Hyphal or pseudohyphal cells may serve as asci, but they do not become swollen or spindle-like. Asci are not borne on ascophores. Species are homothallic or heterothallic.

Sugars may or may not be fermented.

Nitrate is not assimilated.

b. TYPE SPECIES OF THE GENUS

Pichia membranaefaciens Hansen

c. SPECIES ACCEPTED IN THE GENUS

1. *Pichia abadieae* Jacob (1969b)
2. *Pichia acaciae* van der Walt (1966b)
3. *Pichia ambrosiae* van der Walt et Scott (1971a)
4. *Pichia amethionina* Starmer, Phaff, Miranda et Miller (1978b)
   Pichia amethionina Starmer, Phaff, Miranda et Miller var. *amethionina*
   Pichia amethionina Starmer, Phaff, Miranda et Miller var. *pachycereana* Starmer, Phaff, Miranda et Miller (1978b)
5. *Pichia amylophila* Kurtzman, Smiley, Johnson, Wickerham et Fuson (1980a)
6. *Pichia angophorae* Miller et Barker (1968)
7. *Pichia besseyi* Kurtzman et Wickerham (1972)

8. *Pichia bovis* van Uden et do Carmo-Sousa (1957)
9. *Pichia burtonii* Boidin, Pignal, Lehodey, Vey et Abadie (1964)
10. *Pichia cactophila* Starmer, Phaff, Miranda et Miller (1978a)
11. *Pichia carsonii* Phaff et Knapp (1956)
12. *Pichia castillae* Santa María et García Aser (1970)
13. *Pichia chambardii* (Ramirez et Boidin) Phaff (Ramirez & Boidin, 1954)
14. *Pichia delftensis* Beech (1965)
15. *Pichia dispora* (Dekker) Kreger-van Rij (Stelling-Dekker, 1931)
16. *Pichia etchellsii* Kreger-van Rij (1964a)
17. *Pichia farinosa* (Lindner) Hansen (Lindner, 1894)
18. *Pichia fermentans* Lodder (1932)
19. *Pichia fluxuum* (Phaff et Knapp) Kreger-van Rij (Phaff & Knapp, 1956)
20. *Pichia guilliermondii* Wickerham (1966)
21. *Pichia haplophila* Shifrine et Phaff (1956)
22. *Pichia heedii* Phaff, Starmer, Miranda et Miller (1978)
23. *Pichia heimii* Pignal (1970)
24. *Pichia humboldtii* Rodrigues de Miranda et Török (1976)
25. *Pichia kluyveri* Bedford ex Kudriavzev (1960)
26. *Pichia lindneri* Henninger et Windisch (1975a)
27. *Pichia media* Boidin, Pignal, Lehodey, Vey et Abadie (1964)
28. *Pichia membranaefaciens* Hansen (1888b)
29. *Pichia methanolica* Makiguchi (in Kato et al., 1974)
30. *Pichia mississippiensis* Kurtzman, Smiley, Johnson, Wickerham et Fuson (1980a)
31. *Pichia mucosa* Wickerham et Kurtzman (1971)
32. *Pichia naganishii* Kodama (in Kodama & Kyono, 1974a)
33. *Pichia nakazawae* Kodama (1975)
    Pichia nakazawae Kodama var. *nakazawae*
    Pichia nakazawae Kodama var. *akitaensis* Kodama (1975)
34. *Pichia norvegensis* Leask et Yarrow (1976)
35. *Pichia ohmeri* (Etchells et Bell) Kreger-van Rij (Etchells & Bell, 1950a)
36. *Pichia onychis* Yarrow (1965)
37. *Pichia opuntiae* Starmer, Phaff, Miranda, Miller et Barker (1979)
    Pichia opuntiae Starmer, Phaff, Miranda, Miller et Barker var. *opuntiae*
    Pichia opuntiae Starmer, Phaff, Miranda, Miller et Barker var. *thermotolerans* Starmer, Phaff, Miranda, Miller et Barker (1979)
38. *Pichia pastoris* (Guilliermond) Phaff (Guilliermond, 1919)
39. *Pichia philogaea* van der Walt et Johannsen (1975c)
40. *Pichia pijperi* van der Walt et Tscheuschner (1957a)
41. *Pichia pinus* (Holst) Phaff (Holst, 1936)
42. *Pichia quercuum* Phaff et Knapp (1956)
43. *Pichia rabaulensis* Soneda et Uchida (1971)
44. *Pichia rhodanensis* (Ramirez et Boidin) Phaff (Ramirez & Boidin, 1953a)
45. *Pichia saitoi* Kodama, Kyono et Kodama (1962)
46. *Pichia salictaria* Phaff, Miller et Spencer (1964)
47. *Pichia sargentensis* Wickerham et Kurtzman (1971)
48. *Pichia scolyti* (Phaff et Yoneyama) Kreger-van Rij (Phaff & Yoneyama, 1961)

49. *Pichia segobiensis* Santa María et García Aser (1977)
50. *Pichia spartinae* Ahearn, Yarrow et Meyers (1970)
51. *Pichia stipitis* Pignal (1967)
52. *Pichia strasburgensis* (Ramirez et Boidin) Phaff (Ramirez & Boidin, 1953a)
53. *Pichia toletana* (Socias, Ramirez et Peláez) Kreger-van Rij (Socias et al., 1954b)
54. *Pichia trehalophila* Phaff, Miller et Spencer (1964)
55. *Pichia veronae* Kodama (in Kodama & Kyono, 1974b)
56. *Pichia wickerhamii* (van der Walt) Kreger-van Rij (van der Walt, 1959b)

d. KEY TO THE SPECIES OF THE GENUS

1a Maltose assimilated 2
b Maltose not assimilated 31

2a Sucrose assimilated 3
b Sucrose not assimilated 29

3a Galactose assimilated 4
b Galactose not assimilated 17

4a D-Xylose assimilated 5
b D-Xylose not assimilated
P. ohmeri .... p. 348

5a Erythritol assimilated 6
b Erythritol not assimilated 13

6a L-Rhamnose assimilated 7
b L-Rhamnose not assimilated 11

7a Citric acid assimilated 8
b Citric acid not assimilated
P. naganishii .... p. 344

8a Raffinose assimilated 9
b Raffinose not assimilated 10

9a Melibiose assimilated
P. scolyti .... p. 365
b Melibiose not assimilated
P. heimii .... p. 333

10a Lactose assimilated
P. stipitis .... p. 369
b Lactose not assimilated
P. nakazawae var. *nakazawae* .... p. 346

11a Raffinose assimilated
P. burtonii .... p. 314
b Raffinose not assimilated 12

12a Starch assimilated
P. nakazawae var. *akitaensis* .... p. 347
b Starch not assimilated
P. philogaea .... p. 354

13a D-Arabinose assimilated 14
b D-Arabinose not assimilated 15

14a Raffinose assimilated
P. guilliermondii .... p. 329
b Raffinose not assimilated
P. segobiensis .... p. 366

15a L-Rhamnose assimilated
P. strasburgensis .... p. 370
b L-Rhamnose not assimilated 16

16a Starch assimilated
P. carsonii .... p. 317
b Starch not assimilated
P. etchellsii .... p. 323

17a L-Sorbose assimilated 18
b L-Sorbose not assimilated 19

18a Citric acid assimilated
P. spartinae .... p. 367
b Citric acid not assimilated
P. mucosa .... p. 343

19a Erythritol assimilated
P. ambrosiae .... p. 305
b Erythritol not assimilated 20

20a Ribitol assimilated 21
b Ribitol not assimilated 23

21a Raffinose assimilated
P. rabaulensis .... p. 359
b Raffinose not assimilated 22

22a L-Rhamnose assimilated
P. wickerhamii .... p. 374
b L-Rhamnose not assimilated
P. angophorae .... p. 309

23a Growth at 37°C 24
b No growth at 37°C 28

24a Raffinose assimilated
P. onychis .... p. 350
b Raffinose not assimilated 25

25a L-Arabinose assimilated 26
b L-Arabinose not assimilated
P. rhodanensis .... p. 360

26a True hyphae produced 27
b True hyphae not produced
P. bovis .... p. 312

27a Starch assimilated
P. amylophila .... p. 308
b Starch not assimilated
P. mississippiensis .... p. 342

28a Potassium D-gluconate assimilated
P. veronae .... p. 373
b Potassium D-gluconate not assimilated
P. toletana .... p. 371

29a Melibiose assimilated
P. castillae .... p. 319
b Melibiose not assimilated 30

30a Glucose fermented
P. acaciae .... p. 304
b Glucose not fermented
P. media .... p. 337

31a Galactose assimilated 32
b Galactose not assimilated 37

32a Melibiose assimilated
P. abadieae .... p. 302
b Melibiose not assimilated 33

33a D-Ribose assimilated 34
b D-Ribose not assimilated 36

34a Citric acid assimilated
P. farinosa .... p. 325
b Citric acid not assimilated 35

35a Cellobiose assimilated
P. methanolica .... p. 341
b Cellobiose not assimilated
P. haplophila .... p. 330

36a Cellobiose assimilated
P. chambardii .... p. 320
b Cellobiose not assimilated
P. humboldtii .... p. 334

37a Trehalose assimilated 38
b Trehalose not assimilated 43

38a Cellobiose assimilated 39
b Cellobiose not assimilated 40

39a Erythritol assimilated
P. pinus .... p. 356
b Erythritol not assimilated
P. lindneri .... p. 336

40a D-Ribose assimilated
P. trehalophila .... p. 372
b D-Ribose not assimilated 41

41a L-Rhamnose assimilated
P. pastoris .... p. 353
b L-Rhamnose not assimilated 42

42a Pellicles present on liquid media; asci dehiscent
P. saitoi .... p. 361
b Pellicles not present on liquid media; asci persistent
P. dispora .... p. 322

43a Potassium D-gluconate assimilated 44
b Potassium D-gluconate not assimilated 46

44a Cellobiose assimilated
P. salictaria . . . . p. 363
b Cellobiose not assimilated 45

45a Mannitol assimilated
P. amethionina var. *pachycereana* . . . . p. 307
b Mannitol not assimilated
P. amethionina var. *amethionina* . . . . p. 306

46a Glucose fermented 47
b Glucose not fermented 56

47a Mannitol assimilated 48
b Mannitol not assimilated 52

48a Cellobiose assimilated 49
b Cellobiose not assimilated 51

49a D-Xylose assimilated 50
b D-Xylose not assimilated
P. quercuum . . . . p. 358

50a L-Rhamnose assimilated
P. sargentensis . . . . p. 364
b L-Rhamnose not assimilated
P. pijperi . . . . p. 355

51a Ribitol assimilated
P. delftensis . . . . p. 321
b Ribitol not assimilated
P. besseyi . . . . p. 311

52a Cellobiose assimilated
P. norvegensis . . . . p. 347
b Cellobiose not assimilated 53

53a Glucose fermentation weak 54
b Glucose fermentation strong 55

54a Isolated from cacti
P. cactophila . . . . p. 315
b Not isolated from cacti
P. membranaefaciens . . . . p. 339

55a D-Xylose assimilated
P. fermentans .... p. 326
b D-Xylose not assimilated
P. kluyveri .... p. 335

56a Salicin assimilated 57
b Salicin not assimilated 58

57a Citric acid assimilated
P. opuntiae var. *opuntiae* .... p. 351
b Citric acid not assimilated
P. opuntiae var. *thermotolerans* .... p. 352

58a Mannitol assimilated 59
b Mannitol not assimilated 60

59a Growth in yeast nitrogen base with 10% sodium chloride plus 5% glucose
P. delftensis .... p. 321
b No growth in yeast nitrogen base with 10% sodium chloride plus 5% glucose
P. fluxuum .... p. 328

60a Isolated from cacti 61
b Not isolated from cacti
P. membranaefaciens .... p. 339

61a D-Glucosamine assimilated
P. cactophila .... p. 315
b D-Glucosamine not assimilated
P. heedii .... p. 331

Table 18 (see pocket on inside cover)

e. SYSTEMATIC DISCUSSION OF THE SPECIES

1. PICHIA ABADIEAE JACOB (1969b)

Standard description of *Pichia abadieae*

Growth on 5% malt extract agar: After 3 days at 25°C, the cells are spheroidal to ellipsoidal (1.2–4.1) × (2.0–5.4) μm, and occur singly or in pairs. Growth is butyrous and tannish-white in color.

Growth on the surface of assimilation media: Thin climbing pellicles are formed.

Dalmau plate culture on morphology agar: After 7 days at 25°C, growth under the coverglass shows occasional, sparingly branched pseudohyphae, but no true hyphae. Aerobic growth is light tannish-white in color, smooth, glistening and butyrous with a finely serrate margin.

Formation of ascospores: Ascospores were not observed in the present study. Jacob (1969b) reported the asci, which became dehiscent, to produce one to four subspheroidal spores devoid of surface ornamentation. Asci were unconjugated or showed mother–daughter conjugation.

Spores were observed on sterilized carrot slices (Jacob, 1969b).

Fermentation:

Glucose	+ s	Lactose	–
Galactose	+ s	Raffinose	–
Sucrose	–	Trehalose	+ s
Maltose	–		

Assimilation of carbon compounds:

Galactose	+	Raffinose	–	Erythritol	+ w
Sucrose	–	Soluble starch	–	Ribitol	+
Maltose	–	D-Xylose	+	D-Mannitol	+ s
Cellobiose	–	L-Arabinose	+	Succinic acid	+
Trehalose	+	D-Ribose	–	Citric acid	+
Lactose	+ s	L-Rhamnose	+	Inositol	+

Additional carbon compounds tested: L-Sorbose –, melibiose +, melezitose –, inulin –, D-arabinose –, D-glucosamine · HCl +, glucitol +, α-methyl-D-glucoside –, salicin –, potassium D-gluconate + w, DL-lactic acid +

Assimilation of nitrate: –

Growth in vitamin-free medium: –

Growth in 10% sodium chloride plus 5% glucose in yeast nitrogen base: –

Growth at 37°C: –

Comments

Jacob (1969b) reported growth on succinic acid and ethanol to be negative and growth on L-sorbose to be positive. In the present study, growth was obtained on the former two compounds, but not on L-sorbose.

The affinities of *P. abadieae* within the genus *Pichia* are not at all certain, but should the ascospores prove glabrous in the electron microscope, a relationship with *P. carsonii* and *P. etchellsii* would be indicated.

Origin of the strain studied

Tanning liquor, France (1).

Type strain: CBS 6067 (NRRL Y-7499).

2. *PICHIA ACACIAE* VAN DER WALT (1966b)

Standard description of *Pichia acaciae*

Growth on 5% malt extract agar: After 3 days at 25°C, the cells are spheroidal to elongate (1.5–5.5) × (2.5–11) μm, and occur singly, in pairs or in small clusters. Growth is butyrous and tannish-white in color.

Growth on the surface of assimilation media: Pellicles are formed.

Dalmau plate culture on morphology agar: After 7 days at 25°C, growth under the coverglass shows moderately well developed pseudohyphae but no true hyphae. Aerobic growth is butyrous, dull, greyish-white, and has a somewhat wrinkled surface and a lobed margin.

Formation of ascospores: Asci show mother–daughter conjugation, or rarely, conjugation between independent cells. Two to four hat-shaped spores are produced in each ascus, and the spores are released soon after formation.

Spores were observed on YM- and 5% malt extract agar.

Fermentation:

Glucose	+	Lactose	–
Galactose	–	Raffinose	–
Sucrose	–	Trehalose	+
Maltose	+ w		

Assimilation of carbon compounds:

Galactose	+	Raffinose	–	Erythritol	+
Sucrose	–	Soluble starch	–	Ribitol	+
Maltose	+	D-Xylose	– or + w	D-Mannitol	+
Cellobiose	+	L-Arabinose	+	Succinic acid	+
Trehalose	+	D-Ribose	v	Citric acid	+
Lactose	–	L-Rhamnose	–	Inositol	–

Additional carbon compounds tested: L-Sorbose v, melibiose –, melezitose –, inulin –, D-arabinose –, D-glucosamine · HCl –, glucitol +, α-methyl-D-glucoside +, salicin +, potassium D-gluconate + s, DL-lactic acid –

Assimilation of nitrate: –

Growth in vitamin-free medium: –

Growth in 10% sodium chloride plus 5% glucose in yeast nitrogen base: +

Growth at 37°C: +

Comments

Some differences in carbon assimilation for *P. acaciae* were noted among van der Walt's (1966b) original description, Kreger-van Rij's (1970c) study and the present treatment. Consequently, growth on soluble starch and ribitol might be variable rather than positive as given in this study. Additionally, I found the type strain to give a late, weak reaction on D-xylose rather than a negative reaction as previously reported.

P. acaciae is phenotypically quite similar to *P. media,* and the two differ only in that *P. media* is nonfermentative.

Origin of the strains studied

Frass of Buprestidae larvae infesting *Acacia nilotica* var. *kraussiana,* South Africa (1); tunnels of *Sinoxylon ruficorne* in *Combretum apiculatum,* South Africa (1).

Type strain: CBS 5656 (NRRL Y-7117), from frass of Buprestidae larvae.

3. *PICHIA AMBROSIAE* VAN DER WALT ET SCOTT (1971a)

Synonym

Ambrosiozyma ambrosiae (van der Walt et Scott) van der Walt (1972)

Standard description of *Pichia ambrosiae*

Growth on 5% malt extract agar: After 3 days at 25°C, the cells are spheroidal to elongate (2.3–5.5) × (3.9–11.0) μm, and occur singly, in pairs and in clusters. Growth is butyrous and tannish-white in color.

Growth on the surface of assimilation media: Pellicles are not formed.

Dalmau plate culture on morphology agar: After 7 days at 25°C, growth under the coverglass shows moderately branched pseudohyphae and occasional true hyphae. Aerobic growth is tannish-white, glistening and butyrous with an entire to finely lobed margin.

Formation of ascospores: Ascus formation is usually preceded by conjugation between independent cells. The zygote generally divides to form a short chain of cells, and one or sometimes several successive terminal cells serve as asci. The asci, which become dehiscent, usually produce four hat-shaped spores. Van der Walt and Scott (1971a) reported that single-spore isolates form sporogenous colonies thus indicating the species to be homothallic.

Spores were observed on V8 agar.

Fermentation: –

Assimilation of carbon compounds:

Galactose	–	Raffinose	–	Erythritol	+
Sucrose	+	Soluble starch	v	Ribitol	+
Maltose	+	D-Xylose	+	D-Mannitol	+
Cellobiose	+	L-Arabinose	–	Succinic acid	+
Trehalose	+	D-Ribose	+	Citric acid	+
Lactose	–	L-Rhamnose	–	Inositol	–

Additional carbon compounds tested: L-Sorbose –, melibiose –, melezitose +, inulin –, D-arabinose –, D-glucosamine · HCl –, glucitol +, α-methyl-D-glucoside +, salicin +, potassium D-gluconate –, DL-lactic acid – or +w

Assimilation of nitrate: –

Growth in vitamin-free medium: –

Growth in 10% sodium chloride plus 5% glucose in yeast nitrogen base: –

Growth at 37°C: –

Comments

The original description is in good agreement with the present findings. The only differences noted were that one strain failed to assimilate soluble starch and another failed to assimilate DL-lactic acid.

Van der Walt (1972) transferred *P. ambrosiae* to *Ambrosiozyma* when he described that genus. The distinctive feature of *Ambrosiozyma* is the presence of dolipores in hyphal septa. Kreger-van Rij and Veenhuis (1973a) examined the species of *Ambrosiozyma* by transmission electron microscopy and noted all except *A. ambrosiae* to have dolipores. Because of the lack of dolipores, this species has been retained in *Pichia.*

Origin of the strains studied

From tunnels of *Platypus externedentatus* in *Ficus sycomorus,* Natal, South Africa (2).

Type strain: CBS 6003 (NRRL Y-7524).

4. *PICHIA AMETHIONINA* STARMER, PHAFF, MIRANDA ET MILLER (1978b)

This species has two varieties:

Pichia amethionina Starmer, Phaff, Miranda et Miller var. *amethionina*

Pichia amethionina Starmer, Phaff, Miranda et Miller var. *pachycereana* Starmer, Phaff, Miranda et Miller (1978b)

Standard description of *Pichia amethionina* var. *amethionina*

Growth on 5% malt extract agar: After 3 days at 25°C, the cells are ovoidal to elongate (1.2–4.0) × (3.0–9.0) μm, and occur singly, in pairs, in short branched chains and in stellate clusters. Growth is butyrous and tannish-white in color.

Growth on the surface of assimilation media: Pellicles are not formed.

Dalmau plate culture on morphology agar: After 7 days at 25°C, growth under the coverglass shows only a few rudimentary pseudohyphae and no true hyphae. Aerobic growth is white in color, smooth, glistening, and butyrous with an irregular and finely serrate margin.

Formation of ascospores: This species is heterothallic and asci are either unconjugated if formed by diploid cells or conjugated when produced by the pairing of complementary mating types. Each ascus contains one to four hat-shaped spores which are readily released at maturity. Complementary mating types of the variety *amethionina*, derived from the type strain by Starmer et al. (1978b), are NRRL Y-10979 (CBS 6942) and NRRL Y-10980 (CBS 6941). Complementary mating types from the type strain of the variety *pachycereana* are NRRL Y-10982 (CBS 6945) and NRRL Y-10983 (CBS 6944).

Spores were observed on YM- and 5% malt extract agar.

Fermentation: –

Assimilation of carbon compounds:

Galactose	–	Raffinose	–	Erythritol	–
Sucrose	–	Soluble starch	–	Ribitol	–
Maltose	–	D-Xylose	–	D-Mannitol	–
Cellobiose	–	L-Arabinose	–	Succinic acid	+
Trehalose	–	D-Ribose	–	Citric acid	–
Lactose	–	L-Rhamnose	–	Inositol	–

Additional carbon compounds tested: L-Sorbose –, melibiose –, melezitose –, inulin –, D-arabinose –, D-glucosamine · HCl –, glucitol –, α-methyl-D-glucoside –, salicin –, potassium D-gluconate +, DL-lactic acid +

Assimilation of nitrate: –

Growth in vitamin-free medium: –

Growth in 10% sodium chloride plus 5% glucose in yeast nitrogen base: –

Growth at 37°C: +

G + C: 33.0–33.1 mol%, 6 strains of the variety *amethionina*; 32.8–33.6 mol%, 4 strains of the variety *pachycereana* (Starmer et al., 1978b)

Supplementary description of *Pichia amethionina* var. *pachycereana*

The variety *pachycereana* may be separated from the variety *amethionina* by its ability to assimilate D-mannitol.

Comments

Starmer et al. (1978b, 1979) observed no mating between *P. amethionina* and *P. delftensis, P. heedii*, or *P. opuntiae* despite their phenotypic similarity, and it appears that these species are biologically distinct. *P. amethionina* seems unique among the yeasts since all strains require an exogenous source of methionine or cysteine (Starmer et al., 1978b).

Origin of the strains belonging to the variety *amethionina*

From *Rathbunia alamosensis* (1) and *Macherocereus gummosus* (2), Mexico.

Type strain: CBS 6940 (NRRL Y-10978).

Origin of the strains belonging to the variety *pachycereana*

From *Pachycereus pringlei* (2) and *Macherocereus gummosus* (1), Mexico.

Type strain: CBS 6943 (NRRL Y-10981).

5. *PICHIA AMYLOPHILA* KURTZMAN, SMILEY, JOHNSON, WICKERHAM ET FUSON (1980a)

Standard description of *Pichia amylophila*

Growth on 5% malt extract agar: After 3 days at 25°C, the cells are single or rarely clustered, spheroidal, ellipsoidal, or elongate (2.0–5.5) × (3.0–14.0) μm (Fig. 82). Growth is butyrous and tannish-white in color.

Growth on the surface of assimilation media: Pellicles are not formed but rings develop occasionally.

Dalmau plate culture on morphology agar: After 7 days at 25°C, growth under the coverglass shows abundant and highly branched pseudohyphae as well as occasional true hyphae. The true hyphae (diameter: 1.8–2.5 μm) show no evidence of a dolipore septum when viewed under the light microscope. Aerobic growth is cream colored, dull, and moist with a serrate margin and a slightly raised center.

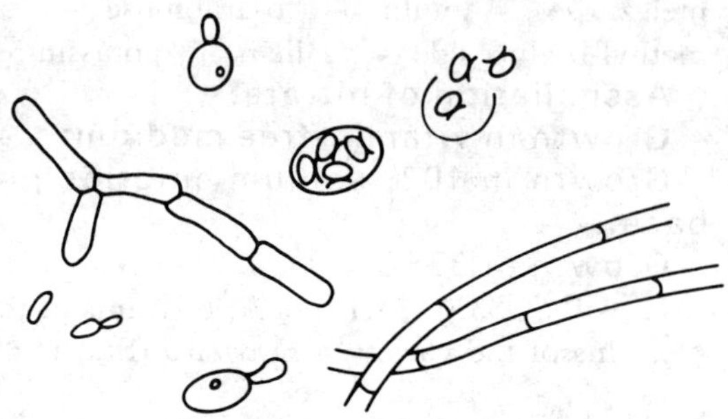

Fig. 82 – *P. amylophila,* CBS 7020
After 1 week on malt extract agar

Formation of ascospores: Diploid cells convert to asci and usually form two hat-shaped spores. Asci are generally free but may form on pseudohyphae where they are sessile or attached by short sterigmata. Upon maturity, the asci become dehiscent. Single-spore isolations showed this species to be heterothallic although most of the ascospores from the parent strain are diploid. Complementary mating types are NRRL YB-1287-82-2 (CBS 7021) and NRRL YB-1287-82-3 (CBS 7022). Asci from this pair predominantly form three or four spores.

Spores were observed on 5% malt extract- and YM agar.

Fermentation:

Glucose	+	Lactose	–
Galactose	–	Raffinose	–
Sucrose	–	Trehalose	–
Maltose	–		

Assimilation of carbon compounds:

Galactose	–	Raffinose	–	Erythritol	–
Sucrose	+	Soluble starch	+	Ribitol	–
Maltose	+	D-Xylose	+	D-Mannitol	+
Cellobiose	+	L-Arabinose	+	Succinic acid	+
Trehalose	+	D-Ribose	–	Citric acid	+
Lactose	–	L-Rhamnose	–	Inositol	–

Additional carbon compounds tested: L-Sorbose –, melibiose –, melezitose +, inulin –, D-arabinose –, D-glucosamine · HCl –, glucitol +, α-methyl-D-glucoside +, salicin −, potassium D-gluconate +, DL-lactic acid +

Assimilation of nitrate: −

Growth in vitamin-free medium: −

Growth in 10% sodium chloride plus 5% glucose in yeast nitrogen base: −

Growth at 37°C: +

G + C: 45.4 mol%, type strain (Kurtzman et al., 1980a)

Comments

P. amylophila is phenotypically quite like *P. mississippiensis* and differs from the latter only in its ability to assimilate soluble starch. Complementary mating types from the two species conjugate but the resulting ascospores are infertile. In a comparison of DNA relatedness, the two taxa showed only 25% base sequence complementarity (Kurtzman et al., 1980a), but this is still higher than the 0–15% commonly found between unrelated species. Thus, the indications are that these species have recently evolved from a common ancestor, and they might be of considerable value for studies of speciation at the genetic and molecular levels.

These two species are rather similar to *P. rhodanensis, P. wickerhamii* and *P. veronae,* which are also heterothallic, but Kurtzman et al. (1980a) found that they gave no mating reaction with these latter three taxa nor did they show significant DNA relatedness.

Origin of the strain studied

Frass from loblolly pine, *Pinus taeda* L., Mississippi, U.S.A.

Type strain: CBS 7020 (NRRL YB-1287).

6. *PICHIA ANGOPHORAE* MILLER ET BARKER (1968)

Standard description of *Pichia angophorae*

Growth on 5% malt extract agar: After 3 days at 25°C, the cells are spheroidal to ellipsoidal (2.0–6.0) × (3.0–7.0) μm, and occur singly, in pairs or in short chains. Growth is butyrous and light tan in color.

Growth on the surface of assimilation media: Thin pellicles are formed.

Dalmau plate culture on morphology agar: After 7 days at 25°C, growth under the coverglass shows well developed pseudohyphae bearing blastospores. True hyphae are not formed. Aerobic growth is light tan in color, moderately glistening, butyrous, and with entire margins.

Formation of ascospores: Miller and Barker (1968) reported ascus development in their original isolates to be preceded by iso- or heterogamic conjugations, or rarely by no conjugation. Two to four hat-shaped spores are formed in each ascus. Asci are dehiscent. Single-spore isolates yielded both sporogenous and asporogenous cultures and the species was described as both homo- and heterothallic (Miller & Barker, 1968). The occurrence of haploid as well as diploid spores could account for these observations. Mating types NRRL Y-7843 (CBS 5831) and NRRL Y-7844 (CBS 5830) originally derived from the type strain readily sporulate following conjugation. The type strain, NRRL Y-7118 (CBS 5823), is now asporogenous, but conjugates and sporulates when paired with mating type NRRL Y-7843.

Spores were observed on 5% malt extract agar.

Fermentation:

Glucose	+	Lactose	–
Galactose	–	Raffinose	–
Sucrose	+	Trehalose	–
Maltose	+		

Assimilation of carbon compounds:

Galactose	–	Raffinose	–	Erythritol	–
Sucrose	+	Soluble starch	–	Ribitol	+
Maltose	+	D-Xylose	+	D-Mannitol	+
Cellobiose	+	L-Arabinose	+	Succinic acid	+
Trehalose	+	D-Ribose	–	Citric acid	+
Lactose	–	L-Rhamnose	–	Inositol	–

Additional carbon compounds tested: L-Sorbose –, melibiose –, melezitose +, inulin –, D-arabinose –, D-glucosamine · HCl +, glucitol +, α-methyl-D-glucoside +, salicin + or w, potassium D-gluconate –, DL-lactic acid –

Assimilation of nitrate: –

Growth in vitamin-free medium: –

Growth in 10% sodium chloride plus 5% glucose in yeast nitrogen base: –

Growth at 37°C: –

Comments

Pichia angophorae is phenotypically similar to *P. spartinae* but differs from the latter by its growth on D-xylose and lack of growth on L-sorbose and at 37°C. No response was noted in mixtures of mating types from the two species (Kurtzman & Ahearn, 1976).

Origin of the strains studied

Gum exudate of *Angophora costata,* Australia (1); Kreger-van Rij (2).

Type strain: CBS 5823 (NRRL Y-7118), from *Angophora costata.*

7. *PICHIA BESSEYI* KURTZMAN ET WICKERHAM (1972)

Standard description of *Pichia besseyi*

Growth on 5% malt extract agar: After 3 days at 25°C, the cells are spheroidal to ellipsoidal (4.3–7.0) × (4.3–8.2) μm. Cells occur singly or occasionally in pairs (Fig. 83). Growth is butyrous and tannish-white in color.

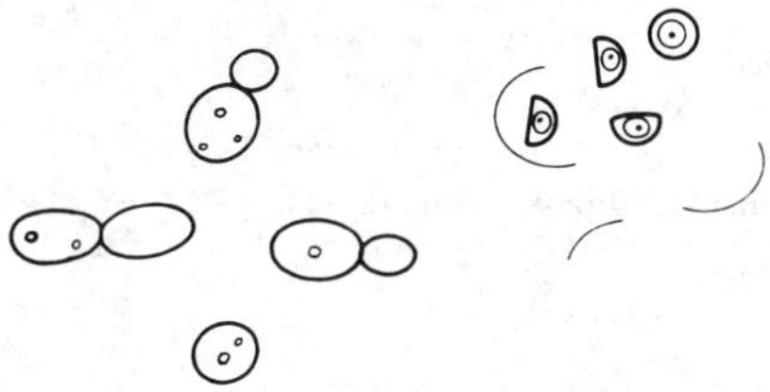

Fig. 83 – *P. besseyi,* CBS 6343
After 1 week on malt extract agar

Growth on the surface of assimilation media: Thin pellicles are formed.

Dalmau plate culture on morphology agar: After 7 days at 25°C, growth under the coverglass shows neither pseudohyphae nor true hyphae, but tree-like outgrowths of pseudohyphae are detected on corn meal agar after 2 weeks. Aerobic growth is dull-glistening and butyrous with an entire margin. The colony color is initially white to cream but becomes pink to brown as ascospores are formed.

Formation of ascospores: Occasionally, asci are formed following conjugation between independent cells, but the majority are unconjugated. Asci usually contain four spores and become dehiscent upon maturity. The shape of the spores is hemispheroidal. The underside of the spores appears concave, and the lower periphery of the spore wall is somewhat thickened. Single-spore isolates from four-spored asci are sporogenous and the species is presumed to be homothallic.

Spores were observed on 5% malt extract-, morphology- and YM agar.

Fermentation:

Glucose	+	Lactose	–
Galactose	–	Raffinose	–
Sucrose	–	Trehalose	–
Maltose	–		

Assimilation of carbon compounds:

Galactose	–	Raffinose	–	Erythritol	–
Sucrose	–	Soluble starch	–	Ribitol	–
Maltose	–	D-Xylose	–	D-Mannitol	+
Cellobiose	–	L-Arabinose	–	Succinic acid	+
Trehalose	–	D-Ribose	–	Citric acid	–
Lactose	–	L-Rhamnose	–	Inositol	–

Additional carbon compounds tested: L-Sorbose –, melibiose –, melezitose –, inulin –, D-arabinose –, D-glucosamine · HCl –, glucitol +, α-methyl-D-glucoside –, salicin –, potassium D-gluconate –, DL-lactic acid +

Assimilation of nitrate: –

Growth in vitamin-free medium: –; thiamine and pyridoxine required

Growth in 10% sodium chloride plus 5% glucose in yeast nitrogen base: –

Growth at 37°C: –

G + C: 35.2 mol%, type strain (Kurtzman, unpublished)

Comments

P. besseyi is rather like *P. membranaefaciens* in that it assimilates a limited number of carbon compounds. However, it assimilates mannitol and glucitol, compounds on which *P. membranaefaciens* is incapable of growth. Likewise, *P. besseyi* assimilates lactic acid, but not ribitol and may by these characters be distinguished from *P. delftensis*.

Origin of the strain studied

Shallow water in a marsh near Bonaventure, Gaspé Peninsula, Quebec, Canada (1).

Type strain: CBS 6343 (NRRL YB-4711).

8. *PICHIA BOVIS* VAN UDEN ET DO CARMO-SOUSA (1957)

Synonym

Zymopichia bovis (van Uden et do Carmo-Sousa) Novák et Zsolt (1961)

Standard description of *Pichia bovis*

Growth on 5% malt extract agar: After 3 days at 25°C, the cells are spheroidal to ovoidal (1.9–5.1) × (2.1–9.2) μm, and occur singly or in pairs. Growth is butyrous and tannish-white in color.

Growth on the surface of assimilation media: Dry climbing pellicles are formed.

Dalmau plate culture on morphology agar: After 7 days at 25°C, growth of pseudohyphae under the coverglass may be quite rudimentary or moderately well developed. True hyphae are not formed. Aerobic growth is tannish-white, smooth, faintly glistening, butyrous, and with an entire margin.

Formation of ascospores: Asci are unconjugated and generally contain no more than two spores. The spores are hat-shaped and liberated from the asci soon after formation.

Spores were observed on YM-, 5% malt extract-, V8-, acetate- and Gorodkowa agar.

Fermentation:

Glucose	+	Lactose	–
Galactose	–	Raffinose	–
Sucrose	–	Trehalose	– or + s
Maltose	–		

Assimilation of carbon compounds:

Galactose	–	Raffinose	–	Erythritol	–
Sucrose	+	Soluble starch	v	Ribitol	–

Maltose	+	D-Xylose	+	D-Mannitol	+
Cellobiose	+	L-Arabinose	+	Succinic acid	+
Trehalose	+	D-Ribose	−	Citric acid	+
Lactose	−	L-Rhamnose	− or +s	Inositol	−

Additional carbon compounds tested: L-Sorbose −, melibiose −, melezitose +, inulin −, D-arabinose −, D-glucosamine · HCl −, glucitol +, α-methyl-D-glucoside +, salicin +, potassium D-gluconate +, DL-lactic acid +

Assimilation of nitrate: −

Growth in vitamin-free medium: −

Growth in 10% sodium chloride plus 5% glucose in yeast nitrogen base: −

Growth at 37°C: + or w

G + C: 42.3 mol%, type strain (Kurtzman et al., unpublished)

Comments

Pichia bovis is phenotypically quite similar to *P. onychis,* but differs from the latter by its inability to ferment sucrose and raffinose and by its failure to assimilate raffinose and inulin.

Origin of the strains studied

Cecum of a cow, van Uden (1); from Santa María (1).

Type strain: CBS 2616 (NRRL YB-4184).

9. *PICHIA BURTONII* BOIDIN, PIGNAL, LEHODEY, VEY ET ABADIE (1964)

Imperfect state: *Candida variabilis* (Lindner) Berkhout; *Trichosporon variabile* (Lindner) Delitsch

Synonyms

Endomycopsis burtonii (Boidin, Pignal, Lehodey, Vey et Abadie) Kreger-van Rij (1970b)
Hyphopichia burtonii (Boidin, Pignal, Lehodey, Vey et Abadie) von Arx et van der Walt (1976)
Monilia variabilis Lindner (1898b)
Oospora variabilis (Lindner) Lindau (1907a)
Candida variabilis (Lindner) Berkhout (1923)
Trichosporon variabile (Lindner) Delitsch (in Lembke, 1943)
Dematium chodati Nechitsch (1904)
Candida chodati (Nechitsch) Berkhout (1923)
Endomycopsis chodatii nom. nud. (Wickerham & Burton, 1952)
? *Sporotrichum carougeaui* Langeron (1922)
Sporotrichum anglicum Castellani (1937b)
Trichosporon behrendii Lodder et Kreger-van Rij (1952)
Fermentotrichon behrendii (Lodder et Kreger-van Rij) Novák et Zsolt (1961)
Candida fibrae Nakase (1971a)
Cladosporium fermentans Goto, Yamakawa et Yokotsuka (1975)

Standard description of *Pichia burtonii*

Growth on 5% malt extract agar: After 3 days at 25°C, the cells are ovoidal to elongate (2.3–5.7) × (6.0–9.0) μm, and occur singly, in pairs and in small clumps. Pseudohyphae and true hyphae may be present. Growth is tannish-yellow in color and may be butyrous to somewhat mycelial.

Growth on the surface of assimilation media: Depending upon the strain, pellicles may be thin and incomplete or quite thick.

Dalmau plate culture on morphology agar: After 7 days at 25°C, growth under the coverglass shows abundant pseudohyphae and true hyphae. Blastospores may be borne on denticles. The denticles occur on other blastospores as well as on hyphae and pseudohyphae. Arthrospores are also produced. Aerobic growth is white to tannish-white in color, dull and sometimes powdery, and usually fringed by a broad mycelial band.

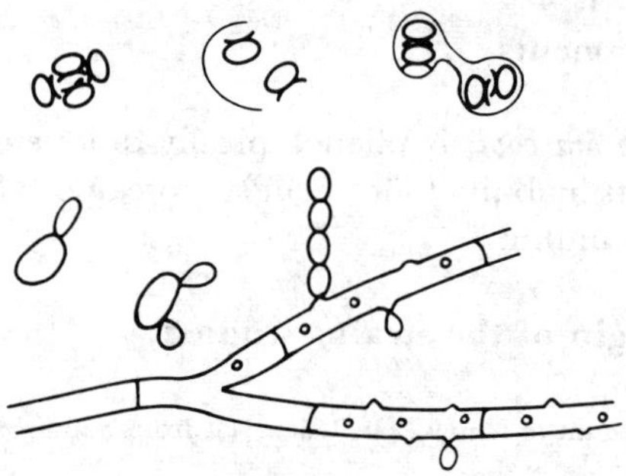

Fig. 84 – *P. burtonii*
CBS 2352 × CBS 2353. After 3 days on malt extract agar

Formation of ascospores: This species is heterothallic. Asci may form directly from diploid cells or result from the pairing of complementary mating types. Each ascus contains one to four hat-shaped spores which are readily released at maturity (Fig. 84). Complementary mating types are NRRL Y-1933 (CBS 2352) and NRRL Y-1938 (CBS 2353).

Spores were observed on 5% malt extract- and YM agar.

Fermentation:

Glucose	+ or s	Lactose	–
Galactose	– or +w	Raffinose	+ or w
Sucrose	+ or s	Trehalose	+ or s
Maltose	+ or w		

Assimilation of carbon compounds:

Galactose	+	Raffinose	+	Erythritol	+
Sucrose	+	Soluble starch	+	Ribitol	+
Maltose	+	D-Xylose	+	D-Mannitol	+
Cellobiose	+	L-Arabinose	v	Succinic acid	+
Trehalose	+	D-Ribose	+	Citric acid	+
Lactose	–	L-Rhamnose	–	Inositol	–

Additional carbon compounds tested: L-Sorbose v, melibiose –, melezitose v, inulin –, D-arabinose –, D-glucosamine · HCl +w or s, glucitol +, α-methyl-D-glucoside +, salicin +, potassium D-gluconate v, DL-lactic acid –

Assimilation of nitrate: –

Growth in vitamin-free medium: +

Growth in 10% sodium chloride plus 5% glucose in yeast nitrogen base: +

Growth at 37°C: – or +w

G + C: 35.5 mol%, one strain (Meyer & Phaff, 1970)

Comments

The presence of budding cells and arthrospores as well as sometimes profuse amounts of true hyphae has prompted the assignment of *P. burtonii* to various genera. In 1952, Wickerham and Burton discovered the ascosporic stage of this species and showed it to be heterothallic. They described it as *Endomycopsis chodatii* but a Latin diagnosis was not provided. Boidin et al. (1964) validly described the species as *P. burtonii,* and in 1970 (b) Kreger-van Rij transferred it to *Endomycopsis.* In 1971, van der Walt and Scott (1971b) showed *Endomycopsis* to be an obligate synonym of *Saccharomycopsis* Schiönning and transferred many of the *Endomycopsis* species to *Saccharomycopsis,* but not *P. burtonii.* Later, von Arx and van der Walt (1976) transferred *P. burtonii* to their newly created genus *Hyphopichia.* Cardinal features of this new genus included heterothallism, presence of septate hyphae, asci formed from conjugating yeast cells and denticulate conidiogenous cells. These workers excluded *P. burtonii* from *Pichia* because it formed expanding septate hyphae and conidia borne on denticles and from *Saccharomycopsis* because it produced asci comprised of pairs of conjugating cells.

In the present treatment, this species is retained in *Pichia* for several reasons. The hyphal septa of *P. burtonii* have a simple closure line typical of other members of *Pichia* with true hyphae (Kreger-van Rij & Veenhuis, 1973a). Placement of this species in *Saccharomycopsis* seems ill-advised because, if one accepts the assignment of *S. lipolytica* to the newly described genus *Yarrowia* (van der Walt & von Arx, 1980), all remaining species of *Saccharomycopsis* are characterized by hyphal septa with plasmodesmata. Finally, although the denticulate conidiogenous cells of *P. burtonii* are quite pronounced, they are not unique to any one taxon and may be found in *Steph. ciferrii, Cephaloascus albidus, S. lipolytica, P. rhodanensis, P. wickerhamii* and *Aureobasidium pullulans.*

Origin of the strains studied

Pollen carried by wild bees in Brazil (1), silage (1), stored rice (6), wheat (2), wheat flour (3), chicken feed (7), fig tree (1), wound on palm tree (1), ragi (1), sputum (1), skin (1), other substrates (6).

Type strain: CBS 2352 (NRRL Y-1933), from pollen.

10. *PICHIA CACTOPHILA* STARMER, PHAFF, MIRANDA ET MILLER (1978a)

Standard description of *Pichia cactophila*

Growth on 5% malt extract agar: After 3 days at 25°C, the cells are ovoidal to elongate (1.2–5.0) × (2.0–8.1) μm, and occur singly, in pairs and in short chains. Growth is butyrous and cream colored.

Growth on the surface of assimilation media: Thin climbing pellicles are formed.

Dalmau plate culture on morphology agar: After 7 days at 25°C, growth under the coverglass shows only rudimentary pseudohyphae and no true hyphae. Aerobic growth is slightly glistening, smooth, butyrous, tannish-white in color, and with irregular and finely serrate margins.

Formation of ascospores: This species has been isolated from nature in the diploid form and cells convert directly to asci producing two to four hat-shaped spores. The spores are liberated at maturity. Single-spore isolates from four-spored asci are asporogenous until paired with the appropriate strain, thus demonstrating this species to be heterothallic (Starmer et al., 1978a). Single-spore isolates from two-spored asci are sporogenous, and, therefore, it appears that these spores are diploid.

Spores were observed on YM agar.

Fermentation:

Glucose	– or + w	Lactose	–
Galactose	–	Raffinose	–
Sucrose	–	Trehalose	–
Maltose	–		

Assimilation of carbon compounds:

Galactose	–	Raffinose	–	Erythritol	–
Sucrose	–	Soluble starch	–	Ribitol	–
Maltose	–	D-Xylose	+ w or s	D-Mannitol	–
Cellobiose	–	L-Arabinose	–	Succinic acid	+
Trehalose	–	D-Ribose	–	Citric acid	v
Lactose	–	L-Rhamnose	–	Inositol	–

Additional carbon compounds tested: L-Sorbose –, melibiose –, melezitose –, inulin –, D-arabinose –, D-glucosamine · HCl +, glucitol –, α-methyl-D-glucoside –, salicin –, potassium D-gluconate –, DL-lactic acid v

Assimilation of nitrate: –

Growth in vitamin-free medium: –

Growth in 10% sodium chloride plus 5% glucose in yeast nitrogen base: v

Growth at 37°C: +

G + C: 36.3–37.5 mol%, 9 strains (Starmer et al., 1978a)

Comments

Pichia cactophila has been isolated from a wide variety of cacti (Starmer et al., 1978a), and this lack of habitat specificity contrasts with the more stringent niche requirements of the other recently isolated cactophilic species, i.e., *P. amethionina, P. heedii,* and *P. opuntiae.* Of the relatively few phenotypic characters available for separating the cactus yeasts, *P. cactophila* may be distinguished from the others by its ability to assimilate D-glucosamine. In addition, the G + C content of this species is nearly 4% higher than that of the other cactus taxa. Another phenotypically similar species is *P. membranaefaciens* which has a G + C content 5% greater than that of *P. cactophila* yet the two presently are separated only by D-glucosamine assimilation, weak or negative in the former species and strong in the latter. Evidently, *P. membranaefaciens* does not inhabit cacti (Starmer et al., 1978a).

The results from the present study were in good agreement with the data presented by Starmer et al. (1978a), except that I found D-xylose assimilation to be late and either weak or strong rather than negative. [**Note in proof:** H.J. Phaff (personal communication) reported *P. cactophila* to form two-spored asci and to be homothallic whereas the four-spored heterothallic strains actually represent a new species to be described as *Pichia pseudocactophila.*]

Origin of the strains studied

Cacti from Mexico (3), cactus from Hawaii, U.S.A. (1).

Type strain: CBS 6926 (NRRL Y-10963) from *Lemaireocerei thurferi,* Baja California, Mexico.

11. PICHIA CARSONII PHAFF ET KNAPP (1956)

Synonyms

Torulaspora carsonii (Phaff et Knapp) van der Walt et Johannsen (1975a)
Debaryomyces vini nom. nud. (Zimmermann, 1938)
Pichia vini nom. nud. (Phaff, 1956)
Pichia vini (Zimmermann) Phaff var. *melibiosi* Santa María (1963a)

Standard description of *Pichia carsonii*

Growth on 5% malt extract agar: After 3 days at 25°C, the cells are spheroidal to elongate (1.4–4.2) × (2.0–6.0) μm, and occur singly, in pairs and in chains. Growth is butyrous and tannish-white in color.

Growth on the surface of assimilation media: Thin to moderately heavy pellicles are formed.

Dalmau plate culture on morphology agar: After 7 days at 25°C, growth under the coverglass shows abundant but sparingly branched pseudohyphae. True hyphae are not produced. Aerobic growth is tannish-white, dull to faintly glistening, butyrous and with an entire or occasionally notched margin.

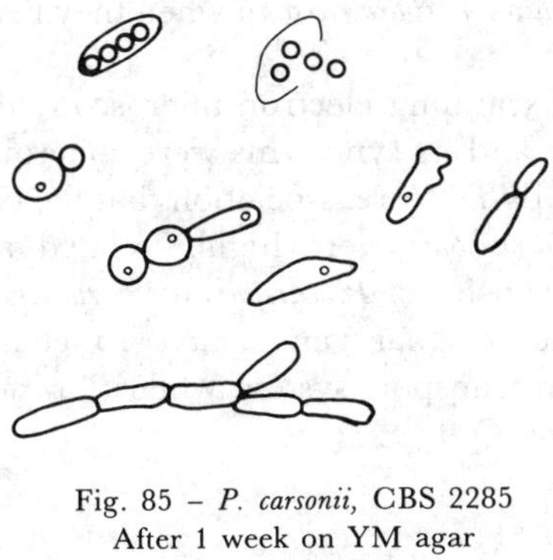

Fig. 85 – *P. carsonii,* CBS 2285
After 1 week on YM agar

Formation of ascospores: Asci may show mother–daughter conjugation or may be unconjugated. One to four smooth spheroidal spores form in each ascus, and asci may be persistent or dehiscent (Fig. 85). Single-spore isolates from four-spored asci are sporogenous, thus indicating the species to be homothallic.

Spores were observed on YM- and morphology agar.

Fermentation: –

Assimilation of carbon compounds:

Galactose	+	Raffinose	v	Erythritol	–
Sucrose	+	Soluble starch	+	Ribitol	+
Maltose	+	D-Xylose	+	D-Mannitol	+
Cellobiose	v	L-Arabinose	v	Succinic acid	+
Trehalose	+	D-Ribose	v	Citric acid	+
Lactose	–	L-Rhamnose	–	Inositol	–

Additional carbon compounds tested: L-Sorbose +, melibiose v, melezitose +, inulin –, D-arabinose –, D-glucosamine · HCl –, glucitol +, α-methyl-D-glucoside +, salicin +, potassium D-gluconate –, DL-lactic acid v

Assimilation of nitrate: –

Growth in vitamin-free medium: –

Growth in 10% sodium chloride plus 5% glucose in yeast nitrogen base: +

Growth at 37°C: – or +w

G + C: 39.6–39.7 mol%, 3 strains (Kurtzman & Smiley, 1979)

Comments

Pichia vini was first isolated by Zimmermann (1938) from spoiled wine and described as *Debaryomyces vini*. Later, Phaff (1956) transferred this species when he emended the genus *Pichia*. Several additional strains were isolated and appeared identical to the original strain except for the presence of dehiscent asci (Kreger-van Rij, 1970c). Santa María (1963a) obtained two similar cultures from 'alpechin' but described them as *P. vini* var. *melibiosi* because they assimilated melibiose and raffinose.

Pichia carsonii, described by Phaff and Knapp (1956), appeared identical with *P. vini* except for its hat-shaped ascospores. Kreger-van Rij (1970c) showed by transmission electron microscopy that the rather small spores of *P. carsonii* were actually spheroidal and practically smooth. Consequently, she considered this species a synonym of *P. vini*. Van der Walt and Johannsen (1975a) pointed out that Zimmermann (1938) failed to provide a Latin diagnosis for *Deb. vini* and that *P. carsonii* was the legitimate name. However, they transferred the species to *Torulaspora* as *T'spora carsonii* when they revised that genus.

Kurtzman and Smiley (1979) demonstrated by scanning electron microscopy that the ascospores from all extant strains of *P. carsonii* and its synonyms were spheroidal and essentially smooth. Furthermore, they showed by DNA reassociation that the type strains of *P. vini, P. vini* var. *melibiosi* and *P. carsonii* were highly related and represented the same species. They rejected the transfer of *P. carsonii* to *Torulaspora,* pointing out that the originally described species of that genus have roughened ascospore walls and coenzyme Q-6 in the electron transport system while *P. carsonii* has smooth-walled ascospores and forms coenzyme Q-9.

Origin of the strains studied

Wine, Germany and South Africa (2); *Quercus kelloggii,* U.S.A. (2); soil, Japan (1); 'alpechin,' Spain (1); *Aulacigaster* sp., U.S.A. (1).

Type strain: CBS 2285 (NRRL YB-4275), from *Quercus kelloggii.*

12. *PICHIA CASTILLAE* SANTA MARÍA ET GARCÍA ASER (1970)

Standard description of *Pichia castillae*

Growth on 5% malt extract agar: After 3 days at 25°C, the cells are ovoidal to ellipsoidal (1.4–3.5) × (2.0–7.0) μm, and occur singly, in pairs and in clusters. Growth is butyrous and tannish-white in color.

Growth on the surface of assimilation media: Thin, slightly climbing pellicles are present.

Dalmau plate culture on morphology agar: After 7 days at 25°C, growth under the coverglass shows only sparse and poorly differentiated pseudohyphae. True hyphae are not present. Aerobic growth is tannish-white in color, somewhat butyrous, and with a smooth, dull to faintly glistening surface. The margin is lobate.

Formation of ascospores: The asci, which show isogamic or heterogamic conjugation, produce two to four hat-shaped spores. Spores are released rapidly following their formation. Santa María and García Aser (1970) reported this species to be homothallic.

Spores were observed on 5% malt extract agar.

Fermentation: –

Assimilation of carbon compounds:

Galactose	+	Raffinose	+	Erythritol	+
Sucrose	–	Soluble starch	–	Ribitol	+
Maltose	+	D-Xylose	+	D-Mannitol	+
Cellobiose	+	L-Arabinose	+	Succinic acid	+
Trehalose	+	D-Ribose	+	Citric acid	+
Lactose	–	L-Rhamnose	–	Inositol	–

Additional carbon compounds tested: L-Sorbose +, melibiose +, melezitose –, inulin –, D-arabinose +, D-glucosamine · HCl –, glucitol +, α-methyl-D-glucoside –, salicin +, potassium D-gluconate +s, DL-lactic acid –

Assimilation of nitrate: –

Growth in vitamin-free medium: –

Growth in 10% sodium chloride plus 5% glucose in yeast nitrogen base: +

Growth at 37°C: +w

Origin of the strain studied

Frass of insects infesting *Gymnocladus canadensis* (1).

Type strain: CBS 6053 (NRRL Y-7501).

13. *PICHIA CHAMBARDII* (RAMIREZ ET BOIDIN) PHAFF (1956)

Synonyms

Saccharomyces chambardi Ramirez et Boidin (1954) (also in Ramirez & Boidin, 1955)
Petasospora chambardi (Ramirez et Boidin) Boidin et Abadie (1954)

Standard description of *Pichia chambardii*

Growth on 5% malt extract agar: After 3 days at 25°C, the cells are ovoidal (1.8–4.1) × (2.6–6.8) μm and occur singly or in pairs. Growth is butyrous and greyish-white in color.

Growth on the surface of assimilation media: Thin pellicles are formed.

Dalmau plate culture on morphology agar: After 7 days at 25°C, growth under the coverglass shows only infrequent undifferentiated pseudohyphal strands. True hyphae are not formed. Aerobic growth is white in color, butyrous, smooth, dull-glistening, and with an occasionally lobed margin.

Formation of ascospores: Asci are unconjugated and form one to four (usually two) hat-shaped spores. The spores are liberated at maturity.

Spores were observed on 5% malt extract agar.

Fermentation: –

Assimilation of carbon compounds:

Galactose	+	Raffinose	–	Erythritol	–
Sucrose	–	Soluble starch	–	Ribitol	–
Maltose	–	D-Xylose	–	D-Mannitol	–
Cellobiose	+	L-Arabinose	–	Succinic acid	+
Trehalose	–	D-Ribose	–	Citric acid	+
Lactose	–	L-Rhamnose	–	Inositol	–

Additional carbon compounds tested: L-Sorbose –, melibiose –, melezitose –, inulin –, D-arabinose –, D-glucosamine · HCl –, glucitol –, α-methyl-D-glucoside –, salicin +, potassium D-gluconate –, DL-lactic acid +

Assimilation of nitrate: –

Growth in vitamin-free medium: –

Growth in 10% sodium chloride plus 5% glucose in yeast nitrogen base: –

Growth at 37°C: –

G + C: 30.2 mol%, one strain (Nakase & Komagata, 1970b)

Origin of the strain studied

Tanning liquor, France (1).

Type strain: CBS 1900 (NRRL Y-2378).

14. *PICHIA DELFTENSIS* BEECH (1965)

Standard description of *Pichia delftensis*

Growth on 5% malt extract agar: After 3 days at 25°C, the cells are ovoidal (2.5–5.3) × (2.7–8.9) μm, and occur singly or in pairs. Growth is butyrous and tannish-white in color.

Growth on the surface of assimilation media: Pellicles are formed.

Dalmau plate culture on morphology agar: After 7 days at 25°C, growth under the coverglass shows only an occasional undifferentiated pseudohyphal outgrowth. True hyphae are not formed. Aerobic growth is white in color, dull to slightly glistening, butyrous and with a finely lobed margin.

Formation of ascospores: Asci may be unconjugated, or show conjugation between independent cells. One to four hat-shaped spores are formed per ascus, and the asci slowly dehisce.

Spores were observed on 5% malt extract agar.

Fermentation:

Glucose	– or +w	Lactose	–
Galactose	–	Raffinose	–
Sucrose	–	Trehalose	–
Maltose	–		

Assimilation of carbon compounds:

Galactose	–	Raffinose	–	Erythritol	–
Sucrose	–	Soluble starch	–	Ribitol	+
Maltose	–	D-Xylose	–	D-Mannitol	+
Cellobiose	–	L-Arabinose	–	Succinic acid	+
Trehalose	–	D-Ribose	–	Citric acid	–
Lactose	–	L-Rhamnose	–	Inositol	–

Additional carbon compounds tested: L-Sorbose –, melibiose –, melezitose –, inulin –, D-arabinose –, D-glucosamine · HCl –, glucitol +, α-methyl-D-glucoside –, salicin –, potassium D-gluconate –, DL-lactic acid –

Assimilation of nitrate: –

Growth in vitamin-free medium: –

Growth in 10% sodium chloride plus 5% glucose in yeast nitrogen base: +

Growth at 37°C: –

Comments

As noted earlier, *P. delftensis* resembles *P. amethionina* and the other species isolated from cacti. It also shows many similarities with *P. membranaefaciens*.

Origin of the strain studied

Naturally fermented apple juice, England (1).

Type strain: CBS 2614 (NRRL Y-7119).

15. *PICHIA DISPORA* (DEKKER) KREGER-VAN RIJ (1970c)

Synonyms

Saccharomyces disporus nom. nud. (Beijerinck, 1908)
Debaryomyces disporus Dekker (Stelling-Dekker, 1931)
Pichia zaruensis Nakase et Komagata (1966)

Standard description of *Pichia dispora*

Growth on 5% malt extract agar: After 3 days at 25°C, the cells are spheroidal to ovoidal (2.5–5.5) × (2.5–6.5) μm and occur singly and in pairs. Growth is butyrous and tannish-white in color.

Growth on the surface of assimilation media: Pellicles are not formed.

Dalmau plate culture on morphology agar: After 7 days at 25°C, growth under the coverglass shows neither hyphae nor pseudohyphae but only an occasional outgrowth of elongated cells. Aerobic growth is tannish-white, butyrous, smooth, glistening, and with an occasionally lobed margin.

Formation of ascospores: Asci may be unconjugated or, more frequently, they result from conjugation between a parent cell and a bud or from the pairing of independent cells. Each ascus contains one or two spheroidal spores which exhibit a barely perceptible equatorial ledge. Asci are persistent.

Spores were observed on 5% malt extract agar.

Fermentation:

Glucose	+	Lactose	–
Galactose	–	Raffinose	–
Sucrose	–	Trehalose	– or +s
Maltose	–		

Assimilation of carbon compounds:

Galactose	–	Raffinose	–	Erythritol	–
Sucrose	–	Soluble starch	–	Ribitol	+
Maltose	–	D-Xylose	–	D-Mannitol	+
Cellobiose	–	L-Arabinose	–	Succinic acid	+
Trehalose	+	D-Ribose	–	Citric acid	–
Lactose	–	L-Rhamnose	–	Inositol	–

Additional carbon compounds tested: L-Sorbose –, melibiose –, melezitose –, inulin –, D-arabinose –, D-glucosamine · HCl –, glucitol +, α-methyl-D-glucoside –, salicin –, potassium D-gluconate –, DL-lactic acid v

Assimilation of nitrate: –

Growth in vitamin-free medium: –

Growth in 10% sodium chloride plus 5% glucose in yeast nitrogen base: –

Growth at 37°C: –

G + C: 37.6 mol%, type strain (Nakase, 1972a)

Comments

Kreger-van Rij (1970c) transferred this species from *Saccharomyces* to *Pichia* because of the presence of a ledge on the spores, but noted that in the transmission electron microscope, the ledge on spores of the type strain appeared not to be circumfluent. Using the scanning electron microscope, Kurtzman and Smiley (1974) confirmed the thin ledge to traverse only about half of the spore circumference.

Observations by scanning electron microscopy showed strains of *P. zaruensis* to form spores with a heavy circumfluent ring (Kurtzman & Smiley, unpublished). However, this has been the only phenotypic difference detected that separates *P. dispora* and *P. zaruensis*. Consequently, I have accepted the decision of Kreger-van Rij (1970c) and Nakase (1972a) to consider *P. zaruensis* as a synonym of *P. dispora*.

Origin of the strains studied

Beijerinck's strain from the exudate of an oak (1); exudate of *Quercus serrata,* Japan (1); forest soil, Japan (1).

Type strain: CBS 794 (NRRL Y-1447), Beijerinck's strain.

16. PICHIA ETCHELLSII KREGER-VAN RIJ (1964a)

Synonyms

Kloeckera faecalis Batista et Silveira (1959a)
Pichia faecalis (Batista et Silveira) Aciole de Queiroz (1970)
Torulaspora etchellsii (Kreger-van Rij) van der Walt et Johannsen (1975a)

Standard description of *Pichia etchellsii*

Growth on 5% malt extract agar: After 3 days at 25°C, the cells are ovoidal to short-elongate (1.8–5.4) × (2.0–8.3) μm, and occur singly or in pairs. Growth is white, dull-glistening and butyrous.

Growth on the surface of assimilation media: Occasional rings are noted, but pellicles do not form.

Dalmau plate culture on morphology agar: After 7 days at 25°C, growth under the coverglass shows moderately well branched pseudohyphae with few blastospores. True hyphae are not present. Aerobic growth on morphology agar is white, dull-glistening and butyrous with a finely striated margin.

Formation of ascospores: Asci show mother–daughter conjugation, or infrequently, conjugation between independent cells. One to four smooth, spheroidal spores are formed per ascus, and the asci may eventually become dehiscent. Single-spore isolates from four-spored asci proved sporogenous, indicating the species to be homothallic.

Spores were observed on YM agar.

Fermentation:

Glucose	+s	Lactose	–
Galactose	–	Raffinose	–
Sucrose	– or +w	Trehalose	–
Maltose	– or +w		

Assimilation of carbon compounds:

Galactose	+	Raffinose	–	Erythritol	–
Sucrose	+	Soluble starch	–	Ribitol	+
Maltose	+	D-Xylose	+	D-Mannitol	+
Cellobiose	+	L-Arabinose	+ or w	Succinic acid	+
Trehalose	+	D-Ribose	–	Citric acid	+
Lactose	–	L-Rhamnose	–	Inositol	–

Additional carbon compounds tested: L-Sorbose +, melibiose –, melezitose +, inulin –, D-arabinose –, D-glucosamine · HCl +, glucitol +, α-methyl-D-glucoside +, salicin v, potassium D-gluconate –, DL-lactic acid v

Assimilation of nitrate: –

Growth in vitamin-free medium: –

Growth in 10% sodium chloride plus 5% glucose in yeast nitrogen base: +

Growth at 37°C: +

G + C: 40.6 mol%, type strain (Price et al., 1978)

Comments

Pichia etchellsii is phenotypically quite similar to *P. carsonii,* differing from the former only by its ability to ferment and by its failure to assimilate starch. However, Kurtzman et al. (unpublished) found only 2% DNA base sequence complementarity between type strains of the two and considered them separate species.

As with *P. carsonii, P. etchellsii* has Q-9 ubiquinone in the electron transport system (Yamada et al., 1973c), and its spores are spheroidal and smooth when viewed under the scanning electron microscope (Kurtzman & Smiley, 1974). For these reasons, its placement in the genus *Torulaspora* by van der Walt and Johannsen (1975a) is rejected.

Origin of the strains studied

Fermenting cucumber brines, U.S.A. (2); human feces (1).

Type strain: CBS 2011 (NRRL Y-7121), one of the strains from cucumber brines.

17. PICHIA FARINOSA (LINDNER) HANSEN (1904)

Synonyms

Saccharomyces farinosus Lindner (1894)
Zygosaccharomyces farinosus (Lindner) Papadakis (1922)
Zygopichia farinosa (Lindner) Klöcker (1924)

Pichia miso nom. nud. (Mogi, 1939b)
Saccharomyces miso (Mogi) Novák et Zsolt (1961)
Zygopichia miso nom. nud. (Mogi, 1942)
Zygosaccharomyces tikumaensis nom. nud. (Mogi, 1942)
Pichia mogii nom. nud. (Ohara & Nonomura, 1954a)
Zygopichia sake Naganishi, (1941a)
Pichia sake (Naganishi) Ohara et Nonomura (1954b)
Zygopichia farinosa (Lindner) Klöcker var. *japonica* nom. nud. (Naganishi 1941b)
Pichia farinosa (Lindner) Hansen var. *japonica* (Naganishi) Ohara et Nonomura (1954b)
? *Pichia minuscula* Soneda (1959a)
Debaryomyces halotolerans Sasaki et Yoshida nom. nud.?

Standard description of *Pichia farinosa*

Growth on 5% malt extract agar: After 3 days at 25°C, the cells are ovoidal to elongate (1.5–5.0) × (3.0–18.0) μm, and occur singly, in pairs and in short chains. Growth is white to occasionally faintly yellowish-white in color.

Growth on the surface of assimilation media: Heavy, dry, climbing pellicles develop.

Dalmau plate culture on morphology agar: After 7 days at 25°C, growth under the coverglass shows pseudohyphae which, depending upon the strain, may be poorly differentiated to moderately branched. True hyphae are not formed. Aerobic growth is white, smooth to rugose and powdery, and generally has lobed margins.

Formation of ascospores: Asci show mother–daughter conjugation and produce one to four spheroidal to subspheroidal spores (Fig. 86). Asci are generally persistent, although free spores are occasionally observed. Single-spore isolates from four-spored asci from the type strain and one additional strain were sporogenous, indicating the species to be homothallic. Kawakami et al. (1961) and Kreger-van Rij (1970c) reported the spores to have a thin, subequatorial ledge visible under the electron microscope.

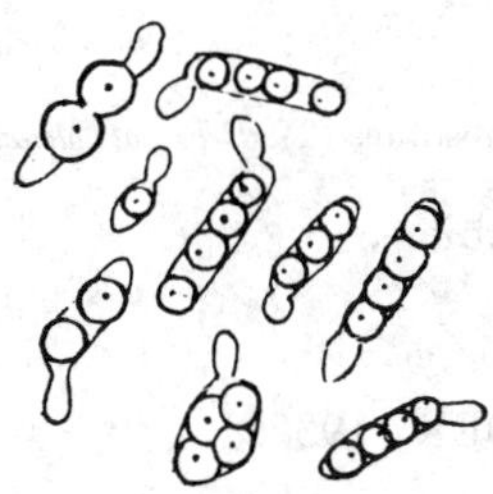

Fig. 86 – *P. farinosa*, CBS 185
After 1 week on V8 agar
(From 'The Yeasts', ed. I, 1952)

Spores were observed on V8-, RG- and YM agar.

Fermentation:

Glucose	+	Lactose	–
Galactose	– or +w	Raffinose	–
Sucrose	–	Trehalose	v
Maltose	–		

Assimilation of carbon compounds:

Galactose	+	Raffinose	–	Erythritol	+
Sucrose	–	Soluble starch	–	Ribitol	+
Maltose	–	D-Xylose	v	D-Mannitol	+
Cellobiose	v	L-Arabinose	– or +w	Succinic acid	v
Trehalose	v	D-Ribose	+	Citric acid	+
Lactose	v	L-Rhamnose	–	Inositol	–

Additional carbon compounds tested: L-Sorbose v, melibiose –, melezitose –, inulin –, D-arabinose –, D-glucosamine · HCl –, glucitol +, α-methyl-D-glucoside –, salicin v, potassium D-gluconate +, DL-lactic acid –

Assimilation of nitrate: –

Growth in vitamin-free medium: v

Growth in 10% sodium chloride plus 5% glucose in yeast nitrogen base: +

Growth at 37°C: +

G + C: 42.5 mol%, type strain (Kurtzman, unpublished)

Comments

A publication describing *Deb. halotolerans* has not been located, but the cultures of this yeast, obtained from the Laboratory of Culture Collection of Microorganisms, Hokkaido University, Japan, appear typical of *P. farinosa.*

Origin of the strains studied

'Jopenbier' (1), miso (2), stored rice (4), wheat flour (1), other substrates (2), *Deb. halotolerans* (2).

Type strain: CBS 185 (NRRL Y-7553) from 'Jopenbier'.

18. *PICHIA FERMENTANS* LODDER (1932)

Imperfect state: *Candida lambica* (Lindner et Genoud) van Uden et Buckley, on the basis of phenotypic similarity.

Synonyms

Saccharomyces dombrowskii Sacchetti (1933)
Pichia dombrowskii Sacchetti
Zymopichia fermentans (Lodder) Novák et Zsolt (1961)
Candida krusei (Cast.) Berkhout var. *transitoria* Saëz (1965) pro parte

Standard description of *Pichia fermentans*

Growth on 5% malt extract agar: After 3 days at 25°C, the cells are ovoidal to ellipsoidal (1.9–6.5) × (4.0–14.4) μm, and occur singly, in pairs and in short chains. Growth is tannish-yellow in color.

Growth on the surface of assimilation media: Thin to thick climbing pellicles are formed.

Dalmau plate culture on morphology agar: After 7 days at 25°C, growth under the coverglass shows abundant and moderately branched pseudohyphae. True hyphae are not present. Aerobic growth is white and dull with the surface wrinkled by convoluted growth. Margins are lobate.

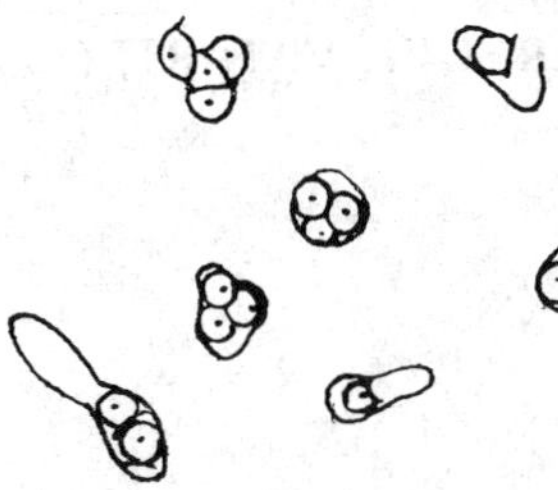

Fig. 87 – *P. fermentans,* CBS 187
After 2 weeks on V8 agar
(From 'The Yeasts', ed. II, 1970)

Formation of ascospores: Asci may be unconjugated or show conjugation between independent cells or between a parent cell and a bud. Two to four spores are formed per ascus, and asci become dehiscent. Spores are hat-shaped but the ledge is not always easily detected under the light microscope (Fig. 87). Single-spore isolates from four-spored asci are sporogenous, indicating the species to be homothallic.

Spores were observed on YM-, ME-, and V8 agar.

Fermentation:

Glucose	+	Lactose	–
Galactose	–	Raffinose	–
Sucrose	–	Trehalose	–
Maltose	–		

Assimilation of carbon compounds:

Galactose	–	Raffinose	–	Erythritol	–
Sucrose	–	Soluble starch	–	Ribitol	–
Maltose	–	D-Xylose	+	D-Mannitol	–
Cellobiose	–	L-Arabinose	–	Succinic acid	+
Trehalose	–	D-Ribose	–	Citric acid	+
Lactose	–	L-Rhamnose	–	Inositol	–

Additional carbon compounds tested: L-Sorbose –, melibiose –, melezitose –, inulin –, D-arabinose –, D-glucosamine · HCl +, glucitol –, α-methyl-D-glucoside –, salicin –, potassium D-gluconate –, DL-lactic acid +

Assimilation of nitrate: –

Growth in vitamin-free medium: –

Growth in 10% sodium chloride plus 5% glucose in yeast nitrogen base: v

Growth at 37°C: +

G + C: 43.1 mol%, type strain (Fuson et al., 1980)

Comments

There is considerable phenotypic similarity between *P. fermentans* and *P. kluyveri*. However, the two may be separated on the basis of D-xylose assimilation by *P. fermentans* (De Camargo & Phaff, 1957). These species also differ by about 14% in G + C content.

Origin of the strains studied

Buttermilk (1), stracchino cheese (1), sputum (1), spoiled orange juice (1), kefyr (1), cecum of cow (1), *Cyanochen cyanopterus* (1), *Cygnus olor* (1).

Type strain: CBS 187 (NRRL Y-1619) from buttermilk.

19. *PICHIA FLUXUUM* (PHAFF ET KNAPP) KREGER-VAN RIJ (1964b)

Synonym

Debaryomyces fluxorum Phaff et Knapp (1956)

Standard description of *Pichia fluxuum*

Growth on 5% malt extract agar: After 3 days at 25°C, the cells are ovoidal to elongate (2.0–6.1) × (3.5–10.5) μm, and occur singly or in pairs. Growth is yellowish-white in color.

Growth on the surface of assimilation media: Dry climbing pellicles are present.

Dalmau plate culture on morphology agar: After 7 days at 25°C, growth under the coverglass shows occasional, sparingly branched pseudohyphae but no true hyphae. Aerobic growth is tannish-white, dull, smooth to partly wrinkled, and with an entire to broadly lobed margin.

Formation of ascospores: Asci are unconjugated and form one or two, rarely three or four, spheroidal, roughened spores (Fig. 88). In addition to the roughening, electron microscopy reveals the spores to have a thin subequatorial ring (Kawakami & Nehira, 1958a; Kreger-van Rij, 1964b; Kurtzman & Smiley, 1974). Asci are persistent.

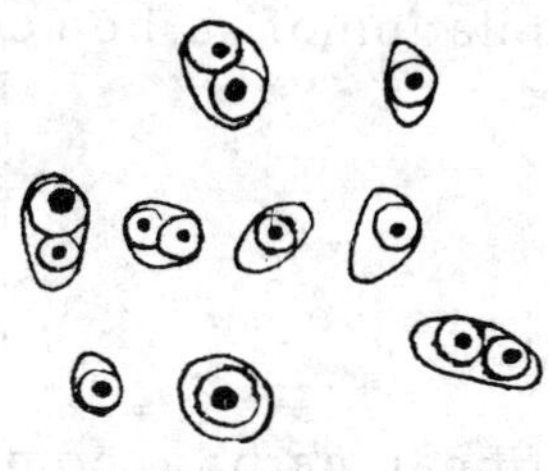

Fig. 88 – *P. fluxuum*, CBS 2287
After 8 days on V8 agar
(From 'The Yeasts', ed. II, 1970)

Spores were observed on YM- and V8 agar.

Fermentation: –

Assimilation of carbon compounds:

Galactose	–	Raffinose	–	Erythritol	–
Sucrose	–	Soluble starch	–	Ribitol	v
Maltose	–	D-Xylose	–	D-Mannitol	+
Cellobiose	–	L-Arabinose	–	Succinic acid	+
Trehalose	–	D-Ribose	–	Citric acid	–
Lactose	–	L-Rhamnose	–	Inositol	–

Additional carbon compounds tested: L-Sorbose –, melibiose –, melezitose –, inulin –, D-arabinose –, D-glucosamine · HCl –, glucitol +, α-methyl-D-glucoside –, salicin –, potassium D-gluconate –, DL-lactic acid v

Assimilation of nitrate: –
Growth in vitamin-free medium: –
Growth in 10% sodium chloride plus 5% glucose in yeast nitrogen base: –
Growth at 37°C: –
G + C: 33.8 mol%, type strain (Price et al., 1978)

Origin of the strains studied

Slime flux of *Quercus kelloggii* (2), slime flux of *Abies concolor* (2), sauerkraut (3).

Type strain: CBS 2287 (NRRL YB-4273), from *Quercus kelloggii.*

20. *PICHIA GUILLIERMONDII* WICKERHAM (1966)

Imperfect state: *Candida guilliermondii* (Cast.) Langeron et Guerra

Synonyms

Candida guilliermondii (Cast.) Langeron et Guerra (1938) pro parte
Endomycopsis guilliermondii nom. nud. (Wickerham & Burton, 1954b)
Myceloblastanon arzti Ota (1924a)
Candida melibiosi Lodder et Kreger-van Rij (1952) pro parte

Standard description of *Pichia guilliermondii*

Growth on 5% malt extract agar: After 3 days at 25°C, the cells are ovoidal to elongate (1.5–4.8) × (2.0–15.0) μm, and occur singly, in pairs, or in short chains. Growth is smooth to wrinkled and tannish-white in color.

Growth on the surface of assimilation media: Generally pellicles are not formed, but rings may be present.

Dalmau plate culture on morphology agar: After 7 days at 25°C, growth under the coverglass shows abundant, well-branched pseudohyphae bearing whorls of blastospores. True hyphae are not present. Aerobic growth is tannish-white, sometimes butyrous, smooth and glistening or dull and rugose. Margins are usually lobate.

Formation of ascospores: This species is heterothallic and only haploid forms have been isolated. Following the pairing of complementary mating types, the resulting asci produce one to four hat-shaped spores. The spores are liberated soon after formation. Complementary mating types are NRRL Y-2075 (CBS 2030) and NRRL Y-2076 (CBS 2031).

Spores were observed on 5% malt extract agar.

Fermentation:

Glucose	+	Lactose	–
Galactose	v	Raffinose	+
Sucrose	+	Trehalose	+
Maltose	–		

Assimilation of carbon compounds:

Galactose	+	Raffinose	+	Erythritol	–
Sucrose	+	Soluble starch	–	Ribitol	+
Maltose	+	D-Xylose	+	D-Mannitol	+
Cellobiose	+	L-Arabinose	+	Succinic acid	+
Trehalose	+	D-Ribose	+	Citric acid	v
Lactose	–	L-Rhamnose	v	Inositol	–

Additional carbon compounds tested: L-Sorbose v, melibiose +, melezitose +, inulin +, D-arabinose +, D-glucosamine · HCl +, glucitol +, α-methyl-D-glucoside +, salicin +, potassium D-gluconate v, DL-lactic acid v

Assimilation of nitrate: –

Growth in vitamin-free medium: –

Growth in 10% sodium chloride plus 5% glucose in yeast nitrogen base: +

Growth at 37°C: +

G + C: 44.4 mol%, type strain (Fuson et al., 1980)

Comments

Wickerham and Burton (1954b) discovered ascosporulation in *C. guilliermondii* var. *guilliermondii* and tentatively named the perfect state *Endomycopsis guilliermondii*. Wickerham (1966) later validated the ascosporic state as *Pichia guilliermondii*. Only a small minority of natural isolates of *C. guilliermondii* are sexually reactive (Wickerham & Burton, 1954b), and Wickerham's (1966) description of *P. guilliermondii* is based on the phenotypic characteristics of those strains having mating competence.

Origin of the strains studied

Frass of *Ulmus americana*, U.S.A. (1); alcohol strain, Argentina (1); frass, Balsam fir, U.S.A. (1); frass *Picea excelsa*, Germany (1); sputum (1); blood (1); atmosphere (1); unknown substrates, received from Castellani (2), Langeron (1), Mackinnon (1).

Type strain: CBS 2030 (NRRL Y-2075), from *Ulmus americana*.

21. *PICHIA HAPLOPHILA* SHIFRINE ET PHAFF (1956)

Standard description of *Pichia haplophila*

Growth on 5% malt extract agar: After 3 days at 25°C, the cells are ovoidal (1.5–4.2) × (2.5–8.0) μm, and occur singly, in pairs or in small clusters. Growth is butyrous and tannish-white in color.

Growth on the surface of assimilation media: Dry, climbing pellicles are present.

Dalmau plate culture on morphology agar: After 7 days at 25°C, growth under the coverglass shows occasional rudimentary pseudohyphae. True hyphae are not formed. Aerobic growth is butyrous with a dull, tannish-white almost powdery surface. The margin is finely lobed.

Formation of ascospores: Asci show conjugation between mother cell and bud or infrequently between independent cells. Two to four hat-shaped spores are formed by each ascus, and they are readily liberated.

Spores were observed on 5% malt extract agar.

Fermentation: –

Assimilation of carbon compounds:

Galactose	+	Raffinose	–	Erythritol	+
Sucrose	–	Soluble starch	–	Ribitol	+
Maltose	–	D-Xylose	+	D-Mannitol	+
Cellobiose	–	L-Arabinose	+	Succinic acid	–
Trehalose	–	D-Ribose	+	Citric acid	–
Lactose	–	L-Rhamnose	–	Inositol	–

Additional carbon compounds tested: L-Sorbose +s, melibiose –, melezitose –, inulin –, D-arabinose – or +s, D-glucosamine · HCl –, glucitol +, α-methyl-D-glucoside –, salicin –, potassium D-gluconate +s, DL-lactic acid –

Assimilation of nitrate: –

Growth in vitamin-free medium: –

Growth in 10% sodium chloride plus 5% glucose in yeast nitrogen base: +s

Growth at 37°C: –

G + C: 41.0 mol%, type strain (Kurtzman et al., unpublished)

Comments

Results of the present study are in good agreement with previous work (Shifrine & Phaff, 1956; Kreger-van Rij, 1970c) except that both strains examined latently assimilated L-sorbose.

Origin of the strains studied

From the bark beetle *Dendroctonus jeffreyi,* U.S.A. (1); orchard soil, Japan (1).

Type strain: CBS 2028 (NRRL Y-7860), from *Dendroctonus jeffreyi.*

22. *PICHIA HEEDII* PHAFF, STARMER, MIRANDA ET MILLER (1978)

Standard description of *Pichia heedii*

Growth on 5% malt extract agar: After 3 days at 25°C, the cells are ovoidal (1.8–4.1) × (3.0–6.8) μm, and occur singly, in pairs or in short chains. Growth is butyrous and cream colored.

Growth on the surface of assimilation media: Thin, climbing pellicles are present.

Dalmau plate culture on morphology agar: After 7 days at 25°C, growth under the coverglass shows rudimentary pseudohyphae but no true hyphae. Aerobic growth is dull, smooth, white in color, butyrous, and has a finely serrate margin.

Formation of ascospores: This species is heterothallic, but only diploid strains have been isolated from nature. Vegetative cells are transformed into asci which produce two to four hat-shaped spores. Asci are dehiscent. Complementary mating types derived from the sporogenous diploid type strain are represented by NRRL Y-10968 (CBS 6931) and NRRL Y-10969 (CBS 6932).

Spores were observed on 5% malt extract agar.

Fermentation: –

Assimilation of carbon compounds:

Galactose	–	Raffinose	–	Erythritol	–
Sucrose	–	Soluble starch	–	Ribitol	–
Maltose	–	D-Xylose	+	D-Mannitol	–
Cellobiose	–	L-Arabinose	–	Succinic acid	+
Trehalose	–	D-Ribose	–	Citric acid	v
Lactose	–	L-Rhamnose	–	Inositol	–

Additional carbon compounds tested: L-Sorbose –, melibiose –, melezitose –, inulin –, D-arabinose –, D-glucosamine · HCl –, glucitol –, α-methyl-D-glucoside –, salicin –, potassium D-gluconate –, DL-lactic acid v

Assimilation of nitrate: –

Growth in vitamin-free medium: –

Growth in 10% sodium chloride plus 5% glucose in yeast nitrogen base: –

Growth at 37°C: +

G + C: 32.3–32.7 mol%, 6 strains (Phaff et al., 1978)

Comments

Phaff et al. (1978) reported *P. heedii* to be isolated only from cacti of the subtribe Pachycereinae and the associated *Drosophila pachea.* It may be separated from *P. amethionina, P. cactophila* and *P. opuntiae* by its strong assimilation of D-xylose. Starmer et al. (1979) observed no mating among *P. heedii, P. amethionina* and *P. opuntiae. P. heedii* may be separated from *P. membranaefaciens* on the basis of its cactus habitat, smaller cell size, comparatively slower release of ascospores and higher maximum growth temperature. The 10% difference in G + C content between these two phenotypically similar species would be expected to preclude conspecificity.

Origin of the strains studied

From *Lophocereus schottii,* Baja California Sur, Mexico (1); *Carnegia gigantea,* Arizona, U.S.A (1).

Type strain: CBS 6930 (NRRL Y-10967).

23. *PICHIA HEIMII* PIGNAL (1970)

Standard description of *Pichia heimii*

Growth on 5% malt extract agar: After 3 days at 25°C, the cells are ovoidal (2.0–4.2) × (3.0–6.0) μm, and occur singly and in pairs. Growth is tannish-white and smooth to somewhat convoluted.

Growth on the surface of assimilation media: Moderately thick pellicles are formed.

Dalmau plate culture on morphology agar: After 7 days at 25°C, growth under the coverglass shows abundant, moderately branched pseudohyphae with large clusters of blastospores. True hyphae are not formed. Infrequently, a pseudohyphal cell will be found which contains up to six endospores. Aerobic growth is tannish-white, glistening, smooth to slightly convoluted, and with a sparingly lobed margin.

Formation of ascospores: Spores were not observed in the present study. Pignal (1970) reported the asci to arise from conjugation between independent cells or, less frequently, from conjugation between a mother cell and a bud. One to four hat-shaped spores were formed in each ascus, and they were liberated soon after formation.

Spores were reported to form on Adams' sporulation medium (Pignal, 1970).

Fermentation:

Glucose	+	Lactose	–
Galactose	+	Raffinose	–
Sucrose	+	Trehalose	+
Maltose	+ w		

Assimilation of carbon compounds:

Galactose	+	Raffinose	+ s	Erythritol	+
Sucrose	+	Soluble starch	–	Ribitol	+
Maltose	+	D-Xylose	+	D-Mannitol	+
Cellobiose	+	L-Arabinose	+	Succinic acid	+
Trehalose	+	D-Ribose	+	Citric acid	+
Lactose	–	L-Rhamnose	+ s	Inositol	–

Additional carbon compounds tested: L-Sorbose –, melibiose –, melezitose +, inulin –, D-arabinose –, D-glucosamine · HCl +, glucitol +, α-methyl-D-glucoside +, salicin +, potassium D-gluconate + s, DL-lactic acid –

Assimilation of nitrate: –

Growth in vitamin-free medium: +

Growth in 10% sodium chloride plus 5% glucose in yeast nitrogen base: +

Growth at 37°C: –

Comments

The results of the present study are in good agreement with the original description except that Pignal (1970) reported a weak fermentation of raffinose and weak growth on L-sorbose, reactions that I found negative.

Origin of the strain studied

Decaying, insect-invaded wood, Equatorial Africa.

Type strain: CBS 6139 (NRRL Y-7502).

24. *PICHIA HUMBOLDTII* RODRIGUES DE MIRANDA ET TÖRÖK (1976)

Imperfect state: *Candida ingens* van der Walt et van Kerken

Standard description of *Pichia humboldtii*

Growth on 5% malt extract agar: After 3 days at 25°C, the cells are ovoidal to elongate (4.5–10.0) × (7.0–16.0) μm, and occur singly, in pairs or in chains. Growth is somewhat dull, lightly wrinkled and tannish-yellow in color.

Growth on the surface of assimilation media: Dry climbing pellicles are formed.

Dalmau plate culture on morphology agar: After 7 days at 25°C, growth under the coverglass shows moderately branched pseudohyphae but no true hyphae. Aerobic growth is tannish-white, dull, wrinkled and with an irregular margin.

Formation of ascospores: Spores were not observed in the present study. Rodrigues de Miranda and Török (1976) induced the type strain of *C. ingens* to sporulate by treatment with *N*-methyl-*N'*-nitro-*N*-nitrosoguanidine. Asci were unconjugated and contained one to four rather large spheroidal spores which were released after formation. Electron microscopy showed the spore walls to be smooth. Heat-treatment of a sporogenous culture gave only sporogenous colonies, indicating the species to be homothallic.

Spores were reported to form on acetate agar.

Fermentation: –

Assimilation of carbon compounds:

Galactose	+	Raffinose	–	Erythritol	–
Sucrose	–	Soluble starch	–	Ribitol	–
Maltose	–	D-Xylose	–	D-Mannitol	–
Cellobiose	–	L-Arabinose	–	Succinic acid	+
Trehalose	–	D-Ribose	–	Citric acid	–
Lactose	–	L-Rhamnose	–	Inositol	–

Additional carbon compounds tested: L-Sorbose –, melibiose –, melezitose –, inulin –, D-arabinose –, D-glucosamine · HCl –, glucitol –, α-methyl-D-glucoside –, salicin –, potassium D-gluconate –, DL-lactic acid +

Assimilation of nitrate: –

Growth in vitamin-free medium: +

Growth in 10% sodium chloride plus 5% glucose in yeast nitrogen base: –

Growth at 37°C: v

Origin of the strain studied

Mutagenized type strain of *C. ingens* (1).

Type strain: CBS 6787 (NRRL Y-10929).

25. PICHIA KLUYVERI BEDFORD EX KUDRIAVZEV (1960)

Synonyms

Pichia kluyveri nom. nud. (Bedford, 1942)
Hansenula kluyveri (Bedford) Kudriavzev (1960)

Standard description of *Pichia kluyveri*

Growth on 5% malt extract agar: After 3 days at 25°C, the cells are ovoidal to elongate (2.0–5.5) × (4.0–11.0) μm, and occur singly, in pairs, or in short chains. Growth is yellowish-tan, dull and with fine wrinkles.

Growth on the surface of assimilation media: Dry, climbing pellicles are formed.

Dalmau plate culture on morphology agar: After 7 days at 25°C, growth under the coverglass shows moderately well branched pseudohyphae, but they produce few blastospores. True hyphae are not formed. Aerobic growth is tannish-white, dull, sometimes almost powdery, and with fine striations radiating from center to colony edge. Margins are entire to finely serrate.

Formation of ascospores: Two to four hat-shaped spores are produced in each ascus, and they are released very soon after formation. Single-spore isolates from four-spored asci gave sporogenous colonies suggesting the species to be homothallic. Spores were observed on acetate agar.

Fermentation:

Glucose	+	Lactose	–
Galactose	–	Raffinose	–
Sucrose	–	Trehalose	–
Maltose	–		

Assimilation of carbon compounds:

Galactose	–	Raffinose	–	Erythritol	–
Sucrose	–	Soluble starch	–	Ribitol	–
Maltose	–	D-Xylose	–	D-Mannitol	–
Cellobiose	–	L-Arabinose	–	Succinic acid	+
Trehalose	–	D-Ribose	–	Citric acid	+ or w
Lactose	–	L-Rhamnose	–	Inositol	–

Additional carbon compounds tested: L-Sorbose –, melibiose –, melezitose –, inulin –, D-arabinose –, D-glucosamine · HCl +, glucitol –, α-methyl-D-glucoside –, salicin –, potassium D-gluconate –, DL-lactic acid + or w

Assimilation of nitrate: –

Growth in vitamin-free medium: –

Growth in 10% sodium chloride plus 5% glucose in yeast nitrogen base: +

Growth at 37°C: +

G + C: 28.3–28.5 mol%, 2 strains (Nakase & Komagata, 1970b)

Origin of the strains studied

Olives, Bedford (1); *P. belgica* var. *microspora* nom. nud.? (1); tropical products (1); fermenting cacao (2).

Type strain: CBS 188 (NRRL Y-11519), from olives.

26. *PICHIA LINDNERI* HENNINGER ET WINDISCH (1975a)

Standard description of *Pichia lindneri*

Growth on 5% malt extract agar: After 3 days at 25°C, the cells are spheroidal to ovoidal (1.5–2.5) × (1.8–4.8) μm and occur singly or in pairs. Growth is tannish-white, glistening and mucoid.

Growth on the surface of assimilation media: Pellicles are not formed.

Dalmau plate culture on morphology agar: After 7 days at 25°C, growth under the coverglass shows neither pseudohyphae nor true hyphae. Aerobic growth is tannish-white, smooth, glistening and butyrous with entire margins.

Formation of ascospores: Sporulation is preceded by conjugation between a parent cell and a bud. Two to four hat-shaped spores are formed in each ascus. Henninger and Windisch (1975a) reported this species to be homothallic.

Spores were observed on 5% malt extract agar.

Fermentation:

Glucose	–, +w, s	Lactose	–
Galactose	–	Raffinose	–
Sucrose	–	Trehalose	–, +w, s
Maltose	–		

Assimilation of carbon compounds:

Galactose	–	Raffinose	–	Erythritol	–
Sucrose	–	Soluble starch	–	Ribitol	+
Maltose	–	D-Xylose	+s	D-Mannitol	+
Cellobiose	+s	L-Arabinose	–	Succinic acid	+s
Trehalose	+	D-Ribose	v	Citric acid	+
Lactose	–	L-Rhamnose	+s	Inositol	–

Additional carbon compounds tested: L-Sorbose –, melibiose –, melezitose –, inulin –, D-arabinose –, D-glucosamine · HCl –, glucitol +, α-methyl-D-glucoside –, salicin +, potassium D-gluconate +s, DL-lactic acid v

Assimilation of nitrate: –

Growth in vitamin-free medium: –

Growth in 10% sodium chloride plus 5% glucose in yeast nitrogen base: +s

Growth at 37°C: –

G + C: 47.7 mol%, type strain (Kurtzman, unpublished)

Comments

Results from the present study are in good agreement with the original description (Henninger & Windisch, 1975a). The weak fermentation shown by *P. lindneri* is also late, and tests terminated before two or three weeks are usually negative.

As shown by the mucoid nature of the cultures, *P. lindneri* produces considerable amounts of an extracellular polysaccharide.

The original spelling of the species name, *lindnerii*, has been treated as an orthographic error and corrected to *lindneri*.

Origin of the strain studied

Forest soil, Germany (1).

Type strain: CBS 6502 (NRRL Y-10948).

27. *PICHIA MEDIA* BOIDIN, PIGNAL, LEHODEY, VEY ET ABADIE (1964)

Standard description of *Pichia media*

Growth on 5% malt extract agar: After 3 days at 25°C, the cells are ovoidal (1.9–3.5) × (3.0–5.5) μm, and occur singly or in pairs. Growth is yellowish-white, smooth, dull-glistening, and butyrous.

Growth on the surface of assimilation media: Pellicles are formed.

Dalmau plate culture on morphology agar: After 7 days at 25°C, growth under the coverglass shows neither pseudohyphae nor true hyphae. Aerobic growth is light tannish-white, smooth, or with minute striations, dull-glistening, butyrous, and with a smooth to finely lobed margin.

Formation of ascospores: Conjugation between parent cell and bud or occasionally between independent cells precedes ascus formation. Asci contain one to four hat-shaped spores which are liberated soon after formation.

Spores were observed on 5% malt extract agar.

Fermentation: –

Assimilation of carbon compounds:

Galactose	+	Raffinose	–	Erythritol	+
Sucrose	–	Soluble starch	– or +w	Ribitol	+
Maltose	+	D-Xylose	+	D-Mannitol	+
Cellobiose	+	L-Arabinose	+	Succinic acid	+
Trehalose	+	D-Ribose	+ or s	Citric acid	+
Lactose	–	L-Rhamnose	–	Inositol	–

Additional carbon compounds tested: L-Sorbose +, melibiose –, melezitose –, inulin –, D-arabinose + or s, D-glucosamine·HCl – or +w, glucitol +, α-methyl-D-glucoside –, salicin +, potassium D-gluconate +s, DL-lactic acid –

Assimilation of nitrate: –
Growth in vitamin-free medium: –
Growth in 10% sodium chloride plus 5% glucose in yeast nitrogen base: +
Growth at 37°C: –

Comments

Kreger-van Rij (1970c) noted some inconsistencies in assimilation between her tests and those of the original description. In turn, I have found some differences with her results. This may reflect differences in the purity of the carbon compounds used or, more likely, variability within *P. media* itself. In contrast to the 1970 study, assimilation of soluble starch was negative or weak, L-arabinose was positive rather than variable, D-ribose was positive but sometimes slow, D-arabinose was positive rather than variable, and galactitol was variable rather than negative.

Origin of the strains studied

Xylophagous insect larvae, France (2).

Type strain: CBS 5521 (NRRL Y-7122).

28. PICHIA MEMBRANAEFACIENS HANSEN (1904)

Imperfect state: *Candida valida* (Leberle) van Uden et Buckley, on the basis of phenotypic similarity.

Synonyms

Saccharomyces membranaefaciens Hansen (1888b)
? *Saccharomyces pyriformis* Ward (1892)
Saccharomyces anomalus Hansen var. *belgicus* Lindner (1895)
? *Pichia hyalospora* (Lindner) Hansen (1904)
Willia belgica Lindner (1905a)
Hansenula belgica (Lindner) H. et P. Sydow (1919)
Endomyces belgica (Lindner) Zender (1925)
Pichia belgica (Lindner) Dekker (Stelling-Dekker, 1931)
Pichia membranaefaciens Hansen var. *belgica* (Lindner) Kodama, Kyono et Kodama (1955)
Saccharomyces mycoderma punctisporus Mélard (1910)
Pichia punctispora (Mélard) Dekker (Stelling-Dekker, 1931)
Pichia alcoholophila Klöcker (1912a)
Pichia calliphorae Klöcker (1912a)
Pichia membranaefaciens Hansen var. *calliphorae* (Klöcker) Dekker (Stelling-Dekker, 1931)
Mycoderma vanlaeriana Lindner et Genoud (1913)
Candida krusei (Cast.) Berkhout var. *vanlaeriana* (Lindner et Genoud) Diddens et Lodder (1942)
Pichia mandshurica Saito (1914)
Pichia membranaefaciens Hansen var. *mandshurica* (Saito) Kodama, Kyono et Kodama (1955)
Zygosaccharomyces chevalieri Guilliermond (1914)
Zygopichia chevalieri (Guilliermond) Klöcker (1924)

Endomyces chodati Zender (1925)
Pichia chodati (Zender) Dekker (Stelling-Dekker, 1931)
Endomyces trumpyi Zender et Bevan (in Zender, 1925)
Pichia chodati (Zender) Dekker var. *trumpyi* (Zender et Bevan) Dekker (Stelling-Dekker, 1931)
Pichia neerlandica Lodder (1932)
Pichia alcoholophila Klöcker var. *naganishii* Lodder (1932)
Pichia derossii nom. nud. (Castelli, 1935)
Zygopichia chiantigiana Castelli (1938)
Mycoderma vini-lafarii Zimmermann (1938)
Pichia membranaefaciens Hansen var. *acidificans* Scrivani (1939)
Pichia chodati (Zender) Dekker var. *fermentans* Mrak, Phaff et Vaughn (1942b)
? *Pichia fermentans* var. *rugosa* nom. nud. (Bedford, 1942)
Pichia silvestris Phaff et Knapp (1956)
Pichia membranaefaciens Hansen var. *sicereum* Santa María (1956c)
Pichia indica nom. nud. (Bahadur, 1957)
Pseudohansenula indica (Bahadur) Novák et Zsolt (1961)
Pichia saccharophila Sasaki et Yoshida (1959)
Pichia tannicola Jacob (1969a)
Pichia farinosa var. *lodderi* Negroni et Fischer nom. nud.?

Standard description of *Pichia membranaefaciens*

Growth on 5% malt extract agar: After 3 days at 25°C, the cells are ovoidal to elongate (1.8–4.5) × (2.5–17.0) μm, and occur singly, in pairs, in chains, or in clusters. Growth is yellowish-tan, dull and smooth or wrinkled.

Growth on the surface of assimilation media: Dry, climbing pellicles are formed.

Dalmau plate culture on morphology agar: After 7 days at 25°C, growth under the coverglass generally shows moderately branched pseudohyphae, although some strains may form none. True hyphae are not produced. Aerobic growth is tannish-white, dull, smooth to wrinkled and with an irregular lobed margin.

Formation of ascospores: Asci may be unconjugated or show conjugation between independent cells or between a parent cell and a bud. One to four spores are usually formed per ascus, but Santa María (1960, 1966) has observed as many as seven. Spores may be spheroidal or hemispheroidal and with or without a ledge (Figs. 89, 90). Spores with ledges may appear nearly hat-shaped or nearly Saturn-shaped. When viewed in the scanning electron microscope, it is not uncommon to find a single strain producing the whole range of spore shapes. Spores are generally liberated from

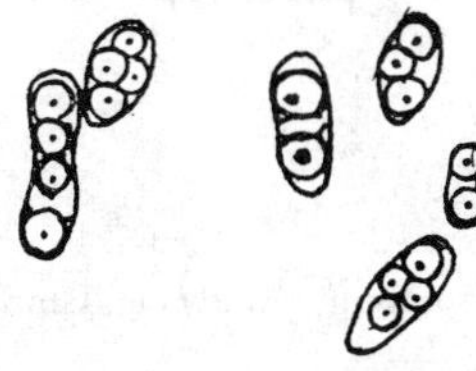

Fig. 89 – *P. membranaefaciens,* CBS 1329
After 8 days on malt extract agar
(From 'The Yeasts', ed. II, 1970)

Fig. 90 – *P. membranaefaciens,* CBS 2100
After 8 days on YM agar
(From 'The Yeasts', ed. II, 1970)

the ascus. Single-spore isolates from four-spored asci show some cultures to be homothallic, but Slooff (1964) and Kreger-van Rij (1970c) clearly demonstrated other strains to be heterothallic. Of further interest was Kreger-van Rij's (1970c) observation of enhanced sporulation in certain self-sporulating strains following their conjugation with one or the other of the mating types. Complementary mating types obtained by Slooff (1964) are CBS 5567 (NRRL Y-6775) and CBS 5568 (NRRL Y-6776).

Spores were observed on YM- and V8 agar.

Fermentation:

Glucose	– or +w	Lactose	–
Galactose	–	Raffinose	–
Sucrose	–	Trehalose	–
Maltose	–		

Assimilation of carbon compounds:

Galactose	–	Raffinose	–	Erythritol	–
Sucrose	–	Soluble starch	–	Ribitol	–
Maltose	–	D-Xylose	v	D-Mannitol	–
Cellobiose	–	L-Arabinose	–	Succinic acid	v
Trehalose	–	D-Ribose	–	Citric acid	v
Lactose	–	L-Rhamnose	–	Inositol	–

Additional carbon compounds tested: L-Sorbose v, melibiose –, melezitose –, inulin –, D-arabinose –, D-glucosamine · HCl v, glucitol –, α-methyl-D-glucoside –, salicin –, potassium D-gluconate –, DL-lactic acid v

Assimilation of nitrate: –

Growth in vitamin-free medium: v

Growth in 10% sodium chloride plus 5% glucose in yeast nitrogen base: + or s

Growth at 37°C: v

G + C: 44.3 mol%, type strain (Fuson et al., 1980)

Comments

Strains of *P. membranaefaciens* show considerable variation in the amount of pseudomycelium produced, in the assimilation of several carbon compounds, in growth at 37°C and in their requirement for vitamins. However, there seems little correlation among the varying characteristics that would allow sub-grouping of the strains, and Kreger-van Rij (1970c) found mating between strains showing differences in one or more of these properties. Consequently, all of this phenotypic variation is considered typical for the species.

Origin of the strains studied

Grape must (1), wine (1), vineyard soil (2), beer (1), raw material for sorghum brandy (1), lambic gueuse (1), feces (1), maize-cattle waste fermentation (3), unknown substrates (9).

Type strain: CBS 107 (NRRL Y-2026).

29. *PICHIA METHANOLICA* MAKIGUCHI (in Kato et al., 1974)

Standard description of *Pichia methanolica*

Growth on 5% malt extract agar: After 3 days at 25°C, the cells are spheroidal to ovoidal (1.9–6.0) × (2.0–8.8) μm, and occur singly, in pairs or in clusters. Growth is cream-colored, smooth, glistening and butyrous.

Growth on the surface of assimilation media: Pellicles are not formed.

Dalmau plate culture on morphology agar: After 7 days at 25°C, growth under the coverglass shows neither pseudohyphae nor true hyphae. Aerobic growth is pale tannish-white, smooth, glistening and butyrous with an entire to lobed margin.

Formation of ascospores: Asci may show parent–bud conjugation or conjugation between independent cells. One to four hat-shaped spores are produced in each ascus, and the spores are freed soon after formation.

Spores were observed on 5% malt extract agar.

Fermentation:

Glucose	+	Lactose	–
Galactose	–	Raffinose	–
Sucrose	–	Trehalose	+
Maltose	–		

Assimilation of carbon compounds:

Galactose	+	Raffinose	–	Erythritol	+
Sucrose	–	Soluble starch	–	Ribitol	+
Maltose	–	D-Xylose	+	D-Mannitol	+
Cellobiose	+	L-Arabinose	+	Succinic acid	+
Trehalose	+	D-Ribose	+	Citric acid	–
Lactose	–	L-Rhamnose	–	Inositol	–

Additional carbon compounds tested: L-Sorbose +, melibiose –, melezitose –, inulin –, D-arabinose +w, D-glucosamine · HCl –, glucitol +, α-methyl-D-glucoside –, salicin +, potassium D-gluconate +w, DL-lactic acid –

Assimilation of nitrate: –

Growth in vitamin-free medium: –

Growth in 10% sodium chloride plus 5% glucose in yeast nitrogen base: –

Growth at 37°C: +

Comments

Some differences were noted in the results obtained from fermentation and assimilation tests in the present study and those reported by Kato et al. (1974). In contrast to the original description, maltose was not fermented; however, L-sorbose and D-arabinose were assimilated, but citric and lactic acids were not utilized. Furthermore, growth was found to be absent in vitamin-free medium. Kato et al. (1974) reported this species to be noteworthy because of its strong assimilation of methanol.

Origin of the strain studied

Soil, Japan (1).

Type strain: CBS 6515 (NRRL Y-7685).

30. *PICHIA MISSISSIPPIENSIS* KURTZMAN, SMILEY, JOHNSON, WICKERHAM ET FUSON (1980a)

Synonym

Candida obtusa (Dietrichson) van Uden et do Carmo-Sousa ex van Uden et Buckley var. *arabinosa* Montrocher (1967)

Standard description of *Pichia mississippiensis*

Growth on 5% malt extract agar: After 3 days at 25°C, the cells, which are quite variable in size, are ellipsoidal to elongate, or infrequently spheroidal (1.1–5.0) × (2.0–12.0) μm, and occur singly, in pairs or in clusters. Growth is butyrous and tannish-white in color.

Growth on the surface of assimilation media: Pellicles are not formed but rings occasionally develop.

Dalmau plate culture on morphology agar: After 7 days at 25°C, growth under the coverglass shows abundant and highly branched pseudohyphae as well as infrequent true hyphae. True hyphae (diameter: 1.7–3.0 μm) show no evidence of a dolipore septum when viewed under the light microscope. Aerobic growth is light tan, glistening, and butyrous; the center is raised and the margin entire.

Formation of ascospores: Asci arise from diploid cells or from the pairing of complementary mating types and form two to four hat-shaped spores. Asci are spheroidal to ellipsoidal and are free or attached to pseudohyphae. Infrequently, a pseudohyphal or hyphal cell may become an ascus. At maturity asci become dehiscent. Complementary mating types of this heterothallic species are NRRL YB-1294-6 (CBS 7024) and NRRL YB-1294-7 (CBS 7025).

Spores were observed on 5% malt extract- and YM agar.

Fermentation:

Glucose	+	Lactose	–
Galactose	–	Raffinose	–
Sucrose	–	Trehalose	–
Maltose	–		

Assimilation of carbon compounds:

Galactose	–	Raffinose	–	Erythritol	–
Sucrose	+	Soluble starch	–	Ribitol	–
Maltose	+	D-Xylose	+	D-Mannitol	+
Cellobiose	+	L-Arabinose	+	Succinic acid	+
Trehalose	+	D-Ribose	–	Citric acid	v
Lactose	–	L-Rhamnose	v	Inositol	–

Additional carbon compounds tested: L-Sorbose −, melibiose −, melezitose +, inulin −, D-arabinose − or +w, D-glucosamine · HCl −, glucitol +, α-methyl-D-glucoside +, salicin v, potassium D-gluconate +, DL-lactic acid +

Assimilation of nitrate: −

Growth in vitamin-free medium: −

Growth in 10% sodium chloride plus 5% glucose in yeast nitrogen base: −

Growth at 37°C: +

G + C: 47.2–48.0 mol%, 3 strains (Kurtzman et al., 1980a)

Comments

The relationship of this species with *P. amylophila* is discussed in the description of the latter species.

Origin of the strains studied

Frass from shortleaf and loblolly pines, Mississippi, U.S.A. (9); laboratory contaminant (1); from a mushroom, France (1).

Type strain: CBS 7023 (NRRL YB-1294), from frass from a loblolly pine.

31. *PICHIA MUCOSA* WICKERHAM ET KURTZMAN (1971)

Standard description of *Pichia mucosa*

Growth on 5% malt extract agar: After 3 days at 25°C, the cells are spheroidal to mostly ellipsoidal (2.5–5.1) × (4.5–8.0) μm and occur singly or in pairs. Growth is tannish-white in color and mucoid.

Growth on the surface of assimilation media: Dry climbing pellicles are formed.

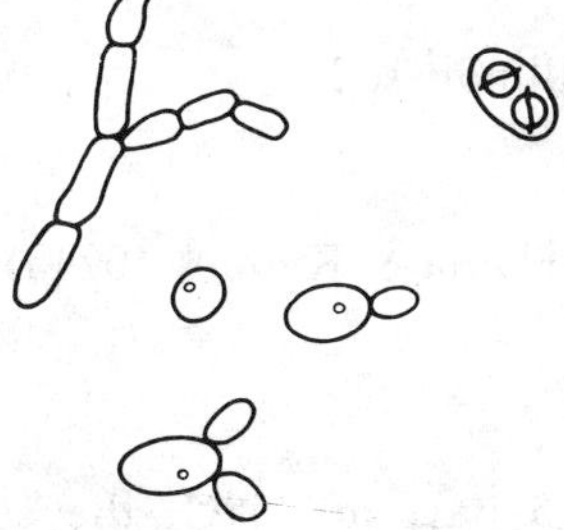

Fig. 91 – *P. mucosa*, CBS 6341
After 1 week on YM agar

Dalmau plate culture on morphology agar: After 7 days at 25°C, growth under the coverglass shows pseudohyphae to be common and highly branched. True hyphae are not present. Aerobic growth is raised, highly glistening, mucoid, grey-white in color and with an entire margin.

Formation of ascospores: The asci are unconjugated and persistent and each contains one or two spores. The spores are spheroidal to occasionally slightly ellipsoidal and have a thick equatorial ring (Fig. 91). Colonies from single ascospore isolates are sporogenous, but since four-spored asci were not available for testing, it is not certain that this species is homothallic.

Spores were observed on 5% malt extract- and YM agar.

Fermentation:

Glucose	+	Lactose	–
Galactose	–	Raffinose	–
Sucrose	–	Trehalose	–
Maltose	–		

Assimilation of carbon compounds:

Galactose	–	Raffinose	–	Erythritol	–
Sucrose	+	Soluble starch	–	Ribitol	–
Maltose	+	D-Xylose	+	D-Mannitol	+ or w
Cellobiose	+	L-Arabinose	–	Succinic acid	v
Trehalose	+	D-Ribose	–	Citric acid	–
Lactose	–	L-Rhamnose	–	Inositol	–

Additional carbon compounds tested: L-Sorbose +, melibiose –, melezitose +, inulin –, D-arabinose v, D-glucosamine · HCl –, glucitol +, α-methyl-D-glucoside +, salicin +, potassium D-gluconate –, DL-lactic acid –

Assimilation of nitrate: –

Growth in vitamin-free medium: –; thiamine, biotin and pyridoxine required for growth

Growth in 10% sodium chloride plus 5% glucose in yeast nitrogen base: +

Growth at 37°C: –

G + C: 33.7–34.2 mol%, 2 strains (Kurtzman et al., unpublished)

Comments

This species is unique among the Saturn-spored members of *Pichia* because it produces substantial amounts of an extracellular polysaccharide. The polysaccharide is neutral and the only sugar present is D-mannose (Seymour et al., 1976).

Origin of the strains studied

Soil from a wooded area, Illinois, U.S.A. (1); soil, South Africa (1).

Type strain: CBS 6341 (NRRL YB-1344), from Illinois soil.

32. *PICHIA NAGANISHII* KODAMA (in Kodama & Kyono, 1974a)

Standard description of *Pichia naganishii*

Growth on 5% malt extract agar: After 3 days at 25°C, the cells are spheroidal to ovoidal (1.8–4.2) × (2.5–6.5) μm, and occur singly, in pairs or in small clusters. Growth is cream-colored, smooth, glistening and butyrous.

Growth on the surface of assimilation media: Thin rings are formed, but no pellicles.

Dalmau plate culture on morphology agar: After 7 days at 25°C, growth under the coverglass shows neither pseudohyphae nor true hyphae. Aerobic growth is faint tan in color, smooth, glistening and butyrous with an entire to finely lobed margin.

Formation of ascospores: Ascus formation may be preceded by conjugation between independent cells or between parent and bud. Two to four hat-shaped spores are produced in each ascus, and they are readily liberated after formation. Single-spore isolates from four-spored asci are sporogenous, indicating the species to be homothallic.

Spores were observed on 5% malt extract- and YM agar.

Fermentation:

Glucose	+	Lactose	–
Galactose	–	Raffinose	–
Sucrose	–	Trehalose	+
Maltose	–		

Assimilation of carbon compounds:

Galactose	+	Raffinose	–	Erythritol	+
Sucrose	+	Soluble starch	–	Ribitol	+
Maltose	+	D-Xylose	+	D-Mannitol	+
Cellobiose	+	L-Arabinose	+	Succinic acid	+
Trehalose	+	D-Ribose	+	Citric acid	–
Lactose	–	L-Rhamnose	+	Inositol	–

Additional carbon compounds tested: L-Sorbose –, melibiose –, melezitose +, inulin –, D-arabinose –, D-glucosamine · HCl +, glucitol +, α-methyl-D-glucoside +w, salicin +, potassium D-gluconate –, DL-lactic acid +

Assimilation of nitrate: –

Growth in vitamin-free medium: –

Growth in 10% sodium chloride plus 5% glucose in yeast nitrogen base: +

Growth at 37°C: +

G + C: 46.1 mol%, type strain (Kodama & Kyono, 1974a)

Comments

The original description of *P. naganishii* (Kodama & Kyono, 1974a) is generally in good agreement with the results obtained in the present study. However, I did find growth on L-sorbose, galactitol and citric acid to be negative rather than positive as originally reported.

Origin of the strain studied

Exudate of *Camellia japonica* L., Japan (1).

Type strain: CBS 6429 (NRRL Y-7654).

33. *PICHIA NAKAZAWAE* KODAMA (1975)

This species has two varieties:
Pichia nakazawae Kodama var. *nakazawae*
Pichia nakazawae Kodama var. *akitaensis* Kodama (1975)

Standard description of *Pichia nakazawae* var. *nakazawae*

Growth on 5% malt extract agar: After 3 days at 25°C, the cells are spheroidal to ellipsoidal (2.0–4.5) × (3.0–9.8) μm, and occur singly, in pairs and in clumps. Growth is cream-colored, smooth, glistening and butyrous.

Growth on the surface of assimilation media: Thin to moderately thick climbing pellicles are produced.

Dalmau plate culture on morphology agar: After 7 days at 25°C, growth under the coverglass shows occasional branching outgrowths of ovoidal cells, but no pseudohyphae or true hyphae. Aerobic growth is tannish-white, smooth or finely striated, glistening, and butyrous with entire to finely lobed margins.

Formation of ascospores: Ascus formation may be preceded by either conjugation between a parent cell and a bud or by conjugation between independent cells. Asci contain two to four hat-shaped spores and these are released soon after formation. Both variaties of *P. nakazawae* were found to be homothallic through examination of single-spore isolates from four-spored asci (Kurtzman, 1984).

Spores were observed on 5% malt extract agar.

Fermentation:

Glucose	+	Lactose	–
Galactose	+	Raffinose	–
Sucrose	–	Trehalose	+
Maltose	+ w		

Assimilation of carbon compounds:

Galactose	+	Raffinose	–	Erythritol	+
Sucrose	+	Soluble starch	+	Ribitol	+
Maltose	+	D-Xylose	+	D-Mannitol	+
Cellobiose	+	L-Arabinose	+	Succinic acid	+
Trehalose	+	D-Ribose	+ s	Citric acid	+
Lactose	–	L-Rhamnose	+	Inositol	–

Additional carbon compounds tested: L-Sorbose + s, melibiose –, melezitose +, inulin –, D-arabinose +, D-glucosamine · HCl +, glucitol +, α-methyl-D-glucoside +, salicin + w, potassium D-gluconate +, DL-lactic acid + w

Assimilation of nitrate: –

Growth in vitamin-free medium: –

Growth in 10% sodium chloride plus 5% glucose in yeast nitrogen base: +

Growth at 37°C: –

G + C: 39.4 mol%, type strain of the variety *nakazawae;* 39.9 mol%, type strain of the variety *akitaensis* (Kurtzman, 1984)

Supplementary description of *Pichia nakazawae* var. *akitaensis*

The variety *akitaensis* differs from the variety *nakazawae* by its failure to ferment galactose, by the presence of a weak and slow sucrose fermentation, and by its inability to assimilate L-rhamnose and lactic acid. The fermentation rate of variety *akitaensis* is quite weak in comparison to the vigorous ability of variety *nakazawae*. Another difference between the two varieties is the formation of moderately well developed pseudohyphae by variety *akitaensis* which causes agar-grown colonies to appear dull and finely convoluted.

Comments

The original description of this species agrees well with the results of the present study except that I found the variety *nakazawae* to slowly assimilate D-ribose and to give a weak and latent reaction on lactic acid. In addition, both varieties assimilated L-sorbose, but neither grew in vitamin-free medium. DNA relatedness between the two varieties was 41%; consequently, the varietal designations were retained (Kurtzman, 1984).

Origin of the strain belonging to the variety *nakazawae*

Exudate of *Quercus myrsinaeforia* Blume, Japan (1).

Type strain: CBS 6700 (NRRL Y-7903).

Origin of the strain belonging to the variety *akitaensis*

Exudate of *Salix* sp., Japan (1).

Type strain: CBS 6701 (NRRL Y-7904).

34. PICHIA NORVEGENSIS LEASK ET YARROW (1976)

Imperfect state: *Candida norvegensis* (Dietrichson) van Uden et Farinha ex van Uden et Buckley, on the basis of phenotypic similarity.

Standard description of *Pichia norvegensis*

Growth on 5% malt extract agar: After 3 days at 25°C, the cells are ovoidal to ellipsoidal $(1.7-4.1) \times (3.0-10.0)$ μm, and occur singly, in pairs or in clusters. Growth is yellowish-white, smooth, glistening and butyrous.

Growth on the surface of assimilation media: Thin pellicles are formed.

Dalmau plate culture on morphology agar: After 7 days at 25°C, growth under the coverglass shows abundant pseudohyphae. Some of the longer pseudohyphal cells are markedly curved. True hyphae are not produced. Aerobic growth is light tan in color, smooth, glistening, butyrous and with an entire margin.

Formation of ascospores: Asci are unconjugated and form one to four hat-shaped spores. Leask and Yarrow (1976) reported that heat-treatment of sporogenous cultures gave only sporogenous colonies, indicating the species to be homothallic. Spores were observed on Fowell's acetate agar.

Fermentation:

Glucose	+ s	Lactose	–
Galactose	–	Raffinose	–
Sucrose	–	Trehalose	–
Maltose	–		

Assimilation of carbon compounds:

Galactose	–	Raffinose	–	Erythritol	–
Sucrose	–	Soluble starch	–	Ribitol	–
Maltose	–	D-Xylose	– or + s	D-Mannitol	–
Cellobiose	+	L-Arabinose	–	Succinic acid	+
Trehalose	–	D-Ribose	–	Citric acid	+ or w
Lactose	–	L-Rhamnose	–	Inositol	–

Additional carbon compounds tested: L-Sorbose –, melibiose –, melezitose –, inulin –, D-arabinose –, D-glucosamine · HCl +, glucitol –, α-methyl-D-glucoside –, salicin v, potassium D-gluconate –, DL-lactic acid + or w

Assimilation of nitrate: –

Growth in vitamin-free medium: –

Growth in 10% sodium chloride plus 5% glucose in yeast nitrogen base: +

Growth at 37°C: –

Origin of the strain studied

Vagina of a pregnant woman (1).

Type strain: CBS 6564 (NRRL Y-7687).

35. *PICHIA OHMERI* (ETCHELLS ET BELL) KREGER-VAN RIJ (1964b)

Synonyms

Endomycopsis ohmeri Etchells et Bell (1950a)
Endomycopsis ohmeri Etchells et Bell var. *minor* Etchells et Bell (1950a)
Candida guilliermondii (Cast.) Langeron et Guerra var. *membranaefaciens* Lodder et Kreger-van Rij (1952)

Standard description of *Pichia ohmeri*

Growth on 5% malt extract agar: After 3 days at 25°C, the cells are ovoidal to cylindroidal (1.7–6.5) × (2.5–25) μm, and occur singly, in pairs and in short chains. Growth is somewhat mycelial and cream-colored.

Growth on the surface of assimilation media: Thin pellicles are formed.

Dalmau plate culture on morphology agar: After 7 days at 25°C, growth under the coverglass shows abundant, moderately branched pseudohyphae. True mycelium was not observed. Aerobic growth is greyish-white, dull and nearly smooth to finely convoluted. Margins are irregularly lobed.

Formation of ascospores: Infrequently, diploid strains are isolated which form unconjugated asci, but this heterothallic species is normally isolated in the haploid form and requires conjugation of complementary mating types before sporulation. Asci form one to four spores and are dehiscent at maturity. Spore shape is determined by the mating types and may be either hat-shaped or spheroidal (Wickerham & Burton, 1954b). Pairing of NRRL Y-2078 (CBS 2037) and NRRL Y-2079 (CBS 568) gives hat-shaped spores (Fig. 92) while the conjugation of NRRL Y-2078 and NRRL Y-2080 (CBS 2038) produces spheroidal spores.

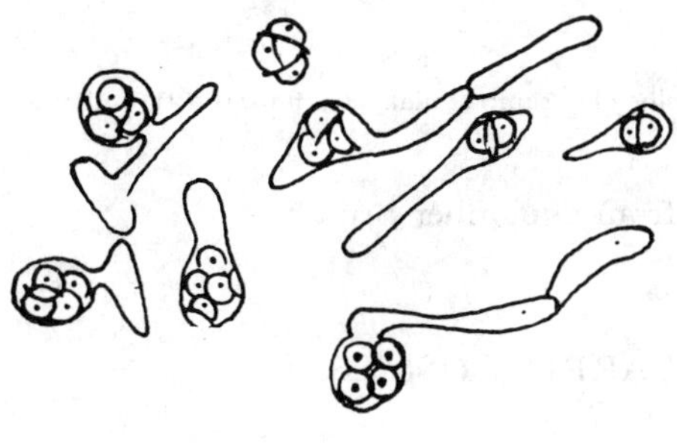

Fig. 92 – *P. ohmeri*
CBS 568 × CBS 2037. After 5 days on malt extract agar (From 'The Yeasts', ed. II, 1970)

Spores were observed on 5% malt extract agar.

Fermentation:

Glucose	+	Lactose	–
Galactose	+ w or s	Raffinose	+
Sucrose	+	Trehalose	– or + w
Maltose	– or + w		

Assimilation of carbon compounds:

Galactose	+	Raffinose	+	Erythritol	–
Sucrose	+	Soluble starch	–	Ribitol	v
Maltose	+	D-Xylose	–	D-Mannitol	+
Cellobiose	+	L-Arabinose	–	Succinic acid	+
Trehalose	+	D-Ribose	v	Citric acid	+
Lactose	–	L-Rhamnose	–	Inositol	–

Additional carbon compounds tested: L-Sorbose +, melibiose –, melezitose –, inulin v, D-arabinose –, D-glucosamine · HCl +, glucitol +, α-methyl-D-glucoside +, salicin +, potassium D-gluconate v, DL-lactic acid v

Assimilation of nitrate: –

Growth in vitamin-free medium: –

Growth in 10% sodium chloride plus 5% glucose in yeast nitrogen base: +

Growth at 37°C: +

G + C: 44.3–44.7 mol%, 4 strains (Fuson et al., 1980)

Comments

Pichia ohmeri is an unusual species because, depending upon the mating types paired, ascospores may be either hat-shaped or spheroidal (Wickerham & Burton, 1954b).

Preliminary genetic studies suggest that only a few genes determine which spore shape results (Kurtzman, unpublished). These studies also demonstrated ascospore fertility through the F_2 generation, thus indicating the various mating types to be members of the same species. Confirming this premise was the work of Fuson et al. (1980) which showed the *P. ohmeri* mating types to exhibit 98–100% DNA relatedness.

Origin of the strains studied

Cucumber brine (3), pleural fluid (1), gooseberry jelly (1), sambal ulak (1), flowers (2), unknown substrates (3).

Type strain: CBS 5367 (NRRL Y-1932), from cucumber brine.

36. *PICHIA ONYCHIS* YARROW (1965)

Standard description of *Pichia onychis*

Growth on 5% malt extract agar: After 3 days at 25°C, the cells are ovoidal (2.0–4.5) × (2.8–7.8) μm, and occur singly or in pairs. Growth is cream-colored, glistening and butyrous.

Growth on the surface of assimilation media: Thin pellicles may be formed.

Dalmau plate culture on morphology agar: After 7 days at 25°C, growth under the coverglass shows a simple, sparingly branched pseudomycelium, but true hyphae are not formed. Aerobic growth is tannish-white, glistening, butyrous, and with an entire or occasionally lobed margin.

Formation of ascospores: Asci are unconjugated and produce one to four hat-shaped spores. The spores are liberated soon after formation.

Spores were observed on 5% malt extract-, YM- and V8 agar.

Fermentation:

Glucose	+	Lactose	–
Galactose	+	Raffinose	+ w
Sucrose	+	Trehalose	–
Maltose	–		

Assimilation of carbon compounds:

Galactose	–	Raffinose	+	Erythritol	–
Sucrose	+	Soluble starch	–	Ribitol	–
Maltose	+	D-Xylose	+	D-Mannitol	+
Cellobiose	+	L-Arabinose	v	Succinic acid	+
Trehalose	+	D-Ribose	–	Citric acid	+
Lactose	–	L-Rhamnose	–	Inositol	–

Additional carbon compounds tested: L-Sorbose –, melibiose –, melezitose +, inulin +, D-arabinose –, D-glucosamine · HCl –, glucitol +, α-methyl-D-glucoside +, salicin +, potassium D-gluconate +, DL-lactic acid +

Assimilation of nitrate: –

Growth in vitamin-free medium: –
Growth in 10% sodium chloride plus 5% glucose in yeast nitrogen base: – or +s
Growth at 37°C: +
G + C: 42.8 mol%, type strain (Kurtzman et al., unpublished)

Comments

Results from the present study are in good agreement with previous work (Yarrow, 1965; Kreger-van Rij, 1970c) except that growth on inulin was found to be positive.

Origin of the strains studied

Infected human nail (1), corn meal (1), dust (1).

Type strain: CBS 5587 (NRRL Y-7123), from an infected nail.

37. PICHIA OPUNTIAE STARMER, PHAFF, MIRANDA, MILLER ET BARKER (1979)

This species has two varieties:
Pichia opuntiae Starmer, Phaff, Miranda, Miller et Barker var. *opuntiae*
Pichia opuntiae Starmer, Phaff, Miranda, Miller et Barker var. *thermotolerans* Starmer, Phaff, Miranda, Miller et Barker (1979)

Standard description of *Pichia opuntiae* var. *opuntiae*

Growth on 5% malt extract agar: After 3 days at 25°C, the cells are ovoidal to ellipsoidal (1.5–4.5) × (2.5–7.5) µm, and occur singly, in pairs or in short chains. Growth is butyrous and tannish-white in color.
Growth on the surface of assimilation media: Thin rings and incomplete pellicles are formed.
Dalmau plate culture on morphology agar: After 7 days at 25°C, growth under the coverglass shows infrequent rudimentary pseudohyphae and no true hyphae. Aerobic growth is light tannish-white in color, smooth, glistening, butyrous, and with entire to finely lobed margins.
Formation of ascospores: This species is heterothallic, and the majority of strains isolated from nature are asporogenous haploids. Following the pairing of complementary mating types, zygotes usually bud diploid cells which become asci. Each ascus contains two to four hat-shaped spores which are liberated soon after formation. Complementary mating types of *P. opuntiae* var. *opuntiae* are NRRL Y-11707 (CBS 7010) and NRRL Y-11708 (CBS 7011); complementary mating types of *P. opuntiae* var. *thermotolerans* are NRRL Y-11709 (CBS 7012) and NRRL Y-11710 (CBS 7013).
Spores were observed on YM agar.
Fermentation: –

Assimilation of carbon compounds:

Galactose	–	Raffinose	–	Erythritol	–
Sucrose	–	Soluble starch	–	Ribitol	–
Maltose	–	D-Xylose	–	D-Mannitol	v
Cellobiose	v	L-Arabinose	–	Succinic acid	+
Trehalose	–	D-Ribose	–	Citric acid	+
Lactose	–	L-Rhamnose	+ s	Inositol	–

Additional carbon compounds tested: L-Sorbose –, melibiose –, melezitose –, inulin –, D-arabinose –, D-glucosamine · HCl –, glucitol –, α-methyl-D-glucoside –, salicin +, potassium D-gluconate –, DL-lactic acid +

Assimilation of nitrate: –

Growth in vitamin-free medium: –

Growth in 10% sodium chloride plus 5% glucose in yeast nitrogen base: –

Growth at 37°C: –

G + C: 33.4–34.0 mol%, 4 strains of the variety *opuntiae;* 32.9–33.3 mol%, 3 strains of the variety *thermotolerans* (Starmer et al., 1979)

Supplementary description of *Pichia opuntiae* var. *thermotolerans*

The variety *thermotolerans* differs from the variety *opuntiae* by its failure to assimilate citric acid. Additionally, it grows at 37°C while the variety *opuntiae* has a maximum temperature for growth of 30–33°C. In my tests, reactions of the variety *thermotolerans* were variable on L-rhamnose and D-glucitol. All known isolates of the variety *opuntiae* have been from cacti in Australia while isolates of the variety *thermotolerans* seem restricted to cacti from North America (Starmer et al., 1979).

Comments

Starmer et al. (1979) reported *P. opuntiae* not to mate with either *P. amethionina* or *P. heedii,* thus indicating the three to be distinct biological species. The relationship of the two varieties of *P. opuntiae* needs further clarification in view of the reported greater fertility in crosses involving the two varieties than in intravarietal crosses of the variety *thermotolerans* (Starmer et al., 1979).

Origin of the strains belonging to the variety *opuntiae*

From *Opuntia inermis,* Yarrawonga, Australia (2).

Type strain: CBS 7010 (NRRL Y-11707).

Origin of the strains belonging to the variety *thermotolerans*

From *Lophocereus schottii,* Mexico (1); moist soil under *Pachycereus pringlei,* Mexico (1).

Type strain: CBS 7012 (NRRL Y-11709).

38. *PICHIA PASTORIS* (GUILLIERMOND) PHAFF (1956)

Synonyms

Zygosaccharomyces pastoris Guilliermond (1919)
Saccharomyces pastori (Guill.) Lodder et Kreger-van Rij (1952)
Petasospora pastori (Guill.) Boidin et Abadie (1954)
Zygowillia pastori (Guill.) Kudriavzev (1960)
Zymopichia pastori (Guill.) Novák et Zsolt (1961)

Standard description of *Pichia pastoris*

Growth on 5% malt extract agar: After 3 days at 25°C, the cells are spheroidal to ovoidal (2.0–4.0) × (2.2–5.8) μm, and occur singly or in pairs. Growth is cream-colored, dull to faintly glistening and butyrous.

Growth on the surface of assimilation media: Pellicles are not formed.

Dalmau plate culture on morphology agar: After 7 days at 25°C, growth under the coverglass shows neither pseudohyphae nor true hyphae. Aerobic growth is white, dull to faintly glistening and butyrous. Margins may be finely serrate or lobed.

Formation of ascospores: Asci may be unconjugated or show conjugation between parent cell and bud or between independent cells. One to four hat-shaped spores are formed in each ascus and they are soon liberated. Single-spore isolates from four-spored asci are sporogenous, indicating this species to be homothallic.

Spores were observed on 5% malt extract-, V8- and acetate agar.

Fermentation:

Glucose	+	Lactose	–
Galactose	–	Raffinose	–
Sucrose	–	Trehalose	–
Maltose	–		

Assimilation of carbon compounds:

Galactose	–	Raffinose	–	Erythritol	–
Sucrose	–	Soluble starch	–	Ribitol	–
Maltose	–	D-Xylose	–	D-Mannitol	+
Cellobiose	–	L-Arabinose	–	Succinic acid	+
Trehalose	+	D-Ribose	–	Citric acid	– or + s
Lactose	–	L-Rhamnose	+	Inositol	–

Additional carbon compounds tested: L-Sorbose –, melibiose –, melezitose –, inulin –, D-arabinose –, D-glucosamine · HCl –, glucitol +, α-methyl-D-glucoside –, salicin –, potassium D-gluconate –, DL-lactic acid v

Assimilation of nitrate: –

Growth in vitamin-free medium: –

Growth in 10% sodium chloride plus 5% glucose in yeast nitrogen base: –

Growth at 37°C: + or w

G + C: 40.2–41.0 mol%, 2 strains (Nakase & Komagata, 1970b)

Origin of the strains studied

Exudate of a chestnut tree, France (1); slime flux on *Quercus kelloggii*, U.S.A. (1).

Type strain: CBS 704 (NRRL Y-1603), from a chestnut tree.

39. *PICHIA PHILOGAEA* VAN DER WALT ET JOHANNSEN (1975c)

Standard description of *Pichia philogaea*

Growth on 5% malt extract agar: After 3 days at 25°C, the cells are spheroidal, ellipsoidal or cylindroidal (2.8–7.5) × (3.0–10.0) μm, and occur singly, in pairs or in short chains. Growth is butyrous and cream-colored.

Growth on the surface of assimilation media: Thin pellicles are formed.

Dalmau plate culture on morphology agar: After 7 days at 25°C, growth under the coverglass shows pseudohyphae to be common but relatively unbranched. True hyphae are not formed. Aerobic growth is tannish-white, glistening and smooth or finely striated. The margin is entire or occasionally lobed.

Formation of ascospores: Ascus formation is preceded by conjugation between a parent cell and a bud. Asci produce one to four hat-shaped spores which are released at maturity. Van der Walt and Johannsen (1975c) reported that heat-treated ascosporogenous cultures yielded only sporogenous isolates, and the species appears homothallic.

Spores were observed on 5% malt extract- and YM agar.

Fermentation:

Glucose	+	Lactose	–
Galactose	+ w	Raffinose	–
Sucrose	–	Trehalose	+
Maltose	– or + w		

Assimilation of carbon compounds:

Galactose	+	Raffinose	–	Erythritol	+
Sucrose	+	Soluble starch	–	Ribitol	+
Maltose	+	D-Xylose	+	D-Mannitol	+
Cellobiose	+	L-Arabinose	+	Succinic acid	+
Trehalose	+	D-Ribose	+ s	Citric acid	+
Lactose	–	L-Rhamnose	–	Inositol	–

Additional carbon compounds tested: L-Sorbose +, melibiose –, melezitose +, inulin –, D-arabinose +, D-glucosamine · HCl +, glucitol +, α-methyl-D-glucoside + s, salicin –, potassium D-gluconate + s, DL-lactic acid –

Assimilation of nitrate: –

Growth in vitamin-free medium: –

Growth in 10% sodium chloride plus 5% glucose in yeast nitrogen base: +

Growth at 37°C: +

Comments

The results reported in the original description are in generally good agreement with the findings of the present study. Van der Walt and Johannsen (1975c) noted sucrose fermentation to be slow but strong. It was negative in the present tests as was growth on salicin, originally reported as positive.

Pichia philogaea shows considerable resemblance to *P. nakazawae* var. *nakazawae,* but differs by its inability to assimilate soluble starch, L-rhamnose and salicin.

Origin of the strains studied

Grassland soil, South Africa (2).

Type strain: CBS 6696 (NRRL Y-7813).

40. *PICHIA PIJPERI* VAN DER WALT ET TSCHEUSCHNER (1957a)

Synonym

Hanseniaspora pijperi (van der Walt et Tscheuschner) Ditlevsen et Hjort (1964)

Standard description of *Pichia pijperi*

Growth on 5% malt extract agar: After 3 days at 25°C, the cells are ovoidal to cylindroidal (3.0–5.8) × (5.0–20.0) μm, and occur singly, in pairs or in short chains. Growth is butyrous and cream-colored.

Growth on the surface of assimilation media: Dry, climbing pellicles are formed.

Dalmau plate culture on morphology agar: After 7 days at 25°C, growth under the coverglass shows poorly to moderately well developed pseudohyphae but no true hyphae. Aerobic growth is light tan, dull, generally smooth, and with an entire or finely lobed margin.

Formation of ascospores: Asci are unconjugated and produce one to four hat-shaped spores. The spores are readily released at maturity.

Spores were observed on 5% malt extract agar.

Fermentation:

Glucose	+	Lactose	–
Galactose	–	Raffinose	–
Sucrose	–	Trehalose	–
Maltose	–		

Assimilation of carbon compounds:

Galactose	–	Raffinose	–	Erythritol	–
Sucrose	–	Soluble starch	–	Ribitol	–
Maltose	–	D-Xylose	+	D-Mannitol	+
Cellobiose	+	L-Arabinose	–	Succinic acid	+
Trehalose	–	D-Ribose	–	Citric acid	+
Lactose	–	L-Rhamnose	–	Inositol	–

Additional carbon compounds tested: L-Sorbose +, melibiose −, melezitose −, inulin −, D-arabinose −, D-glucosamine · HCl −, glucitol +, α-methyl-D-glucoside −, salicin +, potassium D-gluconate −, DL-lactic acid +

Assimilation of nitrate: −

Growth in vitamin-free medium: −

Growth in 10% sodium chloride plus 5% glucose in yeast nitrogen base: −

Growth at 37°C: +w

G + C: 42.5 mol%, type strain (Fuson et al., 1980)

Comments

Despite the appearance of bipolar budding by *P. pijperi,* Kreger-van Rij (1970c) reported that bud scars observed under the transmission electron microscope were not the typical multiple scars of *Hanseniaspora,* but were like those seen in other species of *Pichia.*

Origin of the strains studied

Buttermilk (1), *Drosophila melanogaster* (1).

Type strain: CBS 2887 (NRRL YB-4309), from buttermilk.

41. *PICHIA PINUS* (HOLST) PHAFF (1956)

Synonyms

Zygosaccharomyces pini Holst (1936)
Petasospora pini (Holst) Boidin et Abadie (1954)
Saccharomyces pini (Holst) Shifrine et Phaff (1956)
Zygowillia pini (Holst) Kudriavzev (1960)

Standard description of *Pichia pinus*

Growth on 5% malt extract agar: After 3 days at 25°C, the cells are spheroidal to ovoidal (1.6–4.2) × (2.2–6.0) μm, and occur singly, in pairs and occasionally in small clusters. Growth is tannish-white, smooth, dull to glistening, and butyrous to somewhat mucoid.

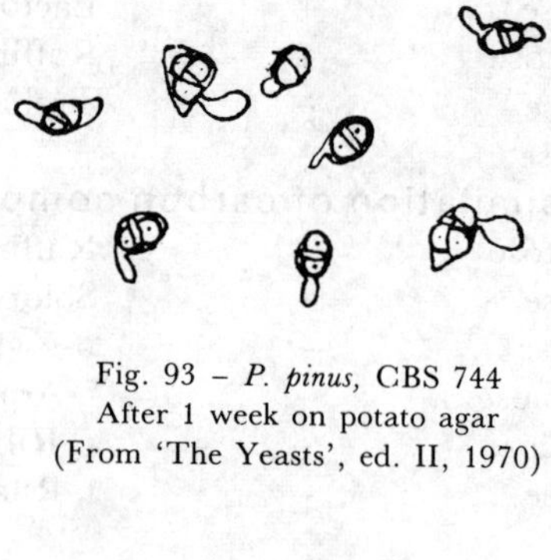

Fig. 93 – *P. pinus,* CBS 744
After 1 week on potato agar
(From 'The Yeasts', ed. II, 1970)

Growth on the surface of assimilation media: Pellicles are not formed.

Dalmau plate culture on morphology agar: After 7 days at

25°C, growth under the coverglass shows either no pseudomycelium or only a few poorly developed pseudohyphae. True hyphae are not formed. Aerobic growth is tannish-white, smooth to finely striated, dull to glistening, and butyrous. Margins range from entire to lobed.

Formation of ascospores: Asci may be unconjugated or show conjugation between a parent cell and a bud or between independent cells. Asci contain two to four hat-shaped spores which are liberated soon after formation (Fig. 93).

Spores were observed on 5% malt extract-, V8- and corn meal agar.

Fermentation:

Glucose	v	Lactose	–
Galactose	–	Raffinose	–
Sucrose	–	Trehalose	v
Maltose	–		

Assimilation of carbon compounds:

Galactose	–	Raffinose	–	Erythritol	+
Sucrose	–	Soluble starch	–	Ribitol	+
Maltose	–	D-Xylose	v	D-Mannitol	+
Cellobiose	+	L-Arabinose	v	Succinic acid	v
Trehalose	+	D-Ribose	v	Citric acid	v
Lactose	–	L-Rhamnose	v	Inositol	–

Additional carbon compounds tested: L-Sorbose v, melibiose –, melezitose –, inulin –, D-arabinose v, D-glucosamine · HCl –, glucitol +, α-methyl-D-glucoside –, salicin +, potassium D-gluconate – or +w, DL-lactic acid –

Assimilation of nitrate: –

Growth in vitamin-free medium: – or +w

Growth in 10% sodium chloride plus 5% glucose in yeast nitrogen base: v

Growth at 37°C: v

G + C: 43.0 mol%, type strain (Fuson et al., 1980)

Comments

As noted earlier, *P. pinus* and *H. glucozyma* show great phenotypic similarity, and Kreger-van Rij (1964b) reported some strains of *P. pinus* to assimilate nitrite. However, Kurtzman et al. (unpublished) found the type strains of the two species to have only 3% DNA relatedness.

Origin of the strains studied

From the bark beetles *Dendroctonus brevicomis* (1), *D. pseudotsugae* (2), *D. frontalis* (1), larvae of *D. monticolae* (1), Curculionidae larvae (1), *Hylurgonotus brunneus* (1), *Araucaria araucana* (1), frass in *Pinus echinata* (1), frass in *Pinus taeda* (1), frass from *Pinus* sp. (2), frass from *Pinus contorta* (4), frass from *Pinus ponderosa* (1), frass from *Abies firma* (1), bark of a dead birch (1).

Type strain: CBS 744 (NRRL Y-11528), from *Dendroctonus brevicomis*.

42. *PICHIA QUERCUUM* PHAFF ET KNAPP (1956)

Synonym

Zymopichia quercibus (Phaff et Knapp) Novák et Zsolt (1961)

Standard description of *Pichia quercuum*

Growth on 5% malt extract agar: After 3 days at 25°C, the cells are spheroidal to ovoidal (1.8–5.0) × (2.0–5.0) μm, and occur singly or in pairs. Growth is tannish-yellow, smooth, glistening and butyrous.

Growth on the surface of assimilation media: Pellicles are not formed.

Dalmau plate culture on morphology agar: After 7 days at 25°C, growth under the coverglass shows very poorly differentiated pseudohyphae and no true hyphae. Aerobic growth is greyish-white, smooth, glistening, and butyrous with an entire to finely lobed margin.

Formation of ascospores: Asci are generally conjugated with the pairing between parent cell and bud or between independent cells. One to four hat-shaped spores are produced in each ascus, and they are liberated soon after formation. Spores were observed on 5% malt extract agar.

Fermentation:

Glucose	+w	Lactose	–
Galactose	–	Raffinose	–
Sucrose	–	Trehalose	–
Maltose	–		

Assimilation of carbon compounds:

Galactose	–	Raffinose	–	Erythritol	–
Sucrose	–	Soluble starch	–	Ribitol	–
Maltose	–	D-Xylose	–	D-Mannitol	+
Cellobiose	+	L-Arabinose	–	Succinic acid	+
Trehalose	–	D-Ribose	–	Citric acid	+
Lactose	–	L-Rhamnose	–	Inositol	–

Additional carbon compounds tested: L-Sorbose –, melibiose –, melezitose –, inulin –, D-arabinose –, D-glucosamine · HCl –, glucitol +, α-methyl-D-glucoside –, salicin +, potassium D-gluconate –, DL-lactic acid +

Assimilation of nitrate: –

Growth in vitamin-free medium: –

Growth in 10% sodium chloride plus 5% glucose in yeast nitrogen base: –

Growth at 37°C: +w

G + C: 40.0 mol%, type strain (Nakase & Komagata, 1970b)

Origin of the strain studied

From a slime flux of *Quercus kelloggii* (1).

Type strain: CBS 2283 (NRRL YB-4281).

43. *PICHIA RABAULENSIS* SONEDA ET UCHIDA (1971)

Standard description of *Pichia rabaulensis*

Growth on 5% malt extract agar: After 3 days at 25°C, the cells are ovoidal to elongate (2.0–5.2) × (3.0–9.0) μm, and occur singly, in pairs or occasionally in short chains. Growth is tannish-white, dull and finely wrinkled.

Growth on the surface of assimilation media: Heavy, dry, climbing pellicles are formed.

Dalmau plate culture on morphology agar: After 7 days at 25°C, growth under the coverglass shows abundant well-branched pseudohyphae, but relatively few blastospores are produced. True hyphae are not formed. Aerobic growth is tannish-white, dull, and almost powdery, and with fine striations. Margins may be entire, serrate or lobed.

Formation of ascospores: Asci are unconjugated and produce one to four hat-shaped spores. The spores are liberated soon after formation.

Spores were observed on 5% malt extract- and YM agar.

Fermentation:

Glucose	+	Lactose	–
Galactose	–	Raffinose	+
Sucrose	+	Trehalose	+ w
Maltose	–		

Assimilation of carbon compounds:

Galactose	–	Raffinose	+	Erythritol	–
Sucrose	+	Soluble starch	–	Ribitol	+
Maltose	+	D-Xylose	+	D-Mannitol	+
Cellobiose	+	L-Arabinose	+	Succinic acid	+
Trehalose	+	D-Ribose	–	Citric acid	+
Lactose	–	L-Rhamnose	+	Inositol	–

Additional carbon compounds tested: L-Sorbose –, melibiose –, melezitose +, inulin –, D-arabinose –, D-glucosamine · HCl –, glucitol +, α-methyl-D-glucoside +, salicin +, potassium D-gluconate + s, DL-lactic acid +

Assimilation of nitrate: –

Growth in vitamin-free medium: –

Growth in 10% sodium chloride plus 5% glucose in yeast nitrogen base: +

Growth at 37°C: +

Comments

The present findings are in good agreement with the original description of *P. rabaulensis*. Soneda and Uchida (1971) reported growth on galactose to be positive while the response to erythritol was variable. In the present study, growth on both of these compounds was negative.

Origin of the strain studied

Feces of an African snail, Rabaul, New Britain (1).

Type strain: CBS 6797 (NRRL Y-7945).

44. *PICHIA RHODANENSIS* (RAMIREZ ET BOIDIN) PHAFF (1956)

Synonyms

Saccharomyces rhodanensis Ramirez et Boidin (1953a) (also in Ramirez & Boidin, 1953b)
Petasospora rhodanensis (Ramirez et Boidin) Boidin et Abadie (1954)
Zymopichia rhodanensis (Ramirez et Boidin) Novák et Zsolt (1961)
Endomycopsis balearica Socias, Ramirez et Genestar (1954a)
Endomycopsis fibuligera (Lindner) Dekker var. *energica* Verona et Zardetto de Toledo (1959a)

Standard description of *Pichia rhodanensis*

Growth on 5% malt extract agar: After 3 days at 25°C, the cells are ovoidal to elongate (2.6–6.0) × (4.0–10.0) μm, and occur singly, in pairs or in chains. Growth is light tan, dull or glistening, and smooth or wrinkled, depending upon the abundance of pseudohyphae.

Growth on the surface of assimilation media: Pellicles are sometimes present.

Dalmau plate culture on morphology agar: After 7 days at 25°C, growth under the coverglass shows abundant branched pseudohyphae with occasional verticillate clusters of blastospores. True hyphae are produced by some strains but not abundantly. Aerobic growth is light tan and may be dull or glistening and butyrous or somewhat tough and wrinkled, depending upon the abundance of hyphae and pseudohyphae. Colony margins are usually entire.

Formation of ascospores: This heterothallic species is frequently isolated from nature in the diploid form, and the unconjugated asci produce one to four hat-shaped spores. The spores are liberated soon after formation. Complementary mating types are represented by NRRL YB-651 sm-14 (CBS 4839) and NRRL Y-7854 Re-1 (CBS 7102).

Spores were observed on 5% malt extract agar.

Fermentation:

Glucose	+	Lactose	–
Galactose	–	Raffinose	–
Sucrose	– or +w	Trehalose	–
Maltose	– or +w		

Assimilation of carbon compounds:

Galactose	–	Raffinose	–	Erythritol	–
Sucrose	+	Soluble starch	–	Ribitol	–
Maltose	+	D-Xylose	+	D-Mannitol	+
Cellobiose	+	L-Arabinose	–	Succinic acid	+
Trehalose	+	D-Ribose	–	Citric acid	+
Lactose	–	L-Rhamnose	+	Inositol	–

Additional carbon compounds tested: L-Sorbose −, melibiose −, melezitose +, inulin −, D-arabinose −, D-glucosamine · HCl −, glucitol +, α-methyl-D-glucoside v, salicin v, potassium D-gluconate +, DL-lactic acid +

Assimilation of nitrate: −

Growth in vitamin-free medium: −

Growth in 10% sodium chloride plus 5% glucose in yeast nitrogen base: −

Growth at 37°C: +

G + C: 51.8–51.9 mol%, 2 strains (Kurtzman et al., 1980a)

Comments

As noted earlier, *P. rhodanensis* is phenotypically similar to *P. amylophila, P. mississippiensis, P. wickerhamii* and *P. veronae,* but none of these heterothallic species shows any mating response when paired with *P. rhodanensis.* Furthermore, *P. rhodanensis* shows little DNA relatedness with these other species (Kurtzman et al., 1980a). *P. toletana,* which is homothallic, differs from *P. rhodanensis* only by its weak fermentation of glucose and by its failure to grow at 37°C. However, the 12% difference in nuclear G + C content of these two taxa would preclude conspecificity.

Origin of the strains studied

Tanning liquor (4), stained leather (1), fermenting canaigre (1), palm nuts (1).

Type strain: CBS 5518 (NRRL Y-7854), from tanning liquor.

45. *PICHIA SAITOI* KODAMA, KYONO ET KODAMA (1962)

Synonym

Pichia krusei nom. nud. (Tsuchiya et al., 1967)

Standard description of *Pichia saitoi*

Growth on 5% malt extract agar: After 3 days at 25°C, the cells are ovoidal (2.3–4.5) × (2.8–7.5) μm, and occur singly and in pairs. Growth is butyrous and tannish-yellow in color.

Growth on the surface of assimilation media: Climbing pellicles are present.

Dalmau plate culture on morphology agar: After 7 days at 25°C, growth under the coverglass shows neither pseudohyphae nor true hyphae. Aerobic growth is light tan, smooth, glistening, butyrous and with a lobate margin.

Formation of ascospores: Asci are generally unconjugated but occasional mother–daughter conjugations occur. Two to four Saturn-shaped spores are produced in each ascus, but the ledge on the spores is sometimes difficult to discern under the light microscope (Fig. 94). The spores are released soon after formation. This species appears to be homothallic. Single-spore isolates from four-spored asci of the type strain gave approximately two-thirds sporogenous colonies and one-third asporogenous colonies. Mixtures of asporogenous cultures showed neither conjugations nor spores. Single-spore isolations from three additional four-spored strains gave only sporogenous cultures.

Spores were observed on YM- and 5% malt extract agar.

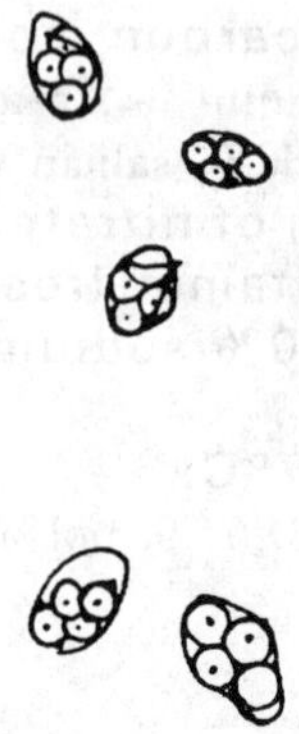

Fig. 94 – *P. saitoi,* CBS 4910
After 1 week on V8 agar
(From 'The Yeasts', ed. II, 1970)

Fermentation:

Glucose	+	Lactose	–
Galactose	–	Raffinose	–
Sucrose	–	Trehalose	–
Maltose	–		

Assimilation of carbon compounds:

Galactose	–	Raffinose	–	Erythritol	–
Sucrose	–	Soluble starch	–	Ribitol	v
Maltose	–	D-Xylose	–	D-Mannitol	+
Cellobiose	–	L-Arabinose	–	Succinic acid	+
Trehalose	+	D-Ribose	–	Citric acid	–
Lactose	–	L-Rhamnose	–	Inositol	–

Additional carbon compounds tested: L-Sorbose –, melibiose –, melezitose –, inulin –, D-arabinose –, D-glucosamine · HCl –, glucitol +, α-methyl-D-glucoside –, salicin –, potassium D-gluconate –, DL-lactic acid +

Assimilation of nitrate: –

Growth in vitamin-free medium: –

Growth in 10% sodium chloride plus 5% glucose in yeast nitrogen base: –

Growth at 37°C: –

G + C: 32.2–33.4 mol%, 4 strains (Nakase, 1972a)

Comments

In an extensive study of *P. saitoi* and *P. dispora* (including *P. zaruensis*), Nakase (1972a) determined the G + C content of the two species to differ by about 4%, therefore strongly suggesting the two taxa to be separate species. Some of the *P. saitoi* strains, isolated since the species was originally described, were found to assimilate ribitol, thus removing the only difference on standard carbon assimilation tests that separated this species from *P. dispora*. However, Nakase pointed out that strains belonging to *P. saitoi* formed thin, smooth, creeping pellicles on liquid media, while strains of *P. dispora* formed no pellicle. In addition, asci of the former dehisce quickly while those of the latter are persistent.

Kurtzman et al. (unpublished) determined that *P. saitoi* showed only a few percent DNA relatedness with *P. dispora,* and consequently represents a distinct species as suggested by Nakase.

Origin of the strains studied

Exudate of *Carpinus* sp. (1), from Ahearn (1), from Kodama (6).

Type strain: CBS 4910 (NRRL Y-6671), from *Carpinus* sp.

46. *PICHIA SALICTARIA* PHAFF, MILLER ET SPENCER (1964)

Standard description of *Pichia salictaria*

Growth on 5% malt extract agar: After 3 days at 25°C, the cells are ovoidal (1.7–4.5) × (2.3–7.2) μm, and occur singly or in pairs. Growth is tannish-yellow, smooth, dull-glistening and butyrous.

Growth on the surface of assimilation media: Heavy, climbing pellicles are formed.

Dalmau plate culture on morphology agar: After 7 days at 25°C, growth under the coverglass has neither pseudohyphae nor true hyphae. Aerobic growth is tannish-white, smooth, dull-glistening, and butyrous. Margins are generally lobed.

Formation of ascospores: Asci may be unconjugated or show conjugation between parent cell and a bud, or between independent cells. One to four hat-shaped spores are formed per ascus and they are quickly liberated.

Spores were observed on 5% malt extract agar.

Fermentation: –

Assimilation of carbon compounds:

Galactose	–	Raffinose	–	Erythritol	–
Sucrose	–	Soluble starch	–	Ribitol	–
Maltose	–	D-Xylose	+	D-Mannitol	+
Cellobiose	+	L-Arabinose	–	Succinic acid	+
Trehalose	–	D-Ribose	–	Citric acid	+
Lactose	–	L-Rhamnose	+	Inositol	–

Additional carbon compounds tested: L-Sorbose –, melibiose –, melezitose –, inulin –, D-arabinose –, D-glucosamine · HCl –, glucitol +, α-methyl-D-glucoside –, salicin +, potassium D-gluconate +, DL-lactic acid +

Assimilation of nitrate: –

Growth in vitamin-free medium: –

Growth in 10% sodium chloride plus 5% glucose in yeast nitrogen base: –

Growth at 37°C: +

G + C: 37.8 mol%, type strain (Nakase & Komagata, 1970b)

Origin of the strains studied

Slime flux of willows (2), human toe (1).

Type strain: CBS 5456 (NRRL Y-6780).

47. *PICHIA SARGENTENSIS* WICKERHAM ET KURTZMAN (1971)

Standard description of *Pichia sargentensis*

Growth on 5% malt extract agar: After 3 days at 25°C, the cells are spheroidal to ellipsoidal (2.8–5.0) × (5.0–9.5) μm, and occur singly, in pairs or occasionally in short chains. Growth is butyrous and tannish-white in color.

Growth on the surface of assimilation media: Pellicles are formed.

Dalmau plate culture on morphology agar: After 7 days at 25°C, growth under the coverglass shows sparsely developed pseudohyphae and no true hyphae. Aerobic growth is slightly raised, dull glistening, butyrous, white in color and with an irregular margin.

Formation of ascospores: Asci may be conjugated or unconjugated and are persistent. Each ascus contains one to four spores which are spheroidal with a thin equatorial to subequatorial ring. Single-spore isolates from four-spored asci are sporogenous, and the species is presumed to be homothallic.

Spores were observed on 5% malt extract- and YM agar.

Fermentation:

Glucose	+	Lactose	–
Galactose	–	Raffinose	–
Sucrose	–	Trehalose	–
Maltose	–		

Assimilation of carbon compounds:

Galactose	–	Raffinose	–	Erythritol	–
Sucrose	–	Soluble starch	–	Ribitol	–
Maltose	–	D-Xylose	+	D-Mannitol	+
Cellobiose	+	L-Arabinose	–	Succinic acid	+
Trehalose	–	D-Ribose	–	Citric acid	–
Lactose	–	L-Rhamnose	+	Inositol	–

Additional carbon compounds tested: L-Sorbose –, melibiose –, melezitose –, inulin –, D-arabinose –, D-glucosamine · HCl –, glucitol +, α-methyl-D-glucoside –, salicin +, potassium D-gluconate –, DL-lactic acid +

Assimilation of nitrate: –

Growth in vitamin-free medium: –; thiamine and biotin are required for growth

Growth in 10% sodium chloride plus 5% glucose in yeast nitrogen base: –

Growth at 37°C: +

G + C: 44.1 mol%, type strain (Kurtzman et al., unpublished)

Comments

Recently, Kurtzman et al. (unpublished) discovered *P. sargentensis* and *H. mrakii* to share 68% DNA base sequence complementarity. These data suggest the two taxa to be conspecific and the ramifications of this finding are discussed under section f., Comments on the genus.

Origin of the strain studied

Outlet of a small lake at Camp Sargent near Peterborough, New Hampshire, U.S.A.

Type strain: CBS 6342 (NRRL YB-4139).

48. PICHIA SCOLYTI (PHAFF ET YONEYAMA) KREGER-VAN RIJ (1964b)

Synonym

Endomycopsis scolyti Phaff et Yoneyama (1961)

Standard description of *Pichia scolyti*

Growth on 5% malt extract agar: After 3 days at 25°C, the cells are ovoidal to ellipsoidal (1.2–3.2) × (2.0–7.8) μm, and occur singly, in pairs and in chains. Growth is white to cream-colored, dull to glistening, and frequently finely wrinkled.

Growth on the surface of assimilation media: Rings and, occasionally, thin pellicles may form.

Dalmau plate culture on morphology agar: After 7 days at 25°C, growth under the coverglass shows abundant moderately branched pseudohyphae with occasional dense clusters of blastospores. Some blastospores have an elongated triangular appearance. True hyphae may also be present. Aerobic growth is white to tannish-white, dull or glistening, frequently finely wrinkled, and occasionally papillate. Margins are irregularly serrate.

Formation of ascospores: This species is heterothallic and usually isolated from nature in the haploid form although at least one diploid has been recovered as a natural isolate. Many of the haploids mate poorly. Asci produce one to four hat-shaped spores which are liberated soon after formation. The original complementary mating types of Phaff and Yoneyama (1961) are NRRL Y-5512 (CBS 4802) and NRRL Y-5513 (CBS 4803). Additional mating types, isolated by Wickerham from frass obtained in Ontario, are NRRL YB-1403-1 (CBS 5731) and NRRL YB-1403-2 (CBS 5732). These latter haploids seem to have greater mating competence.

Spores were observed on 5% malt extract agar.

Fermentation:

Glucose	+ w or s	Lactose	–
Galactose	+ w or s	Raffinose	–
Sucrose	– or + w	Trehalose	+ w or s
Maltose	– or + w		

Assimilation of carbon compounds:

Galactose	+	Raffinose	+	Erythritol	+
Sucrose	+	Soluble starch	–	Ribitol	+
Maltose	+	D-Xylose	+	D-Mannitol	+
Cellobiose	+	L-Arabinose	+	Succinic acid	+
Trehalose	+	D-Ribose	+	Citric acid	+
Lactose	v	L-Rhamnose	+	Inositol	–

Additional carbon compounds tested: L-Sorbose –, melibiose +, melezitose +, inulin –, D-arabinose v, D-glucosamine · HCl v, glucitol +, α-methyl-D-glucoside +, salicin +, potassium D-gluconate +s, DL-lactic acid –

Assimilation of nitrate: –

Growth in vitamin-free medium: –

Growth in 10% sodium chloride plus 5% glucose in yeast nitrogen base: v

Growth at 37°C: v

G + C: 40.1 mol%, type strain (Fuson et al., 1980)

Origin of the strains studied

From frass of *Scolytus ventralis* in *Abies concolor,* California (3); frass in *Abies balsamea,* Ontario (1); from *Dendroctonus pseudotsugae,* Washington (2).

Type strain: CBS 4802 (NRRL Y-5512), from frass of *Scolytus ventralis.*

49. *PICHIA SEGOBIENSIS* SANTA MARÍA ET GARCÍA ASER (1977)

Standard description of *Pichia segobiensis*

Growth on 5% malt extract agar: After 3 days at 25°C, the cells are spheroidal to ovoidal (2.0–4.5) × (2.5–6.3) μm, and occur singly and in pairs. Growth is tannish-white, smooth, glistening, and butyrous.

Growth on the surface of assimilation media: Rings and, occasionally, thin pellicles are formed.

Dalmau plate culture on morphology agar: After 7 days at 25°C, growth under the coverglass shows infrequent and poorly differentiated pseudohyphae. True hyphae are not formed. Aerobic growth is light tannish-white, smooth to occasionally slightly papillate, moderately glistening, butyrous, and with an entire to sparingly lobed margin.

Formation of ascospores: Conjugation between a parent cell and a bud or between independent cells may precede ascus formation. One, but usually two hat-shaped spores are formed in each ascus, and they are readily liberated.

Spores were observed on 5% malt extract agar.

Fermentation:

Glucose	+	Lactose	–
Galactose	+	Raffinose	–
Sucrose	–	Trehalose	+w
Maltose	–		

Assimilation of carbon compounds:

Galactose	+	Raffinose	–	Erythritol	–
Sucrose	+	Soluble starch	–	Ribitol	+
Maltose	+	D-Xylose	+	D-Mannitol	+
Cellobiose	+	L-Arabinose	–	Succinic acid	+
Trehalose	+	D-Ribose	+	Citric acid	+
Lactose	–	L-Rhamnose	–	Inositol	–

Additional carbon compounds tested: L-Sorbose +, melibiose –, melezitose –, inulin –, D-arabinose +s, D-glucosamine · HCl +, glucitol +, α-methyl-D-glucoside +, salicin +, potassium D-gluconate +, DL-lactic acid –

Assimilation of nitrate: –

Growth in vitamin-free medium: –

Growth in 10% sodium chloride plus 5% glucose in yeast nitrogen base: –

Growth at 37°C: –

Comments

Results from the present study are in good agreement with the original description of *P. segobiensis*. The only difference noted was that Santa María and García Aser (1977) reported growth on D-arabinose to be negative, while I found it slow but positive.

Origin of the strains studied

From an adult larva of *Calcophora mariana massiliensis* in *Pinus silvestris* (1), insect frass in the bark of *P. silvestris* (1), larva of *Rhagium inquisitor* (1).

Type strain: CBS 6857 (NRRL Y-11571), from a larva of *Calcophora mariana massiliensis*.

50. *PICHIA SPARTINAE* AHEARN, YARROW ET MEYERS (1970)

Standard description of *Pichia spartinae*

Growth on 5% malt extract agar: After 3 days at 25°C, the cells are spheroidal to ellipsoidal (2.5–5.3) × (3.3–6.5) μm, and occur singly, in pairs or occasionally in short chains. Growth is butyrous and tannish-white in color.

Growth on the surface of assimilation media: Thin rings are formed and occasionally pellicles are present.

Dalmau plate culture on morphology agar: After 7 days at 25°C, growth under the coverglass shows only occasional pseudohyphae, but these are moderately branched and may bear blastospores. True hyphae are not formed. Aerobic growth is light tan, smooth, glistening and butyrous. Margins are entire to lobed.

Formation of ascospores: Asci are unconjugated, or conjugated if formed from the pairing of complementary mating types. Occasionally one, but usually two, hat-shaped spores form in one of the conjugants while the other remains empty (Fig. 95). Asci become dehiscent at maturity. Complementary mating types of this heterothallic species are NRRL Y-7665-1 (CBS 6666) and NRRL Y-7665-3 (CBS 6669). The type strain, NRRL Y-7322 (CBS 6059), is now asporogenous, but conjugates and sporulates when paired with mating type NRRL Y-7665-3.

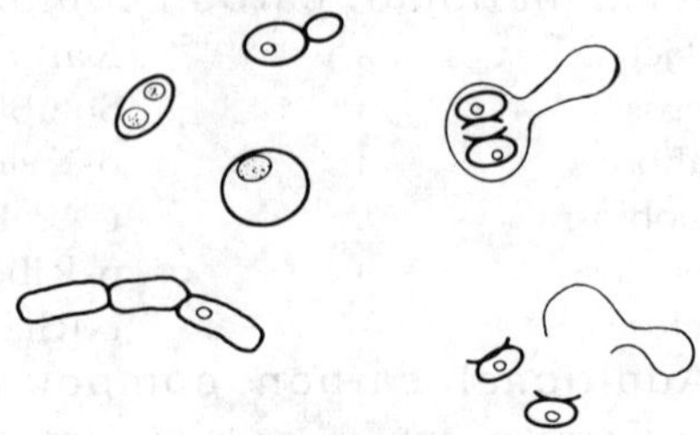

Fig. 95 – *P. spartinae* CBS 6666 × CBS 6669. After 1 week on malt extract agar

Spores were observed on 5% malt extract agar.

Fermentation:

Glucose	+	Lactose	–
Galactose	–	Raffinose	–
Sucrose	+ or w	Trehalose	–
Maltose	– or + w		

Assimilation of carbon compounds:

Galactose	–	Raffinose	–	Erythritol	–
Sucrose	+	Soluble starch	–	Ribitol	+
Maltose	+	D-Xylose	–	D-Mannitol	+
Cellobiose	+	L-Arabinose	–	Succinic acid	+
Trehalose	+	D-Ribose	–	Citric acid	+
Lactose	–	L-Rhamnose	–	Inositol	–

Additional carbon compounds tested: L-Sorbose + or s, melibiose –, melezitose +, inulin –, D-arabinose –, D-glucosamine · HCl v, glucitol +, α-methyl-D-glucoside +, salicin +, potassium D-gluconate v, DL-lactic acid – or + w

Assimilation of nitrate: –

Growth in vitamin-free medium: –

Growth in 10% sodium chloride plus 5% glucose in yeast nitrogen base: +

Growth at 37°C: +

Comments

In the original description, ascospores of *P. spartinae* were reported to be spheroidal or infrequently oval with a slight ledge (Ahearn et al., 1970). Later, Kurtzman and Ahearn (1976) showed a high incidence of malformed spores among strains of *P. spartinae,* and that mature spores were hat-shaped rather than spheroidal.

Pichia spartinae is phenotypically similar to several other heterothallic species of *Pichia* and to certain species of *Candida.* However, there was no mating response between *P. spartinae* and *P. angophorae, P. rhodanensis, P. wickerhamii, P. veronae, C. freyschussii, C. maritima, C. obtusa,* or *C. oregonensis* (Kurtzman & Ahearn, 1976).

Origin of the strains studied

From oyster grass (*Spartina alterniflora* Loisel) marshes, Barataria Bay, Louisiana, U.S.A. (7).

Type strain: CBS 6059 (NRRL Y-7322).

51. PICHIA STIPITIS PIGNAL (1967)

Standard description of *Pichia stipitis*

Growth on 5% malt extract agar: After 3 days at 25°C, the cells are spheroidal to ovoidal (2.2–4.5) × (2.5–6.0) μm, and occur singly or in pairs. Growth is cream-colored, dull-glistening, smooth to faintly wrinkled and butyrous.

Growth on the surface of assimilation media: Rings and incomplete pellicles may form.

Dalmau plate culture on morphology agar: After 7 days at 25°C, growth under the coverglass shows abundant and well branched pseudohyphae with occasional blastospores. True hyphae are not formed. Aerobic growth is light tan, smooth to finely wrinkled, dull-glistening, butyrous, and with an entire to lobed margin.

Formation of ascospores: Preceding ascus formation there may be conjugation between a parent cell and a bud or between independent cells or the ascus may be unconjugated but with a protuberance. Asci produce two hat-shaped spores and these are released soon after formation.

Spores were observed on 5% malt extract agar.

Fermentation:

Glucose	+ s	Lactose	–
Galactose	+ s	Raffinose	–
Sucrose	–	Trehalose	v
Maltose	+ s		

Assimilation of carbon compounds:

Galactose	+	Raffinose	–	Erythritol	+
Sucrose	+	Soluble starch	+	Ribitol	+
Maltose	+	D-Xylose	+	D-Mannitol	+
Cellobiose	+	L-Arabinose	+	Succinic acid	+
Trehalose	+	D-Ribose	+	Citric acid	+
Lactose	+	L-Rhamnose	+	Inositol	–

Additional carbon compounds tested: L-Sorbose – or + s, melibiose –, melezitose +, inulin –, D-arabinose + s, D-glucosamine · HCl +, glucitol +, α-methyl-D-glucoside +, salicin +, potassium D-gluconate v, DL-lactic acid +

Assimilation of nitrate: –

Growth in vitamin-free medium: – or + w

Growth in 10% sodium chloride plus 5% glucose in yeast nitrogen base: –

Growth at 37°C: +

Comments

Results of the present study are in good agreement with the original description. Pignal (1967) noted the phenotypic similarity between *P. stipitis* and *P. scolyti,* and on standard tests the two species differ only in their assimilation of raffinose and melibiose. Kreger-van Rij (1970c) noted the resemblance between *P. stipitis* and *C. shehatae* and suggested that they might represent the same taxon.

Origin of the strains studied

From larvae of *Cetonia* sp., *Dorcus parallelipipedus* and *Laphria* (4); from J. Santa María (1).

Type strain: CBS 5773 (NRRL Y-7124), from insect larvae.

52. *PICHIA STRASBURGENSIS* (RAMIREZ ET BOIDIN) PHAFF (1956)

Synonyms

Saccharomyces strasburgensis Ramirez et Boidin (1953a) (also in Ramirez & Boidin, 1953b)
Petasospora strasburgensis (Ramirez et Boidin) Boidin et Abadie (1954)
Zymopichia strasburgensis (Ramirez et Boidin) Novák et Zsolt (1961)

Standard description of *Pichia strasburgensis*

Growth on 5% malt extract agar: After 3 days at 25°C, the cells are ovoidal to ellipsoidal (1.5–4.5) × (2.9–14.0) μm, and occur singly or in pairs. Growth is cream-colored, smooth, faintly glistening and butyrous.

Growth on the surface of assimilation media: Pellicles are not formed.

Dalmau plate culture on morphology agar: After 7 days at 25°C, growth under the coverglass shows pseudohyphae but no true hyphae. Some strains produce only occasional, sparingly branched pseudohyphae while others form considerable pseudomycelium that may be well branched and bear clusters of blastospores. Aerobic growth is pale tannish-white, smooth to finely wrinkled, dull-glistening and butyrous. Margins may be serrate or lobed.

Formation of ascospores: Asci form without prior conjugation and contain one to four hat-shaped spores which are readily liberated. Kreger-van Rij (1964b) reported this species to be heterothallic, and complementary mating types are CBS 2039 (NRRL Y-11980) and CBS 5005 (NRRL Y-11981).

Spores were observed on 5% malt extract agar.

Fermentation:

Glucose	+	Lactose	–
Galactose	– or +w	Raffinose	+w
Sucrose	+	Trehalose	–
Maltose	–		

Assimilation of carbon compounds:

Galactose	+	Raffinose	+	Erythritol	–
Sucrose	+	Soluble starch	–	Ribitol	+ s
Maltose	+	D-Xylose	+	D-Mannitol	+ s
Cellobiose	+	L-Arabinose	+	Succinic acid	+
Trehalose	+	D-Ribose	– or + w	Citric acid	+
Lactose	–	L-Rhamnose	+	Inositol	–

Additional carbon compounds tested: L-Sorbose –, melibiose –, melezitose +, inulin –, D-arabinose –, D-glucosamine · HCl –, glucitol + s, α-methyl-D-glucoside +, salicin +, potassium D-gluconate + s, DL-lactic acid +

Assimilation of nitrate: –

Growth in vitamin-free medium: –

Growth in 10% sodium chloride plus 5% glucose in yeast nitrogen base: +

Growth at 37°C: +

G + C: 40.6 mol%, type strain (Fuson et al., 1980)

Origin of the strains studied

Stained leather, from Ramirez (2), and from Boidin (1).

Type strain: CBS 2939 (NRRL Y-2383), from stained leather.

53. *PICHIA TOLETANA* (SOCIAS, RAMIREZ ET PELÁEZ) KREGER-VAN RIJ (1964b)

Synonyms

Debaryomyces toletanus Socias, Ramirez et Peláez (1954b)
Zymodebaryomyces toletanus (Socias, Ramirez et Peláez) Novák et Zsolt (1961)
Pichia xylosa Phaff, Miller et Shifrine (1956)
Zymopichia xylosa (Phaff, Miller et Shifrine) Novák et Zsolt (1961)

Standard description of *Pichia toletana*

Growth on 5% malt extract agar: After 3 days at 25°C, the cells are ovoidal (1.5–5.0) × (3.0–9.7) μm, and occur singly and in pairs. Growth is yellowish-white smooth to wrinkled and dull-glistening.

Growth on the surface of assimilation media: Thin pellicles are formed by some strains.

Dalmau plate culture on morphology agar: After 7 days at 25°C, growth under the coverglass shows abundant, branched pseudohyphae with blastospores in a verticillate arrangement. True hyphae are not formed. Aerobic growth is tannish-white, smooth to wrinkled, dull-glistening and butyrous. Margins are usually lobed.

Formation of ascospores: Conjugation of a parent cell and a bud usually precedes ascus formation. Asci contain one to four hat-shaped spores and these may be quickly or slowly liberated. Single-spore isolates from four-spored asci are sporogenous, indicating the species to be homothallic.

Spores were observed on 5% malt extract agar.

Fermentation:

Glucose	+ w or s	Lactose	–
Galactose	–	Raffinose	–
Sucrose	–	Trehalose	–
Maltose	–		

Assimilation of carbon compounds:

Galactose	–	Raffinose	–	Erythritol	–
Sucrose	+	Soluble starch	–	Ribitol	–
Maltose	+	D-Xylose	+	D-Mannitol	+
Cellobiose	+	L-Arabinose	–	Succinic acid	+
Trehalose	+	D-Ribose	–	Citric acid	+
Lactose	–	L-Rhamnose	v	Inositol	–

Additional carbon compounds tested: L-Sorbose –, melibiose –, melezitose v, inulin –, D-arabinose –, D-glucosamine · HCl –, glucitol +, α-methyl-D-glucoside v, salicin + or w, potassium D-gluconate –, DL-lactic acid v

Assimilation of nitrate: –

Growth in vitamin-free medium: –

Growth in 10% sodium chloride plus 5% glucose in yeast nitrogen base: –

Growth at 37°C: –

G + C: 40.2 mol%, type strain (Kurtzman et al., unpublished)

Origin of the strains studied

Tanning fluid (1), *Drosophila miranda* (1), frass of dead *Pinus* sp. (1), diseased *Abies nobilis* (1).

Type strain: CBS 2504 (NRRL YB-4247), from tanning fluid.

54. PICHIA TREHALOPHILA PHAFF, MILLER ET SPENCER (1964)

Standard description of *Pichia trehalophila*

Growth on 5% malt extract agar: After 3 days at 25°C, the cells are spheroidal to ellipsoidal (2.1–6.8) × (3.5–7.5) μm, and occur singly and in pairs. Growth is tannish-yellow, smooth, glistening and butyrous.

Growth on the surface of assimilation media: Pellicles are not formed.

Dalmau plate culture on morphology agar: After 7 days at 25°C, growth under the coverglass shows neither pseudohyphae nor true hyphae. Aerobic growth is tannish-white, smooth, glistening and butyrous. Margins are entire.

Formation of ascospores: Ascus formation may be preceded by conjugation between a parent cell and a bud or between independent cells. Cells with protuberances are also found in sporulating cultures. Two to four hat-shaped spores are formed per ascus and they are quickly liberated.

Spores were observed on 5% malt extract agar.

Fermentation:

Glucose	+	Lactose	–
Galactose	–	Raffinose	–
Sucrose	–	Trehalose	– or +w
Maltose	–		

Assimilation of carbon compounds:

Galactose	–	Raffinose	–	Erythritol	+
Sucrose	–	Soluble starch	–	Ribitol	+
Maltose	–	D-Xylose	+	D-Mannitol	+
Cellobiose	–	L-Arabinose	+	Succinic acid	+
Trehalose	+	D-Ribose	+	Citric acid	–
Lactose	–	L-Rhamnose	–	Inositol	–

Additional carbon compounds tested: L-Sorbose +, melibiose –, melezitose –, inulin –, D-arabinose – or +s, D-glucosamine · HCl –, glucitol +, α-methyl-D-glucoside –, salicin –, potassium D-gluconate –, DL-lactic acid –

Assimilation of nitrate: –

Growth in vitamin-free medium: –

Growth in 10% sodium chloride plus 5% glucose in yeast nitrogen base: –

Growth at 37°C: –

G + C: 34.6 mol%, type strain (Nakase & Komagata, 1970b)

Origin of the strains studied

Slime flux of *Populus trichocarpa* in California (2).

Type strain: CBS 5361 (NRRL Y-6781).

55. PICHIA VERONAE KODAMA (in Kodama & Kyono, 1974b)

Standard description of *Pichia veronae*

Growth on 5% malt extract agar: After 3 days at 25°C, the cells are spheroidal to ellipsoidal (2.5–6.5) × (4.9–8.0) μm, and occur singly or in pairs. Growth is cream colored, faintly glistening and butyrous.

Growth on the surface of assimilation media: Thin climbing pellicles are present.

Dalmau plate culture on morphology agar: After 7 days at 25°C, growth under the coverglass shows well branched pseudohyphae with abundant blastospores. True hyphae are not formed. Aerobic growth is white, smooth, faintly glistening and butyrous. Margins may be entire or finely lobed.

Formation of ascospores: Asci are unconjugated and produce two to four hat-shaped spores. The spores are liberated soon after formation. This species is heterothallic and complementary mating types, derived as single-spore isolates, are NRRL Y-7818-1 (CBS 6895) and NRRL Y-7818-10 (CBS 6896).

Spores were observed on 5% malt extract agar.

Fermentation:

Glucose	+	Lactose	–
Galactose	–	Raffinose	–
Sucrose	+	Trehalose	–
Maltose	–		

Assimilation of carbon compounds:

Galactose	–	Raffinose	+ w	Erythritol	–
Sucrose	+	Soluble starch	–	Ribitol	–
Maltose	+	D-Xylose	+	D-Mannitol	+
Cellobiose	+	L-Arabinose	+ s	Succinic acid	+
Trehalose	+	D-Ribose	–	Citric acid	+
Lactose	–	L-Rhamnose	+	Inositol	–

Additional carbon compounds tested: L-Sorbose –, melibiose –, melezitose +, inulin –, D-arabinose –, D-glucosamine · HCl –, glucitol +, α-methyl-D-glucoside +, salicin + w, potassium D-gluconate +, DL-lactic acid +

Assimilation of nitrate: –

Growth in vitamin-free medium: –

Growth in 10% sodium chloride plus 5% glucose in yeast nitrogen base: –

Growth at 37°C: –

G + C: 49.6 mol%, type strain (Kurtzman et al., unpublished)

Comments

The results obtained in the present study correspond well with the original description. Exceptions concern raffinose and L-arabinose on which Kodama (Kodama & Kyono, 1974b) found no growth, but which I found to be latently utilized.

No mating response has been observed between *P. veronae* and the phenotypically similar species *P. rhodanensis, P. wickerhamii, P. amylophila* and *P. mississippiensis.*

Origin of the strain studied

Exudate of *Aesculus turbinata,* Japan (1).

Type strain: CBS 6591 (LKB-325, NRRL Y-7818).

56. *PICHIA WICKERHAMII* (VAN DER WALT) KREGER-VAN RIJ (1964b)

Synonym

Endomycopsis wickerhamii van der Walt (1959b)

Standard description of *Pichia wickerhamii*

Growth on 5% malt extract agar: After 3 days at 25°C, the cells are ovoidal to elongated (1.7–5.0) × (2.5–12.0) μm, and occur singly, in pairs and in chains. Growth is tannish-yellow, somewhat wrinkled, and dull to faintly glistening.

Growth on the surface of assimilation media: Pellicles are present.

Dalmau plate culture on morphology agar: After 7 days at 25°C, growth under the coverglass shows abundant well branched pseudohyphae bearing blastospores that may be single or in small verticillate clusters. The presence of true hyphae is questionable. All hyphae examined seemed to have slight indentations at the septa rather than parallel walls as would be expected of true hyphae. Aerobic growth is light tannish-white, usually dull, wrinkled and somewhat mycelial. Margins are generally entire.

Formation of ascospores: Asci are unconjugated when formed by diploid strains of this heterothallic species, or conjugated if produced by the pairing of complementary mating types. Two to four hat-shaped spores form in each ascus and they are readily liberated. Complementary mating types derived as single-spore isolates from the type strain are NRRL Y-2435-9 (CBS 7100) and NRRL Y-2435-10 (CBS 7101).

Spores were observed on 5% malt extract agar.

Fermentation:

Glucose	+	Lactose	–
Galactose	–	Raffinose	–
Sucrose	–	Trehalose	–
Maltose	–		

Assimilation of carbon compounds:

Galactose	–	Raffinose	–	Erythritol	–
Sucrose	+	Soluble starch	–	Ribitol	+
Maltose	+	D-Xylose	+	D-Mannitol	+
Cellobiose	+	L-Arabinose	–	Succinic acid	+
Trehalose	+	D-Ribose	–	Citric acid	+
Lactose	–	L-Rhamnose	+	Inositol	–

Additional carbon compounds tested: L-Sorbose –, melibiose –, melezitose v, inulin –, D-arabinose –, D-glucosamine · HCl –, glucitol +, α-methyl-D-glucoside +, salicin +, potassium D-gluconate +, DL-lactic acid +

Assimilation of nitrate: –

Growth in vitamin-free medium: –

Growth in 10% sodium chloride plus 5% glucose in yeast nitrogen base: +

Growth at 37°C: +

G + C: 46.5 mol%, type strain (Kurtzman et al., 1980a)

Comments

In comparisons of phenotypically similar species, *P. wickerhamii* failed to show any mating response with *P. amylophila, P. mississippiensis, P. rhodanensis* or *P. veronae.* Furthermore, *P. wickerhamii* showed only minimal DNA relatedness with these other species (Kurtzman et al., 1980a).

Origin of the strains studied

Frass of Pyralidae larvae in *Encephalartos* sp. (3).

Type strain: CBS 4107 (NRRL Y-2435).

f. COMMENTS ON THE GENUS

Since Hansen established the genus *Pichia* in 1904, the definition of this taxon has changed considerably to accommodate the ever increasing number of new species assigned to it. A historical account of the genus, named in honor of Professor P. Pichi, may be found in the treatments of Lodder and Kreger-van Rij (1952) and Kreger-van Rij (1970c). Many aspects of the taxonomic evolution of *Pichia,* such as ploidy, the presence or absence of a mating system, the type of conjugation, if any, preceding ascus formation, presence of pellicles and the occurrence of gaseous fermentation, were discussed in detail by Kreger-van Rij (1970c), and only some of the more recent developments will be considered here.

Owing to recent electron microscopic and molecular studies, many of the round-spored species have been removed from *Pichia* since the last major treatment of this genus (Kreger-van Rij, 1970c). *P. terricola, P. kudriavzevii* and the more recently described *P. scutulata* were transferred to the genus *Issatchenkia* because their spores proved roughened, but unlike *Debaryomyces* or *Torulaspora,* they had unconjugated asci and coenzyme Q-7 in their electron transport system (Kurtzman et al., 1980b). Additionally, Kurtzman and Smiley (1974) showed by scanning electron microscopy that the spores of *P. polymorpha* and *P. pseudopolymorpha* were roughened and typical of *Debaryomyces.* Other similarities included parent–bud conjugation and the presence of coenzyme Q-9 (Yamada & Kondo, 1972a; Yamada et al., 1973c). Once species relationships were clarified by the DNA studies of Price et al. (1978), these two taxa were transferred to *Debaryomyces* (Price & Phaff, 1979).

The remaining round-spored species have been reluctantly retained in *Pichia* until additional studies can be made. *P. carsonii* and *P. etchellsii,* with glabrous spores and coenzyme Q-9, appear related, but their affinities with the less well characterized *P. humboldtii* and *P. abadieae* remain obscure. *P. ohmeri,* which has coenzyme Q-9 and either hat-shaped or smooth spheroidal spores, seems more akin to the exclusively hat-spored *P. guilliermondii* and *P. scolyti.* However, these latter species also form coenzyme Q-9, and a relationship with *P. carsonii* and *P. etchellsii* may be a distant possibility.

The genera *Pichia* and *Hansenula* show many similarities and have been separated from each other principally on the basis of the nitrate assimilation reaction as proposed by Stelling-Dekker (1931). In recent years it has been recognized that only a few genes are responsible for nitrate reduction, but owing to the failure to detect nitrate variable strains among species in either genus, *Hansenula* and *Pichia* have been maintained.

Recently, Kurtzman et al. (unpublished) discovered *H. mrakii* and *P. sargentensis* to share 68% DNA base sequence complementarity. While this level of DNA relatedness is at the lower limit of that presently accepted between strains of a species (Johnson, 1973; Price et al., 1978), it is clearly much higher than that found between species within a genus, and for this reason these two taxa are being considered conspecific. Consequently, use of the nitrate reaction as a criterion for delimiting either species or genera can no longer be considered valid. This is in keeping with the finding that *Sterigmatomyces halophilus* is nitrate variable (Kurtzman et al., 1984) as well as with the earlier acceptance that species of *Dekkera* are variable in their assimilation of this compound (van der Walt, 1970b).

Discarding the nitrate reaction requires significant changes in current taxonomic schemes. Because of the recency of the finding precipitating these changes, there was insufficient time to incorporate them into the present treatment. However, it is anticipated that generic rearrangements will be made along the following lines. Saturn-spored species presently assigned to *Hansenula* and *Pichia* will be transferred to *Williopsis* Zender. In addition to ascospore shape, these species all show a similar mode of ascus formation and are found in the same general habitats. The remaining species of *Hansenula,* all hat-spored, will be placed in *Pichia* because this latter genus has priority. *Pichia,* even as presently defined, appears to be polyphyletic in its origin and eventually will need to be split into several genera. The points of future scission are presently scarcely perceived because of the paucity of molecular data that seem necessary for defining the anticipated genera.

The majority of species assigned to *Pichia* (and *Hansenula*) do not form true hyphae, and there has been some uncertainty whether those that do should be assigned to this genus. Kreger-van Rij and Veenhuis (1973a) showed marked differences in the ultrastructure of hyphal septa in various taxa of ascomycetous yeasts, and these differences have been useful in the definition of genera. As a result of the ultrastructure work, the mycelial species now accepted in *Pichia* as well as in *Hansenula* have been restricted to those with septa showing only the central closure line.

It seems apparent that the genus *Pichia* represents an amalgam of phylogenetic lines that will not be resolved until extensive studies at the molecular level can be undertaken. Consequently, the present treatment represents only a rather pragmatic attempt to bring order to an apparently diverse assemblage of species.

g. SPECIES NOT ACCEPTED IN THE GENUS

Species for which no strains were available:

Pichia moniliformis Nishimura (1910)
Pichia ovaria Nishimura (1910)
Pichia californica (Seifert) Guilliermond (1912)
Pichia megalospora Kuraishi (1958b)
Pichia orientalis (Beijerinck) Guilliermond (1912)
Pichia radaisii (Lutz) Guilliermond (1912)
Pichia tamarindorum (Seifert) Guilliermond (1912)
Pichia taurica (Seifert) Guilliermond (1912)

Pichia (Zygopichia) uvarum Verona et Luchetti (1941)
Pichia xylopsoci van der Walt et Scott (1971c)
Pichia methanothermo Minami et Yamamura (in Minami et al., 1978)

Newly described species received too late for inclusion in this treatment:
Pichia petrophilum Mu (in Mu & Gao, 1979)
Pichia sorbitophila Rodrigues de Miranda, Appel et Seyfarth (1980)
Pichia cellobiosa Lee et Komagata (1980)
Pichia inositovora Golubev et Blagodatskaya (in Golubev et al., 1981)

Genus 22. *Saccharomyces* Meyen ex Reess

by D. Yarrow

a. DIAGNOSIS OF THE GENUS

Vegetative reproduction by multilateral budding. Cells globose, ellipsoidal or cylindrical. Pseudohyphae may be formed but not septate hyphae.

The vegetative phase is predominantly diploid (or of higher ploidy), conjugation occurring on or soon after germination of the ascospores, diploid ascospores may be formed.

Ascospores globose to short ellipsoidal, with a smooth wall, usually one to four, occasionally more, per ascus. Asci persistent.

Vigorous fermentation.

Starch-like compounds are not produced.

No growth with nitrate as sole source of nitrogen.

b. TYPE SPECIES OF THE GENUS

Saccharomyces cerevisiae Meyen ex Hansen

c. SPECIES ACCEPTED IN THE GENUS

1. *Saccharomyces cerevisiae* Meyen ex Hansen (1883)
2. *Saccharomyces dairensis* Naganishi (1917)
3. *Saccharomyces exiguus* Reess ex Hansen (1888a)
4. *Saccharomyces kluyveri* Phaff, Miller et Shifrine (1956)
5. *Saccharomyces servazzii* Capriotti (1967)
6. *Saccharomyces telluris* van der Walt (1957)
7. *Saccharomyces unisporus* Jörgensen (1909)

d. KEY TO THE SPECIES OF THE GENUS

Based on results of growth tests read on the 4th day of incubation. Assimilation of nitrogen sources tested on auxanogram plates, of carbon sources on auxanogram plates or in liquid shaken cultures; growth in presence of cycloheximide and vitamin requirements in liquid shaken cultures.

1a Ethylamine and cadaverine assimilated 2
b Ethylamine and cadaverine not assimilated 3

2a No growth in presence of 1000 ppm of cycloheximide, sucrose assimilated
Sacch. kluyveri .... p. 388
b Growth in presence of 1000 ppm of cycloheximide, sucrose not assimilated
Sacch. unisporus .... p. 392

3a Growth in presence of 100 ppm of cycloheximide 4
b No growth in presence of 100 ppm of cycloheximide 6

4a Sucrose not assimilated 5
b Sucrose assimilated
Sacch. exiguus .... p. 387

5a Growth in presence of 1000 ppm of cycloheximide
Sacch. servazzii .... p. 389
b No growth in presence of 1000 ppm of cycloheximide
Sacch. dairensis .... p. 386

6a Ascospores smooth 7
b Ascospores rough
Sacch. telluris .... p. 391

7a Cells small (2.5–5.5) × (3.5–7.0) μm, maltose not assimilated 4
b Cells large (3.0–10.0) × (4.5–21.0) μm, maltose may be assimilated
Sacch. cerevisiae .... p. 383

Table 19

Some properties of the species of the genera *Saccharomyces*, *Torulaspora* and *Zygosaccharomyces*

	Assimilation									Growth in presence of cycloheximide		Growth at 37°C	Inositol required	Growth in presence of 1% acetic acid	Average G + C (mol%)
	Ga	Su	Ma	Ra	Me	Ce	Ety	Cad	Lys	100 ppm	1000 ppm				
Saccharomyces															
cerevisiae	v	v	v	v	v	–	–	–	–	–	–	v			39.5
dairensis	+	–	–	–	–	–	–	–	–	+(–)	–	+ w/ –			37.9
exiguus	+	+	–	+ /s	–	–	–/+w	–	–	+(–)	–	–			34.0
kluyveri	+	+	+ /s	+	+	v	+	+	+	–	–	+			40.0
servazzii	+	–	–	–	–	–	–	–	–	+	+	v			34.7
telluris	–	–	–	–	–	–	–	–	–	–	–	+			33.0
unisporus	+	–	–	–	–	–	+	+	+	+	+	– / + w			32.5
Torulaspora															
delbrueckii	v	v	v	v	v	–	–(+)	–	+	–	–	–			43.3
globosa	–	+	–	+	–	–	–	–	+	+	–	+			46.6
pretoriensis	+	+	+	+	–	–	–	–	+	– / + s	–	+			46.1
Zygosaccharomyces															
bailii	v	v	–	–(+)	–	–	+	+	+	– / + s	–	v	–	+	43.7
bisporus	v	–	–	–	–	–	+	+	+	–	–	–	–	+	44.0
cidri	+	+	v	+	+	–	+	+	+	+	+	v	+	–	42.0
fermentati	+	+	+	+	–	+	+	+	+	+	+	+	+	–	42.3
florentinus	v	+	v	+	+	–	+	+	+	+	+	–	–	–	42.6
microellipsoides	+	+	–	+	+	–	+	+	+	–	–	–	–	–	40.2
mrakii	+	+	–	+	+	–	+	+	+	+	+	–	–	–	41.0
rouxii	v	v	v	–	–	–	+	+	+	–	–	v	–	–	40.1

Ga = galactose, Su = sucrose, Ma = maltose, Ra = raffinose, Me = melibiose, Ce = cellobiose, Ety = ethylamine·HCl, Cad = cadaverine·2 HCl, Lys = L-lysine.

e. SYSTEMATIC DISCUSSION OF THE SPECIES

1. *SACCHAROMYCES CEREVISIAE* MEYEN EX HANSEN (1883)

Synonyms

Saccharomyces ellipsoideus Meyen ex Hansen (1883)
Saccharomyces cerevisiae Meyen ex Hansen var. *ellipsoideus* (Hansen) Dekker (Stelling-Dekker, 1931)
Saccharomyces ilicis Grönlund (1892)
Saccharomyces vini-muntz Kayser (1892)
Saccharomyces vordermanii Went et Prinsen Geerligs (1894)
Saccharomyces bayanus Saccardo (1895)
Saccharomyces willianus Saccardo (1895)
Saccharomyces sake Yabe (1897)
Saccharomyces cratericus Lindner (1898a)
Saccharomyces cerevisiae Meyen ex Hansen var. *cratericus* (Lindner) Lodder (1932)
Saccharomyces uvarum Beijerinck (1898b)
Saccharomyces intermedius Hansen (1904)
Saccharomyces pastorianus Hansen (1904)
Saccharomyces turbidans Hansen (1904)
Saccharomyces cerevisiae Meyen ex Hansen var. *turbidans* (Hansen) Dekker (Stelling-Dekker, 1931)
Saccharomyces validus Hansen (1904)
Saccharomyces brasiliensis Lindner (1905b)
Saccharomyces batatae Saito (1907)
Saccharomyces carlsbergensis Hansen (1908)
Saccharomyces uvarum Beijerinck var. *carlsbergensis* Kudriavzev (1960)
Saccharomyces monacensis Hansen (1908)
Saccharomyces carlsbergensis Hansen var. *monacensis* (Hansen) Dekker (Stelling-Dekker, 1931)
Saccharomyces logos van Laer et Denamur ex Jörgensen (1909)
Saccharomyces multisporus Jörgensen (1909)
Saccharomyces tokyo Nakazawa (1909)
Saccharomyces yedo Nakazawa (1909)
Saccharomyces coreanus Saito (1910)
Saccharomyces anamensis Will et Heinrich (in Will, 1913)
Saccharomyces chevalieri Guilliermond (1914)
Saccharomyces lindneri Guilliermond (1914)
Saccharomyces chevalieri Guilliermond var. *lindneri* (Guilliermond) Dekker (Stelling-Dekker, 1931)
Saccharomyces mandshuricus Saito (1914)
Saccharomyces carlsbergensis Hansen var. *mandshuricus* (Saito) Dekker (Stelling-Dekker, 1931)
Saccharomyces mangini Guilliermond (1914)
Saccharomyces paradoxus Batschinskaya (1914)
Zygosaccharomyces paradoxus (Batschinskaya) Klöcker (1924)
Zygosaccharomycodes paradoxus (Batschinskaya) Nishiwaki (1929)
Saccharomyces marchalianus Kufferath (1920a)
Saccharomyces cerevisiae Meyen ex Hansen var. *marchalianus* (Kufferath) Dekker (Stelling-Dekker, 1931)
Saccharomyces chodati Steiner (1924) [non *Saccharomyces chodati* Schweizer (1921)]
Saccharomyces globosus Osterwalder (1924a)
Saccharomyces heterogenicus Osterwalder (1924a)
Saccharomyces oviformis Osterwalder (1924a)
Saccharomyces tubiformis Osterwalder (1924a)
Saccharomyces intermedius Hansen var. *valdensis* Osterwalder (1924a)
Saccharomyces carlsbergensis Hansen var. *valdensis* (Osterwalder) Dekker (Stelling-Dekker, 1931)

Saccharomyces intermedius Hansen var. *turicensis* Osterwalder (1934)
Saccharomyces carlsbergensis Hansen var. *polymorphus* Dekker (Stelling-Dekker, 1931)
Saccharomyces festinans Ward et Baker (1929)
Saccharomyces cerevisiae Meyen ex Hansen var. *festinans* (Ward et Baker) Dekker (Stelling-Dekker, 1931)
Saccharomyces cerevisiae Meyen ex Hansen var. *pulmonalis* (Redaelli) Dekker (Stelling-Dekker, 1931)
Saccharomyces mangini Guilliermond var. *tetraspora* Beijerinck ex Dekker (Stelling-Dekker, 1931)
Saccharomyces cerevisiae Meyen ex Hansen var. *tetraspora* (Beijerinck ex Dekker) Phaff, Miller et Shifrine (1956)
Saccharomyces elongatus Krumbholz (1932)
Saccharomyces cheresiensis Prostoserdov et Afrikian (1933)
Saccharomyces oviformis Osterwalder var. *cheresiensis* (Prostoserdov et Afrikian) Kudriavzev (1954)
Saccharomyces formosensis Nakazawa (1933)
Saccharomyces mangini Guilliermond var. *casei* Sacchetti (1933)
Saccharomyces cerevisiae Meyen ex Hansen var. *onychophilus* Zach (in Wolfram & Zach, 1934)
Saccharomyces ellipsoideus Hansen var. *umbra* nom. nud. (Castelli, 1935)
Saccharomyces ellipsoideus Hansen var. *major* nom. nud. (Castelli, 1938)
Saccharomyces italicus nom. nud. (Castelli, 1938)
Saccharomyces oviformis Osterwalder var. *bisporus* nom. nud. (Castelli, 1938)
Candida robusta Diddens et Lodder (1942)
Saccharomyces fructuum Lodder et Kreger-van Rij (1952)
Saccharomyces steineri Lodder et Kreger-van Rij (1952)
Saccharomyces capensis van der Walt et Tscheuschner (1956d)
Saccharomyces italicus Castelli var. *melibiosi* van Uden et Assis-Lopes (1957)
Saccharomyces oleaceus Santa María (1958a)
Saccharomyces oleaginosus Santa María (1958a)
Saccharomyces aceti Santa María (1959a)
Saccharomyces oxidans Santa María (1959a)
Saccharomyces uvarum Beijerinck var. *melibiosus* Kudriavzev (1960)
Saccharomyces vini Meyen ex Kudriavzev (1960)
Saccharomyces vini Meyen ex Kudriavzev var. *cartilaginosus* Kudriavzev (1960)
Saccharomyces vini Meyen ex Kudriavzev var. *cerevisiae* Kudriavzev (1960)
Saccharomyces hienipiensis Santa María (1962)
Saccharomyces norbensis Santa María (1963a)
Saccharomyces onubensis Santa María (1964)
Saccharomyces diastaticus Andrews et Gilliland ex van der Walt (1965a)
Saccharomyces inusitatus van der Walt (1965b)
Saccharomyces cerevisiae Meyen ex Hansen var. *terrestris* Jensen (1967)
Saccharomyces hispanica Santa María (1968)
Saccharomyces prostoserdovii Kudriavzev (1960)
Saccharomyces beticus Marcilla ex Santa María (1970)
Saccharomyces cordubensis Santa María (1970)
Saccharomyces gaditensis Santa María (1970)

Standard description of *Saccharomyces cerevisiae*

Growth in malt extract: After 3 days at 25°C the cells may be globose or subglobose (5.0–10.0) × (5.0–12.0) μm, ellipsoidal to cylindrical (3.0–9.5) × (4.5–21.0) μm, elongated cells of more than 30 μm in length may be present, single, in pairs, short chains or clusters (Figs. 96, 97). A sediment and occasionally a ring are present. After one month at 20°C a sediment is present, sometimes also a ring and a film.

Growth on malt agar: After one month at 20°C the streak culture is butyrous,

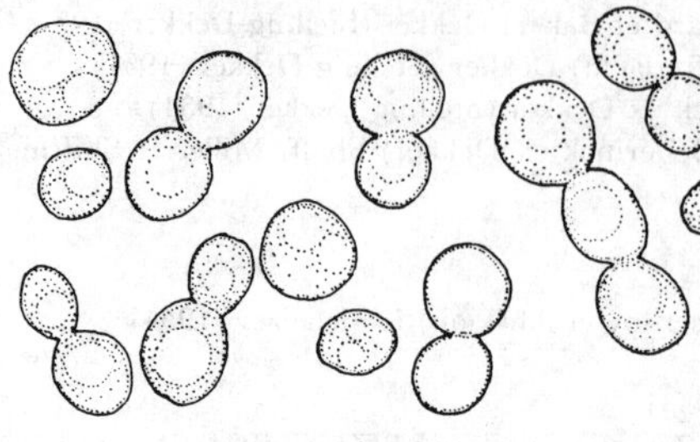

Fig. 96 – *Sacch. cerevisiae,* CBS 1171
After 3 days in malt extract
(From 'The Yeasts', ed. I, 1952)

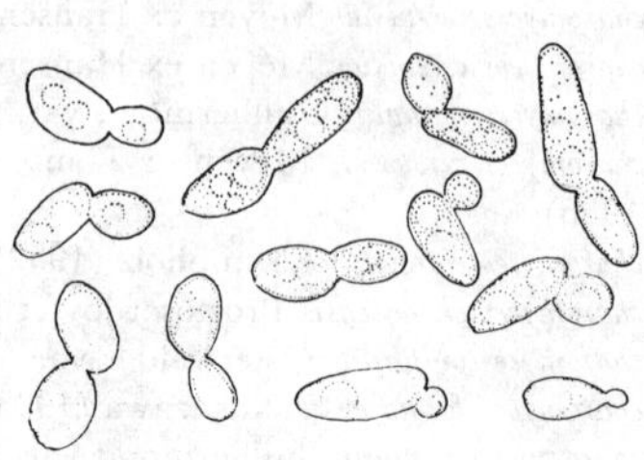

Fig. 97 – *Sacch. cerevisiae,* CBS 380
After 3 days in malt extract
(From 'The Yeasts', ed. I, 1952)

cream to slightly brownish, slightly raised and smooth with light striations or raised and folded, often sectored, glossy or dull.

Slide cultures on corn meal- and morphology agar: Pseudohyphae may be formed, they may be rudimentary or well branched.

Formation of ascospores: Vegetative cells are transformed direct into asci containing one to four, occasionally more, globose to short ellipsoidal ascospores (Fig. 98).

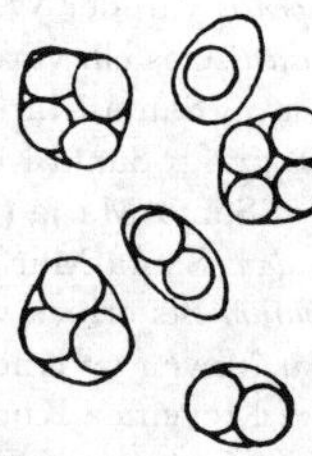

Fig. 98 – *Sacch. cerevisiae*
After 1 week on Gorodkowa agar
(From 'The Yeasts', ed. I, 1952)

Acetate agar was the best medium for inducing sporulation.

Fermentation:

Glucose	+	Maltose	v
Galactose	v	Lactose	–
Sucrose	v		

Assimilation of carbon compounds:

Galactose	v	Raffinose	v	Erythritol	–
Sucrose	v	Soluble starch	v	Ribitol	–
Maltose	v	D-Xylose	–	D-Mannitol	v
Cellobiose	–	L-Arabinose	–	Succinic acid	– or + s
Trehalose	v	D-Ribose	–	Citric acid	–
Lactose	–	L-Rhamnose	–	Inositol	–

Assimilation of nitrate: –

Assimilation of ethylamine · HCl: –

Assimilation of cadaverine · 2 HCl: –

Growth in vitamin-free medium: v

Growth on 50% (w/w) glucose-yeast extract agar: v

Growth at 37° C: v

Growth in the presence of 100 ppm of cycloheximide: –

G + C: 39.0–41.0 mol%, 6 strains (Bicknell & Douglas, 1970); 38.5–40.7 mol%, 44 strains (Nakase & Komagata, 1971d); 38.5–40.7 mol%, 7 strains (Yarrow & Nakase, 1975); 39.0–40.2 mol%, 2 strains

Table 20

Physiological races of *Saccharomyces cerevisiae* given species status in the second edition of 'The Yeasts' (van der Walt, 1970e)

	Fermentation					
	Ga	Su	Ma	Ra	Me	St
Saccharomyces						
aceti	–	–	–	–	–	–
bayanus	–	+	+	+	–	–
capensis	–	+	–	+	–	–
cerevisiae	+	+	+	+	–	–
chevalieri	+	+	–	+	–	–
coreanus	+	+	–	+	+	–
diastaticus	+	+	+	+	–	+
globosus	+	–	–	–	–	–
heterogenicus	–	+	+	–	–	–
hienipiensis	–	–	+	–	+	–
inusitatus	–	+	+	+	+	–
norbensis	–	–	–	–	+	–
oleaceus	+	–	–	+	+	–
oleaginosus	+	–	+	+	+	–
prostoserdovii	–	–	+	–	–	–
steineri	+	+	+	–	–	–
uvarum	+	+	+	+	+	–

Ga = galactose, Su = sucrose, Ma = maltose, Ra = raffinose, Me = melibiose, St = soluble starch.

Comments

Campbell (1972) in a numerical analysis recognized the following ten species at the 90% matching coefficient level: *Sacch. cerevisiae, Sacch. chevalieri, Sacch. coreanus, Sacch. globosus, Sacch. heterogenicus, Sacch. inusitatus, Sacch. italicus, Sacch. oleaginosus, Sacch. prostoserdovii* and *Sacch. uvarum.* Kocková-Kratochvílová (1976) in another numerical study recognized 18 species.

Because of the instability of the characteristics by which they were separated and the absence of reproductive isolation (Scheda & Yarrow, 1966, 1968; Scheda, 1966b; Naumov, 1969a, b, 1972a, b, 1977; Naumov & Yurkevitch, 1970; Neumann, 1972; Yarrow, 1972b) and the high degree of DNA homology (Bicknell & Douglas, 1970; Groot et al., 1975), the species listed in Table 20 with the addition of *Sacch. hispanica* were considered to be merely variant races of *Sacch. cerevisiae* and were listed as synonyms of it by Yarrow & Nakase (1975). *Sacch. gaditensis* and *Sacch. cordubensis,* two species forming a 'flor' on wine and described by Santa María (1970), are interfertile with strains of *Sacch. cerevisiae* and are also reduced to synonymy with it.

Origin of the strains studied

Breweries (18), wine (33), berries (1), palm wine (1), cider (1), ginger wine (1), bili wine (1), slime flux of oak tree (1), cheese (2), pear juice and must (3), apple juice (1), honey (1), grape juice (1), distillery (2), saké-moto (3), grape must (6), yeast cake (1), grapes (2), fruit of *Ilex equifolium* (1), arak (1), sorghum-brandy starter (1), eucalyptus leaves (1), kefyr (1), *Drosophila* (1), soil (4), alpechin (7), olives (5), pulqué (1), soft drinks and fruit juices (9), human skin (1), sputum (11), mouth (1), leg ulcer (1).

Type strain: CBS 1171, isolated by van Wijk from the top yeast used in the Oranjeboom Brewery at Rotterdam, was designated as type by Lodder and Kreger-van Rij (1952).

2. *SACCHAROMYCES DAIRENSIS* NAGANISHI (1917)

Synonym

Saccharomyces castellii Capriotti (1966)

Standard description of *Saccharomyces dairensis*

Growth in malt extract: After 3 days at 25°C, the cells are globose to ellipsoidal (3.0–5.5) × (3.5–7.0) μm, single or in pairs. A sediment is present. After one month at 20°C a sediment is present and sometimes a ring also.

Growth on malt agar: After one month at 20°C the streak culture is butyrous, cream to yellowish, smooth and glossy, the margin is entire or crenate.

Slide cultures on corn meal- and morphology agar: No pseudohyphae are formed.

Formation of ascospores: Vegetative cells are transformed direct into asci containing one to four globose to ellipsoidal ascospores.

Sporulation was observed on acetate agar and on YM agar also in some strains.

Fermentation:

Glucose	+	Maltose	–
Galactose	+	Lactose	–
Sucrose	–		

Assimilation of carbon compounds:

Galactose	+	Raffinose	–	Erythritol	–
Sucrose	–	Soluble starch	–	Ribitol	–
Maltose	–	D-Xylose	–	D-Mannitol	–
Cellobiose	–	L-Arabinose	–	Succinic acid	–
Trehalose	v	D-Ribose	– or + s	Citric acid	–
Lactose	–	L-Rhamnose	–	Inositol	–

Assimilation of nitrate: –

Assimilation of ethylamine · HCl: –

Assimilation of cadaverine · 2 HCl: –
Growth in vitamin-free medium: –
Growth on 50% (w/w) glucose-yeast extract agar: –
Growth at 37°C: +w or –
Growth in the presence of 100 ppm of cycloheximide: + (–); of 1000 ppm: –
G + C: 37.0 mol%, one strain (Bicknell & Douglas, 1970); 38.8 mol%, one strain (Yarrow & Nakase, 1975); 36.6–38.5 mol%, 6 strains

Comments

See under Comments of *Sacch. servazzii.*

Origin of the strains studied

Unknown (1), buttermilk (1), fermenting pickled cucumbers (3), soil (3).

Type strain: CBS 421 isolated by Naganishi.

3. *SACCHAROMYCES EXIGUUS* REESS EX HANSEN (1888a)

Imperfect state: *Candida holmii* (Jörgensen) Meyer et Yarrow

Standard description of *Saccharomyces exiguus*

Growth in malt extract: After 3 days at 25°C, the cells are subglobose to ellipsoidal (2.5–5.0) × (3.5–6.5) μm, single or in pairs (Fig. 99). A sediment is present and occasionally a ring. After one month at 20°C a sediment is present and often also a ring and a film.

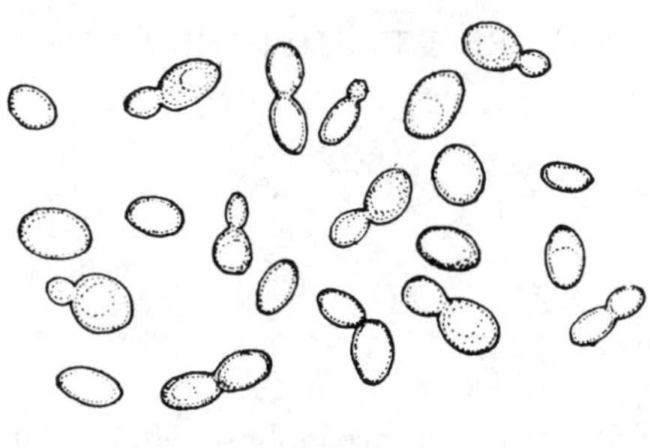

Fig. 99 – *Sacch. exiguus*
After 3 days in malt extract
(From 'The Yeasts', ed. I, 1952)

Growth on malt agar: After one month at 20°C the streak culture is butyrous, cream to tan, smooth with some light striations, flat and rather spreading, glossy, margin usually entire.

Slide cultures on corn meal- and morphology agar: Pseudohyphae are not formed.

Formation of ascospores: Vegetative cells are transformed direct into asci containing one to four globose to ellipsoidal ascospores.

Sporulation was observed on acetate- and V8 agar.

Fermentation:

Glucose	+	Maltose	–
Galactose	+	Lactose	–
Sucrose	+		

Assimilation of carbon compounds:

Galactose	+	Raffinose	+ or s	Erythritol	–
Sucrose	+	Soluble starch	–	Ribitol	–
Maltose	–	D-Xylose	–	D-Mannitol	–
Cellobiose	–	L-Arabinose	–	Succinic acid	– or +s
Trehalose	+	D-Ribose	–	Citric acid	–
Lactose	–	L-Rhamnose	–	Inositol	–

Assimilation of nitrate: –

Assimilation of ethylamine · HCl: – or +w

Assimilation of cadaverine · 2 HCl: –

Growth in vitamin-free medium: – or +w

Growth on 50% (w/w) glucose-yeast extract agar: –

Growth at 37°C: –

Growth in the presence of 100 ppm of cycloheximide: + (–); of 1000 ppm: –

G + C: 33.2–33.7 mol%, 2 strains (Nakase & Komagata, 1971d); 34.9–35.4 mol%, one strain (Yarrow & Nakase, 1975); 32.9–34.9 mol%, 10 strains

Origin of the strains studied

Unknown (2), fermenting cucumber brine (1), soil (1), grapes (2), sauerkraut (3), strawberry (1), sewage slick (1), soft drink (1).

Type strain: CBS 379 from Naganishi was chosen as type by Lodder and Kreger-van Rij (1952).

4. *SACCHAROMYCES KLUYVERI* PHAFF, MILLER ET SHIFRINE (1956)

Synonyms

Saccharomyces smittii Capriotti (1958d)
Saccharomyces silvestris Jensen (1967)

Standard description of *Saccharomyces kluyveri*

Growth in malt extract: After 3 days at 25°C, the cells are globose, ellipsoidal or cylindrical (3.5–7.0) × (4.0–11.0) μm, single, in pairs or in clusters. A sediment and a thin ring are present. After one month at 20°C a sediment and a ring are present.

Growth on malt agar: After one month at 20°C the streak culture is cream to tan, fairly flat and smooth to raised and wrinkled or folded, often sectored, glossy or dull, margin undulating or lobate sometimes fringed with pseudohyphae.

Slide cultures on corn meal- and morphology agar: Pseudohyphae indistinguishable from those of *Saccharomyces cerevisiae* are usually formed.

Formation of ascospores: Vegetative cells are transformed direct into asci containing one to four globose to subglobose ascospores.

Sporulation was observed in all the strains studied on acetate agar.

Fermentation:

Glucose	+	Maltose	– or +w
Galactose	+	Lactose	–
Sucrose	+		

Assimilation of carbon compounds:

Galactose	+	Raffinose	+	Erythritol	–
Sucrose	+	Soluble starch	–	Ribitol	–
Maltose	+ or s	D-Xylose	– or +s	D-Mannitol	v
Cellobiose	v	L-Arabinose	–	Succinic acid	+ or s
Trehalose	+ or s	D-Ribose	–	Citric acid	–
Lactose	–	L-Rhamnose	–	Inositol	–

Assimilation of nitrate: –

Assimilation of ethylamine · HCl: +

Assimilation of cadaverine · 2 HCl: +

Growth in vitamin-free medium: v

Growth on 50% (w/w) glucose-yeast extract agar: –

Growth at 37°C: +

Growth in the presence of 100 ppm of cycloheximide: –

G + C: 40.0 mol%, one strain (Phaff & Meyer, 1969)

Comments

This species closely resembles *Saccharomyces cerevisiae* but Barker and Miller (1969) could not obtain fertile hybrids of the two species. McCullough and Herskowitz (1979) also found the two species to be incompatible for mating. For identification purposes it can readily be distinguished from *Sacch. cerevisiae* by its ability to grow on ethylamine, cadaverine and L-lysine as sole sources of nitrogen.

Origin of the strains studied

Drosophila (3), exudate of trees (3), soil (3), fruit juice (1), single-spore isolates from Wickerham (9).

Type strain: CBS 3082 isolated from *Drosophila pinicola* at Mather.

5. *SACCHAROMYCES SERVAZZII* CAPRIOTTI (1967)

Standard description of *Saccharomyces servazzii*

Growth in malt extract: After 3 days at 25°C, the cells are subglobose to ellipsoidal (3.0–4.5) × (4.0–5.5) μm, single or in pairs. A sediment and a thin ring are present. After one month at 20°C a sediment and a thin ring are present.

Growth on malt agar: After one month at 20°C the streak culture is tan, flat, smooth with faint striations, waxy and glossy, the margin is slightly crenate.

Slide cultures on corn meal- and morphology agar: Pseudohyphae are not formed.

Formation of ascospores: Vegetative cells are transformed direct into asci containing one to four globose to short ellipsoidal ascospores.

Sporulation was observed on acetate agar.

Fermentation:

Glucose	+	Maltose	–
Galactose	+	Lactose	–
Sucrose	–		

Assimilation of carbon compounds:

Galactose	+	Raffinose	–	Erythritol	–
Sucrose	–	Soluble starch	–	Ribitol	–
Maltose	–	D-Xylose	–	D-Mannitol	–
Cellobiose	–	L-Arabinose	–	Succinic acid	–
Trehalose	+ or s	D-Ribose	– or +s	Citric acid	–
Lactose	–	L-Rhamnose	–	Inositol	–

Assimilation of nitrate: –

Assimilation of ethylamine · HCl: –

Assimilation of cadaverine · 2 HCl: –

Growth in vitamin-free medium: –

Growth on 50% (w/w) glucose-yeast extract agar: – or +s

Growth at 37°C: v

Growth in the presence of 100 ppm of cycloheximide: +; **of 1000 ppm:** +

G + C: 33.9 mol%, one strain (Yarrow & Nakase, 1975); 34.4–35.9 mol%, 2 strains

Comments

Saccharomyces servazzii closely resembles *Sacch. dairensis* but the DNA of the strains studied has a lower mol% of guanine plus cytosine. They are also resistant to 1000 ppm of cycloheximide whereas none of the strains of *Sacch. dairensis* is, nor could they be adapted by attempting to cultivate them in increasing concentrations of cycloheximide. Resistance to 1000 ppm of cycloheximide is used in the key for identifying species but more strains need to be isolated and examined before this property can be used with assurance to discriminate between this species and *Sacch. dairensis.*

Both *Sacch. servazzii* and *Sacch. dairensis* are similar to *Sacch. unisporus* in their fermentation and growth patterns, in addition all three require calcium pantothenate and niacin for growth. *Sacch. unisporus* can be identified by its predominantly one-spored asci and its ability to grow with ethylamine, cadaverine and L-lysine as the only source of nitrogen.

Sacch. exiguus differs from all these three species by not requiring calcium pantothenate and niacin.

Origin of the strains studied

Soil (1), unknown (1), both from Finland.

Type strain: CBS 4311. Since a type was not designated for this species by Capriotti (1967) one of the isotypes has been chosen as lectotype.

6. *SACCHAROMYCES TELLURIS* VAN DER WALT (1957)

Imperfect state: *Candida pintolopesii* (van Uden) Meyer et Yarrow

Standard description of *Saccharomyces telluris*

Growth in malt extract: After 3 days at 28°C the cells are globose to ellipsoidal (3.5–6.5) × (4.5–8.5) μm, single, in pairs, short chains or small clusters. Occasionally cylindrical cells which may measure 17 μm in length may be formed. Growth is scant. A sediment has been formed (van der Walt, 1970e). After one month at 20°C a sediment, sometimes also an incomplete ring are present.

Growth on malt agar: After one month at 20°C the streak culture is cream, glossy, smooth with an entire or undulating margin.

Slide cultures on corn meal- and morphology agar: Pseudohyphae may be formed.

Formation of ascospores: Vegetative cells are transformed direct into asci containing one, occasionally two, ascospores. The ascospores are globose or subglobose and vary in appearance from slightly rough to distinctly spiny.

Sporulation was observed on acetate- and Gorodkowa agar.

Fermentation:

Glucose	+	Maltose	–
Galactose	–	Lactose	–
Sucrose	–		

Assimilation of carbon compounds:

Galactose	–	Raffinose	–	Erythritol	–
Sucrose	–	Soluble starch	–	Ribitol	–
Maltose	–	D-Xylose	–	D-Mannitol	–
Cellobiose	–	L-Arabinose	–	Succinic acid	– or +s
Trehalose	–	D-Ribose	–	Citric acid	–
Lactose	–	L-Rhamnose	–	Inositol	–

Assimilation of nitrate: –

Assimilation of ethylamine · HCl: –

Assimilation of cadaverine · 2 HCl: –

Growth in vitamin-free medium: –

Growth on 50% (w/w) glucose-yeast extract agar: –

Growth at 37°C: +

Growth in the presence of 100 ppm of cycloheximide: –

G + C: 32.3–34.0 mol%, 14 strains (Mendonça-Hagler & Phaff, 1975); 32.4 mol%, one strain

Origin of the strains studied

Soil (1), excreta of rodents (1), udder of mouse (1), peritoneal cavity of a guinea pig (1), pigeons (4), poultry (45), unknown (1).

Type strain: CBS 2685, isolated from soil by van der Walt (1957).

7. *SACCHAROMYCES UNISPORUS* JÖRGENSEN (1909)

Synonym

Saccharomyces mongolicus Naganishi (1928)

Standard description of *Saccharomyces unisporus*

Growth in malt extract: After 3 days at 25°C the cells are globose to short ellipsoidal (2.5–4.5) × (3.0–6.0) μm, single or in pairs. A sediment is present. After one month at 20°C a sediment and ring are present.

Growth on malt agar: After one month at 20°C the streak culture is cream, flat, usually glossy, smooth, sometimes with light striations, the margin is entire or undulating.

Slide cultures on corn meal- and morphology agar: Pseudohyphae are not formed.

Formation of ascospores: Vegetative cells are transformed direct into asci containing one, occasionally two, globose to subglobose ascospores.

Sporulation was observed on acetate- and YM agar.

Fermentation:

Glucose	+	Maltose	–
Galactose	+	Lactose	–
Sucrose	–		

Assimilation of carbon compounds:

Galactose	+	Raffinose	–	Erythritol	–
Sucrose	–	Soluble starch	–	Ribitol	–
Maltose	–	D-Xylose	–	D-Mannitol	–
Cellobiose	–	L-Arabinose	–	Succinic acid	–
Trehalose	–	D-Ribose	–	Citric acid	–
Lactose	–	L-Rhamnose	–	Inositol	–

Assimilation of nitrate: –

Assimilation of ethylamine · HCl: +

Assimilation of cadaverine · 2 HCl: +

Growth in vitamin-free medium: –

Growth on 50% (w/w) glucose-yeast extract agar: –

Growth at 37°C: – or +w

Growth in the presence of 100 ppm of cycloheximide: +; of 1000 ppm: +

G + C: 32.4–32.7 mol%, 3 strains (Nakase & Komagata, 1971d)

Origin of the strains studied

Naichin (1), cheese (1), kefyr (3), feces (1), strain patented as *Saccharomyces exiguus* (1).

Type strain: CBS 398, received from the Carlsberg Laboratory in 1924.

f. COMMENTS ON THE GENUS

Stelling-Dekker (1931) reduced *Zygosaccharomyces* Barker to a subgenus of *Saccharomyces*. Lodder and Kreger-van Rij (1952) further extended *Saccharomyces* to include *Torulaspora* Lindner. However, since then, other workers have recognized the former genera, generally separating them by the way the ascus arises. Kudriavzev (1954, 1960) retained *Zygosaccharomyces,* Novák and Zsolt (1961) recognized *Torulaspora*, Capriotti (1958a, b, c) retained both *Zygosaccharomyces* and *Torulaspora*. Van der Walt and Johannsen (1975a) redefined the genus *Torulaspora* and also recognized the genus *Zygosaccharomyces*. Yarrow (von Arx, 1976) provisionally classified some of the *Saccharomyces* species in these genera. Von Arx et al. (1977) renamed several species that had been placed in *Torulaspora* by van der Walt and Johannsen as *Zygosaccharomyces* species. In the present treatment only those species in which the vegetative phase is predominantly diploid are retained in *Saccharomyces*, the haploid species are placed in *Zygosaccharomyces* and *Torulaspora*.

Van der Walt and Johannsen (1975a) also included most of the species of *Debaryomyces* and the species of *Pichia* with globose ascospores in *Torulaspora*. Price et al. (1978) rejected the inclusion of these species in the same genus as the former *Saccharomyces* species on the grounds of differences in DNA base composition, coenzyme Q systems, regulation of arginine metabolism, possible differences in the morphology and the ultrastructure of the ascospores. Kreger-van Rij and Veenhuis (1976b) also argued that differences in the structure of the wall of the ascospore and some physiological properties suffice to separate *Torulaspora* from *Debaryomyces*. Van der Walt and Johannsen (1978) accepted the separation of *Torulaspora* from *Debaryomyces* Lodder et Kreger-van Rij on the criterion of coenzyme Q systems. Accepting these arguments, the following species are here excluded from the genus: *Torulaspora carsonii* (Phaff et Knapp) van der Walt et Johannsen, *T'spora castellii* (Capriotti) van der Walt et Johannsen, *T'spora coudertii* (Saëz) van der Walt et Johannsen, *T'spora etchellsii* (Kreger-van Rij) van der Walt et Johannsen, *T'spora formicaria* (Golubev et Babjeva) van der Walt et Johannsen, *T'spora hansenii* (Zopf) van der Walt et Johannsen, *T'spora melissophila* (van der Walt et van der Klift) van der Walt et Johannsen, *T'spora phaffii* (Capriotti) van der Walt et Johannsen, *T'spora polymorpha* (Klöcker) van der Walt et Johannsen, *T'spora pseudopolymorpha* (Ramirez et Boidin) van der Walt et Johannsen, *T'spora tamarii* (Ohara et Nonomura ex van der Walt et Johannsen) van der Walt et Johannsen, *T'spora vanriji* (van der Walt et Tscheuschner) van der Walt et Johannsen, and *T'spora yarrowii* (Santa María et Aser) van der Walt et Johannsen.

Van der Walt et Johannsen (1975a) distinguished *Zygosaccharomyces* from *Torulaspora* by the following characteristics:

(1) Fusion of independent cells as a rule precedes ascus formation, asci only infrequently arise direct from diploid cells or by somatogamous autogamy. According to the observations of the present author some of the species which were assigned to *Torulaspora* should be placed in *Zygosaccharomyces*. For instance, *Sacch. mrakii* was assigned to *Zygosaccharomyces* by Yarrow (von Arx, 1976) and by Capriotti (1958c) because conjugation was observed to be predominantly between independent cells but, nevertheless, van der Walt and Johannsen placed it in *Torulaspora*. However, in *Zygosacch. rouxii* conjugation generally occurs between individual cells but a few strains were encountered in which, when first examined, conjugation was mainly between a cell and its bud.

(2) Glabrous ascospores. The ascospores of *Torulaspora delbrueckii, T'spora pretoriensis* and *T'spora globosa* often appear rough when observed under the light microscope but rough ascospores could clearly be seen in a few strains of *Zygosacch. rouxii,* as observed by Hjort (1956) also. Although those of *Zygosacch. fermentati* appear to be smooth when seen under the light microscope, Kreger-van Rij (1979) observed some sections of ascospores of the type strain of *Sacch. montanus* van der Walt, CBS 4506, with smooth and some with wrinkled walls. Kurtzman et al. (1975) found that strains of *T'spora pretoriensis* may have either rough or smooth ascospores.

(3) Osmotolerance. Van der Walt and Johannsen themselves remarked that this property does not clearly demarcate *Zygosaccharomyces* from *Torulaspora*. *Zygosacch. rouxii, Zygosacch. bailii* and *Zygosacch. bisporus* can grow on media containing 60% glucose, as well as some strains of *T'spora delbrueckii* and *T'spora pretoriensis*.

(4) All *Zygosaccharomyces* species require an external source of vitamins. However, *Zygosacch. cidri* and *Zygosacch. fermentati* which require inositol, *Zygosacch. florentinus* and *Zygosacch. microellipsoides* which may grow slowly in vitamin-free medium but which grow much better when biotin is added, were all placed in *Torulaspora*.

(5) The preferential fermentation of fructose in a mixture of fructose and glucose, and marked resistance to preservatives. Only three of the eleven species of *Zygosaccharomyces* and *Torulaspora* have been investigated for the first property; of these *Zygosacch. rouxii* and *Zygosacch. bailii* are fructophilic whereas *T'spora delbrueckii* is not (Peynaud & Domercq, 1955). Resistance to preservatives alone does not seem to be an adequate criterion to discriminate between the genera. Both resistant and sensitive species occur in the genus *Issatchenkia, I. orientalis (Candida krusei)* grows in the presence of 1.5% of acetic acid (Pitt, 1974) whereas *I. terricola* is sensitive to 1%.

The genus *Saccharomyces* sensu Lodder and Kreger-van Rij, encompassing both diploid and haploid species, is obviously heterogeneous as is evidenced by the very wide range of the base composition of the DNAs, the estimations of 25 S rRNA complementarity (Bicknell & Douglas, 1970) which suggest considerable evolutionary divergence among the species, and differences in the structure of the wall of the ascospore (Kreger-van Rij & Veenhuis, 1976b; Kreger-van Rij, 1979). Since it is still not possible to assign all the species unequivocally to one of the various genera that have been proposed for them they are provisionally classified as follows:

Saccharomyces containing the diploid species *Sacch. cerevisiae, Sacch. dairensis, Sacch. exiguus, Sacch. kluyveri, Sacch. servazzii* and *Sacch. unisporus*. *Sacch. telluris* is retained temporarily but because of its rough ascospores it probably does not belong here.

Torulaspora containing *T'spora delbrueckii, T'spora globosa* and *T'spora pretoriensis*. In sections of $KMnO_4$-fixed material observed in the electron microscope, the ascospore wall comprises a dark outer layer with under it two lighter layers which may be separated by a very thin dark layer (Kreger-van Rij & Veenhuis, 1976b). On germination the innermost light layer becomes the wall of the vegetative cell. Warts may be absent or present. Conjugation is predominantly between a cell and its bud.

Zygosaccharomyces containing *Zygosacch. bailii, Zygosacch. bisporus, Zygosach. cidri, Zygosacch. fermentati, Zygosacch. florentinus, Zygosacch. microellipsoides, Zygosacch. mrakii* and *Zygosacch. rouxii.* At present three groups can be recognized within this genus based on electron microscopic studies of the spore wall (Kreger-van Rij, 1979) but as half of the species have not yet been studied in sufficient detail to allow them to be assigned to one of these groups they are all retained in one genus.

The first group is the true *Zygosaccharomyces* represented by *Zygosacch. rouxii* which has the type species of the genus, *Zygosacch. barkeri,* among its synonyms. In sections in the electron microscope, the ascospores have a wall made up of a thin dark layer covering a broad light layer, the wall has dark protuberances. Structure and development of the spores are similar to those of the ascospores of *Debaryomyces* species (Kreger-van Rij, 1979). *Zygosacch. rouxii* has the Q-6 ubiquinone in its electron-transport system and it seems inadvisable to classify it with the *Debaryomyces* species which have the Q-9 ubiquinone (Yamada et al., 1976b). Conjugation is between either individual cells or a cell and its bud.

Zygosacch. bailii is the only species of the second group. The wall of the ascospores is smooth and comprises a dark layer over a broad light layer (Kreger-van Rij, 1979). This light layer plus a new layer which is deposited beneath it when the ascospore germinates become the wall of the vegetative cell. Conjugation is between individual cells.

In the third group are *Zygosacch. florentinus* and *Zygosacch. fermentati.* The wall of the ascospores, in sections in the electron microscope, consists of a dark layer which covers a light greyish layer with under it, and sometimes separated from it by a dark line, a very light layer. On germination the light innermost layer forms the new wall of the cell (Kreger-van Rij, 1979). Conjugation is predominantly between individual cells. The ascospores of these species resemble those of *Sacch. cerevisiae* in the structure of the wall (Kreger-van Rij, 1978b). However, the results of Bicknell and Douglas (1970) suggest that the relationship between *Sacch. cerevisiae* and *Zygosacch. florentinus* is very distant. The level of rRNA complementarity between these two species was found to be 85% and suggests a wider evolutionary divergence between *Zygosacch. florentinus* and *Sacch. cerevisiae* than between *Zygosacch. rouxii, Sacch. telluris* and some *Kluyveromyces* species and *Sacch. cerevisiae* where complementarity ranged between 87 and 96%.

Van der Walt (1978) described the genus *Pachytichospora* for *Saccharomyces transvaalensis.*

Cultures of the following species were not available in time to be included: *Saccharomyces astigiensis, Sacch. abuliensis, Sacch. hispalensis* and *Sacch. albasitensis* (Santa María, 1978).

Genus 23. *Saccharomycodes* Hansen

by M.W. Miller and H.J. Phaff

a. DIAGNOSIS OF THE GENUS

Large, diploid, lemon-shaped (apiculate) or elongate cells, which reproduce by bipolar budding on a very broad base (bud-fission). Pseudomycelium, if present, is poorly developed.

The spores are spheroidal with a smooth wall; a narrow ledge may be visible. Asci usually contain four spores, occurring in two pairs; upon germination members of each pair ordinarily conjugate. In some strains spores may germinate without conjugation.

In liquid media a sediment and a ring are formed.

Sugars are fermented; nitrate is not assimilated.

b. TYPE SPECIES OF THE GENUS

Saccharomycodes ludwigii Hansen

c. SPECIES ACCEPTED IN THE GENUS

1. *Saccharomycodes ludwigii* Hansen (1889)

e. SYSTEMATIC DISCUSSION OF THE SPECIES

1. *SACCHAROMYCODES LUDWIGII* HANSEN (1904)

Synonyms

Saccharomyces ludwigii Hansen (1889)
Saccharomycodes ludwigii Hansen var. *vini* Kroemer et Heinrich (1922)
Saccharomycodes bisporus Castelli (1942)
Saënkia bispora (Castelli) Kudriavzev (1960)

Standard description of *Saccharomycodes ludwigii*

Growth in malt extract: After 3 days at 25°C the cells are lemon-shaped with blunt tips, sausage-shaped, curved, broad-oval, or elongated with a swelling in the

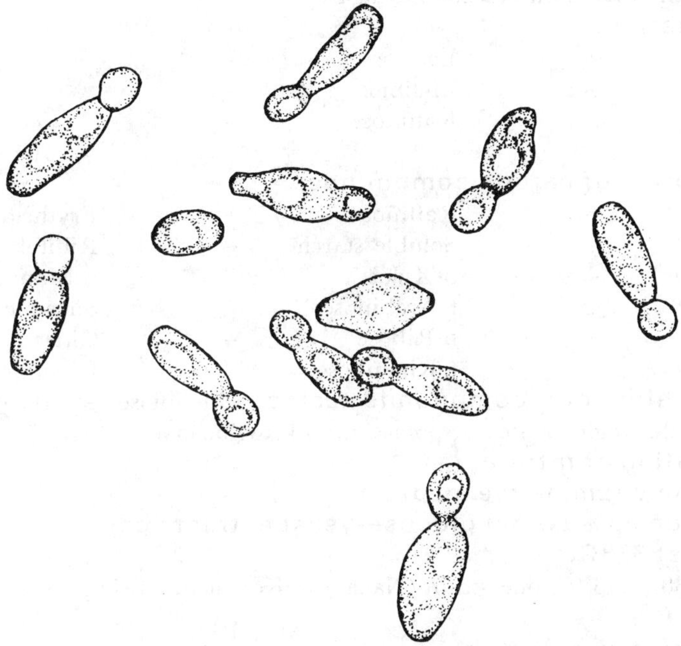

Fig. 100 – *S'codes ludwigii*
After 3 days in malt extract (From 'The Yeasts', ed. II, 1970)

middle, (4–7) × (8–23) μm, and occasionally longer, single or in pairs and sometimes in groups of three. Reproduction occurs by bipolar budding on a broad base, followed by fission (bud-fission). Cells in the process of budding resemble bowling pins (Fig. 100). A sediment and a slight ring are present. After approximately one month a sediment and a ring in evidence.

Growth on malt agar: After 3 weeks the streak culture is cream colored to tan, nearly smooth, semi-glossy, low convex with an irregular border.

Slide culture on potato agar: Aerobically a few short chains of cells. Under the coverslip the pseudomycelium is somewhat better developed consisting of branched chains of elongated cells resembling the 'Mycocandida' type.

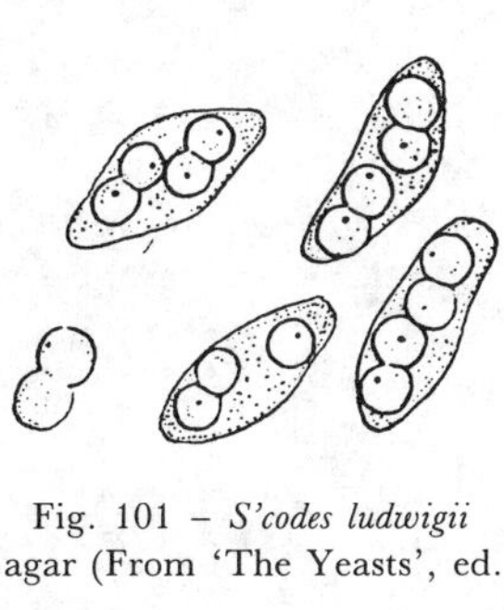

Fig. 101 – *S'codes ludwigii*
On malt agar (From 'The Yeasts', ed. I, 1952)

Formation of ascospores: The spores are spheroidal, smooth with a narrow subequatorial ledge which is visible in sections in the electron microscope (Kreger-van Rij, 1969b); usually there are four spores per ascus in two groups of two (Fig. 101). Ascus walls do not rupture or lyse when the spores are mature. For germination of spores see section a., Diagnosis of the genus.

Sporulation is good on Gorodkowa agar.

Fermentation:

Glucose	+	Lactose	–
Galactose	–	Melibiose	–
Sucrose	+	Raffinose	+
Maltose	–		

Assimilation of carbon compounds:

Galactose	–	Raffinose	+	Erythritol	–
Sucrose	+	Soluble starch	–	Ribitol	–
Maltose	–	D-Xylose	–	D-Mannitol	–
Cellobiose	+	L-Arabinose	–	Succinic acid	–
Trehalose	–	D-Ribose	–	Citric acid	–
Lactose	–	L-Rhamnose	–	Inositol	–

Additional carbon compounds tested: Melibiose –, glucono-δ-lactone –, calcium 2-ketogluconate –, potassium 5-ketogluconate –

Assimilation of nitrate: –

Growth in vitamin-free medium: –

Growth on 50% (w/w) glucose-yeast extract agar: –

Growth at 37°C: v

G + C: 38.3 mol%, one strain (Nakase & Komagata, 1971g)

Origin of the strains studied

Carlsberg Laboratory, Copenhagen (2 diploids, 3 short-celled haploids, 3 long-celled haploids and one strain of *Saccharomycodes bisporus* Castelli); Kluyver, Delft (one diploid strain); CBS (one strain of *Saccharomycodes bisporus*); slime fluxes *Quercus gambelii* (1), *Quercus prinus* (= *Quercus michauxii*) (1).

Type strain: CBS 821 received from Kluyver.

Genus 24. *Saccharomycopsis* Schiönning

by N.J.W. Kreger-van Rij

a. DIAGNOSIS OF THE GENUS

Abundant development of true mycelium with blastospores; in addition, budding yeast cells and pseudomycelium. Arthrospores may occur. The septa of the hyphae have plasmodesmata, with the exception of *S. lipolytica.* *

Asci are formed on the mycelial hyphae. Ascospores are spherical, hat- or Saturn-shaped, smooth or warty; the ledge may be irregular.

b. TYPE SPECIES OF THE GENUS

Saccharomycopsis capsularis Schiönning

c. SPECIES ACCEPTED IN THE GENUS

1. *Saccharomycopsis capsularis* Schiönning (1903)
2. *Saccharomycopsis crataegensis* Kurtzman et Wickerham (1973)
3. *Saccharomycopsis fibuligera* (Lindner) Klöcker (Lindner, 1907)
4. *Saccharomycopsis lipolytica* (Wickerham, Kurtzman et Herman) Yarrow (Wickerham et al., 1970b)
5. *Saccharomycopsis malanga* (Dwidjoseputro) Kurtzman, Vesonder et Smiley (in Dwidjoseputro & Wolf, 1970)
6. *Saccharomycopsis synnaedendra* Scott et van der Walt (van der Walt & Scott, 1971d)
7. *Saccharomycopsis vini* (Kreger-van Rij) van der Walt et Scott (Kreger-van Rij, 1964a)

d. KEY TO THE SPECIES OF THE GENUS

1a	Fermentation absent	2
b	Glucose fermented	4

* See note on p. 413.

2a Starch assimilated
S. synnaedendra .... p. 410
b Starch not assimilated 3

3a Erythritol assimilated
S. lipolytica .... p. 406
b Erythritol not assimilated 7

4a Starch assimilated 5
b Starch not assimilated 7

5a Sucrose assimilated
S. fibuligera .... p. 404
b Sucrose not assimilated 6

6a Ascospores spherical without or with an indistinct ledge
S. capsularis .... p. 401
b Ascospores with a distinct ledge
S. malanga .... p. 409

7a Xylose assimilated
S. crataegensis .... p. 403
b Xylose not assimilated
S. vini .... p. 411

Table 21

Key characters of the species in the genus *Saccharomycopsis*

	Fermentation			Assimilation					Shape of the spores	homo hetero	37°C
	G	Su	Ma	Su	Ma	Er	St	X			
S. capsularis	+ s	–	+ s	–	+	+	+	–	spherical with or without ledge	homo	+w/–
S. fibuligera	+ w,s	+ w,s	+ w,s	+	+	v	+	–	hat-shaped	homo	+
S. malanga	+ s	–	+vw/–	–	+	v	+	–(+)	hat-shaped	homo	+
S. synnaedendra	–			–	+	+	+	v	hat-shaped	homo	–
S. vini	+ w,s	+w,s/–	–	+(–)	v	–	–	–	Saturn-shaped	homo	–
S. crataegensis	+ vw/–	–	–	–	–	–	–	+	elongate with ledge	hetero	–
S. lipolytica	–			–	–	+	–	–	spherical, hat- or crescent-shaped	hetero	–

G = glucose, Su = sucrose, Ma = maltose, Er = erythritol, St = soluble starch, X = D-xylose, homo/hetero = homo- or heterothallic.

e. SYSTEMATIC DISCUSSION OF THE SPECIES

1. *SACCHAROMYCOPSIS CAPSULARIS* SCHIÖNNING (1903)

Synonyms

Endomyces capsularis (Schiönning) Guilliermond (1909)
Williopsis capsularis (Schiönning) Zender (1925)
Endomycopsis capsularis (Schiönning) Dekker (Stelling-Dekker, 1931)
Prosaccharomyces capsularis (Schiönning) Novák et Zsolt (1961)

Standard description of *Saccharomycopsis capsularis*

Growth in malt extract: After 2 days at 25°C budding cells are oval to elongate, (2.5–7) × (7–23) μm, and longer (Fig. 102). Buds are formed on a narrow or a broad base. Hyphae with cross walls may be present in addition to pseudomycelium. A sediment and dull, white islets or a pellicle are formed. After one month at 17°C a sediment and a thick, dull, wrinkled pellicle are present.

Growth on malt agar: After one month at 17°C the streak culture is white, yellowish-white or brown, dull, tough, coarsely or delicately wrinkled, partly hairy.

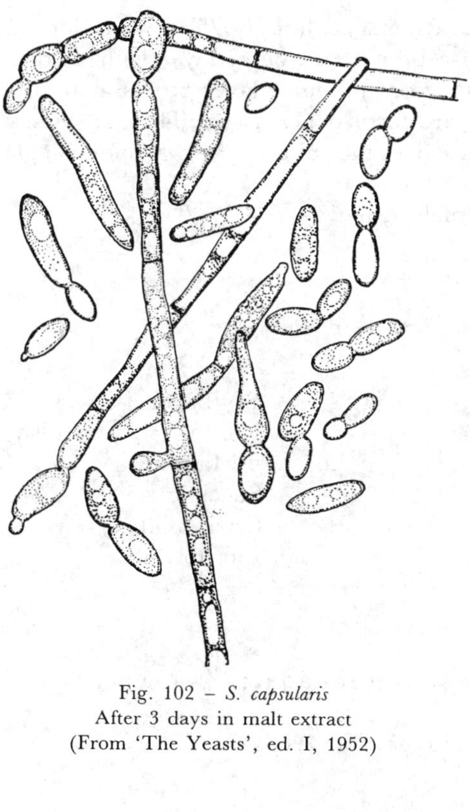

Fig. 102 – *S. capsularis*
After 3 days in malt extract
(From 'The Yeasts', ed. I, 1952)

Slide cultures on potato- and corn meal agar: True, branched mycelium is formed. Blastospores are oval to elongate; they occur in chains at the end of the hyphae, or singly or in short chains along the hyphae (Fig. 103). Pseudomycelium may also be present. The septa have plasmodesmata.

Formation of ascospores: Conjugation between two yeast cells or blastospores, or between hyphal cells precedes ascus formation (Kreger-van Rij & Veenhuis, 1975b). After fusion, a new wall is formed between the two cells. Both cells may become asci, or buds are formed on them which turn into asci. The asci are situated at the end of the hyphae, occasionally in chains, or are intercalary. The spores are spherical or slightly flattened globes, and may have a subequatorial ledge which is not always distinctly visible with the light microscope (Fig. 104). The spores of one of the strains studied (CBS 5638) had a small ledge during their develop-

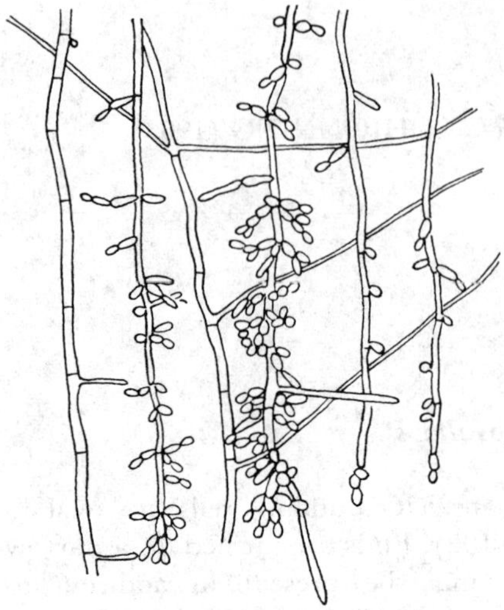

Fig. 103 – *S. capsularis*
After 10 days on potato agar
(From 'The Yeasts', ed. II, 1970)

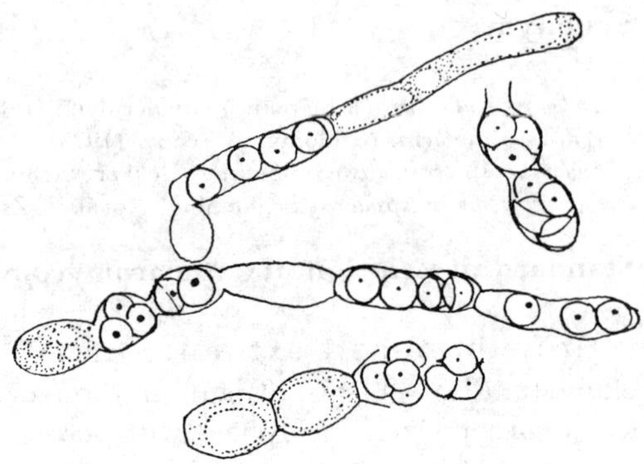

Fig. 104 – *S. capsularis*
After 14 days on V8 agar
(From 'The Yeasts', ed. II, 1970)

ment which disappeared later on (Kreger-van Rij & Veenhuis, 1975b). In sections in the electron microscope, the spore wall has a broad dark outer layer with a crack (Kawakami & Nehira, 1958b). Upon germination the outer layer bursts along the crack. Two to four, generally four, spores are formed per ascus. They are easily liberated from it. The presence of many spores gives the culture a brown color. Single spores gave cultures which sporulated again.

Spores were observed on V8-, YM- and malt agar.

Fermentation:

Glucose	+ s	Maltose	+ s
Galactose	–	Lactose	–
Sucrose	–		

Assimilation of carbon compounds:

Galactose	–	Raffinose	–	Erythritol	+
Sucrose	–	Soluble starch	+	Ribitol	+ (w)
Maltose	+	D-Xylose	–	D-Mannitol	+
Cellobiose	+	L-Arabinose	–	Succinic acid	+
Trehalose	+	D-Ribose	v	Citric acid	v
Lactose	–	L-Rhamnose	–	Inositol	–

Splitting of arbutin: +

Assimilation of nitrate: –

Growth in vitamin-free medium: –

Growth on 50% (w/w) glucose-yeast extract agar: + (vw) or –

Growth at 37°C: + w or –

G + C: 40 mol%, one strain (Nakase & Komagata, 1968a)

Origin of the strains studied

Unknown (3), pollen-pabulum of *Xylocopa caffra* in South Africa (1).

Type strain: CBS 2519, received in Delft from Schnegg (Germany).

2. *SACCHAROMYCOPSIS CRATAEGENSIS* KURTZMAN ET WICKERHAM (1973)

Standard description of *Saccharomycopsis crataegensis*

Growth in malt extract: After 2 days at 25°C loose yeast cells are short-oval to long-oval, (2–3.5) × (3.5–10) μm, single or in pairs. They may be very scarce. Besides, septate hyphae and arthrospores are present. A thin sediment and a thick dry wrinkled pellicle are formed. After one month at 20°C a pellicle and a sediment are present.

Growth on malt agar: After one month at 20°C the streak culture is white and powdery, dull, tough and slightly bumpy. The edge is fringed with mycelium.

Slide cultures on potato- and corn meal agar: Septate branched mycelium is abundantly formed. The cross walls have plasmodesmata. The hyphae may split up into arthrospores. Short-oval to long-oval blastospores occur along the hyphae, often on small projections. Aerial mycelium may be present. Pseudomycelium is occasionally present.

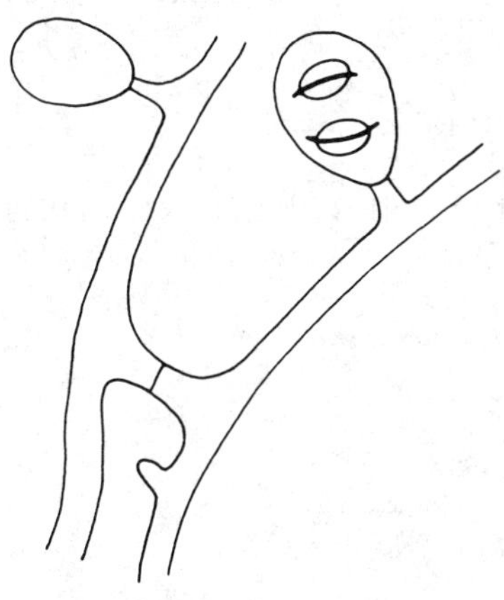

Fig. 105 – *S. crataegensis*
CBS 6447 × CBS 6448. Conjugated hyphal cells and asci on potato agar

Formation of ascospores: Conjugation between hyphal cells of opposite mating type precedes ascus formation. After fusion, the connected cells are separated again by a cross wall. One or both cells may directly change into asci, or buds are formed on them, sometimes on projections, which become asci. Kurtzman and Wickerham (1973) who studied the ascospores with the scanning electron microscope describe them as: elongate, ellipsoidal, frequently slightly flattened with a median longitudinal ledge and wartlike protuberances on the surface. The ledge is distinct under the light microscope (Fig. 105). Two spores are formed per ascus. They may come free after a long time. The presence of many spores gives the culture a red-brown appearance.

Conjugation and sporulation were observed in a mixture of cells of opposite mating type on YM agar.

Fermentation:

Glucose	+ vw or –	Maltose	–
Galactose	–	Lactose	–
Sucrose	–		

Assimilation of carbon compounds:

Galactose	–	Raffinose	–	Erythritol	–
Sucrose	–	Soluble starch	–	Ribitol	v
Maltose	–	D-Xylose	+	D-Mannitol	+
Cellobiose	–	L-Arabinose	+ w or –	Succinic acid	+
Trehalose	v	D-Ribose	+	Citric acid	–
Lactose	–	L-Rhamnose	–	Inositol	–

Splitting of arbutin: –

Assimilation of nitrate: –

Growth in vitamin-free medium: –

Growth on 50% (w/w) glucose-yeast extract agar: –

Growth at 37°C: –

Origin of the strains studied

Fallen hawthorne (*Crataegus* sp.) fruits (2).

Type strain: CBS 6447 (NRRL Y-5902); isotype: CBS 6448 (NRRL Y-5904); the two strains are of opposite mating type.

3. *SACCHAROMYCOPSIS FIBULIGERA* (LINDNER) KLÖCKER (1924)

Synonyms

Endomyces fibuligera Lindner (1907)
Endomycopsis fibuligera (Lindner) Dekker (Stelling-Dekker, 1931)
Pichia fibuligera (Lindner) Boidin, Pignal, Lehodey, Vey et Abadie (1964)
Endomyces lindneri Saito (1913)
Saccharomycopsis lindneri (Saito) Klöcker (1924)
Endomycopsis fibuligera (Lindner) Dekker var. *lindneri* Dekker (Stelling-Dekker, 1931)
Endomyces hordei Saito (1914)
Saccharomycopsis hordei (Saito) Klöcker (1924)
Endomycopsis fibuligera (Lindner) Dekker var. *hordei* (Saito) Dekker (Stelling-Dekker, 1931)
Candida lactosa Dwidjoseputro et Wolf (1970) (see also Kurtzman et al., 1973b)

Standard description of *Saccharomycopsis fibuligera*

Growth in malt extract: After 2 days at 25°C budding yeast cells are short-oval, oval or long-oval, (4–8) × (6–18) μm; branched septate mycelium is also present (Fig.106). The ratio between yeast cells and hyphae varies with the strains. Buds are on a narrow or broad base. A loose sediment is formed. After one month at 17°C a thick, tough, yellow mass is present.

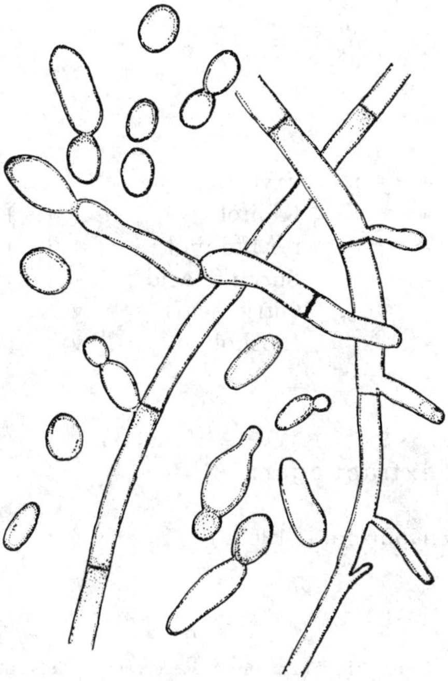

Fig. 106 – *S. fibuligera*
After 3 days in malt extract
(From 'The Yeasts', ed. I, 1952)

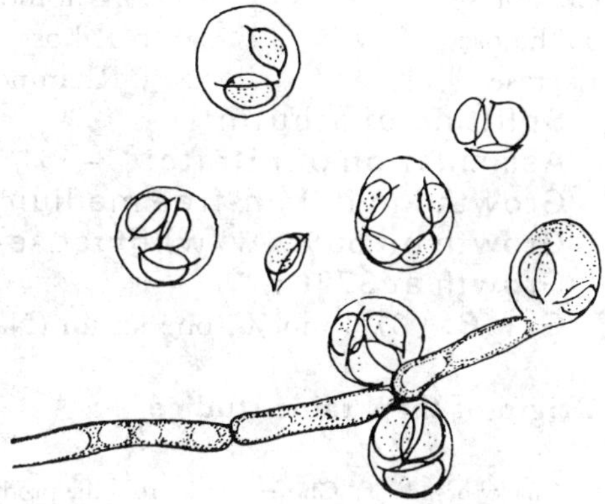

Fig. 107 – *S. fibuligera*
After 3 weeks in water
(From 'The Yeasts', ed. I, 1952)

Growth on malt agar: After one month at 17°C the streak culture is yellow-brown, tough, raised and more or less wrinkled, shiny or partly or entirely hairy.

Slide cultures on potato- and corn meal agar: Branched, septate mycelium is formed, occasionally breaking up into arthrospores. Spherical or oval blastospores are present at the end of the hyphae or along the threads, often in clusters or small chains on side branches. Pseudomycelium may also be present. The septa have plasmodesmata.

Formation of ascospores: Two adjacent cells in a hypha may fuse via protuberances near the cross wall. After fusion a new wall is formed to separate the cells. Both may turn into asci, or buds are formed on them which become asci. The asci are spherical or oval; they are situated at the end of the mycelial hyphae or alongside them, often in clusters. The asci may become detached from the hyphae. The spores are hat-shaped; two to four are formed per ascus (Fig. 107). They are easily liberated. The presence of many spores gives the culture a brown appearance. Single spores give cultures which sporulate again.

Spores were observed on potato- and corn meal agar. Material from a young malt agar culture brought into water may also sporulate.

Fermentation:

Glucose	+ w or s	Maltose	+ w or s
Galactose	–	Lactose	–
Sucrose	+ w or s	Raffinose	v

Assimilation of carbon compounds:

Galactose	–	Raffinose	+ (–)	Erythritol	v
Sucrose	+	Soluble starch	+	Ribitol	+ (–)
Maltose	+	D-Xylose	–	D-Mannitol	+ (–)
Cellobiose	+	L-Arabinose	–	Succinic acid	+ (–)
Trehalose	v	D-Ribose	–	Citric acid	v
Lactose	–	L-Rhamnose	–	Inositol	v

Splitting of arbutin: +

Assimilation of nitrate: –

Growth in vitamin-free medium: –

Growth on 50% (w/w) glucose-yeast extract agar: –

Growth at 37°C: +

G + C: 37.5 mol%, one strain (Nakase & Komagata, 1968a)

Origin of the strains studied

'Chalky bread' (1), Chinese yeast from the production of 'Hoang chui' (1), compressed yeast (1), leaven for the preparation of corn meal bread (1), ragi (3), tape (2), macaroni (1), unknown (7).

Type strain: CBS 2521, the original strain isolated by Lindner from 'chalky bread'.

4. *SACCHAROMYCOPSIS LIPOLYTICA* (WICKERHAM, KURTZMAN ET HERMAN) YARROW (1972a)

Imperfect state: *Candida lipolytica* (Harrison) Diddens et Lodder

Synonyms

Endomycopsis lipolytica Wickerham, Kurtzman et Herman (1970b)
Mycotorula lipolytica Harrison (1928)
Candida lipolytica (Harrison) Diddens et Lodder (1942)
Azymoprocandida lipolytica (Harrison) Novák et Zsolt (1961)
Monilia cornealis Nannizzi (in Bencini & Federici, 1928)
Proteomyces cornealis (Nannizzi) Dodge (1935)
Candida olea van Rij et Verona (1949)
Candida paralipolytica Yamada et Ota (1963)
Candida lipolytica (Harrison) Diddens et Lodder var. *thermotolerans* Blagodatskaja et Kocková-Kratochvílová (1973)
Candida pseudolipolytica Blagodatskaja et Kocková-Kratochvílová (1973)

Standard description of *Saccharomycopsis lipolytica*

Growth in malt extract: After 3 days at 25°C cells are short-oval to elongate, (2–4.5) × (4–22) μm. Budding is on a narrow or a broad base. Besides, pseudomycelium, true mycelium and arthrospores may be present. A sediment and a pellicle may be formed. After one month at 17°C a sediment and a pellicle are present.

Growth on malt agar: After one month at 17°C the streak culture is cream-colored, moist or mat, soft or tough, delicately wrinkled or with coarse folds.

Slide cultures on potato- and corn meal agar: Abundant formation of pseudomycelium and branched, septate, true mycelium (Fig. 108). Arthrospores may be present. Blastospores are arranged in small chains or in verticils on the hyphae. The septa have no plasmodesmata or dolipore, but a closure line, i.e. the remnants of plasmalemma after centripetal closure during development of the septum, or a narrow central pore without Woronin bodies. The edge of the septum around the pore is not tapered.

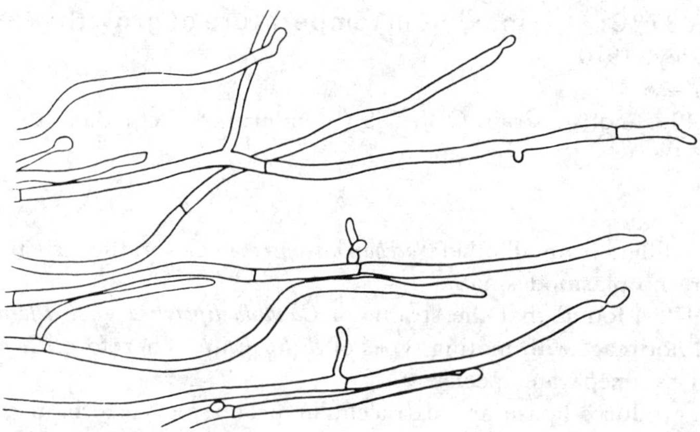

Fig. 108 – *S. lipolytica*
Dalmau plate culture on corn meal agar (From 'The Yeasts', ed. II, 1970)

Formation of ascospores: Fusion of yeast cells of opposite mating type precedes ascus formation. The zygote grows out to mycelium and asci are formed at the end of the hyphae or on short side branches. They are spherical to oval and contain one to four ascospores which are easily liberated. According to Wickerham et al. (1970a), very seldom fusion of yeast cells directly leads to ascus formation without the formation of hyphae. These authors describe the shape of the spores as follows: spherical with a rough or smooth wall; hat-shaped with a narrow or broad ledge; with a ledge grown out at one side which looks like a protuberance; and crescents with protuberances at one or both sides. Weber (1979) examined the ultrastructure of ascospore formation.

Spores were observed on YM agar in a diploid strain (CBS 6124), and after mixing strains of opposite mating type on this medium and on V8 agar.

Fermentation: –

Assimilation of carbon compounds:

Galactose	– (+)	Raffinose	–	Erythritol	+
Sucrose	–	Soluble starch	–	Ribitol	– (+)
Maltose	–	D-Xylose	–	D-Mannitol	+
Cellobiose	–	L-Arabinose	–	Succinic acid	+
Trehalose	–	D-Ribose	– (+)	Citric acid	+
Lactose	–	L-Rhamnose	–	Inositol	–

Additional carbon compounds tested: *n*-Hexadecane +, *n*-decane + (Markovetz & Kallio, 1964; Scheda & Bos, 1966)

Splitting of arbutin: –

Assimilation of nitrate: –

Growth in vitamin-free medium: –

Growth on 50% (w/w) glucose-yeast extract agar: –

Growth at 37°C: –; **maximum temperature of growth:** 33–37°C (van Uden & Buckley, 1970)

Urease: +

G + C: 49.6 mol%, strain CBS 599 (Stenderup & Leth Bak, 1968)

Comments

*S. lipolytica** differs from all other *Saccharomycopsis* species in the structure of the septum: there are no plasmodesmata.

Yarrow (1972a) found that the strains of *Candida lipolytica* var. *deformans (Candida deformans)* did not react with mating types of *S. lipolytica.* Therefore, *Candida deformans* is maintained as a separate species.

S. lipolytica produces lipase and extracellular protease (Wickerham et al., 1970a).

Origin of the strains studied

Material from corn-processing plant (2), stale margarine (2), olives (2), human cornea (1), soil (4), lung tissue (2), cow (1), unknown (1).

Type strain: CBS 6124 (NRRL YB-423), a bisexual strain isolated from steep water in a corn-processing plant.

5. *SACCHAROMYCOPSIS MALANGA* (DWIDJOSEPUTRO) KURTZMAN, VESONDER ET SMILEY (1974b)

Synonym

Hansenula malanga Dwidjoseputro (Dwidjoseputro & Wolf, 1970)

*See note on p. 413.

Standard description of *Saccharomycopsis malanga*

Growth in malt extract: After 2 days at 25°C budding yeast cells which are spherical, oval or elongate and measure (3–6) × (6–12) μm, and septate hyphae with a diameter of 2–4.5 μm are present. The ratio between yeast cells and hyphae varies with the strains; in old strains very few yeast cells were found. A loose sediment is formed. After one month at 20°C a thick, lumpy sediment and, occasionally, fragments of a ring are present.

Growth on malt agar: After one month at 20°C the streak culture is cream colored, tough, dull or dull-shiny and delicately wrinkled. The edge is fringed with mycelium.

Slide cultures on potato- and corn meal agar: Septate branched mycelium with spherical, oval, pip-shaped or elongate blastospores is abundantly formed. The blastospores occur at the end of the hyphae or alongside them. The septa have plasmodesmata. Pseudomycelium may be formed.

Formation of ascospores: Conjugation between hyphal cells or blastospores on the same or in different hyphae precedes ascus formation. Spherical asci are formed on the zygote with two hat-shaped spores. The spores are liberated from the ascus. Single spores gave cultures which sporulated again (Kurtzman et al., 1974b).

Spores were observed on YM-, malt- and corn meal agar, and in cells grown in malt extract and afterwards incubated in water.

Fermentation:

Glucose	+ s	Maltose	+ vw or –
Galactose	–	Lactose	–
Sucrose	–		

Assimilation of carbon compounds: (9 strains, Kurtzman et al., 1974b)

Galactose	–	Raffinose	–	Erythritol	v
Sucrose	–	Soluble starch	+	Ribitol	+
Maltose	+	D-Xylose	– (+)	D-Mannitol	+
Cellobiose	+	L-Arabinose	–	Succinic acid	+
Trehalose	v	D-Ribose	+	Citric acid	–
Lactose	–	L-Rhamnose	–	Inositol	–

Splitting of arbutin: +

Assimilation of nitrate: –

Growth in vitamin-free medium: –

Growth on 50% (w/w) glucose-yeast extract agar: + vw or –

Growth at 37°C: +

Comments

Kurtzman et al. (1974b) found that *S. malanga* produced 3-D-hydroxypalmitic acid in liquid cultures in the form of needle-shaped crystals. *S. fibuligera* which shows some resemblance to *S. malanga* did not produce this compound.

Origin of the strains studied

Ragi-tape (1), Chinese yeast, a commercial starter for fermented rice (1).

Type strain: CBS 6267 (NRRL Y-7175), the strain originally described as *Hansenula malanga*.

6. *SACCHAROMYCOPSIS SYNNAEDENDRA* SCOTT ET VAN DER WALT (VAN DER WALT & SCOTT, 1971d)

Synonyms

Botryoascus synnaedendrus (Scott et van der Walt) von Arx (1972)
Pichia crossotarsi Batra (1971)
Pichia microspora Batra (1971)

Standard description of *Saccharomycopsis synnaedendra*

Growth in malt extract: After 3 days at 25°C yeast cells are oval to ellipsoidal, (1.5–3) × (2–5) μm; single or in pairs. Buds may be formed from a short neck on the mother cell; the latter may show multiple scars. Septate branched mycelium is also present; the hyphae have a diameter of 2–3 μm. The ratio between yeast cells and hyphae varies with the strains. Pseudomycelium may be present. A sediment, islets and an incomplete ring are formed. Growth is slow. After one month at 17°C a sediment and a thick wet pellicle which easily drops are present.

Growth on malt agar: After one month at 17°C the streak culture is white to yellowish-brown, dull, tough, raised and hairy. The margin is fringed with mycelium.

Slide cultures on potato- and corn meal agar: Branched, septate mycelium with spherical or oval blastospores is formed. It may split up into arthrospores. The hyphae may be dentate. The blastospores occur terminally or laterally on the hyphae, singly or in chains. Pseudomycelium may be present. The septa have plasmodesmata.

Formation of ascospores: Conjugation between hyphal cells and blastospores was observed. Probably, a wall is formed between the fused cells before they turn into asci and asci are formed on them as buds. The asci are spherical, oval or elongate; they contain four hat-shaped spores which are liberated (Fig. 109). The presence of many spores gives the culture a pink to reddish appearance.

Spores were observed on V8-, malt-, potato- and corn meal agar.

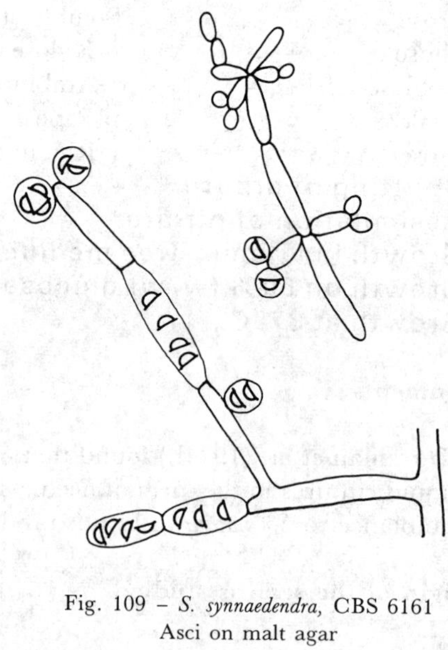

Fig. 109 – *S. synnaedendra*, CBS 6161
Asci on malt agar

Fermentation: –

Assimilation of carbon compounds:

Galactose	–	Raffinose	–	Erythritol	+
Sucrose	–	Soluble starch	+	Ribitol	+
Maltose	+	D-Xylose	v	D-Mannitol	+
Cellobiose	–	L-Arabinose	–	Succinic acid	+
Trehalose	v	D-Ribose	+	Citric acid	+
Lactose	–	L-Rhamnose	+	Inositol	+

Splitting of arbutin: –

Assimilation of nitrate: –

Growth in vitamin-free medium: –

Growth on 50% (w/w) glucose-yeast extract agar: +s

Growth at 37°C: –

Comments

Pichia crossotarsi and *Pichia microspora,* both described by Batra (1971) are considered to be synonyms of *Saccharomycopsis synnaedendra.* Batra found conjugation of ascospores before ascus formation in the strain of *Pichia microspora.* Von Arx (1972) identified a strain labelled *Pichia crossotarsi* with *Ambrosiozyma monospora.* However, the strain which I received from Dr. Batra in 1972 as *Pichia crossotarsi* agreed with the original description, which is distinctly different from *A. monospora.* Unfortunately, at present no living culture of *Pichia crossotarsi* as originally described is available and only fixed material for electron microscopy of the original strain shows that it is not identical with *A. monospora.*

Von Arx (1972) failed to find loose yeast cells in *S. synnaedendra* and, therefore, described a new genus *Botryoascus* for this species which I do not accept.

Origin of the strains studied

Tunnels of *Platypus externedentatus* in *Tabernaemontana ventricosa* in Natal (1), material from tunnels of *Xyleborus* species on *Rapanea melanophloeos* in Natal (1), *Crossotarsus wollastoni* infesting *Dipterocarpus indicus* in India (2).

Type strain: CBS 6161, from *Tabernaemontana ventricosa.*

7. *SACCHAROMYCOPSIS VINI* (KREGER-VAN RIJ) VAN DER WALT ET SCOTT (1971b)

Synonym

Endomycopsis vini Kreger-van Rij (1964a)

Standard description of *Saccharomycopsis vini*

Growth in malt extract: After 2 days at 25°C budding yeast cells which are oval, long-oval and elongate, (3–5) × (8–17) μm and branched mycelium, in varying ratios, are formed. In some strains no yeast cells are present in this medium. A sediment, a ring which easily drops and a thin, dull creeping pellicle or islets are formed. After one month at 17°C a flocculent sediment and a thick pellicle which easily drops and fills up the liquid, are present.

Growth on malt agar: After one month at 17°C the streak culture is yellowish-white, tough, flat or raised, more or less coarsely wrinkled, occasionally hairy. The edge is fringed with mycelium.

Slide cultures on potato- and corn meal agar: True branched mycelium is formed with spherical or oval blastospores in chains or in small clusters. In addition, pseudomycelium may be present. The septa contain plasmodesmata.

Formation of ascospores: Conjugation between adjacent cells in a hypha, between blastospores, or between cells in different hyphae occurs. After fusion, the two cells are separated again by a wall. The fused cells may turn into asci, or buds, sometimes on small stalks, are formed on them which become asci. The asci are spherical or oval; they contain one to four, generally four spores (Fig. 110). The spores are spherical with an equatorial ledge and a second, smaller subequatorial ledge (Kreger-van Rij, 1969a; Kurtzman & Wickerham, 1973). Single spores gave cultures which sporulated again (Kurtzman & Wickerham, 1973).

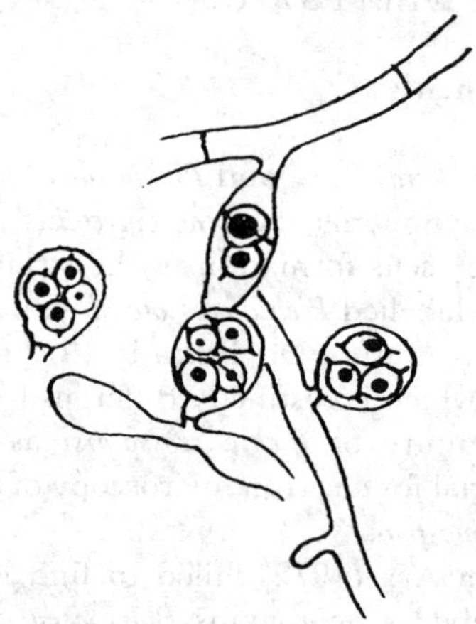

Fig. 110 – *S. vini*
After 6 days on potato agar
(From 'The Yeasts', ed. II, 1970)

Spores were observed on corn meal agar.

Fermentation:

Glucose	+ w or s	Maltose	–
Galactose	–	Lactose	–
Sucrose	+ w or s or –	Raffinose	v

Assimilation of carbon compounds:

Galactose	–	Raffinose	+ (–)	Erythritol	–
Sucrose	+ (–)	Soluble starch	–	Ribitol	v
Maltose	v	D-Xylose	–	D-Mannitol	+ (–)
Cellobiose	–	L-Arabinose	–	Succinic acid	v
Trehalose	+ (–)	D-Ribose	–	Citric acid	–
Lactose	–	L-Rhamnose	–	Inositol	–

Splitting of arbutin: –

Assimilation of nitrate: –

Growth in vitamin-free medium: –

Growth on 50% (w/w) glucose-yeast extract agar: v

Growth at 37°C: –

Origin of the strains studied

Grapes in Brazil (4) and Chile (2).

Type strain: CBS 4110, one of the strains from Chile.

f. COMMENTS ON THE GENUS

The genus *Saccharomycopsis,* invalidly named *Endomycopsis* by Stelling-Dekker (1931), is primarily characterized by the production of both mycelium and budding yeast cells. A number of features differentiate the genus from other ascomycetous yeast genera with hyphae and yeast cells.

1. The structure of the septum. All *Saccharomycopsis* species have plasmodesmata with the exception of *S. lipolytica* (Kreger-van Rij & Veenhuis, 1973a). Although the structure of the septum is probably an important taxonomic criterium, *S. lipolytica* is preliminarily maintained in the genus*. Species with a septum with a dolipore are classified in *Ambrosiozyma.* Mycelial species with a septum without a dolipore or plasmodesmata are classified in *Pichia, Hansenula* and *Arthroascus.*

2. The shape of the ascospores varies considerably in *Saccharomycopsis.* However, species in which the spores are sickle-shaped and have appendages at the tips are classified in the genus *Guilliermondella* and species with hemispherical spores with a very thick wall are in the genus *Stephanoascus.* The asci in the latter genus have a small vegetative cell at the top which is not found in *Saccharomycopsis* species.

The strains of all *Saccharomycopsis* species are nearly always haploid isolates. Asci are formed on the hyphae after cell fusion. In most species the diploid stage consists of the zygote and the buds formed on it which turn into asci.

Special cell contacts are found in *Saccharomycopsis crataegensis, S. malanga* and *S. synnaedendra* (Kreger-van Rij & Veenhuis, 1973b, 1976a). This contact between hyphal cells via denticles leads to submicroscopic changes in the wall of the contacted cell. Dead cells may be penetrated by denticles grown out to stalks.

---

*Van der Walt and von Arx (1980) have now classified *S. lipolytica* in a new genus *Yarrowia.*

Genus 25. *Schizosaccharomyces* Lindner

by D. Yarrow

a. DIAGNOSIS OF THE GENUS (SLOOFF, 1970b)

Cells globose to cylindrical, reproducing by fission; true septate hyphae may develop and break up into arthrospores.

Asci are produced usually by somatic conjugation of vegetative cells. Ascospores may be liberated at an early stage. Spores in the asci are globose or short ellipsoidal, seldom cylindrical. They develop by swelling; during maturation the shape may change to reniform (*Schizosaccharomyces japonicus* var. *versatilis*). The ascospores may infrequently appear to be slightly roughened when observed under the light microscope.

Fermentative ability present.

Nitrate not assimilated.

Positive starch test in Wickerham's liquid medium.

b. TYPE SPECIES OF THE GENUS

Schizosaccharomyces pombe Lindner

c. SPECIES ACCEPTED IN THE GENUS

1. *Schizosaccharomyces japonicus* Yukawa et Maki (1931)
   Schizosaccharomyces japonicus Yukawa et Maki var. *japonicus*
   Schizosaccharomyces japonicus Yukawa et Maki var. *versatilis* (Wickerham et Duprat) Slooff (Wickerham & Duprat, 1945)
2. *Schizosaccharomyces malidevorans* Rankine et Fornachon (1964)
3. *Schizosaccharomyces octosporus* Beijerinck (1894)
4. *Schizosaccharomyces pombe* Lindner (1893)

d. KEY TO THE SPECIES OF THE GENUS

1a Usually four spores per ascus 2
b Usually six to eight spores per ascus 3

2a Maltose fermented
Schiz. pombe . . . . p. 421
b Maltose not fermented
Schiz. malidevorans . . . . p. 418

3a Raffinose and melibiose fermented, true septate hyphae present
Schiz. japonicus . . . . p. 415
b Raffinose and melibiose not fermented, no true septate hyphae
Schiz. octosporus . . . . p. 419

e. SYSTEMATIC DISCUSSION OF THE SPECIES

1. SCHIZOSACCHAROMYCES JAPONICUS YUKAWA ET MAKI (1931)

This species has two varieties:
Schizosaccharomyces japonicus Yukawa et Maki var. *japonicus*
Schizosaccharomyces japonicus Yukawa et Maki var. *versatilis* (Wickerham et Duprat) Slooff (1970b)

Synonym of the variety *japonicus*

Octosporomyces japonicus (Yukawa et Maki) Kudriavzev (1960)

Synonym of the variety *versatilis*

Schizosaccharomyces versatilis nom. nud. (Wickerham & Duprat, 1945)

Standard description of *Schizosaccharomyces japonicus* var. *japonicus* and var. *versatilis*

Growth in malt extract: After 3 days at 25°C the cells are globose, subglobose, ellipsoidal or occasionally cylindrical (5.0–9.0) × (6.0–14.0) μm, single, in pairs or small groups (Fig. 111). Ascospores may be present. A sediment has been formed. After one month at 20°C a sediment is present.

Growth on malt agar: After one month at 20°C the streak is greyish to light brown, glistening, slightly raised in the center with irregular striations, the margin is fringed with hyphae.

Slide cultures on potato- and corn meal agar: True septate hyphae and arthrospores are formed (Fig. 112). Ascospores may be present.

Formation of ascospores: Conjugation between vegetative cells generally precedes the formation of asci containing six to eight, sometimes less, globose to ellipsoidal ascospores. The asci are evanescent, the released ascospores may cohere in chains or clumps. The mature ascospores of the variety *versatilis* become reniform to allantoid (Fig. 114), and give an amyloid reaction with Lugol's iodine; those of the variety *japonicus* do not change shape (Fig. 113) nor stain with Lugol's iodine.

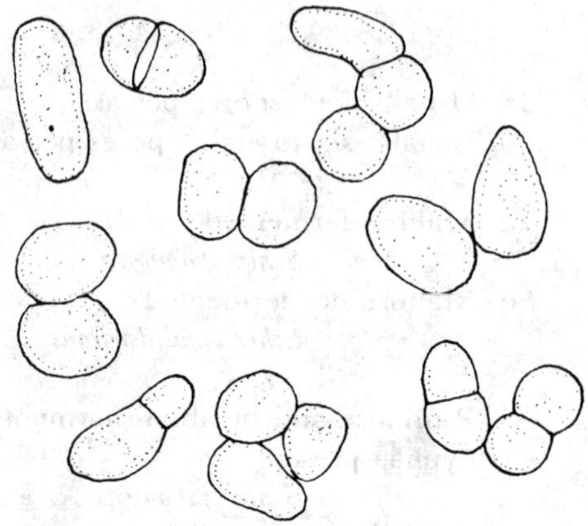

Fig. 111 – *Schiz. japonicus* var. *japonicus*
After 2 days in malt extract
(From 'The Yeasts', ed. II, 1970)

Sporulation was observed on malt-, potato- and corn meal agar, and in malt extract.

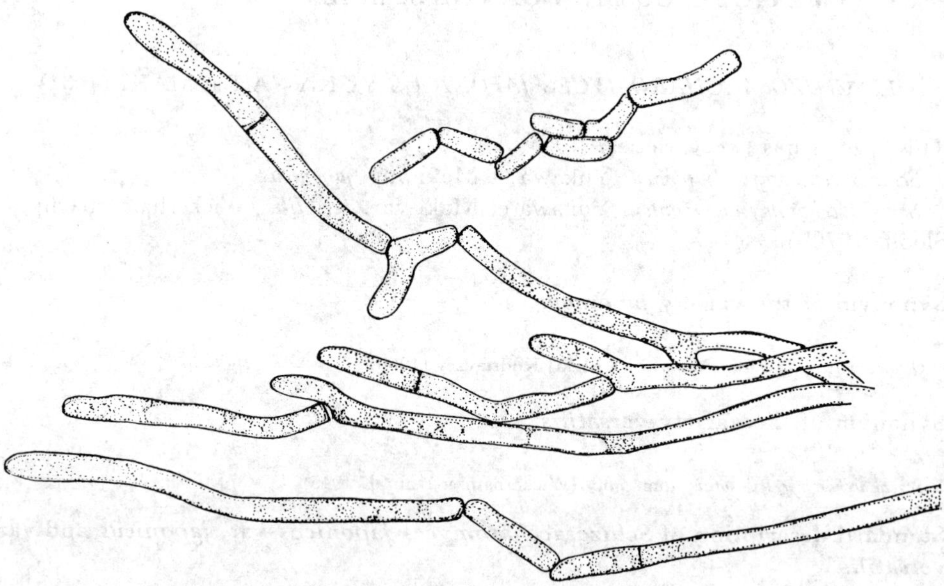

Fig. 112 – *Schiz. japonicus* var. *japonicus*
After 7 days on potato agar (From 'The Yeasts', ed. I, 1952)

Fermentation:

Glucose	+	Lactose	–
Galactose	–	Melibiose	+
Sucrose	+	Raffinose	+ 2/3
Maltose	+		

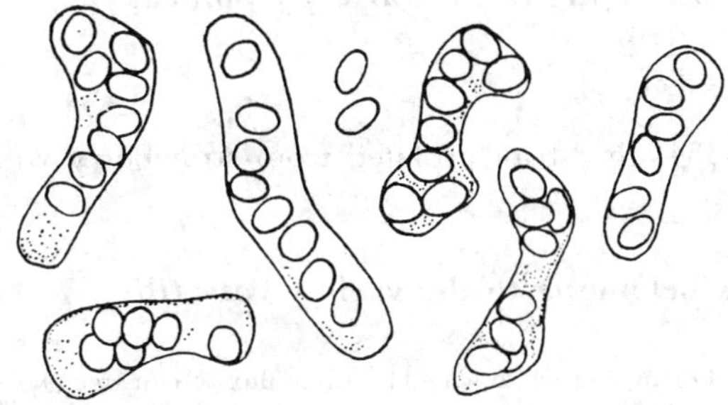

Fig. 113 – *Schiz. japonicus* var. *japonicus*
On malt agar (From 'The Yeasts', ed. I, 1952)

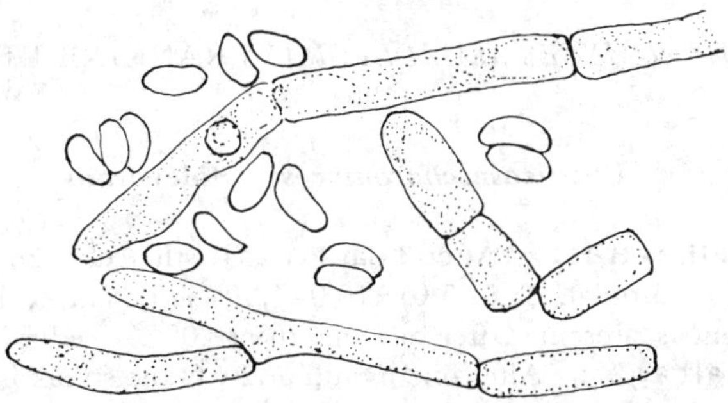

Fig. 114 – *Schiz. japonicus* var. *versatilis*
After 7 days on potato agar (From 'The Yeasts', ed. II, 1970)

Assimilation of carbon compounds:

Galactose	–	Raffinose	+	Erythritol	–
Sucrose	+	Soluble starch	–	Ribitol	–
Maltose	+	D-Xylose	–	D-Mannitol	–
Cellobiose	–	L-Arabinose	–	Succinic acid	–
Trehalose	–	D-Ribose	–	Citric acid	–
Lactose	–	L-Rhamnose	–	Inositol	–

Assimilation of nitrate: –
Growth in vitamin-free medium: –
Growth on 50% (w/w) glucose-yeast extract agar: –
Growth at 37°C: +

Comments

The only differences between the two varieties are the shape and iodine reaction of the mature ascospores.

Bulder (1963) reported that the strains of this species are respiratory deficient.

Origin of the strain belonging to the variety *japonicus*

Strawberry wine (1).

Type strain: CBS 354, the strain isolated from strawberry wine by Yukawa and Maki (1931).

Origin of the strains belonging to the variety *versatilis*

Canned grape juice (5), Portuguese white wine (1), slime flux of elm tree (1).

Type strain: CBS 103, isolated from canned grape juice by Wickerham and Duprat (1945).

2. *SCHIZOSACCHAROMYCES MALIDEVORANS* RANKINE ET FORNACHON (1964)

Standard description of *Schizosaccharomyces malidevorans*

Growth in malt extract: After 3 days at 25°C the cells are subglobose, ellipsoidal or oblong ellipsoidal, (3.5–5.0) × (6.0–12.0) μm, single, in pairs or small groups. A sediment is present. After one month at 20°C a sediment is present.

Growth on malt agar: After one month at 20°C the streak is cream to greyish brown, dull, only slightly raised, the margin is sinuous to lobate.

Slide cultures on potato- and corn meal agar: No hyphae are formed. Ascospores were observed.

Formation of ascospores: Conjugation of vegetative cells generally precedes ascus formation. The asci contain two to four globose to ellipsoidal ascospores and are evanescent; the loose ascospores may cohere in small groups.

Ascospores were observed on potato- and corn meal agar.

Fermentation:

Glucose	+	Lactose	–
Galactose	–	Melibiose	–
Sucrose	+	Raffinose	+ 1/3
Maltose	–		

Assimilation of carbon compounds:

Galactose	–	Raffinose	+	Erythritol	–
Sucrose	+	Soluble starch	–	Ribitol	–
Maltose	–	D-Xylose	–	D-Mannitol	–
Cellobiose	–	L-Arabinose	–	Succinic acid	–
Trehalose	–	D-Ribose	–	Citric acid	–
Lactose	–	L-Rhamnose	–	Inositol	–

Assimilation of nitrate: –

Growth in vitamin-free medium: –

Growth on 50% (w/w) glucose-yeast extract agar: +

Growth at 37°C: +

Comments

This species is able to ferment malic acid very strongly (Rankine & Fornachon, 1964; Slooff, 1970b).

Origin of the strain studied

Grapes (1).

Type strain: CBS 5557, the strain isolated from Listan grapes by Rankine and Fornachon was designated as the type by Slooff (1970b).

3. *SCHIZOSACCHAROMYCES OCTOSPORUS* BEIJERINCK (1894)

Synonyms

Octosporomyces octosporus (Beijerinck) Kudriavzev (1960)
Schizosaccharomyces slooffiae Kumbhojkar (1972)

Standard description of *Schizosaccharomyces octosporus*

Growth in malt extract: After 3 days at 25°C the cells are subglobose (4.5–10.0) μm to cylindrical (4.5–7.0) × (9.0–22.0) μm, single, in pairs or small groups (Fig. 115). A sediment is present. After one month at 20°C a sediment is present.

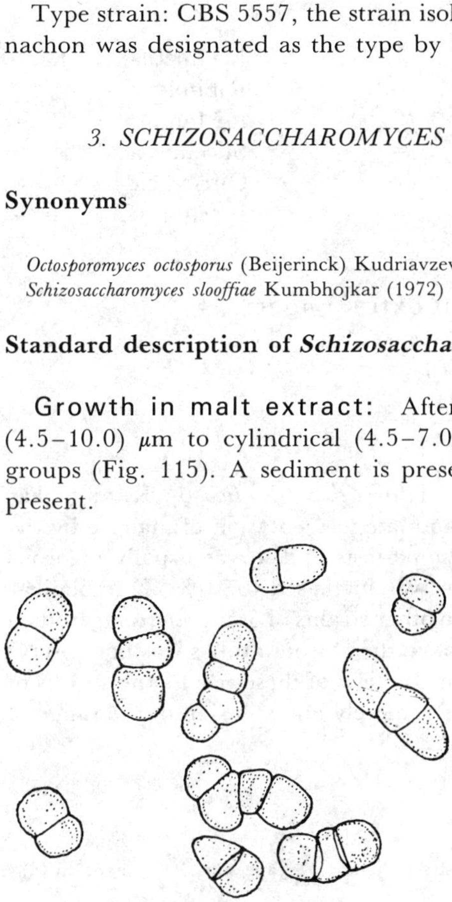

Fig. 115 – *Schiz. octosporus*
After 3 days in malt extract
(From 'The Yeasts', ed. II, 1970)

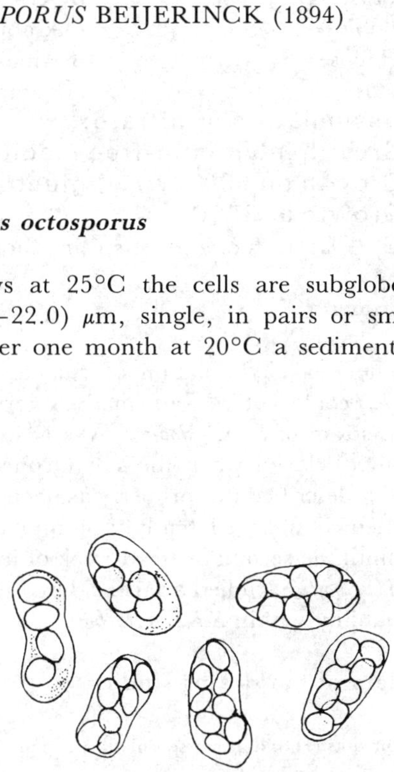

Fig. 116 – *Schiz. octosporus*
After 6 days on malt agar
(From 'The Yeasts', ed. I, 1952)

Growth on malt agar: After one month at 20°C the streak is grey to greyish brown, dull, raised irregularly and striated; the margin is entire, sinuous or lobate.

Slide cultures on potato- and corn meal agar: Hyphae were not observed. Ascospores may be present.

Formation of ascospores: Conjugation of cells usually precedes the formation of asci containing six to eight, sometimes less, globose to short ellipsoidal ascospores (Fig. 116). Conjugation may also occur between ascospores.

Sporulation was observed on malt- and potato agar.

Fermentation:

Glucose	+	Lactose	–
Galactose	–	Melibiose	–
Sucrose	– (+)	Raffinose	–
Maltose	+ or s (–)		

Assimilation of carbon compounds:

Galactose	–	Raffinose	–	Erythritol	–
Sucrose	v	Soluble starch	–	Ribitol	–
Maltose	+ (–)	D-Xylose	–	D-Mannitol	–
Cellobiose	–	L-Arabinose	–	Succinic acid	–
Trehalose	v	D-Ribose	–	Citric acid	–
Lactose	–	L-Rhamnose	–	Inositol	–

Assimilation of nitrate: –

Growth in vitamin-free medium: –

Growth on 50% (w/w) glucose-yeast extract agar: +

Growth at 37°C: +

G + C: 42 mol%, one strain (Storck, 1966)

Comments

The weak assimilation of galactose reported for *Schiz. slooffiae* by Kumbhojkar (1972) could not be confirmed. A very weak and late fermentation of maltose by the type strain of *Schiz. slooffiae* was observed in some tests and it was usually vigorous though delayed when the sugar concentration was increased to 10%. Kumbhojkar (1972) described this property as negative. Two other strains of *Schiz. octosporus* studied fermented maltose latently, accumulation of gas in the Durham tubes was not observed until the second or third week of incubation. In view of these results the strains of *Schiz. slooffiae* studied are considered to represent merely an extension of the range of variability within *Schiz. octosporus*.

Origin of the strains studied

Currants (1) and a non-sporulating variant from this strain, honey (5), cane sugar (1), unknown (3).

Type strain: CBS 371, a strain isolated by Beijerinck from Greek currants, was chosen as the type by Lodder and Kreger-van Rij (1952).

4. *SCHIZOSACCHAROMYCES POMBE* LINDNER (1893)

Synonyms

Schizosaccharomyces vordermani Wehmer (1906)
Schizosaccharomyces mellacei Jörgensen (1909)
Schizosaccharomyces formosensis Nakazawa (1914)
Schizosaccharomyces formosensis Nakazawa var. *akoensis* Nakazawa (1914)
Schizosaccharomyces formosensis Nakazawa var. *tapaniensis* Nakazawa (1914)
Schizosaccharomyces santawensis Nakazawa (1914)
Schizosaccharomyces pinan Nakazawa (1919)
Schizosaccharomyces taito Nakazawa (1919)
Schizosaccharomyces liquefaciens Osterwalder (1924b)
Schizosaccharomyces acidodevoratus Tschalenko (1941)
Schizosaccharomyces pombe Lindner var. *acidodevoratus* (Tschalenko) Dittrich (1964)

Standard description of *Schizosaccharomyces pombe*

Growth in malt extract: After 3 days at 25°C the cells are globose, ellipsoidal or cylindrical (3.0–5.0) × (5.0–15.0–24.0) μm, single, in pairs or small groups (Fig. 117). A sediment is present. After one month at 20°C a sediment is present and sometimes a thin ring.

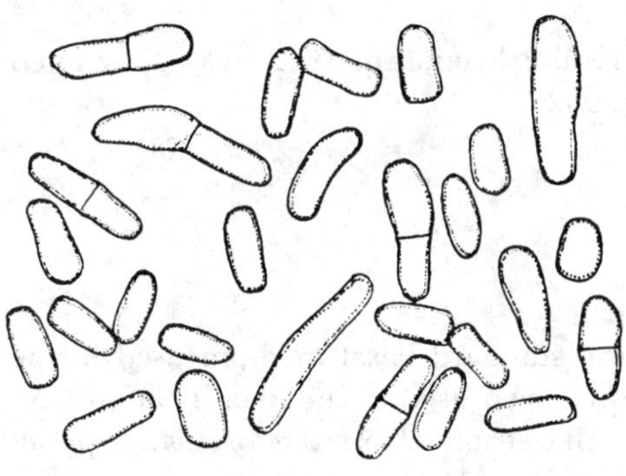

Fig. 117 – *Schiz. pombe*
After 3 days in malt extract
(From 'The Yeasts', ed. I, 1952)

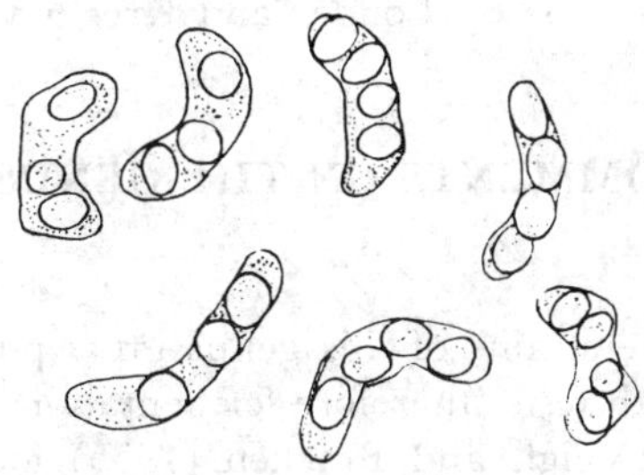

Fig. 118 – *Schiz. pombe*
On malt agar
(From 'The Yeasts', ed. I, 1952)

Growth on malt agar: After one month at 20°C the streak is brownish, dull or glistening, slightly raised, striated, the margin is entire or sinuous.

Slide cultures on potato- and corn meal agar: A few short chains of cells may be observed. Asci may be present.

Formation of ascospores: Conjugation of vegetative cells precedes the formation of evanescent asci containing two to four globose to ellipsoidal ascospores (Fig. 118). The released ascospores may cohere in small groups. Haploid heterothallic strains may be encountered.

Sporulation was observed on malt-, potato- and corn meal agar.

Fermentation:

Glucose	+	Lactose	–
Galactose	–	Melibiose	–
Sucrose	+	Raffinose	+ 1/3
Maltose	+		

Assimilation of carbon compounds:

Galactose	–	Raffinose	+	Erythritol	–
Sucrose	+	Soluble starch	–	Ribitol	–
Maltose	+	D-Xylose	–	D-Mannitol	–
Cellobiose	–	L-Arabinose	–	Succinic acid	–
Trehalose	–	D-Ribose	–	Citric acid	–
Lactose	–	L-Rhamnose	–	Inositol	–

Assimilation of nitrate: –

Growth in vitamin-free medium: –

Growth on 50% (w/w) glucose-yeast extract agar: +

Growth at 37°C: +

G + C: 42 mol%, one strain (Rost & Venner, 1964)

Origin of the strains studied

Arak (1), molasses (7), grape juice (1), brewer's yeast (1), cane sugar (1), palm wine (1), apple (1), Bantu beer (1), unknown (8).

Type strain: CBS 356, received from the Král collection in 1922, was designated as the type by Lodder and Kreger-van Rij (1952).

f. COMMENTS ON THE GENUS

The strains of this genus grow poorly in the standard basal medium used in the growth tests; therefore fermentation results have been used in the identification key.

McVeigh and Bracken (1955) found that the strain of *Schizosaccharomyces pombe* NRRL Y-658 required inositol, biotin, pantothenic acid, thiamine and pyridoxine for complete growth and that another strain required niacin also. They also found that adenine and uracil had a stimulatory effect on early growth. Johnson (1967) reported that in a strain of this species half the progeny of the final generation before the stationary phase lyse unless either 0.2 mg/ml asparagine or 5 mg/ml KH_2PO_4 or both are added to the standard basal medium. Northam and Norris (1951) found that a strain of *Schizosaccharomyces octosporus* grew satisfactorily in a mineral salts medium containing asparagine and ammonium sulphate provided that inositol, pantothenic acid, nicotinic acid, biotin, histidine, methionine and adenine were added. From these results it seems probable that better growth of the *Schizosaccharomyces* species might be obtained if the basal medium were supplemented with adenine, asparagine and additional vitamins.

Genus 26. *Schwanniomyces* Klöcker

by H.J. Phaff and M.W. Miller

a. DIAGNOSIS OF THE GENUS

Cells ovoidal, globose or egg-shaped, occasionally elongate or cylindrical. Vegetative reproduction by multilateral budding. A pseudomycelium is either not formed or it is rudimentary and not characteristic.

Cells are haploid. Before sporulation the nucleus divides, which is followed by fusion. The diploid nucleus ordinarily undergoes reduction division in a special open bud-like structure – the meiosis bud – and this is followed by the formation of one or, more rarely, two ascospores in the mother cell. Alternatively, the nucleus in the mother cell may divide, after which one of the daughter nuclei moves into a small bud which is temporarily separated from the mother cell by a septum. The septum then dissolves, the two nuclei fuse, and meiosis follows. Rarely, diploidization is accomplished by the fusion of two individual cells. The spores are globose, have a warty surface, an equatorial ledge and a pronounced lipid globule. They somewhat resemble the appearance of a walnut.

In liquid media a sediment and a ring are formed. The single species is fermentative.

Nitrate is not assimilated.

b. TYPE SPECIES OF THE GENUS

Schwanniomyces occidentalis Klöcker

c. SPECIES ACCEPTED IN THE GENUS

1. *Schwanniomyces occidentalis* Klöcker (1909a)
   Schwanniomyces occidentalis Klöcker var. *occidentalis*
   Schwanniomyces occidentalis Klöcker var. *persoonii* (van der Walt) Phaff et Miller var. nov. (van der Walt, 1962)

d. KEY TO THE TWO VARIETIES OF *SCHWANNIOMYCES OCCIDENTALIS*

1a Xylose and *n*-alkanes assimilated
Schw. occidentalis var. *occidentalis* .... p. 424
b Xylose and *n*-alkanes not assimilated
Schw. occidentalis var. *persoonii* .... p. 425

e. SYSTEMATIC DISCUSSION OF THE SPECIES

1. SCHWANNIOMYCES OCCIDENTALIS KLÖCKER (1909a)

This species has two varieties:
Schwanniomyces occidentalis Klöcker var. *occidentalis*
Schwanniomyces occidentalis Klöcker var. *persoonii* (van der Walt) Phaff et Miller var. nov.

Synonyms of the variety *occidentalis*

Schwanniomyces castellii Capriotti (1957)
Schwanniomyces alluvius Phaff, Miller et Cooke (1960)

Synonym of the variety *persoonii*

Schwanniomyces persoonii van der Walt (1962), Antonie van Leeuwenhoek *28,* p. 82 (basionym)

Standard description of *Schwanniomyces occidentalis* var. *occidentalis*

Growth in malt extract: After 3 days at room temperature the cells are ovoidal, globose, or sometimes egg-shaped, (3.5–8.0) × (4.5–9.0) μm, occasionally up to 11 μm long; single, in pairs or in short chains or clusters (Fig. 119). A thin ring and a little sediment are present. After 3 weeks a ring and a moderate amount of sediment have formed.

Growth on malt agar: The streak culture after one month is cream colored, smooth, semi-glossy; the texture is pasty to soft; the cross section is broadly convex and the border entire.

Slide culture on corn meal agar: Pseudomycelium is absent aerobically and anaerobically.

Formation of ascospores: The details of the sporulation process and spore morphology are given under the generic diagnosis. There is usually one spore per ascus, occasionally two (Fig. 120).

Sporulation is best after approximately 5–7 days on yeast autolysate agar with 2% glucose.

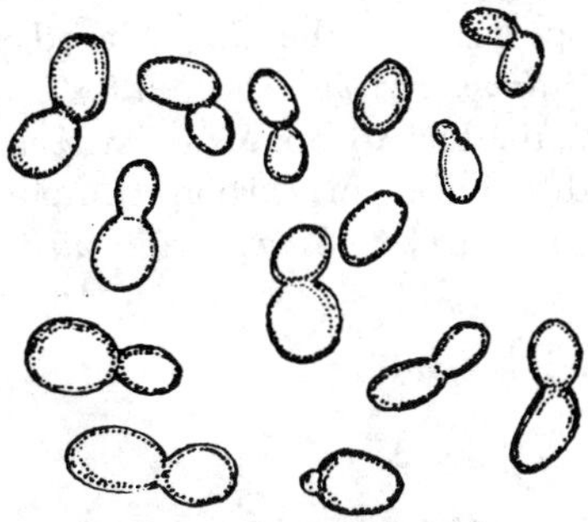

Fig. 119 – *Schw. occidentalis*
After 3 days in malt extract
(From 'The Yeasts', ed. I, 1952)

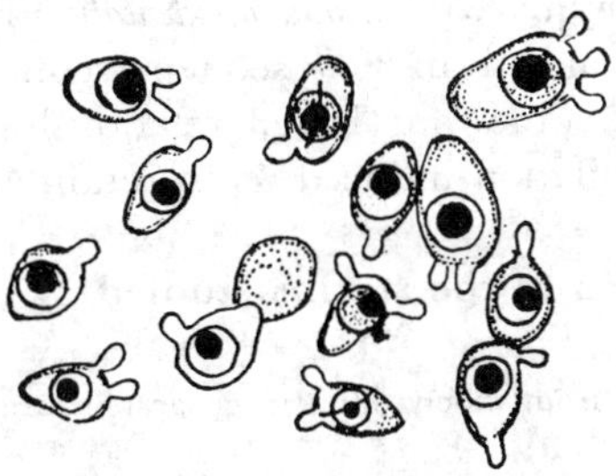

Fig. 120 – *Schw. occidentalis*
After 4 weeks on gypsum block
(From 'The Yeasts', ed. I, 1952)

Fermentation:

Glucose	+	Lactose	–
Galactose	+ s	Melibiose	– or + vw
Sucrose	+	Raffinose	+ or w
Maltose	+ s or –		

Assimilation of carbon compounds:

Galactose	+	Raffinose	+	Erythritol	–
Sucrose	+	Soluble starch	+	Ribitol	+ s or –
Maltose	+	D-Xylose	+	D-Mannitol	+ or s
Cellobiose	v	L-Arabinose	v	Succinic acid	+ s or –
Trehalose	+	D-Ribose	–	Citric acid	+ s or –
Lactose	– (+)	L-Rhamnose	–	Inositol	–

Additional carbon compound tested: *n*-Hexadecane +
Assimilation of nitrate: –
Growth in vitamin-free medium: –
Growth on 50% (w/w) glucose-yeast extract agar: –
Growth at 37°C: + w or –
G + C: 35.2–35.3 mol%, 3 strains (Price et al., 1978)

Supplementary description of *Schwanniomyces occidentalis* var. *persoonii*

Assimilation of carbon compounds:

Galactose	–
D-Xylose	–
n-Hexadecane	–

G + C: 35.2–35.4 mol%, 2 strains (Price et al., 1978)

Comments

Price et al. (1978) carried out DNA/DNA reassociation experiments among the various strains of *Schwanniomyces*. Polynucleotide sequence relationships between reference DNA from *Schw. occidentalis* and DNA from *Schw. castellii* and *Schw. alluvius* were at the 97% level or greater and the last two species were therefore placed into synonymy with *Schw. occidentalis*. The type strain of *Schw. persoonii* showed 83% and a second strain 80% sequence complementarity with the DNA from *Schw. occidentalis*. This somewhat lesser relationship and several differences in carbon compound assimilation induced us to retain *Schw. persoonii* as a variety of *Schw. occidentalis*.

Origin of the strains studied

Soil from widely different geographic areas (12).

Type strain of *Schw. occidentalis* var. *occidentalis*: CBS 1153, a strain received by the CBS from the Carlsberg Laboratory in 1924. It had been isolated by Klöcker from soil of the island of St. Thomas in the West Indies.

Type strain of *Schw. occidentalis* var. *persoonii*: CBS 2169, isolated by van der Walt from soil in the Transvaal, South Africa.

f. COMMENTS ON THE GENUS

The genus *Schwanniomyces* Klöcker can be distinguished from other ascomycetous yeast genera by the unique shape of the ascospores. Kurtzman et al. (1972) have illustrated the spore topography by scanning electron microscopy. Ferreira and Phaff (1959) described the life cycle of *Schwanniomyces* based on nuclear staining of cells in the process of meiosis. They presented evidence that meiosis took place in an open bud-like structure, the meiosis bud, and that the haploid nuclei moved back into the mother cell, where one, or more rarely two, developed into ascospores. Kreger-van Rij (1977a) restudied the life cycle and concluded that conjugation took place between a small daughter cell and the mother cell by lysis of the connecting septum and that meiosis probably occurred in the mother cell. It is possible that both types of life cycles mentioned above occur in *Schwanniomyces*. Kreger-van Rij (1977a) also determined by electron microscopy of ultrathin sections details of spore wall development and internal structure.

Genus 27. *Sporopachydermia* Rodrigues de Miranda

by L. Rodrigues de Miranda

a. DIAGNOSIS OF THE GENUS

Cells ovoidal, ellipsoidal to elongate, occasionally lunate. Reproduction by multilateral budding. Capsules are not formed. On solid medium the growth is butyrous, not mucous. Pseudomycelium or true mycelium are not formed.

Ascospores are round, oval or elliptical with a very thick cell wall without protuberances. In the ascus a very refractile substance is produced between the spores. The spores are liberated from the ascus at maturity and are kept together by that refractile material.

The ability to ferment sugars is lacking.

Inositol as sole carbon source is assimilated.

Nitrate is not assimilated, nitrite may be weakly assimilated.

A characteristic offensive smell is produced on various media.

Extracellular iodophilic amyloid material absent.

b. TYPE SPECIES OF THE GENUS

Sporopachydermia lactativora Rodrigues de Miranda

c. SPECIES ACCEPTED IN THE GENUS

1. *Sporopachydermia cereana* Rodrigues de Miranda (1978)
2. *Sporopachydermia lactativora* Rodrigues de Miranda (1978)

d. KEY TO THE SPECIES OF THE GENUS

1a Ascus formation preceded by conjugation of cells of opposite sexes, ascospores round, erythritol not assimilated
Sporopachydermia lactativora .... p. 429

b Ascus formation directly from diploid cells, ascospores long-ovoidal, erythritol assimilated
Sporopachydermia cereana .... p. 428

e. SYSTEMATIC DISCUSSION OF THE SPECIES

1. *SPOROPACHYDERMIA CEREANA* RODRIGUES DE MIRANDA (1978)

Synonym

Cryptococcus cereanus Phaff, Miller, Miranda, Reed et Starmer (1974)

Standard description of *Sporopachydermia cereana*

Growth in malt extract: After 3–5 days cells are ellipsoidal to elongate, occasionally lunate or curved, (2–4) × (4–13.5) μm, single, in pairs or in small groups of three or four cells. Budding occurs on broad shoulders. A slight ring and little sediment are formed. After one month there is a ring and a scant sediment but no pellicle.

Growth on malt agar: After one month at 20°C the streak culture is cream colored, smooth, glossy, butyrous with entire borders.

Slide culture on potato agar: A rudimentary pseudomycelium is sometimes formed under the coverslip, true mycelium absent.

Formation of ascospores: Vegetative cells are directly transferred into asci, especially the biggest cells which are often lunate or curved like sausages. The asci usually contain one to four long-ovoidal to ellipsoidal spores with a very thick smooth wall (Fig. 121). The spores are liberated from the ascus, that ruptures at maturation.

Spores were observed on Difco malt extract agar. Two strains were studied; ascospores were formed only by one strain.

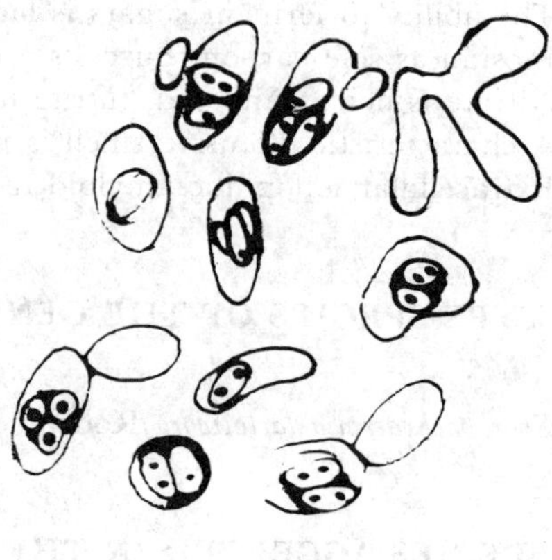

Fig. 121 – *Sporop. cereana* On Difco malt extract agar

Fermentation: –

Assimilation of carbon compounds:

Galactose	–	Raffinose	–	Erythritol	+
Sucrose	–	Soluble starch	+	Ribitol	+
Maltose	–	D-Xylose	+	D-Mannitol	+
Cellobiose	+	L-Arabinose	+ w or –	Succinic acid	+
Trehalose	+	D-Ribose	+	Citric acid	–
Lactose	–	L-Rhamnose	–	Inositol	+

Additional carbon compounds tested: Sodium glucuronate + w, glucono-δ-lactone –, melibiose –

Assimilation of nitrate: –

Assimilation of nitrite: + w

Growth in vitamin-free medium: –
Growth on 50% (w/w) glucose-yeast extract agar: –
Growth at 37°C: +
Starch formation: –
Production of offensively smelling material: +
G + C: 49.58 mol%, one strain (Phaff et al., 1974)

Origin of the strains studied

Rot pockets in cacti species of the genus *Cereus* (2).

Type strain: CBS 6644, isolated by Phaff et al. (1974) from *Cereus giganteus.*

2. *SPOROPACHYDERMIA LACTATIVORA* RODRIGUES DE MIRANDA (1978)

Synonym

Cryptococcus lactativorus Fell et Phaff (1967)

Standard description of *Sporopachydermia lactativora*

Growth in malt extract: After 5 days at 25°C cells are ovoidal, (2.0–3.5) × (3.5–5.5) μm, single, in pairs or in short chains. There is no ring or pellicle and very little sediment. After one month a moderate sediment is formed.

Growth on malt agar: After one month the streak is yellowish white to tan, smooth with faint transverse striations, glossy, butyrous with entire borders.

Slide culture on potato agar: Pseudomycelium or mycelium is not formed.

Formation of ascospores: Conjugation of cells of opposite mating type precedes ascus formation. The spores are round, with a very thick wall and an oil droplet inside; one to four per ascus (Fig. 122). The spore wall is smooth without protuberances; around the spores a highly refractile material is produced. After the liberation of the spores at maturation, the spores are kept together by this material.

Sporulation in a mixture of cells of opposite mating type on Difco malt extract agar was poor; cells of a diploid culture on malt extract agar sporulated more readily.

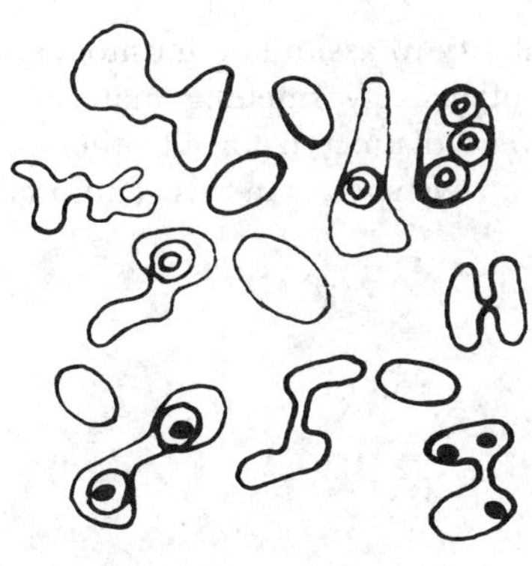

Fig. 122 – *Sporop. lactativora*
On Difco malt extract agar

Fermentation: –

Assimilation of carbon compounds:

Galactose	–	Raffinose	–	Erythritol	–
Sucrose	–	Soluble starch	–	Ribitol	+
Maltose	–	D-Xylose	+	D-Mannitol	–
Cellobiose	–	L-Arabinose	+	Succinic acid	+w
Trehalose	–	D-Ribose	+	Citric acid	–
Lactose	–	L-Rhamnose	–	Inositol	+

Additional carbon compounds tested: Sodium glucuronate –, glucono-δ-lactone –, melibiose –

Assimilation of nitrate: –

Assimilation of nitrite: +w or –

Growth in vitamin-free medium: –

Growth on 50% (w/w) glucose-yeast extract agar: –

Growth at 37°C: +

Starch formation: –

Production of offensively smelling material: +

G + C: 46.3 mol%, one strain (Phaff et al., 1974)

Origin of the strains studied

Antarctic seawater: CBS 5771 and 5772 (Fell & Phaff, 1967); waste lagoon of asphalt plant, Georgia, U.S.A: CBS 6058; mouth of a patient: CBS 6192, isolated in 1970. Strains CBS 5771, 5772 and 6058 are of mating type α, strain CBS 6192 is of mating type *a*.

Type strain: CBS 6989, a diploid strain, selected from a mixture of conjugating cells of strains CBS 5771, 5772 and 6192.

f. COMMENTS ON THE GENUS

The unusually thick wall of the ascospores, the ability to assimilate inositol as sole carbon source and the production of the typical offensively smelling material are characteristics which make the genus *Sporopachydermia* distinct from all other yeast genera. The difference in shape of the ascospores of the two species in this genus makes them easily distinguishable.

Genus 28. *Stephanoascus* Smith, van der Walt et Johannsen

by N.J.W. Kreger-van Rij

a. DIAGNOSIS OF THE GENUS

Formation of budding yeast cells, pseudomycelium and branched septate mycelium with blastospores. Hyphal septa with plasmodesmata.

Asci are formed after fusion of two hyphal cells which grow out to a short hypha with two cross walls. Between these a round ascus develops with a small apical cell attached to it. Immature ascospores are hat-shaped; when mature, they are hemispherical with a thick wall at the convex side.

b. TYPE SPECIES OF THE GENUS

Stephanoascus ciferrii Smith, van der Walt et Johannsen

c. SPECIES ACCEPTED IN THE GENUS

1. *Stephanoascus ciferrii* Smith, van der Walt et Johannsen (1976)

e. SYSTEMATIC DISCUSSION OF THE SPECIES

1. STEPHANOASCUS CIFERRII SMITH, VAN DER WALT ET JOHANNSEN (1976)

Imperfect state: *Candida ciferrii* Kreger-van Rij

Synonym

Candida ciferrii Kreger-van Rij (1965)

Standard description of *Stephanoascus ciferrii*

Growth in malt extract: After 2 days at 25°C budding cells are short-oval to long-oval (2.5–6) × (3–11) μm, often arranged in ramified chains (Fig. 123). True branched mycelium with oval blastospores may be present. A sediment and, occasionally, some islets are formed. After one month at 17°C a loose sediment, a broad ring and, occasionally, islets or a thick dull lumpy pellicle are present.

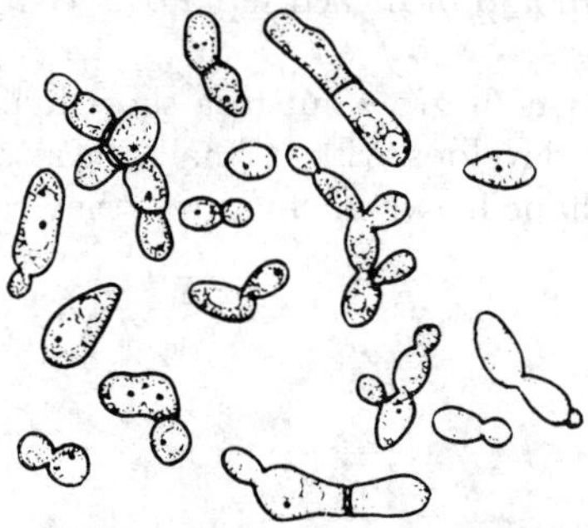

Fig. 123 – *Steph. ciferrii*
In glucose-yeast extract-peptone water
(From 'The Yeasts', ed. II, 1970)

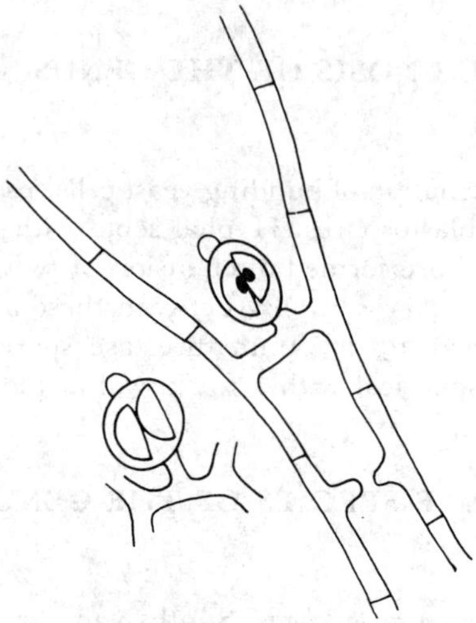

Fig. 124 – *Steph. ciferrii*
CBS 5295 × CBS 6699. Conjugating hyphal cells and asci on corn meal agar

Growth on malt agar: After one month at 17°C the streak culture is yellow to white, tough, raised, coarsely wrinkled and generally hairy.

Slide cultures on potato- and corn meal agar: Branched, septate mycelium is abundantly formed. Short-oval or long-oval blastospores in small chains are arranged alongside the hyphae. Buds arise on the hyphae often on short projections. Aerial mycelium may occur. Pseudomycelium may be present.

Formation of ascospores: Conjugation of hyphal cells of opposite mating type precedes ascus formation. The ascus is spherical and develops as described in the diagnosis of the genus. It contains one to four spores (Fig. 124). Immature spores have a narrow ledge at the base. During maturation, the spore wall extends considerably, but not at the base. Mature spores are hemispherical with a thick wall at the convex side and the ledge no longer visible with the light microscope (Kreger-van Rij, 1980, Fig. 1d).

Spores were observed on V8-, potato- and corn meal agar.

Fermentation: –

Assimilation of carbon compounds:

Galactose	+	Raffinose	+	Erythritol	+
Sucrose	+	Soluble starch	v	Ribitol	+
Maltose	+	D-Xylose	+	D-Mannitol	+

Cellobiose	+	L-Arabinose	+	Succinic acid	+
Trehalose	+	D-Ribose	+	Citric acid	+ or w
Lactose	–	L-Rhamnose	v	Inositol	+

Splitting of arbutin: +
Assimilation of nitrate: –
Growth in vitamin-free medium: –
Growth on 50% (w/w) glucose-yeast extract agar: –
Growth at 37°C: +

Origin of the strains studied

Neck of a cow (1), wood in a cow-shed (1), throat of a pig (1), bovine placenta (1), unknown human origin (1), soil near Pretoria, South Africa (1).

Type strain: Smith et al. (1976) indicated as holotype a dried sporulating mixture of the strains of opposite mating type CBS 6669 and CBS 5295. CBS 6699 and CBS 5295 are the living isotypes.

Genus 29. *Torulaspora* Lindner

by D. Yarrow

a. DIAGNOSIS OF THE GENUS

Vegetative reproduction by multilateral budding. Cells globose or ellipsoidal. Pseudohyphae may be formed but not septate hyphae.

The vegetative phase is predominantly haploid, conjugation occurring usually between a cell and its bud, less frequently between independent cells, just before the ascus is formed. Ascospores globose or ellipsoidal, with either a rough or a smooth wall, one to four per ascus. Asci persistent.

Vigorous fermentation.

Starch-like compounds not produced.

No growth with nitrate as sole source of nitrogen.

b. TYPE SPECIES OF THE GENUS

Torulaspora delbrueckii (Lindner) Lindner

c. SPECIES ACCEPTED IN THE GENUS

1. *Torulaspora delbrueckii* (Lindner) Lindner (Lindner, 1895)
2. *Torulaspora globosa* (Klöcker) van der Walt et Johannsen (Klöcker, 1909a)
3. *Torulaspora pretoriensis* (van der Walt et Tscheuschner) van der Walt et Johannsen (van der Walt & Tscheuschner, 1956c)

d. KEY TO THE SPECIES OF THE GENUS

1a Good growth at 37°C ... 2

b No growth at 37°C

T'spora delbrueckii .... p. 436

See Table 19, p. 381.

e. SYSTEMATIC DISCUSSION OF THE SPECIES

1. TORULASPORA DELBRUECKII (LINDNER) LINDNER (1904)

Imperfect state: *Candida colliculosa* (Hartmann) Meyer et Yarrow

Synonyms

Saccharomyces delbrueckii Lindner (1895)
Zygosaccharomyces delbrueckii (Lindner) Krumbholz (1933)
Debaryomyces delbrueckii (Lindner) Kudriavzev (1954)
Zymodebaryomyces delbrueckii (Lindner) Novák et Zsolt (1961)
Torula colliculosa Hartmann (1903)
Torulopsis colliculosa (Hartmann) Saccardo (1906)
Cryptococcus colliculosus (Hartmann) Skinner (1950)
Candida colliculosa (Hartmann) Meyer et Yarrow (Yarrow & Meyer, 1978)
Torulaspora rosei Guilliermond (1913)
Saccharomyces rosei (Guilliermond) Lodder et Kreger-van Rij (1952)
Debaryomyces rosei (Guilliermond) Kudriavzev (1954)
Zymodebaryomyces rosei (Guilliermond) Novák et Zsolt (1961)
Torulaspora fermentati Saito (1923)
Zygosaccharomyces fermentati (Saito) Krumbholz (1933) [non *Zygosaccharomyces fermentati* Naganishi (1928)]
Saccharomyces fermentati (Saito) Lodder et Kreger-van Rij (1952)
Zygosaccharomyces mongolicus Saito (1923)
Saccharomyces delbrueckii Lindner var. *mongolicus* (Saito) Lodder et Kreger-van Rij (1952)
Torulaspora mongolica (Saito) van der Walt et Johannsen (1975a)
Saccharomyces torulosus Osterwalder (1924a)
Saccharomyces chevalieri Guilliermond var. *torulosus* Dekker (Stelling-Dekker, 1931)
Zygosaccharomyces globiformis Kroemer et Krumbholz (1931)
Torulopsis stellata (Kroemer et Krumbholz) Lodder var. *cambresieri* Lodder et Kreger-van Rij (1952)
Torulaspora nilssoni Capriotti (1958b)
Saccharomyces nilssoni (Capriotti) Santa María (1963b)
Debaryomyces nilssoni (Capriotti) Kodama, Kyono, Naganishi et Takahara (1964c)
Schwanniomyces hominis Batista, Vieira et Coêlho (1959)
Debaryomyces dekkeri Mrak, Phaff, Vaughn et Hansen ex Kudriavzev (1960)
Zymodebaryomyces dekkeri (Mrak, Phaff, Vaughn et Hansen ex Kudriavzev) Novák et Zsolt (1961)
Saccharomyces inconspicuus van der Walt (1965a)
Torulaspora inconspicua (van der Walt) van der Walt et Johannsen (1975a)
Saccharomyces vafer van der Walt (1965a)
Torulaspora vafer (van der Walt) van der Walt et Johannsen (1975a)
Torulaspora benedictae Capriotti (1969)
Saccharomyces saitoanus van der Walt (1970e)
Saccharomyces microellipsodes Osterwalder var. *osmophilus* van der Walt (1970e)
Saccharomyces florenzani Balloni, Materassi et Margheri (1971)

Standard description of *Torulaspora delbrueckii*

Growth in malt extract: After 3 days at 25°C the cells are globose or subglobose, occasionally ellipsoidal, (2.5–6.5) × (2.5–7.0) μm, single or in pairs (Fig. 125). A sediment is present. After one month at 20°C a sediment and usually a ring are present.

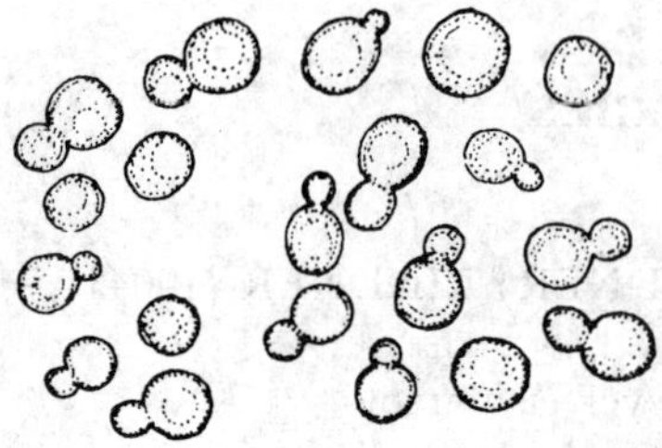

Fig. 125 – *T'spora delbrueckii*
After 3 days in malt extract
(From 'The Yeasts', ed. I, 1952)

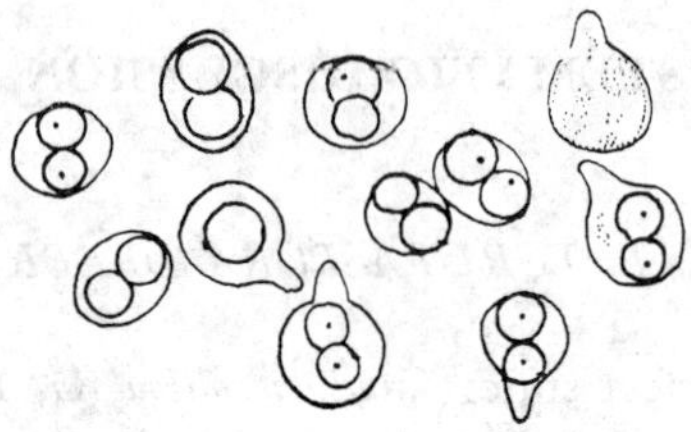

Fig. 126 – *T'spora delbrueckii*
After 6 days on Gorodkowa agar
(From 'The Yeasts', ed. I, 1952)

Growth on malt agar: After one month at 20°C the streak culture is off-white to slightly brownish, smooth and glossy, sometimes lightly striated, with an entire or undulating margin.

Slide cultures on corn meal- and morphology agar: Short chains or bushy clusters of cells may be formed.

Formation of ascospores: Conjugation between a cell and its bud usually precedes the formation of asci containing one or two, rarely up to four, usually distinctly rough ascospores. Cells with protuberances which may resemble conjugation tubes are also formed (Fig. 126).

Acetate-, YM-, V8- and Gorodkowa agar are suitable media for inducing sporulation.

Fermentation:

Glucose	+	Maltose	v
Galactose	v	Lactose	–
Sucrose	v		

Assimilation of carbon compounds:

Galactose	v	Raffinose	v	Erythritol	–
Sucrose	v	Soluble starch	–	Ribitol	– (+s)
Maltose	v	D-Xylose	– or + s	D-Mannitol	+ (–)
Cellobiose	–	L-Arabinose	–	Succinic acid	– (+)
Trehalose	+	D-Ribose	–	Citric acid	–
Lactose	–	L-Rhamnose	–	Inositol	–

Assimilation of nitrate: –

Assimilation of ethylamine · HCl: – (+)

Assimilation of cadaverine · 2 HCl: –

Growth in vitamin-free medium: + s or –

Growth on 50% (w/w) glucose-yeast extract agar: –

Growth at 37°C: –

Growth in the presence of 100 ppm of cycloheximide: –

G + C: 46.3 mol%, one strain (Phaff & Meyer, 1969); 42.2–43.4 mol%, 13 strains (Nakase & Komagata, 1971d); 44.6 mol%, one strain (Yarrow & Nakase, 1975); 43.4–43.9 mol%, 9 strains (Price et al., 1978); 42.9–43.9 mol%, 3 strains

Comments

DNA reassociation experiments have shown *Saccharomyces florenzani, Torulaspora fermentati, T'spora inconspicua, Sacch. microellipsoides* var. *osmophilus, T'spora rosei, Sacch. saitoanus, T'spora vafer* and *T'spora nilssoni* to be synonymous with *T'spora delbrueckii* (Price et al., 1978).

Some strains of *T'spora delbrueckii* can grow with nitrite as source of nitrogen. When tested on auxanogram plates very few strains of this species grow with ethylamine as source of nitrogen, none with cadaverine but all do so with L-lysine. The same pattern of growth was found for *T'spora pretoriensis* and *T'spora globosa.* All the species of *Zygosaccharomyces* grow with all three sources of nitrogen.

Origin of the strains studied

Rhagi (1), gum in sugar factory (1), raw cane sugar (1), grapes (6), grape juice (1), grape must (1), wine (1), kaoliang mash (1), Manchurian sorghum brandy (1), Indonesian sambal (1), dates (1), sour figs (1), soft drinks and fruit juices (12), liver or spleen of wild rodent (1), sheep's milk (2), soil (4), fermenting cucumber brine (2), Mexican colonche (1), unknown (20).

Type strain: CBS 1146 was designated as the type by van der Walt (1970e).

2. *TORULASPORA GLOBOSA* (KLÖCKER) VAN DER WALT ET JOHANNSEN (1975a)

Synonyms

Debaryomyces globosus Klöcker (1909a)
Zymodebaryomyces globosus (Klöcker) Novák et Zsolt (1961)
Saccharomyces kloeckerianus van der Walt (1971)

Standard description of *Torulaspora globosa*

Growth in malt extract: After 3 days at 25°C the cells are globose or subglobose, (3.0–6.0) × (3.0–6.5) μm, single, in pairs or small groups. A sediment is present. After one month at 20°C a sediment is present and usually an incomplete ring also.

Growth on malt agar: After one month at 20°C the streak culture is off-white to dark cream, smooth and glossy, with an entire or lobate margin.

Slide cultures on corn meal- and morphology agar: Pseudohyphae are not formed.

Formation of ascospores: Conjugation of individual cells or rarely of a cell and its bud precedes the formation of asci which contain usually two globose rough ascospores.

Sporulation was observed on YM agar either in the strains individually or when mixed with the compatible mating type.

Fermentation:

Glucose	+	Maltose	–
Galactose	–	Lactose	–
Sucrose	+		

Assimilation of carbon compounds:

Galactose	–	Raffinose	+	Erythritol	–
Sucrose	+	Soluble starch	–	Ribitol	–
Maltose	–	D-Xylose	–	D-Mannitol	v
Cellobiose	–	L-Arabinose	–	Succinic acid	–
Trehalose	– or + s	D-Ribose	–	Citric acid	–
Lactose	–	L-Rhamnose	–	Inositol	–

Assimilation of nitrate: –

Assimilation of ethylamine·HCl: –

Assimilation of cadaverine·2 HCl: –

Growth in vitamin-free medium: +

Growth on 50% (w/w) glucose-yeast extract agar: +

Growth at 37°C: +

Growth in the presence of 100 ppm of cycloheximide: +; **of 1000 ppm:** –

G + C: 45.4 mol%, one strain (Nakase & Komagata, 1971d); 47.0–47.5 mol%, 2 strains (Price et al., 1978)

Origin of the strains studied

Soil (3), human feces (1), feces of hippopotamus (1), unknown (3).

Type strain: CBS 764, received from the Carlsberg Laboratory in 1924 was designated as type by van der Walt (1970e) who by error cited the number of this strain as CBS 765.

3. *TORULASPORA PRETORIENSIS* (VAN DER WALT ET TSCHEUSCHNER) VAN DER WALT ET JOHANNSEN (1975a)

Synonyms

Saccharomyces pretoriensis van der Walt et Tscheuschner (1956c)
Torulaspora franciscae Capriotti (1958a)
Debaryomyces franciscae (Capriotti) Kodama, Kyono, Naganishi et Takahara (1964b)

Standard description of *Torulaspora pretoriensis*

Growth in malt extract: After 3 days at 25°C the cells are globose or subglobose, (3.0–6.0) × (3.5–7.0) μm, single or in pairs. A sediment is present. After one month at 20°C a ring and a sediment are present.

Growth on malt agar: After one month at 20°C the streak culture is cream, smooth and glossy, with an entire or undulating margin.

Slide cultures on corn meal- and morphology agar: Rudimentary pseudohyphae may be formed.

Formation of ascospores: Conjugation between a cell and its bud usually precedes the formation of asci containing one or two, occasionally up to four, ascospores. The spores are usually globose and may be rough or smooth. In one strain they are either ellipsoidal or somewhat cylindrical and curved.

Sporulation was observed on acetate- and YM agar.

Fermentation:

Glucose	+	Maltose	+ or s
Galactose	+ or s	Lactose	–
Sucrose	+		

Assimilation of carbon compounds:

Galactose	+	Raffinose	+	Erythritol	–
Sucrose	+	Soluble starch	–	Ribitol	–
Maltose	+	D-Xylose	v	D-Mannitol	+
Cellobiose	–	L-Arabinose	–	Succinic acid	–
Trehalose	+	D-Ribose	–	Citric acid	–
Lactose	–	L-Rhamnose	–	Inositol	–

Assimilation of nitrate: –

Assimilation of ethylamine · HCl: –

Assimilation of cadaverine · 2 HCl: –

Growth in vitamin-free medium: +

Growth on 50% (w/w) glucose-yeast extract agar: +

Growth at 37°C: +

Growth in the presence of 100 ppm of cycloheximide: – or + s; of 1000 ppm: –

G + C: 43.9 mol%, one strain (Phaff & Meyer, 1969); 45.6 mol%, one strain (Nakase & Komagata, 1971d); 46.3–46.4 mol%, 2 strains (Price et al., 1978)

Origin of the strains studied

Soil (4), sap of citrus tree (1).

Type strain: CBS 2187, isolated from soil in South Africa.

f. COMMENTS ON THE GENUS

See under *Saccharomyces,* section f., Comments on the genus, p. 393.

Genus 30. *Wickerhamia* Soneda

by H.J. Phaff and M.W. Miller

a. DIAGNOSIS OF THE GENUS

Cells are oval to elongate or apiculate, single, in pairs or very short chains, reproducing mainly by bipolar budding on a relatively broad base.

The ascospores are formed without prior conjugation of the vegetative cells. The spores are cap-shaped; the crown deflects to one side of a sinuous brim, giving the appearance of a sporting cap; usually only one or two spores per ascus but occasionally there may be many (up to 16 have been reported); the asci rupture when the spores are mature.

In liquid media a sediment and a ring are formed.

Sugars are fermented; nitrate is not assimilated; inositol and pantothenic acid are not required for growth.

b. TYPE SPECIES OF THE GENUS

Wickerhamia fluorescens Soneda

c. SPECIES ACCEPTED IN THE GENUS

1. *Wickerhamia fluorescens* Soneda (1960)

e. SYSTEMATIC DISCUSSION OF THE SPECIES

1. WICKERHAMIA FLUORESCENS SONEDA (1960)

Synonym

Kloeckera fluorescens Soneda (1959a)

Standard description of *Wickerhamia fluorescens*

Growth in malt extract: After 2 or 3 days at 25°C cells are apiculate with blunt ends, ovoidal, elongate (sometimes sausage-shaped or slightly curved), or in the shape of bowling pins, (4–7) × (8–15–22) μm. They occur singly, in pairs, and occasionally in groups of three; reproduction by bipolar budding on a broad base (Fig. 127); a sediment and a thin ring begin to form. After 3–4 weeks there are a sediment, ring, and sometimes a few islets, but no pellicle.

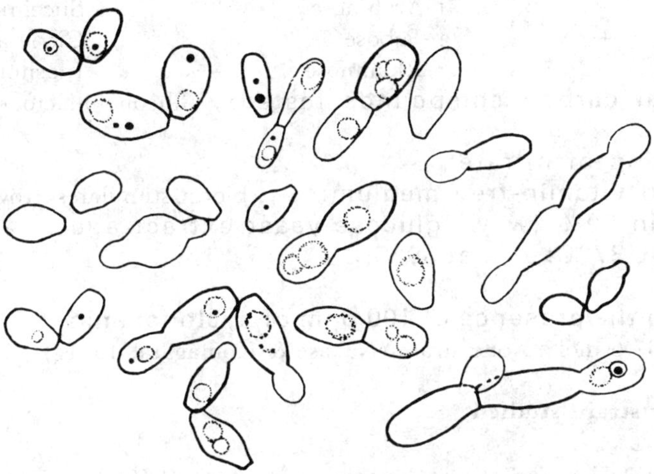

Fig. 127 – *W. fluorescens*
After 8 days in malt extract (From 'The Yeasts', ed. II, 1970)

Growth on malt agar: After 3–4 weeks the streak culture is grey to slightly yellowish, smooth, semiglossy, texture pasty, cross section low convex to flat, border entire. The medium may become yellowish due to excretion of riboflavin, which shows strong fluorescence under ultraviolet light.

Dalmau plate culture on potato agar: Pseudomycelium is not formed.

Formation of ascospores: Spores are formed without immediately preceding conjugation of vegetative cells. The ascospores resemble a sporting cap; the crown is deflected to one side of a sinuous brim; they contain a single, conspicuous refractile lipid globule (Fig. 128). In our study only one and rarely two spores per ascus have been observed, although Soneda reported up to 16 per ascus. Upon maturity asci rupture along an equatorial plane, releasing the spores.

Sporulation is best on corn meal agar, oat meal agar, or potato agar.

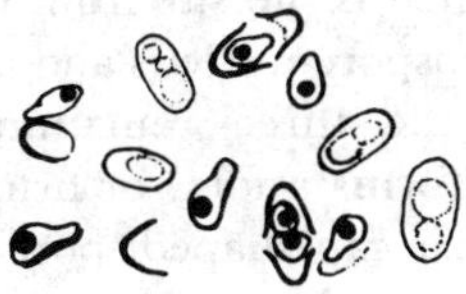

Fig. 128 – *W. fluorescens*. Asci and spores (From 'The Yeasts', ed. II, 1970)

Fermentation:

Glucose	+	Lactose	–
Galactose	+	Melibiose	–
Sucrose	+	Raffinose	+
Maltose	–	Trehalose	–

Assimilation of carbon compounds:

Galactose	+	Raffinose	+	Erythritol	–
Sucrose	+	Soluble starch	–	Ribitol	+
Maltose	–	D-Xylose	–	D-Mannitol	+
Cellobiose	–	L-Arabinose	–	Succinic acid	+
Trehalose	+ vw	D-Ribose	–	Citric acid	+
Lactose	–	L-Rhamnose	–	Inositol	–

Additional carbon compounds tested: Glucono-δ-lactone +, 2-keto-gluconate +

Assimilation of nitrate: –

Growth in vitamin-free medium: –; biotin stimulates growth

Growth on 50% (w/w) glucose-yeast extract agar: +

Growth at 37°C: –; at 30°C: +

Urease: –

Growth in the presence of 100 ppm of cycloheximide: +

G + C: 37.6 mol%, one strain (Nakase & Komagata, 1971g)

Origin of the strain studied

Only one strain of this species is known. It was isolated by Soneda (1959a) from dung of a wild squirrel in Japan.

Type strain: CBS 4565.

f. COMMENTS ON THE GENUS

Mitotic division of the rather large nucleus of *Wickerhamia fluorescens* under phase contrast and in stained preparations has been followed in detail by Matile et al. (1969). Meiosis and ascospore delimitation were studied by Rooney and Moens (1973a,b) with the use of electron microscopy of ultrathin sections. The nuclear behavior is similar to that occurring in *Saccharomyces cerevisiae* in that the nuclear envelope remains intact during the two meiotic divisions and a four-lobed nucleus is formed. One of the four nuclear lobes has a complex spindle-pole body, which is the site from where the prospore wall develops that eventually surrounds the ascospore nucleus and associated cytoplasm. With few exceptions (two-spored asci) the three remaining nuclei degenerate. The outer membrane of the prospore wall forms a fold, or brim, on the outside of the spore, ultimately resulting in an asymmetric cap-shaped spore. Because in the above-cited work and in our own laboratory multispored asci were never observed, it is possible that Soneda's (1960) observation of multispored asci has been based on viewing clumps or aggregates of numerous dehisced spores.

Genus 31. *Wickerhamiella* van der Walt

by J.P. van der Walt

a. DIAGNOSIS OF THE GENUS

Vegetative cells spheroidal, ovoid or ellipsoidal, reproducing by budding.

Asci conjugated, as a rule unispored, dehiscing terminally, ejecting the ascospores, forming persistent saccate structures. Ascospores oblong with obtuse ends, rugose.

Coenzyme Q-9 system present.

Nitrate utilized as sole source of nitrogen.

b. TYPE SPECIES OF THE GENUS

Wickerhamiella domercqii van der Walt

c. SPECIES ACCEPTED IN THE GENUS

1. *Wickerhamiella domercqii* van der Walt (in van der Walt & Liebenberg, 1973a)

e. SYSTEMATIC DISCUSSION OF THE SPECIES

1. WICKERHAMIELLA DOMERCQII VAN DER WALT (in van der Walt & Liebenberg, 1973a)

Imperfect state: *Candida domercqii* (van der Walt et van Kerken) Meyer et Yarrow

Synonyms

Torulopsis domercqii van der Walt et van Kerken (1960b)
Candida domercqii (van der Walt et van Kerken) Meyer et Yarrow (Yarrow & Meyer, 1978)
Torulopsis sacchari Shehata (1960) nom. illegit. publ. sine diagn. latine

Standard description of *Wickerhamiella domercqii*

Growth on morphology agar: After 6 days at 28°C the cells are spheroidal, ovoid or ellipsoidal, reproducing by budding, and occurring singly, in pairs or short chains or small clusters (Fig. 129). The cells of the type strain measure (1.0–1.5) × (2.5–3.5) μm and those of the second strain (CBS 4733), (1.5–3.5) × (3.0–5.0) μm. The colonies are butyrous, cream-colored, raised, shiny to somewhat dull, smooth and with entire margins.

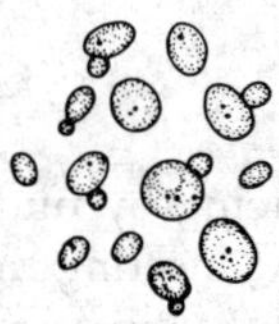

Fig. 129 – *Wick. domercqii*
After 6 days on morphology agar

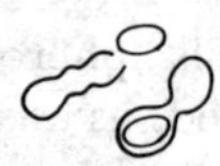

Fig. 130 – *Wick. domercqii*
After 8 days on potato agar

Growth on the surface of assimilation media: Pellicle formation does not occur.

Dalmau plate culture on morphology agar: No pseudomycelium is formed.

Formation of ascospores: Conjugation immediately precedes ascus formation. As a rule one ascospore is produced per ascus. The asci dehisce terminally in the vicinity of the ascospores, ejecting the solitary ascospores, giving rise to persistent saccate structures (Fig. 130). By light microscopy the ascospores appear to be oblong or ellipsoidal, and glabrous. When examined in ultrathin sections by TEM, the ascospores are oblong with obtuse ends and rugose.

Sporulation is rather sparse and is best observed on potato agar.

Fermentation: –

Assimilation of carbon compounds:

Galactose	v	Raffinose	–	Erythritol	–
Sucrose	+w or –	Soluble starch	–	Ribitol	–
Maltose	–	D-Xylose	–	D-Mannitol	+
Cellobiose	–	L-Arabinose	–	Succinic acid	+
Trehalose	–	D-Ribose	+	Citric acid	+
Lactose	–	L-Rhamnose	–	Inositol	–

Assimilation of nitrate: +

Growth in vitamin-free medium: –

Growth in 10% sodium chloride plus 5% glucose in yeast nitrogen base: +

Growth at 37°C: –

Origin of the strains studied

Wine vat (1) (van der Walt & van Kerken, 1960b); effluent from a cane sugar factory (1) (Shehata, 1960).

Type strain: CBS 4351, isolated by van der Walt and van Kerken from the surface of a wine vat.

Genus 32. *Wingea* van der Walt

by J.P. van der Walt

a. DIAGNOSIS OF THE GENUS

Vegetative cells spheroidal to ellipsoidal, reproducing by budding.
Ascus formation preceded by the generation of cells bearing copulatory protuberances.
Asci conjugated or unconjugated, persistent, one- to four-spored.
Ascospores lentiform, glabrous, light brown.
Coenzyme Q-9 system present.
Nitrate not utilized.

b. TYPE SPECIES OF THE GENUS

Wingea robertsii (van der Walt) van der Walt

c. SPECIES ACCEPTED IN THE GENUS

1. *Wingea robertsii* (van der Walt) van der Walt (van der Walt 1959a)

e. SYSTEMATIC DISCUSSION OF THE SPECIES

1. *WINGEA ROBERTSII* (VAN DER WALT) VAN DER WALT (1971)

Synonyms

Pichia robertsii van der Walt (1959a)
Wingea robertsii (van der Walt) van der Walt (1967) nom. inval.

Standard description of *Wingea robertsii* (van der Walt, 1970f)

Growth in malt extract: After 3 days at 28°C the cells are spheroidal to ellipsoidal, (2.5–7.5) × (3.0–8.0) μm, reproducing by budding, and occurring singly, in pairs or occasionally in clusters. Large cells measuring 10 × 10 μm may occasionally be formed (Fig. 131). A sediment, slight ring and occasionally a thin, dull pellicle are formed.

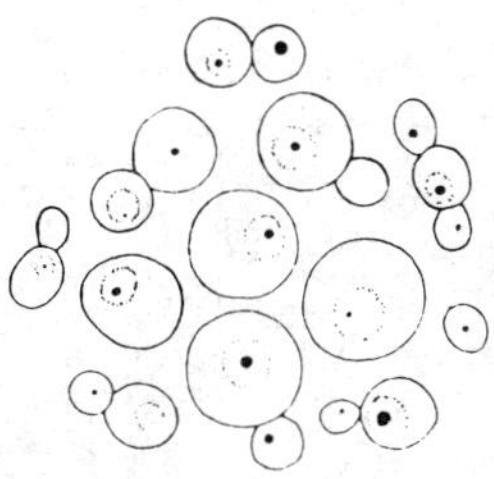

Fig. 131 – *Wi. robertsii*
After 3 days in malt extract
(From 'The Yeasts', ed. II, 1970)

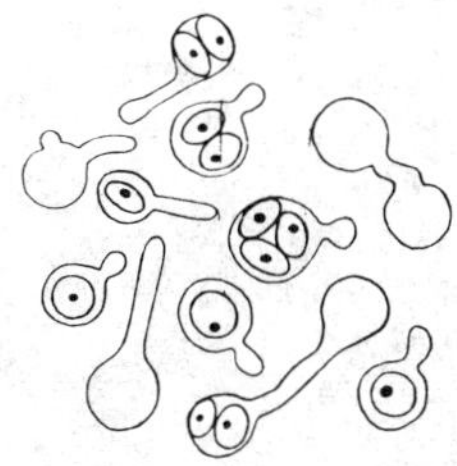

Fig. 132 – *Wi. robertsii*
After 1 week on YM agar
(From 'The Yeasts', ed. II, 1970)

Growth on malt agar: After one month at room temperature the streak culture is butyrous, cream to brownish-cream, occasionally with light brown to chocolate-brown, actively sporulating sectors, dull to somewhat shiny, flat, rather spreading, smooth to verruculose, with transverse striations along the margin. The margin is undulating to lobate.

Dalmau plate culture on corn meal agar: No pseudomycelium is formed.

Formation of ascospores: Cells bearing protuberances are formed. Such cells are either directly converted into unconjugated asci or may fuse by means of the protuberances giving rise to conjugated asci; one to four ascospores per ascus. The asci are persistent. The mature ascospores containing lipid globules are lentiform, light brown in color, glabrous (Fig. 132).

Sporulation is observed on malt-, YM-, V8-, Gorodkowa- and acetate agar.

Fermentation:

Glucose	+	Maltose	+
Galactose	–	Lactose	–
Sucrose	+	Raffinose	+

Assimilation of carbon compounds:

Galactose	+	Raffinose	+	Erythritol	+
Sucrose	+	Soluble starch	+ w	Ribitol	+
Maltose	+	D-Xylose	+	D-Mannitol	+
Cellobiose	+	L-Arabinose	+	Succinic acid	+
Trehalose	+	D-Ribose	v	Citric acid	+
Lactose	–	L-Rhamnose	+	Inositol	–

Assimilation of nitrate: –

Growth in vitamin-free medium: – or + w

Growth on 50% (w/w) glucose-yeast extract agar: +

Growth at 37°C: + w or –

G + C: 42.7 mol% (Price et al., 1978)

Origin of the strains studied

Larval feeds of *Xylocopa caffra* (4) (van der Walt, 1959a); soil (4).

Type strain: CBS 2934, isolated by van der Walt from the larval feed of *Xylocopa caffra*.

Genus 33. *Zygosaccharomyces* Barker

by D. Yarrow

a. DIAGNOSIS OF THE GENUS

Vegetative reproduction by multilateral budding. Cells globose, ellipsoidal and cylindrical. Pseudohyphae may be formed, septate hyphae absent.

The vegetative state is predominantly haploid. Conjugation is usually between individual cells, occasionally between a cell and its bud, just before the formation of persistent asci. Ascospores are globose to ellipsoidal, one to four per ascus.

Vigorous fermentation.

Starch-like compounds are not produced.

No growth with nitrate as sole source of nitrogen.

b. TYPE SPECIES OF THE GENUS

Zygosaccharomyces barkeri Saccardo et Sydow

c. SPECIES ACCEPTED IN THE GENUS

1. *Zygosaccharomyces bailii* (Lindner) Guilliermond (Lindner, 1895)
2. *Zygosaccharomyces bisporus* Naganishi (1917)
3. *Zygosaccharomyces cidri* (Legakis) Yarrow (Legakis, 1961)
4. *Zygosaccharomyces fermentati* Naganishi (1928)
5. *Zygosaccharomyces florentinus* Castelli ex Kudriavzev (1960)
6. *Zygosaccharomyces microellipsoides* (Osterwalder) Yarrow (Osterwalder, 1924a)
7. *Zygosaccharomyces mrakii* Capriotti (1958c)
8. *Zygosaccharomyces rouxii* (Boutroux) Yarrow (Boutroux, 1884)

d. KEY TO THE SPECIES OF THE GENUS

Based on the results of the growth tests read on the 4th day of incubation. Growth with nitrogen sources tested on auxanogram plates, with carbon sources on auxanogram plates or in liquid shaken cultures. Resistance to cycloheximide and requirement for inositol as a growth factor in liquid shaken cultures.

1a Growth in the presence of 1000 ppm of cycloheximide 2
b No growth in the presence of 1000 ppm of cycloheximide 5

2a Cellobiose assimilated
Zygosacch. fermentati .... p. 457
b Cellobiose not assimilated 3

3a Inositol required as vitamin for growth
Zygosacch. cidri .... p. 455
b Inositol not required as vitamin for growth 4

4a Culture does not become pink, maltose usually assimilated
Zygosacch. florentinus .... p. 458
b Culture becomes pink, maltose is not assimilated
Zygosacch. mrakii .... p. 461

5a Melibiose assimilated
Zygosacch. microellipsoides .... p. 460
b Melibiose not assimilated 6

6a Growth in the presence of 1% acetic acid, maltose is not assimilated 7
b No growth in the presence of 1% acetic acid, maltose may be assimilated
Zygosacch. rouxii .... p. 463

7a Cells $(2.5-4.5)\times(3.5-7.5)$ μm, sucrose not assimilated
Zygosacch. bisporus .... p. 453
b Cells larger, $(3.5-7.0)\times(5.5-14.0)$ μm, sucrose may be assimilated
Zygosacch. bailii .... p. 451

See Table 19, p. 381.

e. SYSTEMATIC DISCUSSION OF THE SPECIES

1. *ZYGOSACCHAROMYCES BAILII* (LINDNER) GUILLIERMOND (1912)

Synonyms

Saccharomyces bailii Lindner (1895)
Zygosaccharomyces mandshuricus Saito (1914)
Zygosaccharomyces acidifaciens Nickerson (1943)
Saccharomyces acidifaciens (Nickerson) Lodder et Kreger-van Rij (1952)
Saccharomyces elegans Lodder et Kreger-van Rij (1952)
Saccharomyces elegans Lodder et Kreger-van Rij var. *intermedia* nom. nud. (Verona & Zardetta de Toledo, 1959b)

Standard description of *Zygosaccharomyces bailii*

Growth in malt extract: After 3 days at 25°C the cells are ovoidal, ellipsoidal or cylindrical, (3.5–7.0) × (5.5–14.0) μm, in pairs, short chains or clusters (Fig. 133). A sediment and sometimes a ring are present. After one month at 20°C a sediment and a ring are present.

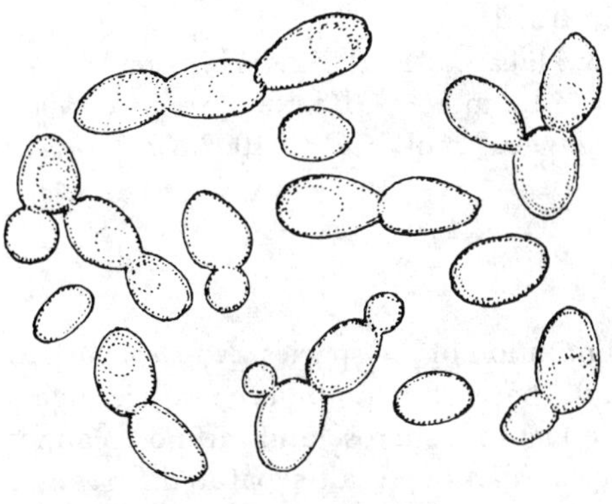

Fig. 133 – *Zygosacch. bailii*
After 3 days in malt extract
(From 'The Yeasts', ed. I, 1952)

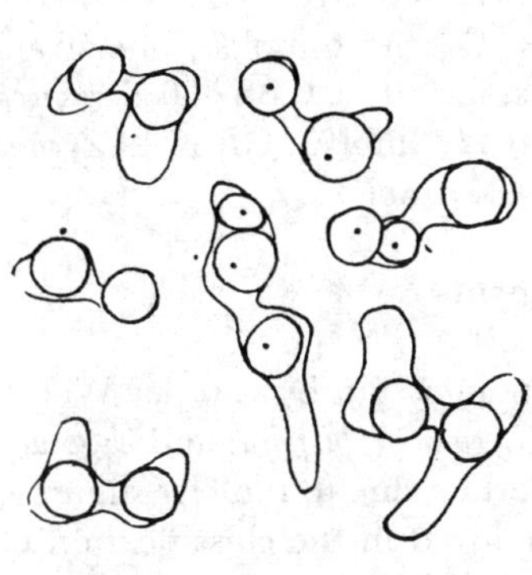

Fig. 134 – *Zygosacch. bailii*
After 1 week on V8 agar
(From 'The Yeasts', ed. I, 1952)

Growth on malt agar: After one month at 20°C the streak culture is white to dark cream, usually smooth and glossy, rarely rough and dull, the margin is sinuous.

Slide cultures on corn meal- and morphology agar: Pseudohyphae may be formed. They consist of chains of ellipsoidal or cylindrical cells which may be branched and bear small verticils of ellipsoidal blastospores. Ascospores may be present.

Formation of ascospores: Conjugation of individual cells precedes the formation of asci containing two to four globose to ellipsoidal, smooth ascospores (Fig. 134).

Sporulation was observed on malt- and YM agar, also on V8- and corn meal agar in some strains.

Fermentation:

Glucose	+	Maltose	–
Galactose	–	Lactose	–
Sucrose	v		

Assimilation of carbon compounds:

Galactose	v	Raffinose	– (+)	Erythritol	–
Sucrose	v	Soluble starch	–	Ribitol	+ (–)
Maltose	–	D-Xylose	–	D-Mannitol	+ (–)
Cellobiose	–	L-Arabinose	–	Succinic acid	–
Trehalose	v	D-Ribose	–	Citric acid	–
Lactose	–	L-Rhamnose	–	Inositol	–

Additional carbon compound tested: Melibiose –

Assimilation of nitrate: –

Assimilation of ethylamine · HCl: +

Assimilation of cadaverine · 2 HCl: +

Growth in vitamin-free medium: –; **in inositol-free medium:** +

Growth on 50% (w/w) glucose-yeast extract agar: +

Growth at 37°C: v

Growth in the presence of 100 ppm of cycloheximide: – or +s; **of 1000 ppm:** –

Growth in the presence of 1% acetic acid: +

G + C: 44.6 mol%, one strain (Yarrow & Nakase, 1975); 42.7–44.6 mol%, 22 strains, including CBS 680 *Zygosacch. bailii* 42.7–43.2 mol%; CBS 685 *Zygosacch. mandshuricus* 43.9 mol%; CBS 749 *Zygosacch. acidifaciens* 43.2 mol%; CBS 1097 *Sacch. elegans* 43.7–44.6 mol%

Comments

As pointed out by van der Walt (1970e) delimitation of the species *Zygosaccharomyces bailii, Zygosacch. bisporus* and *Zygosacch. rouxii* on the basis of their ability to utilize sugars is uncertain due to the few sugars used and the fact that sucrose and raffinose cannot be employed in the classification of these species. Van der Walt separated *Zygosacch. bailii* and *Zygosacch. bisporus* from *Zygosacch. rouxii* by the ability to use maltose but also reported that this ability had been acquired by some strains. The acquisition of the ability to use maltose by several strains which did not do so when first isolated, and which consequently had been identified as *Zygosacch. bailii* or *Zygosacch. bisporus,* was also observed during the present study. Moreover several maltose-negative strains were observed to conjugate and sporulate when mixed with one of the mating types of *Zygosacch. rouxii*; the percentage of guanine plus cytosine in the DNA of these strains also fell within the range of this species. From this it was deduced that the ability to use maltose is not a reliable enough criterion to be used in the classification of these species.

Zygosacch. bailii has been reported to be very resistant to preservatives such as benzoic acid and acetic acid (e.g. Sand, 1973; Pitt, 1974). This has been shown to be primarily due to an energy-dependent excretory mechanism (Warth, 1977). Therefore growth of the strains studied was tested for 14 days on media containing 400 mg/l benzoic acid or 300 mg/l sorbic acid at pH 3.5, or 1% acetic acid, in the hope that this might provide a simple and rapid identification test for these species. *Zygosacch. rouxii* (including the type of *Zygosacch. bisporus* var. *mellis* and of *Zygosacch. bailii* var. *osmophilus*) failed to grow on any of the three media. All the strains of *Zygosacch. bailii* (including the type of *Zygosacch. acidifaciens* and of *Sacch. elegans*) tolerated all three acids at these concentrations. The strains of *Zygosacch. bisporus* did not grow with these concentrations of benzoic acid and sorbic acid but they tolerated 1% acetic acid.

The classification of these species requires considerable further study using techniques such as DNA/DNA hybridization, however, for the time being the following species will be accepted: *Zygosaccharomyces bailii, Zygosacch. bisporus* and *Zygosacch. rouxii*. As most (80%) of the strains of *Zygosacch. bailii* and *Zygosacch. bisporus* grew on a medium containing 10% glucose, 1% tryptone, 1% yeast extract, 2% agar and 1% acetic acid within 3 days, and all did so within 6 days, and as this can be more conveniently employed than the determination of the base composition of the DNA, it is used here in the key for their routine identification.

Contrary to the findings of van der Walt (1970e) all the strains of *Zygosacch. bailii* were found to be able to grow on media containing 60% (w/w) glucose, albeit rather slowly in many cases.

Origin of the strains studied

Sorghum brandy (1), sour red wine (2), perry (1), mayonnaise (2), apple juice (1), vinegar (1), wine (3), pickles (2), *Myoporum* sp. (1), Worcestershire sauce (1), cloudy wine (1), feces (1), grape must (3), concentrated fruit juice (3), soft drinks (12), unknown (6).

Type strain: CBS 680, isolated in the Institute of Brewing, Tokyo, and received from H. Naganishi, was designated as type by Lodder and Kreger-van Rij (1952).

2. *ZYGOSACCHAROMYCES BISPORUS* NAGANISHI (1917)

Synonym

Saccharomyces bisporus (Naganishi) Lodder et Kreger-van Rij (1952)
[nec *Zygosaccharomyces bisporus* Anderson (1917)]

Standard description of *Zygosaccharomyces bisporus*

Growth in malt extract: After 3 days at 25°C the cells are ovoidal to ellipsoidal, (2.5–4.5) × (3.5–7.5) μm, single or in pairs (Fig. 135). A sediment is present and sometimes an incomplete ring also. After one month at 20°C a sediment and usually a ring are present.

Growth on malt agar: After one month at 20°C the streak culture is cream to tan, smooth and glossy, the margin is entire.

Slide cultures on corn meal- and morphology agar: A few short chains of ellipsoidal cells may be formed.

Formation of ascospores: Conjugation of individual cells precedes the for-

mation of asci containing one, usually two, globose, smooth ascospores (Fig. 136).
Sporulation was observed on corn meal- and malt agar.

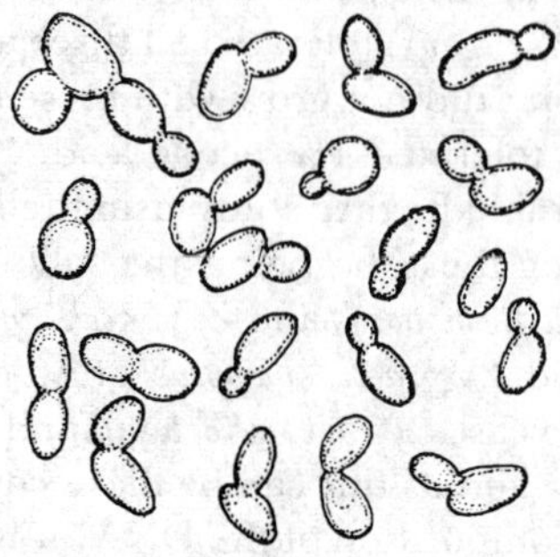

Fig. 135 – *Zygosacch. bisporus*
After 3 days in malt extract
(From 'The Yeasts', ed. I, 1952)

Fig. 136 – *Zygosacch. bisporus,* CBS 1083
After 3 days on corn meal agar

Fermentation:

Glucose	+	Maltose	–
Galactose	–	Lactose	–
Sucrose	–		

Assimilation of carbon compounds:

Galactose	v	Raffinose	–	Erythritol	–
Sucrose	–	Soluble starch	–	Ribitol	v
Maltose	–	D-Xylose	–	D-Mannitol	v
Cellobiose	–	L-Arabinose	–	Succinic acid	–
Trehalose	–	D-Ribose	–	Citric acid	–
Lactose	–	L-Rhamnose	–	Inositol	–

Additional carbon compound tested: Melibiose –

Assimilation of nitrate: –

Assimilation of ethylamine · HCl: +

Assimilation of cadaverine · 2 HCl: +

Growth in vitamin-free medium: – or +w; **in inositol-free medium:** +

Growth on 50% (w/w) glucose-yeast extract agar: +

Growth at 37°C: –

Growth in the presence of 100 ppm of cycloheximide: –

Growth in the presence of 1% acetic acid: +

G + C: 43.7–44.1 mol%, 2 strains (Nakase & Komagata, 1971d); 44.0 mol%, one strain (Price et al., 1978); 43.9–44.6 mol%, 3 strains

Comments

Contrary to the report of van der Walt (1970e) all the strains of this species were found to be able to grow on media containing 60% (w/w) glucose.

The variety *mellis* recognized by van der Walt is considered synonymous with *Zygosacch. rouxii* (see also Comments under *Zygosacch. bailii*).

Origin of the strains studied

Tea beer (2), fermenting pickled cucumbers (1), unknown (1).

Type strain: CBS 702, received from K. Saito in 1923, was designated as the type by Lodder and Kreger-van Rij (1952).

3. *ZYGOSACCHAROMYCES CIDRI* (LEGAKIS) YARROW (in von Arx et al., 1977)

Synonyms

Saccharomyces cidri Legakis (1961)
Torulaspora cidri (Legakis) van der Walt et Johannsen (1975a)
Saccharomyces malacitensis nom. nud. (Santa María, 1960)
Saccharomyces nilssonii (Capriotti) Santa María var. *malacitensis* Santa María (1963b)
Saccharomyces amurcae van der Walt (1970e)
Torulaspora amurcae (van der Walt) van der Walt et Johannsen (1975a)

Standard description of *Zygosaccharomyces cidri*

Growth in malt extract: After 3 days at 25°C the cells are subglobose to ellipsoidal, (2.0–6.5) × (2.5–8.0) μm, single, in pairs or small clusters. A sediment is present. After one month at 20°C a sediment and ring are present.

Growth on malt agar: After one month at 20°C the streak culture is cream, smooth and glossy, occasionally lightly striated near the margin which may be entire or slightly lobate.

Slide cultures on corn meal- and morphology agar: A few short chains of elongated cells may be formed.

Formation of ascospores: Conjugation of individual cells usually precedes the formation of asci containing two to four globose, smooth ascospores (Fig. 137).

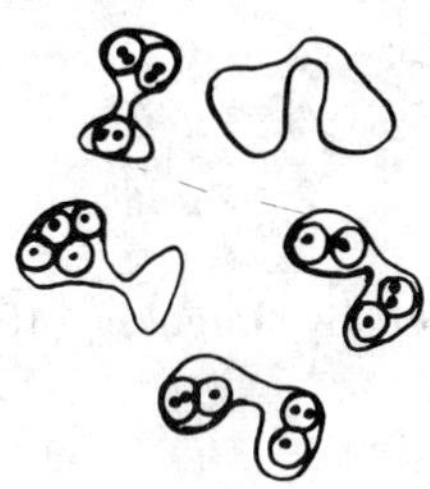

Fig. 137 – *Zygosacch. cidri*, CBS 5666
After 6 days on malt agar

Sporulation was observed on malt agar.

Fermentation:

Glucose	+	Maltose	v
Galactose	+	Lactose	–
Sucrose	+		

Assimilation of carbon compounds:

Galactose	+	Raffinose	+	Erythritol	–
Sucrose	+	Soluble starch	–	Ribitol	+ or s
Maltose	v	D-Xylose	+ or s	D-Mannitol	+
Cellobiose	–	L-Arabinose	–	Succinic acid	+
Trehalose	+	D-Ribose	–	Citric acid	v
Lactose	–	L-Rhamnose	–	Inositol	–

Additional carbon compound tested: Melibiose +

Assimilation of nitrate: –

Assimilation of ethylamine · HCl: +

Assimilation of cadaverine · 2 HCl: +

Growth in vitamin-free medium: – ; **in inositol-free medium:** –

Growth on 50% (w/w) glucose-yeast extract agar: +

Growth at 37°C: v

Growth in the presence of 100 ppm of cycloheximide: + ; **of 1000 ppm:** +

Growth in the presence of 1% acetic acid: –

G + C: 41.7 mol%, 2 strains (Yarrow & Nakase, 1975); 42.6 mol%, one strain (Price et al., 1978); 41.9 mol%, one strain

Comments

Van der Walt (1970e) separated *Saccharomyces amurcae* and *Zygosaccharomyces cidri* by their ability to use maltose. Yarrow and Nakase (1975) considered the separation of these two species on the basis of a single, unstable property to be unjustified and reduced *Sacch. amurcae* to synonymy with *Zygosacch. cidri.*

Zygosacch. cidri and *Zygosacch. fermentati* both require an external supply of the vitamin inositol in order to grow, they are the only species of this genus to do so.

Origin of the strains studied

Cider (3), alpechin (6), rotting bark of *Araucaria* sp. (1).

Type strain: CBS 4575, a strain isolated from cider by Legakis (1961), was designated as type by van der Walt (1970e).

4. *ZYGOSACCHAROMYCES FERMENTATI* NAGANISHI (1928)

Synonyms

[non *Zygosaccharomyces fermentati* (Saito) Krumbholz (1933)]
Debaryomyces mandshuricus Naganishi (1924)
Zymodebaryomyces mandshuricus (Naganishi) Novák et Zsolt (1961)
Torulaspora manchurica (Naganishi) van der Walt et Johannsen (1975a)
Saccharomyces montanus Phaff, Miller et Shifrine (1956)
Saccharomyces montanus nom. nud. (van der Walt, 1970e)

Standard description of *Zygosaccharomyces fermentati*

Growth in malt extract: After 3 days at 25°C the cells are ellipsoidal to cylindrical, (2.0–6.0) × (3.0–8.0) μm, single or in pairs. A sediment is present. After one month at 20°C a sediment and a ring are present.

Growth on malt agar: After one month at 20°C the streak culture is white or off-white, smooth and glossy, the margin is entire or undulating.

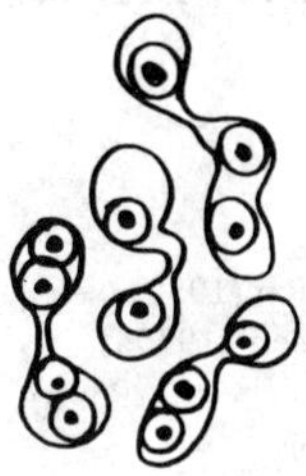

Fig. 138 – *Zygosacch. fermentati*, CBS 6544
After 3 days on malt agar

Slide cultures on corn meal- and morphology agar: Pseudohyphae are formed and consist of chains of cylindrical cells sometimes with small verticils of ellipsoidal blastospores.

Formation of ascospores: Conjugation of individual cells usually precedes the formation of asci containing two to four globose, apparently smooth, ascospores (Fig. 138). The cultures are often highly agglutinative.

Sporulation was observed on malt agar.

Fermentation:

Glucose	+	Maltose	+
Galactose	+ or s	Lactose	–
Sucrose	+		

Assimilation of carbon compounds:

Galactose	+	Raffinose	+	Erythritol	–
Sucrose	+	Soluble starch	–	Ribitol	– or + s
Maltose	+	D-Xylose	v	D-Mannitol	+ (–)
Cellobiose	+	L-Arabinose	–	Succinic acid	+
Trehalose	+	D-Ribose	–	Citric acid	–
Lactose	–	L-Rhamnose	–	Inositol	–

Additional carbon compound tested: Melibiose –

Assimilation of nitrate: –

Assimilation of ethylamine · HCl: +

Assimilation of cadaverine · 2 HCl: +

Growth in vitamin-free medium: –; **in inositol-free medium:** –

Growth on 50% (w/w) glucose-yeast extract agar: + or w

Growth at 37°C: +

Growth in the presence of 100 ppm of cycloheximide: +; **of 1000 ppm:** +

Growth in the presence of 1% acetic acid: –

G + C: 41.5 mol%, one strain (Yarrow & Nakase, 1975); 43.4 mol%, one strain (Price et al., 1978); 41.7–42.5 mol%, 5 strains

Origin of the strains studied

Sediment of peppermint (1), *Drosophila* (1), soft drinks and fruit juices (11), unknown (1).

Type strain: CBS 707, isolated from sediment of peppermint by H. Naganishi (1928).

5. *ZYGOSACCHAROMYCES FLORENTINUS* CASTELLI EX KUDRIAVZEV (1960)

Synonyms

Saccharomyces florentinus (Castelli ex Kudriavzev) Lodder et Kreger-van Rij (1952) (see also Castelli, 1938)
Torulaspora florentina (Castelli ex Kudriavzev) van der Walt et Johannsen (1975a)
Zygosaccharomyces eupagycus Sacchetti ex Kudriavzev (1960) (see also Sacchetti, 1939b)
Saccharomyces eupagycus (Sacchetti ex Kudriavzev) van der Walt (1970e)
Torulaspora eupagyca (Sacchetti ex Kudriavzev) van der Walt et Johannsen (1975a)

Standard description of *Zygosaccharomyces florentinus*

Growth in malt extract: After 3 days at 25°C the cells are globose or subglobose $(3.0-6.5) \times (3.0-7.0)$ μm, single or in pairs (Fig. 139). A sediment is present and sometimes an incomplete ring. After one month at 20°C a sediment and a ring are present.

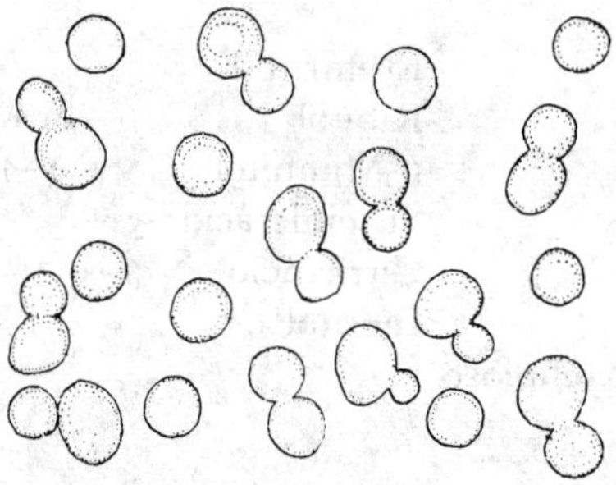

Fig. 139 – *Zygosacch. florentinus*
After 3 days in malt extract
(From 'The Yeasts', ed. I, 1952)

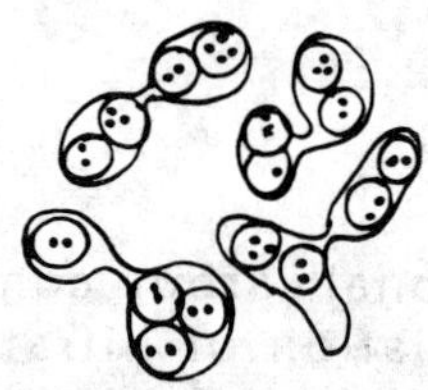

Fig. 140 – *Zygosacch. florentinus,* CBS 6759
After 6 days on corn meal agar

Growth on malt agar: After one month at 20°C the streak culture is white to dark cream, smooth and glossy, the margin may have a short fringe of pseudohyphae.

Slide cultures on corn meal- and morphology agar: Pseudohyphae are formed abundantly and consist of cylindrical cells with small verticils of blastospores. Ascosporulation often occurs under the edge of the coverslip on corn meal agar, conjugation frequently being between cells of adjacent hyphae and resulting in ladder-like formations.

Formation of ascospores: Conjugation of individual cells usually precedes the formation of asci containing two to four globose, smooth ascospores (Fig. 140). Infrequently, conjugation between a cell and its bud or the direct transformation of a cell into an ascus may be observed.

Sporulation was observed on corn meal agar, sometimes on malt-, YM- and V8 agar also.

Fermentation:

Glucose	+	Maltose	+ (–)
Galactose	v	Lactose	–
Sucrose	+ or s		

Assimilation of carbon compounds:

Galactose	+ (–)	Raffinose	+	Erythritol	–
Sucrose	+	Soluble starch	–	Ribitol	–
Maltose	+ (–)	D-Xylose	– or + s	D-Mannitol	+ or s
Cellobiose	–	L-Arabinose	–	Succinic acid	+ or s
Trehalose	+	D-Ribose	–	Citric acid	–
Lactose	–	L-Rhamnose	–	Inositol	–

Additional carbon compound tested: Melibiose +

Assimilation of nitrate: –

Assimilation of ethylamine · HCl: +

Assimilation of cadaverine · 2 HCl: +

Growth in vitamin-free medium: + s or – ; in inositol-free medium: +

Growth on 50% (w/w) glucose-yeast extract agar: +

Growth at 37°C: –

Growth in the presence of 100 ppm of cycloheximide: + ; of 1000 ppm: +

Growth in the presence of 1% acetic acid: –

G + C: 40.5 mol%, one strain (Nakase & Komagata, 1971d); 42.4 mol%, 2 strains (Price et al., 1978); 42.0–43.7 mol%, 3 strains

Comments

The DNA reassociation experiments of Price et al. (1978) showed *Zygosaccharomyces eupagycus* to be conspecific with *Zygosacch. florentinus.*

Origin of the strains studied

Grape must (9), soil (1), *Drosophila* sp. (1), trees and shrubs (6), soft drinks and fruit juices (46).

Type strain: CBS 746, isolated from grape must by Castelli, was indicated as type by Lodder et Kreger-van Rij (1952).

6. *ZYGOSACCHAROMYCES MICROELLIPSOIDES* (OSTERWALDER) YARROW (in von Arx et al., 1977)

Synonyms

Saccharomyces microellipsodes Osterwalder (1924a)
Torulaspora microellipsodes (Osterwalder) van der Walt et Johannsen (1975a)

Standard description of *Zygosaccharomyces microellipsoides*

Growth in malt extract: After 3 days at 25°C the cells are globose, subglobose or ellipsoidal (2.5–6.0)×(2.5–6.5) μm, single or in pairs (Fig. 141). A sediment is present. After one month at 20°C a sediment and usually a ring are present.

Growth on malt agar: After one month at 20°C the streak culture is greyish white to cream, smooth and glossy, sometimes lightly striated, the margin is entire.

Slide cultures on corn meal- and morphology agar: Pseudohyphae are not formed.

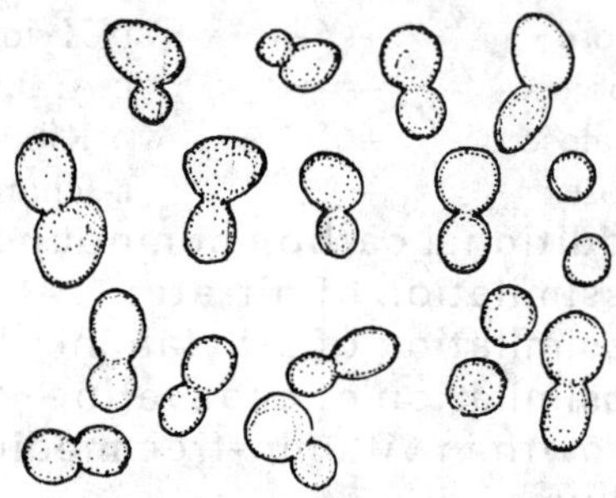

Fig. 141 – *Zygosacch. microellipsoides* After 3 days in malt extract (From 'The Yeasts', ed. I, 1952)

Formation of ascospores: Conjugation of individual cells usually precedes the formation of asci containing two to four globose, smooth ascospores.

Sporulation was observed on corn meal agar, though only after 11 weeks in the case of the type strain. Some strains also sporulated on malt agar and YM agar.

Fermentation:

Glucose	+	Maltose	–
Galactose	+	Lactose	–
Sucrose	+		

Assimilation of carbon compounds:

Galactose	+	Raffinose	+	Erythritol	–
Sucrose	+	Soluble starch	–	Ribitol	–
Maltose	–	D-Xylose	+ s or –	D-Mannitol	v
Cellobiose	–	L-Arabinose	–	Succinic acid	v
Trehalose	– or + s	D-Ribose	–	Citric acid	–
Lactose	–	L-Rhamnose	–	Inositol	–

Additional carbon compound tested: Melibiose +

Assimilation of nitrate: –

Assimilation of ethylamine · HCl: +

Assimilation of cadaverine · 2 HCl: +

Growth in vitamin-free medium: – or + s; **in inositol-free medium:** +

Growth on 50% (w/w) glucose-yeast extract agar: + or w
Growth at 37°C: –
Growth in the presence of 100 ppm of cycloheximide: –
Growth in the presence of 1% acetic acid: –
G + C: 39.3–39.5 mol%, 2 strains (Nakase & Komagata, 1971d); 40.3 mol%, one strain (Price et al., 1978); 39.8–41.0 mol%, 5 strains

Comments

Because of differences in the way the ascus was formed, i.e. conjugation between individual cells in the case of *Zygosacch. microellipsoides* var. *microellipsoides* and between a cell and its bud in the case of the variety *osmophilus,* Yarrow and Nakase (1975) placed the variety *osmophilus* in the group containing *Torulaspora rosei* and its variants. Price et al. (1978) reduced *Zygosacch. microellipsoides* var. *osmophilus* to synonymy with *Torulaspora delbrueckii.*

Origin of the strains studied

Apple juice (1), black currants (1), tea beer (2), exudate of trees (2), soft drinks (5).

Type strain: CBS 427, isolated from apple juice by Osterwalder.

7. *ZYGOSACCHAROMYCES MRAKII* CAPRIOTTI (1958c)

Synonyms

Saccharomyces mrakii (Capriotti) van der Walt (1970e)
Torulaspora mrakii (Capriotti) van der Walt et Johannsen (1975a)

Standard description of *Zygosaccharomyces mrakii*

Growth in malt extract: After 3 days at 25°C the cells are globose to ellipsoidal (3.0–6.5) × (4.0–8.0) μm, single, in pairs or small groups. A sediment is present. After one month at 20°C a sediment and a ring are present.

Growth on malt agar: After one month at 20°C the streak culture is cream to pale pink, smooth and glossy, the margin is undulating. The pink pigment diffuses into the agar around the streak.

Slide cultures on corn meal- and morphology agar: Primitive pseudohyphae may be formed.

Formation of ascospores: Conjugation of individual cells usually precedes the formation of asci containing two to four globose, smooth ascospores.

Sporulation was observed on malt-, potato- and rice agar.

Fermentation:

Glucose	+	Maltose	–
Galactose	+	Lactose	–
Sucrose	+		

Assimilation of carbon compounds:

Galactose	+	Raffinose	+	Erythritol	–
Sucrose	+	Soluble starch	–	Ribitol	–
Maltose	–	D-Xylose	– or + s	D-Mannitol	+
Cellobiose	–	L-Arabinose	–	Succinic acid	+
Trehalose	–	D-Ribose	–	Citric acid	–
Lactose	–	L-Rhamnose	–	Inositol	–

Additional carbon compound tested: Melibiose +

Assimilation of nitrate: –

Assimilation of ethylamine · HCl: +

Assimilation of cadaverine · 2 HCl: +

Growth in vitamin-free medium: v; **in inositol-free medium:** +

Growth on 50% (w/w) glucose-yeast extract agar: + s

Growth at 37°C: –

Growth in the presence of 100 ppm of cycloheximide: +; **of 1000 ppm:** +

Growth in the presence of 1% acetic acid: –

G + C: 41.2–41.3 mol%, 2 strains (Price et al., 1978); 40.5 mol%, one strain

Comments

The high proportion of 48.5 mol% guanine plus cytosine reported by Yarrow and Nakase (1975) has not been confirmed by later determinations on the DNA of the same strain and must therefore be considered erroneous.

This species closely resembles *Zygosacch. florentinus* but Price et al. (1978) found very little reassociation of the DNAs of the type strains.

Origin of the strains studied

Silage (2).

Type strain: CBS 4218, isolated from silage by Capriotti, was designated as type by van der Walt (1970e).

8. ZYGOSACCHAROMYCES ROUXII (BOUTROUX) YARROW (in von Arx et al., 1977)

Imperfect state: *Candida mogii* Vidal-Leiria

Synonyms

Saccharomyces rouxii Boutroux (1884)
Zygosaccharomyces barkeri Saccardo et Sydow (1902)
Saccharomyces soya Saito (1907)
Zygosaccharomyces soya (Saito) Takahashi et Yukawa (1915)

Zygosaccharomyces japonicus Saito var. *soya* (Saito) Dekker (Stelling-Dekker, 1931)
Zygosaccharomyces japonicus Saito (1909)
Zygopichia japonica (Saito) Klöcker (1924)
Zygosaccharomyces mellis-acidi von Richter (1912)
Zygosaccharomyces major Takahashi et Yukawa (1915)
Zygosaccharomyces salsus Takahashi et Yukawa (1915)
Zygopichia salsa (Takahashi et Yukawa) Klöcker (1924)
Zygosaccharomyces nadsonii Guilliermond (1918)
Zygosaccharomyces dairensis Naganishi (1928)
Zygosaccharomyces mellis Fabian et Quinet (1928)
Saccharomyces mellis (Fabian et Quinet) Lodder et Kreger-van Rij (1952)
Saccharomyces bisporus (Naganishi) Lodder et Kreger-van Rij var. *mellis* (Fabian et Quinet) van der Walt (1970e)
Zygosaccharomyces vini Naganishi (1928)
Zygosaccharomyces cavarae Rodio var. *beauverie* Beauverie (1929)
Zygosaccharomyces nussbaumeri Lochhead et Heron (1929)
Zygosaccharomyces richteri Lochhead et Heron (1929)
Zygosaccharomyces amoeboideus Kroemer et Krumbholz (in Krumbholz, 1931)
Zygosaccharomyces cavarae Rodio var. *amoeboideus* Lodder (1932)
Zygosaccharomyces gracilis Karamboloff et Krumbholz (1931)
Zygosaccharomyces polymorphus Kroemer et Krumbholz (in Krumbholz, 1931)
Saccharomyces rouxii Boutroux var. *polymorphus* (Kroemer et Krumbholz) Lodder et Kreger-van Rij (1952)
Zygosaccharomyces variabilis Kroemer et Krumbholz (in Krumbholz, 1931)
Zygosaccharomyces rugosus Lochhead et Farrell (in Lochhead, 1942)
Zygosaccharomyces nectarophilus Lochhead et Farrell (in Lochhead, 1942)
Zygosaccharomyces gracilis Karamboloff et Krumbholz subsp. *italicus* Sacchetti (1932d)
Zygosaccharomyces citrus Lodder (1932)
Zygosaccharomyces major Takahashi et Yukawa var. *threntensis* Lodder (1932)
Zygosaccharomyces felsineus Sacchetti (1932d)
Zygosaccharomyces perspicillatus Sacchetti (1932c)
Zygosaccharomyces ravennatis Sacchetti (1932c)
Zygosaccharomyces nukamiso nom. nud. (Otani, 1939)
Saccharomycodes mestris nom. nud. (Marcilla & Feduchy, 1943)
Zygosaccharomyces halomembranis Etchells et Bell (1950a)
Saccharomyces rouxii Boutroux var. *halomembranis* (Etchells et Bell) Onishi (1957)
Torulopsis osloensis nom. nud. (Dietrichson, 1954)
Saccharomyces osmophilus nom. nud. (Barre & Galzy, 1960)
Kluyveromyces osmophilus Kreger-van Rij (1966)
Torulopsis mogii Vidal-Leiria (1966a)
Saccharomyces bailii Lindner var. *osmophilus* van der Walt (1970e)

Standard description of *Zygosaccharomyces rouxii*

Growth in malt extract: After 3 days at 25°C the cells are subglobose, ovoidal or ellipsoidal (2.5–7.0) × (4.0–9.5) μm, elongate cells may occasionally be present, single, in pairs or small groups. A sediment and sometimes a thin ring, islets or a film are present. After one month at 20°C a sediment and a ring, sometimes also a film or islets, are present.

Growth on malt agar: After one month at 20°C the streak culture is cream to tan, smooth, crateriform, wrinkled or folded, rather flat or raised, glossy or dull, with an entire, lobate or ragged margin.

Slide cultures on corn meal- and morphology agar: Rudimentary pseudohyphae may be formed.

Formation of ascospores: Conjugation of individual cells usually precedes the formation of asci containing one to four globose to short-ellipsoidal ascospores which may appear to be either rough or smooth. Conjugation between a cell and its bud may occur and predominates in some strains, sometimes diploid cells are transformed direct into asci.

Sporulation was observed on YM agar + 2% NaCl, V8-, Gorodkowa-, potato-, malt- or corn meal agar in strains either alone or when mixed with the compatible mating type. Most strains soon lose their ability to reproduce sexually when maintained in the laboratory.

Fermentation:

Glucose	+ (s)	Maltose	v
Galactose	–	Lactose	–
Sucrose	+ s or –		

Assimilation of carbon compounds:

Galactose	v	Raffinose	–	Erythritol	–
Sucrose	v	Soluble starch	–	Ribitol	v
Maltose	v	D-Xylose	–	D-Mannitol	v
Cellobiose	–	L-Arabinose	–	Succinic acid	– (+)
Trehalose	v	D-Ribose	+ s	Citric acid	–
Lactose	–	L-Rhamnose	–	Inositol	–

Additional carbon compound tested: Melibiose –

Assimilation of nitrate: –

Assimilation of ethylamine · HCl: +

Assimilation of cadaverine · 2 HCl: +

Growth in vitamin-free medium: v; **in inositol-free medium:** +

Growth on 50% (w/w) glucose-yeast extract agar: +

Growth at 37°C: v

Growth in the presence of 100 ppm of cycloheximide: –

Growth in the presence of 1% acetic acid: –

G + C: 39.8–40.0 mol%, 4 strains (Nakase & Komagata, 1971c, d); 40.0 mol%, one strain (Bicknell & Douglas, 1970); 39.0–41.2 mol%, 10 strains, including CBS 732 *Zygosacch. rouxii* 40.0 mol%; CBS 736 *Zygosacch. mellis* 40.0 mol%; CBS 4021 *Zygosacch. bailii* var. *osmophilus* 41.0 mol%

Comments

See *Zygosaccharomyces bailii.*

Origin of the strains studied

Concentrated grape must (2), soya (4), syrup of orange peel (1), apple jelly (2), wine (3), confectionary (1), date (1), strawberry juice (1), grapes (4), lemonade (1), honey (9), maple syrup (1), molasses (2), ginger cake (4), urine (1), sputum (1), pickled cucumbers (1), marron glacé (1), juice of salted beans (1), spots on shelf in provision store (1), nukamiso pickles (1), marmelade (1), miso (8), glacé fruit (1), sugar (1), bee food (1), orange juice concentrate (1), black-currant drink (2), unknown (7).

Type strain: CBS 732, the type strain of *Zygosaccharomyces gracilis* Karamboloff et Krumbholz subsp. *italicus* Sacchetti isolated from grape must, was designated as type by Lodder and Kreger-van Rij (1952).

f. COMMENTS ON THE GENUS

See under *Saccharomyces,* section f., Comments on the genus, p. 393.

Chapter IV

Discussion of the genera belonging to the basidiosporogenous yeasts

The basidiosporogenous yeasts include three groups:
A. Filobasidiaceae, the yeast-like genera
B. Teliospore-forming yeasts
C. Tremellales with a yeast phase

A. Filobasidiaceae, the yeast-like genera

Introduction and key to the genera

by K.J. Kwon-Chung

Three genera (*Filobasidium, Filobasidiella, Chionosphaera*) are included in the family Filobasidiaceae. The family is characterized by the formation of slender non-septate basidia with terminal sessile basidiospores; thick-walled probasidia (teliospores) absent; blastospores present. Because the metabasidia, basidiospores and life cycle resemble the smut fungi, Olive (1968) placed the family Filobasidiaceae in the order Ustilaginales.

Donk proposed to place Filobasidiaceae and other basidiomycetous yeasts in Brefeld's hemibasidiomycetes (Donk, 1973) without assigning their orders. He questioned whether they are basidiomycetous but temporarily grouped them together as 'the saprobic Hemibasidiomycetes'. His concept of hemibasidiomycetes included smuts (Ustilaginales), bunts (Tilletiales) and Graphiolales to separate them from the traditional concept of heterobasidiomycetes. Cox (1976) proposed to move Filobasidiaceae from Ustilaginales to the Aphyllophorales of the homobasidiomycetes. His reason was that the metabasidia produced by the members of the family are holobasidia rather than phragmobasidia and the basidiospores are terminal. Subdivision of basidiomycetes into hetero- and homobasidiomycetes, which was based on the morphology of basidia (Rogers, 1934, 1971), is no longer widely accepted. Ainsworth (1973) treated the basidiomycetes (Basidiomycotina) as a subdivision of Eumycota with three classes (Teliomycetes, Hymenomycetes, Gasteromycetes) which are grouped on the basis of teliospore formation and status of the basidiocarp. Since members of Filobasidiaceae produce neither teliospores nor a well developed basidiocarp, they do not fit satisfactorily into any of the three classes. Although a *Chionosphaera* species is reported to produce fasciculate small fruiting bodies, these are inconsistent and not comparable to the basidiocarps of Hymenomycetes or Gasteromycetes. However, the relationship of Filobasidiaceae with some members of Teliomycetes, as well as some members of Phragmobasidiomycetidae of Hymenomycetes, is clear. The non-septate metabasidia that produce apical sessile spores are strongly reminiscent of the Tilletiaceae of Teliomycetes. The life cycle involving indefinite generations of budding yeasts in the haploid state is commonly found among the members of Teliomycetes and some members of Phragmobasidiomycetidae of Hymenomycetes. The chemical structure of the extracellular polysaccharide of *Filobasidiella* species (Bhattacharjee et al., 1978, 1979) is similar to *Tremella mesenterica* (Fraser et al., 1973). Species of *Filobasidium* and *Filobasidiella* produce enzymes such as extracellular deoxyribonuclease and urease which are produced by other basidiomycetous yeasts and the *Tremella* species (Cazin et al., 1969). The ultrastructure of the septal complex shows that the members of Filobasidiaceae lack typical parenthesomes of the dolipore. The dolipore in *Filobasidium capsuligenum* (Moore & Kreger-van Rij, 1972) is identical to that in a *Tremella* species, which is significantly different from the Holobasidiomycetidae. Our opinion is that Filobasidiaceae bridges the gap between Teliomycetes and Phragmobasidiomycetidae of Hymenomycetes. The unique manner of basidiospore formation of *Filobasidiella* does not fit into any of the previously referred orders, such as Ustilaginales or Aphyllophorales. Jülich (1981) established the order Filobasidiales to accommodate this family.

Recently Crane and Schoknecht (1978) described the genus *Rogersomyces* in the Filobasidiaceae to accommodate a new aquatic fungus, *Rogersomyces okefenokeensis*. The genus does not fit satisfactorily within the Filobasidiaceae because its basidiospores are produced on the sterigmata and the fungus lacks a yeast phase.

KEY TO THE GENERA OF FAMILY FILOBASIDIACEAE

1a Basidia arranged in synnemata
Chionosphaera . . . . p. 470

b Basidia produced on the loose hyphae 2

2a Basidiospores produced in chains
Filobasidiella . . . . p. 472

b Basidiospores not produced in chains
Filobasidium . . . . p. 483

A discussion of the three genera in alphabetical order follows.

Genus 1. *Chionosphaera* Cox

by K.J. Kwon-Chung

a. DIAGNOSIS OF THE GENUS

Cox (1976) described a monotypical genus *Chionosphaera* in Filobasidiaceae to accommodate a yeast which produces a fruiting body resembling the basidiocarp of homobasidiomycetes. The species *Chionosphaera apobasidialis* was isolated from dead

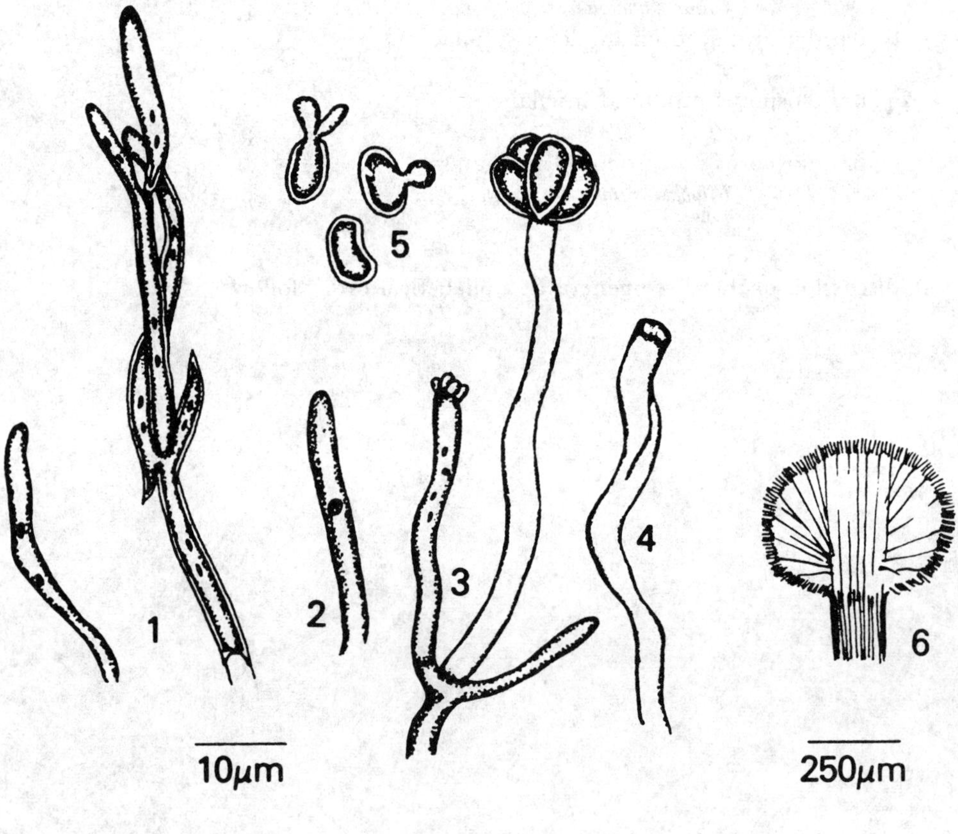

Fig. 142 – *Chionosphaera apobasidialis*
1. Dikaryotic hyphae and early stage of basidium formation 2. Diploidization in young basidium 3. Basidia with basidiospores 4. Empty basidium after release of basidiospores 5. Germinating basidiospores 6. A basidiocarp as it occurs in nature

limbs of *Quercus macrocarpa* and *Carpinus caroliniana* by Cox. Unfortunately, the cultures were not deposited in any culture collection centers and the type material was not deposited in any herbaria. All of the isolates Cox studied were lost. The following generic diagnosis is from the original article by Cox and it is hoped that the fungus will be reisolated from nature.

The fungus may occur in one or two phases. The haploid phase consists of thin-walled oval to oblong uninucleate budding yeast cells with a white mucoid colony. Hyphae or ballistospores are not produced. Pseudomycelium may be produced after several weeks of growth. The dikaryotic phase consists of hyaline hyphae which develop after conjugation between haploid cells of opposite mating types. Clamp connections are not present. Karyogamy and meiosis seem to take place in a clavate basidium (Fig. 142). Fruiting bodies are produced in culture only when *Chionosphaera* is grown in association with a species of *Cladosporium* or other appropriate filamentous hyphomycetes. The fruiting body is small, stalked, with an enlarged upper portion covered with a hymenium of basidia.

f. COMMENTS ON THE GENUS

The characteristics of fusion of yeast cells by conjugation tubes, dikaryotic hyphae without clamp connections and haustorium-like structures in *Chionosphaera apobasidialis* are similar to those reported for *Cryptococcus laurentii* (Kuff.) Skinner (Kurtzman, 1973). Elongated non-septate basidia, with a flower-like arrangement of basidiospores of *Ch. apobasidialis* resemble those of *Filobasidium* species. The major difference between the two genera is the formation of fasciculate fruiting structures by *Chionosphaera,* which is absent in *Filobasidium*. The degree of fasciculation in *Chionosphaera* varies greatly depending on the culture condition. The fasciculate, globose fruiting body is not produced in pure culture and is produced only in the presence of some dematiaceous hyphomycetes such as *Cladosporium* or *Chloridium* spp. Most fruiting bodies of *Chionosphaera apobasidialis* in nature are compact structures less than 1 mm tall, having sturdy stripes with a globose upper portion which resembles the pileus of homobasidiomycetes. Some fruiting structures produced in culture media or on tree limbs placed in a moist chamber are loosely interwoven fascicles or coremia bearing scanty basidial structures. Basidial structures with highly branched fragile synnemata are found on limbs covered by forest litter. In culture, a nutritious medium, e.g., full-strength potato dextrose agar, enhances the degree of fasciculation as well as the formation of globose fructifications just as in the presence of *Cladosporium* or *Chloridium* spp. Although the relationship between *Chionosphaera apobasidialis* and the hyphomycetes is not clear, *Filobasidium floriforme* has a similar type of relationship with *Alternaria* species in which *F. floriforme* parasitizes *Alternaria* hyphae (Olive, 1968).

Genus 2. *Filobasidiella* Kwon-Chung

by K.J. Kwon-Chung and J.W. Fell

a. DIAGNOSIS OF THE GENUS

Asexual reproduction by budding of globose, oval or apiculate cells. Pseudo- or true mycelium lacking in haploid cultures. Growth on solid media is white or cream colored, commonly mucous. Visible carotenoid pigments lacking. Cells are capsulated and the degree of capsulation depends on the environmental factors.

Sexual reproduction is primarily heterothallic. There are a few homothallic strains and genetic control of their sexual pattern is incompletely understood. In these isolates, hyphae with clamp connections develop from uninucleated yeast cells in the apparent absence of mating. Dikaryons may be established throughout the hyphae or restricted to the terminal cells only. Slender nonseptate basidia (holometabasidia) with terminal basidiospores are produced laterally or terminally on the hyphae. Uninucleate basidiospores bud basipetally from four spots on the apex of the basidium forming four long chains of basidiospores. The basidiospores produce homothallic colonies but during many generations of cloning, the strains become heterothallic.

In heterothallic isolates, dikaryotic phase develops following the conjugation of two haploid cells of α and a mating types. Dikaryotic hyphae with clamp connections are produced from the conjugants. The hyphal cross wall contains a typical dolipore but lacks the parenthesomes (Kwon-Chung & Popkin, 1976). A slender holometabasidium with an abruptly expanded apex arises laterally or terminally from the dikaryotic hyphae. Karyogamy takes place in the metabasidium and the zygote nucleus undergoes meiosis in the apical area. Mitosis follows meiosis and uninucleate basidiospores bud basipetally from four spots on the apex of the metabasidium to form four chains of basidiospores. The nuclear genotype of the spores from a simple chain is not identical (Kwon-Chung, 1980), indicating that the nucleus is distributed randomly in the budded spores.

b. TYPE SPECIES OF THE GENUS

Filobasidiella neoformans Kwon-Chung

c. SPECIES ACCEPTED IN THE GENUS

1. *Filobasidiella neoformans* Kwon-Chung (1975)
   Filobasidiella neoformans Kwon-Chung var. *neoformans*
   Filobasidiella neoformans Kwon-Chung var. *bacillispora* (Kwon-Chung) Kwon-Chung (Kwon-Chung, 1976a)

d. KEY TO THE TWO VARIETIES OF *FILOBASIDIELLA NEOFORMANS*

1a Basidiospores spherical, oblong, elliptical, or cylindrical and finely rough, (–)-malic acid not assimilated or slowly and canavanineglycine-bromthymol blue (CGB) agar negative within 2–5 days
Fil. neoformans var. *neoformans* p. 474

b Basidiospores smooth, bacillary form with or without curvature, (–)-malic acid assimilated and CGB agar blue within 2–5 days
Fil. neoformans var. *bacillispora* .... p. 477

e. SYSTEMATIC DISCUSSION OF THE SPECIES

1. *FILOBASIDIELLA NEOFORMANS* KWON-CHUNG (1975)

This species has two varieties:
Filobasidiella neoformans Kwon-Chung var. *neoformans*
Filobasidiella neoformans Kwon-Chung var. *bacillispora* Kwon-Chung (in Kwon-Chung et al., 1982a)

Imperfect state of *Filobasidiella neoformans* var. *neoformans: Cryptococcus neoformans* (Sanf.) Vuillemin var. *neoformans*
Imperfect state of *Filobasidiella neoformans* var. *bacillispora: Cryptococcus neoformans* (Sanf.) Vuillemin var. *gattii* Vanbreuseghem et Takashio

Synonyms of the variety *neoformans*

Saccharomyces neoformans Sanfelice (1895 a,b)
Cryptococcus neoformans (Sanf.) Vuillemin (1901)
Torula neoformans (Sanf.) Weiss (1902)
Blastomyces neoformans (Sanf.) Arzt (1924) (see also Lodder, 1934)
Torulopsis neoformans (Sanf.) Redaelli (1931)
Debaryomyces neoformans (Sanf.) Redaelli, Ciferri et Giordano (1937)
Cryptococcus hominis Vuillemin (1901)

Atelosaccharomyces hominis (Vuill.) Verdun (1912)
Torulopsis hominis (Vuill.) Redaelli (1931)
Debaryomyces hominis (Vuill.) Todd et Herrmann (1936)
Saccharomyces hominis Costantin (1901)
Torula histolytica Stoddard et Cutler (1916)
Torulopsis histolytica (Stod. et Cutl.) Castellani et Jacono (1933)
Torula nasalis Harrison (1928)
Cryptococcus nasalis (Harrison) Dodge (1935)
Cryptococcus psicrophylicus Niño (1934)

Synonyms of the variety *bacillispora*

Cryptococcus bacillisporus Kwon-Chung et Bennett (in Kwon-Chung et al., 1978)
Filobasidiella bacillispora Kwon-Chung (1976a)

Standard description of *Filobasidiella neoformans* var. *neoformans*

Growth in malt extract: After 3 days at 25°C, haploid cells are mostly globose, or oval, occurring singly or with buds. The cell size ranges from 3–8 μm but is mostly 3–5 μm in diameter. Slight sediment is present. A thin ring may be present. The sediment increases as the incubation continues.

Growth on malt agar: After 3–5 days at 25°C, haploid cells are globose to oval measuring 2.5–10 μm in diameter. Colony is white to cream colored with smooth, mucoid texture; the margin is entire. In streak culture, the highly mucous growth runs to the bottom of the slant within one week.

Dalmau plate culture on corn meal agar: Pseudomycelium is absent in haploid cells.

Growth on agar medium containing phenolic compounds: Haploid colonies on media containing substances such as extract of niger seed *(Guizotia abyssinica)* (Staib, 1962; Shields & Ajello, 1966), caffeic acid (Pulverer & Korth, 1971; Hopfer & Blank, 1975; Hopfer & Gröschel, 1975; Healy et al., 1977; Fleming et al., 1977), chlorogenic acid (Shaw & Kapica, 1972) or L-DOPA (Chaskes & Tyndall, 1975) become brown, dark brown to black usually within one to 5 days depending on the media and isolates. Some strains may require 8 days to produce the characteristic pigment on niger seed agar.

Growth on CGB agar: The color of the medium does not change within 2–5 days at 25°C.

Life cycle: Diagram 1 illustrates the life history of *Fil. neoformans* var. *neoformans*. Most isolates are heterothallic but a few strains are homothallic, producing basidiospores without apparent mating. The mating in heterothallic strains is controlled by a bipolar incompatibility system. Two alleles (*α* and *a*) at one locus have been identified. Both heterothallic and homothallic strains produce a mycelium from yeast phases. The heterothallic strains produce dikaryotic hyphae with clamp connections when haploid yeast cells of *α* and *a* mating types are mixed on appropriate media such as malt extract-, V8 juice- and hay infusion agar and the homothallic isolates produce hyphae with clamp connections from a single uninucleate cell in the absence of mating. The nuclear states of the hyphae produced by the homothallic strains are also

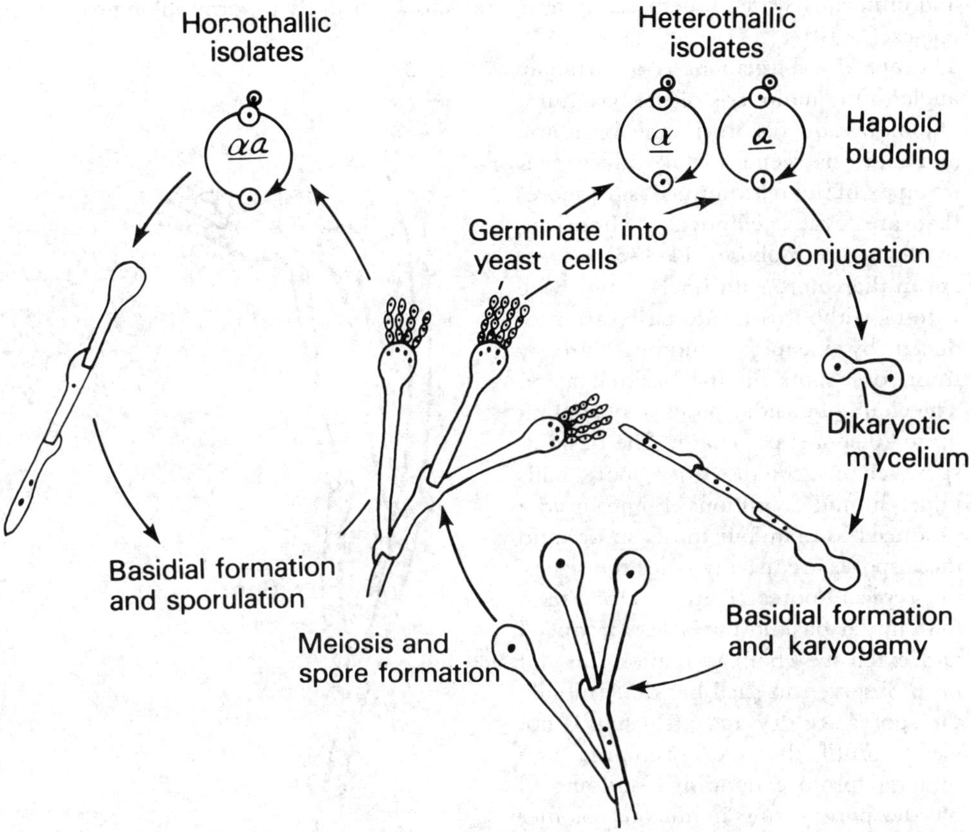

Diagram 1
Life cycle of *Fil. neoformans*

dikaryotic but in some isolates a dikaryon was observed only in terminal cells (Erke, 1976). It is not clear whether the dikaryotic terminal cells represent young growing tips of hyphae or are destined to be the basidiophores. The life cycle discussed below is compiled from gross morphological observations, nuclear stainings and genetic analysis (Kwon-Chung, 1976b, 1980).

Heterothallic strains: Cells of α and a mating types conjugate when mixed on malt extract agar, V8 juice agar, hay infusion agar or sucrose-yeast agar at 25–37°C. Dikaryotic hyphae with clamp connections develop within 3–5 days from the conjugants. After one week, mycelial growth is evident to the naked eye around the margin of colony. The hyphae are hyaline and average 3 μm wide, with a clamp connection at each septum. Regular hyphal branching occurs at the clamps and also irregular thin, 1–1.5 μm, branches termed haustorioid branches originate from clamps. Blastospores bud laterally and terminally from the hyphae and these resemble haploid yeast cells in their size and shape. Mature basidia arise, within one to 2 weeks, singly or in a group laterally or terminally on the hyphae. Basidia are non-septate, slender and uniformly wide in the basal area but expanded abruptly to form oblong to clavate apices (4–10) × (6–13) μm (Fig. 143). The apical enlargement is often at an angle to the main axis of the basidium. Diploidization occurs in the basidium and meisosis seems to take place near the apex of the basidium. Basidiospores that are ovate, elliptical, cylindrical, pyriform to globose, (1.8–3) × (2–5) μm in diameter, with finely rough wall with or without truncate base, are produced by basipetal budding directly from four spots on the basidial apex. The chain of basidiospores may contain up to 20 spores per chain. The basidiospore chains collapse into spore balls under humid conditions. Some basidia produce less than four initial spores and these spores are usually twice the size of the typical spores. Rarely, more than four initial basidiospores are produced and extensive chain formation has not been observed in such basidia. Initially the spores are dry and a capsule is not visible until the spores enlarge and undergo profuse budding. Budding of basidiospores results in mucous colonies of yeast cells representing α, a, or infrequently αa (self-fertile) mating types.

Cytological (Kwon-Chung, 1976b) and genetical studies (Kwon-Chung, 1980) indicate that the small four nuclei of first meiotic product remain in the basidium and each nucleus divides

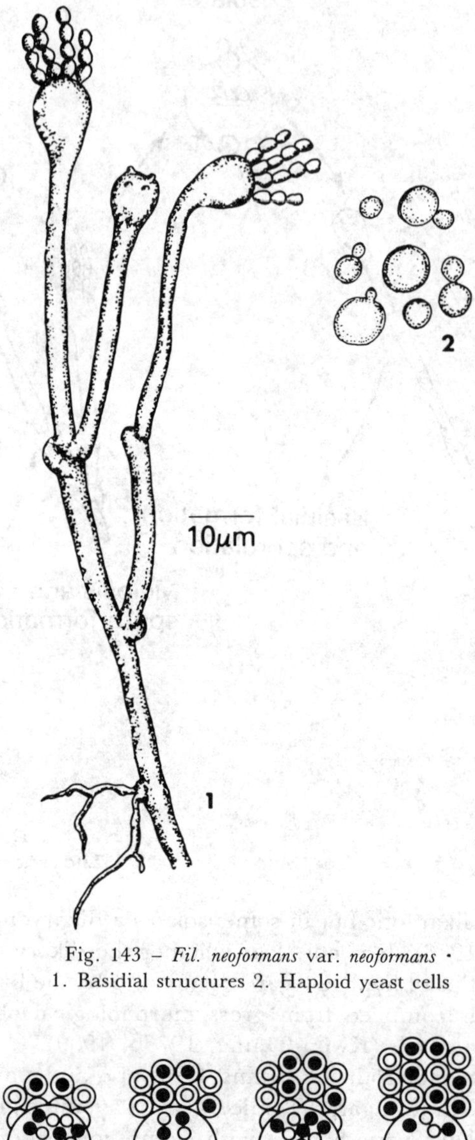

Fig. 143 – *Fil. neoformans* var. *neoformans* · 1. Basidial structures 2. Haploid yeast cells

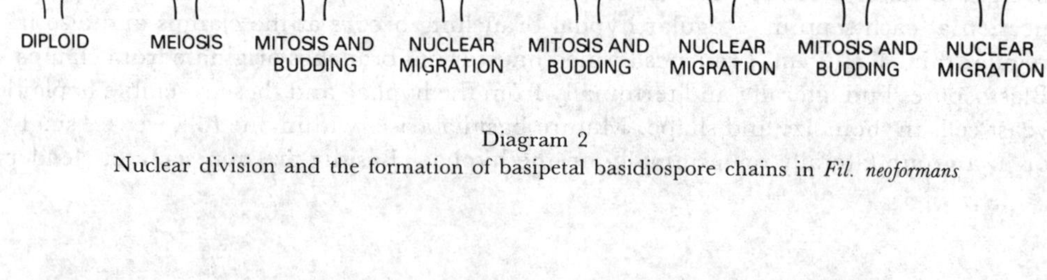

Diagram 2
Nuclear division and the formation of basipetal basidiospore chains in *Fil. neoformans*

mitotically. Daughter nuclei migrate into basidiospores while mother nuclei divide repeatedly during basipetal budding to form uninucleate basidiospores in chains. The spore chains show random distribution of nuclei and, as a result, the spores from a single chain are not genetically identical (Diagram 2).

Homothallic strains: Shadomy (1970) demonstrated hyphal formation in 11 single cultures (Lurie & Shadomy, 1971) and three of the hyphae-forming isolates completed a sexual life cycle without apparent mating (Erke, 1976). The isolates were from human cases of cryptococcosis. The isolates first grew in yeast form and produced hyphae with clamp connections either from single yeast cells or from a pair of yeast cells connected by a clamp connection. Dikaryons were restricted to terminal cells in some strains and not restricted in others. The morphological aspects of basidium and basidiospores were the same as in the mated cultures of heterothallic strains. Colonies developed from single basidiospores were homothallic but yeast cells cloned from such strains became markedly weaker in sporulation in subsequent generations and eventually behaved as one mating type (Kwon-Chung, 1978). Erke (1976) described these self-fertile isolates simply as homothallic but their ploidy is unclear.

Fermentation: –

Assimilation of carbon compounds:

Galactose	+	Raffinose	+ or w	Erythritol	v
Sucrose	+	Soluble starch	+ w or –	Ribitol	v
Maltose	+	D-Xylose	+	D-Mannitol	+
Cellobiose	+ or w	L-Arabinose	+ or s	Succinic acid	v
Trehalose	+	D-Ribose	+ or w	Citric acid	v
Lactose	–	L-Rhamnose	+	Inositol	+

Additional carbon compounds tested: (–)-Malic acid – or + s

Assimilation of nitrate: –

Growth in vitamin-free medium: –

Growth on 50% (w/w) glucose-yeast extract agar: –

Growth at 37°C: +

G + C: 53.2–56.7 mol%, 3 strains (Aulakh et al., 1981)

Natural habitat: pigeon excreta, soil

Supplementary description of *Filobasidiella neoformans* var. *bacillispora*

This variety tends to produce oval to lemon-shaped, (3–5) × (3–7) μm, haploid cells more commonly at 25°C and elongated cells at 37°C. The growth rate is the same at 25°C in the two varieties, but at 37°C var. *bacillispora* is slower than the var. *neoformans*.

In contrast to *Fil. neoformans* var. *neoformans, Fil. neoformans* var. *bacillispora* changes the color of CGB medium into blue-green within 2–5 days and assimilates (–)-malate within 2 days at 25°C.

The life cycle of var. *bacillispora* resembles that of *Fil. neoformans* except that 'haustorioid hyphae' have not been observed. The basidiospores are bacillary (1–1.8) × (3–8) μm, and smooth (Fig. 144). The sexual cycle is produced most abundantly on V8 juice with 4% agar (pH 7) at 25°C in the dark.

The natural habitat of this variety is unknown.
G + C: 56.3–57.2 mol%, 2 strains (Aulakh et al., 1981)

Origin of the strains belonging to the variety *neoformans*

CBS 132 (type strain of *Cryptococcus neoformans*) from peach juice (Sanfelice, 1894), pigeon excreta (4), human (9), single basidiospore isolates (2).

Type strains: CBS 6886 (*a* type) isolated from pigeon excreta in Denmark and CBS 6885 (α type) isolated from a human bone lesion.

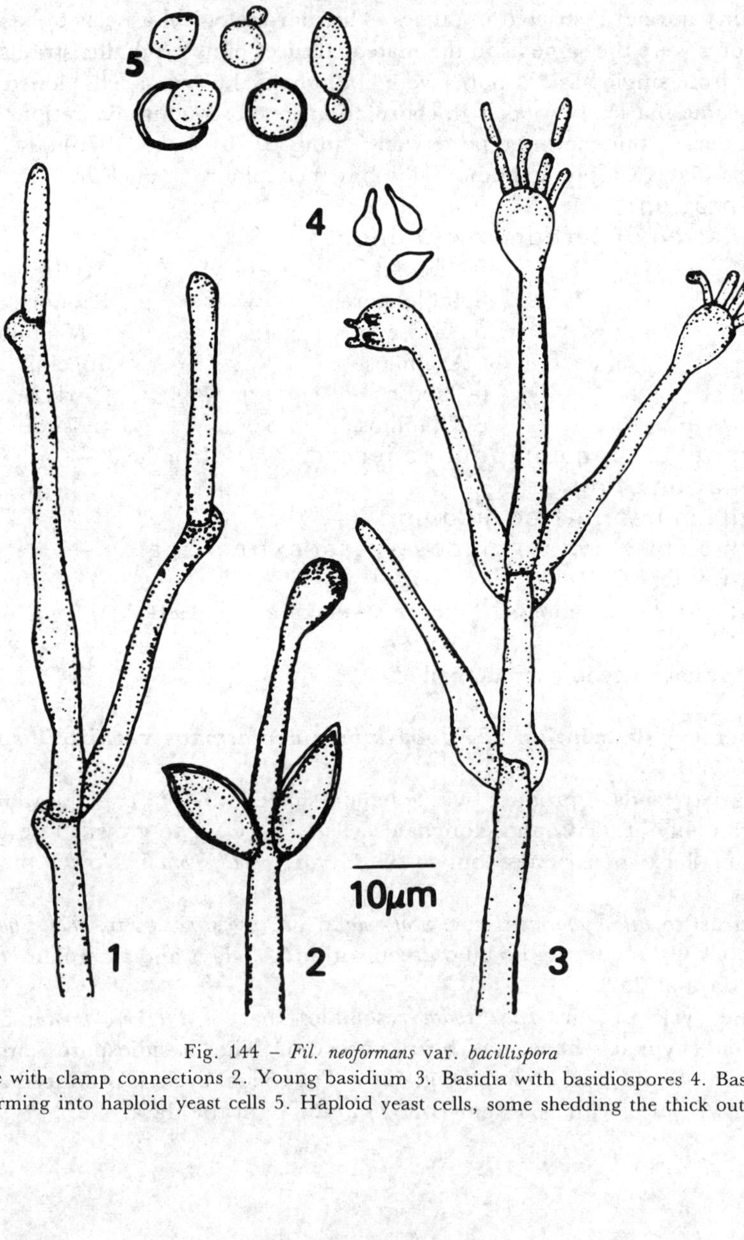

Fig. 144 – *Fil. neoformans* var. *bacillispora*
1. Hyphae with clamp connections 2. Young basidium 3. Basidia with basidiospores 4. Basidiospores transforming into haploid yeast cells 5. Haploid yeast cells, some shedding the thick outer wall

Origin of the strains belonging to the variety *bacillispora*

Eleven strains from human sources. CBS 6992 (serotype B) and CBS 6993 (serotype C) were originally received from Dr. Evans.

Type strains: CBS 6956 (α mating type, serotype B, from sputum) and CBS 6955 (*a* mating type, serotype C, from spinal fluid).

f. COMMENTS ON THE GENUS

Cryptococcus neoformans has been regarded as a homogeneous species except for the existence of four serotypic variations (Evans, 1949, 1950; Wilson et al., 1968) among the isolates. In recent years, evidence of heterogeneity has become increasingly clear within the species. *Cr. neoformans* was considered to contain two perfect states, *Fil. neoformans* (Kwon-Chung, 1975) and *Fil. bacillispora* (Kwon-Chung, 1976a). The *Fil. neoformans* state was produced by the cross between two mating types of serotype A or D isolates, while the *Fil. bacillispora* state was observed when crosses were made among serotype B or C isolates. Intercrossing between A,D and B,C isolates usually produced basidia with two types of spores, the species of *Fil. neoformans* type and the *Fil. bacillispora* type, but in some cases predominantly of the *Fil. bacillispora* type (Rogers et al., 1980). The spores produced by the intercrossings between the type cultures of the two species and many other pairs were sterile. However, CBS 6991 (*a* type, serotype D) produced viable basidiospores (30%) when crossed with CBS 6956 (α type, serotype B). The viable basidiospores produced 1:1 ratio of α and *a* mating type colonies but all were serotype D. All the α type F_1 strains assimilated L-malate while *a* type strains did not. Since the 70% of the basidiospores were sterile, it was not possible to analyze whether the viable progeny were recombinants or whether the D factor was a dominant cytoplasmic factor. Further analyses are required to evaluate these phenomena. The haploid states of *Fil. neoformans* and *Fil. bacillispora* also showed differences in their epidemiology, ecology and biochemistry. While serotype A was widespread throughout the world, D was rare in the United States but common in Europe. B and C types were rare causes of infection except in Southern California (Bennett et al., 1977). Also, the natural reservoir for serotypes A and D seemed to be distinct from that of B and C. Serotypes A and D were associated with pigeon droppings, while the ecological niche of B and C remains unknown. Pigeons seem to be a carrier of *Cr. neoformans* since pigeons caged in different parts of the world contain the fungus in their crops. We have tested the serotype of 11 isolates obtained by Vanbreuseghem from pigeon crops (caged pigeons) and they were all serotype A. The strains of *Fil. bacillispora* and *Cr. neoformans* serotypes B and C have only been isolated from clinical specimens. The two natural isolates reported as serotype C by Denton and Disalvo (1968) and Walter and Coffee (1968) were subsequently determined as serotype A. Although isolates of *Cr. neoformans* are frequently obtained from excreta and the soil contaminated with pigeon droppings, these substrata may not be the primary ecological niche of *Fil. neoformans.* Since the closely related fungus, *Filobasidium floriforme,* has been isolated from dead florets of the large plume grass, *Erianthus giganteus* (Olive, 1968) and *Hibiscus siriaca* (ATCC catalog 1976), effort should be directed to the isolation of *Fil. neoformans* from dead plants. Survey of mating types (Kwon-Chung et al., 1978) revealed that the ratio of α and *a* type was 43:1 among 88 soil and pigeon isolates and 51:1 among 152 clinical isolates of serotypes A and D. About 18% of soil and pigeon isolates and 13% of clinical isolates did not cross with the tester strains. Other workers (Bowman & Ahearn, 1978) have shown even higher percentage of sterile strains. Unequal distribution of α and *a* types was also noted among 31 strains of serotypes B and C.

Biochemical differences noted between serotype A,D and B,C included assimilation of (–)-malic acid and creatinine metabolism (Bennett et al., 1978). Isolates of *Fil. bacillispora* (100%) assimilated (–)-malic acid as sole source of carbon and growth could be detected within 48 hours. Some strains (7%) of serotype B which were unsuccessful in mating, failed to assimilate (–)-malic acid within 5 days. Most (86%) of *Fil. neoformans* did not assimilate the dicarboxylic acid and growth could be detected only after 5 days in the remaining isolates (14%). The strains of serotype B,C utilized creatinine better than A and D types and produced more ammonia. The formation of ammonia resulted in the higher pH of creatinine-dextrose-bromthymol blue agar (CDB) and the color indicator turned blue. This simple one step technique distinguished isolates of serotype B,C from A,D with 97% accuracy by 48 hours. Based on these findings, serotypes B and C of *Cr. neoformans* (haploid state) were described as a separate species, *Cr. bacillisporus* (Kwon-Chung et al. 1978). The CDB medium was later modified by Kwon-Chung et al. (1982b) to obtain higher efficiency in separating the two groups. The modified medium, canavanine-glycine-bromthymol blue (CGB) agar, contains glycine as the sole source of carbon and nitrogen. All B and C serotype isolates assimilated glycine as the sole source of carbon and nitrogen, while most isolates of A and D type did not. L-Canavanine was added to the medium to repress the growth of some A or D serotype isolates which utilize glycine.

Polacheck and Kwon-Chung (1980) demonstrated that creatinine metabolism involves a single step resulting in the production of methylhydantoin and ammonia in both species. The enzyme responsible for this step was identified as creatinine deiminase. A difference was shown in the regulatory mechanism for the synthesis of creatinine deiminase between *Cr. neoformans* and *Cr. bacillisporus.* The enzyme was induced by creatinine in both species but only in *Cr. neoformans* ammonia repressed the enzyme synthesis. This difference explained the biochemical basis of the CDB agar test for the identification of the two species.

The G + C content of *Cr. neoformans* was reported to be 43–45.9 mol% by Erke and Schneidau (1973), 51.1 mol% by Storck et al. (1969) and 49–49.8 mol% by Nakase and Komagata (1971b). Serotypes of the isolates were not determined in these studies. Aulakh et al. (1981) studied the DNA base composition and sequence homology using the type cultures of *Fil. neoformans* and *Fil. bacillispora.* They showed that G + C contents ranged from 53.2 to 56.7 mol% for *Fil. neoformans* and 56.3–57.2 mol% for *Fil. bacillispora* which are significantly higher than those previously reported for *Cr. neoformans.* The lower values may have been due to the contamination with RNA, protein or polysaccharides. Another cause of erroneously low G + C content may have been the use of sheared DNA since Aulakh et al. showed that the G + C mol% ranges of 42.2–45.9 can be obtained with sheared DNA. For DNA hybridizations, all possible homologous and heterologous combinations were used. The DNAs of serotypes A and D of *Fil. neoformans* demonstrated relatedness of 87.7–93.5%. DNAs of serotypes B and C of *Fil. bacillispora* showed 88.5% relatedness. DNAs of *Fil. neoformans* were significantly different from those of *Fil. bacillispora* in that relatedness values of only 55.2–63% were observed in cross hybridization experiments. This DNA relatedness value is higher than generally accepted values which characterize different species.

Vanbreuseghem and Takashio (1970) reported an atypical strain of *Cr. neoformans* which they considered as a variety of the species, *Cr. neoformans* var. *gattii*. The variety produced elongated or bacilliform cells and round cells in the mouse brain.

The biochemical and serological characteristics of the type strain of the variety were identical to those of *Cr. bacillisporus*. It was serotype B, (–)-malate assimilating, CGB positive, and crossed with *a* type of *Fil. neoformans* as well as *Fil. bacillispora*. The morphology of basidiospores that was produced by *Cr. neoformans* var. *gattii* (α type) × *Fil. bacillispora* (*a* type) was typical of the *Fil. bacillispora* type. The cross between *Cr. neoformans* var. *gattii* and *Fil. neoformans* produced basidia with two types of spores: the species of the *Fil. neoformans* type and the *Fil. bacillispora* type. Although the germination rate of the basidiospores produced by *Cr. neoformans* var. *gattii* × *Fil. neoformans* was less than 50%, recombinants were found among the F_1 progeny. Some F_1 progeny of serotype D assimilated (–)-malate, indicating that the (–)-malate marker segregates independently from the serotype marker. These results together with the relatively high value of DNA sequence homology between the two species indicate that they should be classified as two varieties of the same species. By such decision, *Fil. bacillispora* becomes *Fil. neoformans* var. *bacillispora,* but the proper variety name for *Cr. bacillisporus* should be *Cr. neoformans* var. *gattii* since the name has the priority over *Cr. bacillisporus*.

Because of the medical importance of these species, various methods for rapid identification were developed. Staib (1962) found that *Cr. neoformans* was the only tested species of yeast to produce dark brown colonies when grown on an agar medium containing the extract of the bird seed *Guizotia abyssinica*. Shields and Ajello (1966) developed a medium for selective isolation of *Cr. neoformans* by incorporating an extract of *G. abyssinica* into an agar medium containing glucose, creatinine, chloroamphenicol and diphenyl. In 1971, Pulverer and Korth found that various *o*- and *p*-diphenols and *o*-triphenols gave the similar results of the *G. abyssinica*. Strachan et al. (1971) determined that hydroxyl groups in the 3- and/or 4-position of the phenyl ring were common to all compounds which give rise to dark brown pigment. Shaw and Kapica (1972) reported that 3,4-dihydroxyphenylalanine phenoloxidase found in the cells of *Cr. neoformans* catalyzes the melanin synthesis from *o*-diphenols.

According to our experience, the melanin formation is a simple and reliable screening criterion for the two varieties of *Cr. neoformans*. Niger seed agar, chlorogenic acid agar or DOPA agar are satisfactory media for the formation of melanin pigment within 1–5 days by the two variaties. Intensity of the pigment may be slightly weaker in the *Cr. neoformans* var. *gattii* compared with the *Cr. neoformans* var. *neoformans*. The concentration of carbon source in the medium plays a decisive role for the rapid formation of the pigment (Nurudeen & Ahearn, 1979). The medium containing 1% or higher concentration of glucose delays the formation of melanin compared with the medium containing 0.1 or 0.05%. Most isolates of the two varieties produce brown colonies within 1–3 days on niger seed agar with 0.1% glucose. Total elimination of glucose from niger seed agar allowed 69 of 80 isolates to develop the pigment within 12 hours (Palawal & Randawa, 1978). For the primary isolation of the two species from clinical specimens, soil or pigeon excreta, the use of niger seed agar with 0.1% glucose (see p. 52) is recommended.

Niger seed agar is inexpensive to make and the seed extract has a long shelf life compared with various chemically defined substrates of phenoloxidase. Since some other yeast species may produce light brown pigment on the media with the phenolic compounds, the screening technique should also include growth at 37°C, presence of urease, absence of pseudomycelium and utilization of inositol as the sole source of carbon. The good growth at 37°C separates the two varieties from other species of *Cryptococcus* which produce light brown pigment on niger seed agar. Inositol utilization and the urease test separate them from many species of *Candida* which are isolated from clinical specimens. The two varieties of *Cr. neoformans* can be distinguished by the formation of blue-green color by *Cr. neoformans* var. *gattii* on CGB agar (Kwon-Chung et al., 1982b) within 24–48 hours while such color is not formed by *Cr. neoformans* var. *neoformans*.

Most isolates of the two varieties are sensitive to 12 μg per ml or less of cycloheximide and, therefore, agar media containing cycloheximide such as mycosel agar should not be used for isolation. Some isolates grow very slowly at 37°C and the primary isolation tubes or plates should be incubated at 25–35°C.

The mechanism of spore formation similar to *Filobasidiella* species is not found in the members of Ustilaginales (Fischer & Holton, 1957) and the authors are not aware of any comparable situation in basidiomycetes in general. Species of *Entyloma, Tilletia, Urocystis* and *Doassansia* of Ustilaginales produce terminal basidiospores like *Filobasidiella* species but long basipetal spore chains are not produced by the members of these genera. Species of *Schroeteria* (Ustilaginales) produce spores in chains but they are produced by acropetal budding (Zundel, 1953).

Cytology of spore formation in *Fil. neoformans* resembles partly that of *Ustilago* species (Alexopoulos & Mims, 1979). In *Ustilago maydis,* the nuclei of the first meiotic product divide mitotically and daughter nuclei migrate into the basidiospores (sporidia) while mother nuclei remain in the promycelium. Unlike the *Filobasidiella* species, however, the four mother nuclei are separated from each other by septa of promycelium and basipetal chains of sporidia are not produced. The establishment of the order Filobasidiales (Jülich, 1981), therefore, seems to be justified.

Recently, Malloch et al. (1978) described a new species *Filobasidiella arachnophila.* The fungus was described as homothallic and lacks clamp connections and a yeast phase. Although morphology of the fruiting body is similar to *Fil. neoformans,* there was no indication that the sporulation was through meiosis. The fungus may belong to the Fungi imperfecti.

Genus 3. *Filobasidium* Olive

by K.J. Kwon-Chung and J.W. Fell

a. DIAGNOSIS OF THE GENUS

Asexual reproduction by budding of globose, oval, elliptical to elongated cells. Pseudomycelium and true mycelium may be produced. Growth on solid media is white, cream colored, or slightly pinkish, often mucous.

Sexual reproduction is heterothallic. Although the sexual incompatibility systems are not fully understood, the three heterothallic species seem to be controlled by a bipolar system. A dikaryotic mycelium is produced from conjugated yeast cells (Diagram 3). Occasional, single strains have been shown to produce a dikaryon. A clamp connection is present at each cross wall, and the septum contains typical dolipores with or without parenthesomes. The parenthesomes resemble those found in *Tremella* species, rather than those found in homobasidiomycetes.

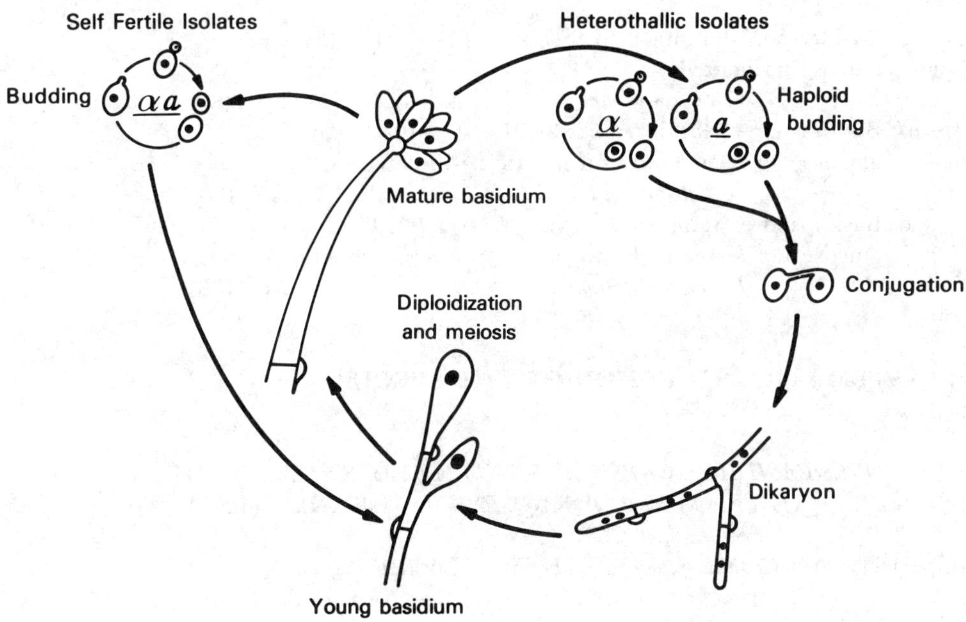

Diagram 3
Life cycle of *F. capsuligenum*

The mycelium produces lateral or terminal holometabasidia. Karyogamy seems to take place in the metabasidium. Sessile basidiospores are produced terminally on the apex of metabasidium in petal-like whorl, giving the basidium a flower-like appearance in apical view.

b. TYPE SPECIES OF THE GENUS

Filobasidium floriforme Olive

c. SPECIES ACCEPTED IN THE GENUS

1. *Filobasidium capsuligenum* (Fell, Statzell, Hunter et Phaff) Rodrigues de Miranda (Fell et al., 1969)
2. *Filobasidium floriforme* Olive (1968)
3. *Filobasidium uniguttulatum* Kwon-Chung (1977)

d. KEY TO THE SPECIES OF THE GENUS

1a Basidiospores elliptical to fusiform, (4–6) × (9–13) μm, glucose fermented
F. capsuligenum .... p. 485

b Basidiospores elliptical to oval, (6–9) × (10–17) μm, glucose not fermented and nitrate assimilated
F. floriforme .... p. 486

c Basidiospores reniform to oval, (3–5) × (4–9) μm, glucose not fermented and nitrate not assimilated
F. uniguttulatum .... p. 488

e. SYSTEMATIC DISCUSSION OF THE SPECIES

1. *FILOBASIDIUM CAPSULIGENUM* (FELL, STATZELL, HUNTER ET PHAFF) RODRIGUES DE MIRANDA (1972)

Imperfect state: *Candida japonica* Diddens et Lodder

Synonyms

Leucosporidium capsuligenum Fell, Statzell, Hunter et Phaff (1969)

Torulopsis capsuligenus van der Walt et van Kerken (1961b)
Candida capsuligena (van der Walt et van Kerken) van Uden et Buckley (1970)
Torulopsis albus Saito et Ota (1934)
Candida japonica Diddens et Lodder (1942)

Standard description of *Filobasidium capsuligenum*

Growth in malt extract: After 3 days at 25°C, the haploid cells are subglobose, oval, elliptical to elongated, (2.5–10.7) × (4.0–17.8) μm, single or in pairs. A thin ring and a little sediment are formed. CBS 4736 produced a heavier film than the remaining two strains tested. After one month, there is a moderate to thick slimy ring and a heavy sediment.

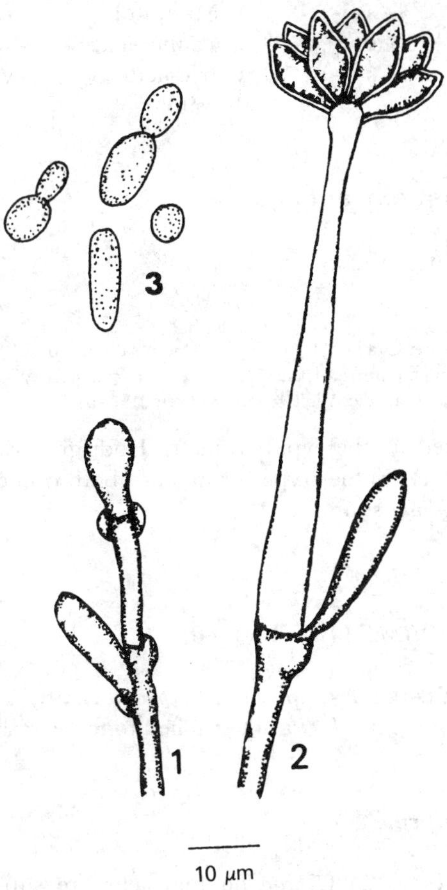

Fig. 145 – *F. capsuligenum*
1. Early stage of basidial formation
2. Mature basidial structures
3. Haploid yeast cells

Growth on malt agar: After 3 days, the cells are similar in size and shape to those formed in malt extract except that elongated pseudomycelial cells occur. The streak culture is white to cream colored, and the texture varies from dry to mucous. The colony margin is entire and smooth or with tufts of mycelium. After 3 weeks, mating type α and *a* strains show differences in colony morphology. Strains of α type are mucous, glistening, cream colored and surface raised with smooth margin; mycelium lacking. Colonies of mating type *a* have a soft, moist or dry texture and are not mucous. The margin of the colony is undulate and heavily myceliated. The mycelium consists of branched, septate hyphae.

Dalmau plate culture on corn meal agar: Pseudomycelium is absent in mating type α, while true mycelium is present in type *a*. Occasionally, a few short chains of undifferentiated, globose cells may be present.

Life cycle: On corn meal agar, conjugation occurs between two mating types within 12 hours of mixing at room temperature. Dikaryotic hyphae with clamp connections grow out from a conjugant pair within 48 hours. After 3 weeks, slender, holometabasidia measuring (3.0–6) × (70–140) μm are produced laterally or terminally on the dikaryotic mycelium; five to nine sessile basidiospores, elliptical to fusiform, (4–6) × (9.5–13) μm, bud on the apices of the metabasidia (Fig. 145). Occasional, single isolates may produce mycelium and basidial structures without apparent mating.

Fermentation:

Glucose	+	Maltose	+ s (–)
Galactose	–	Lactose	–
Sucrose	–		

Assimilation of carbon compounds:

Galactose	+	Raffinose	–	Erythritol	–
Sucrose	+	Soluble starch	+	Ribitol	+
Maltose	+	D-Xylose	+ s (–)	D-Mannitol	+
Cellobiose	+ or s	L-Arabinose	– or + w	Succinic acid	+ w
Trehalose	+	D-Ribose	+ s (–)	Citric acid	+ w
Lactose	–	L-Rhamnose	–	Inositol	+

Assimilation of nitrate: –

Growth in vitamin-free medium: –

Growth on 50% (w/w) glucose-yeast extract agar: –

Growth at 37°C: – or + w

Origin of the strains studied

CBS 1906 (*a* type) described as *Torulopsis albus* by Saito and Ota (1934) and isolated from sake-moto; CBS 4381 (*a* type) from cider by Beech, Bristol; CBS 4736, type culture of *Leucosporidium capsuligenum* (α type) from a wine cellar and described as *Torulopsis capsuligenus* (van der Walt & van Kerken, 1961b).

Type strains: CBS 1906 (*a*) was designated as the type strain by Rodrigues de Miranda (1972). We designate CBS 4736 (α) as another type strain since both α and *a* strains are required to produce the *Filobasidium* state.

2. *FILOBASIDIUM FLORIFORME* OLIVE (1968)

Imperfect state: Haploid state of *F. floriforme* fits the description of *Cryptococcus albidus*. However, the type culture and many strains of *Cr. albidus* varieties failed to mate with α or *a* type of *F. floriforme*.

Standard description of *Filobasidium floriforme*

Growth in malt extract: After 3 days at 25°C, the haploid cells are subglobose, oval, elliptical to long-oval, (2.5–6) × (3–8) μm, single, in pairs or in small clusters. Small amount of sediment is present. After one month, a moderate ring and sediment are present.

Growth on malt agar: Cell morphology is similar to that in malt extract. The

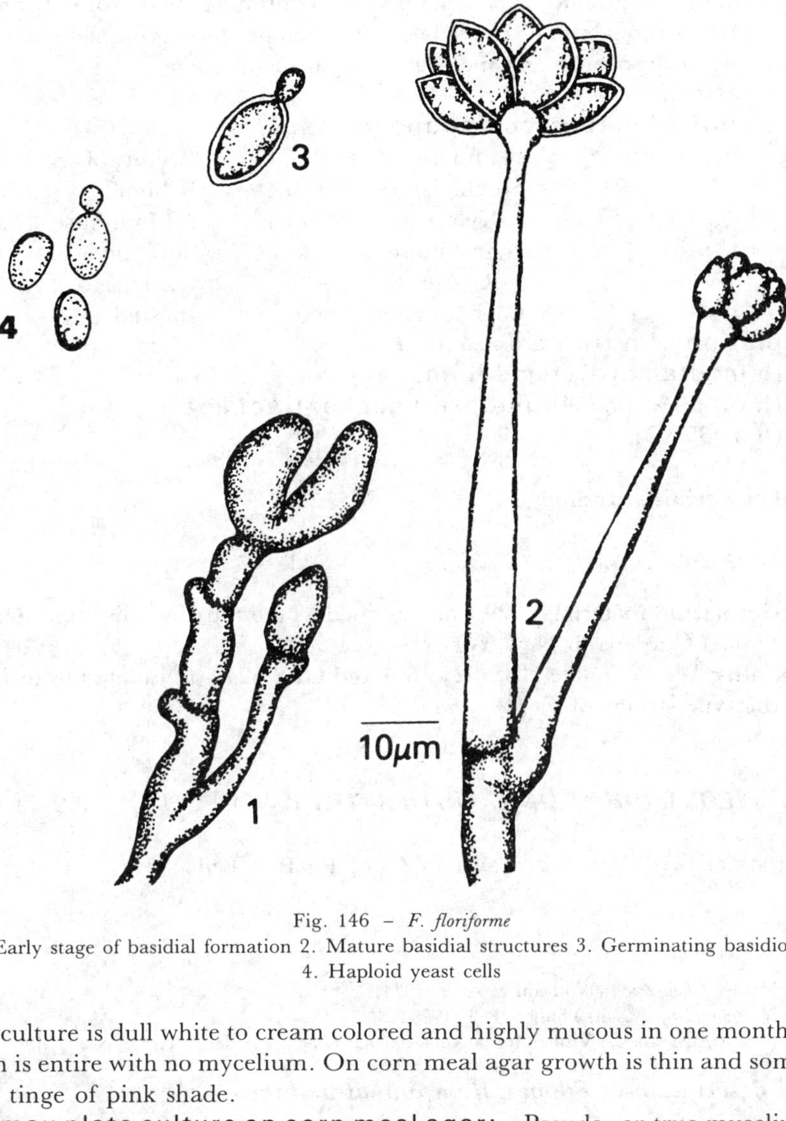

Fig. 146 – *F. floriforme*
1. Early stage of basidial formation 2. Mature basidial structures 3. Germinating basidiospore 4. Haploid yeast cells

streak culture is dull white to cream colored and highly mucous in one month. Colony margin is entire with no mycelium. On corn meal agar growth is thin and some strains bear a tinge of pink shade.

Dalmau plate culture on corn meal agar: Pseudo- or true mycelium is absent in haploid state.

Life cycle: Conjugation occurs when cells of two compatible mating types are mixed on V8 juice- or hay infusion agar. Soon after conjugation, dikaryotic hyphae with clamp connections are produced from the conjugant. The hyphal septum contains a dolipore but no parenthesomes. Long slender holometabasidia, (3.2–7) × (57–212) μm, arise mostly from branching clusters of hyphae. The thin-walled young metabasidia are slightly broader than the hyphae bearing them and contain a denser cytoplasm. Six to eight elliptical to oval basidiospores, (6–9) ×

(10.5–16.5) μm, are formed directly from the expanded apex (Fig. 146) of the basidia. Basidiospores produce blastospores which continue to bud profusely and form yeast colonies. Occasional isolates have been shown to produce a complete sexual state without mating. Karyogamy and meiosis occur in metabasidia.

Fermentation: –

Assimilation of carbon compounds:

Galactose	+ or s	Raffinose	–	Erythritol	–
Sucrose	+	Soluble starch	+ or s	Ribitol	+ s (–)
Maltose	+	D-Xylose	+ or s	D-Mannitol	+ s (–)
Cellobiose	+	L-Arabinose	+	Succinic acid	+ s
Trehalose	+	D-Ribose	–	Citric acid	+ s
Lactose	+	L-Rhamnose	+	Inositol	+

Assimilation of nitrate: +

Growth in vitamin-free medium: –

Growth on 50% (w/w) glucose-yeast extract agar: –

Growth at 37°C: – or + w

Origin of the strains studied

Plume grass (*Erianthus gigantus*) (3), *Hibiscus siriaca* (2).

Type: Herbarium material SC29 was designed as the type by Olive and deposited at the Botanical Gardens in New York.

Type strains: We designate CBS 6241 (*α*) and CBS 6242 (*a*) isolated from *Hibiscus siriaca* as the type strains of *F. floriforme*.

3. *FILOBASIDIUM UNIGUTTULATUM* KWON-CHUNG (1977)

Imperfect state: *Cryptococcus uniguttulatus* (Zach) Phaff et Fell

Synonyms

Eutorulopsis uniguttulata Zach (Wolfram & Zach, 1934)
Cryptococcus uniguttulatus (Zach) Phaff et Fell (1970)
Cryptococcus neoformans (Sanf.) Vuillemin var. *uniguttulatus* (Zach) Lodder et Kreger-van Rij (1952)

Standard description of *Filobasidium uniguttulatum*

Growth in malt extract: After 3 days at 25°C, the haploid cells are globose to oval, single, in pairs or in small clusters, (3.0–5.5) × (3.5–7.0) μm, rarely up to 9 μm long. A thin ring and a little sediment may be present. After one month, there is a moderate ring and a fair amount of sediment.

Growth on malt agar: Shape and size of the cells are similar to those grown in malt extract. After one month at 25°C, the streak culture is white to cream colored with entire margin. Colony texture is smooth, soft and semiglossy.

Dalmau plate culture on corn meal agar: Pseudo- or true mycelium is not

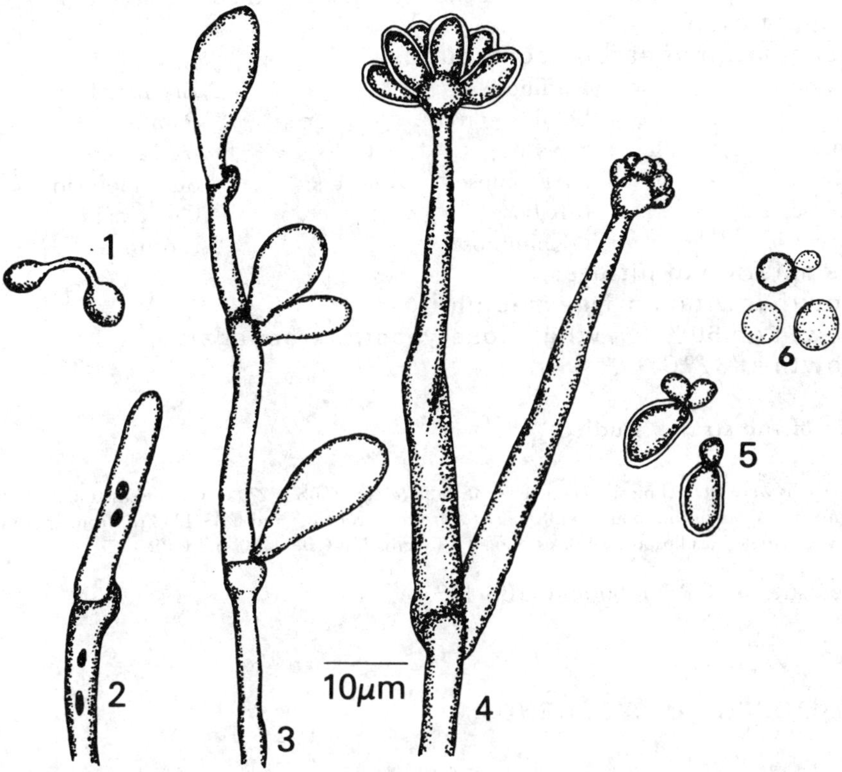

Fig. 147 – *F. uniguttulatum*
1. Conjugating pair of haploid cells 2. Dikaryotic hypha with a clamp connection 3. Early stage of basidial formation 4. Mature basidial structures 5. Germinating basidiospores 6. Haploid yeast cells

formed in the haploid phase.

Life cycle: On V8 juice- or hay infusion agar, cells produce conjugation tubes, (1–5) × (1.5–20) μm within 2–3 days after mixing the two compatible mating types. The conjugation tube is usually sinuous and narrower than the dikaryotic hyphae which arise from the conjugated cells. The dikaryotic hyphae bear clamp connections at each septum. Within one week to 10 days, mature metabasidia appear in the margin of mated cultures. Young metabasidia are thin-walled and broader than the hyphae from which they originate. They develop in a group of two to four or singly on the side of hyphae at the clamp or terminally on the hyphae (Fig. 147). Blastospores are occasionally produced laterally on the hyphae. Metabasidia are elongated (4–6) × (30–90) μm; they taper apically and expand abruptly to produce a globose apex (3–5 μm diameter) with four to ten sessile terminal basidiospores. The basidiospores bud directly from the apex. The spores are oval to kidney-shaped (3–5) × (4–9) μm. Infrequently, haustorioid branches are produced from the clamp of the dikaryotic hyphae. The liberated basidiospores reproduce by budding and form yeast colonies. Although a detailed cytological study was not carried out, analysis of single basidiospores indicated that meiosis occurred in the metabasidium.

Fermentation: –

Assimilation of carbon compounds:

Galactose	– or + w	Raffinose	v	Erythritol	–
Sucrose	+	Soluble starch	+ w or s	Ribitol	–
Maltose	+	D-Xylose	+	D-Mannitol	+ s
Cellobiose	–	L-Arabinose	+ or s	Succinic acid	–
Trehalose	+ s	D-Ribose	–	Citric acid	–
Lactose	–	L-Rhamnose	–	Inositol	+

Assimilation of nitrate: –

Growth in vitamin-free medium: –

Growth on 50% (w/w) glucose-yeast extract agar: –

Growth at 37°C: –

Origin of the strains studied

Five strains were from clinical specimens: CBS 1730 (*a*) and CBS 1727 (α) from diseased nails; CBS 2770 (α) from white spots in the mouth; CBS 2994 (α) from a bronchus and CBS 4257 (α) from sputum. Five strains were single basidiospore cultures obtained by crossing CBS 1730 and CBS 1727.

Type strains: CBS 1730 and CBS 1727.

f. COMMENTS ON THE GENUS

Fell et al. (1969) described the perfect state of *Candida japonica (Candida capsuligena)* as *Leucosporidium capsuligenum.* Basidiospores and basidia were not mentioned but terminal teliospores connected to the hyphae by clamp connections were described. These were later found to be the initial step of metabasidium (Rodrigues de Miranda, 1972). This author observed the complete sexual life cycle of the species and redescribed the species as *F. capsuligenum.*

Physiologically, *F. capsuligenum* differs from other members of the genus by the ability to ferment glucose and maltose. Ultrastructurally, *F. capsuligenum* differs from *F. floriforme* by having cone-shaped vesicular parenthesomes, while the latter species lacks the parenthesome (Moore & Kreger-van Rij, 1972). Such parenthesomes were observed also in *Tremella* species (Khan, 1976). *F. uniguttulatum* differs from *F. floriforme* by the inability to assimilate nitrate and lactose. *Cryptococcus uniguttulatus,* the asexual state of *F. uniguttulatum,* was first described as *Eutorulopsis uniguttulata* by Zach (Wolfram & Zach, 1934). Lodder and Kreger-van Rij (1952) redescribed this yeast as *Cr. neoformans* var. *uniguttulatus* on the basis of morphological and physiological similarities with *Cr. neoformans* but with reduced capsule formation. In 1961, Kreger-van Rij recognized that isolates of *Cr. neoformans* assimilated galactitol as sole carbon source and were able to grow at 37°C whereas the variety was not able to do so. Additional difference in the creatinine utilization was found between the species and variety by Staib (1963). In 1970, Phaff and Fell re-established the species status for the variety as *Cr. uniguttulatus*. With the discovery of the *Filobasidium* state, it is clear now that *Cr. uniguttulatus* is phylogenetically closer to *Candida japonica* and *Cr. albidus* than to *Cr. neoformans*. Physiological and morphological characteristics of *Cr. albidus* fit well to the asexual state of *F. floriforme*. However, we failed to demonstrate the successful mating between the type strains of the *Cr. albidus* varieties with the tester strains of *F. floriforme*.

B. Teliospore-forming yeasts

Introduction and key to the genera

by J.W. Fell

The initial evidence for a basidiomycetous ancestry for certain yeasts was based on morphology. The presence of ballistospores in *Sporobolomyces* and *Bullera* provided an obvious linkage between yeasts and the basidiomycetes, however arguments were expressed both for and against the theory that ballistospores are a form of basidiospore (Lodder & Kreger-van Rij, 1952). More convincing evidence of a basidiomycete–yeast relationship was provided by biochemical analyses, such as the nucleotide composition of DNA (Storck, 1966) and the components of the extracellular heteropolysaccharides (Slodki et al., 1966). The first definitive observation of sexuality in *Rhodotorula* was documented by Banno (1967) who observed a basidiomycetous life cycle in strains of *Rhodotorula glutinis*. Consequently he described the sexual genus *Rhodosporidium*. Previously Nyland (1949) had described a ballistosporogenous yeast with clamp connections and thick-walled chlamydospores as the genus *Sporidiobolus,* a genus that could include the imperfect stages of certain species of *Sporobolomyces, Rhodotorula* and *Bullera*. The importance of Nyland's observations to yeast systematics was probably overlooked as the genus was not mentioned in Lodder and Kreger-van Rij's (1952) review. With Banno's publication the significance of *Sporidiobolus* became apparent and Phaff (1970e) presented a detailed discussion of the genus and recommended a second species to be included in the genus.

With the advent of the 2nd edition of 'The Yeasts', descriptions of new species of basidiomycetes included an additional species of *Rhodosporidium* and the genus *Leucosporidium* with six species. The latter genus consists of some of the non-carotenoid pigmented yeasts of the genus *Candida*. In the following chapters of the present edition, the genus *Rhodosporidium* includes nine species; *Leucosporidium* has five species (one of the species, *Leu. capsuligenum,* was transferred to *Filobasidium*); *Sporidiobolus* has four species. Among these recent discoveries was the observation by Bandoni et al. (1971; 1975) of mating between species of *Sporobolomyces* which resulted in the description of the sexual state of *Sporobolomyces salmonicolor* (Fell & Tallman, 1981).

The life cycles, and in some cases the morphology of these yeasts, are similar to the higher basidiomycetous fungi such as the smuts (Fischer & Holton, 1957) and the mushrooms (Raper et al., 1972). In describing the life cycles, we have attempted to adhere to the nomenclature used for other basidiomycetous fungi (Ainsworth, 1971). The entire sexual reproductive structure is termed the basidium, and is subdivided into the probasidium (or teliospore) where karyogamy takes place, and the metabasidium (or promycelium) where meiosis usually occurs. The metabasidium is divided into two or four cells, a phragmometabasidium, or is one-celled, a holometabasidium. There is considerable variation in the morphology of the metabasidium. The holometabasidium may be short and directly attached to the teliospore (Fig. 156, *Sporid. johnsonii*); separated from the teliospore by a stalk (Fig. 158, *Sporid. salmonicolor*); or elongate with an enlarged terminus (Diagram 5, *Rhodosp. bisporidii*). Similarly, the phragmometabasidia vary morphologically and they include those that attach directly to the teliospore (Fig. 154, *Rhodosp. sphaerocarpum*), attach by a stalk (Fig. 157, *Sporid. ruinenii*) or by a true hypha (*Leu. scottii*).

Basidiospores of most species develop directly on the metabasidium, however, the metabasidia of *Rhodosp. bisporidii* and *Rhodosp. capitatum* produce primary basidiospores which remain attached to the metabasidia. The basidiospores are produced on and shed from these primary basidiospores.

Our use of the term teliospore is not universally accepted and has received some criticism. Some authors refer to the probasidium as a chlamydospore. However, the Ainsworth (1971) definition denotes the chlamydospore as an asexual spore and the teliospore as the spore of the Ustilaginales (smuts) and Uredinales (rusts) from which the basidium is produced. Moore (1972, 1980) argues that similarities between the Ustilaginales and Uredinales are superficial and that teliospore formation, location of meiosis and formation of basidiospores in the Uredinales and Ustilaginales are distinctly different. Therefore, Moore developed the term ustospore for the probasidium in the smuts and basidiomycetous yeasts. Further studies are required to clarify the interrelationships of life cycles in the yeasts to other basidiomycetes and to ascertain the appropriate terminology.

The life cycles of the basidiomycetous yeasts can include homothallic and heterothallic systems. In the latter situation, compatible mating types conjugate and develop a dikaryotic mycelium with clamp connections at the septa. Karyogamy takes place in the teliospore which germinates to a 1–4-celled metabasidium where meiosis occurs. Lateral or terminal haploid basidiospores develop to complete the life cycle. The heterothallic incompatibility systems are biallelic unifactorial (also termed bipolar), multiple allelic unifactorial or bifactorial (tetrapolar). The biallelic unifactorial systems (Diagram 10) are characterized by a single locus with two alleles which results in two mating types (*A1, A2*). Banno's species *Rhodosporidium toruloides* is in this category as are *Rhodosp. diobovatum, Rhodosp. malvinellum, Rhodosp. sphaerocarpum, Leucosporidium antarcticum, Sporidiobolus salmonicolor* and *Sporid. pararoseus*. Multiple allelic unifactorial systems (Diagram 8) consist of a single locus with three or more alleles and consequently a minimum of three distinct mating types (*A1, A2, A3*). When intermated (*A1* × *A2, A1* × *A3, A2* × *A3*) all combinations give complete life cycles. *Rhodosporidium infirmominiatum* is the only yeast species decribed with this system.

Bifactorial incompatibility systems (Diagram 7) consist of two loci; mating competence is determined by *A* and *B* compatibility factors. There is a minimum of four mating types (*A1B1, A2B2, A1B2, A2B1*). Reactions of dissimilar *A* and *B* factors (*A1B1* × *A2B2* and *A1B2* × *A2B1*) result in complete life cycles with fertile progeny. Our designation of the factors as *A* and *B* is in keeping with the accepted factor descriptions for other basidiomycetes. In fungi such as *Schizophyllum commune,* the *A* factor controls nuclear division and clamp cell formation. The *B* factor controls nuclear migration, first during establishment of the dikaryotic condition and secondly by clamp cell fusion (Casselton, 1977). Whether or not the *A* and *B* factors of the basidiomycetous yeasts have the same functions has not been determined. However, we have some evidence to indicate that such is the case. In two bifactorial species, *Rhodosporidium dacryoidum* and *Rhodosp. bisporidii,* interactions of dissimilar *A* and *B* factors follow the typical life cycle process: conjugation, dikaryotic mycelium with clamp connections and teliospores. Interactions of strains with similar *A* and dissimilar *B* factors result in conjugation with poorly formed, twisted mycelium, neither clamps nor teliospores develop. Dissimilar *B* factor responses, analogous to those in *Rhodosporidium* spp., have been studied in detail for several basidiomycetes such as *Coprinus lagopus, Cyathus stercoreus* and *Schizophyllum commune* (Raper, 1978). In addition, we (Fell et al., 1969) observed dissimilar *A* factor behavior in a strain of *Rhodosp. dacryoidum.* This strain, derived from a *A1B1* × *A2B2* mating, developed extensive mycelium, teliospores and pseudoclamp connections. Nuclear stains demonstrated that yeast cells were uninucleate and that they germinated to produce hyphae with binucleate terminal cells. The preceding hyphal cells were uninucleate with the second nucleus trapped in the clamp. This behavior is similar to the $A{\neq}B=$ heterokaryons of *Coprinus lagopus* and *Schizophyllum commune* (Raper, 1978). Both *A* and *B* factors consist of two linked loci, designated α and β, which are separable by crossing over, i.e., a cross of $A\alpha_1\beta_1$ strain with an $A\alpha_2\beta_2$ strain can result in recombinant progeny $A\alpha_1\beta_2$ and $A\alpha_2\beta_1$. In fungi such as *Schizophyllum commune,* 288 different *A* and 81 *B* specificities have been estimated for a total of 23,328 mating types (Raper, 1978). Linked loci resulting in 5 *A* and 3 *B* factors have been described in *Leu. scottii* (Fell & Statzell Tallman, 1982).

In addition to the heterothallic life cycles, homothallism is present among these yeasts. In homothallism a binucleate hypha develops from a diploid cell in the absence of mating. Clamp connections are present and karyogamy takes place in the teliospore. During the life cycles of *Sporid. johnsonii* and *Sporid. ruinenii* reduction division does not occur in the basidium, resulting in diploid basidiospores. Consequently heterothallism has not been observed in either *Sporidiobolus* species. In contrast the diploid cell of the *Rhodosporidium toruloides* homothallic cycle originates from a failure of meiosis in the heterothallic cycle. The progeny from *Rhodosp. toruloides* homothallism are either homothallic or heterothallic.

Strains of several yeasts (*Rhodosp. capitatum, Leu. scottii, Sporid. pararoseus,* etc.) produce teliospores in the absence of mating. Rather than the previously discussed life cycles, the hyphal cells remain uninucleate and clamp connections are lacking. Not all of the species have been studied in sufficient detail to determine if teliospore formation is the result of homothallism or homokaryotic fruiting. In discussing other basidiomycetes, Ullrich (personal communication) points out that strains that sporulate from unclamped uninucleate hyphae are homokaryotic fruiters while homothallic strains initiate as uninucleate cells that become dikaryons prior to fruiting. If this generalization is applicable to the yeasts, then it would appear that teliospore formation in many of the yeast species is due to a homokaryotic process. The most available information is for strains of *Leu. scottii* which appear to be homokaryotic fruiters. In addition to the formation of teliospores on uninucleate hyphae, the homokaryotic strains will mate with normal heterothallic fruiting strains for the formation of a heterothallic life cycle.

Nyland (1949) originally placed *Sporidiobolus* among the Heterobasidiomycetes, a group which is presently considered to correspond to the Phragmobasidiomycetidae (McNabb, 1973). This is a group of fungi with a well-developed basidiocarp that includes Tremellales with a yeast phase (*Tremella* and *Sirobasidium*). The teliospore forming fungi, specifically the rusts and smuts, are placed in a separate class, Teliomycetes (Ainsworth, 1973) or subclass Teliomycetidae (Alexopolous & Mims, 1979), depending on how the fungi are categorized. The Teliomycetes are subdivided into the two orders Uredinales and Ustilaginales. There seems to be little doubt that the teliospore-forming yeasts have a greater affinity to the smuts than to any other group of basidiomycetes. Classification of the yeasts within the Ustilaginales is, however, difficult with the present information. The two families of Ustilaginales, the Ustilaginaceae and the Tilletiaceae were described by Tulasne and Tulasne in 1847. While both families are still accepted, Duran (1973) presented arguments for and against retention of this taxonomy. Family separation is based on the presence of a phragmometabasidium (Ustilaginaceae) or a holometabasidium (Tilletiaceae). Placement of the yeasts within this classification is difficult due to the range of metabasidial morphology within the individual genera *Leucosporidium, Rhodosporidium* and *Sporidiobolus*. Use of the metabasidial morphology would, for example, separate *Sporidiobolus johnsonii* and *Sporid. ruinenii* into different families. An alternative is the formation of a separate family for the yeast-like Ustilaginales. This order consists of angiosperm parasites, and the yeast-like genera would appear to be the only saprophytes, however yeast-like growth on artificial media has been observed with several of the parasitic species (Fischer & Holton, 1957). Unfortunately the knowledge of the systematic interrelationships between the teliospore-forming yeasts and the smuts is lacking.

In considering the taxonomic placement of these yeasts at least two extremes are available. At the one end is the taxonomic scheme presented by Moore (1980) in which the smuts and yeasts are placed in the division Ustomycota and then separated into classes (Ustomycetes and Sporidiomycetes). Further separation of the yeasts into families is based on presence or absence of ballistospores (Sporidiobolaceae and Sporidiaceae). This approach has the advantage of presenting a target for further deliberations. The other extreme is to heed the caution of Donk (1972) who believes that basidiomycete taxonomy is being developed by specialists of narrow groups who lack the broad knowledge and viewpoints to develop adequate decisions. The reaction to this philosophy is to avoid any finite taxonomic pronouncements. We have chosen this conservative approach and we have not classified the yeasts below the level of the Ustilaginales. This decision is rationalized by the optimistic outlook that the next edition of The Yeasts will contain sufficient information to resolve these problems. Among the more important determinations will be the relationships of the yeasts with the smuts, the Filobasidiaceae and the Tremellales.

KEY TO THE GENERA OF THE TELIOSPORE-FORMING YEASTS

1a Ballistospores formed
Sporidiobolus . . . . p. 532
b Ballistospores not formed 2

2a Colony color distinctly yellow to red due to the presence of carotenoid pigments
Rhodosporidium . . . . p. 509
b Distinct carotenoid pigments absent
Leucosporidium . . . . p. 496

A discussion of the three genera in alphabetical order follows.

Genus 1. *Leucosporidium* Fell, Statzell, Hunter et Phaff

by J.W. Fell and A. Statzell Tallman

a. DIAGNOSIS OF THE GENUS

Growth on solid media is white to cream-colored, often mucous. Visible carotenoid pigments are not present. Both yeast and mycelial phases may occur. In the yeast phase the cells are oval to elongate and reproduce by budding; a pseudomycelium may form. Formation of the mycelial phase is usually the result of sexual reproduction. True mycelium with or without clamp connections will form. Terminal or intercalary teliospores, $(3-15) \times (5-16)$ μm, are produced on the mycelium.

Sexual mechanisms, although not completely investigated, include unifactorial (bipolar) and bifactorial (tetrapolar) systems. In these heterothallic systems a dikaryotic mycelium results from the conjugation of a compatible mating pair. A clamp connection is usually present at each cross wall. Intercalary and terminal teliospores develop. Karyogamy takes place in the teliospore (probasidium) which germinates to produce a metabasidium (promycelium) that may be one-celled (holometabasidium) or septate with two to four cells (phragmometabasidium). Meiosis takes place during germination and the resulting haploid basidiospores are terminal on the holometabasidium or lateral and terminal on the phragmometabasidium.

Some strains produce teliospores in the absence of mating. The yeast cells are uninucleate and produce uninucleate teliospore-bearing hyphae without clamp connections. Teliospore germination results in phragmo- or holometabasidia, the type of metabasidium varying with species. Studies of *Leu. scottii* indicate that these strains are homokaryotic fruiters similar to those in other basidiomycetes (Stahl & Esser, 1976).

Ballistospores are not present in either the yeast or mycelial phases. Urease is positive.

b. TYPE SPECIES OF THE GENUS

Leucosporidium scottii Fell, Statzell, Hunter et Phaff

c. SPECIES ACCEPTED IN THE GENUS

1. *Leucosporidium antarcticum* Fell, Statzell, Hunter et Phaff (1969)

2. *Leucosporidium frigidum* Fell, Statzell, Hunter et Phaff (1969)
3. *Leucosporidium gelidum* Fell, Statzell, Hunter et Phaff (1969)
4. *Leucosporidium nivalis* Fell, Statzell, Hunter et Phaff (1969)
5. *Leucosporidium scottii* Fell, Statzell, Hunter et Phaff (1969)
6. *Leucosporidium stokesii* Fell, Statzell, Hunter et Phaff (1969)

d. KEY TO THE SPECIES OF THE GENUS

1a Melibiose assimilated 2
b Melibiose not assimilated 3

2a Maltose assimilated
Leu. gelidum . . . . p. 501
b Maltose not assimilated
Leu. nivalis . . . . p. 502

3a Rhamnose assimilated 4
b Rhamnose not assimilated 5

4a Glucose fermented
Leu. stokesii . . . . p. 507
b Glucose not fermented
Leu. scottii . . . . p. 504

5a Cellobiose assimilated
Leu. frigidum . . . . p. 499
b Cellobiose not assimilated
Leu. antarcticum . . . . p. 498

Table 22

Key characters of the species in the genus *Leucosporidium*

	Assimilation				Fermentation
	Me	Ma	Ce	Rh	G
Leu. nivalis	+	−	+	−	+
Leu. gelidum	+	+	+	+	+
Leu. stokesii	−	+	+	+	+
Leu. frigidum	−	−	+	−	+
Leu. scottii	−	+	+	+	−
Leu. antarcticum	−	v	−	−	−

Me = melibiose, Ma = maltose, Ce = cellobiose, Rh = L-rhamnose, G = glucose.

e. SYSTEMATIC DISCUSSION OF THE SPECIES

1. *LEUCOSPORIDIUM ANTARCTICUM* FELL, STATZELL, HUNTER ET PHAFF (1969)

Standard description of *Leucosporidium antarcticum*

Growth in malt extract: After 3 days at 12°C the cells of self-sporulating strains are (2–4) × (5–22) μm, while heterothallic cells of normal fruiting strains are (3–6) × (3–10) μm. Both types of cells are single, in pairs and short chains. There is a light sediment. After one month there is a light ring and a heavy sediment.

Growth on malt agar: After 3 days at 12°C the cells are approximately the same size as in malt extract. The cells may have a large capsule and refractile inclusions. Growth is cream-colored, smooth, mucous, runny and glistening; the border is entire. Pseudo- and true mycelium may be present. There is little change after one month.

Dalmau plate culture on corn meal agar: Self-sporulating strains have two types of mycelium: short branched chains of pseudomycelium and long branched chains of true mycelium. After 3 weeks teliospores are present which measure (4–6) × (5–8) μm; they occur terminally, singly and occasionally in pairs. They increase in size and at 7 weeks are (5–8) × (5–10) μm, terminal, intercalary, single and in short chains of two to three spores. The teliospores usually have one or two bud-like structures that may represent either vestigial clamp connections or undeveloped sister teliospores. Heterothallic strains form true mycelium and *'Candida'* type pseudomycelium.

Life cycle: Self-sporulating strains: A septate uninucleate hypha with teliospores, but lacking clamp connections, develops directly from a single cell in the absence of mating. Teliospores, in a culture grown on corn meal agar (5°C), were germinated by soaking in distilled water and then transferring to 2% agar. Best results included a total growth-soak period of 3 months. The duration of the individual steps varied, for example, growth on corn meal agar for 2 weeks followed by soaking for 10 weeks or growth for 10 weeks with soaking for 2 weeks. Germination then occurred within one week on 2% agar. When the various steps were at 5°C, the cells produced much less interfering polysaccharides than at 12°C. The metabasidium was four-celled with lateral and terminal basidiospores. Ploidy of the basidiospores has not been determined.

Heterothallism: Compatible mating strains *A1* and *A2* mixed on corn meal agar formed round to ovoidal (7–10) × (8–13) μm teliospores on clamp bearing mycelium after 3 weeks at 12°C. Teliospore germination occurred after soaking 2-month-old teliospores in distilled water for one month followed by one week on 2% agar. Basidiospores developed laterally and terminally on four-celled metabasidia. Six basidiospores from two teliospores were examined and all progeny were self-sporulating.

Neither self-sporulating nor heterothallic strains of *Leu. antarcticum* mated with strains of other *Leucosporidium* species.

Fermentation: –

Assimilation of carbon compounds:

Galactose	v	Raffinose	v	Erythritol	–
Sucrose	v	Soluble starch	–	Ribitol	–
Maltose	v	D-Xylose	v	D-Mannitol	v
Cellobiose	–	L-Arabinose	–	Succinic acid	–
Trehalose	v	D-Ribose	–	Citric acid	–
Lactose	–	L-Rhamnose	–	Inositol	–

Additional carbon compounds tested: Melezitose –, melibiose –, α-methyl-D-glucoside –

Splitting of arbutin: –

Assimilation of nitrate: +

Growth in vitamin-free medium: +

Growth on 50% (w/w) glucose-yeast extract agar: –

Growth at 17°C: +w; at 19°C: –

Starch formation: –

Urease: +

Origin of the strains studied

Antarctic marine waters near the Antarctic Peninsula and in the Indo-Pacific region of the Antarctic (46) (Fell et al., 1969; Fell, 1976).

Type strain: CBS 5942, a self-sporulating strain. Allotype strains: ATCC 26995 (mating type *A1*), ATCC 26996 (mating type *A2*).

2. *LEUCOSPORIDIUM FRIGIDUM* FELL, STATZELL, HUNTER ET PHAFF (1969)

Imperfect state: *Candida frigida* di Menna

Synonym

Candida frigida di Menna (1966a)

Standard description of *Leucosporidium frigidum*

Growth in malt extract: After 3 days at 12°C the cells are single, in pairs and short chains, (3–8) × (4–14) μm; there is a slight ring and a sediment. After one month there is a thick myceliated ring and a heavy sediment.

Growth on malt agar: After 3 days at 12°C the cells are (2–5) × (4–14) μm. Cells are not capsulated. The streak is smooth, dull, cream-colored, the texture is soft. The border is entire and myceliated. There is true mycelium, with single to numerous blastospores that are often in short chains. Teliospores are present, clamp connections are not produced.

Dalmau plate culture on corn meal agar: After 7 days at 12°C, both true, branched, septate mycelium with clusters of blastospores and branched pseudomycelium are formed. Teliospores are terminal and intercalary on the mycelium, either single, in pairs, or short chains. After 3 weeks teliospores are spheroidal to ovoidal, (4–7) × (4–8) μm. After 9 weeks teliospores are (5–8) × (5–11) μm.

Life cycle: Mycelium and teliospores develop directly from a single cell. Clamp connections are not produced. Teliospores germinate to a one- or two-celled metabasidium 8–12 μm in length. Germination was induced at 5°C by growing the culture on corn meal agar for 5 weeks, followed by soaking in distilled water for 7–12 weeks with transfer to 2% agar where germination was observed after 2 weeks. Basidiospores are terminal on one-celled metabasidia; lateral and terminal on two-celled metabasidia (Fig. 148). Nuclear stains indicate that the organism is uninucleate. The nucleus migrates to the teliospore and subsequently into the metabasidium. The nucleus divides and a transverse septum separates the two nuclei resulting in the two-celled metabasidium. Basidiospores form and the nucleus divides repeatedly to provide a nucleus for each new basidiospore.

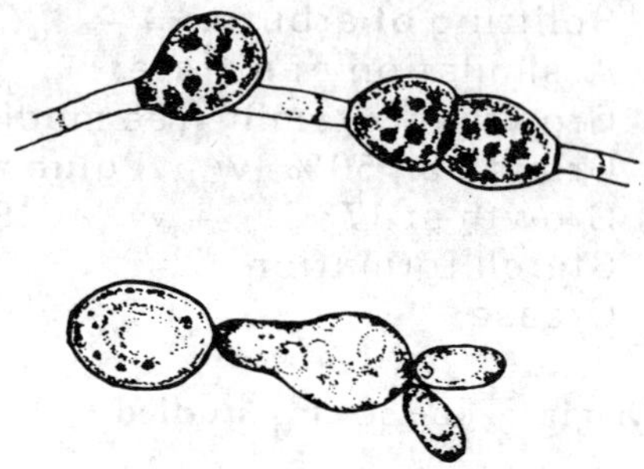

Fig. 148 – *Leu. frigidum*
Intercalary teliospores; teliospore with one-celled metabasidium and terminal sporidia
On corn meal agar
(From 'The Yeasts', ed. II, 1970)

The two isolates examined did not mate with each other or with strains of the other species in the genus.

Fermentation:

Glucose	+ s	Maltose	–
Galactose	+ w,s	Lactose	–
Sucrose	+ s	Raffinose	v

Assimilation of carbon compounds:

Galactose	+	Raffinose	+	Erythritol	–
Sucrose	+	Soluble starch	–	Ribitol	+
Maltose	–	D-Xylose	+	D-Mannitol	+
Cellobiose	+	L-Arabinose	+	Succinic acid	+
Trehalose	+ (s)	D-Ribose	–	Citric acid	+
Lactose	+	L-Rhamnose	–	Inositol	v

Additional carbon compounds tested: Melezitose –, melibiose –, α-methyl-D-glucoside –

Splitting of arbutin: +

Assimilation of nitrate: +

Vitamins required for growth: biotin, thiamine

Growth on 50% (w/w) glucose-yeast extract agar: –

Growth at 17°C: + w; **at 19°C:** –

Starch formation: + (pH independent)

Urease: +

Origin of the strains studied

CBS 5916 from Antarctic snow (Sinclair & Stokes, 1965); CBS 5270 from Antarctic soil (di Menna, 1966a,b). The two isolates differ by their type of metabasidia: CBS 5270 is one-celled, CBS 5916 is two-celled.

Type strain: CBS 5270.

3. *LEUCOSPORIDIUM GELIDUM* FELL, STATZELL, HUNTER ET PHAFF (1969)

Imperfect state: *Candida gelida* di Menna

Synonym

Candida gelida di Menna (1966a)

Standard description of *Leucosporidium gelidum*

Growth in malt extract: After 3 days at 12°C the cells are single and in pairs (2–6) × (5–9) μm. A light sediment is formed. After one month there is a light sediment and a thick myceliated ring.

Growth on malt agar: After 3 days at 12°C the cells measure (2–6) × (4–13) μm. After one month the growth is soft, smooth, semi-glistening and raised. The periphery is heavily myceliated, lobate and wrinkled. The mycelium is true and branched with regularly occurring septa. Few blastospores are formed. Teliospores are terminal and intercalary, either single, in pairs, or in short chains, (5–11) × (7–14) μm in size.

Dalmau plate culture on corn meal agar: After 3 days at 12°C pseudomycelium and true, branched, septate mycelium with occasional clusters of intercalary teliospores that are single and in pairs. After 3 weeks the teliospores are spheroidal to slightly ovoidal (4–5) × (5–6) μm.

Life cycle: A true hypha develops from a single cell. Teliospores form on the hypha; clamp connections are not produced. Germination (12°C) was rare. Teliospores from one month old cultures (malt agar) were soaked 3–4 days in distilled water, followed by transfer to 2% agar. Teliospores germinated after 3–7 days. Basidiospores formed terminally on one- or two-celled metabasidia (Fig. 149).

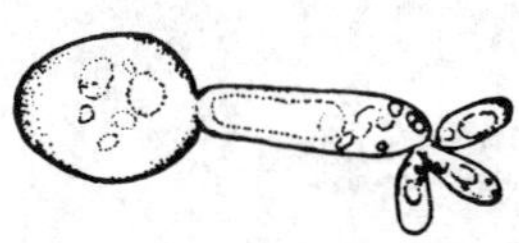

Fig. 149 – *Leu. gelidum*, CBS 5272
Teliospore, one-celled promycelium, terminal sporidia (From 'The Yeasts', ed. II, 1970)

The strain examined did not mate with strains of the other species in the genus.

Fermentation:

Glucose	+ s	Maltose	v
Galactose	+ w	Lactose	–
Sucrose	+	Raffinose	– (+ w)

Assimilation of carbon compounds:

Galactose	+	Raffinose	+	Erythritol	–
Sucrose	+	Soluble starch	+	Ribitol	+
Maltose	+	D-Xylose	+	D-Mannitol	+
Cellobiose	+	L-Arabinose	+	Succinic acid	+ (w)
Trehalose	+	D-Ribose	v	Citric acid	+ (w)
Lactose	–	L-Rhamnose	+	Inositol	v

Additional carbon compounds tested: Melezitose +, melibiose +, α-methyl-D-glucoside v

Splitting of arbutin: +

Assimilation of nitrate: +

Vitamins required for growth: biotin, thiamine

Growth on 50% (w/w) glucose-yeast extract agar: –

Growth at 17°C: + ; **at 19°C:** + w

Starch formation: + (pH independent)

Urease: +

Origin of the strain studied

Antarctic soil (1) (di Menna, 1966a,b).

Type strain: CBS 5272.

4. *LEUCOSPORIDIUM NIVALIS* FELL, STATZELL, HUNTER ET PHAFF (1969)

Imperfect state: *Candida nivalis* di Menna

Synonym

Candida nivalis di Menna (1966a)

Standard description of *Leucosporidium nivalis*

Growth in malt extract: After 3 days at 12°C the cells are single and in pairs, (3–7) × (5–9) μm; there is a light sediment and ring. After one month there is a heavy sediment, a moderate ring and in some cases a film or islets.

Growth on malt agar: The cells are similar to those in malt extract, they are not capsulated. The growth is smooth, semi-glossy, soft, cream-colored, raised, and the border entire with tufts of hyphal development that consist of true, branched, septate hyphae with blastoconidia, as well as branched pseudomycelium. Teliospores are not abundantly produced; they are single, in pairs, and short chains.

Dalmau plate culture on corn meal agar: After one week at 12°C there is true mycelium with branched pseudomycelium and clusters of blastoconidia. Teliospores may be found after 3 weeks at 12°C but they are more abundant after 2–3 months. The 3-week-old teliospores are spheroidal to ovoidal, (4–5) × (5–6) μm, by 9 weeks the teliospores are oval to elongate, (4–11) × (5–16) μm (Fig. 150). They are intercalary and occasionally terminal.

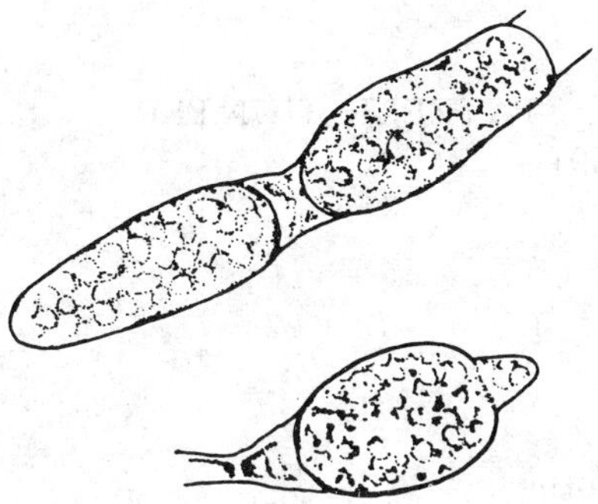

Fig. 150 – *Leu. nivalis,* CBS 5266
Oval and elongate teliospores. After 9 weeks on corn meal agar (From 'The Yeasts', ed. II, 1970)

Life cycle: Mycelium with teliospores but lacking clamp connections develops from a single cell in the absence of mating. Nuclear staining has not been attempted; however, this type of life history is typical of the self-sporulating yeasts of this genus. Germination of the teliospores has not been observed, even through extensive testing with various periods of growing followed by soaking in distilled water. The strain examined did not mate with the strains of other species in the genus.

Fermentation:

Glucose	+ (s)	Maltose	–
Galactose	+ w	Lactose	–
Sucrose	+	Raffinose	+ s or w

Assimilation of carbon compounds:

Galactose	+	Raffinose	+	Erythritol	–
Sucrose	+	Soluble starch	–	Ribitol	+
Maltose	–	D-Xylose	+	D-Mannitol	+
Cellobiose	+	L-Arabinose	+	Succinic acid	+ w,s
Trehalose	+ w	D-Ribose	+ s	Citric acid	v
Lactose	–	L-Rhamnose	–	Inositol	+ (s)

Additional carbon compounds tested: Melezitose –, melibiose +, α-methyl-D-glucoside –

Splitting of arbutin: +

Assimilation of nitrate: +

Vitamins required for growth: biotin, thiamine

Growth on 50% (w/w) glucose-yeast extract agar: –

Growth at 17°C: +; **at 19°C:** + w

Starch formation: + on malt agar; **in liquid glucose-yeast nitrogen base:** –

Urease: +

Origin of the strain studied

Antarctic soil (1) (di Menna, 1966a,b).

Type strain: CBS 5266.

5. *LEUCOSPORIDIUM SCOTTII* FELL, STATZELL, HUNTER ET PHAFF (1969)

Imperfect state: *Candida scottii* Diddens et Lodder

Synonyms

Candida scottii Diddens et Lodder (1942)
Azymocandida scottii (Diddens et Lodder) Novák et Zsolt (1961)

Standard description of *Leucosporidium scottii*

Growth in malt extract: After 3 days at room temperature cells are ovoidal to elongate, (1–7) × (4–16) μm, usually single or in pairs. Reproduction is by apical budding. Strands of pseudomycelium are present. Sediment is moderate. After one month there is a ring, some small islets and a heavy sediment.

Growth on malt agar: After 3 days at room temperature the cells are ovoidal to elongate (1–4) × (3–12) μm. Some strands of true and pseudomycelium are evident. After one month the growth is cream-colored, extremely mucous with dense, extensive mycelium at the periphery.

Dalmau plate culture on corn meal agar: After 3–5 days at room temperature pseudomycelium of the *'Candida'* type begins to form. After one month the pseudomycelium bears large clusters of blastoconidia. True mycelium with endospores is also present.

Life cycle (Diagram 4): *Leu. scottii* has a multiple allelic bifactorial sexual incompatibility system. The mating competency is controlled by two factors designated *A* and *B*, which results in a minimum of four mating types: *A1B1, A2B2, A1B2, A2B1.* Present information suggests that each of the factors consists of two linked loci, designated α and β, which are separable by crossing over, i.e. a cross of $A\alpha_1\beta_1$ strain with $A\alpha_2\beta_2$ strain can result in recombinant progeny $A\alpha_1\beta_2$ and $A\alpha_2\beta_1$. Presently 5 *A* and 3 *B* factors have been identified (Fell & Tallman, 1982). When compatible strains are mixed on corn meal agar they conjugate at room temperature within 30 hours at a frequency of less than 1%. A binucleate hypha develops from one of the conjugate cells (Fig. 151). After 72 hours the hyphae develop extensively with clamp connections at each septum. After 4–7 days intercalary and terminal teliospores develop with one or more clamp structures between the spore and the hypha. The teliospores are spherical, thick-walled, have a granular content and range in size from (7–15) × (7–16) μm. Karyogamy in the teliospores precedes germination. After 5 days the teliospore germinates and the diploid nucleus migrates into the metabasidium. Meiosis occurs in the metabasidium which partitions into four cells. Each metabasidial

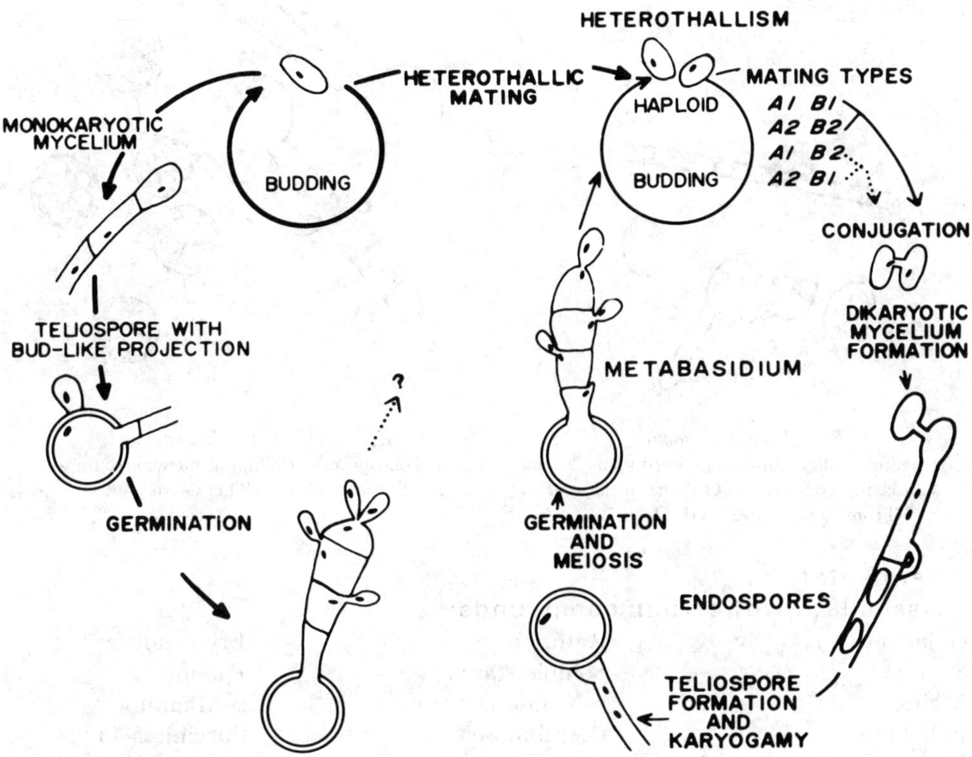

Diagram 4
Life cycle of *Leu. scottii*

cell contains one nucleus which reproduces mitotically resulting in one nucleus in the cell and one in the developing basidiospore. Basidiospores form near the septa and terminally on the distal metabasidial cell (Fig. 152). Budding of the basidiospores results in mucous yeast cells representing the mating types.

Some strains are capable of homokaryotic formation of teliospores. A hypha with uninucleate cells develops from a uninucleate yeast cell. The hyphae do not have clamp connections, although incomplete clamp-like structures may be present. The homokaryotic fruiters will mate with normal fruiting strains resulting in binucleate clamped hyphae with teliospores. Progeny of matings of homokaryotic × normal fruiters are 50% normal fruiters and 50% homokaryotic fruiters. Similar results have been obtained with normal × normal fruiters. This suggests that there are one or more separate genes for fruiting that may be repressed in some heterothallic strains. This gene may not be associated with a specific mating type as suggested by the mating of homokaryotic *A3B1* × normal *A2B2* → normal *A3B1* progeny (Fell & Tallman, 1982).

Strains of *Leu. scottii* do not mate with strains of other species in the genus.

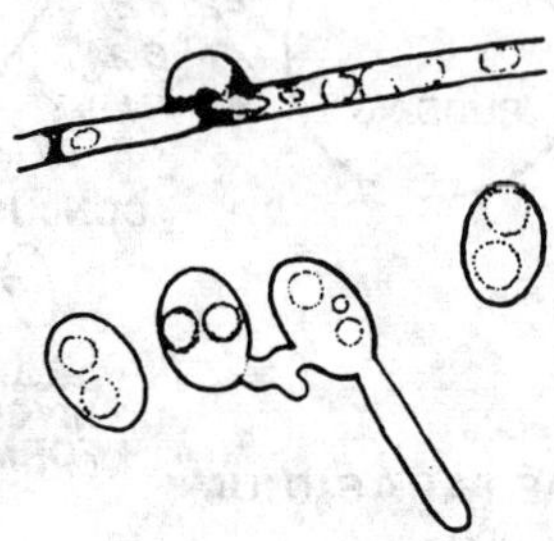

Fig. 151 – *Leu. scottii*
Conjugating cells with developing hypha; hypha with clamp connection. On corn meal agar (From 'The Yeasts', ed. II, 1970)

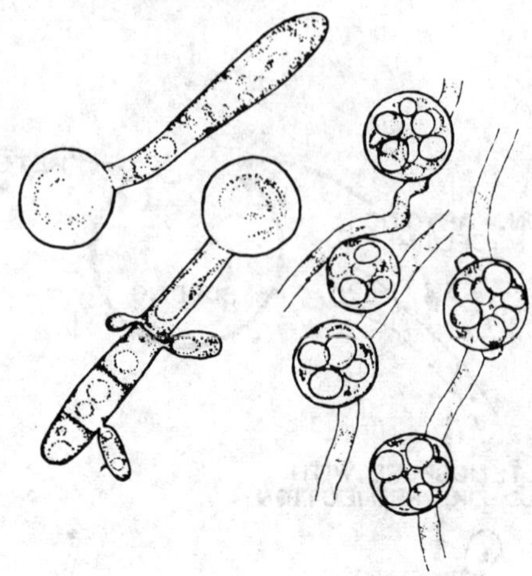

Fig. 152 – *Leu. scottii*
Teliospores, developing metabasidium and basidiospores (From 'The Yeasts', ed. II, 1970)

Fermentation: –

Assimilation of carbon compounds:

Galactose	v	Raffinose	+	Erythritol	–
Sucrose	+	Soluble starch	–	Ribitol	v
Maltose	+	D-Xylose	+	D-Mannitol	+
Cellobiose	+	L-Arabinose	v	Succinic acid	–
Trehalose	+	D-Ribose	v	Citric acid	–
Lactose	v	L-Rhamnose	+	Inositol	–

Additional carbon compounds tested: Melezitose +, melibiose –, α-methyl-D-glucoside +

Splitting of arbutin: +

Assimilation of nitrate: +

Growth in vitamin-free medium: +

Growth on 50% (w/w) glucose-yeast extract agar: –

Maximum temperature of growth: 15–30°C

Starch formation: –

Urease: +

Origin of the strains studied

65 strains were examined: Southern Pacific Ocean (59) (Fell, 1976); Antarctic soil and seawater received from di Menna (3); soil in Australia (1) (Lodder & Kreger-van Rij, 1952); unreported sources (2).

Type strain: CBS 5930 (mating type *A2B2*) from Antarctic seawater. The allotypes are: CBS 5931 (mating type *A1B1*) from Antarctic seawater; CBS 6562 (*A1B2*) and CBS 6561 (*A2B1*) are progeny from *A1B1* × *A2B2* matings (Fell, 1974); CBS 5932 (a self-sporulating strain) from Antarctic seawater.

6. *LEUCOSPORIDIUM STOKESII* FELL, STATZELL, HUNTER ET PHAFF (1969)

Standard description of *Leucosporidium stokesii*

Growth in malt extract: After 3 days at 12°C the cells are (3–6) × (4–11) μm. There is a very light ring, and a moderate sediment. After one month there is little development of the ring; a light film and a heavy sediment form.

Growth on malt agar: After 3 days at 12°C the cells are (2–5) × (3–12) μm. After one month the streak culture is light cream-colored to whitish, with a wrinkled raised surface, or smooth with warty irregularities. The appearance is glistening and the texture soft. The edge is undulate and heavily myceliated. The mycelium is true, branching, with regularly occurring septa. Blastoconidia occur irregularly and sparsely along the mycelium. Teliospores (5–12) × (6–13) μm are abundant, occurring singly, in pairs and chains of three.

Dalmau plate culture on corn meal agar: At 12°C there is a formation of true, branched, septate mycelium and pseudomycelium with abundant clusters of blastospores. After 3 weeks the teliospores are oval to obpyriform, (3–5) × (5–8) μm and usually terminal. After 9 weeks teliospores occur mostly intercalary, singly, in pairs, and in short chains. The teliospores are fewer in number and not as well developed as those on malt agar.

Life cycle: Hyphae and teliospores develop directly from a single cell without mating. Clamp connections are not produced. Teliospore germination resulted in one- to three-celled metabasidia with lateral and terminal basidiospores. Ploidy of the basidiospores has not been determined.

The strain examined did not mate with strains of the other species in the genus.

Fermentation:

Glucose	+ s	Maltose	+ (w)
Galactose	+ w,s	Lactose	–
Sucrose	+	Raffinose	+ (w)

Assimilation of carbon compounds:

Galactose	+	Raffinose	+	Erythritol	–
Sucrose	+	Soluble starch	+ (w)	Ribitol	+
Maltose	+	D-Xylose	+	D-Mannitol	+
Cellobiose	+	L-Arabinose	+	Succinic acid	–
Trehalose	+	D-Ribose	+ w,s	Citric acid	–
Lactose	–	L-Rhamnose	+	Inositol	v

Additional carbon compounds tested: Melezitose +, melibiose –, α-methyl-D-glucoside –

Splitting of arbutin: +

Assimilation of nitrate: +

Vitamins required for growth: biotin, thiamine

Growth on 50% (w/w) glucose-yeast extract agar: –

Growth at 17°C: +; **at 19°C:** + w

Starch formation: + (pH independent)

Urease: +

Origin of the strain studied

Antarctic soil (1) (Sinclair & Stokes, 1965).

Type strain: CBS 5917.

f. COMMENTS ON THE GENUS

Most strains of the six species presently included in *Leucosporidium* are low temperature organisms, with maximal growth temperatures of less than 20°C, although some strains of *Leu. scottii* will grow at 30°C. All species have been reported from Antarctic regions, from snow, terrestrial soils and marine waters (Fell et al., 1969). Additionally, *Leu. gelidum* was isolated from Greenland soils (di Menna, 1966b) and *Leu. scottii* was found in New Zealand, Australia and Canada on sources such as chilled beef (Lodder & Kreger-van Rij, 1952), soils (di Menna, 1965a) and apples (Clark et al., 1954).

Leu. antarcticum and *Leu. scottii* have heterothallic life cycles. *Leu. scottii* has a multiple allelic bifactorial system in which 5 *A* and 3 *B* factors have been identified. *Leu. antarcticum* appears to have a biallelic unifactorial system, although the species has not been intensively studied. All species are capable of developing teliospores in the absence of mating. The hyphae are uninucleate and lack clamp connections. Studies of *Leu. scottii* indicate that this is homokaryotic fruiting rather than a form of homothallism. While the mechanisms of teliospore formation in the other *Leucosporidium* species have not been examined, this type of homokaryotic fruiting is relatively abundant among other basidiomycetes (Stahl & Esser, 1976).

Species separation among species that lack a heterothallic life cycle (*Leu. frigidum, Leu. gelidum, Leu. nivalis* and *Leu. stokesii*) is based on differing carbon assimilation responses. This results in a separation of species which may be artificial and should be tested by the complementarity of the DNA base sequences.

Genus 2. *Rhodosporidium* Banno

by J.W. Fell and A. Statzell Tallman

a. DIAGNOSIS OF THE GENUS

Colony color on solid media includes shades of yellow, orange and red due to the presence of carotenoid pigments. Yeast and mycelial phases may occur. In the yeast phase the cells are oval to elongate and reproduce by budding; a pseudomycelium may form. Formation of the mycelial phase is usually the result of sexual reproduction. True mycelium with or without clamp connections is formed. Terminal or intercalary teliospores $(4-20) \times (5-22)$ μm are produced on the mycelium.

Sexual mechanisms, although not completely understood, are both homothallic and heterothallic. Heterothallism includes unifactorial biallelic (e.g. *Rhodosp. toruloides*), unifactorial multiple allelic (*Rhodosp. infirmominiatum*) and bifactorial (*Rhodosp. dacryoidum*) systems. In all heterothallic systems a dikaryotic mycelium results from the conjugation of a compatible mating pair. A clamp connection is usually present at each cross wall. Intercalary and terminal teliospores develop. Karyogamy takes place in the teliospore (probasidium) which germinates to produce a metabasidium (promycelium) that may be one-celled (holometabasidium) or septate with two to four cells (phragmometabasidium). Meiosis takes place during germination and the resulting haploid basidiospores are terminal on the holometabasidium and lateral and terminal on the phragmometabasidium.

Homothallism, or self-fertility, consists of the development of hyphae with teliospores from a single yeast cell in the apparent absence of mating. The yeast phase is diploid, reduction division takes place with the formation of a dikaryotic mycelium with clamp connections. Karyogamy occurs in the teliospore. Germination of the teliospore results in a metabasidium with diploid basidiospores.

Strains of several species (for example *Rhodosp. capitatum* and *Rhodosp. infirmominiatum*) are self-sporulating; in contrast to the homothallic life cycle, the hyphal cells of these strains remain uninucleate and clamp connections are lacking. Teliospore formation has not been studied in detail, although current information suggests that the process is homokaryotic fruiting rather than a sexual cycle.

Ballistospores are not present in any of the phases.

Sugars are not fermented.

Urease is positive.

b. TYPE SPECIES OF THE GENUS

Rhodosporidium toruloides Banno

c. SPECIES ACCEPTED IN THE GENUS

1. *Rhodosporidium bisporidii* Fell, Hunter et Tallman (1973)
2. *Rhodosporidium capitatum* Fell, Hunter et Tallman (1973)
3. *Rhodosporidium dacryoidum* Fell, Hunter et Tallman (1973)
4. *Rhodosporidium diobovatum* Newell et Hunter (1970)
5. *Rhodosporidium infirmominiatum* Fell, Hunter et Tallman (1973)
6. *Rhodosporidium malvinellum* Fell et Hunter (in Fell, 1970b)
7. *Rhodosporidium paludigenum* Fell et Tallman (1980a)
8. *Rhodosporidium sphaerocarpum* Newell et Fell (1970)
9. *Rhodosporidium toruloides* Banno (1967)

d. KEY TO THE SPECIES OF THE GENUS

1a Nitrate assimilated ... 2
b Nitrate not assimilated
Rhodosp. dacryoidum ... p. 516

2a Inositol assimilated ... 3
b Inositol not assimilated ... 5

3a Teliospores formed after mating of heterothallic strains ... 4
b Teliospores formed in the absence of mating
Rhodosp. capitatum ... p. 514

4a Melibiose assimilated
Rhodosp. bisporidii ... p. 511
b Melibiose not assimilated
Rhodosp. infirmominiatum ... p. 520

5a Melezitose assimilated ... 7
b Melezitose not assimilated ... 6

6a Colony color salmon orange
Rhodosp. paludigenum . . . . p. 524
b Colony color white to mauve
Rhodosp. malvinellum . . . . p. 523

7a Teliospores highly angular
Rhodosp. toruloides . . . . p. 528
b Teliospores not highly angular 8

8a Teliospores cleft, diobovate, or diclavate
Rhodosp. diobovatum . . . . p. 519
b Teliospores globose
Rhodosp. sphaerocarpum . . . . p. 525

Table 23

Key characters of the species in the genus *Rhodosporidium*

	Assimilation				St prod	Spore shape	Sex
	Me	Mz	I	NO_3			
Rhodosp. malvinellum	–	–	–	+	–	SG	H
Rhodosp. paludigenum	–	–	–	+	–	G	H
Rhodosp. sphaerocarpum	–	+	–	+	–	G	H, SS
Rhodosp. toruloides	–	+	–	+	–	A	H, SS
Rhodosp. diobovatum	–	+	–	+	–	G–C	H
Rhodosp. infirmominiatum	–	+	+	+	+	SG–O	HM
Rhodosp. bisporidii	+	+	+	+	+	G–SG	HT
Rhodosp. capitatum	–	+	+	+	+	G–O	SS
Rhodosp. dacryoidum	–	v	–	–	–	SG–D	HT

Me = melibiose, Mz = melezitose, I = inositol, NO_3 = nitrate, St prod = starch production, Sex = sexuality; A = angular, C = cleft, D = dacryoid, G = globose, H = heterothallic, M = multiple allelic bipolar, O = oval, SG = subglobose, SS = self-sporulating, T = tetrapolar.

e. SYSTEMATIC DISCUSSION OF THE SPECIES

1. *RHODOSPORIDIUM BISPORIDII* FELL, HUNTER ET TALLMAN (1973)

(as *Rhodosporidium bisporidiis* Fell, Hunter et Tallman)

Standard description of *Rhodosporidium bisporidii*

Growth in malt extract: After 3 days at 12°C the haploid cells are subglobose to elongate (3–5) × (5–11) μm, usually single or in pairs. Budding may take place on a short neck, there is a light sediment. After one month there is a light film on the surface, and a moderate sediment.

Growth on malt agar: After 3 days at 12°C the haploid cell morphology is similar to that in malt extract. The growth is orient and grenadine pink, smooth, and glistening; the border entire. After one month the color is grenadine pink and the growth mucous and slightly runny, the edge entire. At 24°C there is a development of large round to subglobose $(5-11)\times(5-11)$ μm granulated cells that produce dark brown areas in the colony.

Dalmau plate culture on corn meal agar: After 2 weeks at 12°C and 24°C a rudimentary pseudomycelium may be present.

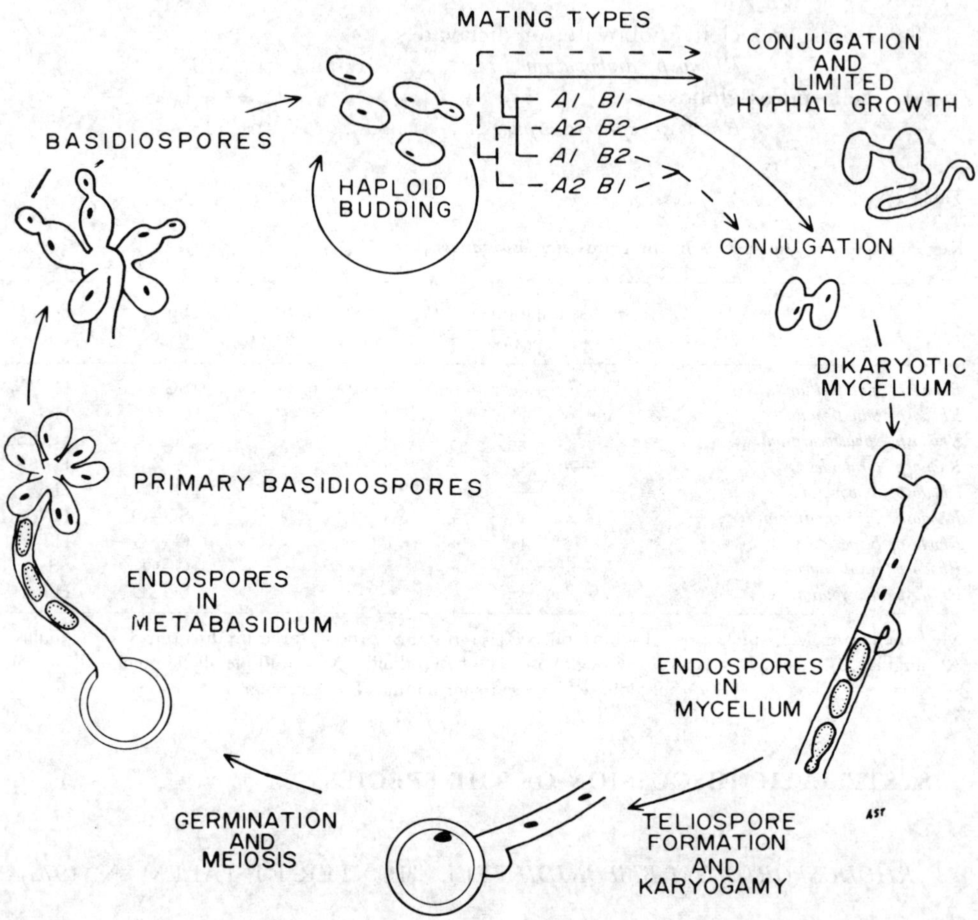

Diagram 5
Life cycle of *Rhodosp. bisporidii*

Life cycle (Diagram 5): The species is heterothallic (bifactorial). Conjugation between mating types *A1B1* × *A2B2* and *A2B1* × *A1B2* takes place within 4 days after mixing 2-day-old stock cultures on corn meal agar at 12°C; the results can be erratic and it may be necessary to repeat the test to obtain a mating reaction. Within the 4-day period mycelium forms with hook-to-peg clamp connections. Conidia do not develop on the mycelium. The teliospores are globose to subglobose (6–11) × (7–11) μm, granulated, intercalary, single, and rarely in pairs. Hook-to-peg clamps connect the teliospores to the hypha. After one month the teliospores are subglobose (8–12) × (10–14) μm. Nuclear stains indicate that the mycelium is binucleate and the teliospores are uninucleate. Teliospores germinate at 12°C and 24°C after soaking 2–3-week-old cultures in distilled water for about 2 months before transferring them to 2% agar plates. Within 3 days after this final transfer, the teliospores germinate to produce a long metabasidium (1–2) × (19–140) μm with zero to one and rarely two to three septa and a terminal swelling 2–4 μm wide. Primary sporidia, which are subglobose or oval (3–4) × (4–8) μm, develop at or near the apex in numbers up to 10; they are attached to the metabasidium either individually or in pairs. One to several basidiospores develop from the primary sporidia. The basidiospores detach and reproduce by budding to develop typical yeast colonies. The primary sporidia remain attached to the metabasidium throughout the process of basidiospore formation. Basidiospores are heterothallic and all four mating types develop. There are some variations in this germination sequence; primary sporidia may be along the sides of the basidium, and in some rare cases, the metabasidium will be branched, and septate, with two 'basidial heads'. The primary sporidia do not always produce basidiospores, i.e., the primary sporidia may germinate to a hypha; the subsequent development of this hypha has not been observed.

Apparently karyogamy can also occur in the hyphae in the absence of a teliospore, as basidia develop directly from thick-walled hyphae as observed in *Rhodosp. capitatum*. Endospores may be present during teliospore germination in numbers of one to six in the basidium and one to four in the mycelium.

Mixes of *A1B1* × *A1B2* and *A2B2* × *A2B1* result in conjugate pairs and twisted, poorly formed hyphae, but no clamps or teliospores, whereas *A1B2* × *A2B2* and *A2B1* × *A1B1* do not indicate any mating reactions.

Conjugation has not been observed when the mating types were mixed with strains of the other species in the genus.

Fermentation: –

Assimilation of carbon compounds:

Galactose	+	Raffinose	+	Erythritol	–
Sucrose	+	Soluble starch	+	Ribitol	v
Maltose	+	D-Xylose	+	D-Mannitol	+
Cellobiose	+	L-Arabinose	+	Succinic acid	+
Trehalose	+	D-Ribose	+ w	Citric acid	+
Lactose	–	L-Rhamnose	+	Inositol	+

Additional carbon compounds tested: Melezitose +, melibiose +, α-methyl-D-glucoside –

Splitting of arbutin: +

Assimilation of nitrate: +

Vitamins required for growth: thiamine; stimulation by biotin

Growth on 50% (w/w) glucose-yeast extract agar: –

Growth at 30°C: + w; **at 33°C:** –

Starch formation: +

Urease: +

Origin of the strains studied

Marine sources of the southern Indian and Pacific Oceans (9) (Fell et al., 1973).

Type strain: CBS 6346 (mating type *A1B1*). Allotype strains: CBS 6347 (mating type *A2B2*); CBS 6348 (*A1B2*) and CBS 6349 (*A2B1*) are progeny from a mating of CBS 6346 and CBS 6347.

2. *RHODOSPORIDIUM CAPITATUM* FELL, HUNTER ET TALLMAN (1973)

Imperfect state: Strains of *Rhodosp. capitatum* fit the description of *Cryptococcus infirmo-miniatus.*

Standard description of *Rhodosporidium capitatum*

Growth in malt extract: At 12°C after 3 days the cells occur singly and in pairs, either subglobose, oval or elongate (1–5) × (2–14) μm. The sediment is light. After one month, there is a moderate sediment, a light film-like ring. At 24°C after 3 days, the cell morphology is similar to that at 12°C, also a few strands of hyphae with immature teliospores are present; after one month there is a heavy sediment and clumps of entwined mycelium; a ring is absent or is in quantities ranging from light to heavy.

Growth on malt agar: After 3 days at 12°C the cells are morphologically similar to those in malt extract. At one week mycelial formation is sparse to abundant, the mycelium is true, branched, septate with intercalary, lateral, and terminal teliospores. The teliospores are brown pigmented, round to subglobose (6–20) × (6–20) μm. They are single, double, and in multiples of 12 or more. Conidia are not formed on the hyphae. The color of the colony is light salmon orange, pinkish- or orange-cinnamon; the growth is mucous; the edge is entire with areas of extensive radiating mycelium. After one month the colony is usually mucous and runny, although some strains may be convoluted. The colors are salmon, light salmon, orange and bittersweet pink. The edge is myceliated, often with dark brown sections of teliospores. There is some increase in the amount of mycelium and teliospores and in the size of the teliospores. Endospores are rare in the mycelium after 2.5 months.

Dalmau plate culture on corn meal agar: After 7 days at 12°C and 24°C there is an extensive true mycelium with brown pigmented teliospores that are single, double, and in groups of 12 or more, intercalary, terminal, and lateral. Their shape is round to oval (5–15) × (5–15) μm. Young developing teliospores are also present, ranging from hyphal swellings to a size of 7 × 10 μm; they lack the heavy cell wall and the large globular inclusions of the developed spores. Conidia do not form on the mycelium. After one month, the teliospores have increased slightly in size. Some teliospores have a basal projection of 1 × 3 μm, that attaches to the hyphae. Endospores are rare in the hyphae at 7 days; they are more numerous after one month, generally less than six per mycelial cell.

Life cycle (Diagram 6): Uninucleate hyphae and teliospores develop direct-

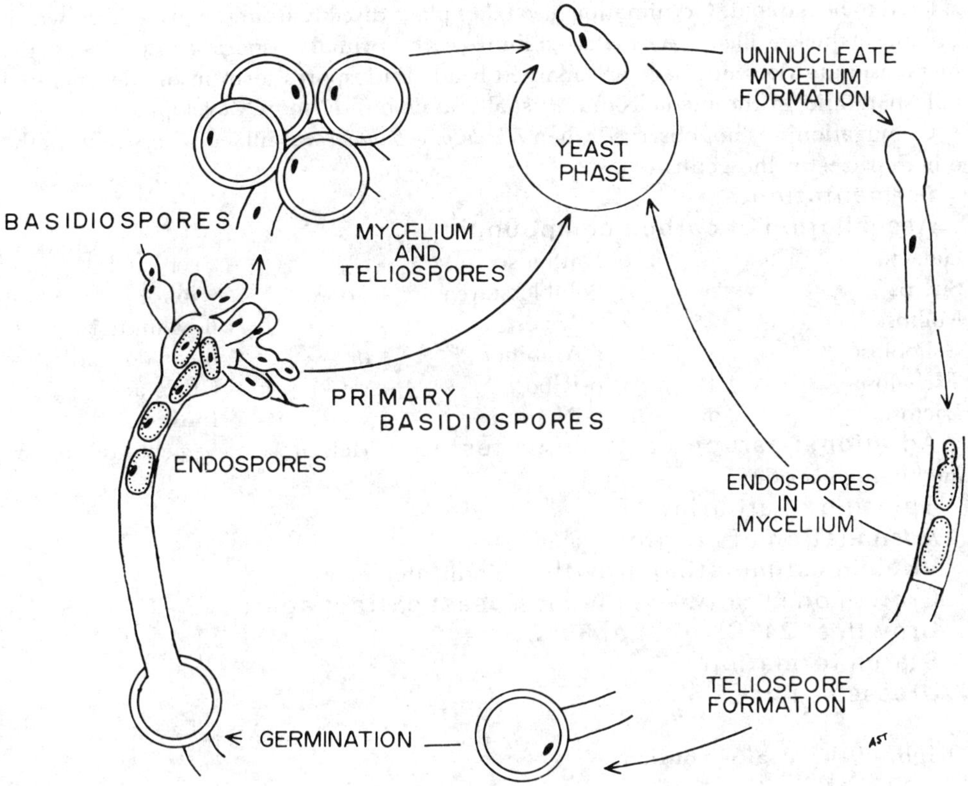

Diagram 6
Life cycle of *Rhodosp. capitatum*

ly from a single cell without mating. Clamp connections are not produced. Germination occurs at 12°C by growing the culture on malt agar for 3 weeks, followed by soaking teliospores in distilled water for 11 weeks, and then transferring to 2% agar. Germination was after 3 days on 2% agar. The metabasidium is non-septate, rarely with one septum, varies in length from 8–80 μm, width of 1–3 μm with a terminal swelling of 5–8 μm. Two to 10 primary sporidia develop on the capitate terminus. The primary sporidia are single and in pairs, subglobose (4–6) × (5–7) μm with one to two basidiospores. Instead of basidiospore production, the primary sporidia may germinate to septate mycelium. After basidiospore development the primary sporidia degenerate and remain attached to the metabasidium. There are also small germinating teliospores (3–5) × (5–7) μm with metabasidia 6–17 μm long and rarely more than two primary sporidia.

Some of the progeny obtained by micromanipulation of basidiospores were morphologically distinct from the parental culture; they were rugose, dry, and mainly composed of masses of teliospores. Yeast cells developed after several weeks in culture.

The teliospores germinated to produce metabasidia typical of the species. Other progeny were yeast-like and similar to the parent culture. None of the progeny were considered to be haploid. Germination also takes place directly from the mycelium, which becomes thick-walled. A metabasidium with primary sporidia develops; the metabasidium does not have an enlarged head. Endospores form in the degenerated teliospore and in the metabasidium usually after basidiospore development.

Conjugation was not observed when *Rhodosp. capitatum* was mixed with strains of the other species in the genus.

Fermentation: –

Assimilation of carbon compounds:

Galactose	+ (s)	Raffinose	+	Erythritol	–
Sucrose	+	Soluble starch	+ or w	Ribitol	–
Maltose	+	D-Xylose	+	D-Mannitol	+
Cellobiose	+	L-Arabinose	+ or w	Succinic acid	+
Trehalose	+	D-Ribose	+ w (–)	Citric acid	+
Lactose	+ w or s	L-Rhamnose	+ w (–)	Inositol	+

Additional carbon compounds tested: Melezitose +, melibiose –, α-methyl-D-glucoside –

Splitting of arbutin: –

Assimilation of nitrate: +

Vitamins stimulating growth: biotin, thiamine

Growth on 50% (w/w) glucose-yeast extract agar: –

Growth at 24°C: +; **at 30°C:** –

Starch formation: +

Urease: +

Origin of the strains studied

Marine sources of the southern Indian Ocean and Pacific Ocean (13) (Fell, 1976) and south Florida (1) (Ahearn & Roth, 1966).

Type strain: CBS 6358 from a zooplankton sample in the southern Indian Ocean.

3. *RHODOSPORIDIUM DACRYOIDUM* FELL, HUNTER ET TALLMAN (1973)

Imperfect state: Haploid strains of *Rhodosp. dacryoidum* fit the description of *Rhodotorula minuta,* however the type strain of *Rh. minuta* does not mate with the mating types of *Rhodosp. dacryoidum.*

Standard description of *Rhodosporidium dacryoidum*

Growth in malt extract: After 3 days at 24°C the haplophase cells are oval (1–5) × (3–9) μm, single, in pairs and short chains; budding may occur on a short stalk. A ring and film are not present; there is a light sediment. After one month there is a light to heavy ring and moderate sediment.

Growth on malt agar: After 7 days at 24°C the growth is carrot red, smooth, creamy, glistening, and slightly raised; the border is entire. Cell morphology is similar to that in malt extract. After one month the growth is slightly darker in color, glistening, creamy to pasty; occasionally the edge is slightly wrinkled.

Dalmau plate culture on corn meal agar: After one week at 24°C mycelial growth may be absent or there may be long thin true mycelium or short primitive pseudomycelium.

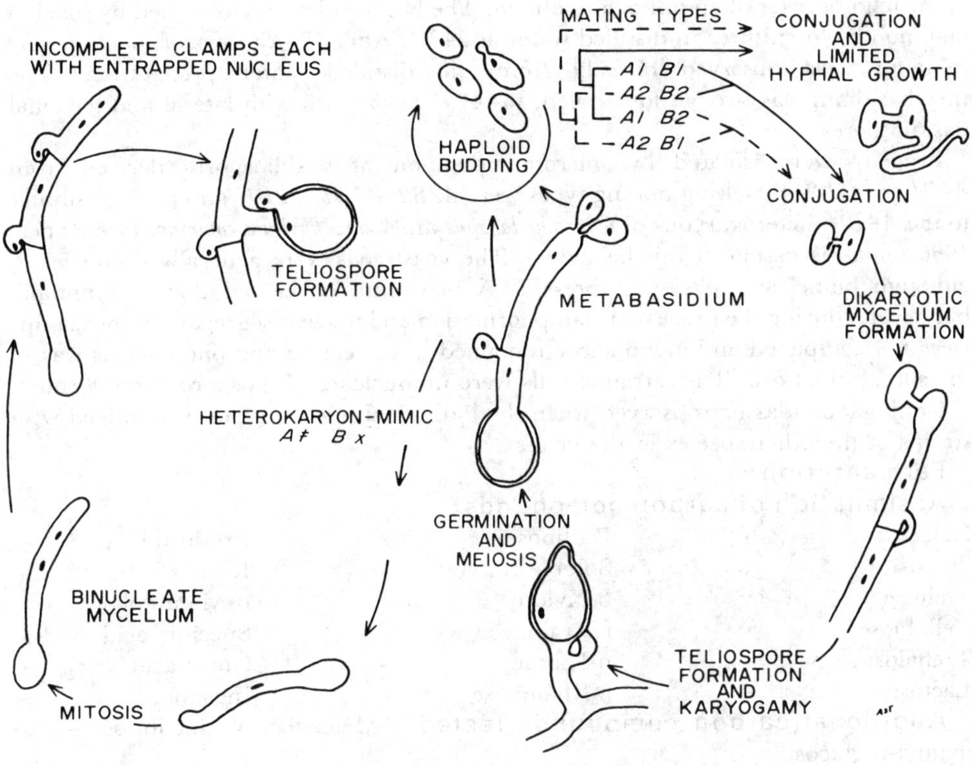

Diagram 7
Life cycle of *Rhodosp. dacryoidum*

Life cycle (Diagram 7): The species is heterothallic (bifactorial). Mating pairs with dissimilar *A* and *B* factors (*A1B1* × *A2B2, A1B2* × *A2B1*) produce the complete life cycle, while mixing of strains with similar *A* and dissimilar *B* factors results in mating and formation of short, distorted, true hyphae lacking both clamp connections and teliospores; there is no mating reaction with pairs having similar *B* factors. Conjugation between mating types with opposite *A* and *B* factors is usually within 24 hours of mixing on corn meal agar at 24°C; this may be variable, mating may not occur for one week or longer. After one week binucleate mycelium is sparse to abundant with hook-to-peg clamp connections. Conidia have not been observed on the hyphae.

Teliospores are lateral or occasionally terminal on the hyphae. The lateral teliospores either develop directly from the hypha on a short sporophore or from clamp connections. After one week the teliospores are subglobose and slightly dacryoid (4–10) × (5–14) μm, or more pronounced dacryoid with a subglobose base (4–8) × (7–10) μm and apical tips measuring (0.3–3) × (0.8–6) μm. After one month most teliospores exceed 6 × 11 μm. The apical tip may be wider than long (1–2) × (2–3) μm and many have an elongate basal tip (1–3) × (5–19) μm. Basal clamp connections join the teliospore to the hypha or to the sporophore. Germination of the teliospores is often difficult to obtain. The best results were obtained by soaking one month old cultures in distilled water at 24°C. After 10–25 days of soaking, germination was observed directly from the distilled water preparation. The metabasidium was two- to four-celled, (3–4) × (24–59) μm with lateral and terminal basidiospores.

Progeny were isolated by micromanipulation of basidiospores derived from *A1B1* × *A2B2*. Resulting mating types were *A2B2, A1B1, A2B1,* and a strain similar to the *A≠B* = heterokaryons of *Coprinus lagopus* and *Schizophyllum commune* (see Raper, 1966 for a discussion of this behavior). The yeast cells were generally uninucleate, although binucleate cells were present. A mycelium developed that was initially binucleate; during the process of clamp formation and nuclear segregation, the clamps were not completed and one nucleus remained in the clamp and one nucleus was in the subterminal cell. The terminal cells were uninucleate. Teliospores were formed.

Conjugation was not observed when the individual mating types were mixed with strains of the other species in the genus.

Fermentation: –

Assimilation of carbon compounds:

Galactose	+	Raffinose	–	Erythritol	–
Sucrose	v	Soluble starch	–	Ribitol	–
Maltose	v	D-Xylose	–	D-Mannitol	v
Cellobiose	v	L-Arabinose	–	Succinic acid	+
Trehalose	+	D-Ribose	–	Citric acid	+
Lactose	–	L-Rhamnose	v	Inositol	–

Additional carbon compounds tested: Melezitose v, melibiose –, α-methyl-D-glucoside –

Splitting of arbutin: v

Assimilation of nitrate: –

Growth in vitamin-free medium: –; weak growth with thiamine or with *p*-aminobenzoic acid

Growth on 50% (w/w) glucose-yeast extract agar: –

Growth at 30°C: +; **at 33°C:** –

Starch formation: –

Urease: +

Origin of the strains studied

Marine waters of the southern Indian and Pacific Oceans (10) (Fell et al., 1973).

Type strain: CBS 6353 (mating type *A1B1*). Allotype strains: CBS 6354 (mating type *A2B2*); CBS 6355 (mating type *A2B1*); CBS 6356 (mating type *A1B2*). CBS 6357 is an *A≠B =* strain from a cross of CBS 6353 and CBS 6354.

4. *RHODOSPORIDIUM DIOBOVATUM* NEWELL ET HUNTER (1970)

Imperfect state: Strains of *Rhodosp. diobovatum* fit the description of *Rhodotorula glutinis*, although the type strain of *Rh. glutinis* does not mate with either mating type of *Rhodosp. diobovatum*.

Standard description of *Rhodosporidium diobovatum*

Growth in malt extract: After 3 days at 22°C, the haplophase cells are round to oval (1–6) × (2–9) μm, single and in pairs. A slight sediment is formed. After one month, a heavy sediment and a thin to moderate film of cells; large round cells (to 9 μm diameter) with large oil globules are present.

Growth on malt agar: Cell morphology is similar to that in malt extract. The streak colony is coral red, raised, mucous, the border entire.

Dalmau plate culture on corn meal agar: Some strains produce only budding cells, whereas other strains produce, in addition to budding cells, a primitive pseudomycelium or fine, true hyphae (diameter 2 μm) bearing blastoconidia with dimensions similar to the yeast-phase cells. These blastoconidia are often formed at septa, on structures homologous to clamp connections, but aberrant and incomplete. Each primary hypha is formed by a single yeast-phase cell.

Life cycle: The species is heterothallic, a biallelic (*A1* and *A2*) system of sexual incompatibility factors exists. Mixing of yeast-phase strains of *A1* and *A2* on corn meal agar results in conjugation within 24 hours. Plasmogamy takes place, and binucleate hyphae (diameter 3 μm) are formed from one of the conjugant cells, or from the conjugation tube joining them. Clamp connections are hook-to-peg or peg-to-peg clamps. After 3 days of hyphal growth, unicellular, round to obovate, diobovate, or diclavate (6–11) × (6–22) μm heavy-walled terminal teliospores form. The teliospores have clamp connections at the basal septum and are binucleate at first, but become uninucleate at maturity. In addition to teliospores, blastoconidia form on the binucleate hyphae. Additional incompatibility factors may exist as several isolates have been examined which mate with *A2* strains to produce mycelium but fail to develop teliospores. Germination of the teliospores occurs after 5–10 days of soaking. A club-shaped phragmometabasidium, (3–5) × (13–22) μm develops from the teliospore with lateral and terminal basidiospores which are either *A1* or *A2* mating types (Fig. 153).

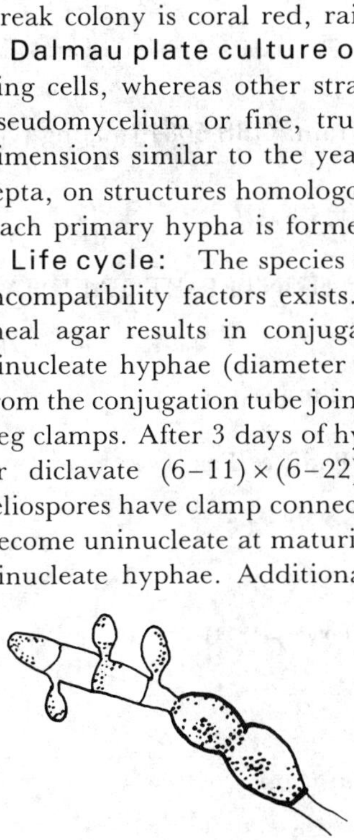

Fig. 153 – *Rhodosp. diobovatum*
Four-celled metabasidium with lateral basidiospores on a diobovate teliospore

Fermentation: –

Assimilation of carbon compounds:

Galactose	+	Raffinose	+	Erythritol	–
Sucrose	+	Soluble starch	+ w	Ribitol	+
Maltose	+	D-Xylose	+	D-Mannitol	+ s or –
Cellobiose	+	L-Arabinose	+	Succinic acid	+ or w
Trehalose	+	D-Ribose	+	Citric acid	+ or w
Lactose	–	L-Rhamnose	–	Inositol	–

Additional carbon compounds tested: Melezitose +, melibiose –, α-methyl-D-glucoside +

Splitting of arbutin: +

Assimilation of nitrate: +

Growth in vitamin-free medium: +

Growth on 50% (w/w) glucose-yeast extract agar: –

Maximum temperature of growth: 35°C

Starch formation: –

Urease: +

G + C: 67.3 mol% (Nakase & Komagata, 1971g)

Origin of the strains studied

Marine sources in southern Florida (14) (Newell & Hunter, 1970). The species has been reported from plants and soils (Kvasnikov et al., 1975; Babjeva & Kratintsev, 1974).

Type strain: CBS 6085 (mating type *A1*). Allotype strain: CBS 6084 (mating type *A2*).

5. *RHODOSPORIDIUM INFIRMOMINIATUM* FELL, HUNTER ET TALLMAN (1973)

(as *Rhodosporidium infirmo-miniatum* Fell, Hunter et Tallman)

Imperfect state: *Cryptococcus infirmo-miniatus* (Okunuki) Phaff et Fell

Synonyms

Torula infirmo-miniata Okunuki (1931)
Rhodotorula glutinis (Fres.) Harrison var. *infirmo-miniata* (Okunuki) Lodder (1934)
Rhodotorula infirmo-miniata (Okunuki) Hasegawa et Banno (1964)
Cryptococcus infirmo-miniatus (Okunuki) Phaff et Fell (1970)
Rhodotorula sinensis Lee (1974)

Standard description of *Rhodosporidium infirmominiatum*

Growth in malt extract: At 12°C after 3 days, the cells are single, in pairs and short chains, subglobose and elongate (1–5) × (3–10) μm. There is a light sediment. After one month a slight film and ring have formed.

Growth on malt agar: After 3 days at 12°C, the cells are subglobose (2–5) × (3–9) μm. The growth is orient to grenadine pink in color, smooth, glistening, and slightly raised; the border is entire and not myceliated. After one month, the colony is usually mucous, runny, glistening, the color grenadine pink, the edge is lighter in color and entire; some strains are dull, dry, grenadine in color with the lighter edge entire. The cells are oval and subglobose (1–5) × (3–11) μm. In addition, there are round to subglobose (2–8) × (5–8) μm granulated cells, with cell walls that appear to be thicker than the typical vegetative cells. The cell walls of the granulated cells are frequently shed from the parent cells. These cells reproduce by budding or formation of mycelium.

Dalmau plate culture on corn meal agar: Mycelium formation is absent at 12°C.

Life cycle (Diagram 8): This species is heterothallic with a multiple allelic unifactorial incompatibility system with three known mating types: *A1, A2* and *A3.*

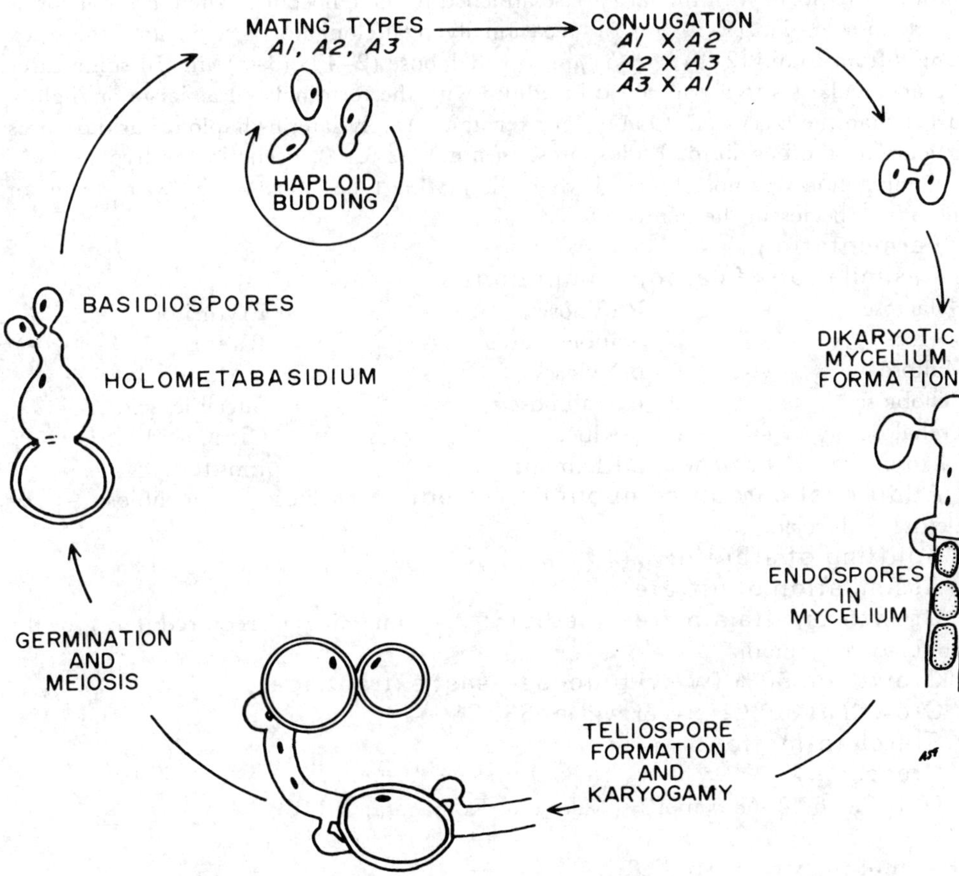

Diagram 8
Life cycle of *Rhodosp. infirmominiatum*

Mating results are similar with mixes of *A1* × *A2, A1* × *A3,* and *A2* × *A3*. Conjugation takes place within 36 hours of mixing on corn meal agar. There is some strain variation in the rate and abundance of mycelium and teliospore formation; usually after 3 days the mycelium is sparse and after one week mycelium and teliospores are abundant. Conidia, on the hyphae, are rare. Clamp connections, which are infrequent, are hook-to-peg. Two morphological types of teliospores form. Some are terminal and intercalary, single and double, round (7–8 μm diameter) to subglobose (5–9) × (7–11) μm with numerous granular bodies; becoming slightly larger (7–10) × (7–11) μm after one month. The second type of teliospore has a thick cell wall, a large globular inclusion, brown pigment, oval shape (9–11) × (14–15) μm or subglobose (8–17) × (8–18) μm, and occurs in groups to 12. The difference in teliospores may be attributed to age, the heavy-walled spores being older.

Germination of the teliospores was obtained by growing the dikaryotic culture for 17 days on corn meal agar followed by soaking the teliospores in distilled water for 20 weeks with transfer to 2% agar for 3 days. The metabasidia are variable in shape; most are pyriform with the large base attached to the teliospore. The basal portion is subglobose (5–7) × (7–9) μm or occasionally oval-elongate (4 × 8) μm, the apex ranges from round (2–3) × (2–3) μm to subglobose (2–4) × (3–5) μm. In some cases the apex enlarges to a two-celled basidium with the terminal cell as large or slightly larger than the basal cell. One to four terminal, rarely lateral, haploid basidiospores develop on the basidium. Endospores, in numbers to 10, occur in the hyphae.

Conjugation was not observed when the mating types were mixed with strains of the other species in the genus.

Fermentation: –

Assimilation of carbon compounds:

Galactose	+	Raffinose	+	Erythritol	–
Sucrose	+	Soluble starch	+	Ribitol	– (+)
Maltose	+	D-Xylose	+	D-Mannitol	–
Cellobiose	+	L-Arabinose	+	Succinic acid	+
Trehalose	+	D-Ribose	+ or s (–)	Citric acid	+
Lactose	+ s or w	L-Rhamnose	v	Inositol	+

Additional carbon compounds tested: Melezitose +, melibiose –, α-methyl-D-glucoside –

Splitting of arbutin: +

Assimilation of nitrate: +

Growth in vitamin-free medium: –; thiamine is required for growth; stimulation by biotin

Growth on 50% (w/w) glucose-yeast extract agar: –

Growth at 30°C: + or w; **at 33°C:** –

Starch formation: +

Urease: +

G + C: 67.8–68.5 mol% (Nakase & Komagata, 1971b)

Origin of the strains studied

Marine waters of the Antarctic, southern Indian and southern Pacific Oceans (14), human source (1) (Fell et al., 1973); air in Tokyo (1) (Okunuki, 1931); diseased pears in China (1) (Lee, 1974).

Type strain: CBS 323 (mating type *A1*), the type strain of *Cryptococcus infirmo-miniatus*. Allotype strains: CBS 6350 (mating type *A2*) and CBS 6352 (mating type *A3*), both from marine sources.

6. *RHODOSPORIDIUM MALVINELLUM* FELL ET HUNTER (in Fell, 1970b)

Imperfect state: Strains of *Rhodosp. malvinellum* fit the description of *Rhodotorula graminis* except for a distinct difference in colony colors. Haploid strains of *Rhodosp. malvinellum* do not mate with the type of strain of *Rh. graminis*.

Standard description of *Rhodosporidium malvinellum*

Growth in malt extract: After 3 days at 12°C the haplophase cells measure (2–7) × (3–7) μm; they are single, occasionally in pairs with budding on short necks. A light sediment is formed. After one month there is a moderate sediment.

Growth on malt agar: After 3 days at 12°C the cell sizes are as above. The streak is creamy white, smooth and semi-glistening. The periphery is entire and the cross section is raised. At 2 weeks the streak is light mauve and at one month the color has intensified.

Dalmau plate culture on corn meal agar: Pseudomycelium is generally absent, occasionally short branched chains of cells are present.

Life cycle: The species is heterothallic (biallelic, unifactorial); when compatible mating types are mixed on corn meal agar at 12°C, conjugation and binucleate mycelium with clamp connections are present within 3 days. Teliospores are rare after 7 weeks and abundant after 3 months. The teliospores are subglobose (5–10) × (7–12) μm. They develop singly and are attached to the hyphae by a short stalk. Germination has only been observed once. The phragmometabasidium is two-celled with lateral and terminal basidiospores.

Fermentation: –

Assimilation of carbon compounds:

Galactose	+ or s	Raffinose	+	Erythritol	–
Sucrose	+	Soluble starch	–	Ribitol	v
Maltose	+	D-Xylose	+ s	D-Mannitol	+
Cellobiose	v	L-Arabinose	v	Succinic acid	+
Trehalose	+	D-Ribose	–	Citric acid	–
Lactose	–	L-Rhamnose	+ or s	Inositol	–

Additional carbon compounds tested: Melezitose –, melibiose –, α-methyl-D-glucoside –

Assimilation of nitrate: +

Growth in vitamin-free medium: –; thiamine is required for growth; in some strains *p*-aminobenzoic acid stimulates growth in the presence of thiamine

Growth on 50% (w/w) glucose-yeast extract agar: –

Growth at 21°C: +; **at 24°C:** –

Starch formation: –

Urease: v

G + C: 50.5 mol% (Nakase & Komagata, 1971g)

Origin of the strains studied

Marine waters of the southern Pacific, Indian and Antarctic Oceans (17) (Fell, 1970b).

Type strain: CBS 6082 (mating type *A1*). Allotype strain: CBS 6083 (mating type *A2*).

7. *RHODOSPORIDIUM PALUDIGENUM* FELL ET TALLMAN (1980a)

Imperfect state: Haploid strains of *Rhodosp. paludigenum* fit the description of *Rhodotorula graminis.* The type strain of *Rh. graminis* does not mate with either mating type of *Rhodosp. paludigenum.*

Standard description of *Rhodosporidium paludigenum*

Growth in malt extract: After 3 days at 25°C the cells are oval to elongate (2–4) × (3–11) μm. They occur singly and in pairs. There is a light sediment. After one month there is a heavy ring with light to heavy sediment.

Growth on malt agar: After 3 days at 25°C the cell morphology is similar to that in malt extract. The one month old streak culture is salmon orange, smooth, glistening, slightly raised or rugose and the border entire.

Dalmau plate culture on corn meal agar: After 6 days at 25°C there are short chains of budding cells, and short lengths of true mycelium may be present.

Life cycle: The species is heterothallic. When opposite mating types were mixed on corn meal agar (prepared with 20 $^0/_{00}$ seawater) and incubated at 22–25°C, true hyphae bearing spheroidal (13–17 μm diameter) teliospores developed within 6 days. The teliospores were in clusters of two to four, intercalary and rarely terminal on the hyphae. Teliospore germination after 4 weeks on corn meal agar, resulted in two- to four-celled metabasidia. Basidiospores were of the two parental types, indicating a unifactorial sexual system. *Rhodosp. paludigenum* does not mate with the other species included in the genus.

Fermentation: –

Assimilation of carbon compounds:

Galactose	+	Raffinose	+	Erythritol	–
Sucrose	+	Soluble starch	–	Ribitol	+
Maltose	+ s	D-Xylose	+ s	D-Mannitol	+
Cellobiose	+	L-Arabinose	+ s	Succinic acid	+ w
Trehalose	+ s	D-Ribose	+ s	Citric acid	+ w
Lactose	–	L-Rhamnose	+ s	Inositol	–

Additional carbon compounds tested: Melezitose –, melibiose –, α-methyl-D-glucoside –

Assimilation of nitrate: +

Growth in vitamin-free medium: +

Growth on 50% (w/w) glucose-yeast extract agar: –
Growth at 37°C: –
Starch formation: –
Urease: +

Origin of the strains studied

Intertidal marsh and swamp waters in South Florida (3) (Fell & Tallman, 1980a).

Type strain: CBS 6566 (mating type *A2*). Allotype strain: CBS 6567 (mating type *A1*).

8. *RHODOSPORIDIUM SPHAEROCARPUM* NEWELL ET FELL (1970)

Imperfect state: Strains of *Rhodosp. sphaerocarpum* fit the description of *Rhodotorula glutinis*. The type strain of *Rh. glutinis* does not mate with mating types of *Rhodosp. sphaerocarpum*.

Synonym

Rhodotorula glutinis (Fres.) Harrison var. *salinaria* Hirosawa et Takada (1969)

Standard description of *Rhodosporidium sphaerocarpum*

Growth in malt extract: After 3 days at 20°C the cells are spheroidal to ellipsoidal, (3–5) × (5–7) μm, single or in pairs. Often daughter cells are budded from very short necks < 0.5 μm long. There is an orange to pink-orange sediment. After one month the sediment becomes thick and a weak ring has formed; spheroidal cells 7 × 7 μm, may be present.

Growth on malt agar: Cell morphology at 20°C is similar to that in malt extract. Streak culture at 3 days is smooth, glossy, orange to pink-orange; texture is highly mucous; cross section is raised, spreading; border is entire. After one month the culture becomes extremely mucous; if kept at 5°C for one month, the culture does not become as mucous as at 20°C, and develops a yellow pigmentation.

Dalmau plate culture on corn meal agar: Some strains produce no mycelium, while some produce, from individual cells, short, fine, primary hyphae (diameter 1–3 μm) with one or two septa; other strains develop extensive, larger (diameter 3.4–5.4 μm) septate hyphae bearing large, spherical, thick-walled, terminal or intercalary teliospores.

Life cycle (Diagram 9): *Rhodosp. sphaerocarpum* has a biallelic unifactorial mating system. Mixing of haploid mating types *A1* and *A2* on corn meal agar at room temperature results in conjugation followed by hyphal development from the conjugation tube, or from one of the two conjugate cells. The mycelium is dikaryotic; clamp connections form at the septa and thick-walled teliospores develop terminally and intercalary on the hyphae (Fig. 154). The teliospores are at first binucleate and even-

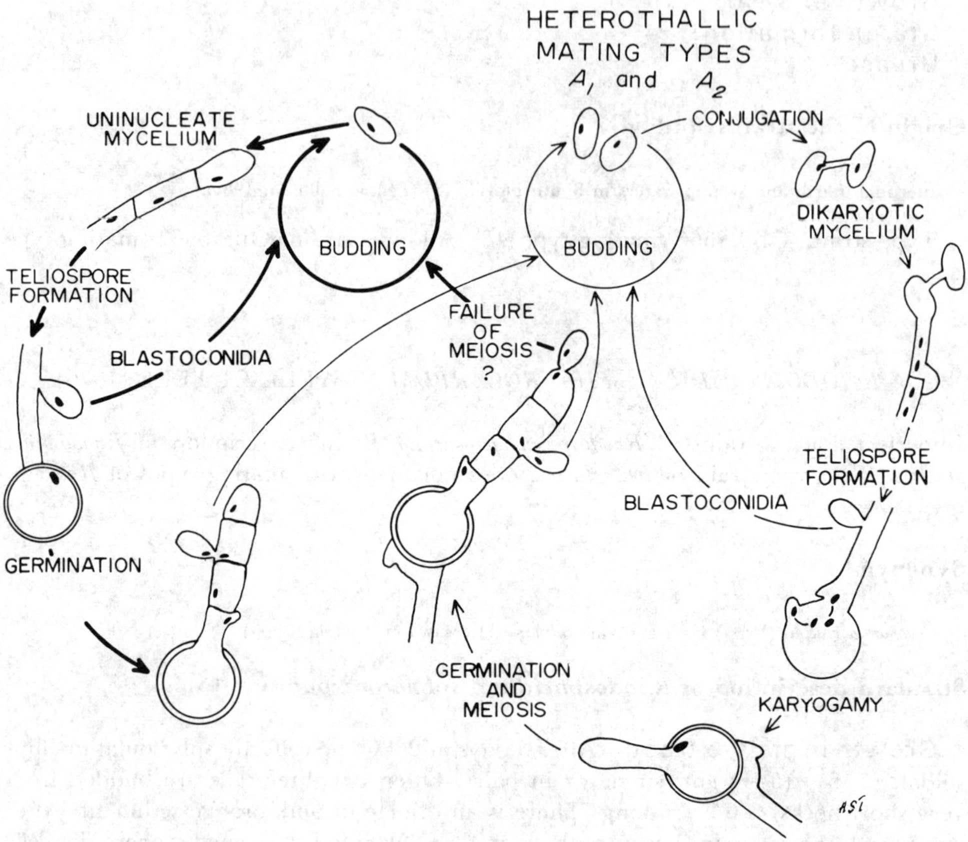

Diagram 9
Life cycle of *Rhodosp. sphaerocarpum*

tually become uninucleate. They are spheroidal, smooth, and dark in color at maturity, occurring singly, in pairs, and in branched chains of as many as 12 or more. When teliospores are present in large quantities the culture becomes dark brown. The teliospores germinate if placed in distilled water at room temperature for one month followed by transfer to malt agar at the same temperature. The teliospore germinates to a four-celled metabasidium, each cell capable of repeated mitotic divisions for the formation of several basidiospores (Fig. 154). The metabasidium eventually breaks up into individual cells. The basidiospores are haploid; a haplo-lethal effect occurs in some strains: examination of progeny from germination of teliospores indicated that all colonies were of mating type *A1*; apparently mating type *A2* did not survive.

Some strains produce teliospores in the absence of mating. The yeast cells are uninucleate and produce uninucleate teliopsore-bearing hyphae without clamp connections. The mycelial formation is much less extensive than the dikaryotic mycelium produced by the conjugating strains. Teliospores of the self-sporulating strains ger-

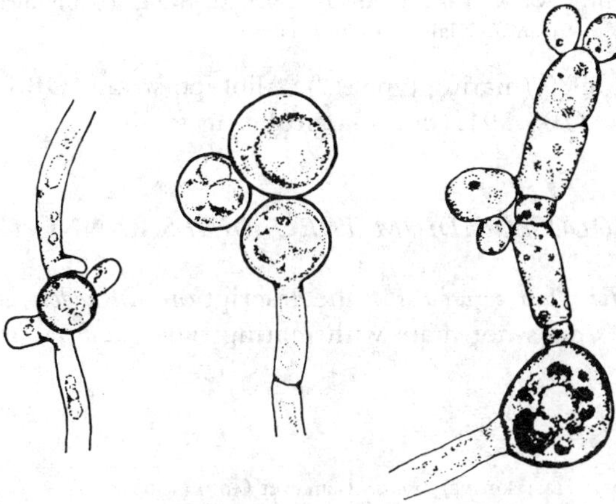

Fig. 154 – *Rhodosp. sphaerocarpum*
Intercalary teliospore with clamp connection; terminal teliospores without clamp connections; teliospore with phragmometabasidium and basidiospores (From 'The Yeasts', ed. II, 1970)

minate under the same conditions as described above for the heterothallic strains to produce phragmometabasidia with haploid basidiospores. Although Newell and Fell (1970) postulated that the haploid cells were the result of a meiotic division in the basidium, it is likely that these strains are homokaryotic fruiters similar to those of *Leu. scottii.* Studies have not been undertaken to confirm either possibility.

Fermentation: –

Assimilation of carbon compounds:

Galactose	+	Raffinose	+	Erythritol	–
Sucrose	+	Soluble starch	–	Ribitol	+
Maltose	+	D-Xylose	+	D-Mannitol	v
Cellobiose	+	L-Arabinose	+	Succinic acid	–
Trehalose	+	D-Ribose	+	Citric acid	–
Lactose	–	L-Rhamnose	–	Inositol	–

Additional carbon compounds tested: Melezitose +, melibiose –, α-methyl-D-glucoside +

Splitting of arbutin: +

Assimilation of nitrate: +

Growth in vitamin-free medium: –; thiamine is required for growth

Growth on 50% (w/w) glucose-yeast extract agar: –

Growth at 27°C: +; at 30°C: +w; at 33°C: –

Starch formation: –

Urease: +

G + C: 64.9–65.4 mol% (Nakase & Komagata, 1971g)

Origin of the strains studied

Marine sources near the Antarctic (Palmer) Peninsula (33) and from the Caribbean Sea (4) (Fell et al.,

1973); salt water (1) (Hirosawa & Takada, 1969). Buck (personal communication) reports *Rhodosp. sphaerocarpum* is prevalent in freshwater lakes in Connecticut.

Type strain: CBS 5939 (mating type *A2*). Allotype strain: CBS 5940 (mating type *A1*). Paratype strain: CBS 5941, a self-sporulating strain.

9. *RHODOSPORIDIUM TORULOIDES* BANNO (1967)

Imperfect strains of *Rhodosp. toruloides* fit the description of *Rhodotorula glutinis*. The type culture of *Rh. glutinis* does not mate with mating types of *Rhodosp. toruloides*.

Synonym

Rhodotorula glutinis (Fres.) Harrison var. *rufusa* Iizuka et Goto (1965)

Standard description of *Rhodosporidium toruloides*

Growth in malt extract: After 4 days at room temperature heterothallic haploid cells are spheroidal to elongate (2–7) × (4–12) μm. The homothallic cells are ovoidal to obtuse oblong (3–7) × (6–30) μm and may form true mycelium and teliospores.

Growth on malt agar: After 4 days at room temperature the cells are similar in size to those in malt extract. The streak culture is pink to orange, the border is entire, the surface is smooth and glistening, the texture is soft to slightly mucous. After one month the color becomes scarlet. The homothallic cultures have dark brown areas consisting of true mycelium and teliospores.

Dalmau plate culture on corn meal agar: After 2 weeks at room temperature pseudomycelium of heterothallic strains may be absent or range from short branched chains of pseudomycelium to abundant clusters of pseudomycelium. The homothallic strains have extensive true mycelium with clamps and teliospores.

Life cycle (Diagram 10): The species is homothallic and heterothallic. Heterothallism: On malt agar at room temperature cells of compatible mating types (*A1* and *A2*) conjugate within 24 hours and plasmogamy occurs. From one cell of the conjugate pair a septate hypha develops with clamp connections at the septa. On this hypha terminal swellings develop, which eventually become highly angular, irregularly shaped, thick-walled teliospores with clamp connections at their bases (Fig. 155). After 1–2 weeks the spores germinate to produce club-shaped metabasidia with one to three septa. Basidiospores bud laterally and terminally. Some of the basidiospores are heterothallic and capable of mating and repeating the life cycle.

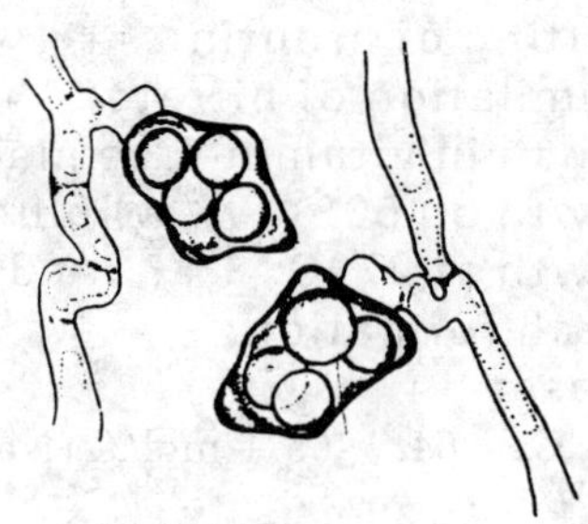

Fig. 155 – *Rhodosp. toruloides*
Teliospores, clamp connections
(From 'The Yeasts', ed. II, 1970)

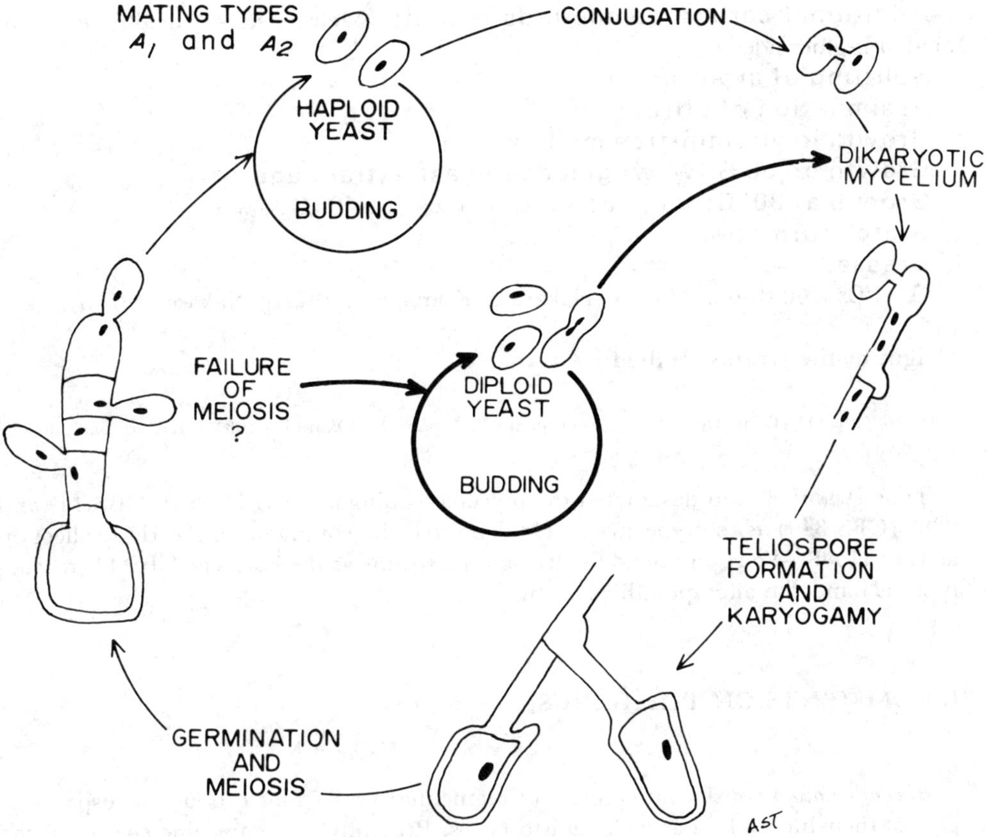

Diagram 10
Life cycle of *Rhodosp. toruloides*

Homothallism: In addition to the heterothallic basidiospores, homothallic basidiospores are produced on the metabasidium. These are uninucleate cells that develop directly into a dikaryotic mycelium with clamp connections and teliospores. The teliospores germinate to produce metabasidia with both heterothallic and homothallic basidiospores.

The heterothallic and homothallic basidiospores are uninucleate while the hyphae with clamp connections, produced by both heterothallic and homothallic phases, is binucleate. The teliospores are initially binucleate, and become uninucleate through karyogamy. At germination the diploid nucleus undergoes reduction division into the four haploid nuclei of the metabasidium.

Fermentation: –

Assimilation of carbon compounds:

Galactose	+ s	Raffinose	+	Erythritol	–
Sucrose	+	Soluble starch	–	Ribitol	+
Maltose	+	D-Xylose	+	D-Mannitol	+
Cellobiose	– or + s	L-Arabinose	+	Succinic acid	+ w
Trehalose	+	D-Ribose	+	Citric acid	+ w
Lactose	–	L-Rhamnose	–	Inositol	–

Additional carbon compounds tested: Melezitose +, melibiose −, α-methyl-D-glucoside +

Splitting of arbutin: +

Assimilation of nitrate: +

Growth in vitamin-free medium: +

Growth on 50% (w/w) glucose-yeast extract agar: −

Growth at 30°C: +; **at 35°C:** +w; **at 40°C:** +w

Starch formation: −

Urease: +

G + C: 60.0–61.0 mol% (Nakase & Komagata, 1971b; Nakase, 1972b)

Origin of the strains studied

Wood pulp (1) (Rennerfelt, 1937); atmosphere in Japan (1) (Okunuki, 1931); marine mammal (2); unknown (2).

Type strain: Banno designated the mycelial conjugate of IFO 0559 (CBS 14) and 0880 (CBS 349) as the type strain. He deposited the conjugate in the IFO collection as IFO 11588. We recommend the living type strains as the holotype CBS 14 (mating type *A1*) and the allotype CBS 349 (mating type *A2*).

f. COMMENTS ON THE GENUS

Rhodosporidium consists of carotenoid pigmented yeasts that produce teliospores as part of their life cycle, but lack ballistospores. Presently there are nine species in the genus which are separated on their physiological capacities, sexual incompatibilities and teliospore morphology (Table 23). There are five nitrate positive, inositol negative species; three nitrate positive, inositol positive species and one nitrate negative species. Among the nitrate positive, inositol negative species, two of the species, *Rhodosp. malvinellum* and *Rhodosp. paludigenum,* do not assimilate melezitose; their separation is based on colony color (*Rhodosp. malvinellum* is white to mauve, *Rhodosp. paludigenum* is salmon orange). The carbon spectrum is similar to that described for *Rhodotorula graminis,* however, strains of those two *Rhodosporidium* species are not sexually compatible with strains of *Rh. graminis.* Strains of the remaining nitrate positive, inositol negative species (*Rhodosp. sphaerocarpum, Rhodosp. toruloides* and *Rhodosp. diobovatum*) have the same assimilation pattern as *Rh. glutinis* but do not mate with the type strains of *Rh. glutinis.* Separation of these three *Rhodosporidium* species requires an examination of the teliospore shapes. The inositol positive species, *Rhodosp. infirmominiatum, Rhodosp. bisporidii* and *Rhodosp. capitatum,* have assimilation patterns similar to that of *Cryptococcus infirmo-miniatus,* although *Rhodosp. bisporidii* differs by the ability to assimilate melibiose. *Rhodosp. dacryoidum* is the only described species that does not assimilate nitrate as a sole source of nitrogen.

With the exception of *Rhodosp. capitatum,* all *Rhodosporidium* spp. have heterothallic life cycles. *Rhodosp. toruloides,* in addition to the heterothallic cycle, has a homothallic life cycle which progresses from a diploid yeast cell to dikaryotic hyphae with clamp connections and teliospores. In contrast, strains of *Rhodosp. capitatum* and *Rhodosp. sphaerocarpum* develop uninucleate hyphae with teliospores. Studies have not determined whether this is a homothallic life cycle or an asexual fruiting development. However, this type of development is typically found in homokaryotic fruiting strains of *Schizophyllum commune* and *Polyporus ciliatus* (Esser et al., 1979; Leonard & Raper, 1969; Stahl & Esser, 1976). Among the heterothallic species, *Rhodosp. bisporidii* and *Rhodosp. dacryoidum* have bifactorial incompatibility systems, *Rhodosp. infirmominiatum* has a unifactorial multiple allelic system, while the incompatibility systems of the remaining species are unifactorial biallelic.

Metabasidial morphology varies in the genus. The metabasidia may be one-celled with terminal basidiospores (*Rhodosp. infirmominiatum*); one-celled with an enlarged terminus and primary basidiospores (*Rhodosp. bisporidii* and *Rhodosp. capitatum*) or septate with lateral and terminal basidiospores (*Rhodosp. dacryoidum, Rhodosp. diobovatum, Rhodosp. malvinellum, Rhodosp. paludigenum, Rhodosp. sphaerocarpum* and *Rhodosp. toruloides*).

Potentially the genus consists of considerably more species than presently recognized. The perfect state of only one of the red pigmented imperfect species (e.g. that includes an imperfect type strain) has been discovered (*Cryptococcus infirmo-miniatus*), indicating that the perfect state of all *Rhodotorula* species, as represented by the type strains, have yet to be determined. With present techniques, elucidation of the life cycles is difficult. Low percentages of mating and teliospore germination are two of the obstacles. Abe et al. (1975) considerably improved the mating frequency to 89% in *Rhodosp. toruloides* by use of a specific ratio of the two mating types in a shaken liquid culture medium. The present teliospore germination technique of a preliminary distilled water soak followed by transfer to solid media, although eventually successful for most species, is time and labor intensive. Therefore studies on the physiology of germination would be of considerable value.

Genus 3. *Sporidiobolus* Nyland

by J.W. Fell and A. Statzell Tallman

a. DIAGNOSIS OF THE GENUS

Colony color on solid media includes shades of pink and red due to the presence of carotenoid pigments. Vegetative reproduction is by budding, formation of pseudomycelium and true mycelium, and by ballistosporulation. The ballistospores develop on simple and branched sterigmata that arise from yeast and mycelial cells.

Sexual mechanisms are homothallic and heterothallic. In the heterothallic system a dikaryotic mycelium results from the conjugation of a compatible mating pair. A clamp connection is usually present at each cross wall. Intercalary and terminal teliospores develop. Karyogamy takes place in the teliospore (probasidium) which germinates to produce a metabasidium (promycelium). Meiosis takes place during germination with the formation of haploid basidiospores.

Homothallism consists of the development of hyphae with teliospores from a single yeast cell in the absence of mating. The yeast phase is diploid, reduction division takes place with the formation of a dikaryotic hypha with clamp connections. Karyogamy occurs in the teliospore. Germination of the teliospore results in a metabasidium with diploid basidiospores.

A strain of *Sporid. pararoseus* produces teliospores on uninucleated hyphae that lack clamp connections. This may be a result of homokaryotic fruiting rather than homothallism.

The metabasidia produced from homothallic and heterothallic systems are one-celled (holometabasidia) or septate with two or more cells (phragmometabasidia). The basidiospores are terminal on the holometabasidium and lateral and terminal on the phragmometabasidium. In some species the metabasidium is connected to the teliospore by an aseptate stalk that ranges in length from 3–26 μm.

Sugars are not fermented; urease is positive.

b. TYPE SPECIES OF THE GENUS

Sporidiobolus johnsonii Nyland

c. SPECIES ACCEPTED IN THE GENUS

1. *Sporidiobolus johnsonii* Nyland (1949)
2. *Sporidiobolus pararoseus* Fell et Tallman (1981)
3. *Sporidiobolus ruinenii* Holzschu, Tredick et Phaff (1981)
4. *Sporidiobolus salmonicolor* Fell et Tallman (1981)

d. KEY TO THE SPECIES OF THE GENUS

1a Nitrate assimilated ... 2
b Nitrate not assimilated
Sporid. pararoseus ... p. 535

2a Homothallic formation of teliospores ... 3
b Heterothallic formation of teliospores
Sporid. salmonicolor ... p. 538

3a Raffinose assimilated
Sporid. ruinenii ... p. 537
b Raffinose not assimilated
Sporid. johnsonii ... p. 533

See Table 24, p. 534.

e. SYSTEMATIC DISCUSSION OF THE SPECIES

1. SPORIDIOBOLUS JOHNSONII NYLAND (1949)

Standard description of *Sporidiobolus johnsonii*

Growth in malt extract: After 3 days at room temperature the cells are ovoidal and elongate (5–7) × (13–27) μm, single, in pairs and in short chains. The cells are budding and may have simple or branched sterigmata bearing kidney-shaped ballistospores (4–7) × (8–14) μm. Short hyphal strands develop from single cells. A thick, dull, corrugated grenadine pink pellicle and a light sediment are present. After one month there is a moderate sediment and a strongly wrinkled pellicle. There is an extensive formation of mycelium with clamp connections and teliospores.

Growth on malt agar: After 5 days the streak culture is flesh colored with a convoluted surface; the appearance is dull; the texture dry and powdery; the periphery has a hyphal fringe. After one month the colony is peach in color. By 3 weeks there is an extensive formation of mycelium and teliospores.

Dalmau plate culture on corn meal agar: After 5 days at room temperature there are budding yeast cells, cells with simple and branched sterigmata bearing bal-

Table 24

Key characters of the ballistospore-forming yeasts

	Assimilation							30°C	Myc	Tel
	NO_3	I	La	Ra	Ma	Ce	LAr			
Bullera alba	−	+	+	+	+	+	+	−	−	−
B. dendrophila	−	+ s	+	−	+	+	+	−/+w	−	−
B. grandispora	−		−		+					
B. piricola	+	+	+ /w	+	+	+ w	+	−	−	−
B. singularis	−	−	+	−	−	+	−	−	−	−
B. tsugae	+	−	+	−	+	+	−	−	−	−
Fibulobasidium inconspicuum	−	+	+ w	+	+	+	+	+	−	−
Sirobasidium magnum	−	+	+	−	−	+	v	+	−	−
Sporidiobolus johnsonii	+	−	−	−	+	+	−	+	+	HM
Sporid. pararoseus	−	−	−	+	+	+	−	v	v	HT, HM
Sporid. ruinenii	+	−	−	+	+ s	+	+ (s)	+	+	HM
Sporid. salmonicolor	+	−	−	v	− (+)	− (+)	v	+ (−)	v	HT
Sporobolomyces albo-rubescens	−	−	−	+	+	−	+	+	−	−
Sp. gracilis	−	−	−	−	−	+ s	−	−	−	−
Sp. holsaticus	+	−	−	v	+	+ /s	+ s	+	+	−
Sp. puniceus	+	+ s	−	+	+ s	+	+	−	+	−
Sp. roseus	+	−	−	+	+	+ (−)	v	−	−	−

Myc = mycelium, Tel = teliospores, NO_3 = nitrate, I = inositol, La = lactose, Ra = raffinose, Ma = maltose, Ce = cellobiose, LAr = L-arabinose; HM = homothallic, HT = heterothallic.

listospores, and extensive true mycelium with clamp connections and spherical teliospores of 10–15 μm diameter.

Formation of ballistospores: Asymmetric, reniform, (2–4) × (5–10) μm ballistospores form on simple or branched sterigmata after 3 days on malt agar.

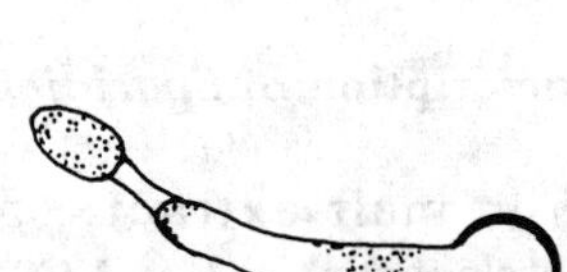

Fig. 156 – *Sporid. johnsonii* Teliospore with one-celled metabasidium and terminal basidiospore

Life cycle: As described by Laffin and Cutter (1959a,b) the species is homothallic. The yeast phase is diploid, reduction division occurs with the formation of a dikaryotic hyphae with clamp connections. Karyogamy occurs in the teliospore. Germination of the teliospore results in a holometabasidium with a terminal diploid basidiospore (Fig. 156). To obtain germination, the cultures were grown on malt agar (22°C) for 4 weeks and then the teliospores were transferred to 2% agar where they germinated after 10 days of incubation.

Fermentation: –

Assimilation of carbon compounds:

Galactose	– (+w)	Raffinose	–	Erythritol	–
Sucrose	+	Soluble starch	+ (s)	Ribitol	+ s
Maltose	+	D-Xylose	+ w	D-Mannitol	+
Cellobiose	+	L-Arabinose	–	Succinic acid	+
Trehalose	+	D-Ribose	v	Citric acid	+
Lactose	–	L-Rhamnose	–	Inositol	–

Additional carbon compounds tested: Melibiose –, melezitose +, α-methyl-D-glucoside +, galactitol –

Assimilation of nitrate: +

Growth in vitamin-free medium: +

Growth on 50% (w/w) glucose-yeast extract agar: –

Growth at 37°C: +

Starch formation: –

Urease: +

Origin of the strains studied

Pustule of *Phragmidium rubi-idaei* (DC.) Karst. on a living leaf of *Rubus idaeus* L. in the State of Washington, U.S.A. (1) (Nyland, 1949).

Type strain: CBS 5470 (Nyland, 1949).

2. *SPORIDIOBOLUS PARAROSEUS* FELL ET TALLMAN (1981)

Imperfect state: *Sporobolomyces shibatanus* (Okunuki) Verona et Cif.

Synonyms

Torula shibatana Okunuki (1931)
Sporobolomyces shibatanus (Okunuki) Verona et Cif. (1938)
Sporobolomyces pararoseus nom. nud. (Olson & Hammer, 1937)
Sporobolomyces marcillae Santa María (1958b)
Prosporobolomyces marcillae (Santa María) Novák et Zsolt (1961)
Sporobolomyces ruber Yamazaki et Fujii (1950)
Sporobolomyces carnicolor Yamazaki et Fujii (1950)
Sporobolomyces japonica Iizuka et Goto (1965)

Standard description of *Sporidiobolus pararoseus*

Growth in malt extract: After 3 days at room temperature there is considerable variation in cell shape; the cells may be ovoidal, spherical, elongate or allantoid (3–9) × (4–21) μm. They occur singly, in pairs, small clusters and short chains. A thin ring and light to moderate sediment are present. After one month a thick corrugated pellicle may form and the sediment may be heavy.

Growth on malt agar: After one month at room temperature the streak culture is pale salmon, grenadine, coral red or peach red in color with a rugose or smooth surface which is glossy or semidull. The texture is soft. The cross section is low convex to flat and the border entire, occasionally with a mycelial fringe.

Dalmau plate culture on corn meal agar: There is considerable strain variation; pseudo- and true hyphae are absent in some strains, whereas other strains will produce light to extensive amounts of pseudo- and/or true hyphae.

Formation of ballistospores: Some strains form asymmetric ballistospores (2–5) × (5–12) μm (1–5 or more/sterigma) on corn meal agar. Phaff (1970c) found a suitable medium for other strains for ballistospore formation that consisted of yeast nitrogen base agar with either ribose, glycerol or glucono-δ-lactone.

Life cycle: When compatible mating types were mixed on corn meal agar at 22°C conjugation was followed by formation of hyphae with hook-to-peg clamps and teliospores. The latter are elliptic-fusiform or spherical (9–13) × (13–24) μm, occurring singly, in pairs or rarely in chains of three. After 35 days on corn meal agar at 22°C the teliospores germinated on the corn meal agar by forming branched hyphae with buds. Continued growth of either the buds or hyphae was not successful. Two mating types were found suggesting a unifactorial mating system with two alleles.

One strain (CBS 5541) produced extensive true mycelium with elliptic-fusiform to spherical teliospores (10–12) × (12–20) μm. The teliospores were single and in chains, located intercalarily and terminally on the hyphae and on short lateral branches. Clamp connections were not formed. This strain did not mate with either the *A1* or *A2* mating types of *Sporid. pararoseus.* The relationship of this strain with the heterothallic strains requires study by determination of G + C contents.

Fermentation: –

Assimilation of carbon compounds:

Galactose	+ or s	Raffinose	+	Erythritol	–
Sucrose	+	Soluble starch	v	Ribitol	v
Maltose	+	D-Xylose	v	D-Mannitol	+
Cellobiose	+	L-Arabinose	–	Succinic acid	v
Trehalose	+	D-Ribose	v	Citric acid	v
Lactose	–	L-Rhamnose	– (+)	Inositol	–

Additional carbon compounds tested: Melibiose –, melezitose +, α-methyl-D-glucoside +, galactitol –

Assimilation of nitrate: –

Growth in vitamin-free medium: +

Growth on 50% (w/w) glucose-yeast extract agar: –

Growth at 37°C: –; **at 30°C:** v

Starch formation: –

Urease: +

G + C: 51.5 mol% (Storck et al., 1969)

Comments

Two strains of *Sporobolomyces shibatanus,* CBS 4217 and 2637, did not mate with either of the mating types of *Sporidiobolus pararoseus.* Storck et al. (1969) found that the G + C contents of the type strain of *Sp. pararoseus* (CBS 484) is 51.5%, whereas the G + C content of CBS 4217 is 55.0% and of CBS 2637 is 60.5%. Apparently these two strains belong to one or more species distinct from *Sporid. pararoseus.*

Origin of the strains studied

Air in U.S.A (2) (Olson & Hammer, 1937); flower (1); soil (1) (Okunuki, 1931); seawater (1).

Type strain: CBS 484, type strain of *Sporobolomyces pararoseus* (Olson & Hammer, 1937), mating type *A1*. Allotype strain: CBS 499 (Olson & Hammer, 1937), mating type *A2*.

3. *SPORIDIOBOLUS RUINENII* HOLZSCHU, TREDICK ET PHAFF (1981)

Standard description of *Sporidiobolus ruinenii*

Growth in malt extract: After 3 days at room temperature the cells are cylindrical or ovoidal (3–9) × (7–13) μm budding or single. Some cells have sterigmata with several bud scars. Short hyphal strands develop from single cells. A light ring and a moderate pellicle may be present. Hyphae with clamp connections and spherical teliospores (4–20 μm in diameter) develop.

Growth on malt agar: After 3 days the streak culture is peach colored. The surface is smooth, the border is entire with tufts of mycelium in some strains; clamp connections are present on the mycelium and teliospores may occur. After one month there is a more extensive development of hyphae and teliospores and the colony is peach red.

Dalmau plate culture on corn meal agar: After 6 days at room temperature true and pseudomycelium may or may not be present. When present the mycelium has clamp connections and spherical teliospores that resemble those on malt agar.

Formation of ballistospores: Asymmetric ballistospores (2–7) × (6–10) μm form on most growth media.

Life cycle: The species is homothallic. Hyphae with clamp connections and teliospores develop from single cells in the absence of mating. Teliospores from a 1.5-week-old corn meal agar streak culture, which was soaked for 12 weeks in distilled water, germinated within 3 days of plating onto 2% agar. Teliospores from a 5-week-old corn meal agar culture were transferred to distilled water where they germinated after 4 weeks of incubation. All procedures were at 18°C. The metabasidia were two- and four-celled with lateral and terminal basidiospores (Fig. 157). The metabasidium connects to the teliospore by an aseptate cell that ranges in length from 3–26 μm. Ploidy of the basidiospores has not been determined.

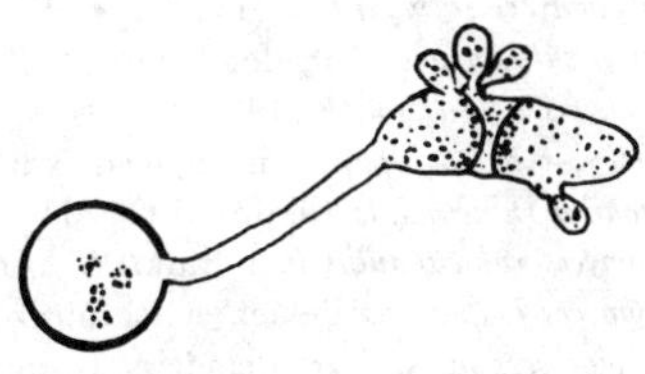

Fig. 157 – *Sporid. ruinenii*
Two-celled metabasidium attached by a stalk to the teliospore; the basidiospores are lateral

Fermentation: –

Assimilation of carbon compounds:

Galactose	+ s	Raffinose	+	Erythritol	–
Sucrose	+	Soluble starch	–	Ribitol	+
Maltose	+ s	D-Xylose	+	D-Mannitol	+
Cellobiose	+	L-Arabinose	+ (s)	Succinic acid	–
Trehalose	+	D-Ribose	+ s	Citric acid	–
Lactose	–	L-Rhamnose	+ s	Inositol	–

Additional carbon compounds tested: Melibiose –, melezitose –, α-methyl-D-glucoside –, galactitol +

Assimilation of nitrate: +

Growth in vitamin-free medium: +

Growth on 50% (w/w) glucose-yeast extract agar: – or + w

Growth at 37°C: –; **at 30°C:** +

Starch formation: –

Urease: +

Origin of the strains studied

Phyllosphere of tropical foliage of *Malphigia coccigera* at Bogor, Indonesia (4) (Ruinen, 1963).

Type strain: CBS 5001 (Ruinen, 1963).

4. *SPORIDIOBOLUS SALMONICOLOR* FELL ET TALLMAN (1981)

Imperfect state: *Sporobolomyces salmonicolor* (Fischer et Brebeck) Kluyver et van Niel

Synonyms

? *Rhodomyces kochii* von Wettstein (1885)
Blastoderma salmonicolor Fischer et Brebeck (1894)
Sporobolomyces salmonicolor (Fischer et Brebeck) Kluyver et van Niel (1925)
Prosporobolomyces salmonicolor (Fischer et Brebeck) Novák et Zsolt (1961)
Aessosporon salmonicolor van der Walt (1970h)
Sporobolomyces odorus Derx (1930)
Sporobolomyces photographus Cif. et Red. var. *odorus* Verona et Cif. (1938)
Pseudomonilia rubicundula Okunuki (1931)
Sporobolomyces rubicundulus (Okunuki) Verona et Cif. (1938)
Sporobolomyces hispanicus Peláez et Ramirez (1956a)
Prosporobolomyces hispanicus (Peláez et Ramirez) Novák et Zsolt (1961)
Sporobolomyces coralliformis Tubaki (1958)
Sporobolomyces salmonicolor (Fischer et Brebeck) Kluyver et van Niel var. *fisherii* Misra et Randhawa (1976)

Standard description of *Sporidiobolus salmonicolor*

Growth in malt extract: After 7 days at room temperature there may be a light to thick ring; when present the pellicle will range from a thin film, or small islets, to a thick corrugated pellicle. Film and pellicle colors are off white, light salmon, flesh and peach. After one month the ring and pellicle have increased in size.

Growth on malt agar: There is considerable strain variation in cell and colony morphology. After 7 days at room temperature, the cells are oval to elongate (2–12) × (3–35) μm, single, in pairs; pseudo- and true hyphae may occur. Cells and hyphae may have sterigmata up to 50 μm or more in length. The streak cultures are shades of flesh, salmon and peach. The edges may be entire or heavily myceliated. The surface is smooth, powdery or deeply corrugated. The texture varies from pasty to tough. After one month there is little change in the cell size, while the streak cultures become deeper in color.

Dalmau plate culture on corn meal agar: There is considerable strain variation, ranging from an absence of pseudo- and true hyphae to extensive growth of both types of hyphae.

Formation of ballistospores: Most strains form kidney-shaped (3–5) × (5–10) μm ballistospores on corn meal agar. Some strains produce ballistospores on malt agar.

Life cycle: The species is heterothallic. Conjugation of opposite mating types on corn meal agar at 22°C results in the formation of true mycelium with clamp connections at the septa. As noted by Bandoni et al. (1975) the mating reactions are not consistent, due possibly to such factors as types of media, age and condition of the strains. Two mating types have been recorded, indicating that the species is unifactorial with two alleles. The teliospores are round (10–14) μm or ovoidal (8–10) × (10–17) μm, terminal, intercalary and lateral on hyphal branches. The teliospores germinate after 18 weeks on corn meal agar at 22°C. The metabasidium (Fig. 158) is one-celled

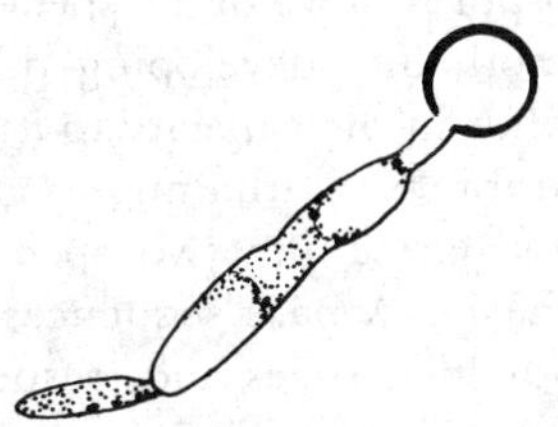

Fig. 158 – *Sporid. salmonicolor*
One-celled metabasidium attached by a stalk to the teliospore; the basidiospore is terminal

Fig. 159 – *Sporid. salmonicolor*
Hyphal formation from a metabasidium

(5–7) × (13–47) μm and is borne on a connecting cell (1–3) × (3–13) μm. One to two basidiospores form terminally on the metabasidium. Hyphae may develop from the metabasidium (Fig. 159). Endospores may be present in the metabasidium.

Fermentation: –

Assimilation of carbon compounds:

Galactose	v	Raffinose	v	Erythritol	–
Sucrose	+	Soluble starch	– (+w)	Ribitol	+
Maltose	– (+)	D-Xylose	+ (s)	D-Mannitol	+
Cellobiose	– (+)	L-Arabinose	v	Succinic acid	– or + w
Trehalose	+	D-Ribose	+ (s)	Citric acid	–
Lactose	–	L-Rhamnose	–	Inositol	–

Additional carbon compounds tested: Melibiose – , melezitose – (+), α-methyl-D-glucoside – (+), galactitol –

Assimilation of nitrate: +

Growth in vitamin-free medium: +

Growth on 50% (w/w) glucose-yeast extract agar: –

Growth at 37°C: – (+); **at 30°C:** + (–)

Starch formation: –

Urease: +

G + C: 63.5–65.0 mol% (Storck et al., 1969)

Origin of the strains studied

Tanning fluid (1) (Peláez & Ramirez, 1956a); seawater (2); aerial contaminants (7) (including those from Bandoni et al., 1971); mixed culture (1) (Kluyver & van Niel, 1925); human sources (2) (Misra & Randhawa, 1976; van der Walt, 1970h; plants (2) (Tubaki, 1958; Derx, 1930).

Type strain: CBS 490, type strain of *Sporobolomyces salmonicolor* (Kluyver & van Niel, 1925), mating type *A1*. Allotype strain: CBS 2630 (mating type *A2*).

f. COMMENTS ON THE GENUS

The genus *Sporidiobolus* is distinguished by the presence of carotenoid pigments, the production of ballistospores and the formation of teliospores. Two of the species of the genus, *Sporid. ruinenii* and *Sporid. johnsonii,* are homothallic, developing dikaryotic hyphae with clamp connections. These two species differ in the capacity to assimilate several carbon compounds (raffinose, melezitose, L-arabinose, L-rhamnose, galactitol and α-methyl-D-glucoside). Holzschu et al. (1981) reported the two species were distinct through studies of the complementarity of the DNA base sequences.

Sporid. salmonicolor and *Sporid. pararoseus* are heterothallic species and, respectively, the perfect states of *Sporobolomyces salmonicolor* and *Sp. shibatanus.* The two species are taxonomically separated by the nitrate assimilation test. Homokaryotic fruiting was observed in a strain, CBS 5541, provisionally included in *Sporid. pararoseus.* However, the relationship between the homokaryotic fruiting and heterothallic strains has not been determined.

Sporid. johnsonii and *Sporid. salmonicolor* have similar carbon and nitrogen assimilation patterns, which suggests a single species consisting of homothallic and heterothallic forms such as observed in *Rhodosporidium toruloides.* Holzschu et al. (1981) found sufficient differences in the n-deoxyribonucleic acid (DNA) base composition and mitochondrial DNA to indicate that *Sporid. johnsonii* and *Sporid. salmonicolor* are distinct species.

Teliospore germination and basidial morphology differ within the genus. *Sporid. johnsonii* teliospores germinate directly to a holometabasidium with terminal basidiospores. The metabasidium of *Sporid. ruinenii* and *Sporid. salmonicolor* is connected to the teliospore by a stalk. The metabasidium of *Sporid. salmonicolor* is one-celled, while the metabasidium of *Sporid. ruinenii* is two- to four-celled. *Sporid. pararoseus* teliospores germinate to branched hyphae, a phenomenon that requires further study.

C. Tremellales with a yeast phase (Sirobasidiaceae and Tremellaceae)

Introduction, list of species and key to the species

by R.J. Bandoni

INTRODUCTION

Knowledge that budding growth occurs in some of the Tremellales dates from the time of Brefeld (1888). Brefeld used this feature, and others, in his attempt to clarify the distinctions between *Exidia* and *Tremella.* He illustrated germination of the basidiospores of several Tremellas and clearly showed the yeast-like development that takes place. Brefeld was followed in this approach to taxonomy by Möller (1895), who studied some of the tropical species of *Tremella* and *Sirobasidium.* Möller noted that the production of a yeast phase was similar in *Tremella* and *Sirobasidium* and suggested that the latter group were the forerunners of the Tremellas. Unfortunately, few later students of the Tremellales have attempted to extend observations on basidiospore germination and development of the haploid stage. The presence of dimorphism in these life histories is of fundamental importance in the taxonomy of the families involved. However, the number of species cultured is small and the information so far obtained cannot therefore be used in the classification of these fungi.

The basic life history patterns in *Tremella* and *Sirobasidium* are essentially as illustrated for *Tremella globospora* Reid by Brough (1974) and in *Sirobasidium magnum* Boedijn by Flegel (1976). In both genera, the haploid phase is a yeast and no mycelium develops until compatible cells are mixed. Conjugation occurs through the development of conjugation tubes of indeterminate length; following fusion of compatible tubes, a dikaryotic mycelium develops. Conidia also are produced on the dikaryotic hyphae; under some conditions, these hyphae may also produce haploid budding cells again. If conditions are suitable, basidiocarps, basidia, and basidiospores are formed by the dikaryotic mycelium. Dikaryotic conidia are produced in the hymenium of many species; these conidia may either precede or accompany the basidia.

The primary basidiospores of *Sirobasidium* are not abstricted; they are deciduous and probably are homologous to the 'epibasidia' or 'protosterigmata' of *Tremella* (Bandoni, 1957). These primary basidiospores germinate by the production of buds or, more commonly, by the production of a ballistospore. This ballistospore is almost certainly the homologue of the *Tremella* basidiospore. The basidiospores of *Tremella* and ballistospores of *Sirobasidium* germinate in one of three ways. First, these spores can produce a sterigma and ballistospore. Second, isolated spores can germinate by budding, leading to the development of a haploid yeast phase. Finally, if two compatible spores are close to one another and conditions are favorable, conjugation tubes may develop and, following fusion, a dikaryotic mycelium is produced.

Sirobasidium, Tremella, Holtermannia and *Fibulobasidium* yeasts resemble certain asporogenous yeasts, especially species of *Cryptococcus* (Kobayashi & Tsubaki, 1965; Slodki et al., 1966). Flegel (1976) observed ballistospore production in cultures of

Sirobasidium magnum, suggesting a possible relationship to the genus *Bullera.* The yeast colonies in all of these genera are whitish, mucoid, and often glistening; the cells of most species are heavily encapsulated. The copious extracellular polysaccharides produced by some species are similar to those formed in species of *Cryptococcus* (Slodki et al., 1966; Slodki, 1966). Similarities have also been found in the cell wall chemistry of *Tremella* and *Cryptococcus* species (Reid & Bartnicki-Garcia, 1976). Although basidia of *Filobasidiella* and *Tremella* are of very different form, their septal pore ultrastructure is remarkably similar (Kwon-Chung & Popkin, 1976; Moore, 1978; Bezerra & Kimbrough, 1978; Khan & Kimbrough, 1980).

Most tremellaceous and sirobasidiaceous yeasts continue budding growth indefinitely under laboratory conditions. When compatible mating strains are mixed on appropriate media (Bandoni, 1963, 1965; Kobayashi & Tsubaki, 1965; Brough, 1974; Flegel, 1976), budding ceases, conjugation occurs, and the dikaryotic phase is initiated. The change in growth form is regulated by diffusible hormones, or erogens, one of which has now been isolated and characterized (Reid 1974; Sakagami et al., 1978a,b, 1979, 1980). This hormone, designated as Tremerogen A-10, is a peptide similar to Rhodotoucine A^{10}, an erogen produced by one mating strain of *Rhodosporidium toruloides* Banno (Sakagami et al., 1980).

In tremellaceous and sirobasidiaceous species examined up to now, the compatibility system is more complex than in most other fungi with yeast phases (Bandoni, 1963, 1979; Brough, 1974; Flegel, 1976). In this system, a modified tetrapolar one, *A* and *a* alleles regulate development of conjugation tubes. Establishment of a dikaryotic mycelium is dependent upon a set of multiple factors or alleles, the *B*'s. Conjugation tubes are produced when *A* and *a* type cells are mixed; a dikaryotic mycelium is then produced only if the mating strains have compatible *B* factors, e.g., *B1* × *B2*.

A study of mating in *Tremella fuciformis* Berk. by Kobayashi and Tsubaki (1965) gave results suggesting the same type of modified tetrapolarity outlined above. Chen and Hou (1978) have discussed the biology of this species and methods used in cultivating it in the Orient. Most monospore isolates of Hawaiian and southern U.S.A. collections of this species appear to be homothallic (Bandoni, unpublished). That is, after a short period of cultivation, dikaryotic hyphae arise in monosporous yeast isolates. These dikaryotic hyphae are homodikaryotic, i.e., the species is facultatively homothallic (Lempke, 1969) as in some homobasidiomycetes. Cultures of *Trem. mesenterica* maintained over a period of 15 years remained entirely in the yeast phase. However, analyses of mating types before and after this period demonstrated instability of the *B* factors or alleles with numerous mating type changes (Bandoni, unpublished). Mating has been studied in few of the known species of dimorphic tremellaceous and sirobasidiaceous fungi; no study to date has elucidated the nature of the *B* factors or alleles.

Although *Tremella fuciformis* is cultivated commercially on wood, many of the dimorphic tremellaceous and sirobasidiaceous fungi occur on fructifications of ascomycetes or on other basidiomycetes. Relationships of these fungi to their 'hosts' has received little attention (Bandoni, 1961; Brough, 1974; Bezerra & Kimbrough, 1978).

Once established, the dikaryotic mycelium may produce basidiocarps if conditions are suitable (Brough, 1974; Flegel, 1976; Chen & Hou, 1978). In *Tremella globospora* and *Sirobasidium magnum,* basidiocarps are formed directly and there is no preliminary

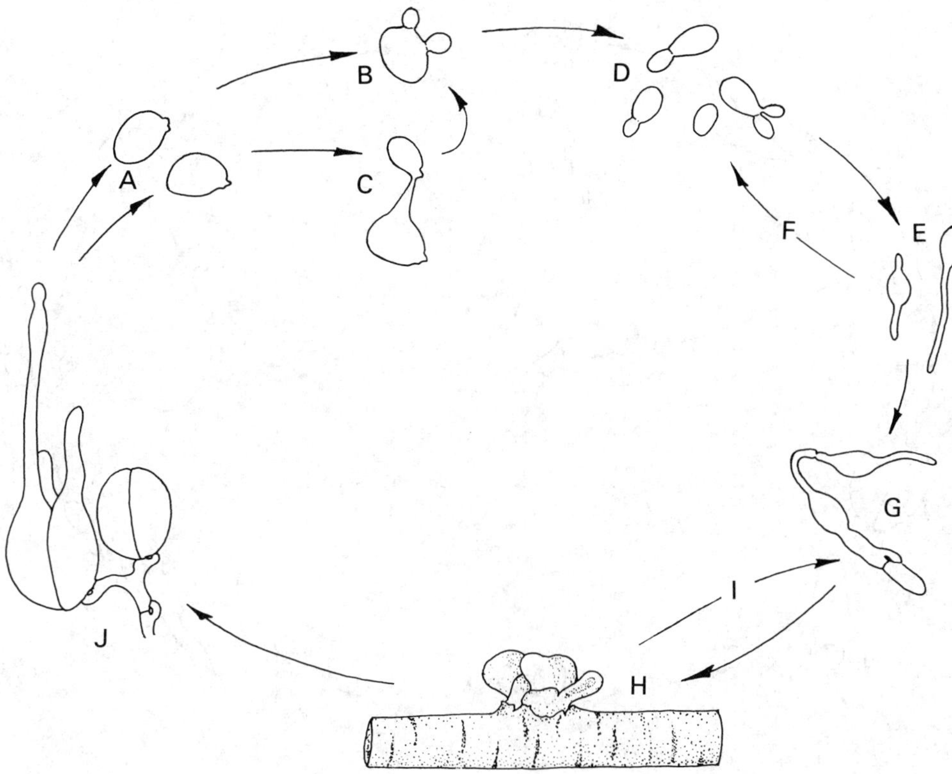

Diagram 11 *Tremella* type life history

A, basidiospores germinate by budding (B) or by repetition (C). Repetitive spores may also germinate by budding. Haploid yeast cells (D) conjugate when compatible pairs are associated (E); those failing to conjugate may form conidia and revert to the yeast phase (F). Conjugation (G) leads to the development of a dikaryotic mycelium and, eventually, to basidiocarp formation (H). Dikaryotic conidia are produced in young basidiocarps of some species; they give rise to dikaryons directly (I). Later, basidia (J) develop in basidiocarps and produce violently abstricted basidiospores.

or concurrent spread of mycelium in the medium. Basidiocarps of *Tremella* and *Sirobasidium* species are irregular in form, most being globose or composed of flattened globes. Those of *Holtermannia* consist of clusters of cornute globes. All are gelatinous and range in size from less than 1 mm to more than 100 mm in diameter; all have amphigenous hymenia. Although hymenial conidia are present in many species, the hymenium typically lacks all types of sterile structures, e.g., cystidia, paraphysoids, setae, etc. Basidia of *Sirobasidium* are catenulate, cells of the fertile hyphae developing into basidia basipetally. Terminal cells function as basidia in *Holtermannia, Tremella* and *Fibulobasidium,* but several basidia are typically attached to one another laterally in *Fibulobasidium.* This attachment is the result of successive clamp formation, the loops of the clamps expanding to form basidia. Mature tremellaceous and sirobasidiaceous basidia typically are four-celled, but two-celled basidia are characteristic of some species. Each cell produces an elongate structure, the protosterigma or epibasidium; the latter is deciduous in *Sirobasidium* and *Fibulobasidium,*

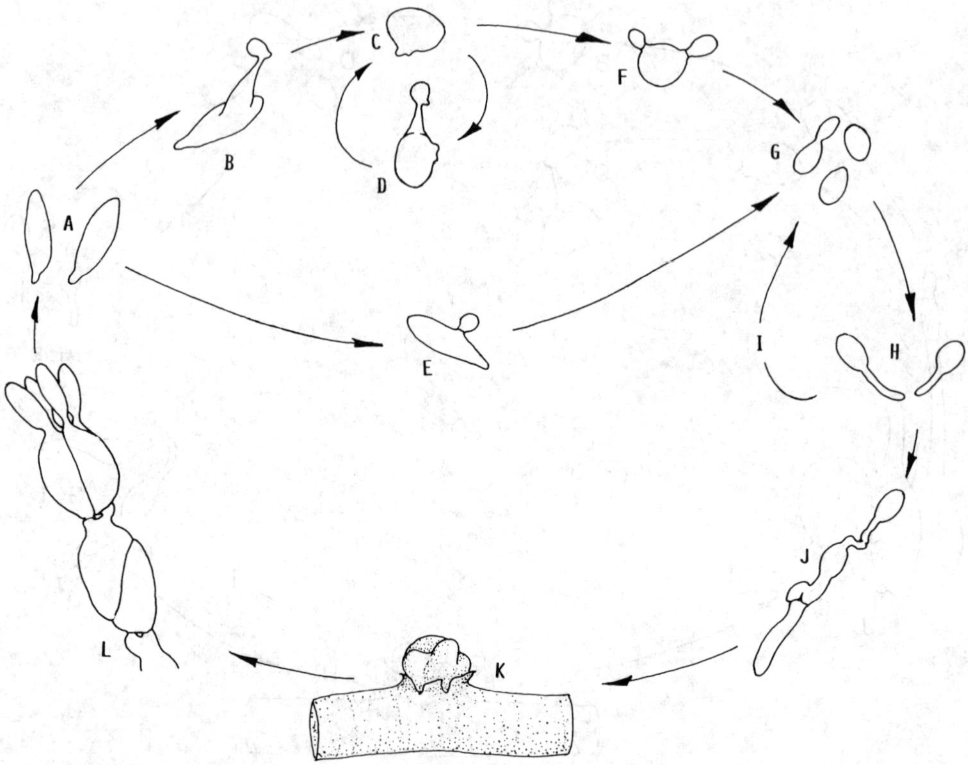

Diagram 12 *Sirobasidium* type life history

A, passively released basidiospores may germinate by the production of a ballistospore (B) or by budding directly (E). Ballistospores (C) may also germinate by formation of a ballistospore, i.e., by repetition (D) or by budding directly (F). Yeast cells (G) conjugate (H) and give rise to a dikaryotic mycelium (J). The dikaryotic mycelium eventually produces a basidiocarp (K) in which chains of basidia (L) give rise to basidiospores.

but is retained on the basidium in *Holtermannia* and *Tremella.* A ballistospore, the basidiospore, is produced on a sterigma borne on the protosterigma.

The life histories of the *Tremella* and *Sirobasidium* types are summarized in Diagrams 11 and 12.

Since there have been few cultural studies of fungi in the Tremellales, the extent of dimorphism within the group is not known. Haploid yeast phases are known to occur in a few auriculariaceous fungi, but these have not as yet been studied. In this treatment, only a few common temperate and tropical species are described. For convenience, yeasts belonging to these groups are referred to by their appropriate basidiomycete names. The same species also are mentioned in keys and descriptions for species of *Cryptococcus, Bullera* and *Candida,* depending upon features of the yeast phase. No descriptions are provided for basidiocarps of the species included; such descriptions are available in taxonomic works dealing with these fungi, e.g., Bandoni, 1958; Bourdot & Galzin, 1927; Kobayashi, 1939; Lowy, 1971; Martin, 1952; Pilat, 1957.

THE YEAST PHASE OF THE FOLLOWING SPECIES WILL BE DESCRIBED:

Sirobasidiaceae

1. *Fibulobasidium inconspicuum* Bandoni
2. *Sirobasidium magnum* Boedijn

Tremellaceae

3. *Holtermannia corniformis* Kobayashi
4. *Tremella aurantia* Schweinitz ex Fries
5. *Tremella brasiliensis* (Möller) Lloyd
6. *Tremella coalescens* Olive
7. *Tremella encephala* Persoon
8. *Tremella foliacea* (Pers. ex S.F. Gray) Persoon
9. *Tremella fuciformis* Berkeley
10. *Tremella globospora* Reid
11. *Tremella mesenterica* Retz. ex Fries
12. *Tremella subanomala* Coker

KEY TO THE YEAST PHASE OF SPECIES OF THE FAMILIES SIROBASIDIACEAE AND TREMELLACEAE

1a Formation of ballistospores 2
b No formation of ballistospores 3

2a Sucrose assimilated
Fibulobasidium inconspicuum .... p. 546
b Sucrose not assimilated
Sirobasidium magnum .... p. 547

3a Inositol assimilated 4
b Inositol not assimilated 7

4a Sucrose assimilated 5
b Sucrose not assimilated 6

5a Erythritol assimilated
Tremella aurantia .... p. 548
b Erythritol not assimilated
Tremella encephala .... p. 551
Tremella subanomala .... p. 554

6a Erythritol assimilated
Tremella foliacea . . . . p. 552
b Erythritol not assimilated
Holtermannia corniformis . . . . p. 548

7a Sucrose assimilated 8
b Sucrose not assimilated 9

8a Melibiose assimilated
Tremella fuciformis . . . . p. 552
b Melibiose not assimilated
Tremella globospora . . . . p. 553

9a Maltose assimilated
Tremella brasiliensis . . . . p. 549
b Maltose not assimilated 10

10a Citric acid assimilated
Tremella coalescens . . . . p. 550
b Citric acid not assimilated
Tremella mesenterica . . . . p. 554

See Table 24 (p. 534) and Table 37 (p. 849)

SYSTEMATIC DISCUSSION OF THE YEAST PHASE OF THE SPECIES OF THE SIROBASIDIACEAE AND TREMELLACEAE

by L. Rodrigues de Miranda and R.J. Bandoni

1. *FIBULOBASIDIUM INCONSPICUUM* BANDONI (1979)

Standard description of *Fibulobasidium inconspicuum*

Growth on malt agar: After 3 days at 25°C cells are spherical to short-oval, (2–5.5) × (2.5–5.5) μm; single or in pairs.

After one month at 20°C the streak culture is grey-brownish yellow, pasty, partly wrinkled and partly smooth with a lobate edge.

Slide culture on potato agar: No pseudomycelium.

Formation of ballistospores: Ballistospores are fusiform to elliptic, (1.5–3.5) × (2–8) μm.

Fermentation: –

Assimilation of carbon compounds:

Galactose	+	Raffinose	+	Erythritol	+ w
Sucrose	+	Soluble starch	+ w	Ribitol	+
Maltose	+	D-Xylose	+	D-Mannitol	+
Cellobiose	+	L-Arabinose	+	Succinic acid	+ w
Trehalose	+	D-Ribose	+	Citric acid	+ w
Lactose	+ w	L-Rhamnose	+	Inositol	+

Additional carbon compounds tested: Melibiose – , sodium glucuronate + , glucono-δ-lactone +

Assimilation of nitrate: –

Assimilation of nitrite: –

Growth in vitamin-free medium: –

Growth at 37°C: + w; at 30°C: +

Starch formation: +

Origin of the strain studied

Spore fall from fruiting body on *Quercus virginiana,* collected New Orleans, Louisiana, U.S.A. by R.J. Bandoni.

2. *SIROBASIDIUM MAGNUM* BOEDIJN (1934)

Standard description of *Sirobasidium magnum*

Growth on malt agar: After 3 days at 25°C cells are spherical to short-oval, (2–4.5) × (2.5–6) μm; single or in pairs.

After one month at 20°C, the streak culture is yellowish or yellowish-white, smooth or with a finely reticulate border, shiny or mat in the middle of the streak, with a smooth or undulating edge.

Slide culture on potato agar: No pseudomycelium.

Formation of ballistospores: Ballistospores are formed on sterigmata; they are spherical, ovoidal or apiculate, (3.5–5) × (4–9) μm. They were observed on corn meal agar.

Fermentation: –

Assimilation of carbon compounds:

Galactose	+w or –	Raffinose	–	Erythritol	v
Sucrose	–	Soluble starch	–	Ribitol	+
Maltose	–	D-Xylose	+	D-Mannitol	+
Cellobiose	+	L-Arabinose	v	Succinic acid	+ or w
Trehalose	+	D-Ribose	+ w	Citric acid	+ w
Lactose	+	L-Rhamnose	+ or w	Inositol	+

Additional carbon compounds tested: Melibiose – , sodium glucuronate + or w, glucono-δ-lactone v

Assimilation of nitrate: –

Assimilation of nitrite: –

Growth in vitamin-free medium: –

Growth at 37°C: – ; at 30°C: +

Starch formation: +

Origin of the strains studied

Five strains isolated from basidiospores by R.J. Bandoni. One of the strains, 11-1 (CBS 6803) came from a fruiting body collected in Shimoda, Japan by T. Flegel; the other strains also came from material collected by Flegel.

3. *HOLTERMANNIA CORNIFORMIS* KOBAYASHI (1937)

Standard description of *Holtermannia corniformis*

Growth on malt agar: After 3 days at 25°C cells are short-oval to long-oval, a few elongate, (2.5–5) × (5–15) μm, and longer; buds are formed at or near the poles, sometimes on a broad base.

After one month at 20°C the streak culture is yellowish-white, smooth, shiny and mucous.

Slide culture on potato agar: Pseudo- and true mycelium absent, or a few hyphal cells, occasionally with cross walls may be present.

Fermentation: –

Assimilation of carbon compounds:

Galactose	+	Raffinose	–	Erythritol	–
Sucrose	–	Soluble starch	+w	Ribitol	+
Maltose	+	D-Xylose	+	D-Mannitol	+
Cellobiose	+	L-Arabinose	–	Succinic acid	+
Trehalose	+	D-Ribose	+	Citric acid	+
Lactose	–	L-Rhamnose	+	Inositol	+

Additional carbon compounds tested: Melibiose –, sodium glucuronate +w, glucono-δ-lactone +

Assimilation of nitrate: –

Assimilation of nitrite: –

Growth in vitamin-free medium: +

Growth at 37°C: –; at 30°C: –; at 25°C: +

Starch formation: +

Color reaction with DBB: +

Origin of the strains studied

Specimen collected and identified by K. Tsubaki on dead wood, Yakeyama Towada National Park, Aomori Pref., Japan. Culture isolated by R.J. Bandoni from a basidiospore.

4. *TREMELLA AURANTIA* SCHWEINITZ EX FRIES (1822)

Standard description of *Tremella aurantia*

Growth on malt agar: After 3 days at 25°C cells are short-oval, (2.5–4.5) × (3–5.5) μm; single or in pairs. Buds are formed on a rather broad base.

After one month at 20°C the streak culture is yellowish-brown, smooth, shiny, slimy (running to the bottom of the tube).

Slide culture on potato agar: No pseudomycelium.

Fermentation: –

Assimilation of carbon compounds:

Galactose	+	Raffinose	+	Erythritol	+ w
Sucrose	+	Soluble starch	+	Ribitol	+
Maltose	+	D-Xylose	+	D-Mannitol	+
Cellobiose	+	L-Arabinose	+	Succinic acid	+
Trehalose	+	D-Ribose	+	Citric acid	+ w
Lactose	+	L-Rhamnose	+	Inositol	+

Additional carbon compounds tested: Melibiose + w, sodium glucuronate +, glucono-δ-lactone +

Assimilation of nitrate: –

Assimilation of nitrite: –

Growth in vitamin-free medium: –

Growth at 37°C: –; **at 30°C:** +

Starch formation: +

Comments

The yeast form of *Tremella aurantia* agrees with *Cryptococcus laurentii,* as was also mentioned by Kurtzman (1973). However, this author found no reactions between the mating types of *Cr. laurentii* var. *laurentii* and two strains of *Tremella aurantia.* Baptist and Kurtzman (1976) showed that there were distinct differences in the enzyme pattern of both species.

Origin of the strain studied

Isolated from fruiting body on *Alnus rubra* Vancouver, B.C., by R.J. Bandoni.

5. *TREMELLA BRASILIENSIS* (MÖLLER) LLOYD (1922)

Standard description of *Tremella brasiliensis*

Growth on malt agar: After 3 days at 25°C cells are short-oval, (1.5–4) × (2.5–5.5) μm; single or in pairs.

After one month at 20°C the streak culture is whitish-brown, smooth, shiny, somewhat slimy, with a smooth edge.

Slide culture on potato agar: No pseudomycelium, but small chains of cells.

Fermentation: –

Assimilation of carbon compounds:

Galactose	+	Raffinose	–	Erythritol	+ w
Sucrose	–	Soluble starch	–	Ribitol	+
Maltose	+	D-Xylose	+	D-Mannitol	+

Cellobiose	+	L-Arabinose	–	Succinic acid	+
Trehalose	+	D-Ribose	–	Citric acid	+
Lactose	+	L-Rhamnose	–	Inositol	–

Additional carbon compounds tested: Melibiose –, melezitose –, sodium glucuronate –, glucono-δ-lactone –

Assimilation of nitrate: –

Assimilation of nitrite: –

Growth in vitamin-free medium: –

Growth at 37°C: + w; at 30°C: +

Starch formation: + w

Urease: +

Color reaction with DBB: +

Origin of the strain studied

From fruiting body on angiosperm wood, Costa Rica, collected by J. Kuijt; culture isolated from a basidiospore by R.J. Bandoni.

6. *TREMELLA COALESCENS* OLIVE (1951)

Standard description of *Tremella coalescens*

Growth on malt agar: After 3 days at 25°C the cells are short-oval to oval, (1.5–3.5) × (3–6) μm; single or in pairs.

After one month at 20°C the streak culture is whitish-brown, smooth, shiny, somewhat slimy with a smooth edge.

Slide culture on potato agar: No pseudomycelium.

Fermentation: –

Assimilation of carbon compounds:

Galactose	–	Raffinose	–	Erythritol	–
Sucrose	–	Soluble starch	–	Ribitol	+
Maltose	–	D-Xylose	+ w	D-Mannitol	+
Cellobiose	–	L-Arabinose	–	Succinic acid	+ w
Trehalose	+	D-Ribose	–	Citric acid	+ w
Lactose	–	L-Rhamnose	–	Inositol	–

Additional carbon compounds tested: Melibiose –, melezitose –, sodium glucuronate –, glucono-δ-lactone –

Assimilation of nitrate: –

Assimilation of nitrite: –

Growth in vitamin-free medium: –

Growth at 37°C: –; at 30°C: +

Starch formation: +

Urease: +

Color reaction with DBB: +

Origin of the strain studied

From fruiting body on angiosperm in swamp, Garyville, Louisiana, U.S.A., collected by R.J. Bandoni; isolated from a basidiospore by Bandoni.

7. *TREMELLA ENCEPHALA* PERSOON (1822)

Standard description of *Tremella encephala*

Growth on malt agar: After 3 days at 25°C the cells are spherical to short-oval, (1.8–7) × (2.5–9) μm; single or in pairs.

After one month at 20°C the streak culture is yellowish-brown, smooth, shiny, slimy.

Slide culture on potato agar: No pseudomycelium.

Fermentation: –

Assimilation of carbon compounds:

Galactose	+	Raffinose	+	Erythritol	–
Sucrose	+	Soluble starch	+	Ribitol	+
Maltose	+	D-Xylose	+	D-Mannitol	+
Cellobiose	+	L-Arabinose	+	Succinic acid	+
Trehalose	+	D-Ribose	+	Citric acid	+
Lactose	+	L-Rhamnose	+	Inositol	+

Additional carbon compounds tested: Melibiose +, sodium glucuronate +, glucono-δ-lactone +

Assimilation of nitrate: –

Assimilation of nitrite: –

Growth in vitamin-free medium: +w

Growth at 37°C: –; **at 30°C:** +

Starch formation: +

Comments

The yeast form of *Tremella encephala* resembles *Cryptococcus laurentii,* but Baptist and Kurtzman (1976) found distinct differences in the enzyme pattern of both species, also between *Tremella encephala* and *Tremella aurantia.*

Origin of the strains studied

Two strains isolated from basidiospores by R.J. Bandoni. They came from fruiting bodies associated with *Stereum sanguineolentum* on conifer wood, Vancouver, B.C., collected by Bandoni.

8. *TREMELLLA FOLIACEA* (PERS. EX S.F. GRAY) PERSOON (1822)

Standard description of *Tremella foliacea*

Growth on malt agar: After 3 days at 25°C cells are spherical to short-oval, (2–8) × (3–8.5) μm; single or in pairs.

After one month at 20°C the streak culture is whitish-brown, smooth, shiny, slightly slimy, with a smooth margin.

Slide culture on potato agar: Pseudomycelium is not formed.

Fermentation: –

Assimilation of carbon compounds:

Galactose	–	Raffinose	–	Erythritol	+ w
Sucrose	–	Soluble starch	–	Ribitol	+
Maltose	+ w	D-Xylose	+	D-Mannitol	+
Cellobiose	–	L-Arabinose	+ w	Succinic acid	+ w
Trehalose	+	D-Ribose	+ w	Citric acid	+ w
Lactose	–	L-Rhamnose	+	Inositol	+

Additional carbon compounds tested: Melibiose – , sodium glucuronate + , glucono-δ-lactone +

Assimilation of nitrate: –

Assimilation of nitrite: –

Growth in vitamin-free medium: –

Growth at 37°C: + w; **at 30°C:** +

Starch formation: +

Origin of the strain studied

From fruiting body on log of *Acer macrophyllum,* Vancouver, B.C., collected by R.J. Bandoni; strain isolated from basidiospore by Bandoni.

9. *TREMELLA FUCIFORMIS* BERKELEY (1856)

Standard description of *Tremella fuciformis*

Growth on malt agar: After 3 days at 25°C cells are spherical, short-oval and oval, (2–4.5) × (2.5–7) μm; single or in pairs.

After one month at 20°C the streak culture is yellowish-brown, smooth, shiny and slimy (running to the bottom of the tube).

Slide culture on potato agar: No pseudomycelium.

Fermentation: –

Assimilation of carbon compounds:

Galactose	+	Raffinose	+	Erythritol	+
Sucrose	+	Soluble starch	v	Ribitol	+
Maltose	+	D-Xylose	+	D-Mannitol	+
Cellobiose	+	L-Arabinose	+	Succinic acid	+ or w

Trehalose	+	D-Ribose	+	Citric acid	+w
Lactose	v	L-Rhamnose	–	Inositol	–

Additional carbon compounds tested: Melibiose +w, melezitose +, sodium glucuronate +, glucono-δ-lactone +

Assimilation of nitrate: –

Assimilation of nitrite: –

Growth in vitamin-free medium: –

Growth at 37°C: +w; at 30°C: +

Starch formation: v

Urease: +

Color reaction with DBB: +

Origin of the strains studied

Fruiting body of a commercially cultivated strain from Taiwan (1); branch of *Quercus virginiana*, Audubon Park, New Orleans (1). Both strains isolated from basidiospores by R.J. Bandoni.

10. *TREMELLA GLOBOSPORA* REID (1970)

Standard description of *Tremella globospora*

Growth on malt agar: After 3 days at 25°C budding yeast cells and true hyphae are present. The yeast cells are spherical to short-oval, (3–5.5) × (4–8) μm; single or in pairs. The hyphae have a diameter of 2 μm.

After one month at 20°C the streak culture is yellowish-white to brown, smooth, shiny, very slimy (running to the bottom of the tube).

Slide culture on potato agar: No pseudomycelium is formed.

Fermentation: –

Assimilation of carbon compounds:

Galactose	+	Raffinose	+	Erythritol	+
Sucrose	+	Soluble starch	+w	Ribitol	+
Maltose	+	D-Xylose	+	D-Mannitol	+
Cellobiose	+	L-Arabinose	+s	Succinic acid	+
Trehalose	+	D-Ribose	+	Citric acid	+w
Lactose	–	L-Rhamnose	–	Inositol	–

Additional carbon compounds tested: Melibiose –, melezitose +, sodium glucuronate –, glucono-δ-lactone +w

Assimilation of nitrate: –

Assimilation of nitrite: +

Growth in vitamin-free medium: –

Growth at 37°C: –; at 30°C: +w

Starch formation: +w

Urease: +

Color reaction with DBB: +

Origin of the strain studied

From fruiting body on *Diaporthe* on *Cornus nuttalli;* isolated by Brough (1).

11. *TREMELLA MESENTERICA* RETZ. EX FRIES (1822)

Standard description of *Tremella mesenterica*

Growth on malt agar: After 3 days at 25°C cells are spherical to short-oval, sometimes cylindrical, (2–5) × (2.5–6) μm and up to 16 μm long; single, in pairs or in short chains.

After one month at 20°C the streak culture is yellowish-white to brown, smooth, shiny, somewhat slimy.

Slide culture on potato agar: No pseudomycelium.

Fermentation: –

Assimilation of carbon compounds:

Galactose	–	Raffinose	–	Erythritol	–
Sucrose	–	Soluble starch	–	Ribitol	+
Maltose	–	D-Xylose	+	D-Mannitol	+
Cellobiose	–	L-Arabinose	–	Succinic acid	v
Trehalose	+	D-Ribose	–	Citric acid	–
Lactose	–	L-Rhamnose	–	Inositol	–

Additional carbon compounds tested: Melibiose –, melezitose –, sodium glucuronate –, glucono-δ-lactone –

Assimilation of nitrate: –

Assimilation of nitrite: –

Growth in vitamin-free medium: –

Growth at 37°C: +w or –; at 30°C: +

Starch formation: v

Urease: +

Color reaction with DBB: +

Origin of the strains studied

From fruiting bodies on *Alnus rubra,* Vancouver, B.C., collected by R.J. Bandoni (2); from fruiting body on angiosperm wood, Costa Rica, collected by J. Kuijt (1); all strains isolated from basidiospores by R.J. Bandoni.

12. *TREMELLA SUBANOMALA* COKER (1920)

Standard description of *Tremella subanomala*

Growth on malt agar: After 2 days at 25°C cells are spherical, short-oval, to oval, (2–4.5) × (2.5–8) μm; single or in pairs.

After one month at 20°C the streak culture is yellowish-brown, smooth, shiny, slimy (running to the bottom of the tube).

Slide culture on potato agar: No pseudomycelium.

Fermentation: –

Assimilation of carbon compounds:

Galactose	+	Raffinose	+ w	Erythritol	–
Sucrose	+	Soluble starch	+	Ribitol	+
Maltose	+	D-Xylose	+	D-Mannitol	+
Cellobiose	+	L-Arabinose	+	Succinic acid	+ w
Trehalose	+	D-Ribose	+	Citric acid	+ w
Lactose	+	L-Rhamnose	+	Inositol	+

Additional carbon compounds tested: Melibiose + w, sodium glucuronate +, glucono-δ-lactone +

Assimilation of nitrate: –

Assimilation of nitrite: –

Growth in vitamin-free medium: –

Growth at 37°C: –; **at 30°C:** +

Starch formation: +

Origin of the strain studied

From fruiting body on stromatic ascomycete on *Salix,* Vancouver, B.C., collected by R.J. Bandoni; isolated from a basidiospore by R.J. Bandoni.

f. COMMENTS

Of the Tremellales discussed, the two ballistosporogenous species: *Fibulobasidium inconspicuum* and *Sirobasidium magnum,* are also in the key to the species of the genus *Bullera* and in the key to all ballistosporogenous species (see Chapter VI).

Holtermannia corniformis is in the *Cryptococcus* and the *Candida* key. The inositol-positive *Tremella* species *Trem. aurantia, Trem. encephala, Trem. foliacea* and *Trem. subanomala* are in the *Cryptococcus* key. The inositol-negative species *Trem. brasiliensis, Trem. coalescens, Trem. fuciformis, Trem. globospora* and *Trem. mesenterica* are in the *Candida* key.

Chapter V

Discussion of the genera belonging to the imperfect yeasts

Genus 1. *Aciculoconidium* King et Jong

by M.Th. Smith

a. DIAGNOSIS OF THE GENUS

Budding cells, mycelium with blastoconidia. Budding cells ovoid or ellipsoid, single, in short chains or clusters; mycelium consisting of branched septate hyphae. Terminal needle-shaped conidia are formed rounded at the base and pointed at the top.

Ascospores and asexual endospores not produced.

Fermentation present.

b. TYPE SPECIES OF THE GENUS

Aciculoconidium aculeatum (Phaff, Miller et Shifrine) King et Jong

c. SPECIES ACCEPTED IN THE GENUS

1. *Aciculoconidium aculeatum* (Phaff, Miller et Shifrine) King et Jong (Phaff et al., 1956)

e. SYSTEMATIC DISCUSSION OF THE SPECIES

1. ACICULOCONIDIUM ACULEATUM (PHAFF, MILLER ET SHIFRINE) KING ET JONG (1976a)

Synonym

Trichosporon aculeatum Phaff, Miller et Shifrine (1956)

Standard description of *Aciculoconidium aculeatum*

Growth in glucose-yeast extract-peptone water: After 3 days at 25°C budding cells and mycelium are formed. Budding cells are ovoid or ellipsoid

(3.5–8.5) × (4.0–15.0) μm, single, in chains or small clusters. Mycelium with thin and broad hyphae, septate. Needle-shaped cells, rounded at the base and pointed at the other end arise frequently from the blastospores from which they easily break off (Fig. 160). An irregular ring and a mycelial sediment have formed. After one month at 25°C, mycelial growth throughout the whole flask is present.

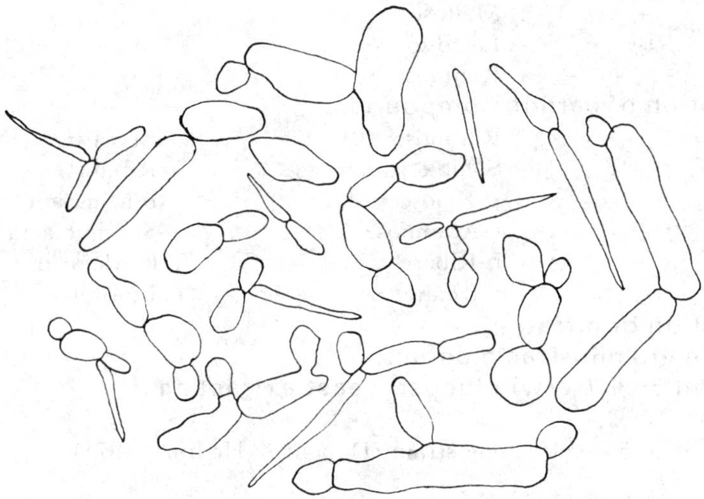

Fig. 160 – *Ac. aculeatum*
After 3 days in glucose-yeast extract-peptone water

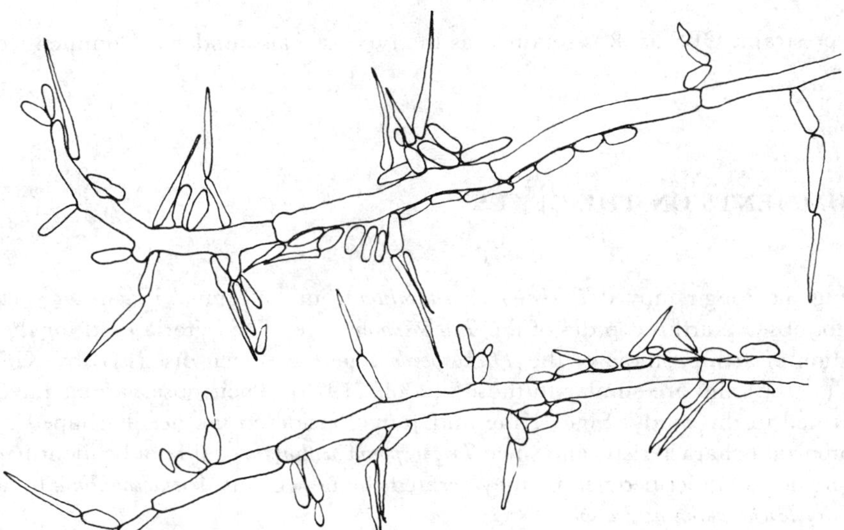

Fig. 161 – *Ac. aculeatum*
Dalmau plate culture on corn meal agar

Growth on glucose-yeast extract-peptone agar: After one month the streak culture is cream colored, smooth to slightly wrinkled, glossy to dull, tough.

Slide culture on corn meal agar: Abundant true mycelium with blastospores, borne singly or successively in chains, terminated by production of needle-shaped cells (0.5–2.0) × (6.0–21.5) μm which are rounded at the base and pointed at the other end (Fig. 161). The mycelium may disarticulate.

Fermentation:

Glucose	+ vw, s	Maltose	–
Galactose	–	Lactose	–
Sucrose	– or + vw	Raffinose	–

Assimilation of carbon compounds:

Galactose	–	Raffinose	–	Erythritol	–
Sucrose	+	Soluble starch	+	Ribitol	+ (w)
Maltose	+	D-Xylose	–	D-Mannitol	+
Cellobiose	+	L-Arabinose	–	Succinic acid	+
Trehalose	+	D-Ribose	–	Citric acid	+
Lactose	–	L-Rhamnose	–	Inositol	–

Assimilation of nitrate: –

Growth in vitamin-free medium: –

Growth on 50% (w/w) glucose-yeast extract agar: –

Growth at 37°C: –

G + C: 54.9–55 mol%, one strain (Dupont & Hedrick, 1971)

Origin of the strains studied

Drosophila occidentalis (1), *Drosophila pinicola* (1).

Type strain: CBS 5578 is retained as the type; see also under f. Comments on the genus.

f. COMMENTS ON THE GENUS

King and Jong removed *Trichosporon aculeatum* from the genus *Trichosporon* since they did not observe arthroconidia of the *Trichosporon* type. The criteria used for the identification of arthroconidia of the *Trichosporon* type have been discussed by King and Jong (1976b) and are similar to those of Cole (1975). Their observations have been confirmed in this study. Since King and Jong considered the needle-shaped conidia an important characteristic and since *Trichosporon aculeatum* could not be identified with any previously described genus, they created the new genus *Aciculoconidium* to accommodate *Trichosporon aculeatum*.

Some comments on the type strain designated by King and Jong (1976a) have to be made. Both authors have stated in their publication that two subcultures of the type strain isolated from *Drosophila pinicola* have been examined. However, the culture chosen by Phaff et al. (1956) as the type and labelled in the original publication as F-145A has been isolated from *Drosophila occidentalis* (Phaff, personal communication). To this culture the number CBS 5578 has been assigned. The culture ATCC 1550 = NRRL YB-4297 considered by King and Jong as the type strain is a subculture of the strain isolated by Phaff from *Drosophila pinicola* (Kurtzman, personal communication). Since subcultures of the authentic strain still exist, the strain CBS 5578 = ATCC 28680 has to be retained as the type.

The needle-shaped cells of *Aciculoconidium aculeatum* resemble those of *Mastigomyces phillippovi* (Imšenecky & Kriss, 1933). However, according to the original description, the latter does not form mycelium. Since no strains of this species were available during this study, the eventual relationship of *M. phillippovi* to *Aciculoconidium aculeatum* could not be examined and discussed.

Genus 2. *Brettanomyces* Kufferath et van Laer

by J.P. van der Walt

a. DIAGNOSIS OF THE GENUS

Cells spheroidal, subglobose to ellipsoidal, frequently ogival, cylindrical to elongate, reproducing by budding. Pseudomycelium and branched, one-celled, non-septate mycelium (blastese) may be formed.

In malt extract and on malt agar slow growth, cultures as a rule short-lived. Production of a characteristic aroma. Under aerobic conditions usually a vigorous production of acetic acid from glucose.

Fermentation usually stimulated by molecular oxygen (Custers effect).

Coenzyme Q-9 system present.

Extraneous vitamin source required.

b. TYPE SPECIES OF THE GENUS

Brettanomyces bruxellensis Kufferath et van Laer

c. SPECIES ACCEPTED IN THE GENUS

1. *Brettanomyces abstinens* Yarrow et Ahearn (1971)
2. *Brettanomyces anomalus* Custers (1940)
3. *Brettanomyces bruxellensis* Kufferath et van Laer (1921)
4. *Brettanomyces claussenii* Custers (1940)
5. *Brettanomyces custersianus* van der Walt (1961)
6. *Brettanomyces custersii* Florenzano (1950)
7. *Brettanomyces intermedius* (Krumbholz et Tauschanoff) van der Walt et van Kerken (Krumbholz & Tauschanoff, 1933)
8. *Brettanomyces lambicus* Kufferath et van Laer (1921)
9. *Brettanomyces naardenensis* Kolfschoten et Yarrow (1970)

d. KEY TO THE SPECIES OF THE GENUS

1a Lactose fermented 2
b Lactose not fermented 3

2a Abundant formation of blastese i.e. a branched, one-celled non-septate mycelium
Br. anomalus . . . . p. 566
b Blastese formation absent
Br. claussenii . . . . p. 569

3a Lactose assimilated, occasionally slowly
Br. custersii . . . . p. 571
b Lactose not assimilated 4

4a Cellobiose assimilated 5
b Cellobiose not assimilated 7

5a Sucrose assimilated
Br. intermedius . . . . p. 572
b Sucrose not assimilated 6

6a Trehalose assimilated
Br. naardenensis . . . . p. 574
b Trehalose not assimilated
Br. abstinens . . . . p. 565

7a Maltose and sucrose fermented 8
b Maltose and sucrose not fermented
Br. custersianus . . . . p. 570

8a Galactose assimilated
Br. lambicus . . . . p. 573
b Galactose not assimilated
Br. bruxellensis . . . . p. 567

Table 25

Key characters of the species in the genus *Brettanomyces*

Species	Fermentation						Assimilation							
	G	Ga	Su	Ma	La	Ra	Ga	Su	Ma	Ce	Tr	La	Ra	NO_3
Br. abstinens	+	+ s	–	–	–	–	+	–	–	+	–	–	–	+
Br. anomalus	+	+	+	v	+	v	+	+	+	+	+	+	+ s	+ s
Br. bruxellensis	+	–	+	+	–	–	–	+	+	–	+	–	+ w	v
Br. claussenii	+	+	+	+ w	+	v	+	+	+	+	+	+	+ w	+ (s)
Br. custersianus	+ s	–	–	–	–	–	– / + w	+ w	– / + w	–	+	–	–	–
Br. custersii	+	+ s	+	+	–	+ s	+	+	+	+	+	+ s	+ s	+
Br. intermedius	+	+ s	+	+ s	–	– / + w	+	+	+	+	+	–	v	v
Br. lambicus	+	– / + w	+	+	–	–	+ s	+	+	–	+	–	–	+
Br. naardenensis	+ s	– / + w	–	–	–	–	+ s	–	+	+	+	–	–	–

G = glucose, Ga = galactose, Su = sucrose, Ma = maltose, Ce = cellobiose, Tr = trehalose, La = lactose, Ra = raffinose, NO_3 = nitrate.

e. SYSTEMATIC DISCUSSION OF THE SPECIES

1. *BRETTANOMYCES ABSTINENS* YARROW ET AHEARN (1971)

Standard description of *Brettanomyces abstinens*

Growth in malt extract: After 5 days at 28°C the cells are ovoid, ellipsoidal, frequently ogival, cylindrical to elongate, (2.0–5.5) × (3.5–13.0–28) μm, reproducing by budding, and occur singly, in pairs or short chains. Pseudomycelium is abundantly formed. A rather coherent sediment is formed and a very slight ring. After 10 days at room temperature a sediment, ring and some islets are present.

A characteristic aroma is produced.

Growth on malt agar: After 6 weeks at room temperature the streak culture is rather coherent, raised, restricted, creamish-brown, somewhat shiny and crispulate. The margin is lobiform.

Growth on malt agar 2% calcium carbonate: After 6 weeks at room temperature the streak culture is butyrous, somewhat shiny, brownish-cream, pulvinate, crispulate or rugose along the center, but smooth towards the margin. The margin is undulating and fringed with pseudomycelium. Acid is produced.

Dalmau plate culture on corn meal agar: Pseudomycelium which branches and bears blastospores is abundantly produced. Filamentous cells are more prominent under aerobic conditions.

Fermentation:

Glucose	+	Maltose	–
Galactose	+ s	Lactose	–
Sucrose	–		

Assimilation of carbon compounds:

Galactose	+	Raffinose	–	Erythritol	–
Sucrose	–	Soluble starch	–	Ribitol	–
Maltose	–	D-Xylose	–	D-Mannitol	–
Cellobiose	+	L-Arabinose	–	Succinic acid	–
Trehalose	–	D-Ribose	–	Citric acid	–
Lactose	–	L-Rhamnose	–	Inositol	–

Assimilation of nitrate: +

Growth in vitamin-free medium: –

Growth on 50% (w/w) glucose-yeast extract agar: –

Growth at 37°C: +

Acid production from glucose (according to the standard method): +

Origin of the strain studied

Ginger ale (1) (Yarrow & Ahearn, 1971).

Type strain: CBS 6055, isolated by Yarrow and Ahearn.

2. *BRETTANOMYCES ANOMALUS* CUSTERS (1940)

Synonyms

Monilia vini Osterwalder (1912)
Oospora vini (Osterwalder) Janke (1924)
Brettanomyces cidri Legakis (1961)
Brettanomyces dublinensis Gilliland (1962)

Standard description of *Brettanomyces anomalus* (van der Walt, 1970g)

Growth in malt extract: After 5 days at 28°C the cells are ellipsoidal, frequently ogival, cylindrical to elongate, (2.0–5.5) × (3.0–19.0) μm, reproducing by budding, and occur singly or in pairs or short chains or clusters. Filamentous, interlaced, frequently branching, non-septate cells which resemble true mycelium, are also formed. These filamentous cells may measure from 200–400 μm in length. A floccose to mucoid sediment is formed. Pellicle formation is absent. After 10 days at room temperature islets or a thin pellicle and a thin, incomplete ring may be formed.

A characteristic aroma is produced.

Growth on malt agar: After 6 weeks at room temperature the streak culture is cream-colored to brown, shiny to dull, smooth or verrucose to crispulate, usually raised in the center. The margin be entire, undulating, lobiform or with fronds.

Growth on malt agar 2% calcium carbonate: After 6 weeks at room temperature the streak culture is coherent, creamish-brown to brown, somewhat shiny to dull, smooth or verrucose, plicate or rugose, usually raised in the center and somewhat spreading at the periphery. The margin may be grained, coherent, lobiform with fronds or fimbriate. Acid is formed.

Dalmau plate culture on corn meal agar: There is a very characteristic and abundant development of blastese, i.e. slender, non-septate, filamentous cells which often branch and resemble a ramified, one-celled mycelium (Fig. 162). Blastospores are usually formed along the filamentous cells but are not arranged in characteristic clusters or verticils.

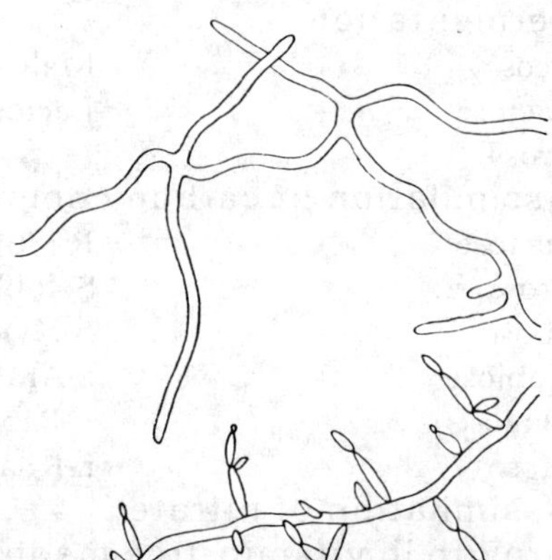

Fig. 162 – *Br. anomalus*
After 10 days on corn meal agar
(From 'The Yeasts', ed. II, 1970)

Fermentation:

Glucose	+	Maltose	− or + w
Galactose	+	Lactose	+
Sucrose	+ s	Raffinose	− or + w

Assimilation of carbon compounds:

Galactose	+	Raffinose	+ s	Erythritol	–
Sucrose	+	Soluble starch	–	Ribitol	v
Maltose	+ s or w	D-Xylose	–	D-Mannitol	–
Cellobiose	+	L-Arabinose	–	Succinic acid	v
Trehalose	+	D-Ribose	v	Citric acid	–
Lactose	+	L-Rhamnose	–	Inositol	–

Assimilation of nitrate: + s

Growth in vitamin-free medium: –

Growth on 50% (w/w) glucose-yeast extract agar: –

Growth at 37°C: +

Acid production from glucose (according to the standard method): +

Origin of the strains studied

Bottled stout beer (1) (Custers, 1940); cider (4) (Beech, 1959; Legakis, 1961); beer (1) (Gilliland, 1962).

Type strain: CBS 77, isolated by Custers from bottled stout.

3. *BRETTANOMYCES BRUXELLENSIS* KUFFERATH ET VAN LAER (1921)

Perfect state: *Dekkera bruxellensis* van der Walt

Synonyms

Brettanomyces bruxellensis Kufferath et van Laer var. *nonmembranaefaciens* Custers (1940)
? *Brettanomyces bruxellensis* Kufferath et van Laer var. *lentus* Custers (1940)

Standard description of *Brettanomyces bruxellensis* (van der Walt, 1970g)

Growth in malt extract: After 5 days at 28°C the cells are occasionally spheroidal, usually ellipsoidal, frequently ogival, or cylindrical to elongate, (2.0–6.5) × (4.0–22.0) μm, reproducing by budding, and occur singly, in pairs, short chains or small clusters (Fig. 163). Pseudomycelium is usually produced. A sediment is formed. Occasionally a thin ring or islets or a thin pellicle may be present. After 10 days at room temperature the sediment is usually floccose to somewhat mucoid. A thin ring, islets or a thin pellicle may be formed.

A characteristic aroma usually develops.

Growth on malt agar: After 6 weeks at room temperature the culture is usually restricted, cream-colored to light brown, occasionaly somewhat salmon-colored usually shiny, smooth to crispulate and somewhat raised. The margin may be entire, undulating or lobiform.

Growth on malt agar 2% calcium carbonate: After 6 weeks at room temperature the streak culture is usually restricted, cream-colored, usually shiny, smooth or pitted to crispulate and somewhat raised. The margin may be entire, undulating to lobiform or occasionally fringed with pseudomycelium. Acid is produced.

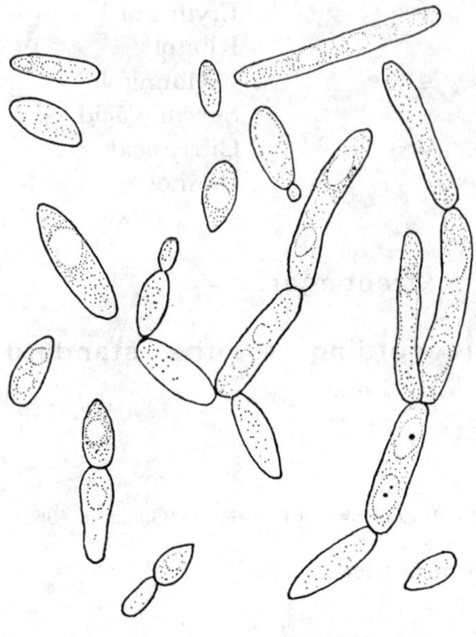

Fig. 163 – *Br. bruxellensis*
After 5 days in malt extract
(From 'The Yeasts', ed. II, 1970)

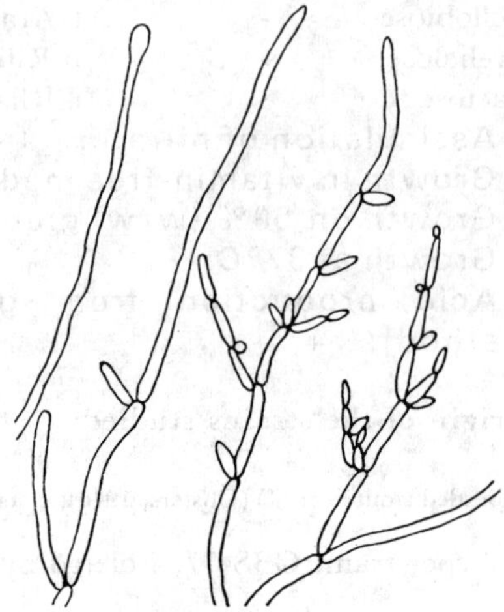

Fig. 164 – *Br. bruxellensis*
After 10 days on corn meal agar
(From 'The Yeasts', ed. II, 1970)

Dalmau plate culture on corn meal agar: Pseudomycelium is usually abundantly produced under both aerobic and anaerobic conditions. It consists of filamentous cells with or without blastospores. The blastospores may be arranged in chains or small, slightly branched verticils. Under anaerobic conditions filamentous cells not infrequently terminate in a vesicle (Fig. 164).

Fermentation:

Glucose	+	Maltose	+
Galactose	–	Lactose	–
Sucrose	+	Raffinose	–

Assimilation of carbon compounds:

Galactose	–	Raffinose	+ w	Erythritol	–
Sucrose	+	Soluble starch	–	Ribitol	v
Maltose	+	D-Xylose	–	D-Mannitol	–
Cellobiose	–	L-Arabinose	–	Succinic acid	v
Trehalose	+	D-Ribose	v	Citric acid	–
Lactose	–	L-Rhamnose	–	Inositol	–

Assimilation of nitrate: v

Growth in vitamin-free medium: –

Growth on 50% (w/w) glucose-yeast extract agar: –

Growth at 37°C: +

Acid production from glucose (according to the standard method): +

Origin of the strains studied

Lambic beer (1) (Kufferath, 1920b); porter beer (1); Skandinavisk Bryggeri Laboratory, Copenhagen (1); National Collection of Yeast Cultures, Nutfield (1).

Type strain: CBS 72, isolated by Kufferath from Belgian lambic beer.

4. BRETTANOMYCES CLAUSSENII CUSTERS (1940)

Synonym

Brettanomyces claussenii Custers var. *sablieri* Legakis (1961) nom. illegit. public. sine diagn. latine

Standard description of *Brettanomyces claussenii* (van der Walt, 1970g)

Growth in malt extract: After 5 days at 28°C the cells are ellipsoidal, frequently ogival, cylindrical to elongate, (2.0–6.5) × (3.5–18.0) μm, reproducing by budding, and occur singly, in pairs or short chains. Filamentous cells are also formed. A sediment is formed. After 10 days at room temperature a sediment and occasionally a slight ring are present.

A characteristic aroma is produced.

Growth on malt agar: After 6 weeks at room temperature the streak culture is cream-colored to brown, shiny, smooth or with some structure, or somewhat pulvinate. The margin is entire to undulating.

Growth on malt agar 2% calcium carbonate: After 6 weeks at room temperature the streak culture is cream-colored to brown, shiny, smooth or with some structure and somewhat raised in the center. The margin is entire to undulating. Acid is produced.

Dalmau plate culture on corn meal agar: Pseudomycelium is formed but not very abundantly. It is usually more abundant in anaerobic areas than in aerobic areas. The pseudomycelium is rudimentary and consists of chains or ramified chains of elongated cells.

Fermentation:

Glucose	+	Maltose	+ w
Galactose	+	Lactose	+
Sucrose	+	Raffinose	+ (–)

Assimilation of carbon compounds:

Galactose	+	Raffinose	+ w	Erythritol	–
Sucrose	+	Soluble starch	–	Ribitol	v
Maltose	+	D-Xylose	–	D-Mannitol	–
Cellobiose	+	L-Arabinose	–	Succinic acid	v
Trehalose	+	D-Ribose	v	Citric acid	–
Lactose	+	L-Rhamnose	–	Inositol	–

Assimilation of nitrate: + (s)

Growth in vitamin-free medium: –

Growth on 50% (w/w) glucose-yeast extract agar: –

Growth at 37°C: +

Acid production from glucose (according to the standard method): +

Origin of the strains studied

Beer (4) (Schiönning, 1908; Zaake, 1960); cider (4) (Legakis, 1961); sherry vat (1) (van der Walt & van Kerken, 1961a).

Type strain: CBS 76, isolated by Schiönning from beer.

5. *BRETTANOMYCES CUSTERSIANUS* VAN DER WALT (1961)

Standard description of *Brettanomyces custersianus* (van der Walt, 1970g)

Growth in malt extract: After 5 days at 28°C the cells are ellipsoidal, frequently ogival, cylindrical to elongate, (1.5–3.5) × (2.0–19.0) μm, reproducing by budding, and occur singly, in pairs or in short, occasionally somewhat branched chains. Much pseudomycelium is produced. A floccose sediment is formed. A thin, dull, pellicle, islets or a slight, frequently incomplete ring may be present. After 10 days at room temperature a dry, creeping pellicle or islets are usually present. Occasionally the pellicle may be rather thick and crispulate. The sediment is floccose to somewhat mucoid. An incomplete ring may be present.

There is a very slight production of a characteristic aroma.

Growth on malt agar: After 6 weeks at room temperature the streak culture is raised, cream-colored to light brown, shiny to somewhat dull, seldom smooth, usually verrucose, crispulate or plicate. The margin is undulating to lobiform.

Growth on malt agar 2% calcium carbonate: After 6 weeks at room temperature the streak culture is raised, cream colored to light brown, shiny to dull, verrucose, crispulate or rugose. The margin is undulating to lobiform. Acid production is slight.

Dalmau plate culture on corn meal agar: Pseudomycelium is abundantly produced. It consists of filamentous cells with no or few blastospores. Under anaerobic conditions the filamentous cells may bear blastospores more abundantly. The blastospores may be arranged in short chains or weakly branched verticils (Fig. 165).

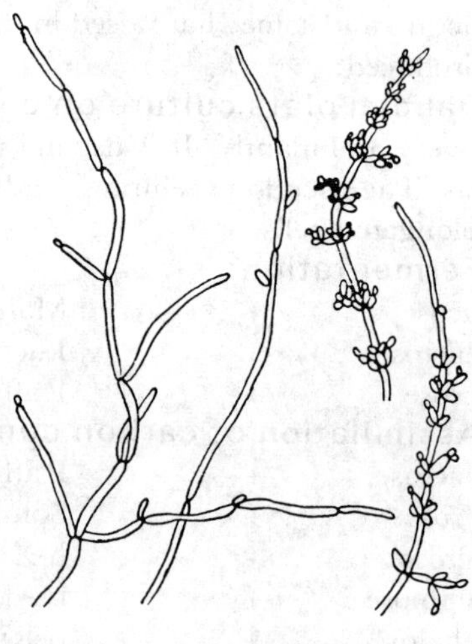

Fig. 165 – *Br. custersianus*
After 10 days on corn meal agar
(From 'The Yeasts', ed. II, 1970)

Fermentation:

Glucose	+ s	Maltose	–
Galactose	–	Lactose	–
Sucrose	–		

Assimilation of carbon compounds:

Galactose	– or + w	Raffinose	–	Erythritol	–
Sucrose	+ w	Soluble starch	–	Ribitol	v
Maltose	– or + w	D-Xylose	–	D-Mannitol	–
Cellobiose	–	L-Arabinose	–	Succinic acid	+
Trehalose	+	D-Ribose	v	Citric acid	–
Lactose	–	L-Rhamnose	–	Inositol	–

Assimilation of nitrate: –

Growth in vitamin-free medium: –

Growth on 50% (w/w) glucose-yeast extract agar: –

Growth at 37°C: v

Acid production from glucose (according to the standard method): + vw

Origin of the strains studied

Contaminated equipment in sorghum beer breweries (14) (van der Walt, 1961).

Type strain: CBS 4805, isolated by van der Walt.

6. *BRETTANOMYCES CUSTERSII* FLORENZANO (1950)

Standard description of *Brettanomyces custersii* (van der Walt, 1970g)

Growth in malt extract: After 5 days at 28°C the cells are globose, usually ellipsoidal, frequently ogival, cylindrical to elongate, (2.0–7.0) × (3.0–17.0) μm, reproducing by budding, and occur singly, in pairs, short chains or small clusters. Pseudomycelium is also formed. A sediment is formed. After 10 days at room temperature a coherent to somewhat floccose sediment is present.

A characteristic aroma is produced.

Growth on malt agar: After 6 weeks at room temperature the streak culture is raised, restricted, creamish-brown, shiny and smooth. The margin is entire to lobiform.

Growth on malt agar 2% calcium carbonate: After 6 weeks at room temperature the streak culture is raised, cream-colored, shiny and smooth. The margin is entire to lobiform. Acid is produced.

Dalmau plate culture on corn meal agar: Pseudomycelium is abundantly produced. It consists of filamentous cells with or without blastospore formation. The blastospores may be arranged in short chains or slightly branched verticils.

Fermentation:

Glucose	+	Maltose	+
Galactose	+ s	Lactose	–
Sucrose	+	Raffinose	+ s

Assimilation of carbon compounds:

Galactose	+	Raffinose	+ s	Erythritol	–
Sucrose	+	Soluble starch	–	Ribitol	v
Maltose	+	D-Xylose	–	D-Mannitol	–
Cellobiose	+	L-Arabinose	–	Succinic acid	+ s
Trehalose	+	D-Ribose	+ s	Citric acid	–
Lactose	+ s	L-Rhamnose	–	Inositol	–

Assimilation of nitrate: +

Growth in vitamin-free medium: –

Growth on 50% (w/w) glucose-yeast extract agar: –

Growth at 37°C: +

Acid production from glucose (according to the standard method): +

Origin of the strain studied

Contaminated brewing equipment in a sorghum beer brewery (1).

Type strain (neotype): CBS 5512, isolated by van der Walt.

7. *BRETTANOMYCES INTERMEDIUS* (KRUMBHOLZ ET TAUSCHANOFF) VAN DER WALT ET VAN KERKEN (van der Walt, 1971)

Perfect state: *Dekkera intermedia* van der Walt

Synonyms

Mycotorula intermedia Krumbholz et Tauschanoff (1933)
Brettanomyces intermedius (Krumbholz et Tauschanoff) van der Walt et van Kerken (1959) nom. inval.
Brettanomyces patavinus Florenzano (1951) nom. illegit. public. sine diagn. latine
Brettanomyces bruxellensis (Kufferath et van Laer) var. *vini* Barret, Bidan et André (1955) nom. illegit. public. sine diagn. latine
Brettanomyces vini Peynaud et Domercq (1956) nom. illegit. public. sine diagn. latine
Brettanomyces schanderlii Peynaud et Domercq (1956) nom. illegit. public. sine diagn. latine

Standard description of *Brettanomyces intermedius* (van der Walt, 1970g)

Growth in malt extract: After 5 days at 28°C the cells are occasionally globose, usually ellipsoidal, frequently ogival, or cylindrical to elongate, $(2.0-7.0) \times (3.5-18.0)$ μm, reproducing by budding, and occur singly, in pairs, short chains or small clusters. A sediment is formed. A slight ring, islets or a thin, dull pellicle may be formed. After 10 days at room temperature a sediment is present which may be dusty, coherent, floccose or occasionally rather mucoid. A ring, islets or a dull, creeping pellicle may be present.

A characteristic aroma is usually produced.

Growth on malt agar: After 6 weeks at room temperature the streak culture is rather restricted, raised, cream-colored to light brown, shiny to somewhat dull, smooth, pitted or crispulate. The margin may be entire, undulating or lobiform, occasionally with some structure.

Growth on malt agar 2% calcium carbonate: After 6 weeks at room temperature the streak culture is usually raised, cream-colored to light brown, shiny to dull, smooth, finely pitted, crispulate or occasionally somewhat rugose. The margin may be entire, undulating, lobiform with some structure or occasionally somewhat fimbriate. Acid is produced.

Dalmau plate culture on corn meal agar: Pseudomycelium is usually abundantly produced and often better developed under anaerobic conditions. It consists of filamentous cells with or without blastospores. The blastospores may be arranged in chains or in somewhat branched verticils.

Fermentation:

Glucose	+	Maltose	+ s
Galactose	+ s	Lactose	–
Sucrose	+	Raffinose	– or + w

Assimilation of carbon compounds:

Galactose	+	Raffinose	v	Erythritol	–
Sucrose	+	Soluble starch	–	Ribitol	v
Maltose	+ or s	D-Xylose	–	D-Mannitol	–
Cellobiose	+	L-Arabinose	–	Succinic acid	v
Trehalose	+	D-Ribose	v	Citric acid	–
Lactose	–	L-Rhamnose	–	Inositol	–

Assimilation of nitrate: v

Growth in vitamin-free medium: –

Growth on 50% (w/w) glucose-yeast extract agar: –

Growth at 37°C: +

Acid production from glucose (according to the standard method): +

Origin of the strains studied

Grape must (2) (Krumbholz & Tauschanoff, 1933; van der Walt & van Kerken, 1961a); wines from Arbois and Mâconnais (2) (Barret et al., 1955); wines from the Gironde and Medoc (2) (Peynaud & Domercq, 1956); South African wines (7) (van der Walt & van Kerken, 1958, 1959); winery equipment (1) (van der Walt & van Kerken, 1961a); contaminated equipment in a sorghum beer brewery (3) (van der Walt, 1961).

Type strain: CBS 73, isolated by Krumbholz and Tauschanoff from grape must.

8. BRETTANOMYCES LAMBICUS KUFFERATH ET VAN LAER (1921)

Standard description of *Brettanomyces lambicus* (van der Walt, 1970g)

Growth in malt extract: After 5 days at 28°C the cells are occasionally spheroidal, usually ellipsoidal, frequently ogival, or cylindrical to elongate, (2.0–7.0) × (3.0–20.0) μm, reproducing by budding, and occur singly, in pairs, short chains or small clusters. Pseudomycelium is usually present. A sediment is formed. A slight ring and a thin, dull, creeping pellicle may also be formed. After 10 days at room temperature a flocculent to coherent sediment is present. A slight ring and a thin, dull, white, creeping pellicle may also be present.

A characteristic aroma is usually produced.

Growth on malt agar: After 6 weeks at room temperature the streak culture is restricted, cream- or salmon-colored to light brown, shiny or dull, usually smooth or somewhat verrucose and raised. The margin may be entire, undulating or lobiform.

Growth on malt agar 2% calcium carbonate: After 6 weeks at room temperature the streak culture is usually restricted, cream-colored, shiny, smooth or somewhat pitted. The margin may be entire to undulating. Acid is produced.

Dalmau plate culture on corn meal agar: Pseudomycelium is abundantly formed both aerobically and anaerobically. It consists of filamentous cells with or without blastospores. The blastospores may be arranged in chains or in somewhat branched verticils.

Fermentation:

Glucose	+	Maltose	+
Galactose	– or +w	Lactose	–
Sucrose	+	Raffinose	–

Assimilation of carbon compounds:

Galactose	+s	Raffinose	–	Erythritol	–
Sucrose	+	Soluble starch	–	Ribitol	–
Maltose	+	D-Xylose	–	D-Mannitol	–
Cellobiose	–	L-Arabinose	–	Succinic acid	v
Trehalose	+	D-Ribose	+s	Citric acid	–
Lactose	–	L-Rhamnose	–	Inositol	–

Assimilation of nitrate: +

Growth in vitamin-free medium: –

Growth on 50% (w/w) glucose-yeast extract agar: –

Growth at 37°C: +

Acid production from glucose (according to the standard method): +

Origin of the strains studied

Belgian lambic beer (1) (Custers, 1940); South African grape must (1) (van der Walt & van Kerken, 1961a).

Type strain: CBS 75, isolated by Custers from lambic beer.

9. *BRETTANOMYCES NAARDENENSIS* KOLFSCHOTEN ET YARROW (1970)

Standard description of *Brettanomyces naardenensis*

Growth in malt extract: After 5 days at 28°C the cells are predominantly cylindrical to elongate, rarely ellipsoidal or ogival, (1.5–4.5) × (4.0–15.0–30) μm, reproducing by budding, and occur singly, in pairs or short chains or small clusters. Pseudomycelium is abundantly formed. A floccose sediment is formed and occasionally a slight, incomplete ring and islets. After 10 days a floccose sediment, islets or pellicle and a ring are present.

The formation of a characteristic aroma is detectable.

Growth on malt agar: After 6 weeks at room temperature the streak culture is butyrous, brownish-cream to light brown, somewhat shiny to dull, pulvinate, crispulate or rugose and somewhat restricted. The margin is lobate and fringed with pseudomycelium.

Growth on malt agar 2% calcium carbonate: After 6 weeks at room temperature the streak culture is rather coherent, creamish-brown to light brown, somewhat shiny to rather dull, pulvinate to verruculose. The margin is lobate. Acid production is not very pronounced.

Dalmau plate culture on corn meal agar: A ramified pseudomycelium bearing few or no blastospores is produced. It is better developed under anaerobic conditions.

Fermentation:

Glucose	+ s	Maltose	–
Galactose	– or + w	Lactose	–
Sucrose	–		

Assimilation of carbon compounds:

Galactose	+ s	Raffinose	–	Erythritol	–
Sucrose	–	Soluble starch	+ s	Ribitol	v
Maltose	+	D-Xylose	+	D-Mannitol	+
Cellobiose	+	L-Arabinose	–	Succinic acid	+
Trehalose	+	D-Ribose	–	Citric acid	–
Lactose	–	L-Rhamnose	v	Inositol	–

Assimilation of nitrate: –

Growth in vitamin-free medium: –

Growth on 50% (w/w) glucose-yeast extract agar: –

Growth at 37°C: –

Acid production from glucose (according to the standard method): +

Origin of the strains studied

Soft drinks (2) (Kolfschoten & Yarrow, 1970); carbonated water (1) (Kolfschoten & Yarrow, 1970).

Type strain: CBS 6042, isolated from aerated lemonade (Kolfschoten & Yarrow, 1970).

f. COMMENTS ON THE GENUS

For a historical and more detailed account of the genus *Brettanomyces,* reference is made to the monograph by Custers (1940) and the relevant chapter of the second edition of "The Yeasts" (van der Walt, 1970g).

Typical of this genus (and its perfect state *Dekkera*) is its marked production of acetic acid on glucose containing media. This accounts for the fact that cultures on media such as malt extract agar which do not contain calcium carbonate as neutralizing agent, are characteristically short-lived. Acid production is, however, less marked in the weakly fermentative species, e.g., *Br. custersianus*. Another unusual feature of the genus is that most of its members display a pronounced Custers effect, i.e., the stimulation of fermentation by molecular oxygen and acetoin (Scheffers, 1966; Scheffers & Wikén, 1969).

The genus *Brettanomyces* displays a rather restricted habitat specificity and is primarily recovered from products of the fermentation industries, from materials obtained from their premises and to a lesser extent from bottled 'soft drinks'. On the basis of resistance to cycloheximide and sorbic acid, van der Walt and van Kerken (1961a) devised a selective medium from the recovery of species of the genus.

As nuclear genome comparison studies (mol% G + C and DNA–DNA complementarity) have not yet been applied to the genus, species delimitation in *Brettanomyces* is still based on phenotypic differentiation. The results of such studies could lead to a revision of the number of species currently accepted. For example, *Br. naardenensis* by virtue of its ability to utilize mannitol, D-xylose, soluble starch and its variable utilization of L-rhamnose, appears to be an anomaly within the genus.

Genus 3. *Bullera* Derx

by L. Rodrigues de Miranda

a. DIAGNOSIS OF THE GENUS

Organisms which reproduce by budding and by the formation of ballistospores which are usually rotationally symmetrical, and which may be globose, obovoid, turbinate, apiculate, or ampulliform. Mycelium or pseudomycelium may be formed; cells are spheroidal, ovoidal, elliptical or cylindrical. The growth on malt agar is cream colored to slightly yellowish or brownish.

Fermentation is absent.

b. TYPE SPECIES OF THE GENUS

Bullera alba (Hanna) Derx

c. SPECIES ACCEPTED IN THE GENUS

1. *Bullera alba* (Hanna) Derx (Hanna, in Bisby et al., 1929)
2. *Bullera dendrophila* van der Walt et Scott (1970)
3. *Bullera grandispora* Derx (1930)
4. *Bullera piricola* Stadelmann (1975)
5. *Bullera singularis* (Phaff et do Carmo-Sousa) Rodrigues de Miranda comb. nov. (Phaff & do Carmo-Sousa, 1962)
6. *Bullera tsugae* Phaff et do Carmo-Sousa (1962)

d. KEY TO THE SPECIES OF THE GENUS

1a	Nitrate assimilated	2
b	Nitrate not assimilated	3

2a Inositol assimilated
B. piricola .... p. 581
b Inositol not assimilated
B. tsugae .... p. 583

3a Lactose assimilated 4
b Lactose not assimilated
B. grandispora .... p. 581

4a Inositol assimilated 5
b Inositol not assimilated
B. singularis .... p. 583

5a Raffinose assimilated 6
b Raffinose not assimilated 7

6a Growth at 30°C
Fibulobasidium inconspicuum .... p. 546
b No growth at 30°C
B. alba .... p. 578

7a Maltose assimilated
B. dendrophila .... p. 579
b Maltose not assimilated
Sirobasidium magnum .... p. 547

See Table 24, p. 534

e. SYSTEMATIC DISCUSSION OF THE SPECIES

1. *BULLERA ALBA* (HANNA) DERX (1948)

Synonym

Sporobolomyces albus Hanna (in Bisby et al., 1929)

Standard description of *Bullera alba*

Growth in malt extract: After 3 days at room temperature the cells are ovoidal, (2–4) × (3.7–6.5) μm, single, in pairs and sometimes in short chains or clusters. A thin ring and a little sediment are formed, but not a pellicle. After one month, a moderate to heavy, slimy, yellowish-brown ring, a heavy sediment and, sometimes, a thin pellicle are formed.

Growth on malt agar: After one month the streak culture is brownish cream

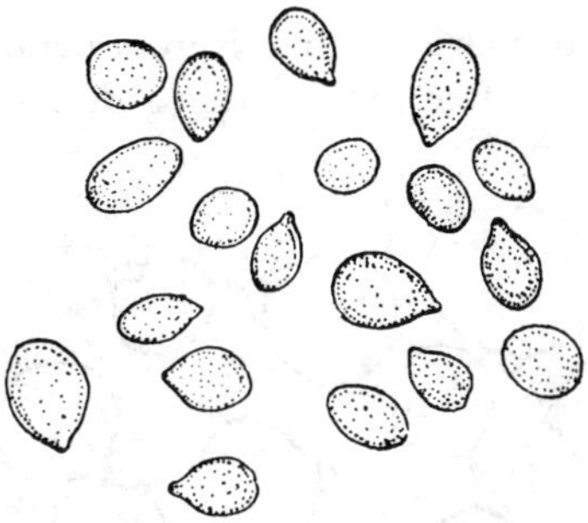

Fig. 166 – *B. alba*
Ballistospores (From 'The Yeasts', ed. I, 1952)

colored sometimes yellowish or pinkish, smooth and glossy, mucous or pasty; in some strains the growth runs to the bottom of the slant, in others the culture is flat or raised with an entire border.

Slide culture on potato agar: In most strains pseudomycelium is absent, but in some strains a rudimentary pseudomycelium may be formed in old cultures.

Formation of ballistospores: Five of the six strains studied formed ballistospores on corn meal agar. The spores are apiculate to globose, (3–8) × (3.5–7.5) μm (Fig. 166). They are borne on sterigmata of variable length.

Fermentation: –

Assimilation of carbon compounds:

Galactose	+	Raffinose	+	Erythritol	v
Sucrose	+	Soluble starch	+ or w	Ribitol	+ (–)
Maltose	+	D-Xylose	+	D-Mannitol	+
Cellobiose	+	L-Arabinose	+	Succinic acid	+
Trehalose	+	D-Ribose	+ or w	Citric acid	+ or w
Lactose	+	L-Rhamnose	+	Inositol	+

Additional carbon compounds tested: Sodium glucuronate + w, glucono-δ-lactone + or w

Assimilation of nitrate: –

Assimilation of nitrite: + w or –

Growth in vitamin-free medium: –

Growth on 50% (w/w) glucose-yeast extract agar: –

Growth at 37°C: – ; **at 30°C:** – ; **at 25°C:** +

Starch formation: +

G + C: 54.4 mol%, one strain (Nakase & Komagata, 1971g)

Origin of the strains studied

Air in a dairy plant (2), rusted straw (1), *Zelkowa* spec. (1), frass of *Sinoxylon ruficorne* in *Lonchocarpus capassa* (1), grass (1).

Type strain: CBS 501, isolated from air in a dairy plant and received from Hammer.

2. *BULLERA DENDROPHILA* VAN DER WALT ET SCOTT (1970)

Standard description of *Bullera dendrophila*

Growth in malt extract: After 3 days at 25°C cells are ellipsoidal

(4.0–8.0) × (5.0–13.0) μm, single, in pairs or in short chains. Reproduction by unilateral or bipolar budding. A sediment is formed. After one month a sediment and a slight ring are present.

Growth on malt agar: After one month at 20°C, the streak culture is mucoid to butyrous, greyish-white, shiny and smooth. The margin is entire to undulating.

Slide culture on potato agar: No pseudomycelium.

Formation of ballistospores: Ballistospores are produced on corn meal agar at room temperature. The ballistospores are subglobose to ellipsoidal or ampulliform, (4.5–9.0) × (5.5–10.5) μm (Fig. 167), borne on sterigmata of variable length.

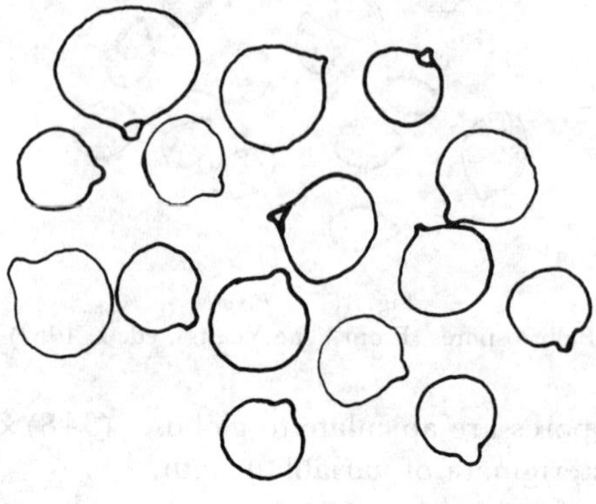

Fig. 167 – *B. dendrophila* Ballistospores

Fermentation: –

Assimilation of carbon compounds:

Galactose	+	Raffinose	–	Erythritol	–
Sucrose	+	Soluble starch	+ w or s	Ribitol	+
Maltose	+	D-Xylose	+	D-Mannitol	+
Cellobiose	+	L-Arabinose	+	Succinic acid	+
Trehalose	+	D-Ribose	+	Citric acid	+ w
Lactose	+	L-Rhamnose	+ or s	Inositol	+ s

Additional carbon compounds tested: Sodium glucuronate + w, glucono-δ-lactone +

Assimilation of nitrate: –

Assimilation of nitrite: –

Growth in vitamin-free medium: –

Growth on 50% (w/w) glucose-yeast extract agar: –

Growth at 37°C: – ; **at 30°C:** – or + w; **at 25°C:** +

Starch formation: +

Comments

Selected isolates from strain CBS 6074 have been described by van der Walt (1973) as *Aessosporon dendrophilum.* As long as this state has not been confirmed by other scientists, this species is retained in the genus *Bullera.*

Origin of the strain studied

Frass of Bupristidae larvae in *Dichrostachys cinera* (1).

Type strain: CBS 6074, isolated from frass of Bupristidae larvae by Scott.

3. *BULLERA GRANDISPORA* DERX (1930)

Original description

Derx (1930) isolated this organism from a chestnut leaf infected with mildew. He gave the following description:

On malt agar the cells measure (4–5) × (12–14) μm. The young culture is dirty white; later on it turns yellowish. It is pasty, smooth, slightly mucilaginous. The non-mucilaginous form is very similar and not very stable.

The ballistospores are symmetrical, without a lateral dent, apiculate, measuring 6 × (9.3–10) μm.

Glucose, sucrose, maltose, peptone and asparagin (very weakly) are assimilated, but not lactose, ethanol or inorganic nitrogen compounds.

Since no strains of *Bullera grandispora* have been preserved up to the present time this species could not be studied. Lodder and Kreger-van Rij (1952) also referred to the original description in their monograph.

4. *BULLERA PIRICOLA* STADELMANN (1975)

Standard description of *Bullera piricola*

Growth in malt extract: After 3 days at 20°C, cells short-oval, oval to cylindrical (4.0–8.6) × (6.4–13.6) μm (Fig. 168), single, in pairs or in short chains. An incomplete ring and little sediment are formed. After one month at room temperature

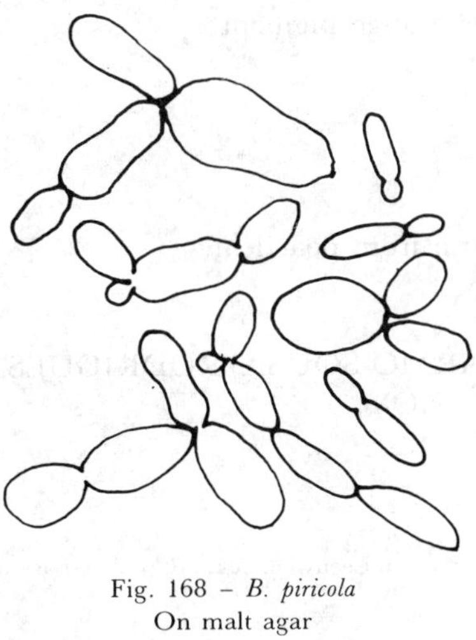

Fig. 168 – *B. piricola*
On malt agar

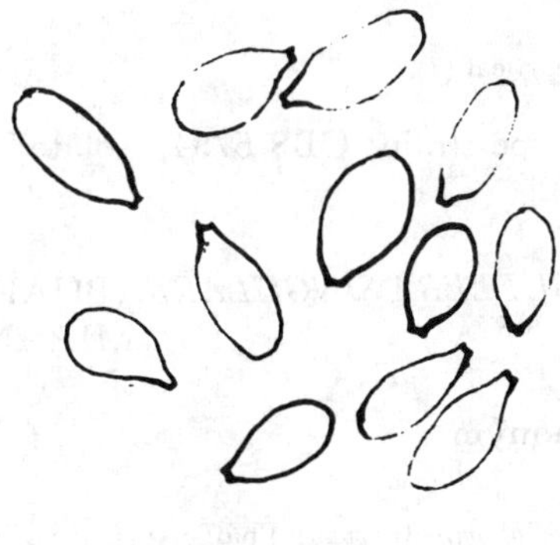

Fig. 169 – *B. piricola*
Ballistospores

a ring and heavy sediment are produced. Some islets may be present.

Growth on malt agar: After one month the streak culture is smooth, glossy, so mucoid that the growth runs to the bottom of the slant, or butyrous. The culture is brownish or pinkish cream colored with an entire, or somewhat lobated margin.

Slide culture on potato agar: Short pseudohyphae consisting of a few elongated cells are present. Septa are also observed.

Formation of ballistospores: Ballistospores are either rotationally symmetrical, ovoid to cylindrical, with an obtuse and a tapering end, or else bilaterally symmetrical, kidncy-shaped, (3.7–7.8) × (8.7–17.0) μm (Fig. 169). The spores are ejected from sterigmata of variable length with or without a septum.

Fermentation: –

Assimilation of carbon compounds:

Galactose	+	Raffinose	+	Erythritol	–
Sucrose	+	Soluble starch	+ w	Ribitol	+
Maltose	+	D-Xylose	+	D-Mannitol	+
Cellobiose	+ w	L-Arabinose	+	Succinic acid	+
Trehalose	+	D-Ribose	– or + w	Citric acid	+
Lactose	+ or w	L-Rhamnose	–	Inositol	+

Additional carbon compounds tested: Sodium glucuronate + w, glucono-δ-lactone – or + w

Assimilation of nitrate: +

Growth in vitamin-free medium: –

Growth on 50% (w/w) glucose-yeast extract agar: –

Growth at 37°C: –; **at 30°C:** –; **at 25°C:** + w or –; **at 22°C:** +

Starch formation: + w

Comments

Although the ballistospores are sometimes kidney-shaped as in *Sporobolomyces,* this species is placed in *Bullera* because of the production of ovoidal, rotationally symmetrical ballistospores and the absence of red or orange pigment.

Origin of the strains studied

Pear leaf (3).

Type strain: CBS 6754, isolated by Stadelmann from pear leaf.

5. *BULLERA SINGULARIS* (PHAFF ET DO CARMO-SOUSA) RODRIGUES DE MIRANDA COMB. NOV.

Synonym

Sporobolomyces singularis Phaff et do Carmo-Sousa (1962) Antonie van Leeuwenhoek, *28*, p. 205 (basionym)

Standard description of *Bullera singularis*

Growth in malt extract: After 3 days at 20°C the cells are ovoidal to slightly elongate (2.5–3.5) × (4.5–7.0) μm, single, in pairs or in short branched chains. A ring forms gradually and a little sediment. After one month at room temperature a moderate ring and sediment are formed.

Growth on malt agar: After one month at 20°C the streak culture is creamy white, highly glossy, smooth with transverse striations, the texture is pasty to slightly mucous; the border is irregular to entire.

Slide culture on potato agar: A few short branched chains may be formed.

Formation of ballistospores: Ballistospores could not be observed in the single available strain in the present study, but Phaff and do Carmo-Sousa (1962) observed ballistospores on corn meal agar. The spores were described as sickle- to kidney-shaped and were apparently bilaterally symmetrical.

Fermentation: –

Assimilation of carbon compounds:

Galactose	–	Raffinose	–	Erythritol	–
Sucrose	–	Soluble starch	–	Ribitol	v
Maltose	–	D-Xylose	–	D-Mannitol	+
Cellobiose	+	L-Arabinose	–	Succinic acid	+ w
Trehalose	+	D-Ribose	–	Citric acid	+ w
Lactose	+	L-Rhamnose	–	Inositol	–

Additional carbon compounds tested: Sodium glucuronate + w, glucono-δ-lactone + w

Assimilation of nitrate: –

Assimilation of nitrite: –

Growth in vitamin-free medium: –

Growth on 50% (w/w) glucose-yeast extract agar: –

Growth at 37°C: – ; at 30°C: – ; at 25°C: +

Starch formation: –

Origin of the strain studied

Frass of *Scolytis tsugae* in *Tsuga heterophylla* (Raf.) Sargent.

Type strain: CBS 5109, isolated by Phaff and do Carmo-Sousa.

6. *BULLERA TSUGAE* PHAFF ET DO CARMO-SOUSA (1962)

Standard description of *Bullera tsugae*

Growth in malt extract: After 3 days at 25°C cells are globose to ovoidal, (2.5–8.0) × (3.5–11.0) μm, single, in pairs or in short chains. An incomplete ring is formed. After one month a ring and sediment are present.

Growth on malt agar: After one month at 20°C the streak culture is greyish-yellow to buff, smooth, highly glistening, slimy with the growth running to the bottom of the slant; the margin is entire.

Slide culture on potato agar: No pseudomycelium.

Formation of ballistospores: In the strain studied ballistospores could be observed neither on malt extract agar nor on corn meal agar. Johri (personal communication), however, described a strain that formed ballistospores abundantly on corn meal agar. The spores in this strain were ovoid, apiculate, (3.0–5.0) × (5.5–6.5) μm.

Fermentation: –

Assimilation of carbon compounds:

Galactose	–	Raffinose	–	Erythritol	–
Sucrose	+	Soluble starch	–	Ribitol	+ w
Maltose	+	D-Xylose	+ w	D-Mannitol	+
Cellobiose	+	L-Arabinose	–	Succinic acid	+ w
Trehalose	+	D-Ribose	+ w	Citric acid	+ w
Lactose	+	L-Rhamnose	–	Inositol	–

Additional carbon compounds tested: Sodium glucuronate + w, glucono-δ-lactone +

Assimilation of nitrate: +

Growth in vitamin-free medium: + w

Growth on 50% (w/w) glucose-yeast extract agar: –

Growth at 37°C: –; **at 30°C:** –; **at 25°C:** +

Starch formation: –

Origin of the strain studied

Frass underneath the bark of *Tsuga heterophylla* (1).

Type strain: CBS 5038, the original strain of *B. tsugae* isolated by Phaff and do Carmo-Sousa.

f. COMMENTS ON THE GENUS

The genera *Bullera, Sporobolomyces* and *Sporidiobolus* are distinguishable from other yeast genera by their production of ballistospores. *Sporidiobolus* species form true mycelium with clamp connections and thick-walled chlamydospores, which are not produced in *Bullera* or *Sporobolomyces.*

The ballistospores in *Bullera* are usually rotationally symmetrical, in *Sporobolomyces* usually bilaterally symmetrical. *Bullera* spores are obovoid, almost globose, turbinate, apiculate or ampulliform, whereas those of *Sporobolomyces* are mostly reniform or allantoid.

Investigations of the cell wall composition (von Arx & Weijman, 1979) have demonstrated that the cell walls of *Bullera* species contain xylose whereas the cell walls of the *Sporobolomyces* species lack this substance. *Sporobolomyces* species produce an orange, red or sometimes purple pigment, whereas the *Bullera* species are cream colored, yellowish or brownish.

Genus 4. *Candida* Berkhout

by Sally A. Meyer, D.G. Ahearn and D. Yarrow

a. DIAGNOSIS OF THE GENUS

Cells vary in shape: globose, ovoid, cylindrical to elongate, rarely apiculate, ogival, triangular or flask-shaped. Reproduction usually by multilateral budding. Pseudomycelium absent, rudimentary or well developed; true mycelium present in some species. Chlamydospores may be present; ascospores, teliospores, ballistospores and arthrospores absent.

Visible pigmentation due to carotenoid pigments absent.

Extracellular polysaccharides may be formed and may give a green to purple color reaction with iodine.

Inositol assimilation variable. Inositol positive strains form pseudomycelium.

Fermentation absent or present.

b. TYPE SPECIES OF THE GENUS

Candida vulgaris Berkhout [= *Candida tropicalis* (Castellani) Berkhout] is the type species of the genus (van Uden & Buckley, 1970).

c. SPECIES ACCEPTED IN THE GENUS

1. *Candida aaseri* Dietrichson ex van Uden et Buckley (1970)
2. *Candida albicans* (Robin) Berkhout (Robin, 1853)
3. *Candida amylolenta* van der Walt, Scott et van der Klift (1972)
4. *Candida anatomiae* (Zwillenberg) Meyer et Yarrow (Zwillenberg, 1966)
5. *Candida antarctica* (Goto, Sugiyama et Iizuka) Kurtzman, Smiley, Johnson et Hoffman (Goto et al., 1969)
6. *Candida apicola* (Hajsig) Meyer et Yarrow (Hajsig, 1958)
7. *Candida apis* (Lavie ex van Uden et Vidal-Leiria) Meyer et Yarrow (van Uden & Vidal-Leiria, 1970)
   Candida apis (Lavie ex van Uden et Vidal-Leiria) Meyer et Yarrow var. *apis*
   Candida apis (Lavie ex van Uden et Vidal-Leiria) Meyer et Yarrow var. *galacta* Golubev et Babjeva (1977)

8. *Candida aquatica* Jones et Slooff (1966)
9. *Candida atlantica* (Siepmann) Meyer et Simione (Siepmann & Höhnk, 1962)
10. *Candida atmospherica* Santa María (1959c)
11. *Candida auriculariae* (Nakase) Meyer et Yarrow (Nakase, 1971a)
12. *Candida auringiensis* Santa María (1978)
13. *Candida austromarina* (Fell et Hunter) Meyer et Yarrow (Fell & Hunter, 1974)
14. *Candida azyma* (van der Walt, Johannsen et Yarrow) Meyer et Yarrow (van der Walt et al., 1978)
15. *Candida bacarum* (Buhagiar) Meyer et Yarrow (Buhagiar, 1975)
16. *Candida beechii* Buckley et van Uden (1968)
17. *Candida berthetii* Boidin, Pignal, Mermier et Arpin (1963)
18. *Candida bimundalis* Wickerham et Santa María var. *bimundalis* (in Wickerham, 1965b) see *Hansenula bimundalis*

    Candida bimundalis Wickerham et Santa María var. *americana* Wickerham (1965b) see *Hansenula americana*
19. *Candida blankii* Buckley et van Uden (1968)
20. *Candida bogoriensis* Deinema (1961)
21. *Candida boidinii* Ramirez (1953)
22. *Candida boleticola* Nakase (1971b)
23. *Candida bombi* Montrocher (1967)
24. *Candida bombicola* (Spencer, Gorin et Tulloch) Meyer et Yarrow (Spencer et al., 1970)
25. *Candida buffonii* (Ramirez) van Uden et Buckley (Ramirez, 1957)
26. *Candida buinensis* Soneda et Uchida (1971)
27. *Candida butyri* Nakase (1971b)
28. *Candida cacaoi* Buckley et van Uden (1968)
29. *Candida cantarellii* (van der Walt et van Kerken) Meyer et Yarrow (van der Walt & van Kerken, 1961b)
30. *Candida cariosilignicola* Lee et Komagata (1980)
31. *Candida castellii* (Capriotti) Meyer et Yarrow (Capriotti, 1961d)
32. *Candida catenulata* Diddens et Lodder (1942)
33. *Candida chilensis* Grinbergs et Yarrow (1970a)
34. *Candida chiropterorum* Grose et Marinkelle (1968)
35. *Candida ciferrii* Kreger-van Rij (1965) see *Stephanoascus ciferrii*
36. *Candida citrea* Nakase (1971c)
37. *Candida colliculosa* (Hartmann) Meyer et Yarrow (Hartmann, 1903) see *Torulaspora delbrueckii*
38. *Candida conglobata* (Redaelli) van Uden et Buckley (Redaelli, 1925)
39. *Candida curiosa* Komagata et Nakase (1965)
40. *Candida curvata* (Diddens et Lodder) Lodder et Kreger-van Rij (Diddens & Lodder, 1942)
41. *Candida dattila* (Kluyver) Meyer et Yarrow (Kluyver, 1914) see *Kluyveromyces thermotolerans*
42. *Candida deformans* (Zach) Langeron et Guerra (in Wolfram & Zach, 1934)
43. *Candida dendrica* (van der Walt, van der Klift et Scott) Meyer et Yarrow (van der Walt et al., 1971b)

44. *Candida dendronema* van der Walt, van der Klift et Scott (van der Walt et al., 1971a)
45. *Candida diddensiae* (Phaff, Mrak et Williams) Fell et Meyer (Phaff et al., 1952)
46. *Candida diffluens* Ruinen (1963)
47. *Candida diversa* Ohara, Nonomura et Yunome ex van Uden et Buckley (1970)
48. *Candida domercqii* (van der Walt et van Kerken) Meyer et Yarrow (van der Walt & van Kerken, 1960b) see *Wickerhamiella domercqii*
49. *Candida edax* van der Walt et Nel (1968)
50. *Candida entomaea* van der Walt, Scott et van der Klift (1972)
51. *Candida entomophila* Scott, van der Walt et van der Klift (van der Walt et al., 1971a)
52. *Candida ergatensis* Santa María (1971)
53. *Candida ernobii* (Lodder et Kreger-van Rij) Meyer et Yarrow (Lodder & Kreger-van Rij, 1952)
54. *Candida etchellsii* (Lodder et Kreger-van Rij) Meyer et Yarrow (Lodder & Kreger-van Rij, 1952)
55. *Candida fabianii* Kodama, Kyono, Iida et Onoyama (1964a) see *Hansenula fabianii*
56. *Candida famata* (Harrison) Meyer et Yarrow (Harrison, 1928)
57. *Candida fennica* (Sonck et Yarrow) Meyer et Ahearn (Sonck & Yarrow, 1969)
58. *Candida fluviatilis* Hedrick (1976)
59. *Candida foliorum* Ruinen (1963)
60. *Candida fragariorum* (Barnett et Buhagiar) Meyer et Yarrow (Barnett & Buhagiar, 1971)
61. *Candida freyschussii* Buckley et van Uden (1968)
62. *Candida friedrichii* van Uden et Windisch (1968)
63. *Candida frigida* di Menna (1966) see *Leucosporidium frigidum*
64. *Candida fructus* (Nakase) Meyer et Yarrow (Nakase, 1971a)
65. *Candida gelida* di Menna (1966) see *Leucosporidium gelidum*
66. *Candida geochares* (van der Walt, Johannsen et Yarrow) Meyer et Yarrow (van der Walt et al., 1978)
67. *Candida glabrata* (Anderson) Meyer et Yarrow (Anderson, 1917)
68. *Candida glaebosa* Komagata et Nakase (1965)
69. *Candida globosa* Yarrow et Meyer (1978) see *Citeromyces matritensis*
70. *Candida graminis* Rodrigues de Miranda et Diem (1974)
71. *Candida gropengiesseri* (Harrison) Meyer et Yarrow (Harrison, 1928)
72. *Candida guilliermondii* (Castellani) Langeron et Guerra (Castellani, 1912b)
73. *Candida haemulonii* (van Uden et Kolipinski) Meyer et Yarrow (van Uden & Kolipinski, 1962)
74. *Candida halonitratophila* (Onishi ex van Uden et Vidal-Leiria) Meyer et Yarrow (van Uden & Vidal-Leiria, 1970)
75. *Candida halophila* Yarrow et Meyer (1978)
76. *Candida hellenica* (Verona et Picci) King et Jong (Verona & Picci, 1958)
77. *Candida holmii* (Jörgensen) Meyer et Yarrow (Jörgensen, 1909)
78. *Candida homilentoma* van der Walt et Nakase (1973)
79. *Candida humicola* (Daszewska) Diddens et Lodder (Daszewska, 1912)
80. *Candida humilis* (Nel et van der Walt) Meyer et Yarrow (Nel & van der Walt, 1968)

81. *Candida hydrocarbofumarica* Yamada et al. ex Ramirez (1974)
82. *Candida hylophila* van der Walt, van der Klift et Scott (van der Walt et al., 1971a)
83. *Candida incommunis* Ohara, Nonomura et Yamazaki (1965)
84. *Candida inconspicua* (Lodder et Kreger-van Rij) Meyer et Yarrow (Lodder & Kreger-van Rij, 1952)
85. *Candida ingeniosa* (di Menna) Meyer et Yarrow (di Menna, 1958b)
86. *Candida ingens* van der Walt et van Kerken (1961c)
87. *Candida inositophila* Nakase (1975)
88. *Candida insectalens* (Scott, van der Walt et van der Klift) Meyer et Yarrow (van der Walt et al., 1971b)
89. *Candida insectamans* Scott, van der Walt et van der Klift (van der Walt et al., 1972)
90. *Candida insectorum* Scott, van der Walt et van der Klift (van der Walt et al., 1972)
91. *Candida intermedia* (Ciferri et Ashford) Langeron et Guerra (Ciferri & Ashford, 1929)
92. *Candida ishiwadae* Sugiyama et Goto (1969)
93. *Candida japonica* Diddens et Lodder (1942) see *Filobasidium capsuligenum*
94. *Candida javanica* Ruinen (1963)
95. *Candida karawaiewii* Yarrow et Meyer (1978)
96. *Candida kefyr* (Beijerinck) van Uden et Buckley (Beijerinck, 1889b)
97. *Candida krissii* Goto, Yamasato et Iizuka (1974)
98. *Candida kruisii* (Kocková-Kratochvílová et Ondrušová) Meyer et Yarrow (Kocková-Kratochvílová & Ondrušová, 1971)
99. *Candida krusei* (Castellani) Berkhout (Castellani, 1910)
100. *Candida lactiscondensi* (Hammer) Meyer et Yarrow (Hammer, 1919)
101. *Candida lambica* (Lindner et Genoud) van Uden et Buckley (Lindner & Genoud, 1913)
102. *Candida lipolytica* (Harrison) Diddens et Lodder (Harrison, 1928) see *Saccharomycopsis lipolytica*
103. *Candida lusitaniae* van Uden et do Carmo-Sousa (1959) see *Clavispora lusitaniae*
104. *Candida magnoliae* (Lodder et Kreger-van Rij) Meyer et Yarrow (Lodder & Kreger-van Rij, 1952)
105. *Candida maltosa* Komagata, Nakase et Katsuya (1964)
106. *Candida mannitofaciens* (Onishi et Suzuki) Meyer et Yarrow (Onishi & Suzuki, 1969)
107. *Candida marina* van Uden et Zobell (1962)
108. *Candida maris* (van Uden et Zobell) Meyer et Yarrow (van Uden & Zobell, 1962)
109. *Candida maritima* (Siepmann) van Uden et Buckley (in Siepmann & Höhnk, 1962)
110. *Candida melibiosica* Buckley et van Uden (1968)
111. *Candida melinii* Diddens et Lodder (1942)
112. *Candida membranaefaciens* (Lodder et Kreger-van Rij) Wickerham et Burton (Lodder & Kreger-van Rij, 1952)
113. *Candida mesenterica* (Geiger) Diddens et Lodder (Geiger, 1910)
114. *Candida methanolovescens* (Oki et Kuono) Meyer et Yarrow (in Oki et al., 1972)
115. *Candida methanosorbosa* (Abe et Yokote) Barnett, Payne et Yarrow (in Yokote et al., 1974)

116. *Candida milleri* Yarrow (1978)
117. *Candida mogii* Vidal-Leiria (1967)
118. *Candida molischiana* (Zikes) Meyer et Yarrow (Zikes, 1911)
119. *Candida mucilagina* Phaff, Starmer, Miranda et Miller (1980)
120. *Candida multis-gemmis* (Buhagiar) Meyer et Yarrow (Buhagiar, 1975)
121. *Candida musae* (Nakase) Meyer et Yarrow (Nakase, 1971a)
122. *Candida muscorum* di Menna (1958a)
123. *Candida naeodendra* van der Walt, Johannsen et Nakase (1973)
124. *Candida nemodendra* (van der Walt, van der Klift et Scott) Meyer et Yarrow (van der Walt et al., 1971b)
125. *Candida nitrativorans* van der Walt, Scott et van der Klift (1972) see *Hansenula sydowiorum*
126. *Candida nitratophila* (Shifrine et Phaff) Meyer et Yarrow (Shifrine & Phaff, 1956)
127. *Candida nivalis* di Menna (1966) see *Leucosporidium nivalis*
128. *Candida nodaensis* Yarrow et Meyer (1978)
129. *Candida norvegensis* (Dietrichson) van Uden et Farinha ex van Uden et Buckley (1970) see *Pichia norvegensis*
130. *Candida norvegica* (Reiersöl) Meyer et Yarrow (Reiersöl, 1958)
131. *Candida oleophila* Montrocher (1967)
132. *Candida oregonensis* Phaff et do Carmo-Sousa (1962)
133. *Candida parapsilosis* (Ashford) Langeron et Talice (Ashford, 1928)
134. *Candida pararugosa* Nakase, Komagata et Fukazawa (1978)
135. *Candida pelliculosa* Redaelli (1925) see *Hansenula anomala*
136. *Candida peltata* (Yarrow) Meyer, Ahearn et Yarrow (Yarrow, 1968)
137. *Candida philyla* (van der Walt, van der Klift et Scott) Meyer et Yarrow (van der Walt et al., 1971b)
138. *Candida pignaliae* (Jacob) Meyer et Yarrow (Jacob, 1970)
139. *Candida pintolopesii* (van Uden) Meyer et Yarrow (van Uden, 1952)
    Candida pintolopesii (van Uden) Meyer et Yarrow var. *pintolopesii*
    Candida pintolopesii (van Uden) Meyer et Yarrow var. *slooffiae* (van Uden et do Carmo-Sousa) Mendonça-Hagler et Phaff ex Meyer, Ahearn et Yarrow (van Uden & do Carmo-Sousa, 1957a)
140. *Candida pinus* (Lodder et Kreger-van Rij) Meyer et Yarrow (Lodder & Kreger-van Rij, 1952)
141. *Candida polymorpha* Ohara et Nonomura ex Meyer et Ahearn (1983)
142. *Candida pseudointermedia* Nakase, Komagata et Fukazawa (1976)
143. *Candida psychrophila* (Goto, Sugiyama et Iizuka) Meyer et Yarrow (Goto et al., 1969)
144. *Candida pulcherrima* (Lindner) Windisch (Lindner, 1901) see *Metschnikowia pulcherrima*
145. *Candida pustula* (Buhagiar) Meyer et Yarrow (Buhagiar, 1975)
146. *Candida quercitrusa* (van Uden et do Carmo-Sousa) Meyer et Phaff (van Uden & do Carmo-Sousa, 1959)
147. *Candida quercuum* Nakase (1971b)
148. *Candida reukaufii* (Grüss) Diddens et Lodder (Grüss, 1918) see *Metschnikowia reukaufii*

149. *Candida rhagii* (Diddens et Lodder) Jurzitza, Kühlwein et Kreger-van Rij (Diddens & Lodder, 1942)
150. *Candida rugopelliculosa* Nakase (1971c)
151. *Candida rugosa* (Anderson) Diddens et Lodder (Anderson, 1917)
152. *Candida sake* (Saito et Ota) van Uden et Buckley (Saito & Ota, 1934)
153. *Candida salmanticensis* (Santa María) van Uden et Buckley (Santa María, 1963a)
154. *Candida santamariae* Montrocher (1967)
155. *Candida savonica* Sonck (1974)
156. *Candida schatavii* (Kocková-Kratochvílová et Ondrušová) Meyer et Yarrow (Kocková-Kratochvílová & Ondrušová, 1971)
157. *Candida scottii* Diddens et Lodder (1942) see *Leucosporidium scottii*
158. *Candida shehatae* Buckley et van Uden (1967)
159. *Candida silvae* Vidal-Leiria et van Uden (1963)
160. *Candida silvanorum* van der Walt, van der Klift et Scott (van der Walt et al., 1971a)
161. *Candida silvatica* (van der Walt, van der Klift et Scott) Meyer et Yarrow (van der Walt et al., 1971b)
162. *Candida silvicola* Shifrine et Phaff (1956) see *Hansenula holstii*
163. *Candida silvicultrix* van der Walt, Scott et van der Klift (1972)
164. *Candida solani* Lodder et Kreger-van Rij (1952)
165. *Candida sonckii* Hopsu-Havu, Tunnela et Yarrow (1978)
166. *Candida sonorensis* (Miller, Phaff, Miranda, Heed et Starmer) Meyer et Yarrow (Miller et al., 1976a)
167. *Candida sorbophila* (Nakase) Meyer et Yarrow (Nakase, 1975)
168. *Candida sorbosa* Hedrick et Burke ex van Uden et Buckley (1970) see *Issatchenkia occidentalis*
169. *Candida sorboxylosa* Nakase (1971c)
170. *Candida spandovensis* (Henninger et Windisch) Meyer et Yarrow (Henninger & Windisch, 1976b)
171. *Candida sphaerica* (Hammer et Cordes) Meyer et Yarrow (Hammer & Cordes, 1920) see *Kluyveromyces marxianus* var. *lactis*
172. *Candida steatolytica* Yarrow (1969a)
173. *Candida stellata* (Kroemer et Krumbholz) Meyer et Yarrow (in Krumbholz, 1931)
174. *Candida succiphila* Lee et Komagata (1980)
175. *Candida suecica* Rodrigues de Miranda et Norkrans (1968)
176. *Candida tannotolerans* (Jacob) Meyer et Yarrow (Jacob, 1970)
177. *Candida tenuis* Diddens et Lodder (1942)
178. *Candida tepae* Grinbergs (1967)
179. *Candida terebra* Sugiyama et Goto (1969)
180. *Candida torresii* (van Uden et Zobell) Meyer et Yarrow (van Uden & Zobell, 1962)
181. *Candida tropicalis* (Castellani) Berkhout (Castellani, 1910)
182. *Candida tsukubaensis* Onishi (1972)
183. *Candida utilis* (Henneberg) Lodder et Kreger-van Rij (Henneberg, 1926)
184. *Candida valdiviana* Grinbergs et Yarrow (1970a)
185. *Candida valida* (Leberle) van Uden et Buckley (Leberle, 1909)
186. *Candida vanderwaltii* (Vidal-Leiria) Meyer et Yarrow (Vidal-Leira, 1966b)

187. *Candida variabilis* (Lindner) Berkhout (Lindner, 1898) see *Pichia burtonii*
188. *Candida vartiovaarai* (Capriotti) van Uden et Buckley (Capriotti, 1961e)
189. *Candida veronae* Florenzano ex van Uden et Buckley (1970)
190. *Candida versatilis* (Etchells et Bell) Meyer et Yarrow (Etchells & Bell, 1950b)
191. *Candida vinaria* Ohara, Nonomura et Yunome ex Smith (1973)
192. *Candida vini* (Desmazières ex Lodder) van Uden et Buckley (Lodder, 1934)
193. *Candida viswanathii* Sandhu et Randhawa (1962)
194. *Candida wickerhamii* (Capriotti) Meyer et Yarrow (Capriotti, 1958g)
195. *Candida xestobii* Yarrow et Meyer (1978)
196. *Candida zeylanoides* (Castellani) Langeron et Guerra (Castellani, 1920)

d. KEYS TO THE SPECIES IN THE GENUS

For convenience in identification of the large number of species included in the genus *Candida,* the genus has been divided into 10 groups based on physiological properties. The groups are neither mutually exclusive nor necessarily composed of natural clusters of species, although Group I presumably includes only basidiomycetous anamorphs, whereas the remaining groups contain the anascosporogenous forms. Because of the variable nature of some species, they occur in more than one group and some species occur at multiple levels in a key. Use of a single characteristic often makes separation of some species difficult, thus the complete description must be consulted to verify the identification.

CANDIDA GROUP KEY

1a	DBB positive	Group I	p. 593	
b	DBB negative			2
2a	Inositol assimilated	Group II	p. 595	
b	Inositol not assimilated			3
3a	Nitrate assimilated	Group III	p. 596	
b	Nitrate not assimilated			4
4a	Erythritol assimilated			5
b	Erythritol not assimilated			6
5a	Raffinose assimilated	Group IV	p. 598	
b	Raffinose not assimilated	Group V	p. 599	
6a	Maltose assimilated			7
b	Maltose not assimilated			8

7a	Raffinose assimilated	Group VI	p. 601
b	Raffinose not assimilated	Group VII	p. 602
8a	Mannitol assimilated		9
b	Mannitol not assimilated	Group VIII	p. 604
9a	Galactose assimilated	Group IX	p. 605
b	Galactose not assimilated	Group X	p. 606

ABBREVIATIONS USED IN THE KEYS AND TABLE

For the abbreviations of the generic names see p. XI.

In Table 36, the indications weak and slow have been omitted; absence of fermentation of glucose (FG –) means that the other sugars are not fermented either.

Arb	splitting of arbutin
Ce	cellobiose
Cit	citric acid
Co-Q	coenzyme Q number
DAr	D-arabinose
DBB	color reaction with DBB
Er	erythritol
F	fermentation
G	glucose
Ga	galactose
Gal	galactitol
G + C	mol% guanine + cytosine
Glu	D-glucitol
Gly	glycerol
I	inositol
La	lactose
Lac	DL-lactic acid
LAr	L-arabinose
Ma	maltose
Man	D-mannitol
Me	melibiose
Mz	melezitose
NO_3	nitrate
Osm	50% glucose-yeast extract agar
Pell	pellicle formation after 3 days
PH	pseudomycelium
Psm	pseudomycelium
Ra	raffinose
Rh	L-rhamnose
Ri	D-ribose
Rib	ribitol
Sa	salicin
Sor	L-sorbose
St	soluble starch
Stpr	starch production
Su	sucrose
Suc	succinic acid
TH	true mycelium
Tr	trehalose
True myc	true mycelium
Ur	urease
Vit	growth in vitamin-free medium
X	D-xylose
20, 25, 37, 40	growth at 20, 25, 37, 40°C

GROUP I

DBB +

1	Nitrate +	2
	Nitrate –	22
2	Me +	3
	Me –	7
3	Er +	
	C. antarctica	p. 615
	Er –	4
4	Ma +	5
	Ma –	
	C. nivalis (see *Leu. nivalis)*	p. 502
5	La +	
	C. aquatica	p. 620
	La –	6
6	Su +	
	C. gelida (see *Leu. gelidium)*	p. 501
	Su –	
	C. javanica	p. 719
7(2)	Er +	8
	Er –	11
8	Ra +	9
	Ra –	10
9	La +	
	C. tsukubaensis	p. 821
	La –	
	C. bacarum	p. 629
10	Ce +	
	C. graminis	p. 690
	Ce –	
	C. diffluens	p. 664
11(7)	Vit +	12
	Vit –	17
12	Ce +	13
	Ce –	16
13	Ra +	14
	Ra –	
	C. foliorum	p. 680
14	Rh +	
	C. scottii (see *Leu. scottii)*	p. 504
	Rh –	15
15	St +	
	C. ingeniosa	p. 709
	St –	
	C. fragariorum	p. 681
16(12)	37°C +	
	C. foliorum	p. 680
	37°C –	
	Leu. antarcticum	p. 498
17(11)	Ma +	18
	Ma –	19
18	La +	
	C. muscorum	p. 754
	La –	
	Leu. stokesii	p. 507
19	Su +	20
	Su –	21

20 20°C +
C. curiosa p. 655
20°C −
C. frigida (see *Leu. frigidum)* p. 499

21 Ce +
C. pustula p. 778
Ce −
C. sonckii p. 802

22(1) I + 23
I − 29

23 Su + 24
Su − 28

24 Er + 25
Er − 27

25 Me + 26
Me −
C. curvata p. 657

26 Pell +
C. humicola p. 702
Pell −
C. amylolenta p. 613

27 La +
C. podzolica (Cr. laurentii) p. 843
La −
C. japonica (see *F. capsuligenum*) p. 485

28(23) Ma +
Holt. corniformis p. 548
Ma −
C. marina p. 735

29(22) Su + 30
Su − 33

30 X + 31
X −
C. auriculariae p. 624

31 Me + 32
Me −
Trem. globospora p. 553

32 True myc + , Psm +
C. humicola p. 702
True myc − , Psm −
Trem. fuciformis p. 552

33(29) Ma + 34
Ma − 35

34 La +
Trem. brasiliensis p. 549
La −
C. bogoriensis p. 634

35 Ce +
Rh. fujisanensis p. 905
Ce − 36

36 Tr + 37
Tr −
C. hylophila p. 705

37 Vit +
C. philyla p. 769
Vit − 38

38 Cit +
Trem. coalescens p. 550
Cit −
Trem. mesenterica p. 554

Table 26

Species in Group I

C. amylolenta	*C. fragariorum*	*C. muscorum*	*Leu. stokesii*
C. antarctica	*C. frigida*	*C. nivalis*	*Rh. fujisanensis*
C. aquatica	*C. gelida*	*C. philyla*	*Trem. brasiliensis*
C. auriculariae	*C. graminis*	*C. podzolica*	*Trem. coalescens*
C. bacarum	*C. humicola*	*C. pustula*	*Trem. fuciformis*
C. bogoriensis	*C. hylophila*	*C. scottii*	*Trem. globospora*
C. curiosa	*C. ingeniosa*	*C. sonckii*	*Trem. mesenterica*
C. curvata	*C. japonica*	*C. tsukubaensis*	
C. diffluens	*C. javanica*	*Holt. corniformis*	
C. foliorum	*C. marina*	*Leu. antarcticum*	

GROUP II

DBB − inositol +

1	Nitrate +		2
	Nitrate −		4
2	Er +		3
	Er −		
	C. valdiviana	p. 823	
3	37°C +		
	C. edax	p. 667	
	37°C −		
	C. incommunis	p. 707	
4(1)	Er +		5
	Er −		9
5	Ma +		6
	Ma −		
	C. auringiensis	p. 625	
6	True myc +		7
	True myc −		8
7	Suc +		
	C. ciferrii (see *Steph. ciferrii*)	p. 432	
	Suc −		
	C. chiropterorum	p. 651	
8	Pell +		
	C. hydrocarbofumarica	p. 704	
	Pell −		
	C. blankii	p. 633	
9(4)	Cit +		
	C. steatolytica	p. 807	
	Cit −		10
10	FMa +		
	C. inositophila	p. 712	
	FMa −		
	C. hellenica	p. 698	

Table 27

Species in Group II

C. auringiensis	*C. edax*	*C. inositophila*
C. blankii	*C. hellenica*	*C. steatolytica*
C. chiropterorum	*C. hydrocarbofumarica*	*C. valdiviana*
C. ciferrii	*C. incommunis*	

GROUP III

DBB – inositol – nitrate +

1 FG + 2
FG – 33

2 Ce + 3
Ce – 22

3 Ma + 4
Ma – 18

4 Rh + 5
Rh – 11

5 Vit + 6
Vit – 7

6 Er +
C. nitrativorans (see *H. sydowiorum*) p. 210
Er –
C. ernobii p. 673

7 St + 8
St – 10

8 FMa +
C. ishiwadae p. 718
FMa – 9

9 Su +
C. silvicola (see *H. holstii*) p. 192
Su –
C. molischiana p. 750

10(7) 37°C +
C. bimundalis var. *bimundalis* (see *H. bimundalis*) p. 178
37°C –
C. bimundalis var. *americana* (see *H. americana*) p. 171

11(4) Er + 12
Er – 14

12 Vit +
C. pelliculosa (see *H. anomala*) p. 174
Vit – 13

13 Su +
C. chilensis p. 650
Su –
C. molischiana p. 750

14(11) 37°C + 15
37°C – 16

15 St +
C. fabianii (see *H. fabianii*) p. 189
St –
C. utilis p. 822

16 Mz +
C. vartiovaarai p. 827
Mz – 17

17 Glu +
C. versatilis p. 831
Glu –
C. mannitofaciens p. 734

18(3) Ga + 19
Ga – 21

19 X + 20
X –
C. halophila p. 697

20 37°C +
C. methanosorbosa p. 746
37°C –
C. wickerhamii p. 837

21(18) 37°C +
C. berthetii p. 631
37°C –
C. norvegica p. 760

22(2) Er + 23
Er – 24

23 Rh +
C. cariosilignicola p. 646
Rh –
C. boidinii p. 635

24(22) Rh +
C. nitratophila p. 758
Rh – 25

25 Ma + 26
Ma – 30

26 Ga + 27
Ga –
C. globosa (see *Cit. matritensis)* p. 118

27 FGa +
C. versatilis p. 831
FGa – 28

28 FMa +
C. etchellsii p. 674
FMa – 29

29 Su +
C. multis-gemmis p. 752
Su –
C. nodaensis p. 759

30(25) Su + 31
Su – 32

31 FRa +
C. lactiscondensi p. 728
FRa –
C. magnoliae p. 731

32 Tr +
C. pignaliae p. 770
Tr –
C. halonitratophila p. 695

33(1) X + 34
X –
C. domercqii (see *Wick. domercqii)* p. 444

34 Ga +
C. vanderwaltii p. 826
Ga – 35

35 37°C +
C. melinii p. 741
37°C – 36

36 Mz +
C. buffonii p. 640
Mz –
C. norvegica p. 760

Table 28

Species in Group III

C. berthetii	*C. halonitratophila*	*C. nodaensis*
C. bimundalis var. *bimundalis*	*C. halophila*	*C. norvegica*
C. bimundalis var. *americana*	*C. ishiwadae*	*C. pelliculosa*
C. boidinii	*C. lactiscondensi*	*C. pignaliae*
C. buffonii	*C. magnoliae*	*C. silvicola*
C. cariosilignicola	*C. mannitofaciens*	*C. utilis*
C. chilensis	*C. melinii*	*C. vanderwaltii*
C. domercqii	*C. methanosorbosa*	*C. variovaarai*
C. ernobii	*C. molischiana*	*C. versatilis*
C. etchellsii	*C. multis-gemmis*	*C. wickerhamii*
C. fabianii	*C. nitrativorans*	
C. globosa	*C. nitratophila*	

GROUP IV

DBB – inositol – nitrate – erythritol + raffinose +

1	Rh +		2
	Rh –		4
2	Lac +		
	C. famata	p. 675	
	Lac –		3
3	FG + slow		
	C. insectorum	p. 715	
	FG + not slow		
	C. silvanorum	p. 797	
4(1)	La +		5
	La –		7
5	True myc +		6
	True myc –		
	C. famata	p. 675	
6	FMa +		
	C. fennica	p. 677	
	FMa –		
	C. entomophila	p. 670	
7(4)	Gal +		8
	Gal –		10
8	FG + strong		9
	FG + weak or –		
	C. famata	p. 675	
9	FMe +		
	C. friedrichii	p. 683	
	FMe –		
	C. membranaefaciens	p. 742	
10(7)	True myc +		
	C. variabilis (see *P. burtonii*)	p. 314	
	C. fennica	p. 677	
	True myc –		11
11	FG + strong		12
	FG + weak or –		
	C. famata	p. 675	
12	St +		
	C. silvicultrix	p. 799	
	St –		
	C. rhagii	p. 782	

Table 29

Species in Group IV

C. entomophila	*C. insectorum*	*C. silvicultrix*
C. famata	*C. membranaefaciens*	*C. variabilis*
C. fennica	*C. rhagii*	
C. friedrichii	*C. silvanorum*	

GROUP V

DBB – inositol – nitrate – erythritol + raffinose –

1	25°C +		2
	25°C –		
	C. psychrophila	p. 777	
2	Ma +		3
	Ma –		21
3	Gal +		4
	Gal –		8
4	La +		5
	La –		6
5	St +		
	C. shehatae	p. 794	
	St –		
	C. entomaea	p. 668	
6	37°C +		7
	37° –		
	C. dendronema	p. 661	
7	Psm +		
	C. terebra	p. 816	
	Psm –		
	C. peltata	p. 768	
8(3)	Ga +		9
	Ga –		
	C. mesenterica	p. 744	
9	37°C +		10
	37°C –		17
10	FG +		11
	FG –		
	C. aaseri	p. 608	
11	Rh +		12
	Rh –		16
12	La +		13
	La –		14
13	FMa +		
	C. shehatae	p. 794	
	FMa –		
	C. tenuis	p. 813	
14	Vit +		
	C. homilentoma	p. 700	
	C. tenuis	p. 813	
	Vit –		15
15	St +		
	C. naeodendra	p. 756	
	St –		
	C. veronae	p. 829	
16(11)	La +		
	C. shehatae	p. 794	
	La –		
	C. butyri	p. 642	
	C. diddensiae	p. 662	
17(9)	Su +		18
	Su –		
	C. polymorpha	p. 774	

18	Rh +		19
	Rh –		20
19	FGa +		
	C. tenuis	p. 813	
	FGa –		
	C. atlantica	p. 621	
20	La +		
	C. ergatensis	p. 672	
	La –		
	C. atmospherica	p. 623	
21(2)	FG +		22
	FG –		27
22	Ce +		23
	Ce –		24
23	37°C +		
	C. cacaoi	p. 643	
	37°C –		
	C. conglobata	p. 654	
24	La +		
	C. succiphila	p. 810	
	La –		25
25	Vit +		26
	Vit –		
	C. boleticola	p. 637	
26	Ga +		
	C. schatavii	p. 793	
	Ga –		
	C. cantarellii	p. 645	
27(21)	True myc +		28
	True myc –		29
28	Ce +		
	C. deformans	p. 659	
	Ce –		
	C. lipolytica (see *S. lipolytica*)	p. 406	
29	Ce +		
	C. nemodendra	p. 757	
	Ce –		
	C. pinus	p. 773	

Table 30

Species in Group V

C. aaseri	*C. diddensiae*	*C. polymorpha*
C. atlantica	*C. entomaea*	*C. psychrophila*
C. atmospherica	*C. ergatensis*	*C. schatavii*
C. boleticola	*C. homilentoma*	*C. shehatae*
C. butyri	*C. lipolytica*	*C. succiphila*
C. cacaoi	*C. mesenterica*	*C. tenuis*
C. cantarellii	*C. naeodendra*	*C. terebra*
C. conglobata	*C. nemodendra*	*C. veronae*
C. deformans	*C. peltata*	
C. dendronema	*C. pinus*	

GROUP VI

DBB − inositol − nitrate − erythritol − maltose + raffinose +

1	Ga +	2
	Ga −	18
2	Man +	3
	Man − *C. xestobii*	p. 838
3	FG +	4
	FG − *C. famata*	p. 675
	C. glaebosa	p. 688
4	Ce +	5
	Ce −	15
5	Mz +	6
	Mz − *P. ohmeri*	p. 348
6	Me +	7
	Me −	11
7	37°C +	8
	37°C −	10
8	LAr +	9
	LAr − *C. melibiosica*	p. 739
9	FG + weak, slow or − *C. famata*	p. 675
	FG +, not weak or slow *C. guilliermondii*	p. 692
10(7)	FMa + *C. salmanticensis*	p. 789
	FMa − *C. famata*	p. 675
11(6)	La +	12
	La −	14
12	FLa + *C. sphaerica* (see *K. marxianus* var. *lactis*)	p. 241
	FLa −	13
13	FG + weak, slow or − *C. famata*	p. 675
	FG +, not weak or slow *C. intermedia*	p. 716
14(11)	FGa + *C. pseudointermedia*	p. 776
	FGa − *C. famata*	p. 675
15(4)	Mz +	16
	Mz −	17
16	X + *C. haemulonii*	p. 694
	X − *C. dattila* (see *K. thermotolerans*)	p. 248
17	LAr + *C. mogii*	p. 748
	LAr − *C. colliculosa* (see *T'spora delbrueckii*)	p. 436
18(1)	Ce + *C. maritima*	p. 738
	Ce − *C. colliculosa* (see *T'spora delbrueckii*)	p. 436

Table 31

Species in Group VI

C. colliculosa	*C. haemulonii*	*C. pseudointermedia*
C. dattila	*C. intermedia*	*C. salmanticensis*
C. famata	*C. maritima*	*C. sphaerica*
C. glaebosa	*C. melibiosica*	*C. xestobii*
C. guilliermondii	*C. mogii*	*P. ohmeri*

GROUP VII

DBB – inositol – nitrate – erythritol – maltose + raffinose –

1	Ga +		2
	Ga –		23
2	St +		3
	St –		9
3	La +		4
	La –		5
4	FGa +		
	C. shehatae	p. 794	
	FGa –		
	C. fluviatilis	p. 679	
5(3)	FMa +		6
	FMa –		8
6	FSu +		
	C. tropicalis	p. 819	
	FSu –		7
7	Ce +		
	C. viswanathii	p. 835	
	Ce –		
	C. albicans	p. 611	
8(5)	Ce +		
	C. kruisii	p. 724	
	Ce –		
	C. catenulata	p. 648	
9(2)	37°C +		10
	37°C –		17
10	Ce +		11
	Ce –		15
11	FG +		12
	FG –		
	C. mucilagina	p. 751	
12	FLa +		
	C. sphaerica (see *K. marxianus* var. *lactis)*	p. 241	
	FLa –		13
13	Rh +		
	C. lusitaniae (see *Cl. lusitaniae)*	p. 121	
	Rh –		14
14	FMa +		
	C. maltosa	p. 733	
	FMa –		
	C. pulcherrima (see *M. pulcherrima)*	p. 273	
15(10)	FG +		16
	FG –		
	C. azyma	p. 628	
16	LAr +		
	C. parapsilosis	p. 765	
	LAr –		
	C. quercitrusa	p. 780	

17(9)	FG +	18
	FG –	
	C. tepae p. 815	
18	La +	
	C. xestobii p. 838	
	La –	19
19	Ce +	20
	Ce –	22
20	Pell +	
	C. oleophila p. 762	
	Pell –	21
21	DAr +	
	C. buinensis p. 641	
	DAr –	
	C. pulcherrima (see *M. pulcherrima)* p. 273	
	C. reukaufii (see *M. reukaufii)* p. 275	
	C. sake p. 787	
22(19)	Psm +	
	C. sake p. 787	
	Psm –	
	C. colliculosa (see *T'spora delbrueckii)* p. 436	
23(1)	37°C +	24
	37°C –	27
24	Su +	25
	Su –	
	C. insectamans p. 714	
25	Rh +	26
	Rh –	
	C. quercuum p. 781	
26	Rib +	
	C. lusitaniae (see *Cl. lusitaniae)* p. 121	
	Rib –	
	C. freyschussii p. 682	
27(23)	Man +	28
	Man –	
	C. solani p. 801	
28	Mz +	29
	Mz –	30
29	Ce +	
	C. oregonensis p. 763	
	Ce –	
	C. musae p. 753	
30	Psm +	
	C. suecica p. 811	
	Psm –	
	C. colliculosa (see *T'spora delbrueckii)* p. 436	

Table 32

Species in Group VII

C. albicans	*C. insectamans*	*C. oregonensis*	*C. shehatae*
C. azyma	*C. kruisii*	*C. parapsilosis*	*C. solani*
C. buinensis	*C. lusitaniae*	*C. pulcherrima*	*C. sphaerica*
C. catenulata	*C. maltosa*	*C. quercitrusa*	*C. suecica*
C. colliculosa	*C. mucilaginosa*	*C. quercuum*	*C. tepae*
C. fluviatilis	*C. musae*	*C. reukaufii*	*C. tropicalis*
C. freyschussii	*C. oleophila*	*C. sake*	*C. xestobii*

GROUP VIII

DBB – inositol – nitrate – erythritol – maltose – D-mannitol –

1 Ga + 2
Ga – 8

2 Vit +
C. ingens p. 710
Vit – 3

3 FG + 4
FG – 7

4 Su + 5
Su – 6

5 La +
C. kefyr p. 722
La –
C. holmii p. 699
C. milleri p. 747

6 Tr +
C. humilis p. 703
Tr –
C. tannotolerans p. 812

7(3) Sor +
C. vinaria p. 832
Sor –
C. austromarina p. 627

8(1) FSu +
C. stellata p. 809
FSu – 9

9 Tr + 10
Tr – 11

10 FTr +
C. glabrata p. 687
FTr –
C. castellii p. 647

11 Ce + 12
Ce – 13

12 37°C +
C. norvegensis (see *P. norvegensis*) p. 347
37°C –
C. dendrica p. 660

13 X + 14
X – 16

14 Sor +
C. sorboxylosa p. 805
Sor – 15

15 Pell +
C. lambica p. 730
Pell –
C. citrea p. 653

16(13) 37°C + 17
37°C – 22

17 25°C + 18
25°C –
C. pintolopesii var. *slooffiae* p. 772

18 Vit + 19
Vit – 21

19 Gly + 20
Gly –
C. rugopelliculosa p. 784

20 40°C +
C. krusei p. 726
40°C –
C. sorbosa (see *I. occidentalis*) p. 216

21(18) FG + *C. pintolopesii* var. *pintolopesii* p. 771
FG − *C. inconspicua* p. 708

22(16) FG + slow *C. valida* p. 825
FG + not slow *C. citrea* p. 653

Table 33

Species in Group VIII

C. austromarina	*C. ingens*	*C. rugopelliculosa*
C. castellii	*C. kefyr*	*C. sorbosa*
C. citrea	*C. krusei*	*C. sorboxylosa*
C. dendrica	*C. lambica*	*C. stellata*
C. glabrata	*C. milleri*	*C. tannotolerans*
C. holmii	*C. norvegensis*	*C. valida*
C. humilis	*C. pintolopesii* var. *pintolopesii*	*C. vinaria*
C. inconspicua	*C. pintolopesii* var. *slooffiae*	

GROUP IX

DBB − inositol − nitrate − erythrithol − maltose − D-mannitol + galactose +

1 La + 2
La − 3

2 Mz + *C. sphaerica* (see *K. marxianus* var. *lactis)* p. 241
Mz − *C. kefyr* p. 722

3 Su + 4
Su − 11

4 FG + 5
FG − 9

5 37°C + 6
37°C − 7

6 Cit + *C. geochares* p. 686
Cit − *C. bombicola* p. 639

7 Ce + *C. gropengiesseri* p. 691
Ce − 8

8 LAr + *C. spandovensis* p. 806
LAr − *C. colliculosa* (see *T'spora delbrueckii)* p. 436

9(4) 37°C + *C. pararugosa* p. 766
37°C − 10

10 Ra + *C. apis* var. *apis* p. 618
Ra − *C. apis* var. *galacta* p. 619

11(3) 37°C + 12
37°C − 15

12	Tr +		13
	Tr –		14
13	FG +		
	C. torresii	p. 817	
	FG –		
	C. insectalens	p. 713	
14	Lac +		
	C. rugosa	p. 785	
	Lac –		
	C. sorbophila	p. 804	
15(11)	Cit +		16
	Cit –		18
16	FGa +		
	C. savonica	p. 792	
	FGa –		17
17	FG + weak or –		
	C. krissii	p. 723	
	C. zeylanoides	p. 839	
	FG + not weak		
	C. santamariae	p. 790	
18(15)	FG +		19
	FG –		
	C. vinaria	p. 832	
19	Ce +		
	C. torresii	p. 817	
	Ce –		
	C. colliculosa (see *T'spora delbrueckii)*	p. 436	

Table 34

Species in Group IX

C. apis var. *apis*	*C. kefyr*	*C. spandovensis*
C. apis var. *galacta*	*C. krissii*	*C. sphaerica*
C. bombicola	*C. pararugosa*	*C. torresii*
C. colliculosa	*C. rugosa*	*C. vinaria*
C. geochares	*C. santamariae*	*C. zeylanoides*
C. gropengiesseri	*C. savonica*	
C. insectalens	*C. sorbophila*	

GROUP X

DBB – inositol – nitrate – erythritol – maltose – D-mannitol + galactose –

1	Tr +		2
	Tr –		10
2	37°C +		3
	37°C –		5
3	Ce +		
	C. methanolovescens	p. 745	
	Ce –		4
4	FG +		
	C. bombi	p. 638	
	FG –		
	C. silvatica	p. 798	
5(2)	Rh +		
	C. anatomiae	p. 614	
	Rh –		6

6	FG + weak or –	
	C. krissii	p. 723
	C. zeylanoides	p. 839
	FG + not weak	7
7	Ce +	8
	Ce –	9
8	Ri +	
	C. beechii	p. 630
	Ri –	
	C. santamariae	p. 790
9	Cit +	
	C. fructus	p. 685
	Cit –	
	C. colliculosa (see *T'spora delbrueckii)*	p. 436
10(1)	FSu +	11
	FSu –	12
11	Suc +	
	C. bombi	p. 638
	Suc –	
	C. apicola	p. 617
12	Sor +	
	C. krissii	p. 723
	Sor –	13
13	Ce +	
	C. sonorensis	p. 803
	Ce –	14
14	Rh +	
	C. maris	p. 736
	Rh –	15
15	FG + weak or –	16
	FG + not weak	
	C. diversa	p. 665
16	Psm +	
	C. silvae	p. 796
	C. vini	p. 834
	Psm –	
	C. karawaiewii	p. 720

Table 35

Species in Group X

C. anatomiae	*C. fructus*	*C. silvae*
C. apicola	*C. karawaiewii*	*C. silvatica*
C. beechii	*C. krissii*	*C. sonorensis*
C. bombi	*C. maris*	*C. vini*
C. colliculosa	*C. methanolovescens*	*C. zeylanoides*
C. diversa	*C. santamariae*	

Table 36 *(see pocket on inside cover)*

e. SYSTEMATIC DISCUSSION OF THE SPECIES

1. *CANDIDA AASERI* DIETRICHSON EX VAN UDEN ET BUCKLEY (1970)

Synonym

Azymocandida aaseri nom. nud. (Novák & Zsolt, 1961)

Standard description of *Candida aaseri* (van Uden & Buckley, 1970)

Growth in glucose-yeast extract-peptone water: After 3 days at 25°C the cells are elongate and of irregular shape, (2–3) × (5–12) μm (Fig. 170). A delicate, powdery pellicle may be formed.

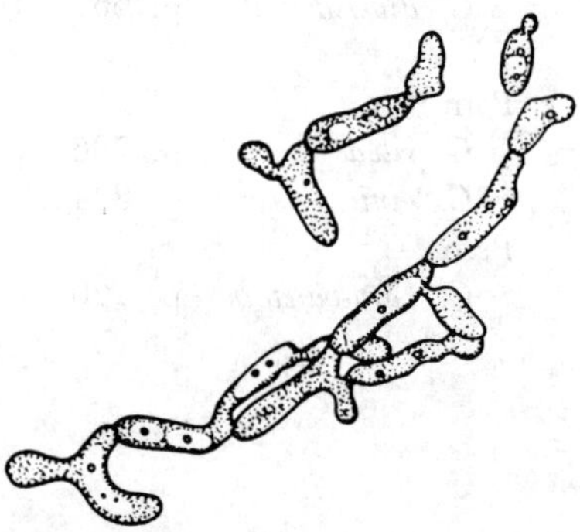

Fig. 170 – *C. aaseri*
In glucose-yeast extract-peptone water
(From 'The Yeasts' ed. II, 1970)

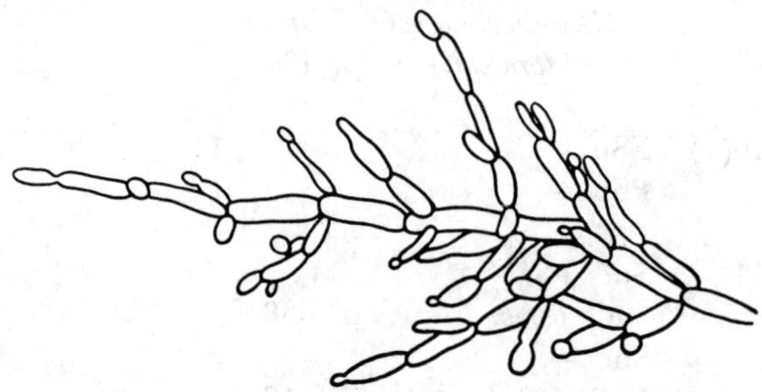

Fig. 171 – *C. aaseri*
Dalmau plate culture on corn meal agar
(From 'The Yeasts' ed. II, 1970)

Growth on glucose-yeast extract-peptone agar: After one month at 25°C the streak culture is cream-colored, semidull and soft; the surface has a delicate, punctilated structure.

Dalmau plate culture on corn meal agar: A primitive pseudomycelium is abundantly formed. It consists of short pseudohyphae and cylindrical cells in a tree-like arrangement (Fig. 171).

Fermentation: –

Assimilation of carbon compounds:

Galactose	+	Melezitose	+	Ribitol	+
L-Sorbose	+	Soluble starch	–	Galactitol	–
Sucrose	+	D-Xylose	+	D-Mannitol	+
Maltose	+	L-Arabinose	+	D-Glucitol	+
Cellobiose	+	D-Arabinose	–	Salicin	+
Trehalose	+	D-Ribose	+	DL-Lactic acid	–
Lactose	+ s	L-Rhamnose	–	Succinic acid	+
Melibiose	–	Glycerol	+	Citric acid	+
Raffinose	–	Erythritol	+	Inositol	–

Splitting of arbutin: +

Assimilation of nitrate: –
Growth in vitamin-free medium: –
Growth at 37°C: +
Color reaction with DBB: –
G + C: 37.1 mol%, type strain (Meyer & Phaff, 1972)

Origin of the strain studied

From sputum (Dietrichson, 1954) (1).

Type strain: CBS 1913.

2. *CANDIDA ALBICANS* (ROBIN) BERKHOUT (1923)

Synonyms

Oidium albicans Robin (1853)
Saccharomyces albicans (Robin) Reess (1877)
Dematium albicans (Robin) Laurent (1889)
Monilia albicans (Robin) Zopf (1890)
Parasaccharomyces albicans (Robin) Froilano de Mello et Gonzaga Fernandes (1918)
Myceloblastanon albicans (Robin) Ota (1928)
Mycotorula albicans (Robin) Langeron et Talice (1932)
Syringospora albicans (Robin) Dodge (1935)
Procandida albicans (Robin) Novák et Zsolt (1961)
Syringospora robinii Quinquaud (1868)
Saccharomyces tumefaciens-albus Foulerton (1900)
Monilia tumefaciens-alba (Foulerton) Ota (1924c)
Myceloblastanon tumefaciens-album (Foulerton) Ota (1928)
Endomyces pinoyi Cast. (1912a)
Monilia pinoyi (Cast.) Cast. et Chalmers (1913)
Myceloblastanon pinoyi (Cast.) Ota (1928)
Blastodendrion pinoyi (Cast.) Langeron et Talice (1932)
Mycotorula pinoyi (Cast.) Saggese (1934)
Monilia decolorans Cast. et Low (1913)
Myceloblastanon decolorans (Cast. et Low) Ota (1928)
Castellania decolorans (Cast. et Low) Dodge (1935)
Endomyces pulmonalis Cast. (1913)
Monilia pulmonalis (Cast.) Cast. et Chalmers (1913)
Candida pulmonalis (Cast.) Basgal (1931)
Castellania pulmonalis (Cast.) Dodge (1935)
Monilia psilosis Ashf. (1917)
Myceloblastanon psilose (Ashf.) Ota (1928)
Candida psilosis (Ashf.) de Almeida (1933)
Syringospora psilosis (Ashf.) Dodge (1935)
Parasaccharomyces ashfordi Anderson (1917)
Myceloblastanon ashfordi (Anderson) Ota (1928)
Monilia alba Cast. et Chalmers (1919)
Castellania alba (Cast. et Chalmers) Dodge (1935)
Monilia metalondinensis Cast. et Chalmers var. *alba* (Cast. et Chalmers) Cast. (1937a)
Monilia metalondinensis Cast. et Chalmers (1919)

Candida metalondinensis (Cast. et Chalmers) Berkhout (1923)
Myceloblastanon metalondinense (Cast. et Chalmers) Ota (1928)
Castellania metalondinensis (Cast. et Chalmers) Dodge (1935)
Monilia nabarroi Cast. et Chalmers (1919)
Myceloblastanon nabarroi (Cast. et Chalmers) Ota (1928)
Castellania nabarroi (Cast. et Chalmers) Dodge (1935)
Monilia pinoyi (Cast.) Cast. et Chalmers var. *nabarroi* (Cast. et Chalmers) Cast. (1937a)
Monilia pseudolondinensis Cast. et Chalmers (1919)
Castellania pseudolondinensis (Cast. et Chalmers) Dodge (1935)
Monilia metalondinensis Cast. et Chalmers var. *pseudolondinensis* (Cast. et Chalmers) Cast. (1937a)
Monilia pseudolondinoides Cast. et Chalmers (1919)
Castellania pseudolondinoides (Cast. et Chalmers) Dodge (1935)
Monilia pseudometalondinensis Cast. et Chalmers (1919)
Castellania pseudometalondinensis (Cast. et Chalmers) Dodge (1935)
Cryptococcus copellii Neveu-Lemaire (1921)
Myceloblastanon copellii (Neveu-Lemaire) Ota (1928)
Torulopsis copellii (Neveu-Lemaire) de Almeida (1933)
Castellania copellii (Neveu-Lemaire) Dodge (1935)
Monilia butantanensis Gomes (1924a,b)
Candida butantanensis (Gomes) Langeron et Talice (1932)
Parendomyces butantanensis (Gomes) Dodge (1935)
Myceloblastanon cutaneum Ota (1924a)
Monilia cutanea (Ota) Nann. (1934)
[non *Monilia cutanea* (de Beurm., Gougerot et Vaucher) Cast. et Chalmers (1913) and synonyms]
Blastodendrion cutaneum (Ota) Dodge (1935)
Myceloblastanon gruetzii Ota (1924a)
Oidiomyces unguium nom. nud. (Ota, 1924c)
Onychomyces unguium nom. nud.
Myceloblastanon favrei Ota (1925)
Cryptococcus favrei (Ota) Poll. et Nann. (1930)
Blastodendrion favrei (Ota) Langeron et Talice (1932)
Candida favrei (Ota) de Almeida (1933)
Monilia richmondi Shaw (1926)
Castellania richmondi (Shaw) Dodge (1935)
Monilia aldoi Pereira Filho (1927)
Mycotoruloides aldoi (Pereira Filho) Langeron et Talice (1932)
Candida aldoi (Pereira Filho) Cast. et Jacono (1933)
Monilia periunguealis Niño (1930)
Parendomyces periunguealis (Niño) Dodge (1935)
Mycotorula periunguealis Niño (1938)
Monilia alvarezsotoi Mazza et Niño (in Mazza et al., 1931)
Zymonema alvarezsotoi (Mazza et Niño) Dodge (1935)
Mycotorula alvarezsotoi (Mazza et Niño) Niño (1938)
Blastodendrion erectum Langeron et Talice (1932)
Mycotoruloides ovalis Langeron et Talice (1932)
Mycotoruloides triadis Langeron et Talice (1932)
Monilia triadis (Langeron et Talice) Cast. (1937a)
Candida triadis (Langeron et Talice) Langeron et Guerra (1938)
Blastodendrion oosporoides Zach (in Wolfram & Zach, 1933)
Parasaccharomyces oosporoides (Zach) Dodge (1935)
Cryptococcus pinoyisimilis Cast. (1933)
Blastodendrion pinoyisimilis (Cast.) Cast. et Jacono (1933)
Candida pinoyisimilis (Cast.) Cast. et Jacono (1933)
Monilia pinoyisimilis (Cast.) Cast. et Jacono (1933)
Mycocandida pinoyisimilis (Cast.) Red. et Cif. (Ciferri & Redaelli, 1935)
Candida desidiosa Cif. et Red. (1935)

Mycoderma desidiosum (Cif. et Red.) Dodge et Moore (1936)
Candida mycotoruloidea Red. et Cif. (Ciferri & Redaelli, 1935)
Parasaccharomyces colardi Dodge (1935)
Syringospora negroni Dodge (1935)
Monilia stellatoidea Jones et Martin (1938)
Candida stellatoidea (Jones et Martin) Langeron et Guerra (1939a)
Candida albicans (Robin) Berkhout var. *stellatoidea* (Jones et Martin) Diddens et Lodder (1942)
Procandida stellatoidea (Jones et Martin) Novák et Zsolt (1961)
Syringospora stellatoidea van der Walt (1970k)
Mycotorula albicans (Robin) Langeron et Talice var. *vuillemini* (Landrieu) Red., Cif. et Cavallero (1939)
Candida truncata nom. nud. (Vanbreuseghem, 1948)
Candida intestinalis Batista et Silveira (1959c)
Candida biliaria Batista et Silveira (1959c)
Candida genitalis Batista et Silveira (1962)
Candida claussenii Lodder et Kreger van Rij (1952)
Candida albicans (Robin) Berkhout var. *claussenii* (Lodder et Kreger van Rij) Verona et Montemartini (1959)
Syringospora claussenii van der Walt (1970k)
Procandida langeronii nod. num. (Novák & Zsolt, 1961)
Candida langeronii Dietrichson ex van Uden et Buckley (1970)
Candida nouvelii Saëz (1973)

Standard description of *Candida albicans*

Growth in glucose-yeast extract-peptone water: After 3 days at 25°C the cells are usually globose or short-ovoid, sometimes long-oval to elongate, (2.9–7.2) × (2.9–14.4) μm, and occur singly, budding, in short chains and in clusters (Fig. 172). A sediment is present. After one month a ring and a sediment are present.

Growth on glucose-yeast extract-peptone agar: After one month at 25°C the streak culture is cream-colored, glistening, or somewhat waxy, soft and usually smooth. Some strains may become wrinkled and have a mycelial border.

Dalmau plate culture on corn meal agar: Pseudomycelium is abundantly formed and true mycelium may be found. Generally, pseudomycelium of the "*Mycotorula*" type is present – grapelike clusters of blastospores occurring along the

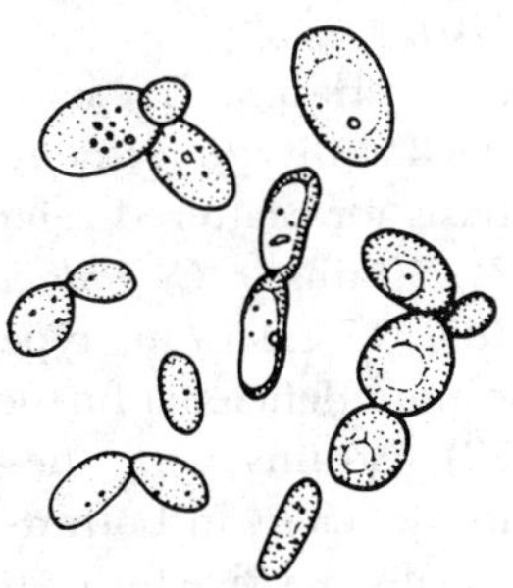

Fig. 172 – *C. albicans*
In glucose-yeast extract-peptone water
(From 'The Yeasts' ed. II, 1970)

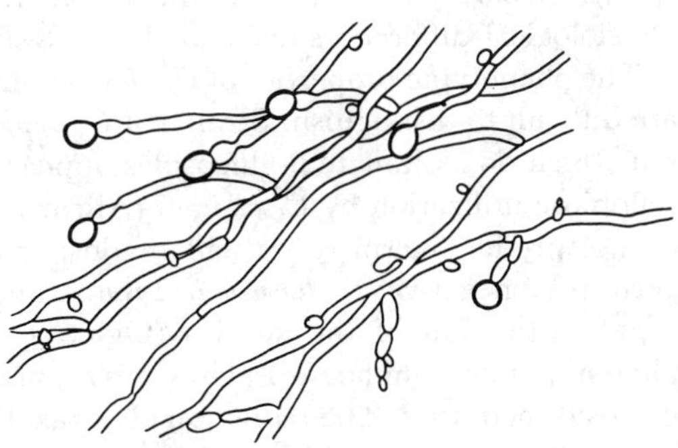

Fig. 173 – *C. albicans*
Dalmau plate culture on corn meal agar
(From 'The Yeasts' ed. II, 1970)

pseudohyphal chains. Also, pseudomycelium with branched chains of cells and blastospores occurring at irregular intervals may be observed (Fig. 173). Chlamydospores may be formed.

In sera, other tissue fluids and various defined media incubated at 37°C: Yeast cells of typical strains germinate to produce "germ tubes", thin filaments that lack a constriction at their point of origin.

Fermentation:

Glucose	+	Maltose	+
Galactose	v	Lactose	–
Sucrose	– (+)		

Assimilation of carbon compounds:

Galactose	+	Melezitose	v	Ribitol	– or + s
L-Sorbose	v	Soluble starch	+	Galactitol	–
Sucrose	+(–)	D-Xylose	+	D-Mannitol	+
Maltose	+	L-Arabinose	v	D-Glucitol	+(–)
Cellobiose	–	D-Arabinose	– or + s	Salicin	–
Trehalose	+	D-Ribose	– or + s	DL-Lactic acid	v
Lactose	–	L-Rhamnose	–	Succinic acid	v
Melibiose	–	Glycerol	v	Citric acid	v
Raffinose	–	Erythritol	–	Inositol	–

Splitting of arbutin: –
Assimilation of nitrate: –
Growth in vitamin-free medium: –
Growth at 37°C: +
Color reaction with DBB: –
G + C: 34.3–35.6 mol%, 15 strains (Meyer, unpublished)

Comments

Using DNA reassociation techniques Meyer (1979) found a significant degree of DNA relatedness between the type strains of *C. albicans, C. stellatoidea, C. claussenii* and *C. langeronii* and recommended that these species be reduced to synonymy. Previously, species separation was determined on the basis of a few morphological and physiological differences (see van Uden & Buckley, 1970).

The phenotypic properties of *C. albicans* and *C. tropicalis* overlap so that some strains are difficult to distinguish. Strains of *C. tropicalis* which lack ability to ferment sucrose but which do assimilate β-glucosides appear to be the basis for scattered reference to cellobiose utilization by *C. albicans* (Ahearn et al., 1977). Strains of *C. albicans* which do not produce germ tubes, chlamydospores or both (e.g. *C. claussenii,* type strain, germ tube negative; *C. langeronii,* type strain, chlamydospore deficient) further cloud rapid distinction of the species (Ogletree et al., 1978). Strains with these latter phenotypes are similar to isolates of *C. tropicalis* that are deficient in both α- and β-glucoside activities. The reduction of tetrazolium dyes, usually positive for *C. tropicalis,* has been used as an aid in distinguishing such strains (Schlitzer & Ahearn, 1982). Pseudohyphal cells that resemble germ tubes are frequently produced by *C. tropicalis* (Bowman & Ahearn, 1975). Also, germ tube production has been reported for rare oral isolates of *C. tropicalis* (Martin & White, 1981). These latter isolates have not been subjected to DNA complementarity studies with *C. albicans*. Strains of *C. tropicalis* which assimilate sucrose and β-glucosides but do not ferment sucrose are phenotypically similar to *C. viswanathii*.

Origin of the strains studied

Skin disorder (1), blood (14), seawater (12), sputum (8), mouths (6), vagina (7), animal feces (1), miscellaneous clinical (3), unknown (1).

Type strain: CBS 562, isolated in Uruguay from a skin disorder (MacKinnon, 1936).

3. *CANDIDA AMYLOLENTA* VAN DER WALT, SCOTT ET VAN DER KLIFT (1972)

Synonym

Cryptococcus amylolentus (van der Walt, Scott et van der Klift) Golubev (1981)

Standard description of *Candida amylolenta*

Growth in glucose-yeast extract-peptone water: After 3 days at 25°C the cells are spheroidal, ovoidal or ellipsoidal (2.8–5.6) × (3.9–8.4) μm (Fig. 174).

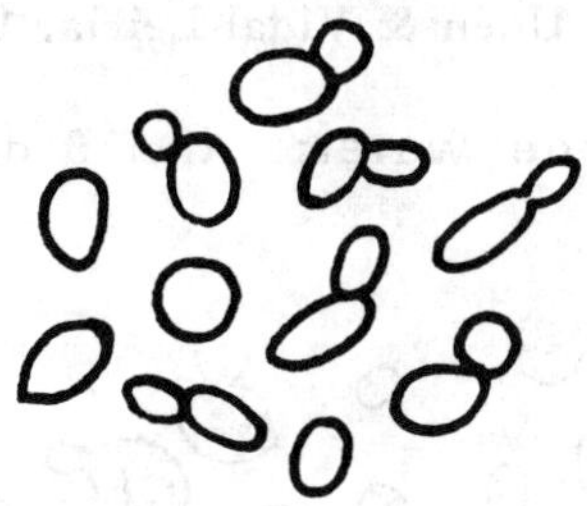

Fig. 174 – *C. amylolenta*
In glucose-yeast extract-peptone water

Growth on glucose-yeast extract-peptone agar: After one month at 25°C the streak culture is white to cream-colored, dull and smooth. The margin is entire to undulating.

Dalmau plate culture on corn meal agar: The pseudomycelium consists of ramified chains of ovoid to elongate cells, bearing clusters of short-ovoid blastospores in verticillate positions.

Fermentation: –

Assimilation of carbon compounds:

Galactose	+	Melezitose	+ or s	Ribitol	+
L-Sorbose	v	Soluble starch	+	Galactitol	v
Sucrose	+	D-Xylose	+	D-Mannitol	+
Maltose	+	L-Arabinose	+	D-Glucitol	+
Cellobiose	+	D-Arabinose	+ or s	Salicin	+
Trehalose	+	D-Ribose	+	DL-Lactic acid	–
Lactose	–	L-Rhamnose	+	Succinic acid	+
Melibiose	+ or s	Glycerol	+	Citric acid	+
Raffinose	+	Erythritol	+	Inositol	+

Splitting of arbutin: +
Assimilation of nitrate: –
Growth in vitamin-free medium: –
Growth on 50% (w/w) glucose-yeast extract agar: +
Growth at 37°C: v
Starch formation: +
Urease: +
Color reaction with DBB: +
G + C: 55.1 mol%, type strain (Meyer, Stenderup & Christiansen, unpublished)

Origin of the strains studied

Frass of the tunnels of *Enneadesmus forficulus* Fairm. in *Dombeya rotundifolia* (Hochst.) Harv. near Grobersdal in the Transvaal (1); frass from the tunnels of *Sinoxylon ruficorne* Fahr. infesting *Dichrostachys cinerea* L. (Wright et Arn.) near Grobersdal in the Transvaal (1).

Type strain: CBS 6039, from frass of the tunnels of *Enneadesmus forficulus*.

4. *CANDIDA ANATOMIAE* (ZWILLENBERG) MEYER ET YARROW (Yarrow & Meyer, 1978)

Synonym

Torulopsis anatomiae Zwillenberg (1966)

Standard description of *Candida anatomiae* (van Uden & Vidal-Leiria, 1970)

Growth in glucose-yeast extract-peptone water: After 3 days at 25°C the cells are spheroidal to short-ovoidal, (2–6) × (3–8) μm, single, in pairs or in short chains (Fig. 175). Elongate cells may occur. A ring may be present.

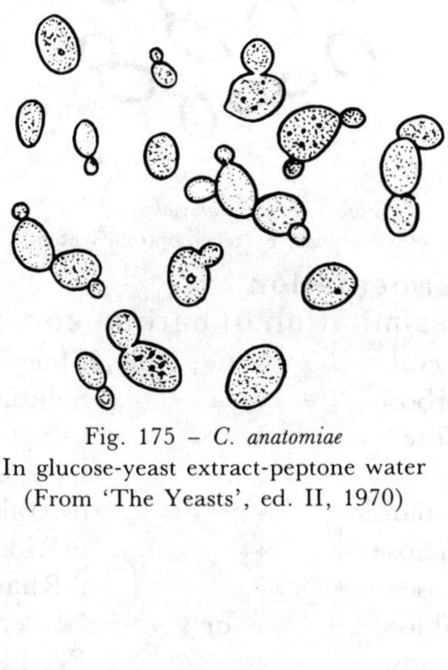

Fig. 175 – *C. anatomiae*
In glucose-yeast extract-peptone water
(From 'The Yeasts', ed. II, 1970)

Growth on glucose-yeast extract-peptone agar: After one month at 25°C the streak culture is whitish cream-colored, glossy, soft and smooth.

Dalmau plate culture on corn meal agar: A rudimentary pseudomycelium consisting of branched chains of elongate cells may be formed in small amounts.

Fermentation:

Glucose	+	Maltose	–
Galactose	–	Lactose	–
Sucrose	–		

Assimilation of carbon compounds:

Galactose	–	Melezitose	–	Ribitol	–
L-Sorbose	–	Soluble starch	–	Galactitol	–
Sucrose	–	D-Xylose	–	D-Mannitol	+
Maltose	–	L-Arabinose	–	D-Glucitol	–
Cellobiose	+	D-Arabinose	–	Salicin	+
Trehalose	+	D-Ribose	+ w	DL-Lactic acid	–
Lactose	–	L-Rhamnose	+	Succinic acid	v
Melibiose	–	Glycerol	–	Citric acid	v
Raffinose	–	Erythritol	–	Inositol	–

Splitting of arbutin: +

Assimilation of nitrate: –

Growth in vitamin-free medium: –

Growth at 37°C: –

Color reaction with DBB: –

G + C: 39.0 mol%, type strain (Meyer, Stenderup & Christiansen, unpublished)

Origin of the strain studied

Formalin (1%, w/v) solution (1).

Type strain: CBS 5547.

5. *CANDIDA ANTARCTICA* (GOTO, SUGIYAMA ET IIZUKA) KURTZMAN, SMILEY, JOHNSON ET HOFFMAN (1983)

Synonyms

Sporobolomyces antarcticus Goto, Sugiyama et Iizuka (1969)
Trichosporon oryzae Ito, Iizuka et Sato (1974)
Sterigmatomyces aphidis Henninger et Windisch (1975b)

Standard description of *Candida antarctica*

Growth in glucose-yeast extract-peptone water: After 3 days at 25°C the cells are elongate, cylindrical, (3.2–3.3) × (7.2–10.8) μm, slender, often pointed at one or both ends, occurring singly, budding and in chains. Long cells, some curved and of irregular shapes are also present (2.8–3.3) × (19.6–30.8) μm (Fig. 176). A ring and sediment are formed. After one month a thick pellicle is present along with a dense sediment.

Growth on glucose-yeast extract-peptone agar: After one month at 25°C the streak culture is brownish, wrinkled, crusty, with an irregular border and with mycelial growth.

Dalmau plate culture on corn meal agar: Extensive mycelial development consists of septate hyphae which are long, branched, often wavy and in some strains fragmented. Chains of cylindrical blastospores are present in some strains while others lack blastospores. Endospores may be present.

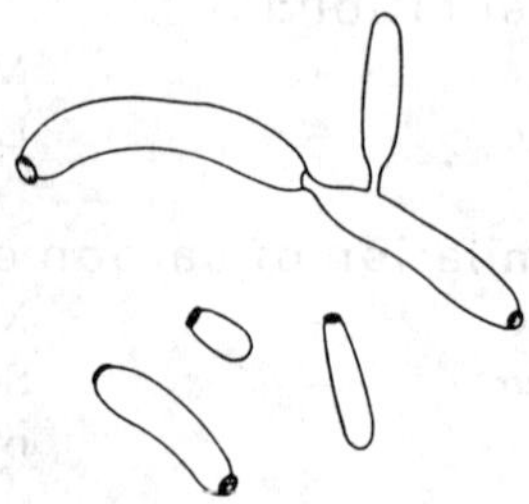

Fig. 176 – *C. antarctica*
In glucose-yeast extract-peptone water

Fermentation: –

Assimilation of carbon compounds:

Galactose	+	Melezitose	+	Ribitol	v
L-Sorbose	+	Soluble starch	+	Galactitol	–
Sucrose	+	D-Xylose	+	D-Mannitol	+
Maltose	+	L-Arabinose	+	D-Glucitol	+
Cellobiose	+ or s	D-Arabinose	+	Salicin	v
Trehalose	+	D-Ribose	+	DL-Lactic acid	+
Lactose	+	L-Rhamnose	v	Succinic acid	+
Melibiose	+	Glycerol	+	Citric acid	+
Raffinose	+	Erythritol	+	Inositol	+

Splitting of arbutin: +

Assimilation of nitrate: +

Growth in vitamin-free medium: +

Growth on 50% (w/w) glucose-yeast extract agar: –

Growth at 37°C: +

Urease: +

Color reaction with DBB: +

Acid formation: + (strong)

G + C: 61.77–61.96 mol% (Kurtzman et al. 1983).

Comments

Based on morphological and physiological properties, Jong and King (1977) concluded that *Sterigmatomyces aphidis* and *Trichosporon oryzae* were conspecific. They also noted the similarity of these strains to *Candida edax* and proposed that the three species represented a single taxon. *C. edax* was the earliest described species and had priority, whereas *St. aphidis* and *Tr. oryzae* were considered synonyms. Later, Kurtzman et al. (1983) reported on the DNA relatedness of *St. aphidis, Tr. oryzae* and the morphological and physiological similar *Sporobolomyces antarcticus*. They found sufficient DNA relatedness among these strains to consider them synonymous. Since the epithet *antarcticus* has priority, the new combination, *Candida antarctica* was proposed and *St. aphidis* and *Tr. oryzae* were made synonyms of *C. antarctica*. The lower GC contents of *C. edax*, 46.8% (Stenderup et al., 1972), compared to that of the strains representing *C. antarctica* precludes a significant degree of genetic relatedness among these species.

Origin of the strains studied

Secretions of Aphidadae on leaves of *Solanum pseudocapsicum* (1); unpolished rice in Japan (1).

Type strain: CBS 5955 from sediment in Lake Vanda, Antarctica.

6. *CANDIDA APICOLA* (HAJSIG) MEYER ET YARROW
(Yarrow & Meyer, 1978)

Synonyms

Torulopsis apicola Hajsig (1958)
Torulopsis bacillaris (Kroemer et Krumbholz) Lodder var. *obesa* nom. nud. (Ohara et al., 1960a)

Standard description of *Candida apicola* (van Uden & Vidal-Leiria, 1970)

Growth in glucose-yeast extract-peptone water: After 3 days at 25°C the cells are spherical to ovoid, (2–5) × (3–6) μm (Fig. 178). A ring and a pellicle may be present.

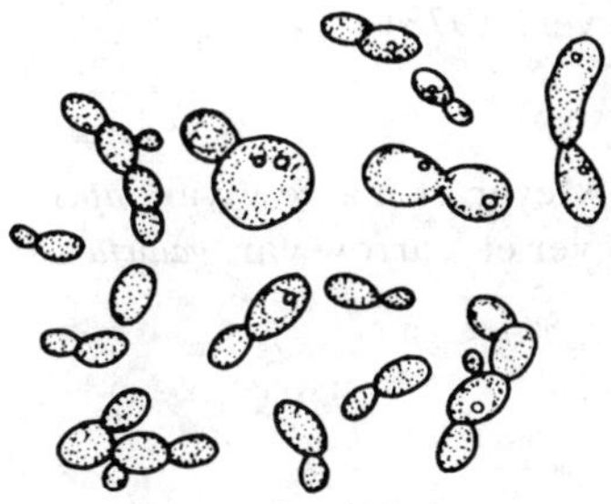

Fig. 178 – *C. apicola*
In glucose-yeast extract-peptone water
(From 'The Yeasts' ed. II, 1970)

Growth on glucose-yeast extract-peptone agar: After one month at 25°C the streak culture is yellowish to cream-colored, glossy, soft and smooth.

Dalmau plate culture on corn meal agar: No pseudomycelium is formed.

Fermentation:

Glucose	+ or s	Maltose	–
Galactose	–	Lactose	–
Sucrose	+	Raffinose	+ or w

Assimilation of carbon compounds:

Galactose	–	Melezitose	–	Ribitol	–
L-Sorbose	+	Soluble starch	–	Galactitol	–
Sucrose	+	D-Xylose	–	D-Mannitol	+
Maltose	–	L-Arabinose	–	D-Glucitol	+
Cellobiose	–	D-Arabinose	–	Salicin	–
Trehalose	–	D-Ribose	–	DL-Lactic acid	–
Lactose	–	L-Rhamnose	–	Succinic acid	–
Melibiose	–	Glycerol	+	Citric acid	–
Raffinose	+	Erythritol	–	Inositol	–

Additional carbon compound tested: D-Glucosamine –
Splitting of arbutin: –
Assimilation of nitrate: –
Growth in vitamin-free medium: v
Growth at 37°C: +
Color reaction with DBB: –
G + C: 46.3 mol%, type strain (Stenderup et al., 1972)

Origin of the strains studied

Intestinal tract of a bee in Yugoslavia (1); grape must in Japan (1); a yeast preparation used against mite infestation of bees in France (1); from stored sugar in The Netherlands (1).

Type strain: CBS 2868, from the intestinal tract of a bee (Hajsig, 1958).

7. *CANDIDA APIS* (LAVIE EX VAN UDEN ET VIDAL-LEIRIA) MEYER ET YARROW (Yarrow & Meyer, 1978)

This species has two varieties:
Candida apis (Lavie ex van Uden et Vidal-Leiria) Meyer et Yarrow var. *apis*
Candida apis (Lavie ex van Uden et Vidal-Leiria) Meyer et Yarrow var. *galacta* Golubev et Babjeva (1977)

Synonyms of the variety *apis*

Paratorulopsis apis nom. nud. (Novák & Zsolt, 1961)
Torulopsis apis Lavie ex van Uden et Vidal-Leiria (1970)

Synonym of the variety *galacta*

Torulopsis apis Lavie ex van Uden et Vidal-Leiria var. *galacta* Golubev et Babjeva (1977)

Standard description of *Candida apis* var. *apis*

Growth in glucose-yeast extract-peptone water: After 3 days at 25°C the cells are spherical, 1.5–3.5 μm, to slight ovoid and occur singly, in pairs and in short chains and small clusters (Fig. 179). A ring and sometimes a pellicle are present.

Growth on glucose-yeast extract-peptone agar: After one month at 25°C the streak culture is cream-colored, beige, butyrous, soft and smooth. The margin is entire.

Dalmau plate culture on corn meal agar: Pseudomycelium is absent.

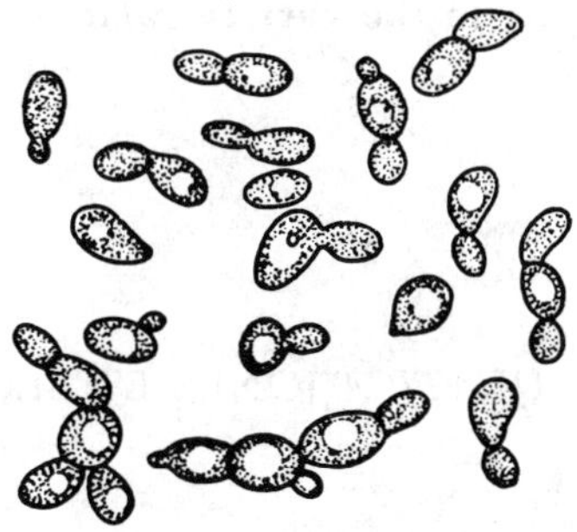

Fig. 179 – *C. apis* var. *apis*
In glucose-yeast extract-peptone water (From 'The Yeasts', ed. II, 1970)

Fermentation: –

Assimilation of carbon compounds:

Galactose	+ w	Melezitose	–	Ribitol	–
L-Sorbose	–	Soluble starch	–	Galactitol	–
Sucrose	+	D-Xylose	–	D-Mannitol	+
Maltose	–	L-Arabinose	–	D-Glucitol	+
Cellobiose	–	D-Arabinose	–	Salicin	–
Trehalose	–	D-Ribose	–	DL-Lactic acid	–
Lactose	–	L-Rhamnose	–	Succinic acid	–
Melibiose	–	Glycerol	+	Citric acid	–
Raffinose	+	Erythritol	–	Inositol	–

Splitting of arbutin: –
Assimilation of nitrate: –
Growth in vitamin-free medium: v
Growth at 37°C: –
Urease: –
Color reaction with DBB: –
G + C: 47.7 mol%, type strain (Meyer, Stenderup & Christiansen, unpublished)

Supplementary description of *Candida apis* var. *galacta*:

Assimilation of raffinose: –

Comments

The original description reported a difference of 2.8 mol% G + C between *C. apis* var. *apis* (46.0%) and var. *galacta* (48.8%).

Origin of the strain belonging to the variety *apis*

Trachea of a bee (Lavie, 1954) (1).

Type strain: CBS 2674.

Origin of the strain belonging to the variety *galacta*

Cocoon of red ants (1).

Type strain CBS 6939.

8. *CANDIDA AQUATICA* JONES ET SLOOFF (1966)

Synonym

Vanrija aquatica (Jones et Slooff) Moore (1980)

Standard description of *Candida aquatica*

Growth in glucose-yeast extract-peptone water: After 3 days at 25°C the cells are ovoid or elongate, $(3-7) \times (7-12)$ μm. Often the budding is polar. Repetitive polar budding may result in the formation of cell groups with a star-like arrangement (Fig. 180) (van Uden & Buckley, 1970).

Growth on glucose-yeast extract-peptone agar: After one month at 25°C the streak culture is yellowish cream-colored, semidull, soft, slightly wrinkled (van Uden & Buckley, 1970).

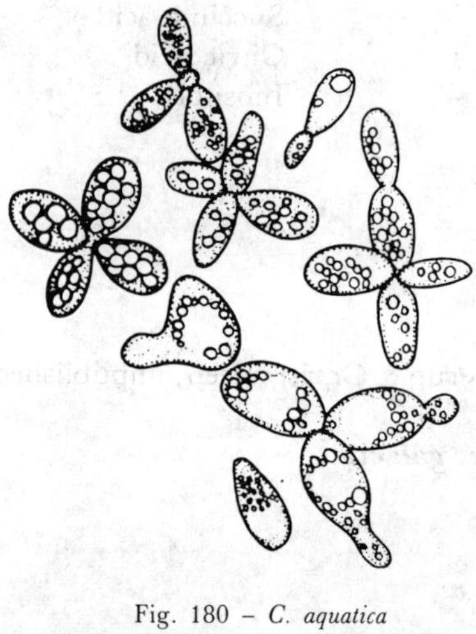

Fig. 180 – *C. aquatica*
In glucose-yeast extract-peptone water
(From 'The Yeasts' ed. II, 1970)

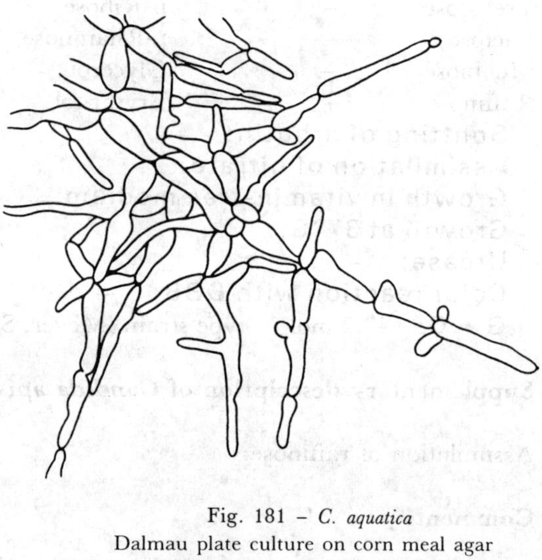

Fig. 181 – *C. aquatica*
Dalmau plate culture on corn meal agar
(From 'The Yeasts' ed. II, 1970)

Dalmau plate culture on corn meal agar: The pseudomycelium consists of branched chains of ovoid or elongate cells. The elongate cells may be flask-, dumbbell- or club-shaped or display other bizarre forms (Fig. 181) (van Uden & Buckley, 1970).

Fermentation:

Glucose	+ w	Maltose	v
Galactose	–	Lactose	–
Sucrose	v	Raffinose	v

Assimilation of carbon compounds:

Galactose	+	Melezitose	+	Ribitol	+
L-Sorbose	+	Soluble starch	+	Galactitol	–
Sucrose	+	D-Xylose	+	D-Mannitol	+
Maltose	+	L-Arabinose	+	D-Glucitol	+
Cellobiose	+	D-Arabinose	–	Salicin	+
Trehalose	+	D-Ribose	–	DL-Lactic acid	–
Lactose	+	L-Rhamnose	–	Succinic acid	+ s
Melibiose	+	Glycerol	v	Citric acid	+ s
Raffinose	+	Erythritol	–	Inositol	–

Splitting of arbutin: +
Assimilation of nitrate: +
Growth in vitamin-free medium: – or + s
Growth at 37°C: –
Starch formation: +
Urease: +
Color reaction with DBB: +
G + C: 58.4 mol%, type strain (Meyer, Stenderup & Christiansen, unpublished)

Origin of the strain studied

Scums on a freshwater lake in England (Jones & Slooff, 1966) (1).

Type strain: CBS 5443.

9. *CANDIDA ATLANTICA* (SIEPMANN) MEYER ET SIMIONE (1983)

Synonym

Trichosporon atlanticum Siepmann (in Siepmann & Höhnk, 1962)

Standard description of *Candida atlantica*

Growth in glucose-yeast extract-peptone water: After 3 days at 25°C the cells are globose to ovoid, (2.3–4.6) × (2.3–5.7) μm, and occur singly, budding and in short chains and clusters. Irregular-shaped cells and pseudomycelium are also present. A sediment exists. After one month a thin film is evident as well as a sediment.

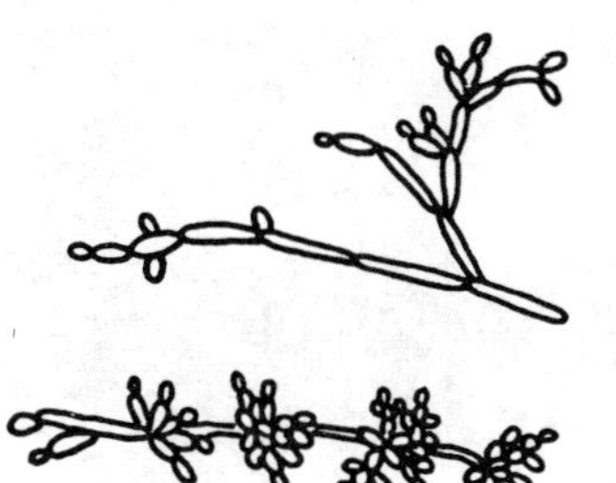

Fig. 182 – *C. atlantica*
Dalmau plate culture on corn meal agar

Growth on glucose-yeast extract-peptone agar: After one month at 25°C the streak culture is cream-colored, somewhat glistening, primarily smooth and soft with some "craters" in the central area of the streak. The margin is entire.

Dalmau plate culture on corn meal agar: Pseudomycelium consists of long, wavy hyphae with few blastospores (Fig. 182).

Fermentation:

Glucose	+ s or w	Maltose	–
Galactose	–	Lactose	–
Sucrose	–		

Assimilation of carbon compounds:

Galactose	+	Melezitose	+	Ribitol	+ or s
L-Sorbose	+ or s	Soluble starch	–	Galactitol	–
Sucrose	+	D-Xylose	+	D-Mannitol	+
Maltose	+	L-Arabinose	+	D-Glucitol	+
Cellobiose	–	D-Arabinose	+	Salicin	+
Trehalose	+	D-Ribose	+	DL-Lactic acid	+
Lactose	–	L-Rhamnose	+	Succinic acid	–
Melibiose	–	Glycerol	+	Citric acid	–
Raffinose	–	Erythritol	+	Inositol	–

Splitting of arbutin: +

Assimilation of nitrate: –

Growth in vitamin-free medium: –

Growth at 37°C: –

Urease: –

Color reaction with DBB: –

G + C: 38.5 mol%, type strain (Meyer & Simione, 1983)

Comments

Trichosporon atlanticum was placed in synonymy with *C. diddensiae* because of the lack of arthrospores and its physiological and morphological similarity to *C. diddensiae* (van Uden & Buckley, 1970). DNA reassociation studies (Meyer & Simione, 1978; Meyer, unpublished) revealed insignificant DNA homology between the type strains of these species requiring the removal of *Tr. atlanticum* from *C. diddensiae* and the recognition of it as a separate species. The proper combination was provided by Meyer and Simione (1983).

Origin of the strain studied

Seawater (1).

Type strain: CBS 5263.

10. CANDIDA ATMOSPHERICA SANTA MARÍA (1959c)

Standard description of *Candida atmospherica*

Growth on glucose-yeast extract-peptone water: After 3 days at 25°C the cells are globose, ovoid to cylindrical (Fig. 183).

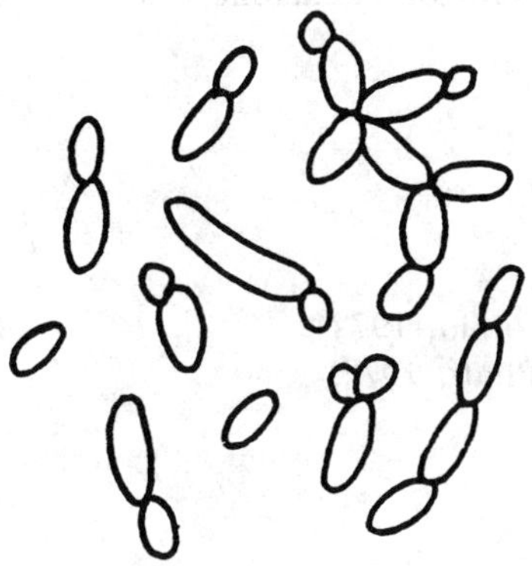

Fig. 183 – *C. atmospherica*
In glucose-yeast extract-peptone water

Growth in glucose-yeast extract-peptone agar: After one month at 25°C the streak culture is white or cream colored.

Dalmau plate culture on corn meal agar: The pseudomycelium consists of branched chains of cells, giving a wavy appearance and having few blastospores (Fig. 184).

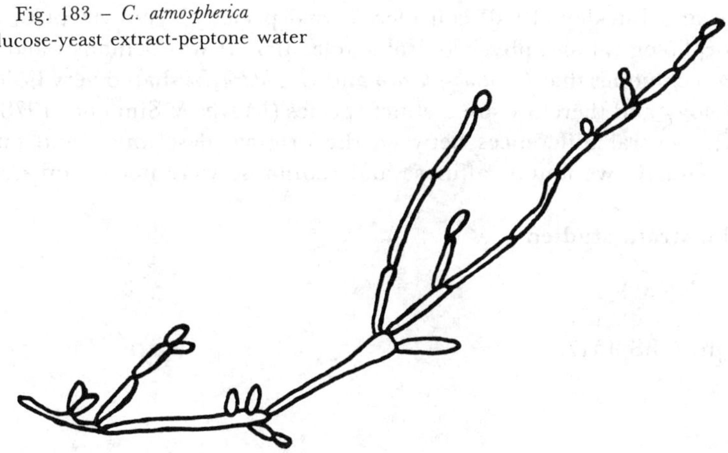

Fig. 184 – *C. atmospherica*
Dalmau plate culture on corn meal agar

Fermentation:

Glucose	+	Maltose	–
Galactose	–	Lactose	–
Sucrose	–		

Assimilation of carbon compounds:

Galactose	+	Melezitose	+	Ribitol	+
L-Sorbose	+ s	Soluble starch	–	Galactitol	–
Sucrose	+	D-Xylose	+	D-Mannitol	+
Maltose	+	L-Arabinose	+	D-Glucitol	+

Cellobiose	+	D-Arabinose	+	Salicin	+ s
Trehalose	+	D-Ribose	+	DL-Lactic acid	−
Lactose	−	L-Rhamnose	−	Succinic acid	+
Melibiose	−	Glycerol	+	Citric acid	+
Raffinose	−	Erythritol	+	Inositol	−

Additional carbon compound tested: Glucono-δ-lactone + s
Splitting of arbutin: +
Assimilation of nitrate: −
Growth in vitamin-free medium: −
Growth at 37°C: −
Urease: −
Color reaction with DBB: −
Coenzyme Q system: Q 9 (Yamada & Kondo, 1972)
G + C: 39.8 mol%, type strain (Meyer & Phaff, 1972)

Comments

Van Uden and Buckley (1970) considered this species a synonym of *C. diddensiae* based on morphological and physiological similarities. It was demonstrated by DNA reassociation techniques that *C. atmospherica* and *C. diddensiae* shared very little base sequence homology and therefore are distinct species (Meyer & Simione, 1978; Meyer, unpublished). Several differences between the original description and our results were noted, namely, we found raffinose and rhamnose were not assimilated.

Origin of the strain studied

Air (1).

Type strain: CBS 4547.

11. *CANDIDA AURICULARIAE* (NAKASE) MEYER ET YARROW (Yarrow & Meyer, 1978)

Synonym

Torulopsis auriculariae Nakase (1971a)

Standard description of *Candida auriculariae*

Growth in YM broth: After 3 days at 25°C the cells are round to oval, (3–4.5) × (4–7) μm, and occur singly, budding or in clusters (Fig. 185). A sediment

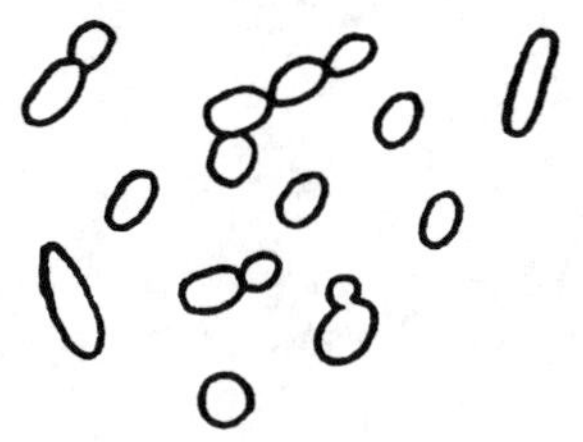

Fig. 185 – *C. auriculariae*
In glucose-yeast extract-peptone water

is formed. After one month a ring is also present (Nakase, 1971a).

Growth on glucose-yeast extract-peptone agar: After one month at 25°C the streak culture is cream-colored, mucoid, glistening, moist, convex and entire.

Dalmau plate culture on corn meal agar: Pseudomycelium absent.

Fermentation: –

Assimilation of carbon compounds:

Galactose	–	Melezitose	+	Ribitol	–
L-Sorbose	–	Soluble starch	–	Galactitol	–
Sucrose	+	D-Xylose	–	D-Mannitol	+
Maltose	+ s	L-Arabinose	–	D-Glucitol	+ s
Cellobiose	–	D-Arabinose	–	Salicin	–
Trehalose	+ s	D-Ribose	–	DL-Lactic acid	–
Lactose	–	L-Rhamnose	–	Succinic acid	+
Melibiose	–	Glycerol	–	Citric acid	–
Raffinose	–	Erythritol	–	Inositol	–

Splitting of arbutin: –

Assimilation of nitrate: –

Growth in vitamin-free medium: –

Growth at 37°C: –

Urease: +

Color reaction with DBB: +

G + C: 54.4 mol% (Nakase, 1971a)

Origin of the strain studied

Fruiting body of *Auricularia auricula-judae,* in Zaruga-take (Mt. Zaru), Hamanashi Prefecture (1).

Type strain: CBS 6379.

12. *CANDIDA AURINGIENSIS* SANTA MARÍA (1978)

Standard description of *Candida auringiensis*

Growth in glucose-yeast extract-peptone water: After 3 days at 25°C cells are globose to oval with some teardrop-shaped, (2.0–4.0) × (2.5–5.0) μm, and occur singly and in pairs (Fig. 186). A sediment is present.

Growth on glucose-yeast extract-peptone agar: After one month at 25°C the streak culture is white creamy, smooth and shiny. The margin is mostly entire with an occasional tuft of mycelial growth.

Dalmau plate culture on corn meal agar: Pseudohyphae are formed.

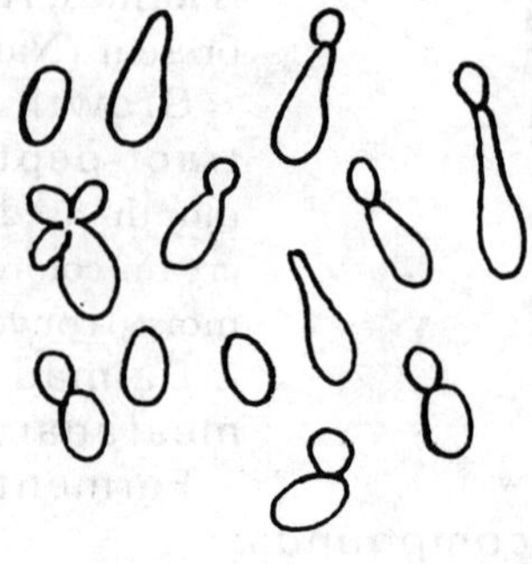

Fig. 186 – *C. auringiensis*
In glucose-yeast extract-peptone water

Fermentation:

Glucose	+	Maltose	–
Galactose	+	Lactose	+ s
Sucrose	–		

Assimilation of carbon compounds:

Galactose	+	Melezitose	–	Ribitol	+
L-Sorbose	+	Soluble starch	–	Galactitol	–
Sucrose	+	D-Xylose	+	D-Mannitol	+
Maltose	–	L-Arabinose	+	D-Glucitol	+
Cellobiose	+	D-Arabinose	– or + s	Salicin	+
Trehalose	+	D-Ribose	– or + s	DL-Lactic acid	–
Lactose	+	L-Rhamnose	–	Succinic acid	+
Melibiose	–	Glycerol	+	Citric acid	–
Raffinose	–	Erythritol	+	Inositol	+

Additional carbon compound tested: D-Glucosamine·HCl +

Splitting of arbutin: +

Assimilation of nitrate: –

Growth in vitamin-free medium: –

Growth at 37°C: +

Urease: –

Color reaction with DBB: –

Origin of the strains studied

Alpechin in Spain (3).

Type strain: CBS 6913.

13. *CANDIDA AUSTROMARINA* (FELL ET HUNTER) MEYER ET YARROW (Yarrow & Meyer, 1978)

Synonym

Torulopsis austromarina Fell & Hunter (1974)

Standard description of *Candida austromarina* (Fell & Hunter, 1974)

Growth in malt extract: After 3 days at 12°C the cells are globose, 1.3–7.4 μm, to ovoid (3.4–4.0) × (4.0–7.4) μm, and occur singly or in pairs. The buds may be on short stalks. A sediment is present, but no ring or pellicle.

Growth on malt agar: After one month at 12°C the streak culture is white, smooth, soft, with an entire margin.

Dalmau plate culture on corn meal agar: Pseudomycelium absent.

Fermentation: –

Assimilation of carbon compounds:

Galactose	+	Melezitose	–	Ribitol	–
L-Sorbose	–	Soluble starch	–	Galactitol	–
Sucrose	–	D-Xylose	–	D-Mannitol	–
Maltose	–	L-Arabinose	–	D-Glucitol	–
Cellobiose	–	D-Arabinose	–	Salicin	–
Trehalose	v	D-Ribose	–	DL-Lactic acid	–
Lactose	–	L-Rhamnose	–	Succinic acid	+
Melibiose	–	Glycerol	+ s	Citric acid	–
Raffinose	–	Erythritol	–	Inositol	–

Splitting of arbutin: –

Assimilation of nitrate: –

Growth in vitamin-free medium: –

Growth at 37°C: –

Color reaction with DBB: –

G + C: 39.0 mol%, 2 strains (Meyer, unpublished)

Origin of the strains studied

Antarctic Ocean (2).

Type strain: CBS 6179, isolated from Antarctic Ocean.

14. *CANDIDA AZYMA* (VAN DER WALT, JOHANNSEN ET YARROW) MEYER ET YARROW (Yarrow & Meyer, 1978)

Synonym

Torulopsis azyma van der Walt, Johannsen et Yarrow (1978)

Standard description of *Candida azyma*

Growth in glucose-yeast extract-peptone water: After 3 days at 25°C the cells are ovoid to ellipsoidal, occasionally somewhat cylindrical, (1.5–3.5) × (2.0–6.0) μm, occurring singly, in pairs or short chains (Fig. 187). A sediment and ring are formed. After one month at room temperature, a ring, incomplete, sinking pellicle and sediment are present (van der Walt et al., 1978).

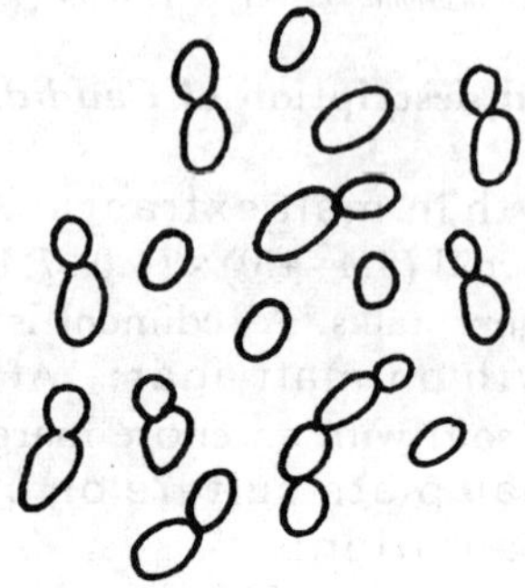

Fig. 187 – *C. azyma*
In glucose-yeast extract-peptone water

Growth on glucose-yeast extract-peptone agar: After one month at 25°C the streak culture is slightly beige to cream-colored, smooth, moist, and shiny. The margins may be entire to undulating.

Dalmau plate culture on corn meal agar: The pseudomycelium consists of branched chains of ovoid cells.

Fermentation: –

Assimilation of carbon compounds:

Galactose	+	Melezitose	+	Ribitol	v
L-Sorbose	+	Soluble starch	–	Galactitol	v
Sucrose	+	D-Xylose	v	D-Mannitol	+ or s
Maltose	+	L-Arabinose	v	D-Glucitol	+ or s
Cellobiose	–	D-Arabinose	v	Salicin	–
Trehalose	+	D-Ribose	–	DL-Lactic acid	–
Lactose	–	L-Rhamnose	–	Succinic acid	+
Melibiose	–	Glycerol	+ or s	Citric acid	–
Raffinose	–	Erythritol	–	Inositol	–

Splitting of arbutin: –

Assimilation of nitrate: –

Growth in vitamin-free medium: –

Growth on 50% (w/w) glucose-yeast extract agar: + (van der Walt et al., 1978)

Growth at 37°C: +

Color reaction with DBB: – (van der Walt et al., 1978)

G + C: 54.4 mol%

Origin of the strain studied

Grassland soil near Nelspruit (1).

Type strain: CBS 6826.

15. *CANDIDA BACARUM* (BUHAGIAR) MEYER ET YARROW (Yarrow & Meyer, 1978)

Synonym

Torulopsis bacarum Buhagiar (1975)

Standard description of *Candida bacarum*

Growth in glucose-yeast extract-peptone water: After 4 days at 25°C the cells are ovoid and elongate, (1.4–2.9) × (3.6–5.1) μm.

Growth on glucose-yeast extract-peptone agar: After one month at 25°C the streak culture is brownish or cream-colored, smooth, convex, with an entire to slightly undulating margin.

Dalmau plate culture on corn meal agar: The pseudomycelium is poorly developed, consisting of ramified chains of ovoid cells.

Fermentation: –

Assimilation of carbon compounds:

Galactose	–	Melezitose	+	Ribitol	+ s
L-Sorbose	v	Soluble starch	+	Galactitol	–
Sucrose	+	D-Xylose	+	D-Mannitol	+ s
Maltose	+	L-Arabinose	+	D-Glucitol	+
Cellobiose	+ s	D-Arabinose	+	Salicin	+ s
Trehalose	+	D-Ribose	+	DL-Lactic acid	+ s
Lactose	–	L-Rhamnose	–	Succinic acid	+
Melibiose	–	Glycerol	+	Citric acid	+ s
Raffinose	+	Erythritol	+	Inositol	–

Splitting of arbutin: +

Assimilation of nitrate: +

Growth in vitamin-free medium: –

Growth on 50% (w/w) glucose-yeast extract agar: –

Growth at 37°C: –

Urease: +

Color reaction with DBB: +

G + C: 52.4 mol%, type strain

Origin of the strain studied

Black currants (1).

Type strain: CBS 6526, isolated from black currants.

16. *CANDIDA BEECHII* BUCKLEY ET VAN UDEN (1968)

Standard description of *Candida beechii*

Growth in glucose-yeast extract-peptone water: After 3 days at 25°C the cells are ovoid to cylindrical, (1.5–4.5) × (4.5–11) μm; they occur mostly in groups; much longer cylindrical cells also occur (Fig. 188). A ring is present (van Uden & Buckley, 1970).

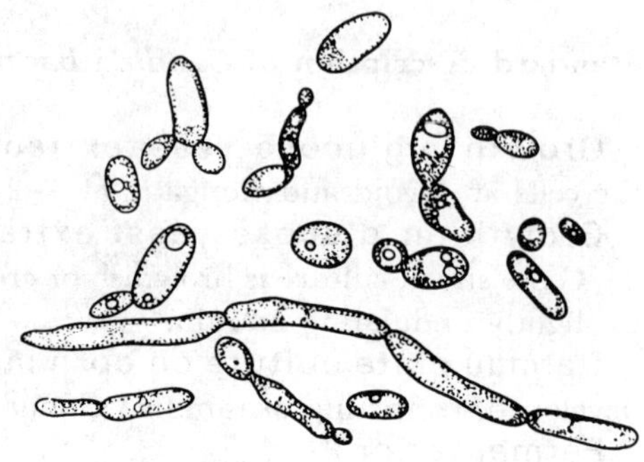

Fig. 188 – *C. beechii*
In glucose-yeast extract-peptone water
(From 'The Yeasts' ed. II, 1970)

Growth on glucose-yeast extract-peptone agar: After one month at 25°C the streak culture is cream-colored, delicately wrinkled, lacy, moist and soft. There is a mycelial edge.

Dalmau plate culture on corn meal agar: Pseudomycelium is abundantly formed. It consists of branched formations of long pseudohyphae. True mycelium also occurs. The pseudomycelium is generally bare of blastospores (Fig. 189); in some parts stalagmoid, oval or bacilliform blastoconidia are formed (van Uden & Buckley, 1970).

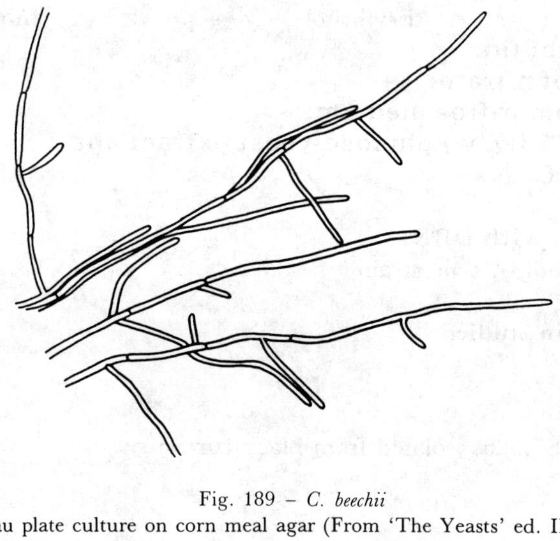

Fig. 189 – *C. beechii*
Dalmau plate culture on corn meal agar (From 'The Yeasts' ed. II, 1970)

Fermentation:

Glucose	+	Maltose	–
Galactose	–	Lactose	–
Sucrose	–	Trehalose	+

Assimilation of carbon compounds:

Galactose	–	Melezitose	–	Ribitol	+
L-Sorbose	+	Soluble starch	–	Galactitol	–
Sucrose	–	D-Xylose	+ s	D-Mannitol	+
Maltose	–	L-Arabinose	–	D-Glucitol	+
Cellobiose	+	D-Arabinose	–	Salicin	+
Trehalose	+	D-Ribose	+ s	DL-Lactic acid	+ s
Lactose	–	L-Rhamnose	–	Succinic acid	+
Melibiose	–	Glycerol	+	Citric acid	+
Raffinose	–	Erythritol	–	Inositol	–

Additional carbon compound tested: Methanol –

Splitting of arbutin: +

Assimilation of nitrate: –

Growth in vitamin-free medium: –

Growth at 37°C: –

Color reaction with DBB: –

G + C: 41.2 mol%, type strain (Meyer, Stenderup & Christiansen, unpublished)

Origin of the strain studied

Cider in England (1).

Type strain: CBS 4261.

17. CANDIDA BERTHETII BOIDIN, PIGNAL, MERMIER ET ARPIN (1963)

Standard description of *Candida berthetii*

Growth in glucose-yeast extract-peptone water: After 3 days at 25°C the cells are globose to ovoid, (3.0–5.5) × (4–9.5) μm; sausage-shaped cells also occur (Fig. 190). A dry pellicle is present (van Uden & Buckley, 1970).

Growth on glucose-yeast extract-peptone agar: After one month at 25°C the streak culture is greyish cream-colored, semidull, soft, almost smooth (van Uden & Buckley, 1970).

Dalmau plate culture on corn meal agar: The pseudomycelium consists of short pseudohyphae in a tree-like arrangement. Chains and clusters of globose blastospores occur in pleural and terminal positions in some parts of the pseudomycelium (Fig. 191). Other parts may be bare of blastospores (van Uden & Buckley, 1970).

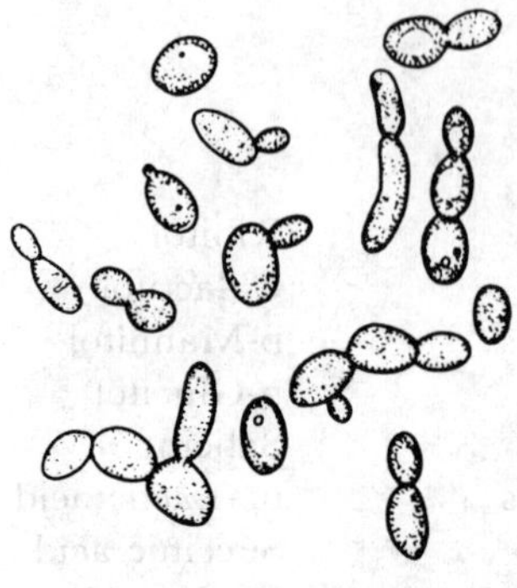

Fig. 190 – *C. berthetii*
In glucose-yeast extract-peptone water
(From 'The Yeasts' ed. II, 1970)

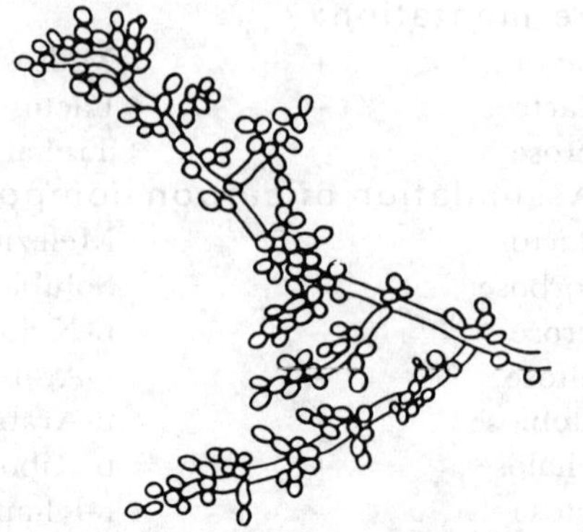

Fig. 191 – *C. berthetii*
Dalmau plate culture on corn meal agar
(From 'The Yeasts' ed. II, 1970)

Fermentation:

Glucose	+ s	Maltose	–
Galactose	–	Lactose	–
Sucrose	–	Trehalose	–

Assimilation of carbon compounds:

Galactose	–	Melezitose	–	Ribitol	–
L-Sorbose	–	Soluble starch	–	Galactitol	–
Sucrose	–	D-Xylose	–	D-Mannitol	–
Maltose	–	L-Arabinose	–	D-Glucitol	–
Cellobiose	+	D-Arabinose	–	Salicin	+
Trehalose	–	D-Ribose	–	DL-Lactic acid	–
Lactose	–	L-Rhamnose	–	Succinic acid	+ w
Melibiose	–	Glycerol	+	Citric acid	+ w
Raffinose	–	Erythritol	–	Inositol	–

Splitting of arbutin: +
Assimilation of nitrate: +
Growth in vitamin-free medium: +
Growth at 37°C: +
Color reaction with DBB: –
G + C: 39.8 mol%, type strain (Stenderup et al., 1972)

Origin of the strains studied

Tree gum in the Cameroons (Boidin et al., 1963) (1), insect tunnels (3).

Type strain: CBS 5452.

18a. CANDIDA BIMUNDALIS WICKERHAM ET SANTA MARÍA VAR. *BIMUNDALIS*

see *Hansenula bimundalis* p. 178.

18b. CANDIDA BIMUNDALIS WICKERHAM ET SANTA MARÍA VAR. *AMERICANA* WICKERHAM

see *Hansenula americana* p. 171.

19. CANDIDA BLANKII BUCKLEY ET VAN UDEN (1968)

Standard description of *Candida blankii*

Growth in glucose-yeast extract-peptone water: After 3 days at 25°C the cells are ovoid to elongate, (2.5–4) × (4.5–13) μm. The elongate cells are often irregularly curved (Fig. 192) (van Uden & Buckley, 1970).

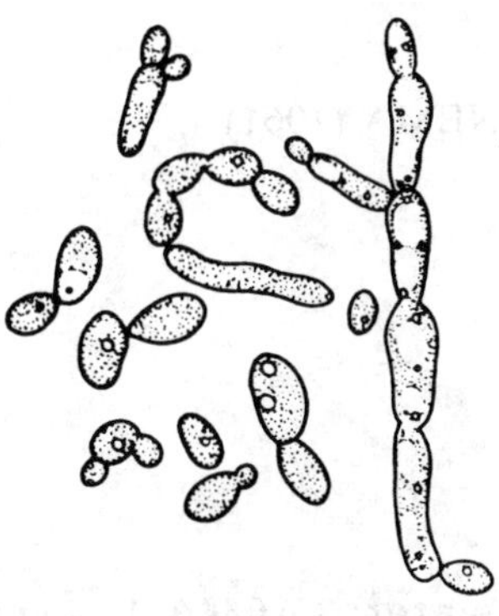

Fig. 192 – *C. blankii*
In glucose-yeast extract-peptone water
(From 'The Yeasts' ed. II, 1970)

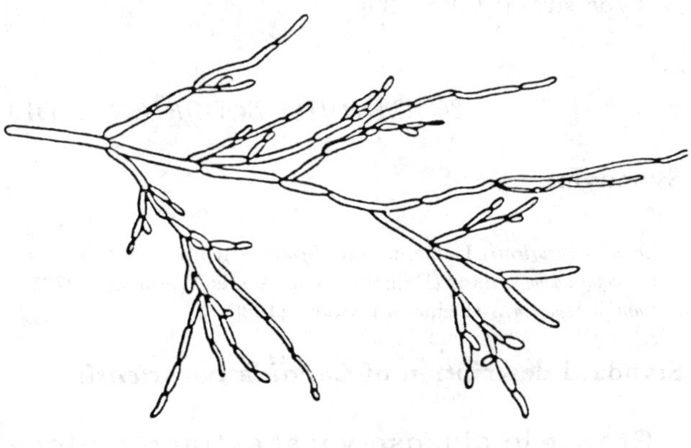

Fig. 193 – *C. blankii*
Dalmau plate culture on corn meal agar
(From 'The Yeasts' ed. II, 1970)

Growth on glucose-yeast extract-peptone agar: After one month at 25°C the streak culture is whitish cream-colored, semidull, soft to membranous, wrinkled (van Uden & Buckley, 1970).

Dalmau plate culture on corn meal agar: The pseudomycelium consists of branched formations of pseudohyphae bearing chains of ovoid cells (Fig. 193) (van Uden & Buckley, 1970).

Fermentation:

Glucose	– or + s	Maltose	–
Galactose	–	Lactose	–
Sucrose	–		

Assimilation of carbon compounds:

Galactose	+	Melezitose	+	Ribitol	+
L-Sorbose	+	Soluble starch	+	Galactitol	+
Sucrose	+	D-Xylose	+	D-Mannitol	+
Maltose	+	L-Arabinose	+	D-Glucitol	+
Cellobiose	+	D-Arabinose	–	Salicin	+
Trehalose	+	D-Ribose	–	DL-Lactic acid	–
Lactose	+	L-Rhamnose	+	Succinic acid	+ w

Melibiose	–	Glycerol	+	Citric acid	+
Raffinose	+	Erythritol	+	Inositol	+ s

Splitting of arbutin: +
Assimilation of nitrate: –
Growth in vitamin-free medium: +
Growth at 37°C: +
Color reaction with DBB: –
G + C: 55.1 mol%, type strain (Meyer, Stenderup & Christiansen, unpublished)

Origin of the strain studied

Organs of a mink in Canada (1).

Type strain: CBS 1898.

20. *CANDIDA BOGORIENSIS* DEINEMA (1961)

Synonyms

Candida bogoriensis Deinema var. *lipolytica* Ruinen (1963)
Rhodotorula bogoriensis (Deinema) von Arx et Weijman (1979)
Vanrija bogoriensis (Deinema) Moore (1980)

Standard description of *Candida bogoriensis*

Growth in glucose-yeast extract-peptone water: After 3 days at 25°C the cells are long-ovoid to sausage-shaped, (1.5–4) × (7–14) μm (Fig. 194) (van Uden & Buckley, 1970).

Growth on glucose-yeast extract-peptone agar: After one month at 25°C the streak culture is yellowish-brown, mucoid, soft and smooth (van Uden & Buckley, 1970).

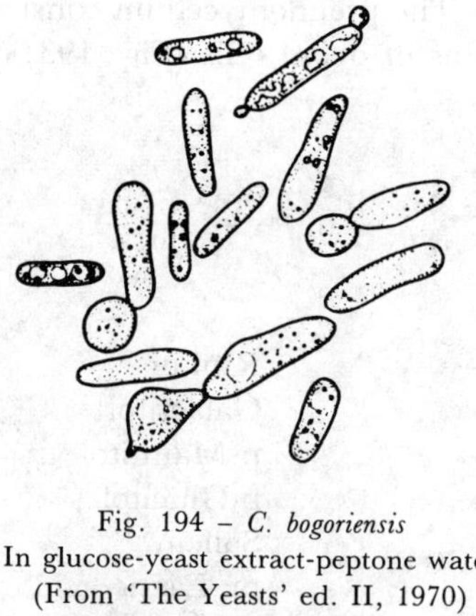

Fig. 194 – *C. bogoriensis*
In glucose-yeast extract-peptone water
(From 'The Yeasts' ed. II, 1970)

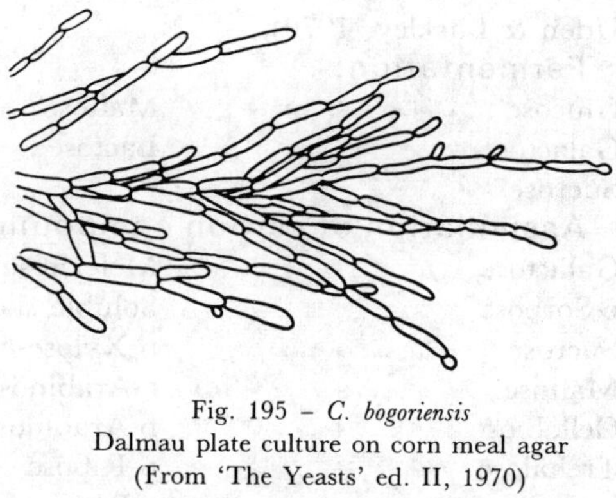

Fig. 195 – *C. bogoriensis*
Dalmau plate culture on corn meal agar
(From 'The Yeasts' ed. II, 1970)

Dalmau plate culture on corn meal agar: The pseudomycelium consists of branched chains of short pseudohyphae (Fig. 195) (van Uden & Buckley, 1970).

Fermentation: –

Assimilation of carbon compounds:

Galactose	– or +s	Melezitose	+	Ribitol	+
L-Sorbose	– or +s	Soluble starch	+	Galactitol	–
Sucrose	–	D-Xylose	+	D-Mannitol	+
Maltose	+	L-Arabinose	+	D-Glucitol	+
Cellobiose	+	D-Arabinose	+s	Salicin	+s
Trehalose	+	D-Ribose	+	DL-Lactic acid	–
Lactose	–	L-Rhamnose	–	Succinic acid	+
Melibiose	–	Glycerol	+s	Citric acid	+
Raffinose	–	Erythritol	–	Inositol	–

Splitting of arbutin: +

Assimilation of nitrate: –

Growth in vitamin-free medium: +s

Growth at 37°C: –

Urease: +

Color reaction with DBB: +

Coenzyme Q system: Q 10 (Yamada & Kondo, 1972)

G + C: 57.3 mol%, type strain (Meyer & Phaff, 1972)

Comments

In restricting *Candida* to the anamorphs of Saccharomycetales, von Arx and Weijman (1979) transferred *C. bogoriensis* which is basidiomycetous to the genus *Rhodotorula* based on the type of conidiogenesis and the lack of xylose in cell hydrolysates.

Origin of the strain studied

Surface of leaves in Indonesia (1).

Type strain: CBS 4101.

21. *CANDIDA BOIDINII* RAMIREZ (1953)

Synonyms

Candida koshuensis Yokotsuka et Goto (1955)
Candida olivarium Santa María (1958a)
Candida methanolica Oki et Kuonu (in Oki et al., 1972)
Candida queretana Herrera et Ulloa (1978)

Standard description of *Candida boidinii*

Growth in glucose-yeast extract-peptone water: After 3 days at 25°C the cells are long-ovoid to cylindrical, often slightly curved, (1.5–3.5) × (7–12) μm (Fig. 196). A creeping, powdery pellicle is present (van Uden & Buckley, 1970).

Growth on glucose-yeast extract-peptone agar: After one month at 25°C the streak culture is yellowish cream-colored, moist to dull, soft, delicately wrinkled (van Uden & Buckley, 1970).

Dalmau plate culture on corn meal agar: The pseudomycelium consists of ramified chains of coarse, short pseudohyphae and of more slender pseudohyphal tree-like arrangements bearing verticillated chains and groups of ovoid blastospores (Fig. 197) (van Uden & Buckley, 1970).

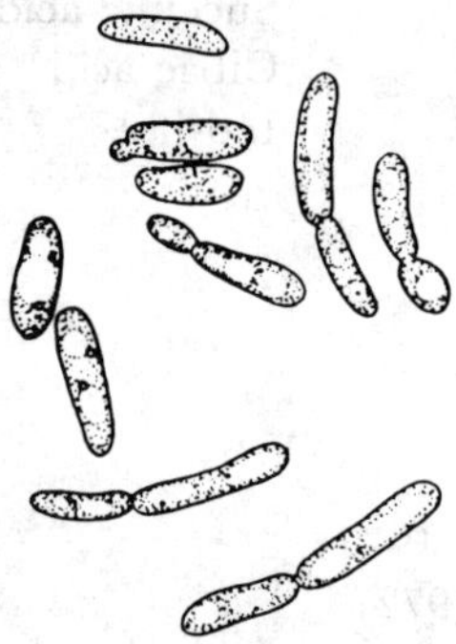

Fig. 196 – *C. boidinii*
In glucose-yeast extract-peptone water
(From 'The Yeasts' ed. II, 1970)

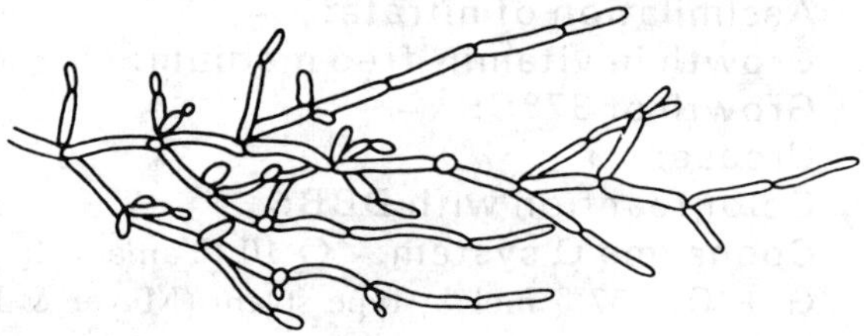

Fig. 197 – *C. boidinii*
Dalmau plate culture on corn meal agar
(From 'The Yeasts' ed. II, 1970)

Fermentation:

Glucose	+	Maltose	–
Galactose	–	Lactose	–
Sucrose	–		

Assimilation of carbon compounds:

Galactose	+	Melezitose	–	Ribitol	+
L-Sorbose	–	Soluble starch	–	Galactitol	–
Sucrose	–	D-Xylose	+	D-Mannitol	+
Maltose	–	L-Arabinose	+ s or –	D-Glucitol	+
Cellobiose	–	D-Arabinose	–	Salicin	–
Trehalose	–	D-Ribose	+	DL-Lactic acid	+ s
Lactose	–	L-Rhamnose	–	Succinic acid	–
Melibiose	–	Glycerol	+	Citric acid	–
Raffinose	–	Erythritol	+	Inositol	–

Additional carbon compound tested: Methanol +

Splitting of arbutin: –

Assimilation of nitrate: +

Growth in vitamin-free medium: + w or –

Growth at 37°C: v

Urease: –

Color reaction with DBB: –

G + C: 33.0 mol%, type strain (Meyer & Phaff, 1972)

Origin of the strains studied

Tanning liquor in France (1); fermenting olives in Spain (Santa María, 1958a) (1); soil in South Africa (1); seawater (1); wine (1); ginger ale (1); ward floor (1); washed bottles (1); tepache de queretara in Mexico (1).

Type strain: CBS 2428, from tanning liquor (Ramirez, 1953).

22. *CANDIDA BOLETICOLA* NAKASE (1971b)

Standard description of *Candida boleticola*

Growth in glucose-yeast extract-peptone water: After 3 days at 25°C the cells are long-oval, elongate to cylindrical, (3.6–4.3) × (5.1–6.5) μm, and occur singly, in pairs and short chains. A pellicle is present.

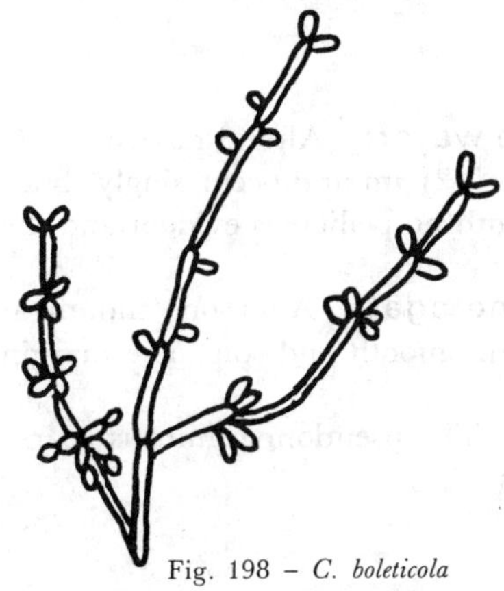

Fig. 198 – *C. boleticola*
Dalmau plate culture on corn meal agar

Growth on glucose-yeast extract-peptone agar: After one month at 25°C the streak culture is cream-colored, dry, wrinkled, dull and fringed with mycelium.

Dalmau plate culture on corn meal agar: Pseudomycelium consists of branched chains of long cells bearing chains of blastospores, mostly in terminal positions (Fig. 198).

Fermentation:

Glucose	+	Maltose	–
Galactose	–	Lactose	–
Sucrose	–		

Assimilation of carbon compounds:

Galactose	+	Melezitose	–	Ribitol	+
L-Sorbose	– or + s	Soluble starch	–	Galactitol	–
Sucrose	–	D-Xylose	+	D-Mannitol	+
Maltose	–	L-Arabinose	–	D-Glucitol	+
Cellobiose	–	D-Arabinose	– or + s	Salicin	+ or s
Trehalose	+	D-Ribose	+	DL-Lactic acid	–
Lactose	–	L-Rhamnose	–	Succinic acid	+
Melibiose	–	Glycerol	+	Citric acid	+ s
Raffinose	–	Erythritol	+	Inositol	–

Additional carbon compound tested: Methanol –

Splitting of arbutin: +

Assimilation of nitrate: –

Growth in vitamin-free medium: –

Growth on 50% (w/w) glucose-yeast extract agar: –
Growth at 37°C: –
Color reaction with DBB: –
G + C: 43.2–44.1 mol%, 3 strains (Nakase, 1971b)

Origin of the strain studied

Mushroom (1).

Type strain: CBS 6420.

23. *CANDIDA BOMBI* MONTROCHER (1967)

Standard description of *Candida bombi*

Growth in glucose-yeast extract-peptone water: After 3 days at 25°C the cells are globose to short-ovoid, (2.9–3.6) × (5.8–7.2) μm and occur singly, budding and in small clusters and chains. After one month no pellicle is evident and the sediment is floccose.

Growth on glucose-yeast extract-peptone agar: After one month at 20°C the streak culture is cream-colored to light tan, smooth and soft. The margin is entire to slightly irregular.

Dalmau plate culture on corn meal agar: The pseudomycelium is sparse and consists of a few branched chains of ovoid cells.

Fermentation:

Glucose	+	Maltose	–
Galactose	–	Lactose	–
Sucrose	+	Raffinose	+

Assimilation:

Galactose	–	Melezitose	–	Ribitol	–
L-Sorbose	+	Soluble starch	–	Galactitol	–
Sucrose	+	D-Xylose	–	D-Mannitol	+ s
Maltose	–	L-Arabinose	–	D-Glucitol	+
Cellobiose	–	D-Arabinose	–	Salicin	–
Trehalose	+ s or –	D-Ribose	–	DL-Lactic acid	–
Lactose	–	L-Rhamnose	–	Succinic acid	+
Melibiose	–	Glycerol	+	Citric acid	+ or s
Raffinose	+	Erythritol	–	Inositol	–

Splitting of arbutin: –
Assimilation of nitrate: –
Growth in vitamin-free medium: v
Growth at 37°C: +
Urease: –
Color reaction with DBB: –
G + C: 48.3 mol% (Meyer & Phaff, 1972)

Origin of the strain studied

Bee (1).

Type strain: CBS 5836, from a bee.

24. *CANDIDA BOMBICOLA* (SPENCER, GORIN ET TULLOCH) MEYER ET YARROW (Yarrow & Meyer, 1978)

Synonym

Torulopsis bombicola Spencer, Gorin et Tulloch (1970)

Standard description of *Candida bombicola*

Growth in glucose-yeast extract-peptone water: After 3 days at 25°C the cells are spheroidal to oval, (1.0–2.0) × (1.7–2.6) μm. A ring but no pellicle is formed (Spencer et al., 1970).

Growth on glucose-yeast extract-peptone agar: After one month at room temperature the streak culture is cream-colored to grey, smooth, shiny, convex with an entire to undulating margin.

Dalmau plate culture on corn meal agar: Pseudomycelium is either absent or consists of elongated cells forming branched chains.

Fermentation:

Glucose	+	Maltose	–
Galactose	–	Lactose	–
Sucrose	+	Raffinose	v

Assimilation of carbon compounds:

Galactose	+ s	Melezitose	–	Ribitol	–
L-Sorbose	+ s	Soluble starch	–	Galactitol	–
Sucrose	+	D-Xylose	– or + s	D-Mannitol	+
Maltose	–	L-Arabinose	–	D-Glucitol	+ s
Cellobiose	–	D-Arabinose	–	Salicin	–
Trehalose	–	D-Ribose	–	DL-Lactic acid	–
Lactose	–	L-Rhamnose	–	Succinic acid	+ s
Melibiose	–	Glycerol	+	Citric acid	–
Raffinose	v	Erythritol	–	Inositol	–

Splitting of arbutin: –

Assimilation of nitrate: –

Growth in vitamin-free medium: –

Growth at 37°C: +

Urease: –

Color reaction with DBB: –

G + C: 49.8 mol%, type strain (Stenderup et al., 1972)

Origin of the strain studied

Bumble bee (1).

Type strain: CBS 6009.

25. *CANDIDA BUFFONII* (RAMIREZ) VAN UDEN ET BUCKLEY (1970)

Synonyms

Torulopsis buffoni Ramirez (1957)
Paratorulopsis buffoni (Ramirez) Novák et Zsolt (1961)

Standard description of *Candida buffonii*

Growth in glucose-yeast extract-peptone water: After 3 days at 25°C the cells are elongate, (2–3.5) × (10–22) μm (Fig. 199). A ring may be formed (van Uden & Buckley, 1970).

Growth on glucose-yeast extract-peptone agar: After one month at 25°C the streak culture is greyish-brown, glistening, soft and smooth. On potato agar the streak culture is mucoid (van Uden & Buckley, 1970).

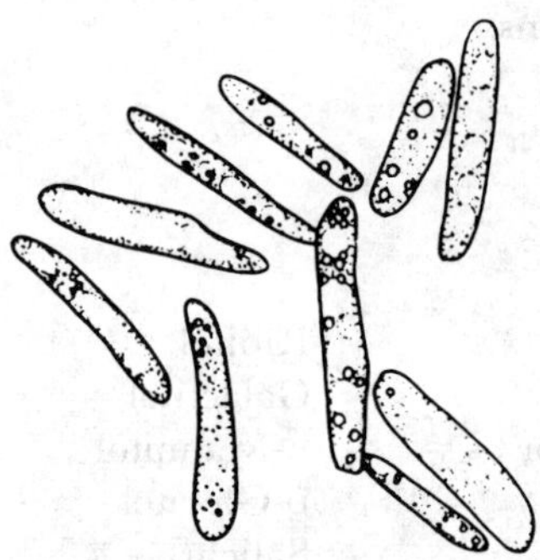

Fig. 199 – *C. buffonii*
In glucose-yeast extract-peptone water
(From 'The Yeasts' ed. II, 1970)

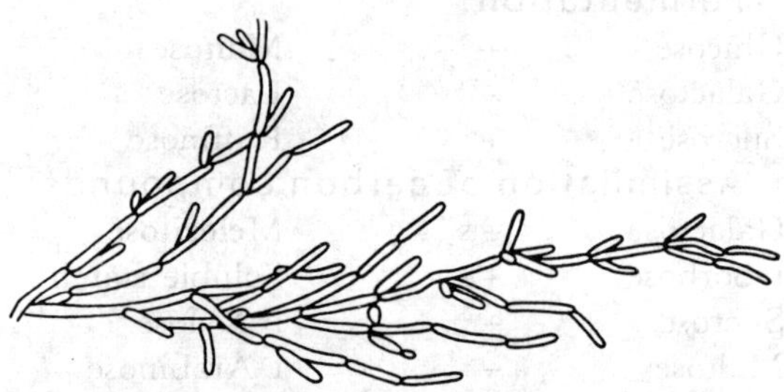

Fig. 200 – *C. buffonii*
Dalmau plate culture on corn meal agar
(From 'The Yeasts' ed. II, 1970)

Dalmau plate culture on corn meal agar: Pseudomycelium is abundantly formed. It consists of branched chains of elongate cells (Fig. 200) (van Uden & Buckley, 1970).

Fermentation: –

Assimilation of carbon compounds:

Galactose	–	Melezitose	+	Ribitol	+
L-Sorbose	+ s	Soluble starch	+	Galactitol	–
Sucrose	–	D-Xylose	+ s	D-Mannitol	+

Maltose	+	L-Arabinose	+ s	D-Glucitol	+
Cellobiose	+	D-Arabinose	–	Salicin	+ s
Trehalose	+	D-Ribose	–	DL-Lactic acid	–
Lactose	–	L-Rhamnose	–	Succinic acid	–
Melibiose	–	Glycerol	+ s	Citric acid	– or + s
Raffinose	–	Erythritol	–	Inositol	–

Splitting of arbutin: + s

Assimilation of nitrate: +

Growth in vitamin-free medium: –

Growth at 37°C: –

Color reaction with DBB: –

G + C: 50.8 mol%, type strain (Meyer, Stenderup & Christiansen, unpublished)

Origin of the strain studied

Carpophore of *Boletus edulis* var. *alba* (Ramirez, 1957) (1).

Type strain: CBS 2838.

26. *CANDIDA BUINENSIS* SONEDA ET UCHIDA (1971)

Standard description of *Candida buinensis*

Growth in glucose-yeast extract-peptone water: After 3 days at 25°C the cells are spheroidal to ovoidal, elongated and cylindrical, (2.0–3.5) × (3.5–6.0) μm, and occur singly, in pairs, clusters and short chains (Fig. 201).

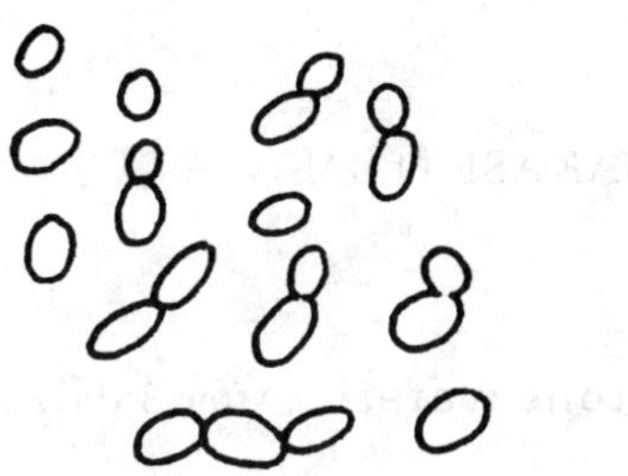

Fig. 201 – *C. buinensis*
In glucose-yeast extract-peptone water

Growth on glucose-yeast extract-peptone agar: After one month at 25°C the streak culture is white to cream-colored, creamy and smooth, convex; margin is entire.

Dalmau plate culture on corn meal agar: The pseudomycelium consists of ramified chains of elongated cells bearing blastospores.

Fermentation:

Glucose	+	Maltose	– or + vw
Galactose	+ s	Lactose	–
Sucrose	–		

Assimilation of carbon compounds:

Galactose	+	Melezitose	+	Ribitol	–
L-Sorbose	+	Soluble starch	–	Galactitol	–

Sucrose	+	D-Xylose	+	D-Mannitol	+
Maltose	+	L-Arabinose	– or + s	D-Glucitol	+ s
Cellobiose	+	D-Arabinose	+	Salicin	+
Trehalose	+	D-Ribose	+ s	DL-Lactic acid	–
Lactose	–	L-Rhamnose	–	Succinic acid	+
Melibiose	–	Glycerol	+ s	Citric acid	+
Raffinose	–	Erythritol	–	Inositol	–

Additional carbon compound tested: 2-Ketogluconate –
Splitting of arbutin: +
Assimilation of nitrate: –
Growth in vitamin-free medium: –
Growth at 37°C: –
Color reaction with DBB: –
G + C: 31.5 mol%, type strain

Comments

Original description reports fermentation of glucose absent; assimilation of D-ribose –, succinic acid –, D-mannitol –, D-glucitol –, salicin –, L-arabinose +. These results could not be confirmed. *C. buinensis* is physiologically similar to *C. sake,* but may be differentiated by growth on D-arabinose.

Origin of the strain studied

Gelatinous materials of tree ferm collected in Buin, 1970 (1).

Type strain: CBS 6796.

27. *CANDIDA BUTYRI* NAKASE (1971b)

Standard description of *Candida butyri*

Growth in glucose-yeast extract-peptone water: After 3 days at 25°C the cells are globose to ovoid, (3.6–5.1) × (3.6–7.2) μm, and occur singly, in pairs and chains (Fig. 202). Pseudomycelium present.

Growth on glucose-yeast extract-peptone agar: After one month at 25°C the streak culture is white to cream-colored, shiny, smooth and wrinkled, fringed with mycelium.

Dalmau plate culture on corn meal agar: Abundant pseudomycelium consists of ramified chains of cells

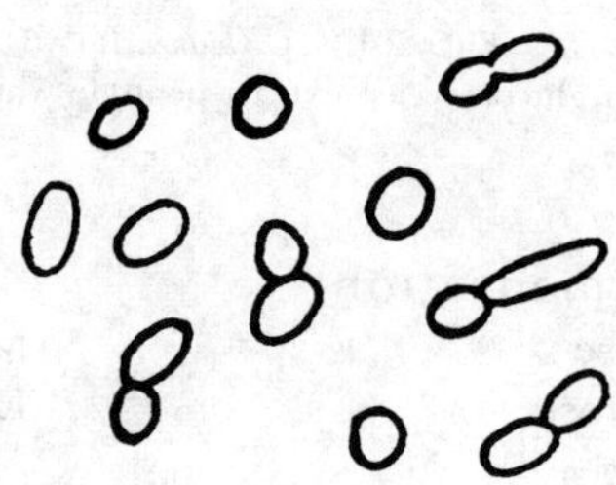

Fig. 202 – *C. butyri*
In glucose-yeast extract-peptone water

bearing groups of blastospores in verticils.

Fermentation:

Glucose	+	Maltose	–
Galactose	+ s or –	Lactose	–
Sucrose	–		

Assimilation of carbon compounds:

Galactose	+	Melezitose	– or + s	Ribitol	+
L-Sorbose	+ s	Soluble starch	–	Galactitol	–
Sucrose	+ or s	D-Xylose	+	D-Mannitol	+
Maltose	+ or s	L-Arabinose	+	D-Glucitol	+
Cellobiose	+ or s	D-Arabinose	+	Salicin	+ or s
Trehalose	v	D-Ribose	+ or s	DL-Lactic acid	–
Lactose	–	L-Rhamnose	–	Succinic acid	+
Melibiose	– or + s	Glycerol	+	Citric acid	+
Raffinose	–	Erythritol	+	Inositol	–

Splitting of arbutin: +

Assimilation of nitrate: –

Growth in vitamin-free medium: –

Growth at 37°C: +

Starch formation: –

Color reaction with DBB: –

G + C: 34.4–34.9 mol%, 2 strains (Nakase, 1971b)

Origin of the strain studied

Butter (1).

Type strain: CBS 6421.

28. CANDIDA CACAOI BUCKLEY ET VAN UDEN (1968)

Standard description of *Candida cacaoi*

Growth in glucose-yeast extract-peptone water: After 3 days at 25°C the cells are short-ovoid to long-ovoid, (2.5–4) × (3–7) μm; they often occur in chains and groups (Fig. 203). A thin, dry pellicle is formed (van Uden & Buckley, 1970).

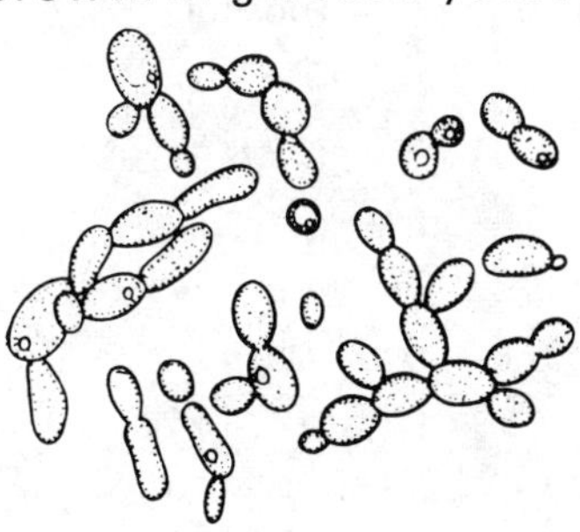

Fig. 203 – *C. cacaoi*
In glucose-yeast extract-peptone water
(From 'The Yeasts' ed. II, 1970)

Growth on glucose-yeast extract-peptone agar: After one month at 25°C the streak culture is whitish, dull, soft and smooth (van Uden & Buckley, 1970).

Dalmau plate culture on corn

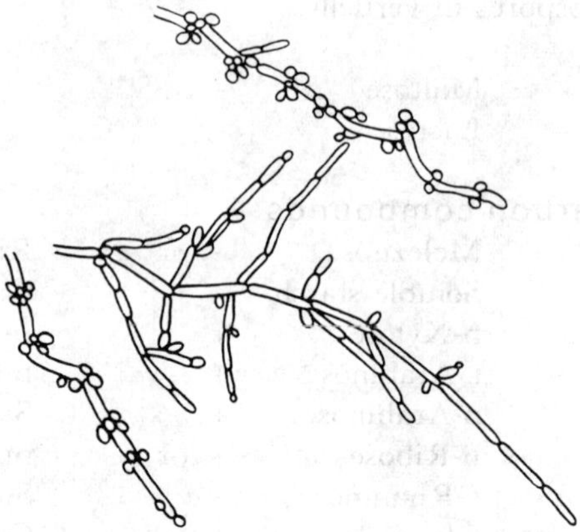

Fig. 204 – *C. cacaoi*
Dalmau plate culture on corn meal agar (From 'The Yeasts' ed. II, 1970)

meal agar: The pseudomycelium consists of ramified chains of pseudohyphae bearing chains and groups of roundish blastospores in verticillated positions (Fig. 204) (van Uden & Buckley, 1970).

Fermentation:

Glucose	+	Maltose	–
Galactose	–	Lactose	–
Sucrose	–	Trehalose	+

Assimilation of carbon compounds:

Galactose	+	Melezitose	–	Ribitol	+
L-Sorbose	–	Soluble starch	–	Galactitol	–
Sucrose	–	D-Xylose	+	D-Mannitol	+
Maltose	–	L-Arabinose	v	D-Glucitol	+
Cellobiose	+	D-Arabinose	–	Salicin	+
Trehalose	+	D-Ribose	+ s	DL-Lactic acid	–
Lactose	–	L-Rhamnose	–	Succinic acid	+
Melibiose	–	Glycerol	+	Citric acid	–
Raffinose	–	Erythritol	+	Inositol	–

Splitting of arbutin: +
Assimilation of nitrate: –
Growth in vitamin-free medium: +
Growth at 37°C: +
Color reaction with DBB: –
G + C: 41.7 mol% (Meyer & Phaff, 1972)

Origin of the strain studied

Fermenting cacao in Trinidad (1).

Type strain: CBS 2020.

29. *CANDIDA CANTARELLII* (VAN DER WALT ET VAN KERKEN) MEYER ET YARROW (Yarrow & Meyer, 1978)

Synonyms

Torulopsis cantarellii van der Walt et van Kerken (1961b)
Torulopsis vinacea nom. nud. (Ohara et al., 1964)

Standard description of *Candida cantarellii*

Growth in glucose-yeast extract-peptone water: After 3 days at 25°C the cells are globose, short-ovoid to ovoid, (3.0–5.0) × (4.5–6.5) μm (Fig. 205). Only a sediment is formed (van Uden & Vidal-Leiria, 1970).

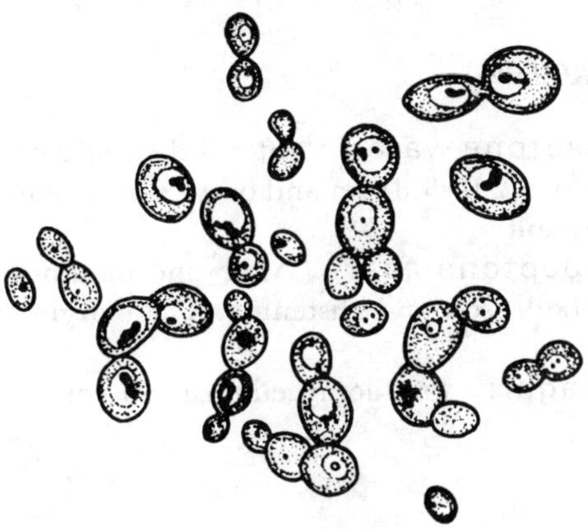

Fig. 205 – *C. cantarellii*
In glucose-yeast extract-peptone water
(From 'The Yeasts' ed. II, 1970)

Growth on glucose-yeast extract-peptone agar: After one month at 25°C the streak culture is cream-colored, butyrous, dull, soft and wrinkled (van Uden & Vidal-Leiria, 1970).

Dalmau plate culture on corn meal agar: No pseudomycelium is formed (van Uden & Vidal-Leiria, 1970).

Fermentation:

Glucose	+	Maltose	–
Galactose	–	Lactose	–
Sucrose	–	Trehalose	+ w

Assimilation of carbon compounds:

Galactose	–	Melezitose	–	Ribitol	+
L-Sorbose	+	Soluble starch	–	Galactitol	–
Sucrose	–	D-Xylose	–	D-Mannitol	+
Maltose	–	L-Arabinose	–	D-Glucitol	+
Cellobiose	–	D-Arabinose	–	Salicin	–
Trehalose	+	D-Ribose	–	DL-Lactic acid	+ s
Lactose	–	L-Rhamnose	–	Succinic acid	+ w
Melibiose	–	Glycerol	+	Citric acid	–
Raffinose	–	Erythritol	+	Inositol	–

Splitting of arbutin: –
Assimilation of nitrate: –
Growth in vitamin-free medium: +
Growth at 37°C: v

Color reaction with DBB: –
G + C: 41.9 mol%, type strain (Meyer, Stenderup & Christiansen, unpublished)

Origin of the strains studied (van Uden & Vidal-Leiria, 1970)

Grape must in South Africa (van der Walt & van Kerken, 1961b) (1); grape must in Japan (Ohara et al., 1964) (3).

Type strain: CBS 4878, from grape must in South Africa.

30. *CANDIDA CARIOSILIGNICOLA* LEE ET KOMAGATA (1980)

Standard description of *Candida cariosilignicola*

Growth in glucose-yeast extract-peptone water: After 3 days at 25°C the cells are globose to slightly ovoid, $(1.5–3.5) \times (2.0–3.8)$ μm and occur singly, budding and in small clusters. A sediment is present.

Growth on glucose-yeast extract-peptone agar: After one month at 25°C the streak culture is white, creamy, smooth, soft and glistening. The margin is entire.

Dalmau plate culture on corn meal agar: Pseudomycelium is absent.

Fermentation:

Glucose	+	Maltose	–
Galactose	–	Lactose	–
Sucrose	+	Raffinose	+

Assimilation of carbon compounds:

Galactose	–	Melezitose	–	Ribitol	+
L-Sorbose	+	Soluble starch	–	Galactitol	+ s, w
Sucrose	+	D-Xylose	+	D-Mannitol	+
Maltose	–	L-Arabinose	+	D-Glucitol	+
Cellobiose	–	D-Arabinose	+	Salicin	–
Trehalose	+	D-Ribose	+	DL-Lactic acid	–
Lactose	–	L-Rhamnose	+	Succinic acid	–
Melibiose	–	Glycerol	+	Citric acid	+
Raffinose	+	Erythritol	+	Inositol	–

Additional carbon compound tested: Methanol +
Splitting of arbutin: –
Assimilation of nitrate: +
Growth in vitamin-free medium: –
Growth at 37°C: +
Urease: –
Color reaction with DBB: –
Coenzyme Q system: Q 7 (Lee & Komagata, 1980)
G + C: 35.1 mol%, type strain (Lee & Komagata, 1980)

Origin of the strain studied

Decayed timber in Yamanashi Prefecture, Japan (1).

Type strain: CBS 8001.

31. CANDIDA CASTELLII (CAPRIOTTI) MEYER ET YARROW (Yarrow & Meyer, 1978)

Synonym

Torulopsis castellii Capriotti (1961d)

Standard description of *Candida castellii*

Growth in glucose-yeast extract-peptone water: After 3 days at 25°C the cells are ovoid to cylindrical, (1.5–2.5) × (3.5–5) μm (Fig. 206). Only a sediment is formed (van Uden & Buckley, 1970).

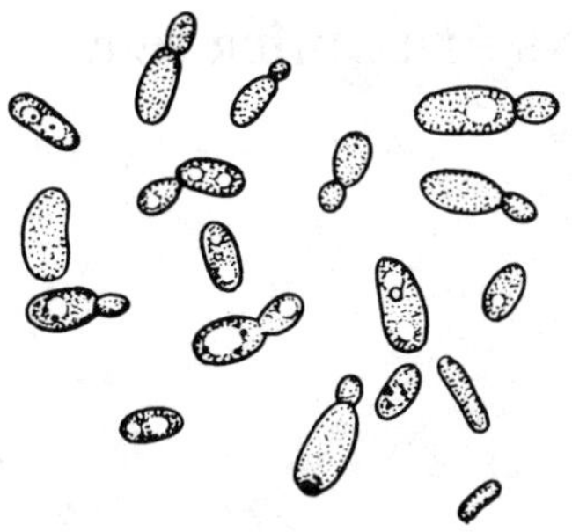

Fig. 206 – *C castellii*
In glucose-yeast extract-peptone water
(From 'The Yeasts' ed. II, 1970)

Growth on glucose-yeast extract-peptone agar: After one month at 25°C the streak culture is cream-colored, glossy, soft and smooth (van Uden & Buckley, 1970).

Dalmau plate culture on corn meal agar: No pseudomycelium is formed.

Fermentation:

Glucose	+	Maltose	–
Galactose	–	Lactose	–
Sucrose	–	Trehalose	–

Assimilation of carbon compounds:

Galactose	–	Melezitose	–	Ribitol	–
L-Sorbose	–	Soluble starch	–	Galactitol	–
Sucrose	–	D-Xylose	–	D-Mannitol	–
Maltose	–	L-Arabinose	–	D-Glucitol	–
Cellobiose	–	D-Arabinose	–	Salicin	–
Trehalose	+	D-Ribose	–	DL-Lactic acid	–
Lactose	–	L-Rhamnose	–	Succinic acid	–
Melibiose	–	Glycerol	+	Citric acid	–
Raffinose	–	Erythritol	–	Inositol	–

Splitting of arbutin: –

Assimilation of nitrate: –

Growth in vitamin-free medium: –
Growth at 37°C: +
Color reaction with DBB: –
G + C: 43.3 mol%, type strain (Meyer, Stenderup & Christiansen, unpublished)

Comments

Candida castellii can be differentiated from *C. glabrata* on vitamin requirements: inositol is required by *C. castellii* and not by *C. glabrata* and pyridoxin is required by *C. glabrata* and not by *C. castellii*.

Origin of the strains studied

Soil in Finland (1); vagina (1).

Type strain: CBS 4332.

32. *CANDIDA CATENULATA* DIDDENS ET LODDER (1942)

Synonyms

Monilia rugosa nom. nud.
Blastodendrion brumptii nom. nud. (Guerra, 1935)
Candida brumptii nom. nud. (Langeron & Guerra, 1938)
Candida ravautii nom. nud. (Langeron & Guerra, 1938)

Standard description of *Candida catenulata*

Growth in glucose-yeast extract-peptone water: After 3 days at 25°C the cells are ovoid or elongate, often sausage-like or pear-like in shape, (1.5–5.5) × (2.9–8.8) μm, and occur singly, budding and in short chains (Fig. 207). Pseudomycelial cells are present. A pellicle is not present, although islets appear and growth is creeping up the sides of the flask.

Growth on glucose-yeast extract-peptone agar: After one month at 25°C the streak culture is greyish to cream-colored, semidull, soft and wrinkled.

Dalmau plate culture on corn meal agar: The pseudomycelium consists of ramified chains of short pseudohyphae or long-ovoid cells. The

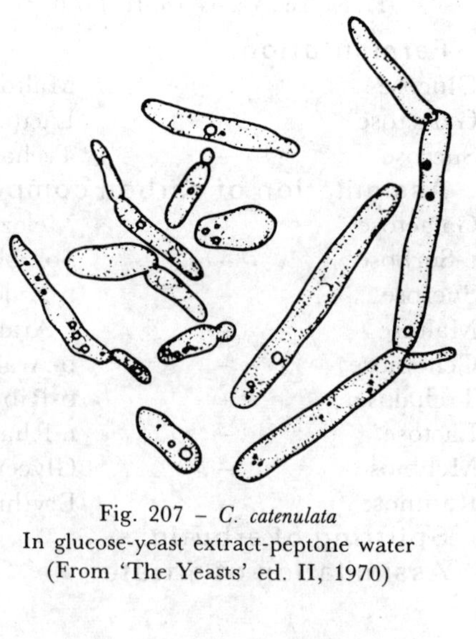

Fig. 207 – *C. catenulata*
In glucose-yeast extract-peptone water
(From 'The Yeasts' ed. II, 1970)

pseudohyphae are often curved or wavy and may display swollen parts (Fig. 208).

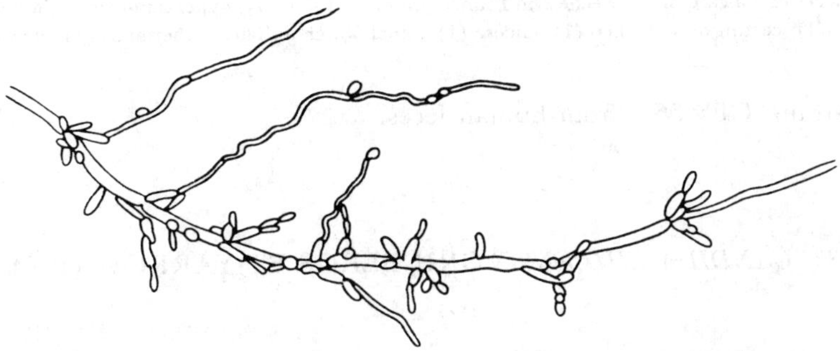

Fig. 208 – *C. catenulata*
Dalmau plate culture on corn meal agar (From 'The Yeasts' ed. II, 1970)

Fermentation:

Glucose	v	Maltose	–
Galactose	–	Lactose	–
Sucrose	–		

Assimilation of carbon compounds:

Galactose	+	Melezitose	–	Ribitol	–
L-Sorbose	–	Soluble starch	+	Galactitol	–
Sucrose	–	D-Xylose	+	D-Mannitol	+
Maltose	+	L-Arabinose	–	D-Glucitol	+
Cellobiose	–	D-Arabinose	–	Salicin	–
Trehalose	–	D-Ribose	–	DL-Lactic acid	+
Lactose	–	L-Rhamnose	–	Succinic acid	v
Melibiose	–	Glycerol	+	Citric acid	v
Raffinose	–	Erythritol	–	Inositol	–

Splitting of arbutin: –
Assimilation of nitrate: –
Growth in vitamin-free medium: +
Growth at 37°C: v
Urease: –
Color reaction with DBB: –
Coenzyme Q system: Q 9 (Yamada & Kondo, 1972)
G + C: 54.5 mol%, type strain (Stenderup & Bak, 1968)

Comments

DNA reassociation studies revealed *C. catenulata, C. brumptii* and *C. ravautii* to be conspecific (Simione & Meyer, 1978). Since both *C. brumptii* and *C. ravautii* were described in 1935 without a Latin description, and *C. catenulata* was described properly in 1942, *C. catenulata* has priority and *C. brumptii* and *C. ravautii* are placed in synonymy.

Origin of the strains studied

Human feces (1); a case of "perlèche" in France (Guerra, 1935) (1); hyperkeratosis of the foot in France (*C. ravautii*) (1); cecum of a chicken (1); cheese (1); canal water in The Netherlands (1); skin (3); mastitis milk (1).

Type strain: CBS 565, from human feces.

33. *CANDIDA CHILENSIS* GRINBERGS ET YARROW (1970a)

Standard description of *Candida chilensis*

Growth in glucose-yeast extract-peptone water: After 3 days at 25°C the cells are short-ovoid to elongate, (3.6–5.8) × (4.3–7.2) μm, pseudomycelium is formed (Fig. 209). After one month a pellicle is formed and the cell suspension is mucoid.

Growth on malt agar: After one month at 20°C the streak culture is brownish, cream-colored to beige, shiny, smooth, and butyrous. The margin is entire to slightly fringed.

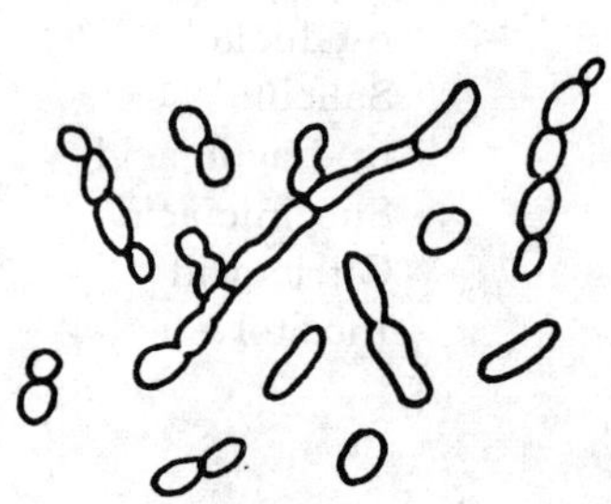

Fig. 209 – *C. chilensis*
In glucose-yeast extract-peptone water

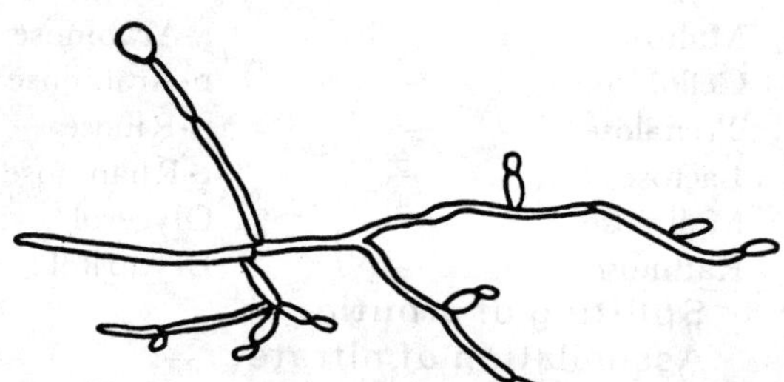

Fig. 210 – *C. chilensis*
Dalmau plate culture on corn meal agar

Dalmau plate culture on corn meal agar: The pseudomycelium consists of long wavy pseudohyphae bearing short chains and clusters of blastospores (Fig. 210).

Fermentation:

Glucose	+ or s	Maltose	–
Galactose	–	Lactose	–
Sucrose	–		

Assimilation of carbon compounds:

Galactose	+	Melezitose	+	Ribitol	+
L-Sorbose	+	Soluble starch	–	Galactitol	–
Sucrose	+	D-Xylose	+	D-Mannitol	+
Maltose	+	L-Arabinose	–	D-Glucitol	+
Cellobiose	+	D-Arabinose	+	Salicin	+
Trehalose	+	D-Ribose	+	DL-Lactic acid	+
Lactose	+ s	L-Rhamnose	–	Succinic acid	+
Melibiose	–	Glycerol	+	Citric acid	+
Raffinose	–	Erythritol	+	Inositol	–

Splitting of arbutin: +

Assimilation of nitrate: +

Growth in vitamin-free medium: –

Growth at 37°C: –

Color reaction with DBB: –

G + C: 44.0 mol%, type strain (Meyer, Stenderup & Christiansen, unpublished)

Origin of the strains studied

Rotten tree stump (2).

Type strain: CBS 5719.

34. CANDIDA CHIROPTERORUM GROSE ET MARINKELLE (1968)

Standard description of *Candida chiropterorum*

Growth in glucose-yeast extract-peptone water: After 3 days at 25°C the cells are spheroidal to ovoidal, (2.2–2.8) × (4.4–7.0) μm, and occur in branched chains. Few single or budding cells are evident. After one month a thick sediment is present.

Growth on glucose-yeast extract-peptone agar: After one month at 25°C the streak culture is whitish to cream-colored; hairy projections may be present giving a rough feathery appearance; the cross section is convex and the margin is fringed with mycelium.

Dalmau plate culture on corn meal agar: True mycelium is abundantly formed and consists of ramified chains of elongate cells bearing clusters of blastospores in verticillate positions (Fig. 211).

Fermentation: –

Assimilation of carbon compounds:

Galactose	+	Melezitose	+ s	Ribitol	+
L-Sorbose	+	Soluble starch	+	Galactitol	+
Sucrose	+	D-Xylose	+	D-Mannitol	+
Maltose	+	L-Arabinose	+ s	D-Glucitol	+
Cellobiose	+	D-Arabinose	+ s	Salicin	+

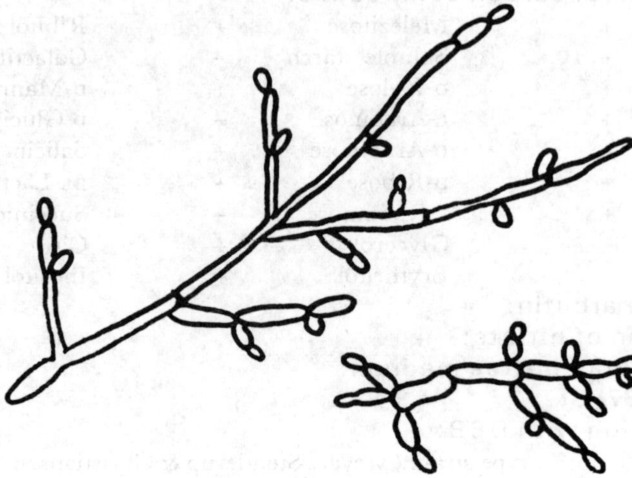

Fig. 211 – *C. chiropterorum*
Dalmau plate culture on corn meal agar

Trehalose	+	D-Ribose	+	DL-Lactic acid	+ s
Lactose	–	L-Rhamnose	+	Succinic acid	–
Melibiose	–	Glycerol	+	Citric acid	+
Raffinose	+ s	Erythritol	+	Inositol	+

Splitting of arbutin: +
Assimilation of nitrate: –
Growth in vitamin-free medium: –
Growth on 50% (w/w) glucose-yeast extract agar: + ; on 60%: –
Growth at 37°C: +
Color reaction with DBB: –
G + C: 48.0 mol%, type strain (Meyer, Stenderup & Christiansen, unpublished)

Comments

In the original description it was stated that lactose was assimilated and melezitose was not. These reactions could not be confirmed.

Origin of the strain studied

Liver of *Mormoops megalophylla* (1).

Type strain: CBS 6064.

35. CANDIDA CIFERRII KREGER-VAN RIJ

see *Stephanoascus ciferrii* p. 431.

36. *CANDIDA CITREA* NAKASE (1971c)

Standard description of *Candida citrea*

Growth in glucose-yeast extract-peptone water: After 4 days at 25°C the cells are globose to ovoid to elongate, (3.4–5.6) × (4.5–11.2) μm, and occur singly and in pairs (Fig. 212). Pseudomycelium is present.

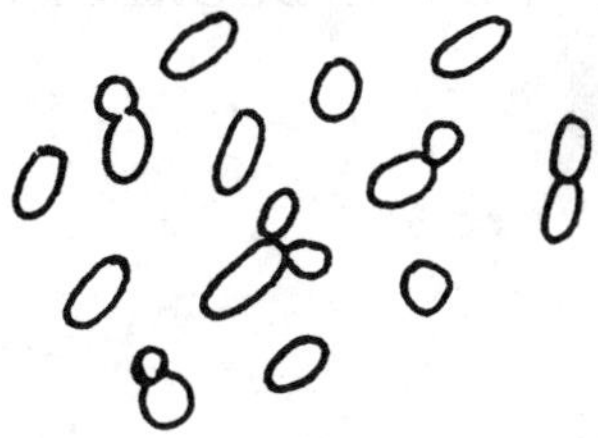

Fig. 212 – *C. citrea*
In glucose-yeast extract-peptone water

Growth on glucose-yeast extract-peptone agar: After one month at 25°C the streak culture is white, flat, dry, smooth, butyrous and entire.

Dalmau plate culture on corn meal agar: The pseudomycelium consists of elongate cells bearing chains and clusters of ovoid to somewhat triangular blastospores.

Fermentation:

Glucose	+	Maltose	–
Galactose	–	Lactose	–
Sucrose	–		

Assimilation of carbon compounds:

Galactose	–	Melezitose	–	Ribitol	–
L-Sorbose	–	Soluble starch	–	Galactitol	–
Sucrose	–	D-Xylose	– or + s	D-Mannitol	–
Maltose	–	L-Arabinose	–	D-Glucitol	–
Cellobiose	–	D-Arabinose	–	Salicin	–
Trehalose	–	D-Ribose	–	DL-Lactic acid	–
Lactose	–	L-Rhamnose	–	Succinic acid	+
Melibiose	–	Glycerol	+	Citric acid	v
Raffinose	–	Erythritol	–	Inositol	–

Additional carbon compound tested: Glucono-δ-lactone +
Splitting of arbutin: –
Assimilation of nitrate: –
Growth in vitamin-free medium: –
Growth at 37°C: –
Color reaction with DBB: –
G + C: 32.9–33.2 mol%, 3 strains (Nakase, 1971c)

Origin of the strains studied

Lemon (2); banana (1).

Type strain: CBS 6374, from lemon.

37. *CANDIDA COLLICULOSA* (HARTMANN) MEYER ET YARROW

see *Torulaspora delbrueckii* p. 435.

38. *CANDIDA CONGLOBATA* (REDAELLI) VAN UDEN ET BUCKLEY (1970)

Synonym

Torulopsis conglobata Redaelli (1925)

Standard description of *Candida conglobata*

Growth in glucose-yeast extract-peptone water: After 3 days at 25°C the cells are ovoid to cylindrical, (3–5) × (7–20) μm (Fig. 213). A dry pellicle may be present (van Uden & Buckley, 1970).

Growth on glucose-yeast extract-peptone agar: After one month at 25°C the streak culture is yellowish to cream-colored, semiglistening, somewhat membranous, delicately wrinkled (van Uden & Buckley, 1970).

Dalmau plate culture on corn meal agar: The pseudomycelium consists of long, delicate pseudohyphae which branch on wide angles; blastospores are scarce (Fig. 214) (van Uden & Buckley, 1970).

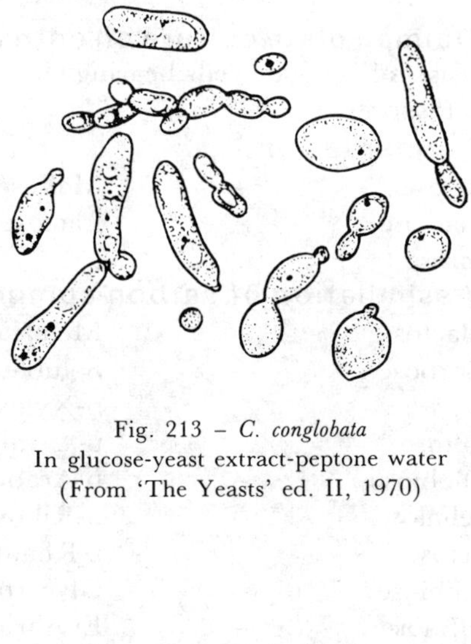

Fig. 213 – *C. conglobata*
In glucose-yeast extract-peptone water
(From 'The Yeasts' ed. II, 1970)

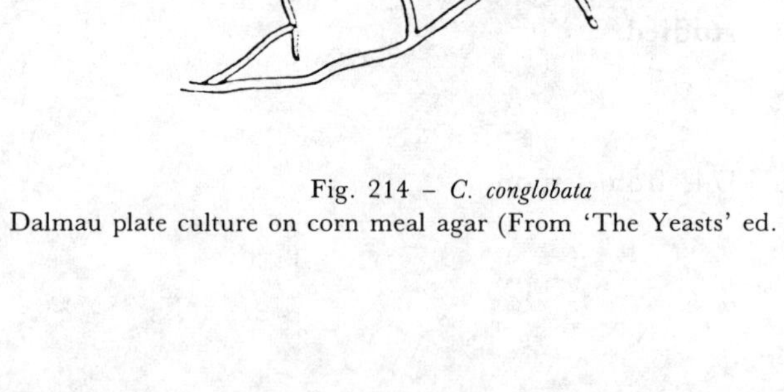

Fig. 214 – *C. conglobata*
Dalmau plate culture on corn meal agar (From 'The Yeasts' ed. II, 1970)

Fermentation:

Glucose	+	Lactose	–
Galactose	+ w	Trehalose	+
Sucrose	–	Cellobiose	v
Maltose	–		

Assimilation of carbon compounds:

Galactose	+	Melezitose	–	Ribitol	+
L-Sorbose	+	Soluble starch	–	Galactitol	–
Sucrose	–	D-Xylose	+	D-Mannitol	+
Maltose	–	L-Arabinose	+	D-Glucitol	+
Cellobiose	+ or w	D-Arabinose	+ s	Salicin	+
Trehalose	+	D-Ribose	+	DL-Lactic acid	v
Lactose	–	L-Rhamnose	–	Succinic acid	+
Melibiose	–	Glycerol	+	Citric acid	–
Raffinose	–	Erythritol	+	Inositol	–

Additional carbon compound tested: Methanol –

Splitting of arbutin: +

Assimilation of nitrate: –

Growth in vitamin-free medium: –

Growth at 37°C: –

Color reaction with DBB: –

G + C: 39.2–40.9 mol%, 2 strains (Meyer, Stenderup & Christiansen, unpublished)

Origin of the strains studied

Human origin (1); unknown (1).

Type strain: CBS 2018, isolated by Redaelli (1925).

39. *CANDIDA CURIOSA* KOMAGATA ET NAKASE (1965)

Standard description of *Candida curiosa*

Growth in malt extract: After 3 days at 15°C the cells are short-oval to oval, (2.6–6.5) × (3–9) μm, and occur singly or in pairs (Fig. 215). Occasionally septate cells are observed. Only a sediment is formed. After a month at 10°C a ring is also present (Komagata & Nakase, 1965).

Growth on malt agar: After one month at 10°C the streak culture is yellowish white to light brownish grey, raised, dull, mucoid and smooth or wrinkled. The margin may be fringed with mycelium (Komagata & Nakase, 1965).

Dalmau plate culture on corn meal agar: Pseudo- and true mycelium are abundantly developed. Blastospores are short-oval to oval and occur singly or in chains or in verticils (Fig. 216) (Komagata & Nakase, 1965).

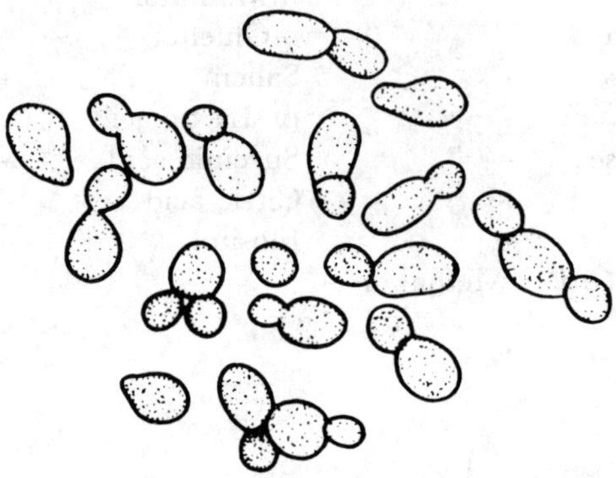

Fig. 215 – *C. curiosa*
In glucose-yeast extract-peptone water
(From 'The Yeasts' ed. II, 1970)

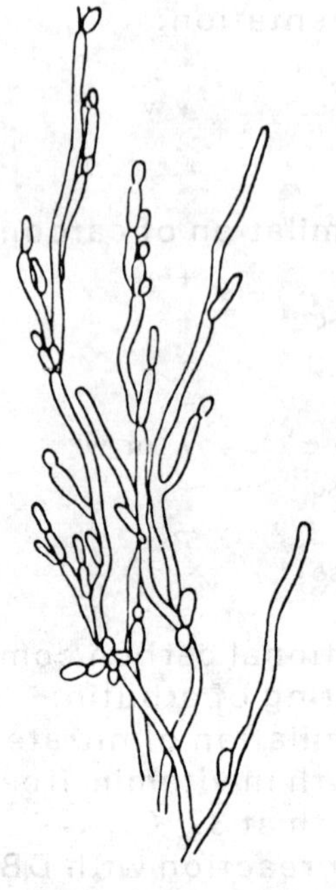

Fig. 216 – *C. curiosa*
Dalmau plate culture on corn meal agar
(From 'The Yeasts', ed. II. 1970)

Fermentation:

Glucose	+	Lactose	–
Galactose	+ s or –	Raffinose	+ s
Sucrose	+	Melibiose	–
Maltose	–	Trehalose	–

Assimilation of carbon compounds:

Galactose	+	Melezitose	–	Ribitol	+
L-Sorbose	+ s	Soluble starch	–	Galactitol	–
Sucrose	+	D-Xylose	+ s	D-Mannitol	+
Maltose	–	L-Arabinose	+	D-Glucitol	+
Cellobiose	+	D-Arabinose	+ s	Salicin	+ s
Trehalose	+	D-Ribose	–	DL-Lactic acid	–
Lactose	+ s or –	L-Rhamnose	v	Succinic acid	+ s
Melibiose	–	Glycerol	+ s or –	Citric acid	–
Raffinose	+	Erythritol	–	Inositol	–

Splitting of arbutin: +
Assimilation of nitrate: +
Growth in vitamin-free medium: –
Growth at 37°C: – ; **25°C:** –
Starch formation: +
Urease: +
Color reaction with DBB: +

Coenzyme Q system: Q 8 (Yamada & Kondo, 1972)
G + C: 52.9 mol% (Meyer, Stenderup & Christiansen, unpublished)

Comments

Van Uden & Buckley (1970) recognized *C. nivalis* as a synonym of *C. curiosa.* Since Fell et al. (1969) observed the formation of teliospores in a strain of *C. nivalis*, and described it as a new species of *Leucosporidium (Leu. nivalis), C. nivalis* is no longer considered a synonym of *C. curiosa.* No sexual state of *C. curiosa* has thus far been reported.

Origin of the strain studied

Frozen food in Japan (1).

Type strain: CBS 5688.

40. CANDIDA CURVATA (DIDDENS ET LODDER) LODDER ET KREGER-VAN RIJ (1952)

Synonyms

Candida heveanensis (Groenewege) Diddens et Lodder var. *curvata* Diddens et Lodder (1942)
Azymocandida curvata (Diddens et Lodder) Novák et Zsolt (1961)
Apiotrichum curvatum (Diddens et Lodder) von Arx et Weijman (1979)
Vanrija curvata (Diddens et Lodder) Moore (1980)

Standard description of *Candida curvata* (van Uden & Buckley, 1970)

Growth in glucose-yeast extract-peptone water: After 3 days at 25°C the cells are ovoid to sausage-shaped, often curved, (2–3) × (6–10) μm, many cells show lateral budding on a rather broad base (Fig. 217). A thin, dull, creeping pellicle is usually formed.

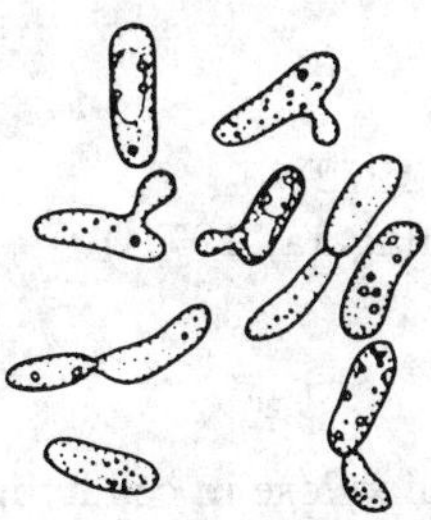

Fig. 217 – *C. curvata*
In glucose-yeast extract-peptone water
(From 'The Yeasts' ed. II, 1970)

Growth on glucose-yeast extract-peptone agar: After one month at 25°C the streak culture is either brownish, mucoid, soft and smooth or yellowish, moist to dull, soft and wrinkled.

Dalmau plate culture on corn meal agar: The pseudomycelium consists either of branched chains of short pseudohyphae or long, wavy pseudohyphae bearing a few blastospores (Fig. 218).

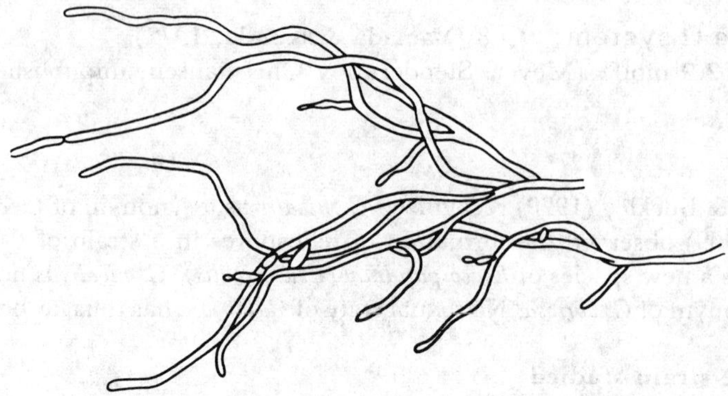

Fig. 218 – *C. curvata*
Dalmau plate culture on corn meal agar (From 'The Yeasts' ed. II, 1970)

Fermentation: –

Assimilation of carbon compounds:

Galactose	+	Melezitose	v	Ribitol	+ or w
L-Sorbose	v	Soluble starch	+	Galactitol	–
Sucrose	+	D-Xylose	+	D-Mannitol	v
Maltose	+ or w	L-Arabinose	v	D-Glucitol	v
Cellobiose	+	D-Arabinose	–	Salicin	+
Trehalose	+	D-Ribose	+	DL-Lactic acid	v
Lactose	+	L-Rhamnose	+ or w	Succinic acid	v
Melibiose	–	Glycerol	+	Citric acid	v
Raffinose	+	Erythritol	+ or w	Inositol	+ or w

Splitting of arbutin: +
Assimilation of nitrate: –
Growth in vitamin-free medium: + w
Growth at 37°C: v
Urease: +
Color reaction with DBB: +
Coenzyme Q system: Q 10 (Yamada & Kondo, 1972)
G + C: 59.8–60.3 mol%, 2 strains (Nakase & Komagata, 1971f)

Comments

Because of the type of conidiogenesis and the presence of xylose in cell hydrolysates, von Arx and Weijman (1979) transferred *C. curvata* to *Apiotrichum* as *Apiotrichum curvatum*.

Origin of the strains studied (van Uden & Buckley, 1970)

Human sputum (2); human feces (1); human urine (2); tissue culture medium (1); uterus of a cow (1); uncertain origin (3).

Type strain: CBS 570, isolated in The Netherlands from sputum.

41. *CANDIDA DATTILA* (KLUYVER) MEYER ET YARROW

see *Kluyveromyces thermotolerans* p. 247.

42. *CANDIDA DEFORMANS* (ZACH) LANGERON ET GUERRA (1938)

Synonyms

Pseudomonilia deformans Zach (in Wolfram & Zach, 1934)
Candida lipolytica (Harrison) Diddens et Lodder var. *deformans* (Zach) van Uden et Buckley (1970)

Standard description of *Candida deformans*

Growth in glucose-yeast extract-peptone water: After 3 days at 25°C the cells are oval to elongate, (3.6–6.5) × (5.1–10.8) μm, and occur singly, budding and in chains. Pseudomycelium is abundant. No distinct pellicle is evident and the sediment is floccose.

Growth on glucose-yeast extract-peptone agar: After one month at 25°C the streak culture is white, folded and spreading throughout the tube; very tough and difficult to pick up.

Dalmau plate culture on corn meal agar: Branched pseudohyphae and septate hyphae are present. Blastospores are arranged singly, in short chains and clusters.

Fermentation: –

Assimilation of carbon compounds:

Galactose	–	Melezitose	–	Ribitol	+ w, s
L-Sorbose	+	Soluble starch	–	Galactitol	–
Sucrose	–	D-Xylose	–	D-Mannitol	+
Maltose	–	L-Arabinose	–	D-Glucitol	+
Cellobiose	+	D-Arabinose	–	Salicin	+
Trehalose	–	D-Ribose	+ w, s	DL-Lactic acid	+
Lactose	–	L-Rhamnose	–	Succinic acid	+ or w
Melibiose	–	Glycerol	+	Citric acid	+
Raffinose	–	Erythritol	+	Inositol	–

Splitting of arbutin: –
Assimilation of nitrate: –
Growth in vitamin-free medium: –
Growth at 37°C: –
Urease: +
Color reaction with DBB: –

Comments

Recently, Yarrow observed ascospores in mixtures prepared from the type strain of *C. deformans* and a fresh isolate of *Saccharomycopsis lipolytica*. Therefore, *C. deformans* may be considered a synonym of *S. lipolytica*.

Origin of the strain studied

Nail (1).

Type strain: CBS 2071.

43. *CANDIDA DENDRICA* (VAN DER WALT, VAN DER KLIFT ET SCOTT) MEYER ET YARROW (Yarrow & Meyer, 1978)

Synonym

Torulopsis dendrica van der Walt, van der Klift et Scott (van der Walt et al., 1971b)

Standard description of *Candida dendrica*

Growth in glucose-yeast extract-peptone water: After 3 days at 25°C the cells are spheroidal to ovoid, ellipsoidal and cylindrical, (2.0–5.5) × (3.0–6.5) μm, and occur in pairs, short chains and clusters (Fig. 219). A sediment is formed. After one month at 20°C only a sediment is present.

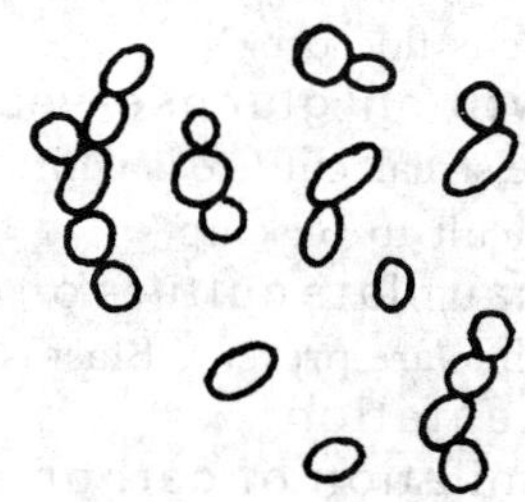

Fig. 219 – *C. dendrica*
In glucose-yeast extract-peptone water

Growth on glucose-yeast extract-peptone agar: After one month at 25°C the streak culture is smooth, shiny, cream-colored. The margin is lobate to undulating.

Dalmau plate culture on corn meal agar: No pseudomycelium is produced.

Fermentation:

Glucose	+ s or vs	Maltose	–
Galactose	–	Lactose	–
Sucrose	–		

Assimilation of carbon compounds:

Galactose	–	Melezitose	–	Ribitol	–
L-Sorbose	–	Soluble starch	–	Galactitol	–
Sucrose	–	D-Xylose	–	D-Mannitol	–
Maltose	–	L-Arabinose	–	D-Glucitol	–
Cellobiose	+ or s	D-Arabinose	–	Salicin	+
Trehalose	–	D-Ribose	–	DL-Lactic acid	–
Lactose	–	L-Rhamnose	–	Succinic acid	+
Melibiose	–	Glycerol	+	Citric acid	+
Raffinose	–	Erythritol	–	Inositol	–

Splitting of arbutin: +

Assimilation of nitrate: –

Growth in vitamin-free medium: –
Growth at 37°C: –
Color reaction with DBB: –
G + C: 43 mol%, type strain (Meyer, Stenderup & Christiansen, unpublished)

Origin of the strain studied

Frass of Cerambycid larvae infesting *Cryptocarya latifolia* Sond. near Lake Sibaya (1).

Type strain: CBS 6151.

44. *CANDIDA DENDRONEMA* VAN DER WALT, VAN DER KLIFT ET SCOTT (van der Walt et al., 1971a)

Standard description of *Candida dendronema*

Growth in glucose-yeast extract-peptone water: After 3 days at 25°C the cells are various shapes: spheroidal, ovoid, elongated, curved, teardrop, and triangular,(2.2–5.1) × (3.6–5.1) μm, and occur singly, budding and in clusters or short chains (Fig. 220). Pseudomycelium is present. No pellicle is formed. After one month a thick sediment and ring are present. Growth is floccose or clumpy.

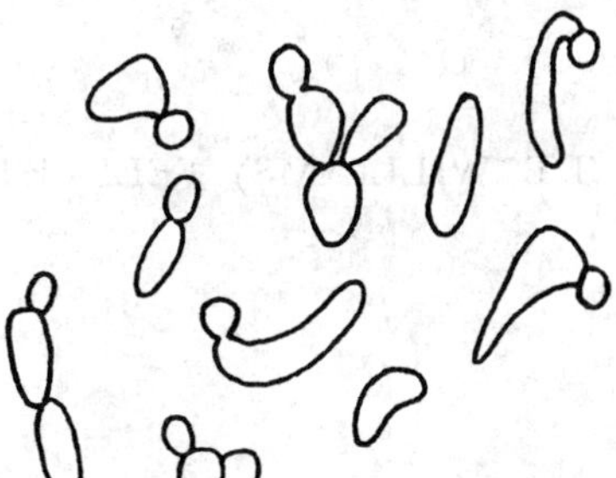

Fig. 220 – *C. dendronema*
In glucose-yeast extract-peptone water

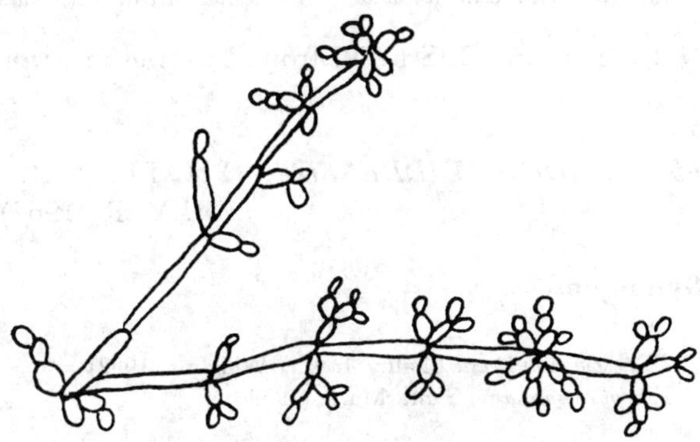

Fig. 221 – *C. dendronema*
Dalmau plate culture on corn meal agar

Growth on glucose-yeast extract-peptone agar: After one month at 25°C the streak culture is cream-colored, raised, wrinkled, lacy-appearing with some pseudomycelium evident.

Dalmau plate culture on corn meal agar: Pseudomycelium consists of ramified chains of cells, some curved, with round, triangular and teardrop-shaped blastospores arranged in grape-like clusters (Fig. 221).

Fermentation:

Glucose	+ or s	Maltose	–
Galactose	+ s	Lactose	–
Sucrose	– or + s		

Assimilation of carbon compounds:

Galactose	+	Melezitose	–	Ribitol	+
L-Sorbose	+	Soluble starch	– or +s	Galactitol	+
Sucrose	+	D-Xylose	+	D-Mannitol	+
Maltose	+	L-Arabinose	+	D-Glucitol	+
Cellobiose	+	D-Arabinose	+	Salicin	+
Trehalose	+	D-Ribose	+	DL-Lactic acid	–
Lactose	–	L-Rhamnose	+	Succinic acid	+
Melibiose	–	Glycerol	+	Citric acid	+
Raffinose	–	Erythritol	+	Inositol	–

Additional carbon compounds tested: Hexadecane +, tetradecane +s
Splitting of arbutin: +
Assimilation of nitrate: –
Growth in vitamin-free medium: –
Growth on 50% (w/w) glucose-yeast extract agar: +; **on 60%:** –
Growth at 37°C: –
Color reaction with DBB: –
G + C: 38.9 mol%, type strain (Meyer, Stenderup & Christiansen, unpublished)

Origin of the strains studied

Frass of Cerambycid larvae infesting *Diospyros inhacaensis* F. White near Lake Sibaya, Natal (1); frass of Cerambycid larvae in *Acacia karro* Hayne near Mtunzini, Natal (1).

Type strain: CBS 6270, from *Diospyros inhacaensis*.

45. *CANDIDA DIDDENSIAE* (PHAFF, MRAK ET WILLIAMS) FELL ET MEYER (1967)

Synonyms

Trichosporon diddensii Phaff, Mrak et Williams (1952)
Torulopsis saccharini Santa María (1959b)

Standard description of *Candida diddensiae* (van Uden & Buckley, 1970)

Growth in glucose-yeast extract-peptone water: After 3 days at 25°C the cells are generally oval, (2.3–4.6) × (2.5–5.7) μm, and occur singly, in pairs and in short chains and clusters (Fig. 222). Cells of other shapes may also be present: triangular, elongate to filamentous (Fig. 223). Pseudomycelium is formed. After one month a thin filmy pellicle and a smooth sediment are present.

Growth on glucose-yeast extract-peptone agar: After one month at 25°C the streak culture is cream-colored, off-white, lacy to wrinkled with aerial mycelium as well as a mycelial border. Individual colonies may appear flat, doughnut-like with sunken centers and a raised rim.

Dalmau plate culture on corn meal agar: Pseudomycelium consists of

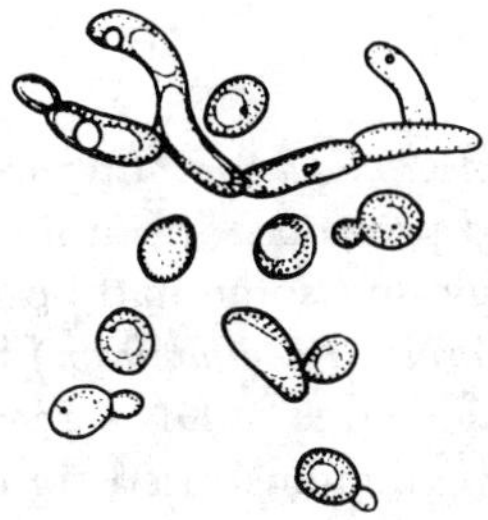

Fig. 222 – *C. diddensiae*
In glucose-yeast extract-peptone water
(From 'The Yeasts' ed. II, 1970)

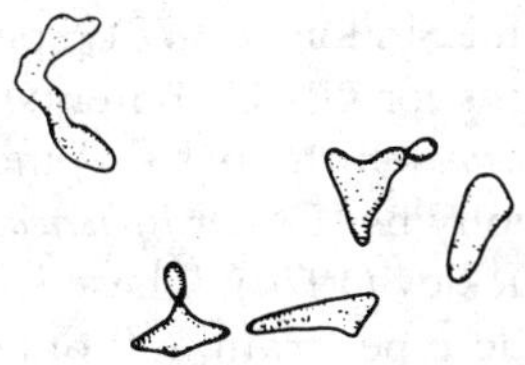

Fig. 223 – *C. diddensiae*
In glucose-yeast extract-peptone water
(From 'The Yeasts' ed. II, 1970)

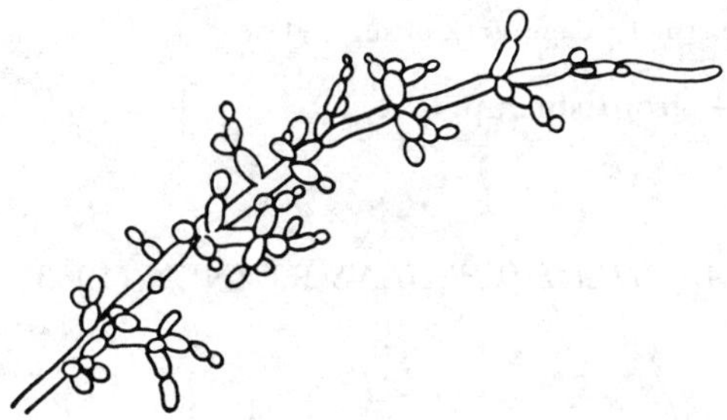

Fig. 224 – *C. diddensiae*
Dalmau plate culture on corn meal agar (From 'The Yeasts' ed. II, 1970)

branched chains of elongate cells with clusters of blastospores at intervals giving a wavy appearance (Fig. 224).

Fermentation:

Glucose	+ or s	Maltose	– or +w
Galactose	– or +s	Lactose	–
Sucrose	– or +w		

Assimilation of carbon compounds:

Galactose	+	Melezitose	+	Ribitol	+
L-Sorbose	–	Soluble starch	–	Galactitol	–
Sucrose	+	D-Xylose	+	D-Mannitol	+
Maltose	+	L-Arabinose	+	D-Glucitol	+
Cellobiose	+	D-Arabinose	– or +s	Salicin	+
Trehalose	+	D-Ribose	+ or s	DL-Lactic acid	–
Lactose	–	L-Rhamnose	–	Succinic acid	+
Melibiose	–	Glycerol	+	Citric acid	+
Raffinose	–	Erythritol	+	Inositol	–

Additional carbon compound tested: Glucono-δ-lactone +

Splitting of arbutin: +

Assimilation of nitrate: –

Growth in vitamin-free medium: –

Growth at 37°C: +

Color reaction with DBB: –

G + C: 38.8 mol%, type strain (Meyer & Phaff, 1972)

Comments

Meyer and Simione (1978) reported the heterogeneity of *C. diddensiae* after examining 10 strains for G + C contents. The strains studied had previously been identified as *C. diddensiae* on the basis of morphology and physiology and some had been considered synonyms (*C. atmospherica, C. polymorpha, T. saccharini, Tr. atlanticum*) by van Uden & Buckley (1970). These strains were studied for the degree of DNA reassociation with the type strain and only one strain (*T. saccharini*) demonstrated significant (91%) DNA relatedness to be considered a member of the species *C. diddensiae.*

Origin of the strains studied

Shrimp from the Gulf of Mexico (1); damp sack of sugar (1).

Type strain: CBS 2214, from shrimp.

46. *CANDIDA DIFFLUENS* RUINEN (1963)

Synonyms

Rhodotorula diffluens (Ruinen) von Arx et Weijman (1979)
Vanrija diffluens (Ruinen) Moore (1980)

Standard description of *Candida diffluens*

Growth in glucose-yeast extract-peptone water: After 3 days at 25°C the cells are subglobose, ovoid, or slightly irregular, (3–5) × (5–11) μm (Fig. 225). The cells are encapsulated (van Uden & Buckley, 1970).

Growth on glucose-yeast extract-peptone agar: After one month at 25°C the streak culture is brownish cream-colored, glistening, soft, smooth and

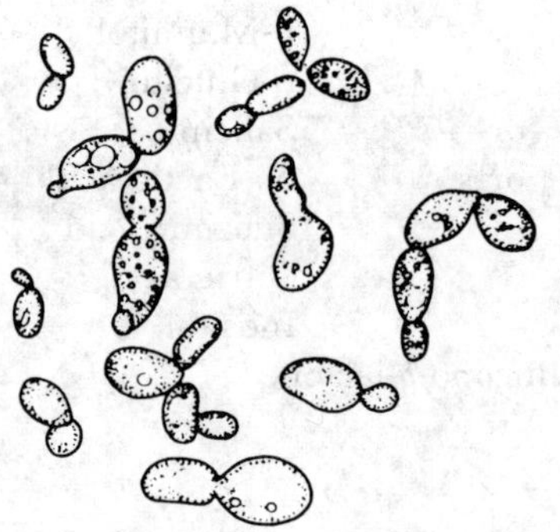

Fig. 225 – *C. diffluens*
In glucose-yeast extract-peptone water
(From 'The Yeasts' ed. II, 1970)

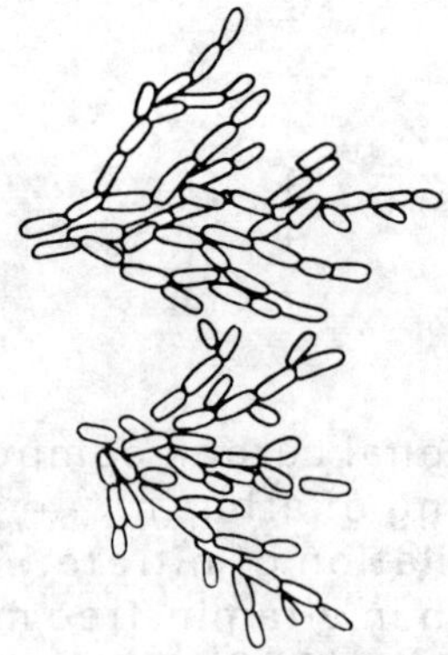

Fig. 226 – *C. diffluens*
Dalmau plate culture on corn meal agar
(From 'The Yeasts' ed. II, 1970)

mucoid (van Uden & Buckley, 1970).

Dalmau plate culture on corn meal agar: The pseudomycelium is either very primitive consisting of branched chains of cylindrical cells (Fig. 226), or displays long pseudohyphal cells bearing spindle-shaped blastospores (van Uden & Buckley, 1970).

Fermentation: –

Assimilation of carbon compounds:

Galactose	+ s	Melezitose	+	Ribitol	v
L-Sorbose	+ s	Soluble starch	–	Galactitol	–
Sucrose	+	D-Xylose	–	D-Mannitol	+ s
Maltose	+	L-Arabinose	–	D-Glucitol	+
Cellobiose	–	D-Arabinose	–	Salicin	v
Trehalose	+	D-Ribose	–	DL-Lactic acid	–
Lactose	–	L-Rhamnose	–	Succinic acid	+ s
Melibiose	–	Glycerol	+ s	Citric acid	+ s
Raffinose	–	Erythritol	+	Inositol	–

Splitting of arbutin: v

Assimilation of nitrate: +

Growth in vitamin-free medium: +

Growth at 37°C: –

Urease: +

Color reaction with DBB: +

Coenzyme Q system: Q 10 (Yamada & Kondo, 1972)

G + C: 60.5 mol%, one strain (Nakase & Komagata, 1971f)

Comments

Based on the type of conidiogenesis and the absence of xylose in cell hydrolysates von Arx and Weijman (1979) transferred this species to *Rhodotorula*.

Origin of the strain studied

Leaves of *Tillandsia usneoides* in Dutch Guyana (Ruinen, 1963) (1).

Type strain: CBS 5233.

47. *CANDIDA DIVERSA* OHARA, NONOMURA ET YUNOME EX VAN UDEN ET BUCKLEY (1970)

Synonyms

Candida fimetaria Soneda var. *diversa* nom. nud. (Ohara et al., 1960b)
Torulopsis arnaudii Capriotti et Fatichenti (1969)

Standard description of *Candida diversa*

Growth in glucose-yeast extract-peptone water: After 3 days at 25°C two types of cells occur: small cylindrical ones, (1.5–2) × (3.5–6) μm, and larger ovoid ones, (4–8) × (5–10) μm (Fig. 227) (van Uden & Buckley, 1970).

Growth on malt agar: After one month at 25°C the streak culture is greyish cream-colored, semidull, soft and smooth (van Uden & Buckley, 1970).

Dalmau plate culture on corn meal agar: The pseudomycelium consists of branched chains of cylindrical cells, sometimes with clusters of blastospores along the pseudohyphae (Fig. 228) (van Uden & Buckley, 1970).

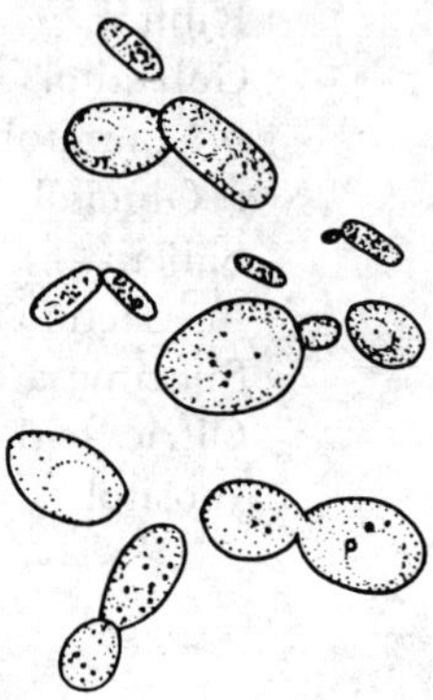

Fig. 227 – *C. diversa*
In glucose-yeast extract-peptone water
(From 'The Yeasts' ed. II, 1970)

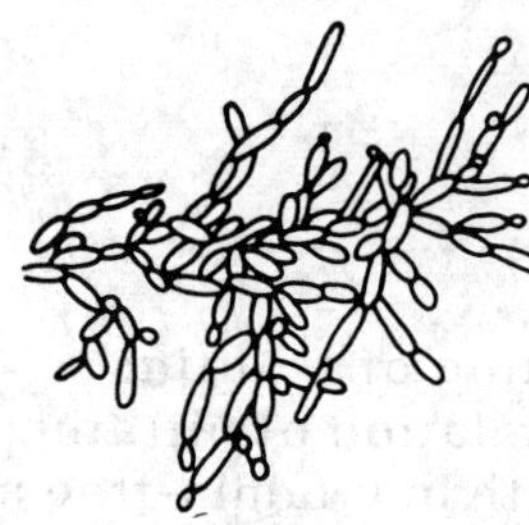

Fig. 228 – *C. diversa*
Dalmau plate culture on corn meal agar
(From 'The Yeasts' ed. II, 1970)

Fermentation:

Glucose	+	Maltose	–
Galactose	–	Lactose	–
Sucrose	–		

Assimilation of carbon compounds:

Galactose	–	Melezitose	–	Ribitol	+
L-Sorbose	–	Soluble starch	–	Galactitol	–
Sucrose	–	D-Xylose	v	D-Mannitol	+
Maltose	–	L-Arabinose	–	D-Glucitol	+
Cellobiose	–	D-Arabinose	–	Salicin	–
Trehalose	–	D-Ribose	–	DL-Lactic acid	–
Lactose	–	L-Rhamnose	–	Succinic acid	+
Melibiose	–	Glycerol	–	Citric acid	+
Raffinose	–	Erythritol	–	Inositol	–

Additional carbon compound tested: Methanol –

Splitting of arbutin: –

Assimilation of nitrate: –

Growth in vitamin-free medium: –

Growth at 37°C: v

Color reaction with DBB: –

Coenzyme Q system: Q 7 (Yamada & Kondo, 1972)

G + C: 36.0–36.6 mol%, 3 strains including type strain (36.6%) (Meyer, Stenderup & Christiansen, unpublished)

Origin of the strains studied

Grape must in Japan (Ohara et al., 1960b) (1); grape must (Capriotti & Fatichenti, 1969) (2).

Type strain: CBS 4074.

48. CANDIDA DOMERCQII (VAN DER WALT ET VAN KERKEN) MEYER ET YARROW

see *Wickerhamiella domercqii* p. 443.

49. CANDIDA EDAX VAN DER WALT ET NEL (1968)

Standard description of *Candida edax* (van der Walt & Nel, 1968)

Growth in glucose-yeast extract-peptone water: After 3 days at 28°C the cells are spheroidal, ellipsoidal to cylindrical, (2.0–5.5) × (3.0–17.5) μm, and occur singly, in pairs and occasionally in short chains (Fig. 229). Pseudomycelium is produced. A sediment and a thin, dry, creeping pellicle are formed. After one month at room temperature, a sediment and a thin dry pellicle are present.

Growth on glucose-yeast extract-peptone agar: After one month at room temperature, the culture is yellowish-cream, butyrous, flat, slightly shiny, finely pitted at the center, but smooth to finely striated along the margin. The margin may be entire, undulating or lobiform.

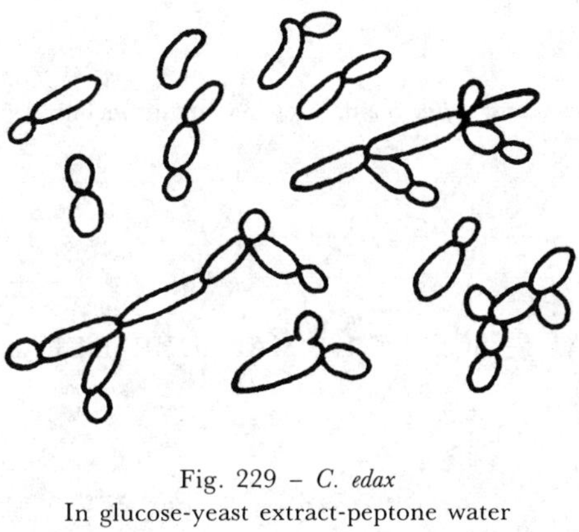

Fig. 229 – *C. edax*
In glucose-yeast extract-peptone water

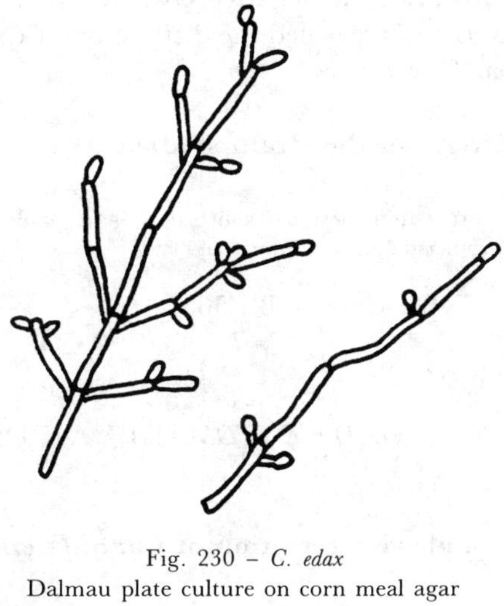

Fig. 230 – *C. edax*
Dalmau plate culture on corn meal agar

Dalmau plate culture on corn meal agar: Pseudomycelium is abundantly formed aerobically, but very sparsely in the areas under the coverslips. Blastospores may occur singly, in pairs, short chains or clusters along the pseudomycelium. Sections of the pseudomycelium may be coiled (Fig. 230). True mycelium is also formed. Aerial pseudomycelium may be present.

Fermentation: –

Assimilation of carbon compounds:

Galactose	+	Melezitose	+	Ribitol	+
L-Sorbose	+	Soluble starch	+ s	Galactitol	+ s
Sucrose	+	D-Xylose	+	D-Mannitol	+
Maltose	+	L-Arabinose	+	D-Glucitol	+
Cellobiose	+	D-Arabinose	+	Salicin	+
Trehalose	+	D-Ribose	+ or s	DL-Lactic acid	+
Lactose	+ s	L-Rhamnose	+ or s	Succinic acid	+ s
Melibiose	+	Glycerol	+	Citric acid	+ s
Raffinose	+	Erythritol	+	Inositol	+

Splitting of arbutin: +

Assimilation of nitrate: +

Growth in vitamin-free medium: +

Growth at 37°C: +

Urease: –

Color reaction with DBB: –

G + C: 46.8 mol%, type strain (Stenderup et al., 1972)

Comments

Based on morphological and physiological similarities, Jong and King (1977) concluded that *C. edax, Sterigmatomyces aphidis* and *Trichosporon oryzae* were conspecific. However, the negative DBB reaction and the lower G + C content (46.8% vs. 62%) of *C. edax* revealed the distinction of *C. edax* as a separate species. See also Comments on *C. antarctica*.

Origin of the strain studied

Frass from deserted subcortical insect galleries in *Sclerocarya caffra* Sond. growing in the vicinity of Groblersdal in the Transvaal (1).

Type strain: CBS 5657.

50. *CANDIDA ENTOMAEA* VAN DER WALT, SCOTT ET VAN DER KLIFT (1972)

Standard description of *Candida entomaea*

Growth in glucose-yeast extract-peptone water: After 3 days at 25°C the cells are mostly oval; cylindrical to elongate cells also occur, (1.8–2.9) × (2.9–5.8) μm. They occur singly, in pairs, small clusters and short chains (Fig. 231).

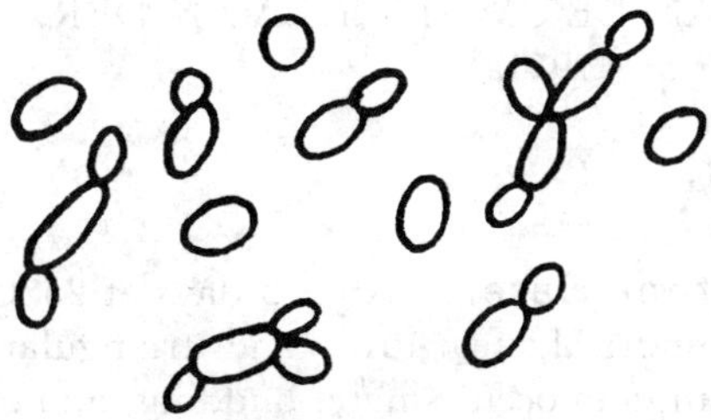

Fig. 231 – *C. entomaea*
In glucose-yeast extract-peptone water

Pseudomycelium is formed. A sediment is present. After a month a floccose sediment and a dull creeping pellicle are evident.

Growth on glucose-yeast extract-peptone agar: After one month at room temperature the streak culture is cream-colored, butyrous, flat, smooth with some structure along the center and somewhat shiny. The margin is lobate (van der Walt et al., 1972).

Dalmau plate culture on corn meal agar: Ramified pseudomycelium with small clusters of blastospores at verticillate positions is produced.

Fermentation:

Glucose	+	Maltose	v
Galactose	+ s	Lactose	–
Sucrose	+ s or –		

Assimilation of carbon compounds:

Galactose	+	Melezitose	+	Ribitol	+
L-Sorbose	–	Soluble starch	–	Galactitol	+ s
Sucrose	+	D-Xylose	+	D-Mannitol	+
Maltose	+	L-Arabinose	+	D-Glucitol	+
Cellobiose	+	D-Arabinose	+ s	Salicin	+
Trehalose	+	D-Ribose	+	DL-Lactic acid	+ s
Lactose	+ or s	L-Rhamnose	+	Succinic acid	+
Melibiose	–	Glycerol	+	Citric acid	+
Raffinose	–	Erythritol	+	Inositol	–

Splitting of arbutin: +

Assimilation of nitrate: –

Growth in vitamin-free medium: –

Growth on 50% (w/w) glucose-yeast extract agar: +

Growth at 37°C: +

Color reaction with DBB: –

G+C: 43.5 mol%, type strain (Meyer, Stenderup & Christiansen, unpublished)

Origin of the strain studied

Material from the tunnel of an unknown pinhole borer in *Pinus radiata* D. Don near Alexandria in the Southern Cape Province, South Africa (1).

Type strain: CBS 6306.

51. *CANDIDA ENTOMOPHILA* SCOTT, VAN DER WALT ET VAN DER KLIFT (van der Walt et al., 1971a)

Standard description of *Candida entomophila*

Growth in glucose-yeast extract-peptone water: After 3 days at 25°C the cells are spheroidal, oval, long-oval, cylindrical, tear-drop and triangular, $(2.8-4.2) \times (7.0-8.4)$ and $(5.6-7.0) \times (14-21)$ μm, and occur singly, budding, and in chains (Fig. 232). Spheroidal to ellipsoidal asexual endospores are formed in hyphal elements. The growth is floccose and a light sediment is formed. After one month at 25°C a viscous sediment is present.

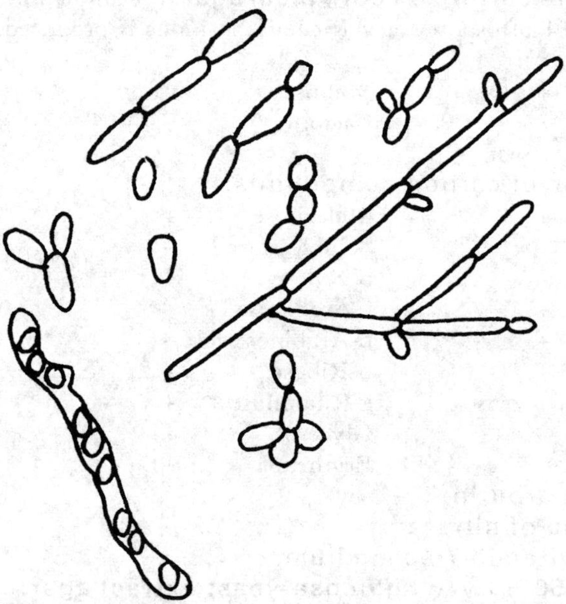

Fig. 232 – *C. entomophila*
In glucose-yeast extract-peptone water

Growth on glucose-yeast extract-peptone agar: After one month at 25°C the streak culture is cream-colored, semi-glossy to dull with smooth and folded areas; mycelial growth is prominent.

Dalmau plate culture on corn meal agar: Ramified pseudomycelium and true mycelium are produced. Small clusters and short chains of blastospores occur in verticils (Fig. 233). Some blastospores are triangular. Often the mycelial strands are long, straight, occasionally branched, with few blastospores present. A wavy appearance is evident.

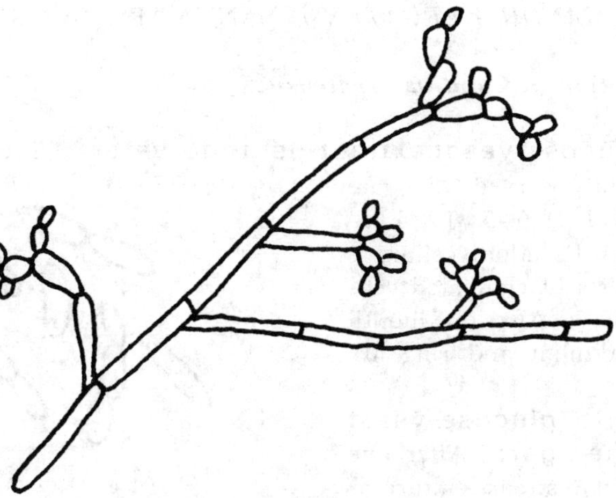

Fig. 233 – *C. entomophila*
Dalmau plate culture on corn meal agar

Fermentation:

Glucose	+	Lactose	–
Galactose	+ s or –	Raffinose	v
Sucrose	+ s	Melibiose	+ s or –
Maltose	–		

Assimilation of carbon compounds:

Galactose	+	Melezitose	+	Ribitol	+
L-Sorbose	+ s	Soluble starch	–	Galactitol	–
Sucrose	+	D-Xylose	+	D-Mannitol	+
Maltose	+	L-Arabinose	+	D-Glucitol	+
Cellobiose	+	D-Arabinose	+ s	Salicin	+
Trehalose	+	D-Ribose	+	DL-Lactic acid	–
Lactose	+	L-Rhamnose	–	Succinic acid	+
Melibiose	+	Glycerol	+	Citric acid	+
Raffinose	+	Erythritol	+	Inositol	–

Splitting of arbutin: +
Assimilation of nitrate: –
Growth in vitamin-free medium: –
Growth on 50% (w/w) glucose-yeast extract agar: +
Starch formation: –
Growth at 37°C: +
Color reaction with DBB: –
G + C: 56.3 mol%, type strain (Meyer, Stenderup & Christiansen, unpublished)

Origin of the strains studied

Material from the tunnels of *Crossotarsus externedentatus* Fairm. in *Ficus sycomorus* L. near St. Lucia and in *Tabernaemontana ventricosa* Hochst. ex A.Dc. (2).

Type strain: CBS 6160, from *Ficus sycomorus* L.

52. *CANDIDA ERGATENSIS* SANTA MARÍA (1971)

Standard description of *Candida ergatensis*

Growth in glucose-yeast extract-peptone water: After 3 days at 25°C the cells are slightly curved, elongate and sausage-shaped, (3.6–5.8) × (5.8–7.2) μm (Fig. 234). Pseudomycelium is present and consists of elongated cells with oval blastospores. After one month a flaky, thick sediment and ring are present.

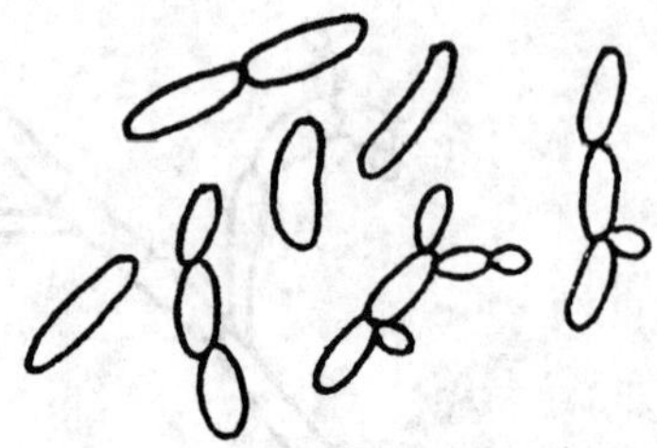

Fig. 234 – *C. ergatensis*
In glucose-yeast extract-peptone water

Growth on glucose-yeast extract-peptone agar: After one month at 25°C the streak culture is cream-colored, shiny, raised, folded at periphery and fringed with mycelium.

Dalmau plate culture on corn meal agar: The pseudomycelium consists of long slender, often wavy, branched strands with chains and clusters of blastospores.

Fermentation:

Glucose	+	Maltose	–
Galactose	+ s	Lactose	–
Sucrose	–		

Assimilation of carbon compounds:

Galactose	+	Melezitose	–	Ribitol	+
L-Sorbose	+	Soluble starch	–	Galactitol	–
Sucrose	+	D-Xylose	+	D-Mannitol	+
Maltose	+	L-Arabinose	+ s or –	D-Glucitol	+
Cellobiose	+	D-Arabinose	+ s or –	Salicin	+
Trehalose	+	D-Ribose	+ s or –	DL-Lactic acid	–
Lactose	+	L-Rhamnose	–	Succinic acid	+
Melibiose	–	Glycerol	+	Citric acid	+
Raffinose	–	Erythritol	+	Inositol	–

Splitting of arbutin: +

Assimilation of nitrate: –

Growth in vitamin-free medium: –

Growth on 50% (w/w) glucose-yeast extract agar: –

Growth at 37°C: –

Urease: –

Color reaction with DBB: –

G + C: 36.5 mol%, type strain (Meyer, Stenderup & Christiansen, unpublished)

Origin of the strain studied

Ergates faber (1).

Type strain: CBS 6248.

53. *CANDIDA ERNOBII* (LODDER ET KREGER-VAN RIJ) MEYER ET YARROW (Yarrow & Meyer, 1978)

Synonym

Torulopsis ernobii Lodder et Kreger-van Rij (1952)

Standard description of *Candida ernobii*

Growth in glucose-yeast extract-peptone water: After 3 days at 25°C the cells are globose or nearly so, (3–5) × (3–6) μm (Fig. 235). A ring may be present (van Uden & Vidal-Leiria, 1970).

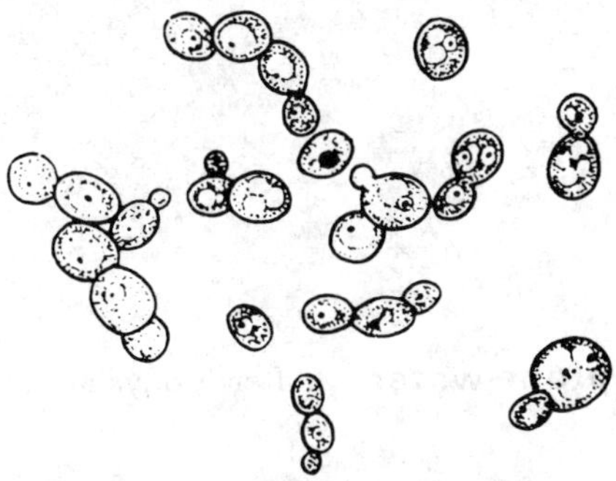

Fig. 235 – *C. ernobii*
In glucose-yeast extract-peptone water
(From 'The Yeasts' ed. II, 1970)

Growth on glucose-yeast extract-peptone agar: After one month at 25°C the streak culture is brownish to greyish-white, almost dull, soft, smooth or slightly wrinkled (van Uden & Vidal-Leiria, 1970).

Dalmau plate culture on corn meal agar: Chains of roundish cells may be present. One strain formed pseudomycelium bearing verticillated globose blastospores on potato agar (van Uden & Vidal-Leiria, 1970).

Fermentation:

Glucose	+	Maltose	–
Galactose	–	Lactose	–
Sucrose	–		

Assimilation of carbon compounds:

Galactose	–	Melezitose	+ s	Ribitol	+
L-Sorbose	v	Soluble starch	+	Galactitol	–
Sucrose	+	D-Xylose	+	D-Mannitol	+
Maltose	+	L-Arabinose	–	D-Glucitol	+
Cellobiose	+	D-Arabinose	–	Salicin	+
Trehalose	–	D-Ribose	–	DL-Lactic acid	–
Lactose	–	L-Rhamnose	+	Succinic acid	v
Melibiose	–	Glycerol	+	Citric acid	v
Raffinose	–	Erythritol	–	Inositol	–

Splitting of arbutin: +
Assimilation of nitrate: +
Growth in vitamin-free medium: + s
Growth at 37°C: –
Color reaction with DBB: –
Coenzyme Q system: Q 8 (Yamada & Kondo, 1972)

G + C: 36.1 mol%, type strain (Nakase & Komagata, 1971c)

Origin of the strains studied

Ernobius, isolated by Müller (1934) (1), and by Jurzitza (1962) (2).

Type strain: CBS 1737, isolated by Müller.

54. CANDIDA ETCHELLSII (LODDER ET KREGER-VAN RIJ) MEYER ET YARROW (Yarrow & Meyer, 1978)

Synonyms

Torulopsis etchellsii Lodder et Kreger-van Rij (1952)
Brettanomyces sphaericus Etchells et Bell (1950b)
Torulopsis citrus Recca et Mrak (1952)

Standard description of *Candida etchellsii*

Growth in glucose-yeast extract-peptone water: After 3 days at 25°C the cells are globose to short-ovoid, (2–4) × (2.5–5) μm (Fig. 236). A ring is present (van Uden & Buckley, 1970).

Growth on glucose-yeast extract-peptone agar: After one month at 25°C the streak culture is greyish cream-colored, sometimes with a brownish tinge, glossy, soft, smooth or striped at the margin (van Uden & Buckley, 1970).

Dalmau plate culture on corn meal agar: No pseudomycelium is formed.

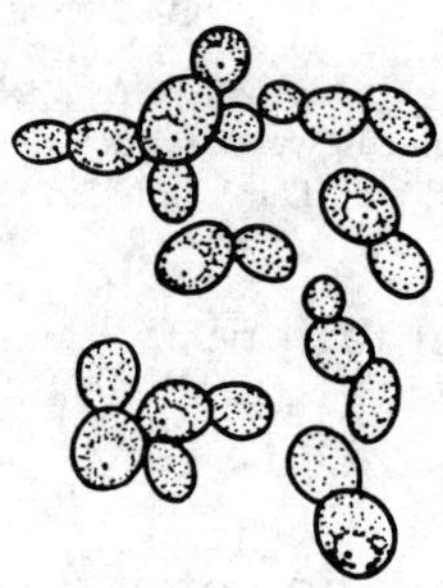

Fig. 236 – *C. etchellsii*
In glucose-yeast extract-peptone water
(From 'The Yeasts' ed. II, 1970)

Fermentation:

Glucose	+	Maltose	+
Galactose	–	Lactose	–
Sucrose	–		

Assimilation of carbon compounds:

Galactose	+	Melezitose	–	Ribitol	–
L-Sorbose	v	Soluble starch	–	Galactitol	–
Sucrose	–	D-Xylose	–	D-Mannitol	v
Maltose	+	L-Arabinose	–	D-Glucitol	+
Cellobiose	–	D-Arabinose	–	Salicin	v
Trehalose	–	D-Ribose	v	DL-Lactic acid	v
Lactose	–	L-Rhamnose	–	Succinic acid	v

Melibiose	–	Glycerol	+	Citric acid	–
Raffinose	–	Erythritol	–	Inositol	–

Splitting of arbutin: –
Assimilation of nitrate: +
Growth in vitamin-free medium: v
Growth at 37°C: –
Color reaction with DBB: –
Coenzyme Q system: Q 9 (Yamada & Kondo, 1972)
G + C: 52.0–52.7 mol%, 4 strains (Nakase & Komagata, 1971c)

Origin of the strains studied

Cucumber brine in USA (1); concentrated lemon in USA (1); sputum (2).

Type strain: CBS 1750, from cucumber brine (Etchells & Bell, 1950b).

55. *CANDIDA FABIANII* KODAMA, KYONO, IIDA ET ONOYAMA

see *Hansenula fabianii* p. 188.

56. *CANDIDA FAMATA* (HARRISON) MEYER ET YARROW (Yarrow & Meyer, 1978)

Perfect state: *Debaryomyces hansenii* (Zopf) Lodder et Kreger-van Rij

Synonyms

Torula candida Saito (1922)
Torulopsis candida (Saito) Lodder (1934)
Cryptococcus candidus (Saito) Skinner (1947a)
Mycotorula famata Harrison (1928)
Geotrichum famatum (Harrison) Dodge (1935)
Torulopsis famata (Harrison) Lodder et Kreger-van Rij (1952)
Candida famata (Harrison) Novák et Zsolt (1961)
Cryptococcus minor Pollacci et Nannizzi (in Nannizzi, 1934)
Torulopsis minor (Pollacci et Nannizzi) Lodder (1934)
Parendomyces minor (Pollacci et Nannizzi) Dodge (1935)
Blastodendrion flareri Ciferri et Redaelli (1935)
Candida flareri (Ciferri et Redaelli) Langeron et Guerra (1938)

Standard description of *Candida famata*

Growth in glucose-yeast extract-peptone water: After 3 days at 25°C cells are almost spherical to ovoid. The size of the cells varies with the strain from (2.5–4) × (3–5) μm to (4–7) × (5–8.5) μm (Fig. 237). A thin pellicle may be formed (van Uden & Vidal-Leiria, 1970).

Growth on glucose-yeast extract-peptone agar: After one month at 25°C the streak culture is greyish-white to cream-colored, semi-dull to dull, soft, smooth, or partly striped, warty or slightly wrinkled (van Uden & Vidal-Leiria, 1970).

Dalmau plate culture on corn meal agar: No pseudomycelium is formed, though short branched chains of round-oval cells are produced by some strains (van Uden & Vidal-Leiria, 1970).

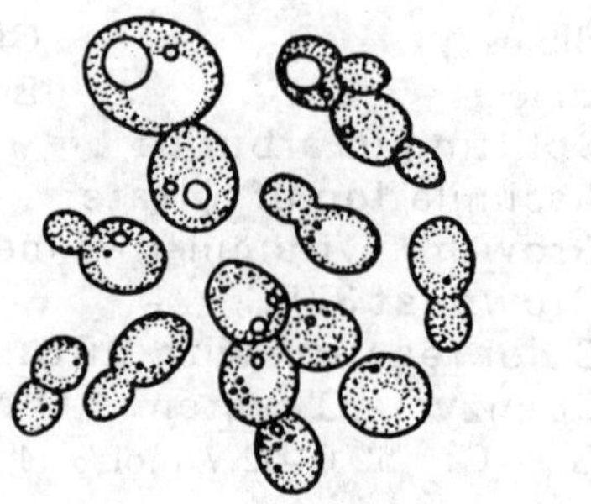

Fig. 237 – *C. famata*
In glucose-yeast extract-peptone water
(From 'The Yeasts' ed. II, 1970)

Fermentation:

Glucose	v	Lactose	–
Galactose	–	Raffinose	v
Sucrose	v	Trehalose	v
Maltose	–		

Assimilation of carbon compounds:

Galactose	+	Melezitose	+	Ribitol	+
L-Sorbose	v	Soluble starch	v	Galactitol	v
Sucrose	+	D-Xylose	+	D-Mannitol	+
Maltose	+	L-Arabinose	+	D-Glucitol	+
Cellobiose	+	D-Arabinose	v	Salicin	+
Trehalose	+	D-Ribose	v	DL-Lactic acid	+
Lactose	v	L-Rhamnose	v	Succinic acid	+
Melibiose	v	Glycerol	+	Citric acid	v
Raffinose	+	Erythritol	v	Inositol	–

Splitting of arbutin: +

Assimilation of nitrate: –

Growth in vitamin-free medium: v

Growth at 37°C: v

Color reaction with DBB: –

Coenzyme Q system: Q 9 (Yamada & Kondo, 1972)

G + C: 36.9 mol%, type strain (Meyer, Stenderup & Christiansen, unpublished)

Comments

Upon transfer of the *Torulopsis* species to the genus *Candida* (Yarrow & Meyer, 1978), the species, *Torulopsis candida,* would have become the new combination *Candida candida.* Since this is a tautonym and therefore illegitimate under Article 23 of the International Code of Botanical Nomenclature, the oldest synonym, *Mycotorula famata* Harrison, is the species epithet used for the new combination, *Candida famata.* Although Novák and Zsolt (1961) had previously recommended this combination, it was not validly published according to Article 33 of the International Code of Botanical Nomenclature.

Origin of the strains studied

Authentic strain of *Torula candida* from air in Japan (Saito, 1922) (1); authentic strain of *Mycotorula famata* from a wound of a hand in England (Harrison, 1928) (1); authentic strain of *Cryptococcus minor* from scales of a case of psoriasis in Italy (Nannizzi, 1934) (1); authentic strain of *Blastodendrion flareri* from an erythematous-squamous lesion in Italy (Ciferri & Redaelli, 1935) (1); a strain labelled *Debaryomyces mucosus* from pus in France (Hufschmitt et al., 1931) (1); an apple in Italy (1); tobacco in The Netherlands (1); frozen beef in Australia (1); an insect in The Netherlands (1); skin lesions in a pig in England (1); fermenting butter in The Netherlands (1); baker's yeast in Portugal; human origin (probably) (3).

Type strain: CBS 940, from air.

57. *CANDIDA FENNICA* (SONCK ET YARROW) MEYER ET AHEARN (1983)

N.J.W. Kreger-van Rij classifies this species in the genus *Trichosporon* as *Tr. fennicum* and *Tr. melibiosaceum*, see p. 948 and p. 953.

Synonyms

Trichosporon fennicum Sonck et Yarrow (1969)
Trichosporon melibiosaceum Scott et van der Walt (1970b)

Standard description of *Candida fennica*

Growth in glucose-yeast extract-peptone water: After 3 days at 25°C the cells are round to oval and elongate, (2.2–6.5) × (2.9–6.5) μm. Pseudomycelium may be present (Fig. 238). The sediment is somewhat dispersed or clumpy and the liquid may appear viscous.

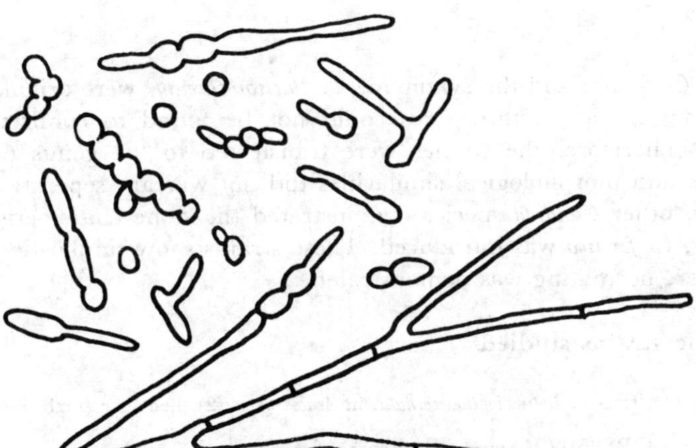

Fig. 238 – *C. fennica*
In glucose-yeast extract-peptone water

Growth in glucose-yeast extract-peptone agar: After one month at 25°C the streak culture is white to grey with both dry and mucoid areas. The center is folded and the border is mycelial.

Dalmau plate culture on corn meal agar: The pseudomycelium consists of long wavy branched chains of cells with short chains of round to oval blastospores. True mycelium is also evident with cells which possess knobby protuberances.

Fermentation:

Glucose	+	Maltose	+
Galactose	+	Lactose	– or +s
Sucrose	+	Raffinose	– or +s

Assimilation of carbon compounds:

Galactose	+	Melezitose	v	Ribitol	+
L-Sorbose	+s or –	Soluble starch	+ or s	Galactitol	–
Sucrose	+	D-Xylose	+s	D-Mannitol	+
Maltose	+	L-Arabinose	v	D-Glucitol	+
Cellobiose	v	D-Arabinose	– or +s	Salicin	+ or s
Trehalose	+	D-Ribose	+ or s	DL-Lactic acid	– or +s
Lactose	v	L-Rhamnose	–	Succinic acid	+
Melibiose	v	Glycerol	+ or s	Citric acid	+ or s
Raffinose	+ or s	Erythritol	+	Inositol	–

Additional carbon compounds tested: Inulin –, methanol –

Splitting of arbutin: +

Assimilation of nitrate: –

Growth in vitamin-free medium: + or s

Growth at 37°C: v

Urease: –

Color reaction with DBB: –

Comments

Although *C. fennica* and the synonym, *Tr. melibiosaceum,* were originally described as *Trichosporon* species, arthrospores could not be found to confirm this generic designation. Therefore, the species were transferred to the genus *Candida*. Their physiological and morphological similarities did not warrant separate species status and since no other *Candida* species demonstrated the same characteristics, the new combination, *C. fennica* was introduced. These strains show similarities to *Pichia burtonii;* however, no mating was demonstrated.

Origin of the strains studied

Anus of a cat (1); frass of *Botrichioplites cornutus* in *Acacia karoo* (2); hens (1); birch trees (17).

Type strain: CBS 5928, from the anus of a cat.

58. *CANDIDA FLUVIATILIS* HEDRICK (1976)

Standard description of *Candida fluviatilis*

Growth in glucose-yeast extract-peptone water: After 3 days at 25°C the cells are round, oval to elongate and cylindrical, (2.9–5.7) × (3.6–6.5) μm; they occur singly, in pairs or short chains. After one month, a granular sediment is present.

Growth on glucose-yeast extract-peptone agar: After one month at 25°C the streak culture is light-cream colored, semi-glossy, smooth and creamy. The margin is entire.

Dalmau plate culture on corn meal agar: "*Candida*" type pseudomycelium, with branching tree-like limbs; the cells are about equal length. No true mycelium is formed. The pseudohyphae are formed only beneath the coverglass (Hedrick, 1976).

Fermentation:

Glucose	– or +w, s	Maltose	– or +w
Galactose	–	Lactose	–
Sucrose	–	Raffinose	–

Assimilation of carbon compounds:

Galactose	+	Melezitose	+	Ribitol	+
L-Sorbose	–	Soluble starch	+	Galactitol	–
Sucrose	+	D-Xylose	+	D-Mannitol	+
Maltose	+	L-Arabinose	–	D-Glucitol	+
Cellobiose	+	D-Arabinose	–	Salicin	+
Trehalose	+	D-Ribose	–	DL-Lactic acid	+
Lactose	+	L-Rhamnose	–	Succinic acid	+
Melibiose	–	Glycerol	+	Citric acid	+
Raffinose	–	Erythritol	–	Inositol	–

Splitting of arbutin: +
Assimilation of nitrate: –
Growth in vitamin-free medium: +
Growth at 37°C: +
Urease: –
Color reaction with DBB: –
G + C: 40 mol%, type strain

Origin of the strains studied

Greater Calumet River near Gary, Indiana, USA (2).

Type strain: CBS 6776.

59. *CANDIDA FOLIORUM* RUINEN (1963)

Standard description of *Candida foliorum*

Growth in glucose-yeast extract-peptone water: After 3 days at 25°C the cells are slender, long-ovoid to cylindrical, (1.5–4) × (4–12) μm (Fig. 239). They are encapsulated. A ring may be present (van Uden & Buckley, 1970).

Growth on glucose-yeast extract-peptone agar: After one month at 25°C the streak culture is yellowish, glistening, mucous, smooth and entire (van Uden & Buckley, 1970).

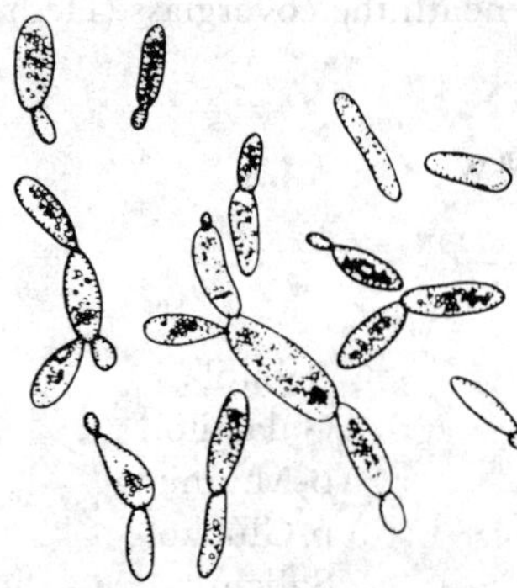

Fig. 239 – *C. foliorum*
In glucose-yeast extract-peptone water
(From 'The Yeasts' ed. II, 1970)

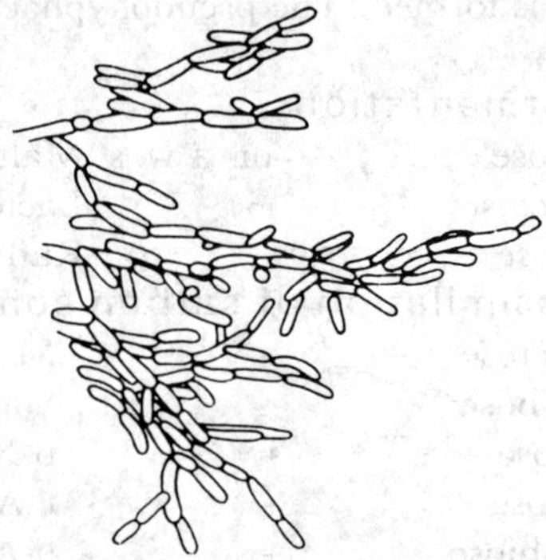

Fig. 240 – *C. foliorum*
Dalmau plate culture on corn meal agar
(From 'The Yeasts' ed. II, 1970)

Dalmau plate culture on corn meal agar: The pseudomycelium is primitive and consists of ramified chains of cylindrical cells (Fig. 240) (van Uden & Buckley, 1970).

Fermentation: –

Assimilation of carbon compounds:

Galactose	–	Melezitose	–	Ribitol	v
L-Sorbose	–	Soluble starch	–	Galactitol	–
Sucrose	–	D-Xylose	+ or s	D-Mannitol	+
Maltose	–	L-Arabinose	v	D-Glucitol	+
Cellobiose	+ w, s	D-Arabinose	v	Salicin	–
Trehalose	+	D-Ribose	v	DL-Lactic acid	+ s
Lactose	–	L-Rhamnose	–	Succinic acid	+ s
Melibiose	–	Glycerol	+ s	Citric acid	+ s
Raffinose	–	Erythritol	–	Inositol	–

Splitting of arbutin: –

Assimilation of nitrate: +

Growth in vitamin-free medium: + s

Growth at 37°C: +

Urease: +
Color reaction with DBB: +

Origin of the strain studied

Leaves of *Drymoglossum pilloselliodes* in Indonesia (Ruinen, 1963) (1).

Type strain: CBS 5234.

60. *CANDIDA FRAGARIORUM* (BARNETT ET BUHAGIAR) MEYER ET YARROW (Yarrow & Meyer, 1978)

Synonym

Torulopsis fragaria Barnett et Buhagiar (1971)

Standard description of *Candida fragariorum*

Growth in glucose-yeast extract-peptone water: After 3 days at 25°C the cells are oval or cylindrical. In defined medium with glucose as the sole carbon source, the cells measure (1.8–6.1) × (3.4–12.4) μm (Barnett & Buhagiar, 1971).

Growth on glucose-yeast extract-peptone agar: After one month at 25°C the streak culture is beige, smooth, shiny and flat with an entire or slight irregular edge.

Dalmau plate culture on corn meal agar: Short chains of cells are observed in the streak.

Fermentation: –

Assimilation of carbon compounds:

Galactose	v	Melezitose	+	Ribitol	v
L-Sorbose	+	Soluble starch	–	Galactitol	–
Sucrose	+	D-Xylose	+	D-Mannitol	+
Maltose	+	L-Arabinose	+ or s	D-Glucitol	+
Cellobiose	+	D-Arabinose	+ or s	Salicin	+
Trehalose	+	D-Ribose	v	DL-Lactic acid	v
Lactose	+	L-Rhamnose	–	Succinic acid	+
Melibiose	–	Glycerol	v	Citric acid	+
Raffinose	+	Erythritol	–	Inositol	–

Splitting of arbutin: +
Assimilation of nitrate: +
Growth in vitamin-free medium: +
Growth at 37°C: –
Urease: +
Color reaction with DBB: +

Origin of the strains studied

Fresh strawberries (4).

Type strain: CBS 6254.

61. *CANDIDA FREYSCHUSSII* BUCKLEY ET VAN UDEN (1968)

Standard description of *Candida freyschussii*

Growth in glucose-yeast extract-peptone water: After 3 days at 25°C the cells are ovoid, (3–5) × (4–7) μm (Fig. 241) (van Uden & Buckley, 1970).

Growth on glucose-yeast extract-peptone agar: After one month at 25°C the streak culture is greyish cream-colored, glistening, soft, smooth or slightly punctilated (van Uden & Buckley, 1970).

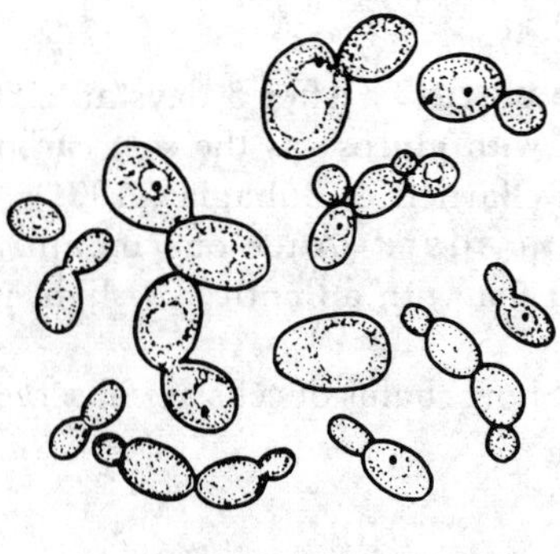

Fig. 241 – *C. freyschussii*
In glucose-yeast extract-peptone water
(From 'The Yeasts' ed. II, 1970)

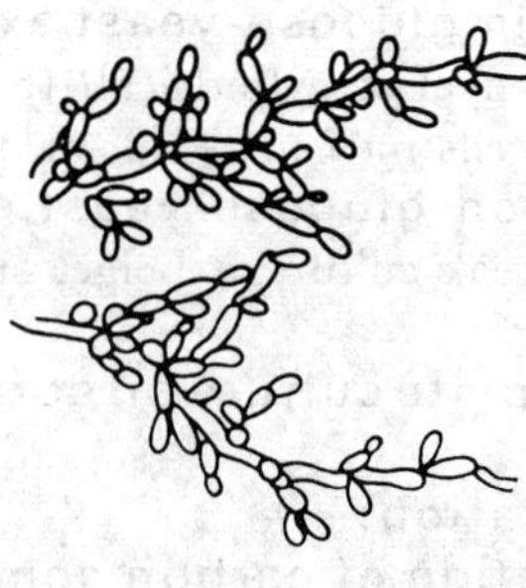

Fig. 242 – *C. freyschussii*
Dalmau plate culture on corn meal agar
(From 'The Yeasts' ed. II, 1970)

Dalmau plate culture on corn meal agar: The pseudomycelium is either primitive consisting of dense chains of ovoid cells or formed by more openly branched chains of long-ovoid cells (Fig. 242) (van Uden & Buckley, 1970).

Fermentation:

Glucose	+	Maltose	–
Galactose	–	Lactose	–
Sucrose	–	Cellobiose	+

Assimilation of carbon compounds:

Galactose	–	Melezitose	+	Ribitol	–
L-Sorbose	–	Soluble starch	–	Galactitol	–
Sucrose	+	D-Xylose	+	D-Mannitol	+
Maltose	+	L-Arabinose	–	D-Glucitol	+
Cellobiose	+	D-Arabinose	–	Salicin	+
Trehalose	– or +s	D-Ribose	–	DL-Lactic acid	+

Lactose	–	L-Rhamnose	+	Succinic acid	+
Melibiose	–	Glycerol	+	Citric acid	+
Raffinose	–	Erythritol	–	Inositol	–

Splitting of arbutin: +
Assimilation of nitrate: –
Growth in vitamin-free medium: –
Growth at 37°C: +
Color reaction with DBB: –
G + C: 45.3 mol%, one strain (Meyer, Stenderup & Christiansen, unpublished)

Origin of the strains studied

Wood pulp in Sweden (2).

Type strain: CBS 2162.

62. *CANDIDA FRIEDRICHII* VAN UDEN ET WINDISCH (1968)

Standard description of *Candida friedrichii*

Growth in glucose-yeast extract-peptone water: After 3 days at 25°C the cells are ovoid or sausage-shaped, often curved in an irregular fashion, (1.5–3) × (7–12) μm; much longer cells may occur (Fig. 243). A thin pellicle is present (van Uden & Buckley, 1970).

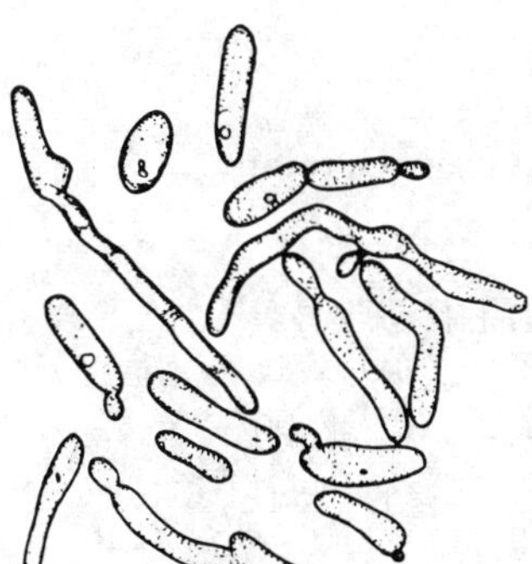

Fig. 243 – *C. friedrichii*
In glucose-yeast extract-peptone water
(From 'The Yeasts' ed. II, 1970)

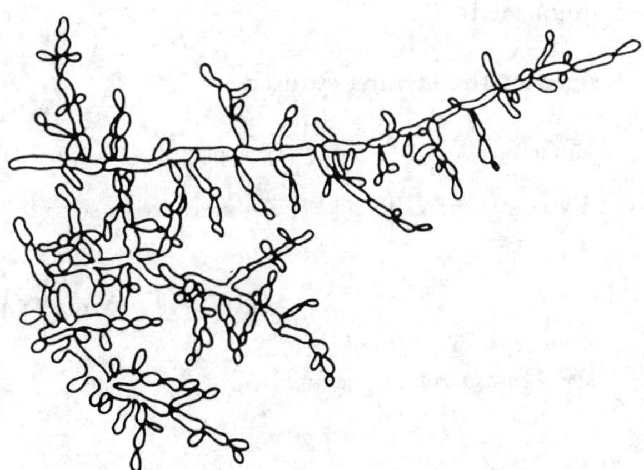

Fig. 244 – *C. friedrichii*
Dalmau plate culture on corn meal agar
(From 'The Yeasts', ed. II, 1970)

Growth on glucose-yeast extract-peptone agar: After one month at 25°C the streak culture is cream-colored, dull, soft, delicately wrinkled (van Uden & Buckley, 1970).

Dalmau plate culture on corn meal agar: The pseudomycelium consists of branched chains of short or medium-sized pseudohyphae, bearing short chains of ovoid and stalagmoid blastospores. The pseudohyphae are often curved, flask-shaped or stalagmoid (Fig. 244) (van Uden & Buckley, 1970).

Fermentation:

Glucose	+	Lactose	–
Galactose	+ or s	Raffinose	v
Sucrose	–	Melibiose	+
Maltose	–		

Assimilation of carbon compounds:

Galactose	+	Melezitose	+	Ribitol	+
L-Sorbose	+	Soluble starch	–	Galactitol	+
Sucrose	+	D-Xylose	+	D-Mannitol	+
Maltose	+	L-Arabinose	+	D-Glucitol	+
Cellobiose	+	D-Arabinose	+	Salicin	+
Trehalose	+	D-Ribose	+	DL-Lactic acid	v
Lactose	–	L-Rhamnose	–	Succinic acid	+
Melibiose	+	Glycerol	+	Citric acid	+
Raffinose	+	Erythritol	+	Inositol	–

Splitting of arbutin: +

Assimilation of nitrate: –

Growth in vitamin-free medium: –

Growth at 37°C: –

Color reaction with DBB: –

G+C: 32.9–33.7 mol%, 2 strains (Meyer, Stenderup & Christiansen, unpublished)

Origin of the strains studied

Sorbitol solution in Germany (1); apple jelly (1); cheese (1).

Type strain: CBS 4114, isolated from sorbitol solution in Germany.

63. *CANDIDA FRIGIDA* DI MENNA

see *Leucosporidium frigidum* p. 499.

64. *CANDIDA FRUCTUS* (NAKASE) MEYER ET YARROW (Yarrow & Meyer, 1978)

Synonym

Torulopsis fructus Nakase (1971a)

Standard description of *Candida fructus*

Growth in glucose-yeast extract-peptone water: After 3 days at 25°C cells are short-oval, (2–5.5) × (2.5–7) μm, and occur singly, in pairs and in clusters (Fig. 245). After one month a heavy sediment is present and no pellicle is formed.
and no pellicle is formed.

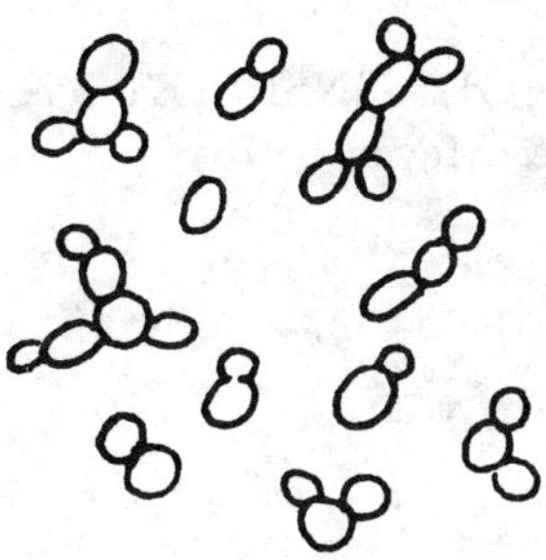

Fig. 245 – *C. fructus*
In glucose-yeast extract-peptone water

Growth on glucose-yeast extract-peptone agar: After one month at 25°C the streak culture is smooth, glistening, soft, white to yellowish-white. The margin is entire.

Dalmau plate culture on corn meal agar: Chains of oval cells occur.

Fermentation:

Glucose	+	Maltose	–
Galactose	–	Lactose	–
Sucrose	–		

Assimilation of carbon compounds:

Galactose	–	Melezitose	–	Ribitol	+
L-Sorbose	+	Soluble starch	–	Galactitol	–
Sucrose	–	D-Xylose	+	D-Mannitol	+ s
Maltose	–	L-Arabinose	–	D-Glucitol	+
Cellobiose	–	D-Arabinose	– or + s	Salicin	–
Trehalose	+	D-Ribose	– or + s	DL-Lactic acid	–
Lactose	–	L-Rhamnose	–	Succinic acid	+
Melibiose	–	Glycerol	+	Citric acid	+
Raffinose	–	Erythritol	–	Inositol	–

Additional carbon compound tested: Methanol –

Splitting of arbutin: –

Assimilation of nitrate: –

Growth in vitamin-free medium: –

Growth at 37°C: –

Color reaction with DBB: –

G + C: 48.5 mol%; 48.8 mol%, type strain (Nakase, 1971a)

Origin of the strain studied

Banana from a food market, Tokyo (1).

Type strain: CBS 6380.

65. *CANDIDA GELIDA* DI MENNA

see *Leucosporidium gelidum* p. 501.

66. *CANDIDA GEOCHARES* (VAN DER WALT, JOHANNSEN ET YARROW) MEYER ET YARROW (Yarrow & Meyer, 1978)

Synonym

Torulopsis geochares van der Walt, Johannsen et Yarrow (1978)

Standard description of *Candida geochares*

Growth in glucose-yeast extract-peptone water: After 3 days at 25°C the cells are spheroidal to slightly ovoid, (2.2–4.3) × (2.9–4.3) μm, occurring singly, in pairs, short chains, and clusters (Fig. 246). A sediment is formed.

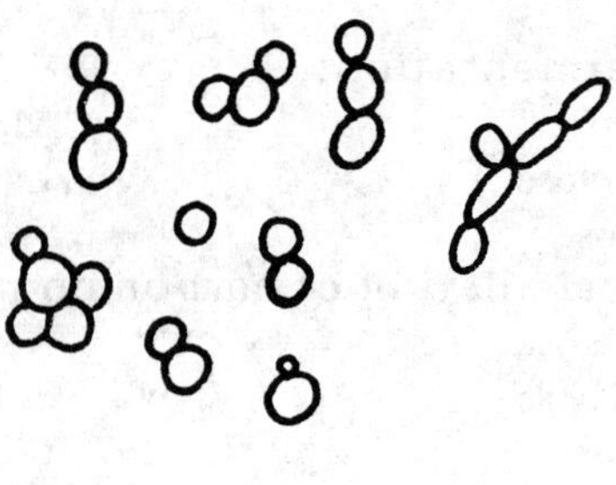

Fig. 246 – *C. geochares*
In glucose-yeast extract-peptone water

Growth on glucose-yeast extract-peptone agar: After one month at 25°C the streak culture is beige to cream-colored and shiny. The margin is entire to slightly undulating.

Dalmau plate culture on corn meal agar: No pseudomycelium is present.

Fermentation:

Glucose	+ s	Maltose	–
Galactose	–	Lactose	–
Sucrose	+ s		

Assimilation of carbon compounds:

Galactose	+ s	Melezitose	–	Ribitol	– or + s
L-Sorbose	+	Soluble starch	–	Galactitol	–
Sucrose	+ or s	D-Xylose	–	D-Mannitol	+
Maltose	–	L-Arabinose	–	D-Glucitol	+
Cellobiose	v	D-Arabinose	–	Salicin	+
Trehalose	–	D-Ribose	– or + s	DL-Lactic acid	–
Lactose	–	L-Rhamnose	–	Succinic acid	+ s
Melibiose	–	Glycerol	+	Citric acid	+
Raffinose	–	Erythritol	–	Inositol	–

Splitting of arbutin: +
Assimilation of nitrate: –
Growth in vitamin-free medium: –
Growth at 37°C: +
Urease: –
Color reaction with DBB: –
G + C: 54.1 mol%, type strain (van der Walt et al., 1978)

Origin of the strain studied

Grassland soil collected near Groblersdal (1).

Type strain: CBS 6870.

67. *CANDIDA GLABRATA* (ANDERSON) MEYER ET YARROW
(Yarrow & Meyer, 1978)

Synonyms

Cryptococcus glabratus Anderson (1917)
Torulopsis glabrata (Anderson) Lodder et de Vries (1938)

Standard description of *Candida glabrata*

Growth in glucose-yeast extract-peptone water: After 3 days at 25°C the cells are ovoid, (2.5–4.5) × (4–6) μm (Fig. 247). After one month at 25°C some strains may have formed a ring, a thin pellicle and a sediment. Most strains only form a sediment (van Uden & Vidal-Leiria, 1970).

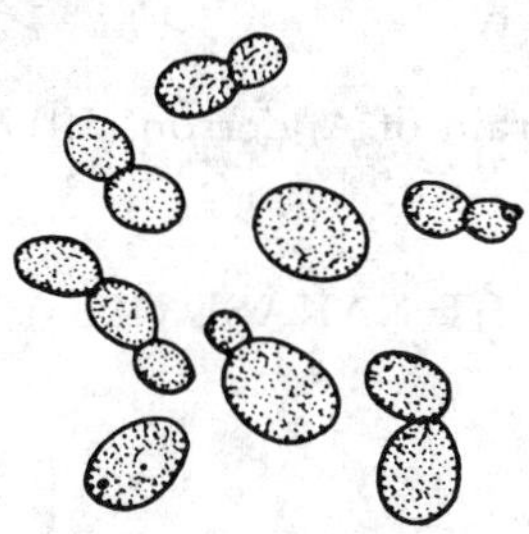

Fig. 247 – *C. glabrata*
In glucose-yeast extract-peptone water
(From 'The Yeasts' ed. II, 1970)

Growth on glucose-yeast extract-peptone agar: After one month at 25°C the streak culture is cream-colored, soft, glossy and smooth (van Uden & Vidal-Leiria, 1970).

Dalmau plate culture on corn meal agar: No pseudomycelium is formed. Some strains may form a few branched chains of ovoid cells (van Uden & Vidal-Leiria, 1970).

Fermentation:

Glucose	+	Maltose	–
Galactose	–	Lactose	–
Sucrose	–	Trehalose	+ or w

Assimilation of carbon compounds:

Galactose	–	Melezitose	–	Ribitol	–
L-Sorbose	–	Soluble starch	–	Galactitol	–
Sucrose	–	D-Xylose	–	D-Mannitol	–
Maltose	–	L-Arabinose	–	D-Glucitol	–
Cellobiose	–	D-Arabinose	–	Salicin	–
Trehalose	+	D-Ribose	–	DL-Lactic acid	–
Lactose	–	L-Rhamnose	–	Succinic acid	–
Melibiose	–	Glycerol	v	Citric acid	–
Raffinose	–	Erythritol	–	Inositol	–

Splitting of arbutin: –
Assimilation of nitrate: –
Growth in vitamin-free medium: –
Growth at 37°C: +
Color reaction with DBB: –
Coenzyme Q system: Q 6 (Yamada & Kondo, 1972)
G + C: 39.6–40.2 mol%, 5 strains (Mendonça-Hagler & Phaff, 1975)

Comments

C. glabrata can be differentiated from *C. castellii* on vitamin requirements (see *C. castellii*.

Origin of the strains studied

Human feces (Anderson, 1917) (1); human vulval ulcer in The Netherlands (1); human oral cavities in The Netherlands (2); concentrated orange juice, USA (1); urine of diabetics in The Netherlands (2); baker's yeast (1); vagina (1); sorghum malt (1); frass of Cossidae larvae (1).

Type strain: CBS 138, the presumed authentic strain of Anderson (1917).

68. *CANDIDA GLAEBOSA* KOMAGATA ET NAKASE (1965)

Standard description of *Candida glaebosa*

Growth in glucose-yeast extract-peptone water: After 7 days at 15°C the cells are short-ovoid, (2–3) × (3–5) μm, single, in pairs, or clumped (Fig. 248). A granular sediment, a ring and islets are present (van Uden & Buckley, 1970).

Growth on glucose-yeast extract-peptone agar: After one month at 25°C the streak culture is whitish, dull, soft fragile and delicately wrinkled (van Uden & Buckley, 1970).

Dalmau plate culture on corn meal agar: A primitive pseudomycelium is formed. It consists of ramified chains of cylindrical cells (Fig. 249) (van Uden & Buckley, 1970).

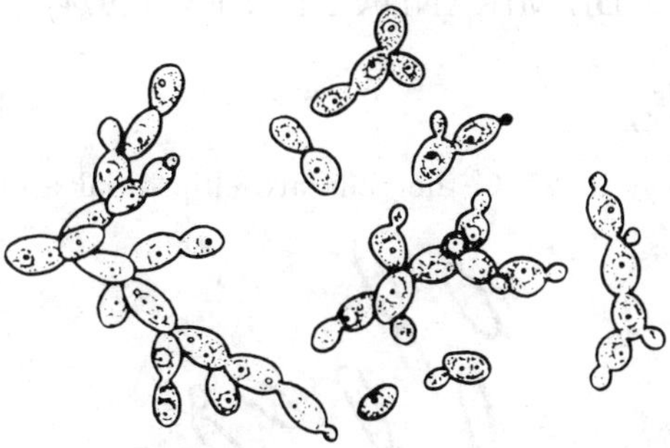

Fig. 248 – *C. glaebosa*
In glucose-yeast extract-peptone water
(From 'The Yeasts' ed. II, 1970)

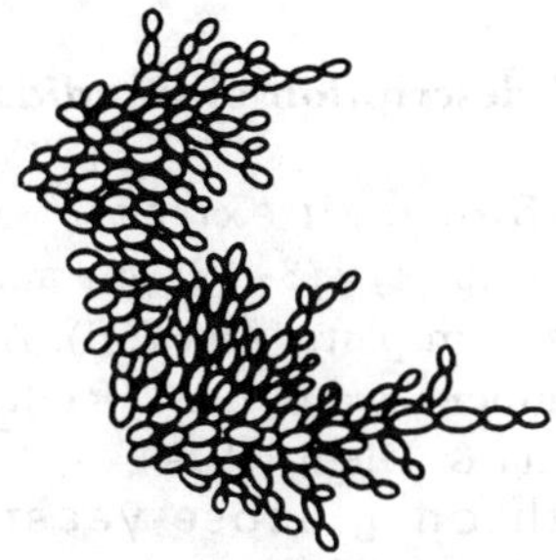

Fig. 249 – *C. glaebosa*
Dalmau plate culture on corn meal agar
(From 'The Yeasts' ed. II, 1970)

Fermentation: –

Assimilation of carbon compounds:

Galactose	+	Melezitose	v	Ribitol	+
L-Sorbose	+	Soluble starch	+ w	Galactitol	–
Sucrose	+	D-Xylose	+	D-Mannitol	+
Maltose	+	L-Arabinose	+ w, s	D-Glucitol	+
Cellobiose	+	D-Arabinose	v	Salicin	+
Trehalose	+	D-Ribose	–	DL-Lactic acid	+
Lactose	+	L-Rhamnose	–	Succinic acid	+
Melibiose	+	Glycerol	+	Citric acid	+
Raffinose	+	Erythritol	–	Inositol	–

Splitting of arbutin: +

Assimilation of nitrate: –

Growth in vitamin-free medium: –

Growth at 37°C: – ; at 25°C: +

Color reaction with DBB: –

Coenzyme Q system: Q 9 (Yamada & Kondo, 1972)

G + C: 42.4 mol%, type strain (Meyer, Stenderup & Christiansen, unpublished)

Origin of the strain studied

Frozen cuttle-fish in Japan (Komagata & Nakase, 1965) (1).

Type strain: CBS 5691.

69. *CANDIDA GLOBOSA* YARROW ET MEYER

see *Citeromyces matritensis* p. 117.

70. CANDIDA GRAMINIS RODRIGUES DE MIRANDA ET DIEM (1974)

Standard description of *Candida graminis*

Growth in malt extract: After 3 days at 25°C the cells are ellipsoidal and elongate, (1.5–4) × (3–12) μm and occur singly or in pairs (Fig. 250). A ring and a sediment are present (Rodrigues de Miranda & Diem, 1974).

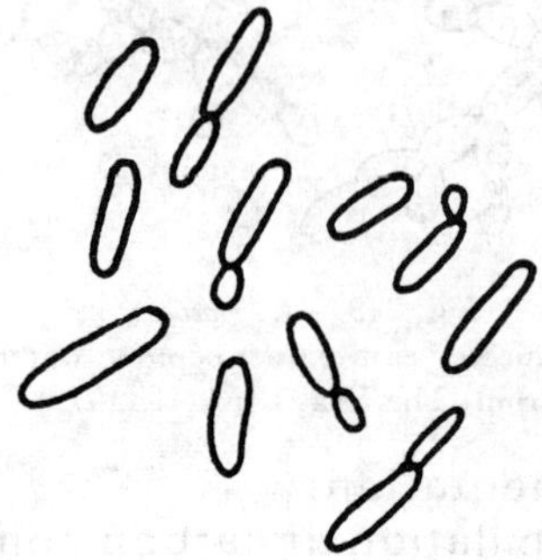

Fig. 250 – *C. graminis*
In glucose-yeast extract-peptone water

Growth on glucose-yeast extract-peptone agar: After one month at 25°C the streak culture is beige, smooth, creamy to slightly mucoid with an entire margin.

Dalmau plate culture on corn meal agar: Pseudomycelium is sparse and consists of branched chains of elongated cells.

Fermentation: –

Assimilation of carbon compounds:

Galactose	+ s	Melezitose	+	Ribitol	+
L-Sorbose	+	Soluble starch	v	Galactitol	–
Sucrose	+	D-Xylose	– or + s	D-Mannitol	+
Maltose	+	L-Arabinose	–	D-Glucitol	+
Cellobiose	+	D-Arabinose	– or + s	Salicin	+
Trehalose	+	D-Ribose	+ or s	DL-Lactic acid	+
Lactose	+ s	L-Rhamnose	–	Succinic acid	+
Melibiose	–	Glycerol	+	Citric acid	+
Raffinose	–	Erythritol	+	Inositol	–

Splitting of arbutin: +

Assimilation of nitrate: +

Growth in vitamin-free medium: –

Growth at 37°C: –

Urease: +

Color reaction with DBB: +

Origin of the strain studied

Barley leaves (1).

Type strain: CBS 6403.

71. *CANDIDA GROPENGIESSERI* (HARRISON) MEYER ET YARROW (Yarrow & Meyer, 1978)

Synonyms

Torula gropengiesseri Harrison (1928)
Torulopsis gropengiesseri (Harrison) Lodder (1934)
Cryptococcus gropengiesseri (Harrison) Skinner (1947a)

Standard description of *Candida gropengiesseri*

Growth in glucose-yeast extract-peptone water: After 3 days at 25°C the cells are ovoid to cylindrical, (1.5–2.5) × (3.5–6) μm (Fig. 251). A ring is present (van Uden & Vidal-Leiria, 1970).

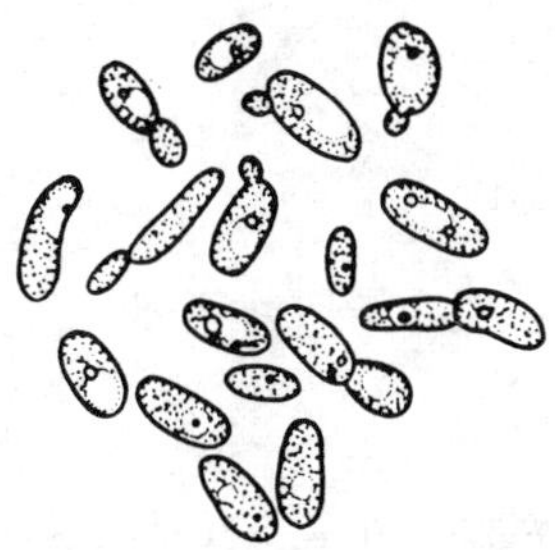

Fig. 251 – *C. gropengiesseri*
In glucose-yeast extract-peptone water
(From 'The Yeasts' ed. II, 1970)

Growth on glucose-yeast extract-peptone agar: After one month at 25°C the streak culture is greyish cream-colored, glistening, soft and smooth (van Uden & Vidal-Leiria, 1970).

Dalmau plate culture on corn meal agar: No pseudomycelium is formed; short chains of ovoid cells may occur (van Uden & Vidal-Leiria, 1970).

Fermentation:

Glucose	+	Maltose	–
Galactose	–	Lactose	–
Sucrose	+		

Assimilation of carbon compounds:

Galactose	+	Melezitose	–	Ribitol	–
L-Sorbose	–	Soluble starch	–	Galactitol	–
Sucrose	+	D-Xylose	–	D-Mannitol	+
Maltose	–	L-Arabinose	–	D-Glucitol	+
Cellobiose	+	D-Arabinose	–	Salicin	+
Trehalose	–	D-Ribose	–	DL-Lactic acid	–
Lactose	–	L-Rhamnose	–	Succinic acid	–
Melibiose	–	Glycerol	+	Citric acid	–
Raffinose	+	Erythritol	–	Inositol	–

Splitting of arbutin: +
Assimilation of nitrate: –
Growth in vitamin-free medium: +w
Growth at 37°C: –
Color reaction with DBB: –
Coenzyme Q system: Q 9 (Yamada & Kondo, 1972)
G + C: 55.7 mol%, one strain (Meyer, Stenderup & Christiansen, unpublished); 57.1 mol% (Nakase & Komagata, 1971c)

Origin of the strain studied

An insect (Gropengiesser, 1925) (1).

Type strain: CBS 156.

72. *CANDIDA GUILLIERMONDII* (CASTELLANI) LANGERON ET GUERRA (1938)

Perfect state: *Pichia guilliermondii* Wickerham

Synonyms

Endomyces guilliermondi Castellani (1912b)
Monilia guilliermondi (Castellani) Castellani et Chalmers (1913)
Myceloblastanon guilliermondi (Castellani) Ota (1928)
Mycotorula guilliermondi (Cast.) Langeron et Guerra (1935)
Blastodendrion guilliermondi (Cast.) Guerra (1935)
Castellania guilliermondi (Cast.) Dodge (1935)
Monilia pseudoguilliermondi Castellani et Chalmers (1919)
Castellania pseudoguilliermondi (Cast. et Chalmers) Dodge (1935)
Monilia guilliermondi (Cast.) Castellani et Chalmers var. *pseudoguilliermondi* (Cast. et Chalmers) Castellani (1937a)
Torula fermentati Saito (1922)
Myceloblastanon arzti Ota (1924b)
Blastodendrion arzti Ota (1924b)
Myceloblastanon krausi Ota (1924b)
Blastodendrion krausi Ota (1924b)
Mycotorula krausi (Ota) Red. et Cif. (1947)
Microanthomyces alpinus Grüss (1926b, 1927)
Candida melibiosi Lodder et Kreger-van Rij (1952)
Trichosporon appendiculare Batista et Silveira (1959c)
Torulopsis carpophila nom. nud. (Miller & Mrak, 1953)
Candida guilliermondii (Cast.) Langeron et Guerra var. *carpophila* Phaff et Miller (1961)
Torulopsis kestonii Scarr et Rose (1966)
Candida kestonii (Scarr et Rose) Meyer et Yarrow (Yarrow & Meyer, 1978)
Candida guilliermondii var. *japonica* Sugiyama et Goto (1969)
Candida parapsilosis (Ashf.) Lang. et Talice var. *tuxtlensis* Herrera et al. (1973)

Standard description of *Candida guilliermondii*

Growth in glucose-yeast extract-peptone water: After 3 days at 25°C the cells are short-ovoid, (2–4.5) × (2.5–7) μm; small cylindrical cells also occur (Fig. 252). A ring and islets may be present (van Uden & Buckley, 1970).

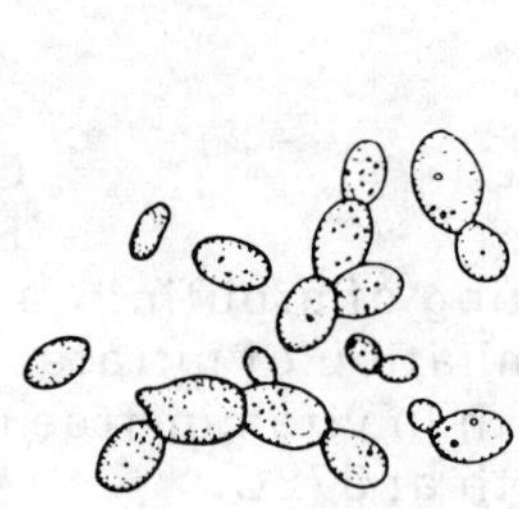

Fig. 252 – *C. guilliermondii*
In glucose-yeast extract-peptone water
(From 'The Yeasts', ed. II, 1970)

Growth on glucose-yeast extract-peptone agar: After one

month at 25°C the streak culture is yellowish cream-colored, soft, glistening and smooth or dull and wrinkled.

Dalmau plate culture on corn meal agar: Pseudomycelium differs among strains. Some strains exhibit sparce pseudomycelial development consisting of branched chains of cells while others demonstrate extensive growth (Figs. 253 and 254). Grape-like clusters of blastospores along the pseudohyphae or branched pseudohyphae with ovoid blastospores in short chains are frequently observed.

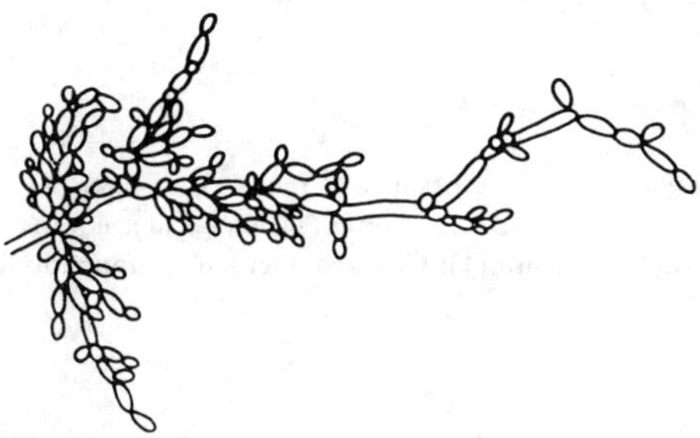

Fig. 253 – *C. guilliermondii*
Dalmau plate culture on corn meal agar
(From 'The Yeasts' ed. II, 1970)

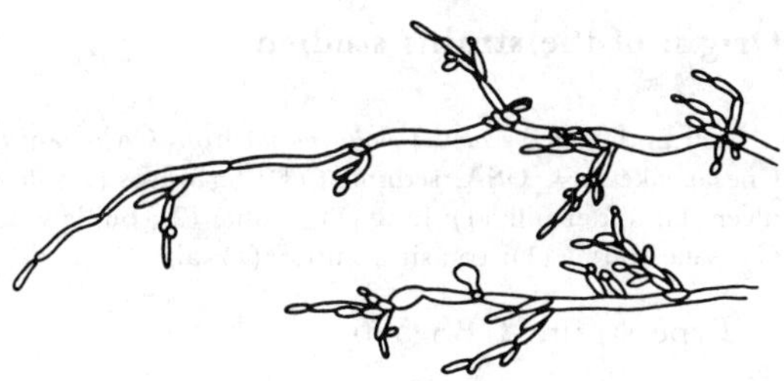

Fig. 254 – *C. guilliermondii*
Dalmau plate culture on corn meal agar
(From 'The Yeasts' ed. II, 1970)

Fermentation:

Glucose	+	Maltose	–
Galactose	v	Lactose	–
Sucrose	+	Raffinose	+ or w

Assimilation of carbon compounds:

Galactose	+	Melezitose	+	Ribitol	+
L-Sorbose	+ s	Soluble starch	–	Galactitol	+ or s
Sucrose	+	D-Xylose	+	D-Mannitol	+
Maltose	+	L-Arabinose	+	D-Glucitol	+
Cellobiose	+	D-Arabinose	+ or s	Salicin	+
Trehalose	+	D-Ribose	v	DL-Lactic acid	v
Lactose	–	L-Rhamnose	v	Succinic acid	+
Melibiose	+	Glycerol	+	Citric acid	+
Raffinose	+	Erythritol	–	Inositol	–

Splitting of arbutin: +
Assimilation of nitrate: –
Growth in vitamin-free medium: –
Growth at 37°C: +
Color reaction with DBB: –
Coenzyme Q system: Q 9 (Yamada & Kondo, 1972)
G + C: 44.1–44.4 mol%, 2 strains (Nakase & Komagata, 1971f)

Comments

The differences between *C. guilliermondii* var. *guilliermondii* and the variety *carpophila* appear too insignificant at this time to warrant the recognition of the variety *carpophila*, since strains of the variety *guilliermondii* indicated variable results on several of the carbon sources (e.g., L-rhamnose and galactitol utilization; galactose fermentation) originally used in establishing the variety *carpophila*. Furthermore, Meyer and Phaff (1972) showed a significant degree of DNA relatedness between the type strain of *C. guilliermondii* var. *guilliermondii* and two strains of the variety *carpophila*.

Origin of the strains studied

A strain labelled *Monilia guilliermondi* from Castellani (1); pigeon (2); Chesapeake Bay, USA, water (6); Chesapeake Bay, USA, sediment (1); fig wasps (1); flower (1); washed beer bottle (1); blood (1); equine ulcer (1); buttermilk (1); lung (1); leather (2); bull's semen (1); sputum (1); Cossidae larvae (1); appendix (1); sake starter (1); trypsin solution (1); air (1).

Type strain: CBS 566.

73. *CANDIDA HAEMULONII* (VAN UDEN ET KOLIPINSKI) MEYER ET YARROW (Yarrow & Meyer, 1978)

Synonym

Torulopsis haemulonii van Uden et Kolipinski (1962)

Standard description of *Candida haemulonii*

Growth in glucose-yeast extract-peptone water: After 3 days at 25°C the cells are ovoid to globose, rarely slender, (3–5) × (3–6.5) μm (Fig. 255). A ring is present (van Uden & Vidal-Leiria, 1970).

Growth on glucose-yeast extract-peptone agar: After one month at 25°C the streak culture is cream-colored, glistening, soft and smooth (van Uden & Vidal-Leiria, 1970).

Dalmau plate culture on corn meal agar: No pseudomycelium is formed; a few short chains of globose cells may be present (van Uden & Vidal-Leiria, 1970).

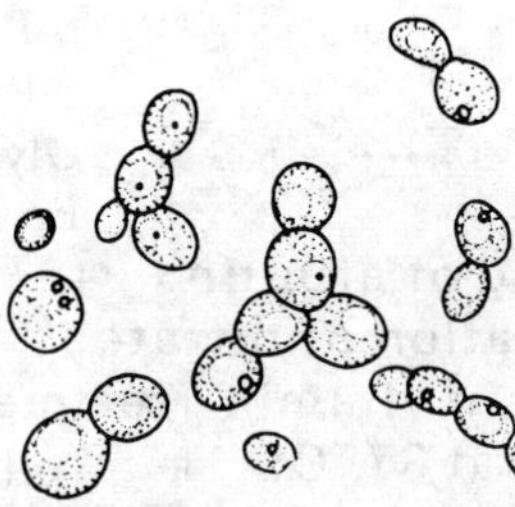

Fig. 255 – *C. haemulonii*
In glucose-yeast extract-peptone water
(From 'The Yeasts' ed. II, 1970)

Fermentation:

Glucose	+	Lactose	–
Galactose	–	Raffinose	+ w(s)
Sucrose	+	Trehalose	+
Maltose	–		

Assimilation of carbon compounds:

Galactose	+	Melezitose	+	Ribitol	+
L-Sorbose	–	Soluble starch	–	Galactitol	+ s
Sucrose	+	D-Xylose	+	D-Mannitol	+
Maltose	+	L-Arabinose	+	D-Glucitol	+
Cellobiose	–	D-Arabinose	–	Salicin	–
Trehalose	+	D-Ribose	–	DL-Lactic acid	–
Lactose	–	L-Rhamnose	+	Succinic acid	+
Melibiose	–	Glycerol	+	Citric acid	+
Raffinose	+ w(s)	Erythritol	–	Inositol	–

Splitting of arbutin: –

Assimilation of nitrate: –

Growth in vitamin-free medium: + s

Growth at 37°C: +

Color reaction with DBB: –

Coenzyme Q system: Q 9 (Yamada & Kondo, 1972)

G + C: 45.9–46.1 mol%, 2 strains (Nakase & Komagata, 1971c); 47.8 mol%, type strain (Meyer, Stenderup & Christiansen, unpublished)

Origin of the strains studied

Gut contents of *Haemulon sciurus* (blue-striped grunt), Florida (van Uden & Kolipinski, 1962) (1); seawater off Portugal (van Uden & Kolipinski, 1962) (1).

Type strain: CBS 5149, from gut contents of *Haemulon sciurus.*

74. CANDIDA HALONITRATOPHILA (ONISHI EX VAN UDEN ET VIDAL-LEIRIA) MEYER ET YARROW (Yarrow & Meyer, 1978)

Synonym

Torulopsis halonitratophila Onishi ex van Uden et Vidal-Leiria (1970)

Standard description of *Candida halonitratophila*

Growth in glucose-yeast extract-peptone water: After 3 days at 25°C the cells are globose to ovoid, (2–4) × (3–5) μm (Fig. 256); they frequently occur in groups. A ring may be present (van Uden & Vidal-Leiria, 1970).

Growth on glucose-yeast extract-peptone agar: After one month at 25°C the streak culture is cream-colored, glossy, soft and smooth (van Uden & Vidal-Leiria, 1970).

Dalmau plate culture on corn meal agar: No pseudomycelium is formed. Chains of globose and ovoid cells may occur (van Uden & Vidal-Leiria, 1970).

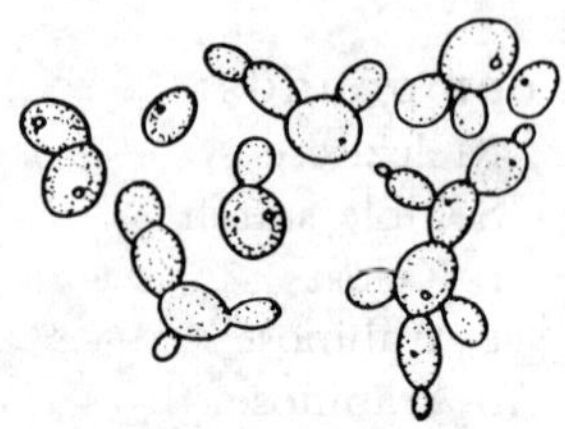

Fig. 256 – *C. halonitratophila*
In glucose-yeast extract-peptone water + 10% NaCl (w/v)
(From 'The Yeasts' ed. II, 1970)

Fermentation [NaCl was added to a final concentration of 10% (w/v)]:

Glucose	+ s	Maltose	–
Galactose	–	Lactose	–
Sucrose	–		

Assimilation of carbon compounds [NaCl was added to a final concentration of 10% (w/v)]:

Galactose	v	Melezitose	–	Ribitol	–
L-Sorbose	–	Soluble starch	–	Galactitol	–
Sucrose	–	D-Xylose	–	D-Mannitol	+
Maltose	–	L-Arabinose	–	D-Glucitol	v
Cellobiose	–	D-Arabinose	–	Salicin	–
Trehalose	–	D-Ribose	–	DL-Lactic acid	v
Lactose	–	L-Rhamnose	–	Succinic acid	– or + s
Melibiose	–	Glycerol	+	Citric acid	–
Raffinose	–	Erythritol	–	Inositol	–

Assimilation of nitrate: +
Growth in vitamin-free medium: –
Growth at 37°C: –
Growth on 50% glucose-peptone agar: + ; **on 60%:** + s
Urease: –
Color reaction with DBB: –
G + C: 52.4 mol%, type strain (Meyer, Stenderup & Christiansen, unpublished)

Origin of the strains studied

Soy mash, Japan (Onishi, 1960) (2).

Type strain: CBS 5240.

75. *CANDIDA HALOPHILA* YARROW ET MEYER (1978)

Synonym

Torulopsis halophilus nom. nud. (Onishi, 1957)

Standard description of *Candida halophila*

Growth in glucose-yeast extract-peptone water: After 3 days at 25°C the cells are globose to subglobose, (2.0–4.0) × (2.0–5.0) μm and occur singly and budding. A sediment is present and sometimes a ring.

Growth on glucose-yeast extract-peptone agar: After one month at 25°C the streak culture is white, smooth, and waxy in appearance.

Dalmau plate culture on corn meal agar: Pseudomycelium is absent.

Fermentation:

Glucose	+	Maltose	–
Galactose	+	Lactose	–
Sucrose	+ s		

Assimilation of carbon compounds:

Galactose	+ s	Melezitose	–	Ribitol	–
L-Sorbose	–	Soluble starch	–	Galactitol	–
Sucrose	+ s	D-Xylose	–	D-Mannitol	+
Maltose	–	L-Arabinose	+ s	D-Glucitol	–
Cellobiose	+ or s	D-Arabinose	–	Salicin	+
Trehalose	+ s	D-Ribose	–	DL-Lactic acid	–
Lactose	–	L-Rhamnose	–	Succinic acid	+ s
Melibiose	–	Glycerol	+ s	Citric acid	+ s
Raffinose	–	Erythritol	–	Inositol	–

Splitting of arbutin: +

Assimilation of nitrate: +

Growth in vitamin-free medium: –

Growth at 37°C: –

Urease: –

Color reaction with DBB: –

Comments

This species was originally described without a Latin description (Onishi, 1957). Yarrow and Meyer (1978) provided a Latin description at the same time they proposed to amend the diagnosis of the genus *Candida* to include nonhyphal species.

Origin of the strain studied

Soy mash (1).

Type strain: CBS 4019.

76. *CANDIDA HELLENICA* (VERONA ET PICCI) KING ET JONG (1977)

Synonym

Trichosporon hellenicum Verona et Picci (1958)

Standard description of *Candida hellenica*

Growth in glucose-yeast extract-peptone water: After 3 days at 25°C a pseudomycelium of long branched and curved cells is present. No single cells are available to measure. After 3 days on a rotary shaker the cells are elongated, sausage-shaped, (2.1–2.8) × (7.0–8.4) μm to pseudomycelial, often curved. Long branched pseudomycelium is evident. A thick, flaky sediment is present and the liquid is somewhat viscous. After one month the sediment is very thick and the liquid is clear.

Growth on glucose-yeast extract-peptone agar: After one month at 25°C the streak culture is white, dull, mycelial bordered with slight folds near the center, otherwise flat.

Dalmau plate culture on corn meal agar: Septate mycelium is present as well as branched chains of slender, often curved pseudohyphae, bearing ovoid blastospores.

Fermentation:

Glucose	+	Maltose	–
Galactose	+ w or –	Lactose	–
Sucrose	+	Raffinose	–

(A thick crusty pellicle is present in all tubes.)

Assimilation of carbon compounds:

Galactose	+	Melezitose	– or + s	Ribitol	+ s or w
L-Sorbose	+	Soluble starch	+	Galactitol	+
Sucrose	+	D-Xylose	+	D-Mannitol	+
Maltose	+	L-Arabinose	+	D-Glucitol	+
Cellobiose	+ or w	D-Arabinose	–	Salicin	+
Trehalose	+	D-Ribose	–	DL-Lactic acid	–
Lactose	–	L-Rhamnose	– or + w	Succinic acid	–
Melibiose	–	Glycerol	+	Citric acid	–
Raffinose	+ or w	Erythritol	–	Inositol	+ or s

Additional carbon compounds tested: Inulin –, D-glucosamine · HCl + s, methanol –, α-methyl-D-glucoside –

Splitting of arbutin: +

Assimilation of nitrate: –

Growth in vitamin-free medium: –

Growth at 37°C: +

Urease: –

Color reaction with DBB: –

G + C: 43.7 mol%, type strain

Comments

In their study of the genus *Trichosporon*, King and Jong (1977) noted that *Trichosporon hellenicum* failed to produce arthrospores and recommended that this species be transferred to the genus *Candida*. No known species of *Candida* with the same properties as this yeast was found to exist, thus it was designated the new combination, *C. hellenica*.

Origin of the strain studied

Grape must (1).

Type strain: CBS 4099.

77. *CANDIDA HOLMII* (JÖRGENSEN) MEYER ET YARROW (Yarrow & Meyer, 1978)

Perfect state: *Saccharomyces exiguus* Reess ex Hansen

Synonyms

Torula holmii Jörgensen (1909)
Torulopsis holmii (Jörgensen) Lodder (1934)
Cryptococcus holmii (Jörgensen) Skinner (1947a)
Torulopsis holmii (Jörgensen) Skinner var. *acidi-lactici* Ramirez et Sierra (1956)

Standard description of *Candida holmii*

Growth in glucose-yeast extract-peptone water: After 3 days at 25°C the cells are ovoid to cylindrical, (2.5–3) × (6–8) μm; also globose cells (3–5) μm occur; they predominate in some strains (Fig. 257). Only a sediment is formed (van Uden & Vidal-Leiria, 1970).

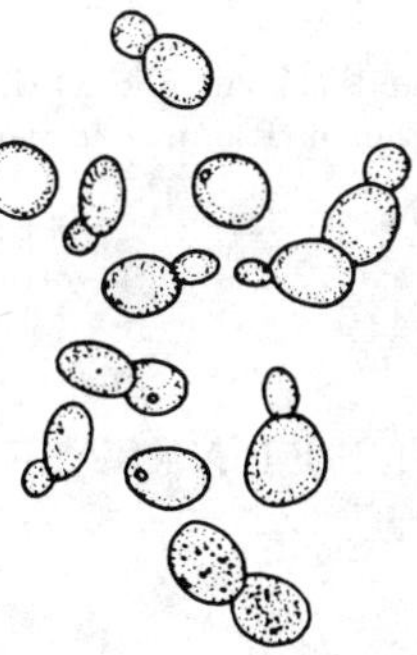

Fig. 257 – *C. holmii*
In glucose-yeast extract-peptone water
(From 'The Yeasts' ed. II, 1970)

Growth on glucose-yeast extract-peptone agar: After one month at 25°C the streak culture is greyish to cream-colored, glossy, soft, smooth with striped margins (van Uden & Vidal-Leiria, 1970).

Dalmau plate culture on corn meal agar: No pseudomycelium is formed. A few chains of ovoid cells may be present (van Uden & Vidal-Leiria, 1970).

Fermentation:

Glucose	+	Lactose	–
Galactose	+	Raffinose	v
Sucrose	+	Trehalose	+
Maltose	–		

Assimilation of carbon compounds:

Galactose	+	Melezitose	–	Ribitol	–
L-Sorbose	–	Soluble starch	–	Galactitol	–
Sucrose	+	D-Xylose	–	D-Mannitol	–
Maltose	–	L-Arabinose	–	D-Glucitol	–
Cellobiose	–	D-Arabinose	–	Salicin	–
Trehalose	+	D-Ribose	–	DL-Lactic acid	v
Lactose	–	L-Rhamnose	–	Succinic acid	–
Melibiose	–	Glycerol	v	Citric acid	–
Raffinose	+	Erythritol	–	Inositol	–

Splitting of arbutin: –

Assimilation of nitrate: –

Growth in vitamin-free medium: –

Growth at 37°C: v

Color reaction with DBB: –

Coenzyme Q system: Q 6 (Yamada & Kondo, 1972)

G+C: 32.9–35.1 mol%, 6 strains (Yarrow, 1978)

Comments

C. holmii is physiologically similar to *C. milleri* but has a G + C value more than 10 percent lower than *C. milleri* (46.8%). Yarrow (1978) found that *C. holmii* grew well in the absence of niacin and calcium pantothenate whereas two of the three strains of *C. milleri* which he studied did not grow in the absence of niacin and all three were stimulated by calcium pantothenate.

Origin of the strains studied

Buttermilk in The Netherlands (1); human sputum in The Netherlands (1); cucumber brine in USA (2); the authentic strain of *T. holmii* var. *acidi-lactici* from "alpechin" in Spain (Ramirez & Sierra, 1956) (1); unknown origin (2).

Type strain: CBS 135, from buttermilk.

78. *CANDIDA HOMILENTOMA* VAN DER WALT ET NAKASE (1973)

Standard description of *Candida homilentoma*

Growth in glucose-yeast extract-peptone water: After 3 days at 25°C the cells are spheroidal, ellipsoidal, ovoid, to cylindrical, (2.2–3.6) × (3.6–7.2) μm, and occur singly, in pairs, chains and clusters (Fig. 258). Pseudomycelium is also pre-

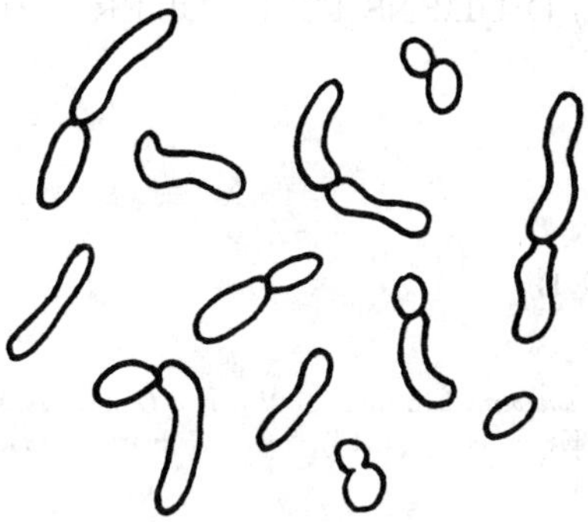

Fig. 258 – *C. homilentoma*
In glucose-yeast extract-peptone water

sent. A sediment and ring are present. After one month the culture is very viscous.

Growth on glucose-yeast extract-peptone agar: After one month at 25°C the streak culture is flat, cream-colored, dry and smooth with a mycelial border.

Dalmau plate culture on corn meal agar: Long, slender, branched, wavy pseudomycelium is abundant. Few blastospores are present.

Fermentation:

Glucose	+	Maltose	– or +s
Galactose	+	Lactose	–
Sucrose	–		

Assimilation of carbon compounds:

Galactose	+	Melezitose	+	Ribitol	+
L-Sorbose	– or +w	Soluble starch	+	Galactitol	–
Sucrose	+	D-Xylose	+	D-Mannitol	+
Maltose	+	L-Arabinose	+	D-Glucitol	+
Cellobiose	+	D-Arabinose	v	Salicin	+
Trehalose	+	D-Ribose	v	DL-Lactic acid	–
Lactose	–	L-Rhamnose	+	Succinic acid	+
Melibiose	–	Glycerol	+	Citric acid	+
Raffinose	–	Erythritol	+	Inositol	–

Splitting of arbutin: +

Assimilation of nitrate: –

Growth in vitamin-free medium: +

Growth on 50% (w/w) glucose-yeast extract agar: +

Growth at 37°C: +

Urease: –

Color reaction with DBB: –

G + C: 48.3–49.0 mol%, 2 strains (van der Walt & Nakase, 1973)

Origin of the strains studied

Tunnel of *Xylion adustus* Fabr. infesting *Ficus sycomorus* L. in the Dukuduku Forest near St. Lucia, Northern Natal (1); tunnels of *Sinoxylon ruficorne* Fabr. infesting *Combretum apiculatum* Sond. in the vicinity of Nylstroom, Northern Transvaal (1).

Type strain: CBS 6312, from *Combretum apiculatum*.

79. CANDIDA HUMICOLA (DASZEWSKA) DIDDENS ET LODDER (1942)

Synonyms

Torula humicola Daszewska (1912)
Mycotorula humicola (Daszewska) Harrison (1928)
Azymoprocandida humicola (Daszewska) Novák et Zsolt (1961)
Vanrija humicola (Daszewska) Moore (1980)
Apiotrichum porosum Stautz (1931)
Candida suaveolens Langeron et Guerra 1938 [nec *Sachsia suaveolens* Lindner (1895); nec *Oospora suaveolens* (Lindner) Lindau (1907b); nec *Oidium suaveolens* Krzemecki (1913); nec *Cylindrium suaveolens* (Krzemecki) Burns (1933)]
Sporobolomyces albidus Ramirez (1957)
Nadsonia slovaca Kocková-Kratochvílová et Svobodová-Poláková (1959)
Apiotrichum humicola (Daszewska) v. Arx et Weijman (1979)

Standard description of *Candida humicola*

Growth in glucose-yeast extract-peptone water: After 3 days at 25°C the cells are variable in form and size. They may be short-ovoid, drop-like, lemon-shaped, long-ovoid to cylindrical and measure (3.5–5) × (8–30) μm (Fig. 259). A pellicle is present (van Uden & Buckley, 1970).

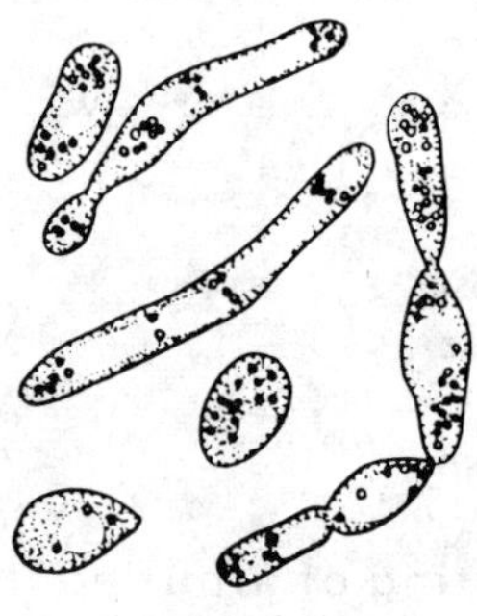

Fig. 259 – *C. humicola*
In glucose-yeast extract-peptone water
(From 'The Yeasts' ed. II, 1970)

Growth on glucose-yeast extract-peptone agar: After one month at 25°C the streak culture is yellow, glistening, smooth to densely wrinkled, mucous where smooth, membranous where wrinkled, sometimes slightly powdery on the surface (van Uden & Buckley, 1970).

Dalmau plate culture on corn meal agar: Pseudomycelium and true mycelium are abundantly formed. Hyphae and pseudohyphae often are wavy in appearance. Characteristically, branching is along narrow angles resulting in branches that often are almost parallel to the main axes. The blastospores are not well differentiated and all kinds of transitions occur between hyphal swellings, short pseudohyphae and oval yeast cells (Fig. 260) (van Uden & Buckley, 1970).

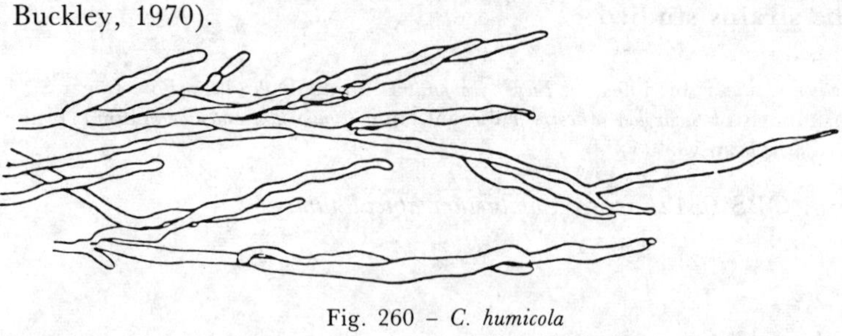

Fig. 260 – *C. humicola*
Dalmau plate culture on corn meal agar (From 'The Yeasts', ed. II, 1970)

Fermentation: –

Assimilation of carbon compounds:

Galactose	v	Melezitose	v	Ribitol	v
L-Sorbose	v	Soluble starch	v	Galactitol	v
Sucrose	+	D-Xylose	+	D-Mannitol	+
Maltose	+	L-Arabinose	v	D-Glucitol	v
Cellobiose	v	D-Arabinose	v	Salicin	+
Trehalose	+	D-Ribose	+	DL-Lactic acid	v
Lactose	v	L-Rhamnose	v	Succinic acid	v
Melibiose	+	Glycerol	v	Citric acid	v
Raffinose	v	Erythritol	+	Inositol	v

Splitting of arbutin: +

Assimilation of nitrate: –

Growth in vitamin-free medium: v

Growth at 37°C: –

Starch formation: +

Urease: +

Color reaction with DBB: +

Coenzyme Q system: Q 9 (type strain); Q 9 and Q 10 for 3 other strains (Yamada & Kondo, 1972)

G + C: 58.8–63.2 mol%, 3 strains (Nakase & Komagata, 1971f)

Comments

Von Arx and Weijman (1979) transferred *C. humicola* to *Apiotrichum* because of its heterobasidiomycetous properties, primarily sympodial conidiogenesis and the presence of xylose in cell hydrolysates.

Origin of the strains studied

Soil (1); culture contaminant (1); mushrooms (2); decaying toadstools (2); skin (2); water (1).

Type strain: CBS 571, isolated in Switzerland by Daszewska (1912) from heath soil.

80. *CANDIDA HUMILIS* (NEL ET VAN DER WALT) MEYER ET YARROW (Yarrow & Meyer, 1978)

Synonym

Torulopsis humilis Nel et van der Walt (1968)

Standard description of *Candida humilis*

Growth in glucose-yeast extract-peptone water: After 2 days at 25°C the cells are globose to ovoid, (3.6–6.5) × (4.3–7.2) μm and occur singly, in pairs and short chains. A sediment is present. After one month at 20°C a sediment and thin ring are present.

Growth on glucose-yeast extract-peptone agar: After one month at 25°C the streak culture is cream-colored, moist, glistening and smooth; the margin is entire or slightly undulating.

Dalmau plate culture on corn meal agar: No pseudomycelium is formed.

Fermentation:

Glucose	+	Maltose	–
Galactose	+	Lactose	–
Sucrose	–	Trehalose	+

Assimilation of carbon compounds:

Galactose	+	Melezitose	–	Ribitol	–
L-Sorbose	–	Soluble starch	–	Galactitol	–
Sucrose	–	D-Xylose	–	D-Mannitol	–
Maltose	–	L-Arabinose	–	D-Glucitol	–
Cellobiose	–	D-Arabinose	–	Salicin	–
Trehalose	+	D-Ribose	–	DL-Lactic acid	+
Lactose	–	L-Rhamnose	–	Succinic acid	–
Melibiose	–	Glycerol	+	Citric acid	–
Raffinose	–	Erythritol	–	Inositol	–

Splitting of arbutin: –

Assimilation of nitrate: –

Growth in vitamin-free medium: –

Growth at 37°C: –

Color reaction with DBB: –

G + C: 47.1 mol%, type strain (Meyer, Stenderup & Christiansen, unpublished)

Origin of the strain studied

Bantu beer (1).

Type strain: CBS 5658.

81. CANDIDA HYDROCARBOFUMARICA YAMADA ET AL. EX RAMIREZ (1974)

Synonym

Candida hydrocarbofumarica nom. nud. (Yamada et al., 1970)

Standard description of *Candida hydrocarbofumarica*

Growth in YM broth: After 3 days at 25°C the cells are spheroidal to ovoid, (2–5) × (2–7) μm, and occur singly, in pairs or short chains. A thin, smooth pellicle and a sediment are formed (Yamada et al., 1970).

Growth on YM agar: After one month at 20°C the streak culture is cream-colored, smooth or slightly wrinkled, dull, soft, with a ciliate margin (Yamada et al., 1970).

Dalmau plate culture on corn meal agar: Pseudomycelium consists of branched chains of cells with blastospores formed in verticils.

Fermentation: –

Assimilation of carbon compounds:

Galactose	+	Melezitose	+	Ribitol	+ or s
L-Sorbose	+	Soluble starch	+	Galactitol	+ or s
Sucrose	+	D-Xylose	+	D-Mannitol	+
Maltose	+	L-Arabinose	+	D-Glucitol	+
Cellobiose	+	D-Arabinose	– or +s	Salicin	+
Trehalose	+	D-Ribose	– or +s	DL-Lactic acid	–
Lactose	+ or s	L-Rhamnose	+	Succinic acid	+
Melibiose	–	Glycerol	+s	Citric acid	+
Raffinose	+s	Erythritol	+	Inositol	+

Splitting of arbutin: +

Assimilation of nitrate: –

Growth in vitamin-free medium: +

Growth at 37°C: +

Urease: –

Color reaction with DBB: –

Coenzyme Q system: Q 9 (Yamada & Kondo, 1972)

G + C: 56.6 mol%, one strain (Meyer, Stenderup & Christiansen, unpublished)

Comments

Yamada et al. (1970) described *C. hydrocarbofumarica,* a new yeast species, which showed strong ability to produce fumaric acid from *n*-paraffins. They reported the fermentation of glucose, galactose, sucrose, maltose (weak), lactose (weak) and trehalose. These results could not be confirmed.

Origin of the strain studied

Soil (1).

Type strain: CBS 6734.

82. *CANDIDA HYLOPHILA* VAN DER WALT, VAN DER KLIFT ET SCOTT (van der Walt et al., 1971a)

Standard description of *Candida hylophila*

Growth in glucose-yeast extract-peptone water: After 3 days at 25°C the cells are spheroidal, ovoid, ellipsoidal or cylindrical, (1.0–3.0) × (1.5–9.0) μm, occasionally triangular or irregular, and occur singly, in pairs, short chains and small clusters (Fig. 261). Pseudomycelium is present. After one month at room temperature a granular to floccose sediment, islets and an incomplete ring are present (van der Walt et al., 1971a).

Growth on glucose-yeast extract-peptone agar: After one month at 25°C the streak culture is brownish-cream, moist, raised, lacy with a fringed margin.

Dalmau plate culture on corn meal agar: The pseudomycelium consists of long slender pseudohyphae often curved bearing few blastospores. Van der Walt et al. (1971a) reported that the blastospores were often triangular and occurred in clusters or short chains.

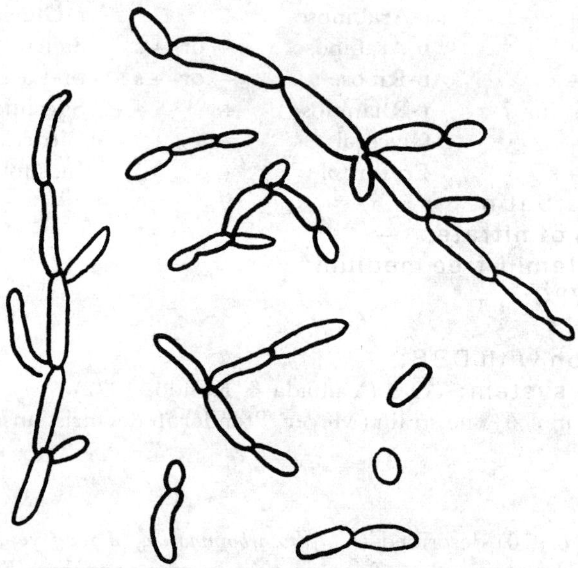

Fig. 261 – *C. hylophila*
In glucose-yeast extract-peptone water

Fermentation: –

Assimilation of carbon compounds:

Galactose	–	Melezitose	–	Ribitol	–
L-Sorbose	v	Soluble starch	–	Galactitol	–
Sucrose	–	D-Xylose	–	D-Mannitol	+ or s
Maltose	–	L-Arabinose	–	D-Glucitol	+ or s
Cellobiose	–	D-Arabinose	–	Salicin	–
Trehalose	–	D-Ribose	– or + s	DL-Lactic acid	+
Lactose	–	L-Rhamnose	–	Succinic acid	+
Melibiose	–	Glycerol	+	Citric acid	–
Raffinose	–	Erythritol	–	Inositol	–

Additional carbon compound tested: Methanol –

Splitting of arbutin: –

Assimilation of nitrate: –

Growth in vitamin-free medium: –

Growth on 50% (w/w) glucose-yeast extract agar: –

Growth at 37°C: –
Starch formation: –
Urease: +
Color reaction with DBB: +

Origin of the strain studied

Tunnels of *Xyleborus aemulus* Woll. in *Rapanea melanophloeos* L. growing near Knysna, Cape Province in South Africa (1).

Type strain: CBS 6226.

83. CANDIDA INCOMMUNIS OHARA, NONOMURA ET YAMAZAKI (1965)

Standard description of *Candida incommunis*

Growth in glucose-yeast extract-peptone water: After 3 days at 25°C the cells are cylindrical, (2–3) × (10–17) μm, single or in short chains (Fig. 262). A sediment and a thin pellicle are present. After one month at 17°C, a slimy, thick pellicle is formed (Ohara et al., 1965).

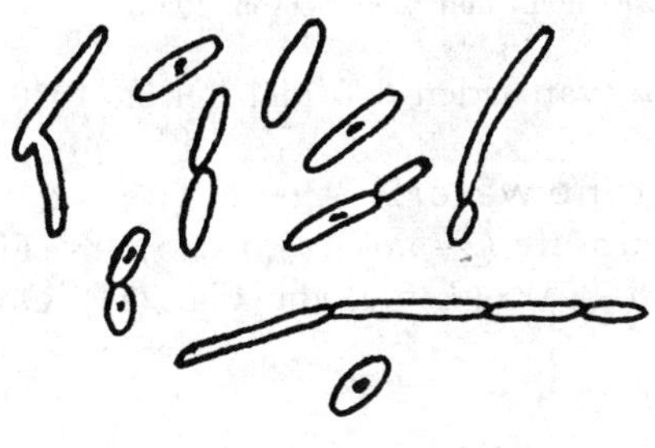

Fig. 262 – *C. incommunis*
In glucose-yeast extract-peptone water

Growth on glucose-yeast extract-peptone agar: After one month at 25°C the streak culture is light tan with a smooth periphery and wrinkled central area. Some pseudomycelium is present at the margin.

Dalmau plate culture on corn meal agar: Abundant pseudomycelial development consisting of long, branching cylindrical cells. A "wavy" appearance is evident.

Fermentation:

Glucose	– or +w, s	Maltose	–
Galactose	–	Lactose	–
Sucrose	–		

Assimilation of carbon compounds:

Galactose	–	Melezitose	–	Ribitol	–
L-Sorbose	+	Soluble starch	–	Galactitol	–
Sucrose	+	D-Xylose	v	D-Mannitol	+ s
Maltose	+	L-Arabinose	–	D-Glucitol	+ s
Cellobiose	+ s	D-Arabinose	–	Salicin	+
Trehalose	+	D-Ribose	–	DL-Lactic acid	–
Lactose	–	L-Rhamnose	–	Succinic acid	+
Melibiose	–	Glycerol	+	Citric acid	+
Raffinose	–	Erythritol	+	Inositol	+

Splitting of arbutin: +
Assimilation of nitrate: +
Growth in vitamin-free medium: –
Growth at 37°C: –
Urease: –
Color reaction with DBB: –
G + C: 48.9 mol%, type strain (Meyer, Stenderup & Christiansen, unpublished)

Origin of the strain studied

Grape must produced in Yamanashi Prefecture, Japan (1).

Type strain: CBS 5604.

84. CANDIDA INCONSPICUA (LODDER ET KREGER-VAN RIJ) MEYER ET YARROW (Yarrow & Meyer, 1978)

Synonyms

Torulopsis inconspicua Lodder et Kreger-van Rij (1952)
Torulopsis inconspicua Lodder et Kreger-van Rij var. *filiforme* nom. nud. (Dietrichson, 1954)

Standard description of *Candida inconspicua* (van Uden & Vidal-Leiria, 1970)

Growth in glucose-yeast extract-peptone water: After 3 days at 25°C the cells are ovoid. In some strains the cells measure (3–5.5) × (5.5–7.5) μm (Fig. 263); in others, they are smaller, measuring (1.5–3.5) × (3–6.5) μm (Fig. 264). Only a sediment is formed.

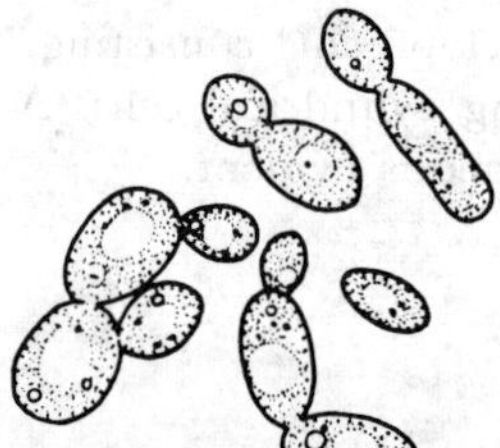

Fig. 263 – *C. inconspicua*
In glucose-yeast extract-peptone water
(From 'The Yeasts' ed. II, 1970)

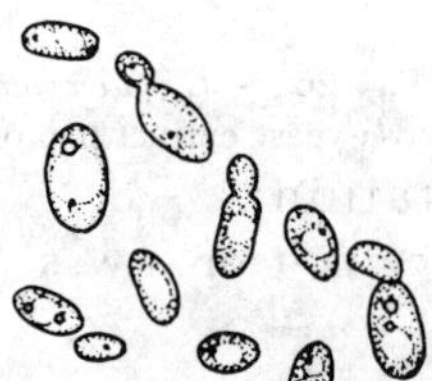

Fig. 264 – *C. inconspicua*
In glucose-yeast extract-peptone water
(From 'The Yeasts'. ed. II, 1970)

Growth on glucose-yeast extract-peptone agar: After one month at 25°C the streak culture is greyish-white, semidull, soft, smooth or striped.

Dalmau plate culture on corn meal agar: A very primitive pseudomycelium consisting of branched chains of cylindrical or ovoid cells may be formed.

Fermentation: –

Assimilation of carbon compounds:

Galactose	–	Melezitose	–	Ribitol	–
L-Sorbose	–	Soluble starch	–	Galactitol	–
Sucrose	–	D-Xylose	–	D-Mannitol	–
Maltose	–	L-Arabinose	–	D-Glucitol	–
Cellobiose	–	D-Arabinose	–	Salicin	–
Trehalose	–	D-Ribose	–	DL-Lactic acid	+
Lactose	–	L-Rhamnose	–	Succinic acid	+ or w
Melibiose	–	Glycerol	+	Citric acid	–
Raffinose	–	Erythritol	–	Inositol	–

Assimilation of nitrate: –

Growth in vitamin-free medium: –

Growth at 37°C: +

Color reaction with DBB: –

Coenzyme Q system: Q 8 (Yamada & Kondo, 1972)

G + C: 36.7 mol%, type strain (Meyer, Stenderup & Christiansen, unpublished)

Origin of the strains studied

Sputum (4), one in The Netherlands, two in Germany, and one, the authentic strain of *T. inconspicua* var. *filiforme,* in Norway (Dietrichson, 1954); human tongue (1); unknown origin (1).

Type strain: CBS 180, in The Netherlands from sputum.

85. *CANDIDA INGENIOSA* (DI MENNA) MEYER ET YARROW (Yarrow & Meyer, 1978)

Synonyms

Torulopsis ingeniosa di Menna (1958b)
Rhodotorula ingeniosa (di Menna) von Arx et Weijman (1979)
Vanrija ingeniosa (di Menna) Moore (1980)

Standard description of *Candida ingeniosa*

Growth in glucose-yeast extract-peptone water: After 3 days at 25°C the cells are ovoid to cylindrical, single or in pairs, (2–3.5) × (3.5–8) μm (Fig. 265). A thin capsule may be present. After one month a ring is present (van Uden & Vidal-Leiria, 1970).

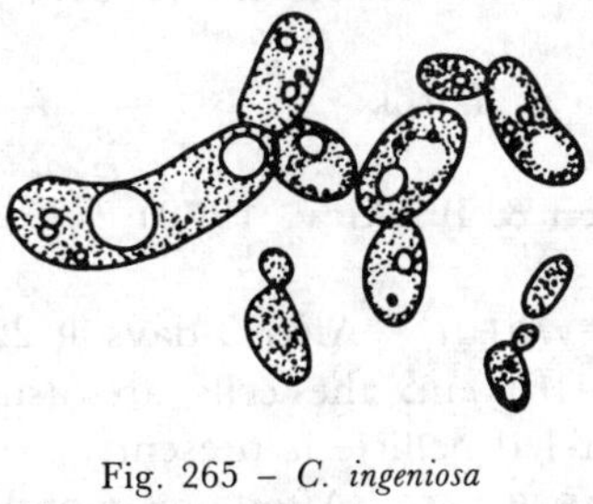

Fig. 265 – *C. ingeniosa*
In glucose-yeast extract-peptone water
(From 'The Yeasts', ed. II, 1970)

Growth on glucose-yeast extract-peptone agar: After one month at 25°C the streak culture is yellowish, mucoid, glistening, soft and smooth (van Uden & Vidal-Leiria, 1970).

Dalmau plate culture on corn meal agar: Pseudomycelium is absent.
Fermentation: –
Assimilation of carbon compounds:

Galactose	+ w	Melezitose	+	Ribitol	–
L-Sorbose	+	Soluble starch	+	Galactitol	–
Sucrose	+	D-Xylose	+ s	D-Mannitol	+
Maltose	+	L-Arabinose	–	D-Glucitol	+
Cellobiose	+	D-Arabinose	–	Salicin	+
Trehalose	+	D-Ribose	+ s	DL-Lactic acid	+ w
Lactose	+ w	L-Rhamnose	–	Succinic acid	+ w
Melibiose	–	Glycerol	+	Citric acid	+
Raffinose	+	Erythritol	–	Inositol	–

Splitting of arbutin: +
Assimilation of nitrate: +
Growth in vitamin-free medium: +
Growth at 37°C: v
Urease: +
Color reaction with DBB: +
Coenzyme Q system: Q 10 (Yamada & Kondo, 1972)
G + C: 55.6 mol%, type strain (Nakase & Komagata, 1971c)

Comments

Several carbon assimilation results were found to differ from those reported by van Uden & Vidal-Leiria (1970). They found L-sorbose, trehalose, xylose, D-ribose, glycerol, D-mannitol, D-glucitol and citric acid to be negative. Based on the type of conidiogenesis and the lack of xylose in cell hydrolysates, von Arx and Weijman (1979) transferred *C. ingeniosa* to the genus *Rhodotorula.*

Origin of the strain studied

Pasture grass leaves in New Zealand (di Menna, 1958b) (1).

Type strain: CBS 4240.

86. *CANDIDA INGENS* VAN DER WALT ET VAN KERKEN (1961c)

Perfect state: *Pichia humboldtii* Rodrigues de Miranda et Török

Standard description of *Candida ingens* (van Uden & Buckley, 1970)

Growth in glucose-yeast extract-peptone water: After 3 days at 25°C the cells are long-ovoid to cylindrical, (4–8) × (8–16) μm; the cells are usually arranged in ramified chains (Fig. 266). A dull, wrinkled pellicle is present.

Growth on glucose-yeast extract-peptone agar: After one month at

25°C the streak culture is yellowish to cream-colored, semidull, soft, with a punctilated to warty surface.

Dalmau plate culture on corn meal agar: The pseudomycelium consists of ramified chains of coarse pseudohyphae (Fig. 267).

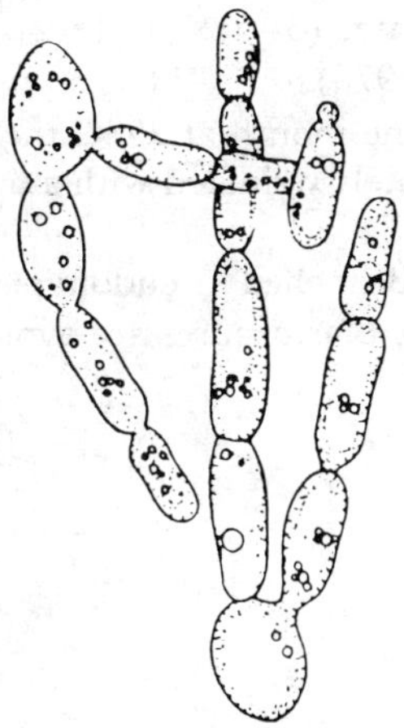

Fig. 266 – *C. ingens*
In glucose-yeast extract-peptone water
(From 'The Yeasts' ed. II, 1970)

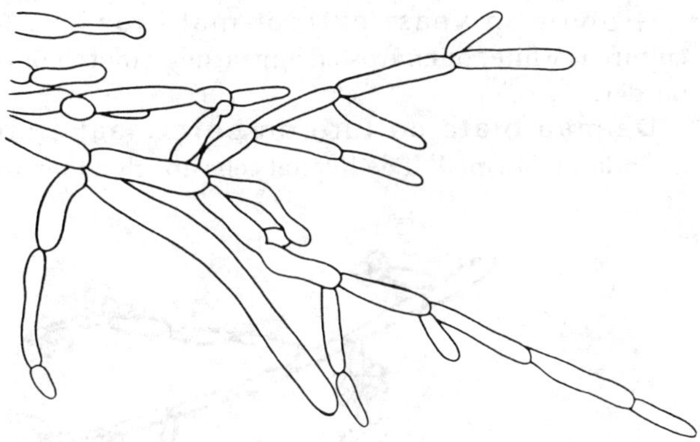

Fig. 267 – *C. ingens*
Dalmau plate culture on corn meal agar
(From 'The Yeasts' ed. II, 1970)

Fermentation: –

Assimilation of carbon compounds:

Galactose	+	Melezitose	–	Ribitol	–
L-Sorbose	v	Soluble starch	–	Galactitol	–
Sucrose	–	D-Xylose	–	D-Mannitol	–
Maltose	–	L-Arabinose	–	D-Glucitol	–
Cellobiose	–	D-Arabinose	–	Salicin	–
Trehalose	–	D-Ribose	–	DL-Lactic acid	v
Lactose	–	L-Rhamnose	–	Succinic acid	v
Melibiose	–	Glycerol	+	Citric acid	–
Raffinose	–	Erythritol	–	Inositol	–

Assimilation of nitrate: –

Growth in vitamin-free medium: +

Growth at 37°C: v

Urease: +

Color reaction with DBB: –

G + C: 39.5 mol%, type strain (Stenderup et al., 1972)

Origin of the strains studied

Brewery material (4); industrial sewage (2).

Type strain: CBS 4603, isolated in South Africa from brewery material.

87. *CANDIDA INOSITOPHILA* NAKASE (1975)

Standard description of *Candida inositophila*

Growth in yeast extract-malt extract broth: After 3 days at 25°C, growth is poor. The cells are oval, long-oval or elongate, (3–5) × (7–15.2) μm. A wrinkled pellicle and a sediment are formed (Nakase, 1975).

Growth on yeast extract-malt agar: After one month at 25°C the streak culture is white, often frosted appearing, smooth or delicately wrinkled with a mycelial border.

Dalmau plate culture on corn meal agar: Branched pseudohyphae are abundantly formed. The hyphal cells are elongate and the blastospores are ovoid (Fig. 268).

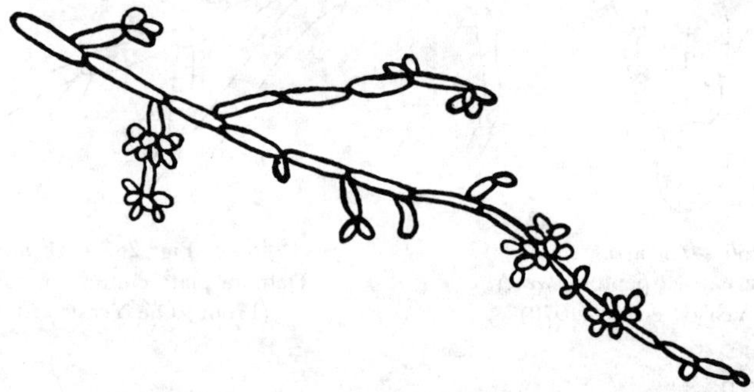

Fig. 268 – *C. inositophila*
Dalmau plate culture on corn meal agar

Fermentation:

Glucose	+	Maltose	+ s
Galactose	v	Lactose	–
Sucrose	+ s	Raffinose	v

Assimilation of carbon compounds:

Galactose	+	Melezitose	+	Ribitol	+
L-Sorbose	+	Soluble starch	+	Galactitol	+ s
Sucrose	+	D-Xylose	+	D-Mannitol	+
Maltose	+	L-Arabinose	+	D-Glucitol	+
Cellobiose	+	D-Arabinose	–	Salicin	+
Trehalose	+	D-Ribose	+ s or –	DL-Lactic acid	–
Lactose	–	L-Rhamnose	+	Succinic acid	+
Melibiose	–	Glycerol	+	Citric acid	–
Raffinose	+	Erythritol	–	Inositol	+

Splitting of arbutin: +

Assimilation of nitrate: –

Growth in vitamin-free medium: –

Growth at 37°C: +

Urease: –
Color reaction with DBB: –
G + C: 43.9 mol%, type strain (Nakase, 1975)

Comments

Candida inositophila is similar to *C. steatolytica*; they are differentiated on utilization of citric acid.

Origin of the strain studied

Washings of ion-exchange resin in a guanosine monophosphate manufacturing plant (1).

Type strain: CBS 6736.

88. CANDIDA INSECTALENS (SCOTT, VAN DER WALT ET VAN DER KLIFT) MEYER ET YARROW (Yarrow & Meyer, 1978)

Synonym

Torulopsis insectalens Scott, van der Walt et van der Klift (van der Walt et al., 1971b)

Standard description of *Candida insectalens*

Growth in glucose-yeast extract-peptone water: After 3 days at 25°C the cells are predominantly spheroidal 2.9 × 2.9 μm and occur singly, in pairs, small clusters or short chains. Some cells are ovoidal to ellipsoidal. A sediment is present. After one month at 20°C a sediment is present.

Growth on glucose-yeast extract-peptone agar: After one month at 25°C the streak culture is yellowish-white, smooth, convex, with an entire to undulating margin.

Dalmau plate culture on corn meal agar: No pseudomycelium is formed.

Fermentation: –

Assimilation of carbon compounds:

Galactose	+ s	Melezitose	–	Ribitol	+ s
L-Sorbose	–	Soluble starch	–	Galactitol	–
Sucrose	–	D-Xylose	–	D-Mannitol	+
Maltose	–	L-Arabinose	–	D-Glucitol	+
Cellobiose	+ s	D-Arabinose	v	Salicin	+
Trehalose	+	D-Ribose	–	DL-Lactic acid	+ s
Lactose	–	L-Rhamnose	–	Succinic acid	+
Melibiose	–	Glycerol	–	Citric acid	–
Raffinose	–	Erythritol	–	Inositol	–

Additional carbon compound tested: Methanol –

Splitting of arbutin: + or s

Assimilation of nitrate: –
Growth in vitamin-free medium: –
Growth on 50% (w/w) glucose-yeast extract agar: –
Growth at 37°C: +
Color reaction with DBB: –
G + C: 44.9 mol%, type strain (Stenderup et al., 1972)

Origin of the strains studied

Tunnels of *Crossotarus externedentatus* Fairm. infecting *Ficus sycomorus* L. near St. Lucia (1); insect tunnels in *Cryptocarya latifolia* (1).

Type strain: CBS 6036 from *Ficus sycomorus.*

89. *CANDIDA INSECTAMANS* SCOTT, VAN DER WALT ET VAN DER KLIFT (van der Walt et al., 1972)

Standard description of *Candida insectamans*

Growth in glucose-yeast extract-peptone water: After 3 days at 25°C the cells are globose to oval, long-oval to cylindrical, (1.5–6.5) × (2.5–12.0) μm (Fig. 269) and occur singly, in pairs and in small clusters. Pseudomycelium is present. After one month a thick floccose sediment and ring are formed (van der Walt et al., 1972).

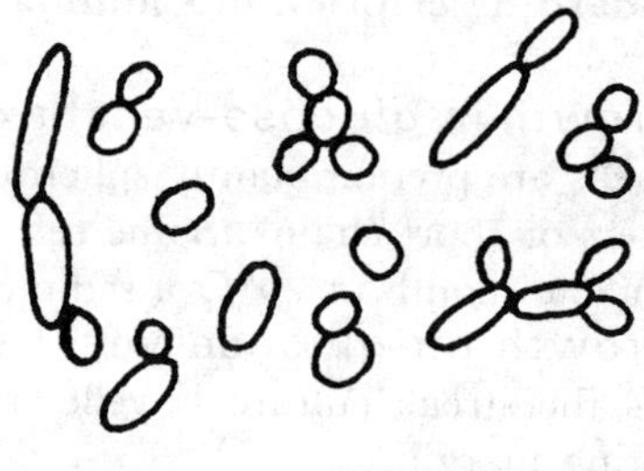

Fig. 269 – *C. insectamans*
In glucose-yeast extract-peptone water

Growth on glucose-yeast extract-peptone agar: After one month at 25°C the streak culture is cream-colored to tan with a slightly raised, wrinkled center, smooth near periphery and fringed with pseudomycelium.

Dalmau plate culture on corn meal agar: Extensive, long, slender ramified pseudomycelium is formed with few blastospores in small clusters or short chains.

Fermentation:

Glucose	+ w or s	Maltose	–
Galactose	–	Lactose	–
Sucrose	–		

Assimilation of carbon compounds:

Galactose	–	Melezitose	– or + s	Ribitol	+
L-Sorbose	–	Soluble starch	+	Galactitol	–
Sucrose	–	D-Xylose	+	D-Mannitol	+
Maltose	+	L-Arabinose	–	D-Glucitol	+

Cellobiose	+	D-Arabinose	–	Salicin	+
Trehalose	+	D-Ribose	+ or s	DL-Lactic acid	–
Lactose	–	L-Rhamnose	–	Succinic acid	+ s
Melibiose	–	Glycerol	– or + s	Citric acid	+
Raffinose	–	Erythritol	–	Inositol	–

Splitting of arbutin: +
Assimilation of nitrate: –
Growth in vitamin-free medium: –
Growth on 50% (w/w) glucose-yeast extract agar: –
Growth at 37°C: +
Color reaction with DBB: –
G + C: 35.9 mol%, type strain (Meyer, Stenderup & Christiansen, unpublished)

Origin of the strain studied

Frass of Buprestid larvae infesting a moribund specimen of *Acacia nilotica* L. var. *kraussiana* in the vicinity of Groblersdal in the Transvaal (1).

Type strain: CBS 6033.

90. *CANDIDA INSECTORUM* SCOTT, VAN DER WALT ET VAN DER KLIFT (van der Walt et al., 1972)

Standard description of *Candida insectorum*

Growth in glucose-yeast extract-peptone water: After 3 days at 25°C the cells are globose, oval to long-oval, cylindrical to elongate, (1.5–5.0) × (2.5–9.5–19) μm, and occur singly, in pairs, clusters, and chains (Fig. 270). Pseudomycelium is present. After one month a floccose sediment is present and occasionally an incomplete ring (van der Walt et al., 1972).

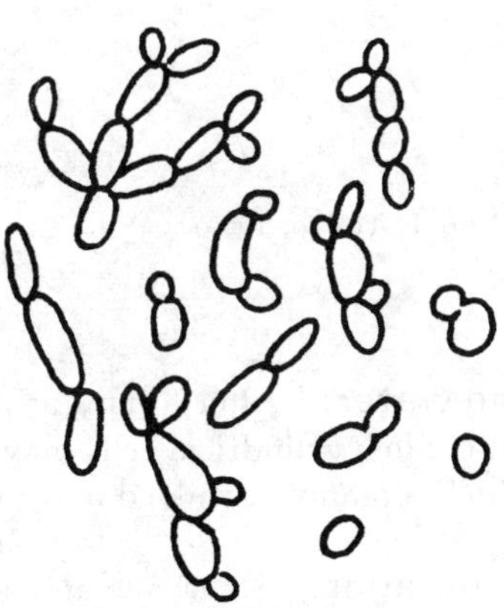

Fig. 270 *C. insectorum*
In glucose-yeast extract-peptone water

Growth on glucose-yeast extract-peptone agar: After one month at 25°C the streak culture is yellowish-white to brownish, creamy, flat and glistening. The center area may be rough and the border lobate and fringed with pseudomycelium.

Dalmau plate culture on corn meal agar: A ramified pseudomycelium is formed with blastospores in small clusters and short chains.

Fermentation:

Glucose	+ s	Maltose	–
Galactose	+ s	Lactose	–
Sucrose	–		

Assimilation of carbon compounds:

Galactose	+	Melezitose	+	Ribitol	+
L-Sorbose	+ s	Soluble starch	v	Galactitol	–
Sucrose	+	D-Xylose	+	D-Mannitol	+
Maltose	+	L-Arabinose	+	D-Glucitol	+
Cellobiose	+	D-Arabinose	+ s	Salicin	+
Trehalose	+	D-Ribose	+	DL-Lactic acid	–
Lactose	+ s or –	L-Rhamnose	+	Succinic acid	+
Melibiose	+ s or –	Glycerol	+	Citric acid	+
Raffinose	+	Erythritol	+	Inositol	–

Splitting of arbutin: +

Assimilation of nitrate: –

Growth in vitamin-free medium: -

Growth at 37°C: v

Urease: –

Color reaction with DBB: –

G + C: 39.5 mol%, type strain (Stenderup et al., 1972)

Origin of the strain studied

Frass of *Phoracantha recurva* Newman infesting *Eucalyptus maculata* Hook near Kwa-Moonami in Natal (1).

Type strain: CBS 6213.

91. *CANDIDA INTERMEDIA* (CIFERRI ET ASHFORD) LANGERON ET GUERRA (1938)

Synonyms

Blastodendrion intermedium Cif. et Ashf. (1929)
Candida intermedia var. *ethanophila* nom. nud. (Verona & Zardetto de Toledo, 1954)

Standard description of *Candida intermedia*

Growth in glucose-yeast extract-peptone water: After 3 days at 25°C the cells are ovoid, (2.5–5) × (4.5–8) μm, usually in chains; cylindrical cells may also occur (Fig. 271) (van Uden & Buckley, 1970). A thick, creamy, wrinkled and folded pellicle is formed.

Growth on glucose-yeast extract-peptone agar: After one month at 25°C the streak culture is whitish to cream-colored, somewhat dull, soft, smooth or delicately wrinkled (van Uden & Buckley, 1970).

Dalmau plate culture on corn meal agar: The pseudomycelium consists of

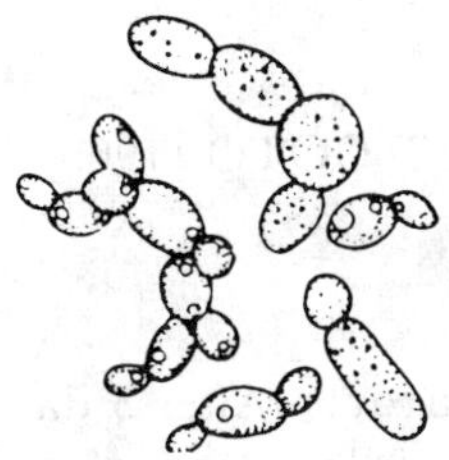

Fig. 271 – *C. intermedia*
In glucose-yeast extract-peptone water
(From 'The Yeasts' ed. II, 1970)

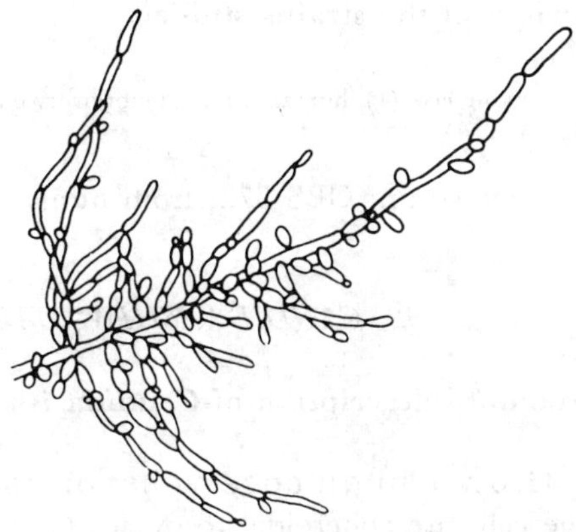

Fig. 272 – *C. intermedia*
Dalmau plate culture on corn meal agar
(From 'The Yeasts' ed. II, 1970)

short pseudohyphae arranged in a tree-like manner, bearing ovoid blastospores (Fig. 272) (van Uden & Buckley, 1970).

Fermentation:

Glucose	+	Lactose	–
Galactose	+ or s	Raffinose	v
Sucrose	+	Trehalose	v
Maltose	v		

Assimilation of carbon compounds:

Galactose	+	Melezitose	+	Ribitol	– or + s
L-Sorbose	+	Soluble starch	+	Galactitol	–
Sucrose	+	D-Xylose	+	D-Mannitol	+
Maltose	+	L-Arabinose	v	D-Glucitol	+
Cellobiose	+	D-Arabinose	–	Salicin	+
Trehalose	+	D-Ribose	– or + s	DL-Lactic acid	–
Lactose	+	L-Rhamnose	v	Succinic acid	v
Melibiose	–	Glycerol	–	Citric acid	v
Raffinose	+	Erythritol	–	Inositol	–

Splitting of arbutin: +
Assimilation of nitrate: –
Growth in vitamin-free medium: –
Growth on 50% (w/w) glucose-yeast extract agar: –
Growth at 37°C: – or + w
Urease: –
Color reaction with DBB: –
Coenzyme Q system: Q 9 (Yamada & Kondo, 1972)
G + C: 44.4 mol%, type strain (Nakase & Komagata, 1971f)

Origin of the strains studied

Human feces (1); human throat (1); brewery equipment (2); condensed milk (1); unknown (2); grape (1); seawater (1).

Type strain: CBS 572, from human feces.

92. *CANDIDA ISHIWADAE* SUGIYAMA ET GOTO (1969)

Standard description of *Candida ishiwadae*

Growth in glucose-yeast extract-peptone water: After 3 days at 25°C the cells are spheroidal to ovoid, (2.9–5.1) × (3.6–6.5) μm and occur singly and in pairs. A sediment is present. After one month at 20°C a ring and sediment are present.

Growth on glucose-yeast extract-peptone agar: After one month at 25°C the streak culture is cream to white, mucoid, glossy, smooth and the margin is entire.

Dalmau plate culture on corn meal agar: Branched chains of cylindrical cells, somewhat curved, are present.

Fermentation:

Glucose	+	Maltose	+
Galactose	+w or –	Lactose	–
Sucrose	+w or –		

Assimilation of carbon compounds:

Galactose	v	Melezitose	+	Ribitol	+ or s
L-Sorbose	– or +s	Soluble starch	+	Galactitol	v
Sucrose	+	D-Xylose	+ or w	D-Mannitol	+
Maltose	+	L-Arabinose	+s or –	D-Glucitol	+
Cellobiose	+	D-Arabinose	+	Salicin	+
Trehalose	+	D-Ribose	+	DL-Lactic acid	+
Lactose	–	L-Rhamnose	+	Succinic acid	+
Melibiose	–	Glycerol	+	Citric acid	+
Raffinose	–	Erythritol	+ or s	Inositol	–

Splitting of arbutin: +

Assimilation of nitrate: +

Growth in vitamin-free medium: –

Growth at 37°C: –

Color reaction with DBB: –

G + C: 36.7 mol%, type strain (Meyer, Stenderup & Christiansen, unpublished)

Origin of the strain studied

Soil (1).

Type strain: CBS 6022.

93. *CANDIDA JAPONICA* DIDDENS ET LODDER

see *Filobasidium capsuligenum* p. 484.

94. *CANDIDA JAVANICA* RUINEN (1963)

Synonym

Rhodotorula javanica (Ruinen) von Arx et Weijman (1979)

Standard description of *Candida javanica* (van Uden & Buckley, 1970)

Growth in glucose-yeast extract-peptone water: After 3 days at 25°C the cells are long-oval to spindle-shaped (1.5–3) × (7–24) μm (Fig. 273). The cells are encapsulated; a thin ring may be present.

Growth on glucose-yeast extract-peptone agar: After one month at 25°C the streak culture is yellowish, glistening, slightly mucoid, soft and almost smooth.

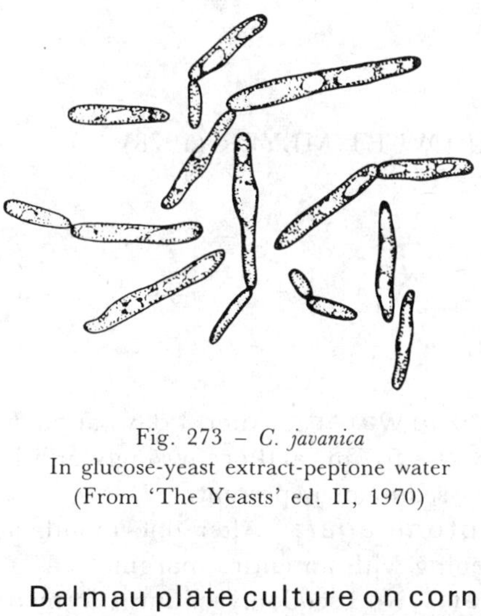

Fig. 273 – *C. javanica*
In glucose-yeast extract-peptone water
(From 'The Yeasts' ed. II, 1970)

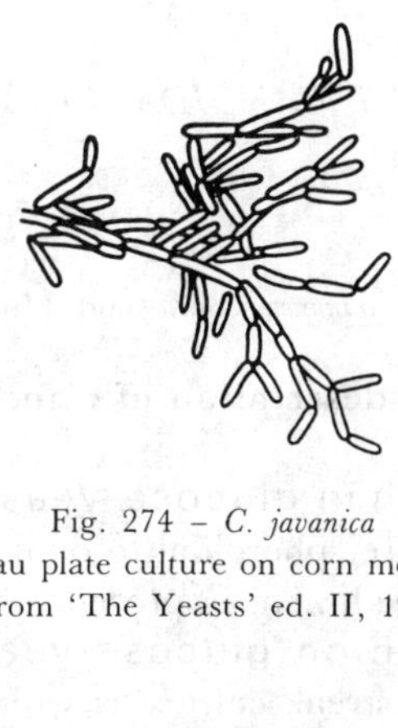

Fig. 274 – *C. javanica*
Dalmau plate culture on corn meal agar
(From 'The Yeasts' ed. II, 1970)

Dalmau plate culture on corn meal agar: The pseudomycelium consists of branched chains of spindle-shaped cells in a tree-like arrangement (Fig. 274).

Fermentation: –

Assimilation of carbon compounds:

Galactose	+	Melezitose	–	Ribitol	+
L-Sorbose	+	Soluble starch	+	Galactitol	+
Sucrose	–	D-Xylose	+	D-Mannitol	+
Maltose	+	L-Arabinose	+	D-Glucitol	+
Cellobiose	+	D-Arabinose +		Salicin	+
Trehalose	+	D-Ribose	+	DL-Lactic acid	–

Lactose	–	L-Rhamnose	–	Succinic acid	–
Melibiose	+	Glycerol	+	Citric acid	–
Raffinose	+	Erythritol	–	Inositol	–

Additional carbon compound tested: α-Methyl-D-glucoside –
Splitting of arbutin: +
Assimilation of nitrate: +
Growth in vitamin-free medium: +s
Growth at 37°C: –
Urease: +
Color reaction with DBB: +

Comments

Von Arx and Weijman (1979) transferred this species to *Rhodotorula* on the basis of basipetal cell formation and lack of xylose in cell hydrolysates.

Origin of the strain studied

Leaf of *Ixora coccinea* in Indonesia (Ruinen, 1963) (1).

Type strain: CBS 5236.

95. *CANDIDA KARAWAIEWII* YARROW ET MEYER (1978)

Synonym

Torulopsis karawaiewii nom. nud. (Jurzitza, 1970)

Standard description of *Candida karawaiewii*

Growth in glucose-yeast extract-peptone water: After 3 days at 25°C the cells are spheroidal to ovoidal (2.5–4.0) × (3.0–4.5) μm, and occur singly and in pairs. A sediment is present. After one month a sediment is present.

Growth on glucose-yeast extract-peptone agar: After one month at 25°C the streak culture is white, smooth, glistening with an entire margin.

Dalmau plate culture on corn meal agar: No pseudomycelium is formed.

Fermentation:

Glucose	+s	Maltose	–
Galactose	–	Lactose	–
Sucrose	–		

Assimilation of carbon compounds:

Galactose	–	Melezitose	–	Ribitol	+s or –
L-Sorbose	–	Soluble starch	–	Galactitol	–
Sucrose	–	D-Xylose	–	D-Mannitol	+s
Maltose	–	L-Arabinose	–	D-Glucitol	+ or s

Cellobiose	–	D-Arabinose	–	Salicin	v
Trehalose	–	D-Ribose	–	DL-Lactic acid	–
Lactose	–	L-Rhamnose	–	Succinic acid	v
Melibiose	–	Glycerol	+ s	Citric acid	v
Raffinose	–	Erythritol	–	Inositol	–

Additional carbon compound tested: Methanol –
Splitting of arbutin: +
Assimilation of nitrate: –
Growth in vitamin-free medium: –
Growth on 50% (w/w) glucose-yeast extract agar: –
Growth at 37°C: –
Color reaction with DBB: –
G + C: 37.8 mol%, type strain (Meyer, Stenderup & Christiansen, unpublished)

Origin of the strains studied

Larvae of *Ernobius* species (1); larvae of *Ernobius abietis* (2).

Type strain: CBS 5214.

96. *CANDIDA KEFYR* (BEIJERINCK) VAN UDEN ET BUCKLEY (1970)

Perfect state: *Kluyveromyces marxianus* (Hansen) van der Walt var. *marxianus*

Synonyms

Saccharomyces kefyr Beijerinck (1889b)
Mycotorula kefyr (Beijerinck) Harrison (1928)
Torulopsis kefyr (Beijerinck) Lodder (1934)
Cryptococcus kefyr (Beijerinck) Skinner (1947a)
Mycotorula lactis Trüper (1928)
Mycotorula lactosa Harrison (1928)
Candida pseudotropicalis (Cast.) Basgal var. *lactosa* (Harrisson) Diddens et Lodder (1942)
Endomyces pseudotropicalis Cast. (1911)
Monilia pseudotropicalis (Cast.) Cast. et Chalmers (1913)
Atelosaccharomyces pseudotropicalis (Cast.) Froilano de Mello et Gonzaga Fernandes (1918)
Myceloblastanon pseudotropicalis (Cast.) Ota (1928)
Mycocandida pseudotropicalis (Cast.) Cif. et Red. (1935)
Castellania pseudotropicalis (Cast.) Dodge (1935)
Monilia pseudotropicalis (Cast.) Cast. et Chalmers var. *pseudotropicalis* Cast. (1937a)
Monilia pseudotropicalis (Cast.) Cast. et Chalmers var. *metapseudotropicalis* Cast. (1937a)
Mycotorula pseudotropicalis (Cast.) Red. et Cif. (1947)
Candida pseudotropicalis (Cast.) Basgal (1931)
Torula cremoris Hammer et Cordes (1920)
Candida mortifera Red. (1925)
Mycocandida mortifera (Red.) Langeron et Talice (1932)
Monilia mortifera (Red.) Martin, Jones, Yao et Lee (1937)
Candida mortifera Red. var. α Red. (1925)
Blastodendrion procerum Zach (in Wolfram & Zach, 1934)

Monilia macedoniensis Cast. et Chalmers (1919)
Candida macedoniensis (Cast. et Chalmers) Berkhout (1923)
Myceloblastanon macedoniense (Cast. et Chalmers) Ota (1928)
Blastodendrion macedoniensis (Cast. et Chalmers) Langeron et Guerra (1935)
Mycotorula macedoniensis (Cast. et Chalmers) Red. et Cif. (Ciferri & Redaelli, 1935)
Castellania macedoniensis (Cast. et Chalmers) Dodge (1935)
Mycotoruloides macedoniensis (Cast. et Chalmers) Dodge et Moore (1936)
Monilia macedoniensoides Cast. et Taylor (1925)
Castellania macedoniensoides (Cast. et Taylor) Dodge (1935)
Monilia macedoniensis Cast. et Chalmers var. *macedoniensoides* (Cast. et Taylor) Cast. (1937a)
Saccharomyces fragrans Beijerinck (1895)

Standard description of *Candida kefyr*

Growth in glucose-yeast extract-peptone water: After 3 days at 25°C the cells are short-ovoid to long-ovoid, (2.5–9) × (5–14) μm; much longer cells may also occur (van Uden & Buckley, 1970).

Growth on glucose-yeast extract-peptone agar: After one month at 25°C the streak culture is cream-colored to yellowish, semidull, soft, smooth to finely reticulate (van Uden & Buckley, 1970).

Dalmau plate culture on corn meal agar: Pseudomycelium is abundant in some strains, scarce or almost absent in others. It consists of branched pseudohyphae bearing chains of blastospores, often in a verticillated position. Other parts of the pseudomycelium may consist of bare, wavy elongate cells (van Uden & Buckley, 1970).

Fermentation:

Glucose	+	Lactose	v
Galactose	+	Raffinose	v
Sucrose	+	Inulin	+
Maltose	–		

Assimilation of carbon compounds:

Galactose	+	Melezitose	–	Ribitol	–
L-Sorbose	–	Soluble starch	–	Galactitol	–
Sucrose	+	D-Xylose	v	D-Mannitol	v
Maltose	–	L-Arabinose	v	D-Glucitol	v
Cellobiose	v	D-Arabinose	–	Salicin	v
Trehalose	v	D-Ribose	v	DL-Lactic acid	+
Lactose	+	L-Rhamnose	–	Succinic acid	v
Melibiose	–	Glycerol	v	Citric acid	v
Raffinose	+	Erythritol	–	Inositol	–

Splitting of arbutin: +

Assimilation of nitrate: –

Growth in vitamin-free medium: –

Growth at 37°C: +

Color reaction with DBB: –
Coenzyme Q system: Q 6 (Yamada & Kondo, 1972)
G + C: 41.3 mol% (Stenderup & Bak, 1968)

Comments

Johannsen & van der Walt (1978) demonstrated the conspecificity between *Kluyveromyces* species by cross mating. *K. fragilis* and *K. marxianus* mated and produced viable progeny. Since these species had been considered perfect forms of *C. pseudotropicalis* and *C. macedoniensis,* respectively, these *Candida* species are reduced to synonymy. Stenderup and Bak (1968) had previously shown significant genetic relatedness between *C. pseudotropicalis* and *K. fragilis* by employing DNA–RNA reassociation techniques. *C. kefyr* is also considered to be synonymous because of similar physiological and morphological properties and G + C percent. Furthermore, crosses between *C. kefyr* and *K. marxianus* var. *lactis* gave ascigerous recombinants (van der Walt, unpublished). *C. kefyr* is the epithet of priority.

Origin of the strains studied

Strain received from Castellani as *Monilia pseudotropicalis (1);* kefir grains in The Netherlands (1); buttermilk in The Netherlands and described as *Mycotorula lactosa* (Harrison, 1928) (1); milk (1); fermenting cream (1); lung (1); nail (1); cheese (1); yoghurt (1); fermented milk (1); unknown origin, sent by Joekes (London), studied by Berkhout (1); compressed yeast (2); canned apples (1).

Type strain: CBS 834, from kefir grains.

97. *CANDIDA KRISSII* GOTO, YAMASATO ET IIZUKA (1974)

Standard description of *Candida krissii*

Growth in glucose-yeast extract-peptone water: After 3 days at 25°C cells are mostly oval to long-oval, some cylindrical, (2.3–5.7) × (4.6– 10.3) μm, others are curved. They occur singly, in pairs and chains (Fig. 275). Pseudomycelium is present. A sediment and a ring are formed. After one month a sediment, ring and film are present.

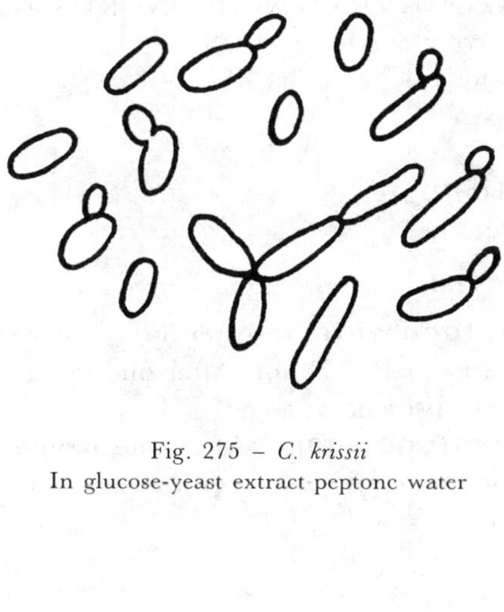

Fig. 275 – *C. krissii*
In glucose-yeast extract peptone water

Growth on glucose-yeast extract-peptone agar: After one month at 25°C the streak culture is smooth, creamy, glistening, off-white with an entire or undulating border.

Dalmau plate culture on corn meal agar: Abundant pseudomycelium of the "*Mycotoruloides*" type is formed. Extensive development occurs with clusters and short chains of ovoid blastospores appearing along the hyphal strand.

Fermentation:

Glucose	– or +w	Maltose	–
Galactose	–	Lactose	–
Sucrose	–		

Assimilation of carbon compounds:

Galactose	+s or –	Melezitose	–	Ribitol	+s or –
L-Sorbose	+s	Soluble starch	–	Galactitol	–
Sucrose	–	D-Xylose	–	D-Mannitol	+
Maltose	–	L-Arabinose	–	D-Glucitol	+
Cellobiose	+s or –	D-Arabinose	–	Salicin	– or +s
Trehalose	+s or –	D-Ribose	–	DL-Lactic acid	–
Lactose	–	L-Rhamnose	–	Succinic acid	+
Melibiose	–	Glycerol	+	Citric acid	+
Raffinose	–	Erythritol	–	Inositol	–

Splitting of arbutin: +s or –

Assimilation of nitrate: –

Growth in vitamin-free medium: –

Growth on 50% (w/w) glucose-yeast extract agar: +

Growth at 37°C: –

Color reaction with DBB: –

G + C: 43.4 mol%, type strain (Meyer, unpublished)

Origin of the strain studied

Seawater (1).

Type strain: CBS 6519, from seawater.

98. *CANDIDA KRUISII* (KOCKOVÁ-KRATOCHVÍLOVÁ ET ONDRUŠOVÁ) MEYER ET YARROW (Yarrow & Meyer, 1978)

Synonym

Torulopsis kruisii Kocková-Kratochvílová et Ondrušová (1971)

Standard description of *Candida kruisii*

Growth in glucose-yeast extract-peptone water: After 3 days at 25°C the cells are long-ovoid to cylindrical, $(2.9–3.6) \times (5.1–8.7)$ μm. After one month at 25°C a sediment, incomplete ring and islets are present.

Growth on glucose-yeast extract-peptone agar: After one month at 25°C the streak culture is cream-colored, smooth, semi-glossy, with areas that are folded and dry-appearing. The margin is entire.

Dalmau plate culture on corn meal agar: The pseudomycelium consists of ramified chains of elongated cells bearing chains of blastospores in verticils (Fig. 276).

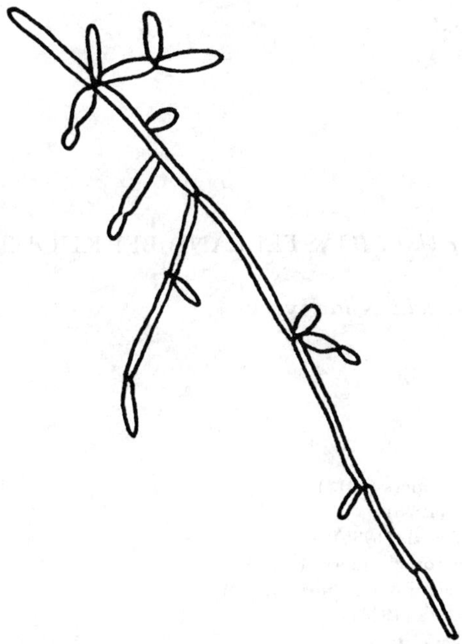

Fig. 276 – *C. kruisii*
Dalmau plate culture on corn meal agar

Fermentation:

Glucose	+	Maltose	–
Galactose	+ or s	Lactose	–
Sucrose	–		

Assimilation of carbon compounds:

Galactose	+	Melezitose	+ s	Ribitol	+
L-Sorbose	+ s or –	Soluble starch	+	Galactitol	–
Sucrose	+	D-Xylose	+	D-Mannitol	+
Maltose	+	L-Arabinose	+ s or –	D-Glucitol	+
Cellobiose	+	D-Arabinose	–	Salicin	+
Trehalose	+	D-Ribose	+ s or –	DL-Lactic acid	–
Lactose	–	L-Rhamnose	–	Succinic acid	+
Melibiose	–	Glycerol	+	Citric acid	+
Raffinose	–	Erythritol	–	Inositol	–

Splitting of arbutin: +

Assimilation of nitrate: –

Growth in vitamin-free medium: v

Growth on 50% (w/w) glucose-yeast extract agar: –

Growth at 37°C: v

Color reaction with DBB: –
G + C: 44.6 mol%, type strain

Origin of the strain studied

Boletus purpureus (1).

Type strain: CBS 6451.

99. *CANDIDA KRUSEI* (CASTELLANI) BERKHOUT (1923)

Perfect state: *Issatchenkia orientalis* Kudriavzev

Synonyms

Saccharomyces krusei Cast. (1910)
Endomyces krusei Cast. (1912b)
Monilia krusei (Cast.) Cast. et Chalmers (1913)
Myceloblastanon krusei (Cast.) Ota (1928)
Trichosporon krusei (Cast.) Cif. et Red. (1935)
Mycotoruloides krusei (Cast.) Langeron et Guerra (1935)
Monilia parakrusei Cast. (in Castellani & Chalmers, 1919)
Myceloblastanon parakrusei (Cast.) Ota (1928)
Castellania parakrusei (Cast.) Dodge (1935)
Mycoderma chevalieri Guilliermond (1914)
Candida chevalieri (Guill.) Westerdijk (in CBS catalogue 1933)
Mycoderma monosa Anderson (1917)
Mycoderma bordetii Kuff. (1920a)
Monilia inexpectata Mazza, Niño et Egües (1930)
Mycocandida inexpectata (Mazza, Niño et Egües) Talice et Mackinnon (1934)
Pseudomonilia inexpectata (Mazza, Niño et Egües) Dodge (1935)
Trichosporon dendriticum nom. nud. (Ciferri & Redaelli, 1935)
Candida dendritica (Cif. et Red.) Dodge et Moore (1936)
Monilia krusoides Cast. (1937a)
Pseudomycoderma miso nom. nud. (Mogi, 1940)
Candida castellanii van Uden et Assis-Lopes (1953b)
Candida tamarindi nom. nud. (Lewis & Johar, 1955)
Procandida tamarindi (Lewis et Johar) Novák et Zsolt (1961)
Candida lobata Batista et Silveira (1959c)
Candida soosii Novák (1964)
Candida requinyii Szép et Novák (1963)
Candida brassicae Amano, Goto et Kagami (1975)

Standard description of *Candida krusei*

Growth in glucose-yeast extract-peptone water: After 3 days at 25°C the cells are ovoid, elongate and cylindrical, (2.2–5.6) × (4.3–15.2) μm and occur singly, budding and in chains (Fig. 277). Long cells, some curved cells, and pseudomycelium are also present. A dry, wrinkled, creeping pellicle is present.

Growth on glucose-yeast extract-peptone agar: After one month at

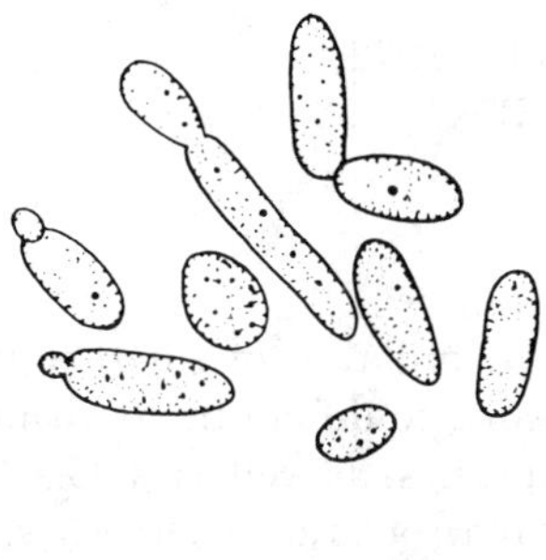

Fig. 277 – *C. krusei*
In glucose-yeast extract-peptone water
(From 'The Yeasts' ed. II, 1970)

25°C the streak culture is off-white, dull, soft, smooth with some wrinkling and an irregular margin, fringed with pseudomycelium.

Dalmau plate culture on corn meal agar: Pseudomycelium is extensive and consists of chains of elongated cells, sometimes slender, with frequent branching. Clusters and chains of blastospores occur along the pseudohyphae (Fig. 278).

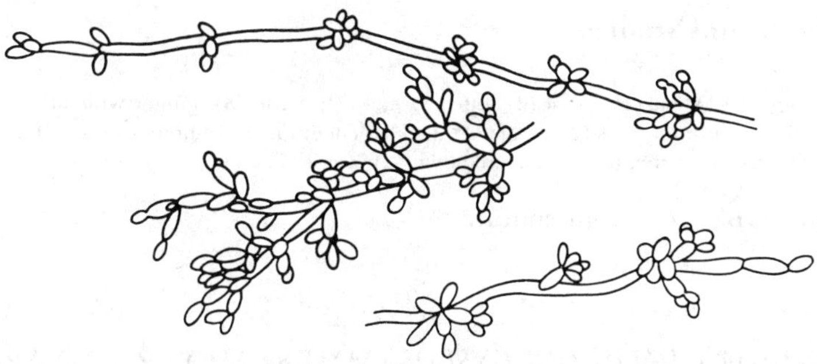

Fig. 278 – *C. krusei*
Dalmau plate culture on corn meal agar (From 'The Yeasts' ed. II, 1970)

Fermentation:

Glucose	+	Maltose	–
Galactose	–	Lactose	–
Sucrose	–		

Assimilation of carbon compounds:

Galactose	–	Melezitose	–	Ribitol	–
L-Sorbose	v	Soluble starch	–	Galactitol	–
Sucrose	–	D-Xylose	–	D-Mannitol	–
Maltose	–	L-Arabinose	–	D-Glucitol	–
Cellobiose	–	D-Arabinose	–	Salicin	–
Trehalose	–	D-Ribose	–	DL-Lactic acid	+
Lactose	–	L-Rhamnose	–	Succinic acid	+
Melibiose	–	Glycerol	+	Citric acid	v
Raffinose	–	Erythritol	–	Inositol	–

Splitting of arbutin: –
Assimilation of nitrate: –
Growth in vitamin-free medium: +
Growth on 50% (w/w) glucose-yeast extract agar: –
Growth at 37°C: +
Urease: v

Color reaction with DBB: –
Coenzyme Q system: Q 7 (Yamada & Kondo, 1972)
G + C: 39.6 mol%, type strain (Stenderup & Bak, 1968)

Comments

The original description of *C. brassicae* reported assimilation of sucrose, raffinose (weak), xylose (weak), and maltose (weak). These results could not be confirmed. The similarity of the morphological and physiological properties, as well as % G + C (39.8) indicated its synonymy with *C. krusei*. Using DNA reassociation techniques, Kurtzman et al. (1980b) showed that *C. requinyii* is conspecific with *C. krusei* and that *C. krusei* is the imperfect state of *Issatchenkia orientalis*.

Origin of the strains studied

Sputum (1); yoghurt (1); baker's yeast (2); infected nail (1); cacao (3); ginger wine (1); air (1); case of bronchocandidiasis (1); human feces (2); silage (1); pickled cucumbers (2); homare-miso (1); tamarind (1); marmalade (1); industrial fermentation (1); unknown (2); yeast cake (1).

Type strain: CBS 573, from sputum.

100. CANDIDA LACTISCONDENSI (HAMMER) MEYER ET YARROW (Yarrow & Meyer, 1978)

(as *Candida lactis-condensi* (Hammer) Meyer et Yarrow)

Synonyms

Torula lactis-condensi Hammer (1919)
Torulopsis lactis-condensi (Hammer) Lodder et Kreger-van Rij (1952)
Torulopsis caroliniana Etchells et Bell (1950b)

Standard description of *Candida lactiscondensi*

Growth in glucose-yeast extract-peptone water: After 3 days at 25°C the cells are ovoid, (1.5–3) × (2.5–5) µm (Fig. 279). A ring may be present (van Uden & Vidal-Leiria, 1970).

Growth on glucose-yeast extract-peptone agar: After one month at 25°C the streak culture is cream-colored, glossy, soft and smooth. The cultures are often poorly developed and die easily (van Uden & Vidal-Leiria, 1970).

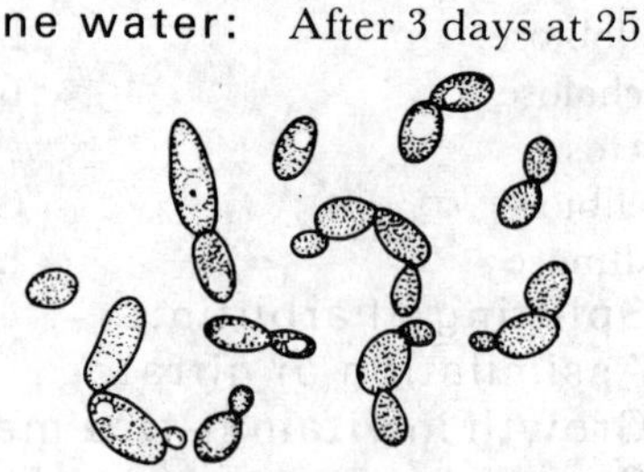

Fig. 279 – *C. lactiscondensi*
In glucose-yeast extract-peptone water
(From 'The Yeasts' ed. II, 1970)

Dalmau plate culture on corn meal agar: No pseudomycelium is formed.

Fermentation:

Glucose	+	Maltose	–
Galactose	–	Lactose	–
Sucrose	+	Raffinose	+

Assimilation of carbon compounds:

Galactose	–	Melezitose	–	Ribitol	–
L-Sorbose	–	Soluble starch	–	Galactitol	–
Sucrose	+	D-Xylose	–	D-Mannitol	–
Maltose	–	L-Arabinose	–	D-Glucitol	–
Cellobiose	–	D-Arabinose	–	Salicin	–
Trehalose	–	D-Ribose	–	DL-Lactic acid	–
Lactose	–	L-Rhamnose	–	Succinic acid	–
Melibiose	–	Glycerol	–	Citric acid	–
Raffinose	+ s	Erythritol	–	Inositol	–

Splitting of arbutin: –

Assimilation of nitrate: +

Growth in vitamin-free medium: v

Growth at 37°C: –

Color reaction with DBB: –

Coenzyme Q system: Q 8 (Yamada & Kondo, 1972)

G + C: 43.2 mol%, type strain (Meyer, Stenderup & Christiansen, unpublished)

Origin of the strains studied

Sweetened condensed milk in U.S.A. (Hammer, 1919) (1); the authentic strain of *Torulopsis caroliniana*, cucumber brine in U.S.A. (Etchells & Bell, 1950b) (1); pickled onions in South Africa (1).

Type strain: CBS 52, from sweetened condensed milk.

101. *CANDIDA LAMBICA* (LINDNER ET GENOUD) VAN UDEN ET BUCKLEY (1970)

Perfect state: *Pichia fermentans* Lodder, on the basis of phenotypic and % G + C similarity.

Synonyms

Mycoderma lambica Lindner et Genoud (1913)
Torula monosa Kluyver (1914)
Mycotorula monosa (Kluyver) Harrison (1928)
Candida monosa (Kluyver) Diddens et Lodder (1942)
Candida fimetaria Soneda (1959b)

Standard description of *Candida lambica*

Growth in glucose-yeast extract-peptone water: After 3 days at 25°C the cells are ovoid, (3.6–5.8) × (7.2–11.6) μm and occur singly, budding and in short chains and small clusters (Fig. 280). A pellicle is formed. After one month a heavy sediment is present.

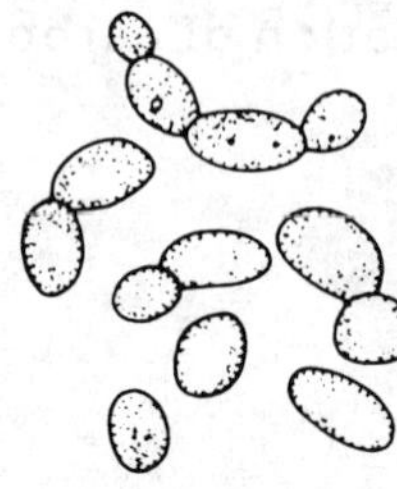

Fig. 280 – *C. lambica*
In glucose-yeast extract-peptone water (From 'The Yeasts' ed. II, 1970)

Growth on glucose-yeast extract-peptone agar: After one month at 25°C the streak culture is cream-colored, soft, smooth or finely wrinkled, glistening in the smooth strains, somewhat dull in the wrinkled strains (van Uden & Buckley, 1970).

Dalmau plate culture on corn meal agar: Pseudomycelium is abundantly formed. Branched chains of elongated cells, sometimes long and slender, with clusters and short chains or round to oval blastospores are present (Fig. 281).

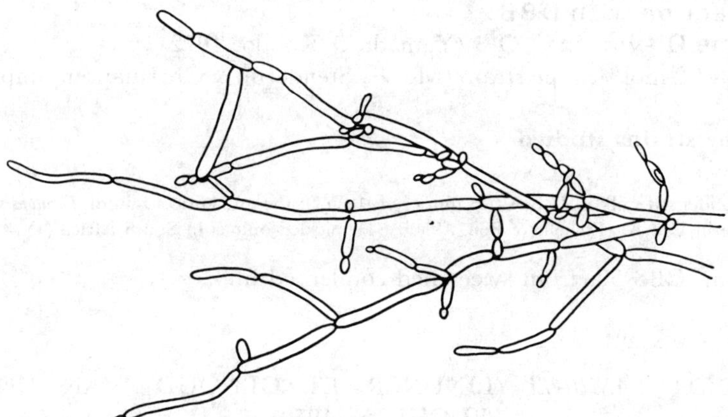

Fig. 281 – *C. lambica*
Dalmau plate culture on corn meal agar (From 'The Yeasts' ed. II, 1970)

Fermentation:

Glucose	+	Maltose	–
Galactose	–	Lactose	–
Sucrose	–		

Assimilation of carbon compounds:

Galactose	–	Melezitose	–	Ribitol	–
L-Sorbose	–	Soluble starch	–	Galactitol	–
Sucrose	–	D-Xylose	+	D-Mannitol	–
Maltose	–	L-Arabinose	–	D-Glucitol	–
Cellobiose	–	D-Arabinose	–	Salicin	–
Trehalose	–	D-Ribose	–	DL-Lactic acid	+
Lactose	–	L-Rhamnose	–	Succinic acid	+

Melibiose	–	Glycerol	+	Citric acid	+
Raffinose	–	Erythritol	–	Inositol	–

Splitting of arbutin: –
Assimilation of nitrate: –
Growth in vitamin-free medium: –
Growth on 50% (w/w) glucose-yeast extract agar: –
Growth at 37°C: +w
Urease: –
Color reaction with DBB: –
G + C: 42.4–43.2 mol% (Nakase & Komagata, 1971f)

Origin of the strains studied

Lambic beer (1), brewer's yeast (1), animal dung (2), potato flour (1), Chesapeake Bay water (18), unknown (2), cattle fodder (1).

Type strain: CBS 1876, from lambic beer.

102. CANDIDA LIPOLYTICA (HARRISON) DIDDENS ET LODDER

see *Saccharomycopsis lipolytica* p. 406

103. CANDIDA LUSITANIAE VAN UDEN ET DO CARMO-SOUSA

see *Clavispora lusitaniae* p. 120

104. CANDIDA MAGNOLIAE (LODDER ET KREGER-VAN RIJ) MEYER ET YARROW (Yarrow & Meyer, 1978)

Synonym

Torulopsis magnoliae Lodder et Kreger-van Rij (1952)

Standard description of *Candida magnoliae*

Growth in glucose-yeast extract-peptone water: After 3 days at 25°C the cells are ovoid to globose, (1.5–5) × (2–5) μm (Fig. 282). A ring is present. Some strains also form a pellicle or islets (van Uden & Vidal-Leiria, 1970).

Growth on glucose-yeast extract-peptone agar: After one month at 25°C the streak culture is yellowish to cream-colored, semidull to glossy, soft, almost smooth (van Uden & Vidal-Leiria, 1970).

Dalmau plate culture on corn meal agar: No pseudomycelium is formed; a few short chains of globose or ovoid cells may be present (van Uden & Vidal-Leiria, 1970).

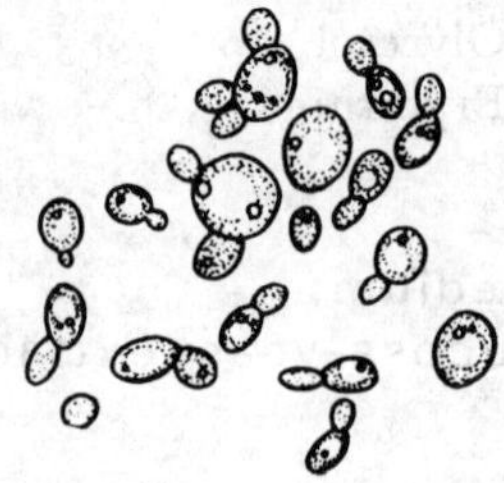

Fig. 282 – *C. magnoliae*
In glucose-yeast extract-peptone water (From 'The Yeasts' ed. II, 1970)

Fermentation:

Glucose	+	Maltose	–
Galactose	–	Lactose	–
Sucrose	+	Raffinose	–

Assimilation of carbon compounds:

Galactose	v	Melezitose	–	Ribitol	–
L-Sorbose	v	Soluble starch	–	Galactitol	–
Sucrose	+	D-Xylose	–	D-Mannitol	+
Maltose	–	L-Arabinose	–	D-Glucitol	+
Cellobiose	–	D-Arabinose	–	Salicin	v
Trehalose	–	D-Ribose	v	DL-Lactic acid	–
Lactose	–	L-Rhamnose	–	Succinic acid	v
Melibiose	–	Glycerol	+	Citric acid	–
Raffinose	v	Erythritol	–	Inositol	–

Splitting of arbutin: v
Assimilation of nitrate: +
Growth in vitamin-free medium: v
Growth at 37°C: v
Color reaction with DBB: –
Coenzyme Q system: Q 9 (Yamada & Kondo, 1972)
G + C: 60.0 mol%, 2 strains (Nakase & Komagata, 1971c)

Comments

Van der Walt & Johannsen (1973) reported that *T. magnoliae* produced a dangeardien and asigned the species to the genus *Entelexis.* We did not study the sexual cycle of this organism and, therefore, treat it as a *Candida* species.

Origin of the strains studied

Flowers (2); gut of a bee (3); concentrated orange juice in South Africa (1); human throat (1); human sputum in The Netherlands (1); unknown origin (1).

Type strain: CBS 166, in The Netherlands from a flower.

105. *CANDIDA MALTOSA* KOMAGATA, NAKASE ET KATSUYA (1964)

Synonyms

Candida cloacae Komagata, Nakase et Katsuya (1964)
Candida subtropicalis Nakase, Fukazawa et Tsuchiya (1972)
Candida novellus nom. nud.

Standard description of *Candida maltosa*

Growth in glucose-yeast extract-peptone water: After 3 days at 25°C cells are globose to short-oval, (3.6–7.2) × (3.6–7.9) μm, occur singly, in pairs, short chains and clusters (Fig. 283). Large oval cells are also present as well as pseudomycelium. A sediment is formed. After one month a sediment and ring are present.

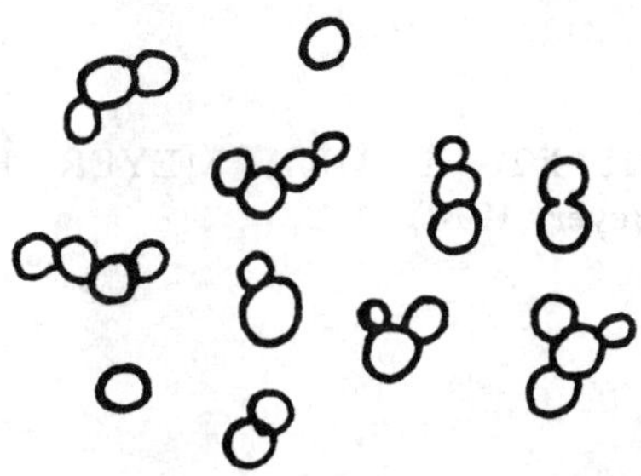

Fig. 283 – *C. maltosa*
In glucose-yeast extract-peptone water

Growth on glucose-yeast extract-peptone agar: After one month at 25°C the streak culture is creamy, smooth or folded with a mycelial border.

Dalmau plate culture on corn meal agar: A well-developed pseudomycelium is formed with blastospores occurring singly, in short chains and clusters.

Fermentation:

Glucose	+	Maltose	+ w or s
Galactose	v	Lactose	–
Sucrose	+	Trehalose	+

Assimilation of carbon compounds:

Galactose	+	Melezitose	+	Ribitol	+ or s
L-Sorbose	+ s	Soluble starch	–	Galactitol	–
Sucrose	+	D-Xylose	+	D-Mannitol	+
Maltose	+	L-Arabinose	–	D-Glucitol	+
Cellobiose	+ or s	D-Arabinose	–	Salicin	+
Trehalose	+	D-Ribose	–	DL-Lactic acid	– or + s
Lactose	–	L-Rhamnose	–	Succinic acid	+
Melibiose	–	Glycerol	+	Citric acid	v
Raffinose	–	Erythritol	–	Inositol	–

Splitting of arbutin: +
Assimilation of nitrate: –
Growth in vitamin-free medium: –
Growth at 37°C: +
Color reaction with DBB: –

Coenzyme Q system: Q 9 (Yamada & Kondo, 1972)
G + C: 35.6–36.8 mol%, 6 strains (Meyer et al., 1975)

Comments

DNA reassociation studies (Meyer et al., 1975) revealed *C. maltosa, C. cloacae, C. subtropicalis* and *C. novellus* nom. nud. to be conspecific and a species separate from *C. sake.* The epithet, *C. maltosa,* has priority. Therefore the other organisms were placed in synonymy.

Origin of the strains studied

Adhesives in a monosodium glutamate tank (1); mud (1); air (1); soil (2); oil slick (1).

Type strain: CBS 5611, from adhesives in a monosodium glutamate tank.

106. *CANDIDA MANNITOFACIENS* (ONISHI ET SUZUKI) MEYER ET YARROW (Yarrow & Meyer, 1978)

Synonym

Torulopsis mannitofaciens Onishi et Suzuki (1969)

Standard description of *Candida mannitofaciens*

Growth in glucose-yeast extract-peptone water: After 3 days at 25°C cells are globose to subglobose, 3.0–4.5 μm and occur singly and in pairs. A sediment is formed. After one month at 20°C a ring and sediment are present.

Growth on glucose-yeast extract-peptone agar: After one month at 25°C the streak culture is white, smooth, soft and glossy. The margin is entire.

Dalmau plate culture on corn meal agar: Pseudomycelium is absent.

Fermentation:

Glucose	+	Maltose	+ s
Galactose	+ s	Lactose	–
Sucrose	+ s	Raffinose	–

Assimilation of carbon compounds:

Galactose	+	Melezitose	–	Ribitol	–
L-Sorbose	–	Soluble starch	–	Galactitol	–
Sucrose	+	D-Xylose	+ s	D-Mannitol	+
Maltose	+	L-Arabinose	+	D-Glucitol	–
Cellobiose	+	D-Arabinose	–	Salicin	+ s
Trehalose	+	D-Ribose	+ s	DL-Lactic acid	–
Lactose	–	L-Rhamnose	–	Succinic acid	+ s
Melibiose	+ s	Glycerol	+	Citric acid	–
Raffinose	+ s	Erythritol	–	Inositol	–

Splitting of arbutin: –
Assimilation of nitrate: +
Growth in vitamin-free medium: –
Growth at 37°C: –
Urease: –
Color reaction with DBB: –
G + C: 47.5 mol% (Meyer, Stenderup & Christiansen, unpublished)

Origin of the strain studied

Soy mash (1).

Type strain: CBS 5981.

107. CANDIDA MARINA VAN UDEN ET ZOBELL (1962)

Standard description of *Candida marina*

Growth in glucose-yeast extract-peptone water: After 3 days at 25°C the cells are ovoid, (3–4.5) × (6–8) μm (Fig. 284). A dull pellicle is present (van Uden & Buckley, 1970).

Growth on glucose-yeast extract-peptone agar: After one month at 25°C the streak culture is brownish-yellow, dull, soft, smooth (van Uden & Buckley, 1970).

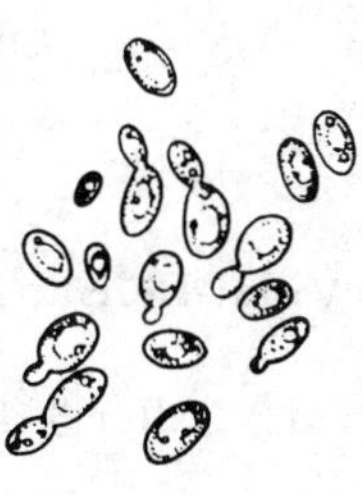

Fig. 284 – *C. marina*
In glucose-yeast extract-peptone water
(From 'The Yeasts' ed. II, 1970)

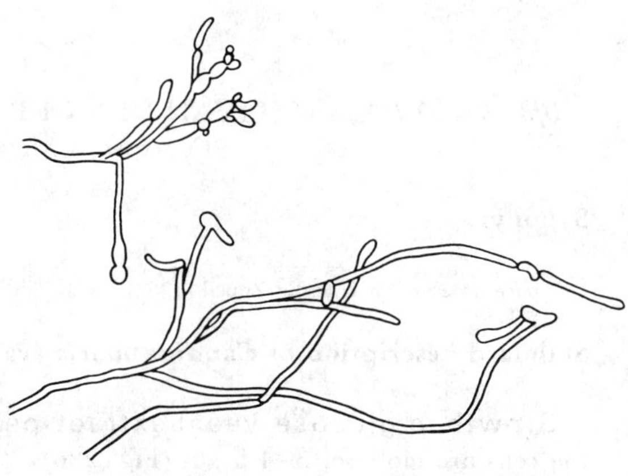

Fig. 285 – *C. marina*
Dalmau plate culture on corn meal agar
(From 'The Yeasts' ed. II, 1970)

Dalmau plate culture on corn meal agar: The pseudomycelium consists of long wavy pseudohyphae bearing, mostly in terminal positions, globose and ovoid, often curved cells (Fig. 285) (van Uden & Buckley, 1970).

Fermentation: –

Assimilation of carbon compounds:

Galactose	–	Melezitose	–	Ribitol	+
L-Sorbose	+	Soluble starch	–	Galactitol	–
Sucrose	–	D-Xylose	+	D-Mannitol	+
Maltose	–	L-Arabinose	+	D-Glucitol	+
Cellobiose	–	D-Arabinose	+ s	Salicin	+ s
Trehalose	+	D-Ribose	+	DL-Lactic acid	–
Lactose	–	L-Rhamnose	+	Succinic acid	–
Melibiose	–	Glycerol	+	Citric acid	–
Raffinose	–	Erythritol	+	Inositol	+

Splitting of arbutin: +

Assimilation of nitrate: –

Growth in vitamin-free medium: –

Growth at 37°C: –

Urease: +

Color reaction with DBB: +

Coenzyme Q system: Q 10 (Yamada & Kondo, 1972)

G + C: 64.0 mol%, type strain (Stenderup et al., 1972); 61.2 mol%, one strain (Nakase & Komagata, 1971f)

Origin of the strain studied

Seawater over the Great Barrier Reef (van Uden & Zobell, 1962) (1).

Type strain: CBS 5235.

108. CANDIDA MARIS (VAN UDEN ET ZOBELL) MEYER ET YARROW (Yarrow & Meyer, 1978)

Synonym

Torulopsis maris van Uden et Zobell (1962)

Standard description of *Candida maris* (van Uden & Vidal-Leiria, 1970)

Growth in glucose-yeast extract-peptone water: After 3 days at 25°C the cells are globose, 3–4.5 μm (Fig. 286).

Growth on glucose-yeast extract-peptone agar: After one month at 25°C the streak culture is cream-colored, glossy, soft and smooth.

Dalmau plate culture on corn meal agar: No pseudomycelium is formed.

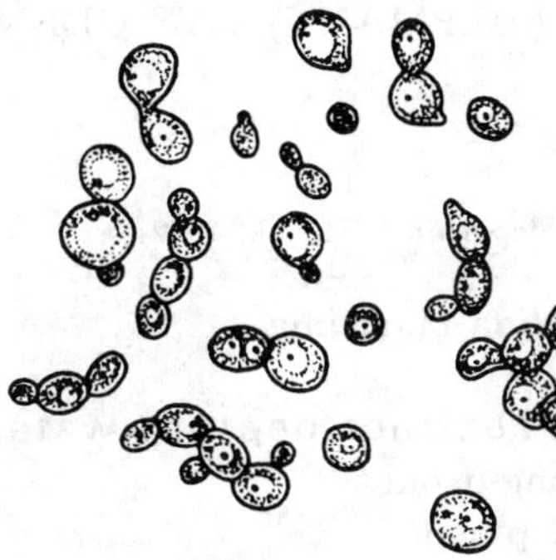

Fig. 286 – *C. maris*
In glucose-yeast extract-peptone water (From 'The Yeasts' ed. II, 1970)

Fermentation: –

Assimilation of carbon compounds:

Galactose	–	Melezitose	–	Ribitol	+
L-Sorbose	–	Soluble starch	–	Galactitol	–
Sucrose	–	D-Xylose	+	D-Mannitol	+
Maltose	–	L-Arabinose	–	D-Glucitol	+
Cellobiose	–	D-Arabinose	–	Salicin	–
Trehalose	–	D-Ribose	–	DL-Lactic acid	–
Lactose	–	L-Rhamnose	+	Succinic acid	+
Melibiose	–	Glycerol	+	Citric acid	–
Raffinose	–	Erythritol	–	Inositol	–

Splitting of arbutin: –
Assimilation of nitrate: –
Growth in vitamin-free medium: + s
Growth at 37°C: –
Color reaction with DBB: –
Coenzyme Q system: Q 7 (Yamada & Kondo, 1972)
G + C: 50.9 mol%, type strain (Meyer, Stenderup & Christiansen, unpublished)

Origin of the strain studied

Seawater (van Uden & Zobell, 1962) (1).

Type strain: CBS 5151.

109. CANDIDA MARITIMA (SIEPMANN) VAN UDEN ET BUCKLEY (1970)

Synonym

Trichosporon maritima Siepmann (in Siepmann & Höhnk, 1962)

Standard description of *Candida maritima*

Growth in glucose-yeast extract-peptone water: After 3 days at 25°C the cells are globose to long-ovoid (2–5) × (2.5–9) μm, elongated pseudohyphal cells may also occur (Fig. 287). A ring may be present (van Uden & Buckley, 1970).

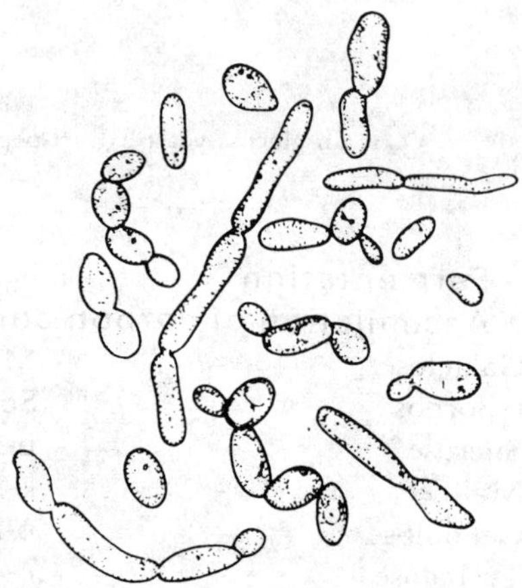

Fig. 287 – *C. maritima*
In glucose-yeast extract-peptone water
(From 'The Yeasts' ed. II, 1970)

Growth on glucose-yeast extract-peptone agar: After one month at 25°C the streak culture is greyish to cream-colored, glistening to somewhat dull, soft, smooth, or wrinkled (van Uden & Buckley, 1970).

Dalmau plate culture on corn meal agar: The pseudomycelium consists of ramified chains of pseudohyphae, often long and wavy, bearing pleural blastoconidia, with branched chains of globose and ovoid blastospores (Fig. 288) (van Uden & Buckley, 1970).

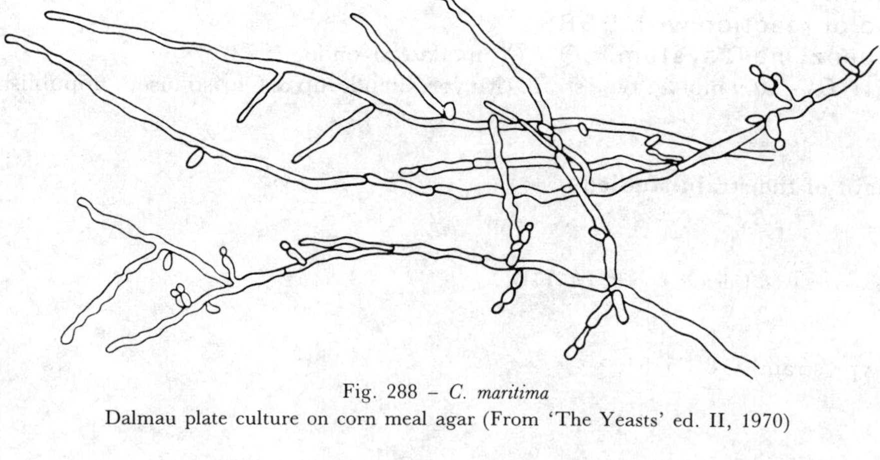

Fig. 288 – *C. maritima*
Dalmau plate culture on corn meal agar (From 'The Yeasts' ed. II, 1970)

Fermentation:

Glucose	+ s	Maltose	–
Galactose	–	Lactose	–
Sucrose	+ s or –	Raffinose	+ s

Assimilation of carbon compounds:

Galactose	–	Melezitose	+	Ribitol	–
L-Sorbose	–	Soluble starch	v	Galactitol	–
Sucrose	+	D-Xylose	+	D-Mannitol	+
Maltose	+	L-Arabinose	–	D-Glucitol	+
Cellobiose	+	D-Arabinose	–	Salicin	+
Trehalose	+	D-Ribose	+	DL-Lactic acid	+
Lactose	–	L-Rhamnose	+	Succinic acid	+
Melibiose	–	Glycerol	+	Citric acid	v
Raffinose	+	Erythritol	–	Inositol	–

Splitting of arbutin: +

Assimilation of nitrate: –

Growth in vitamin-free medium: –

Growth on 50% (w/w) glucose-yeast extract agar: –

Growth at 37°C: –

Urease: –

Color reaction with DBB: –

G + C: 45.4 mol%

Origin of the strains studied

Seawater in the North Atlantic (Siepmann & Höhnk, 1962) (1), tunnels of ambrosia beetles (1), larvae of *Rhyphenes mailei* (1), tanning fluid (2).

Type strain: CBS 5107, from seawater was chosen as type by van Uden and Buckley (1970).

110. CANDIDA MELIBIOSICA BUCKLEY ET VAN UDEN (1968)

Synonyms

Candida parapsilosis (Ashf.) Langeron et Talice var. *hokkai* nom. nud. (Goto & Yokotsuka, 1962)
Torulopsis navarrensis Moriyon et Ramirez (1974)
Candida navarrensis (Moriyon et Ramirez) Meyer et Yarrow (Yarrow & Meyer, 1978)
Torulopsis pampelonensis Ramirez et Martinez (1978)
Candida pampelonensis (Ramirez et Martinez) Meyer et Yarrow (Yarrow & Meyer, 1978)

Standard description of *Candida melibiosica*

Growth in glucose-yeast extract-peptone water: After 3 days at 25°C the cells are globose to ovoid, (2.2–4.3) × (2.2–6.5) μm (Fig. 289). A thin pellicle is present.

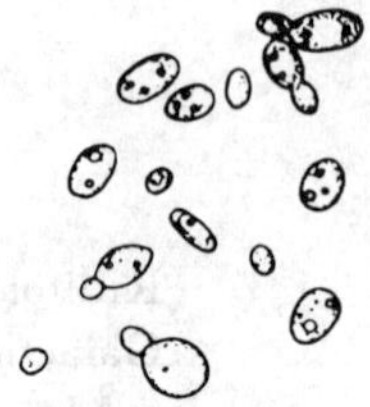

Fig. 289 – *C. melibiosica*
In glucose-yeast extract-peptone water
(From 'The Yeasts' ed. II, 1970)

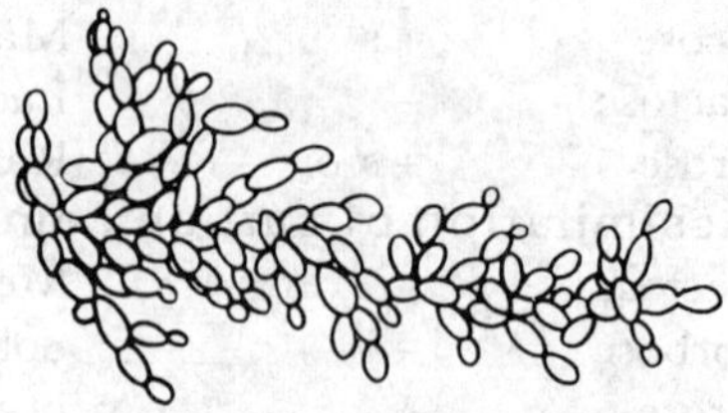

Fig. 290 – *C. melibiosica*
Dalmau plate culture on corn meal agar
(From 'The Yeasts' ed. II, 1970)

Growth on glucose-yeast extract-peptone agar: After one month at 25°C the streak culture is yellowish, creamy, soft, and smooth with an entire margin.

Dalmau plate culture on corn meal agar: Branched chains of oval to elongate cells bearing many blastospores give a bushy feathery appearance at low power magnification (Fig. 290).

Fermentation:

Glucose	+	Lactose	–
Galactose	+ or s	Raffinose	+ s
Sucrose	– or + s	Melibiose	+ s
Maltose	–	Trehalose	+

Assimilation of carbon compounds:

Galactose	+	Melezitose	+	Ribitol	+ or s
L-Sorbose	+	Soluble starch	–	Galactitol	–
Sucrose	+	D-Xylose	+	D-Mannitol	+
Maltose	+	L-Arabinose	–	D-Glucitol	+
Cellobiose	+	D-Arabinose	–	Salicin	+
Trehalose	+	D-Ribose	+ s	DL-Lactic acid	–
Lactose	–	L-Rhamnose	–	Succinic acid	+
Melibiose	+	Glycerol	+ s	Citric acid	+
Raffinose	+	Erythritol	–	Inositol	–

Splitting of arbutin: +
Assimilation of nitrate: –
Growth in vitamin-free medium: + s
Growth at 37°C: +
Color reaction with DBB: –
G + C: 54.5–55.5 mol%, 3 strains (Meyer, unpublished)

Origin of the strains studied

Sputum (1); soil (1) and leaves (1) in a beech forest in Navarra, Spain.

Type strain: CBS 5814, from sputum.

111. CANDIDA MELINII DIDDENS ET LODDER (1942)

Perfect state: *Hansenula canadensis* Wickerham

Standard description of *Candida melinii*

Growth in glucose-yeast extract-peptone water: After 3 days at 25°C the cells are ovoid, (2–4) × (3–7) μm, single or in pairs (Fig. 291). Also long slender cells occur (van Uden & Buckley, 1970).

Growth on glucose-yeast extract-peptone agar: After one month at 25°C the streak culture is greyish to cream-colored, semidull, soft, smooth (van Uden & Buckley, 1970).

Dalmau plate culture on corn meal agar: The pseudomycelium consists of ramified chains of coarse pseudohyphae bearing clusters and chains of globose and ovoid blastospores in verticillated positions (Fig. 292) (van Uden & Buckley, 1970).

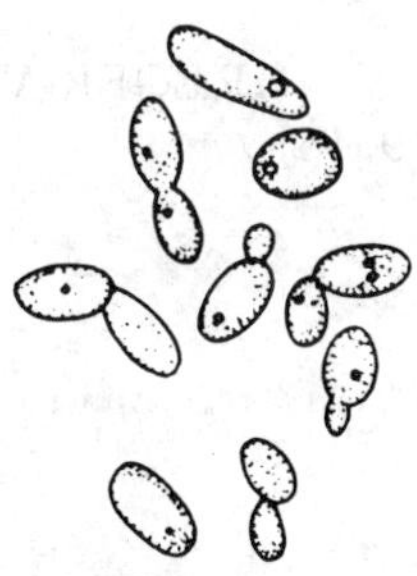

Fig. 291 – *C. melinii*
In glucose-yeast extract-peptone water
(From 'The Yeasts' ed. II, 1970)

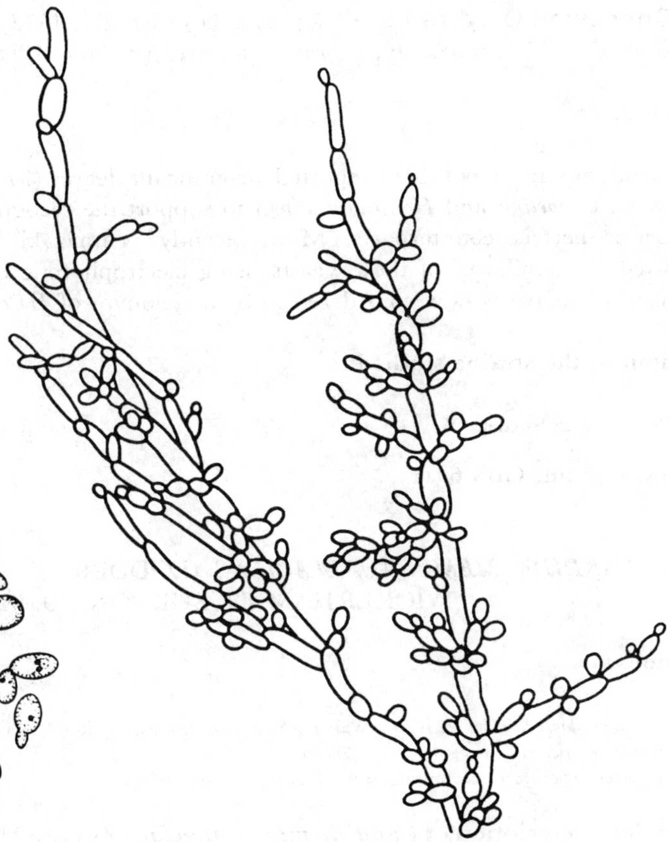

Fig. 292 – *C. melinii*
Dalmau plate culture on corn meal agar
(From 'The Yeasts, ed. II, 1970)

Fermentation: –

Assimilation of carbon compounds:

Galactose	–	Melezitose	+	Ribitol	–
L-Sorbose	–	Soluble starch	–	Galactitol	–
Sucrose	+	D-Xylose	+	D-Mannitol	+
Maltose	+	L-Arabinose	–	D-Glucitol	+
Cellobiose	+	D-Arabinose	–	Salicin	+
Trehalose	+	D-Ribose	–	DL-Lactic acid	+ s
Lactose	–	L-Rhamnose	+	Succinic acid	+
Melibiose	–	Glycerol	+	Citric acid	+ or w
Raffinose	–	Erythritol	–	Inositol	–

Splitting of arbutin: +

Assimilation of nitrate: +

Growth in vitamin-free medium: v

Growth at 37°C: +

Color reaction with DBB: –

Coenzyme Q system: Q 7 (Yamada & Kondo, 1972)

G + C: 40.9 mol%, type strain (Stenderup & Bak, 1968)

Comments

Stenderup and Bak (1968) reported a significant degree (97%) of DNA relatedness between *C. melinii* and *Hansenula wingei* to support the recognition of these yeasts as imperfect-perfect counterparts. More recently, Yamazaki and Komagata (1982) showed the similarity of these yeasts using electrophoretic comparisons of ten enzymes. *H. wingei* is now considered to be a synonym of *H. canadensis.*

Origin of the strains studied

Wood pulp in Sweden (2).

Type strain: CBS 601.

112. CANDIDA MEMBRANAEFACIENS (LODDER ET KREGER-VAN RIJ) WICKERHAM ET BURTON (1954b)

Synonyms

Candida melibiosi Lodder et Kreger-van Rij var. *membranaefaciens* Lodder et Kreger-van Rij (1952)
Candida majoricensis Genestar Serra (1956)
Procandida majoricensis (Genestar Serra) Novák et Zsolt (1961)

Standard description of *Candida membranaefaciens* (van Uden & Buckley, 1970)

Growth in glucose-yeast extract-peptone water: After 3 days at 25°C the cells are almost spherical to short-ovoid, (3–6) × (3.5–7) μm (Fig. 293); they frequently occur in groups. Islets may be present.

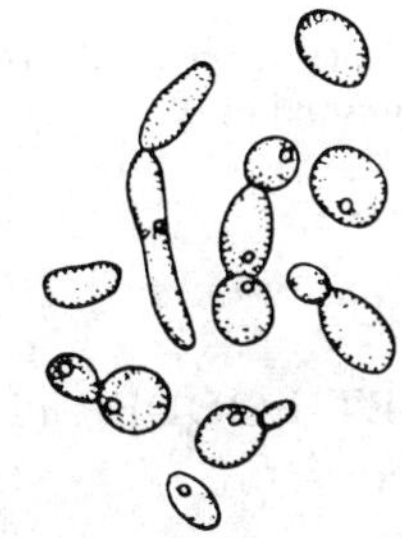

Fig. 293 – *C. membranaefaciens*
In glucose-yeast extract-peptone water
(From 'The Yeasts' ed. II, 1970)

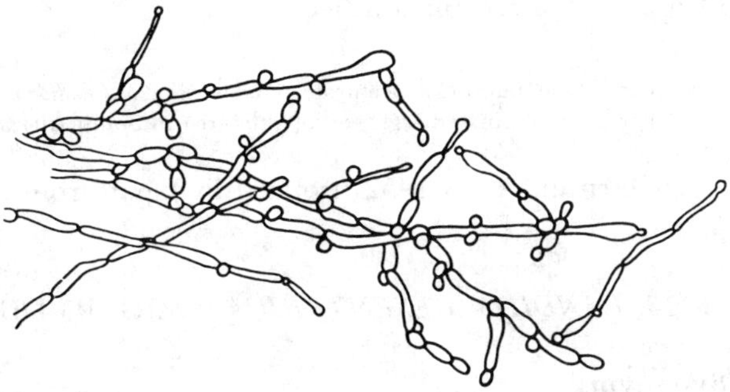

Fig. 294 – *C. membranaefaciens*
Dalmau plate culture on corn meal agar
(From 'The Yeasts' ed. II, 1970)

Growth on glucose-yeast extract-peptone agar: After one month at 25°C the streak culture is cream-colored, dull-glistening to glistening, soft, partly to wholly wrinkled.

Dalmau plate culture on corn meal agar: The pseudomycelium consists of ramified chains of long pseudohyphae bearing chains and clusters of roundish blastospores in verticillated positions; the pseudohyphae may be terminally swollen, club-like or curved (Fig. 294).

Fermentation:

Glucose	+	Lactose	–
Galactose	v	Raffinose	v
Sucrose	v	Trehalose	v
Maltose	–		

Assimilation of carbon compounds:

Galactose	+	Melezitose	+	Ribitol	+
L-Sorbose	+	Soluble starch	–	Galactitol	+
Sucrose	+	D-Xylose	+	D-Mannitol	+
Maltose	+	L-Arabinose	+	D-Glucitol	+
Cellobiose	+	D-Arabinose	+	Salicin	+
Trehalose	+	D-Ribose	+	DL-Lactic acid	v
Lactose	–	L-Rhamnose	–	Succinic acid	v
Melibiose	+	Glycerol	+	Citric acid	v
Raffinose	+	Erythritol	+	Inositol	–

Additional carbon compound tested: Inulin +

Splitting of arbutin: v

Assimilation of nitrate: –

Growth in vitamin-free medium: +s

Growth at 37°C: v

Color reaction with DBB: –

G + C: 33.4 mol%, one strain (Stenderup et al., 1972)

Origin of the strains studied

Scale insect (1); tanning concentrate and described as *Candida majoricensis* (Genestar Serra, 1956) (1); ulcer of a leper probably in India (1); eucalyptus tree in South Africa (1); human origin (3).

Type strain: CBS 1952, isolated in Japan from human urine.

113. CANDIDA MESENTERICA (GEIGER) DIDDENS ET LODDER (1942)

Synonyms

Pseudomonilia mesenterica Geiger (1910)
Azymoprocandida mesenterica (Geiger) Novák et Zsolt (1961)
Pseudomonilia albomarginata Geiger (1910)
Candida albomarginata (Geiger) Windisch (1953)

Standard description of *Candida mesenterica*

Growth in glucose-yeast extract-peptone water: After 3 days at 20°C the cells are ovoid to elongate, (2–4.5) × (6–12) μm; much longer pseudohyphal and hyphal cells also occur; the cells often occur in groups (Fig. 295). A pellicle or islets may be present (van Uden & Buckley, 1970).

Growth on glucose-yeast extract-peptone agar: After one month at 25°C the streak culture is cream-colored to whitish, dull, membranous, warty or folded, often with hairy projections, sometimes partially or wholly covered with a short velvet-like aerial mycelium (van Uden & Buckley, 1970).

Dalmau plate culture on corn meal agar: Pseudomycelium is abundantly formed; it is much branched and often bare of blastospores. The blastospores occur as irregular verticils or singly; they are often rod-like and somewhat curved (Fig. 296). True, septate hyphae also occur (van Uden & Buckley, 1970).

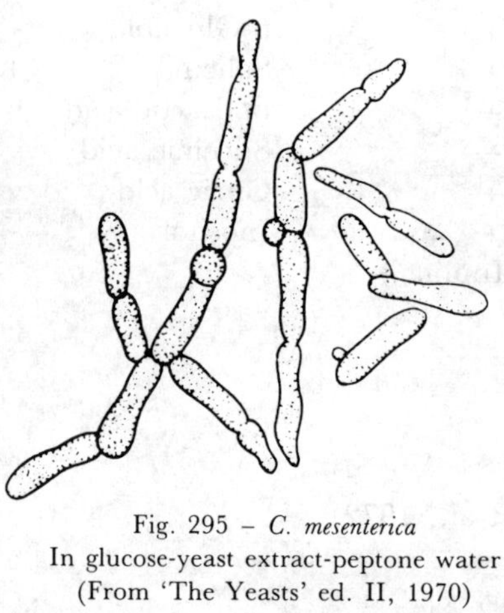

Fig. 295 – *C. mesenterica*
In glucose-yeast extract-peptone water
(From 'The Yeasts' ed. II, 1970)

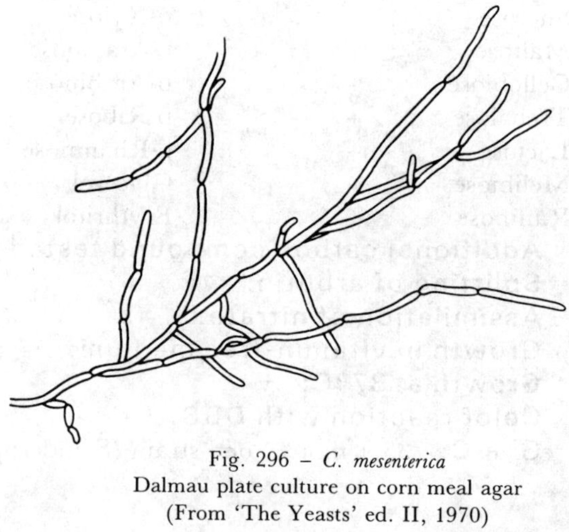

Fig. 296 – *C. mesenterica*
Dalmau plate culture on corn meal agar
(From 'The Yeasts' ed. II, 1970)

Fermentation: –

Assimilation of carbon compounds:

Galactose	–	Melezitose	v	Ribitol	v
L-Sorbose	v	Soluble starch	–	Galactitol	–
Sucrose	+	D-Xylose	–	D-Mannitol	v
Maltose	+	L-Arabinose	–	D-Glucitol	v
Cellobiose	v	D-Arabinose	–	Salicin	v
Trehalose	+	D-Ribose	v	DL-Lactic acid	+ s
Lactose	–	L-Rhamnose	–	Succinic acid	v
Melibiose	–	Glycerol	+	Citric acid	v
Raffinose	–	Erythritol	+	Inositol	–

Splitting of arbutin: v

Assimilation of nitrate: –

Growth in vitamin-free medium: + s

Growth at 37°C: –

Color reaction with DBB: –

G + C: 47.6 mol%, one strain (Meyer & Phaff, 1972)

Origin of the strains studied

Beer conduit tubes (5); whey (1).

Type strain: CBS 602, isolated in Germany from beer conduit tubes and described as *Pseudomonilia mesenterica.*

114. CANDIDA METHANOLOVESCENS (OKI ET KUONO) MEYER ET YARROW (Yarrow & Meyer, 1978)

Synonym

Torulopsis methanolovescens Oki et Kuono (in Oki et al., 1972)

Standard description of *Candida methanolovescens* (Barnett et al., 1983)

Growth in glucose-yeast extract-peptone water: After 3 days at 25°C the cells are globose to short-oval, (2.5–4.5) × (3.5–5.5) μm, and occur singly, in pairs and in threes. A sediment is present.

Growth on glucose-yeast extract-peptone agar: After one month at 25°C the streak culture is white, smooth and entire.

Dalmau plate culture on corn meal agar: No pseudomycelium is present.

Fermentation:

Glucose	+ s	Maltose	–
Galactose	–	Lactose	–
Sucrose	–		

Assimilation of carbon compounds:

Galactose	–	Melezitose	–	Ribitol	+
L-Sorbose	–	Soluble starch	–	Galactitol	–
Sucrose	–	D-Xylose	+ s	D-Mannitol	+
Maltose	–	L-Arabinose	–	D-Glucitol	+
Cellobiose	+	D-Arabinose	–	Salicin	+
Trehalose	+	D-Ribose	v	DL-Lactic acid	–
Lactose	–	L-Rhamnose	+	Succinic acid	+ s
Melibiose	–	Glycerol	+	Citric acid	+
Raffinose	–	Erythritol	–	Inositol	–

Additional carbon compound tested: Methanol +

Splitting of arbutin: +

Assimilation of nitrate: –

Growth in vitamin-free medium: –

Growth on 50% (w/w) glucose-yeast extract agar: –

Growth at 37°C: + or w

Urease: –

Color reaction with DBB: –

Coenzyme Q system: Q 7

G + C: 49.7 mol%

Origin of the strain studied

Flower of *Rhododendron indici* in Japan (1).

Type strain: CBS 6511.

115. CANDIDA METHANOSORBOSA (ABE ET YOKOTE) BARNETT, PAYNE ET YARROW (1983)

Synonyms

Torulopsis methanosorbosa Abe et Yokote (in Yokote et al., 1974)
Torulopsis nagoyaensis Asai et Makiguchi (in Asai et al., 1976)
Candida nagoyaensis (Asai et Makiguchi) Meyer et Yarrow (Yarrow & Meyer, 1978)

Standard description of *Candida methanosorbosa*

Growth in glucose-yeast extract-peptone water: After 3 days at 25°C the cells are globose to ovoid (3.0–5.0) × (3.5–5.5) μm and occur singly and in pairs.

Growth on glucose-yeast extract-peptone agar: After one month at 25°C the streak culture is cream-colored, smooth and glossy. The margin is entire.

Dalmau plate culture on corn meal agar: Pseudomycelium is absent.

Fermentation:

Glucose	+	Maltose	–
Galactose	+ s	Lactose	–
Sucrose	–	Trehalose	+

Assimilation of carbon compounds:

Galactose	+	Melezitose	–	Ribitol	+
L-Sorbose	+	Soluble starch	–	Galactitol	–
Sucrose	–	D-Xylose	+	D-Mannitol	+
Maltose	–	L-Arabinose	+	D-Glucitol	+
Cellobiose	+	D-Arabinose	–	Salicin	+
Trehalose	+	D-Ribose	+	DL-Lactic acid	+
Lactose	–	L-Rhamnose	+	Succinic acid	+
Melibiose	–	Glycerol	+	Citric acid	+
Raffinose	–	Erythritol	–	Inositol	–

Additional carbon compounds tested: Glucono-δ-lactone + , methanol +

Splitting of arbutin: +

Assimilation of nitrate: +

Growth in vitamin-free medium: –

Growth at 37°C: +

Color reaction with DBB: –

Origin of the strain studied

Sewage in Japan (1).

Type strain: CBS 7029.

116. *CANDIDA MILLERI* YARROW (1978)

Standard description of *Candida milleri*

Growth in glucose-yeast extract-peptone water: After 3 days at 25°C the cells are globose to ovoid, (2.2–4.3) × (2.2–5.7) μm and occur singly and in pairs, budding. After one month there is a fair amount of sediment and no pellicle.

Growth on glucose-yeast extract-peptone agar: After one month at 25°C the streak culture is smooth, glistening cream-colored and entire.

Dalmau plate culture on corn meal agar: No pseudomycelium is formed.

Fermentation:

Glucose	+	Maltose	–
Galactose	+	Lactose	–
Sucrose	+	Raffinose	+

Assimilation of carbon compounds:

Galactose	+	Melezitose	–	Ribitol	–
L-Sorbose	–	Soluble starch	–	Galactitol	–
Sucrose	+	D-Xylose	–	D-Mannitol	–
Maltose	–	L-Arabinose	–	D-Glucitol	–
Cellobiose	–	D-Arabinose	–	Salicin	–
Trehalose	+	D-Ribose	–	DL-Lactic acid	– or +s
Lactose	–	L-Rhamnose	–	Succinic acid	+w or –
Melibiose	–	Glycerol	v	Citric acid	–
Raffinose	+	Erythritol	–	Inositol	–

Splitting of arbutin: –
Assimilation of nitrate: –
Growth in vitamin-free medium: –
Growth on 50% (w/w) glucose-yeast extract agar: –
Growth at 37°C: –
Color reaction with DBB: –
G + C: 46.1–46.8 mol%, 3 strains (Yarrow, 1978)

Comments

See *Candida holmii*, Comments.

Origin of the strains studied

Sourdough (2); alpechin (1).

Type strain: CBS 6897, from San Francisco sourdough.

117. *CANDIDA MOGII* VIDAL-LEIRIA (1967)

Synonyms

Torulopsis miso nom. nud. (Mogi, 1938)
Torulopsis miso α var. 1 nom. nud. (Mogi, 1938)
Torulopsis miso α var. 2 nom. nud. (Mogi, 1942)

Standard description of *Candida mogii*

Growth in glucose-yeast extract-peptone water: After 3 days at 25°C the cells are ovoid, measuring (2.5–3) × (3–5) μm, single, in pairs or little chains (Fig. 297). Some pseudomycelium may also occur. A ring is present (van Uden & Buckley, 1970).

Growth on glucose-yeast extract-peptone agar: After one month at 25°C the streak culture is cream-colored, glistening to dull, soft, smooth or wrinkled (van Uden & Buckley, 1970).

Dalmau plate culture on corn meal agar: The pseudomycelium consists of formations of pseudohyphae bearing single ovoid blastospores and chains of blastospores in pleural positions (Fig. 298) (van Uden & Buckley, 1970).

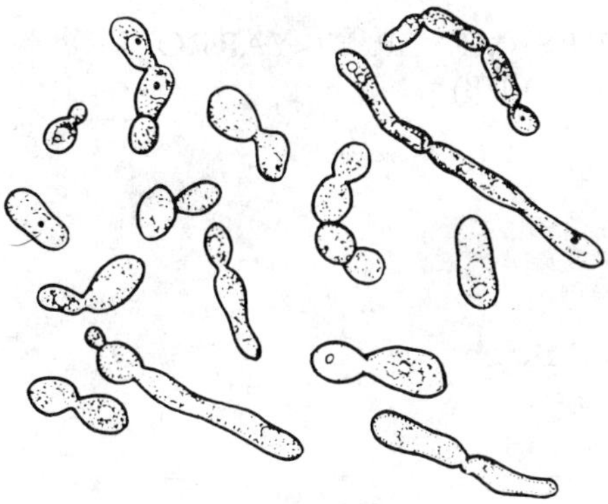

Fig. 297 – *C. mogii*
In glucose-yeast extract-peptone water
(From 'The Yeasts' ed. II, 1970)

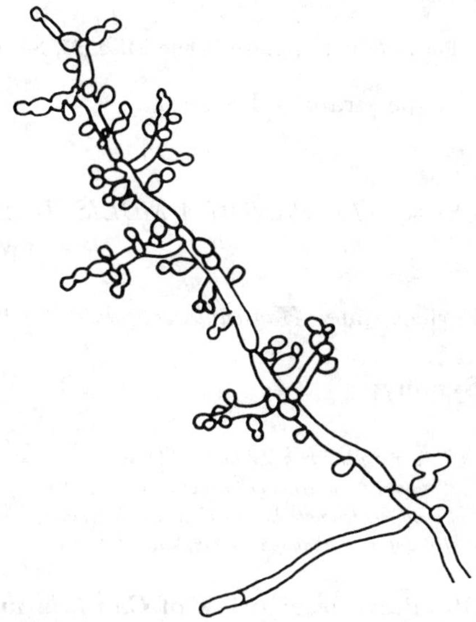

Fig 298 – *C. mogii*
Dalmau plate culture on corn meal agar
(From 'The Yeasts', ed. II, 1970)

Fermentation:

Glucose	+	Lactose	–
Galactose	–	Raffinose	–
Sucrose	+	Trehalose	+
Maltose	+ s or –		

Assimilation of carbon compounds:

Galactose	+	Melezitose	–	Ribitol	+ s or –
L-Sorbose	–	Soluble starch	+ s or –	Galactitol	+ s or –
Sucrose	+	D-Xylose	+	D-Mannitol	+
Maltose	+	L-Arabinose	+	D-Glucitol	+ s or –
Cellobiose	–	D-Arabinose	+ s or –	Salicin	–
Trehalose	+	D-Ribose	+	DL-Lactic acid	–
Lactose	–	L-Rhamnose	–	Succinic acid	+
Melibiose	–	Glycerol	+	Citric acid	+
Raffinose	+	Erythritol	–	Inositol	–

Additional carbon compound tested: α-Methyl-D-glucoside + s or –

Splitting of arbutin: –

Assimilation of nitrate: –

Growth in vitamin-free medium: –

Growth on 50% (w/w) glucose-yeast extract agar: +

Growth at 37°C: +

Urease: –

Color reaction with DBB: –

G + C: 46.1 mol% (Nakase & Komagata, 1971f)

Origin of the strains studied

Fuchu miso (3), Saiko White Miso (1), Siendal Miso (1), all of them isolated by Mogi.

Type strain: CBS 2032.

118. CANDIDA MOLISCHIANA (ZIKES) MEYER ET YARROW (Yarrow & Meyer, 1978)

Perfect state: *Hansenula capsulata* Wickerham

Synonyms

Torula molischiana Zikes (1911)
Torulopsis molischiana (Zikes) Lodder (1934)
Cryptococcus molischianus (Zikes) Skinner (1947a)
Torulopsis methanophiles Urakami (1975)

Standard description of *Candida molischiana*

Growth in glucose-yeast extract-peptone water: After 3 days at 25°C the cells are short-ovoid to nearly spherical, (2–5) × (3–6) μm, single, in pairs or in short chains. Elongate cells sometimes occur (Fig. 299). Besides a sediment, an incomplete ring may be present. The cells may or may not be encapsulated (van Uden & Vidal-Leiria, 1970).

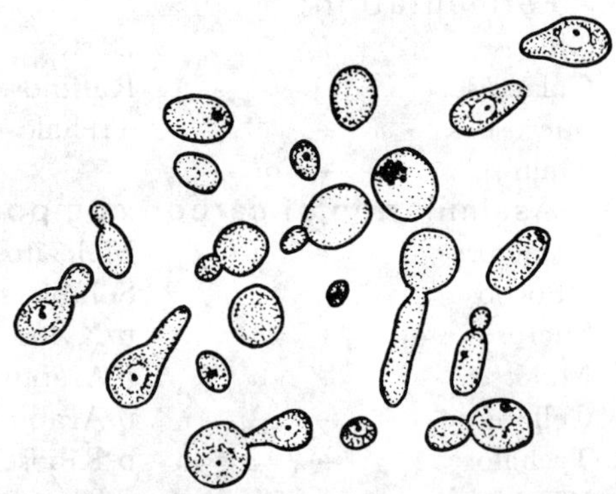

Fig. 299 – *C. molischiana*
In glucose-yeast extract-peptone water
(From 'The Yeasts' ed. II, 1970)

Growth on glucose-yeast extract-peptone agar: After one month at 25°C the streak culture is greyish-white, mucoid, soft and smooth. Capsule-less strains have a butyrous streak culture (van Uden & Vidal-Leiria, 1970).

Dalmau plate culture on corn meal agar: No pseudomycelium is formed (van Uden & Vidal-Leiria, 1970).

Fermentation:

Glucose	+	Lactose	–
Galactose	–	Trehalose	+
Sucrose	–	Cellobiose	+
Maltose	–		

Assimilation of carbon compounds:

Galactose	–	Melezitose	+	Ribitol	+
L-Sorbose	–	Soluble starch	+	Galactitol	–
Sucrose	–	D-Xylose	+	D-Mannitol	+
Maltose	+	L-Arabinose	+ w	D-Glucitol	+
Cellobiose	+	D-Arabinose	+	Salicin	+
Trehalose	+	D-Ribose	+	DL-Lactic acid	–
Lactose	–	L-Rhamnose	v	Succinic acid	v
Melibiose	–	Glycerol	+	Citric acid	–
Raffinose	–	Erythritol	+	Inositol	–

Splitting of arbutin: +
Assimilation of nitrate: +
Growth in vitamin-free medium: –
Growth at 37°C: +
Color reaction with DBB: –
G + C: 50.0 mol%, type strain (Meyer, Stenderup & Christiansen, unpublished)

Comments

Our results agreed with those of van Uden and Vidal-Leiria (1970) except for the utilization of rhamnose and succinic acid, both of which gave variable results.

Origin of the strains studied (van Uden & Vidal-Leiria, 1970)

Unknown source obtained by the CBS from the American Type Culture Collection in 1933 (1); sawmill (1); soil (1).

Type strain: CBS 136.

119. *CANDIDA MUCILAGINA* PHAFF, STARMER, MIRANDA ET MILLER (1980)

Standard description of *Candida mucilagina*

Growth in glucose-yeast extract-peptone water: After 3 days at 25°C the cells are spheroidal, sometimes ovoidal, (3.2–7.9) × (3.9–7.9) μm, single, in pairs, and sometimes in groups of three, surrounded by a mucilaginous capsule; pellicle, ring and visible sediment are lacking. After 21 days a thin ring and moderate amount of sediment are present.

Growth on glucose-yeast extract-peptone agar: After one month at 25°C the streak culture is watery and very slimy. The growth runs to the bottom of the tube.

Dalmau plate culture on corn meal agar: Pseudohyphae and true hyphae are lacking.

Fermentation: –

Assimilation of carbon compounds:

Galactose	+	Melezitose	+	Ribitol	+ or s
L-Sorbose	+ or s	Soluble starch	–	Galactitol	–
Sucrose	v	D-Xylose	+	D-Mannitol	v
Maltose	v	L-Arabinose	+	D-Glucitol	v
Cellobiose	+	D-Arabinose	v	Salicin	+
Trehalose	–	D-Ribose	v	DL-Lactic acid	v
Lactose	+ s or –	L-Rhamnose	v	Succinic acid	+
Melibiose	–	Glycerol	+	Citric acid	+
Raffinose	–	Erythritol	–	Inositol	–

Additional carbon compounds tested: Inulin –, methanol –, hexadecane –

Assimilation of nitrate: –

Growth in vitamin-free medium: –

Growth on 50% (w/w) glucose-yeast extract agar: –

Growth at 37°C: +

Starch formation: +

Urease: –

Color reaction with DBB: –

Coenzyme Q system: Q 8 (Phaff et al., 1980)

G + C: 43.2–44.0 mol%, 5 strains (Phaff et al., 1980)

Comments

Superficially, *C. mucilagina* resembles species of the genus *Lipomyces*.

Origin of the strains studied

Rotting tissue of *Stenocereus gummosis* (agria) near Todos Santos, Baja California Sur, Mexico (11); rotting cladodes of *Opuntia inermis* in New South Wales and Queensland, Australia (29).

Type strain: CBS 7071.

120. CANDIDA MULTIS-GEMMIS (BUHAGIAR) MEYER ET YARROW (Yarrow & Meyer, 1978)

Synonym

Torulopsis multis-gemmis Buhagiar (1975)

Standard description of *Candida multis-gemmis*

Growth in glucose-yeast extract-peptone water: After 4 days at 25°C the cells are short-ovoid to globose, 2.9–5.8 μm and occur singly, in pairs, chains or clusters.

Growth on glucose-yeast extract-peptone agar: After one month at

25°C the streak culture is white to cream-colored, semi-glossy, mat, smooth and entire.

Dalmau plate culture on corn meal agar: Pseudomycelium is absent or consists of short chains of a few budding cells.

Fermentation:

Glucose	+ s	Maltose	–
Galactose	–	Lactose	–
Sucrose	–	Raffinose	–

Assimilation of carbon compounds:

Galactose	+	Melezitose	+	Ribitol	+ s
L-Sorbose	+	Soluble starch	–	Galactitol	–
Sucrose	+	D-Xylose	+	D-Mannitol	v
Maltose	+	L-Arabinose	+	D-Glucitol	+ or s
Cellobiose	–	D-Arabinose	–	Salicin	–
Trehalose	+	D-Ribose	–	DL-Lactic acid	–
Lactose	–	L-Rhamnose	–	Succinic acid	+
Melibiose	–	Glycerol	+	Citric acid	+
Raffinose	–	Erythritol	–	Inositol	–

Splitting of arbutin: –

Assimilation of nitrate: +

Growth in vitamin-free medium: –

Growth on 50% (w/w) glucose-yeast extract agar: –

Growth at 37°C: –

Urease: –

Color reaction with DBB: –

G + C: 35.4 mol%, type strain

Origin of the strains studied

Raspberries (2).

Type strain: CBS 6524.

121. *CANDIDA MUSAE* (NAKASE) MEYER ET YARROW (Yarrow & Meyer, 1978)

Synonym

Torulopsis musae Nakase (1971a)

Standard description of *Candida musae*

Growth in glucose-yeast extract-peptone water: After 3 days at 25°C cells are round to oval, (2–5) × (2.5–7.0) μm, and occur singly, in pairs, chains and clusters. After one month at 20°C no pellicle is present, but a heavy sediment is evident.

Growth on glucose-yeast extract-peptone agar: After one month at 25°C the streak culture is yellowish-white, smooth, soft, shining, with an entire margin.

Dalmau plate culture on corn meal agar: Chains of budding cells are formed.

Fermentation:

Glucose	+	Maltose	–
Galactose	–	Lactose	–
Sucrose	–	Trehalose	+ s

Assimilation of carbon compounds:

Galactose	–	Melezitose	+	Ribitol	+ s
L-Sorbose	+	Soluble starch	–	Galactitol	–
Sucrose	+	D-Xylose	+ or s	D-Mannitol	+
Maltose	+	L-Arabinose	–	D-Glucitol	+
Cellobiose	–	D-Arabinose	–	Salicin	–
Trehalose	+	D-Ribose	+ s or –	DL-Lactic acid	–
Lactose	–	L-Rhamnose	–	Succinic acid	+
Melibiose	–	Glycerol	+ w	Citric acid	+
Raffinose	–	Erythritol	–	Inositol	–

Additional carbon compounds tested: Glucono-δ-lactone +, 2-keto-gluconate +

Splitting of arbutin: + or s

Assimilation of nitrate: –

Growth in vitamin-free medium: –

Growth on 50% (w/w) glucose-yeast extract agar: –

Growth at 37°C: –

Color reaction with DBB: –

G + C: 47.8 mol% (Nakase, 1971a)

Origin of the strain studied

Banana from a food market in Tokyo (1).

Type strain: CBS 6381.

122. *CANDIDA MUSCORUM* DI MENNA (1958a)

Synonyms

Azymocandida muscorum (di Menna) Novák et Zsolt (1961)
Rhodotorula muscorum (di Menna) von Arx et Weijman (1979)

Standard description of *Candida muscorum*

Growth in glucose-yeast extract-peptone water: After 3 days at 25°C the cells are ovoid to irregularly cylindrical, (2–4) × (6–20) μm (Fig. 300) (van Uden & Buckley, 1970).

Growth on glucose-yeast extract-peptone agar: After one month at 25°C the streak culture is yellowish, cream-colored, mucoid, soft, smooth; warty outgrowths may be present (van Uden & Buckley, 1970).

Dalmau plate culture on corn meal agar: The pseudomycelium consists of ramified chains of coarse pseudohyphae, bearing few oval blastospores (Fig. 301) (van Uden & Buckley, 1970).

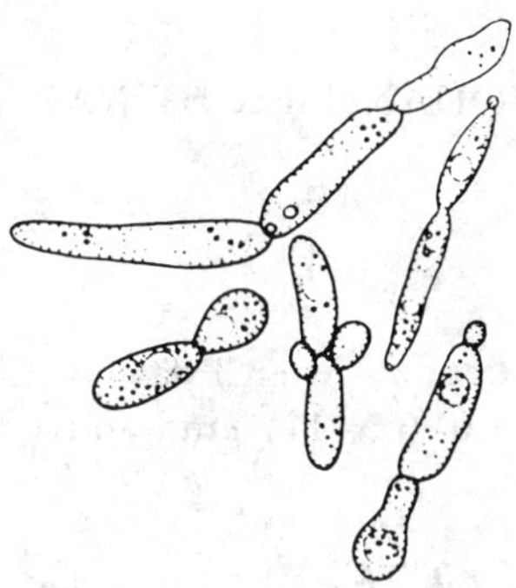

Fig. 300 – *C. muscorum*
In glucose-yeast extract-peptone water
(From 'The Yeasts' ed. II, 1970)

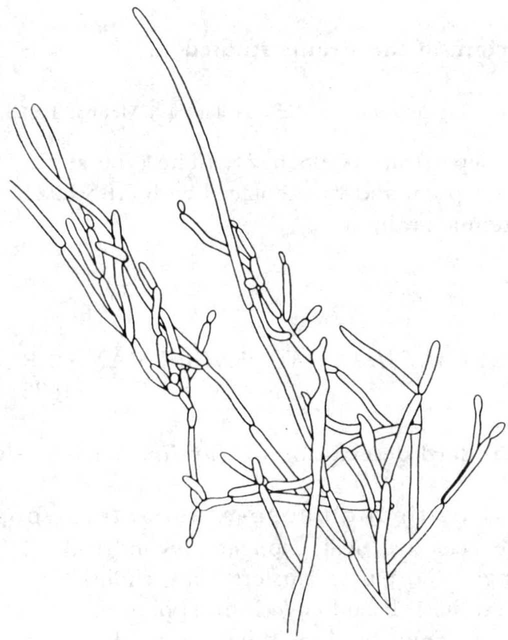

Fig. 301 – *C. muscorum*
Dalmau plate culture on corn meal agar
(From 'The Yeasts' ed. II, 1970)

Fermentation: –

Assimilation of carbon compounds:

Galactose	+	Melezitose	+	Ribitol	+ s
L-Sorbose	+	Soluble starch	–	Galactitol	+ s
Sucrose	+	D-Xylose	+	D-Mannitol	+
Maltose	+	L-Arabinose	–	D-Glucitol	+
Cellobiose	+	D-Arabinose	+ or s	Salicin	+
Trehalose	+	D-Ribose	+ or s	DL-Lactic acid	–
Lactose	+	L-Rhamnose	–	Succinic acid	+
Melibiose	–	Glycerol	+	Citric acid	+ s
Raffinose	+	Erythritol	–	Inositol	–

Splitting of arbutin: +

Assimilation of nitrate: +

Growth in vitamin-free medium: –

Growth at 37°C: –

Color reaction with DBB: +

G + C: 46.1 mol%, type strain (Meyer, Stenderup & Christiansen, unpublished)

Comments

Based on the type of conidiogenesis and the lack of xylose in cell hydrolysates, von Arx and Weijman (1979) transferred *C. muscorum* to the genus *Rhodotorula*.

Origin of the strains studied

Rotting sphagnum in New Zealand (di Menna, 1958a) (1); barley leaf (1).

Type strain: CBS 6921. (The type strain, CBS 2912, no longer agreed with the description and was replaced with CBS 6921. Both cultures were derived from the di Menna strains.)

123. *CANDIDA NAEODENDRA* VAN DER WALT, JOHANNSEN ET NAKASE (1973)

Standard description of *Candida naeodendra*

Growth in glucose-yeast extract-peptone water: After 3 days at 25°C the cells are oval, elongate, cylindrical, (1.5–5.0) × (2.0–6.5–11) μm, and occur singly, in pairs, clusters and chains (Fig. 302). Pseudomycelium is present. An incomplete ring, thin, small islets and floccose sediment are formed. After one month an incomplete ring and rather floccose sediment are present (van der Walt et al., 1973).

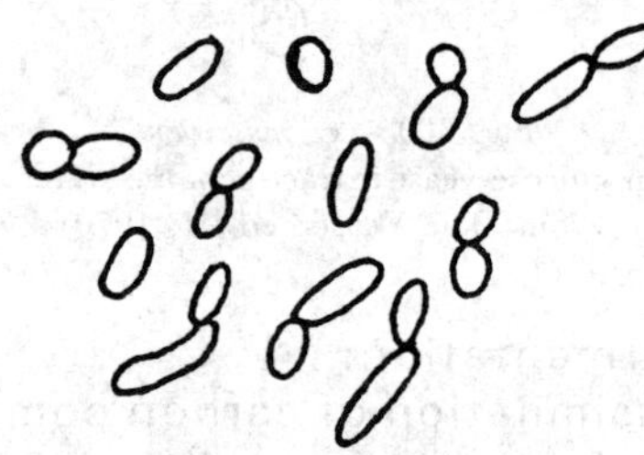

Fig. 302 – *C. naeodendra*
In glucose-yeast extract-peptone water

Growth on glucose-yeast extract-peptone agar: After one month at 25°C the streak culture is cream-colored, smooth, and fringed with mycelium (van der Walt et al., 1973).

Dalmau plate culture on corn meal agar: A ramified pseudomycelium is abundantly produced. Some blastospores are triangular, usually in short chains or small clusters. Van der Walt et al. (1973) reported the presence of aerial chains of blastoconidia.

Fermentation:

Glucose	+ or s	Maltose	–
Galactose	+ s	Lactose	–
Sucrose	–		

Assimilation of carbon compounds:

Galactose	+	Melezitose	+	Ribitol	+
L-Sorbose	+ or s	Soluble starch	+	Galactitol	–
Sucrose	+	D-Xylose	+	D-Mannitol	+
Maltose	+	L-Arabinose	+	D-Glucitol	+
Cellobiose	+	D-Arabinose	+ s	Salicin	+
Trehalose	+	D-Ribose	+	DL-Lactic acid	–
Lactose	–	L-Rhamnose	+	Succinic acid	+
Melibiose	–	Glycerol	+	Citric acid	+
Raffinose	–	Erythritol	+	Inositol	–

Splitting of arbutin: +

Assimilation of nitrate: –

Growth in vitamin-free medium: –

Growth on 50% (w/w) glucose-yeast extract agar: +

Growth at 37°C: +

Color reaction with DBB: –

Fat splitting: +

G + C: 38.3 mol% (van der Walt et al., 1973)

Origin of the strain studied

Frass from galleries of Buprestid larvae infesting *Acacia nilotica* L. var. *Kraussiana* (Benth.) Brenan in the vicinity of Groblersdal in the Transvaal (1).

Type strain: CBS 6032.

124. CANDIDA NEMODENDRA (VAN DER WALT, VAN DER KLIFT ET SCOTT) MEYER ET YARROW (Yarrow & Meyer, 1978)

Synonym

Torulopsis nemodendra van der Walt, van der Klift et Scott (van der Walt et al., 1971b)

Standard description of *Candida nemodendra*

Growth in glucose-yeast extract-peptone water: After 3 days at 25°C the cells are spheroidal 1.5–4.3 μm, and occur singly and budding. A sediment is present. After one month a sediment and a ring are present.

Growth on glucose-yeast extract-peptone agar: After one month at 25°C the streak culture is off-white, smooth, shiny, and the margin is entire.

Dalmau plate culture on corn meal agar: No pseudomycelium is formed.

Fermentation: –

Assimilation of carbon compounds:

Galactose	+	Melezitose	–	Ribitol	+
L-Sorbose	+	Soluble starch	–	Galactitol	+
Sucrose	–	D-Xylose	+	D-Mannitol	+
Maltose	–	L-Arabinose	+	D-Glucitol	+

Cellobiose	+	D-Arabinose	+ s	Salicin	+
Trehalose	+	D-Ribose	+	DL-Lactic acid	–
Lactose	–	L-Rhamnose	–	Succinic acid	+
Melibiose	–	Glycerol	+	Citric acid	–
Raffinose	–	Erythritol	+	Inositol	–

Additional carbon compound tested: Methanol +
Splitting of arbutin: +
Assimilation of nitrate: –
Growth in vitamin-free medium: –
Growth on 50% (w/w) glucose-yeast extract agar: –
Growth at 37°C: v
Color reaction with DBB: –
G + C: 42.4 mol%, type strain (Meyer, Stenderup & Christiansen, unpublished)

Origin of the strains studied

Material from the tunnels of *Xyleborus aemulus* Woll. in *Rapanea melanopholoeos* L. Mez near Knysna in the Cape Province, South Africa (2).

Type strain: CBS 6280.

125. CANDIDA NITRATIVORANS VAN DER WALT, SCOTT ET VAN DER KLIFT

see *Hansenula sydowiorum* p. 210

126. CANDIDA NITRATOPHILA (SHIFRINE ET PHAFF) MEYER ET YARROW (Yarrow & Meyer, 1978)

Synonym

Torulopsis nitratophila Shifrine et Phaff (1956)

Standard description of *Candida nitratophila* (van Uden & Vidal-Leiria, 1970)

Growth in glucose-yeast extract-peptone water: After 3 days at 25°C the cells are ovoid, (2–3.5) × (2.5–5) μm (Fig. 303). Only a sediment is formed.

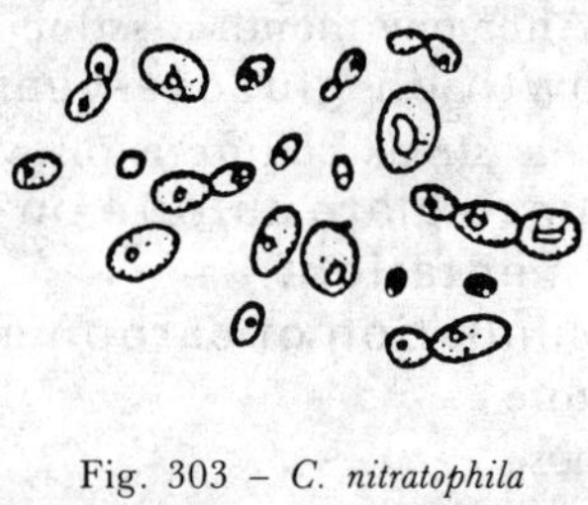

Fig. 303 – *C. nitratophila*
In glucose-yeast extract-peptone water
(From 'The Yeasts', ed. II, 1970)

Growth on glucose-yeast extract-peptone agar: After one month at 25°C the streak culture is greyish-white, glossy, soft, almost smooth.

Dalmau plate culture on corn meal agar: No pseudomycelium is formed; branched chains of oval cells occur.

Fermentation:

Glucose	+ s	Maltose	–
Galactose	–	Lactose	–
Sucrose	–	Trehalose	+

Assimilation of carbon compounds:

Galactose	+	Melezitose	–	Ribitol	+
L-Sorbose	–	Soluble starch	–	Galactitol	–
Sucrose	–	D-Xylose	+	D-Mannitol	+
Maltose	–	L-Arabinose	+	D-Glucitol	+
Cellobiose	–	D-Arabinose	–	Salicin	–
Trehalose	+	D-Ribose	+	DL-Lactic acid	–
Lactose	–	L-Rhamnose	+	Succinic acid	+ s
Melibiose	–	Glycerol	+	Citric acid	–
Raffinose	–	Erythritol	–	Inositol	–

Splitting of arbutin: –

Assimilation of nitrate: +

Growth in vitamin-free medium: –

Growth at 37°C: v

Color reaction with DBB: –

G + C: 36.6 mol%, one strain (Nakase & Komagata, 1971c)

Origin of the strain studied

Bark beetle (Shifrine & Phaff, 1956) (1).

Type strain: CBS 2027.

127. *CANDIDA NIVALIS* DI MENNA

see *Leucosporidium nivalis* p. 502.

128. *CANDIDA NODAENSIS* YARROW ET MEYER (1978)

Synonym

Torulopsis nodaensis nom. nud. (Onishi, 1957)

Standard description of *Candida nodaensis*

Growth in glucose-yeast extract-peptone water: After 3 days at 25°C the cells are globose to oval (2.0–3.0) × (3.0–4.5) μm and occur singly and budding.

Growth on glucose-yeast extract-peptone agar: After one month at 20°C the streak culture is off-white to cream-colored, smooth, soft and dull. The margin is entire.

Dalmau plate culture on corn meal agar: Pseudomycelium is absent.

Fermentation:

Glucose	+	Maltose	–
Galactose	–	Lactose	–
Sucrose	–		

Assimilation of carbon compounds:

Galactose	+	Melezitose	–	Ribitol	–
L-Sorbose	+	Soluble starch	–	Galactitol	–
Sucrose	–	D-Xylose	–	D-Mannitol	+
Maltose	+ s	L-Arabinose	–	D-Glucitol	+ s
Cellobiose	–	D-Arabinose	–	Salicin	–
Trehalose	–	D-Ribose	–	DL-Lactic acid	–
Lactose	–	L-Rhamnose	–	Succinic acid	–
Melibiose	–	Glycerol	+	Citric acid	+
Raffinose	–	Erythritol	–	Inositol	–

Splitting of arbutin: –

Assimilation of nitrate: +

Growth in vitamin-free medium: –

Growth at 37°C: –

Urease: –

Color reaction with DBB: –

G + C: 53 mol% (Yarrow & Meyer, 1978)

Origin of the strain studied

Soy mash (1).

Type strain: CBS 3094.

129. CANDIDA NORVEGENSIS (DIETRICHSON) VAN UDEN ET FARINHA EX VAN UDEN ET BUCKLEY

see *Pichia norvegensis* p. 347

130. CANDIDA NORVEGICA (REIERSÖL) MEYER ET YARROW (Yarrow & Meyer, 1978)

Synonyms

Torulopsis norvegica Reiersöl (1958)
Paratorulopsis norvegica (Reiersöl) Novák et Zsolt (1961)
Torulopsis vanzylii van der Walt et van Kerken (1961b)

Standard description of *Candida norvegica*

Growth in glucose-yeast extract-peptone water: After 3 days at 25°C

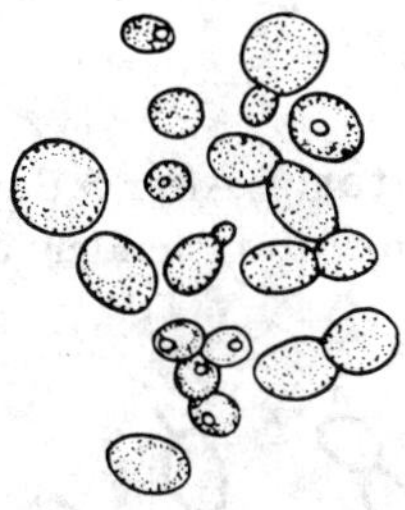

Fig. 304 – *C. norvegica*
In glucose-yeast extract-peptone water
(From 'The Yeasts' ed. II, 1970)

the cells are ovoid to globose, (2–6) × (3–6) μm (Fig. 304). Only a sediment is formed (van Uden & Vidal-Leiria, 1970).

Growth on glucose-yeast extract-peptone agar: After one month at 25°C the streak culture is cream-colored, glossy, soft, smooth or finely punctilated (van Uden & Vidal-Leiria, 1970).

Dalmau plate culture on corn meal agar: A rudimentary pseudomycelium of branched chains of cells is formed (van Uden & Vidal-Leira, 1970).

Fermentation:

Glucose	v	Maltose	–
Galactose	–	Lactose	–
Sucrose	–		

Assimilation of carbon compounds:

Galactose	–	Melezitose	–	Ribitol	–
L-Sorbose	–	Soluble starch	–	Galactitol	–
Sucrose	–	D-Xylose	+	D-Mannitol	+
Maltose	–	L-Arabinose	v	D-Glucitol	+
Cellobiose	+	D-Arabinose	–	Salicin	+
Trehalose	–	D-Ribose	–	DL-Lactic acid	+
Lactose	–	L-Rhamnose	v	Succinic acid	v
Melibiose	–	Glycerol	+	Citric acid	v
Raffinose	–	Erythritol	–	Inositol	–

Splitting of arbutin: +

Assimilation of nitrate: +

Growth in vitamin-free medium: + s

Growth at 37°C: –

Color reaction with DBB: –

G + C: 40.7 mol%, type strain (Meyer, Stenderup & Christiansen, unpublished)

Origin of the strains studied

Human sputum in Norway (Reiersöl, 1958) (1); authentic strain of *T. vanzylii* from the floor of a wine cellar in South Africa (van der Walt & van Kerken, 1961b) (1); apple juice in Switzerland (1); cider, England (2); a beer bottle washing machine in Sweden (2); Chesapeake Bay water (3); Chesapeake Bay sediment (2).

Type strain: CBS 4239, from human sputum.

131. CANDIDA OLEOPHILA MONTROCHER (1967)

Standard description of *Candida oleophila*

Growth in glucose-yeast extract-peptone water: After 3 days at 25°C the cells are oval to elongate, (2.9–4.3) × (4.3–6.5) μm and slender cylindrical, (1.4–3.6) × (4.3–9.4) μm. Some cells are curved (Fig. 305). Cells occur singly, in pairs, in short branched and unbranched chains. Pseudomycelium is present. A thin, delicately wrinkled, waxy pellicle and a fluffy sediment are present. After one month the pellicle is complete, mucoid to waxy and slightly folded.

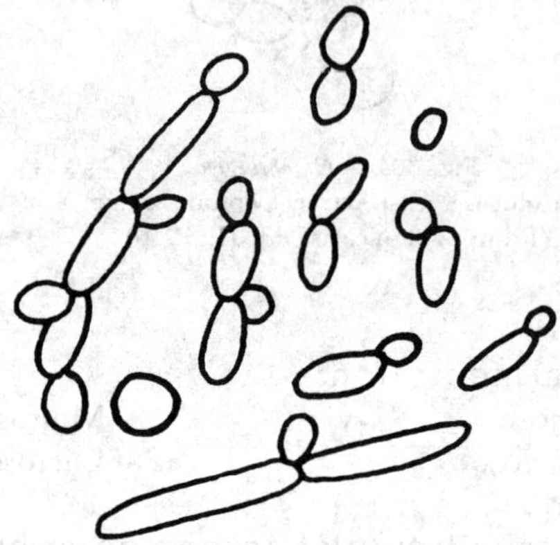

Fig. 305 – *C. oleophila*
In glucose-yeast extract-peptone water

Growth on glucose-yeast extract-peptone agar: After one month at 25°C the streak culture is dull, cream-colored, lacy, with raised growth radiating from the center. A lobed margin is evident with occasional tufts of mycelial growth.

Dalmau plate culture on corn meal agar: The pseudomycelium consists of branched chains of cells with elongated blastospores.

Fermentation:

Glucose	+	Maltose	– or + s
Galactose	+ or s	Lactose	–
Sucrose	+ s or –		

Assimilation of carbon compounds:

Galactose	+	Melezitose	+	Ribitol	+ s
L-Sorbose	+	Soluble starch	–	Galactitol	–
Sucrose	+	D-Xylose	+	D-Mannitol	+
Maltose	+	L-Arabinose	– or + s	D-Glucitol	+
Cellobiose	+	D-Arabinose	v	Salicin	+
Trehalose	+	D-Ribose	+ s	DL-Lactic acid	+ s
Lactose	–	L-Rhamnose	–	Succinic acid	+
Melibiose	–	Glycerol	+	Citric acid	+
Raffinose	–	Erythritol	–	Inositol	–

Additional carbon compound tested: 2-Ketogluconate +

Splitting of arbutin: +

Assimilation of nitrate: –

Growth in vitamin-free medium: –

Growth at 37°C: –

Color reaction with DBB: –

Growth in the presence of 100 ppm cycloheximide: +

Coenzyme Q system: Q 9 (Yamada & Kondo, 1972)
G + C: 40.7–41.5 mol%, 3 strains (Meyer & Simione, unpublished)

Comments

A second strain designated as *Candida oleophila* Iizuka was published without a Latin description four years after Montrocher's valid description appeared. These organisms are unrelated. DNA reassociation studies demonstrated the difference between *C. oleophila* Montrocher and *C. sake.* Two additional strains previously identified as *C. sake* also showed significant reassociation with *C. oleophila* type strain to be considered representatives of the species (Meyer & Simione, unpublished).

Origin of the strains studied

Olives (1); tonic water (1); cider (1); flower (1).

Type strain: CBS 2219, from olives.

132. CANDIDA OREGONENSIS PHAFF ET DO CARMO-SOUSA (1962)

Synonym

Candida obtusa (Dietrichson) van Uden et do Carmo-Sousa var. *oregonensis* (Phaff et do Carmo-Sousa) Montrocher (1967)

Standard description of *Candida oregonensis* (van Uden & Buckley, 1970)

Growth in glucose-yeast extract-peptone water: After 3 days at 25°C cells are short-ovoid, (2–4) × (2.5–5) μm, often in chains and groups (Fig. 306).

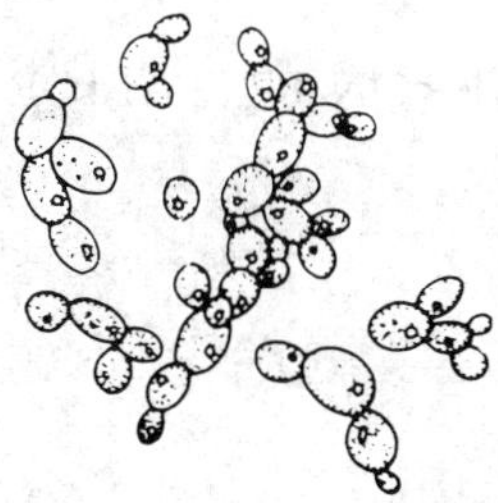

Fig. 306 – *C. oregonensis*
In glucose-yeast extract-peptone water
(From 'The Yeasts' ed. II, 1970)

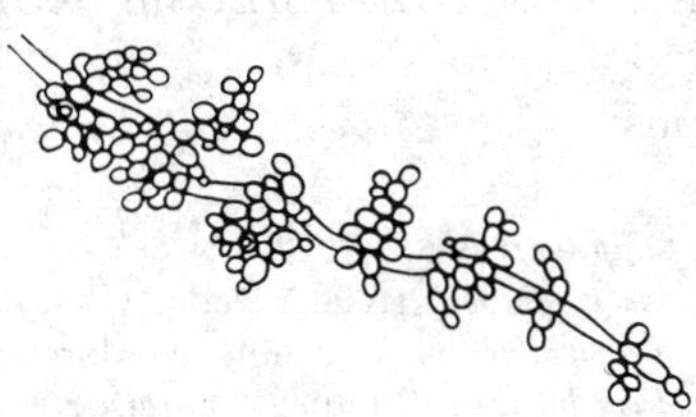

Fig. 307 – *C. oregonensis*
Dalmau plate culture on corn meal agar
(From 'The Yeasts' ed. II, 1970)

Growth on glucose-yeast extract-peptone agar: After one month at 25°C the streak culture is greyish to cream-colored, glistening, soft, smooth or slightly reticulated.

Dalmau plate culture on corn meal agar: Pseudomycelium is abundantly formed. The branched chains of pseudohyphae bear chains or clusters of globose blastospores in verticils or irregularly arranged along the hyphal strand (Fig. 307).

Fermentation:

Glucose	+ or s	Maltose	– or + s
Galactose	–	Lactose	–
Sucrose	–		

Assimilation of carbon compounds:

Galactose	–	Melezitose	+	Ribitol	+ or s
L-Sorbose	–	Soluble starch	+ or s	Galactitol	–
Sucrose	+	D-Xylose	+	D-Mannitol	+
Maltose	+	L-Arabinose	–	D-Glucitol	+
Cellobiose	+	D-Arabinose	– or + s	Salicin	+
Trehalose	+	D-Ribose	– or + s	DL-Lactic acid	–
Lactose	–	L-Rhamnose	+	Succinic acid	+
Melibiose	–	Glycerol	–	Citric acid	–
Raffinose	–	Erythritol	–	Inositol	–

Splitting of arbutin: +

Assimilation of nitrate: –

Growth in vitamin-free medium: –

Growth at 37°C: –

Color reaction with DBB: –

Coenzyme Q system: Q 9 (Yamada & Kondo, 1972)

G + C: 48.1 mol% (Meyer & Phaff, 1972)

Origin of the strains studied

Insect frass (Phaff & do Carmo-Sousa, 1962) (1); alpechin (1).

Type strain: CBS 5036, from insect frass.

133. *CANDIDA PARAPSILOSIS* (ASHFORD) LANGERON ET TALICE (1932)

Synonyms

Monilia parapsilosis Ashf. (1928)
Mycocandida parapsilosis (Ashf.) Dodge (1935)
Monilia onychophila Poll. et Nann. (in Marengo, 1926)
Blastodendrion intestinale Mattlet var. *epidermicum* Cif. et Alfonseca (1931)
Blastodendrion globosum Zach (in Wolfram & Zach, 1933)
Schizoblastosporion globosum (Zach) Dodge (1935)
Blastodendrion gracile Zach (in Wolfram & Zach, 1933)
Schizoblastosporion gracile (Zach) Dodge (1935)
Candida parapsilosis Ashf. var. *intermedia* van Rij et Verona (1949)
Saccharomyces verticillatus nom. nud. (Dietrichson, 1954)
Saccharomyces vossii nom. nud. (Dietrichson, 1954)

Standard description of *Candida parapsilosis*

Growth in glucose-yeast extract-peptone water: After 3 days at 25°C the cells are oval, elliptical and elongated, (2.9–4.3) × (3.6–7.2) μm, and occur singly, in pairs and in short chains and clusters. Pseudomycelium also occurs (Fig. 308). A sediment is present. After one month a sediment and a ring are present.

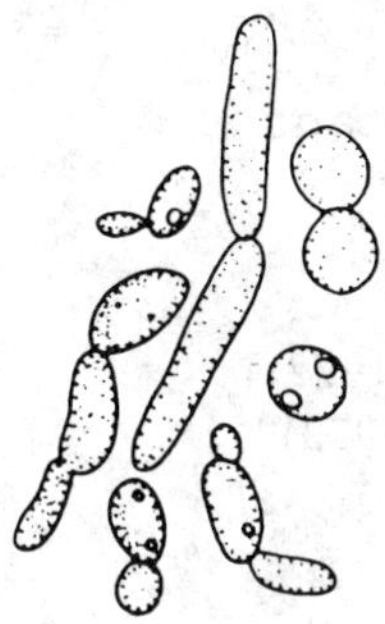

Fig. 308 – *C. parapsilosis*
In glucose-yeast extract-peptone water
(From 'The Yeasts' ed. II, 1970)

Growth on glucose-yeast extract-peptone agar: After one month at 25°C the streak culture is cream-colored to yellowish, glistening and soft; most strains are smooth, others are partly or wholly wrinkled (van Uden & Buckley, 1970).

Dalmau plate culture on corn meal agar: Pseudomycelium consists of branched chains of elongated cells with chains and clusters of round to oval blastospores forming at intervals along the hyphae (Fig. 309). Rare strains form practically no hyphae.

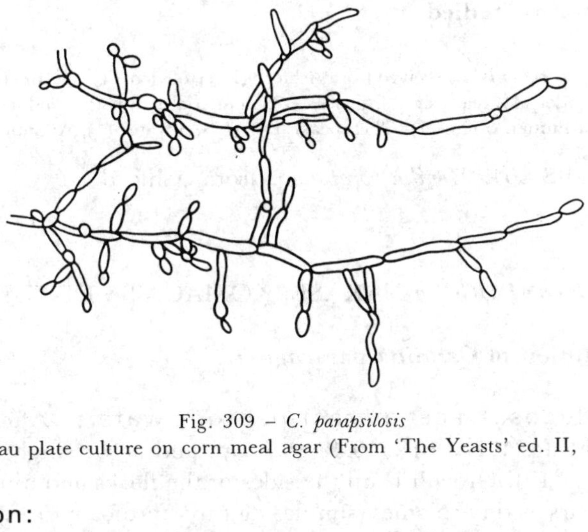

Fig. 309 – *C. parapsilosis*
Dalmau plate culture on corn meal agar (From 'The Yeasts' ed. II, 1970)

Fermentation:

Glucose	+	Maltose	– or +w
Galactose	v	Lactose	–
Sucrose	– or +w		

Assimilation of carbon compounds:

Galactose	+	Melezitose	–	Ribitol	+ or s
L-Sorbose	v	Soluble starch	–	Galactitol	–
Sucrose	+	D-Xylose	+	D-Mannitol	+

Maltose	+	L-Arabinose	+	D-Glucitol	+
Cellobiose	–	D-Arabinose	–	Salicin	–
Trehalose	+	D-Ribose	v	DL-Lactic acid	v
Lactose	–	L-Rhamnose	–	Succinic acid	v
Melibiose	–	Glycerol	+	Citric acid	v
Raffinose	–	Erythritol	–	Inositol	–

Splitting of arbutin: –
Assimilation of nitrate: –
Growth in vitamin-free medium: – or +w
Growth on 50% (w/w) glucose-yeast extract agar: +
Growth at 37°C: +
Urease: –
Color reaction with DBB: –
Coenzyme Q system: Q 9 (Yamada & Kondo, 1972)
G + C: 40.5 mol%, type strain (Meyer & Phaff, 1972)

Comments

C. parapsilosis is no longer considered the imperfect form of *Lodderomyces elongisporus*. Meyer and Phaff (1972) demonstrated the lack of genetic relatedness between the type strains of these species using DNA reassociation techniques.

Origin of the strains studied

Strain labelled *Monilia parapsilosis* received from Ashford (Puerto Rico) (1), sputum (1), olives (1), infected nail (4), skin (2), bladder (1), tea beer (1), pickles (2), industrial strain for production of citric acid (1), unknown, possibly of human origin (4), Chesapeake Bay (USA) water (1), Atlantic Ocean (11).

Type strain: CBS 604, *Monilia parapsilosis* from Ashford.

134. CANDIDA PARARUGOSA NAKASE, KOMAGATA ET FUKAZAWA (1978)

Standard description of *Candida pararugosa*

Growth in glucose-yeast extract-peptone water: After 3 days at 25°C the cells are elongate (1.4–3.6) × (10.1–21.7) μm and occur singly, budding and in short chains (Fig. 310). Growth is up the sides of the flasks and a pellicle is present. The pellicle appears as tiny colonies (simulating tiny styrofoam discs) covering the entire surface of the medium. After one month the pellicle is folded, rough and thick. A light sediment is present.

Growth on glucose-yeast extract-peptone agar: After one month at 25°C the streak culture is off-white to beige, raised, wrinkled, and rough with a mycelial border.

Dalmau plate culture on corn meal agar: Pseudomycelium is abundantly formed. It consists of branched chains of elongate cells.

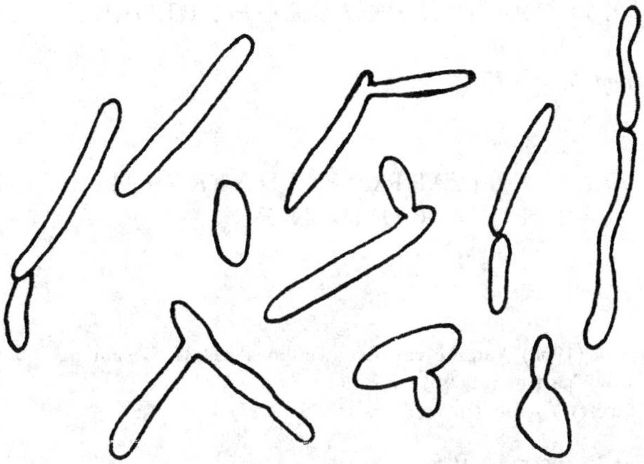

Fig. 310 – *C. pararugosa*
In glucose-yeast extract-peptone water

Fermentation: –

Assimilation of carbon compounds:

Galactose	+	Melezitose	–	Ribitol	–
L-Sorbose	+	Soluble starch	–	Galactitol	–
Sucrose	+	D-Xylose	–	D-Mannitol	+
Maltose	–	L-Arabinose	+ or s	D-Glucitol	+
Cellobiose	–	D-Arabinose	+ s	Salicin	–
Trehalose	–	D-Ribose	–	DL-Lactic acid	+ s
Lactose	–	L-Rhamnose	–	Succinic acid	+
Melibiose	–	Glycerol	+	Citric acid	–
Raffinose	–	Erythritol	–	Inositol	–

Splitting of arbutin: –

Assimilation of nitrate: –

Growth in vitamin-free medium: –

Growth at 37°C: +

Color reaction with DBB: –

G + C: 48.3 mol%, type strain (Nakase et al., 1978)

Comments

The results of some of the carbon assimilation tests differed from those reported in the original description. No growth was reported on D-arabinose, glycerol, D-mannitol, D-glucitol, lactate and succinate.

Origin of the strain studied

Human feces (1).

Type strain: CBS 1010.

135. CANDIDA PELLICULOSA REDAELLI

see *Hansenula anomala* p.173

136. CANDIDA PELTATA (YARROW) MEYER, AHEARN ET YARROW COMB. NOV.

Synonyms

Torulopsis peltata Yarrow (1968) Antonie van Leeuwenhoek *34*, 81 (basionym)
Selenotila peltata (Yarrow) Yarrow (1969b)
Selenozyma peltata (Yarrow) Yarrow (in von Arx et al., 1977)

Standard description of *Candida peltata*

Growth in glucose-yeast extract-peptone water: After 3 days at 25°C the cells are spherical to ovate, or lunate, (2.2–3.5) × (2.5–4.8) μm, single or in pairs. A sediment is present. After one month at 25°C a ring and thin pellicle may be present.

Growth on glucose-yeast extract-peptone agar: After one month at 25°C the streak culture is greyish-white, either dull with reticulate markings or glistening and mucoid.

Slide cultures on potato agar and corn meal agar: Neither septate nor pseudohyphae are formed.

Fermentation:

Glucose	+	Maltose	– or +s
Galactose	+s	Lactose	–
Sucrose	– or +s		

Assimilation of carbon compounds:

Galactose	+	Melezitose	+	Ribitol	+
L-Sorbose	+	Soluble starch	+s	Galactitol	+
Sucrose	+	D-Xylose	+	D-Mannitol	+
Maltose	+	L-Arabinose	+	D-Glucitol	+
Cellobiose	+	D-Arabinose	+	Salicin	+
Trehalose	+	D-Ribose	+	DL-Lactic acid	+
Lactose	–	L-Rhamnose	+	Succinic acid	+
Melibiose	–	Glycerol	+	Citric acid	+
Raffinose	–	Erythritol	+	Inositol	–

Splitting of arbutin: +

Assimilation of nitrate: –

Growth in vitamin-free medium: –

Growth at 37°C: +

Color reaction with DBB: –

Comments

The finding by Golubev (1977) that the perfect state of *Selenozyma intestinalis,* the type species of the genus *Selenozyma,* belongs in the genus *Metschnikowia* indicates that the shape of the cells, striking though it may be, is not an adequate criterion for recognizing a separate genus. Therefore, *Selenozyma peltata* is transferred to *Candida.*

Origin of the strains studied

Bovine mastitis (1) and a mucoid variant of this strain (1).

Type strain: CBS 5576, isolated from bovine mastitis (Yarrow, 1968).

137. *CANDIDA PHILYLA* (VAN DER WALT, VAN DER KLIFT ET SCOTT) MEYER ET YARROW (Yarrow & Meyer, 1978)

Synonyms

Torulopsis philyla van der Walt, van der Klift et Scott (van der Walt et al., 1971b)
Vanrija philyla (van der Walt, van der Klift et Scott) Moore (1980)

Standard description of *Candida philyla*

Growth in glucose-yeast extract-peptone water: After 3 days at 25°C the cells are ellipsoidal to ovoidal, (2.5–3.6) × (3.6–5.7) μm and occur singly or in pairs. A thin ring and sediment are present. After one month at 20°C a ring and sediment are present.

Growth on glucose-yeast extract-peptone agar: After one month at 25°C the streak culture is cream-colored, glistening, smooth, mucoid, and the margin is entire.

Dalmau plate culture on corn meal agar: No pseudomycelium is formed.

Fermentation: –

Assimilation of carbon compounds:

Galactose	–	Melezitose	–	Ribitol	+ s
L-Sorbose	–	Soluble starch	–	Galactitol	–
Sucrose	–	D-Xylose	– or + s	D-Mannitol	+
Maltose	–	L-Arabinose	–	D-Glucitol	+
Cellobiose	–	D-Arabinose	–	Salicin	–
Trehalose	+	D-Ribose	–	DL-Lactic acid	–
Lactose	–	L-Rhamnose	–	Succinic acid	+
Melibiose	–	Glycerol	–	Citric acid	+
Raffinose	–	Erythritol	–	Inositol	–

Splitting of arbutin: –

Assimilation of nitrate: –

Growth in vitamin-free medium: +

Growth at 37°C: –

Starch formation: –
Urease: +
Color reaction with DBB: +
G + C: 62.9 mol%, type strain (Meyer, Stenderup & Christiansen, unpublished)

Origin of the strain studied

Tunnels of *Xyleborus ferrugineus* F. infecting *Harpephyllum caffrum* Bernh. ex Krass near Mtunzini, Natal (1).

Type strain: CBS 6272.

138. CANDIDA PIGNALIAE (JACOB) MEYER ET YARROW (Yarrow & Meyer, 1978)

Synonym

Torulopsis pignaliae Jacob (1970)

Standard description of *Candida pignaliae*

Growth in glucose-yeast extract-peptone water: After 3 days at 25°C the cells are spheroidal, 3.6–7.2 μm and occur singly, in pairs, short chains and clusters. After one month there is a small sediment, no pellicle or ring.

Growth on glucose-yeast extract-peptone agar: After one month at 25°C the streak culture is white to cream-colored, moist, shiny convex and the margin is entire.

Dalmau plate culture on corn meal agar: Pseudomycelium absent.

Fermentation:

Glucose	+	Maltose	–
Galactose	–	Lactose	–
Sucrose	–	Raffinose	–

Assimilation of carbon compounds:

Galactose	–	Melezitose	–	Ribitol	+ or s
L-Sorbose	– or + s	Soluble starch	–	Galactitol	–
Sucrose	–	D-Xylose	+ or s	D-Mannitol	+
Maltose	–	L-Arabinose	+ or s	D-Glucitol	+
Cellobiose	–	D-Arabinose	v	Salicin	–
Trehalose	+	D-Ribose	–	DL-Lactic acid	–
Lactose	–	L-Rhamnose	–	Succinic acid	+
Melibiose	–	Glycerol	+ or s	Citric acid	–
Raffinose	–	Erythritol	–	Inositol	–

Additional carbon compound tested: Methanol +
Splitting of arbutin: –
Assimilation of nitrate: +
Growth in vitamin-free medium: –

Growth at 37°C: –

Color reaction with DBB: –

G + C: 42.7 mol%, type strain (Meyer, Stenderup & Christiansen, unpublished)

Origin of the strain studied

Tanning liquor (1).

Type strain: CBS 6071.

139. CANDIDA PINTOLOPESII (VAN UDEN) MEYER ET YARROW (Yarrow & Meyer, 1978)

This species has two varieties:

Candida pintolopesii (van Uden) Meyer et Yarrow var. *pintolopesii*

Candida pintolopesii (van Uden) Meyer et Yarrow var. *slooffiae* (van Uden et do Carmo-Sousa) Mendonça-Hagler et Phaff ex Meyer, Ahearn et Yarrow var. nov.

Perfect state: *Saccharomyces telluris* van der Walt

Synonyms of the variety *pintolopesii*

Torulopsis pintolopesii van Uden (1952)
Torulopsis bovina van Uden et do Carmo-Sousa (1957)
Candida bovina (van Uden et do Carmo-Sousa) van Uden et Vidal-Leiria (1970)

Synonym of the variety *slooffiae*

Candida slooffii van Uden et do Carmo-Sousa (1957)

Standard description of *Candida pintolopesii* var. *pintolopesii*

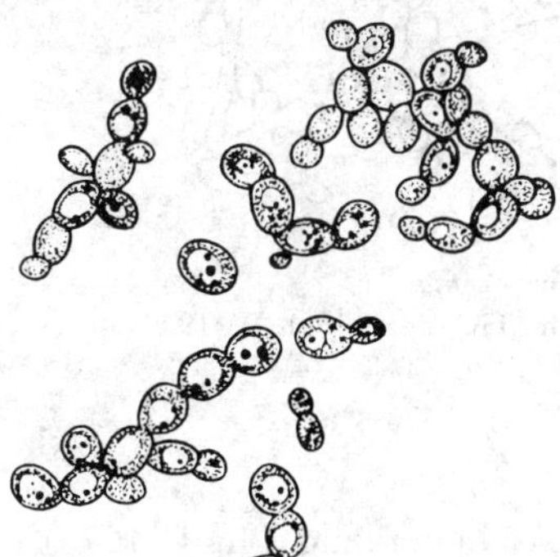

Fig. 311 – *C. pintolopesii* var. *pintolopesii* In glucose-yeast extract-peptone water (From 'The Yeasts', ed. II, 1970)

Growth in glucose-yeast extract-peptone water: After 24 hours at 37°C the cells are ovoid, (2.5–6) × (3–9) μm, single, in pairs and in groups (Fig. 311). Only a sediment is formed.

Growth on glucose-yeast extract-peptone agar: After one month at 37°C the streak culture is cream-colored, brownish, semidull to glossy, soft and smooth.

Dalmau plate culture on corn meal agar: Pseudomycelium may be absent or some strains may form branched chains of ovoid to elongate cells.

Fermentation:

Glucose	+	Maltose	–
Galactose	–	Lactose	–
Sucrose	–		

Assimilation of carbon compounds:

Galactose	–	Melezitose	–	Ribitol	–
L-Sorbose	–	Soluble starch	–	Galactitol	–
Sucrose	–	D-Xylose	–	D-Mannitol	–
Maltose	–	L-Arabinose	–	D-Glucitol	–
Cellobiose	–	D-Arabinose	–	Salicin	–
Trehalose	–	D-Ribose	–	DL-Lactic acid	v
Lactose	–	L-Rhamnose	–	Succinic acid	–
Melibiose	–	Glycerol	–	Citric acid	–
Raffinose	–	Erythritol	–	Inositol	–

Splitting of arbutin: –

Assimilation of nitrate: –

Growth in vitamin-free medium: –

Growth at 37°C: +; at 25°C: +

Color reaction with DBB: –

G + C: 32.4–33.9 mol%, 10 strains (Mendonça-Hagler & Phaff, 1975)

Supplementary description of *Candida pintolopesii* var. *slooffiae*

The variety differs from the species by having a minimal growth temperature of 28°C, by requiring inositol in Wickerham's assimilation medium and by forming more well-developed pseudomycelium (Fig. 312) (Mendonça-Hagler & Phaff, 1975).

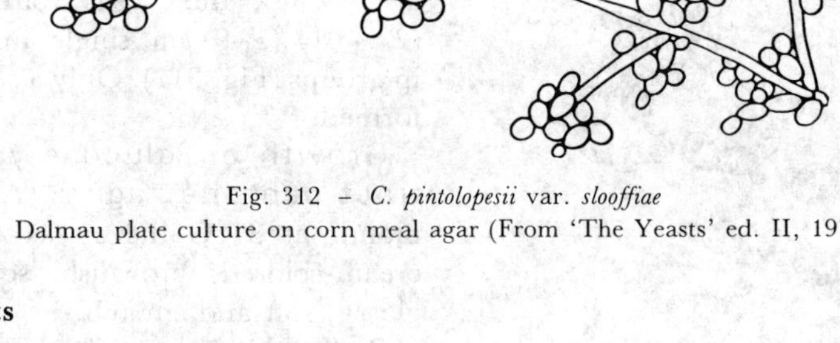

Fig. 312 – *C. pintolopesii* var. *slooffiae*
Dalmau plate culture on corn meal agar (From 'The Yeasts' ed. II, 1970)

Comments

By DNA reassociation experiments Mendonça-Hagler and Phaff (1975) showed that significant genetic relatedness existed between strains of *C. pintolopesii, C. bovina, C. slooffii* and *Saccharomyces telluris,* and that they represented a single taxon. *C. pinto-*

lopesii is the proper epithet and the imperfect state of *Sacch. telluris*. *C. bovina* is a synonym of *C. pintolopesii*, and *C. slooffii* because of its morphological and physiological differences is considered a variety.

Origin of the strains belonging to the variety *pintolopesii*

Liver and spleen of white mice in Portugal (van Uden, 1952) (3); peritoneal cavity of a rat in Uruguay (Artagaveytia-Allende, 1953) (1); peritoneal cavity of a mouse inoculated with soil in USA (1); cecal contents of a rat in Israel (Aschner et al., 1954) (1); intestinal contents of a chinchilla in USA (1); intestinal contents of pigeons in USA (4); feces of a rat (1); an unspecified domestic animal in Germany (1); origin unknown (1); bovine cecal contents in Portugal (1); cecal contents of a horse (1); cecal contents of a pig (1); diseased crop of a turkey (1); digestive tract of an unspecified domestic animal in Germany (1).

Type strain: CBS 1787, from the liver of a white mouse.

Origin of the strains belonging to the variety *slooffiae*

Digestive tract of horses (van Uden & do Carmo-Sousa, 1957) (6); digestive tract of a pig (van Uden et al., 1958) (1).

Type strain: CBS 2419, isolated from a horse.

140. *CANDIDA PINUS* (LODDER ET KREGER-VAN RIJ) MEYER ET YARROW
(Yarrow & Meyer, 1978)

Synonyms

Torulopsis pinus Lodder et Kreger-van Rij (1952)
Paratorulopsis pinus (Lodder et Kreger-van Rij) Novák et Zsolt (1961)

Standard description of *Candida pinus*

Growth in glucose-yeast extract-peptone water: After 3 days at 25°C the cells are globose, 3–4 μm (Fig. 313). Only a sediment is formed (van Uden & Vidal-Leiria, 1970).

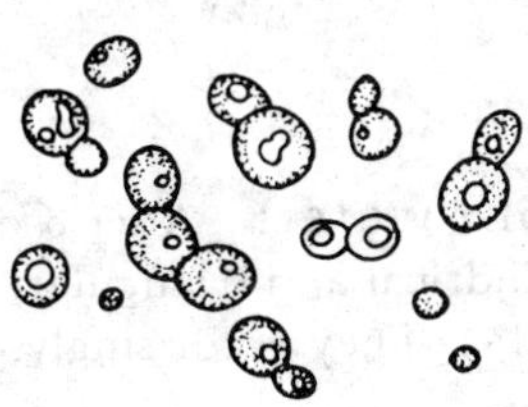

Fig. 313 – *C. pinus*
In glucose-yeast extract-peptone water
(From 'The Yeasts', ed. II, 1970)

Growth on glucose-yeast extract-peptone agar: After one month at 25°C the streak culture is yellowish, glossy, soft and mucous, smooth (van Uden & Vidal-Leiria, 1970).

Dalmau plate culture on corn meal agar: No pseudomycelium is formed.

Fermentation: –

Assimilation of carbon compounds:

Galactose	–	Melezitose	–	Ribitol	+
L-Sorbose	–	Soluble starch	–	Galactitol	–
Sucrose	–	D-Xylose	+	D-Mannitol	+
Maltose	–	L-Arabinose	–	D-Glucitol	+
Cellobiose	–	D-Arabinose	–	Salicin	–
Trehalose	+	D-Ribose	+	DL-Lactic acid	–
Lactose	–	L-Rhamnose	–	Succinic acid	–
Melibiose	–	Glycerol	+	Citric acid	+
Raffinose	–	Erythritol	+	Inositol	–

Splitting of arbutin: –

Assimilation of nitrate: –

Growth in vitamin-free medium: –

Growth at 37°C: –

Color reaction with DBB: –

Coenzyme Q system: Q 7 (Yamada & Kondo, 1972)

G + C: 37.3 mol%, one strain (Nakase & Komagata, 1971c)

Origin of the strain studied

Pine tree in Sweden (1).

Type strain: CBS 970.

141. *CANDIDA POLYMORPHA* OHARA ET NONOMURA EX MEYER ET AHEARN (1983)

Standard description of *Candida polymorpha*

Growth in glucose-yeast extract-peptone water: After 3 days at 25°C the cells are diverse in shape. Globose, ovoid, cylindrical and triangular cells are present as well as others with irregular shapes (Fig. 314). They occur singly, in pairs and in short chains. A sediment is present.

Growth on glucose-yeast extract-peptone agar: After one month at 25°C the streak culture is white, delicately wrinkled, lacy, and soft, the margin is lobed.

Dalmau plate culture on corn meal agar: The pseudomycelium consists of branched chains of cells bearing ovoid blastospores in short chains.

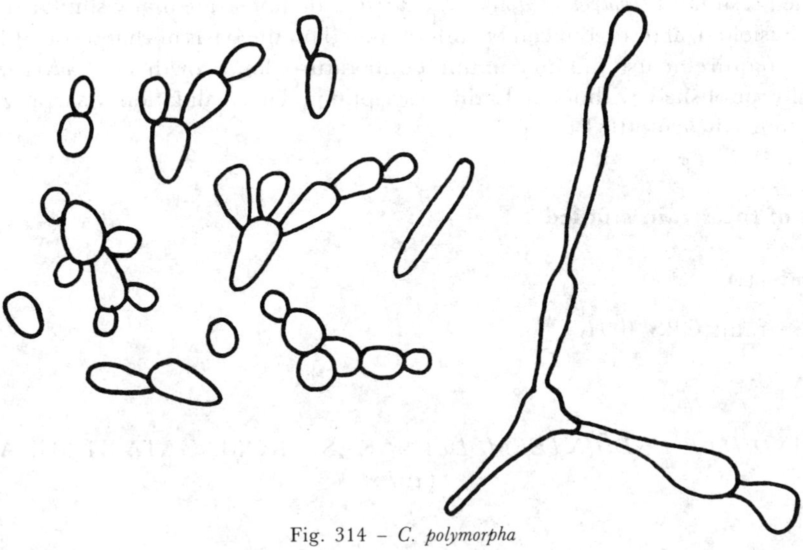

Fig. 314 – *C. polymorpha*
In glucose-yeast extract-peptone water

Fermentation:

Glucose	– or +w	Maltose	–
Galactose	–	Lactose	–
Sucrose	–		

Assimilation of carbon compounds:

Galactose	+	Melezitose	+	Ribitol	+
L-Sorbose	+	Soluble starch	–	Galactitol	–
Sucrose	–	D-Xylose	+	D-Mannitol	+s
Maltose	+	L-Arabinose	+	D-Glucitol	+s
Cellobiose	+	D-Arabinose	–	Salicin	+
Trehalose	+s	D-Ribose	+	DL-Lactic acid	–
Lactose	–	L-Rhamnose	+	Succinic acid	+
Melibiose	–	Glycerol	+	Citric acid	+s
Raffinose	–	Erythritol	+	Inositol	–

Additional carbon compound tested: Glucono-δ-lactone –
Splitting of arbutin: +
Assimilation of nitrate: –
Growth in vitamin-free medium: –
Growth at 37°C: –
Urease: –
Color reaction with DBB: –
G + C: 36.6 mol%, type strain (Meyer & Phaff, 1972)

Comments

Van Uden & Buckley (1970) considered *C. polymorpha* a synonym of *C. diddensiae.* DNA reassociation experiments demonstrated insignificant genome homology for these yeasts to be representatives of the same species (Meyer & Simione, 1978 and unpublished results). *C. polymorpha* and *C. diddensiae* demonstrate many similarities in the usual physiological tests, but can be differentiated on the basis of rhamnose utilization, vitamin requirements and maximum temperature for growth. *C. polymorpha* was originally published without a Latin description. This validation was provided by Meyer and Ahearn (1983).

Origin of the strain studied

Soya mash (1).

Type strain: CBS 4094.

142. CANDIDA PSEUDOINTERMEDIA NAKASE, KOMAGATA ET FUKAZAWA (1976)

Standard description of *Candida pseudointermedia*

Growth in glucose-yeast extract-peptone water: After 3 days at 25°C cells are globose, short-oval to long-oval, (3.6–5.1) × (4.3–7.2) µm, and occur singly, in pairs or in chains (Fig. 315). Pseudomycelial growth is evident. A pellicle is present. After one month a sediment and pellicle are present.

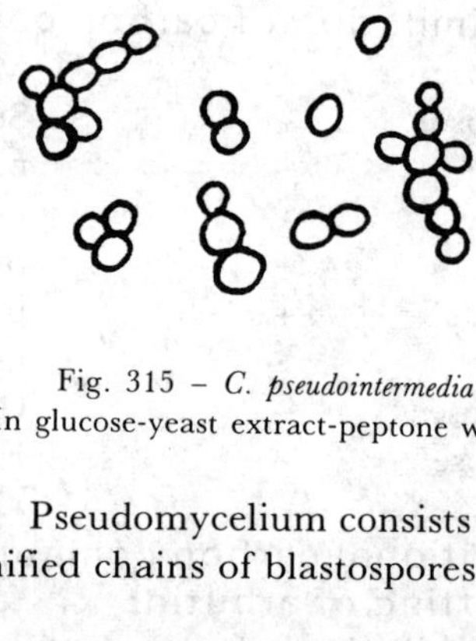

Fig. 315 – *C. pseudointermedia*
In glucose-yeast extract-peptone water

Growth on glucose-yeast extract-peptone agar: After one month at 25°C the streak culture is white, dry, dull, yellowish, raised, mostly smooth with some wrinkled areas and fringed with mycelium.

Dalmau plate culture on corn meal agar: Pseudomycelium consists of tree-like branched chains of pseudohyphae bearing ramified chains of blastospores in verticillate positions.

Fermentation:

Glucose	+	Maltose	+ w or s
Galactose	+	Lactose	–
Sucrose	+	Raffinose	–

Assimilation of carbon compounds:

Galactose	+	Melezitose	+	Ribitol	+ or s
L-Sorbose	+	Soluble starch	+	Galactitol	v
Sucrose	+	D-Xylose	+ or s	D-Mannitol	+
Maltose	+	L-Arabinose	− or +s	D-Glucitol	+
Cellobiose	+	D-Arabinose	v	Salicin	+
Trehalose	+	D-Ribose	+ or s	DL-Lactic acid	−
Lactose	−	L-Rhamnose	+s or vw	Succinic acid	+
Melibiose	−	Glycerol	v	Citric acid	+
Raffinose	+	Erythritol	−	Inositol	−

Additional carbon compound tested: Inulin −
Splitting of arbutin: +
Assimilation of nitrate: −
Growth in vitamin-free medium: −
Growth on 50% (w/w) glucose-yeast extract agar: −
Growth at 37°C: v
Color reaction with DBB: −
G + C: 44.4–44.9 mol%, 2 strains (Nakase et al., 1976)

Comments

The original description reported the ability to ferment raffinose. This could not be confirmed.

Origin of the strain studied

"Kamaboko", a fish-paste product in Japan (1).

Type strain: CBS 6918.

143. CANDIDA PSYCHROPHILA (GOTO, SUGIYAMA ET IIZUKA) MEYER ET YARROW (Yarrow & Meyer, 1978)

Synonym

Torulopsis psychrophila Goto, Sugiyama et Iizuka (1969)

Standard description of *Candida psychrophila*

Growth in malt extract: After 3 days at 10°C the cells are mostly globose, 2.9–4.3 μm and few are slightly ovoidal. They occur mostly in small clusters of four to six cells. Single and budding cells are also present. A sediment is formed.

Growth on malt agar: After one month at 10°C the streak culture is creamy, shiny, mucoid, raised, and entire (Goto et al., 1969).

Dalmau plate culture on corn meal agar: Pseudomycelium is absent. However, Goto et al. (1969) reported primitive pseudomycelium in the original description.

Fermentation (at 15°C): –

Assimilation of carbon compounds (at 15°C):

Galactose	+	Melezitose	–	Ribitol	+
L-Sorbose	+ s or –	Soluble starch	–	Galactitol	–
Sucrose	–	D-Xylose	+	D-Mannitol	+
Maltose	–	L-Arabinose	+	D-Glucitol	+
Cellobiose	–	D-Arabinose	–	Salicin	–
Trehalose	+	D-Ribose	–	DL-Lactic acid	–
Lactose	–	L-Rhamnose	–	Succinic acid	–
Melibiose	–	Glycerol	+	Citric acid	–
Raffinose	–	Erythritol	+	Inositol	–

Splitting of arbutin: –

Assimilation of nitrate: –

Growth in vitamin-free medium: –

Growth at 37°C: –; **at 25°C:** –; **at 15°C:** +

Color reaction with DBB: –

G + C: 35.9 mol%

Origin of the strain studied

Penguin dung on Ross Island, Antarctica (1).

Type strain: CBS 5956.

144. *CANDIDA PULCHERRIMA* (LINDNER) WINDISCH

see *Metschnikowia pulcherrima* p. 273

145. *CANDIDA PUSTULA* (BUHAGIAR) MEYER ET YARROW

(Yarrow & Meyer, 1978)

Synonym

Torulopsis pustula Buhagiar (1975)

Standard description of *Candida pustula*

Growth in glucose-yeast extract-peptone water: After 3 days at 25°C the cells are oval, long-oval to cylindrical, and pear-shaped, (3.4–5.6) × (8.4–10.6) μm (Fig. 316). After one month a ring and sediment are present.

Growth on glucose-yeast extract-peptone agar: After one month at 25°C the streak is white to tan-white, mucoid, glossy, smooth, convex with an entire margin.

Dalmau plate culture on corn meal agar: Branched chains of long, ovoid cells are formed with short chains of blastospores.

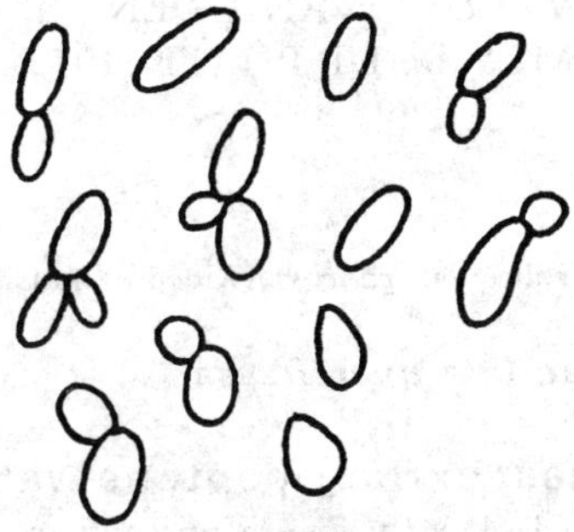

Fig. 316 – *C. pustula*
In glucose-yeast extract-peptone water

Fermentation: –

Assimilation of carbon compounds:

Galactose	– or + s	Melezitose	–	Ribitol	+
L-Sorbose	+	Soluble starch	–	Galactitol	–
Sucrose	–	D-Xylose	+	D-Mannitol	+
Maltose	–	L-Arabinose	+	D-Glucitol	+
Cellobiose	+	D-Arabinose	+	Salicin	+
Trehalose	+	D-Ribose	+ s	DL-Lactic acid	–
Lactose	–	L-Rhamnose	–	Succinic acid	+
Melibiose	–	Glycerol	+ s	Citric acid	+
Raffinose	–	Erythritol	–	Inositol	–

Splitting of arbutin: +

Assimilation of nitrate: +

Growth in vitamin-free medium: –

Growth on 50% (w/w) glucose-yeast extract agar: –

Growth at 37°C: –

Color reaction with DBB: +

Urease: +

G + C: 51.0 mol%, type strain

Origin of the strains studied

Black currants (2).

Type strain: CBS 6527.

146. *CANDIDA QUERCITRUSA* (VAN UDEN ET DO CARMO-SOUSA) MEYER ET PHAFF (1972)

Synonym

Candida parapsilosis Langeron et Talice var. *querci* van Uden et do Carmo-Sousa (1959)

Standard description of *Candida quercitrusa*

Growth in glucose-yeast extract-peptone water: After 3 days at 25°C the cells are globose to oval (3–4) × (5–7) μm and occur singly, in pairs and in short chains and clusters. Long-oval and cylindrical cells are also present.

Growth on glucose-yeast extract-peptone agar: After one month at 25°C the streak culture is white, creamy, soft, smooth and delicately wrinkled.

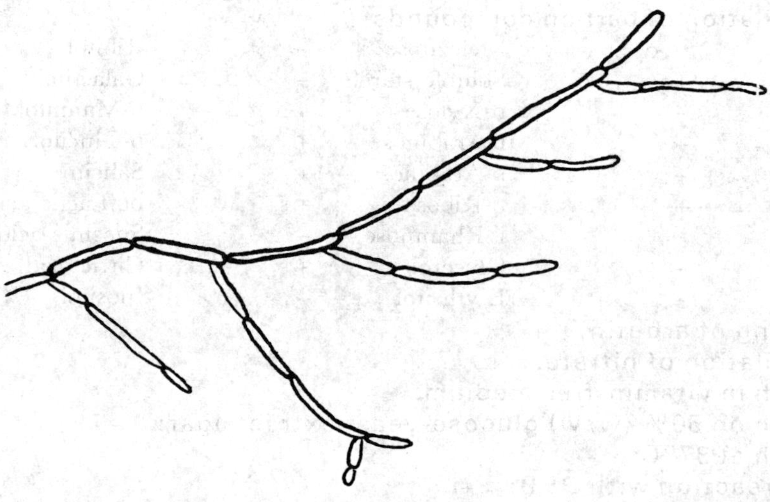

Fig. 317 – *C. quercitrusa*
Dalmau plate culture on corn meal agar

Dalmau plate culture on corn meal agar: Pseudomycelium consists of ramified chains of oval to elongate cells (Fig. 317).

Fermentation:

Glucose	+	Maltose	–
Galactose	–	Lactose	–
Sucrose	–		

Assimilation of carbon compounds:

Galactose	+	Melezitose	+	Ribitol	–
L-Sorbose	+	Soluble starch	–	Galactitol	–
Sucrose	+	D-Xylose	+	D-Mannitol	+

Maltose	+	L-Arabinose	−	D-Glucitol	+
Cellobiose	−	D-Arabinose	−	Salicin	−
Trehalose	+	D-Ribose	−	DL-Lactic acid	−
Lactose	−	L-Rhamnose	−	Succinic acid	−
Melibiose	−	Glycerol	+	Citric acid	−
Raffinose	−	Erythritol	−	Inositol	−

Additional carbon compound tested: D-Glucosamine · HCl +
Splitting of arbutin: −
Assimilation of nitrate: −
Growth in vitamin-free medium: −
Growth at 37°C: +
Urease: −
Color reaction with DBB: −
G + C: 41.5 mol%, type strain (Meyer & Phaff, 1972)

Comments

Meyer and Phaff (1972) reported that the type strains of *C. parapsilosis* var. *querci* and *C. parapsilosis* demonstrated insignificant DNA reassociation to be considered members of the same species and recommended that *C. parapsilosis* var. *querci* be elevated to the level of species. Physiological differences include inability to grow in L-arabinose and at 37°C.

Origin of the strain studied

Insect frass on an oak tree (1).

Type strain: CBS 4214.

147. *CANDIDA QUERCUUM* NAKASE (1971b)

Standard description of *Candida quercuum*

Growth in glucose-yeast extract-peptone water: After 3 days at 25°C the cells are ovoid to elongate, (1.4–3.6) × (2.9–5.8) μm, and occur singly and budding. Primitive pseudomycelium is present.

Growth on glucose-yeast extract-peptone agar: After one month at 25°C the streak culture is beige-white, dry, dull, smooth and entire.

Dalmau plate culture on corn meal agar: The pseudomycelium consists of ramified chains of long pseudohyphae bearing round to ovoid blastospores, some in verticillate positions.

Fermentation:

Glucose	+ w, s	Maltose	−
Galactose	−	Lactose	−
Sucrose	−		

Assimilation of carbon compounds:

Galactose	–	Melezitose	+	Ribitol	–
L-Sorbose	–	Soluble starch	–	Galactitol	–
Sucrose	+	D-Xylose	+	D-Mannitol	+
Maltose	+	L-Arabinose	–	D-Glucitol	+
Cellobiose	+	D-Arabinose	–	Salicin	+
Trehalose	+	D-Ribose	–	DL-Lactic acid	+
Lactose	–	L-Rhamnose	–	Succinic acid	+
Melibiose	–	Glycerol	+	Citric acid	+
Raffinose	–	Erythritol	–	Inositol	–

Splitting of arbutin: +
Assimilation of nitrate: –
Growth in vitamin-free medium: –
Growth on 50% (w/w) glucose-yeast extract agar: –
Growth at 37°C: +
Color reaction with DBB: –
G + C: 38.3 mol%, type strain (Nakase, 1971b)

Origin of the strain studied

An oak tree (1).

Type strain: CBS 6422.

148. CANDIDA REUKAUFII (GRÜSS) DIDDENS ET LODDER

see *Metschnikowia reukaufii* p. 275

149. CANDIDA RHAGII (DIDDENS ET LODDER) JURZITZA, KÜHLWEIN ET KREGER-VAN RIJ (1960)

Synonym

Candida tropicalis (Cast.) Berkh. var. *rhagii* Diddens et Lodder (1942)

Standard description of *Candida rhagii*

Growth in glucose-yeast extract-peptone water: After 3 days at 25°C the cells are globose to short-ovoid, (2.5–4) × (3.5–5) μm (Fig. 318); cylindrical and flask-shaped cells, measuring (2–3.5) × (6–10) μm, occur in some strains (Fig. 319) (van Uden & Buckley, 1970).

Growth on glucose-yeast extract-peptone agar: After one month at 25°C the streak culture is cream-colored, semidull or glistening, soft, smooth or slightly warty (van Uden & Buckley, 1970).

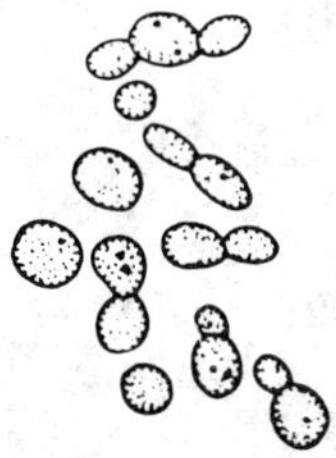

Fig. 318 – *C. rhagii*
In glucose-yeast extract-peptone water
(From 'The Yeasts' ed. II, 1970)

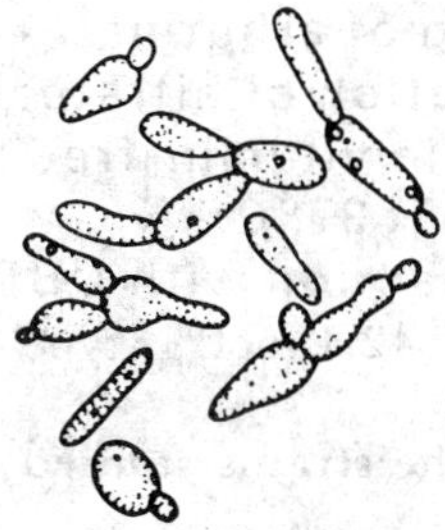

Fig. 319 – *C. rhagii*
In glucose-yeast extract-peptone water
(From 'The Yeasts' ed. II, 1970)

Dalmau plate culture on corn meal agar: The pseudomycelium consists of ramified chains of long pseudohyphae bearing groups of oval blastospores in verticillated positions, as well as single or paired blastospores in other sites (Fig. 320) (van Uden & Buckley, 1970).

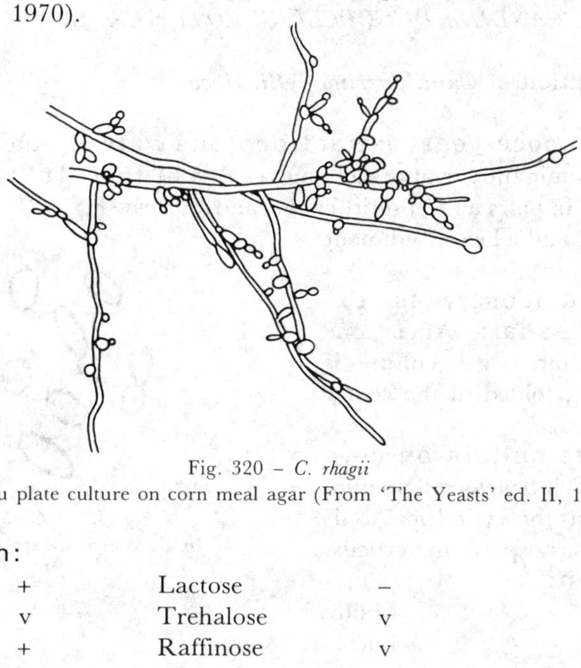

Fig. 320 – *C. rhagii*
Dalmau plate culture on corn meal agar (From 'The Yeasts' ed. II, 1970)

Fermentation:

Glucose	+	Lactose	–
Galactose	v	Trehalose	v
Sucrose	+	Raffinose	v
Maltose	–		

Assimilation of carbon compounds:

Galactose	+	Melezitose	+	Ribitol	+
L-Sorbose	–	Soluble starch	–	Galactitol	–
Sucrose	+	D-Xylose	+	D-Mannitol	+
Maltose	+	L-Arabinose	v	D-Glucitol	+
Cellobiose	+	D-Arabinose	–	Salicin	+
Trehalose	+	D-Ribose	+	DL-Lactic acid	v
Lactose	–	L-Rhamnose	–	Succinic acid	v
Melibiose	–	Glycerol	+	Citric acid	–
Raffinose	+	Erythritol	+	Inositol	–

Splitting of arbutin: +
Assimilation of nitrate: –
Growth in vitamin-free medium: v
Growth at 37°C: v
Color reaction with DBB: –
G + C: 42.4 mol%, type strain (Meyer & Phaff, 1972)

Origin of the strains studied

Insects: *Harpium (Rhagium) inquisitor* (Jurzitza et al., 1960) (2); *Rhagium* sp. (Müller, 1934) (1); *Gaurotes virginea* (1).

Type strain: CBS 4237, isolated from *Harpium (Rhagium) inquisitor* in Germany (Jurzitza et al., 1960).

150. CANDIDA RUGOPELLICULOSA NAKASE (1971c)

Standard description of *Candida rugopelliculosa*

Growth in glucose-yeast extract-peptone water: After 3 days at 25°C the cells are predominantly oval to elongate, (2.8–5.6) × (5.6–11.2) μm, some round, and occur singly, in pairs and in short chains and clusters (Fig. 321). A thick, wrinkled, folded pellicle and a heavy sediment are formed.

Growth on glucose-yeast extract-peptone agar: After one month at 25°C the streak culture is white to pale tan, folded in the center with a mycelial edge.

Dalmau plate culture on corn meal agar: Well-developed pseudomycelium consists of cylindrical cells with clusters of blastospores in verticils.

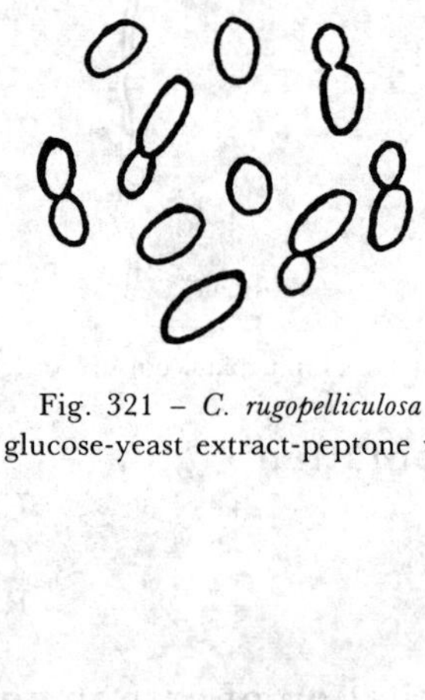

Fig. 321 – *C. rugopelliculosa*
In glucose-yeast extract-peptone water

Fermentation:

Glucose	+	Maltose	–
Galactose	–	Lactose	–
Sucrose	–		

Assimilation of carbon compounds:

Galactose	–	Melezitose	–	Ribitol	–
L-Sorbose	–	Soluble starch	–	Galactitol	–
Sucrose	–	D-Xylose	–	D-Mannitol	–
Maltose	–	L-Arabinose	–	D-Glucitol	–
Cellobiose	–	D-Arabinose	–	Salicin	–
Trehalose	–	D-Ribose	–	DL-Lactic acid	+
Lactose	–	L-Rhamnose	–	Succinic acid	+
Melibiose	–	Glycerol	–	Citric acid	–
Raffinose	–	Erythritol	–	Inositol	–

Additional carbon compound tested: D-Glucosamine · HCl +
Splitting of arbutin: –
Assimilation of nitrate: –
Growth in vitamin-free medium: +
Growth on 50% (w/w) glucose-yeast extract agar: –
Growth at 37°C: +
Color reaction with DBB: –
G + C: 30 mol%, one strain (Nakase, 1971c)

Comments

Physiologically similar to *C. krusei* and *C. citrea*, *C. rugopelliculosa* differs from them by its inability to assimilate glycerol; also, from *C. citrea* by its ability to assimilate D-glucosamine, to grow at 37°C and in vitamin-free medium. Furthermore, *C. rugopelliculosa* has a G+C% lower than *C. krusei* (40%) and *C. citrea* (33%).

Origin of the strain studied

Soybean protein factory (1).

Type strain: CBS 6377.

151. CANDIDA RUGOSA (ANDERSON) DIDDENS ET LODDER (1942)

Synonyms

Mycoderma rugosa Anderson (1917)
Azymocandida rugosa (Anderson) Novák et Zsolt (1961)
Candida rugosa (Anderson) Diddens et Lodder var. *elegans* nom. nud. (Dietrichson, 1954)

Standard description of *Candida rugosa*

Growth in glucose-yeast extract-peptone water: After 3 days at 25°C the cells are ovoid to elongate, sometimes sausage-shaped or curved, (2.5–4) × (7–15) μm (Fig. 322). Islets or a pellicle are usually present (van Uden & Buckley, 1970).

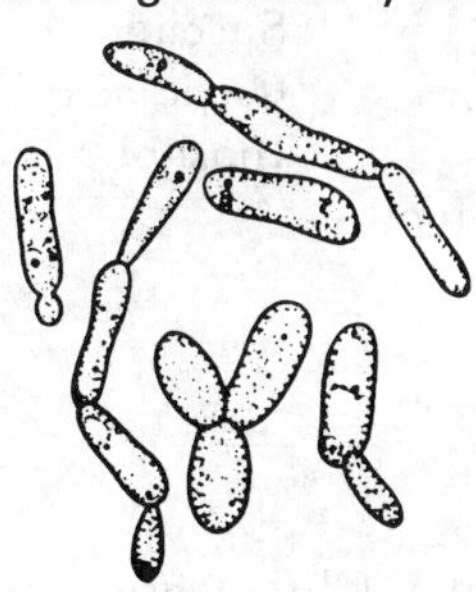

Fig. 322 – *C. rugosa*
In glucose-yeast extract-peptone water
(From 'The Yeasts' ed. II, 1970)

Growth on glucose-yeast extract-peptone agar: After one month at 25°C the streak culture is variable in appearance. The color ranges from dirty white to greyish-yellow; the surface is wrinkled in most strains but may be partly or wholly smooth in others. The wrinkled strains are dull with a moist appearance; the partly or wholly smooth strains are dull-glistening (van Uden & Buckley, 1970).

Dalmau plate culture on corn meal agar: A primitive much branched pseudomycelium is abundantly formed. It may exclusively consist of short pseudohyphae. Blastospores, if formed at all, are scarce (Fig. 323) (van Uden & Buckley, 1970).

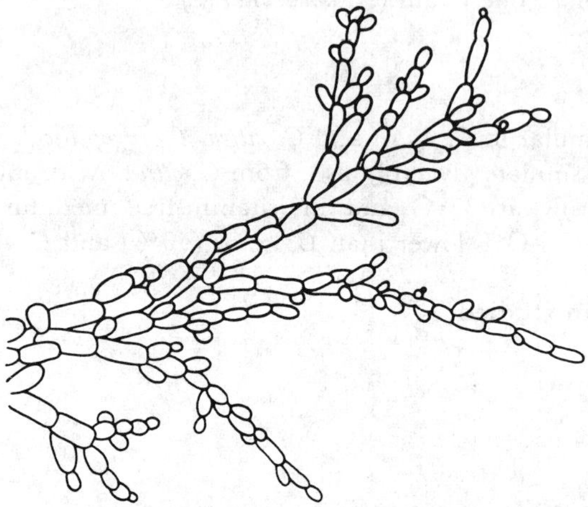

Fig. 323 – *C. rugosa*
Dalmau plate culture on corn meal agar (From 'The Yeasts' ed. II, 1970)

Fermentation: –

Assimilation of carbon compounds:

Galactose	+	Melezitose	–	Ribitol	v
L-Sorbose	v	Soluble starch	–	Galactitol	–
Sucrose	–	D-Xylose	v	D-Mannitol	+
Maltose	–	L-Arabinose	v	D-Glucitol	+
Cellobiose	–	D-Arabinose	v	Salicin	–
Trehalose	–	D-Ribose	–	DL-Lactic acid	+
Lactose	–	L-Rhamnose	–	Succinic acid	+
Melibiose	–	Glycerol	v	Citric acid	v
Raffinose	–	Erythritol	–	Inositol	–

Additional carbon compound tested: Methanol –

Splitting of arbutin: –

Assimilation of nitrate: –

Growth in vitamin-free medium: –

Growth at 37°C: +

Color reaction with DBB: –

G + C: 52.1 mol%, type strain (Meyer, Stenderup & Christiansen, unpublished); 48.3–50.7 mol%, 3 strains (Nakase & Komagata, 1971f)

Origin of the strains studied (van Uden & Buckley, 1970)

Human feces (2); stale butter (3); a margarine factory (1); bovine droppings (1); seawater off San Diego, California (1); sputum (1); uncertain origin (2).

Type strain: CBS 613, isolated from human feces (Anderson, 1917).

152. CANDIDA SAKE (SAITO ET OTA) VAN UDEN ET BUCKLEY (1970)

Synonyms

Torula lambica Kufferath (1925)
Mycotorula lambica Harrison (1928) [nec *Mycoderma lambica* Lindner et Genoud (1913)]
Candida tropicalis (Cast.) Berkhout var. *lambica* (Harrison) Diddens et Lodder (1942)
Eutorulopsis sake Saito et Ota (1934)
Torulopsis sake (Saito et Ota) Lodder et Kreger-van Rij (1952)
Candida natalensis van der Walt et Tscheuschner (1957a)
Candida vanriji Capriotti (1958f)
Candida salmonicola Komagata et Nakase (1965)
Candida australis Goto, Sugiyama et Iizuka (1969)

Standard description of *Candida sake*

Growth in glucose-yeast extract-peptone water: After 3 days at 25°C the cells vary in size and shape. Globose to slightly oval cells predominate. Others are long oval to cylindrical. They occur singly, in pairs and in short chains. Pseudomycelium is evident. After one month a pellicle and sediment may be present.

Growth on glucose-yeast extract-peptone agar: After one month at 25°C the streak culture is off-white to cream-colored, glistening, soft, smooth or lacy to rough and folded.

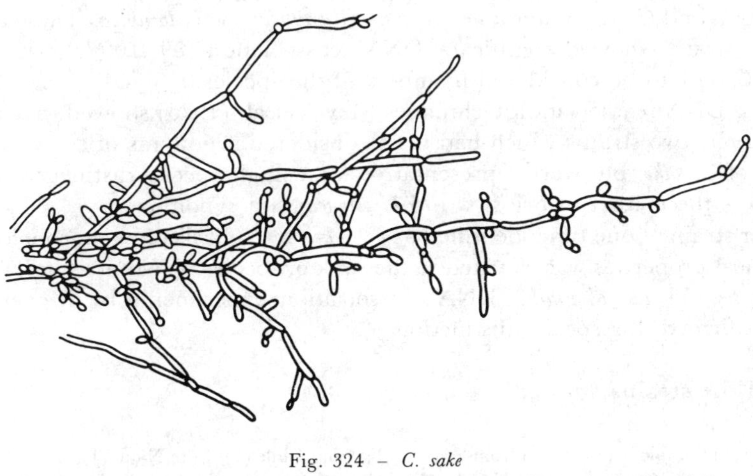

Fig. 324 – *C. sake*
Dalmau plate culture on corn meal agar (From 'The Yeasts' ed. II, 1970)

Dalmau plate culture on corn meal agar: The pseudomycelium varies from branched chains of cells with a few blastospores to well-developed pseudohyphae with many blastospores and long, curved cells with few blastospores, giving a wavy appearance (Fig. 324).

Fermentation:

Glucose	+	Maltose	v
Galactose	+ s or w	Lactose	–
Sucrose	v		

Assimilation of carbon compounds:

Galactose	+	Melezitose	v	Ribitol	v
L-Sorbose	+	Soluble starch	–	Galactitol	–
Sucrose	+	D-Xylose	+	D-Mannitol	+
Maltose	+	L-Arabinose	–	D-Glucitol	+
Cellobiose	v	D-Arabinose	–	Salicin	+ or s
Trehalose	+	D-Ribose	– or + s	DL-Lactic acid	v
Lactose	–	L-Rhamnose	–	Succinic acid	+
Melibiose	–	Glycerol	+ or s	Citric acid	v
Raffinose	–	Erythritol	–	Inositol	–

Splitting of arbutin: + or s

Assimilation of nitrate: –

Growth in vitamin-free medium: v

Growth at 37°C: –

Color reaction with DBB: –

Growth in the presence of 100 ppm cycloheximide: –

Coenzyme Q system: Q 9 (Yamada & Kondo, 1972)

G + C: 37.8–41.0 mol%, 10 strains (Meyer et al., 1975)

Comments

Strains of *C. sake* include the previously described species *C. australis, C. natalensis, C. salmonicola* and *C. vanriji* and one variety, *C. tropicalis* var. *lambica.* These and seven additional strains showed significant DNA reassociation (89–100%) with the type strain of *C. sake* to be considered members of the species.

By using DNA reassociation techniques Meyer et al. (1975) showed that *C. maltosa* and *C. cloacae,* two strains which had been considered synonyms of *C. sake,* were not members of *C. sake,* but were representatives of a single species distinct from *C. sake. C. maltosa* is the epithet of priority and *C. cloacae* is a synonym.

Another strain at one time identified as *C. sake* on the basis of its morphological and physiological properties was elevated to the rank of species by Montrocher (1967) and described as *Candida oleophila.* DNA reassociation experiments by Meyer (unpublished) confirmed this species distinction.

Origin of the strains studied

Sake-moto (1), lambic beer (1), soil from Sweden (1), from Chile (1), from Natal (1); frozen salmon stick (1), water from the Chesapeake Bay, U.S.A. (1), sediment from the Chesapeake Bay, U.S.A. (1), Lake Bonney (1), stream (1), grape juice (1), beech sap (1), unknown source (1).

Type strain: CBS 159, isolated in Japan from "sake-moto" yeast and described as *Eutorulopsis sake* (Saito & Ota, 1934).

153. CANDIDA SALMANTICENSIS (SANTA MARÍA) VAN UDEN ET BUCKLEY (1970)

Synonym

Torulopsis salmanticensis Santa María (1963a)

Standard description of *Candida salmanticensis*

Growth in glucose-yeast extract-peptone water: After 3 days at 25°C the cells are ovoid, (2.5–5) × (5–9) μm; occurring mostly in groups (Fig. 325) (van Uden & Buckley, 1970).

Growth on glucose-yeast extract-peptone agar: After one month at 25°C the streak culture is whitish, cream-colored, semidull, soft and smooth (van Uden & Buckley, 1970).

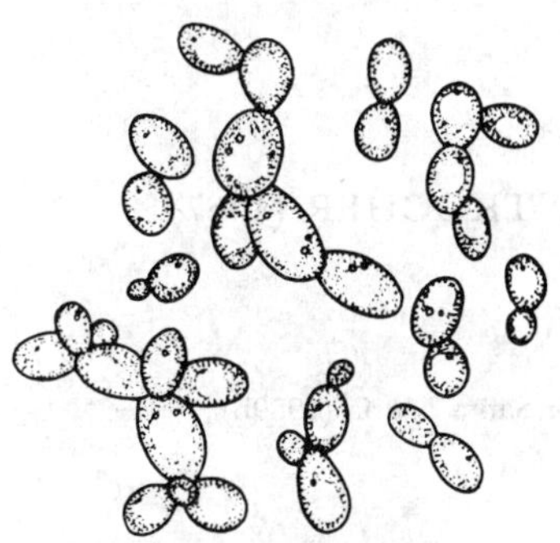

Fig. 325 – *C. salmanticensis*
In glucose-yeast extract-peptone water
(From 'The Yeasts' ed. II, 1970)

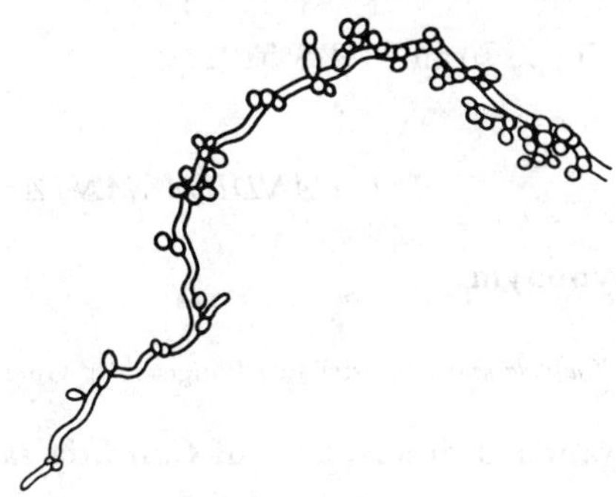

Fig. 326 – *C. salmanticensis*
Dalmau plate culture on corn meal agar
(From 'The Yeasts' ed. II, 1970)

Dalmau plate culture on corn meal agar: The pseudomycelium consists of a sparsely branched formation of wavy pseudohyphae bearing chains and clusters of ovoid blastospores (Fig. 326) (van Uden & Buckley, 1970).

Fermentation:

Glucose	+	Lactose	–	Cellobiose	–
Galactose	+ s	Raffinose	+	Melezitose	+
Sucrose	+	Melibiose	+		
Maltose	+	Trehalose	+		

Assimilation of carbon compounds:

Galactose	+	Melezitose	+	Ribitol	+ s
L-Sorbose	+	Soluble starch	+ s	Galactitol	–
Sucrose	+	D-Xylose	+	D-Mannitol	+
Maltose	+	L-Arabinose	–	D-Glucitol	+
Cellobiose	+	D-Arabinose	+	Salicin	+
Trehalose	+	D-Ribose	–	DL-Lactic acid	+
Lactose	+ s or –	L-Rhamnose	–	Succinic acid	+
Melibiose	+	Glycerol	–	Citric acid	+ s
Raffinose	+	Erythritol	–	Inositol	–

Additional carbon compounds tested: 2-Ketogluconate +, D-glucosamine · HCl +

Splitting of arbutin: +

Assimilation of nitrate: –

Growth in vitamin-free medium: –

Growth at 37°C: v

Urease: –

Color reaction with DBB: –

G + C: 45.5 mol%, type strain (Meyer, Stenderup & Christiansen, unpublished)

Origin of the strain studied

Alpechin (Santa María, 1963a) (1).

Type strain: CBS 5121.

154. *CANDIDA SANTAMARIAE* MONTROCHER (1967)

Synonym

Candida krusei (Castellani) Langeron et Guerra var. *saccharicola* Santa María (1959b)

Standard description of *Candida santamariae*

Growth in glucose-yeast extract-peptone water: After 3 days at 25°C the cells are ovoid, long-ovoid to elongate, (2.8–6.7) × (5.1–11.2) μm occurring in groups or in chains. Some small ovoid and cylindrical cells are also present (Fig. 327).

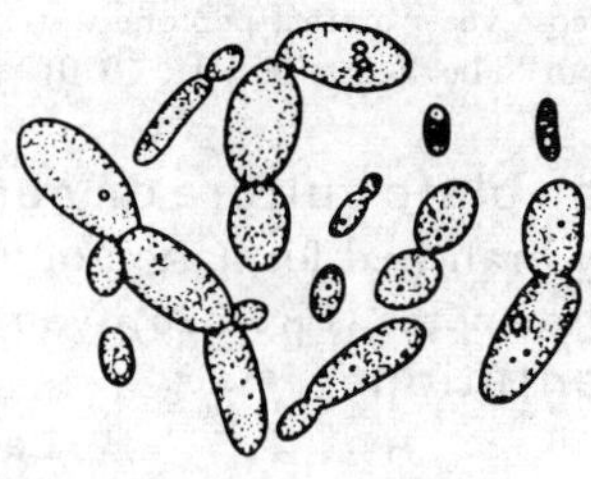

Fig. 327 – *C. santamariae*
In glucose-yeast extract-peptone water
(From 'The Yeasts' ed. II, 1970)

Growth on glucose-yeast extract-peptone agar: After one month at 25°C the streak culture is tan, creamy, glistening, soft, slightly wrinkled, and the margin is fringed with mycelium.

Dalmau plate culture on corn meal agar: The pseudomycelium consists of long slender branched chains of cylindrical, often slightly curved, cells with single and short-chained ovoid blastospores occurring laterally (Fig. 328).

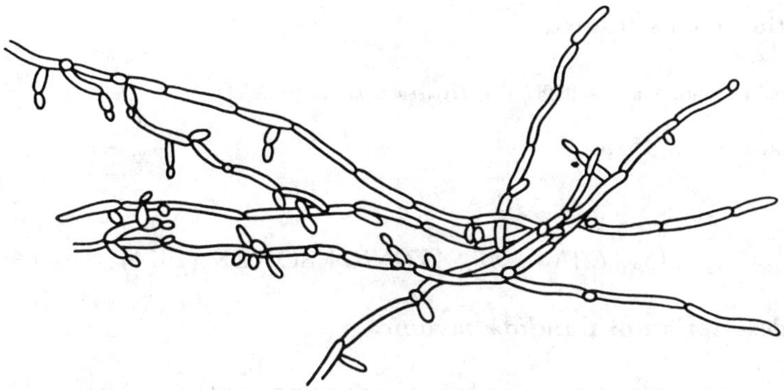

Fig. 328 – *C. santamariae*
Dalmau plate culture on corn meal agar (From 'The Yeasts' ed. II, 1970)

Fermentation:

Glucose	+	Maltose	–
Galactose	–	Lactose	–
Sucrose	–	Trehalose	+

Assimilation of carbon compounds:

Galactose	+w or –	Melezitose	–	Ribitol	+
L-Sorbose	+	Soluble starch	–	Galactitol	–
Sucrose	–	D-Xylose	– or +s	D-Mannitol	+
Maltose	–	L-Arabinose	–	D-Glucitol	+
Cellobiose	+ or s	D-Arabinose	–	Salicin	v
Trehalose	+	D-Ribose	–	DL-Lactic acid	+w or –
Lactose	–	L-Rhamnose	–	Succinic acid	+
Melibiose	–	Glycerol	+	Citric acid	+
Raffinose	–	Erythritol	–	Inositol	–

Additional carbon compound tested: Methanol –

Splitting of arbutin: +

Assimilation of nitrate: –

Growth in vitamin-free medium: –

Growth at 37°C: –

Color reaction with DBB: –

Coenzyme Q system: Q 9 (Yamada & Kondo, 1972)

G + C: 38.8 mol%, type strain (Meyer, Stenderup & Christiansen, unpublished)

Comments

Montrocher (1967) described the variety *C. santamariae* var. *membranaefaciens* based on differences from the species in sucrose utilization, galactose fermentation and pellicle formation. Since these differences could not be confirmed and the GC percent of *C. santamariae* var. *membranaefaciens* was higher (46%) than that of *C. santamariae* the relationship of these strains is questionable and further study is necessary.

Origin of the strains studied

Sugar in Spain (Santa María, 1959b) (1); *Hymenochaeta rubiginosa* (1).

Type strain: CBS 4515.

155. CANDIDA SAVONICA SONCK (1974)

Standard description of *Candida savonica*

Growth in glucose-yeast extract-peptone water: After 3 days at 28°C the cells are various shapes: spheroidal, ovoid or ellipsoidal, (2.8–8.4) × (5.6–11.2) μm and cylindrical 3.5 × 12 μm and occur singly, in pairs and occasionally in short chains (Fig. 329). Pseudomycelium is evident. After one month at room temperature only a sediment is present.

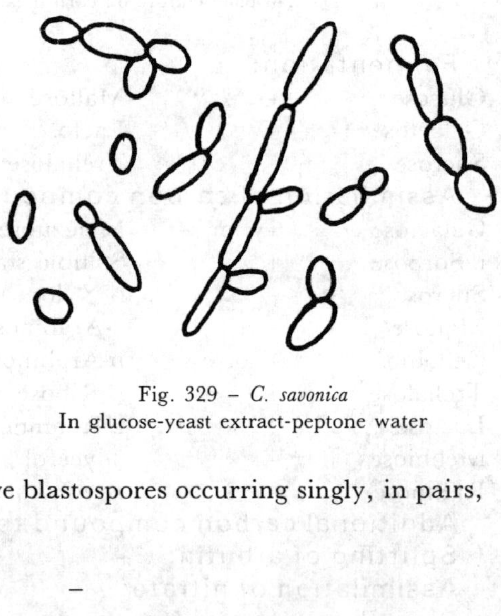

Fig. 329 – *C. savonica*
In glucose-yeast extract-peptone water

Growth on glucose-yeast extract-peptone agar: After one month at 25°C the streak culture is greenish-white, butyrous, granular, slightly shiny with a clearly visible striated border of pseudomycelium.

Dalmau plate culture on corn meal agar: Extensive pseudomycelium is formed. Branched pseudohyphae have blastospores occurring singly, in pairs, short chains or clusters.

Fermentation:

Glucose	+	Maltose	–
Galactose	+	Lactose	–
Sucrose	–		

Assimilation of carbon compounds:

Galactose	+	Melezitose	–	Ribitol	+
L-Sorbose	–	Soluble starch	–	Galactitol	–
Sucrose	–	D-Xylose	+ s	D-Mannitol	+
Maltose	–	L-Arabinose	–	D-Glucitol	+
Cellobiose	+	D-Arabinose	– or + s	Salicin	+
Trehalose	+	D-Ribose	–	DL-Lactic acid	–

Lactose	–	L-Rhamnose	–	Succinic acid	+
Melibiose	–	Glycerol	+	Citric acid	+
Raffinose	–	Erythritol	–	Inositol	–

Splitting of arbutin: +
Assimilation of nitrate: –
Growth in vitamin-free medium: +w or –
Growth at 37°C: –
Urease: –
Color reaction with DBB: –
G + C: 48 mol%, type strain

Origin of the strain studied

Wood in Finland (1).

Type strain: CBS 6563.

156. *CANDIDA SCHATAVII* (KOCKOVÁ-KRATOCHVÍLOVÁ ET ONDRUŠOVÁ) MEYER ET YARROW (Yarrow & Meyer, 1978)

Synonym

Torulopsis schatavii Kocková-Kratochvílová et Ondrušová (1971)

Standard description of *Candida schatavii*

Growth in glucose-yeast extract-peptone water: After 3 days at 25°C the cells are globose to ovoid, (2.9–5.8) × (3.6–7.2) μm, and elongate and occur singly, in pairs and chains (Fig. 330). Pseudomycelial development is evident.

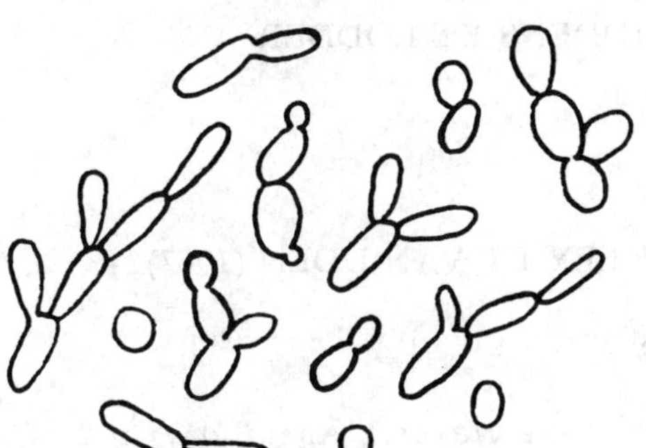

Fig. 330 – *C. schatavii*
In glucose-yeast extract-peptone water

Growth on glucose-yeast extract-peptone agar: After one month at 25°C the streak culture is white to cream-colored, smooth, moist, glistening, with a fringed margin.

Dalmau plate culture on corn meal agar: The pseudomycelium consists of a network of thin branching pseudohyphae bearing a few ovoid blastospores.

Fermentation:

Glucose	+s	Maltose	–
Galactose	+w or –	Lactose	–
Sucrose	–	Raffinose	–

Assimilation of carbon compounds:

Galactose	+	Melezitose	–	Ribitol	+
L-Sorbose	–	Soluble starch	–	Galactitol	–
Sucrose	–	D-Xylose	+s or –	D-Mannitol	+
Maltose	–	L-Arabinose	–	D-Glucitol	+
Cellobiose	–	D-Arabinose	– or +s	Salicin	–
Trehalose	+	D-Ribose	+	DL-Lactic acid	–
Lactose	–	L-Rhamnose	–	Succinic acid	+
Melibiose	–	Glycerol	+	Citric acid	+
Raffinose	–	Erythritol	+	Inositol	–

Additional carbon compound tested: Methanol –

Splitting of arbutin: +s or –

Assimilation of nitrate: –

Growth in vitamin-free medium: +

Growth at 37°C: –

Color reaction with DBB: –

G + C: 44.6 mol%, type strain

Origin of the strain studied

Fomitopsis pinicola (1).

Type strain: CBS 6452.

157. *CANDIDA SCOTTII* DIDDENS ET LODDER

see *Leucosporidium scottii* p. 504

158. *CANDIDA SHEHATAE* BUCKLEY ET VAN UDEN (1967)

Standard description of *Candida shehatae*

Growth in glucose-yeast extract-peptone water: After 3 days at 25°C the cells are ovoid, (2–4.5) × (3.5–6) μm; cylindrical and pseudohyphal cells may also occur (Fig. 331) (van Uden & Buckley, 1970).

Growth on glucose-yeast extract-peptone agar: After one month at 25°C the streak culture is cream-colored, glistening, soft and smooth or slightly reticulated (van Uden & Buckley, 1970).

Dalmau plate culture on corn meal agar: The pseudomycelium consists of ramified chains of curved pseudohyphae bearing blastospores in varying amounts in

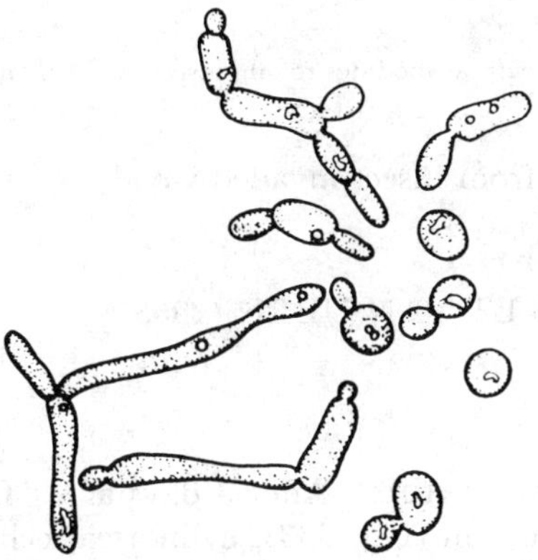

Fig. 331 – *C. shehatae*
In glucose-yeast extract-peptone water
(From 'The Yeasts' ed. II, 1970)

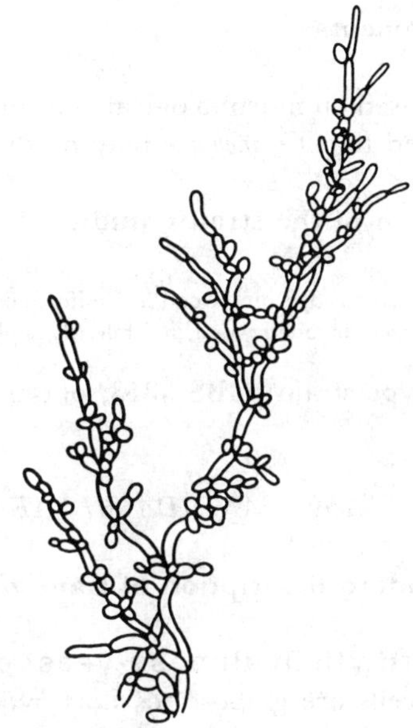

Fig. 332 – *C. shehatae*
Dalmau plate culture on corn meal agar
(From 'The Yeast' ed. II, 1970)

more or less verticillated positions (Fig. 332) (van Uden & Buckley, 1970).

Fermentation:

Glucose	+	Maltose	+
Galactose	+	Lactose	–
Sucrose	–	Trehalose	+

Assimilation of carbon compounds:

Galactose	+	Melezitose	+	Ribitol	+
L-Sorbose	+	Soluble starch	+	Galactitol	–
Sucrose	+	D-Xylose	+	D-Mannitol	+
Maltose	+	L-Arabinose	v	D-Glucitol	+
Cellobiose	+	D-Arabinose	v	Salicin	+
Trehalose	+	D-Ribose	+	DL-Lactic acid	+
Lactose	+	L-Rhamnose	v	Succinic acid	+
Melibiose	–	Glycerol	+	Citric acid	+
Raffinose	–	Erythritol	v	Inositol	–

Splitting of arbutin: +
Assimilation of nitrate: –
Growth in vitamin-free medium: v
Growth at 37°C: +
Color reaction with DBB: –
Coenzyme Q system: Q 9 (Yamada & Kondo, 1972)
G + C: 41.2 mol%, one strain (Nakase & Komagata 1971f)

Comments

Based on morphological and physiological similarities, Kreger-van Rij (1970c) suggested that *C. shehatae* may be the imperfect state of *Pichia stipitis.*

Origin of the strains studied

Insect-invaded pine wood in California (2); *Leptura cerambyciformis,* a wood-destroying insect in Germany (1); insect-invaded wood in Chile (2); soil in South Africa (1).

Type strain: CBS 5813, obtained in California from insect-invaded wood.

159. CANDIDA SILVAE VIDAL-LEIRIA ET VAN UDEN (1963)

Standard description of *Candida silvae*

Growth in glucose-yeast extract-peptone water: After 3 days at 25°C the cells are globose to short-ovoid, (1.5–4) × (2–6) μm (Fig. 333); cylindrical cells measuring up to 12 μm in length may occur. A thin pellicle is sometimes present (van Uden & Buckley, 1970).

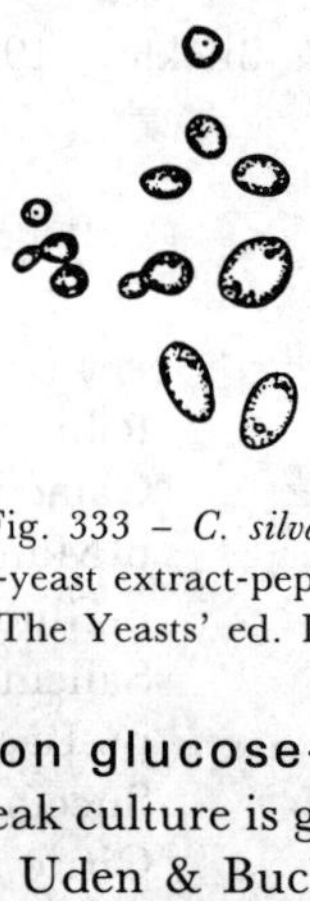

Fig. 333 – *C. silvae*
In glucose-yeast extract-peptone water
(From 'The Yeasts' ed. II, 1970)

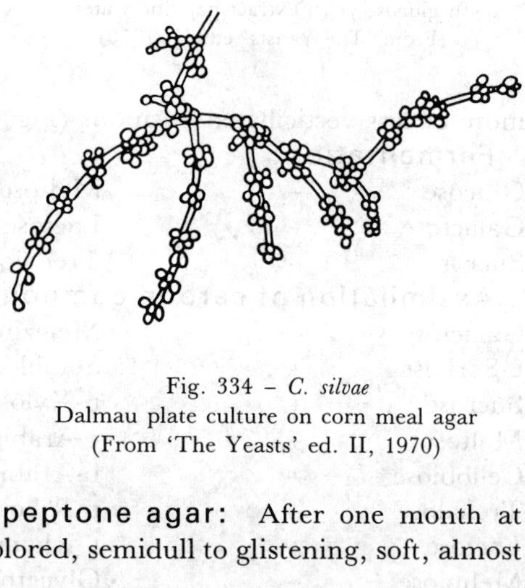

Fig. 334 – *C. silvae*
Dalmau plate culture on corn meal agar
(From 'The Yeasts' ed. II, 1970)

Growth on glucose-yeast extract-peptone agar: After one month at 25°C the streak culture is greyish to cream-colored, semidull to glistening, soft, almost smooth (van Uden & Buckley, 1970).

Dalmau plate culture on corn meal agar: The pseudomycelium consists of ramified chains of thin pseudohyphae, bearing clusters of roundish blastospores in verticillated positions (Fig. 334); a primitive formation of short pseudohyphae may also occur (van Uden & Buckley, 1970).

Fermentation:

Glucose	+ w or –	Maltose	–
Galactose	–	Lactose	–
Sucrose	–		

Assimilation of carbon compounds:

Galactose	–	Melezitose	–	Ribitol	+ or s
L-Sorbose	–	Soluble starch	–	Galactitol	–
Sucrose	–	D-Xylose	–	D-Mannitol	+
Maltose	–	L-Arabinose	–	D-Glucitol	+
Cellobiose	–	D-Arabinose	v	Salicin	–
Trehalose	–	D-Ribose	–	DL-Lactic acid	v
Lactose	–	L-Rhamnose	–	Succinic acid	+
Melibiose	–	Glycerol	+	Citric acid	v
Raffinose	–	Erythritol	–	Inositol	–

Additional carbon compound tested: Methanol –
Splitting of arbutin: –
Assimilation of nitrate: –
Growth in vitamin-free medium: + s
Growth at 37°C: v
Color reaction with DBB: –
G + C: 44 mol%, one strain (Meyer, Stenderup & Christiansen, unpublished)

Origin of the strains studied (van Uden & Buckley, 1970)

Digestive tract of a horse in Portugal (2); sputum (1); vaginal discharge (1).

Type strain: CBS 5498, in Portugal from the digestive tract of a horse.

160. CANDIDA SILVANORUM VAN DER WALT, VAN DER KLIFT ET SCOTT (van der Walt et al., 1971a)

Standard description of *Candida silvanorum* (van der Walt et al., 1971a)

Growth in glucose-yeast extract-peptone water: After 3 days at 25°C the cells are globose, oval, long-oval, (2.2–4.3) × (2.9–6.5) μm, and a few triangular to irregular shaped cells (Fig. 335). Pseudomycelium, curved and branched, is present. A sediment is present. After one month a sediment and ring are present.

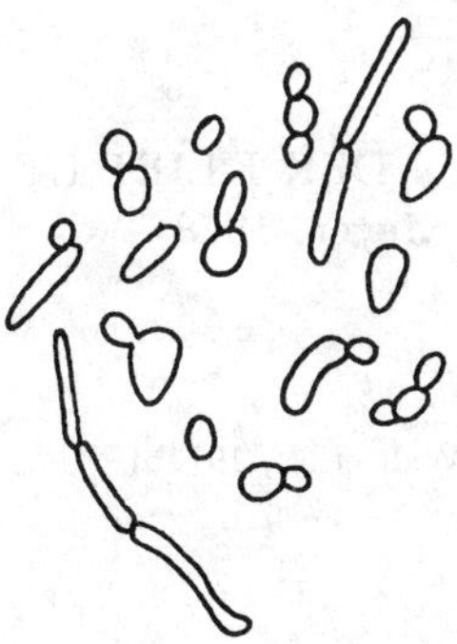

Fig. 335 – *C. silvanorum*
In glucose-yeast extract-peptone water

Growth on glucose-yeast extract-peptone agar: After one month at 20°C the streak culture is cream-colored, wrinkled, somewhat shiny, and fringed with pseudomycelium.

Dalmau plate culture on corn meal agar: Pseudomycelium is branched and abundantly produced. Long slender cells, often curved, have a few, usually triangular, blastospores occurring singly, in small clusters or short chains.

Fermentation:

Glucose	+	Maltose	– or + s
Galactose	+ or s	Lactose	–
Sucrose	– or + w,s	Raffinose	– or + s

Assimilation of carbon compounds:

Galactose	+	Melezitose	+	Ribitol	+
L-Sorbose	+ s	Soluble starch	+	Galactitol	–
Sucrose	+	D-Xylose	+	D-Mannitol	+
Maltose	+	L-Arabinose	+	D-Glucitol	+
Cellobiose	+	D-Arabinose	v	Salicin	+
Trehalose	+	D-Ribose	+ s	DL-Lactic acid	–
Lactose	–	L-Rhamnose	+	Succinic acid	+
Melibiose	+	Glycerol	+	Citric acid	+
Raffinose	+	Erythritol	+	Inositol	–

Additional carbon compound tested: Inulin –

Splitting of arbutin: +

Assimilation of nitrate: –

Growth in vitamin-free medium: –

Growth on 50% (w/w) glucose-yeast extract agar: +

Growth at 37°C: +

Color reaction with DBB: –

G + C: 41.7 mol%, type strain (Meyer, Stenderup & Christiansen, unpublished)

Origin of the strain studied

Material from the tunnels of *Xylion adustus* Fahr. in *Peltophorum africanum* (1).

Type strain: CBS 6274.

161. CANDIDA SILVATICA (VAN DER WALT, VAN DER KLIFT ET SCOTT) MEYER ET YARROW (Yarrow & Meyer, 1978)

Synonym

Torulopsis silvatica van der Walt, van der Klift et Scott (van der Walt et al., 1971b)

Standard description of *Candida silvatica*

Growth in glucose-yeast extract-peptone water: After 3 days at 25°C cells are spheroidal to ellipsoidal, (1.5–3.0) × (2.5–4.0) μm, and occur singly, in pairs, small clusters or chains. After one month at room temperature only a sediment is present (van der Walt et al., 1971b).

Growth on glucose-yeast extract-peptone agar: After one month at 25°C the streak culture is cream-colored, smooth, shiny with an undulating margin.

Dalmau plate culture on corn meal agar: No pseudomycelium is formed.

Fermentation: –

Assimilation of carbon compounds:

Galactose	–	Melezitose	–	Ribitol	+
L-Sorbose	–	Soluble starch	–	Galactitol	–
Sucrose	–	D-Xylose	–	D-Mannitol	+
Maltose	–	L-Arabinose	–	D-Glucitol	+
Cellobiose	–	D-Arabinose	–	Salicin	–
Trehalose	+	D-Ribose	–	DL-Lactic acid	+
Lactose	–	L-Rhamnose	–	Succinic acid	+
Melibiose	–	Glycerol	+	Citric acid	–
Raffinose	–	Erythritol	–	Inositol	–

Additional carbon compound tested: Methanol –

Splitting of arbutin: –

Assimilation of nitrate: –

Growth in vitamin-free medium: –

Growth on 50% (w/w) glucose-yeast extract agar: –

Growth at 37°C: +

Color reaction with DBB: –

G + C: 55.6 mol%, type strain (Meyer, Stenderup & Christiansen, unpublished)

Origin of the strain studied

Tunnels of *Crossotarsus externedentatus* Fairmi in *Ficus sycomorus* near St. Lucia, Natal (1).

Type strain: CBS 6277.

162. *CANDIDA SILVICOLA* SHIFRINE ET PHAFF

see *Hansenula holstii* p. 192

163. *CANDIDA SILVICULTRIX* VAN DER WALT, SCOTT ET VAN DER KLIFT (1972)

Standard description of *Candida silvicultrix*

Growth in glucose-yeast extract-peptone water: After 3 days at 25°C the cells are round or oval, (2.2–4.3) × (2.2–5.1) μm (Fig. 336), and occur singly, in pairs, short chains and small clusters. Pseudomycelium is present. After one month a sediment is present, but rings and pellicle are not.

Growth on glucose-yeast extract-peptone agar: After one month at 25°C the streak culture is creamy-white, smooth, convex and shiny. The margin is entire or fringed with mycelium.

Dalmau plate culture on corn meal agar: Abundant pseudomycelium with dense clusters of blastospores arranged at verticillate positions is formed.

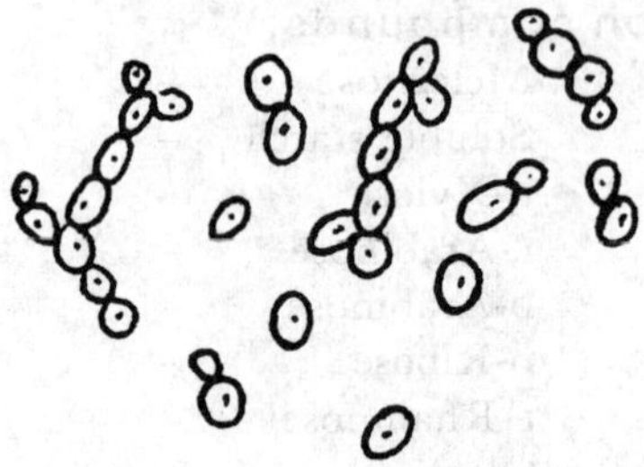

Fig. 336 – *C. silvicultrix*
In glucose-yeast extract-peptone water

Fermentation:

Glucose	+	Lactose	–
Galactose	+	Raffinose	+
Sucrose	+	Melibiose	+
Maltose	– or +s		

Assimilation of carbon compounds:

Galactose	+	Melezitose	+	Ribitol	+
L-Sorbose	–	Soluble starch	v	Galactitol	–
Sucrose	+	D-Xylose	+	D-Mannitol	+
Maltose	+	L-Arabinose	+	D-Glucitol	+
Cellobiose	+	D-Arabinose	+	Salicin	+
Trehalose	+	D-Ribose	+	DL-Lactic acid	+
Lactose	–	L-Rhamnose	–	Succinic acid	+
Melibiose	+	Glycerol	+	Citric acid	+
Raffinose	+	Erythritol	+	Inositol	–

Additional carbon compound tested: Inulin –

Splitting of arbutin: +

Assimilation of nitrate: –

Growth in vitamin-free medium: +

Growth on 50% (w/w) glucose-yeast extract agar: +

Growth at 37°C: +

Color reaction with DBB: –

G + C: 36.6 mol%, type strain

Origin of the strain studied

Material from the tunnels of *Sinoxylon ruficorne* Fahr. infesting *Acacia karroo* Hayne near Winburg in the Orange Free State (1).

Type strain: CBS 6269.

164. *CANDIDA SOLANI* LODDER ET KREGER-VAN RIJ (1952)

Standard description of *Candida solani*

Growth in glucose-yeast extract-peptone water: After 3 days at 25°C the cells are short-ovoid to long-ovoid, sometimes curved, $(2.5-5) \times (4-10)$ μm (Fig. 337). A thin, delicately wrinkled pellicle is present (van Uden & Buckley, 1970).

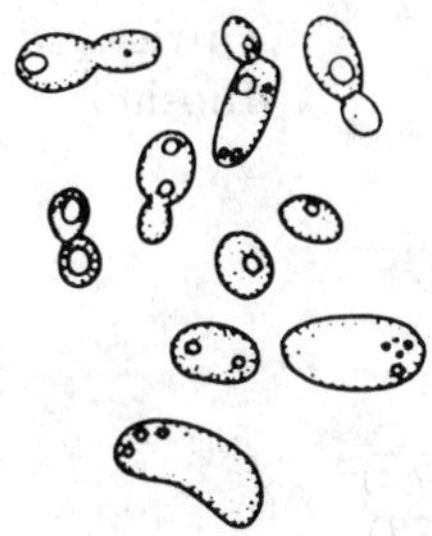

Fig. 337 – *C. solani*
In glucose-yeast extract-peptone water
(From 'The Yeasts' ed. II, 1970)

Growth on glucose-yeast extract-peptone agar: After one month at 25°C the streak culture is greyish-white, dull, soft; the surface is finely pointed as if covered by flour (van Uden & Buckley, 1970).

Dalmau plate culture on corn meal agar: The pseudomycelium consists of ramified chains of short pseudohyphae bearing chains of oval and cylindrical blastospores (Fig. 338) (van Uden & Buckley, 1970).

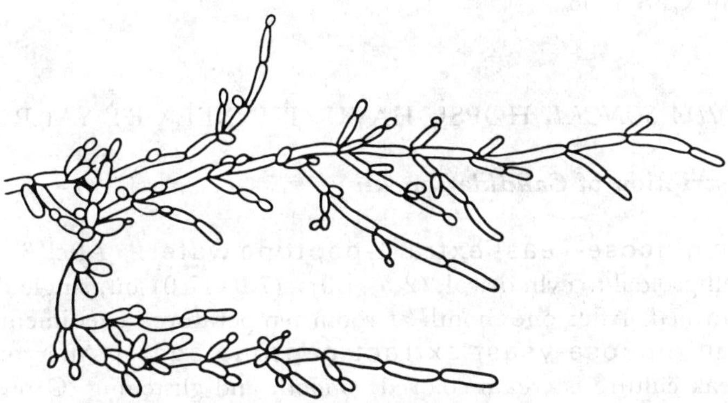

Fig. 338 – *C. solani*
Dalmau plate culture on corn meal agar (From 'The Yeasts' ed. II, 1970)

Fermentation:

Glucose	+	Maltose	–
Galactose	–	Lactose	–
Sucrose	–		

Assimilation of carbon compounds:

Galactose	–	Melezitose	+	Ribitol	–
L-Sorbose	+	Soluble starch	–	Galactitol	–
Sucrose	+	D-Xylose	+	D-Mannitol	–
Maltose	+	L-Arabinose	–	D-Glucitol	–
Cellobiose	+	D-Arabinose	–	Salicin	+
Trehalose	+	D-Ribose	–	DL-Lactic acid	+
Lactose	–	L-Rhamnose	–	Succinic acid	+
Melibiose	–	Glycerol	+	Citric acid	–
Raffinose	–	Erythritol	–	Inositol	–

Splitting of arbutin: +
Assimilation of nitrate: –
Growth in vitamin-free medium: –
Growth at 37°C: –
Color reaction with DBB: –
Coenzyme Q system: Q7 (Yamada & Kondo, 1972)
G + C: 41.9 mol%, type strain (Meyer & Phaff, 1972)

Origin of the strain studied

Material in a potato flour mill in The Netherlands (1).

Type strain: CBS 1908.

165. CANDIDA SONCKII HOPSU-HAVU, TUNNELA ET YARROW (1978)

Standard description of *Candida sonckii*

Growth in glucose-yeast extract-peptone water: After 3 days at 25°C the cells are ellipsoidal to cylindrical, (2.5–4.0) × (7.0–12.0) μm, single or in pairs. A sediment is formed. After one month at room temperature, a sediment is present.
Growth on glucose-yeast extract-peptone agar: After one month at 25°C the streak culture is cream-colored, smooth and glistening. Growth is slow.
Dalmau plate culture on corn meal agar: Pseudomycelium is not formed.
Fermentation: –
Assimilation of carbon compounds:

Galactose	+	Melezitose	–	Ribitol	+
L-Sorbose	+ or s	Soluble starch	–	Galactitol	–
Sucrose	–	D-Xylose	+	D-Mannitol	+
Maltose	–	L-Arabinose	v	D-Glucitol	+
Cellobiose	–	D-Arabinose	–	Salicin	–
Trehalose	+	D-Ribose	–	DL-Lactic acid	–
Lactose	–	L-Rhamnose	–	Succinic acid	+
Melibiose	–	Glycerol	+ or s	Citric acid	–
Raffinose	–	Erythritol	–	Inositol	–

Additional carbon compounds tested: α-Methyl-D-glucoside –, inulin –
Splitting of arbutin: –
Assimilation of nitrate: +
Growth in vitamin-free medium: –
Growth on 50% (w/w) glucose-yeast extract agar: –
Growth at 37°C: –
Color reaction with DBB: +
G + C: 44.1 mol% (Hopsu-Havu et al., 1978)

Origin of the strain studied

Bathing pool (1).

Type strain: CBS 6713.

166. *CANDIDA SONORENSIS* (MILLER, PHAFF, MIRANDA, HEED ET STARMER) MEYER ET YARROW (Yarrow & Meyer, 1978)

Synonym

Torulopsis sonorensis Miller, Phaff, Miranda, Heed et Starmer (1976a)

Standard description of *Candida sonorensis*

Growth in glucose-yeast extract-peptone water: After 3 days at 25°C the cells are spheroidal to elongate, (3.4–5.6)×(4.5–8.4) μm, and occur singly, in pairs, short chains or small clusters. A sediment is present. After one month a sediment, ring and film are present.

Growth on glucose-yeast extract-peptone agar: After one month at 25°C the streak culture is cream-colored, smooth, semi-glossy, soft and convex; the border is entire or slightly lobed.

Dalmau plate culture on corn meal agar: No pseudomycelium is formed or branched chains of cells may be present.

Fermentation:

Glucose	+	Maltose	–
Galactose	–	Lactose	–
Sucrose	–		

Assimilation of carbon compounds:

Galactose	–	Melezitose	–	Ribitol	+
L-Sorbose	–	Soluble starch	–	Galactitol	–
Sucrose	–	D-Xylose	+	D-Mannitol	+
Maltose	–	L-Arabinose	+	D-Glucitol	+
Cellobiose	+	D-Arabinose	–	Salicin	+
Trehalose	–	D-Ribose	+s or –	DL-Lactic acid	– or +s
Lactose	–	L-Rhamnose	–	Succinic acid	+ or w

Melibiose	–	Glycerol	+s	Citric acid	–
Raffinose	–	Erythritol	–	Inositol	–

Additional carbon compound tested: Methanol +
Splitting of arbutin: +
Assimilation of nitrate: –
Growth in vitamin-free medium: –
Growth on 50% (w/w) glucose- yeast extract agar: –
Growth at 37°C: +
Color reaction with DBB: –
G + C: 36.1 mol%, type strain (Miller et al., 1976a)

Origin of the strain studied

Lamaireocereus thurberi (organ-pipe cactus) at Caborca, Sonora, Mexico (1).

Type strain: CBS 6792.

167. *CANDIDA SORBOPHILA* (NAKASE) MEYER ET YARROW (Yarrow & Meyer, 1978)

Synonym

Torulopsis sorbophila Nakase (1975)

Standard description of *Candida sorbophila*

Growth in glucose-yeast extract-peptone water: After 3 days at 25°C the cells are globose to ovoid, (2.2–4.5) × (2.2–5.0) μm, and occur singly, in pairs, and small clusters. A sediment is present. After one month a sediment and ring are present.

Growth on glucose-yeast extract-peptone agar: After one month at 25°C the streak culture is yellowish-white, glistening, convex, smooth, with an entire margin.

Dalmau plate culture on corn meal agar: A sparse pseudomycelium consisting of ramified chains of ovoid cells may be formed.

Fermentation: –

Assimilation of carbon compounds:

Galactose	+	Melezitose	–	Ribitol	–
L-Sorbose	+	Soluble starch	–	Galactitol	–
Sucrose	–	D-Xylose	+ s or –	D-Mannitol	+
Maltose	–	L-Arabinose	v	D-Glucitol	+
Cellobiose	–	D-Arabinose	–	Salicin	–
Trehalose	–	D-Ribose	–	DL-Lactic acid	–
Lactose	–	L-Rhamnose	–	Succinic acid	+
Melibiose	–	Glycerol	+ s	Citric acid	–
Raffinose	–	Erythritol	–	Inositol	–

Additional carbon compound tested: Methanol –
Splitting of arbutin: –
Assimilation of nitrate: –
Growth in vitamin-free medium: –
Growth at 37°C: + or w
Color reaction with DBB: –
G + C: 50.0 mol%, one strain (Nakase, 1975)

Origin of the strain studied

Washings of ion-exchange resin of guanosine monophosphate plant (1).

Type strain: CBS 6739.

168. CANDIDA SORBOSA HEDRICK ET BURKE EX VAN UDEN ET BUCKLEY

see *Issatchenkia occidentalis* p. 215

169. CANDIDA SORBOXYLOSA NAKASE (1971c)

Standard description of *Candida sorboxylosa*

Growth in glucose-yeast extract-peptone water: After 3 days at 25°C cells are short-oval to long-oval or elongate, (2.8–5.6) × (3.9–8.4) μm, and occur singly, in pairs and in chains (Fig. 339). A sediment is present. After one month a sediment and film are present.

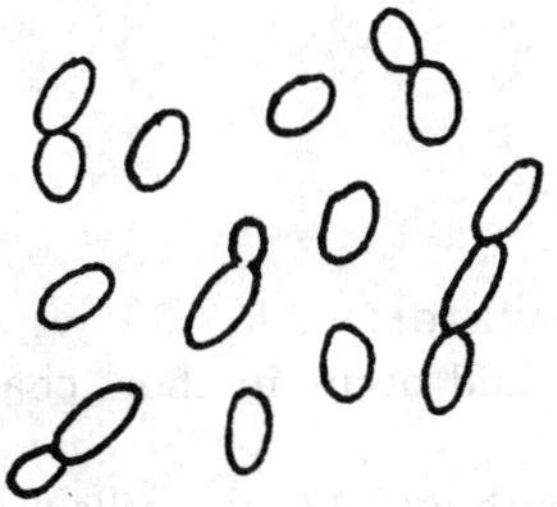

Fig. 339 – *C. sorboxylosa*
In glucose-yeast extract-peptone water

Growth on glucose-yeast extract-peptone agar: After one month at 25°C the streak culture is off-white, smooth, dull, raised, soft to butyrous and has an entire to erose margin.

Dalmau plate culture on corn meal agar: The pseudomycelium consists of short pseudohyphae bearing chains of spheroidal to ovoid blastospores in verticils.

Fermentation:

Glucose	– or + w	Maltose	–
Galactose	–	Lactose	–
Sucrose	–		

Assimilation of carbon compounds:

Galactose	–	Melezitose	–	Ribitol	–
L-Sorbose	+ or s	Soluble starch	–	Galactitol	–
Sucrose	–	D-Xylose	+	D-Mannitol	–
Maltose	–	L-Arabinose	–	D-Glucitol	–
Cellobiose	–	D-Arabinose	–	Salicin	–
Trehalose	–	D-Ribose	–	DL-Lactic acid	+
Lactose	–	L-Rhamnose	–	Succinic acid	+
Melibiose	–	Glycerol	+	Citric acid	+
Raffinose	–	Erythritol	–	Inositol	–

Splitting of arbutin: –
Assimilation of nitrate: –
Growth in vitamin-free medium: –
Growth on 50% (w/w) glucose-yeast extract agar: –
Growth at 37°C: v
Color reaction with DBB: –
G + C: 39.8 mol%, 2 strains (Nakase, 1971c)

Origin of the strains studied

Pineapple (1); fermenting cacao (1); palm wine (1); insect tunnels in felled timber (1).

Type strain: CBS 6378, from pineapple.

170. CANDIDA SPANDOVENSIS (HENNINGER ET WINDISCH) MEYER ET YARROW (Yarrow & Meyer, 1978)

Synonym

Torulopsis spandovensis Henninger et Windisch (1976b)

Standard description of *Candida spandovensis*

Growth in glucose-yeast extract-peptone water: After 3 days at 25°C the cells are round, 2.9–5.1 μm, bud multilaterally and occur in short chains and clusters.

Growth on malt extract agar: After one month at 25°C the cells are round or short-oval, 2.8–6.0 μm, show multilateral budding and occur singly, occasionally in pairs and in small clusters. The streak culture is butyrous, cream-colored, flat, shiny and smooth. The margin is entire. A slight fruit odor is evident (Henninger & Windisch, 1976b).

Dalmau plate culture on corn meal agar: The pseudomycelium consists of chains of globose to ovoid, and teardrop-shaped cells bearing small roundish blastospores in verticillated positions.

Fermentation:

Glucose	+	Maltose	–
Galactose	+ or s	Lactose	–
Sucrose	+ s	Raffinose	+ s or –

Assimilation of carbon compounds:

Galactose	+	Melezitose	–	Ribitol	+
L-Sorbose	+ or s	Soluble starch	–	Galactitol	+
Sucrose	+	D-Xylose	+	D-Mannitol	+
Maltose	–	L-Arabinose	+	D-Glucitol	+
Cellobiose	–	D-Arabinose	–	Salicin	–
Trehalose	–	D-Ribose	–	DL-Lactic acid	+ s
Lactose	–	L-Rhamnose	–	Succinic acid	+
Melibiose	–	Glycerol	+	Citric acid	+ s
Raffinose	+	Erythritol	–	Inositol	–

Splitting of arbutin: –
Assimilation of nitrate: –
Growth in vitamin-free medium: –
Growth on 50% (w/w) glucose-yeast extract agar: +
Growth at 37°C: –
Urease: –
Color reaction with DBB: –
G + C: 52.7 mol%, type strain

Comments

We were unable to confirm growth at 37°C.

Origin of the strains studied

German Pilsener beer (2).

Type strain: CBS 6875.

171. CANDIDA SPHAERICA (HAMMER ET CORDES) MEYER ET YARROW

see *Kluyveromyces marxianus* var. *lactis* p. 240

172. CANDIDA STEATOLYTICA YARROW (1969a)

Standard description of *Candida steatolytica*

Growth in malt extract: After 3 days at 25°C the cells are oval to cylindrical, (2–4) × (4.5–10) μm; mycelial cells up to 30 μm in length may also be present (Fig. 340). A ring and islets are formed. After one month at room temperature a thick, off-white pellicle has developed (Yarrow, 1969a).

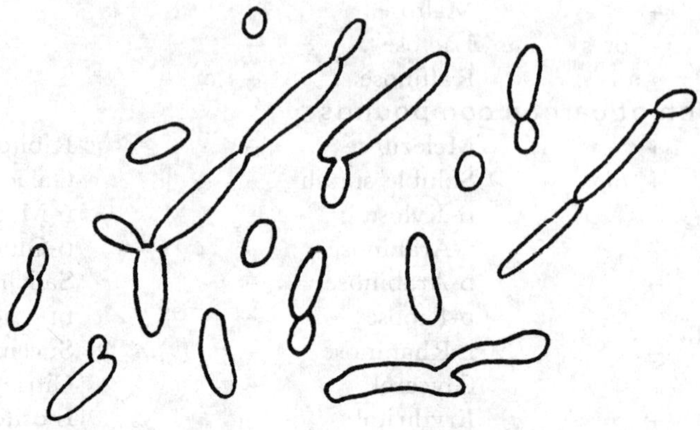

Fig. 340 – *C. steatolytica*
In glucose-yeast extract-peptone water

Growth on glucose-yeast extract-peptone agar: After one month at 25°C the streak culture is beige, shiny, creamy, soft and mostly smooth with slight irregularity at the base and center of the streak. An extensive mycelial border is evident.

Dalmau plate culture on corn meal agar: Both pseudo- and true hyphae are abundantly formed with blastospores arranged in verticils.

Fermentation:

Glucose	+	Maltose	+s or –
Galactose	+ or s	Lactose	–
Sucrose	+ or s	Raffinose	+s or –

Assimilation of carbon compounds:

Galactose	+	Melezitose	+ or s	Ribitol	+
L-Sorbose	+	Soluble starch	+	Galactitol	+ or s
Sucrose	+	D-Xylose	+	D-Mannitol	+
Maltose	+	L-Arabinose	+	D-Glucitol	+
Cellobiose	+	D-Arabinose	+s or –	Salicin	+
Trehalose	+	D-Ribose	+s	DL-Lactic acid	+s or –
Lactose	+s or –	L-Rhamnose	+	Succinic acid	+
Melibiose	–	Glycerol	+	Citric acid	+
Raffinose	+ or s	Erythritol	–	Inositol	+

Additional carbon compounds tested: 5-Ketogluconate +, inulin –

Splitting of arbutin: +

Assimilation of nitrate: –

Growth in vitamin-free medium: –

Growth on 50% (w/w) glucose-yeast extract agar: v

Growth at 37°C: +

Urease: –

Color reaction with DBB: –

G + C: 43.4–44.9 mol%, 2 strains

Comments

Four strains previously identified as *C. steatolytica* on the basis of morphological and physiological properties were examined for DNA base composition and degree of DNA reassociation with the type strain. The GC contents ranged from 43.4 to 55.6% and only one strain showed significant reassociation (98%) to be considered a member of the species. The other strains remain unidentified (Meyer & Yarrow, unpublished). *C. steatolytica* is similar to *C. inositophila,* but can be distinguished by citrate utilization and the formation of true hyphae.

Origin of the strains studied

Mastitic bovine udder (1), grape must (1), soft drink factory (1).

Type strain: CBS 5839.

173. *CANDIDA STELLATA* (KROEMER ET KRUMBHOLZ) MEYER ET YARROW (Yarrow & Meyer, 1978)

Synonyms

Saccharomyces stellatus Kroemer et Krumbholz (in Krumbholz, 1931)
Torulopsis stellata (Kroemer et Krumbholz) Lodder (1932)
Cryptococcus stellatus (Kroemer et Krumbholz) Skinner (1947a)
Saccharomyces bacillaris Kroemer et Krumbholz (1931)
Torulopsis bacillaris (Kroemer et Krumbholz) Lodder (1932)
Cryptococcus bacillaris (Kroemer et Krumbholz) Skinner (1947a)
Brettanomyces italicus Verona et Florenzano (1947)

Standard description of *Candida stellata* (van Uden & Vidal-Leiria, 1970)

Growth in glucose-yeast extract-peptone water: After 3 days at 25°C cells are ovoid, (2–3) × (4–6) μm (Fig. 341). They are often arranged in groups which may have a star-like configuration. Only a sediment is formed.

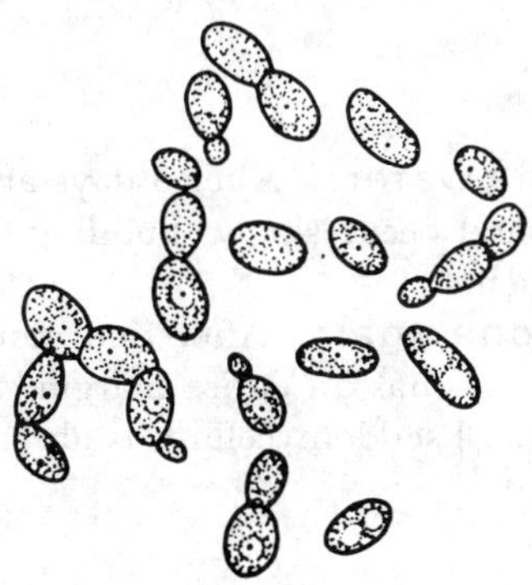

Fig. 341 – *C. stellata*
In glucose-yeast extract-peptone water
(From 'The Yeasts', ed. II, 1970)

Growth on glucose-yeast extract-peptone agar: After one month at 25°C the streak culture is greyish-white to brownish, glossy, soft and smooth. The cultures die off easily, probably due to acid formation.

Dalmau plate culture on corn meal agar: No pseudomycelium is formed.

Fermentation:

Glucose	+	Maltose	–
Galactose	–	Lactose	–
Sucrose	+	Raffinose	+ or s

Assimilation of carbon compounds:

Galactose	–	Melezitose	–	Ribitol	–
L-Sorbose	–	Soluble starch	–	Galactitol	–
Sucrose	+	D-Xylose	–	D-Mannitol	–
Maltose	–	L-Arabinose	–	D-Glucitol	–
Cellobiose	–	D-Arabinose	–	Salicin	–
Trehalose	–	D-Ribose	–	DL-Lactic acid	–
Lactose	–	L-Rhamnose	–	Succinic acid	–
Melibiose	–	Glycerol	–	Citric acid	–
Raffinose	+	Erythritol	–	Inositol	–

Splitting of arbutin: –
Assimilation of nitrate: –
Growth in vitamin-free medium: –
Growth at 37°C: v
Color reaction with DBB: –
Coenzyme Q system: Q6 and Q8 (Yamada & Kondo, 1972)
G + C: 42.0 mol%, one strain (Nakase & Komagata, 1971c)

Origin of the strains studied (van Uden & Vidal-Leiria, 1970)

Overripe grapes in Germany (2), grapes in Italy (1), grapes in France (1); *Drosophila pinicola* in USA (1); soil in Italy (1); "tea beer" in The Netherlands (1).

Type strain: CBS 157, from grapes and originally described as *Saccharomyces stellatus* (Krumbholz, 1931).

174. CANDIDA SUCCIPHILA LEE ET KOMAGATA (1980)

Standard description of *Candida succiphila*

Growth in glucose-yeast extract-peptone water: After 3 days at 25°C the cells are predominantly globose, 2.3–4.4 μm and occur singly, budding and in short chains and clusters. A light sediment is present.
Growth on glucose-yeast extract-peptone agar: After one month at 25°C the streak culture is cream-colored, smooth and has an entire margin.
Dalmau plate culture on corn meal agar: Pseudomycelium is absent.
Fermentation:

Glucose	+	Maltose	–
Galactose	+	Lactose	–
Sucrose	–		

Assimilation of carbon compounds:

Galactose	+	Melezitose	–	Ribitol	+
L-Sorbose	+	Soluble starch	–	Galactitol	+ s or w
Sucrose	–	D-Xylose	+	D-Mannitol	+
Maltose	–	L-Arabinose	+	D-Glucitol	+
Cellobiose	–	D-Arabinose	+	Salicin	+
Trehalose	+	D-Ribose	+	DL-Lactic acid	–
Lactose	+	L-Rhamnose	+	Succinic acid	–
Melibiose	–	Glycerol	+	Citric acid	–
Raffinose	–	Erythritol	+	Inositol	–

Additional carbon compound tested: Methanol +
Splitting of arbutin: –
Assimilation of nitrate: –
Growth in vitamin-free medium: – or + w
Growth at 37°C: +
Urease: –
Color reaction with DBB: –
Coenzyme Q system: Q7 (Lee & Komagata, 1980)
G + C: 40.9 mol%, type strain (Lee & Komagata, 1980)

Origin of the strain studied

Peach tree sap in Yamanashi Prefecture, Japan (1).

Type strain: CBS 8003.

175. *CANDIDA SUECICA* RODRIGUES DE MIRANDA ET NORKRANS (1968)

Standard description of *Candida suecica*

Growth in malt extract: After 3 days at 25°C the cells are oval to long-oval or cylindrical, (2.5–7) × (3–18) μm, single, in pairs and short chains. Pseudomycelial development is evident. After a month a yellowish-white, dull, wrinkled, somewhat creeping pellicle is formed (Rodrigues de Miranda & Norkrans, 1968).

Growth on malt agar: After one month at 25°C the streak culture is whitish-grey, partially smooth and partially wrinkled, soft with a lobed margin (Rodrigues de Miranda & Norkrans, 1968).

Dalmau plate culture on corn meal agar: Pseudomycelium is abundant, long slender, branched chains, with blastospores in verticils (Rodrigues de Miranda & Norkrans, 1968).

Fermentation:

Glucose	+s or –	Maltose	–
Galactose	–	Lactose	–
Sucrose	–	Trehalose	–

Assimilation of carbon compounds:

Galactose	–	Melezitose	–	Ribitol	+
L-Sorbose	+	Soluble starch	–	Galactitol	–
Sucrose	+	D-Xylose	+	D-Mannitol	+
Maltose	+	L-Arabinose	–	D-Glucitol	+
Cellobiose	+s or –	D-Arabinose	–	Salicin	– or +s
Trehalose	+	D-Ribose	–	DL-Lactic acid	–
Lactose	–	L-Rhamnose	–	Succinic acid	–
Melibiose	–	Glycerol	–	Citric acid	+w, s
Raffinose	–	Erythritol	–	Inositol	–

Additional carbon compounds tested: Glucono-δ-lactone +, 2-ketogluconate +

Splitting of arbutin: – or +w

Assimilation of nitrate: –

Growth in vitamin-free medium: –

Growth on 50% (w/w) glucose-yeast extract agar: –

Growth at 37°C: –

Color reaction with DBB: –

G + C: 42.7 mol%, type strain (Meyer, Stenderup & Christiansen, unpublished)

Origin of the strain studied

Water sample of Kungsbackafjorden (west coast of Sweden) (1).

Type strain: CBS 5724.

176. CANDIDA TANNOTOLERANS (JACOB) MEYER ET YARROW (Yarrow & Meyer, 1978)

Synonym

Torulopsis tannotolerans Jacob (1970)

Standard description of *Candida tannotolerans*

Growth in glucose-yeast extract-peptone water: After 3 days at 25°C the cells are spheroidal to ovoid, (2.9–3.6) × (3.6–5.1) μm, and occur singly, in pairs and short chains. A sediment is present. After one month a sediment is present.

Growth on glucose-yeast extract-peptone agar: After one month at 25°C the streak culture is white to cream-colored, smooth, flat and the margin is entire.

Dalmau plate culture on corn meal agar: No pseudomycelium is formed.

Fermentation:

Glucose	+	Maltose	–
Galactose	+	Lactose	–
Sucrose	–	Raffinose	–

Assimilation of carbon compounds:

Galactose	+	Melezitose	–	Ribitol	–
L-Sorbose	–	Soluble starch	–	Galactitol	–
Sucrose	–	D-Xylose	+ s	D-Mannitol	–
Maltose	–	L-Arabinose	–	D-Glucitol	–
Cellobiose	–	D-Arabinose	–	Salicin	–
Trehalose	–	D-Ribose	–	DL-Lactic acid	–
Lactose	–	L-Rhamnose	–	Succinic acid	–
Melibiose	–	Glycerol	+	Citric acid	–
Raffinose	–	Erythritol	–	Inositol	–

Splitting of arbutin: –

Assimilation of nitrate: –

Growth in vitamin-free medium: –

Growth on 50% (w/w) glucose-yeast extract agar: –

Growth at 37°C: –

Color reaction with DBB: –

G + C: 33.5 mol%, type strain (Meyer, Stenderup & Christiansen, unpublished)

Origin of the strain studied

Tanning liquor (1).

Type strain: CBS 6070.

177. CANDIDA TENUIS DIDDENS ET LODDER (1942)

Standard description of *Candida tenuis*

Growth in glucose-yeast extract-peptone water: After 3 days at 25°C the cells are ovoid to elongate, (1.5–3.5) × (4–11) μm (Fig. 342). They often occur in groups. A ring may be present (van Uden & Buckley, 1970).

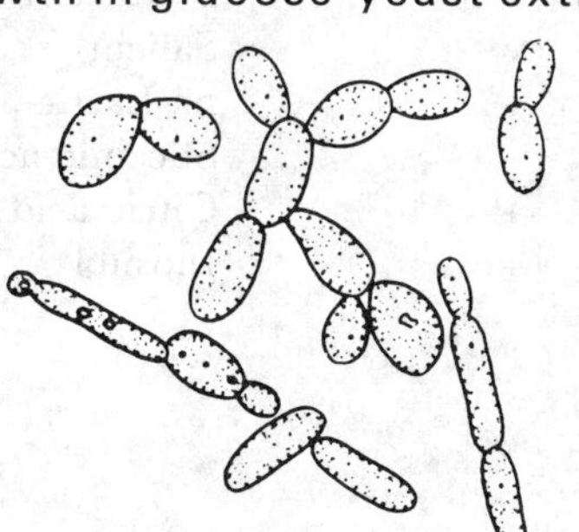

Fig. 342 – *C. tenuis*
In glucose-yeast extract-peptone water
(From 'The Yeasts', ed. II, 1970)

Growth on glucose-yeast extract-peptone agar: After one month at 25°C the streak culture is greyish to cream-colored; most strains are semidull, soft and smooth or partly wrinkled; other strains are soft to membranous and wrinkled all over the surface (van Uden & Buckley, 1970).

Dalmau plate culture on corn meal agar: Pseudomycelium is abundantly formed. The pseudohyphae are thin, often wavy. Blastospore formation varies. In some strains the pseudomycelium consists solely of bare pseudohyphae (Fig. 343); in other strains pleural blastoconidia and chains of verticillated blastospores occur (Fig. 344); occasionally clusters of verticillated blastospores are seen (van Uden & Buckley, 1970).

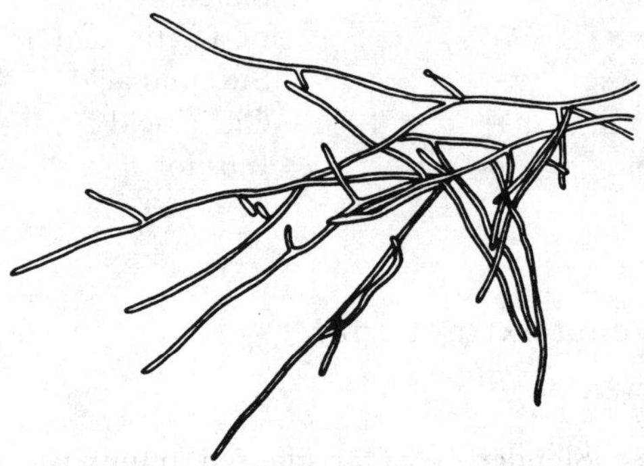

Fig. 343 – *C. tenuis*
Dalmau plate culture on corn meal agar
(From 'The Yeasts' ed. II, 1970)

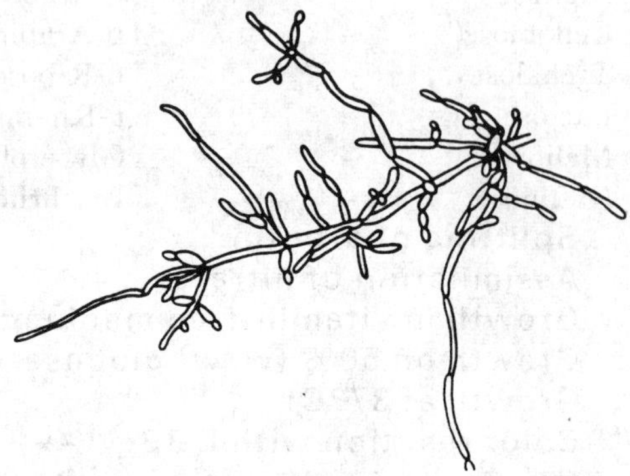

Fig. 344 – *C. tenuis*
Dalmau plate culture on corn meal agar
(From 'The Yeasts' ed. II, 1970)

Fermentation:

Glucose	+	Maltose	–
Galactose	+ or w	Lactose	–
Sucrose	–	Trehalose	v

Assimilation of carbon compounds:

Galactose	+	Melezitose	+	Ribitol	+
L-Sorbose	v	Soluble starch	v	Galactitol	–
Sucrose	+	D-Xylose	+	D-Mannitol	+
Maltose	+	L-Arabinose	v	D-Glucitol	+
Cellobiose	+	D-Arabinose	v	Salicin	+
Trehalose	+	D-Ribose	+	DL-Lactic acid	v
Lactose	v	L-Rhamnose	+	Succinic acid	v
Melibiose	–	Glycerol	+	Citric acid	v
Raffinose	–	Erythritol	+	Inositol	–

Splitting of arbutin: +
Assimilation of nitrate: –
Growth in vitamin-free medium: +s
Growth at 37°C: v
Color reaction with DBB: –
Coenzyme Q system: Q 9 (Yamada & Kondo, 1972)
G + C: 44.0 mol%, one strain (Stenderup & Bak, 1968)

Origin of the strains studied (van Uden & Buckley, 1970)

Chesapeake Bay (U.S.A.) water (1); bark beetles (U.S.A.) (3); beetles (Germany): *Harpium mordax* (1), *Rhagium sycophanta* (1), *Leptura rubra* (1), *Rhagium bifasciatum* (1); larva of *Mycroplophorus magallanicus* in Chile (1); bovine mastitis in Norway (1); human abscess (The Netherlands) (1); soil in South Africa (1).

Type strain: CBS 615, in the U.S.A. from bark beetles.

178. CANDIDA TEPAE GRINBERGS (1967)

Standard description of *Candida tepae*

Growth in glucose-yeast extract-peptone water: After 3 days at 25°C the cells are globose to ovoid, (2.2–4.3) × (3.6–7.2) μm and occur singly, in pairs, chains and clusters (Fig. 345). A sparse sediment and ring are present.

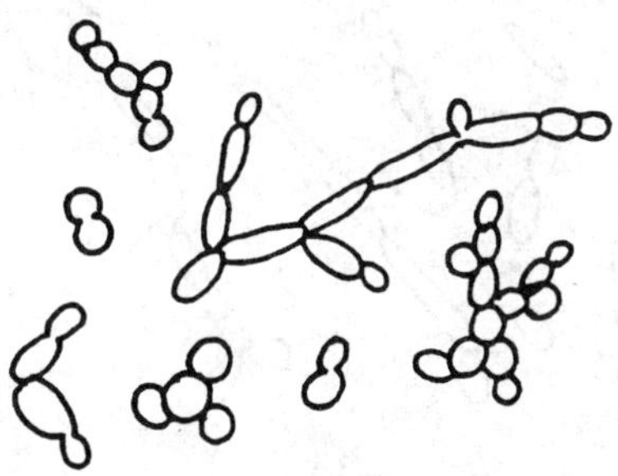

Fig. 345 – *C. tepae*
In glucose-yeast extract-peptone water

Growth on glucose-yeast extract-peptone agar: After one month at 25°C the streak culture is off-white, creamy, smooth, butyrous, shiny with an entire margin.

Dalmau plate culture on corn meal agar: The pseudomycelium consists of short branches of chains of oval cells giving a "bushy" appearance.

Fermentation: –

Assimilation of carbon compounds:

Galactose	+ s	Melezitose	–	Ribitol	v
L-Sorbose	+ s	Soluble starch	–	Galactitol	–
Sucrose	+	D-Xylose	+	D-Mannitol	+ or s
Maltose	+	L-Arabinose	–	D-Glucitol	+ or s
Cellobiose	+	D-Arabinose	– or + s	Salicin	+ s
Trehalose	+ s	D-Ribose	–	DL-Lactic acid	+ or s
Lactose	–	L-Rhamnose	–	Succinic acid	+ or s
Melibiose	–	Glycerol	+	Citric acid	+ or s
Raffinose	–	Erythritol	–	Inositol	–

Splitting of arbutin: +
Assimilation of nitrate: –
Growth in vitamin-free medium: –
Growth at 37°C: –
Urease: –
Color reaction with DBB: –
G + C: 41.0 mol%, type strain (Meyer, Stenderup & Christiansen, unpublished)

Origin of the strain studied

Rotten *Laurelia phyllipiana* (1).

Type strain: CBS 5115.

179. *CANDIDA TEREBRA* SUGIYAMA ET GOTO (1969)

Standard description of *Candida terebra*

Growth in glucose-yeast extract-peptone water: After 3 days at 25°C the cells are elongate, allantoid, (2.8–4.5) × (6.7–10.1) μm, and occur singly, in pairs and short chains (Fig. 346). Pseudohyphae are present. A sediment is formed. After one month a thick sediment and film are present.

Growth on glucose-yeast extract-peptone agar: After one month at 25°C the streak culture is cream-colored, smooth, dry and fringed with pseudomycelium.

Dalmau plate culture on corn meal agar: The pseudomycelium consists of extensive, long, delicate pseudohyphae; blastospores are scarce.

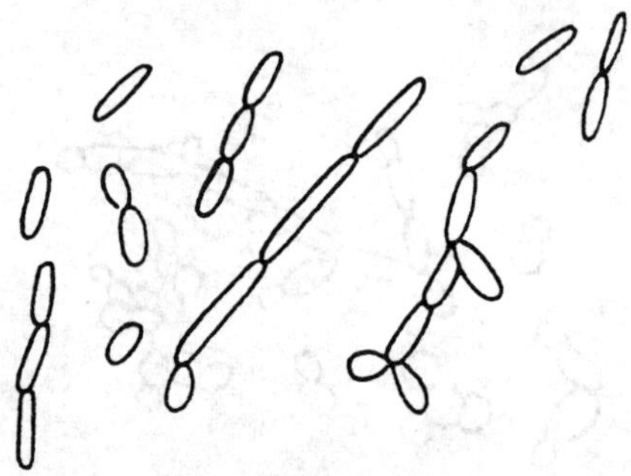

Fig. 346 – *C. terebra*
In glucose-yeast extract-peptone water

Fermentation:

Glucose	+ s	Maltose	– or + s
Galactose	+ s	Lactose	–
Sucrose	– or + s	Raffinose	–

Assimilation of carbon compounds:

Galactose	+	Melezitose	+	Ribitol	+
L-Sorbose	v	Soluble starch	–	Galactitol	+ s
Sucrose	+	D-Xylose	+	D-Mannitol	+
Maltose	+	L-Arabinose	+	D-Glucitol	+
Cellobiose	+	D-Arabinose	+	Salicin	+
Trehalose	+	D-Ribose	+ s	DL-Lactic acid	–
Lactose	–	L-Rhamnose	+	Succinic acid	+
Melibiose	–	Glycerol	+	Citric acid	+ s
Raffinose	–	Erythritol	+ or s	Inositol	–

Splitting of arbutin: +

Assimilation of nitrate: –

Growth in vitamin-free medium: –

Growth on 50% (w/w) glucose-yeast extract agar: –

Growth at 37°C: +

Color reaction with DBB: –

G + C: 43.6 mol%, type strain (Meyer, Stenderup & Christiansen, unpublished)

Origin of the strain studied

Soil (1).

Type strain: CBS 6023.

180. CANDIDA TORRESII (VAN UDEN ET ZOBELL) MEYER ET YARROW (Yarrow & Meyer, 1978)

Synonym

Torulopsis torresii van Uden et Zobell (1962)

Standard description of *Candida torresii*

Growth in glucose-yeast extract-peptone water: After 3 days at 25°C two types of cells are present. The larger ones are ovoid, (3–4) × (4.4–5) μm; the smaller ones are cylindrical, (1–2) × (3–4.5) μm (Fig. 347). Only a sediment is formed (van Uden & Vidal-Leiria, 1970).

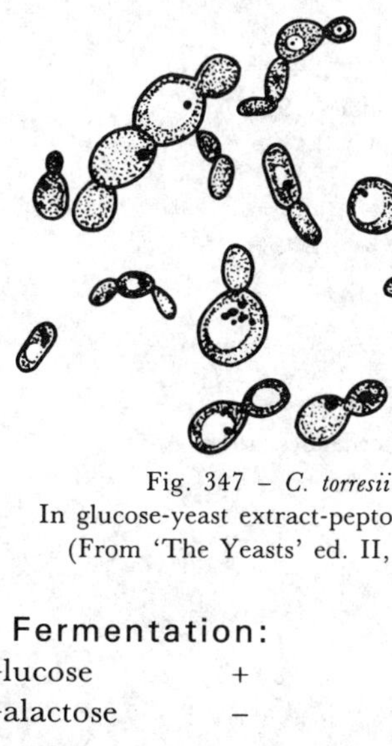

Fig. 347 – *C. torresii*
In glucose-yeast extract-peptone water
(From 'The Yeasts' ed. II, 1970)

Growth on glucose-yeast extract-peptone agar: After one month at 25°C the streak culture is cream-colored, glossy, soft and smooth (van Uden & Vidal-Leiria, 1970).

Dalmau plate culture on corn meal agar: No pseudomycelium is formed.

Fermentation:

Glucose	+	Maltose	–
Galactose	–	Lactose	–
Sucrose	–	Trehalose	+

Assimilation of carbon compounds:

Galactose	+	Melezitose	–	Ribitol	+
L-Sorbose	+	Soluble starch	–	Galactitol	–
Sucrose	–	D-Xylose	+	D-Mannitol	+
Maltose	–	L-Arabinose	+	D-Glucitol	+
Cellobiose	+	D-Arabinose	–	Salicin	+
Trehalose	+	D-Ribose	–	DL-Lactic acid	–
Lactose	–	L-Rhamnose	–	Succinic acid	+
Melibiose	–	Glycerol	+	Citric acid	–
Raffinose	–	Erythritol	–	Inositol	–

Splitting of arbutin: +
Assimilation of nitrate: –
Growth in vitamin-free medium: –
Growth at 37°C: v
Color reaction with DBB: –
Coenzyme Q system: Q9 (Yamada & Kondo, 1972)
G + C: 51.7 mol%, type strain (Meyer, Stenderup & Christiansen, unpublished)

Origin of the strain studied

Seawater (1).

Type strain: CBS 5152.

181. CANDIDA TROPICALIS (CASTELLANI) BERKHOUT (1923)

Synonyms

Oidium tropicale Cast. (1910)
Endomyces tropicalis Cast. (1911)
Monilia tropicalis (Cast.) Cast. et Chalmers (1913)
Atelosaccharomyces tropicalis (Cast.) Froilano de Mello et Gonzaga Fernandes (1918)
Myceloblastanon tropicale (Cast.) Ota (1928)
Castellania tropicalis (Cast.) Dodge (1935)
Procandida tropicalis (Cast.) Novák et Zsolt (1961)
Monilia candida Bon. sensu Hansen (Bonorden, 1851; Hansen, 1886)
Myceloblastanon candidum (Bon. sensu Hansen) Ota (1928)
Monilia bonordenii Vuill. (1911)
Candida bonordenii (Vuill.) Basgal (1931)
Endomyces paratropicalis Cast. (1911)
Monilia paratropicalis (Cast.) Cast. et Chalmers (1913)
Atelosaccharomyces paratropicalis (Cast.) Froilano de Mello et Gonzaga Fernandes (1918)
Myceloblastanon paratropicale (Cast.) Ota (1928)
Candida paratropicalis (Cast.) Basgal (1931)
Castellania paratropicalis (Cast.) Dodge (1935)
Endomyces bronchialis Cast. (1912b)
Monilia bronchialis (Cast.) Cast. et Chalmers (1913)
Myceloblastanon bronchiale (Cast.) Ota (1928)
Candida bronchialis (Cast.) Basgal (1931)
Castellania bronchialis (Cast.) Dodge (1935)
Endomyces entericus Cast. (1912b)
Monilia enterica (Cast.) Cast. et Chalmers (1913)
Myceloblastanon entericum (Cast.) Ota (1928)
Candida enterica (Cast.) de Almeida (1933)
Castellania enterica (Cast.) Dodge (1935)
Endomyces faecalis Cast. (1912b)
Monilia faecalis (Cast.) Cast. et Chalmers (1913)
Myceloblastanon faecale (Cast.) Ota (1928)
Castellania faecalis (Cast.) Dodge (1935)
Endomyces insolitus Cast. (1912b)
Monilia insolita (Cast.) Cast. et Chalmers (1913)

Myceloblastanon insolitum (Cast.) Ota (1928)
Candida insolita (Cast.) Basgal (1931)
Castellania insolita (Cast.) Dodge (1935)
Endomyces niveus Cast. (1912b)
Monilia nivea (Cast.) Cast. et Chalmers (1913)
Myceloblastanon niveum (Cast.) Ota (1928)
Candida nivea (Cast.) Basgal (1931)
Castellania nivea (Cast.) Dodge (1935)
Monilia metatropicalis Cast. et Chalmers (1919)
Castellania metatropicalis (Cast. et Chalmers) Dodge (1935)
Candida vulgaris Berkhout (1923)
Geotrichoides vulgaris (Berkhout) Langeron et Talice (1932)
Mycotorula interdigitalis Red. (1931)
Mycotorula dimorpha Red. et Cif. (Ciferri & Redaelli, (1935)
Syringospora dimorpha (Red. et Cif.) Dodge et Moore (1936)
Mycotorula trimorpha Red. et Cif. (Ciferri & Redaelli, 1935)
Mycotoruloides trimorpha (Red. et Cif.) Dodge et Moore (1936)
Pseudomonilia miso Mogi (1942)
Mycotorula japonica Yamaguchi (1943)
Monilia kefyr nom. nud.
Candida kefyr nom. nud. [nec *Candida kefyr* (Beijerinck) van Uden et Buckley (1970)]
Geotrichoides kefyr nom. nud. (Langeron & Talice, 1932)
Oidium breve nom. nud.
Candida benhamii Novák et Vitéz (1964)
Candida paratropicalis Baker, Salkin, Pincus et D'Amato (1981)

Standard description of *Candida tropicalis*

Growth in glucose-yeast extract-peptone water: After 3 days at 25°C the cells are globose, short-ovoid to long ovoid, (4.3–7.2) × (5.8–10.8) μm (Fig. 348). A sediment, a ring and islets are present.

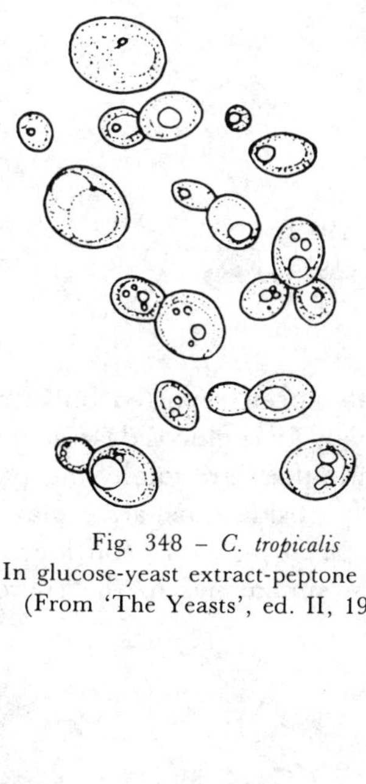

Fig. 348 – *C. tropicalis*
In glucose-yeast extract-peptone water
(From 'The Yeasts', ed. II, 1970)

Growth on glucose-yeast extract-peptone agar: After one month at 25°C the streak culture is cream-colored, off-white to grey, dull, soft, smooth and creamy or wrinkled and tough. A mycelial border is evident.

Dalmau plate culture on corn meal agar: Pseudomycelium is abundantly formed and consists of long, branched pseudohyphae bearing blastospores singly, in short chains and clusters (Fig. 349). True mycelium also may occur.

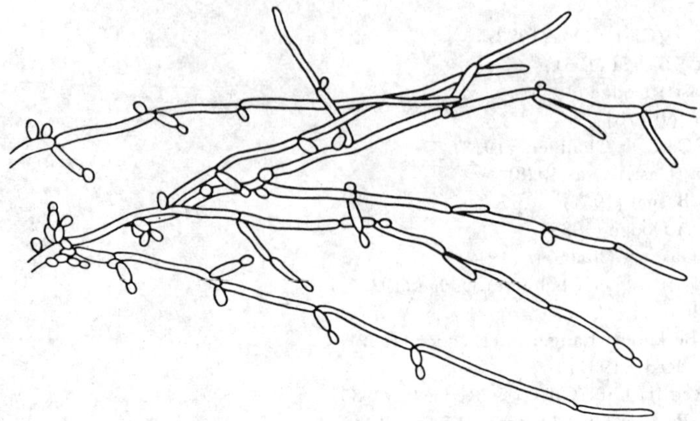

Fig. 349 – *C. tropicalis*
Dalmau plate culture on corn meal agar (From 'The Yeasts', ed. II, 1970)

Fermentation:

Glucose	+	Maltose	+
Galactose	+	Lactose	–
Sucrose	v	Trehalose	+ or s

Assimilation of carbon compounds:

Galactose	+	Melezitose	v	Ribitol	v
L-Sorbose	v	Soluble starch	+	Galactitol	–
Sucrose	v	D-Xylose	+	D-Mannitol	+
Maltose	+	L-Arabinose	+ w or –	D-Glucitol	+
Cellobiose	v	D-Arabinose	–	Salicin	v
Trehalose	+	D-Ribose	v	DL-Lactic acid	v
Lactose	–	L-Rhamnose	–	Succinic acid	+
Melibiose	–	Glycerol	v	Citric acid	v
Raffinose	–	Erythritol	–	Inositol	–

Additional carbon compound tested: Inulin –

Splitting of arbutin: v

Assimilation of nitrate: –

Growth in vitamin-free medium: –

Growth at 37°C: +

Color reaction with DBB: –

Coenzyme Q system: Q 9 (Yamada & Kondo, 1972)

G + C: 35.9–36.1 mol%, 3 strains (Meyer et al., 1975)

Comments

Sucrose-negative isolates of *C. tropicalis* have, to date, been reported only from clinical specimens (Ahearn et al., 1977). These isolates vary in α-glucosidase activity, some show assimilatory but lack fermentative ability and others are totally incapable of α-glucosidase activity. Strains which are initially α-glucosidase negative may acquire this capacity after maintenance in culture. *C. paratropicalis* is considered a synonym based on its morphological and physiological similarities to the sucrose-negative isolates of *C. tropicalis*.

Origin of the strains studied

Human material on Ceylon (1), blood (16), sputum (2), urine (4), fodder (1), oil (1), miscellaneous clinical (5), Chesapeake Bay, U.S.A. (8), soil (1), seawater (1), bronchial washings (1), maxillary sinus (1), unknown (1).

Type strain: CBS 94, received from Castellani; it was probably isolated on Ceylon from human material.

182. CANDIDA TSUKUBAENSIS ONISHI (1972)

Standard description of *Candida tsukubaensis*

Growth in glucose-yeast extract-peptone water: After 3 days at 25°C only a few single elongated cells 2.9 × 8.7 μm are present. The growth is primarily long, irregularly branched pseudomycelium. A granular sediment is evident. After one month a ring is present and the culture appears brownish and viscous.

Growth on malt agar: After one month at 25°C the streak culture is yellowish, slightly folded and wrinkled (giving a dry, cracked appearance) with a fringed margin.

Dalmau plate culture on corn meal agar: True mycelium consists of ramified chains of broad hyphae which often appear fragmented.

Fermentation: –

Assimilation of carbon compounds:

Galactose	+	Melezitose	+	Ribitol	–
L-Sorbose	+	Soluble starch	+	Galactitol	–
Sucrose	+	D-Xylose	+	D-Mannitol	–
Maltose	+	L-Arabinose	+	D-Glucitol	–
Cellobiose	+	D-Arabinose	+ s	Salicin	–
Trehalose	+	D-Ribose	+	DL-Lactic acid	+
Lactose	+	L-Rhamnose	–	Succinic acid	+
Melibiose	–	Glycerol	+	Citric acid	+
Raffinose	+	Erythritol	+	Inositol	+

Splitting of arbutin: +

Assimilation of nitrate: +

Growth in vitamin-free medium: +

Growth at 37°C: +

Urease: +

Color reaction with DBB: +

Origin of the strain studied

Flower on Mount Tsukuba, Japan (1).

Type strain: CBS 6389.

183. CANDIDA UTILIS (HENNEBERG) LODDER ET KREGER-VAN RIJ (1952)

Perfect state: *Hansenula jadinii* (A. et R. Sartory, Weill et Meyer) Wickerham

Synonyms

Torula utilis Henneberg (1926)
Torulopsis utilis (Henneberg) Lodder (1934)
Cryptococcus utilis (Henneberg) Anderson et Skinner (1947)
Candida guilliermondii (Cast.) Langeron et Guerra var. *nitratophila* Diddens et Lodder (1942)
Torulopsis utilis (Henneberg) Lodder var. *major* Thaysen et Morris (1943)

Standard description of *Candida utilis* (van Uden & Buckley, 1970)

Growth in glucose-yeast extract-peptone water: After 3 days at 25°C the cells are ovoid to cylindrical, (3.5–4.5) × (7–13) μm (Fig. 351). A thin ring may be present.

Growth on glucose-yeast extract-peptone agar: After one month at 25°C the streak culture is greyish to cream-colored, semidull or, more rarely, glistening, soft, smooth in most strains, finely punctuated in other strains.

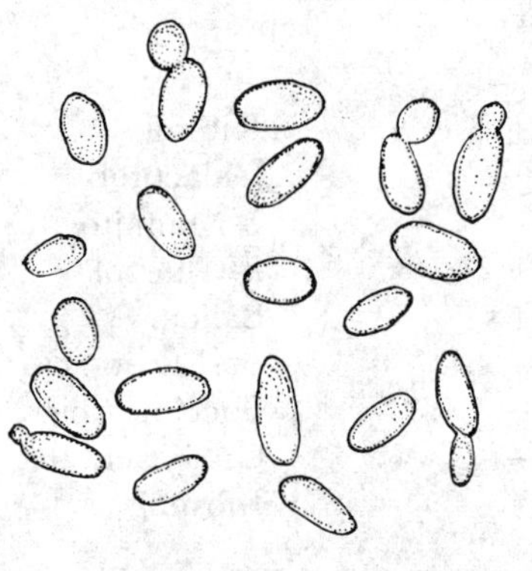

Fig. 351 – *C. utilis*
After 3 days in malt extract
(From 'The Yeasts' ed. II, 1970)

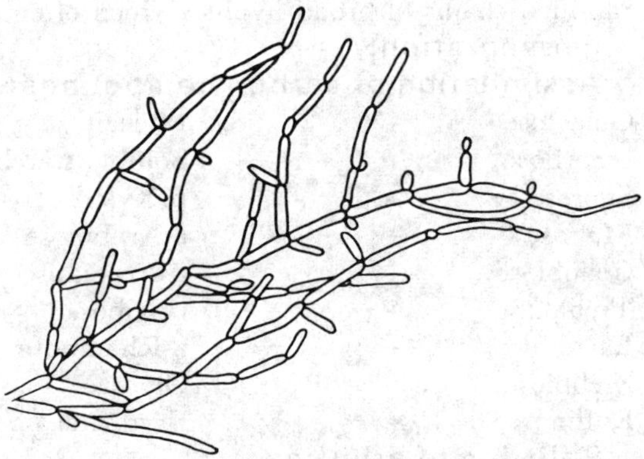

Fig. 352 – *C. utilis*
Dalmau plate culture on corn meal agar
(From 'The Yeasts' ed. II, 1970)

Dalmau plate culture on corn meal agar: A primitive pseudomycelium is abundantly formed. It consists of branched chains of short, coarse pseudohyphae and ovoid cells (Fig. 352).

Fermentation:

Glucose	+	Maltose	–
Galactose	–	Lactose	–
Sucrose	+	Raffinose	+

Assimilation of carbon compounds:

Galactose	–	Melezitose	+	Ribitol	–
L-Sorbose	–	Soluble starch	–	Galactitol	–
Sucrose	+	D-Xylose	+ or w	D-Mannitol	v
Maltose	+	L-Arabinose	–	D-Glucitol	–
Cellobiose	+	D-Arabinose	–	Salicin	+
Trehalose	+ or w	D-Ribose	–	DL-Lactic acid	+
Lactose	–	L-Rhamnose	–	Succinic acid	+ or w
Melibiose	–	Glycerol	+	Citric acid	+
Raffinose	+	Erythritol	–	Inositol	–

Splitting of arbutin: +
Assimilation of nitrate: +
Growth in vitamin-free medium: v
Growth at 37°C: +
Color reaction with DBB: –
Coenzyme Q system: Q7 (Yamada & Kondo, 1972)
G + C: 45.8 mol% (Stenderup & Bak, 1968)

Origin of the strains studied

Yeast factory (1), unknown (8), food yeast (2).

Type strain: CBS 621, isolated by Henneberg (1926) in Germany.

184. CANDIDA VALDIVIANA GRINBERGS ET YARROW (1970a)

Standard description of *Candida valdiviana*

Growth in glucose-yeast extract-peptone water: After 3 days at 25°C the cells are globose, oval, and elongate, (2.2–3.6) × (2.9–5.1) μm and occur singly, budding and in short chains (Fig. 353).

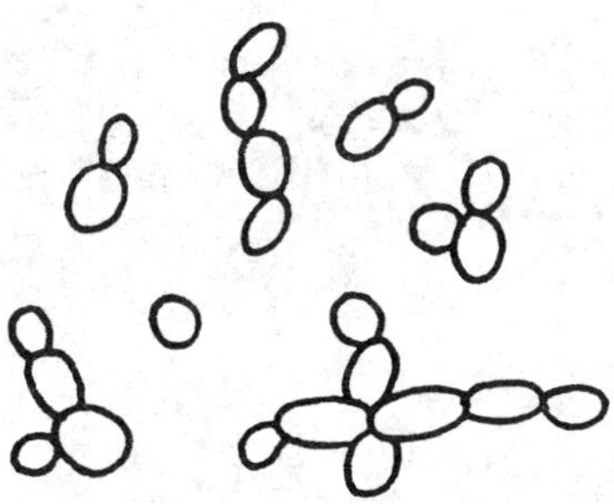

Fig. 353 – *C. valdiviana*
In glucose-yeast extract-peptone water

Growth on glucose-yeast extract-peptone agar: After one month at 25°C the streak culture is white, creamy, glistening, smooth and entire.

Dalmau plate culture on corn meal agar: Extensive pseudomycelium is formed. Branched chains of elongated cells bear chains and clusters of round to oval blastospores.

Fermentation:

Glucose	+s	Maltose	–
Galactose	–	Lactose	–
Sucrose	–		

Assimilation of carbon compounds:

Galactose	+	Melezitose	+ or s	Ribitol	+s or –
L-Sorbose	+	Soluble starch	–	Galactitol	+s or –
Sucrose	+	D-Xylose	+	D-Mannitol	+
Maltose	+	L-Arabinose	+s or –	D-Glucitol	+
Cellobiose	+	D-Arabinose	+s or –	Salicin	+
Trehalose	+	D-Ribose	–	DL-Lactic acid	+s or –
Lactose	v	L-Rhamnose	–	Succinic acid	+
Melibiose	+	Glycerol	+ or s	Citric acid	+
Raffinose	+	Erythritol	–	Inositol	+

Splitting of arbutin: +

Assimilation of nitrate: +

Growth in vitamin-free medium: –

Growth at 37°C: –

Urease: –

Color reaction with DBB: –

G + C: 55.3 mol% (Meyer, Stenderup & Christiansen, unpublished)

Origin of the strains studied

Sputum (1), unknown (1).

Type strain: CBS 5721, from sputum.

185. CANDIDA VALIDA (LEBERLE) VAN UDEN ET BUCKLEY (1970)

Perfect state: *Pichia membranaefaciens* Hansen

Synonyms

Mycoderma valida Leberle (1909)
Mycokluyveria valida (Leberle) Ciferri et Redaelli (1947)
Mycoderma decolorans Will (1910)
Mycokluyveria decolorans (Will) Ciferri et Redaelli (1947)
Mycoderma tannicum Asai (1918)
Mycokluyveria tannica (Asai) Ciferri et Redaelli (1947)
Mycoderma lafarii Janke (1930)
Mycokluyveria lafarii (Janke) Ciferri et Redaelli (1947)
Candida mycoderma (Reess) Lodder et Kreger-van Rij (1952) pro parte
Azymocandida mycoderma (Reess) Novák et Zsolt (1961) pro parte

Standard description of *Candida valida*

Growth in glucose-yeast extract-peptone water: After 3 days at 25°C the cells are ovoid to cylindrical, $(1.4–3.6) \times (3.6–14.4)$ μm and occur singly and in pairs, clusters and chains (Fig. 354). Pseudomycelium is evident. A pellicle is present and growth extends up the sides of the flask. After one month a sediment is present.

Growth on glucose-yeast extract-peptone agar: After one month at 25°C the streak culture is off-white to beige, dull, flat, and delicately wrinkled. The margin is irregular and fringed with mycelium.

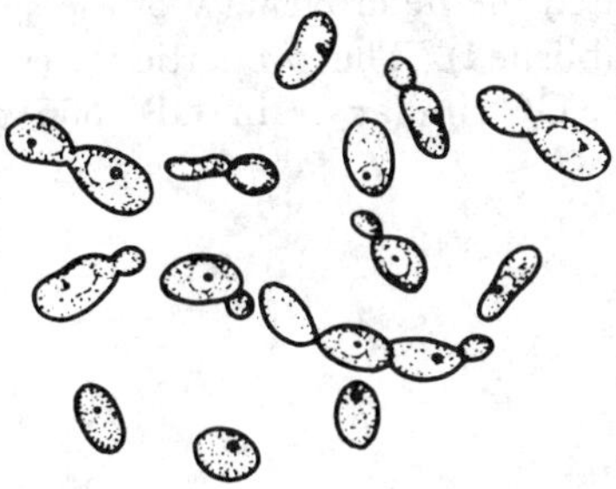

Fig. 354 – *C. valida*
In glucose-yeast extract-peptone water
(From 'The Yeasts' ed. II, 1970)

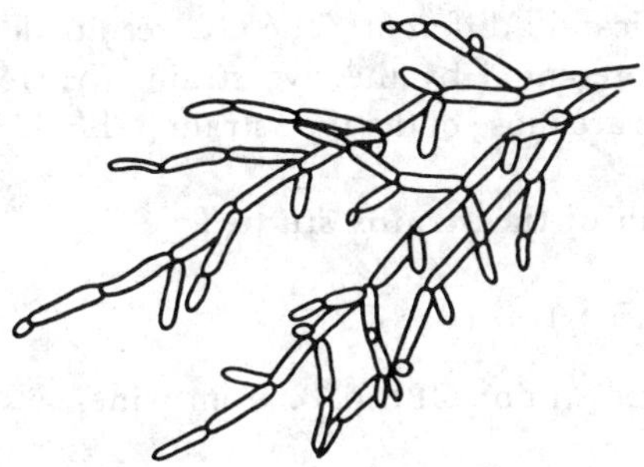

Fig. 355 – *C. valida*
Dalmau plate culture on corn meal agar
(From 'The Yeasts' ed. II, 1970)

Dalmau plate culture on corn meal agar: Abundant development of pseudomycelium consisting of branched chains of ovoid and cylindrical cells. Pseudohyphae may be long, slightly curved with few blastospores, giving a feathery appearance (Fig. 355).

Fermentation:

Glucose	+ s	Maltose	–
Galactose	–	Lactose	–
Sucrose	–		

Assimilation of carbon compounds:

Galactose	–	Melezitose	–	Ribitol	–
L-Sorbose	–	Soluble starch	–	Galactitol	–
Sucrose	–	D-Xylose	–	D-Mannitol	–
Maltose	–	L-Arabinose	–	D-Glucitol	–
Cellobiose	–	D-Arabinose	–	Salicin	–
Trehalose	–	D-Ribose	–	DL-Lactic acid	v
Lactose	–	L-Rhamnose	–	Succinic acid	v
Melibiose	–	Glycerol	+	Citric acid	–
Raffinose	–	Erythritol	–	Inositol	–

Additional carbon compound tested: Glucosamine · HCl +

Splitting of arbutin: –

Assimilation of nitrate: –

Growth in vitamin-free medium: v

Growth at 37°C: –

Color reaction with DBB: –

G + C: 43.2 mol%, type strain (Meyer, Stenderup & Christiansen, unpublished)

Comments

The type strain and five other strains previously identified as *C. valida* were obtained from the CBS and ATCC for DNA studies. The determination of DNA base composition (%GC) by the thermal denaturation method revealed only one strain had a %GC similar to that of the type strain. The values for the remaining four strains were significantly different. The GC results demonstrated the heterogeneity of the species as represented by the five strains (Meyer, unpublished). The properties presented above are those of the type strain (CBS 638) and the GC similar strain (CBS 636) only.

Origin of the strains studied

Wine (1), lager (1).

Type strain: CBS 638, from wine.

186. CANDIDA VANDERWALTII (VIDAL-LEIRIA) MEYER ET YARROW (Yarrow & Meyer, 1978)

Synonym

Torulopsis vanderwaltii Vidal-Leiria (1966b)

Standard description of *Candida vanderwaltii*

Growth in glucose-yeast extract-peptone water: After 3 days at 25°C the cells are ovoid or ellipsoid, (2–3) × (5–6) μm (Fig. 356). A ring and islets may be present (van Uden & Vidal-Leiria, 1970).

Growth on glucose-yeast extract-peptone agar: After one month at 25°C the streak culture is whitish cream-colored, glossy, soft and smooth (van Uden & Vidal-Leiria, 1970).

Dalmau plate culture on corn meal agar: No pseudomycelium is formed; ramified chains of ovoid cells may occur (van Uden & Vidal-Leiria, 1970).

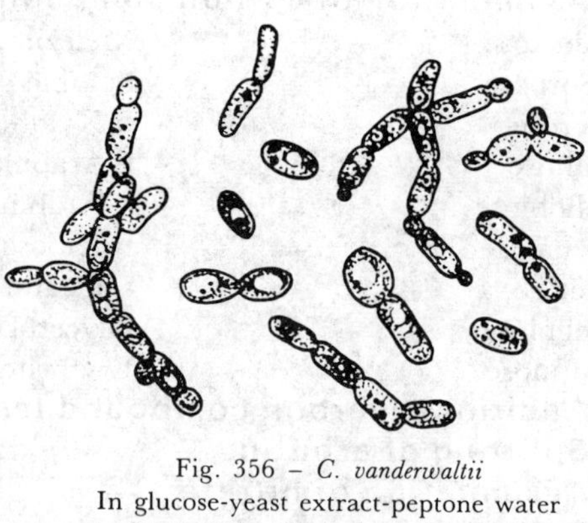

Fig. 356 – *C. vanderwaltii*
In glucose-yeast extract-peptone water
(From 'The Yeasts' ed. II, 1970)

Fermentation: –

Assimilation of carbon compounds:

Galactose	+	Melezitose	–	Ribitol	+
L-Sorbose	+	Soluble starch	–	Galactitol	– or +s
Sucrose	+s or –	D-Xylose	+	D-Mannitol	+
Maltose	–	L-Arabinose	+	D-Glucitol	+
Cellobiose	–	D-Arabinose	+s	Salicin	–
Trehalose	+	D-Ribose	+s	DL-Lactic acid	–
Lactose	–	L-Rhamnose	–	Succinic acid	+
Melibiose	–	Glycerol	+	Citric acid	+
Raffinose	–	Erythritol	–	Inositol	–

Splitting of arbutin: –

Assimilation of nitrate: +

Growth in vitamin-free medium: –

Growth at 37°C: v

Color reaction with DBB: –

G + C: 49.0 mol%, type strain (Meyer, Stenderup & Christiansen, unpublished)

Origin of the strains studied

Winery equipment in South Africa (2).

Type strain: CBS 5524.

187. CANDIDA VARIABILIS (LINDNER) BERKHOUT

see *Pichia burtonii* p. 313

188. CANDIDA VARTIOVAARAI (CAPRIOTTI) VAN UDEN ET BUCKLEY (1970)

Synonym

Torulopsis vartiovaarai Capriotti (1961e)

Standard description of *Candida vartiovaarai*

Growth in glucose-yeast extract-peptone water: After 3 days at 25°C the cells are short-ovoid to long-ovoid, (2.5–4) × (4.5–8) μm (Fig. 357). A pellicle may be present (van Uden & Buckley, 1970).

Growth on glucose-yeast extract-peptone agar: After one month at 25°C the streak culture is greyish cream-colored, semidull, soft and smooth (van Uden & Buckley, 1970).

Dalmau plate culture on corn meal agar: A well differentiated pseudomycelium is formed. It consists of ramified chains of elongate cells bearing clusters of globose and short-ovoid blastospores in verticillated positions (Fig. 358) (van Uden & Buckley, 1970).

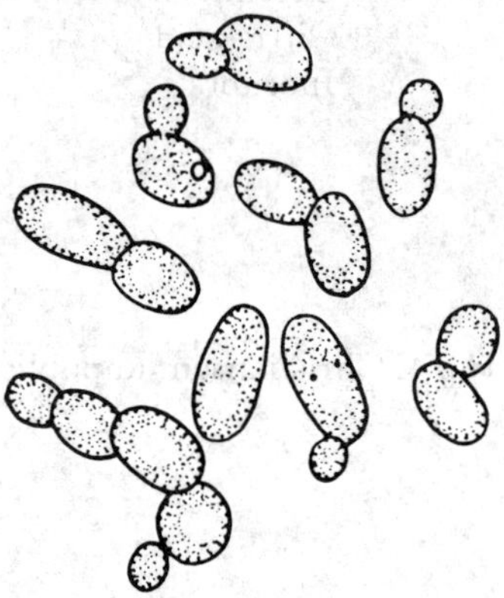

Fig. 357 – *C. vartiovaarai*
In glucose-yeast extract-peptone water
(From 'The Yeasts' ed. II, 1970)

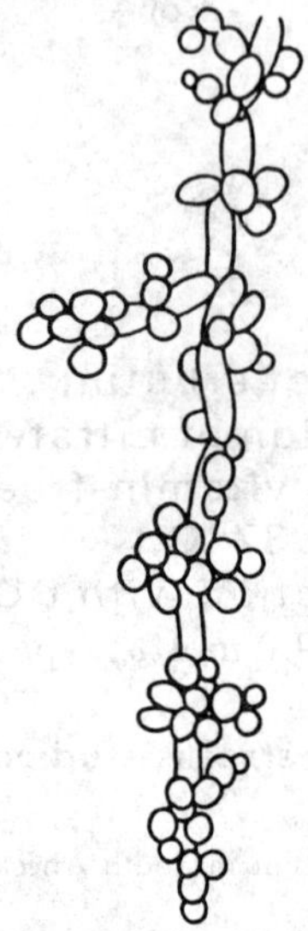

Fig. 358 – *C. vartiovaarai*
Dalmau plate culture on corn meal agar
(From 'The Yeasts' ed. II, 1970)

Fermentation:

Glucose	+	Maltose	–
Galactose	–	Lactose	–
Sucrose	+		

Assimilation of carbon compounds:

Galactose	–	Melezitose	+	Ribitol	–
L-Sorbose	–	Soluble starch	–	Galactitol	–
Sucrose	+	D-Xylose	+	D-Mannitol	+
Maltose	+	L-Arabinose	–	D-Glucitol	+
Cellobiose	+	D-Arabinose	–	Salicin	+
Trehalose	+	D-Ribose	–	DL-Lactic acid	+
Lactose	–	L-Rhamnose	–	Succinic acid	+
Melibiose	–	Glycerol	+	Citric acid	+
Raffinose	–	Erythritol	–	Inositol	–

Splitting of arbutin: +

Assimilation of nitrate: +

Growth in vitamin-free medium: +

Growth at 37°C: –

Color reaction with DBB: –

G + C: 48.8 mol%, type strain (Meyer, Stenderup & Christiansen, unpublished)

Origin of the strains studied

Soil in Finland (Capriotti, 1961e) (2); cider in Great Britain (1); water (1).

Type strain: CBS 4289, isolated in Finland from soil (Capriotti, 1961e).

189. CANDIDA VERONAE FLORENZANO EX VAN UDEN ET BUCKLEY (1970)

Synonym

Trichosporon veronae nom. nud. (Florenzano, 1953)

Standard description of *Candida veronae*

Growth in glucose-yeast extract-peptone water: After 3 days at 25°C the cells are long-ovoid to cylindrical, (2–5) × (4–10) μm. They often occur in groups. Long pseudohyphal cells and pseudomycelial formation may also occur (Fig. 359). A ring is formed (van Uden & Buckley, 1970).

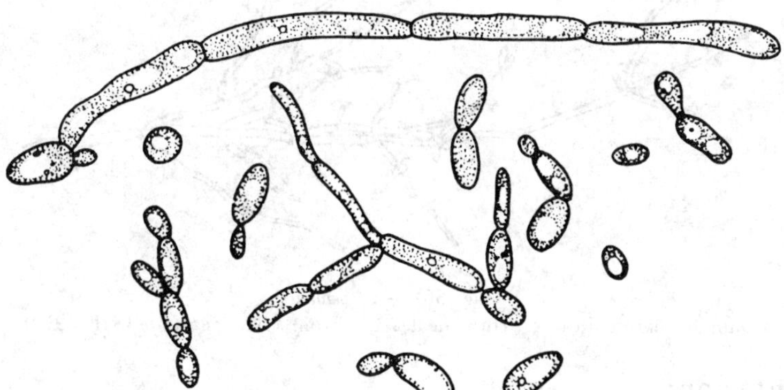

Fig. 359 – *C. veronae*
In glucose-yeast extract-peptone water (From 'The Yeasts' ed. II, 1970)

Growth on glucose-yeast extract-peptone agar: After one month at 25°C the streak culture is yellowish cream-colored, semidull and soft; the surface has a delicate, punctilated structure (van Uden & Buckley, 1970).

Dalmau plate culture on corn meal agar: The pseudomycelium is of two types. Long, wavy pseudohyphae may branch in an open, irregular way (Fig. 360), or the pseudohyphae may have a tree-like arrangement (Fig. 361). In both forms ovoid blastospores occur scarcely and may form short chains, either pleurally or terminally (van Uden & Buckley, 1970).

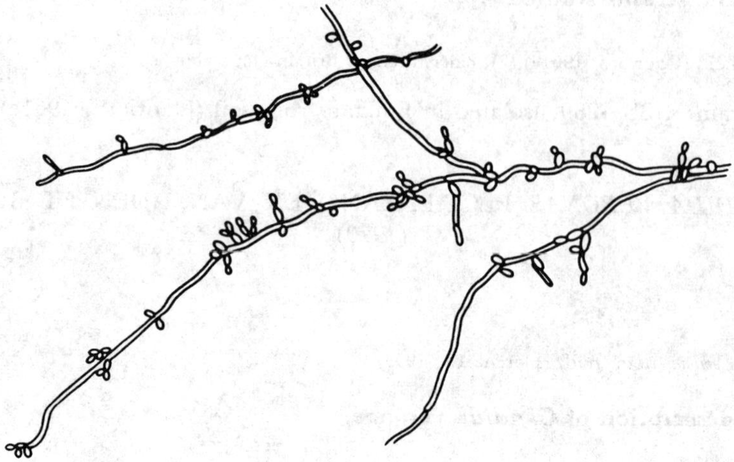

Fig. 360 – *C. veronae*
Dalmau plate culture on corn meal agar (From 'The Yeasts' ed. II, 1970)

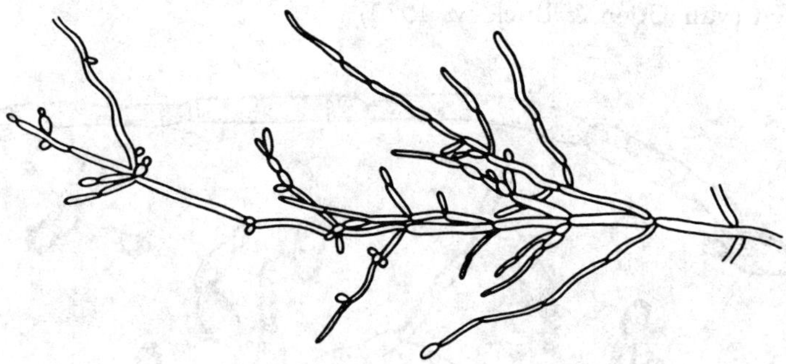

Fig. 361 – *C. veronae*
Dalmau plate culture on corn meal agar (From 'The Yeasts' ed. II, 1970)

Fermentation:

Glucose	+	Maltose	–
Galactose	+ s or –	Lactose	–
Sucrose	–	Trehalose	+ s

Assimilation of carbon compounds:

Galactose	+	Melezitose	+	Ribitol	+
L-Sorbose	–	Soluble starch	–	Galactitol	–
Sucrose	+	D-Xylose	+	D-Mannitol	+
Maltose	+	L-Arabinose	+	D-Glucitol	+
Cellobiose	+	D-Arabinose	+ s	Salicin	+ s
Trehalose	+	D-Ribose	+	DL-Lactic acid	–
Lactose	–	L-Rhamnose	+	Succinic acid	+
Melibiose	–	Glycerol	+	Citric acid	+
Raffinose	–	Erythritol	+	Inositol	–

Splitting of arbutin: +s
Assimilation of nitrate: –
Growth in vitamin-free medium: –
Growth at 37°C: +
Color reaction with DBB: –
G + C: 43.7 mol%, type strain (Meyer & Phaff, 1972)

Origin of the strain studied

Wine must in Italy (1).

Type strain: CBS 5815.

190. *CANDIDA VERSATILIS* (ETCHELLS ET BELL) MEYER ET YARROW (Yarrow & Meyer, 1978)

Synonyms

Brettanomyces versatilis Etchells et Bell (1950b)
Torulopsis versatilis (Etchells et Bell) Lodder et Kreger-van Rij (1952)
Torulopsis anomala Lodder et Kreger-van Rij (1952)

Standard description of *Candida versatilis*

Growth in glucose-yeast extract-peptone water: After 3 days at 25°C the cells are globose to short-ovoid and ovoid, (2.5–3.5) × (3–4.5) μm (Fig. 362). A ring is present (van Uden & Vidal-Leiria, 1970).

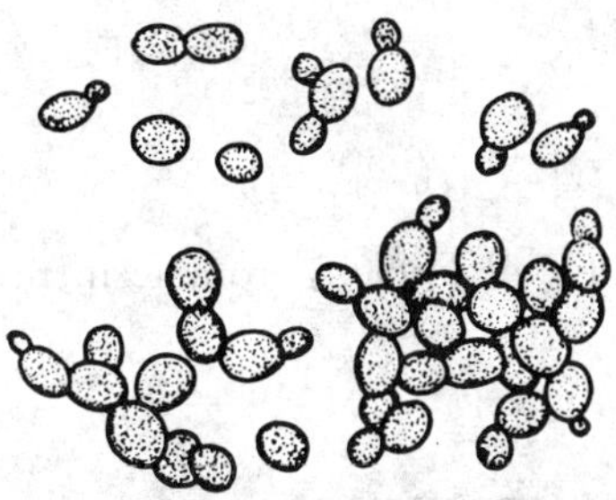

Fig. 362 – *C. versatilis*
In glucose-yeast extract-peptone water
(From 'The Yeasts' ed. II, 1970)

Growth on glucose-yeast extract-peptone agar: After one month at 25°C the streak culture is cream-colored with a brownish tinge, glistening, soft and smooth (van Uden & Vidal-Leiria, 1970).

Dalmau plate culture on corn meal agar: No pseudomycelium is formed. A few short chains of ovoid cells may be present (van Uden & Vidal-Leiria, 1970).

Fermentation:

Glucose	+	Lactose	+w or –
Galactose	+	Raffinose	–
Sucrose	v	Trehalose	+
Maltose	+	Cellobiose	v

Assimilation of carbon compounds:

Galactose	+	Melezitose	–	Ribitol	–
L-Sorbose	–	Soluble starch	–	Galactitol	–
Sucrose	v	D-Xylose	–	D-Mannitol	v
Maltose	+	L-Arabinose	v	D-Glucitol	+
Cellobiose	v	D-Arabinose	–	Salicin	v
Trehalose	+	D-Ribose	v	DL-Lactic acid	–
Lactose	v	L-Rhamnose	–	Succinic acid	–
Melibiose	v	Glycerol	+	Citric acid	–
Raffinose	v	Erythritol	–	Inositol	–

Splitting of arbutin: v

Assimilation of nitrate: +

Growth in vitamin-free medium: +w

Growth at 37°C: –

Color reaction with DBB: –

Coenzyme Q system: Q9 (Yamada & Kondo, 1972)

G + C: 53.7 mol%, type strain (Meyer, Stenderup & Christiansen, unpublished); Nakase & Komagata (1971c) found 45.4–46.6 mol% G + C for 3 strains they studied.

Comments

From the degree of variability demonstrated in the physiological tests plus the results reported on % G + C, it is obvious that *Candida versatilis,* as represented by the strains presently included, is a heterogeneous species.

Origin of the strains studied

Fermenting cucumber brine, U.S.A. (5); a pickle brine tank, U.S.A. (1); sugar (1).

Type strain: CBS 1752, isolated by Etchells and Bell (1950b) from fermenting cucumber brine.

191. CANDIDA VINARIA OHARA, NONOMURA ET YUNOME EX SMITH (1973)

Synonym

Candida vinaria nom. nud. (Ohara et al., 1960b)

Standard description of *Candida vinaria*

Growth in glucose-yeast extract-peptone water: After 3 days at 25°C the cells are globose to ovoid, (2.2–3.6) × (3.6–5.7) μm and elongate. They occur singly and in pairs. After one month the sediment appears floccose.

Growth on glucose-yeast extract-peptone agar: After one month at 25°C the streak culture is off-white, smooth, shiny and entire.

Dalmau plate culture on corn meal agar: A primitive pseudomycelium may be formed.

Fermentation: –

Assimilation of carbon compounds:

Galactose	+	Melezitose	–	Ribitol	–
L-Sorbose	+	Soluble starch	–	Galactitol	–
Sucrose	–	D-Xylose	+ s or –	D-Mannitol	+ s or –
Maltose	–	L-Arabinose	–	D-Glucitol	+ s or –
Cellobiose	–	D-Arabinose	–	Salicin	–
Trehalose	–	D-Ribose	+ s or –	DL-Lactic acid	–
Lactose	–	L-Rhamnose	–	Succinic acid	–
Melibiose	–	Glycerol	+	Citric acid	–
Raffinose	–	Erythritol	–	Inositol	–

Splitting of arbutin: –

Assimilation of nitrate: –

Growth in vitamin-free medium: –

Growth at 37°C: –

Urease: –

Color reaction with DBB: –

Coenzyme Q system: Q9 (Yamada & Kondo, 1972)

G + C: 44.1 mol% (Nakase & Komagata, 1971f)

Comments

C. vinaria was previously considered a synonym of *C. zeylanoides* (van Uden & Buckley, 1970). Although the physiological characteristics of these species are similar, their GC contents are significantly different (*C. vinaria,* 44.1%; *C. zeylanoides,* 57.6%) and indicative of distinct species. These species may be distinguished on β-glucoside utilization. *C. vinaria* lacks the ability to utilize β-glucosides whereas *C. zeylanoides* possesses this capacity.

Origin of the strain studied

Grape must in Japan (1).

Type strain: CBS 4077.

192. CANDIDA VINI (DESMAZIÈRES EX LODDER) VAN UDEN ET BUCKLEY (1970)

Synonyms

Mycoderma vini Desmazières (1823) ex Lodder (1934)
Mycoderma vini Des. var. *paradoxa* nom. nud. (Zimmermann, 1938)
Mycokluyveria vini (Desm.) Ciferri et Redaelli (1947)
Saccharomyces mycoderma Reess (1870)
Candida mycoderma (Reess) Lodder et Kreger-van Rij (1952) pro parte
Azymocandida mycoderma (Reess) Novák et Zsolt (1961) pro parte
Mycoderma cerevisiae Desmazières (1823) ex Leberle (1909)
Mycokluyveria cerevisiae (Desm.) Ciferri et Redaelli (1947)
Mycoderma gallica Leberle (1909)

Standard description of *Candida vini*

Growth in glucose-yeast extract-peptone water: After 3 days at 25°C the cells are ovoid to cylindrical, (2.9–4.3) × (5.8–10.1) μm and occur singly, in pairs, short chains and small clusters (Fig. 363). A light sediment is apparent.

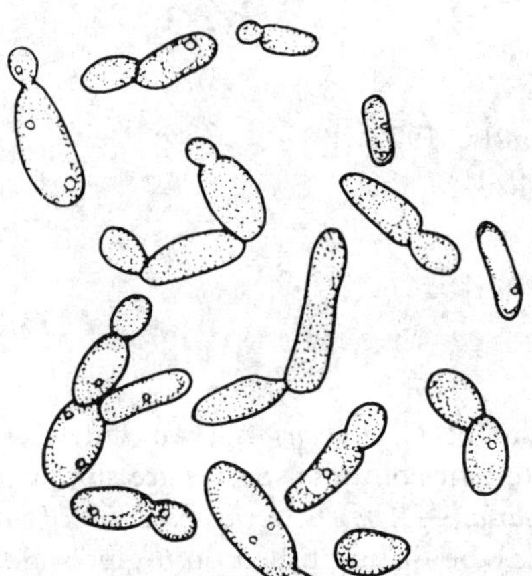

Fig. 363 – *C. vini*
In glucose-yeast extract-peptone water
(From 'The Yeasts' ed. II, 1970)

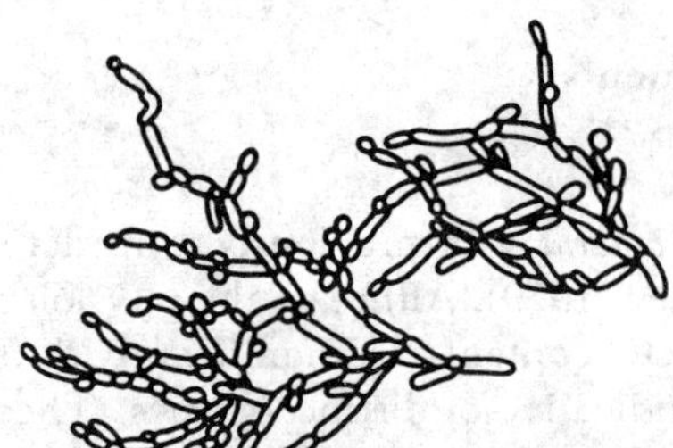

Fig. 364 – *C. vini*
Dalmau plate culture on corn meal agar
(From 'The Yeasts' ed. II, 1970)

Growth on glucose-yeast extract-peptone agar: After one month at 25°C the streak culture is cream-colored, soft, butyrous, smooth, somewhat shiny with a lobed margin.

Dalmau plate culture on corn meal agar: The pseudomycelium consists of branched chains of cylindrical cells that bear ovoidal blastospores. Pseudomycelial cells and blastospores are similar in size and shape (Fig. 364). Short tufts of pseudohyphae growing out of the streak give a "bushy" formation.

Fermentation: –

Assimilation of carbon compounds:

Galactose	–	Melezitose	–	Ribitol	v
L-Sorbose	–	Soluble starch	–	Galactitol	–
Sucrose	–	D-Xylose	–	D-Mannitol	+
Maltose	–	L-Arabinose	–	D-Glucitol	+
Cellobiose	–	D-Arabinose	–	Salicin	–
Trehalose	–	D-Ribose	–	DL-Lactic acid	v
Lactose	–	L-Rhamnose	–	Succinic acid	v
Melibiose	–	Glycerol	v	Citric acid	–
Raffinose	–	Erythritol	–	Inositol	–

Additional carbon compound tested: Methanol –

Splitting of arbutin: –

Assimilation of nitrate: –

Growth in vitamin-free medium: –

Growth at 37°C: –

Color reaction with DBB: –

G + C: 34.1 mol%, type strain (Meyer, Stenderup & Christiansen, unpublished)

Origin of the strains studied

Sour wine, studied by Lodder (1934) and classified as *Mycoderma vini* (1); turbid wines (3), one of these strains described as *Mycoderma vini* var. *paradoxa* by Zimmermann (1938); one, labelled *Mycoderma gallica* from Germany, classified by Lodder (1934) as *Mycoderma cerevisiae;* one, labelled *Mycoderma cerevisiae,* from Denmark; wines in Italy (2); beer in The Netherlands (1).

Type strain: CBS 639 from sour wine.

193. CANDIDA VISWANATHII SANDHU ET RANDHAWA (1962)

Standard description of *Candida viswanathii* (van Uden & Buckley, 1970)

Growth in glucose-yeast extract-peptone water: After 3 days at 25°C the cells are globose, ovoid or cylindrical, (2.5–7) × (4–12) μm (Fig. 365). Pseudomycelial formations may be present (van Uden & Buckley, 1970).

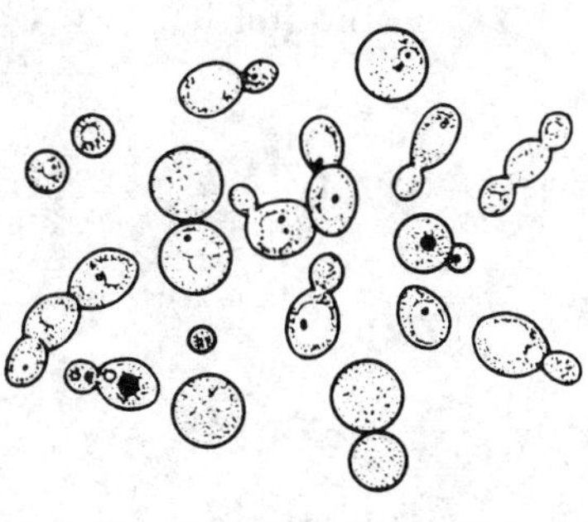

Fig. 365 – *C. viswanathii*
In glucose-yeast extract-peptone water
(From 'The Yeasts' ed. II, 1970)

Growth on glucose-yeast extract-peptone agar: After one month at 25°C the streak culture is cream-colored, semidull, soft to membranous, wrinkled; hairy projections may be present (van Uden & Buckley, 1970).

Dalmau plate culture on corn meal agar: The pseudomycelium consists of long wavy pseudohyphae arranged in irregularly branched formations bearing verticillated, branched chains of ovoid blastospores (Fig. 366) (van Uden & Buckley, 1970).

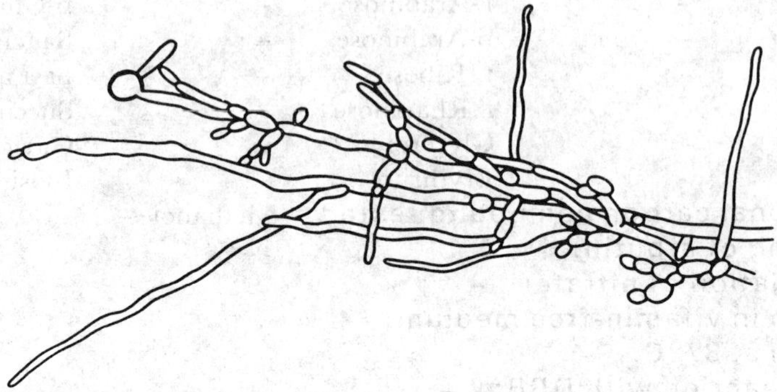

Fig. 366 – *C. viswanathii*
Dalmau plate culture on corn meal agar (From 'The Yeasts' ed. II, 1970)

Fermentation:

Glucose	+	Maltose	+
Galactose	+ w	Lactose	–
Sucrose	+ s or –	Trehalose	+ or s

Assimilation of carbon compounds:

Galactose	+	Melezitose	+	Ribitol	+
L-Sorbose	–	Soluble starch	+	Galactitol	–
Sucrose	+	D-Xylose	+	D-Mannitol	+
Maltose	+	L-Arabinose	+ s	D-Glucitol	+
Cellobiose	+	D-Arabinose	–	Salicin	+ or s
Trehalose	+	D-Ribose	–	DL-Lactic acid	–
Lactose	–	L-Rhamnose	–	Succinic acid	+
Melibiose	–	Glycerol	+	Citric acid	+
Raffinose	–	Erythritol	–	Inositol	–

Splitting of arbutin: +
Assimilation of nitrate: –
Growth in vitamin-free medium: –
Growth at 37°C: +
Color reaction with DBB: –
G + C: 46.3 mol%, type strain (Meyer & Phaff, 1972)

Origin of the strains studied

Cerebrospinal fluid (Viswanathan & Randhawa, 1959) (1); sputum (Sandhu & Randhawa, 1962) (1).

Type strain: CBS 4024, isolated in India from cerebrospinal fluid.

194. *CANDIDA WICKERHAMII* (CAPRIOTTI) MEYER ET YARROW (Yarrow & Meyer, 1978)

Synonym

Torulopsis wickerhamii Capriotti (1958g)

Standard description of *Candida wickerhamii*

Growth in glucose-yeast extract-peptone water: After 3 days at 25°C cells are nearly spherical, 3–6 μm (Fig. 367). Only a sediment is formed (van Uden & Vidal-Leiria, 1970).

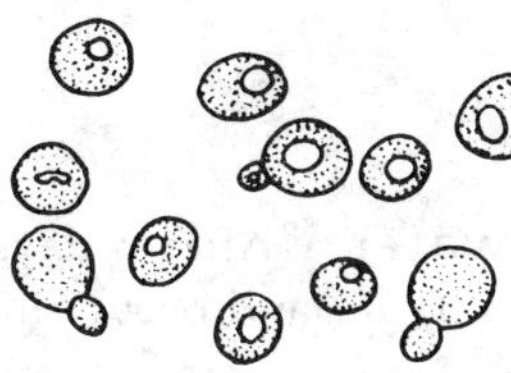

Fig. 367 – *C. wickerhamii*
In glucose-yeast extract-peptone water
(From 'The Yeasts' ed. II, 1970)

Growth on glucose-yeast extract-peptone agar: After one month at 25°C the streak culture is greyish-yellow, butyrous to mucous, semidull to glossy, soft and smooth (van Uden & Vidal-Leiria, 1970).

Dalmau plate culture on corn meal agar: No pseudomycelium is formed.

Fermentation:

Glucose	+	Maltose	–
Galactose	–	Lactose	–
Sucrose	–	Cellobiose	+

Assimilation of carbon compounds:

Galactose	+	Melezitose	–	Ribitol	+
L-Sorbose	–	Soluble starch	–	Galactitol	–
Sucrose	–	D-Xylose	+	D-Mannitol	+
Maltose	–	L-Arabinose	+	D-Glucitol	+
Cellobiose	+	D-Arabinose	+ s	Salicin	+
Trehalose	–	D-Ribose	+	DL-Lactic acid	v
Lactose	–	L-Rhamnose	+	Succinic acid	+ s
Melibiose	–	Glycerol	+	Citric acid	–
Raffinose	–	Erythritol	–	Inositol	–

Splitting of arbutin: +

Assimilation of nitrate: +

Growth in vitamin-free medium: + s

Growth at 37°C: –

Color reaction with DBB: –

G + C: 38.1 mol%, type strain (Meyer, Stenderup & Christiansen, unpublished)

Origin of the strains studied

Silage mixture containing olive husks, molasses and whey in Italy (Capriotti, 1958g) (2).

Type strain: CBS 2928, one of these strains.

195. CANDIDA XESTOBII YARROW ET MEYER (1978)

Synonym

Torulopsi xestobii nom. nud. (Jurzitza, 1970)

Standard description of *Candida xestobii*

Growth in glucose-yeast extract-peptone water: After 3 days at 25°C the cells are round to slightly oval, (3.0–4.5) × (3.5–6.0) μm and occur singly, budding and in pairs.

Growth on glucose-yeast extract-peptone agar: After one month at 25°C the streak culture is tan, smooth, glistening, and entire.

Dalmau plate culture on corn meal agar: Pseudomycelium is absent.

Fermentation:

Glucose	+ s	Maltose	–
Galactose	–	Lactose	–
Sucrose	–		

Assimilation of carbon compounds:

Galactose	+	Melezitose	+ s or –	Ribitol	+ or s
L-Sorbose	+	Soluble starch	–	Galactitol	–
Sucrose	+	D-Xylose	+	D-Mannitol	–
Maltose	+	L-Arabinose	+ s	D-Glucitol	–
Cellobiose	+ s or –	D-Arabinose	–	Salicin	+ s or –
Trehalose	+	D-Ribose	+ s	DL-Lactic acid	–
Lactose	+	L-Rhamnose	–	Succinic acid	+ s
Melibiose	+ s	Glycerol	+	Citric acid	+
Raffinose	+ s or –	Erythritol	–	Inositol	–

Additional carbon compound tested: α-Methyl-D-glucoside + or s

Splitting of arbutin: + s

Assimilation of nitrate: –

Growth in vitamin-free medium: –

Growth on 50% (w/w) glucose-yeast extract agar: –

Growth at 37°C: –

Urease: –

Color reaction with DBB: –

G + C: 48.7 mol%, type strain (Meyer, Stenderup & Christiansen, unpublished)

Origin of the strain studied

Xestobium plumbeum (1).

Type strain: CBS 5975.

196. CANDIDA ZEYLANOIDES (CASTELLANI) LANGERON ET GUERRA (1938)

Synonyms

Monilia zeylanoides Cast. (1920)
Mycotorula zeylanoides (Cast.) Red. et Cif. (Ciferri & Redaelli, 1935)
Parendomyces zeylanoides (Cast.) Dodge (1935)
Pseudomonilia zeylanoides (Cast.) Dodge et Moore (1936)
Azymocandida zeylanoides (Cast.) Novák et Zsolt (1961)
Cryptococcus macroglossiae Cast. (1925)
Monilia macroglossiae Cast. (1925)
Torulopsis macroglossiae (Cast.) Cast. et Jacono (1933)
Mycocandida macroglossiae (Cast.) Red. et Cif. (Ciferri & Redaelli, 1935)
Blastodendrion macroglossiae (Cast.) Langeron et Guerra (1935)
Parendomyces macroglossiae (Cast.) Dodge (1935)
Monilia zeylanoides Cast. var. *macroglossiae* Cast. (1937a)
Cryptococcus uvae Poll. et Nann. (apud Motta, 1926)
Monilia uvae (Poll. et Nann.) Vuillemin (1931)
Torulopsis uvae (Poll. et Nann.) Lodder (1934)
Syringospora uvae (Poll. et Nann.) Dodge (1935)
Blastodendrion canis v. Szilvinyi (1933–'34)
Monilia parazeylanoides Cast. (1937a)
Pichia dubia nom. nud. (Dietrichson, 1954)
Trichosporon piscium Siepmann (in Siepmann & Höhnk, 1962)
Candida iberica Ramirez et Gonzalez (1972)

Standard description of *Candida zeylanoides*

Growth in glucose-yeast extract-peptone water: After 3 days at 25°C the cells are short-ovoid to long-ovoid, (2.9–5.8) × (5.0–10.8) μm and occur singly, in pairs, and short chains (Fig. 368). Much longer cells may occur. A ring may be present. After one month a pellicle may be present.

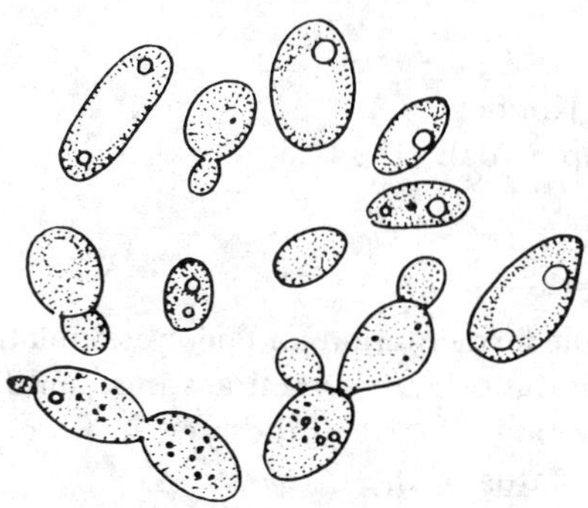

Fig. 368 – *C. zeylanoides*
In glucose-yeast extract-peptone water
(From 'The Yeasts' ed. II, 1970)

Growth on glucose-yeast extract-peptone agar: After one month at 25°C the streak culture is cream-colored to yellow, dull, soft, butyrous, smooth and flat.

Dalmau plate culture on corn meal agar: Pseudomycelium consists of chains of cells, frequently curved, bearing globose, oval or elongate blastospores singly and in small clusters and short chains. Greater blastospore development is present at the proximal end of the pseudohyphal strand (Fig. 369). The pseudomycelium gives a feather-like appearance at low power.

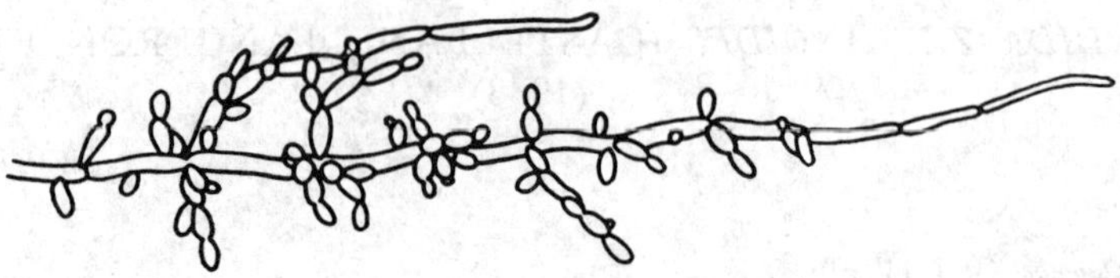

Fig. 369 – *C. zeylanoides*
Dalmau plate culture on corn meal agar (From 'The Yeasts' ed. II, 1970)

Fermentation:

Glucose	– or +w	Maltose	–
Galactose	–	Lactose	–
Sucrose	–	Trehalose	+s or –

Assimilation of carbon compounds:

Galactose	v	Melezitose	–	Ribitol	v
L-Sorbose	+ or s	Soluble starch	–	Galactitol	–
Sucrose	–	D-Xylose	–	D-Mannitol	+
Maltose	–	L-Arabinose	–	D-Glucitol	+
Cellobiose	v	D-Arabinose	–	Salicin	v
Trehalose	+	D-Ribose	–	DL-Lactic acid	–
Lactose	–	L-Rhamnose	–	Succinic acid	+
Melibiose	–	Glycerol	+	Citric acid	+
Raffinose	–	Erythritol	–	Inositol	–

Splitting of arbutin: +
Assimilation of nitrate: –
Growth in vitamin-free medium: v
Growth at 37°C: –
Color reaction with DBB: –
Coenzyme Q system: Q 9 (Yamada & Kondo, 1972)
G + C: 57.6 mol%, type strain (Stenderup & Bak, 1968)

Comments

Strains of *C. tropicalis,* which may give negligible fermentation and poor assimilation reactions upon examination shortly after their isolation, are sometimes misidentified as *C. zeylanoides.* No significant differences were noted between the type strain of *C. iberica* and *C. zeylanoides* to warrant the species status of the *C. iberica.*

Origin of the strains studied

Cr. macroglossiae from blastomycotic macroglossia (1); chilled beef (1); *Cr. uvae* from uvula (1); sputum (1); dog's skin (1); seawater (1); skin (1); fish skin *(Tr. piscium)* (1); red clover (1); feces (2); sausage (1).

Type strain: CBS 619, received from Castellani (London) as *Cryptococcus macroglossiae;* isolated in Ceylon from a case of blastomycotic macroglossia.

f. COMMENTS ON THE GENUS

The genus *Candida,* containing the anamorphs of species of ascomycetous genera as diverse as *Metschnikowia* and *Citeromyces* as well as basidiomycetous genera such as *Leucosporidium* and *Filobasidium,* is an unnatural group of yeasts. The heterogeneity of the genus is emphasized further by the wide range of DNA base composition of 30–66 mol% G + C, and coenzyme Q systems ranging from Q 6 to Q 10 (Yamada & Kondo, 1972). The taxonomic boundaries of *Candida* are broad so that the genus acts as a repository for all the asexual yeasts except those with:
(a) acetic acid production (*Brettanomyces, Eeniella*)
(b) visually detectable red, pink or orange pigments (*Rhodotorula, Phaffia, Oosporidium* and some *Cryptococcus* species)
(c) arthroconidia (*Trichosporon*)
(d) unipolar and bipolar budding on a broad base (*Kloeckera, Malassezia* and *Schizoblastosporion*)
(e) blastoconidia formed on sympodulae (*Sympodiomyces*)
(f) buds formed on stalks (*Sterigmatomyces*)
(g) needle-shaped terminal conidia (*Aciculoconidium*)
(h) triangular cells (*Trigonopsis*)
(i) enteroblastic-basipetal budding usually with mucoid colonies and the ability to grow on inositol as the sole carbon source (*Cryptococcus*)
(j) ballistoconidia (*Bullera* and *Sporobolomyces*)

Van Uden and Buckley (1970), who recognized the imperfect genus *Torulopsis,* pointed out that the ability to form hyphae is an inadequate criterion for separating this genus from *Candida.* While studying the relationships between psychrophobic yeasts by DNA reassociation, Mendonça-Hagler and Phaff (1975) found *Candida slooffii* to be conspecific with *Torulopsis pintolopesii.* They also reached the conclusion that the presence or absence of pseudohyphae is of minor value for differentiating species.

Yarrow and Meyer (1978) called into doubt the legitimacy of the names *Torulopsis* Berlese typified by *Torulopsis rosea* Berlese and *Torulopsis* Lodder et Kreger-van Rij typified by *Torulopsis colliculosa* Hartmann. Agreeing with the arguments of van Uden and Buckley (1970) that there is no sound taxonomic reason to classify filamentous and nonfilamentous species in separate genera, Yarrow and Meyer proposed an amendment to the description of the genus *Candida* to include nonhyphal species.

The presence of lunate and falcate cells, a striking morphological feature, was the differentiating characteristic Yarrow (in von Arx et al., 1977) used to distinguish

Selenozyma from *Candida*. Golubev (1977) induced sporulation in the neotype strain of *Selenozyma intestinalis* and because the asci were club-shaped and the ascospores acicular, the teleomorph was described in *Metschnikowia*. Since the other species of *Metschnikowia* do not have lunate or falcate cells and since their anamorphs are classified in *Candida*, these cell shapes are of no value for generic differentiation. Therefore *Selenozyma peltata* (Yarrow) Yarrow whose teleomorph is unknown was herein transferred to *Candida*.

The separation of *Candida* from the genus *Cryptococcus* is not clear in our opinion. *Cryptococcus* is supposed to differ from *Candida* by the inability to form hyphae and to ferment, combined with the ability to grow on inositol as a sole carbon source. However, there are exceptions. Formation of septate mycelium and pseudohyphae by *Cryptococcus laurentii* was reported by Boidin et al. (1963) in strains isolated in Cameroon, by Rodrigues de Miranda (1974) in strains isolated from soil in Denmark, and observed by one of the present authors in several strains isolated from forest soil in Belgium. These strains are distinguishable from the species recently described as *Candida podzolica* Babjeva and Reshetova (1975). In the description of the genus *Cryptococcus* in this monograph by Rodrigues de Miranda, *Cr. laurentii* is described as having no pseudo- or true hyphae. We recognize *Cr. laurentii* as variable in this characteristic*. The occurrence of hyphal strains of *Cryptococcus neoformans* has also been reported (Shadomy, 1970; Erke & Schneidau, 1973). According to the descriptions given by Phaff and Fell (1970), some strains of *Cryptococcus infirmo-miniatus* and *Cr. macerans* form pseudohyphae. Moreover, the imperfect states of two species of the basidiomycetous *Filobasidium* have been described as *Candida capsuligena* and *Cryptococcus uniguttulatus*. A more natural classification of the imperfect yeasts presently described in *Cryptococcus* and *Candida* could be achieved if only those species with ascomycetous characters are retained in *Candida* and basidiomycetous species such as *Candida capsuligena* are placed in another genus.

Von Arx and Weijman (1979) restricted the genus *Candida* to anamorphs of Saccharomycetales. They transferred some basidiomycetous species of *Candida* with polysaccharides containing xylose to *Apiotrichum* and others with polysaccharides containing fucose to *Rhodotorula*. (*Apiotrichum porosum*, a synonym of *Candida humicola*, was designated the type of the genus). Because von Arx and Weijman (1979) considered the name *Cryptococcus* to be illegitimate, they included former species of both *Candida* and *Cryptococcus* in *Apiotrichum* and *Rhodotorula*. In later studies, Moore (1980) proposed the genus *Vanrija* for the heterobasidiomycetous yeasts then in *Candida* with *Candida humicola* (*Torula humicola* Daszewska) as the type species. Golubev (1981b) classified *Candida humicola*, *C. curvata*, *C. amylolenta*, *C. marina* and *C. podzolica* in *Cryptococcus*, which he amended to include filamentous species. Verona and Florenzano (1980) proposed amending the diagnosis of *Cryptococcus* by adding the statement "DBB positive" in order to distinguish the genus from *Candida*.

In the absence of the teleomorph, the ascomycetous and basidiomycetous nature of a yeast may be discerned by differences in their cell walls. Kreger-van Rij and Veenhuis (1971a) showed the laminar structure of the wall of basidiomycetous yeasts

---

* Since Rodrigues de Miranda does not accept *C. podzolica* in the genus *Cryptococcus*, a description of this species is given under the genus *Candida*, p. 843.

versus two-layered cell wall of the ascomycetes, and Bartnicki-Garcia (1970), Bastide et al. (1979), von Arx and Weijman (1979) and Golubev (1981) reported the difference in the chemical composition of the cell walls of the two groups.

Other characteristics suggested for discrimination between the two groups such as production of extracellular DNAse and urease (Cazin et al., 1969; Sen & Komagata, 1979; Hagler & Ahearn, 1981) and the molar percent G + C of the DNA have been shown to overlap (van der Walt et al., 1974). Thus, until the subdivision of the basidiomycetous anamorphs is clarified, we have chosen to maintain those yeasts not obviously distinguishable by pigmentation or their colony character as a subgroup of the genus *Candida*.

g. SPECIES NOT ACCEPTED IN THE GENUS

Candida podzolica is not accepted by us in the genus *Candida* (see f. Comments on the genus) as we consider it to be identical with *Cryptococcus laurentii*. However, Rodrigues de Miranda does not agree with this reclassification and, in this monograph, does not indicate *C. podzolica* as a synonym of *Cr. laurentii*. Therefore, a description of *Candida podzolica* Babjeva et Reshetova (1975), partly taken over from these authors, follows here.

Standard description of *Candida podzolica*

Growth in malt extract: After 3 days the cells are polymorphic, oval, apiculate to elongate and irregularly spherical, (3.8–4.8) × (4.4–13.8) μm. A sediment and a ring are formed, but no pellicle.

Growth on malt agar: After one month the culture is brownish and markedly folded. Smooth and folded variants may occur.

Growth on potato agar: True and pseudomycelium is formed.

Fermentation: –

Assimilation of carbon compounds:

Galactose	+ s	Melezitose	+	Ribitol	–
L-Sorbose	+	Soluble starch	+	Galactitol	+
Sucrose	+	D-Xylose	+	D-Mannitol	–
Maltose	+	L-Arabinose	+	D-Glucitol	+ s
Cellobiose	+ s	D-Arabinose	+	Salicin	+
Trehalose	+	D-Ribose	+	DL-Lactic acid	+ s
Lactose	+	L-Rhamnose	+	Succinic acid	+
Melibiose	+	Glycerol	–	Citric acid	+
Raffinose	+	Erythritol	–	Inositol	+

Splitting of arbutin: +

Assimilation of nitrate: –

Growth in vitamin-free medium: +

Growth at 37°C: –

Starch formation: +

Urease: +
G + C: 62.8 mol% (Babjeva & Reshetova, 1975)

Type strain: CBS 6819 (BKB Y-1982), isolated from soil.

Species which have not been discussed:

Candida acutus Goto (S. Goto, J. Gen. Appl. Microbiol. *25,* 145–148, 1979)

Candida ethanolica Rybářǒvá et al. (J. Rybářǒvá, F. Stros & A. Kocková-Kratochvílová, Z. Allgem. Mikrobiol. *20,* 279–281, 1980)

Candida fermenticarens van der Walt et Baker (J.P. van der Walt & P.B. Baker, Bothalia *12,* 561–562, 1978)

Candida fusiformata Buhagiar (R.W.M. Buhagiar, J. Gen. Microbiol. *110,* 91–97, 1979)

Candida guilliermondii (Castellani) Langeron et Guerra var. *soya* Onishi et Suzuki (H. Onishi & T. Suzuki, Appl. Microbiol. *18,* 1031–1035, 1969)

Candida methylica Trotsenko et al. (J.A. Trotsenko, S.V. Bykovskaya & N.N. Kirikova, Doklady Akad. Nauk S.S.R. *211,* 1230–1232, 1973)

Candida paludigena Golubev et al. (W.J. Golubev, V.M. Blagodatskaya, S.A. Suetin & S. Trotsenko, Int. J. Syst. Bacteriol. *31,* 91–96, 1981)

Genus 5. *Cryptococcus* Kützing emend. Phaff et Spencer

by L. Rodrigues de Miranda

a. DIAGNOSIS OF THE GENUS

Cells spheroidal, ovoidal, elongate, amoeboid or polymorphic. Reproduction by multilateral budding. Most strains have capsulated cells; extent of capsule formation depends on the medium. Under certain conditions of growth the capsule may contain starch-like compounds, which are released into the medium by many strains.

On solid media most species have a slimy appearance; carotenoid pigments may be produced, but young malt agar cultures of most species are not strongly pigmented. In one species a dark brown to black pigment is formed under certain conditions. Pseudomycelium is either not formed or rudimentary.

Growth is slow in liquid media and initially limited to the formation of a ring; later a sediment is produced. The accumulated growth sometimes changes the entire contents of the flask into a slimy mass.

Ascospores, teliospores or ballistospores are not produced.

Ability to ferment sugars is lacking.

All species assimilate inositol as sole carbon source.

b. TYPE SPECIES OF THE GENUS

Cryptococcus mollis Kützing 1833

c. SPECIES ACCEPTED IN THE GENUS

1. *Cryptococcus albidus* (Saito) Skinner (Saito, 1922)
   Cryptococcus albidus (Saito) Skinner var. *albidus*
   Cryptococcus albidus (Saito) Skinner var. *aerius* (Saito) Phaff et Fell (Saito, 1922)
2. *Cryptococcus ater* Castellani ex Cooke (1966)
3. *Cryptococcus dimennae* Fell et Phaff (1967)
4. *Cryptococcus elinovii* Golubev et Tauson (1979)
5. *Cryptococcus flavus* (Saito) Phaff et Fell (Saito, 1922)
6. *Cryptococcus gastricus* Reiersöl et di Menna (1958)

7. *Cryptococcus heveanensis* (Groenewege) Baptist et Kurtzman (Groenewege, 1921b)
8. *Cryptococcus hungaricus* (Zsolt) Phaff et Fell (Zsolt, 1957)
9. *Cryptococcus infirmo-miniatus* (Okunuki) Phaff et Fell (Okunuki, 1931)
10. *Cryptococcus kuetzingii* Fell et Phaff (1967)
11. *Cryptococcus laurentii* (Kufferath) Skinner (Kufferath, 1920a)
12. *Cryptococcus luteolus* (Saito) Skinner (Saito, 1922)
13. *Cryptococcus macerans* (Frederiksen) Phaff et Fell (Frederiksen, 1956)
14. *Cryptococcus magnus* (Lodder et Kreger-van Rij) Baptist et Kurtzman (Lodder & Kreger-van Rij, 1952)
15. *Cryptococcus melibiosum* (Shifrine et Phaff) Phaff et Fell (Shifrine & Phaff, 1956)
16. *Cryptococcus neoformans* (Sanfelice) Vuillemin (Sanfelice, 1895a,b)
    Cryptococcus neoformans (Sanfelice) Vuillemin var. *neoformans*
    Cryptococcus neoformans (Sanfelice) Vuillemin var. *gattii* Vanbreuseghem et Takashio (1970)
17. *Cryptococcus skinneri* Phaff et do Carmo-Sousa (1962)
18. *Cryptococcus terreus* di Menna (1954b)
19. *Cryptococcus uniguttulatus* (Zach) Phaff et Fell (Wolfram & Zach, 1934)

d. KEY TO THE SPECIES OF THE GENERA *CRYPTOCOCCUS* AND *HOLTERMANNIA*, AND TO *TREMELLA* SPECIES WHICH ASSIMILATE INOSITOL

1a Nitrate assimilated 2
b Nitrate not assimilated 7

2a Sucrose assimilated 3
b Sucrose not assimilated 6

3a Maltose assimilated 4
b Maltose not assimilated
Cr. kuetzingii . . . . p. 859

4a Young growth on malt agar hyaline to slightly pigmented slightly
Cr. albidus . . . . p. 850
b Young growth on malt agar deep red 5

5a Erythritol assimilated
Cr. macerans . . . . p. 862
b Erythritol not assimilated
Cr. infirmo-miniatus . . . . p. 859

6a Melezitose assimilated
Cr. elinovii .... p. 854
b Melezitose not assimilated
Cr. terreus .... p. 869

7a Sucrose strongly assimilated 8
b Sucrose not assimilated 20

8a Maltose assimilated 9
b Maltose not assimilated
Cr. dimennae .... p. 853

9a Young growth on malt agar red or orange
Cr. hungaricus .... p. 858
b Young growth on malt agar hyaline or slightly pigmented 10

10a Lactose strongly assimilated 11
b Lactose assimilation negative or weak 16

11a Melibiose assimilated 12
b Melibiose not assimilated 14

12a Erythritol assimilated 13
b Erythritol not assimilated
Trem. encephala .... p. 551
Trem. subanomala .... p. 554

13a Starch produced
Cr. laurentii .... p. 860
Trem. aurantia .... p. 548
b Starch not produced
Cr. flavus .... p. 854

14a Rhamnose assimilated 15
b Rhamnose not assimilated
Cr. magnus .... p. 863

15a Erythritol assimilated
Cr. heveanensis .... p. 857
b Erythritol not assimilated
Cr. ater .... p. 852

16a Growth at 37°C 17
b No growth at 37°C 18

17a Canavanine-glycine bromthymol blue (CGB) agar blue within 2 to 5 days
Cr. neoformans var. *gattii* .... p. 867
b CGB agar unchanged within 2 to 5 days
Cr. neoformans var. *neoformans* .... p. 866

18a Erythritol assimilated 19
b Erythritol not assimilated
Cr. uniguttulatus .... p. 870

19a Starch produced
Cr. luteolus.......................... p. 861
b Starch not produced
Cr. flavus .... p. 854

20a Maltose assimilated 21
b Maltose not assimilated 23

21a Cellobiose assimilated 22
b Cellobiose not assimilated
Trem. foliacea......................... p. 552

22a Lactose assimilated
Cr. gastricus .... p. 856
b Lactose not assimilated
Holtermannia corniformis .... p. 548

23a Melibiose assimilated
Cr. melibiosum .... p. 864
b Melibiose not assimilated 24

24a Cellobiose assimilated 25
b Cellobiose not assimilated
Sporopachydermia lactativora .... p. 429

25a Rhamnose assimilated
Cr. skinneri .... p. 868
b Rhamnose not assimilated
Sporopachydermia cereana .... p. 428

Key characters of the species in the genus *Cryptococcus* and of *Tremella* and *Holtermannia* species

	Assimilation									Color of culture	St prod	37°C	CGB
	Su	Ma	La	Ce	Me	St	Rh	Er	Mz				
Nitrate-positive species:													
Cr. albidus var. *albidus*	+	+/w	v	+	v	v	v	–		hyaline	+/w	–(+)	
Cr. albidus var. *aerius*	+	+	+	+	+	v	v	–		hyaline	–	–	
Cr. elinovii	–	+	+	+	–	+	+	–	+	hyaline	+	–	
Cr. infirmo-miniatus	+	+	+ s/w	+	–	+	v	–		red	+	–	
Cr. kuetzingii	+	–	v	+	–	–	–	–		hyaline	+	–	
Cr. macerans	+	+/w	+/w	+	+w/–	+s	–(+)	+		red	+	–	
Cr. terreus	–	+/w	+	+	–	+/w	+	–	–	hyaline	+	–	
Nitrate-negative species:													
Cr. ater	+	+	+	+	–	+/w	+	–		hyaline[1]	+	–	
Cr. dimennae	+	–	+	+	–	–	+	–		hyaline	+	–	
Cr. flavus	+	+	+w	+	+	+w	+w	+		sl. yellow	–	–	
Cr. gastricus	–	+	+	+	–	+	+	–		hyaline	+	–	
Cr. heveanensis	+	+	+	+	–	+w	+	+		hyaline	+	–	
Cr. hungaricus	+	+	v	+	v	+w	+/w	–(+)		red	+	–	
Cr. laurentii	+	+	+	+	+	+/w	+(w)	+		hyaline	+	v	
Cr. luteolus	+	+	–/w	+	+	+w	+/w	+		hyaline	+	–	
Cr. magnus	+	+	+	+	–	+w	–	–		hyaline	+w	–	
Cr. melibiosum	–	–	+	+	+	–	–	–		hyaline	+w	+w	
Cr. neoformans var. *neoformans*	+	+	–	+/w	–	v	+	v		hyaline	+/w	+	–
Cr. neoformans var. *gattii*	+	+	–	+/w	–	v	+	v		hyaline	+/w	+	+
Cr. skinneri	–	–	–	+	–	–	+	–/w		hyaline	+w	–	
Cr. uniguttulatus	+	+	–	–	–	+w	+/w	–		hyaline	+/w	–	
Trem. aurantia	+	+	+	+	+w	+	+	+w		hyaline	+	–	
Trem. encephala	+	+	+	+	+	+	+	–		hyaline	+	–	
Trem. foliacea	–	+w	–	–	–	–	+	+w		hyaline	+	+w	
Trem. subanomala	+	+	+	+	+w	+	+	–		hyaline	+	–	
Holterm. corniformis	–	+	–	+	–	+w	+	–		hyaline	+	–	

Su = sucrose, Ma = maltose, La = lactose, Ce = cellobiose, Me = melibiose, St = soluble starch, Rh = L-rhamnose, Er = erythritol, Mz = melezitose, St prod = starch formation, CGB = CGB agar turns blue.

[1] After several weeks growth dark brown or black on certain media.

e. SYSTEMATIC DISCUSSION OF THE SPECIES

1. *CRYPTOCOCCUS ALBIDUS* (SAITO) SKINNER (1947b)

This species has two varieties:
Cryptococcus albidus (Saito) Skinner var. *albidus*
Cryptococcus albidus (Saito) Skinner var. *aerius* (Saito) Phaff et Fell (1970)

Perfect state: The haploid strains of the species *Filobasidium floriforme* have the same assimilation pattern as *Cryptococcus albidus* var. *albidus* and therefore the two species cannot be distinguished by physiological tests. However, not all strains which may be identified with *Cr. albidus* var. *albidus* give mating reactions with the types of *Filobasidium floriforme*.

Synonyms of the variety *albidus*

Torula albida Saito (1922)
Torulopsis albida (Saito) Lodder (1934)
Torula gelatinosa Saito (1922)
Torula alpina nom. nud.
Torulopsis rotundata Redaelli (1925)
Torulopsis albida (Saito) Lodder var. *japonica* Lodder (1934)
Torulopsis liquefaciens Saito et Ota (1934)
Torulopsis nodaensis Saito et Ota (1934)
Torulopsis diffluens Zach (Wolfram & Zach, 1934)
Cryptococcus diffluens (Zach) Lodder et Kreger-van Rij (1952)
Cryptococcus albidus (Saito) Skinner var. *diffluens* (Zach) Phaff et Fell (1970)
Torulopsis diffluens Zach var. *urugaiensis* Artagaveytia-Allende et de Queiroz (1970)
Torulopsis acris var. *granulosa* Marcilla et Feduchy
Torulopsis pseudaeria Zsolt (1958)
Paratorulopsis pseudaeria (Zsolt) Novák et Zsolt (1961)
Hansenula amylofaciens nom. nud. (Dietrichson, 1954)
Cryptococcus mucorugosus nom. nud. (Benham, 1955)
Cryptococcus neoformans (Sanf.) Vuill. var. *innocuous* nom. nud. (Benham, 1955)
Cryptococcus terricolus Pedersen (1958)
Cryptococcus genitalis nom. nud. (Castellani, 1963)
Naganishia globosa Goto (1963)
Cryptococcus albidus (Saito) Skinner var. *ovalis* Sugiyama et Goto (1967)

Synonyms of the variety *aerius*

Torula aeria Saito (1922)
Cryptococcus aerius (Saito) Nannizzi (Pollacci & Nannizzi, 1927)
Torulopsis aeria (Saito) Lodder (1934)
Paratorulopsis aeria (Saito) Novák et Zsolt (1961)

Standard description of *Cryptococcus albidus* var. *albidus*

Growth in malt extract: After 3 days at 25°C the cells are generally globose to ovoidal; single, in pairs and often in little clusters or short chains; occasionally some elongate cells are present. Some strains have rather small cells, e.g. (3–6.1) × (3.5–6.2) μm, but others have much larger cells, e.g. (3.5–8.8) × (5.5–10.2) μm and even up to 14 μm long. Many other strains have intermediate cell sizes. Giant cells occur in some strains. Usually a small amount of sediment is formed. After one month a moderate non-slimy ring, no film or a few moist islands and a moderate to heavy sediment are present.

Growth on malt agar: After one month at 20°C the streak culture is cream-colored to slightly yellowish or tan, mostly smooth, highly glossy, soft and slimy with the border entire or sometimes rough and wrinkled with irregular border.

Slide culture on potato agar: Pseudomycelium absent.

Fermentation: –

Assimilation of carbon compounds:

Galactose	v	Raffinose	+ w (+)	Erythritol	–
Sucrose	+	Soluble starch	v	Ribitol	v
Maltose	+ or w	D-Xylose	+ or w	D-Mannitol	+ (w)
Cellobiose	+	L-Arabinose	+ or w	Succinic acid	+ or w
Trehalose	+ or w	D-Ribose	v	Citric acid	+ or w
Lactose	v	L-Rhamnose	v	Inositol	+

Additional carbon compounds tested: Melibiose v, sodium glucuronate + or w, glucono-δ-lactone + or w

Assimilation of nitrate: +

Growth in vitamin-free medium: – (+); if –, thiamine stimulates growth

Growth on 50% (w/w) glucose-yeast extract agar: –

Growth at 37°C: – (+)

Starch formation: + or w

G + C: 53.7 mol%, strain CBS 142; 54.1–55.1 mol%, strain CBS 160; 50.5 mol%, strain CBS 5106 (Nakase & Komagata, 1971b); 55 mol%, strain CBS 142 (Storck et al., 1969)

Supplementary description of *Cryptococcus albidus* var. *aerius*

This variety differs from the variety *albidus* in being unable to synthesize starch.

The following characteristics are variable in the variety *albidus,* but uniform in the variety *aerius:* it is vitamin independent; the maximum temperature of growth is 30°C or below; it strongly assimilates galactose, lactose, melibiose and D-ribose.

G + C: 55.5 mol% (Nakase & Komagata, 1968b)

Comments

Phaff and Fell (1970) separated the variety *diffluens* from the variety *albidus* by its inability to assimilate lactose. However it appeared to be impossible to make a sharp distinction with this criterium: some strains originally classified in the variety *albidus* assimilated lactose very weakly or not and some strains of the variety *diffluens* assimilated lactose strongly or weakly. Therefore the variety *diffluens* is rejected.

The variety *aerius* is maintained, because it is well distinguishable from the var. *albidus* by the absence of starch production.

Origin of the strains belonging to the variety *albidus*

Air (4), bumble bee (1), skin of dog (1), sputum, lung and bronchial secrete (7), patients with various diseases (4), nails (3), skin (1), saké moto (2), wine (2), dew-retted flax (2), leaves (2), soil (1), blue cheese *(Naganishia globosa)* (1), goat dung (1), washed bottle in brewery (1), purification tank for polluted water (1), unknown (4), a strain labelled *Torulopsis dattila* var. *armeniensis* of unknown origin.

Type strain: CBS 142, isolated by Saito from the air.

Origin of the strains belonging to the variety *aerius*

Air in Tokyo (1), soil in Norway (1), asbestos cement (1).

Type strain: CBS 155, isolated by Saito from the air.

2. *CRYPTOCOCCUS ATER* CASTELLANI EX COOKE (1966)

Synonym

Cryptococcus laurentii (Kuff.) Skinner var. *magnus* Lodder et Kreger-van Rij forma *ater* (Cast.) Cooke (1966)

Standard description of *Cryptococcus ater*

Growth in malt extract: After 3 days at 25°C the cells are ovoidal to globose, (3.7–6.9) × (4.0–8.0) μm, rarely up to 9 μm long; mainly single or in pairs, a few short chains of three and four cells. A thin ring and a little sediment begin to form. After one month there is a moderate, nonslimy ring and much sediment.

Growth on malt agar: After one month the streak culture is cream to tan, smooth, glossy, slimy (running to the bottom of the tube); border is entire. On media low in sugar and rich in organic nitrogen there is a tendency for the growth to turn dark in color.

Slide culture on potato agar: Pseudomycelium absent.

Fermentation: –

Assimilation of carbon compounds:

Galactose	+ (w)	Raffinose	+ w	Erythritol	–
Sucrose	+	Soluble starch	+ or w	Ribitol	+
Maltose	+	D-Xylose	+	D-Mannitol	+
Cellobiose	+	L-Arabinose	+	Succinic acid	+ w
Trehalose	+	D-Ribose	+ w	Citric acid	+ w
Lactose	+	L-Rhamnose	+	Inositol	+

Additional carbon compounds tested: Melibiose –, sodium glucuronate + w, glucono-δ-lactone – or + w

Assimilation of nitrate: –

Assimilation of nitrite: + w

Growth in vitamin-free medium: –; thiamine stimulates growth

Growth on 50% (w/w) glucose-yeast extract agar: –

Growth at 37°C: –; maximum temperature of growth: 28–30°C
Starch formation: +

Origin of the strains studied

Ulcers on leg of patient (3).

Type strain: CBS 4685, the original strain of Castellani (1960) isolated from a multiple ulcer on the leg of a young man.

3. *CRYPTOCOCCUS DIMENNAE* FELL ET PHAFF (1967)

Standard description of *Cryptococcus dimennae*

Growth in malt extract: After 3 days at 25°C the cells are oval or ovoid, occasionally spheroidal, (2.8–6.8) × (3.5–7.2) μm; single or in pairs. No ring or pellicle and very little sediment. After one month there is a thin ring, a thin smooth pellicle and a moderate amount of sediment.

Growth on malt agar: After one month at 20°C the streak culture is tan to pinkish cream-colored; the surface is smooth and glossy, the texture soft and the border entire.

Slide culture on potato agar: Pseudomycelium absent.

Fermentation: –

Assimilation of carbon compounds:

Galactose	+	Raffinose	+	Erythritol	–
Sucrose	+	Soluble starch	–	Ribitol	+
Maltose	–	D-Xylose	+	D-Mannitol	+
Cellobiose	+	L-Arabinose	+	Succinic acid	+ w
Trehalose	+	D-Ribose	+	Citric acid	+ w
Lactose	+	L-Rhamnose	+	Inositol	+

Additional carbon compounds tested: Melibiose –, sodium glucuronate +, glucono-δ-lactone +

Assimilation of nitrate: –

Assimilation of nitrite: +

Growth in vitamin-free medium: –

Growth on 50% (w/w) glucose-yeast extract agar: –

Growth at 37°C: –; at 30°C: +

Starch formation: +

G + C: 53.1 mol% (Nakase & Komagata, 1971b)

Origin of the strain studied

Leaves of pasture plants in New Zealand (1).

Type strain: CBS 5770, isolated by di Menna.

4. *CRYPTOCOCCUS ELINOVII* GOLUBEV ET TAUSON (1979)

Standard description of *Cryptococcus elinovii*

Growth in malt extract: After 3 days at 25°C the cells are subglobose to oval; single, in pairs or short chains, (3.5–7.0) × (4.0–9.0) μm. A little sediment is formed. After one month at 20°C a ring and a moderate to heavy sediment are formed.

Growth on malt agar: After one month at 20°C the streak culture is brownish cream-colored to tan. The surface is smooth, glossy and the texture is mucous so that the growth runs to the bottom of the slant.

Slide culture on potato agar: Pseudomycelium absent.

Fermentation: –

Assimilation of carbon compounds:

Galactose	+	Raffinose	–	Erythritol	–
Sucrose	–	Soluble starch	+	Ribitol	+
Maltose	+	D-Xylose	+	D-Mannitol	+
Cellobiose	+	L-Arabinose	+	Succinic acid	+
Trehalose	+	D-Ribose	+	Citric acid	+
Lactose	+	L-Rhamnose	+	Inositol	+

Additional carbon compounds tested: Melibiose –, melezitose +, sodium glucuronate +, glucono-δ-lactone +

Assimilation of nitrate: +

Growth in vitamin-free medium: +

Growth on 50% (w/w) glucose-yeast extract agar: –

Growth at 37°C: –; **at 30°C:** +

Starch formation: +

G + C: 57–58 mol%, one strain (Golubev & Tauson, 1979)

Origin of the strain studied

Soil on high mountain in U.S.S.R. (1).

Type strain: CBS 7051, isolated by Golubev.

5. *CRYPTOCOCCUS FLAVUS* (SAITO) PHAFF ET FELL (1970)

Synonyms

Torula flava Saito (1922)
Chromotorula flava (Saito) Harrison (1928)
Rhodotorula flava (Saito) Lodder (1934)

Standard description of *Cryptococcus flavus*

Growth in malt extract: After 3 days at 25°C the cells are oval, ovoid or irregular; single or in pairs, (3.2–5.2) × (5.0–8.8) μm (Fig. 370). A thin ring and a little

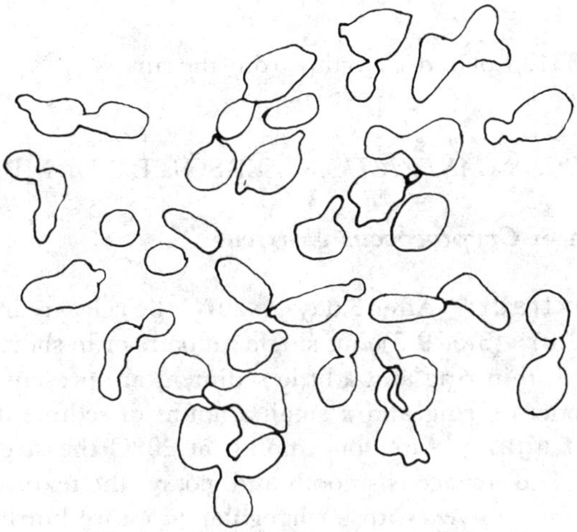

Fig. 370 – *Cr. flavus,* CBS 331
After 3 days in malt extract

sediment begin to form. After one month a moderate yellowish ring and sediment are present.

Growth on malt agar: After one month at 20°C the streak culture is light brown-yellow, semiglossy and nearly smooth with a butyrous texture and entire borders.

Slide culture on potato agar: No pseudomycelium. A few chains of cells may be formed.

Fermentation: –

Assimilation of carbon compounds:

Galactose	+	Raffinose	+	Erythritol	+
Sucrose	+	Soluble starch	+ w	Ribitol	+ w
Maltose	+	D-Xylose	+	D-Mannitol	+
Cellobiose	+	L-Arabinose	+	Succinic acid	+ vw
Trehalose	+	D-Ribose	+	Citric acid	–
Lactose	+ w	L-Rhamnose	+ w	Inositol	+

Additional carbon compounds tested: Melibiose +, sodium glucuronate + w, glucono-δ-lactone +

Assimilation of nitrate: –

Assimilation of nitrite: + w

Growth in vitamin-free medium: –

Growth on 50% (w/w) glucose-yeast extract agar: –

Growth at 37°C: – ; **at 30°C:** +

Starch formation: –

G + C: 55.0 mol% (Storck et al., 1969)

Origin of the strain studied

Air in Tokyo (1).

Type strain: CBS 331, isolated by Saito from the air.

6. *CRYPTOCOCCUS GASTRICUS* REIERSÖL ET DI MENNA (1958)

Standard description of *Cryptococcus gastricus*

Growth in malt extract: After 3 days at 20°C the cells are mainly globose, but also short-oval, (3.0–7.0) × (3.5–9.5) μm; single, in pairs or in short chains. After 3–5 days a partially formed, thin ring and a little sediment are present. After one month at 20°C there is a moderate ring and a small amount of sediment.

Growth on malt agar: After one month at 20°C the streak culture is pale yellow to tan-colored. The surface is smooth and glossy, the texture slightly mucous. The growth is flat to low convex with an irregular to entire border.

Slide culture on potato agar: Pseudomycelium absent; a few chains of cells may be formed under the coverslip.

Fermentation: –

Assimilation of carbon compounds:

Galactose	+	Raffinose	–	Erythritol	–
Sucrose	–	Soluble starch	+	Ribitol	v
Maltose	+	D-Xylose	+	D-Mannitol	+
Cellobiose	+	L-Arabinose	+	Succinic acid	+ w
Trehalose	+	D-Ribose	+ w	Citric acid	–
Lactose	+	L-Rhamnose	+	Inositol	+

Additional carbon compounds tested: Melibiose –, sodium glucuronate +, glucono-δ-lactone +

Assimilation of nitrate: –

Assimilation of nitrite: –

Growth in vitamin-free medium: –

Growth on 50% (w/w) glucose-yeast extract agar: –

Growth at 37°C: –; at 30°C: –; at 25°C: +

Starch formation: +

G + C: 51.0 mol%, strain CBS 1927; 65.5 mol%, strain CBS 2288 (Storck et al., 1969)

Comments

The great difference in G + C % of the two strains investigated makes it likely that they are not of the same species. However the morphological and physiological properties are identical and it is not yet possible to distinguish them with the usual criteria.

Origin of the strains studied

Soil in New Zealand (1), washing of the stomach of a tuberculosis patient (Norway) (1).

Type strain: CBS 1927, isolated by di Menna from soil in New Zealand.

7. *CRYPTOCOCCUS HEVEANENSIS* (GROENEWEGE) BAPTIST ET KURTZMAN (1976)

Synonyms

Torula heveanensis Groenewege (1921b)
Candida heveanensis (Groenewege) Diddens et Lodder (1942)
Torulopsis heveanensis (Groenewege) Mager et Aschner (1947)
Cryptococcus laurentii (Kuff.) Skinner var. *magnus* Lodder et Kreger-van Rij (1952)

Standard description of *Cryptococcus heveanensis*

Growth in malt extract: After 3 days at 25°C the cells are ovoidal to elongate, single, in pairs or in short chains of three or four cells, (2.0–4.5) × (3.0–7.0) μm. A small ring and some sediment are present. After one month a moderate thick ring and a heavy sediment are formed.

Growth on malt agar: After one month the streak culture is smooth, glossy, mucous so that the growth runs to the bottom of the slant, cream-colored with a tan line in the middle. The border is entire.

Slide culture on potato agar: No pseudomycelium.

Fermentation: –

Assimilation of carbon compounds:

Galactose	+	Raffinose	+ w	Erythritol	+
Sucrose	+	Soluble starch	+ w	Ribitol	+
Maltose	+	D-Xylose	+	D-Mannitol	+
Cellobiose	+	L-Arabinose	+	Succinic acid	+
Trehalose	+	D-Ribose	+	Citric acid	+ w
Lactose	+	L-Rhamnose	+	Inositol	+

Additional carbon compounds tested: Melibiose –, sodium glucuronate +, glucono-δ-lactone +

Assimilation of nitrate: –

Assimilation of nitrite: +

Growth in vitamin-free medium: –

Growth on 50% (w/w) glucose-yeast extract agar: –

Growth at 37°C: –; at 30°C: +

Starch formation: +

G + C: 52.0 mol%, one strain (Baptist & Kurtzman, 1976)

Comments

The classification of the original strain of *Torula heveanensis* as the variety *magnus* in the species *Cr. laurentii* is no longer recognized by Baptist and Kurtzman (1976). They described it as a separate species, based on a G + C content of 52.0 mol%, that differs too much from the G + C content value for *Cr. laurentii* var. *laurentii* (59.7 mol%, Baptist & Kurtzman, 1976) and of the value for a second strain in the variety *magnus* (54.1 mol%, Baptist & Kurtzman, 1976) that they described as *Cryptococcus magnus*. *Cr. heveanensis* can be separated from *Cr. laurentii* var. *laurentii* by its inability to assimilate melibiose; from *Cr. magnus* it differs in the assimilation of L-rhamnose, erythritol and ribitol.

Origin of the strain studied

Rubber (1).

Type strain: CBS 569, isolated by Groenewege from rubber.

8. *CRYPTOCOCCUS HUNGARICUS* (ZSOLT) PHAFF ET FELL (1970)

Synonym

Dioszegia hungarica Zsolt (1957)

Standard description of *Cryptococcus hungaricus*

Growth in malt extract: After 3 days at 17°C the cells are globose to ovoidal (4.0–6.3) × (5.0–8.7) μm; single, in pairs or in small clusters; budding is multilateral; occasionally a bud is connected to the mother cell by a very short neck. A slight ring and a little sediment may be present. After one month there is a moderate brownish red ring and some sediment.

Growth on malt agar: After one month at 20°C the streak culture is reddish to slightly orange, semiglossy, smooth, pasty with an entire border.

Slide culture on potato agar: No pseudomycelium.

Fermentation: –

Assimilation of carbon compounds:

Galactose	+	Raffinose	+ or w	Erythritol	– (+)
Sucrose	+	Soluble starch	+ w	Ribitol	v
Maltose	+	D-Xylose	+	D-Mannitol	+
Cellobiose	+	L-Arabinose	+	Succinic acid	+ w
Trehalose	+ or w	D-Ribose	+ or w	Citric acid	+ w
Lactose	v	L-Rhamnose	+ or w	Inositol	+ w

Additional carbon compounds tested: Melibiose v, sodium glucuronate + w, glucono-δ-lactone +

Assimilation of nitrate: –

Assimilation of nitrite: + w

Growth in vitamin-free medium: –
Growth on 50% (w/w) glucose-yeast extract agar: –
Growth at 37°C: –; **at 30°C:** –; **at 25°C:** +
Starch formation: +

Origin of the strains studied

Soil in Hungary (1), flowers of *Lewesia pygmaea* (1), seawater (4), leaves of cereals (1).

Type strain: CBS 4214, isolated by Zsolt from soil in Hungary.

9. *CRYPTOCOCCUS INFIRMO-MINIATUS* (OKUNUKI) PHAFF ET FELL

see *Rhodosporidium infirmominiatum* p. 520

10. *CRYPTOCOCCUS KUETZINGII* FELL ET PHAFF (1967)

Standard description of *Cryptococcus kuetzingii*

Growth in malt extract: After 3 days at 25°C the cells are spheroidal to ovoidal (3.2–5.6) × (3.7–7.2) μm, single or in pairs. A slight ring may be present and a little sediment.

Growth on malt agar: After one month at 20°C the streak culture is cream to tan. The surface is smooth, glistening and the texture mucous, so that the growth runs to the bottom of the slant.

Slide culture on potato agar: Pseudomycelium absent.
Fermentation: –
Assimilation of carbon compounds:

Galactose	–	Raffinose	+ or w	Erythritol	–
Sucrose	+	Soluble starch	–	Ribitol	+
Maltose	–	D-Xylose	+	D-Mannitol	+
Cellobiose	+	L-Arabinose	+	Succinic acid	+ or w
Trehalose	+	D-Ribose	+	Citric acid	+ or w
Lactose	v	L-Rhamnose	–	Inositol	+

Additional carbon compounds tested: Melibiose –, sodium glucuronate + or w, glucono-δ-lactone –
Assimilation of nitrate: +
Growth in vitamin-free medium: –
Growth on 50% (w/w) glucose-yeast extract agar: –
Growth at 37°C: –; **at 30°C:** +
Starch formation: +

Origin of the strains studied

Medlar fruit *(Mespilus)* (1), trachea rinsings of patient (1), air in The Netherlands (1), unknown (1).

Type strain: CBS 1926, isolated by Protti from a medlar fruit.

11. CRYPTOCOCCUS LAURENTII (KUFFERATH) SKINNER (1947b)

Synonyms

Torula laurentii Kufferath (1920a)
Torulopsis laurentii (Kuff.) Lodder (1934)
Torulopsis carnescens Verona et Luchetti (1936)
Torula flavescens Saito (1922)
Torulopsis flavescens (Saito) Lodder (1934)
Cryptococcus flavescens (Saito) Skinner (1947b)
Cryptococcus laurentii (Kufferath) Skinner var. *flavescens* (Saito) Lodder et Kreger-van Rij (1952)
Torula aurea Saito (1922)
Chromotorula aurea (Saito) Harrison (1928)
Rhodotorula aurea (Saito) Lodder (1934)
Rhodotorula peneaus Phaff, Mrak et Williams (1952)
Rhodotorula nitens Mackenzie et Auret (1963)

Standard description of *Cryptococcus laurentii*

Growth in malt extract: After 3 days at 25°C the cells are spheroidal to ovoidal to elongate; single, in pairs or in short chains of three or four cells. The cell size can vary considerably in the different strains. In a number of strains the cells are (2.0–5.5) × (3.0–7.0) μm, in others the width is up to 8.5 μm and the length up to 13 μm. A thin ring, a moderate sediment and sometimes some islets are present. After one month a moderate to thick slimy ring, a heavy sediment and sometimes a pellicle or islets are formed.

Growth on malt agar: After one month at 20°C the streak culture is cream colored, yellowish, pinkish, orange or tan colored. The surface is smooth and glossy, sometimes pasty (not mucous) but in most strains the texture is slimy so that the growth runs to the bottom of the slant. The border is entire to lobate.

Slide culture on potato agar: Pseudomycelium absent or rudimentary.

Fermentation: –

Assimilation of carbon compounds:

Galactose	+	Raffinose	+	Erythritol	+ (–)
Sucrose	+	Soluble starch	+ or w	Ribitol	+
Maltose	+	D-Xylose	+ (w)	D-Mannitol	+
Cellobiose	+	L-Arabinose	+	Succinic acid	v
Trehalose	+	D-Ribose	+	Citric acid	+ or w
Lactose	+	L-Rhamnose	+ (w)	Inositol	+ (w)

Additional carbon compounds tested: Melibiose +, sodium glucuronate + or w, glucono-δ-lactone + (w)

Assimilation of nitrate: –
Assimilation of nitrite: v
Growth in vitamin-free medium: v
Growth on 50% (w/w) glucose-yeast extract agar: –
Growth at 37°C: v; at 30°C: v; at 25°C: +
Starch formation: +
G + C: 51.2–59.0 mol%, 2 strains (Nakase & Komagata, 1971b); 58 mol%, one strain (Storck et al., 1969); 59.7 mol%, one strain (Baptist & Kurtzman, 1976)

Comments

Strains of opposite sex have been isolated and described by Kurtzman (1973). After conjugation the zygote forms true mycelium with chlamydospore-like structures but without clamp connections. The complete life cycle is not known. The strains CBS 6473 and CBS 6474 are of mating type α; CBS 6475 and CBS 6476 of mating type *a*. Corn meal agar is a suitable medium for mating.

Erythritol assimilation was the criterion to distinguish the variety *flavescens* from the variety *laurentii*. As a great number of strains originally classified in the variety *flavescens* assimilate erythritol and a few strains classified in the variety *laurentii* assimilate erythritol weakly, the variety *flavescens* is no longer recognized.

Origin of the strains studied

Palm wine (1), Italian muscatel wine (1), bottle in brewery (1), slime flux of oak (1), tumor (1), bronchial tube of patient (1), leaves of tropical plants (4), insect frass in *Acacia karroo* (1), wheat (2), corn (3), seawater (1), air (3), deep frozen beans (1), shrimp (1), soil (3).

Type strain: CBS 139, isolated by Kufferath from palm wine.

12. *CRYPTOCOCCUS LUTEOLUS* (SAITO) SKINNER (1947b)

Synonyms

Torula luteola Saito (1922)
Chromotorula luteola (Saito) Harrison (1928)
Torulopsis luteola (Saito) Lodder (1934)

Standard description of *Cryptococcus luteolus*

Growth in malt extract: After 3 days at 25°C the cells are ovoidal to elongate, (3.1–6.0) × (5.5–9.0) μm; single, in pairs or short chains. A ring begins to form and a little sediment is present. After one month a moderate ring and a fair amount of sediment are formed, sometimes a thin film.

Growth on malt agar: After one month at 20°C the streak culture is cream-colored to yellowish or tan to grey-brown. The surface is smooth and glossy, the texture slimy so that the growth runs to the bottom of the slant. The border is entire.

Slide culture on potato agar: Pseudomycelium is absent.
Fermentation: –
Assimilation of carbon compounds:

Galactose	+	Raffinose	+	Erythritol	+
Sucrose	+	Soluble starch	+ w	Ribitol	+ or w
Maltose	+	D-Xylose	+	D-Mannitol	+
Cellobiose	+	L-Arabinose	+	Succinic acid	+ w
Trehalose	+	D-Ribose	+	Citric acid	+ or w
Lactose	– or + w	L-Rhamnose	+ or w	Inositol	+

Additional carbon compounds tested: Melibiose +, sodium glucuronate + w, glucono-δ-lactone +
Assimilation of nitrate: –
Assimilation of nitrite: v
Growth in vitamin-free medium: – or + vw
Growth on 50% (w/w) glucose-yeast extract agar: –
Growth at 37°C: –; **at 30°C:** + or w
Starch formation: +
G + C: 50.5 mol%, one strain (Nakase & Komagata, 1971b)

Origin of the strains studied

Air in Tokyo (1), foliage of tropical plants (2).

Type strain: CBS 943, isolated by Saito from the air in Tokyo.

13. *CRYPTOCOCCUS MACERANS* (FREDERIKSEN) PHAFF AND FELL (1970)

Synonym

Rhodotorula macerans Frederiksen (1956)

Standard description of *Cryptococcus macerans*

Growth in malt extract: After 3 days at 25°C the cells are oval or long-oval, single or in pairs, (3.0–4.5) × (5.0–15.0) μm. A thin ring and some sediment are present. After one month a moderate to heavy, slimy reddish ring and a heavy sediment are formed and sometimes some islets.

Growth on malt agar: After one month at 20°C the streak culture is bright coral-red to purplish-pink, surface nearly smooth with some rugose or sectored areas, glossy with entire borders. The texture is pasty.
Slide culture on potato agar: Pseudomycelium absent or rudimentary.
Fermentation: –

Assimilation of carbon compounds:

Galactose	+ or w	Raffinose	+ or w	Erythritol	+
Sucrose	+	Soluble starch	+ s	Ribitol	v
Maltose	+ or w	D-Xylose	+	D-Mannitol	+
Cellobiose	+	L-Arabinose	+	Succinic acid	+ w
Trehalose	+	D-Ribose	+	Citric acid	+ w
Lactose	+ or w	L-Rhamnose	− (+)	Inositol	+

Additional carbon compounds tested: Melibiose + w or −, sodium glucuronate + or w, glucono-δ-lactone + or w

Assimilation of nitrate: +

Growth in vitamin-free medium: −

Growth on 50% (w/w) glucose-yeast extract agar: −

Growth at 37°C: −; **at 30°C:** −; **at 25°C:** +

Starch formation: +

Comments

Mating types have been found by the present author; CBS 2206 and CBS 6532 are of one type (*a*), CBS 2425 and CBS 2426 of the opposite type (*α*). The life cycle is not yet completely known (unpublished).

Origin of the strains studied

Dew-retted flax (3), pharynx of *Axis axis* (2), flowers (2).

Type strain: CBS 2206, isolated by Frederiksen from dew-retted flax.

14. CRYPTOCOCCUS MAGNUS (LODDER ET KREGER-VAN RIJ) BAPTIST ET KURTZMAN (1976)

Synonym

Cryptococcus laurentii (Kuff.) Skinner var. *magnus* Lodder et Kreger-van Rij (1952)

Standard description of *Cryptococcus magnus*

Growth in malt extract: After 3 days at 25°C the cells are ovoidal to elongate (3.5–15.0) × (4.5–45.0) μm; single, in pairs or in short chains of up to six cells. A ring begins to form and a little sediment is present. After one month a moderate ring and a fair amount of sediment are formed.

Growth on malt agar: After one month at 20°C the streak culture is cream-colored to tan. The surface is smooth, highly glossy and the texture slimy, so that the growth runs to the bottom of the slant. The border is entire.

Slide culture on potato agar: Pseudomycelium absent.

Fermentation: −

Assimilation of carbon compounds:

Galactose	+	Raffinose	+ w	Erythritol	–
Sucrose	+	Soluble starch	+ w	Ribitol	–
Maltose	+	D-Xylose	+ w	D-Mannitol	+
Cellobiose	+	L-Arabinose	+	Succinic acid	+ w
Trehalose	+	D-Ribose	+	Citric acid	+ w
Lactose	+	L-Rhamnose	–	Inositol	+

Additional carbon compounds tested: Melibiose –, sodium glucuronate + w, glucono-δ-lactone +

Assimilation of nitrate: –

Assimilation of nitrite: + w

Growth in vitamin-free medium: –

Growth on 50% (w/w) glucose-yeast extract agar: –

Growth at 37°C: –; **at 30°C:** +

Starch formation: + w

G + C: 49.0 mol%, one strain (Storck et al., 1969); 54.1 mol%, one strain (Baptist & Kurtzman, 1976)

Origin of the strain studied

Atmosphere in The Netherlands (1).

Type strain: CBS 140 isolated by Sobels from the atmosphere in The Netherlands.

15. *CRYPTOCOCCUS MELIBIOSUM* (SHIFRINE ET PHAFF) PHAFF ET FELL (1970)

Synonyms

Torulopsis melibiosum Shifrine et Phaff (1956)
Paratorulopsis melibiosi (Shifrine et Phaff) Novák et Zsolt (1961)

Standard description of *Cryptococcus melibiosum*

Growth in malt extract: After 3 days at 25°C the cells are (2.0–5.3) × (2.5–5.3) μm, spheroidal to globose, single or in pairs. A thin ring and a little sediment are present. After one month at 20°C a delicate ring and a moderate sediment are formed.

Growth on malt agar: After one month at 20°C the streak culture is cream-colored, smooth, glistening, rather mucous and flat with an entire border.

Slide culture on potato agar: Pseudomycelium absent.

Fermentation: –

Assimilation of carbon compounds:

Galactose	+	Raffinose	–	Erythritol	–
Sucrose	–	Soluble starch	–	Ribitol	+
Maltose	–	D-Xylose	+	D-Mannitol	–
Cellobiose	+	L-Arabinose	+	Succinic acid	+w
Trehalose	–	D-Ribose	– or +w	Citric acid	–
Lactose	+	L-Rhamnose	–	Inositol	+

Additional carbon compounds tested: Melibiose +, sodium glucuronate –, glucono-δ-lactone +

Assimilation of nitrate: –

Assimilation of nitrite: –

Growth in vitamin-free medium: –

Growth on 50% (w/w) glucose-yeast extract agar: –

Growth at 37°C: +w; **at 30°C:** +

Starch formation: +w

G + C: 61.0 mol%, one strain (Storck et al., 1969)

Comments

Cryptococcus melibiosum has a cell wall of ascomycetous nature in contradistinction to all the other species classified in *Cryptococcus* (Rodrigues de Miranda, 1978).

Origin of the strains studied

Dendroctonus monticolae (bark beetle) in *Pinus ponderosa* (2).

Type strain: CBS 2102, received from Phaff, isolated from *Dendroctonus monticolae.*

16. *CRYPTOCOCCUS NEOFORMANS* (SANFELICE) VUILLEMIN (1901)

This species has two varieties:
Cryptococcus neoformans (Sanfelice) Vuillemin var. *neoformans*
Cryptococcus neoformans (Sanfelice) Vuillemin var. *gattii* Vanbreuseghem et Takashio (1970)

Perfect state of *Cryptococcus neoformans* var. *neoformans*: *Filobasidiella neoformans* Kwon-Chung var. *neoformans*
Perfect state of *Cryptococcus neoformans* var. *gattii*: *Filobasidiella neoformans* Kwon-Chung var. *bacillispora* Kwon-Chung

Synonyms of the variety *neoformans*

Saccharomyces neoformans Sanfelice (1895a,b)
Torula neoformans (Sanf.) Weiss (1902)
Blastomyces neoformans (Sanf.) Arzt (1924) (see also Lodder, 1934)
Torulopsis neoformans (Sanf.) Redaelli (1931)

Debaryomyces neoformans (Sanf.) Redaelli, Ciferri et Giordano (1937)
Saccharomyces lithogenes Sanfelice (1895b)
Cryptococcus lithogenes (Sanf.) Vuillemin (1901)
Blastomyces lithogenes (Sanf.) Sasakawa (1922)
Torulopsis lithogenes (Sanf.) de Almeida (1933)
Cryptococcus hominis Vuillemin (1901) pro parte
Saccharomyces hominis Costantin (1901)
Atelosaccharomyces hominis (Vuill.) Verdun (1912)
Torulopsis hominis (Vuill.) Redaelli (1931)
Debaryomyces hominis (Vuill.) Todd et Herrmann (1936)
Cryptococcus costantini Froilano de Mello et Gonzaga Fernandes (1918)
Torulopsis costantini (F. de Mello et G. Fernandes) de Almeida (1933)
Saccharomyces plimmeri Costantin (1901)
Torula plimmeri (Costantin) Weiss (1902)
Cryptococcus plimmeri (Costantin) Neveu-Lemaire (1912)
Torulopsis plimmeri (Costantin) de Almeida (1933)
Torula klein Weiss (1902)
Cryptococcus kleini (Weiss) Cohn (apud Guégen, 1904)
Atelosaccharomyces busse-buschki de Beurmann et Gougerot (1909)
Saccharomyces blanchardi Guiart (1910)
Atelosaccharomyces breweri Verdun (1912)
Cryptococcus breweri (Verdun) Castellani et Chalmers (1913)
Saccharomyces breweri (Verdun) Neveu-Lemaire (1921)
Torulopsis breweri (Verdun) de Almeida (1933)
Torula histolytica Stoddard et Cutler (1916) pro parte
Torulopsis histolytica (Stod. et Cutl.) Castellani et Jacono (1933)
Cryptococcus cerebriloculosis Freemann et Weidman (1923)
Torula nasalis Harrison (1928)
Cryptococcus nasalis (Harrison) Dodge (1935)
Cryptococcus hondurianus Castellani (in Castellani & Jacono, 1933)
Cryptococcus hominis Vuill. var. *hondurianus* Cast. (in Castellani & Jacono, 1933)
Torulopsis hominis (Vuill.) Red. var. *honduriana* Cast. (in Castellani & Jacono, 1933)
Cryptococcus psicrophylicus Niño (1934)
Torulopsis neoformans (Sanf.) Red. var. *sheppei* Giordano (1935)
Cryptococcus meningitidis Dodge (1935)
Cryptococcus neoformans (Sanf.) Vuillemin var. *gattii* Vanbreuseghem et Takashio (1970) pro parte

Synonym of the variety *gattii*

Cryptococcus bacillisporus Kwon-Chung et Bennett (in Kwon-Chung et al., 1978)

Standard description of *Cryptococcus neoformans* var. *neoformans*

Growth in malt extract: After 3 days at 25°C the cells are spherical, single, in pairs or in small groups, $(3.0-7.0) \times (3.3-7.9)$ μm (Fig. 371). A thin ring is usually present, some sediment and occasionally a moist film. After one month there is a well developed ring and heavy sediment.

Growth on malt agar: After one month at 20°C the streak culture is cream-colored to yellowish-brown. The surface is glossy and smooth and the texture mucous to slimy causing the growth to run the bottem of the slant. In most strains the border is entire.

Slide culture on potato agar: Pseudomycelium absent. Extracellular material is abundantly formed.

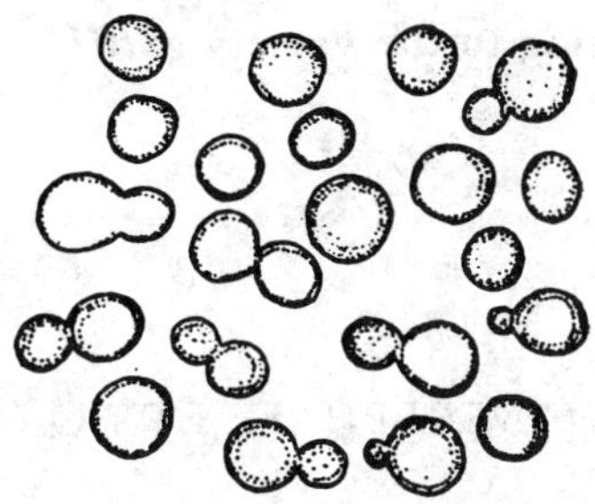

Fig. 371 – *Cr. neoformans* var. *neoformans*
After 3 days in malt extract (From 'The Yeasts' ed. I, 1952)

Fermentation: –

Assimilation of carbon compounds:

Galactose	+	Raffinose	+ or w	Erythritol	v
Sucrose	+	Soluble starch	v	Ribitol	+
Maltose	+	D-Xylose	+	D-Mannitol	+
Cellobiose	+ or w	L-Arabinose	v	Succinic acid	v
Trehalose	+	D-Ribose	+ or w	Citric acid	v
Lactose	–	L-Rhamnose	+	Inositol	+

Additional carbon compounds tested: Melibiose –, sodium glucuronate + or w, glucono-δ-lactone +

Assimilation of nitrate: –

Assimilation of nitrite: –

Growth in vitamin-free medium: –

Growth on 50% (w/w) glucose-yeast extract agar: –

Growth at 37°C: +

Starch formation: + or w

G + C: 49.0–49.8 mol%, one strain (Nakase & Komagata, 1971b); 51.5 mol%, one strain (Storck et al., 1969)

Supplementary description of *Cryptococcus neoformans* var. *gattii*

Canavanine-glycine bromthymol blue (CGB) agar turns blue within 2–5 days; *Cryptococcus neoformans* var. *neoformans* leaves this medium unchanged.

Comments

See *Filobasidiella*, Comments on the genus, p. 479.

Origin of the strains belonging to the variety *neoformans*

Fruit juice (1), ulcus on cheek of patient (1), blastomycosis of the skin of patient (1), spinal fluid (3), tumor (1), sputum (1), lung (1), human patients (7), sick goat (1), mastitis of cow (2), mastitis milk (3), soil via mouse (3), frass in *Tsuga heterophylla* (1), unknown origin (5).

Type strain: CBS 132, isolated by Sanfelice from fermenting fruit juice.

Origin of the strain belonging to the variety *gattii*

Human (1).

Type strain: CBS 6289.

17. CRYPTOCOCCUS SKINNERI PHAFF ET DO CARMO-SOUSA (1962)

Standard description of *Cryptococcus skinneri*

Growth in malt extract: After 3 days at 25°C the cells are ovoidal to globose (2.6–5.2) × (4.5–7.2) μm, single, in pairs or in small groups. Only a little sediment is formed. After one month at 20°C a moderate ring and a small amount of sediment are formed.

Growth on malt agar: After one month the streak culture is pale, smooth, glistening, slightly mucous, raised; a little of the growth runs to the bottom of the slant. The border is entire.

Slide culture on potato agar: Pseudomycelium absent.

Fermentation: –

Assimilation of carbon compounds:

Galactose	+	Raffinose	–	Erythritol	– or +w
Sucrose	–	Soluble starch	–	Ribitol	+
Maltose	–	D-Xylose	+	D-Mannitol	+
Cellobiose	+	L-Arabinose	+	Succinic acid	+w
Trehalose	+	D-Ribose	+	Citric acid	+w
Lactose	–	L-Rhamnose	+	Inositol	+

Additional carbon compounds tested: Melibiose –, sodium glucuronate +, glucono-δ-lactone +

Assimilation of nitrate: –

Assimilation of nitrite: –

Growth in vitamin-free medium: –

Growth on 50% (w/w) glucose-yeast extract agar: –

Growth at 37°C: –; at 30°C: +w

Starch formation: +w

G + C: 53.0 mol%, one strain (Storck et al., 1969)

Origin of the strain studied

Frass in bark of a diseased *Tsuga heterophylla* (1).

Type strain: CBS 5029, isolated by Phaff and do Carmo-Sousa from frass in *Tsuga heterophylla*.

18. CRYPTOCOCCUS TERREUS DI MENNA (1954b)

Synonym

Cryptococcus himalayensis Goto et Sugiyama (1970)

Standard description of *Cryptococcus terreus*

Growth in malt extract: After 3 days at 25°C the cells are mainly globose or short-ovoidal, single, in pairs or short chains, (3.5–6.5) × (4.0–8.0) μm. Sometimes the beginning of a ring is formed and a little sediment is present. After one month at 20°C a moderate ring and a moderate to heavy sediment are formed.

Growth on malt agar: After one month at 20°C the streak culture is brownish cream-colored to yellowish-tan. The surface is smooth, glossy, and the texture is in some strains soft but not mucous, in others mucous so that the growth runs to the bottom of the slant. The border is entire.

Slide culture on potato agar: Pseudomycelium absent.

Fermentation: –

Assimilation of carbon compounds:

Galactose	+	Raffinose	–	Erythritol	–
Sucrose	–	Soluble starch	+ or w	Ribitol	v
Maltose	+ or w	D-Xylose	+	D-Mannitol	+
Cellobiose	+	L-Arabinose	+	Succinic acid	v
Trehalose	+	D-Ribose	+	Citric acid	v
Lactose	+	L-Rhamnose	+	Inositol	+

Additional carbon compounds tested: Melibiose –, melezitose –, sodium glucuronate +, glucono-δ-lactone +

Assimilation of nitrate: +

Growth in vitamin-free medium: +

Growth on 50% (w/w) glucose-yeast extract agar: –

Growth at 37°C: –; **at 30°C:** +

Starch formation: +

G + C: 59.5 mol%, one strain (Storck et al., 1969)

Comments

Cryptococcus himalayensis was described by Goto and Sugiyama (1970) as a separate species, because of its ability to assimilate sucrose, raffinose, and melezitose and its inability to assimilate maltose. Repeated tests have proved the opposite and this species therefore fits well in the standard description for *Cryptococcus terreus.*

Origin of the strains studied

Soil (4).

Type strain: CBS 1895, isolated from soil in New Zealand, obtained from di Menna.

19. *CRYPTOCOCCUS UNIGUTTULATUS* PHAFF ET FELL (1970)

Perfect state: *Filobasidium uniguttulatum* Kwon-Chung

Synonyms

Eutorulopsis uniguttulata Zach (Wolfram & Zach, 1934)
Cryptococcus neoformans (Sanfelice) Vuillemin var. *uniguttulatus* (Zach) Lodder et Kreger-van Rij (1952)

Standard description of *Cryptococcus uniguttulatus*

Growth in malt extract: After 3 days at 25°C the cells are globose to ovoidal, single, in pairs or in small clusters, (3.0–5.2) × (3.5–9.5) μm. An incomplete ring and a little sediment may be present. After one month a moderate ring and a heavy sediment are formed.

Growth on malt agar: After one month at 20°C the streak culture is cream colored to yellowish, smooth, semiglossy, soft but not slimy with an entire border.

Slide culture on potato agar: Pseudomycelium absent.

Fermentation: –

Assimilation of carbon compounds:

Galactose	v	Raffinose	+ or w	Erythritol	–
Sucrose	+	Soluble starch	+ w	Ribitol	v
Maltose	+	D-Xylose	+	D-Mannitol	+
Cellobiose	–	L-Arabinose	+	Succinic acid	+ or w
Trehalose	+ or w	D-Ribose	– or + w	Citric acid	v
Lactose	–	L-Rhamnose	+ or w	Inositol	+

Additional carbon compounds tested: Melibiose –, sodium glucuronate + w, glucono-δ-lactone +

Assimilation of nitrate: –

Assimilation of nitrite: –

Growth in vitamin-free medium: + w or –

Growth on 50% (w/w) glucose-yeast extract agar: –

Growth at 37°C: – ; **at 30°C:** v

Starch formation: + or w

G + C: 51.5–58.0 mol%, 2 strains (Storck et al., 1969)

Origin of the strains studied

Diseased nails (2), throat of patient (1), bronchial tube of patient (1), sputum (1), skin under foot of patient (1).

Type strain: CBS 1730, isolated from nail and obtained from Zach.

f. COMMENTS ON THE GENUS

In 1833 Kützing described *Cryptococcus mollis* and based the description of the genus on this species. The original material, collected by Kützing is deposited in the Rijksherbarium at Leiden University. As *Cryptococcus* species are characterized by a combination of morphological and physiological properties, it is not possible with only dead material available to recognize this species.

Investigation of the original material has shown that a great part of the cells are yeasts of basidiomycetous nature. Therefore *Cryptococcus mollis* can be maintained as the type for the name of the genus (Rodrigues de Miranda & Batenburg-van der Vegte, 1981).

Phaff and Spencer (1969) placed those imperfect species not forming pseudo- or true mycelium but which can assimilate inositol in *Cryptococcus,* and those which lack this ability in *Rhodotorula.* This character does not coincide with a red color of the culture, and *Cryptococcus* includes some red species.

Phaff and Spencer's diagnosis of *Cryptococcus* mainly includes basidiomycetous species, but also ascomycetous ones. For two of the latter, *Cr. lactativorus* and *Cr. cereanus,* the perfect state has been found and they are now classified in the genus *Sporopachydermia.* For a third species, *Cr. melibiosum* – the only ascomycetous one left in the genus – the perfect state is not yet known. It will probably be found as soon as more isolates are available.

Golubev et al. (1974) proposed the use of the assimilation of sodium glucuronate as a character in the diagnosis of *Cryptococcus* instead of inositol. Sodium glucuronate is formed during the first step in the catabolism of inositol, but not all inositol-positive yeasts are sodium glucuronate positive and, alternately, not all glucuronate-positive yeasts can assimilate inositol.

All *Cryptococcus* species assimilate sodium glucuronate with the exception of *Cr. melibiosum,* and exclusion of the ascomycetous species would make the genus more homogeneous. On the other hand, introduction of sodium glucuronate assimilation as a criterion for distinction between the genera *Cryptococcus, Rhodotorula* and *Candida* would bring several *Rhodotorula* and *Candida* species into *Cryptococcus,* among them ascomycetous *Candida* species (unpublished), thus leading to less homogeneity of *Cryptococcus.* This shows that neither inositol nor sodium glucuronate is suitable for distinguishing between ascomycetous and basidiomycetous species. For the present, I propose to retain inositol assimilation as a feature of the genus *Cryptococcus.*

g. SPECIES NOT ACCEPTED IN THE GENUS

The following species have not been accepted in the genus *Cryptococcus* because they do not assimilate inositol.

Cryptococcus bhutanensis Goto et Sugiyama (1970)
Cryptococcus lupi Baharaeen et Vishniac (1982)
Cryptococcus vishniacii Vishniac et Hempfling (1979)

The following species were not included because they came belatedly to the attention of the author.

Cryptococcus phylloplanus Shivas et Rodrigues de Miranda (1983a)

Cryptococcus hinnuleus Shivas et Rodrigues de Miranda (1983a)

Genus 6. *Kloeckera* Janke

by M.Th. Smith

a. DIAGNOSIS OF THE GENUS

Cells are lemon-shaped (apiculate), oval to long-oval or elongate. Vegetative reproduction by bipolar budding in basipetal succession. Pseudomycelium may be absent or present, ranging from primitive to well developed.

Fermentation is present.

Nitrate is not assimilated.

The strains of all species have an absolute requirement for inositol and pantothenate (Phaff, 1970d).

b. TYPE SPECIES OF THE GENUS

Kloeckera apiculata (Reess emend. Klöcker) Janke

c. SPECIES ACCEPTED IN THE GENUS

1. *Kloeckera africana* (Klöcker) Janke (Klöcker, 1912b)
2. *Kloeckera apiculata* (Reess emend. Klöcker) Janke (Reess, 1870)
3. *Kloeckera apis* Lavie ex Smith, Simione et Meyer (1977)
4. *Kloeckera corticis* (Klöcker) Janke (Klöcker, 1912b)
5. *Kloeckera japonica* Saito et Ohtani (1931)
6. *Kloeckera javanica* (Klöcker) Janke (Klöcker, 1912b)

d. KEY TO THE SPECIES OF THE GENUS

1a	Maltose assimilated	2
b	Maltose not assimilated	3

2a Growth at 34°C
Kl. africana .... p. 874
b No growth at 34°C
Kl. corticis .... p. 878

3a Growth at 37°C
Kl. apis .... p. 877
b No growth at 37°C 4

4a Sucrose assimilated
Kl. javanica .... p. 880
b Sucrose not assimilated 5

5a 2-Ketogluconate assimilated
Kl. apiculata .... p. 875
b 2-Ketogluconate not assimilated
Kl. japonica .... p. 879

See Table 12, p. 155

e. SYSTEMATIC DISCUSSION OF THE SPECIES

1. KLOECKERA AFRICANA (KLÖCKER) JANKE (1928)

Perfect state: *Hanseniaspora vineae* van der Walt et Tscheuschner

Synonym

Pseudosaccharomyces africanus Klöcker (1912b)

Standard description of *Kloeckera africana*

Growth in glucose-yeast extract-peptone water: After 2 days at 25°C the cells are apiculate or round to oval or elongate, (2.0–7.5) × (4.0–14.5) μm, mostly single, sometimes in pairs. Reproduction by bipolar budding. A sediment has formed. After one month a very thin ring may be present.

Growth on glucose-yeast extract-peptone agar: After one month the streak culture is white to cream-colored, smooth, glossy; center is raised and periphery is flat.

Slide culture on potato agar: Rudimentary or well developed pseudomycelium with or without clusters of blastospores.

Fermentation:

Glucose	+	Maltose	–
Galactose	–	Lactose	–
Sucrose	–		

Assimilation of carbon compounds:

Galactose	–	Raffinose	–	Erythritol	–
Sucrose	v	Soluble starch	–	Ribitol	–
Maltose	+	D-Xylose	–	D-Mannitol	–
Cellobiose	+	L-Arabinose	–	Succinic acid	–
Trehalose	–	D-Ribose	–	Citric acid	–
Lactose	–	L-Rhamnose	–	Inositol	–

Additional carbon compounds tested: Glycerol –, 2-ketogluconate –

Assimilation of nitrate: –

Growth in vitamin-free medium: –

Growth on 50% (w/w) glucose-yeast extract agar: +

Growth at 37°C: –; **at 34°C:** +

G + C: 38.8–40.5 mol%, 3 strains (Meyer et al., 1978)

Origin of the strains studied

Soil (1), unknown (2).

Type strain: CBS 277, isolated by Klöcker from soil.

2. *KLOECKERA APICULATA* (REESS EMEND. KLÖCKER) JANKE (1928)

Perfect state: *Hanseniaspora uvarum* (Niehaus) Shehata, Mrak et Phaff

Synonyms

Saccharomyces apiculatus Reess (1870)
Pseudosaccharomyces apiculatus (Reess) Klöcker (1912b)
Pseudosaccharomyces germanicus Klöcker (1912b)
Kloeckera germanica (Klöcker) Janke (1928)
Pseudosaccharomyces muelleri Klöcker (1912b)
Kloeckera muelleri (Klöcker) Janke (1928)
Kloeckera lindneri (Klöcker) Janke var. *pelliculosa* Lodder (1934)

Standard description of *Kloeckera apiculata*

Growth in glucose-yeast extract-peptone water: After 2 days at 25°C the cells are apiculate, round to oval or elongate, (1.5–5.0) × (2.5–11.5) μm, single or in pairs (Fig. 372). Reproduction by bipolar budding. A sediment is present. After one month a very thin ring has formed.

Growth on glucose-yeast extract-peptone agar: After one month the streak culture is white to creamy, smooth, glossy, with a raised center and a flat periphery.

Slide culture on potato agar: Pseudomycelium is present (tree-like formation), rudimentary or absent.

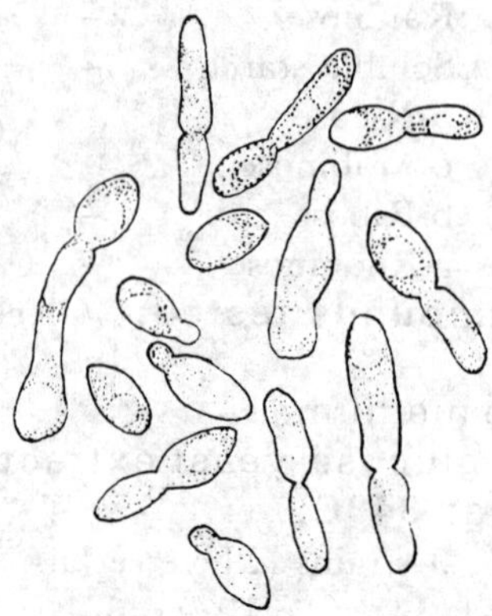

Fig. 372 – *Kl. apiculata*
After 3 days in malt extract (From 'The Yeasts' ed. I, 1952)

Fermentation:

Glucose	+	Maltose	–
Galactose	–	Lactose	–
Sucrose	–		

Assimilation of carbon compounds:

Galactose	–	Raffinose	–	Erythritol	–
Sucrose	–	Soluble starch	–	Ribitol	–
Maltose	–	D-Xylose	–	D-Mannitol	–
Cellobiose	+	L-Arabinose	–	Succinic acid	–
Trehalose	–	D-Ribose	–	Citric acid	–
Lactose	–	L-Rhamnose	–	Inositol	–

Additional carbon compounds tested: Glycerol –, 2-ketogluconate +
Assimilation of nitrate: –
Growth in vitamin-free medium: –
Growth on 50% (w/w) glucose-yeast extract agar: v
Growth at 37°C: –
G + C: 32.7–34.6 mol%, 4 strains (Meyer et al., 1978)

Origin of the strains studied

Soil (3), fruit juice (1), rotting strawberries (1).

Type strain: The strain from Hansen's collection (CBS 104) was designated as the type by Lodder and Kreger-van Rij (1952).

3. *KLOECKERA APIS* LAVIE EX SMITH, SIMIONE ET MEYER (1977)

Perfect state: *Hanseniaspora guilliermondii* Pijper

Standard description of *Kloeckera apis*

Growth in glucose-yeast extract-peptone water: After 2 days at 25°C the cells are apiculate, ovoid or elongate, (2.2–5.8) × (4.5–10.2) μm, occasionally longer, single or in pairs. Reproduction by bipolar budding. A sediment is present. After one month a thin ring has formed.

Growth on glucose-yeast extract-peptone agar: After one month the streak culture is white to cream-colored, smooth, glossy; in cross sections a raised center and a flat periphery.

Slide culture on potato agar: Pseudomycelium is absent.

Fermentation:

Glucose	+	Maltose	–
Galactose	–	Lactose	–
Sucrose	–		

Assimilation of carbon compounds:

Galactose	–	Raffinose	–	Erythritol	–
Sucrose	–	Soluble starch	–	Ribitol	–
Maltose	–	D-Xylose	–	D-Mannitol	–
Cellobiose	+	L-Arabinose	–	Succinic acid	–
Trehalose	–	D-Ribose	–	Citric acid	–
Lactose	–	L-Rhamnose	–	Inositol	–

Additional carbon compounds tested: Glycerol –, 2-ketogluconate +

Assimilation of nitrate: –

Growth in vitamin-free medium: –

Growth on 50% (w/w) glucose-yeast extract agar: +

Growth at 37°C: +

G + C: 33.9 mol%, one strain (Meyer et al., 1978)

Origin of the strain studied

One strain, isolated by Lavie (1954) from a bee, was examined.

Type strain: CBS 2591, isolated by Lavie from a bee.

4. *KLOECKERA CORTICIS* (KLÖCKER) JANKE (1928)

Perfect state: *Hanseniaspora osmophila* (Niehaus) Phaff, Miller et Shifrine

Synonyms

Pseudosaccharomyces corticis Klöcker (1912b)
Pseudosaccharomyces magnus de Rossi (1920)
Kloeckera magna (de Rossi) Janke (1928)
Pseudosaccharomyces santacruzensis Klöcker (1912b)
Kloeckera santacruzensis (Klöcker) Janke (1928)

Standard description of *Kloeckera corticis*

Growth in glucose-yeast extract-peptone water: After 2 days at 25°C the cells are lemon-shaped, oval or long-oval, (2.2–6.0) × (4.5–14.8) μm, single or in pairs. Reproduction by bipolar budding. A sediment is present. After one month a thin ring has formed.

Growth on glucose-yeast extract-peptone agar: The streak culture is white to cream-colored, smooth, glossy; the center is raised and the periphery is flat.

Slide culture on potato agar: Pseudomycelium in a branched (tree-like) formation is formed.

Fermentation:

Glucose	+	Maltose	–
Galactose	–	Lactose	–
Sucrose	–		

Assimilation of carbon compounds:

Galactose	–	Raffinose	–	Erythritol	–
Sucrose	v	Soluble starch	–	Ribitol	–
Maltose	+	D-Xylose	–	D-Mannitol	–
Cellobiose	+	L-Arabinose	–	Succinic acid	–
Trehalose	–	D-Ribose	–	Citric acid	–
Lactose	–	L-Rhamnose	–	Inositol	–

Additional carbon compounds tested: Glycerol –, 2-ketogluconate –

Assimilation of nitrate: –

Growth in vitamin-free medium: –

Growth on 50% (w/w) glucose-yeast extract agar: v

Growth at 37°C: –; **at 34°C:** –

G + C: 39.8–40.5 mol%, 6 strains (Meyer et al., 1978)

Origin of the strains studied

Soil (1), grapes (1), bark and moss (1), cider (1), unknown (3).

Type strain: CBS 106, isolated by Klöcker.

5. *KLOECKERA JAPONICA* SAITO ET OHTANI (1931)

Perfect state: *Hanseniaspora valbyensis* Klöcker

Synonym

Kloeckera corticis (Klöcker) Janke var. *pulquensis* Ulloa et Herrera (1973)

Standard description of *Kloeckera japonica*

Growth in glucose-yeast extract-peptone water: After 2 days at 25°C the cells are apiculate, round to oval or elongate, (2.0–5.5) × (3.0–10.2) μm, single or in pairs. Reproduction by bipolar budding. A sediment is present. After one month a thin ring may have been formed.

Growth on glucose-yeast extract-peptone agar: After one month the streak culture is white to cream-colored, smooth, glossy; center is somewhat raised and periphery is flat.

Slide culture on potato agar: Pseudomycelium is present (tree-like formation), rudimentary or absent.

Fermentation:

Glucose	+	Maltose	–
Galactose	–	Lactose	–
Sucrose	–		

Assimilation of carbon compounds:

Galactose	–	Raffinose	–	Erythritol	–
Sucrose	–	Soluble starch	–	Ribitol	–
Maltose	–	D-Xylose	–	D-Mannitol	–
Cellobiose	+	L-Arabinose	–	Succinic acid	–
Trehalose	–	D-Ribose	–	Citric acid	–
Lactose	–	L-Rhamnose	–	Inositol	–

Additional carbon compounds tested: Glycerol –, 2-ketogluconate –

Assimilation of nitrate: –

Growth in vitamin-free medium: –

Growth on 50% (w/w) glucose-yeast extract agar: –

Growth at 37°C: –

G + C: 28.8–29.3 mol%, 5 strains (Meyer et al., 1978)

Origin of the strains studied

Tree exudate (1), beer (2), pulque (1), unknown (1).

Type strain: CBS 281, isolated by Saito and Ohtani from exudate of tree.

6. *KLOECKERA JAVANICA* (KLÖCKER) JANKE (1928)

Perfect state: *Hanseniaspora occidentalis* Smith

Synonyms

Pseudosaccharomyces javanicus Klöcker (1912b)
Pseudosaccharomyces antillarum Klöcker (1912b)
Kloeckera antillarum (Klöcker) Janke (1928)
Pseudosaccharomyces indicus Klöcker (1912b)
Kloeckera indica (Klöcker) Janke (1928)

Pseudosaccharomyces jensenii Klöcker (1912b)
Kloeckera jensenii (Klöcker) Janke (1928)
Pseudosaccharomyces lafari Klöcker (1912b)
Kloeckera lafarii (Klöcker) Janke (1928)
Kloeckera javanica (Klöcker) Janke var. *lafarii* (Klöcker) Miller et Phaff (1958)
Pseudosaccharomyces willi Klöcker (1912b)
Kloeckera willi (Klöcker) Janke (1928)
Pseudosaccharomyces malaiana Klöcker (1912b)
Kloeckera malaiana (Klöcker) Janke (1928)

Standard description of *Kloeckera javanica*

Growth in glucose-yeast extract-peptone water: After 2 days at 25°C the cells are lemon-shaped, oval, sometimes round, (1.8–6.2) × (3.0–11.0) μm, single or in pairs. Reproduction by bipolar budding. A sediment is formed. After one month a thin ring is present.

Growth on glucose-yeast extract-peptone agar: After one month the culture is white to cream-colored, smooth, glossy; a raised center and a flat periphery.

Slide culture on potato agar: Pseudomycelium is generally lacking, but some strains produce a rudimentary or well developed pseudomycelium.

Fermentation:

Glucose	+	Maltose	–
Galactose	–	Lactose	–
Sucrose	+		

Assimilation of carbon compounds:

Galactose	–	Raffinose	–	Erythritol	–
Sucrose	+	Soluble starch	–	Ribitol	–
Maltose	–	D-Xylose	–	D-Mannitol	–
Cellobiose	+	L-Arabinose	–	Succinic acid	–
Trehalose	–	D-Ribose	–	Citric acid	–
Lactose	–	L-Rhamnose	–	Inositol	–

Additional carbon compounds tested: Glycerol +, 2-ketogluconate –

Assimilation of nitrate: –

Growth in vitamin-free medium: –

Growth on 50% (w/w) glucose-yeast extract agar: –

Growth at 37°C: –

G + C: 34.9–35.6 mol%, 7 strains (Meyer et al., 1978)

Origin of the strains studied

Soil (6), unknown (3).

Type strain: CBS 282, isolated by Klöcker from soil.

f. COMMENTS ON THE GENUS

Kloeckera lindneri (Klöcker) Janke (1928), regarded as a synonym of *Kl. apiculata* by Phaff (1970d), is not considered in this study since the status of *Kl. lindneri* within the genus *Kloeckera* has not yet been established (Meyer et al., 1978).

Will (1916a) described two apiculate yeast species, viz. *Pseudosaccharomyces cerevisiae* and *Pseudosaccharomyces vini*. In a later publication (1916b) however, Will stated that both species could not be considered as apiculates. Janke in 1924 accepted both species as representatives of the apiculate genus *Kloeckera* and made the combinations: *Kloeckera cerevisiae* (Will) Janke and *Kloeckera vini* (Will) Janke. Since authentic cultures are not available nor a definite conclusion from the original description could be made the names have to be considered nomina dubia.

Ogata et al. (1970a,b) published two papers on a methanol-assimilating yeast, tentatively identified and labelled "*Kloeckera* sp. No 2201". This strain has been identified as *Candida boidinii* by Smith (in von Arx, 1976).

Kozlova and Meisel (1975) in a paper on formation of peroxisomes in yeast organisms, mentioned a species *Kloeckera boidinii* nom. nud. Although a subculture of this species has not been examined in this study, the conclusion seems justified, by comparing the paper of Kozlova and Meisel with the publications of Ogata et al., to consider *Kl. boidinii* nom. nud. as a synonym of *Candida boidinii*.

Genus 7. *Malassezia* Baillon

by D. Yarrow and D.G. Ahearn

a. DIAGNOSIS OF THE GENUS

Cells are globose, oblong-ellipsoidal to cylindrical. Reproduction is by budding on a broad base and from the same site at one pole. The extruded bud is separated from the cell by a septum followed by fission. The separation of the bud leaves a scar. Successive scars form a small collar from which later buds emerge.

Well developed hyphae occur infrequently in culture with most isolates but are observed commonly in skin scales.

Growth may be stimulated by natural oil such as olive oil and occurs in the presence of 400 ppm cycloheximide.

Optimum temperature for growth is 35–37°C, weak growth at 25°C.

No fermentation.

b. TYPE SPECIES OF THE GENUS

Malassezia furfur (Robin) Baillon

c. SPECIES ACCEPTED IN THE GENUS

1. *Malassezia furfur* (Robin) Baillon (Robin, 1853)
2. *Malassezia pachydermatis* (Weidman) Dodge (Weidman, 1925)

d. KEY TO THE SPECIES OF THE GENUS

1a Good growth on media such as glucose-peptone-yeast extract agar, YM agar and malt agar at 37°C without addition of oil
Mal. pachydermatis .... p. 884

b No growth or only very poor growth on such media without the addition of olive oil
Mal. furfur (conidial yeast-like forms) ... p. 883

e. SYSTEMATIC DISCUSSION OF THE SPECIES

1. MALASSEZIA FURFUR (ROBIN) BAILLON (1889)

Synonyms

Microsporon furfur Robin (1853)
Sporotrichum furfur (Robin) Saccardo (1886)
Pityrosporum furfur (Robin) Emmons (in Emmons et al., 1970)
Saccharomyces ovalis Bizzozero (1884)
Saccharomyces sphaericus Bizzozero (1884) [non *Saccharomyces sphaericus* von Nägeli (1879)]
Malassezia ovalis (Bizz.) Acton et Panja (1927)
Pityrosporum ovale (Bizz.) Castellani et Chalmers (1913)
Pityrosporum malassezi Sabouraud (1904)
Cryptococcus malassezi (Sab.) Benedek (1930)
Pityrosporum orbiculare Gordon (1951)

Standard description of *Malassezia furfur*

In skin scales short, often curved, rarely branched hyphae 2.5–4.0 μm in diameter predominate. Globose cells up to 8 μm in diameter may be observed.

Growth on glucose-peptone-yeast extract agar overlaid with olive oil: After 3 days at 37°C two forms can be recognized. The ovale form is most abundant with ovoidal, ellipsoidal or cylindrical cells, (1.5–4.5) × (2.0–6.5) μm. The orbiculare form, which is usually sparse, consists predominantly of globose cells measuring 2.5–4.5 μm in diameter. Transition from one form to the other may occur in culture.

The streak culture is cream to yellowish, smooth or lightly wrinkled, glistening or dull, the margin is entire or lobate.

G + C: 62.9 mol%, one strain (CBS 1878)

Comments

Because of their nutritional requirements it is not possible to conduct the usual physiological tests on the cultivatable forms of this species. Salkin and Gordon (1977) reported that, in contrast to ovoidal cells, globose cells did not grow in yeast nitrogen base plus either olive oil or Tween 20. Moreover one out of the six cultures grew on yeast-carbon base with oil and potassium nitrate.

Pityrosporum ovale and *Pit. orbiculare* are considered here to be variant cultural forms of *Malassezia furfur* and are treated as synonyms of it. Burke (1961) observed typical structures of *Mal. furfur* in a patient with a high plasma cortisol level experimentally infected with *Pit. orbiculare.* Moreover she observed hyphal elements in cultures of *Pit. orbiculare.* Dorn and Roehnert (1977) reported that the globose cells of *Pit. orbiculare* converted readily to hyphal elements in a defined medium. Sternberg and Keddie (1961) found a high titre of agglutinating antibodies to *Pit. orbiculare* in the serum of a patient with pityriasis versicolor. Keddie and Shadomy (1963) isolated *Pit. orbiculare* from scales of tinea versicolor in which the presence of *Mal. furfur* had been demonstrated by staining. Keddie (1967) and Emmons et al. (1970) concluded that *Pit. orbiculare* and *Mal. furfur* are synonymous. Slooff (1970c) reviewed the literature on these two genera. Salkin and Gordon (1977) reported changes in the characters of 11 isolates of *Pit. orbiculare* during maintenance in culture. They observed an alteration from the globose *orbiculare* form to the ovoidal and cylindrical *ovale* form and suggested that these forms are stages in the development of *Mal. furfur*. Such changes in cell form were also observed in cultures used in this study.

Origin of the strains studied

Human skin scales (14), eye lid (1), ear of pig (1), ear of horse (1), ear of cow (1), skin of eland (1), intestinal polyp (1).

Type specimen: 7019, obtained from lesions of pityriasis versicolor on the trunk of a fifteen-year-old female by V.K. Hopsu-Havu has been deposited in the herbarium of the Centraalbureau voor Schimmelcultures at Baarn as neotype.

2. *MALASSEZIA PACHYDERMATIS* (WEIDMAN) DODGE (1935)

Synonyms

Pityrosporum pachydermatis Weidman (1925)
Pityrosporum canis Gustafson (1955)

Standard description of *Malassezia pachydermatis*

Growth on glucose-peptone yeast extract agar: After 3 days at 37°C the cells are subglobose to ellipsoidal, (2.5–5.5) × (3.0–6.5) μm (Fig. 373).

The streak culture is cream, smooth, dull, with an entire or slightly lobate margin.

G + C: 56.1 mol%, one strain (CBS 1879)

Comments

The strains of this species do not grow in the liquid media used in the standard assimilation tests. Slooff (1970c) reported that growth occurs on glucose, D-mannitol and D-glucitol, sometimes on glycerol, lactic acid and citric acid also, when tested on auxanogram plates.

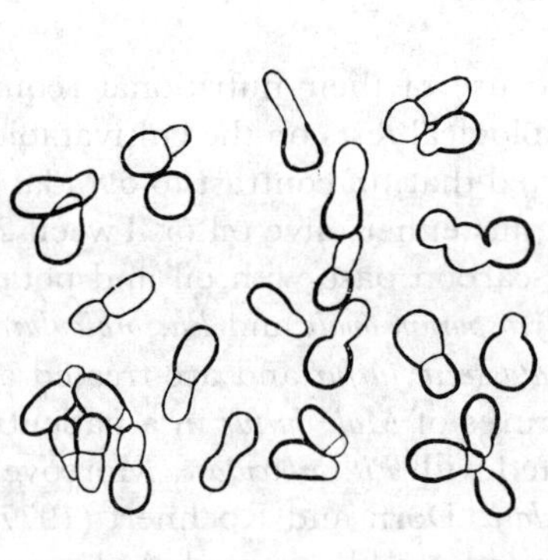

Fig. 373 – *Mal. pachydermatis,* CBS 1879
After 2 weeks on glucose-yeast extract-peptone agar (From 'The Yeasts' ed. II, 1970)

Origin of the strains studied

Otitis externa of dogs (8), inflamed ears of dogs (3), ears of dogs (10), human foot (1), human scalp (1), human eye (1).

Type strain: CBS 1879, isolated from otitis externa of a dog by Gustafson (1955), was designated as the type by Slooff (1970c).

Genus 8. *Oosporidium* Stautz

by M.Th. Smith

a. DIAGNOSIS OF THE GENUS (DO CARMO-SOUSA, 1970b)

Cells of various shapes reproducing vegetatively by multipolar budding on a broad base combined with septation. The cells usually remain attached to each other forming long chains. True septa may be present independently from the broad base budding. No arthrospores.

Ascospores not formed.

Asexual endospores formed by protoplasmic cleavage.

Alcoholic fermentation absent.

Pink or orange-yellow pigments of a non-carotenoid nature present.

b. TYPE SPECIES OF THE GENUS

Oosporidium margaritiferum Stautz

c. SPECIES ACCEPTED IN THE GENUS

1. *Oosporidium margaritiferum* Stautz (1931)

e. SYSTEMATIC DISCUSSION OF THE SPECIES

1. *OOSPORIDIUM MARGARITIFERUM* STAUTZ (1931)

Synonym

Trichosporon margaritiferum (Stautz) Buchwald (1939)

Standard description of *Oosporidium margaritiferum*

Growth in glucose-yeast extract-peptone water: After 3 days at 25°C

no visible growth occurs. After one week a granular or flocculent sediment is present. The cells are globose, ovoid or elongate, (5.0–12.0) × (6.0–12.5) μm. Reproduction is multilateral, usually by budding on a broad base combined with septation. The cells usually remain attached to each other and form long branched chains (Fig. 374). After one month a sediment is present.

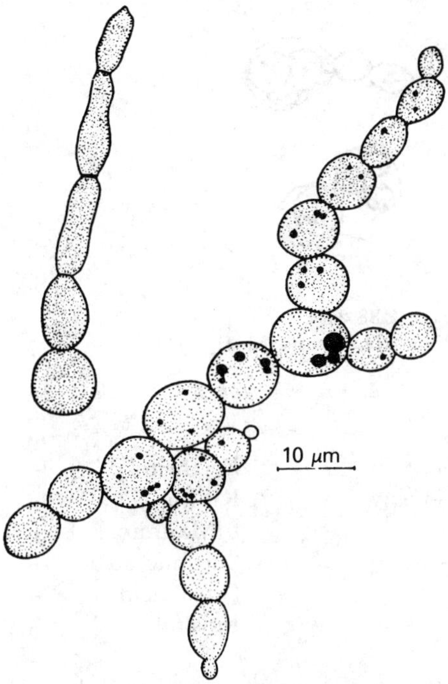

Fig. 374 – *O. margaritiferum,* CBS 2531
After 1 week in glucose-yeast extract-peptone water (From 'The Yeasts' ed. II, 1970)

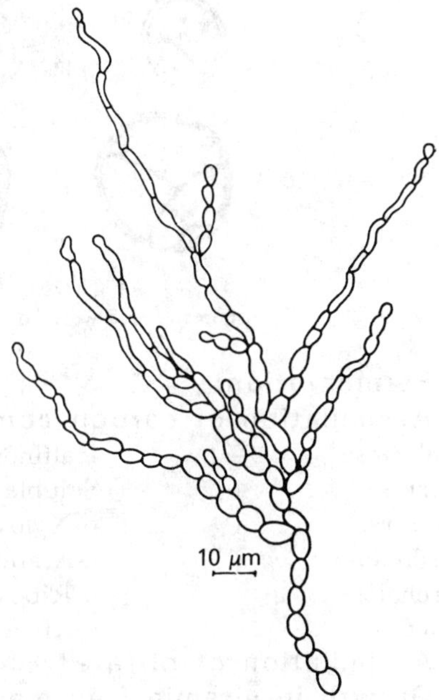

Fig. 375 – *O. margaritiferum,* CBS 2531
Dalmau plate culture on corn meal agar (From 'The Yeasts' ed. II, 1970)

Growth on glucose-yeast extract-peptone agar: After one month at 25°C the streak culture is light pink to light orange, dull, rugose, raised, with a fragile surface. The growth-rate is slow.

Slide culture on corn meal agar: Pseudomycelium is present consisting of branched chains of ovoid, globose or elongate cells. At the periphery cells tend to become more and more elongate (Fig. 375). True septa may be formed.

Formation of asexual spores: Asexual endospores are formed by protoplasmic cleavage (Fig. 376). Cells of variable size and shape may contain one to ten endospores of various sizes, globose or ellipsoidal in shape. The endospores may start germinating inside the mother cell. The endospores were observed on YM agar and Gorodkowa agar both in slants and in slide cultures after 4–6 weeks at 18–20°C, in slide cultures on corn meal- and potato agar. The slants had been inoculated with cultures grown in 15% liquid malt extract for 5–6 days at 25°C on a shaker (do Carmo-Sousa, 1970b).

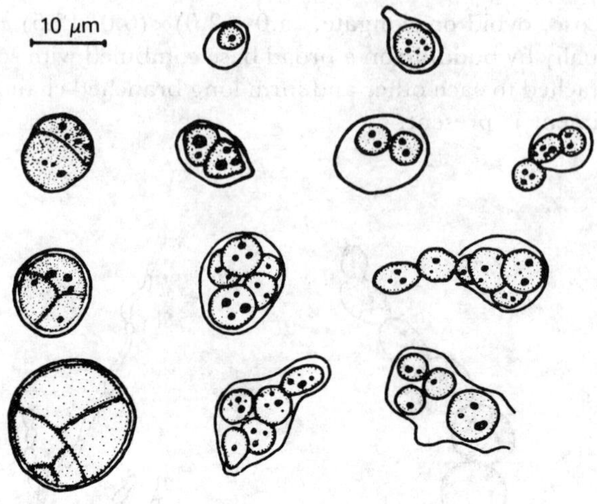

Fig. 376 – *O. margaritiferum,* CBS 2531
Endospores after 5 weeks on YM agar (From 'The Yeasts' ed. II, 1970)

Fermentation: –

Assimilation of carbon compounds*:

Galactose	–	Raffinose	+	Erythritol	–
Sucrose	+	Soluble starch	+ or w	Ribitol	–
Maltose	+	D-Xylose	–	D-Mannitol	+ w
Cellobiose	–	L-Arabinose	–	Succinic acid	+
Trehalose	+	D-Ribose	–	Citric acid	–
Lactose	–	L-Rhamnose	–	Inositol	–

Assimilation of nitrate:** +

Growth in vitamin-free medium: –

Growth on 50% (w/w) glucose-yeast extract agar: –

Maximum temperature of growth: 32–34°C

Origin of the strains studied

Slime flux (2).

Type strain: CBS 2531, isolated by Stautz (1931).

*Tested on yeast nitrogen base-carbon source-agar slants. The liquid medium test is not suitable for this organism.

**Tested on yeast carbon base-potassium nitrate agar slants.

f. COMMENTS ON THE GENUS

The genus *Oosporidium,* introduced by Stautz in 1931, has been considered a synonym of the genus *Trichosporon* by Buchwald (1939), Diddens and Lodder (1942) and Lodder and Kreger-van Rij (1952). However, do Carmo-Sousa (1970b) re-instated the genus.

The genus is separated from the genus *Trichosporon* by the absence of arthrospores and from the genus *Candida* by the ability of endospore formation. Do Carmo-Sousa (1969) studied extensively the endospore formation in the genera *Oosporidium* and *Trichosporon.*

Genus 9. *Phaffia* Miller, Yoneyama et Soneda

by M.W. Miller

a. DIAGNOSIS OF THE GENUS

Vegetative cells are ellipsoidal and occur singly, in pairs, and occasionally in short chains. Carotenoid pigments are synthesized (mainly astaxanthin and a minor proportion of β-carotene) which give cells en masse an orange to salmon-red color. True mycelium is absent, but a rudimentary pseudomycelium may be present. Chlamydospores are formed. A pellicle is formed on liquid media.

Fermentative ability present.

Starch-like compounds are formed (synthesis is pH-independent).

Inositol is not assimilated.

Nitrate is not assimilated.

b. TYPE SPECIES OF THE GENUS

Phaffia rhodozyma Miller, Yoneyama et Soneda

c. SPECIES ACCEPTED IN THE GENUS

1. *Phaffia rhodozyma* Miller, Yoneyama et Soneda (1976b)

e. SYSTEMATIC DISCUSSION OF THE SPECIES

1. PHAFFIA RHODOZYMA MILLER, YONEYAMA ET SONEDA (1976b)

Standard description of *Phaffia rhodozyma*

Growth in malt extract: After 3 days at 20°C, the cells are ellipsoidal, (3.8–7.5) × (5.5–10.5) μm, occurring singly, in pairs, or occasionally in short chains. Budding occurs frequently at the same site. Spheroidal chlamydospores with refractile granules may be formed after several days (Fig. 377). A thin ring and pellicle are form-

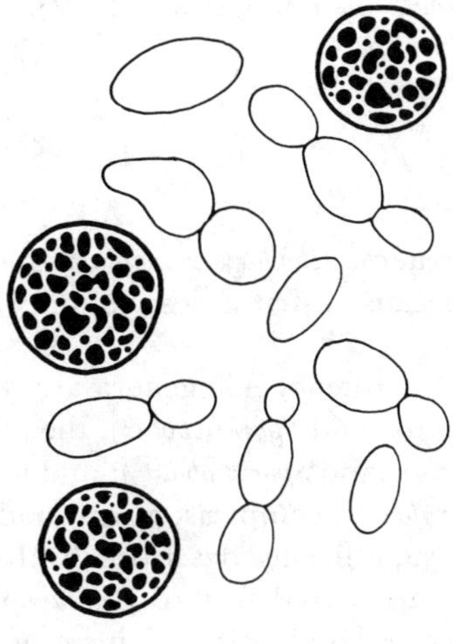

Fig. 377 – *Ph. rhodozyma*
After 3 days on malt extract agar

ed, and there is little sediment. After one month, there are a thick, orange ring, a thin pellicle or islets, and a moderate amount of sediment.

Growth on malt agar: After one month, the streak culture is orange- to salmon-red, punctilate to nearly smooth, glossy to semidull, soft and broadly convex to convex in cross section; the border is lobulate to nearly entire.

Dalmau plate culture on corn meal agar: In the aerobic portion, the border is occasionally irregular; under the coverslip, there is an occasional tuft of cells or rudimentary pseudomycelium consisting of limited numbers of cells in a chain. No true mycelium is formed.

Fermentation:

Glucose	+	Lactose	–
Galactose	–	Melibiose	–
Sucrose	+ s	Raffinose	+ s
Maltose	+ s		

Assimilation of carbon compounds:

Galactose	–	Raffinose	+	Erythritol	–
Sucrose	+	Soluble starch	+ s or –	Ribitol	–
Maltose	+	D-Xylose	+ s	D-Mannitol	+ s
Cellobiose	+	L-Arabinose	+	Succinic acid	+
Trehalose	+	D-Ribose	–	Citric acid	–
Lactose	–	L-Rhamnose	–	Inositol	–

Assimilation of nitrate: –

Growth in vitamin-free medium: –

Growth on 50% (w/w) glucose-yeast extract agar: –

Growth at 37°C: – (maximum ca. 27°C)

Starch formation: + (pH-independent)

Carotenoid pigments: + (ca. 85% is astaxanthin, Andrews et al., 1976)

Urease: +

G + C: 48.3 mol%, one strain (Miller et al., 1976b)

Origin of the strains studied

Slime fluxes of *Cornus brachypoda* (2), *Fagus crenata* (1), *Alnus japonica* (1), *Betula maximowicziana* (1), *B. tauschii* (1), all from Japan, *B. verrucosa* from U.S.S.R. (2), *B. papyrifera* from Alaska (1), *Betula* sp. from Finland (1).

Type strain: CBS 5905, isolated from *Fagus crenata* by Phaff et al. (1972).

f. COMMENTS ON THE GENUS

Ph. rhodozyma was tentatively described under the name *Rhodozyma montanae* by Phaff et al. (1972). This binomial is a nomen nudum, because a Latin description was not provided.

Specific criteria differentiating *Phaffia* from other carotenogenic genera are: the inability to form ballistospores (positive for *Sporobolomyces* and *Sporidiobolus*); the absence of an observed sexual cycle (positive for *Rhodosporidium* and *Sporidiobolus*); and fermentative ability (negative for *Cryptococcus* and *Rhodotorula*). A comparison was made also with *Rhodomyces dendrorheus,* a yeast-like organism superficially described by Ludwig (1891, 1896). Other than color and source of isolation (red flux of *Betula* sp.), no similarity was found. Apparently, Ludwig never isolated this organism in pure culture.

Golubev et al. (1977) isolated 67 additional strains in U.S.S.R. from *Betula verrucosa* fluxes and bark during a seven year survey.

Because of its astaxanthin content *Ph. rhodozyma* is being considered as a dietary pigment source for pen-reared salmonids and crustaceans (Johnson et al., 1977). An enzymatic procedure for the extraction of astaxanthin from the yeast cells was reported by Johnson et al. (1978).

Genus 10. *Rhodotorula* Harrison

by J.W. Fell, A. Statzell Tallman and D.G. Ahearn

a. DIAGNOSIS OF THE GENUS

Cells spheroidal, ovoidal or elongate. Reproduction by multilateral budding. Strains of some species form pseudo- or true hyphae. Ascospores or ballistospores are not formed. Red or yellow carotenoid pigments are synthesized in malt agar cultures. Many strains have a mucous appearance due to capsule formation, but others are pasty or dry and wrinkled.

No assimilation of inositol as sole source of carbon. Starch-like substances are not synthesized. Fermentative ability is lacking.

Some species include homothallic or heterothallic stages of the genus *Rhodosporidium* Banno, or non-ballistosporulating sporobolomycetes.

b. TYPE SPECIES OF THE GENUS

Rhodotorula glutinis (Fresenius) Harrison

c. SPECIES ACCEPTED IN THE GENUS

1. *Rhodotorula acheniorum* (Buhagiar et Barnett) Rodrigues de Miranda (Buhagiar & Barnett, 1973)
2. *Rhodotorula araucariae* Grinbergs et Yarrow (1970b)
3. *Rhodotorula aurantiaca* (Saito) Lodder (Saito, 1922)
4. *Rhodotorula glutinis* (Fresenius) Harrison (Fresenius, 1852)
5. *Rhodotorula graminis* di Menna (1958a)
6. *Rhodotorula lactosa* Hasegawa (1959)
7. *Rhodotorula minuta* (Saito) Harrison (Saito, 1922)
8. *Rhodotorula rubra* (Demme) Lodder (Demme, 1889)

d. KEY TO THE SPECIES OF THE GENUS

1a Nitrate assimilated 2
b Nitrate not assimilated 7

2a Melibiose assimilated 3
b Melibiose not assimilated 4

3a Erythritol assimilated
Rh. acheniorum .... p. 895
b Erythritol not assimilated
Rh. lactosa .... p. 900

4a Raffinose assimilated 5
b Raffinose not assimilated 6

5a Melezitose assimilated
Rh. glutinis .... p. 898
b Melezitose not assimilated
Rh. graminis .... p. 899

6a Melezitose assimilated
Rh. aurantiaca .... p. 897
b Melezitose not assimilated
Rh. araucariae .... p. 896

7a Raffinose assimilated
Rh. rubra .... p. 903
b Raffinose not assimilated
Rh. minuta .... p. 901

Table 38

Key characters of the species in the genus *Rhodotorula*

	Assimilation					
	NO_3	Me	Ma	Mz	Ra	Er
Rh. acheniorum	+	+	+	+	+	+
Rh. araucariae	+	–	–	–	–	–
Rh. aurantiaca	+	–	+	+	–	–
Rh. glutinis	+	–	+	+	+	–
Rh. graminis	+	–	–(+)	–	+	–
Rh. lactosa	+	+	+	+	+	–
Rh. minuta	–	–	v	v	–	v
Rh. rubra	–	–	v	v	+	–

NO_3 = nitrate, Me = melibiose, Ma = maltose, Mz = melezitose, Ra = raffinose, Er = erythritol.

e. SYSTEMATIC DISCUSSION OF THE SPECIES

1. *RHODOTORULA ACHENIORUM* (BUHAGIAR ET BARNETT) RODRIGUES DE MIRANDA (1975)

Synonym

Sterigmatomyces acheniorum Buhagiar et Barnett (1973)

Standard description of *Rhodotorula acheniorum*

Growth in malt extract: After 3 days at 24°C, the cells are ellipsoidal, mostly elongate (1.6–3.9) × (2.4–8.2) μm, and contain conspicuous fat globules. The bud may be borne on a short pedicle. After one month there is a ring and sediment, no pellicle.

Growth on malt agar: After 3 days at 24°C the colony is buff, occasionally pale flesh colored, smooth to rough. After 3 weeks or possibly longer the colony is pale vinaceous with well defined edges.

Dalmau plate culture on corn meal agar: Short chains of budding cells, with sparse pseudomycelium.

Fermentation: –

Assimilation of carbon compounds:

Galactose	+	Raffinose	+	Erythritol	+
Sucrose	+	Soluble starch	–	Ribitol	+ (w)
Maltose	+	D-Xylose	+	D-Mannitol	+
Cellobiose	+	L-Arabinose	+	Succinic acid	+
Trehalose	+	D-Ribose	+	Citric acid	+
Lactose	+ (w)	L-Rhamnose	–	Inositol	–

Additional carbon compounds tested: Melezitose +, melibiose +

Assimilation of nitrate: +

Growth in vitamin-free medium: –

Growth on 50% (w/w) glucose-yeast extract agar: –

Maximum temperature for growth: 30–32°C

Starch formation: –

Urease: +

Origin of the strain studied

Fresh strawberries in England (1) (Buhagiar & Barnett, 1973).

Type strain: CBS 6386.

2. *RHODOTORULA ARAUCARIAE* GRINBERGS ET YARROW (1970b)

Standard description of *Rhodotorula araucariae*

Growth in malt extract: After 3 days at 24°C the cells are oval to subglobose (3.0–6.0) × (3.4–7.4) μm, single or budding. Grinbergs and Yarrow (1970b) report cells to 11 μm. After one month there is a ring and moderate sediment, no pellicle; the cells are the same size range.

Growth on malt agar: After 3 days at 24°C the colony is pale flesh, smooth and glistening. The cross section is raised, the texture soft, the edges entire. After one month the flesh color is more intense.

Dalmau plate culture on corn meal agar: Pseudomycelium absent.

Fermentation: –

Assimilation of carbon compounds:

Galactose	+ s	Raffinose	–	Erythritol	–
Sucrose	–	Soluble starch	–	Ribitol	+
Maltose	–	D-Xylose	+	D-Mannitol	+
Cellobiose	–	L-Arabinose	–	Succinic acid	–
Trehalose	+	D-Ribose	–	Citric acid	–
Lactose	–	L-Rhamnose	–	Inositol	–

Additional carbon compounds tested: Melezitose –, melibiose –

Assimilation of nitrate: +

Growth in vitamin-free medium: +

Growth on 50% (w/w) glucose-yeast extract agar: –

Growth at 37°C: –

Starch formation: –

Urease: +

Origin of the strain studied

Rotting *Araucaria araucana* in Chile (1) (Grinbergs & Yarrow, 1970b).

Type strain: CBS 6031.

3. *RHODOTORULA AURANTIACA* (SAITO) LODDER (1934)

Synonyms

Torula aurantiaca Saito (1922)
Torulopsis aurantiaca (Saito) Cif. et Red. (1925)
Chromotorula aurantiaca (Saito) Harrison (1928)
Rhodotorula glutinis (Fres.) Harrison var. *aurantiaca* (Saito) Hasegawa (1958)
Mycotorula colostri Castelli (1932)
Rhodotorula crocea Shifrine et Phaff (1956)

Standard description of *Rhodotorula aurantiaca*

Growth in malt extract: After 3 days at 18°C the cells are elongate to cylindrical, (3–5) × (6–13) μm, occasionally 16 μm or longer; in pairs or single, sometimes in short chains; little sediment, there may be a thin, partial ring. After one month a thin to well developed orange or red ring and a moderate amount of sediment.

Growth on malt agar: The cell morphology is similar to that in malt extract, but cells may be slightly longer. The streak culture after one month is reddish, orange, or dark yellow, finely wrinkled or smooth, semiglossy, soft, flat to low convex, with an entire to slightly irregular border.

Dalmau plate culture on corn meal agar: Pseudomycelium absent or rudimentary, consisting of short chains of elongate cells.

Fermentation: –

Assimilation of carbon compounds:

Galactose	+	Raffinose	–	Erythritol	–
Sucrose	+	Soluble starch	–	Ribitol	v
Maltose	+	D-Xylose	+	D-Mannitol	+
Cellobiose	v	L-Arabinose	v	Succinic acid	+
Trehalose	v	D-Ribose	–	Citric acid	v
Lactose	–	L-Rhamnose	–	Inositol	–

Additional carbon compounds tested: Melezitose +, melibiose –

Assimilation of nitrate: +

Growth in vitamin-free medium: –; thiamine is required and *p*-aminobenzoic acid by most strains

Growth on 50% (w/w) glucose-yeast extract agar: –

Maximum temperature for growth: 28–38°C

Starch formation: –

Urease: +

G + C: 55.4 mol%; strain *Rh. crocea*: 58.8 mol% (Nakase & Komagata, 1971b)

Origin of the strains studied

21 strains from Japan, Europe, North America, and the Antarctic, Atlantic and Indian Oceans. Sources include: air (1), clinical (1), unknown (1), bark beetle (1), plants and soils (7), marine waters (10).

Type strain: CBS 317 from atmosphere (Saito, 1922).

4. RHODOTORULA GLUTINIS (FRESENIUS) HARRISON (1928)

Perfect states: *Rhodosporidium diobovatum* Newell et Hunter, *Rhodosporidium sphaerocarpum* Newell et Fell, *Rhodosporidium toruloides* Banno

Synonyms

Cryptococcus glutinis Fres. (1852)

Saccharomyces glutinis (Fres.) Cohn (1875)
Torula glutinis (Fres.) Pringsh. et Bilewsky (1911)
Torulopsis glutinis (Fres.) Dodge (1935)
Torula rubescens Saito (1922)
Mycotorula rubescens (Saito) Cif. et Red. (1925)
Rhodotorula rubescens (Saito) Harrison (1928)
Rhodotorula glutinis (Fres.) Harrison var. *rubescens* (Saito) Lodder (1934)
Torula rufula Saito (1922)
Torulopsis rufula (Saito) Cif. et Red. (1925)
Rhodotorula rufula (Saito) Harrison (1928)
Rhodotorula glutinis (Fres.) Harrison var. *rufula* (Saito) Lodder (1934)
Torula rubra Schimon var. α Saito (1922)
Torulopsis bronchialis Cif. et Red. (1925)
Cryptococcus bronchialis (Cif. et Red.) Nann. (1934)
Torulopsis saitoi Cif. et Red. (1925)
Rhodotorula glutinis (Fres.) Harrison var. *saitoi* (Cif. et Red.) Lodder (1934)
Torula miniata Okunuki (1931)
Torula suganii Okunuki (1931)
Rhodotorula suganii (Okunuki) Lodder (1934)
Rhodotorula longissima Lodder (1934)
Rhodotorula gracilis Rennerfelt (1937)
Rhodotorula glutinis (Fres.) Harrison var. *lusitanica* Marcilla, Feduchy et Gomes (1945)
Rhodotorula glutinis (Fres.) Harrison var. *dairenensis* Hasegawa et Banno (1958)
Rhodotorula terrea Sugiyama et Goto (1969)

Standard description of *Rhodotorula glutinis*

Growth in malt extract: After 3 days at 23°C the cells are ovoidal to globose (2.3–5.0) × (4.0–10) μm or more elongate (12–16 μm) and wider (7 μm). There is usually a thin ring and a little sediment. After one month a medium to heavy, pink to orange or salmon-colored ring and a heavy sediment.

Growth on malt agar: The cell morphology is similar to that in malt extract, but the cells may be longer. The streak culture after one month is coral red to salmon-colored or slightly orange; the surface varies from smooth, often with fine transverse striations, to wrinkled; the appearance from highly glossy to semiglossy. The texture from mucous to pasty to slightly tough. The cross section is flat to broad convex and the border is irregular to entire, often with rudimentary pseudomycelium.

Dalmau plate culture on corn meal agar: Pseudomycelium generally absent or rudimentary; in some strains well developed pseudohyphae or true hyphae with large dark pigmented teliospores which indicate self-sporulating strains of *Rhodosporidium.*

Fermentation: –

Assimilation of carbon compounds:

Galactose	+	Raffinose	+	Erythritol	–
Sucrose	+	Soluble starch	v	Ribitol	v
Maltose	+	D-Xylose	+	D-Mannitol	v
Cellobiose	+	L-Arabinose	v	Succinic acid	+
Trehalose	+	D-Ribose	v	Citric acid	v
Lactose	–	L-Rhamnose	v	Inositol	–

Additional carbon compounds tested: Melezitose +, melibiose –

Assimilation of nitrate: +
Growth in vitamin-free medium: + ; thiamine rarely required
Growth on 50% (w/w) glucose-yeast extract agar: –
Growth at 37°C: v
Starch formation: –
Urease: +
G + C: 60.0–61.2 mol%; 66.8–67.8 mol% (Nakase & Komagata, 1971b)

Origin of the strains studied

370 strains from Japan, Sweden, U.S.A., Germany, Spain, Canada, and the Atlantic, Indian and Antarctic Oceans. Sources included: air (6), human patient (1), soil (1), wood pulp (20), brewery water (1), plant leaves (2), boric acid (1), brined pickles (1), olive brine (2), textiles (1), flowers (6), tree slime fluxes (4), marine (316), oil drillings (1), lake trout intestine (1), brine shrimp (2), unknown sources (4).

Type strain: CBS 20 isolated by Pringsheim from atmosphere and originally designated *Torula rubra* var. *a*.

5. *RHODOTORULA GRAMINIS* DI MENNA (1958a)

Synonym

Rhodotorula rosa (Nishiwaki) Goto et Yokotsuka (1962)

Standard description of *Rhodotorula graminis*

Growth in malt extract: After 3 days at 20°C the cells are globose to ovoidal or elongate, (2.5–7) × (4–15) μm; single, in pairs or in short chains or clusters. Oil droplets often in the cells or in the medium. A thin ring and a little sediment are present. After one month a moderate to heavy pink ring, usually a heavy sediment, and occasionally some moist film islets are present.

Growth on malt agar: The cell morphology is similar to that in malt extract. After one month at room temperature the streak culture is coral red, the surface is smooth or rugose to corrugated and glossy; texture soft, occasionally mucous; the cross section is flat to low convex and the border irregular with some rudimentary pseudomycelium.

Dalmau plate culture on corn meal agar: Pseudomycelium absent or rudimentary, consisting of chains of elongate cells. Some strains have a well developed branched pseudomycelium growing into the agar.

Fermentation: –

Assimilation of carbon compounds:

Galactose	+	Raffinose	+	Erythritol	–
Sucrose	+	Soluble starch	–	Ribitol	v
Maltose	– (+)	D-Xylose	+ (w)	D-Mannitol	+ (–)
Cellobiose	v	L-Arabinose	v	Succinic acid	+
Trehalose	+	D-Ribose	v	Citric acid	+w or –
Lactose	–	L-Rhamnose	v	Inositol	–

Additional carbon compounds tested: Melezitose –, melibiose –
Assimilation of nitrate: +
Growth in vitamin-free medium: +
Growth on 50% (w/w) glucose-yeast extract agar: –
Growth at 37°C: –; **at 28–33°C:** +
Starch formation: –
Urease: +
G + C: 70.0 mol% (Storck et al., 1969)

Origin of the strains studied

67 strains from New Zealand, Indonesia, North America, Japan, the Indian and Atlantic Oceans. Sources included: plants (6), marine waters (59), unknown (1), soil (1).

Type strain: CBS 2826 from pasture grass (di Menna, 1958a).

6. *RHODOTORULA LACTOSA* HASEGAWA (1959)

Standard description of *Rhodotorula lactosa*

Growth in malt extract: After 5 days at 20°C the cells are single or in pairs, short-ovoidal to long-ovoidal (2.5–4.0) × (4.5–8.0) μm or longer (11–12.5) μm. There is a thin incomplete ring and a little sediment. Growth is slow. After one month there is a thin to moderate ring and a moderate sediment.

Growth on malt agar: Cells in young cultures are larger than those grown in liquid malt extract. After one month at 20°C the streak culture is pink, smooth, highly glossy, soft but not slimy, convex and little spreading, border entire.

Dalmau plate culture on corn meal agar: Pseudomycelium absent.
Fermentation: –
Assimilation of carbon compounds:

Galactose	–	Raffinose	+	Erythritol	–
Sucrose	+	Soluble starch	–	Ribitol	+ s
Maltose	+	D-Xylose	+	D-Mannitol	+
Cellobiose	+	L-Arabinose	+	Succinic acid	+
Trehalose	+ s	D-Ribose	–	Citric acid	+ s
Lactose	+ s or –	L-Rhamnose	+	Inositol	–

Additional carbon compounds tested: Melezitose +, melibiose +
Assimilation of nitrate: +
Growth in vitamin-free medium: –; thiamine and *p*-aminobenzoic acid are required
Growth on 50% (w/w) glucose-yeast extract agar: –
Growth at 37°C: –; **at 30°C:** +
Starch formation: –
Urease: +
G + C: 57.3 mol% (Nakase & Komagata, 1971b)

Origin of the strains studied

From air near Osaka, Japan (2).

Type strain: CBS 5826.

7. *RHODOTORULA MINUTA* (SAITO) HARRISON (1928)

Synonyms

Torula minuta Saito (1922)
Torulopsis minuta (Saito) Cif. et Red. (1925)
Rhodotorula texensis Phaff et al. var. *minuta* (Saito) Hasegawa, Banno et Yamauchi (1960)
Rhodotorula pallida Lodder (1934)
Rhodotorula marina Phaff, Mrak et Williams (1952)
Rhodotorula texensis Phaff, Mrak et Williams (1952)
Rhodotorula minuta (Saito) Harrison var. *texensis* (Phaff, Mrak et Williams) Phaff et Ahearn (1970)
Rhodotorula tokyoensis Kobayashi (1953)
Rhodotorula laryngis Reiersöl (1955)
Rhodotorula slooffii Novák et Vörös-Felkai (1962)
Rhodotorula zsoltii Galgóczy et Novák (1965)

Standard description of *Rhodotorula minuta*

Growth in malt extract: After 3 days at 25°C the cells are ovoidal to globose (2.3–4.5) × (3.5–6.5) μm, to elongate, single and pairs. There is a thin, pink ring and a little sediment. After one month a moderate orange to red ring and a light sediment are present.

Growth on malt agar: The cell morphology is similar to that in malt extract. After one month at 25°C the streak culture is pink, smooth, glossy, soft but not mucous, the cross section flat to low convex, and the border entire.

Dalmau plate culture on corn meal agar: Pseudomycelium absent.

Fermentation: –

Assimilation of carbon compounds:

Galactose	v	Raffinose	– (+w)	Erythritol	v
Sucrose	+ (–)	Soluble starch	–	Ribitol	+ (–)
Maltose	v	D-Xylose	+ (–)	D-Mannitol	+ (–)
Cellobiose	v	L-Arabinose	+ (–)	Succinic acid	+ or w
Trehalose	+	D-Ribose	v	Citric acid	–
Lactose	v	L-Rhamnose	–	Inositol	–

Additional carbon compounds tested: Melezitose + (–), melibiose –

Assimilation of nitrate: –

Growth in vitamin-free medium: – ; *p*-aminobenzoic acid and thiamine are required

Growth on 50% (w/w) glucose-yeast extract agar: –

Maximal temperature for growth: 28–38°C

Starch formation: –

Urease: +

G + C: 51.0–54.0 mol% (Storck et al., 1969; Nakase & Komagata, 1971b)

Origin of the strains studied

151 strains from Japan, Europe, North America, New Zealand, Antarctica. Sources included: air (1), clinical (23), terrestrial plants and animals (8), marine water, plants and animals (115), pickles (1), unknown sources (3).

Type strain: CBS 319 isolated by Saito from atmosphere in Tokyo.

8. *RHODOTORULA RUBRA* (DEMME) LODDER (1934)

Synonyms

Saccharomyces ruber Demme (1889)
Cryptococcus ruber (Demme) Vuill. (1901)
Torulopsis rubra (Demme) de Almeida (1933)
Torula mucilaginosa Jörg. (1909)
Torulopsis mucilaginosa (Jörg.) Cif. et Red. (1925)
Rhodotorula mucilaginosa (Jörg.) Harrison (1928)
Torula sanguinea Schimon (1911)
Torulopsis sanguinea (Schimon) Cif. et Red. (1925)
Rhodotorula sanguinea (Schimon) Harrison (1928)
Rhodotorula mucilaginosa (Jörg.) Harrison var. *sanguinea* (Schimon) Lodder (1934)
Cryptococcus ludwigi Anderson (1918)
Cryptococcus mena Fontoynont et Boucher (1923)
Torulopsis mena (Font. et Boucher) Dodge (1935)
Blastodendrion carbonei Cif. et Red. (1925)
Rhodotorula mucilaginosa (Jörg.) Harrison var. *carbonei* Lodder (1934)
Eutorulopsis dubia Cif. et Red. (1925)
Mycotorula pulmonalis Cif. et Red. (1925)
Geotrichum pulmonale (Cif. et Red.) Dodge (1935)
Torulopsis sanniei Cif. et Red. (1925)
Cryptococcus sanniei (Cif. et Red.) Nannizzi (1934)
Rhodotorula sanniei (Cif. et Red.) Lodder (1934)
Torulopsis biourgei Cif. et Red. (1925)
Blastodendrion simplex Cif. et Red. (1925)
Cryptococcus rubrorugosus Castellani (1927)
Cryptococcus pararoseus Castellani (1927)
Rhodotorula mucilaginosa (Jörg.) Harrison var. *pararosea* (Cast.) Lodder (1934)
Torula aclotiana (Harrison, 1928)
Rhodotorula aclotiana Harrison (1928)
Cryptococcus corallinus A. et R. Sartory et al. (1930)
Torulopsis nitritophila Cif. et Ashford (Ashford & Ciferri, 1930)
Cryptococcus radiatus A. et R. Sartory et Meyer (1931)
Torula decolans Okunuki (1931)
Torulopsis mannitica Castelli (1932)
Torulopsis aurantia Zach (Wolfram & Zach, 1934)
Rhodotorula mucilaginosa (Jörg.) Harrison var. *plicata* Lodder (1934)
Rhodotorula rubra (Demme) Lodder var. *longa* Lodder (1934)
Rhodotorula rubra (Demme) Lodder var. *curvata* Lodder (1934)
Rhodotorula mucilaginosa (Jörg.) Harrison var. *kentuckyi* Giovannozzi (1948)
Rhodotorula pilimanae Hedrick et Burke (1951)
Rhodotorula vuilleminii Saëz (1967)
Rhodotorula ulzamae Moriyon et Ramirez (1974)
Mycotorula cisnerosi nom. nud.
Mycotorula germanica nom. nud.

Standard description of *Rhodotorula rubra*

Growth in malt extract: After 3 to 5 days at room temperature cells vary from short-ovoidal to elongate; single, pairs, short chains or clusters; width varies 2–5.5 μm. The length of the short celled strains varies 2.5–6.5 μm; strains with intermediate sized cells 4.5–10 μm; long-celled strains 6.5–14 μm and occasionally longer. There is a partial or complete ring and little sediment. After one month there is a moderate to heavy pink or salmon-colored ring, usually no pellicle; sometimes a partial or complete, slimy, moist film and a moderate sediment are present.

Growth on malt agar: The cell morphology is similar to that in malt extract. After one month the streak culture varies from deep coral to pink or salmon-colored; the surface is glistening and usually smooth, sometimes reticulate, rugose or corrugated; the texture varies from soft to mucous; cross section low convex to flat and the border entire, occasionally developing rudimentary pseudomycelium at the thin part of the slant.

Dalmau plate culture on corn meal agar: Pseudomycelium absent or rudimentary, consisting mainly of chains of elongate cells; some strains produce extensive, well developed pseudo- or true mycelium in corn meal agar supplemented with 1% Tween 80.

Fermentation: –

Assimilation of carbon compounds:

Galactose	v	Raffinose	+	Erythritol	–
Sucrose	+	Soluble starch	–	Ribitol	+(–)
Maltose	v	D-Xylose	+	D-Mannitol	v
Cellobiose	v	L-Arabinose	v	Succinic acid	+(w)
Trehalose	+	D-Ribose	v	Citric acid	v
Lactose	–	L-Rhamnose	v	Inositol	–

Additional carbon compounds tested: Melezitose v, melibiose –

Assimilation of nitrate: –

Growth in vitamin-free medium: – (+w); thiamine is required

Growth on 50% (w/w) glucose-yeast extract agar: –

Growth at 37°C: v

Starch formation: –

Urease: +

G + C: 60–63.5 mol% (Nakase & Komagata, 1968a, 1971b; Marmur & Doty, 1962)

Origin of the strains studied

230 strains from Europe, North America, Japan, Hawaii, Atlantic and Indian Oceans. Sources included: clinical (42), beer (1), atmosphere (3), fresh water in a brewery (1), fermenting tobacco (2), jam (1), flowers (8), trout intestine (3), raw sewage (2), marine sources, Atlantic and Indian Oceans (131), fresh water of Florida Everglades (23), insect larvae (1), bird (1), unknown sources (11).

Type strain: CBS 17 from Pollacci in Italy as *Saccharomyces ruber*.

f. COMMENTS ON THE GENUS

We currently accept eight species in the genus. Six of these are nitrate positive, four of which were included in the previous revision of Phaff and Ahearn (1970). The six nitrate-positive species can be differentiated by abilities to assimilate erythritol, melibiose, melezitose, maltose and raffinose (Table 38).

One of the important criteria that influenced our taxonomic decisions in retaining and eliminating species was the mol percent guanine plus cytosine (G + C) content of the DNA. *Rh. glutinis,* the most prevalent species of the genus is a complex of imperfect forms of at least three sexual species (*Rhodosporidium toruloides, Rhodosp. sphaerocarpum* and *Rhodosp. diobovatum*). Of those isolates of *Rh. glutinis* that have been examined for G + C content, there are two groupings: one with a G + C content of 60.0–61.2 mol% that includes imperfect forms of *Rhodosporidium toruloides;* the other group with a G + C of 66.8–67.8 mol% representing *Rhodosporidium diobovatum.* Sexual cycles for some strains in both groups have not been reported.

Rh. graminis has the highest G + C content (70 mol%) reported for the genus. That species is distinguished by the ability to assimilate raffinose, but not melezitose. *Rh. lactosa,* a rarely isolated species, has a G + C content of 57 mol% and the unique ability to assimilate melibose. The type strain of *Rh. aurantiaca* has a G + C content of 55.4 mol% and the ability to utilize maltose and melezitose but not raffinose. For practical purposes we retained the synonymy of *Rh. crocea* with *Rh. aurantiaca. Rh. crocea* differs from *Rh. aurantiaca* by its G + C content of 58.8 mol% and its ability to assimilate 5-ketogluconate, a compound that is difficult to obtain. The G + C contents of *Rh. acheniorum* and *Rh. araucariae* have not been determined. *Rh. acheniorum* is the only nitrate-positive species that assimilates erythritol. *Rh. araucariae* assimilates only a few of the tested compounds and is distinct from other members of the genus (Table 38) by the inability to utilize α- and β-glucosides and α-galactosides.

Only two nitrate-negative species, *Rh. minuta* and *Rh. rubra,* were retained which are separated by raffinose assimilation. *Rh. pallida* and *Rh. marina* were placed in synonymy with *Rh. minuta.* These three species had previously been separated on differences in assimilation patterns of lactose and maltose. Instability among these reactions is common for strains of these yeasts making practical separation difficult. The G + C of the three species is within a narrow range (50–54 mol%) and they all share the essential requirement for *p*-aminobenzoic acid. Similarly the minor differences in sugar assimilations between *Rh. pilimanae* (G + C: 60.7 mol%) and *Rh. rubra* (G + C: 60.0–63.5 mol%) are not sufficient for maintenance of separate species.

Currently in depth biochemical studies are developing an insight into the phylogenetic relationships of species of *Rhodotorula* with species of other imperfect genera. Johnson and Phaff (1978) examined pigment production in *Torulopsis fujisanensis,* a species with a whitish-yellow to brownish pigment. By altering the growth medium, for example employing ethylamine as a sole nitrogen source in yeast carbon base, Johnson and Phaff were able to enhance pigmentation. By chemical extraction and analysis, the authors demonstrated the production of carotenoid pigments in *T. fujisanensis* and transferred the species to the genus *Rhodotorula.* Von Arx and Weijman (1979) separated the anamorphic basidiomycetous yeasts into two groups based on daughter cell formation and chemical composition of intact cell hydrolysates. The authors concluded that pigmentation of the colonies was an inadequate character to distinguish genera. Studies such as these, when expanded and coupled to other biochemical investigations of a broader range of species will undoubtedly aid in clarifying the phylogeny of the anamorphic yeasts. Unfortunately, the use of relatively sophisticated tests for the separation of genera is currently impractical for most laboratories. Until more information is available, we have maintained separation of *Rhodotorula* from the other imperfect genera by the presence of visible carotenoid pigments.

g. SPECIES NOT ACCEPTED IN THE GENUS

The following species was not included because it came belatedly to the attention of the authors:

Rhodotorula armeniaca Shivas et Rodrigues de Miranda (1983b)

The species *Rhodotorula fujisanensis* (Soneda) Johnson et Phaff (1978), synonym: *Torulopsis fujisanensis* Soneda (1959a), has not been accepted by us in the genus *Rhodotorula* (see f. Comments on the genus). *Rh. fujisanensis* occurs in the key to group I of the genus *Candida,* but is not described in that genus. The following description is reproduced from van Uden and Vidal-Leiria (1970).

Standard description of *Rhodotorula fujisanensis*

Growth in glucose-yeast extract-peptone water: After 3 days at 25°C the cells are long-ovoid to cylindrical, (2.5–5) × (7–15) μm. A ring and an incomplete pellicle are present.

Growth on glucose-yeast extract-peptone agar: After one month at 25°C the streak culture is greyish-yellow, semidull, soft, smooth or finely punctate.

Dalmau plate culture on corn meal agar: A rudimentary pseudomycelium is formed that consists of ramified chains of ovoid cells.

Fermentation: –

Assimilation of carbon compounds:

Galactose	+	Melezitose	–	Ribitol	+
L-Sorbose	+	Soluble starch	–	Galactitol	+
Sucrose	–	D-Xylose	+	D-Mannitol	+
Maltose	–	L-Arabinose	+	D-Glucitol	+
Cellobiose	+	D-Arabinose	+	Salicin	+
Trehalose	–	D-Ribose	+	DL-Lactic acid	–
Lactose	–	L-Rhamnose	–	Succinic acid	–
Melibiose	–	Glycerol	+	Citric acid	+ w
Raffinose	–	Erythritol	–	Inositol	–

Assimilation of nitrate: –

Growth in vitamin-free medium: +

Maximum temperature for growth: 27–28°C

Comments

Mixing of cells from strains CBS 6371 and CBS 4551 on potato dextrose agar, followed by incubation at 12–18°C for 18 days can result in the formation of hyphae with clamps and teliospores. The teliospores are globose, 10–26 μm diameter, to occasionally elongate. These results are inconsistent and attempts to induce teliospore germination have been unsuccessful. This species clearly belongs in one of the sexual basidiomycetous genera; however, formal description of the sexual state is delayed pending complete details of the life cycle.

Origin of the strains studied

Hare feces (1), frass (1), leaves (1).

Type strain: CBS 4551, isolated from the feces of a hare (Soneda, 1959a).

Genus 11. *Sarcinosporon* King et Jong

by M.Th. Smith

a. DIAGNOSIS OF THE GENUS

Vegetative growth filamentous or unicellular. Reproduction by budding singly or successively in chains. Sarcina-like agglomerates of cells are formed by septation in different planes of individual blastospores or of cells in filaments.

Ascospores not produced.

Asexual endospores are formed.

Fermentation absent.

b. TYPE SPECIES OF THE GENUS

Sarcinosporon inkin (Oho) King et Jong

c. SPECIES ACCEPTED IN THE GENUS

1. *Sarcinosporon inkin* (Oho) King et Jong (Oho, 1919)

e. SYSTEMATIC DISCUSSION OF THE SPECIES

1. SARCINOSPORON INKIN (OHO) KING ET JONG (1975)

Synonyms

Sarcinomyces inkin Oho (1919)
Trichosporon inkin (Oho) do Carmo-Sousa et van Uden (1967)
Prototheca filamenta Arnold et Ahearn (1972)

Standard description of *Sarcinosporon inkin*

Growth in glucose-yeast extract-peptone water: After 3 days at 25°C

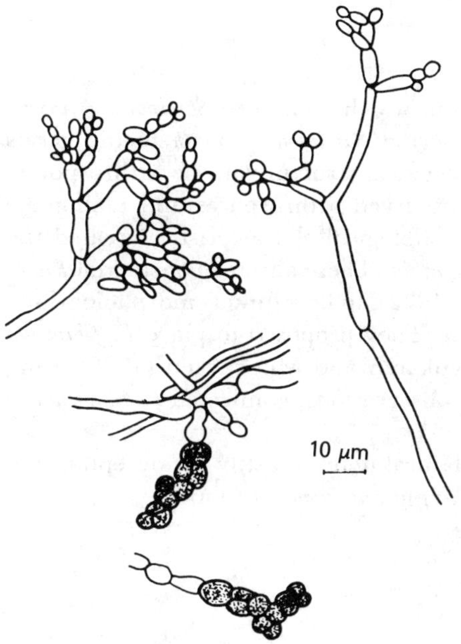

Fig. 378 – *Sar. inkin*
Dalmau plate culture on corn meal agar
(From 'The Yeasts' ed. II, 1970)

budding cells are ovoid or ellipsoidal, single, in pairs or groups. Elongate segments and filaments are present. A wrinkled, creeping pellicle and sediment are present.

Growth on glucose-yeast extract-peptone agar: After one month the streak culture is dry, friable and cream colored with a raised center and radiating furrows.

Slide culture on corn meal agar: True mycelium and blastospores are formed. Blastospores may arise singly or successively, in simple or branched chains. Sarcina-like agglomerates are formed by septation in different planes of blastospores or of cells in filaments (Fig. 378).

Formation of asexual spores: The sarcina-like agglomerates contain at maturity one to several endospores. Endospores are ovoid or ellipsoid, (1.3–6.3) × (1.3–9.0) μm, and are produced on glucose-yeast extract-peptone agar and Harrold's M40Y agar (King & Jong, 1975).

Fermentation: –

Assimilation of carbon compounds:

Galactose	+	Raffinose	–	Erythritol	+
Sucrose	+	Soluble starch	+	Ribitol	–
Maltose	+	D-Xylose	+	D-Mannitol	+(s)
Cellobiose	+	L-Arabinose	–	Succinic acid	–
Trehalose	+	D-Ribose	+	Citric acid	–
Lactose	+	L-Rhamnose	–	Inositol	+

Assimilation of nitrate: –

Growth in vitamin-free medium: –

Growth on 50% (w/w) glucose-yeast extract agar: –

Growth at 37°C: +

G + C: 57.0–57.6 mol%, one strain (Dupont & Hedrick, 1971)

Origin of the strain studied

Skin lesion (1).

Type strain: Dried material of type culture ATCC 18020 (=CBS 5585) on corn meal agar plate deposited by King and Jong (1975) in the herbarium of the National Fungus Collections, Beltsville, Maryland.

f. COMMENTS ON THE GENUS

Oho (1919; Ota, 1926a) described an organism which he named *Sarcinomyces inkin.* Do Carmo-Sousa and van Uden (1967) transferred *Sarcinomyces inkin* to the genus *Trichosporon* since they considered some morphological structures to be arthrospores. However, King and Jong (1975) not having observed arthrospores and taking the sarcina-like agglomerates as an important morphological characteristic, created the genus *Sarcinosporon* to accommodate *Trichosporon inkin.* These authors considered *Prototheca filamenta,* described by Arnold and Ahearn (1972), to be a fungus morphologically and physiologically similar to *Sarcinosporon inkin.* They proposed to place *P. filamenta* in synonymy with *Sarcinosporon inkin.* Nadakavukaren and McCracken (1975) using scanning and transmission electron microscope observations, confirmed *P. filamenta* to be a fungus and not a colorless alga.

In 1976 Jong and King published a nomenclatural note to justify the acceptance of the name *Sarcinosporon* King et Jong over the name *Sarcinomyces* Oho.

Genus 12. *Schizoblastosporion* Ciferri

by N.J.W. Kreger-van Rij

a. DIAGNOSIS OF THE GENUS

Cells reproducing by bipolar budding on a broad base.
Ascospores are not formed.
Fermentation absent.

b. TYPE SPECIES OF THE GENUS

Schizoblastosporion starkeyi-henricii Ciferri

c. SPECIES ACCEPTED IN THE GENUS

1. *Schizoblastosporion starkeyi-henricii* Ciferri (1930)

e. SYSTEMATIC DISCUSSION OF THE SPECIES

1. SCHIZOBLASTOSPORION STARKEYI-HENRICII CIFERRI (1930)

Standard description of *Schizoblastosporion starkeyi-henricii*

Growth in malt extract: After 3 days at 25°C the cells are elliptical, long-elliptical or cylindrical, (2.5–6.8) × (4.5–20) μm; single or in pairs. Buds are formed on a broad base at both poles (Fig. 379). Nonpolar buds occur only rarely. A thin sediment is formed. After one month at 17°C a sediment, a ring and, occasionally, some islets are present.

Growth on malt agar: After one month at 17°C the streak culture is greyish-brown, soft, dull, flat, occasionally with tiny protuberances.

Slide cultures on potato- and corn meal agar: Pseudomycelium is absent, or it consists of small chains of cells, occasionally with side branches.

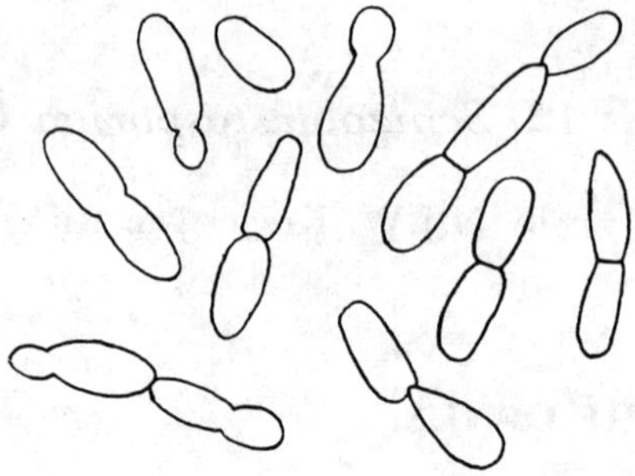

Fig. 379 – *Schizobl. starkeyi-henricii*
After 3 days in malt extract (From 'The Yeasts' ed. II, 1970)

Fermentation: –

Assimilation of carbon compounds:

Galactose	+	Raffinose	–	Erythritol	–
Sucrose	–	Soluble starch	–	Ribitol	–
Maltose	–	D-Xylose	–	D-Mannitol	v
Cellobiose	–	L-Arabinose	–	Succinic acid	+
Trehalose	+ w or –	D-Ribose	–	Citric acid	–
Lactose	–	L-Rhamnose	–	Inositol	–

Splitting of arbutin: –

Assimilation of nitrate: –

Growth in vitamin-free medium: –

Growth on 50% (w/w) glucose-yeast extract agar: –

Growth at 37°C: –

Origin of the strains studied

Soil in Denmark (1) (Lund, 1954); in New Zealand (5) (di Menna, 1957, 1965b); in Germany (1) and in Norway (1) (Roberts, 1960); stomach contents (1).

Type strain: CBS 2159, isolated by Lund from a soil sample from peat bog.

Genus 13. *Sporobolomyces* Kluyver et van Niel

by J.W. Fell and A. Statzell Tallman

a. DIAGNOSIS OF THE GENUS

Colony colors on solid media are shades of pink, red and orange due to the presence of carotenoid pigments. Vegetative reproduction is by budding, by the formation of pseudomycelium and true mycelium, and by the development of ballistospores. The ballistospores are bilaterally symmetrical, usually obovoid, pyriform and reniform; they develop obliquely at the tips of simple and branched sterigmata on yeast and mycelial cells.

Fermentation does not occur; urease is positive.

Species in this genus may represent the imperfect states of *Sporidiobolus* spp.

b. TYPE SPECIES OF THE GENUS

Sporobolomyces roseus Kluyver et van Niel

c. SPECIES ACCEPTED IN THE GENUS

1. *Sporobolomyces albo-rubescens* Derx (1930)
2. *Sporobolomyces gracilis* Derx (1930)
3. *Sporobolomyces holsaticus* Windisch ex Yarrow et Fell (1980)
4. *Sporobolomyces puniceus* (Komagata et Nakase) Ahearn et Yarrow (Komagata & Nakase, 1965)
5. *Sporobolomyces roseus* Kluyver et van Niel (1925)
6. *Sporobolomyces salmonicolor* (Fischer et Brebeck) Kluyver et van Niel (Fischer & Brebeck, 1894), see *Sporidiobolus salmonicolor*
7. *Sporobolomyces shibatanus* (Okunuki) Verona et Ciferri (Okunuki, 1931), see *Sporidiobolus pararoseus*

d. KEY TO THE SPECIES OF THE GENUS

1a Inositol assimilated
Sp. puniceus .... p. 917
b Inositol not assimilated 2

2a Nitrate assimilated 3
b Nitrate not assimilated 5

3a Mycelium present
Sp. holsaticus .... p. 915
see also *Sporidiobolus salmonicolor* ...... p. 538
b Mycelium not present 4

4a Starch assimilated
Sp. roseus .... p. 918
b Starch not assimilated, or slowly and weakly assimilated
Sp. salmonicolor (see *Sporidiobolus salmonicolor*) .... p. 538

5a Maltose assimilated 6
b Maltose not assimilated
Sp. gracilis .... p. 913

6a Cellobiose assimilated
Sp. shibatanus (see *Sporidiobolus pararoseus*) p. 535
b Cellobiose not assimilated
Sp. albo-rubescens .... p. 912

See Table 24, p. 534.

e. SYSTEMATIC DISCUSSION OF THE SPECIES

1. SPOROBOLOMYCES ALBO-RUBESCENS DERX (1930)

Synonym

Sporobolomyces photographus Cif. et Red. var. *albo-rubescens* (Derx) Verona et Cif. (1938)

Standard description of *Sporobolomyces albo-rubescens*

Growth in malt extract: After 3 days at room temperature the cells are ovoidal and spherical (3–8) × (5–10) μm, single and in pairs; reproduction by multilateral budding. There is a light strawberry pink ring and a moderate sediment. After one month the ring remains light and the sediment is moderate to heavy.

Growth on malt agar: The cell morphology is similar to that in malt extract. After one month at room temperature the streak culture is coral pink, smooth, highly glossy, flat and the border is entire.

Dalmau plate culture on corn meal agar: Pseudomycelium or true mycelium are not formed.

Formation of ballistospores: The original description illustrated pyriform ballistospores 4 × (9.5–11) μm that are flattened on one side. The capacity of the type culture to produce ballistospores has been lost as they were not observed by us, Lodder and Kreger-van Rij (1952) or Phaff (1970c). Another strain, CBS 6239, examined by Yarrow (personal communication), formed ballistospores on corn meal agar at room temperature that were oval and slightly flattened on the lateral plane.

Fermentation: –

Assimilation of carbon compounds:

Galactose	+	Raffinose	+	Erythritol	–
Sucrose	+	Soluble starch	–	Ribitol	+
Maltose	+	D-Xylose	+	D-Mannitol	–
Cellobiose	–	L-Arabinose	+	Succinic acid	+ w
Trehalose	+	D-Ribose	+	Citric acid	+ w
Lactose	–	L-Rhamnose	–	Inositol	–

Additional carbon compounds tested: Melezitose +, melibiose –, α-methyl-D-glucoside –

Assimilation of nitrate: –

Growth in vitamin-free medium: –

Growth on 50% (w/w) glucose-yeast extract agar: –

Growth at 37°C: + w; **at 30°C:** +

Starch formation: –

Urease: +

G + C: 63.0 mol% (Storck et al., 1969)

Origin of the strains studied

Smut infected leaves in southern France (1) (Derx, 1930); dialysis equipment (1) (examined by Yarrow).

Type strain: CBS 482 (Derx, 1930).

2. *SPOROBOLOMYCES GRACILIS* DERX (1930)

Standard description of *Sporobolomyces gracilis*

Growth in malt extract: Of the two strains examined, the type strain did not grow in malt extract. Phaff (1970c) examined that strain in liquid yeast nitrogen base-glucose assimilation medium on a shaker after 4 days at 18°C and found the cells were ovoidal to cylindrical (2–4) × (7–16) μm, occurring singly and in pairs, occasionally in clusters of 4–5 cells. The other strain grew in malt extract. After 7 days there was a moderate sediment, light ring, cells occurred singly, in pairs and threes. The cells were oval to elongate (3–5) × (5–9) μm. After one month spherical cells to 16 μm with or without sterigmata were observed.

Growth on malt agar: The cells are similar to those in liquid media. After one month at room temperature the streak culture is scarlet red, smooth, glistening, broadly convex; the border is entire.

Dalmau plate culture on corn meal agar: Pseudomycelium or true mycelium are absent.

Formation of ballistospores: Ballistospores were not observed with the type strain in the present studies nor by Phaff (1970c), although pyriform ballistospores were illustrated by Derx (1930). Another strain (CBS 7028) formed sterigmata after one month in malt extract at room temperature; ballistospores did not develop.

Fermentation: –

Assimilation of carbon compounds:

Galactose	–	Raffinose	–	Erythritol	–
Sucrose	–	Soluble starch	–	Ribitol	+ s
Maltose	–	D-Xylose	+	D-Mannitol	+
Cellobiose	+ s	L-Arabinose	–	Succinic acid	v
Trehalose	+ s	D-Ribose	+ s	Citric acid	v
Lactose	–	L-Rhamnose	–	Inositol	–

Additional carbon compounds tested: Melezitose –, melibiose –, α-methyl-D-glucoside –

Assimilation of nitrate: –

Growth in vitamin-free medium: –

Growth on 50% (w/w) glucose-yeast extract agar: –

Growth at 37°C: –; **at 30°C:** –

Starch formation: –

Urease: +

Origin of the strains studied

Decaying leaves (1) (Derx, 1930); unknown (1).

Type strain: CBS 71 (Derx, 1930).

3. *SPOROBOLOMYCES HOLSATICUS* WINDISCH EX YARROW ET FELL (1980)

Synonyms

Sporobolomyces holsaticus nom. nud. (Windisch, 1949)
Prosporobolomyces holsaticus (Windisch) Novák et Zsolt (1961)

Standard description of *Sporobolomyces holsaticus*

Growth in malt extract: After 7 days at room temperature the cells (Fig. 380) are similar to those on malt agar. There is a thick off-white or peach ring and light sediment. By 2 weeks there is a thick myceliated growth.

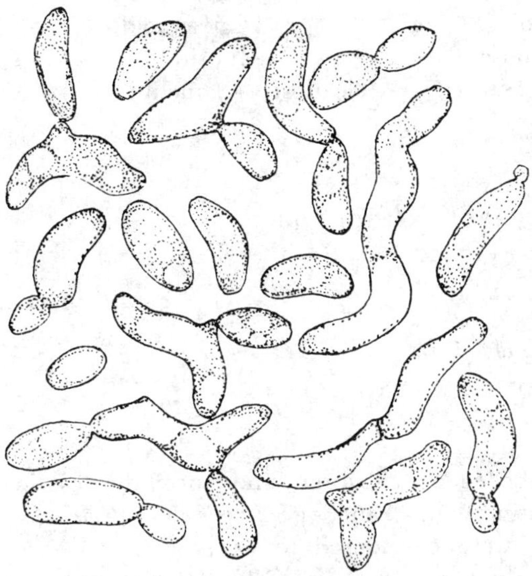

Fig. 380 – *Sp. holsaticus*
After 3 days in malt extract
(From 'The Yeasts' ed. I, 1952)

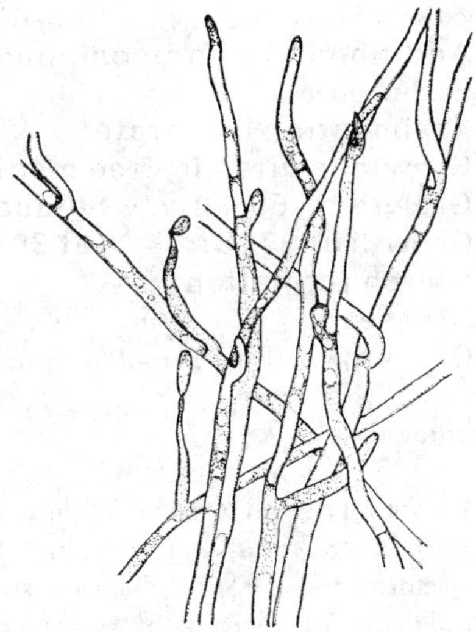

Fig. 381 – *Sp. holsaticus*
Slide culture on potato agar
(From 'The Yeasts' ed. I, 1952)

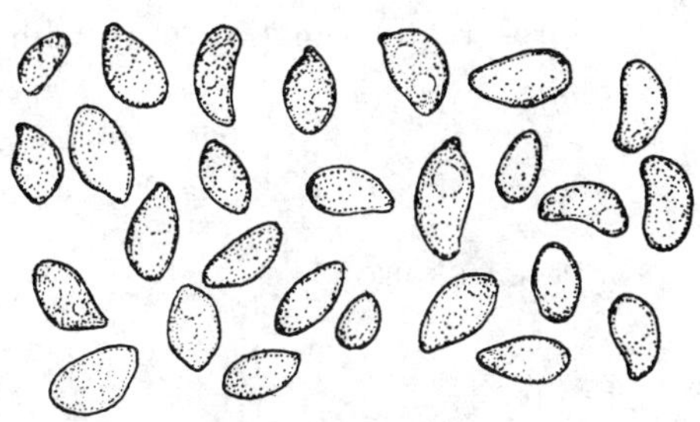

Fig. 382 – *Sp. holsaticus*
Ballistospores
(From 'The Yeasts' ed. I, 1952)

Growth on malt agar: After 2 weeks at room temperature the cells are ovoidal to elongate $(2-4) \times (5-28)$ μm, occurring singly and in pairs. There is an extensive development of true hyphae and occasionally pseudohyphae at the border. The streak culture is light saffron peach or pale salmon becoming flesh color after one month. The smooth surface is dull, the texture is tough, and the cross section is flat to broadly convex.

Dalmau plate culture on corn meal agar: After 4–7 days there is an extensive development of true and occasionally pseudohyphae. True hyphae are sparsely septate and often bear ballistospores or budding cells at the tips (Fig. 381).

Formation of ballistospores: Positive on most media. Obovoid, pyriform and reniform spores are discharged that measure $(2-4) \times (5-9)$ μm (Fig. 382).

Fermentation: –

Assimilation of carbon compounds:

Galactose	+ s	Raffinose	v	Erythritol	–
Sucrose	+	Soluble starch	– or + w	Ribitol	+
Maltose	+	D-Xylose	+	D-Mannitol	+
Cellobiose	+ or s	L-Arabinose	+ s	Succinic acid	+ w
Trehalose	+	D-Ribose	v	Citric acid	–
Lactose	–	L-Rhamnose	–	Inositol	–

Additional carbon compounds tested: Melezitose +, melibiose –, α-methyl-D-glucoside +

Assimilation of nitrate: +

Growth in vitamin-free medium: v

Growth on 50% (w/w) glucose-yeast extract agar: –

Growth at 37°C: –; **at 30°C:** +

Starch formation: –

Urease: +

G + C: 62.0, 64.0 mol% (Storck et al., 1969)

Comments

We found that most strains that were labelled *'Sp. holsaticus'*, and that fit the description of *Sp. holsaticus*, were capable of mating with strains of *Sporidiobolus salmonicolor*. Therefore, those *'Sp. holsaticus'* strains were considered to be strains of *Sporid. salmonicolor*. Two strains of *Sp. holsaticus*, the type strain and CBS 488, did not mate with strains of *Sporid. salmonicolor* which suggests either that these strains of *Sp. holsaticus* have lost their sexual capacity or that they represent the imperfect state of a species that is distinct from *Sporid. salmonicolor* (see Comments on the genus). Because strains of *Sp. holsaticus* and *Sporid. salmonicolor* appear to have similar physiological and morphological characteristics, identification procedures require mating tests with *Sporid. salmonicolor* mating types.

Origin of the strains studied

Fodder yeast (1) (Windisch, 1949); etiolated grass (1) (from Derx, CBS 488).

Type strain: CBS 1522 (Windisch, 1949).

4. *SPOROBOLOMYCES PUNICEUS* (KOMAGATA ET NAKASE) AHEARN ET YARROW (in Yarrow, 1972a)

Synonym

Candida punicea Komagata et Nakase (1965)

Standard description of *Sporobolomyces puniceus*

Growth in malt extract: After 3 days at 18°C the cells are round, oval to elongate (5–10) × (10–20) μm, occurring singly, in pairs and short chains. A sediment is formed. After one month the cells are larger (5–17) × (10–40) μm.

Growth on malt agar: After one month at 18°C the colony is smooth, glistening, slightly raised, the border entire. The color is very pale rose. The cells are similar to those in malt extract, with the addition of short strands of pseudo- and true hyphae.

Dalmau plate culture on corn meal agar: After 2 weeks at 18°C, short strands of pseudo- and true hyphae are present.

Formation of ballistospores: Not observed in the present study. Ahearn et al. (1968) reported the rare production of large kidney-shaped ballistospores on corn meal agar at 15°C and 90% relative humidity. A similar observation was made by Yarrow (1969a).

Fermentation: –

Assimilation of carbon compounds:

Galactose	–	Raffinose	+	Erythritol	–
Sucrose	+	Soluble starch	–	Ribitol	–
Maltose	+ s	D-Xylose	–	D-Mannitol	+
Cellobiose	+	L-Arabinose	+	Succinic acid	–
Trehalose	+ s	D-Ribose	+ s	Citric acid	–
Lactose	–	L-Rhamnose	–	Inositol	+ s

Additional carbon compounds tested: Melezitose + s, melibiose –, α-methyl-D-glucoside –

Assimilation of nitrate: +

Growth in vitamin-free medium: –

Growth on 50% (w/w) glucose-yeast extract agar: –

Growth at 37°C: –; at 30°C: –

Starch formation: +

Urease: +

Origin of the strain studied

Frozen fish in Japan (1) (Komagata & Nakase, 1965).

Type strain: CBS 5689 (Komagata & Nakase, 1965).

5. *SPOROBOLOMYCES ROSEUS* KLUYVER ET VAN NIEL (1925)

Synonyms

Sporobolomyces tenuis Kluyver et van Niel (1925)
Torula photographa nom. nud.
Sporobolomyces photographus Cif. et Red. (1925)
Amphiernia rubra Grüss (1926a,b)
Sporobolomyces salmoneus Derx (1930)

Sporobolomyces salmonicolor (Fischer et Brebeck) Kluyver et van Niel var. *salmoneus* (Derx) Verona et Cif. (1938).
Sporobolomyces salmoneus Derx var. *albus* Derx (1930).
Sporobolomyces salmonicolor (Fischer et Brebeck) Kluyver et van Niel var. *albus* (Derx) Verona et Cif. (1938)
Sporobolomyces pollaccii Verona et Cif. (1938)
Cryptococcus pulverulentus nom. nud.
Sporobolomyces miniatus Yamazaki et Fujii (1950)
Sporobolomyces boleticola Ramirez (1957)
Sporobolomyces roseus Kluyver et van Niel var. *madurae* Janke (1954)

Standard description of *Sporobolomyces roseus*

Growth in malt extract: After 3 days at room temperature the cell morphology (Fig. 383) is similar to that on malt agar. There is a thin ring and a light sediment. A thin film may be present. After one month there is a thin to heavy ring and a moderate to heavy sediment.

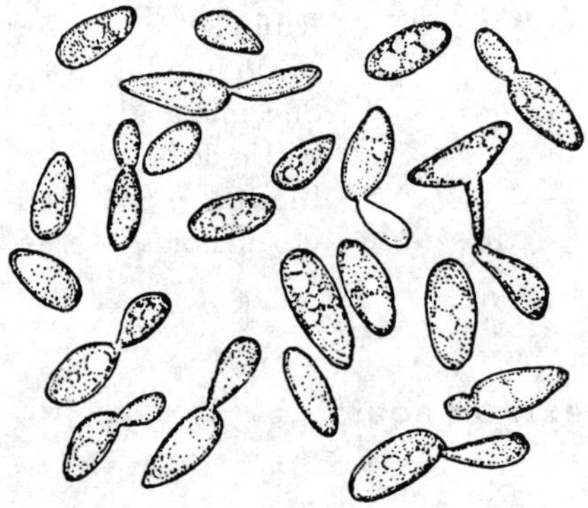

Fig. 383 – *Sp. roseus*
After 5 days in malt extract
(From 'The Yeasts' ed. I, 1952)

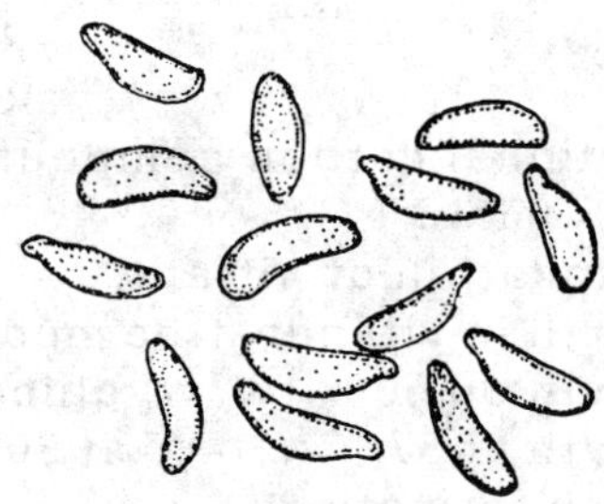

Fig. 384 – *Sp. roseus*
Ballistospores
(From 'The Yeasts' ed. I, 1952)

Growth on malt agar: At room temperature cells are spherical, ovoidal to elongate (3–8) × (5–18) μm. At one month the streak culture may be orange, grenadine or jasper pink. The surface ranges from nearly smooth or slightly rough to finely wrinkled to corrugated. The appearance is dull to glossy and the texture soft; the cross section is flat to raised and the periphery entire or with a fringe of small colonies arising from ballistospores.

Dalmau plate culture on corn meal agar: Pseudo- or true hyphae absent. Short chains of cells may be present.

Formation of ballistospores: On corn meal and malt agars, ballistospores may form abundantly. The ballistospores develop on simple or branched sterigmata that are positioned terminally and laterally on the cells. There may be one to three, occasionally more, sterigmata per cell. Ballistospores are oval, obovoid and reniform (2–4) × (7–13) μm (Fig. 384).

Fermentation: –

Assimilation of carbon compounds:

Galactose	v	Raffinose	+	Erythritol	–
Sucrose	+	Soluble starch	+	Ribitol	v
Maltose	+	D-Xylose	v	D-Mannitol	+
Cellobiose	+(–)	L-Arabinose	v	Succinic acid	v
Trehalose	+	D-Ribose	v	Citric acid	v
Lactose	–	L-Rhamnose	–	Inositol	–

Additional carbon compounds tested: Melezitose +(–), melibiose –, α-methyl-D-glucoside v

Assimilation of nitrate: +

Growth in vitamin-free medium: v

Growth on 50% (w/w) glucose-yeast extract agar: –

Growth at 37°C: –; **at 30°C:** –

Starch formation: –

Urease: +

G + C: 53.5–56.0 mol% (Storck et al., 1969)

Origin of the strains studied

The present study (Fell & Tallman) examined strains from air in dairy (1), blighted potatoes (1) (Kluyver & van Niel, 1925) and an unknown source (1) (strain B, Kluyver & van Niel, 1925). Phaff (1970c) examined 35 strains from beer (2), air (3), plants (18), human sources (2), fresh water (1), insects (5), unknown (4). Our results agreed with those of Phaff.

Type strain: CBS 486 (strain B, Kluyver & van Niel, 1925).

6. *SPOROBOLOMYCES SALMONICOLOR* (FISCHER ET BREBECK) KLUYVER ET VAN NIEL

see *Sporidiobolus salmonicolor* p. 538

7. *SPOROBOLOMYCES SHIBATANUS* (OKUNUKI) VERONA ET CIFERRI

see *Sporidiobolus pararoseus* p. 535

f. COMMENTS ON THE GENUS

The genus *Sporobolomyces* is characterized by species which produce visible carotenoid pigments and develop ballistospores. Seven species are presently recognized in the genus. *Sporobolomyces salmonicolor* is the most commonly isolated species and is the imperfect state of *Sporidiobolus salmonicolor.* Two species, *Sporobolomyces odorus* and *Sp. hispanicus,* that were included in the second edition of "The Yeasts" (1970), have been included as synonyms of *Sporidiobolus salmonicolor. Sp. holsaticus,* based on standard morphological and biochemical criteria, is similar to *Sporid. salmonicolor.* Most strains that we received labelled *Sp. holsaticus* were sexually compatible with strains of *Sporid. salmonicolor.* A notable exception was the type strain (CBS 1522) of *Sp. holsaticus* which did not show a sexual response. Storck et al. (1969) found that CBS 1522 had a G + C content of 64.0%. Several of the strains of *Sp. salmonicolor* that they studied (CBS 2630, 4209, 483, 490, 496), had G + C contents ranging from 63.5 to 65.0. Similarity in G + C contents suggests that *S. holsaticus* may be a synonym of *Sporid. salmonicolor* and that CBS 1522 may have lost its sexual capacity. This can be examined by a study of the complementarity of the DNA base sequences.

Sp. roseus, although morphologically and biochemically quite similar to strains of *Sporid. salmonicolor,* is probably a distinct species. This observation is based on the Storck et al. (1969) report that strains of *Sp. roseus* have a G + C content ranging from 53.5 to 56.0%, in contrast to the considerably higher G + C content of *Sporid. salmonicolor. Sp. roseus* can be distinguished from *Sp. holsaticus* and many strains of *Sporid. salmonicolor* by the absence of true and pseudomycelium. In addition, *Sp. roseus* assimilates soluble starch, a compound which most strains of *Sporid. salmonicolor* and *Sp. holsaticus* do not assimilate, although we have observed, in rare instances, the weak and slow assimilation of soluble starch by strains of *Sporid. salmonicolor* and *Sp. holsaticus. Sp. puniceus* has only been isolated one time. This species is distinctive as the only species in the genus that assimilates inositol and by the very pale rose color of the colony, in contrast to deeper pigmented colors of other species in the genus. Development of ballistospores is difficult to induce.

Sp. gracilis and *Sp. albo-rubescens* also are rare species and the type cultures have lost their ability to produce ballistospores. As such, the yeasts can be identified as species of *Rhodotorula. Sp. gracilis* can be identified as *Rh. minuta,* and *Sp. albo-rubescens* as *Rh. rubra.* The G + C content of *Sp. albo-rubescens* is 63.0% (Storck et al., 1969) and *Rh. rubra* 60–63.5% (Nakase & Komagata, 1971b) which suggests a similarity in the two species. Apparently the G + C of *Sp. gracilis* has not been determined.

Sp. shibatanus is distributed over a wide range of habitats. This species is the imperfect state of *Sporidiobolus pararoseus.* Based on studies of mating and G + C content, not all strains identified as *Sp. shibatanus* are imperfect strains of *Sporid. pararoseus* (see *Sporid. pararoseus* Comments). Therefore future studies may develop a basis for additional imperfect species as a *Sp. shibatanus* species complex.

g. SPECIES NOT INCLUDED IN THE GENUS

The following species were not included because they came belatedly to the attention of the authors:

Sporobolomyces elongatus Shivas et Rodrigues de Miranda (1983b)
Sporobolomyces foliicola Shivas et Rodrigues de Miranda (1983b)

Genus 14. *Sterigmatomyces* Fell

by J.W. Fell, A. Statzell Tallman and C.P. Kurtzman

a. DIAGNOSIS OF THE GENUS

Cells are round to oval and produce one or more sterigmata- or conidiophore-like structures. A conidium develops terminally on the sterigma; when the new cell is mature it may in turn produce sterigmata, often resulting in short branched chains of cells. In some species the cells separate at a septum in the mid-region of the sterigma, otherwise separation is at the distal position on the sterigma. The sterigma is generally 1.5–3.0 μm in length, but may be reduced or may elongate to give the appearance of a short hypha. True mycelium may or may not be produced. Ballistospores are not produced. Strains in this genus may be imperfect states of *Sterigmatosporidium* (Kraepelin & Schulze, 1982).

b. TYPE SPECIES OF THE GENUS

Sterigmatomyces halophilus Fell

c. SPECIES ACCEPTED IN THE GENUS

1. *Sterigmatomyces elviae* Sonck et Yarrow (1969)
2. *Sterigmatomyces halophilus* Fell (1966)
3. *Sterigmatomyces nectairii* Rodrigues de Miranda (1975)
4. *Sterigmatomyces penicillatus* Rodrigues de Miranda (1975)
5. *Sterigmatomyces polyborus* Scott et van der Walt (1970b)
6. *Sterigmatomyces tursiopsis* Kurtzman, Smiley, Johnson et Hoffman (1984)

d. KEY TO THE SPECIES OF THE GENUS

1a Colony color salmon to orange
St. nectairii .... p. 924
b Colony color cream to white 2

2a Sucrose assimilated 3

b Sucrose not assimilated

St. halophilus . . . . p. 923

3a Maltose assimilated 4

b Maltose not assimilated 5

4a α-Methyl-D-glucoside assimilated

St. polyborus . . . . p. 927

b α-Methyl-D-glucoside not assimilated

St. penicillatus . . . . p. 925

5a Rhamnose assimilated

St. tursiopsis . . . . p 928

b Rhamnose not assimilated

St. elviae . . . . p. 922

Table 39

Key characters of the species in the genus *Sterigmatomyces*

	Assimilation										37°C	mol% G+C
	Su	Ma	La	Me	Ra	Ri	Rh	Er	α-G	NO_3		
St. elviae	+	−	+	−	+	+	−	+	−	−	+	52.8
St. halophilus	−	−	−/+s	−	−	+	−	+	−	v	−	55.0
St. nectairii	−	−	−	−	−	−	−	−	−	+	−	52.5
St. penicillatus	+	+	+	−	−	+	+	+	−	−	−	48.9
St. polyborus	+	+	+(s)	+	+s	+	+	+s	+	−	−	49.6
St. tursiopsis	+	−	+	+	+	+	+	+	−	−	−	55.2

Su = sucrose, Ma = maltose, La = lactose, Me = melibiose, Ra = raffinose, Ri = D-ribose, Rh = L-rhamnos Er = erythritol, α-G = α-methyl-D-glucoside, NO_3 = nitrate, 37°C = growth at 37°C, mol% G + C = mol guanine + cytosine in the nuclear DNA (average value; range given with each species description).

e. SYSTEMATIC DISCUSSION OF THE SPECIES

1. *STERIGMATOMYCES ELVIAE* SONCK ET YARROW (1969)

Standard description of *Sterigmatomyces elviae*

Growth in malt extract: After 3 days at 24°C the cells are round to ov (3–6) × (3–10) μm with one to five short (<1 μm) sterigmata on each cell. A light ring an sediment are present. After one month the cells are within the same size range; a moderat ring and sediment are present.

Growth on malt agar: There are two colony types; in one type the streak cultur is smooth, dull, cream colored, edges entire, with cells similar to those described in malt extract. The differences exhibited by the other type are a glistening colony, and ellipsoidal and apiculate cells (2–7) × (3–13) μm with one or rarely more short sterigmata (<0.5 μm), the general appearance is that of a typically budding yeast cell.

Dalmau plate culture on corn meal agar: Pseudo- or true mycelium absent.

Fermentation: –

Assimilation of carbon compounds:

Galactose	–	Raffinose	+	Erythritol	+
Sucrose	+	Soluble starch	–	Ribitol	+
Maltose	–	D-Xylose	+	D-Mannitol	+
Cellobiose	+	L-Arabinose	+	Succinic acid	+ s or w
Trehalose	+ s	D-Ribose	+	Citric acid	+ s or w
Lactose	+	L-Rhamnose	–	Inositol	–

Additional carbon compounds tested: Melibiose –, α-methyl-D-glucoside –, melezitose –

Assimilation of nitrate: –

Growth in vitamin-free medium: +

Growth on 50% (w/w) glucose-yeast extract agar: –

Growth at 37°C: +

Starch formation: –

Urease: +

G + C: 52.8 mol%, type strain (Kurtzman et al., 1984)

Origin of the strain studied

Human source in Finland (1) (Sonck & Yarrow, 1969).

Type strain: CBS 5922.

2. *STERIGMATOMYCES HALOPHILUS* FELL (1966)

Synonyms

Sterigmatomyces halophilus Fell var. *indicus* Fell (1966)
Sterigmatomyces indicus (Fell) Fell (1970a)

Standard description of *Sterigmatomyces halophilus*

Growth in malt extract: After 3 days at room temperature cells are round to oval (2–7) × (3–11) μm with one to ten or more sterigmata, 0.5 × (1–3) μm. New cells initiate as a distal swelling of the sterigma with subsequent enlargement and development. This secondary cell can produce sterigmata with cells, resulting in short branched chains. Cells separate by septal formation in the mid-region of the sterigma. A light film is produced which develops into powdery islets and after one month there is a heavy white pellicle. A sediment is present.

Growth on malt agar: Cell morphology is similar to that in malt extract

(Fig. 385). The cells are varied in size (1–4) × (3–17) μm and some are thin and long, 127 μm. The sterigmata generally are 0.5 × (1–3) μm, some elongate to 26 μm. The width of the elongated sterigma remains about 0.5 μm. The streak is chalky-white, pasty, with a dry granular surface, and the border is irregular to lobate. Colonies may develop that are smooth, glistening, or dull, and with margins entire.

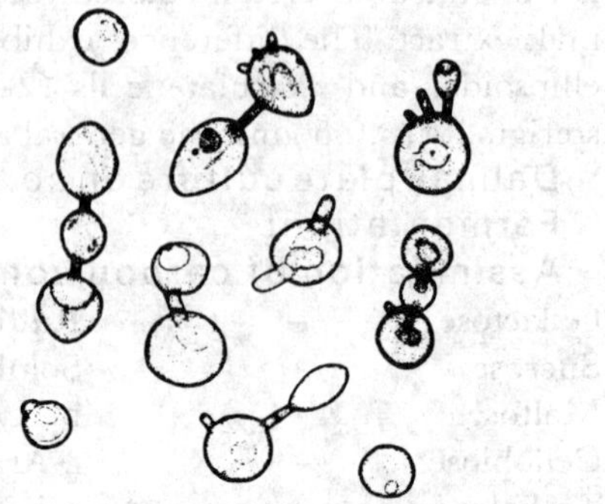

Fig. 385 – *St. halophilus*
After 3 days on malt agar. After Fell (1966)
(From 'The Yeasts' ed. II, 1970)

Dalmau plate culture on corn meal agar: Pseudomycelium or true mycelium absent; short branched chains of cells may be produced.

Fermentation: –

Assimilation of carbon compounds:

Galactose	v	Raffinose	–	Erythritol	+
Sucrose	–	Soluble starch	–	Ribitol	+
Maltose	–	D-Xylose	+	D-Mannitol	+
Cellobiose	v	L-Arabinose	+	Succinic acid	+ or w
Trehalose	+	D-Ribose	+	Citric acid	+ or w
Lactose	– or + s	L-Rhamnose	–	Inositol	–

Additional carbon compounds tested: Melibiose –, α-methyl-D-glucoside –, melezitose –

Assimilation of nitrate: v

Growth in vitamin-free medium: –; thiamine required for growth

Growth on 50% (w/w) glucose-yeast extract agar: v

Growth at 37°C: –; **at 30°C:** v

Starch formation: –

Urease: + (w, s)

G + C: 54.9–55.2 mol%, 4 strains (Kurtzman et al., 1984)

Origin of the strains studied

Marine (8), atmosphere (1), patient (1).

Type strain: CBS 4609 from the atmosphere (Fell, 1966).

3. *STERIGMATOMYCES NECTAIRII* RODRIGUES DE MIRANDA (1975)

Standard description of *Sterigmatomyces nectairii*

Growth in malt extract: After 3 days the cells are round to oval (3–5) × (4–7) μm with 1–10 short (<0.5 μm) conidiophores. Daughter cells develop laterally or terminally on the conidiophore, branching of the conidiophore occurs. Daughter cell separation is at the distal position. Within one month conidiophores are branched and irregularly shaped (0.7–1) × (1–17) μm. A sediment is present.

Growth on malt agar: Cells are similar in size to those on malt agar; colony is cream becoming salmon colored at 5 days and orange by one month. The texture is smooth, the edges entire.

Dalmau plate culture on corn meal agar: Within one month, short, sparse strands of true mycelium are present.

Fermentation: –

Assimilation of carbon compounds:

Galactose	v	Raffinose	–	Erythritol	–
Sucrose	–	Soluble starch	–	Ribitol	+
Maltose	–	D-Xylose	+ or w	D-Mannitol	+
Cellobiose	–	L-Arabinose	–	Succinic acid	+ or w
Trehalose	+	D-Ribose	–	Citric acid	–
Lactose	–	L-Rhamnose	–	Inositol	–

Additional carbon compounds tested: Melibiose –, α-methyl-D-glucoside –, melezitose –

Assimilation of nitrate: +

Growth in vitamin-free medium: –

Growth on 50% (w/w) glucose-yeast extract agar: –

Growth at 37°C: –

Starch formation: –

Urease: +w

G + C: 52.5 mol%, type strain (Kurtzman et al., 1984)

Origin of the strain studied

From cheese in France (1) (Rodrigues de Miranda, 1975).

Type strain: CBS 6405.

4. *STERIGMATOMYCES PENICILLATUS* RODRIGUES DE MIRANDA (1975)

Standard description of *Sterigmatomyces penicillatus*

Growth in malt extract: After 3 days at 24°C the cells (1–6) × (2–11) μm have a variety of shapes that may be spheroidal, ovoidal, obpyriform and apiculate. When attached to short hyphal strands the cells are ventricose. The sterigmata (1–6 per cell) range in size from 0.5 × (0.7–9) μm. After 30 days the majority of the cells are (1–7) × (2–8) μm with as many as 17 sterigmata per cell. The sterigmata range in size from 0.5 × (0.7–35) μm. Daughter cell separation is at the distal position where an adjacent septum was formed.

Growth on malt agar: After 3 days at 24°C cells' sizes are similar to those in malt extract and the sterigmata (1–15 or more per cell) are simple, branched with a length of 40 μm or greater. The streak culture is dull white to pale luteus, pasty, soft, rugose with an irregularly crenulate border. Colonies also develop which are smooth, glistening, soft, mucous, raised, pale luteus to cream-colored with a border that is entire. The cells in these colonies are spheroidal to ovoidal (2–3) × (3–4) μm. Sterigmata are rare in the smooth colony type, thus the cells have the appearance of typical budding yeast cells.

Dalmau plate culture on corn meal agar: Rugose colonies develop branched true mycelium (<2 μm wide); cross walls are rare, sterigmata are lateral, short to elongate, with terminal conidia (Fig. 386). The latter may develop numerous sterigmata with conidia to give the appearance of a penicillate head. Hyphae may also develop in the smooth colonies, however, the majority of the cells are similar to those on malt agar.

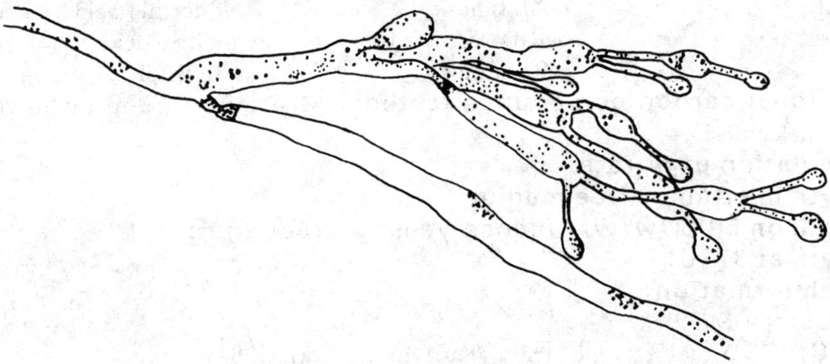

Fig. 386 – *St. penicillatus*
Hypha on corn meal agar; conidia on sterigmata

Fermentation: –

Assimilation of carbon compounds:

Galactose	+	Raffinose	–	Erythritol	+
Sucrose	+	Soluble starch	+	Ribitol	+
Maltose	+	D-Xylose	+	D-Mannitol	+
Cellobiose	+	L-Arabinose	+	Succinic acid	+
Trehalose	+	D-Ribose	+	Citric acid	+
Lactose	+	L-Rhamnose	+	Inositol	+

Additional carbon compounds tested: Melibiose –, α-methyl-D-glucoside –, melezitose +

Assimilation of nitrate: –

Growth in vitamin-free medium: –

Growth on 50% (w/w) glucose-yeast extract agar: –

Growth at 37°C: –

Starch formation: – or +w

Urease: +

G + C: 48.9 mol%, type strain (Kurtzman et al., 1984)

Origin of the strain studied

Contaminated agar plate (1).

Type strain: CBS 5492.

5. *STERIGMATOMYCES POLYBORUS* SCOTT ET VAN DER WALT (1970b)

Standard description of *Sterigmatomyces polyborus*

Growth in malt extract: After 3 days at 24°C the cells are spheroidal to ovoidal (1.3–4.0) × (2.0–4.0) μm. The cells may be single or produce conidia on sterigmata that are <4.0 μm in length. After one month there may be as many as 15 sterigmata per cell. Daughter cells generally separate at the distal end of sterigmata. Islets, a light ring and sediment are present.

Growth on malt agar: After 3 days at 24°C the streak culture is dull, white to pale luteus, powdery, rugose with irregularly crenulate margins. The cells are similar to those in malt extract. Variants are present that form colonies which are glistening, raised, smooth and pale luteus. The margin is entire. The encapsulated cells bear few short (<0.5 μm in length) sterigmata. The cells appear similar to budding cells of the genus *Cryptococcus*.

Dalmau plate culture on corn meal agar: Non-septate hypha-like elongations of the cell occur. Neither pseudo- nor true mycelium is present.

Fermentation: –

Assimilation of carbon compounds:

Galactose	+	Raffinose	+ s	Erythritol	+ s
Sucrose	+	Soluble starch	–	Ribitol	+
Maltose	+	D-Xylose	+	D-Mannitol	+
Cellobiose	+	L-Arabinose	+ s	Succinic acid	+
Trehalose	+	D-Ribose	+	Citric acid	+
Lactose	+ (s)	L-Rhamnose	+	Inositol	+

Additional carbon compounds tested: Melibiose +, α-methyl-D-glucoside +, melezitose +

Assimilation of nitrate: –

Growth in vitamin-free medium: –

Growth on 50% (w/w) glucose-yeast extract agar: –

Growth at 37°C: –

Starch formation: –

Urease: +

G + C: 49.6 mol%, type strain (Kurtzman et al., 1984)

Origin of the strain studied

An insect tunnel in a tree (1) (Scott & van der Walt, 1970b).

Type strain: CBS 6072.

6. *STERIGMATOMYCES TURSIOPSIS* KURTZMAN, SMILEY, JOHNSON ET HOFFMAN (1984)

Standard description of *Sterigmatomyces tursiopsis*

Growth in malt extract: After 3 days at 24°C the cells are spheroidal to ovoidal, sometimes ventricose and irregularly shaped, measuring (1.3–5) × (2–13) μm. Conidia are produced on sterigmata, 1–4 per cell and measuring to 7 μm in length. The conidia may produce additional sterigmata and conidia that result in branched chains of cells. Conidia separate at the distal end of the sterigmata. True mycelium is produced. A ring, powdery islets and slight sediment are present. After one month there are 1–22 sterigmata per cell. Variants (which develop smooth colonies on agar) in malt extract consist of oval (1.3–5) × (2–5) μm encapsulated cells with short (<0.5 μm) sterigmata that appear as typical budding yeast cells. A slight ring and sediment are present.

Growth on malt agar: After 3 days at 24°C the streak culture has a white powdery surface with pale luteus rugoid areas. Below the surface is a glistening, soft, butyrous mass of cells. The margin is crenulate. True mycelium and non-septate filamentous elongations of the cell occur. Variants occur which consist of spheroidal to ovoidal cells (1.3–4.0) × (1.3–4.0) μm. The colonies are cream to pale luteus, with a smooth, glistening, slightly raised surface; the margin is entire. The cells develop on short (<0.5 μm) sterigmata and are encapsulated; pseudo- and true hyphae are present.

Dalmau plate culture on corn meal agar: True mycelium (<2 μm wide) is branched with lateral and terminal sterigmata. Cross walls are rare. Terminal conidia may develop numerous sterigmata with conidia to present the appearance of penicillate heads.

Fermentation: –

Assimilation of carbon compounds:

Galactose	+	Raffinose	+	Erythritol	+
Sucrose	+	Soluble starch	–	Ribitol	+
Maltose	–	D-Xylose	+	D-Mannitol	+
Cellobiose	+	L-Arabinose	+	Succinic acid	+
Trehalose	+	D-Ribose	+	Citric acid	+
Lactose	+	L-Rhamnose	+	Inositol	+

Additional carbon compounds tested: Melibiose +, α-methyl-D-glucoside –, melezitose +w

Assimilation of nitrate: –

Growth in vitamin-free medium: –

Growth on 50% (w/w) glucose-yeast extract agar: –

Growth at 37°C: –

Starch formation: +

Urease: +w

G + C: 55.2 mol%, type strain (Kurtzman et al., 1984)

Origin of the strain studied

Bottlenosed dolphin, *Tursiops truncatus* (1).

Type culture: CBS 6133.

f. COMMENTS ON THE GENUS

The majority of the *Sterigmatomyces* spp. have chalky white colonies and cell replication is via daughter cells produced on sterigmata; only one bud is produced per sterigma. Daughter cells usually detach by a separation in the mid-region of the sterigma in *St. elviae* and *St. halophilus,* but separation is at the distal end of the sterigma for *St. nectairii, St. penicillatus, St. polyborus* and *St. tursiopsis.* In some species variants develop with sterigmata reduced to the extent that they appear as typical budding yeast cells. These colonies are smooth, mucous and runny with cells that may be encapsulated; in contrast to dry, rugose, non-capsulated characteristics of typical sterigma-forming cells and colonies. *St. elviae* is particularly troublesome in this regard as we had a difficult time locating a sub-culture of the type that had retained the sterigma-forming characteristics.

St. nectairii is distinct within the genus: the colonies are salmon to orange in color, mother–daughter cell separation is terminal on the sterigma, and the latter can branch or elongate to produce additional daughter cells. This type of reproduction is analogous to that observed in *Sympodiomyces*. However, the two genera, *Sympodiomyces* and *Sterigmatomyces,* are phylogenetically distinct. *Sympodiomyces* has characteristics of the ascomycetes, while *Sterigmatomyces* appears to be a basidiomycete. The characteristics that are used to depict *Sterigmatomyces* as a basidiomycete are: presence of urease activity, laminar nature of the cell wall (Kreger-van Rij & Veenhuis, 1971a; Kurtzman et al., 1984), and the color reaction with Diazonium Blue B salt (van der Walt & Hopsu Havu, 1976). These characteristics have been demonstrated for all six species including *St. nectairii*.

DNA reassociation studies confirmed the species now accepted in *Sterigmatomyces* to be distinct (Kurtzman et al., 1984). Surprisingly, *St. halophilus* and *St. indicus* exhibited 100% DNA relatedness and must be considered conspecific. These two species had been previously separated on the basis of nitrate utilization, positive in the former and negative in the latter.

Kraepelin and Schulze (1982) described *Sterigmatosporidium* as the sexual state of *Sterigmatomyces*. The genus includes one species, *Sterigmatosporidium polymorphum*, which did not mate with strains of *St. polyborus, St. penicillatus* and *St. elviae*. Further, little DNA complementarity was detected between this new taxon and the six species that we have accepted in *Sterigmatomyces* (Kurtzman et al., 1984).

Genus 15. *Sympodiomyces* Fell et Statzell

by J.W. Fell

a. DIAGNOSIS OF THE GENUS

The streak culture on malt agar is dull creme, the surface is smooth, the texture soft but not mucous. Yeast-like cells are present. A conidiophore develops directly from a yeast cell, producing a terminal conidium. At the side of the conidium a new growing tip of the conidiophore develops forming a new terminal conidium. In this way an elongate conidiophore is formed with successive conidia. The conidiophores possess several scars. Conidial formation is irregular, both in distance between conidia and location on the conidiophore. Branched true mycelium is inconspicuous and restricted in development. Conidia on the mycelium are single on a short peg or sympodial on an elongate conidiophore. Clusters of conidia may develop.

Ascospores, teliospores and ballistospores have not been observed.

b. TYPE SPECIES OF THE GENUS

Sympodiomyces parvus Fell et Statzell

c. SPECIES ACCEPTED IN THE GENUS

1. *Sympodiomyces parvus* Fell et Statzell (1971)

e. SYSTEMATIC DISCUSSION OF THE SPECIES

1. SYMPODIOMYCES PARVUS FELL ET STATZELL (1971)

Standard description of *Sympodiomyces parvus*

Growth in malt extract: After 5 days at 12°C the single cells vary in morphology (Fig. 387); they are round, oval and subglobose $(1.0-1.3)\times(1.3-3.4)$ μm. The cells are single, in pairs and short branched chains. The cells in chains are generally larger (2.6–3.4) × (4.7–5.4) μm and oval in shape. The conidiophores at this age are short, which gives the appearance of typical budding yeast cells, or are elongate to 4 μm. There is a moderate sediment. After one month the conidiophores have elongated to 11 μm in length with numerous bud scars and one or more conidia.

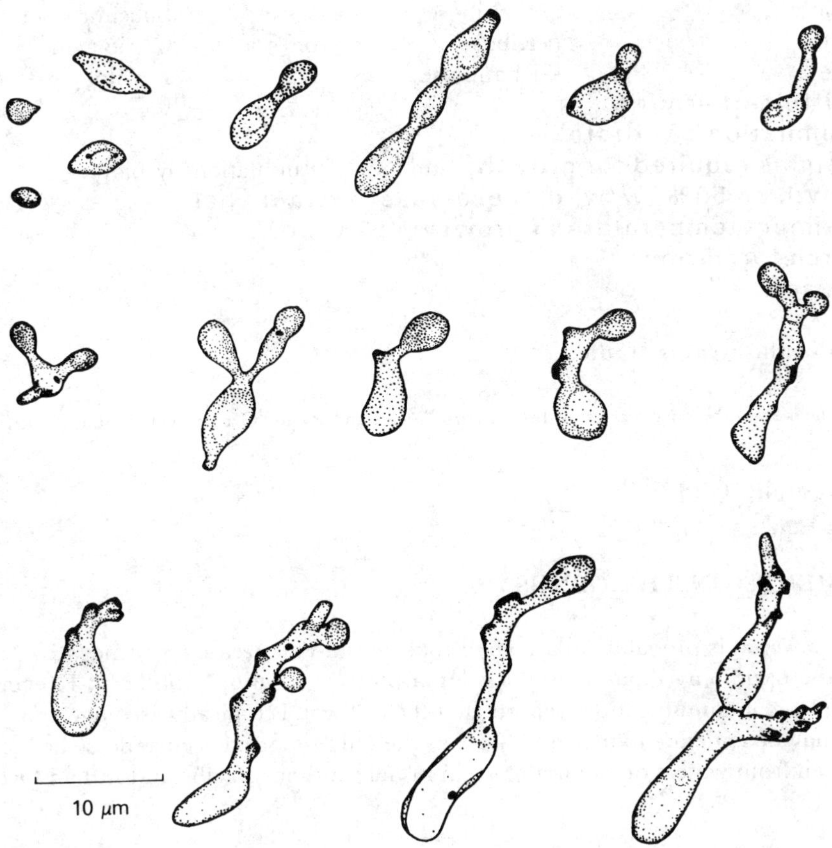

Fig. 387 – *Symp. parvus*
Yeast phase with the sympodial formation of conidia
After 1–4 weeks in malt extract (From Fell & Statzell, 1971)

Growth on malt agar: After one month at 12°C the streak culture is dull creme, the surface is smooth, the texture is soft but not mucous. The growth is convex and the periphery is entire, apart from tufts of non-aerial mycelium at the edges.

Dalmau plate culture on corn meal agar: After one month at 12°C there is true mycelium with clearly demarked septa and clusters of conidia. The latter form much like single cells, i.e. the new bud forms terminally on a conidiophore that elongates sympodially to produce new conidia. The hyphae are approximately 2 μm in width and there are often spheroidal intercalary cells in the range of 5 × 7 μm. Mycelial formation is restricted on corn meal agar and is more extensive, but still in conspicuous, on malt agar.

Fermentation: –

Assimilation of carbon compounds:

Galactose	+	Raffinose	–	Erythritol	+
Sucrose	+ s	Soluble starch	–	Ribitol	+
Maltose	+ s	D-Xylose	+	D-Mannitol	+ s
Cellobiose	+	L-Arabinose	+ s	Succinic acid	–
Trehalose	+ or s	D-Ribose	– or + s	Citric acid	–
Lactose	–	L-Rhamnose	+ or s	Inositol	+ s

Splitting of arbutin: +

Assimilation of nitrate: –

Vitamins required for growth: thiamine; stimulation by biotin

Growth on 50% (w/w) glucose-yeast extract agar: –

Maximum temperature of growth: 25°C

Starch formation: –

Urease: –

Origin of the strains studied

52 isolates from the open waters of the southern Indian and Pacific Oceans (Fell & Statzell, 1971; Fell, 1976).

Type strain: CBS 6147.

COMMENTS ON THE GENUS

Sympodiomyces is probably an ascomycetous yeast. The urease test is negative, a test which the basidiomycetous yeasts develop a positive reaction. Similarly, Kreger-van Rij (personal communication) reports that the cell wall is typically ascomycetous. The genus may be endemic to marine waters as the only reported occurrences (Fell, 1976) have been from waters of Antarctic origin in the southern Pacific and Indian Oceans.

Genus 16. *Trichosporon* Behrend

by N.J.W. Kreger-van Rij

a. DIAGNOSIS OF THE GENUS (DO CARMO-SOUSA, 1970c)

Budding cells of various shapes. Pseudomycelium well-developed or reduced. Septate mycelium and arthrospores always present, usually abundant. Ring and pellicle often formed on liquid media.

Ascospores, basidiospores and teliospores not produced.

Asexual endospores may be formed.

Fermentation present or absent.

b. TYPE SPECIES OF THE GENUS

Trichosporon ovoides Behrend

c. SPECIES ACCEPTED IN THE GENUS

1. *Trichosporon aculeatum* Phaff, Miller et Shifrine (1956)
2. *Trichosporon aquatile* Hedrick et Dupont (1968)
3. *Trichosporon brassicae* Nakase (1971a)
4. *Trichosporon capitatum* Diddens et Lodder (1942)
5. *Trichosporon cutaneum* (de Beurm., Gougerot et Vaucher) Ota (De Beurm. et al., 1909)
6. *Trichosporon eriense* Hedrick et Dupont (1968)
7. *Trichosporon fennicum* Sonck et Yarrow (1969)
8. *Trichosporon fermentans* Diddens et Lodder (1942)
9. *Trichosporon figueirae* Batista et Silveira (1960)
10. *Trichosporon inkin* (Oho) do Carmo-Sousa et van Uden (Oho, 1919)
11. *Trichosporon melibiosaceum* Scott et van der Walt (1970b)
12. *Trichosporon pullulans* (Lindner) Diddens et Lodder (Lindner, 1895)
13. *Trichosporon sericeum* (Stautz) Diddens et Lodder (Stautz, 1931)
14. *Trichosporon terrestre* van der Walt et Johannsen (1975d)
15. *Trichosporon variabile* (Lindner) Delitsch (Lindner, 1898)

d. KEY TO THE SPECIES OF THE GENUS

1a Nitrate assimilated 2
b Nitrate not assimilated 3

2a Growth at 37°C
Tr. terrestre .... p. 959
b No growth at 37°C
Tr. pullulans .... p. 955

3a Glucose fermented 4
b Glucose not fermented 9

4a Maltose assimilated 5
b Maltose not assimilated 8

5a Erythritol assimilated 6
b Erythritol not assimilated
Tr. aculeatum .... p. 936
see also *Aciculoconidium aculeatum* .... p. 558

6a Melibiose assimilated
Tr. melibiosaceum .... p. 953
b Melibiose not assimilated 7

7a Lactose assimilated
Tr. fennicum .... p. 948
b Lactose not assimilated
Tr. variabile (see *Pichia burtonii*) .... p. 314

8a Ribitol assimilated
Tr. fermentans .... p. 949
b Ribitol not assimilated
Tr. eriense .... p. 946

9a Lactose assimilated 10
b Lactose not assimilated 13

10a Sarcina-like groups of cells produced
Tr. inkin .... p. 953
see also *Sarcinosporon inkin* .... p. 906
b No sarcina-like groups of cells produced 11

11a Inositol assimilated
Tr. cutaneum .... p. 942
b Inositol not assimilated 12

12a Growth at 37°C
Tr. figueirae .... p. 952
b No growth at 37°C
Tr. aquatile .... p. 936

13a Maltose assimilated 14
b Maltose not assimilated 15

14a Inositol assimilated
Tr. brassicae .... p. 937
b Inositol not assimilated
Tr. aculeatum .... p. 936
see also *Aciculoconidium aculeatum* .... p. 558

15a Growth at 37°C
Tr. capitatum .... p. 938
b No growth at 37°C 16

16a Xylose assimilated
Tr. eriense .... p. 946
b Xylose not assimilated
Tr. sericeum .... p. 957

Table 40

Key characters of the species in the genus *Trichosporon*

	DBB	ur	F	Assimilation									37°C
				NO_3	Ma	La	Me	Ce	X	Er	Rib	I	
ascomycetous:													
Tr. aculeatum	−	−	+w	−	+	−	−	+	−	−	+(w)	−	−
Tr. fennicum	−	−	+	−	+	+vw	−	v	+	+	+	−	−
Tr. melibiosaceum	−	−	+	−	+	−/+vw	+	+	+	+	+	−	−
Tr. variabile	−	−	+	−	+	−	−	+	+	+	+	−	v
Tr. capitatum	−	−	−	−	−	−		−	−	−	−	−	+
Tr. fermentans	−	−	+	−	−	−		+	+	−	+	−	v
Tr. eriense	−	−	+w/s	−	−	−		+	+	−	−	−	−
Tr. terrestre	−	−	−	+	+	−		+vw/−	+	+	+	+	+
Tr. sericeum	−	+w	−	−	−	−		−	−	−	−	−	−
basidiomycetous:													
Tr. aquatile	−	+	−	−	+	+		+	+	+	−	−	−
Tr. brassicae	+	+	−	−	+	−		+w	+	−	−	+	−
Tr. cutaneum	+	+	−	−	+(−)	+		+(−)	+	v	v	+	v
Tr. figueirae	+	+	−	−	+	+		+	+	+	+w	−	+
Tr. inkin	+	+	−	−	+	+		+	+	+	−	+	+
Tr. pullulans	+	+	−	+	+	+/w		+	v	+	v	v	−

DBB = color reaction with Diazonium Blue B, ur = urease, F = fermentation of glucose, NO_3 = nitrate, Ma = maltose, La = lactose, Me = melibiose, Ce = cellobiose, X = D-xylose, Er = erythritol, Rib = ribitol, I = inositol.

e. SYSTEMATIC DISCUSSION OF THE SPECIES

1. TRICHOSPORON ACULEATUM PHAFF, MILLER ET SHIFRINE (1956)

see *Aciculoconidium aculeatum* p. 558

Comments

King and Jong (1976a) described splitting up of the hyphae in this species as "disarticulation", a type of splitting which they considered to be distinctly different from arthrospore formation e.g. in *Geotrichum candidum.* Smith, in this monograph, follows King and Jong's classification of the species *aculeatum* in the genus *Aciculoconidium.* I have not been able to find differences between the arthrospores of *Tr. aculeatum* and of *Geotrichum candidum,* and have therefore retained the species in the genus *Trichosporon.*

The ultrastructure of the cell wall is ascomycetous. The hyphal septa have a closure line. In sections in the electron microscope the hyphal cells may show as many as six nuclear profiles. These characters have also been observed in cells of *Tr. fennicum, Tr. melibiosaceum* and *Pichia burtonii,* which may point to a relationship.

2. TRICHOSPORON AQUATILE HEDRICK ET DUPONT (1968)

Standard description of *Trichosporon aquatile*

Growth in glucose-yeast extract-peptone water: After 3 days at 25°C branched septate hyphae are present and many arthrospores of all sizes. Hedrick and Dupont (1968) found spherical, oval and elongate cells budding on a broad base and measuring (4–5) × (5–12) μm. A white, wrinkled, dry pellicle and some loose sediment are formed. After one month a dry pellicle and a loose sediment are present.

Growth on glucose-yeast extract-peptone agar: After one month at 20°C the streak culture is white, dry, folded, tough and fringed with mycelium.

Slide cultures on potato- and corn meal agar: Septate, branched hyphae abundant; budding yeast cells and blastospores rare. Arthrospores of various lengths are formed. Aerial mycelium may occur. The ultrastructure of the cell wall is basidiomycetous.

Fermentation: –

Assimilation of carbon compounds:

Galactose	+	Raffinose	–	Erythritol	+
Sucrose	+	Soluble starch	+	Ribitol	–
Maltose	+	D-Xylose	+	D-Mannitol	–
Cellobiose	+	L-Arabinose	+	Succinic acid	+
Trehalose	+	D-Ribose	+	Citric acid	+ w
Lactose	+	L-Rhamnose	–	Inositol	–

Splitting of arbutin: +

Assimilation of nitrate: –

Assimilation of nitrite: –
Growth in vitamin-free medium: –
Growth on 50% (w/w) glucose-yeast extract agar: +
Growth at 37°C: –
Urease: +
Color reaction with DBB: – (on solid medium)
G + C: 55 mol%, type strain (Dupont & Hedrick, 1971)

Origin of the strain studied

Fresh water (1).

Type strain: CBS 5973 (NRRL Y-7051).

3. *TRICHOSPORON BRASSICAE* NAKASE (1971a)

Standard description of *Trichosporon brassicae*

Growth in glucose-yeast extract-peptone water: After 3 days at 25°C long-oval to cylindrical cells, (3–4.5) × (5–20) μm, producing buds on a more or less broad base. Branched hyphae with few septa and arthrospores of various lengths are present. A sediment and a pellicle which easily drops are formed. After one month a sediment and a ring are present.

Growth on glucose-yeast extract-peptone agar: After one month the streak culture is yellowish white, smooth and soft.

Slide cultures on potato- and corn meal agar: Branched, septate hyphae falling apart into arthrospores. Long-oval blastospores often occur in chains on the hyphae. On potato agar, single budding cells may be present. The ultrastructure of the cell wall is basidiomycetous.

Fermentation: –

Assimilation of carbon compounds:

Galactose	+	Raffinose	–	Erythritol	–
Sucrose	+	Soluble starch	–	Ribitol	–
Maltose	+	D-Xylose	+	D-Mannitol	+
Cellobiose	+ w	L-Arabinose	+	Succinic acid	+
Trehalose	+	D-Ribose	+	Citric acid	+
Lactose	–	L-Rhamnose	–	Inositol	+

Splitting of arbutin: +
Assimilation of nitrate: –
Growth in vitamin-free medium: –
Growth on 50% (w/w) glucose-yeast extract agar: + w
Growth at 37°C: –
Urease: +
Color reaction with DBB: +
G + C: 54.9 mol%, type strain (Nakase, 1971a)

Origin of the strain studied

Cabbage (1).

Type strain: CBS 6382, isolated by Nakase from a fresh cabbage from a Tokyo market.

4. *TRICHOSPORON CAPITATUM* DIDDENS ET LODDER (1942)

Synonyms

Endoblastomyces thermophilus nom. nud. (Odinzova, in Kudriavzev, 1960)
Geotrichum linkii nom. nud. (Vörös-Felkai, 1961)
Geotrichum capitatum (Diddens et Lodder) von Arx (in von Arx et al., 1977)
Ascotrichosporon capitatum nom. nud. (Kocková-Kratochvílová et al., 1977b)

Standard description of *Trichosporon capitatum* (do Carmo-Sousa, 1970c)

Growth in glucose-yeast extract-peptone water: After 3 days at 25°C, oval to long-oval cells with buds on a more or less broad base $(3.5-6.7) \times (7.4-12.3)$ μm, septate, branched hyphae and arthrospores (Fig. 388). A thin, creeping, dull, smooth or slightly wrinkled pellicle is formed.

Growth on glucose-yeast extract-peptone agar: After one month at 25°C the streak culture is white to cream colored, wrinkled to slightly coremial, rugose or cottony, glossy to dull, of a soft membranous texture, slightly raised with a straight border fringed with mycelium.

Slide cultures on potato- and corn meal agar: True mycelium and ar-

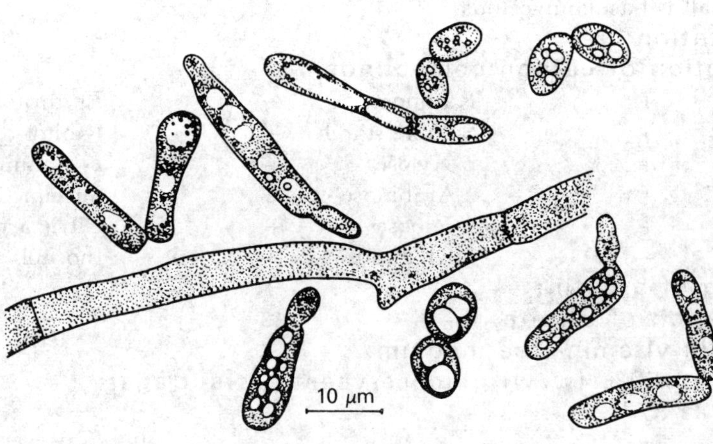

Fig. 388 – *Tr. capitatum*
After 3 days in glucose-yeast extract-peptone water (From 'The Yeasts' ed. II, 1970)

throspores. Blastospores usually in clusters along and at the end of the hyphae (Fig. 389). The ultrastructure of the cell wall is ascomycetous. The hyphal septa have plasmodesmata arranged in a circle.

Formation of asexual spores: Ellipsoidal endospores may form by internal budding in the hyphae, in intercalary or terminal segments, either at the septa or at the tip of tubular ingrowths. They are arranged in rows. Their number per segment varies from 1 to 15 (Fig. 390). The endospores were observed in 3-month-old cultures on glucose-yeast extract-peptone agar at 18–20°C and in slide cultures after 2–3 weeks at 25°C.

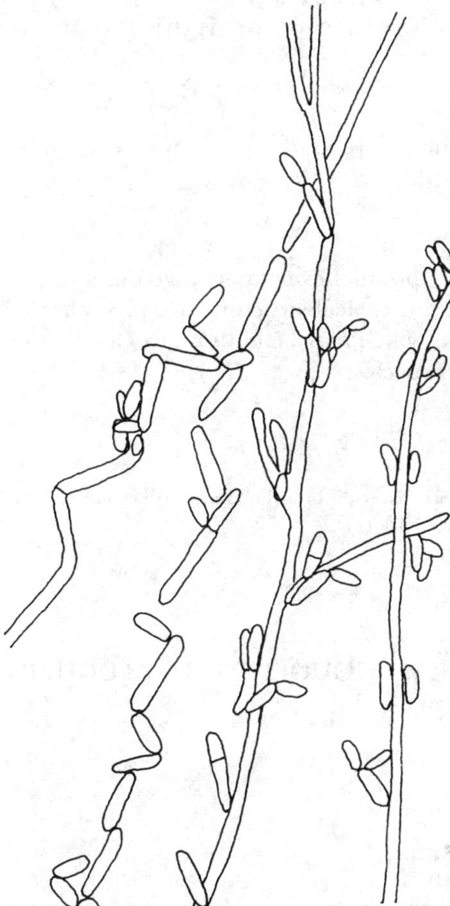

Fig. 389 – *Tr. capitatum*
Slide culture on potato agar
(From 'The Yeasts' ed. I, 1952)

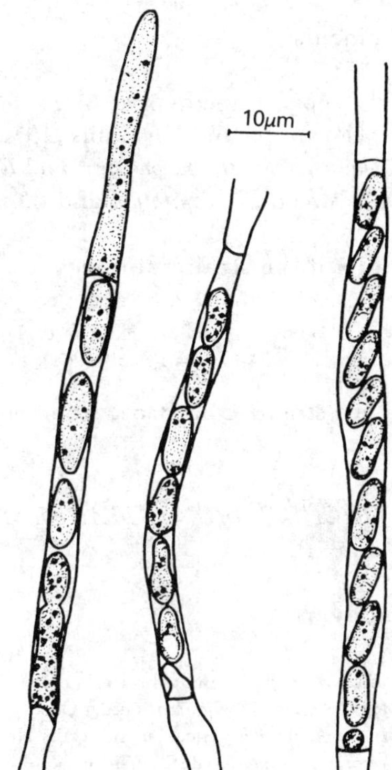

Fig. 390 – *Tr. capitatum*. Endospores after
3 months on glucose-yeast extract-peptone agar
(From 'The Yeasts' ed. II, 1970)

Fermentation: –

Assimilation of carbon compounds:

Galactose	+ w	Raffinose	–	Erythritol	–
Sucrose	–	Soluble starch	–	Ribitol	–
Maltose	–	D-Xylose	–	D-Mannitol	–
Cellobiose	–	L-Arabinose	–	Succinic acid	+
Trehalose	–	D-Ribose	–	Citric acid	+
Lactose	–	L-Rhamnose	–	Inositol	–

Splitting of arbutin: –

Assimilation of nitrate: –

Growth in vitamin-free medium: –

Growth on 50% (w/w) glucose-yeast extract agar: –

Growth at 37°C: +; maximum temperature of growth: 44–46°C

Starch formation: –

Urease: –

Color reaction with DBB: –

G + C: 31.5 mol%, strain CBS 2465 (Guého, 1979)

Comments

The septa in the hyphae of *Tr. capitatum* have plasmodesmata arranged in a circle. Kreger-van Rij and Veenhuis (1972) observed this typical structure also in hyphae of *Endomyces magnusii, E. ovetensis* and *E. tetrasperma,* which may be indicative of a relationship between *Tr. capitatum* and these *Endomyces* species.

Origin of the strains studied

Wood pulp (2), compressed yeast (1), sputa of patients with bronchopulmonary conditions (6) (Dietrichson, 1954; Vörös-Felkai, 1961); unknown, South Africa (1).

Type strain: CBS 2364, from wood pulp.

5. *TRICHOSPORON CUTANEUM* (DE BEURM., GOUGEROT ET VAUCHER) OTA (1926b)

Synonyms

Oidium cutaneum de Beurm. et al. (1909)
Monilia cutanea (de Beurm. et al.) Cast. et Chalmers (1913)
Mycoderma cutaneum (de Beurm. et al.) Neveu-Lemaire (1921)
Geotrichoides cutaneus (de Beurm. et al.) Langeron et Talice (1932)
Geotrichum cutaneum (de Beurm. et al.) de Almeida (1933)
Proteomyces cutaneus (de Beurm. et al.) Dodge (1935)
Basidiotrichosporon cutaneum nom. nud. (Kocková-Kratochvílová et al., 1977b)
? *Oidium moniliaforme* Weigm. et Wolff (1909)
? *Oospora moniliaformis* (Weigm. et Wolff) Berkhout (1923)
Endomyces rotundus Cast. (1912b)

Monilia rotunda (Cast.) Cast. et Chalmers (1913)
Oidium rotundatum (Cast.) Cast. et Chalmers (1919)
Oospora rotundata (Cast.) Berkhout (1923)
Mycoderma rotundatum (Cast.) Brumpt (1927)
Myceloblastanon rotundatum (Cast.) Ota (1928)
Geotrichum rotundatum (Cast.) Cif. et Red. (1929)
Candida rotundata (Cast.) Basgal (1931)
Trichosporon rotundatum (Cast.) Puntoni (1935)
Endomyces rugosus Cast. (1912b) [nec *Mycoderma rugosa* Anderson (1917); nec *Candida rugosa* (And.) Diddens et Lodder (1942)]
Monilia rugosa (Cast.) Cast. et Chalmers (1913)
Hemispora rugosa (Cast.) Cast. et Chalmers (1919)
Parendomyces rugosus (Cast.) Ota (1924a)
Trichosporon rugosum (Cast.) Ota (1926b)
Geotrichum rugosum (Cast.) Dodge (1935)
Parendomyces balzeri Gougerot et Burnier (in Balzer et al., 1912)
Saccharomyces balzeri (Goug. et Burnier) Cast. et Chalmers (1913)
Monilia balzeri (Goug. et Burnier) Neveu-Lemaire (1921)
Trichosporon balzeri (Goug. et Burnier) Ota (1926b)
Trichosporium balzeri (Goug. et Burnier) Bolognesi et Chiurco (1927)
Geotrichoides balzeri (Goug. et Burnier) Langeron et Talice (1932)
Candida balzeri (Goug. et Burnier) de Almeida (1933)
Proteomyces balzeri (Goug. et Burnier) Dodge (1935)
Proteomyces infestans Moses et Vianna (1913)
Sporotrichum infestans (Moses et Vianna) Sartory (1922)
Mycoderma infestans (Moses et Vianna) da Fonseca et de Arêa Leão (1929)
Trichosporon infestans (Moses et Vianna) Cif. et Red. (1935)
Geotrichum infestans (Moses et Vianna) Brumpt (1936)
Trichosporon cutaneum (de Beurm. et al.) Ota var. *infestans* (Moses et Vianna) Diddens et Lodder (1942)
Oidium pulmoneum de Magalhães (1914) [nec *Oidium pulmoneum* Bennett (1842); *Oospora pulmonea* (Bennett) Sacc. (1886); *Mycoderma pulmoneum* (Bennett) Vuill. (1891) (in Brumpt, 1927); *Monilia pulmonea* (Bennett) Cast. et Chalmers (1913); *Geotrichum pulmoneum* (Bennett) Cif. et Red. (1929)]
Neogeotrichum pulmoneum de Magalhães (1932)
? *Oidium brasiliense* de Magalhães (1918) [nec *Zymonema brasiliense* Splendore (1912); *Coccidioides brasiliensis* (Splend.) de Almeida (1929); *Paracoccidioides brasiliensis* (Splend.) de Almeida (1930); *Monilia brasiliensis* (Splend.) Vuill. (1931)]
? *Mycoderma brasiliense* (de Mag.) Neveu-Lemaire (1921)
? *Monilia brasiliensis* (de Mag.) Brumpt (1922)
? *Myceloblastanon brasiliense* (de Mag.) Ota (1928)
? *Geotrichum brasiliense* (de Mag.) Cif. et Red. (1929)
? *Candida brasiliensis* (de Mag.) Basgal (1931)
? *Trichosporon brasiliense* (de Mag.) Puntoni (1935)
Hemispora pararugosa Cast. et al. (1921)
Trichosporon pararugosum (Cast. et al.) Nann. (1934)
Mycoderma pararugosum (Cast. et al.) Dodge (1935)
Parendomyces asteroides Rischin (1921)
Trichosporon asteroides (Rischin) Ota (1926b)
Trichosporon asteroides (Rischin) Bolognesi et Chiurco (1927)
Geotrichoides asteroides (Rischin) Langeron et Talice (1932)
Proteomyces asteroides (Rischin) Dodge (1935) [nec *Monilia asteroides* Cast. (1914); *Oidium asteroides* (Cast.) Cast. et Chalmers (1919); *Mycoderma asteroides* (Cast.) Brumpt (1927); *Geotrichum asteroides* (Cast.) Basgal (1931)]
? *Oospora cerebriformis* Kambayashi (1923)
? *Trichosporon cerebriforme* (Kamb.) Ota (1928)
? *Oospora granulosa* Kambayashi (1923)
? *Trichosporon granulosum* (Kamb.) Ota (1928)

? *Trichosporon equinum* Fambach (1926)
Trichosporon asahii Akagi (1929)
Trichosporon hortai Niño (1933) [nec *Trichosporon hortai* Brumpt (1913)]
Trichosporon humahuaquensis Mazza et Niño (1933)
? *Geotrichoides paludosus* Smit (1934)
Geotrichum amycelicum nom. nud. (Ciferri & Redaelli, 1935)
Hemispora coremiformis Moore (1935)
Proteomyces variabilis Boedijn et Verbunt (1938)
Mycotorula mucinosa Goidànich et al. (1939)
Trichosporon minor nom. nud. (de Arêa Leão, 1940)
? *Geotrichum rotundatum* (Cast.) Cif. et Red. var. *gallicum* nom. nud. (Castellani, 1940)
Trichosporon multisporum nom. nud. (Cochet, 1940)
Trichosporon cutaneum (de Beurm. et al.) Ota var. *multisporum* nom. nud. (Lodder & Kreger-van Rij, 1952)
? *Trichosporon ribeiroi* nom. nud. (de Moraes, 1941)
Geotrichum hirtum nom. nud. (Windisch, 1952)
Trichosporon undulatum nom. nud. (Windisch, 1952)
Trichosporon cutaneum (de Beurm. et al.) Ota var. *peneaus* Phaff et al. (1952)
? *Trichosporon cutaneum* (de Beurm. et al.) Ota var. *curvatum* nom. nud. (Ohara & Nonomura, 1954d)
? *Trichosporon aneurinolyticum* nom. nud. (Yonezawa et al., 1957)
Schizosaccharomyces zambettakesi Ramirez (1957)
Geotrichum zambettakesi Ramirez (1959)
Trichosporon pardi nom. nud. (Fernandez-Baquero & Trespalacios, 1959)
Endomycopsis mali (Lewis) Dekker var. *faecalis* Batista et Silveira (1959b)
Trichosporon loboi Batista et al. (1963)
Geotrichum vanriji Saëz (1964)
Protendomycopsis domschii Windisch (1965)
Endomyces laibachii Windisch (1965)
Trichosporon cutaneum (de Beurm. et al.) Ota var. *antarcticum* Goto, Sugiyama et Iizuka (1969)
Trichosporon jirovecii Fragner (1969)
Trichosporon cutaneum (de Beurm. et al.) Ota var. *jirovecii* (Fragner) Fragner (1970)
Endomycopsis montevidensis de Queiroz (1973)
Trichosporon dulcitum (Berkhout) Weijman (1979)
Trichosporon loubieri (Morenz) Weijman (1979)

Standard description of *Trichosporon cutaneum* (do Carmo-Sousa, 1970c)

Growth in glucose-yeast extract-peptone water: After 3 days at 25°C true mycelium and arthrospores are always present (Fig. 391). Width and length of the arthrospores are variable. Budding cells may be abundant or scarce, or even absent in strains which have been maintained in culture collections for a long time. The yeast cells are spherical to oval, (3.5–7) × (3.5–14) μm. Budding may be on a narrow or a broad base; it is stimulated by growing the cultures on a shaker. A pellicle is often present; it is cream colored or white, smooth, wrinkled, undulating, rugose or folded, thin or thick, often creeping, sometimes tending to sink to the bottom. In the absence of a pellicle, a ring is sometimes formed; it may be incomplete or thick. The sediment is butyrous or flocculent.

Growth on glucose-yeast extract-peptone agar: After one month at 25°C the streak culture is white to yellowish, seldom with a brownish tinge, smooth, wrinkled, rugose or folded, velvety, powdery, hairy or coremial; dull, moist or shiny; of a pasty to tough texture, with an entire or irregular radially folded border, fringed with mycelium. In 3–4-week-old cultures chlamydospores may be observed (van der

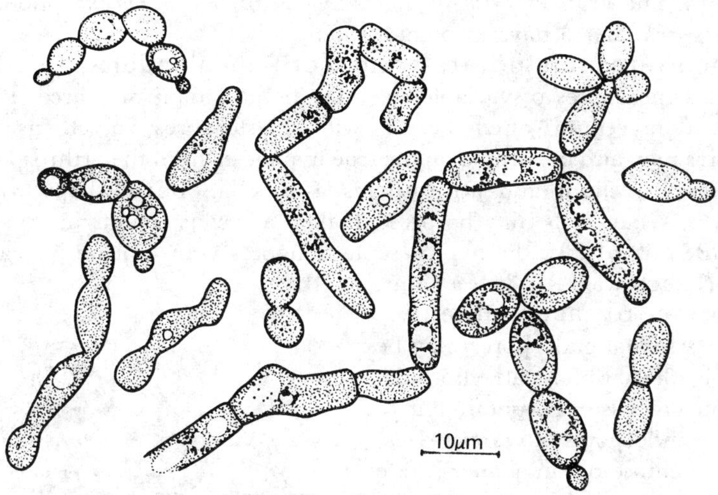

Fig. 391 – *Tr. cutaneum*
After 3 days in glucose-yeast extract-peptone water (From 'The Yeast' ed. II, 1970)

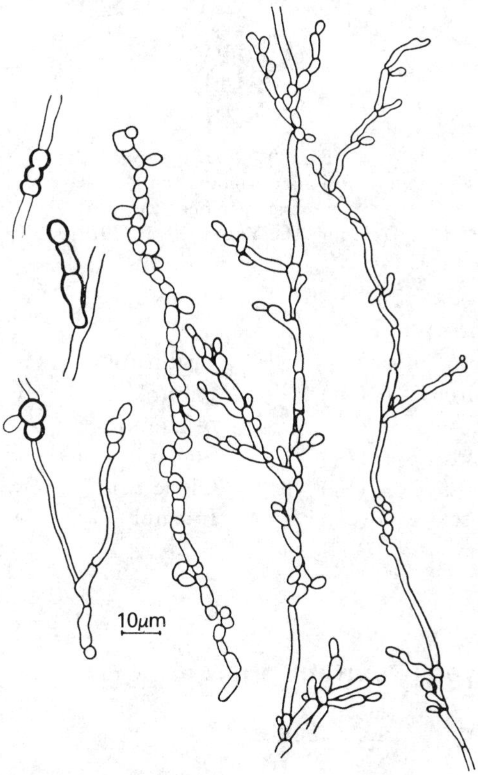

Fig. 392 – *Tr. cutaneum*
Dalmau plate culture on corn meal agar
(From 'The Yeasts' ed. II, 1970)

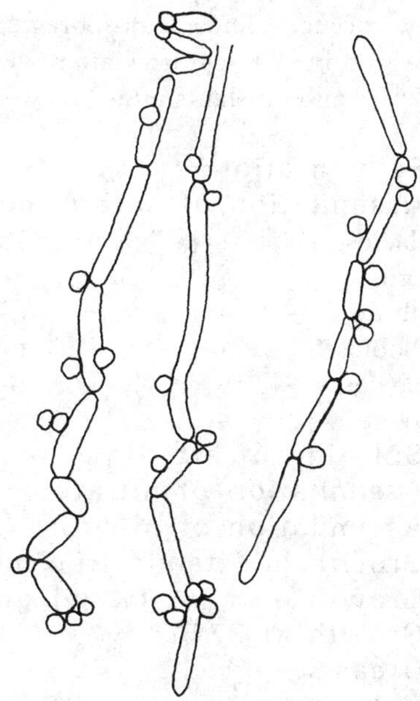

Fig. 393 – *Tr. cutaneum* strain *Tr. multisporum*
Slide culture on potato agar
(From 'The Yeasts' ed. I, 1952)

Walt, 1970i). The strain of *Trichosporon dulcitum* (CBS 508.67) is exceptional in forming chlamydospores after 3 days in potato water.

Slide cultures on potato- and corn meal agar: True mycelium is abundant; arthrospores of variable size may be abundant or scarce (Figs. 392 and 393). Pseudomycelium often occurs with blastospores in chains or clusters. Blastospores may also be formed on the true hyphae or from the arthrospores. In some isolates, the mycelium and arthrospores assume a bizarre shape with very few blastospores. Giant cells may be present; they are conspicuous in the strain of *Tr. loubieri* (CBS 252.61). In the hyphal septa dolipores with a primitive cap have been observed (Kreger-van Rij & Veenhuis, 1971b).

Formation of asexual endospores: Asexual endospores may be formed (Fig. 394). Spherical, ellipsoidal or pear-shaped cells may contain one to six or more endospores of various sizes, spherical or ellipsoidal in shape. These endospores were formed in very recently isolated cultures on glucose-yeast extract-peptone agar after 3 months at 25°C. The capacity of forming endospores by protoplasmic cleavage was apparently lost after a few transfers. Endospores formed by internal budding may occur. The endospores were observed in slide cultures after 3 weeks at 25°C and on slants after one week at 24°C.

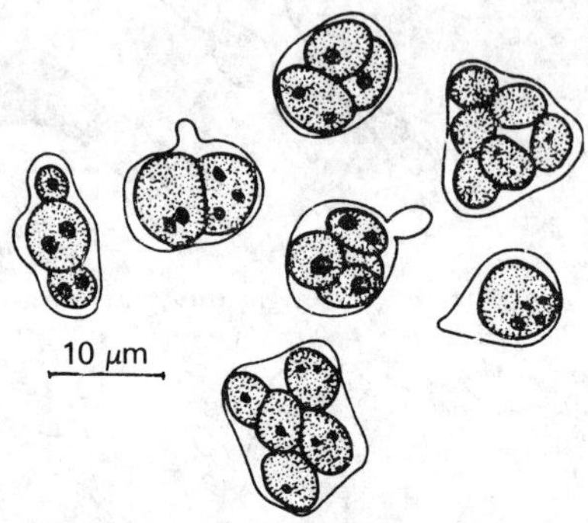

Fig. 394 – *Tr. cutaneum*
Endospores after 3 months on glucose-yeast extract-peptone agar
(From 'The Yeasts' ed. II, 1970)

Fermentation: –

Assimilation of carbon compounds:

Galactose	+ (–)	Raffinose	v	Erythritol	v
Sucrose	+ (–)	Soluble starch	+ (–)	Ribitol	v
Maltose	+ (–)	D-Xylose	+	D-Mannitol	v
Cellobiose	+ (–)	L-Arabinose	+ (–)	Succinic acid	v
Trehalose	v	D-Ribose	+ (–)	Citric acid	v
Lactose	+	L-Rhamnose	v	Inositol	+

Splitting of arbutin: +

Assimilation of nitrate: –

Assimilation of nitrite: v

Growth in vitamin-free medium: –

Growth on 50% (w/w) glucose-yeast extract agar: +w or –

Growth at 37°C: v

Urease: +

Color reaction with DBB: +

G + C: 60 mol%, CBS 5580; 57 mol%, *Tr. loubieri,* CBS 252.61; 57.5 mol%, *Geotrichum vanriji,* CBS 439.64; 60 mol%, *Tr. amycelicum,* CBS 186.38 (all by Guého, 1979)

Comments

The standard description of *Tr. cutaneum* shows that many characters, both morphological and physiological, vary among the strains studied. Cultural conditions may influence or determine the production of yeast cells, hyphae and arthrospores; moreover, freshly isolated, smooth and soft or even mucous strains, when maintained for a longer period on, e.g., malt agar, may change into rough and tough forms which fail to produce budding yeast cells and blastospores under the usual conditions. Variations in the physiological properties, such as the maximum temperature of growth (29–41°C, according to do Carmo-Sousa, 1970c) and the assimilation of various carbon compounds suggest that the standard description includes more than one species.

At present, the species *Tr. figueirae* has been reestablished. It differs from *Tr. cutaneum* by the inability to assimilate inositol. Other species accepted and described since 1970 are *Tr. aquatile* and *Tr. brassicae;* they differ only slightly from *Tr. cutaneum.* New synonyms since 1970 are the following *Trichosporon* species and varieties: *Tr. cutaneum* var. *antarcticum, Tr. jirovecii, Tr. lutetiae, Tr. dulcitum* and *Tr. loubieri.* In contrast with the original description, I found the assimilation of maltose and lactose positive in *Tr. lutetiae.*

A perfect state of *Tr. cutaneum* and of the other basidiomycetous *Trichosporon* species is not yet known. If these species are heterothallic, finding mating types may promote a better differentiation of the imperfect states if suitable indicators appear to be present among the characters of the standard description. For a successful mating reaction, combination of strains which are physiologically as similar as possible and with closely similar G + C mol% may be required.

The species name *cutaneum* is not generally accepted, some authors (do Carmo-Sousa, 1970c) think that *beigelii* has priority. The latter name is older and was given to a yeast isolated from hair in a case of white piedra. *Trichosporon beigelii* has the following synonyms:

Pleurococcus beigelii Küchenmeister et Rabenhorst (in Rabenhorst, 1867)
Hyalococcus beigeli (Küch. et Raben.) Schroeter (1886)
Chlamydotomus beigeli (Küch. et Raben.) Trevisan (1889) (in Saccardo, 1889)
Micrococcus beigeli (Küch. et Raben.) Migula 1900.

There is no type material available of *Tr. beigelii* and Diddens and Lodder (1942) found the original descriptions insufficient for recognizing the species. Therefore, they considered the name *Tr. beigelii* a nomen dubium, just as *Trichosporon ovale* Unna (1896). Instead, they used the name *Tr. cutaneum* and indicated as the type of the species a strain *Geotrichoides cutaneus* (de Beurm. et al.) Langeron et Talice which came from Langeron and was probably the strain studied by Ota. The claim for the validity of *Tr. beigelii* as a species name is ecological; isolates from cases of white piedra were indicated as such. Among the strains described by Diddens and Lodder (1942) and by do Carmo-Sousa (1970c) several came from cases of white piedra. Dr. G.A. de Vries (personal communication) showed that none of them attacked keratin. These strains are not all similar in physiological properties, e.g. the maximum temperature of growth varies among them. Therefore, apart from the question whether ecological data suffice for the determination of a species, a more accurate speciation of the strains of *Tr. cutaneum,* perhaps only possible when the perfect state is found, is necessary before the question of the name of the species may be settled.

The sources of isolation of this species vary considerably, although many of them come from human or animal material.

Origin of the strains studied

White piedra: *Tr. humahuaquensis, Geotrichum hirtum, Oidium brasiliense, Tr. giganteum* (9); axilar piedra: *Tr. minor* (1); skin lesions: *Parendomyces asteroides, Geotrichum amycelicum, Hemispora coremiformis, Proteomyces variabilis, Tr. loboi* (5); nail: *Tr. jirovecii* (1); human clinical specimens: *Geotrichoides cutaneus, Oospora rotundata, Tr. rugosum, Tr. pararugosum, Tr. asahii* (10); a mouse inoculated with pus from human generalized infection: *Proteomyces infestans* (1); feces rat: *Tr. multisporum* (1); brain of pig with meningo-encephalitis (1); cecum of pig (1); rumen of cows (2); intestine of *Cervus eldi siamensis: Tr. lutetiae* (1); cow: *Tr. loubieri* (1); ganglia of livestock (2); cecum of diseased turkey (1); frozen chicken (1); amphibian tissue culture (1); fresh shrimp: *Tr. cutaneum* var. *peneaus* (1); soil: *Protendomycopsis domschii, Endomyces laibachii* (4); industrial sewage (1); water purification tank: *Endomycopsis montevidensis* (1); intertidal mud (1), fresh water (5), wood pulp (1), mushroom: *Schizosaccharomyces zambettakesi* (1), orange peel (2), fodder yeast (1), unknown (5).

Type strain: CBS 2466, strain *Geotrichoides cutaneus* from Baudet, who received it from Langeron.

6. *TRICHOSPORON ERIENSE* HEDRICK ET DUPONT (1968)

Synonym

Geotrichum eriense (Hedrick et Dupont) Weijman (1979)

Standard description of *Trichosporon eriense*

Growth in glucose-yeast extract-peptone water: After 2 days at 25°C branched, septate hyphae with a diameter of 3–4 μm are present. A loose sediment is formed. After one month at room temperature a flocculent sediment is present.

Growth in malt extract (1:10): After 2 days at 25°C, branched hyphae, arthrospores of various lengths and round or elongate budding cells, (3–4) × (4–10) μm, are present.

Growth on glucose-yeast extract-peptone agar: After one month at 20°C the streak culture is white, tough, flat or slightly wrinkled.

Slide culture on corn meal agar: Septate, branched hyphae are formed, under the coverslip with many blastospores occasionally on short protuberances over the whole hypha, single or in short chains, round to pear-shaped. No arthrospores. Aerial hyphae occur. On malt extract (1:100) agar: blastospores and arthrospores are abundant. Single budding cells also occur (Fig. 395).

The ultrastructure of the cell wall is ascomycetous. The hyphal septa have plasmodesmata in a random arrangement.

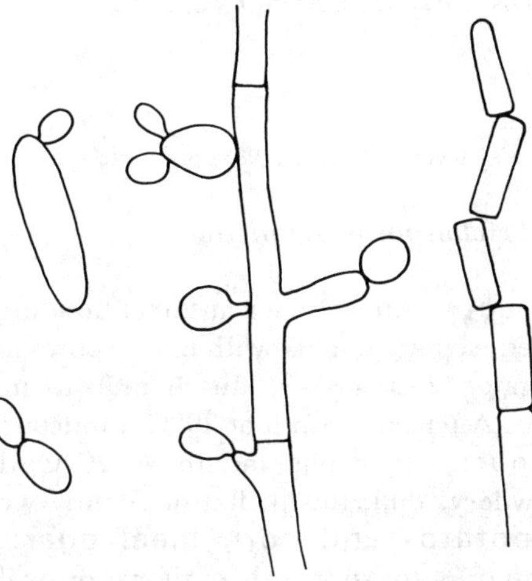

Fig. 395 – *Tr. eriense*
On malt extract (1:100) agar, 1000 ×.
Single budding cells, hyphae with blastospores and arthrospores

Fermentation:

Glucose	+ w or s	Maltose	–
Galactose	–	Lactose	–
Sucrose	–		

Assimilation of carbon compounds:

Galactose	+	Raffinose	–	Erythritol	–
Sucrose	–	Soluble starch	–	Ribitol	–
Maltose	–	D-Xylose	+	D-Mannitol	+
Cellobiose	+	L-Arabinose	–	Succinic acid	+
Trehalose	–	D-Ribose	–	Citric acid	+
Lactose	–	L-Rhamnose	–	Inositol	–

Splitting of arbutin: +
Assimilation of nitrate: –
Growth in vitamin-free medium: –
Growth on 50% (w/w) glucose-yeast extract agar: –
Growth at 37°C: –
Urease: –

Comments

In contrast with the original description, I observed a slow fermentation of glucose.

Origin of the strain studied

Water of Lake Erie, near Cleveland, U.S.A. (1).

Type strain: CBS 5974 (NRRL Y-7052), isolated from fresh water.

7. *TRICHOSPORON FENNICUM* SONCK ET YARROW (1969)

Synonym

Candida fennica (Sonck et Yarrow) Meyer et Ahearn (1983) pro parte

Standard description of *Trichosporon fennicum*

Growth in malt extract: After 2 days at 25°C budding yeast cells, pseudomycelium and true branched, septate hyphae with blastospores are formed. The yeast cells are spherical to elongate, (2–5) × (2–9) μm. A mucous mass partly as a ring, partly submerged is present. After one month at 20°C a mucous mass fills the bottle.

Growth on malt agar: After one month at 20°C the streak culture is yellowish-white, partly powdery, dull, tough, flat or slightly wrinkled.

Slide cultures on potato- and corn meal agar: True mycelium is abundantly formed. Blastospores are spherical, pyriform or oval and occur singly or in small branched chains on the hyphae, occasionally on small elevations. Arthrospores may be present. Most blastospores are formed outside the coverslip. Aerial hyphae occur.

In sections in the electron microscope the hyphal septa have a closure line. The ultrastructure of the cell wall is ascomycetous.

Fermentation:

Glucose	+	Lactose	+ vw or –
Galactose	+ w	Raffinose	– or + vw
Sucrose	+	Melibiose	–
Maltose	+		

Assimilation of carbon compounds:

Galactose	+	Raffinose	+	Erythritol	+
Sucrose	+	Soluble starch	+	Ribitol	+
Maltose	+	D-Xylose	+	D-Mannitol	+
Cellobiose	v	L-Arabinose	+	Succinic acid	+
Trehalose	+	D-Ribose	+	Citric acid	+
Lactose	+ vw	L-Rhamnose	–	Inositol	–

Additional carbon compound tested: Melibiose –

Splitting of arbutin: +

Assimilation of nitrate: –

Growth in vitamin-free medium: +

Growth on 50% (w/w) glucose-yeast extract agar: +

Growth at 37°C: –

Urease: –

Comments

Tr. fennicum closely resembles *Pichia burtonii,* but differs by the fermentation and assimilation of lactose. More important, the type strain of *Tr. fennicum* does not react with the mating types of *P. burtonii.* Therefore, it is retained as a separate species. I observed a very weak assimilation reaction with lactose and, in contrast with the original description, found no growth at 37°C.

Meyer et al. (see p. 678) did not observe arthrospores in the strains of *Tr. fennicum* and *Tr. melibiosaceum* and, therefore transferred *Tr. fennicum* to the genus *Candida.* Moreover, the authors considered the differences between *Tr. fennicum* and *Tr. melibiosaceum* too small for their distinction as separate species and they made *Tr. melibiosaceum* a synonym of *Candida fennica.*

Since I observed arthrospores in the strains of both *Tr. fennicum* and *Tr. melibiosaceum,* I retained these species in the genus *Trichosporon. Pichia burtonii* which is very similar to these species also has arthrospores (see p. 314). The distinction between *Tr. fennicum* and *Tr. melibiosaceum* is very small, but I propose to keep them as separate species until relationships, also with *P. burtonii,* have been clarified by DNA studies.

Origin of the strain studied

Anus of domestic cat in Finland (1).

Type strain: CBS 5928, from a cat.

8. *TRICHOSPORON FERMENTANS* DIDDENS ET LODDER (1942)

Synonyms

Fermentotrichon fermentans (Diddens et Lodder) Novák et Zsolt (1961)
Geotrichum fermentans (Diddens et Lodder) von Arx (in von Arx et al., 1977)

Standard description of *Trichosporon fermentans* (do Carmo-Sousa, 1970c)

Growth in glucose-yeast extract-peptone water: After 3 days at 25°C oval to long-oval budding cells, true mycelium and arthrospores are formed (Fig. 396). The budding cells measure (3.9–7) × (7–17.5) μm. A flocculent sediment is present.

Growth on glucose-yeast extract-peptone agar: After one month at 20°C the streak culture is white to cream-colored, hairy, glossy to dull, slightly raised; the border is fringed with mycelium.

Slide cultures on potato- and corn meal agar: True mycelium, arthrospores and blastospores are formed, the latter singly on the hyphae, in short chains or in clusters at the end of side branches (Fig. 397). The septa in the hyphae have plasmodesmata in a random arrangement. The ultrastructure of the cell wall is ascomycetous.

Formation of asexual endospores: Asexual spores, ellipsoidal or ovoid, 2–3 per hyphal segment may occur (Fig. 398). They were observed very rarely in 3-month-old cultures on glucose-yeast extract-peptone agar at 18–20°C.

Fermentation:

Glucose	+ s	Maltose	–
Galactose	+ w or s	Lactose	–
Sucrose	–		

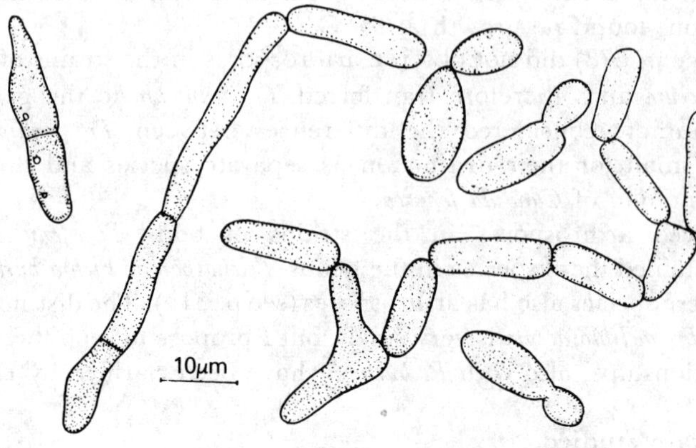

Fig. 396 – *Tr. fermentans*
After 3 days in glucose-yeast extract-peptone water (From 'The Yeasts' ed. II, 1970)

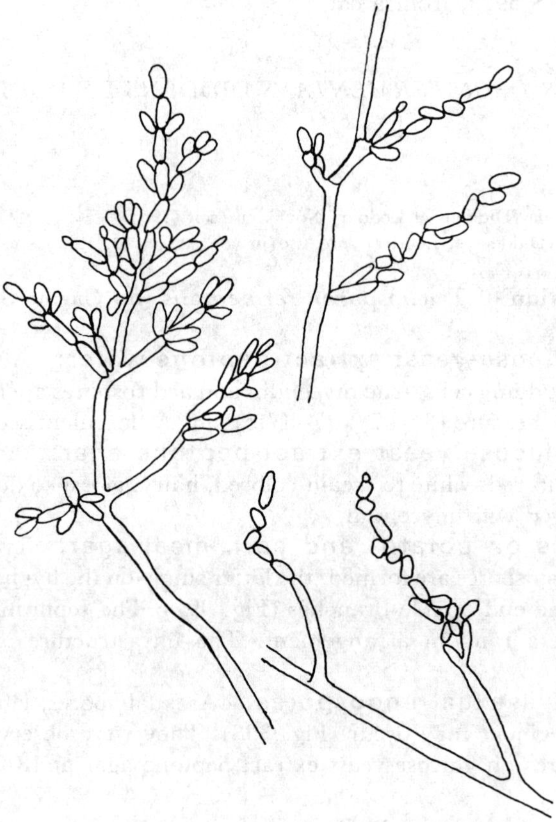

Fig. 397 – *Tr. fermentans*
Slide culture on potato agar (From 'The Yeasts' ed. I, 1952)

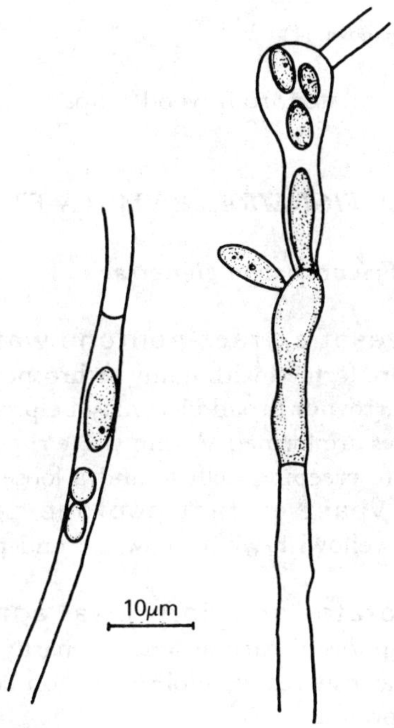

Fig. 398 – *Tr. fermentans*
Endospores after 3 months on glucose-yeast extract-peptone agar
(From 'The Yeasts' ed. II, 1970)

Assimilation of carbon compounds:

Galactose	+	Raffinose	–	Erythritol	–
Sucrose	–	Soluble starch	–	Ribitol	+
Maltose	–	D-Xylose	+	D-Mannitol	+
Cellobiose	+	L-Arabinose	–	Succinic acid	+
Trehalose	–	D-Ribose	–	Citric acid	+
Lactose	–	L-Rhamnose	–	Inositol	–

Splitting of arbutin: +
Assimilation of nitrate: –
Growth on 50% (w/w) glucose-yeast extract agar: –
Growth in vitamin-free medium: +
Maximum temperature of growth: 36–38°C
Urease: –
Starch formation: –
Color reaction with DBB: –
G + C: 42 mol%, type strain (Guého, 1979)

Origin of the strains studied

Wood pulp (1), leather (2), unknown (1).

Type strain: CBS 2529, isolated from wood pulp.

9. *TRICHOSPORON FIGUEIRAE* BATISTA ET SILVEIRA (1960)

Standard description of *Trichosporon figueirae*

Growth in glucose-yeast extract-peptone water: After 2 days at 25°C branched, septate hyphae are formed with many arthrospores, part of them spherical. Yeast cells budding on a narrow or broad base may be present. In some arthrospores cross walls in different planes are formed. A thin white rugose pellicle is present. After one month a thin dull white creeping pellicle and a loose sediment are present.

Growth on glucose-yeast extract-peptone agar: After one month at 20°C the streak culture is yellowish-white or white and powdery, tough, wrinkled, fringed with mycelium.

Slide cultures on potato- and corn meal agar: Abundant, branched, septate hyphae and arthrospores of various lengths, partly spherical. On potato agar a few budding cells and blastospores were observed; on corn meal agar no budding cells or blastospores were present.

Fermentation: –

Assimilation of carbon compounds:

Galactose	+	Raffinose	–	Erythritol	+
Sucrose	–	Soluble starch	–	Ribitol	+w
Maltose	+	D-Xylose	+	D-Mannitol	v
Cellobiose	+	L-Arabinose	+	Succinic acid	+
Trehalose	+w	D-Ribose	+	Citric acid	+
Lactose	+	L-Rhamnose	+	Inositol	–

Splitting of arbutin: +

Assimilation of nitrate: –

Assimilation of nitrite: +

Growth in vitamin-free medium: –

Growth on 50% (w/w) glucose-yeast extract agar: +

Growth at 37°C: +; **at 40°C:** +

Urease: +

Color reaction with DBB: +

Comments

Do Carmo-Sousa (1970c) listed *Tr. figueirae* among the synonyms of *Tr. cutaneum*. The original strain of *Tr. figueirae* was one of the few not assimilating inositol. I have re-established *Tr. figueirae* as a separate species based on this character. The other inositol-negative, nitrate-negative basidiomycetous *Trichosporon* species is *Tr. aquatile*. The two species differ in the ability to grow at 37°C.

Origin of the strains studied

Human feces (1); unknown, received from Dr. H.J. Phaff (1).

Type strain: CBS 4829, from feces.

10. TRICHOSPORON INKIN (OHO) DO CARMO-SOUSA ET VAN UDEN (1967)

see *Sarcinosporon inkin* p. 906

Comments

In the cells of *Tr. inkin* cross walls are formed in different often perpendicular planes yielding sarcina-like structures. King and Jong (1975) indicated these cells as sporangia with endospores and classified the species in a separate genus *Sarcinosporon*. However, formation of septa in different planes is occasionally also observed in cells of *Tr. cutaneum* and *Tr. figueirae*. Therefore, there does not seem to be a reason for separating *Tr. inkin* by this property from the genus *Trichosporon*. Dolipores were observed in the septa of the sarcina-like structures of *Tr. inkin*.

Smith, in this monograph, agrees with King and Jong and describes the species *inkin* as *Sarcinosporon inkin*.

11. TRICHOSPORON MELIBIOSACEUM SCOTT ET VAN DER WALT (1970b)

Synonym

Candida fennica (Sonck et Yarrow) Meyer et Ahearn (1983) pro parte

Standard description of *Trichosporon melibiosaceum*

Growth in malt extract: After 2 days at 25°C budding yeast cells, pseudomycelium, branched, septate mycelium and arthrospores may be formed. The yeast cells are spherical to elongate and measure (2–6) × (2–10) μm; they occur singly, in pairs or in chains. A loose sediment is formed. After one month at room temperature a thick mucous mass fills the bottle.

Growth on malt agar: After one month at 20°C the streak culture is yellowish-white, tough, dull, coarsely wrinkled; the edge is fringed with mycelium.

Slide cultures on potato- and corn meal agar: Abundant formation of true mycelium with spherical to oval blastospores, single or in small branched chains, occasionally on small elevations. The mycelium may break up into arthrospores. Most blasto- and arthrospores occur outside the coverslip. Aerial mycelium may be present.

In sections in the electron microscope, the hyphal septa show a closure line. The ultrastructure of the cell wall is ascomycetous.

Fermentation:

Glucose	+	Lactose	–
Galactose	+ w	Raffinose	+ w
Sucrose	+	Melibiose	– or + vw
Maltose	+		

Assimilation of carbon compounds:

Galactose	+	Raffinose	+	Erythritol	+
Sucrose	+	Soluble starch	+	Ribitol	+
Maltose	+	D-Xylose	+	D-Mannitol	+
Cellobiose	+	L-Arabinose	v	Succinic acid	+
Trehalose	+	D-Ribose	+ w	Citric acid	+
Lactose	– or + vw	L-Rhamnose	–	Inositol	–

Additional carbon compound tested: Melibiose +

Splitting of arbutin: +

Assimilation of nitrate: –

Growth in vitamin-free medium: +

Growth on 50% (w/w) glucose-yeast extract agar: +

Growth at 37°C: –

Urease: –

Comments

Tr. melibiosaceum closely resembles *Pichia burtonii* and only differs in the assimilation of melibiose from this species. The type strain of *Tr. melibiosaceum* could not be found to react with the mating types of *Pichia burtonii* and, therefore, it is considered to be a separate species.

See also Comments *Tr. fennicum.*

Origin of the strain studied

From frass of *Bostrichioplites cornutus,* a bostrichid infesting a moribund specimen of *Acacia karroo* near Nelspruit, Transvaal.

Type strain: CBS 6087, isolated by Scott and van der Walt (1970b).

12. *TRICHOSPORON PULLULANS* (LINDNER) DIDDENS ET LODDER (1942)

Synonyms

Oidium pullulans Lindner (1895)
Oospora pullulans (Lindner) Lindau (1907b)
Basidiotrichosporon pullulans nom. nud. (Kocková-Kratochvílová et al., 1977b)
Oosporidium fuscans Stautz (1931)
Trichosporon fuscans (Stautz) Buchwald (1939)

Standard description of *Trichosporon pullulans* (do Carmo-Sousa, 1970c)

Growth in glucose-yeast extract-peptone water: After 3 days at 20°C budding cells are ovoid to cylindrical, seldom globose, $(3.5-7) \times (5.3-21)$ μm, single or in pairs (Fig. 399). A dull or moist, smooth or rough, creeping pellicle is usually formed, or islets and an incomplete ring. The sediment is scarce or absent.

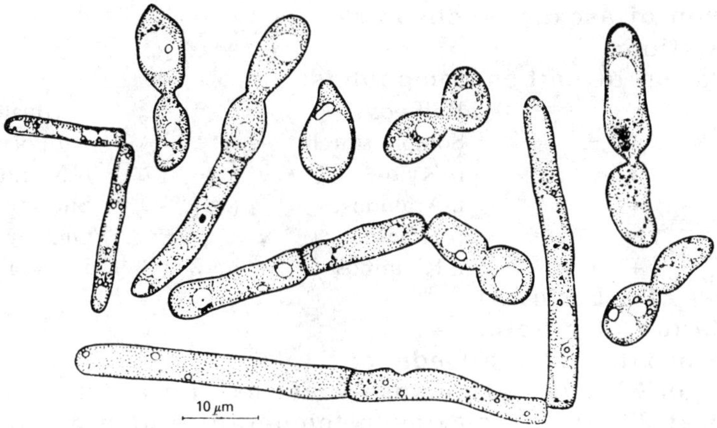

Fig. 399 – *Tr. pullulans*
After 3 days in glucose-yeast extract-peptone water (From 'The Yeasts' ed. II, 1970)

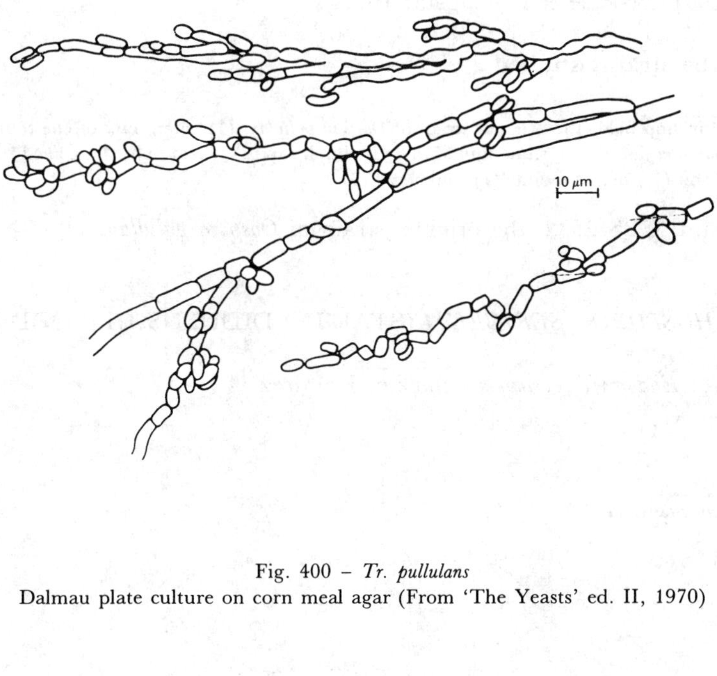

Fig. 400 – *Tr. pullulans*
Dalmau plate culture on corn meal agar (From 'The Yeasts' ed. II, 1970)

Growth on glucose-yeast extract-peptone agar: After one month at 20°C the streak culture is orange-yellow, sometimes with a whitish surface, wrinkled to rugose or velvety, usually dull, rarely glossy, of a pasty or membranous texture, flat or raised, with an entire or undulating border fringed with mycelium.

Slide culture on corn meal agar: True mycelium and arthrospores are abundant; pseudomycelium sparse to abundant with globose to ovoid or elongate blastospores; these arise singly, in pairs, short chains or wreaths (Fig. 400).

Formation of asexual endospores: –

Fermentation: –

Assimilation of carbon compounds:

Galactose	+	Raffinose	+	Erythritol	+
Sucrose	+	Soluble starch	+	Ribitol	v
Maltose	+	D-Xylose	v	D-Mannitol	+
Cellobiose	+	L-Arabinose	+ (–)	Succinic acid	+
Trehalose	+	D-Ribose	v	Citric acid	+
Lactose	+ or w	L-Rhamnose	+ or w	Inositol	v

Splitting of arbutin: +

Assimilation of nitrate: +

Growth in vitamin-free medium: +

Growth on 50% (w/w) glucose-yeast extract agar: –

Growth at 37°C: – : **maximum temperature of growth:** 23–27°C or less

Starch formation: +

Urease: +

Color reaction with DBB: +

G + C: 55 mol%, strain NRRL Y-1485 (Dupont & Hedrick, 1971); 57.1 mol%, strain AJ 4831 (Nakase & Komagata, 1971g)

Origin of the strains studied

Air sample in hop field, the original strain of *Oospora pullulans* (1), slime flux of chestnut, the original strain of *Oosporidium fuscans* (1), slime flux of *Acer* (1), birch sap (2), wood pulp (1), cooled beef (1), butter (1), Antarctic soil (1), beer conduits (1), unknown (3).

Type strain: CBS 2532, the original strain of *Oospora pullulans.*

13. *TRICHOSPORON SERICEUM* (STAUTZ) DIDDENS ET LODDER (1942)

Perfect state: *Endomyces ovetensis* Peláez et Ramirez

Synonym

Oospora sericea Stautz (1931)

Standard description of *Trichosporon sericeum*

Growth in malt extract: After 2 days at 25°C the cells are long-oval to elongate, (4–7) × (15–25) μm; buds are formed on a broad base; besides, mycelial hyphae with cross walls are present (Fig. 401). A sediment and a thin, dull, wrinkled pellicle are formed. After one month at 17°C, a ring and a pellicle which easily drops are present.

Growth on malt agar: After one month at 16°C the streak culture is yellow, dull-shiny, slightly raised, moderately tough, punctuate, with fine wrinkles.

Slide cultures on potato- and corn meal agar: Pseudomycelium consisting of cells formed by budding on a broad base is abundantly produced. This pseudomycelium breaks up easily. In addition, true mycelium is produced. The true

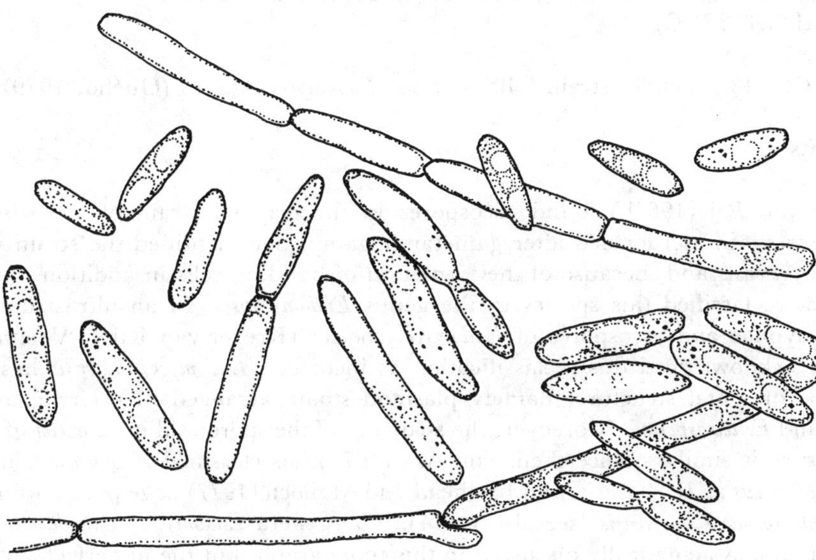

Fig. 401 – *Tr. sericeum*
After 3 days in malt extract (From 'The Yeasts' ed. I, 1952)

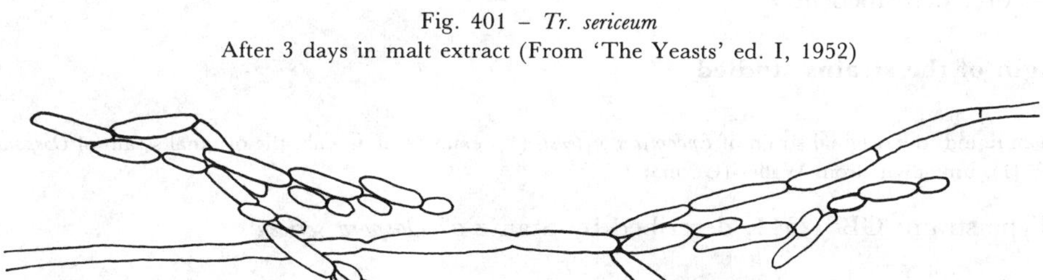

Fig. 402 – *Tr. sericeum*
Slide culture on potato agar (From 'The Yeasts' ed. I, 1952)

mycelium breaks up into arthrospores (Fig. 402). The hyphal septa and the cross wall between mother cell and bud have plasmodesmata arranged in a circle.

Fermentation: –

Assimilation of carbon compounds:

Galactose	+	Raffinose	–	Erythritol	–
Sucrose	–	Soluble starch	–	Ribitol	–
Maltose	–	D-Xylose	–	D-Mannitol	–
Cellobiose	–	L-Arabinose	–	Succinic acid	+
Trehalose	–	D-Ribose	–	Citric acid	–
Lactose	–	L-Rhamnose	–	Inositol	–

Splitting of arbutin: –

Assimilation of nitrate: –

Growth in vitamin-free medium: –

Growth on 50% (w/w) glucose-yeast extract agar: –

Growth at 37°C: –

Urease: +w

G + C: 48.5 mol%, strain CBS 192.55, *Endomyces ovetensis* (Guého, 1979)

Comments

Kreger-van Rij (1964b) found ascospores in the original strain of *Tr. sericeum (Oospora sericea)* in asci formed after gametangiogamy. She identified the strain with *Endomyces ovetensis* and, because of the formation of budding cells in addition to true mycelium, reclassified this species in the genus *Endomycopsis.* In an ultrastructural study of hyphae and ascospores of *Endomyces* species (Kreger-van Rij & Veenhuis, 1972) she acknowledged this reclassification as incorrect. *Endomyces ovetensis* has the same type of septal structure, namely plasmodesmata arranged in a circle, as *E. magnusii* and *E. tetrasperma.* Moreover, the structure of the spore wall in *E. ovetensis* and *E. tetrasperma* is similar. Since then, von Arx (1977) has classified *E. ovetensis* in the genus *Dipodascus* as *D. ovetensis,* and Redhead and Malloch (1977) have placed it in the genus *Zendera* as *Z. ovetensis* (see also p. 14). The genera *Endomyces, Dipodascus* and *Zendera* are not systematically discussed in this monograph, but the imperfect state of *E. ovetensis* conforms to the diagnosis of the genus *Trichosporon* and this species is, therefore, described here.

Origin of the strains studied

Tan liquid, the original strain of *Endomyces ovetensis* (1), exudate of an oak, the original strain of *Oospora sericea* (1), unknown, from Waller (Germany) (1).

Type strain: CBS 2544, described by Stautz as *Oospora sericea.*

14. TRICHOSPORON TERRESTRE VAN DER WALT ET JOHANNSEN (1975d)

Synonym

Geotrichum terrestre (van der Walt et Johannsen) Weijman (1979)

Standard description of *Trichosporon terrestre*

Growth in glucose-yeast extract-peptone water: After 3 days at 25°C branched, septate mycelium is abundantly formed; arthrospores and some pseudomycelium may also be present. Loose budding cells are short-oval to elongate and measure (2–6) × (3–10) μm. The hyphae have a diameter of 2–3.5 μm (Fig. 403). There is a flocculent sediment. After one month at room temperature a loose sediment is present.

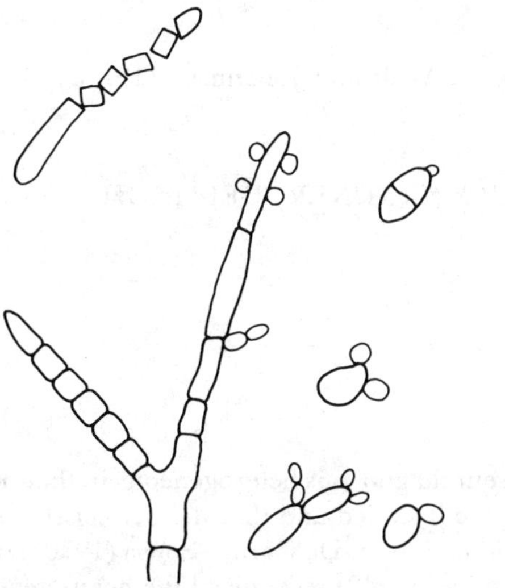

Fig. 403 – *Tr. terrestre*
After 3 days in glucose-yeast extract-peptone water (1000 ×) Single budding cells, true mycelium with blastospores and arthrospores

Growth on glucose-yeast extract-peptone agar: After one month at room temperature the streak culture is cream-colored, dry, soft, raised and delicately wrinkled, with a lobate margin.

Slide cultures on potato- and corn meal agar: Branched, septate mycelium is formed. Blastospores occur on the hyphae near the septa and in between, singly or in chains, sometimes on short protuberances. Arthrospores vary in length, and may be very short and spherical. The septa have plasmodesmata in a random arrangement.

Fermentation: –

Assimilation of carbon compounds:

Galactose	+	Raffinose	+	Erythritol	+
Sucrose	+	Soluble starch	–	Ribitol	+
Maltose	+	D-Xylose	+	D-Mannitol	+
Cellobiose	+ vw or –	L-Arabinose	+	Succinic acid	+
Trehalose	+	D-Ribose	+	Citric acid	+
Lactose	–	L-Rhamnose	+	Inositol	+

Splitting of arbutin: + vw and s or –

Assimilation of nitrate: +

Growth in vitamin-free medium: –

Growth on 50% (w/w) glucose-yeast extract agar: +

Growth at 37°C: +
Urease: –
Color reaction with DBB: –

Comments

Tr. terrestre is the only nitrate-positive ascomycetous *Trichosporon* species.

Origin of the strain studied

Grassland soil near Pretoria, South Africa (1).

Type strain: CBS 6697, isolated by van der Walt and Johannsen (1975d).

15. TRICHOSPORON VARIABILE (LINDNER) DELITSCH

see *Pichia burtonii* p. 313

f. COMMENTS ON THE GENUS

The genus *Trichosporon* defined in the present diagnosis is heterogeneous in that both ascomycetous and basidiomycetous species are included and that the perfect states of the ascomycetous species may belong to various genera. Do Carmo-Sousa (1970c) who adopted this diagnosis from Diddens and Lodder (1942) recognized the heterogeneity and distinguished three groups among the seven species.

Since 1970, a few new *Trichosporon* species have been described, among others by Hedrick and Dupont. These authors (1968) also studied the utilization of amino acids by *Trichosporon* species and applied numerical taxonomy to the ten species they accepted (Dupont & Hedrick, 1971). Since then several species have been transferred to other genera: King and Jong (1975) classified *Tr. inkin* in the new genus *Sarcinosporon,* and (1976a) *Tr. aculeatum* in a new genus *Aciculoconidium;* von Arx (in von Arx et al., 1977) transferred *Tr. capitatum, Tr. fermentans* and *Tr. penicillatum* to the genus *Geotrichum,* and Weijman (1979) classified *Tr. eriense* and *Tr. terrestre* in the genus *Geotrichum.*

It was surmised that *Tr. cutaneum* was of basidiomycetous affinity because Aschner and Cury (1951) found that it could produce starch-like compounds. Moreover, it was urease positive, another feature of many basidiomycetous yeasts. Finding the typical layered basidiomycetous cell wall and dolipores in the septa (Kreger-van Rij & Veenhuis, 1971b) corroborated the supposition that *Tr. cutaneum* is the imperfect state of a basidiomycete. Later on, other characters typical of some basidiomycetes have been added, namely the color reaction with Diazonium Blue B (van der Walt & Hopsu Havu, 1976) and the presence of xylose in the hydrolysate of the cells (Weijman, 1979). These and other new methods, such as the release of spheroplasts from cells by the use of specific enzymes (Bastide et al., 1979) have made it easier to distinguish between ascomycetous and basidiomycetous yeasts in the absence of a perfect state. Some authors have consequently proposed to split the genus *Trichosporon*. Kocková-Kratochvílová et al. (1977b) suggested two new genera: *Ascotrichosporon* and *Basidiotrichosporon,* but these names have not been validly published because of the lack of a Latin diagnosis. Weijman (1979) also suggested a separation of basidiomycetous and ascomycetous species, the former remaining in the genus *Trichosporon*.

The basidiomycetous yeast species in this chapter are: *Tr. aquatile, Tr. brassicae, Tr. cutaneum, Tr. figueirae, Tr. inkin* and *Tr. pullulans*. For none of them a perfect state has yet been found. If the presence of a dolipore in the septum, observed in strains of *Tr. cutaneum, Tr. inkin* and *Tr. pullulans,* is taken as a criterion, they will probably not belong to the teliospore-forming yeasts, but rather to a group including the Filobasidiaceae and Tremellaceae. Finding of mating types will make a better speciation among these species possible, perhaps splitting up *Tr. cutaneum*. As mentioned on p. 953, I do not think that a separation of *Tr. inkin* in the genus *Sarcinosporon* is warranted. The six basidiomycetous species now accepted are probably narrowly related and restriction of the genus *Trichosporon* to the basidiomycetous species will make it more homogeneous.

Of the ascomycetous species: *Tr. aculeatum, Tr. capitatum, Tr. eriense, Tr. fennicum, Tr. fermentans, Tr. melibiosaceum, Tr. terrestre, Tr. sericeum* and *Tr. variabile, Tr. sericeum* has a perfect state in *Endomyces ovetensis* and *Tr. variabile* in *Pichia burtonii*. *Tr. fennicum* and *Tr. melibiosaceum* are very similar to *P. burtonii,* but do not mate with the types of the latter. As mentioned on p. 936, *Tr. acuelatum* shows some resemblance with *P. burtonii*. The perfect state of *Tr. capitatum* may belong to a group of *Endomyces* species (see p. 940). It is uncertain to which genera the perfect state of *Tr. eriense, Tr. fermentans* and *Tr. terrestre* may belong. I agree with the reclassification of *Tr. penicillatum* in *Geotrichum;* it is very similar to *Geotrichum candidum* and I have not been able to find budding cells in the strain studied. Classification of *Tr. eriense, Tr. fermentans* and *Tr. terrestre* in either *Trichosporon* or *Geotrichum* depends on the presence or absence of budding cells. Generally, there is no doubt about this criterion, especially in fresh isolates. In older strains it may be difficult to ascertain budding, as it is sometimes difficult to tell the difference between pseudo- and true mycelium. Buds have a constriction at the base, but this may be narrow as well as broad, and in true mycelium constriction between hyphal cells may occur. Blastospore formation in, e.g., *Tr. eriense* is distinct and on a narrow base, and a redefinition of the genus *Geotrichum* would be necessary to include this species in it. A convenient and practical way out of these problems seems for the present to retain *Trichosporon* in its heterogeneous state. *Trichosporon* differs from *Candida* in the formation of arthrospores, absent in the latter.

Behrend (1890) described the genus *Trichosporon* with the single species *Tr. ovoides,* and, therefore, this is the type of the genus. Diddens and Lodder (1942) considered the original description of this species sufficient for defining the genus, but not for the species and indicated *Tr. ovoides* as a nomen dubium. They designated *Tr. cutaneum* as the type species of the genus. I agree with these authors in their view on the diagnosis of *Tr. ovoides,* but, in accordance with the International Code of Botanical Nomenclature, have to accept *Tr. ovoides* as the type of the genus.

g. SPECIES NOT ACCEPTED IN THE GENUS

Species received too late for inclusion in this treatment:

Trichosporon beemeri Kuttin et Müller (1981)
Trichosporon sporotrichoides (van Oorschot) van Oorschot et de Hoog (1981)

Genus 17. *Trigonopsis* Schachner

by D. Yarrow

a. DIAGNOSIS OF THE GENUS (SLOOFF, 1970d)

Cells triangular or ellipsoidal; tetrahedral and rhombohedral cells are also present. Reproduction occurs by budding from the angles on the triangular and tetrahedral cells; on the ellipsoidal cells budding is multilateral.

Pseudohyphae are not produced.

A film or a thin pellicle develops on liquid media.

The organisms have no fermentative ability.

Nitrate is not assimilated.

b. TYPE SPECIES OF THE GENUS

Trigonopsis variabilis Schachner

c. SPECIES ACCEPTED IN THE GENUS

1. *Trigonopsis variabilis* Schachner (1929)

e. SYSTEMATIC DISCUSSION OF THE SPECIES

1. TRIGONOPSIS VARIABILIS SCHACHNER (1929)

Standard description of *Trigonopsis variabilis*

Growth in malt extract: After 3 days at 25°C some of the cells are triangular or of the shape of a dented tetrahedron, seldom of a rhombohedron, single or in pairs. The ellipsoidal cells measure (2.0–6.0) × (4.0–9.0) μm; the length of the sides of the tetrahedral and triangular cells varies from 3.5 to 7.0 μm (Fig. 404). A sediment and thin ring are present. After one month at 20°C a sediment, ring and sometimes a film are present.

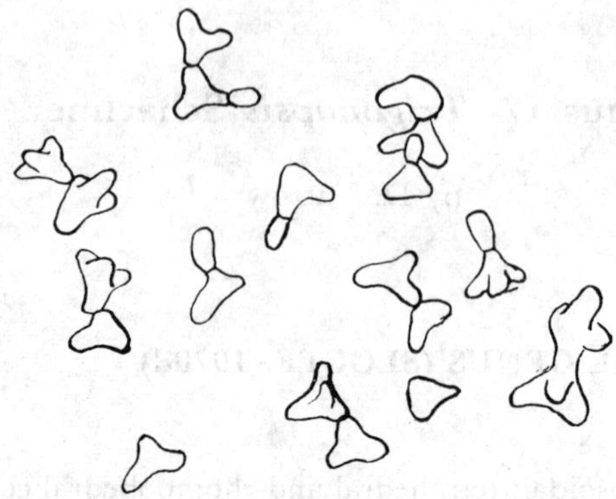

Fig. 404 – *Trig. variabilis,* CBS 1040
After 3 days in malt extract (From 'The Yeasts' ed. II, 1970)

Growth on malt agar: After one month at 20°C the streak culture is flat and cream-colored, the surface smooth and shining.

Slide culture on potato agar: No pseudohyphae; some short chains of cells may have been formed.

Fermentation: –

Assimilation of carbon compounds:

Galactose	+	Raffinose	–	Erythritol	–
Sucrose	–	Soluble starch	–	Ribitol	–
Maltose	–	D-Xylose	– or +w	D-Mannitol	+ or s
Cellobiose	–	L-Arabinose	–	Succinic acid	+
Trehalose	+	D-Ribose	+ or s	Citric acid	+
Lactose	–	L-Rhamnose	–	Inositol	–

Splitting of arbutin: –

Assimilation of nitrate: –

Growth in vitamin-free medium: –

Growth on 50% (w/w) glucose-yeast extract agar: –

Growth at 37°C: +

G + C: 44.0 mol%, one strain (Nakase & Komagata, 1968a)

Comments

The proportion of triangular to ellipsoidal cells formed by strains of this species is influenced by the nitrogen source (Sentheshanmuganathan & Nickerson, 1962; Picci & Verona, 1963; Šašek & Becker, 1969), the carbon source (Matthewson & Barnett, 1974), the growth temperature (Rambelli, 1959; Picci & Verona, 1963; Packer & Bersten, 1977), the presence of the detergents Tween 80 and Span 80 and of short-chain alcohols (Packer & Bersten, 1977).

Origin of the strains studied

Beer (1) and a mutant of it rarely producing triangular cells; grape must (1).

Type strain: CBS 1040, "Rasse I" isolated from beer by Schachner.

Chapter VI

Keys

1. Key to the genera

by N.J.W. Kreger-van Rij

1a Vegetative reproduction exclusively by cross wall formation without constriction
Schizosaccharomyces ... p. 414

b Vegetative reproduction exclusively by cells formed on stalks
Sterigmatomyces ... p. 921

c Vegetative reproduction by the formation of terminal buds on a conidiophore; the conidiophore elongates to develop a succession of conidia
Sympodiomyces ... p. 930

d Other forms of vegetative reproduction ... 2

2a Vegetative reproduction by unipolar budding on a broad base; true mycelium may occur
Malassezia ... p. 882

b Vegetative reproduction by bipolar budding on a broad base ... 3

c Vegetative reproduction by multilateral budding; true mycelium, arthrospores and ballistospores may also occur ... 8

3a Ascospores formed 4
b Ascospores not formed 6

4a Ascospores cap-shaped
Wickerhamia .... p. 440
b Ascospores hat- or helmet-shaped
Hanseniaspora .... p. 154
c Ascospores spherical 5

5a Ascospores smooth, hyaline, conjugating in pairs in the ascus; a narrow ledge is not visible under the light microscope
Saccharomycodes .... p. 396
b Ascospores warty and brown
Nadsonia .... p. 279
c Ascospores smooth or warty, with or without an indistinctly visible ledge, not brown, not conjugating in pairs in the ascus
Hanseniaspora .... p. 154

6a Glucose fermented 7
b Glucose not fermented
Schizoblastosporion .... p. 909

7a Strong acetic acid formation from glucose; on malt agar growth is slow and cells are short-lived
Eeniella *
b No strong acetic acid formation from glucose; growth not slow
Kloeckera .... p. 873

8a Ballistospores formed 9
b Ballistospores not formed 11

9a Teliospores formed
Sporidiobolus .... p. 532
b Teliospores not formed 10

10a Cultures pink, red or orange
Sporobolomyces .... p. 911
b Cultures cream colored to slightly yellow
Bullera, *Fibulobasidium*, *Sirobasidium* } see *Bullera* key .... p. 577

* The genus *Eeniella* was described by Smith et al. (1981) and is not yet presented in this monograph.

11a Some vegetative cells triangular
Trigonopsis .... p. 963
b Vegetative cells not triangular 12

12a Strong acetic acid formation from glucose; on malt agar growth is slow and cells are short-lived 13
b Not this combination of characters 14

13a Ascospores formed
Dekkera .... p. 146
b Ascospores not formed
Brettanomyces .... p. 562

14a Ascospores formed 15
b Ascospores not formed 41

15a Nitrate assimilated 16
b Nitrate not assimilated 19

16a Abundant true mycelium and budding cells; the hyphal septa have a dolipore visible under the light microscope as a small dark dot in the middle of the septum
Ambrosiozyma .... p. 106
b True mycelium present, scarce or absent; without septal dolipores 17

17a Hat-shaped ascospores formed in globose compartment at distal end of tube-shaped ascus
Pachysolen .... p. 289
b Ascus not tube-shaped 18

18a Ascospores spherical with a warty wall
Citeromyces .... p. 117
b Ascospores hat-shaped, hemispherical or Saturn-shaped
Hansenula .... p. 165
c Ascospores oblong with obtuse ends
Wickerhamiella .... p. 443

19a Ascus is sac-like protrusion on a vegetative cell; spores light amber or brown
Lipomyces .... p. 252
b Not this combination of characters 20

20a Abundant true mycelium in addition to budding yeast cells 21
b True mycelium scarce or absent 27

21a The hyphal septa have a dolipore visible under the light microscope as a small dark dot in the middle of the septum
Ambrosiozyma .... p. 106
b Hyphal septa without dolipores 22

22a Asci spindle-shaped, vegetative cells lemon-shaped or elongate, ascospores with a ledge
Arthroascus .... p. 114
b Not this combination of characters 23

23a Asci formed exclusively on the true hyphae 24
b Asci not formed exclusively on the true hyphae 26

24a Asci spherical with a small vegetative apical cell attached to it
Stephanoascus .... p. 431
b Asci without apical cell 25

25a Ascospores lunate; the wall at the two poles is thickened and gives the appearance of appendages
Guilliermondella .... p. 151
b Ascospores not lunate
Saccharomycopsis .... p. 399

26a Nitrate assimilated
Hansenula .... p. 165
b Nitrate not assimilated
Pichia .... p. 295

27a Ascospores needle-shaped, one or two per ascus
Metschnikowia .... p. 266
b Ascospores spindle-shaped 28
c Ascospores of a different shape 29

28a Ascospores with a whip-like appendage
Nematospora .... p. 285
b Ascospores without whip-like appendage; parasitic in intestine of *Drosophila*; not yet in culture
Coccidiascus .... p. 123

29a Ascospores oval to cylindrical; growth only on media with gaseous CO_2 present; occurrence confined to digestive tract of rabbits and certain other rodents
Cyniclomyces .... p. 125
b Not this combination of characters 30

30a Asci dehiscent 31
b Asci persistent 33

31a Ascospores clavate, and warty which may not be visible under the light microscope
Clavispora .... p. 120
b Ascospores spherical or ellipsoidal, thick-walled
Sporopachydermia .... p. 427
c Ascospores hat- or Saturn-shaped
Pichia .... p. 295
d Ascospores reni- or crescentiform
Kluyveromyces .... p. 224
e Ascospores spherical or ellipsoidal 32

32a Fermentation of glucose vigorous
Kluyveromyces .... p. 224
b Fermentation of glucose weak, slow or absent
Pichia .... p. 295

33a Ascospores spherical, warty with an equatorial ledge
Schwanniomyces .... p. 423
b Ascospores lentiform, light brown
Wingea .... p. 446
c Ascospores hat- or Saturn-shaped
Pichia .... p. 295
d Ascospores oblong with obtuse ends; not thick-walled; one, seldom two per ascus
Lodderomyces .... p. 263
e Ascospores spherical or ellipsoidal 34

34a Conjugation immediately preceding ascus formation 35
b No conjugation immediately preceding ascus formation 39

35a Ascospores spherical or ellipsoidal, warty or with ridges 36
b Ascospores spherical and smooth 37

36a Fermentation of glucose vigorous
Torulaspora .... p. 434
b Fermentation of glucose not vigorous; usually slow, weak or absent
Debaryomyces .... p. 130

37a Early formation of a pellicle on liquid media
Pichia .... p. 295
b No early formation of a pellicle on liquid media 38

38a Conjugation usually between individual cells
Zygosaccharomyces .... p. 449
b Conjugation usually between a cell and its bud
Torulaspora .... p. 434

39a Early formation of a pellicle on liquid media
Issatchenkia .... p. 214
b No early formation of a pellicle on liquid media 40

40a Ascospores ellipsoidal, thick-walled
Pachytichospora .... p. 292
b Ascospores spherical or ellipsoidal, not thick-walled
Saccharomyces .... p. 379

41a Teliospores formed 42
b Teliospores not formed 43

42a Cultures yellow, orange or red due to the presence of carotenoid pigments
Rhodosporidium .... p. 509
b Cultures not yellow, orange or red due to the presence of carotenoid pigments
Leucosporidium .... p. 496

43a Basidiospores formed 44
b Basidiospores not formed 46

44a Basidia arranged in synnemata
Chionosphaera .... p. 470
b Basidia not arranged in synnemata 45

45a Basidiospores produced in chains on the apex of the basidium
Filobasidiella .... p. 472
b Basidiospores produced in a whorl on the apex of the basidium
Filobasidium .... p. 483

46a Cultures pink or red due to the presence of carotenoid pigments 47
b Cultures not pink or red due to the presence of carotenoid pigments 50

47a Glucose fermented
Phaffia .... p. 890
b Glucose not fermented 48

48a Inositol assimilated
Cryptococcus .... p. 845
b Inositol not assimilated 49

49a Starch-like compounds formed
Yeast-like forms of *Taphrina**
b Starch-like compounds not formed
Rhodotorula .... p. 893

50a Multilateral budding on a broad base combined with the formation of septa; no arthrospores
Oosporidium .... p. 886
b Multilateral budding, true mycelium and arthrospores 51
c Multilateral budding; true mycelium may be formed, no arthrospores 52

51a Sarcina-like agglomerates of cells formed by septation in different planes
Sarcinosporon .... p. 906
b No formation of sarcina-like agglomerates of cells
Trichosporon .... p. 933

52a Formation of terminal needle-shaped blastospores
Aciculoconidium .... p. 558
b No formation of needle-shaped blastospores 53

53a No formation of pseudomycelium, inositol assimilated
Cryptococcus, *Holtermannia*, *Tremella* } see *Cryptococcus* key .... p. 846
b Not this combination of characters
Candida, *Tremella* } see *Candida* key .... p. 591

* The genus *Taphrina* is not described in this monograph. According to Lodder (1970), *Taphrina* species may grow like yeast cells; they grow slowly and the cultures are pink to pale ochre.

2. Keys to species and genera not requiring characteristics of sexual reproduction

by N.J.W. Kreger-van Rij

The characters used in these keys are mainly: the features of vegetative reproduction, the presence or absence of a pellicle on liquid media and the appearance of the culture on solid media, the results of a limited number of fermentation and assimilation tests, growth without the addition of vitamins and growth at 37°C.

The first key leads directly to certain genera, and in addition to eight groups of species, each with its own key. The eight keys lead to species or occasionally groups of species. The latter is the case if features of sexual reproduction are required for the differentiation of these species, e.g. in the genus *Lipomyces*, or when special key-characters are used for their distinction in the genus, e.g. in the genera *Torulaspora, Saccharomyces* and *Zygosaccharomyces*.

Several of the haploid, heterothallic sporogenous species in the following keys are also found in the key to the species of the genus *Candida*.

The species *Coccidiascus legeri* and *Chionosphaera apobasidialis* were not included in the key because these organisms have not been obtained in pure culture. *Cyniclomyces guttulatus* is omitted due to special conditions required for growth.

Abbreviations used in the keys

Ce	cellobiose	St	soluble starch
Cit	citric acid	Su	sucrose
Er	erythritol	Suc	succinic acid
G	glucose	Tr	trehalose
Ga	galactose	X	D-xylose
I	inositol	F	fermentation,
La	lactose		e.g., FGa = fermentation of galactose
LAr	L-arabinose		
Ma	maltose		Ga = assimilation of galactose
Man	D-mannitol		
Me	melibiose	True myc	true mycelium
Mz	melezitose	Psm	pseudomycelium
Ra	raffinose	Pell	pellicle formation
Rh	L-rhamnose	Vit	growth in vitamin-free medium
Ri	D-ribose		
Rib	ribitol	37°C	growth at 37°C

For the abbreviations of the generic names see p. xi.

1a Vegetative reproduction exclusively by cross wall formation without constriction
Schizosaccharomyces . . . . p. 414

b Vegetative reproduction exclusively by cells formed on stalks
Sterigmatomyces . . . . p. 921

c Vegetative reproduction by the formation of terminal buds on a conidiophore; the conidiophore elongates to develop a succession of buds
Sympodiomyces . . . . p. 930

d Vegetative reproduction by unipolar budding on a broad base; true mycelium may occur
Malassezia . . . . p. 882

e Vegetative reproduction by bipolar budding on a broad base
see key A. . . . . p. 977

f Multilateral budding on a broad base combined with the formation of septa; no arthrospores; cultures light pink to orange-yellow
Oosporidium . . . . p. 886

g Vegetative reproduction by the formation of true mycelium and arthrospores and by multilateral budding
Trichosporon . . . . p. 933

h Other types of vegetative reproduction 2

2a Ballistospores formed
see key 3. . . . . p. 989

b Ballistospores not formed 3

3a Some vegetative cells are triangular
Trigonopsis . . . . p. 963

b Vegetative cells not triangular 4

4a Cultures pink, orange or red due to carotenoid pigments
see key B. . . . . p. 978

b Cultures not colored due to carotenoid pigments 5

5a Strong acetic acid formation from glucose; on malt agar slow growth and cells short-lived
Dekkera . . . . p. 146
Brettanomyces . . . . p. 562

b Not this combination of features 6

6a Nitrate assimilated 7
b Nitrate not assimilated 9

7a Inositol assimilated 8
b Inositol not assimilated
see key C. . . . . p. 979
see *Candida* key . . . . p. 591

8a Pseudo- and/or true mycelium produced
see *Candida* key . . . . p. 591
b Pseudo- and true mycelium not produced
Cryptococcus . . . . p. 845
Filobasidium floriforme . . . . p. 486

9a Inositol assimilated 10
b Inositol not assimilated 15
see also *Candida* key . . . . p. 591

10a Pseudo- and true mycelium produced 11
see also *Candida* key . . . . p. 591
b Pseudo- and true mycelium not produced 14

11a Sucrose assimilated 12
b Sucrose not assimilated 13

12a Glucose fermented
Saccharomycopsis fibuligera . . . . p. 404
b Glucose not fermented
Stephanoascus ciferrii . . . . p. 432

13a Glucose fermented
Pichia abadieae . . . . p. 302
b Glucose not fermented
Saccharomycopsis synnaedendra . . . . p. 410

14a Glucose fermented
Pichia abadieae . . . . p. 302
b Glucose not fermented
Cryptococcus, *Tremella*, *Holtermannia*, *Sporopachydermia* } see *Cryptococcus* key . . . . p. 846
Filobasidiella neoformans . . . . p. 474
Filobasidium uniguttulatum . . . . p. 488

15a Glucose fermented 16
b Glucose not fermented 17

16a Erythritol assimilated
see key D. . . . . p. 981
b Erythritol not assimilated
see key E. . . . . p. 982

17a Erythritol assimilated
see key F. . . . . p. 987
b Erythritol not assimilated
see key G. . . . . p. 988

A. Key to species with vegetative reproduction by bipolar budding on a broad base

1 FG + 2
FG –
N. commutata p. 280
Schizobl. starkeyi-henricii p. 909

2 Ce + 3
Ce – 8

3 37°C + 4
37°C – 5

4 Ra +
S'codes ludwigii p. 397
Ra –
H'spora guilliermondii p. 156
Kl. apis p. 877

5 Ma +
H'spora osmophila p. 158
Kl. corticis p. 878
H'spora vineae p. 162
Kl. africana p. 874
Ma – 6

6 FSu + 7
FSu –
H'spora uvarum p. 159
Kl. apiculata p. 875
H'spora valbyensis p. 161
Kl. japonica p. 879

7 Ra +
S'codes ludwigii p. 397
Ra –
H'spora occidentalis p. 157
Kl. javanica p. 880

8 FSu + 9
FSu –
N. elongata p. 281

9 FMa +
N. fulvescens p. 283
FMa –
W. fluorescens p. 441

B. *Key to species with pink, red or orange cultures due to carotenoid pigments and not forming ballistospores*

Couplet	Test	Species	Page / go to
1	FG +	*Phaffia rhodozyma*	p. 890
	FG –		2
2	Nitrate +		3
	Nitrate –		15
3	I +		4
	I –		7
4	Er +	*Cr. macerans*	p. 862
	Er –		5
5	Man +		6
	Man –	*Rhodosp. infirmominiatum*	p. 520
6	Me +	*Rhodosp. bisporidii*	p. 511
	Me –	*Rhodosp. capitatum*	p. 514
7(3)	Er +	*Rh. acheniorum*	p. 895
	Er –		8
8	Me +	*Rh. lactosa*	p. 900
	Me –		9
9	Mz +		10
	Mz –		12
10	Vit +	*Rhodosp. diobovatum*	p. 519
		Rhodosp. toruloides	p. 528
		Rh. glutinis	p. 898
	Vit –		11
11	Ra +	*Rhodosp. sphaerocarpum*	p. 525
	Ra –	*Rh. aurantiaca*	p. 897
12(9)	Vit +		13
	Vit –		14
13	Ra +	*Rhodosp. paludigenum*	p. 524
		Rh. graminis	p. 899
	Ra –	*Rh. araucariae*	p. 896
14	Ra +	*Rhodosp. malvinellum*	p. 523
	Ra –	*St. nectairii*	p. 924
15(2)	I +	*Cr. hungaricus*	p. 858
	I –		16
16	Ra +	*Rh. rubra*	p. 903
	Ra –		17
17	Cit +	*Rhodosp. dacryoidum*	p. 516
	Cit –	*Rh. minuta*	p. 901

C. *Key to nitrate positive, inositol negative species*

1 FSu + 2
FSu − 17

2 Ma + 3
Ma − 16

3 Er + 4
Er − 9

4 Vit + 5
Vit −
H. subpelliculosa p. 208

5 St + 6
St − 8

6 Rh + 7
Rh −
H. anomala p. 174

7 Me +
H. sydowiorum p. 210
Me −
H. ciferrii p. 185

8 Me +
H. sydowiorum p. 210
Me −
H. lynferdii p. 195

9(3) Vit + 10
Vit − 12

10 Ce + 11
Ce −
Cit. matritensis p. 118

11 Cit +
H. jadinii p. 194
Cit −
H. beijerinckii p. 177

12(9) Rh + 13
Rh − 15

13 37°C +
H. petersonii p. 202
37°C − 14

14 Me +
Leu. gelidum p. 501
Me −
Leu. stokesii p. 507

15 Ce +
H. fabianii p. 189
Ce −
Cit. matritensis p. 118

16(2) 25°C +
H. saturnus p. 206
25°C −
Leu. frigidum p. 499

17(1) Su + 18
Su − 33

18 FG + 19
FG − 28

19 Er + 20
Er − 22

20 LAr +
H. holstii p. 192
LAr − 21

21 True myc and psm +
A. platypodis p. 111
True myc and psm −
H. polymorpha p. 204

22 St +
H. holstii p. 192
St − 23

23 Psm + 24
Psm −
H. californica p. 180

24 Rib + 25
Rib − 26

25 LAr +
H. silvicola p. 207
LAr −
H. beckii p. 175

26 FG + strong 27
FG + weak or −
H. beckii p. 175

27 37°C +
H. bimundalis p. 178
37°C −
H. americana p. 171

28(18) Pell +
H. alni p. 170
Pell − 29

29 Vit + 30
Vit − 31

30 Ce +
Leu. scottii p. 504
Ce −
Leu. antarcticum p. 498

31 Ce + 32
Ce −
Wick. domercqii p. 444

32 37°C + strong
H. canadensis p. 181
37°C + weak or −
H. beckii p. 175

33(17) Er + 34
Er − 39

34 Psm +
H. muscicola p. 199
Psm − 35

35 St +
H. capsulata p. 183
St − 36

36 37°C + 37
37°C −
H. wickerhamii p. 211

37 Ce + 38
Ce −
H. philodendra p. 203

38 FG +
H. glucozyma p. 190
FG −
H. henricii p. 191

39(33) FG + 40
FG − 43

40 Tr +
H. minuta p. 197
Tr − 41

41 Rib +
Pa. tannophilus p. 290
Rib − 42

42 Vit +
H. mrakii p. 198
Vit −
H. dimennae p. 186

43(39) True myc and psm +
Leu. antarcticum p. 498
True myc and psm − 44

44 Ri + 45
Ri −
H. dryadoides p. 187

45 37°C +
H. nonfermentans p. 201
37°C −
Wick. domercqii p. 444

D. Key to nitrate negative, inositol negative, erythritol positive, fermenting species

1	Su +			2
	Su –			21
2	True myc +			3
	True myc –			8
3	Ra +			4
	Ra –			6
4	St +			5
	St –			
		P. scolyti	p. 365	
5	Vit +			
		P. burtonii	p. 314	
	Vit –			
		S. fibuligera	p. 404	
6	LAr +			
		A. cicatricosa	p. 107	
		A. monospora	p. 108	
	LAr –			7
7	37°C +			
		S. fibuligera	p. 404	
	37°C –			
		A. philentoma	p. 110	
8(2)	Ra +			9
	Ra –			18
9	St +			10
	St –			17
10	Rh +			11
	Rh –			15
11	La +			12
	La –			13
12	FG + strong			
		Deb. pseudo-polymorphus	p. 141	
	FG + weak			
		Deb. hansenii	p. 135	
13	FG + strong			
		Wi. robertsii	p. 446	
	FG + weak			14
14	Vit +			
		Deb. vanriji var. *vanriji*	p. 143	
	Vit –			
		Deb. hansenii	p. 135	
15(10)	FG + strong			
		Deb. poly-morphus	p. 139	
	FG + weak			16
16	Vit +			
		Deb. vanriji var. *vanriji*	p. 143	
	Vit –			
		Deb. hansenii	p. 135	
		Deb. marama	p. 137	
17(9)	Vit +			
		P. heimii	p. 333	
	Vit –			
		P. scolyti	p. 365	
18(8)	St +			19
	St –			20
19	La +			
		P. stipitis	p. 369	
	La –			
		P. nakazawae	p. 346	
20	Rh +			
		P. naganishii	p. 344	
	Rh –			
		P. philogaea	p. 354	
21(1)	True myc +			22
	True myc –			23

22 37°C + strong
S. malanga p. 409
37°C + weak or –
S. capsularis p. 401

23 Ma +
P. acaciae p. 304
Ma – 24

24 Ga + 25
Ga – 26

25 Pell +
P. farinosa p. 325
Pell –
P. methanolica p. 341

26 Ce +
P. pinus p. 356
Ce –
P. trehalophila p. 372

E. Key to nitrate negative, inositol negative, erythritol negative, fermenting species

1 Ce + 2
Ce – 55

2 Su + 3
Su – 50

3 La + 4
La – 15

4 St + 5
St – 8

5 Ga + 6
Ga –
Schw. occidentalis var. *persoonii* p. 425

6 FG + strong 7
FG + weak
Deb. hansenii p. 135

7 Pell +
Deb. castellii p. 132
Pell –
Schw. occidentalis var. *occidentalis* p. 424

8(4) Ma + 9
Ma – 11

9 37°C +
K. marxianus var. *lactis* p. 241
37°C – 10

10 FLa +
K. marxianus var. *lactis* p. 241
FLa –
K. aestuarii p. 228

11 Ra + 12
Ra – 14

12 37°C +
K. marxianus var. *bulgaricus* p. 236
K. marxianus var. *lactis* p. 241
K. marxianus var. *marxianus* p. 234
37°C – 13

13 FG + rapid 10
FG + slow
Deb. tamarii p. 142

14 FLa +
K. marxianus var. *lactis* p. 241
FLa –
K. wickerhamii p. 250

15(3) 37°C + 16
37°C − 38

16 Ra + 17
Ra − 29

17 Ga + 18
Ga − 26

18 St + 19
St − 20

19 FG + strong
Schw. occidentalis var. *occidentalis* p. 424
FG + weak
Deb. hansenii p. 135

20 Ri + 21
Ri − 23

21 Pell +
P. ohmeri p. 348
Pell − 22

22 Me +
P. guilliermondii p. 329
Me −
P. strasburgensis p. 370

23 Rh +
P. strasburgensis p. 370
Rh − 24

24 Cit + 25
Cit −
K. marxianus var. *drosophilarum* p. 239
K. marxianus var. *vanudenii* p. 242
Sacch. kluyveri p. 388
Zygosacch. fermentati p. 457

25 X + slow
K. marxianus var. *drosophilarum* p. 239
X −
P. ohmeri p. 348

26(17) True myc +
S. fibuligera p. 404
True myc − 27

27 St +
Schw. occidentalis var. *persoonii* p. 425
St − 28

28 Rh +
P. rabaulensis p. 359
Rh −
P. onychis p. 350

29(16) X + 30
X − 37

30 Rib + 31
Rib − 34

31 Pell +
P. wickerhamii p. 374
Pell − 32

32 Rh +
Cl. lusitaniae p. 121
Rh − 33

33 FG + rapid
M. pulcherrima p. 273
FG + slow
P. etchellsii p. 323

34 LAr + 35
LAr −
P. rhodanensis p. 360

35 Pell +
P. bovis p. 312
Pell − 36

36 St +
P. amylophila p. 308
St −
P. mississippiensis p. 342

37(29) True myc +
S. fibuligera p. 404
True myc −
P. spartinae p. 367

38(15) Vit +
Deb. vanriji var. *yarrowii* p. 144
Vit − 39

39 Ra + 40
Ra − 42

40 St + 19
St − 41

41 Pell +
P. veronae p. 373
Pell −
K. marxianus var. *dobzhanskii* p. 238

42 LAr + 43
LAr − 45

43 FSu + 44
FSu −
P. bovis p. 312

44 Rib +
P. angophorae p. 309
Rib −
P. veronae p. 373

45 Rib + 46
Rib − 48

46 Cit + 47
Cit −
M. lunata p. 272
M. pulcherrima p. 273
M. reukaufii p. 275
M. zobellii p. 276

47 True myc +
Ac. aculeatum p. 558
True myc −
M. pulcherrima p. 273
P. segobiensis p. 366

48 Cit +
P. toletana p. 371
Cit − 49

49 Pell +
P. mucosa p. 343
Pell −
M. bicuspidata p. 269

50(2) True myc +
S. malanga p. 409
True myc − 51

51 Man + 52
Man −
P. norvegensis p. 347

52 Rib +
P. lindneri p. 336
Rib − 53

53 Cit + 54
Cit −
P. sargentensis p. 364

54 Pell +
P. pijperi p. 355
Pell −
P. quercuum p. 358

55(1) True myc + 56
True myc − 59

56 Pell + 57
Pell − 58

57 X +
S. crataegensis p. 403
X –
S. vini p. 411

58 Rib +
G. selenospora p. 151
Rib –
Nem. coryli p. 286

59 Pell + 60
Pell – 74

60 Ra +
K. marxianus var. *drosophilarum* p. 239
Ra – 61

61 Man + 62
Man – 65

62 FMa +
Zygosacch. rouxii p. 463
FMa – 63

63 Tr +
P. saitoi p. 361
Tr – 64

64 Rib +
P. delftensis p. 321
Rib –
P. besseyi p. 311

65(61) Vit + 66
Vit – 70

66 FG + strong 67
FG + weak
P. membranaefaciens p. 339

67 FMa +
Zygosacch. rouxii p. 463
FMa – 68

68 37°C + 69
37°C –
I. scutulata p. 219

69 40°C +
I. orientalis p. 218
40°C –
I. occidentalis p. 216

70 FMa +
Zygosacch. rouxii p. 463
FMa – 71

71 FG + weak 72
FG + slow
I. terricola p. 221
FG + rapid 73

72 From cacti
P. cactophila p. 315
Not from cacti
P. membranaefaciens p. 339

73 X +
P. fermentans p. 326
X –
P. kluyveri p. 335

74(59) Su + 75
Su – 89

75 Ga + 76
Ga – 88

76 St + 77
St – 78

77 X +
Schw. occidentalis var. *occidentalis* p. 424
X –
Sacch. cerevisiae p. 383

78 Culture pink, the pigment diffuses into the agar
Zygosacch. mrakii p. 461
Culture not pink 79

79 37°C + 80
37°C – 84

80 Cit + 81
Cit – 82

81 FG + rapid
K. marxianus var. *drosophilarum* p. 239
Zygosacch. cidri p. 455
FG + slow
Lod. elongisporus p. 263

82 X + 83
X –
K. thermotolerans p. 248
Sacch. cerevisiae p. 383
Sacch. kluyveri p. 388
T'spora pretoriensis p. 439
Zygosacch. bailii p. 451
Zygosacch. rouxii p. 463

83 Suc +
K. marxianus var. *drosophilarum* p. 239
K. marxianus var. *wikenii* p. 244
Sacch. kluyveri p. 388
Zygosacch. cidri p. 455
Suc –
T'spora pretoriensis p. 439

84(79) Ma +
K. thermotolerans p. 248
Sacch. cerevisiae p. 383
T'spora delbrueckii p. 436
Zygosacch. cidri p. 455
Zygosacch. florentinus p. 458
Zygosacch. rouxii p. 463
Ma – 85

85 FSu + 86
FSu – 87

86 FGa +
K. lodderi p. 232
K. polysporus p. 246
Sacch. cerevisiae p. 383
Sacch. exiguus p. 387
T'spora delbrueckii p. 436
Zygosacch. cidri p. 455
Zygosacch. florentinus p. 458
Zygosacch. microellipsoides p. 460
FGa –
Sacch. cerevisiae p. 383
T'spora delbrueckii p. 436
Zygosacch. bailii p. 451
Zygosacch. florentinus p. 458
Zygosacch. rouxii p. 463

87 FGa +
K. africanus p. 229
K. phaffii p. 245
Sacch. cerevisiae p. 383
T'spora delbrueckii p. 436
FGa –
Sacch. cerevisiae p. 383
T'spora delbrueckii p. 436
Zygosacch. bailii p. 451
Zygosacch. rouxii p. 463

88(75) St +
Sacch. cerevisiae p. 383
Schw. occidentalis var. *persoonii* p. 425
St –
K. waltii p. 249
Sacch. cerevisiae p. 383

T'spora delbrueckii p. 436
T'spora globosa p. 437
Zygosacch. bailii p. 451
Zygosacch. florentinus p. 458
Zygosacch. rouxii p. 463

89(74) FGa + 90
FGa − 91

90 Ri +
K. blattae p. 230
Sacch. dairensis p. 386
Sacch. servazzii p. 389
Ri −
K. africanus p. 229
K. phaffii p. 245
P'spora transvaalensis p. 293
Sacch. cerevisiae p. 383
Sacch. dairensis p. 386
Sacch. servazzii p. 389
Sacch. unisporus p. 392
T'spora delbrueckii p. 436

91 Rh +
P. pastoris p. 353
Rh − 92

92 37°C +
K. delphensis p. 231
Sacch. cerevisiae p. 383
Sacch. telluris p. 391
Zygosacch. bailii p. 451
Zygosacch. rouxii p. 463
37°C −
P. dispora p. 322
Sacch. cerevisiae p. 383
T'spora delbrueckii p. 436
Zygosacch. bailii p. 451
Zygosacch. bisporus p. 453
Zygosacch. rouxii p. 463

F. *Key to nitrate negative, inositol negative, erythritol positive, non-fermenting species*

1 Su + 2
Su − 7

2 Psm + 3
Psm − 4

3 37°C +
Deb. vanriji var. *vanriji* p. 143
37°C −
P. ambrosiae p. 305

4 Ra + 5
Ra −
Deb. melissophilus p. 138

5 Pell +
Deb. hansenii p. 135
Deb. marama p. 137
Pell − 6

6 Culture mucoid
L. lipofer p. 257
L. starkeyi p. 259
L. tetrasporus p. 261
Culture not mucoid
Deb. hansenii p. 135

7(1) True myc +
S. lipolytica p. 406
True myc − 8

8 Ce + 9
Ce −
P. haplophila p. 330

9 Pell + 10
Pell − 11

10 Ra +
P. castillae p. 319
Ra −
P. media p. 337

11 Ma +
Deb. coudertii p. 133
Ma −
P. pinus p. 356

G. Key to nitrate negative, inositol negative, erythritol negative, non-fermenting species

1 Su + 2
Su − 6

2 Rib + 3
Rib − 5

3 Vit +
Deb. vanriji var. *yarrowii* p. 144
Vit − 4

4 Psm well developed
P. carsonii p. 317
Psm − or primitive
Deb. hansenii p. 135

5 Ra +
L. kononenkoae p. 256
Ra −
M. krissii p. 271

6(1) Vit + 7
Vit − 8

7 Ga +
P. humboldtii p. 334
Ga −
P. membranaefaciens p. 339

8 True myc + 9
True myc − 10

9 X +
S. crataegensis p. 403
X −
Ar. javanensis p. 115

10 From cacti 11
Not from cacti 13

11 X + 12
X −
P. amethionina p. 306
P. opuntiae p. 351

12 Glucosamine +
P. cactophila p. 315
Glucosamine −
P. heedii p. 331

13(10) 37°C + 14
37°C − 16

14 Ce + 15
Ce −
P. membranaefaciens p. 339

15 Rib +
P. lindneri p. 336
Rib −
P. salictaria p. 363

16 Ga +
P. chambardii p. 320
Ga − 17

17 Ma +
L. anomalus p. 254
Ma − 18

18 Man +
P. delftensis p. 321
P. fluxuum p. 328
Man −
P. membranae-faciens p. 339

3. Key to species forming ballistospores

by J.W. Fell and L. Rodrigues de Miranda

See Table 24, p. 534.

1 Nitrate + 2
Nitrate − 9

2 I + 3
I − 4

3 La +
B. piricola p. 581
La −
Sp. puniceus p. 917

4 La +
B. tsugae p. 583
La − 5

5 Homothallic formation of teliospores 6
No homothallic formation of teliospores 7

6 Ra +
Sporid. ruinenii p. 537
Ra −
Sporid. johnsonii p. 533

7 Heterothallic formation of teliospores
Sporid. salmonicolor p. 538
No heterothallic formation of teliospores 8

8 True myc +
Sp. holsaticus p. 915
True myc −
Sp. roseus p. 918

9 La + 10
La − 14

10 I + 11
I −
B. singularis p. 583

11 Ra + 12
Ra − 13

12 30°C +
Fib. inconspicuum p. 546
30°C −
B. alba p. 578

13 Ma +
B. dendrophila p. 579
Ma −
Sir. magnum p. 547

14 Ma + 15
B. grandispora p. 581
Ma −
Sp. gracilis p. 913

15 Ce +
Sporid. pararoseus p. 535
Ce –
Sp. albo-rubescens p. 912

4. Key to species that produce true mycelium

by N.J.W. Kreger-van Rij

1 Culture pink, red or orange* 2
Culture not pink, red or orange 8

2 Ballistospores + 3
Ballistospores – 6

3 Teliospores +
see *Sporidiobolus* key p. 533
Teliospores – 4

4 Nitrate + 5
Nitrate –
Sp. shibatanus p. 535

5 I +
Sp. puniceus p. 917
I –
Sp. holsaticus p. 915
Sp. salmonicolor p. 538

6(2) Teliospores +
see *Rhodosporidium* key p. 510
Teliospores – 7

7 Nitrate +
St. nectairii p. 924
Nitrate –
Rh. rubra p. 903

8(1) 25°C + 9
25°C –
see *Leucosporidium* key p. 497
C. curiosa p. 655

9 Strong formation of acid from glucose, slow growth
Br. anomalus p. 566
Not this combination of characters 10

10 Nitrate + 11
Nitrate – 33

11 I + 12
I – 17

12 Arthrospores + 13
Arthrospores – 14

13 37°C +
Tr. terrestre p. 959
37°C –
Tr. pullulans p. 955

14 Clamp connections +
F. floriforme p. 486
Clamp connections – 15

* Cultures of *Aureobasidium pullulans* (de Bary) Arnaud, a species which is not presented in this monograph, may be pink and yeast-like and surrounded by true hyphae on YM agar. The colonies may sometimes first turn green and then black. For a description see Ellis (1971).

15 Man + 16
Man –
C. tsukubaensis p. 821

16 Urease +
C. antarctica p. 615
Urease –
C. edax p. 667

17(11) Ce + 18
Ce – 32

18 FG + 19
FG – 29

19 Su + 20
Su –
H. muscicola p. 199

20 Er + 21
Er – 24

21 Vit +
H. ciferrii p. 185
Vit – 22

22 FSu +
H. subpelliculosa p. 208
FSu – 23

23 LAr +
H. holstii p. 192
LAr –
A. platypodis p. 111

24(20) FSu +
H. petersonii p. 202
FSu – 25

25 Ga + 26
Ga – 27

26 St +
H. holstii p. 192
St –
H. silvicola p. 207

27 FG + strong 28
FG + weak
H. beckii p. 175

28 37°C +
H. bimundalis p. 178
37°C –
H. americana p. 171

29(18) Vit + 30
Vit – 31

30 Er +
Tr. pullulans p. 955
Er –
Leu. scottii p. 504

31 37°C + strong
H. canadensis p. 181
37°C + weak or –
H. beckii p. 175

32(17) FG +
H. subpelliculosa p. 208
FG –
O. margaritiferum p. 886

33(10) Conidia produced as single terminal buds on sympodial conidiophores
Symp. parvus p. 930
No sympodial conidiophores 34

34 Cells on sterigmata 35
Cells not on sterigmata 36

35 Ma +
St. penicillatus p. 925
Ma –
St. tursiopsis p. 928

36 Clamp connections + 37
Clamp connections – 39

37 FG +
F. capsuligenum p. 485
FG – 38

38 37°C +
Fil. neoformans p. 474
37°C –
F. uniguttulatum p. 488

39(36) I + 40
I – 52

40 FG + 41
FG – 43

41 X + 42
X –
S. fibuligera p. 404

42 Cit +
C. steatolytica p. 807
Cit –
C. hellenica p. 698

43(40) 37°C + 44
37°C – 47

44 La + 45
La – 46

45 Cells in sarcina-like structures
Sar.(Tr.) inkin p. 906
No sarcina-like structures
Tr. cutaneum p. 942

46 Suc +
Steph. ciferrii p. 432
Suc –
C. chiropterorum p. 651

47(43) La + 48
La – 50

48 Arthrospores +
Tr. cutaneum p. 942
Arthrospores – 49

49 Er +
C. humicola p. 702
Er –
C. podzolica p. 843

50(47) Su + 51
Su –
S. synnaedendra p. 410

51 Arthrospores +
Tr. brassicae p. 937
Arthrospores –
C. humicola p. 702

52(39) Er + 53
Er – 69

53 FG + 54
FG – 63

54 Su + 55
Su – 62

55 X + 56
X –
S. fibuligera p. 404

56 Vit + 57
Vit – 59

57 Me +
Tr. melibiosaceum p. 953
Me – 58

58 La +
Tr. fennicum p. 948
La –
P. burtonii p. 314

59(56) LAr + 60
LAr –
A. philentoma p. 110

60 Ra + 61

Ra –
A. monospora p. 108
A. cicatricosa p. 107

61 Rh +
P. scolyti p. 365
Rh –
C. entomophila p. 670

62(54) Cit +
S. capsularis p. 401
Cit –
S. malanga p. 409

63(53) 37°C +
Tr. figueirae p. 952
37°C – 64

64 Su + 65
Su – 68

65 X + 66
X –
C. mesenterica p. 744

66 Urease + 67
Urease –
P. ambrosiae p. 305

67 Arthrospores +
Tr. aquatile p. 936
Arthrospores –
C. humicola p. 702

68(64) Ce +
C. deformans p. 659
Ce –
S. lipolytica p. 406

69(52) Su + 70
Su – 79

70 X + 71
X – 75

71 St + 72
St – 74

72 FSu +
C. tropicalis p. 819
FSu – 73

73 Ce +
P. amylophila p. 308
Ce –
C. albicans p. 611

74(71) LAr +
P. missis-sippiensis p. 342
LAr –
P. rhodanensis p. 360

75(70) 37°C + 76
37°C – 78

76 Bud formation + 77
Bud formation –
Schiz. japonicus p. 415

77 Ce +
S. fibuligera p. 404
Ce –
Nem. coryli p. 286

78(75) Ce +
Ac. aculeatum p. 558
Ce –
S. vini p. 411

79(69) Ce + 80
Ce – 83

80 Ma +
S. malanga p. 409
Ma – 81

81 Tr +
C. beechii p. 630
Tr – 82

82 Rib +
Tr. fermentans p. 949
Rib –
Tr. eriense p. 946

83(79)	X +		84
	X –		86
84	Ga +		85
	Ga –		
		S. crataegensis	p. 403
85	Rib +		
		G. selenospora	p. 151
	Rib –		
		Geotrichum candidum *	
86(83)	37°C +		
		Tr. capitatum	p. 938
	37°C –		87
87	Ga +		
		Tr. sericeum	p. 957
	Ga –		88
88	Arthrospores +		
		Ar. javanensis	p. 115
	Arthrospores –		
		S. vini	p. 411

5. Key to *Candida* species that do not produce pseudo- or true mycelium

by N.J.W. Kreger-van Rij

1	Nitrate +		2
	Nitrate –		31
2	FG +		3
	FG –		23
3	Ce +		4
	Ce –		13
4	Ma +		5
	Ma –		10
5	Er +		6
	Er –		7
6	Su +		
		C. pelliculosa	p. 768
	Su –		
		C. molischiana	p. 750
7	37°C +		
		C. fabianii	p. 675
	37°C –		8
8	Tr +		9
	Tr –		
		C. ernobii	p. 673
9	X +		
		C. mannitofaciens	p. 734
	X –		
		C. versatilis	p. 831
10(4)	37°C +		
		C. methanosorbosa	p. 746
	37°C –		11
11	Tr +		
		C. halophila	p. 697
	Tr –		12

* *Geotrichum candidum* Link is not described in this monograph. Descriptions are given by Carmichael (1957) and by van Uden and do Carmo-Sousa (1959).

12 Ga +
C. wickerhamii p. 837
Ga −
C. norvegica p. 760

13(3) Ma + 14
Ma − 18

14 X +
C. multis-gemmis p. 752
X − 15

15 Tr + 16
Tr − 17

16 Ga +
C. versatilis p. 831
Ga −
C. globosa p. 689

17 FMa +
C. etchellsii p. 674
FMa −
C. nodaensis p. 759

18(13) Tr + 19
Tr − 21

19 FSu +
C. cariosilignicola p. 646
FSu − 20

20 Ga +
C. nitratophila p. 758
Ga −
C. pignaliae p. 770

21(18) FSu + 22
FSu −
C. halonitratophila p. 695

22 Man +
C. magnoliae p. 731
Man −
C. lactiscondensi p. 728

23(2) Ma + 24
Ma − 27

24 Er + 25
Er − 26

25 LAr +
C. bacarum p. 629
LAr −
C. graminis p. 690

26 St +
C. ingeniosa p. 709
St −
C. fragariorum p. 681

27(23) Tr + 28
Tr − 30

28 Ce +
C. pustula p. 778
Ce − 29

29 Cit +
C. vanderwaltii p. 826
Cit −
C. sonckii p. 802

30(27) Ce +
C. norvegica p. 760
Ce −
C. domercqii p. 667

31(1) FG + 32
FG − 69

32 Ce + 33
Ce − 45

33 I +
C. japonica p. 719
I − 34

34 Su + 35
Su – 41

35 Tr + 36
Tr – 40

36 Man + 37
Man –
C. xestobii p. 838

37 LAr + 38
LAr – 39

38 Ra +
C. famata p. 675
Ra –
C. peltata p. 768

39 La +
C. sphaerica p. 807
La –
C. pulcherrima p. 778

40(35) Ra +
C. gropengiesseri p. 691
Ra –
C. geochares p. 686

41(34) Tr + 42
Tr – 44

42 LAr +
C. torresii p. 817
LAr – 43

43 Rib +
C. methanol-ovescens p. 745
Rib –
C. anatomiae p. 614

44(41) X +
C. sonorensis p. 803
X –
C. dendrica p. 660

45(32) Er + 46
Er – 47

46 LAr +
C. succiphila p. 810
LAr –
C. cantarellii p. 645

47 Ma + 48
Ma – 52

48 Cit + 49
Cit – 51

49 Man + 50
Man –
C. xestobii p. 838

50 Rh +
C. haemulonii p. 694
Rh –
C. musae p. 753

51(48) Mz +
C. dattila p. 659
Mz –
C. colliculosa p. 654

52(47) Tr + 53
Tr – 61

53 Su + 54
Su – 57

54 Ga + 55
Ga – 56

55 Man +
C. colliculosa p. 654
Man –
C. holmii p. 699
C. milleri p. 747

56 Cit +
C. bombi p. 638
Cit –
C. colliculosa p. 654

57(53) 37°C + 58
37°C – 59

58 FTr +
C. glabrata p. 687
FTr –
C. castellii p. 647

59 Cit +
C. fructus p. 685
Cit – 60

60 Man +
C. colliculosa p. 654
Man –
C. humilis p. 703

61(52) Man + 62
Man – 67

62 Ra + 63
Ra – 65

63 Cit +
C. bombi p. 638
Cit – 64

64 Suc +
C. bombicola p. 639
Suc –
C. apicola p. 617

65(62) 37°C + 66
37°C –
C. karawaiewii p. 720

66 Cit +
C. geochares p. 686
Cit –
C. bombicola p. 639

67(61) Su +
C. stellata p. 809
Su – 68

68 37°C +
C. pintolopesii var. *pintolopesii* p. 771
37°C –
C. tannotolerans p. 812

69(31) Ce + 70
Ce – 75

70 X + 71
X –
C. insectalens p. 713

71 Ra +
C. famata p. 675
Ra – 72

72 Vit +
Rh. fujisanensis p. 905
Vit – 73

73 LAr + 74
LAr –
C. tepae p. 815

74 Er +
C. nemodendra p. 757
Er –
C. mucilagina p. 751

75(69) 37°C + 76
37°C – 79

76 Su +
C. azyma p. 628
Su – 77

77 Tr +
C. silvatica p. 798
Tr – 78

78 Man +
C. sorbophila p. 804
Man –
C. inconspicua p. 708

79(75) Er + 80
Er − 81

80 25°C +
C. pinus p. 773
25°C −
C. psychrophila p. 777

81 Su + 82
Su − 83

82 Urease +
C. auriculariae p. 624
Urease −
C. apis p. 618

83 Ga + 84
Ga − 85

84 Suc +
C. austromarina p. 627
Suc −
C. vinaria p. 832

85 Tr +
C. philyla p. 769
Tr −
C. maris p. 736

6. Key to yeasts pathogenic for man and animal

by D.G. Ahearn and R.L. Schlitzer

A. Introduction

Because of the cosmopolitan distribution of yeasts in nature, the fortuitous occurrence of any species in clinical specimens is not surprising. Fortunately, of the more than 500 species described in this monograph, less than 30 are associated commonly with man, and fewer than 20 have been established as the etiological agents of disease (Ahearn & Schlitzer, 1981; Kahanpää, 1971). The associations of most species with man are transient and saprophytic in nature, but for some species and hosts, the relationship is a commensal colonization or parasitism. The balance between colonization and parasitism is dependent mostly on physiological stress of the host.

Malnutrition, pregnancy and diseases associated with impaired host defenses such as diabetes and malignant neoplasms have been historically associated with predisposition to yeast infections (Hart et al., 1969). Indiscriminant use of broad spectrum antibiotics, immunosuppressive and antineoplastic therapy, and postoperative procedures, especially hyperalimentation (Montgomerie & Edwards, 1978), have resulted in a marked increase in the incidence as well as in the severity of yeast infections (Edwards et al., 1978; Wingard et al., 1979; Kiehn et al., 1980). These infections, currently and in the past, are mostly caused by *Candida albicans* and *Cryptococcus neoformans;* however, recent iatrogenic conditions have increased the number of adventitious fungi (i.e., fungi that are unusual or accidental agents of infection of man) (Ahearn, 1978; Brandstetter & Brause, 1980; Quirke et al., 1980). The species of yeasts implicated as pathogens of man and their common habitats are listed in Table 41. This list is restricted in comparison to general lists of medically important (or man- and animal-associated) yeasts because we have included only those species which best fit the criteria of McGinnis et al. (1980): the fungus must be isolated; the fungus must be properly identified; the clinical disease must be compatible with a mycosis; and the fungus must be seen in the clinical and pathological specimens. Of the yeasts which best fit these criteria, *C. albicans, Cr. neoformans* and *C. tropicalis* are the species with the more virulent representatives (Ahearn, 1978). Progressive invasion by other species is rare or questionable, and if such invasion does occur, it is limited to the most seriously debilitated patients. For review and aspects of infections by various yeasts, see Ahearn and Holzschu (1977), Odds (1979), Holzschu et al. (1979b) and Pien et al. (1980).

Most adventitious yeast pathogens are ubiquitous in nature, but a few species, mainly *C. albicans* and *C. glabrata,* can be considered obligatory commensals of man and warm-blooded animals (van Uden, 1962b). In various surveys, from 20 to 80% of normal humans harbor *C. albicans* and a lesser percentage *C. glabrata* within their oral-gastrointestinal tracts (Odds, 1979; Kahanpää, 1971). Both species are found host-free in nature, but these occurrences (particularly that of *C. albicans*) usually have been related to recent contamination by human or animal excrement (van Uden, 1962b; Ahearn, 1973). *C. glabrata* appears to have a greater index of survival in host-free habitats than *C. albicans* (van Uden, 1962b; Ahearn, 1973; Buck, 1978). Unlike the other species, the isolation of *Cr. neoformans* from clinical specimens, even in small numbers, usually indicates disease.

The worldwide prevalence of yeast infections is unknown. Fraser et al. (1979) compared the incidence and morbidity of systemic mycoses in the United States among patients requiring hospitalization in 1970 and 1976 and reported a 72% increase in mortality due to candidiasis (244 deaths) and cryptococcosis (123 deaths). Ahearn (1978) estimated that for the United States in 1976, there were over two million cases of vulvovaginitis and several thousand cases each of systemic candidiasis and cryptococcosis. Over 90% of the candida infections were caused by *C. albicans. Candida tropicalis* probably accounted for about half of the remainder, and all other species combined accounted for less than 5% of the total infections. *Malassezia furfur* (syn. *Pityrosporum ovale*) is frequently encountered as the etiological agent of tinea versicolor, but this syndrome typically is so mild and so seldom treated that it is rarely considered an infection and is not considered in the above epidemiological estimates.

B. Identification of species

A variety of commercial products employing selected miniaturized or dehydrated assimilatory media or serological systems has been developed that provides rapid presumptive identifications of medically important yeasts (Bowman & Ahearn, 1976; Shinoda et al., 1981). A commercial product is available also for the rapid identification of *Cr. neoformans* by detection of phenoloxidase activity (melanin synthesis) with dihydroxyphenylalanine as the substrate (Cooper, 1980). These tests all give quite reasonable accuracy, but some strains deviate from the norm in key characters. For example, *C. albicans* may not readily demonstrate germ tubes and more often chlamydospores; even pseudohyphae may be lacking in freshly isolated strains (Ogletree et al., 1978; Crow et al., 1977; Ahearn & Schlitzer, 1981). Accurate identifications, therefore, may not be possible for a few strains solely with phenotypic properties. An extreme example is seen with the sucrose-negative variant of *C. tropicalis* and *C. viswanathii.* These yeasts have nearly identical assimilation patterns, but about an 8.0% difference in the guanine plus cytosine content of their DNA. A few of the adventitious yeasts have known sexual stages (see Table 8, p. 37), but these stages are seldom observed in routine clinical laboratories. Even in instances when the species is homothallic, which is most often the case with *Sacch. cerevisiae,* ascospores are frequently lacking in strains from clinical specimens. The clinical isolates of *Sacch. cerevisiae* and *K. marxianus* are probably most often of commercial origin and probably aneuploid, thus non-sporulating in culture. In spite of the above obstacles, most yeasts are remarkably stable in their phenotypic properties. A selective key for the identification of the adventitious yeasts and some of the common animal-associated yeasts that are phenotypically similar to the adventitious yeasts is presented below. Steps in the key include multiple characters, so that the key permits a presumptive species identification, i.e., a distinction from most species presented as animal-associated in this monograph. The key does not necessarily preclude the possibility of another species, however, and definitive identification may require additional tests. Key characters of adventitious and selected man- and animal-associated yeasts are given in Table 42.

C. *Key to selected man- and animal-associated yeasts*

1a Growth on Sabouraud's-, malt-, or yeast nitrogen base + glucose agar 2
b No growth on the above media without addition of olive oil .... *Malassezia furfur*
c No growth on yeast nitrogen base + glucose agar, growth on Sabouraud's- or malt agar .... *Malassezia pachydermatis*

2a Germ tubes produced .... *Candida albicans*
b No germ tubes produced 3

3a Hyphae or pseudohyphae absent or scarce 4
b Hyphae or pseudohyphae well developed 14

4a Inositol assimilated, starch synthesized 5
b Inositol not assimilated, starch not synthesized 9

5a Nitrate assimilated 6
b Nitrate not assimilated 7

6a Sucrose assimilated .... *Cryptococcus albidus*
b Sucrose not assimilated .... *Cryptococcus terreus*

7a Lactose assimilated .... *Cryptococcus laurentii*
b Lactose not assimilated 8

8a Dulcitol assimilated; melanin produced	*Cryptococcus neoformans*
b Dulcitol not assimilated; melanin not synthesized ...	*Cryptococcus uniguttulatus*
9a Cultures red or orange; nitrate not assimilated	*Rhodotorula rubra*
b Cultures not red or orange; nitrate assimilated or not assimilated	10
10a Nitrate and methanol assimilated	*Hansenula polymorpha*
b Nitrate and methanol not assimilated.............	11
11a Lactose assimilated	*Candida famata*
b Lactose not assimilated..........................	12
12a Xylose assimilated.............................	*Candida guilliermondii* (see also *C. famata*)
b Xylose not assimilated	13
13a Raffinose assimilated and fermented	see *Saccharomyces cerevisiae*
b Raffinose neither assimilated nor fermented	*Candida glabrata*
14a Arthroconidia produced	15
b Arthroconidia not produced......................	20
15a DBB reaction positive..........................	16
b DBB reaction negative	17
16a Nitrate assimilated............................	*Trichosporon pullulans*
b Nitrate not assimilated	*Trichosporon beigelii* (syn. *Tr. cutaneum*)
17a Xylose assimilated	18
b Xylose not assimilated	*Geotrichum capitatum* (syn. *Tr. capitatum*)
18a Cellobiose assimilated	*Geotrichum fermentans* (syn. *Tr. fermentans*)
b Cellobiose not assimilated	19
19a Arthroconidia in chains within primary filaments ...	*Geotrichum candidum**
b Arthroconidia in branched chains arising on branches from primary filaments..........................	*Geotrichum penicillatum**
20a Lactose fermented	*Candida kefyr* (syn. *C. pseudotropicalis*)
b Lactose not fermented...........................	21

* This species is not discussed in this monograph.

21a Raffinose and melibiose assimilated .... *Candida guilliermondii*
b Raffinose and melibiose not assimilated............ 22

22a Trehalose assimilated 23
b Trehalose not assimilated .... *Candida krusei*

23a Cellobiose assimilated 24
b Cellobiose not assimilated 26

24a Rhamnose assimilated............................. *Candida lusitaniae*
b Rhamnose not assimilated 25

25a TCC reaction positive............................ *Candida tropicalis*
b TCC reaction negative to weak.................... *Candida viswanathii*

26a Maltose fermented, TTC reaction positive or negative 27
b Maltose not fermented, TCC reaction positive (rose to red color) .... *Candida parapsilosis*

27a Sucrose fermented, TCC reaction positive (deep maroon color)................................... *Candida tropicalis*
b Sucrose not fermented, TCC reaction negative 28

28a Maltose assimilated (repeat germ tube test) .... *Candida albicans*
b Maltose not assimilated .... *Candida zeylanoides*

Table 41

Habitats of yeasts pathogenic for man

Species	Association with man	Host free
Candida albicans	mucous membranes (common)	soil, air, water, plants (rare)
Malassezia furfur (syn. *Pityrosporum ovale*)	skin (common)	none
Cryptococcus neoformans	none	pigeon droppings (common)
Candida tropicalis	sputa, urine, feces (common)	organically enriched soil and aquatic habitats (common)
Candida parapsilosis	skin, sputa (common)	soil, water, plants (common)
Candida glabrata (syn. *Torulopsis glabrata*)	urine, sputa, feces (common)	soil, water (rare)
Candida guilliermondii	skin, sputa (common)	soil, water (rare)
Rhodotorula spp. (mainly *Rh. rubra*)	skin (occasional)	soil, water, plants (common)
Geotrichum candidum	sputa (occasional)	sewage, fruits (common)
Candida krusei	sputa (occasional)	soil, water, plants (common)
Candida viswanathii	unknown	unknown
Trichosporon beigelii (syn. *Tr. cutaneum*)	skin and hair (occasional)	soil, water (common)
Saccharomyces cerevisiae	sputa, feces (occasional)	soil, fruits
Candida pseudotropicalis (*K. marxianus*)	sputa, feces (occasional)	dairy products
Candida lusitaniae	sputa, feces (occasional)	animal manure, water
Hansenula polymorpha	unknown	insects, soil

Table 42

Key characters of adventitious and man- and animal-associated yeasts

	TH	PH	Art	Fermentation					
				G	Su	Ma	La	Ra	Tr
C. albicans	−(+)	+	−	+	−	+	−	−	v
C. famata	−	−	−	v	v	−	−	v	+
C. glabrata	−	−	−	+	−	−	−	−	+
C. guilliermondii	−	+(−)	−	+	+	−	−	+	v
C. pseudotropicalis	−	+	−	+	+	−	+	+	−
C. krusei	−	+	−	+	−	−	−	−	−
C. lusitaniae	−	+	−	+	v	−	−	−	+
C. parapsilosis	−	+	−	+	−	−	−	−	v
C. tropicalis	−(+)	+	−	+	v	+	−	−	+
C. viswanathii	−(+)	+	−	+	−	+	−	−	v
Cr. albidus	−(+)	−	−	−	−	−	−	−	−
Cr. laurentii	−(+)	−	−	−	−	−	−	−	−
Cr. neoformans	−(+)	−	−	−	−	−	−	−	−
Cr. terreus	−	−	−	−	−	−	−	−	−
Cr. uniguttulatus	−(+)	−	−	−	−	−	−	−	−
G. candidum	+	−	+	v	−	−	−	−	−
G. capitatum (*Tr. capitatum*)	+	−	+	−	−	−	−	−	−
G. fermentans (*Tr. fermentans*)	+	−	+	+	−	−	−	−	−
G. penicillatum	+	−	+	v	−	−	−	−	−
H. polymorpha	−	−	−	+	−	−	−	−	−
Rh. rubra	−	−	−	−	−	−	−	−	−
Sacch. cerevisiae	−	v	−	+	+	+	−	+	v
Tr. beigelii (*Tr. cutaneum*)	+	+	+	−	−	−	−	−	−
Tr. pullulans	+	+	+	−	−	−	−	−	−

TH = true hyphae, PH = pseudohyphae, Art = arthrospores, G = glucose, Su = sucrose, Ma = maltose, La = lactose, Ra = raffinose, Tr = trehalose, Ce = cellobiose, X = D-xylose, Rh = L-rhamnose,

Assimilation											DBB	37°C	ur	Meth
Su	Ma	La	Ce	Tr	Ra	X	Rh	Dul	I	NO_3				
+(−)	+	−	−	+	−	+(−)	−	−	−	−	−	+	−	−
+	+	v	+	+	+	+	v	v	−	−	−	v	−	−
−	−	−	−	+	−	−	−	−	−	−	−	+	−	−
+	+	−	+	+	+	+	v	+	−	−	−	+	−	−
+	−	+	v	−	+	+	−	−	−	−	−	+	−	−
−	−	−	−	−	−	−	−	−	−	−	−	+	v	−
+	+	−	+	+	−	+	+	−	−	−	−	+	−	−
+	+	−	−	+	−	+	−	−	−	−	−	+	−	−
+(−)	+	−	+(−)	+	−	+	−	−	−	−	−	+	−	−
+	+	−	+	+	−	+	−	−	−	−	−	+	−	−
+	+	v	+	+	+	+	v	v	+	+	+	−(+)	+	−
+	+	+	+	+	+	+	+	+	+	−	+	v	+	−
+	+	−	v	+	v	+	+	+	+(−)	−	+	+	+	−
−	+	+	+	+	−	+	+	v	+	+	+	−	+	−
+	+	−	−	v	v	+	+	−	+	−	+	−	+	−
−	−	−	−	−	−	+	−	−	−	−	−	v	−	−
−	−	−	−	−	−	−	−	−	−	−	−	+	−	−
−	−	−	+	−	−	+	−	−	−	−	−	v	−	−
−	−	−	−	−	−	+	−	−	−	−	−	−	−	−
+	+	−	+	+	−	+	v	−	−	+	−	+	−	+
+	v	−	v	+	+	+	v	v	−	−	+	v	+	−
+	+	−	−	v	+	−	−	−	−	−	−	v	−	−
+	+	+	+	v	v	+	v	v	+	−	+	v	v	−
+	+	+	+	+	+	v	+	−	+	+	+	−	+	−

Dul = dulcitol (= galactitol), I = inositol, NO_3 = nitrate, DBB = color reaction with Diazonium Blue B, 37°C = growth at 37°C, ur = urease, Meth = assimilation of methanol.

Glossary of terms used in this monograph

Ainsworth & Bisby's Dictionary of the Fungi (Ainsworth, 1971) was used for the explanation of many terms in this glossary.

acerose – needle-shaped
acicular – see acerose
acropetal – produced successively in the direction of the apex; the apical part is youngest
aerobic – free oxygen is required for growth; an atmosphere with free oxygen
agglutinate – clump or bind together
allantoid – curved with rounded ends, sausage-shaped
allele – any of the alternative expressions of a gene
allotype – paratype of the opposite sex to the holotype
amphigenous hymenium – spore layer of fruit body formed on two sides
amphiploidy (allopolyploidy) – production of a fertile species by the fusion of the genomes of two separate but usually related species; the new species is reproductively isolated from the parent species
ampulliform – flask-like form; shaped like a Florence flask
anamorph (anamorphic state) – see imperfect state
anastomosis – fusion between hyphae
annellation – formation of ring-like protrusions near the apex of a cell resulting from successive formation of conidia
apex – end of a cell
apical – at the end
apiculate – cell with short projections at both poles, lemon-shaped
apiculum – short projection on a cell
arthroconidium – see arthrospore
arthrospore – an asexual spore resulting from the division of a hypha or a single cell by fission
ascophore – ascus-bearing hypha
ascus – the sac-like cell in which ascospores are produced
autogamy – fusion of nuclei within a cell in the absence of cell fusion
axenic culture – a pure culture

ballistoconidium – see ballistospore
ballistospore – a spore that is forcibly projected
base composition of DNA – molar ratio of guanine + cytosine (G + C) or adenine + thymine (A + T) expressed as the percent that the particular base pair comprises of the whole DNA molecule
basidiocarp – basidia-producing fruit body
basidiophore – basidium-bearing hypha
basidiospore – a (sexual) spore from a basidium
basidium – organ of the perfect state of a basidiomycete bearing basidiospores
basipetal – produced successively in the direction of the base; the apical part is oldest
bifactorial mating system – see tetrapolar incompatibility system
bilaterally symmetrical – with one plane of symmetry
bipolar allelic – two opposite mating types comprising a single locus in a one-locus form
bipolar biallelic – a single incompatibility factor with two alleles resulting in two opposite mating types
bipolar budding – budding from both ends of the cell
bipolar incompatibility system, unifactorial mating system – a system with a single incompatibility factor; bipolar, because of the two incompatibility groups found among the progeny of a cross
bipolar multiallelic – a single incompatibility factor with more than two alleles resulting in more than two mating types, found only in the basidiomycetes
bisexual – two opposite mating types in the genome of a single cell
blastese – formation of hyphae without septa
blastoconidium – see blastospore
blastospore – a vegetative spore formed as a bud on true- or pseudomycelium
botuliform – cylindrical with rounded ends
bud – formation of a new cell from an outgrowth on the mother cell on a narrow base
bud-fission – formation of a new cell from an outgrowth on the mother cell on a broad base, but with a constriction between the two cells
bud-meiosis – meiosis preceding ascus formation which takes place in the bud attached to the mother cell; the latter becomes the ascus

capsule – a hyaline sheath of polysaccharide surrounding the yeast cell
catenulate – in a chain
chlamydospore – a thick-walled, intercalary or terminal asexual spore
clamp connection – a hyphal outgrowth which, at cell division, makes a connection between the resulting two cells by fusion with the lower one, e.g., as found during cell division in many dikaryotic basidiomycetes
clavate – club-shaped
clone – the asexual offspring from a single cell
closure line – the site in the center of the septum where during growth the septum has closed; there is a double layer of plasmalemma visible in a

longitudinal section through the center of the hypha

coenzyme Q number – the number of isoprene units in the side chain of coenzyme Q (ubiquinone)

collarette – cup-shaped structure found at the apex of certain conidiogenous cells such as on the conidiophore of *Phialophora* or after successive budding at the same site on the cells of certain yeasts

compatible mating types – opposite mating types

complementary mating types – see compatible mating types

conidiophore – a specialized hypha on which conidia are produced

conidium (conidiospore) – see blastospore

conjugated ascus – ascus developed from the fusion of two cells

conjugation – fusion of two cells

conspecific – belonging to the same species

coremium – a group of conidiophores joined together

cornute – horned

crescentiform – see lunate

crescent-shaped – see lunate

crispulate – with a wavy margin

cross – to bring together genetic material from different individuals

cross wall – see septum

cystidium – a sterile, conical, prismatic or cylindrical end of a non-vascular hypha in the hymenium in basidiomycetes

dacryoid – pear-shaped; having one end rounded and the other more or less pointed

dactyloid – finger-like

dangeardien – the structure in which karyogamy and meiosis take place

dehiscent, dehiscing – opening when mature

dentate – with tooth-like projections

denticle – a small tooth-like projection

diclavate – club-shaped at both ends

dikaryon – two conjugate nuclei

dikaryotic – binucleate condition, i.e. each cell contains two nuclei

dimorphic – having two forms

diobovate – being inversely ovate at both ends

diplo-haplontic life cycle – the cells reproducing vegetatively may be haploid or diploid

diploid – having a double set (2n) of chromosomes

diplontic life cycle – the vegetative cells are diploid

diplophase – the part of the life cycle in which the cells are diploid

distal – terminal, away from the point of origin

DNA – deoxyribonucleic acid

DNA reassociation – the pairing or reannealing of complementary single strands of DNA to form a double-stranded molecule (duplex)

DNA sequence homology – the extent of homology or relatedness between the sequences of bases on complementary DNA molecules

DNA sequence relatedness – see DNA sequence homology

dolipore – a septal pore with a thickened edge

dormant – in state of rest

ellipsoidopedunculate – stalk-like and ellipsoidal

endospore (vegetative) – an asexual spore formed within a vegetative cell

epibasidium – a structure between a probasidium and basidiospores

epithet – the second (specific) part of a Latin binomial

erose – having delicate tooth-like projections from the edge

evanescent ascus – ascus, breaking down when mature and liberating ascospores

exosporium – the outer layer of the spore wall

falcate – sickle-shaped

fascicle – bundle

fasciculate – growth in fascicles

filament (filamentous) – thread-like hyphal or pseudohyphal cell

fimbriate – with a torn or minutely fringed margin

galeate – hat-shaped

gametangiogamy – fusion of gametangia

gametangium – special cells by which conjugation may take place

G + C content – mol percent guanine + cytosine (G + C) of DNA; see also base composition of DNA

genotype – the genetic constitution of an individual or taxon

glabrous – smooth, not hairy

haploid – having a single set (n) of chromosomes

haplontic life cycle – the vegetative cells are haploid

haustorium – a special hyphal branch, especially one within a living cell of the host, for absorption of food

heterogamous conjugation – conjugation of two cells of different form or size

heterothallic – sexual reproduction is only possible after interaction of different mating types

hirsute – having long hairs

holometabasidium – non-septate metabasidium

holomorph – undivided shape

holotype – material or strain indicated as type by the original author

homokaryon – a strain in which all nuclei are of the same genotype

homokaryotic fruiting – formation of a "teliospore" in the absence of a dikaryophase; the mycelial cells are uninucleate and lack clamp connections

homonym – the same name given to two or more different taxa of the same rank based on different types

homothallic – sexual reproduction without the preceding interaction of complementary mating types

hyaline – having no color

hybrid – progeny of a cross involving genetically different parents

hymenium – spore bearing layer of a fruit body
hypha – one of the filaments of a mycelium

imperfect state – asexual part of a life cycle
incompatibility – inability to sexually reproduce
intercalary cell – a cell between two others
interfertility – fertility between two organisms, i.e. two organisms can mate and form sexual spores
intramatrical – living in the matrix or substratum
isogamous – conjugating cells of the same shape and size
isotype – any specimen or strain of the type collection other than the holotype

karyogamy – the fusion of the two sex nuclei after cell fusion

lateral – at the side
lectotype – a type taken later from the original material or one of the strains on which the original description was based of a taxon for which the author named no type
lenticular – shape of a double convex lens
lentiform – see lenticular
life cycle – the stage or series of stages between one spore form and the subsequent development of this same spore type
lobate – lobed
lobiform – in the shape of a lobe
locus – the site of a gene in a chromosome
lunate – like a new moon

mating – fusion of two cells that are sexually reactive
mating types – cells of opposite types which are able to conjugate
meiosis – reduction division associated with sexual reproduction
meiosis bud – a bud attached to the mother cell which serves as the site of meiosis after fusion of the bud and mother cell nuclei
mesophilic – optimum temperature of growth 20–40°C
metabasidium – the part of the basidium in which meiosis occurs
micropores – see plasmodesmata
mitosis – nuclear division with retention of the original number of chromosomes
multilateral budding – budding at many sites on the cell
multiple allelic – more than two alleles per incompatibility factor; more than two alleles for control of a particular function
multipolar budding – see multilateral budding
mycelium – a mass of hyphae of an individual fungus; see also true- and pseudomycelium

neotype – strain or material designated as type when the material or strain on which the original description was based does not exist
new combination, combinatio nova, comb. nov. – name of a species which has been transferred from one genus to another
nomen conservandum, nom. cons. – the name of a genus or species the conservation of which has been approved by an International Botanical Congress
nomen illegitinum, nom. illegit., illegitimate name – a name which is nomenclaturally superfluous when published, i.e. if the taxon to which it was applied, as circumscribed by its author, included the type of a name or epithet which ought to have been adopted under the rules of the International Code of Botanical Nomenclature
nomen invalidum, nom. inval., invalid name – a name which has not been validly published, i.e. in accordance with the International Code of Botanical Nomenclature
nomen nudum, nom. nud. – a name published without an adequate description, i.e. in accordance with the International Code of Botanical Nomenclature
nuclear segregation – the segregation of genetically different nuclei contained in a heterokaryotic multikaryon
numerical analysis, numerical taxonomy – taxonomic arrangement based upon quantification of the similarities and differences

oblate – flattened at the poles
oblong – twice as long as wide and having somewhat truncate ends
obovate – inversely ovate, narrowest at the base
obovoid – see obovate
obpyriform – pear-shaped but attached at the broad end
obtuse – rounded or blunt
ogival – pointed like a Gothic arch
orbicular – spherical
osmophilic – able to grow in a medium with elevated, i.e. greater than usual, sugar- or salt concentration
osmotic medium – medium with elevated sugar or salt concentration
osmotolerance – resistance to osmotic pressure

paraphysoid – in basidiomycetes: sterile accessory hymenial structures allied to basidia, cystidia or hyphae
parasexuality – recombination of hereditary properties based not on sexual reproduction (meiosis) but on mitotic non-disjunction
paratype – strain other than the holotype on which the description of a species is based
parenthesome – endoplasmic reticulum of a special structure at both sides of the septal dolipore in the basidiomycetes
pedicle – a small stalk-like structure
peduncle – stalk
perfect state – the state of the life cycle in which asco- or basidiospores are formed after nuclear fusion
persistent ascus – ascus not opening or breaking down when mature

phenotype – the totality of characteristics of an individual as a result of the interaction between genotype and environment
phragmometabasidium – a metabasidium divided by septa
pileus – the hymenium-supporting part of the fruit body of a hymenomycete
plasmodesmata – isthmus-like strands of protoplasm connecting adjacent cells; in hyphal septa also indicated as micropores
plasmogamy – fusion of the cytoplasm of two sexual cells
plicate – folded into pleats
primary sporidia – basidiospores which act as sterigmata; they remain attached to the metabasidium and bear the deciduous basidiospores
probasidium – the part or stage of the basidium in which karyogamy occurs
promycelium – see metabasidium
protosterigma – basal, filamentous or inflated part of the sterigma
prototrophic – growth factor-independent
pseudohyphae – see pseudomycelium
pseudomycelium – a series of cells that remain attached to each other; each cell was formed as a bud
psychrophilic – optimum temperature of growth below 20°C
pulcherrima cells – round cells containing a large oil drop as with *Metschnikowia (Candida) pulcherrima*
pulcherrimin – a red iron-containing compound produced by some strains of *Kluyveromyces* spp. and certain other yeasts
pulvinate – cushion-like
pyriform – pear-shaped

reannealing experiments – see DNA reassociation
recumbent – prostrate on the surface from which it extends
reticulate – net-like
rhizoid – a root-like structure
rotational symmetry – with all planes of symmetry through the length axis
rugose – wrinkled

saccate – like a sack or bag
Saturn-shaped – spherical with a ledge around the middle
self-fertile – see homothallic
septum – a cross wall in a hypha
seta – a stiff hair or bristle
sib mating – intercrossing of siblings
somatogamous conjugation – fusion of two whole cells
sphaeropedunculate – large spherical tip on a short stalk
sporidium – see basidiospore
sporophore – a spore-producing or supporting structure
stalagmoid – like a long tear or drop
sterigma – a process from a cell for supporting a spore
sympodium – a conidiophore that extends in growth of a succession of apices; a spore is produced at each apex and the growth continues to the side of the apex; the result is a zigzag appearance
syngamy – sexual reproduction resulting in karyogamy and zygote formation
synnema – see coremium

teleomorph, teleomorphic state – see perfect state
teliospore – a thick-walled spore which is a probasidium
tetrapolar incompatibility system, bifactorial mating system – a system with two incompatibility factors, *A* and *B*, which control mating competence; tetrapolar, because of the four incompatibility groups found among the progeny of a cross
true hyphae – see true mycelium
true mycelium – a mass of hyphal cells with tip growth and cross walls
truncate – ending abruptly as though the end was cut off
turbinate – like a top in form
type material – specimen material on which the name of a taxonomic group, i.e. a species, is based
type species – the species which was designated as the type of a genus
type strain – the preserved, original specimen, or progeny thereof, upon which the description of a taxon is based; for bacteria: a living, pure culture of an organism descended from a strain designated as the nomenclatural type

unifactorial mating system – see bipolar incompatibility system
unisexual – vegetative cells of a single mating type

ventricose – swelling out in the middle or at one side; inflated
verrucose – having small rounded processes or warts
verticil – whorl
vesicle – a bladder-like sac, cell or cavity; a thin-walled body
vestigial clamp connection – hyphal outgrowth which does not fuse with the lower cell; see also clamp connection

Woronin bodies – membrane bound refractive spherical bodies occurring on either side of the pore of an ascomycete septum

zygote – a cell that is the result of the fusion of two sexual cells

Bibliography

Abadie, F. (1967) L'uréase chez les levures. Ann. Inst. Pasteur 113, 791–813.

Abadie, F., Pignal, M.C. & Jacob, J.L. (1963) Les levures à spores verruqueuses. Bull. Soc. Mycol. France 79, 16–70.

Abe, K., Kusaka, I. & Fukui, S. (1975) Morphological change in the early stages of the mating process of *Rhodosporidium toruloides.* J. Bacteriol. 122, 710–718.

Aciole de Queiroz, L. (1970) *Pichia faecalis* nov. comb. Revisión de una cepa de levadura nombrada *Kloeckera faecalis* Batista & Silveira, 1959. Mycopathol. Mycol. Appl. 41, 327–329.

Acton, H.W. & Panja, G. (1927) Seborrhoeic dermatitis or pityriasis capitis: a lesion caused by the *Malassezia ovale.* Ind. Med. Gaz. 62, 603–614.

Adams, A.M. (1949) A convenient method of obtaining ascospores from baker's yeast. Can. J. Res. Sect. C 27, 179–189.

Ahearn, D.G. (1973) Effects of environmental stress on aquatic yeast populations. In: H. Stevenson & R. Colwell (Eds.) Estuarine microbial ecology. Belle W. Baruch Symposia in Marine Science. Univ. South Carolina Press, Columbia, pp. 433–439.

Ahearn, D.G. (1978) Medically important yeasts. Annu. Rev. Microbiol. 32, 59–68.

Ahearn, D.G. & Holzschu, D.L. (1977) The white yeasts as disease agents: historical perspective. In: Pan American Health Organization Proc. 4th Internat. Cong. Mycoses. Washington D.C. PAHO Sci. Publ. No. 356, pp. 119–123.

Ahearn, D.G. & Roth, F.J. (1966) Physiology and ecology of psychrotrophic carotenogenic yeasts. Dev. Ind. Microbiol. 7, 301–309.

Ahearn, D.G. & Schlitzer, R.L. (1981) Yeast infections. In: A. Balows (Ed.) Diagnostic procedures in bacterial, mycotic and parasitic infections. 6th ed. American Public Health Association, Washington D.C., pp. 991–1012.

Ahearn, D.G., Roth, F.J., Fell, J.W. & Meyers, S.P. (1960) Use of shaken cultures in the assimilation test for yeast identification. J. Bacteriol. 79, 369–371.

Ahearn, D.G., Jannach, J.R. & Roth, F.J. (1966) Speciation and densities of yeasts in human urine specimens. Sabouraudia 5, 110–119.

Ahearn, D.G., Meyers, S.P. & Nichols, R.A. (1968) Extracellular proteinases and yeast like fungi. Appl. Microbiol. 16, 1370–1374.

Ahearn, D.G., Yarrow, D. & Meyers, S.P. (1970) *Pichia spartinae* sp. n. from Louisiana marshland habitats. Antonie van Leeuwenhoek 36, 503–508.

Ahearn, D.G., Meyer, S.A., Mitchell, G., Nicholson, M.A. & Ibrahim, A.I. (1977) Sucrose-negative variants of *Candida tropicalis.* J. Clin. Microbiol. 5, 494–496.

Ainsworth, G.C. (1971) Ainsworth & Bisby's Dictionary of the Fungi. 6th edn., Commonwealth Mycological Institute, Kew.

Ainsworth, G.C. (1973) Introduction and keys to higher taxa. In: G.C. Ainsworth, F.K. Sparrow & A.S. Sussman (Eds.), The Fungi, Academic Press, New York–London, pp. 1–7.

Akagi, S. (1929) Jpn. J. Dermatol. Urol. 29, 53 (cited from Lodder & Kreger-van Rij, 1952).

Alexopoulos, C.J. & Mims, C.W. (1979) Introductory Mycology, 3rd ed., Wiley & Sons, Inc., New York.

Allen, L.A. & Thornley, B.D. (1929) The morphology and physiology of two lactose-fermenting yeasts and chemical changes during the ripening of cheese from milk containing them. Ann. Appl. Biol. 16, 578–595.

Almeida, F.P. de (1929) Estudo comparativo do granuloma coccidióidico nos Estados Unidos e no Brasil. An. Fac. Med. São Paulo 4, 91–98.

Almeida, F.P. de (1930) Estudo comparativo do granuloma coccidióidico nos Estados Unidos e no Brasil. Novo genêro para o parasito brasileiro. An. Fac. Med. São Paulo 5, 125–141.

Almeida, F. de (1933) As blastomycoses no Brasil. An. Fac. Med. São Paulo 9, 69–164.

Amano, Y., Goto, S. & Kagami, M. (1975) A strongly ethanol assimilating new yeast *Candida brassicae* nov. sp. J. Ferment. Technol. 53, 311–314.

Anderson, H.W. (1917) Yeast-like fungi of the human intestinal tract. J. Infect. Dis. 21, 341–386.

Anderson, H.W. (1918) The influence of illumination gas and its constituents on certain bacteria and fungi. In: C.A. Ludwig, Am. J. Bot. 5, 1–31.

Anderson, K.W. & Skinner, C.E. (1947) Yeasts in decomposing fleshy fungi. Mycologia 39, 166–170.

Andrews, A.G., Phaff, H.J. & Starr, M.P. (1976) Carotenoids of *Phaffia rhodozyma,* a red-pigmented fermenting yeast. Phytochemistry 15, 1003–1007.

Arêa Leão, A.E. de (1940) Consideracoes sobre os Thallosporados. O gênero *Trichosporon. Trichosporon minor* n. sp. produtor da piedra axilar. Mem. Inst. Oswaldo Cruz 35, 729–745.

Arnold, P. & Ahearn, D.G. (1972) The systematics of the genus *Prototheca* with a description of a new species *P. filamenta.* Mycologia 64, 265–275.

Artagaveytia-Allende, R.C. (1953) Sobre una levadura hallada en el peritoneo de ratones inoculados con diversos materiales. Arch. Soc. Bio. Montevideo 20, 1.

Artagaveytia-Allende, R.C. & Queiroz, L.A. de (1970) *Cryptococcus diffluens* var. *Uruguaiensis* Allende et de Queiroz. Inst. Micol. Univ. Fed. Pernambuco, Publ. 663, pp. 1–9.

Arx, J.A. von (1972) On *Endomyces, Endomycopsis* and related yeast-like fungi. Antonie van Leeuwenhoek 38, 289–309.

Arx, J.A. von (1976) Centraalbureau voor Schimmelcultures Baarn and Delft. Progress Report

1975. Verh. Kon. Ned. Akad. Wetensch., Afd. Natuurk., 2e Reeks 67, 1–21.

Arx, J.A. von (1977) Notes on *Dipodascus, Endomyces* and *Geotrichum* with the description of two new species. Antonie van Leeuwenhoek 43, 333–340.

Arx, J.A. von (1979) Propagation in the yeasts and yeast-like fungi. In: B. Kendrick (Ed.), The whole fungus, Vol. 2. Proc. 2nd Int. Mycol. Conf., Kananaskis, pp. 555–571.

Arx, J.A. von (1981) A mycologist's view of yeasts. In: F.A. Skinner, S.M. Passmore & R.R. Davenport (Eds.), Biology and activities of yeasts, Society for Applied Bacteriology Symposium Series No. 9. Academic Press, London, pp. 53–61.

Arx, J.A. von & Schipper, M.A.A. (1978) The CBS fungus collection. In: D. Perlman (Ed.), Advances in applied microbiology, Vol. 24. Academic Press, London–New York, pp. 215–236.

Arx, J.A. von & Walt, J.P. van der (1976) The ascigerous state of *Candida chodatii.* Antonie van Leeuwenhoek 42, 309–314.

Arx, J.A. von & Weijman, A.C.M. (1979) Conidiation and carbohydrate composition in some *Candida* and *Torulopsis* species. Antonie van Leeuwenhoek 45, 547–555.

Arx, J.A. von, Rodrigues de Miranda, L., Smith, M.Th. & Yarrow, D. (1977) The genera of yeasts and the yeast-like fungi. Studies in Mycology. No. 14, Centraalbureau voor Schimmelcultures, Baarn.

Arzt, L. (1924) Zur Klinik und Pathologie der Sprosspilzerkrankungen. Arch. Dermatol. Syph. 145, 311.

Asai, T. (1918) Physiologische Untersuchungen über eine neue, in der Gerbbrühe gedeihende Kahmhefe. J. Coll. Sci. Imp. Univ. Tokyo 39, 1–42.

Asai, Y., Makiguichi, N., Shimada, M. & Kurimura, Y. (1976) New species of methanol-assimilating yeasts. J. Gen. Appl. Microbiol. 22, 197–202.

Aschner, M. & Cury, A. (1951) Starch production in the genus *Trichosporon.* J. Bacteriol. 62, 350–352.

Aschner, M., Mager, J. & Leibowitz, J. (1945) Production of extracellular starch in cultures of capsulated yeasts. Nature 156, 295.

Aschner, M., Halevy, S. & Awram, D. (1954) The yeast flora of rats on normal diets and on diets supplemented with antibiotics. Bull. Res. Counc. Isr. 2, no. 2.

Ashford, B.K. (1917) The etiology of sprue. Am. J. Med. Sci. 154, 157–176.

Ashford, B.K. (1928) Certain conditions of the gastro-intestinal tract in Porto Rico and their relation to tropical sprue. Am. J. Trop. Med. 8, 507–538.

Ashford, B.K. & Ciferri, R. (1930) A new species of *Torulopsis: Torulopsis nitritophila* Cif. and Ashf. Zentralbl. Bakteriol. Parasitenk., Abt. II, 81, 63–67.

Aulakh, H.S., Straus, S.E. & Kwon-Chung, K.J. (1981) Genetic relatedness of *Filobasidiella neoformans (Cryptococcus neoformans)* and *Filobasidiella bacillispora (Cryptococcus bacillisporus)* as determined by deoxyribonucleic acid base composition and sequence homology studies. Int. J. Syst. Bacteriol. 31, 97–103.

Babjeva, I.P. & Gorin, S.E. (1975) *Lipomyces anomalus* sp. nov. Antonie van Leeuwenhoek 41, 185–191.

Babjeva, I.P. & Kartintsev, A.V. (1974) Occurrence of the yeast *Rhodosporidium diobovatum* on plants. Mosc. Univ. Soil Sci. Bull. (Engl. transl. of Vestn Mosk Univ. Pichuoved) 29 (5/6).

Babjeva, I.P. & Reshetova, I.S. (1975) A new yeast species *Candida podzolica* sp. n. from the soil. Microbiology (transl. of Mikrobiologiya) 44, 294–333.

Bacon, J.S.D., Jones, D., Farmer, V.C. & Webley, D.M. (1968) The occurrence of α-1,3-glucan in *Cryptococcus, Schizosaccharomyces* and *Polysporus* species, and its hydrolysis by a Streptomycete culture filtrate lysing cell walls of *Cryptococcus.* Biochim. Biophys. Acta 158, 313–315.

Bahadur, K. (1957) Isolation and morphological studies of the Dhar yeast. Zentralbl. Bakteriol. Parasitenk., Abt. II, 110, 305–308.

Baharaeen, S. & Vishniac, H.S. (1982) *Cryptococcus lupi* sp. nov. an antarctic basidio-blastomycete. Int. J. Syst. Bact. 32, 229–233.

Baillon, E.H. (1889) Traité de Botanique Médicale Cryptogamique. O. Doin, Paris.

Bak, A.L. (1973) DNA base composition in mycoplasma, bacteria, and yeast. Curr. Top. Microbiol. Immunol. 61, 89–149.

Bak, A.L. & Stenderup, A. (1969) Deoxyribonucleic acid homology in yeasts. Genetic relatedness within the genus *Candida.* J. Gen. Microbiol. 59, 21–30.

Baker, J.G., Salkin, I.F., Pincus, D.H. & D'Amato, R.F. (1981) *Candida paratropicalis,* a new species of *Candida.* Mycotaxon 13, 115–119.

Baker, J.M. (1963) Ambrosia beetles and their fungi, with particular reference to *Platypus cylindrus* Fab. Symposia of the Society for General Microbiology XIII, Symbiotic associations. Cambridge University Press, pp. 232–265.

Baker, J.M. & Kreger-van Rij, N.J.W. (1964) *Endomycopsis platypodis* sp. n. (Ascomycetes): an auxiliary ambrosia fungus of *Platypus cylindrus* Fab. (Col. Platypodidae). Antonie van Leeuwenhoek 30, 433–441.

Balloni, W., Materassi, R. & Margheri, M.C. (1971) *Saccharomyces (Torulaspora) florenzanii* sp. n., a new wine fermenting yeast from Sardegna (Italia). Zentralbl. Bakteriol. Parasitenk., Abt. II, 126, 386–388.

Ballou, C.E. (1974) Some aspects of the structure, immunochemistry, and genetic control of yeast mannans. Adv. Enzymol. 40, 239–270.

Ballou, C.E. (1976) Structure and biosynthesis of the mannan component of the yeast cell envelope. Adv. Microb. Physiol. 14, 93–157.

Balzer, F., Gougerot, H. & Burnier, R. (1912) Ann. Dermatol. Syph. V, 3, 282 (cited from Lodder & Kreger-van Rij, 1952).

Bandoni, R.J. (1957) The spores and basidia of *Sirobasidium.* Mycologia 49, 250–255.

Bandoni, R.J. (1958) Some Tremellaceous fungi in the C.G. Lloyd collection. Lloydia 21, 137–151.

Bandoni, R.J. (1961) The genus *Naematelia.* Am. Midl. Nat. 66, 319–328.

Bandoni, R.J. (1963) Conjugation in *Tremella mesenterica.* Can. J. Bot. 41, 467–474.

Bandoni, R.J. (1965) Secondary control of conjugation in *Tremella mesenterica.* Can. J. Bot. 43, 627–630.

Bandoni, R.J. (1979) *Fibulobasidium:* a new genus in the Sirobasidiaceae. Can. J. Bot. 57, 264–268.

Bandoni, R.J., Lobo, K.J. & Brezden, S.A. (1971) Conjugation and chlamydospores in *Sporobolomyces odorus.* Can. J. Bot. 49, 683–686.

Bandoni, R.J., Johri, B.N. & Reid, S.A. (1975) Mating among isolates of three species of *Sporobolomyces.* Can. J. Bot. 53, 2942–2944.

Banno, I. (1967) Studies on the sexuality of *Rhodotorula.* J. Gen. Appl. Microbiol. 13, 167–196.

Baptist, J.N. & Kurtzman, C.P. (1976) Comparative enzyme patterns in *Cryptococcus laurentii* and its taxonomic varieties. Mycologia 68, 1195–1203.

Baptist, J.N., Shaw, C.R. & Mandel, M. (1971) Comparative zone electrophoresis of enzymes of *Pseudomonas solanacearam* and *Pseudomonas cepacia.* J. Bacteriol. 108, 799–803.

Barker, B.T.P. (1901) A conjugating 'yeast'. Phil. Trans. Roy. Soc. London, Ser. B, 194, 467–485.

Barker, E.R. & Miller, M.W. (1969) Some properties of *Saccharomyces kluyveri.* Antonie van Leeuwenhoek 35, 159–171.

Barnett, J.A. & Buhagiar, R.W.M. (1971) *Torulopsis fragaria* species nova, a yeast from fruit. J. Gen. Microbiol. 67, 233–238.

Barnett, J.A. & Ingram, M. (1955) Technique in the study of yeast assimilation reactions. J. Appl. Microbiol. 18, 131–148.

Barnett, J.A. & Pankhurst, R.J. (1974) A new key to The Yeasts. North-Holland Publ. Co., Amsterdam.

Barnett, J.A., Payne, R.W. & Yarrow, D. (1979) A guide to identifying and classifying yeasts. Cambridge University Press, Cambridge.

Barnett, J.A., Payne, R.W. & Yarrow, D. (1983) Yeasts: characteristics and identification, Cambridge Univ. Press, Cambridge.

Barre, P. & Galzy, P. (1960) Etude et détermination d'une levure osmophile. Ann. Technol. 4, 345–348.

Barret, A., Bidan, P. & André, L. (1955) Sur quelques accidents de vinification dûs à des levures à voile. C.R. Acad. Agric. France 41, 426–431.

Bartnicki-Garcia, S. (1968) Cell wall chemistry, morphogenesis, and taxonomy of fungi. Annu. Rev. Microbiol. 22, 87–108.

Bartnicki-Garcia, S. (1970) Cell wall composition and other biochemical markers in fungal taxonomy. In: J.B. Harborne (Ed.), Phytochemical Phylogeny. Academic Press, New York–London, pp. 81–103.

Bary, A. de (1859) Zur Kenntnisse einiger Agaricinen. Bot. Z. 17, 401–404.

Basgal, W. (1931) Contribuição ao estudo das blastomicoses pulmonares. Thesis, Institute Oswaldo Cruz, Rio de Janeiro.

Bastide, J.M., El Hadi Hadibi & Bastide, M. (1979) Taxonomic significance of yeast sphaeroplast release after enzymic treatment of intact cells. J. Gen. Microbiol. 113, 147–153.

Batista, A.C. & Silveira, J.S. (1959a) *Kloeckera faecalis,* n. sp. isolado de fezes humanas. O. Hospital 56, 505–507.

Batista, A.C. & Silveira, J.S. (1959b) Una variedade nova de *Endomycopsis Mali* (Lewis) Dekker assinalada sôbre fezes humanas. Rev. Ass. Med. Bras. 5, 353–354.

Batista, A.C. & Silveira, J.S. (1959c) Alqumas novas espécies de *Candida* do aparelho digestivo e orgãos anexos do homen. O. Hospital 56, 293–299.

Batista, A.C. & Silveira, J.S. (1960) *Trichosporon figueirae* n. sp. assinalado como fungo intestinal. Mycopathol. Mycol. Appl. 12, 196–198.

Batista, A.C. & Silveira, J.S. (1962) Micose pulmonar e *Candida genitalis* nobis n.sp. Univ. de Recife, Inst. Micol. 170, 3.

Batista, A.C., Vieira, J.R. & Coêlho, R.P. (1959) *Schwanniomyces hominis* n. sp. como causa de lesion epidermica en recien nacido. Rev. Ecuatoriana Hig. Med. Trop. 16, 136–141.

Batista, A.C., Campos, S.T.C. & Oliveira, S. (1963) Instituto de Micologia, Universidade do Recife, Publ. no. 207 (cited from do Carmo-Sousa, 1970c).

Batra, L.R. (1963) Contribution to our knowledge of ambrosia fungi. II. *Endomycopsis fasciculata* nom. nov. (Ascomycetes). Am. J. Bot. 50, 481–487.

Batra, L.R. (1971) Two new Hemiascomycetes: *Pichia crossotarsi* and *P. microspora.* Mycologia 63, 994–1001.

Batra, L.R. (1973) Nematosporaceae (Hemiascomycetidae): Taxonomy, pathogenicity, distribution and vector relations. Tech. Bull. No. 1469, ARS, US Dept. of Agric. Washington, D.C., 71 pp.

Batschinskaya, A.A. (1914) Entwicklungsgeschichte und Kultur des neuen Hefepilzes *Saccharomyces paradoxus.* J. Microbiol. Epidémiol. Immunobiol. 1, 231 (cited by Kudriavzev, 1960).

Beauverie, J. (1929) Sur un *Zygosaccharomyces* de la datte iso-hétérogame. Bull. Soc. Mycol. France 45, 153–170.

Beck, O. (1922) Eine neue *Endomyces*-Art, *Endomyces bisporus.* Ann. Mycol. 20, 219–227.

Bedford, C.L. (1942) A taxonomic study of the genus Hansenula. Mycologia 34, 628–649.

Beech, F.W. (1957) Thesis, University of Bristol.

Beech, F.W. (1959) The yeast flora of apple juice and ciders. J. Appl. Bacteriol. 21, 257–265.

Beech, F.W. (1965) *Pichia delftensis* sp. n. Antonie van Leeuwenhoek 31, 81–83.

Beech, F.W. & Davenport, R.R. (1969) The isolation of non-pathogenic yeasts. In: D.A. Shapton &

G.W. Gould (Eds.), Isolation Methods for Microbiologists. Academic Press, London–New York, pp. 71–88.

Beech, F.W. & Davenport, R.R. (1971) Isolation, purification and maintenance of yeasts. In: J.R. Morris and D.W. Ribbons (Eds.), Methods in Microbiology, Vol. 4, Academic Press, London–New York, pp. 153–182.

Beech, F.W., Carr, J.G. & Codner, R.C. (1955) A multipoint inoculator for plating bacteria or yeasts. J. Gen. Microbiol. 13, 408–410.

Beech, F.W. et al. (1980) Media and methods for growing yeasts: proceedings of a discussion meeting. In: F.A. Skinner, S.M. Passmore & R.R. Davenport (Eds.), Biology and activities of yeasts. Academic Press, London, pp. 259–293.

Behrend, G. (1890) Über Trichomycosis nodosa (Juhel-Rénoy); piedra (Osorio). Berl. Klin. Wochenschr. 27, 464–467.

Beijerinck, M.W. (1889a) L'auxanographie, ou la méthode de l'hydrodiffusion dans la gélatine appliquée aux recherches microbiologiques. Arch. Néerl. Sci. Exactes Nat. 23, 367–372.

Beijerinck, M.W. (1889b) Sur le Kéfir. Arch. Néerl. Sci. Exactes Nat. 23, 428–444.

Beijerinck, M.W. (1892) Zur Ernährungs-Physiologie des Kahmpilzes. Zentralbl. Bakteriol. Parasitenk. 11, 68–75.

Beijerinck, M.W. (1894) *Schizosaccharomyces octosporus,* eine achtsporige Alkoholhefe. Zentralbl. Bakteriol. Parasitenk. 16, 49–58.

Beijerinck, M.W. (1895) Über Nachweis und Verbreitung der Glukase, das Enzym der Maltose. Zentralbl. Bakteriol. Parasitenk., Abt. II, 1, 221–229, 265, 329.

Beijerinck, M.W. (1897) Weitere Beobachtungen über die Octosporushefe. Zentralbl. Bakteriol. Parasitenk., Abt. II, 3, 449–455, 518–525.

Beijerinck, M.W. (1898a) Über Regeneration der Sporenbildung bei Alkoholhefen wo diese Funktion im Verschwinden begriffen ist. Zentralbl. Bakteriol. Parasitenk., Abt. II, 4, 657–663, 721–730.

Beijerinck, M.W. (1898b) Sur la régénération de la faculté de produire des spores chez des levures en voie de la perdre. Arch. Néerl. Sci. Exactes Nat., Ser. II, 2, 269–289.

Beijerinck, M.W. (1908) Die Erscheinung der Flockenbildung oder Agglutination bei Alkoholhefen. Zentralbl. Bakteriol. Parasitenk., Abt. II, 20, 137–157, 641–650.

Beijerinck, M.W. (1912) Mutation bei Mikroben. Folia Microbiol. 1, 1–97; also in Verzamelde Geschriften V, 25, 25–88 (1922).

Bencini, A. & Federici, E. (1928) Su due nuove specie di miceti rinvenuti nella cornea. Atti R. Accad. Fisiocrit. Siena, Ser. X, 3, 743–766.

Benda, I. (1962) *Torulopsis burgeffiana* nov. spec., eine von Weinbeeren isolierte, neue Hefeart. Antonie van Leeuwenhoek 28, 208–214.

Benedek, T. (1930) *Cryptococcus malassezi (Pityrosporon malassezi* Sabouraud 1895). Zentralbl. Bakteriol. Parasitenk., Abt. I, 116, 317–331.

Benham, Rh.W. (1955) The genus *Cryptococcus:* the present status and criteria for the identification of species. Trans. N.Y. Acad. Sci., Ser. II, 17, 418–429.

Bennett, J.E., Kwon-Chung, K.J. & Howard, D.H. (1977) Epidemiologic differences among serotypes of *Cryptococcus neoformans.* Am. J. Epidemiol. 105, 582–586.

Bennett, J.E., Kwon-Chung, K.J. & Theodore, T.S. (1978) Biochemical differences between serotypes of *Cryptococcus neoformans.* Sabouraudia 16, 167–174.

Bennett, J.H. (1842) On the parasitic vegetable structures found growing in living animals. Trans. R. Soc. Edinburgh 15, 277–294.

Berkeley, M.J. (1856) J. Bot. & Kew Misc. 8, 277.

Berkhout, C.M. (1923) De schimmelgeslachten *Monilia, Oidium, Oospora* en *Torula.* Thesis, Utrecht.

Berlese, A.N. (1895) I funghi diversi dai Saccaromiceti e capaci di determinare la fermentazione alcoolica. G. Viticolt. Enol. 3, 52–55.

Bernhardt, E. (1946) Time saving in the preparation of corn meal agar and in the identification of yeast-like fungi. Mycologia 38, 228–229.

Besson, M. (1966) Les membranes des ascospores de levures au microscope électronique. Bull. Soc. Mycol. France 82, 489–503.

Beurmann, L. de & Gougerot, H. (1909) Tribune Méd. 42, 501 (cited from Lodder & Kreger-van Rij, 1952).

Beurmann, L. de, Gougerot, H. & Vaucher (1909) Les exascoses Saccharomycoses (Mycose de Busse-Buschke) et Parasaccharomycoses. Zymonématoses (Mycose de Gilchrist). Parendomycoses et Endomycoses (Muguet). Bull. Mém. Soc. Méd. Hôp. Paris, 28, 250–263.

Bezerra, J.L. & Kimbrough, J.W. (1978) A new species of *Tremella* parasitic on *Rhytidhysterium rufulum.* Can. J. Bot. 56, 3021–3033.

Bhattacharjee, A.K., Kwon-Chung, K.J. & Glaudemans, C.P.J. (1978) On the structure of the polysaccharide from *Cryptococcus neoformans* serotype C. Imm. Chem. 15, 673–679.

Bhattacharjee, A.K., Kwon-Chung, K.J. & Glaudemans, C.P.J. (1979) The structure of the capsular polysaccharide from *Cryptococcus neoformans* serotype D. Carbohydr. Res. 73, 183–192.

Bicknell, J.N. & Douglas, H.C. (1970) Nucleic acid homologies among species of *Saccharomyces.* J. Bacteriol. 101, 505–512.

Bisby, G.R., Buller, A.H.R. & Dearness, J. (1929) The Fungi of Manitoba. Longmans Green & Co., London.

Bizzozero, J. (1884) Über die Mikrophyten der normalen Oberhaut des Menschen. Virchows Arch. Pathol. Anat. 98, 441–459.

Black, S.H. & Gorman, C. (1971) The cytology of *Hansenula.* III Nuclear segregation and envelopment during ascosporogenesis in *Hansenula wingei.* Arch. Mikrobiol. 79, 231–248.

Blagodatskaja, V. & Kocková-Kratochvílová, A. (1973) The heterogeneity of the species *Candida lipolytica, Candida pseudolipolytica* n. sp. and *Candida lipolytica* var. *thermotolerans* n. var. Biológia (Bratislava) 28, 709–716.

Boedijn, K.B. (1934) Bull. Jard. Bot. Buitenz., Ser. III, 13, 266–268.

Boedijn, K.B. & Verbunt, J.A. (1938) Annotations about dermatomycoses in Batavia. Mycopathologia 1, 185–198.

Boidin, J. & Abadie, F. (1954) Les levures des liqueurs tannantes végétales; leur action sur les tanins pyrogalliques. Bull. Soc. Mycol. France 70, 353–383.

Boidin, J. & Adzet, J.-M. (1957) Deux curieuses levures isolées d'extraits tannants d'origine végétale: *Pachysolen* (nov. gen.) *tannophilus* nov. sp. et *P. pelliculatus* nov. sp. Bull. Soc. Mycol. France 73, 331–342.

Boidin, J., Abadie, F., Jacob, J.L. & Pignal, M.C. (1962) Les levures à spores réniformes. Bull. Soc. Mycol. France 78, 155–203.

Boidin, J., Pignal, M.-C., Mermier, G. & Arpin, M. (1963) Quelques levures camerounaises. Cah. Maboké 1, 86–100.

Boidin, J., Pignal, M.-C., Lehodey, Y., Vey, A. & Abadie, F. (1964) Le genre *Pichia* sensu lato (Première Contribution). Bull. Soc. Mycol. France 80, 396–438.

Boidin, J., Abadie, F. & Lehodey, Y. (1965a) Le genre *Pichia* sensu lato (Deuxième Contribution). Bull. Soc. Mycol. France 81, 5–23.

Boidin, J., Pignal, M.-C. & Besson, M. (1965b) Le genre *Pichia* sensu lato (Quatrième Contribution). Bull. Soc. Mycol. France 81, 566–606.

Bolognesi, G. & Chiurco, G.A. (1927) Micosi chirurg. (cited from Lodder & Kreger-van Rij, 1952).

Bonorden, H.F. (1851) Handbuch der allgemeinen Mykologie als Anleitung zum Studium derselben, nebst speziellen Beiträgen zur Vervollkommnung dieses Zweiges der Naturkunde. Schweizerbart, Stuttgart.

Bourdot, H. & Galzin, A. (1927) Contribution à la Flore Mycologique de la France, I. Hyménomycètes de France. Hétérobasidies – Homobasidies Gymnocarpes. Marcel Bry, Sceaux.

Boutroux, L. (1884) Ann. Sci. Nat., Sér. VI, 17 (cited by Saccardo in Sylloge Fungorum).

Bowman, P.I. & Ahearn, D.G. (1975) Evaluation of the Uni-Yeast-Tek kit for the identification of medically important yeasts. J. Clin. Microbiol. 2, 354–358.

Bowman, P.I. & Ahearn, D.G. (1976) Evaluation of commercial systems for the identification of clinical yeast isolates. J. Clin. Microbiol. 4, 49–53.

Bowman, P.I. & Ahearn, D.G. (1978) Ecology of *Cryptococcus neoformans* in Georgia. Proc. IV Int. Conf. Mycoses, Pan Am. Health Org. Sci. Publ. No. 356, 258–268.

Brady, B.L. (1960) Some observations on the freeze-drying of yeasts. In: A.S. Parkes & A.U. Smith (Eds.), Recent Research in Freeze Drying. Blackwell Scientific Publications, Oxford, pp. 243–247.

Brandstetter, R.D. & Brause, B.D. (1980) *Candida parapsilosis* endocarditis. J.Am. Med. Assoc. 243, 1073.

Bréchot, P., Guibert, L. & Croson, M. (1958) La lyophilisation des levures (deuxième partie). Ann. Inst. Pasteur 95, 62–72.

Brefeld, O. (1888) Untersuchungen aus dem Gesamtgebiete der Mykologie. VII. Protobasidiomyceten. Leipzig.

Brefeld, O. (1891) Die Formen der Ascomyceten und ihre Kultur in Nährlösungen. In: Untersuchungen aus dem Gesamtgebiete der Mykologie. Vol. 9, H. Schöningh, Munster, pp. 119–156.

Brenner, D.J. (1973) DNA reassociation in the taxonomy of enteric bacteria. Int. J. Syst. Bacteriol. 23, 298–307.

Brenner, D.J., Fanning, G.R., Rake, A. & Johnson, K.E. (1969) A batch procedure for thermal elution of DNA from hydroxyapatite. Anal. Biochem. 28, 447–459.

Brock, T.D. (1959) Mating reaction in *Hansenula wingei.* Relation of cell surface properties to agglutination. J. Bacteriol. 78, 59–68.

Brock, T.D. (1965) Biochemical and cellular changes occurring during conjugation in *Hansenula wingei.* J. Bacteriol. 90, 1019–1025.

Brough, S. (1974) *Tremella globospora,* in the field and in culture. Can. J. Bot. 52, 1853–1859.

Brumpt, E. (1913) Précis de Parasitologie, 2me éd., Paris.

Brumpt, E. (1922) Précis de Parasitologie, 3me éd., Paris.

Brumpt, E. (1927) Précis de Parasitologie, 4me éd., Paris.

Brumpt, E. (1936) Précis de Parasitologie, 5me éd., Paris.

Buchwald, N.F. (1939) Fungi imperfecti (Deuteromycetes). Kongelige Veterinaer-og Landbohojskole, Kopenhagen.

Buck, J.D. (1978) Comparison of the in situ and in vitro survival of *Candida albicans* in seawater. Microbial Ecol. 4, 291–302.

Buckley, H.R. (1971) Fungi pathogenic for man and animals: 2. The subcutaneous and deep-seated mycoses. In: J.R. Norris & D.W. Ribbons (Eds.), Methods in Microbiology, Vol. 4. Academic Press, London–New York, pp. 461–478.

Buckley, H.R. & Uden, N. van (1967) *Candida shehatae* sp. n., a yeast associated with wood-destroying insects. Mycopathol. Mycol. Appl. 32, 297–301.

Buckley, H.R. & Uden, N. van (1968) Five new *Candida* species. Mycopathol. Mycol. Appl. 36, 257–266.

Buckley, H.R., Campbell, C.K. & Thompson, J.C. (1969) Techniques for the isolation of pathogenic fungi. In: D.A. Shapton & G.W. Gould (Eds.), Isolation Methods for Microbiologists. Academic Press, London–New York, pp. 113–126.

Buecher, E.J. & Phaff, H.J. (1972) Dimorphism in a new isolate of *Saccharomycopsis* Schiönning. Can. J. Microbiol. 18, 901–907.

Buhagiar, R.W.M. (1975) *Torulopsis bacarum, Torulopsis pustula* and *Torulopsis multis-gemmis* sp. nov., three new yeasts from soft fruit. J. Gen. Microbiol. 86, 1–11.

Buhagiar, R.W.M. & Barnett, J.A. (1973) *Sterigmatomyces acheniorum* species nova, a yeast from strawberries. J. Gen. Microbiol. 77, 71–78.

Bulder, C.J.E.A. (1955) Some observations on the lipolytic activity of micro-organisms and a new method for its detection. Antonie van Leeuwenhoek 21, 433–445.

Bulder, C.J.E.A. (1963) On respiratory deficiency in yeasts. Thesis, Delft.

Burke, R.C. (1961) Tinea versicolor: susceptibility factors and experimental infection in human beings. J. Invest. Dermatol. 36, 389–401.

Burns, W.E. (1933) Iowa State Coll. J. Sci. 7, 4.

Butler, E.E. & Petersen, L.J. (1970) Sexual reproduction in *Geotrichum candidum.* Science 169, 481–482.

Butler, E.E. & Petersen, L.J. (1972) *Endomyces geotrichum*, a perfect state of *Geotrichum candidum.* Mycologia 64, 365–374.

Camargo, R. De & Phaff, H.J. (1957) Yeasts occurring in drosophila flies and in fermenting tomato fruits in Northern California. Food Res. 22, 367–372.

Campbell, I. (1972) Numerical analysis of the genera *Saccharomyces* and *Kluyveromyces.* J. Gen. Microbiol. 73, 279–301.

Campbell, I. (1973) Numerical analysis of *Hansenula, Pichia* and related yeast genera. J. Gen. Microbiol. 77, 427–441.

Capriotti, A. (1957) New blastomycetes isolated from soils of Spain I: *Schwanniomyces castellii* nov. spec. Arch. Mikrobiol. 26, 434–438.

Capriotti, A. (1958a) Neue Hefen, isoliert aus Böden Spaniens II: *Torulaspora franciscae* nov. spec. Arch. Mikrobiol. 28, 238–243.

Capriotti, A. (1958b) *Torulaspora nilssoni* nov. spec. A new yeast from Swedish soil. Arch. Mikrobiol. 28, 247–254.

Capriotti, A. (1958c) *Zygosaccharomyces mrakii* nova species. A new yeast from silage. Arch. Mikrobiol. 30, 387–392.

Capriotti, A. (1958d) *Saccharomyces smittii* nov. spec. A new yeast isolated from Dutch, Italian and Swedish soils. Antonie van Leeuwenhoek 24, 215–220.

Capriotti, A. (1958e) *Debaryomyces castellii* nov. spec. (Eine neue Hefe aus Böden Schwedens). Arch. Mikrobiol. 28, 344–348.

Capriotti, A. (1958f) *Candida vanriji* species nova, a new yeast from Swedish soils. Arch. Mikrobiol. 30, 226–230.

Capriotti, A. (1958g) *Torulopsis wickerhamii* nova species. A new yeast from silage. Arch. Mikrobiol. 30, 383–384.

Capriotti, A. (1959) The yeasts of certain soils from Sweden. Lantbruks-Högskol. Ann. 25, 185–220.

Capriotti, A. (1961a) *Debaryomyces cantarellii* nova species, a new yeast from Finnish soil. Arch. Mikrobiol. 39, 123–129.

Capriotti, A. (1961b) *Debaryomyces phaffii* sp. n., a new yeast isolated from Finnish soil. J. Bacteriol. 82, 326–330.

Capriotti, A. (1961c) *Hansenula wickerhamii* sp. n., a new yeast from Finnish soil. J. Bacteriol. 82, 359–360.

Capriotti, A. (1961d) *Torulopsis castellii* sp. nov., a yeast isolated from Finnish soil. J. Gen. Microbiol. 26, 41–43.

Capriotti, A. (1961e) *Torulopsis vartiovaarai* sp. nov. A new yeast isolated from Finnish soil. Can. J. Microbiol. 7, 681–684.

Capriotti, A. (1966) *Saccharomyces castellii* n. sp. Una nuova specie die lievito isolata da un terreno della Finlandia. Studi Sassar, Sez. III, 14, 453–459.

Capriotti, A. (1967) *Saccharomyces servazzii* n. sp. A new yeast from Finland soil. Ann. Microbiol. Enzimol. 17, 79–84.

Capriotti, A. (1969) *Torulaspora benedictae* sp. n. A new yeast from sheep's milk in Sardinia, Italy. Ann. Microbiol. Enzimol. 19, 107–114.

Capriotti, A. & Fatichenti, F. (1969) *Torulopsis arnaudii* sp. n. A new yeast from Sardinian grapemust. Ann. Microbiol. Enzimol. 19, 115–119.

Carmichael, J.W. (1957) *Geotrichum candidum.* Mycologia 49, 820–830.

Carmo-Sousa, L. do (1969) Endospore formation in the genus *Trichosporon.* In: A. Kocková-Kratochvílová (Ed.), Proc. 2nd Symp. Yeasts, 1966, Bratislava, pp. 87–92.

Carmo-Sousa, L. do (1970a) The genus *Nematospora* Peglion. In: J. Lodder (Ed.), The Yeasts, a taxonomic study. North-Holland Publ. Co., Amsterdam, pp. 441–447.

Carmo-Sousa, L. do (1970b) The genus *Oosporidium* Stautz. In: J. Lodder (Ed.), The Yeasts, a taxonomic study. North-Holland Publ. Co., Amsterdam, pp. 1161–1166.

Carmo-Sousa, L. do (1970c) The genus *Trichosporon* Behrend. In: J. Lodder (Ed.), The Yeasts, a taxonomic study. North-Holland Publ. Co., Amsterdam, pp. 1309–1352.

Carmo-Sousa, L. do & Phaff, H.J. (1962) An improved method for the detection of spore discharge in the Sporobolomycetaceae. J. Bacteriol. 83, 434–435.

Carmo-Sousa, L. do & Uden, N. van (1967) Reisolation of *Sarcinomyces inkin* and its transfer to the genus *Trichosporon.* Mycologia 59, 653–657.

Casselton, L.A. (1977) Dikaryon formation in higher basidiomycetes. In: J.E. Smith & D.R. Berry (Eds.), The Filamentous Fungi, Vol. 3, Developmental Mycology. J. Wiley & Sons, pp. 275–297.

Castellani, A.J. (1910) Tropical bronchomycosis. Observations on a new species of Epidermophyton found in tinea cruris. A new intestinal *Spirillum.* Phillipp. J. Sci. B5, 197–210.

Castellani, A.J. (1911) Observations on fungi of the genus *Endomyces* affecting man in the tropics. Zentralbl. Bakteriol. Parasitenk., Abt. I, 58, 236–238.

Castellani, A.J. (1912a) Observations on the fungi found in tropical bronchomycosis. Lancet 1, 13–15.

Castellani, A.J. (1912b) Note on the importance of Hyphomycetes and other fungi in tropical pathology. Br. Med. J. 2, 1208–1212.

Castellani, A.J. (1913) Further observations on the fungi of the genus *Endomyces* found in man. Arch. Parasitol. 16, 184–186.

Castellani, A.J. (1914) J. Trop. Med. Hyg. 17, 305.

Castellani, A.J. (1920) The etiology of thrush. J. Trop. Med. Hyg. 23, 17–22.

Castellani, A.J. (1925) Notes on three new yeast-like organisms and a new bacillus with remarks on the clinical conditions from which they have been isolated: Furunculosis blastomycetica, Stomatitis cryptococcobacillaris. J. Trop. Med. Hyg. 28, 217–223.

Castellani, A.J. (1927) Fungi and fungous diseases. Arch. Dermatol. Syphilol. 16, 383–425.

Castellani, A.J. (1928) Blastomycosis and some other conditions due to yeast-like fungi (budding fungi). Am. J. Trop. Med. 8, 379–422.

Castellani, A.J. (1933) Blastomycosis, a short general account. Med. Press Circ. 136, 438–443.

Castellani, A.J. (1937a) A short general account for medical men of the genus *Monilia.* Trop. Med. Hyg. 40, 293–306.

Castellani, A.J. (1937b) A preliminary report on two pathogenic fungi: *Trichophyton dankaliense* n. sp. and *Sporotrichum anglicum* n. sp. J. Trop. Hyg. 40, 313–318.

Castellani, A.J. (1940) J. Trop. Med. Hyg., March 15 (cited from do Carmo-Sousa, 1970c).

Castellani, A.J. (1960) A capsulated yeast producing black pigment: *Cryptococcus ater* n. sp. J. Trop. Med. Hyg. 63, 27–30.

Castellani, A.J. (1963) Balanoposthitis chronica ulcerative cryptococcica. Dermatol. Trop. 2, 137–141.

Castellani, A.J. & Chalmers, A.J. (1913) Manual of Tropical Medicine. 2nd ed., Baillière, Tindal and Cox, London.

Castellani, A.J. & Chalmers, A.J. (1919) Manual of Tropical Medicine, 3rd ed., London.

Castellani, A.J. & Jacono, I. (1933) Observations on fungi isolated from cases of blastomycosis cutis and blastomycosis pulmonalis in North America and Europe. Remarks on blastomycetin. J. Trop. Med. Hyg. 36, 297–321.

Castellani, A.J. & Low, G.C. (1913) The rôle played by fungi in sprue. J. Trop. Med. Hyg. 16, 33–35.

Castellani, A.J. & Taylor, F.E. (1925) Further observations on vaginal Monilias and vaginal moniliases. J. Trop. Med. Hyg. 28, 242–249.

Castellani, A.J., Douglas, M. & Thompson, T. (1921) Notes on certain forms of bronchitis clinically resembling tuberculosis: bronchohemisporosis, bronchomoniliasis, bronchoanaeromycosis. J. Trop. Med. Hyg. 24, 149–152.

Castelli, T. (1932) Due nuove Torulopsidacee a pigmento rosso rosa. G. Biol. Appl. Ind. Chim. Alim. 2, 1–13.

Castelli, T. (1933) Su alcune *Hansenule* della fermentazione panaria. Arch. Mikrobiol. 4, 514–529.

Castelli, T. (1935) Su alcuni blastomiceti dei mosti umbri. Boll. Ist. Sieroter. Milan. 9, 911–916.

Castelli, T. (1938) Nuovi blastomiceti isolati da mosti del chianti e zone limitrofe. Arch. Mikrobiol. 9, 449–468.

Castelli, T. (1940) Considerazioni sulla *Torulopsis pulcherrima.* Arch. Mikrobiol. 11, 127–136.

Castelli, T. (1942) Una nuova specie di *Saccharomycodes: Saccharomycodes bisporus.* Arch. Mikrobiol. 12, 260–265.

Castelli, T. (1948) I lieviti della fermentazione vinaria nella regione Pugliese. Ric. Sci. 18, 3–31.

Cazin, J., Kozel, T.R., Lupan, D.M. & Burt, W.R. (1969) Extracellular deoxyribonuclease production by yeasts. J. Bacteriol. 100, 760–762.

Chaskes, S. & Tyndall, R.L. (1975) Pigment production by *Cryptococcus neoformans* from *para-* and *ortho*-diphenols: effect of the nitrogen source. J. Clin. Microbiol. 1, 509–514.

Chaskes, S. & Tyndall, R.L. (1978) Pigment production by *Cryptococcus neoformans* and other *Cryptococcus* species from aminophenols and diaminobenzenes. J. Clin. Microbiol. 7, 146–152.

Chatton, E.R. (1913) *Coccidiascus legeri* n.g., n. sp. levure ascosporée parasite des cellules intestinales de *Drosophila funebris* Fabr. C.R. Soc. Biol. 75, 117–120.

Chen, Ping-Chao & Hou, Hsing-Hsiung (1978) *Tremella fuciformis.* In: S.T. Chang & W.A. Hayes (Eds.), The Biology and Cultivation of Edible Mushrooms. Academic Press, New York, pp. 629–643.

Christensen, W.B. (1946) Urea decomposition as a means of differentiating *Proteus* and Paracolon cultures from each other and from *Salmonella* and *Shigella* type. J. Bacteriol. 52, 461–466.

Ciferri, R. (1930) Contribuzioni alla sistematica delle Torulopsidaceae II–XIV. Arch. Protistenk. 71, 405–452.

Ciferri, R. & Alfonseca, J.D. (1931) A new variety of *Blastodendrion intestinale* Mattl. Zentralbl. Bakteriol. Parasitenk., Abt. II, 83, 273–276.

Ciferri, R. & Ashford, B.K. (1929) A new species of *Blastodendrion.* Porto Rico J. Publ. Health Trop. Med. 5, 91–105.

Ciferri, R. & Redaelli, P. (1925) Monografia delle Torulopsidaceae a pigmento rosso. Atti Ist. Bot. R. Univ. Pavia, Ser. I, 2, 147–303.

Ciferri, R. & Redaelli, P. (1929) Studies on the Torulopsidaceae, a trial general systematic classification of the asporogenous ferments. Ann. Mycol. 27, 243–295.

Ciferri, R. & Redaelli, P. (1935) Contribuzione alla sistematica delle Torulopsidaceae. XV-XXXIII. Arch. Mikrobiol. 6, 9–72.

Ciferri, R. & Redaelli, P. (1947) Reisolamento del *Mycoderma glutinis-farinulae* Desm., e proposta del cambiamento del nome generico di *Mycoderma* Leberle in *Mycokluyveria* nobis. Mycopathologia 4, 54–59.

Clark, P.S., Wallace, R.H. & David, J.J. (1954) Yeasts occurring on apples and in apple cider. Can. J. Microbiol. 1, 145–149.

Cochet, G. (1940) Recherches sur la flore mycologique de l'intestin des rongeurs domestiques. Thesis, Alençon.

Cohn, F. (1875) Untersuchungen über Bakterien. Beitr. Biol. Pflanz. (Heft 2), 127–224.

Coker, W.C. (1920) J. Elisha Mitchell Sci. Soc. 35, 148.

Cole, G.T. (1975) The thallic mode of conidiogenesis in the Fungi Imperfecti. Can. J. Bot. 53, 2983–3001.

Cole, G.T. & Samson, R.A. (1979) Patterns of Development in Conidial Fungi. Pitman Publishing Ltd., London.

Conti, S.F. & Brock, T.D. (1965) Electron microscopy of cell fusion in conjugating *Hansenula wingei*. J. Bacteriol. 90, 524–533.

Cooke, W.B. (1966) A note on *Cryptococcus ater* Castellani. Mycopathol. Mycol. Appl. 30, 349–352.

Cooper, B.H. (1980) Clinical laboratory evaluation of a screening medium (CN Screen) for *Cryptococcus neoformans*. J. Clin. Microbiol. 11, 672–674.

Cordro'ch, M. (1937) Nouvelle espèce de levure du genre *Zygosaccharomyces: Z. ashbyii*. Ann. Ferment. 3, 87–104.

Costantin, M.J. (1901) Sur les levures des animaux. Bull. Soc. Mycol. France 17, 145–148.

Cox, D.E. (1976) A new homobasidiomycete with anomalous basidia. Mycologia 68, 481–510.

Crandall, M.A. & Brock, T.D. (1968) Molecular basis of mating in the yeast *Hansenula wingei*. Bacteriol. Rev. 32, 139–163.

Crane, J.L. & Schoknecht, J.D. (1978) *Rogersomyces*, a new genus in the Filobasidiaceae (Homobasidiomycetes) from an aquatic habit. Am. J. Bot. 65, 902–906.

Crosa, J.H., Brenner, J.D. & Falkow, S. (1973) Use of a single-strand specific nuclease for analysis of bacterial and plasmid deoxyribonucleic acid homo- and heteroduplexes. J. Bacteriol. 115, 904–911.

Crow, S.A., Bowman, P.I. & Ahearn, D.G. (1977) Isolation of *Candida albicans* from the North Sea. Appl. Environ. Microbiol. 33, 738–739.

Custers, M.Th.J. (1940) Onderzoekingen over het gistgeslacht *Brettanomyces*. Thesis, Delft.

Dalmau, L.M. (1929) Remarques sur la technique mycologique. Ann. Parasitol. Hum. Comp. 7, 536–545.

Dastur, J.F. & Singh, J. (1930) A new *Nematospora* on cotton bolls in the Central Provinces (India). Ann. Mycol. 28, 291–296.

Daszewska, W. (1912) Étude sur la désagrégation de la cellulose dans la terre de bruyère et la tourbe. Bull. Soc. Bot. Genève, Sér. 2, 4, 255–316.

Davenport, R.R. (1980) An outline guide to media and methods for studying yeasts and yeast-like organisms. In: F.A. Skinner, S.M. Passmore & R.R. Davenport (Eds.), Biology and activities of yeasts. The Society for Applied Bacteriology Symposium Series No. 9. Academic Press, London, pp. 261–263.

Deinema, M.H. (1961) Intra- and extra-cellular lipid production by yeasts. Meded. Landbouwhogeschool, Wageningen 61, 1–54.

Demme, R. (1889) *Saccharomyces ruber*. Ann. Micrographie 2, 555–556.

Denton, J.F. & Disalvo, A.F. (1968) The prevalence of *Cryptococcus neoformans* in various natural habitats. Sabouraudia 6, 213–217.

Derx, H.G. (1930) Etude sur les Sporobolomycètes. Ann. Mycol. 28, 1–23.

Derx, H.G. (1948) *Itersonilia* nouveau genre de Sporobolomycètes à mycélium bouclé. Bull. Bot. Gardens, Buitenzorg, Ser. III, 17, 465–472.

Desmazières, J.B. (1823) Catalogue des Plantes Omises dans les Flores du Nord de la France.

Diddens, H.A. & Lodder, J. (1939) On some sporogenous yeasts and their imperfect stages. Mycopathologia 2, 28–36.

Diddens, H.A. & Lodder, J. (1942) Die anaskosporogenen Hefen, II Hälfte. North-Holland Publ. Co., Amsterdam.

Dietrichson, E. (1954) Étude d'une collection norvégienne de levures. Ann. Parasitol. Hum. Comp. 29, 271–288, 460–498.

Dijken, J.P. van & Harder, W. (1974) Optimal conditions for the enrichment and isolation of methanol-assimilating yeasts. J. Gen. Microbiol. 84, 409–411.

Ditlevsen, E. & Hjort, A. (1964) *Hanseniaspora pijperi* (van der Walt et Tscheuschner) nov. comb. C.R. Trav. Lab. Carlsberg 34, 347–357.

Dittrich, H.H. (1964) Die alkoholische Vergärung der L-Apfelsäure durch *Schizosaccharomyces pombe* var. *acidodevoratus*. Zentralbl. Bakteriol. Parasitenk., Abt. II, 118, 406–421.

Dobzhansky, T. (1976) Organismic and molecular aspects of species formation. In: F.J. Ayala (Ed.), Molecular Evolution. Sinauer Publ., Sunderland, Massachusetts, pp. 95–105.

Dodge, C.W. (1935) Medical Mycology. Mosby, St. Louis.

Dodge, C.W. & Moore, M. (1936) Morphology, physiology and cytology of *Syringospora inexorabilis (Monilia inexorabilis)*. Ann. Missouri Bot. Gard. 23, 129–150.

Dombrowski, W. (1910) Die Hefen in Milch und Milchprodukten. Zentralbl. Bakteriol. Parasitenk., Abt. II, 28, 345–403.

Donk, M.A. (1972) The heterobasidiomycetes: A reconnaissance. I. A restricted emendation. Proc. Kon. Ned. Akad. Wetensch., Ser. C, 75, 365–375.

Donk, M.A. (1973) The heterobasidiomycetes: A reconnaissance. IV. Proc. Kon. Ned. Akad. Wetensch., Ser. C, 76, 109–125.

Dooren de Jong, L.E. den (1927) Een nieuwe, vetvormende gist. Ned. Tijdschr. Hyg. Microbiol. Serol. 1, 136–148.

Dorn, M. & Roehnert, K. (1977) Dimorphism of *Pityrosporum orbiculare* in defined culture medium. J. Invest. Dermatol. 69, 244–248.

Douglas, H.C., Grindall, D.P. & Talbott, H. (1969) Electrophoretic variants of phosphoglucomutase in *Saccharomyces* species. J. Bacteriol. 99, 287–290.

Dupont, P.F. & Hedrick, L.R. (1971) Deoxyribonucleic acid base composition and numerical taxonomy of yeasts in the genus *Trichosporon*. J. Gen. Microbiol. 66, 349–359.

Duran, R. (1973) Ustilaginales. In: G.C. Ainsworth, F.K. Sparrow & A.S. Sussman (Eds.), The Fungi, an advanced treatise. Vol. 4B. Academic Press, New York–London, pp. 281–300.

Dwidjoseputro, D. & Wolf, F.T. (1970) Microbiological studies of Indonesian fermented foodstuffs. Mycopathol. Mycol. Appl. 41, 211–222.

Edwards, J.E., Jr., Lehrer, R.I., Stiehm, E.R., Fischer, T.J. & Young, L.S. (1978) Severe candidal infections: clinical perspective, immune defense mechanisms, and current concepts of therapy. Ann. Intern. Med. 89, 91–106.

Eijkman, C. (1901) Ueber Enzyme bei Bakterien und Schimmelpilzen. Zentralbl. Bakteriol. Parasitenk., Abt. I, 29, 841–848.

Ellis, M.B. (1971) Dematiaceous Hyphomycetes. Commonwealth Mycological Institute, Kew, Surrey, England.

Emmons, C.W., Binford, C.H. & Utz, J.P. (1970) Medical Mycology, 2nd ed., Lea & Febiger, Philadelphia.

Erke, K.H. (1976) Light microscopy of basidia, basidiospores and nuclei in spores and hyphae of *Filobasidiella neoformans (Cryptococcus neoformans).* J. Bacteriol. 128, 445–455.

Erke, K.H. & Schneidau, J.D. (1973) Relationship of some *Cryptococcus neoformans* hyphaeforming strains to standard strains and to other species of yeasts as determined by deoxyribonucleic acid base ratios and homologies. Infect. Immun. 7, 941–948.

Esser, K., Salch, F. & Meinhardt (1979) Genetics of fruit body production in higher basidiomycetes II. Monokaryotic and dikaryotic fruiting in *Schizophyllum commune.* Curr. Genetics, 1, 85–88.

Etchells, J.L. & Bell, T.A. (1950a) Film yeasts on commercial cucumber brines. Food Technol. 4, 77–83.

Etchells, J.L. & Bell, T.A. (1950b) Classification of yeasts from the fermentation of commercially brined cucumbers. Farlowia 4, 87–112.

Evans, E.E. (1949) An immunologic comparison of twelve strains of *Cryptococcus neoformans (Torula histolytica).* Proc. Soc. Exp. Biol. Med. 71, 644–646.

Evans, E.E. (1950) The antigenic composition of *Cryptococcus neoformans.* I. A serologic classification by means of the capsular and agglutination reactions. J. Immunol. 64, 423–430.

Fabian, F.W. & Hall, H.H. (1933) Yeasts found in fermented maple syrup. Zentralbl. Bakteriol. Parasitenk., Abt. II, 89, 31–47.

Fabian, F.W. & Quinet, R.I. (1928) A study of the cause of honey fermentation. Tech. Bull. 92, Agric. Expt. Stat. Michigan State Coll.

Fambach (1926) *Trichosporon equinum.* Z. Infektionskr. 29, 124–142.

Fell, J.W. (1961) A new species of *Saccharomyces* isolated from a subtropical estuary. Antonie van Leeuwenhoek 27, 27–30.

Fell, J.W. (1966) *Sterigmatomyces,* a new fungal genus from marine areas. Antonie van Leeuwenhoek 32, 99–104.

Fell, J.W. (1970a) The genus *Sterigmatomyces* Fell. In: J. Lodder (Ed.), The Yeasts, a taxonomic study. North-Holland Publ. Co., Amsterdam, pp. 1229–1234.

Fell, J.W. (1970b) Yeasts with heterobasidiomycetous life cycles. In: D.G. Ahearn (Ed.), Recent Trends in Yeast Research. Spectrum, Vol. I. Georgia State University, Atlanta, pp. 49–66.

Fell, J.W. (1974) Heterobasidiomycetous yeasts *Leucosporidium* and *Rhodosporidium.* Their systematics and sexual incompatability systems. Trans. Mycol. Soc. Japan 15, 316–323.

Fell, J.W. (1976) Yeasts in oceanic regions. In: E.B. Gareth Jones (Ed.), Recent Advances in Aquatic Mycology. Elek Science, London, pp. 93–124.

Fell, J.W. & Hunter, I.L. (1968) Isolation of heterothallic yeast strains of *Metschnikowia* Kamienski and their mating reactions with *Chlamydozyma* Wickerham spp. Antonie van Leeuwenhoek 34, 365–376.

Fell, J.W. & Hunter, I.L. (1974) *Torulopsis austromarina* sp. nov. A yeast isolated from the Antarctic Ocean. Antonie van Leeuwenhoek 40, 307–310.

Fell, J.W. & Meyer, S.A. (1967) Systematics of yeast species in the *Candida parapsilosis* group. Mycopathol. Mycol. Appl. 32, 177–193.

Fell, J.W. & Phaff, H.J. (1967) Three new yeasts: *Cryptococcus dimennae, Cryptococcus kuetzingii* and *Cryptococcus lactativorus* spp. n. Antonie van Leeuwenhoek 33, 464–472.

Fell, J.W. & Pitt, J.I. (1969) Taxonomy of the yeast genus *Metschnikowia:* a correction and a new variety. J. Bacteriol. 98, 853–854.

Fell, J.W. & Statzell, A.C. (1971) *Sympodiomyces* gen. n., a yeast-like organism from southern marine waters. Antonie van Leeuwenhoek 37, 359–367.

Fell, J.W. & Statzell Tallman, A. (1980a) *Rhodosporidium paludigenum* sp. nov. A basidiomycetous yeast from intertidal waters of South Florida. Int. J. Syst. Bacteriol. 30, 658–659.

Fell, J.W. & Statzell Tallman, A. (1980b) Mating between strains of the yeasts *Aessosporon salmonicolor* and *Sporobolomyces* spp. Int. J. Syst. Bacteriol. 30, 206–207.

Fell, J.W. & Statzell Tallman, A. (1981) Heterothallism in the basidiomycetous yeast genus *Sporidiobolus* Nyland. Curr. Microbiol. 5, 77–82.

Fell, J.W. & Statzell Tallman, A. (1982) Multiple allelic incompatibility factors among bifactorial strains of the yeast *Leucosporidium (Candida) scottii.* Curr. Microbiol. 7, 213–216.

Fell, J.W., Statzell, A., Hunter, I.L. & Phaff, H.J. (1969) *Leucosporidium* gen. nov. The heterobasidiomycetous stage of several yeasts of the genus *Candida.* Antonie van Leeuwenhoek 35, 433–462.

Fell, J.W., Hunter, I.L. & Tallman, A.S. (1973) Marine basidiomycetous yeasts (*Rhodosporidium* spp. n.) with tetrapolar and multiple allelic bipolar mating systems. Can. J. Microbiol. 19, 643–657.

Ferenczy, L. (1981) Microbial protoplast fusion. In: S.W. Glover & D.A. Hopwood (Eds.), Genetics as a Tool in Microbiology. Society for General Microbiology Symposium 31, Cambridge University Press.

Fernandez-Baquero, G. & Trespalacios, F. (1959) Bol. Soc. Cubana Dermatol. Sifilogr. 16, 21 (cited from do Carmo-Sousa, 1970c).

Ferreira, J.D. & Phaff, H.J. (1959) Life cycle and nuclear behavior of a species of the yeast genus *Schwanniomyces.* J. Bacteriol. 78, 352–361.

Filippov, G.S. (1932) Porča fruktovyh konservov, vyzvannaja drozzevym gribkom. Tr. Cent. Nauchno-Issled. biohim. inst. pisc. vkus. prom. Narkomsnaba Sojuza SSR 2, 26–32.

Fiol, J.B. (1967) Les besoins vitaminiques dans les genres *Debaryomyces* Kloecker, *Schwanniomyces* Kloecker, *Hansenula* H. et P. Sydow et *Endomycopsis* Dekker. C.R. Acad. Sci. Paris, Sér. D, 264, 1605–1607.

Fiol, J.B. & Billon-Grand, G. (1978) Osidases, nitrite et nitrate réductases dans les genres *Hanseniaspora* et *Kloeckera.* Mycopathologia 63, 47–51.

Fischer, B. & Brebeck, C. (1894) Zur Morphologie, Biologie, und Systematik der Kahmpilze, der *Monilia candida* Hansen und des Soorerregers. C. Fischer, Jena.

Fischer, G.W. & Holton, C.S. (1957) Biology and control of the smut fungi. The Ronald Press Co., New York.

Flegel, T. (1976) Conjugation and growth of *Sirobasidium magnum* in laboratory culture. Can. J. Bot 54, 411–418.

Flegel, T.W. (1977) Let's call a yeast a yeast. Can. J. Microbiol. 23, 945–946.

Fleming, W.H., Hopkins, J.M. & Land, G.A. (1977) New culture medium for the presumptive identification of *Candida albicans* and *Cryptococcus neoformans.* J. Clin. Microbiol. 5, 236–243.

Florenzano, G. (1950) Due nuove specie dei lieviti asporogeni, isolati da vini (*Trichosporon intermedium* nov. spec. e *Brettanomyces custersii* nov. spec.) Ric. Sci. 20, 1494–1498.

Florenzano, G. (1951) Diffusione e significato enologico dei lieviti *Brettanomyces.* Atti Accad. Ital. Vite Vino 3, 236–249.

Florenzano, G. (1953) In: T. di Enologia, Sansoni Firenze, Vol. 2, pp. 299.

Fonseca, O.O.R. da (1922) Sobre os agentes das blastomycoses europeas. Cyclo sexuado e posiçao do levedo de Hudelo. Bras. Med. 36, 101 (rev. in: Bull. Inst. Pasteur 21, 651).

Fonseca, O.O.R. da & Arêa Leão, A.E. de (1929) Bras. Med. 43, 667 (cited by Lodder & Kreger-van Rij, 1952).

Fontoynont, M. & Boucher, H. (1923) Contribution à l'étude des mycoses de Madagascar. Ann. Dermatol. Syph. Série VI, 4, 318–344.

Foulerton, A.G.R. (1900) On the pathogenic action of Blastomycetes. J. Pathol. Bacteriol. 6, 37–63.

Fowell, R.R. (1952) Sodium acetate agar as a sporulation medium for yeasts. Nature 170, 578.

Fowell, R.R. (1969) Sporulation and hybridization of yeasts. In: A.H. Rose & J.S. Harrison (Eds.), The Yeasts, Vol. 1, Academic Press, London–New York, pp. 303–383.

Fragner, P. (1969) *Trichosporon jirovecii* sp. nov. Ceska Mykol. 23, 160–162.

Fragner, P. (1970) Spannweite der Art *Trichosporon cutaneum.* Ceska Mykol. 24, 153–161.

Fraser, C.G., Jennings, H.J. & Moyna, P. (1973) Structural analysis of an acid polysaccharide from *Tremella mesenterica.* Can. J. Biochem. 51, 219–224.

Fraser, D.W., Ward, J.I., Ajello, L. & Plikaytis, B.D. (1979) Aspergillosis and other systemic mycoses. J. Am. Med. Assoc. 242, 1631–1635.

Frazer, H.L. (1944) Observations on the method of transmission of internal boll disease of cotton by the cotton stainer-bug. Ann. Appl. Biol. 21, 271–290.

Frederiksen, P.S. (1956) A new *Rhodotorula* species, *Rhodotorula macerans* sp. n. isolated from field-retted flax straw. Friesia 5, 234–239.

Freeman, W. & Weidman, F.D. (1923) Cystic blastomycosis of the cerebral gray matter caused by *Torula histolytica* Stoddard and Cutler. Arch. Neurol. Psychiatry 9, 569–603.

Fresenius, G. (1850–63, 1852) Beiträge zur Mykologie. H.L. Brönner, Frankfurt.

Fries, E. (1822) Systema Mycologicum 2, 213.

Froilano de Mello, I. & Gonzaga Fernandes, L. (1918) Ensaio de classificação de fungos pertencentes a classe dos Blastomyces. Arq. Hig. Patol. Exot. 6, 207–316.

Fukazawa, Y., Shinonda, T., Nishikawa, A. & Nakase, T. (1980) Synonymy of *Saccharomyces cerevisiae* Hansen 1883 and *Saccharomyces uvarum* Beijerinck 1898: Significance of cell wall antigens in yeast classification. Int. J. Syst. Bacteriol. 30, 196–205.

Fuson, G.B., Price, C.W. & Phaff, H.J. (1979) Deoxyribonucleic acid sequence relatedness among some members of the yeast genus *Hansenula.* Int. J. Syst. Bacteriol. 29, 64–69.

Fuson, G.B. Price, C.W. & Phaff, H.J. (1980) Deoxyribonucleic acid base sequence relatedness among strains of *Pichia ohmeri* that produce dimorphic ascospores. Int. J. Syst. Bacteriol. 30, 217–219.

Galgóczy, J. & Novák, E.K. (1962) A new yeast *Paratorulopsis banhegyii* n. sp. from human skin. Acta Microbiol. Acad. Sci. Hung. 9, 77–79.

Galgóczy, J. & Novák, E.K. (1965) *Rhodotorula zsoltii* n. sp., a new species of yeasts, and some notes on the taxonomy of the genus *Rhodotorula.* Acta Microbiol. Acad. Sci. Hung. 12, 151–155

Ganesan, A.T. & Swaminathan, M.S. (1958) Staining the nucleus in yeasts. Stain Technol. 33, 115–121.
Gäumann, E. (1964) Die Pilze, II Aufl., Birkhäuser Verlag, Basel & Stuttgart.
Geiger, A. (1910) Beiträge zur Kenntnis der Sprosspilze ohne Sporenbildung. Zentralbl. Bakteriol. Parasitenk., Abt. II, 27, 97–149.
Genestar Serra, R. (1956) Nueva especie de levadura aislada de concentrado de tanino *Candida majoricensis* nov. sp. Microbiol. Esp. 9, 275–280.
Gilliland, R.B. (1962) The *Brettanomyces* II. Taxonomic significance of slow fermentations and description of a new species. J. Inst. Brew. 68, 51–59.
Giordano, A. (1935) Rôle du *Torulopsis neoformans* (Sanfelice) Red. en pathologie humaine. Boll. Sez. Ital. Soc. Intern. Microbiol. 7, 119–123.
Giovannozzi, M. (1939) Studi sulla fermentazione del tabacco. IV nota. Sui blastomiceti del tabacco Kentucky in fermentazione – i lieviti sporigeni. Boll. Tec. Regio Ist. Sper. Colt. Tab. 'Leonardo Angeloni', Scafati, no. 1.
Giovannozzi, M. (1941) Studi sulla fermentazione dei tabacchi. VI nota. I blastomiceti del tabacco in fermentazione – i lieviti asporigeni. Boll. Tec. Regio Ist. Sper. Colt. Tab. 'Leonardo Angeloni' Scafati, no. 3–4.
Giovannozzi, M. (1948) Microflora blastomicetica dei tabacchi. Mycopathologia 4, 260–267.
Goidànich, G.R., Ciferri, R. & Redaelli, P. (1939) Identificazione di alcuni lieviti isolati da pasta di legno. Mycopathologia 2, 48–51.
Golubev, W.I. (1973) *Nadsonia commutata* nov. sp. Mikrobiologiya 42, 1058–1061.
Golubev, W.I. (1977) *Metschnikowia lunata* sp. nov. Antonie van Leeuwenhoek 43, 317–322.
Golubev, W.I. (1981a) Variability in the ability of *Cryptococcus* species to form pseudohyphae. Mikrobiologiya 49, 520–524
Golubev, W.I. (1981b) Novye kombinatsii drozhzhevykh gribov roda *Cryptococcus.* Mikol. Fitopatol. 15, 467–468.
Golubev, V.I. & Babjeva, I.P. (1972) *Debaryomyces formicarius* sp. n. and *Debaryomyces cantarellii* associated with the ants of the group *Formica rufa* L. J. Gen. Appl. Microbiol. 18, 249–254.
Golubev, V.I. & Babjeva, I.P. (1977) *Torulopsis apis* var. *galacta* var. nov. Int. J. Syst. Bacteriol. 27, 165–167.
Golubev, V.I. & Blagodatskaya, V.M. (1978) The significance of the ability to assimilate D-glucuronic acid, D-glucono-δ-lactone and 5-keto-D-gluconate in identification of species of *Candida* Berkhout. Mikrobiologiya 47, 841–848. (in Russian)
Golubev, V.I. & Tauson, E.L. (1979) Heterogeneity of a yeast fungus, *Cryptococcus terreus* di Menna. Mikol. Fitopatol. 13, 462–467.
Golubev, W.I., Okunev, O.N. & Vdavina, N.W. (1974) Assimilation of inositol as criterion in yeast-identification (translated title). Mikrobiologiya 6, 1047–1050.
Golubev, W.I., Babjeva, I.P., Blagodatskaya, V.M. & Reshetova, I.S. (1977) Taxonomy of yeasts isolated from the sap of birch *(Betula verricosus* Ehrh.). Mikrobiologiya 46, 564–569.
Golubev, W.I., Blagodatskaya, V.M., Suetin, S.O. & Trotsenko, R.S. (1981) *Pichia inositovora* and *Candida paludigena,* two new species of yeasts isolated from peat. Int. J. Syst. Bacteriol. 31, 91–96.
Gomes, J.M. (1924a) Ann. Paulistas Med. Cirurg. 12.
Gomes, J.M. (1924b) Mycose bronchopulmonar *Monilia butantanensis* n. sp. Mem. Inst. Butantan (Sao Paulo).
Gordon, M.A. (1951) The lipophilic mycoflora of the skin. I. In vitro culture of *Pityrosporum orbiculare* n. sp. Mycologia 43, 524–535.
Gorin, P.A.J. (1973) The position of phosphate groups in the phosphomannans of *Hansenula capsulata,* as determined by carbon-13 magnetic resonance spectroscopy. Can. J. Chem. 51, 2105–2109.
Gorin, P.A.J. & Spencer, J.F.T. (1968) Galactomannans of *Trichosporon fermentans* and other yeasts; proton magnetic resonance and chemical studies. Can. J. Chem. 46, 2299–2304.
Gorin, P.A.J. & Spencer, J.F.T. (1970) Proton magnetic resonance spectroscopy – an aid in identification and chemotaxonomy of yeasts. Adv. Appl. Microbiol. 13, 25–89.
Gorin, P.A.J., Spencer, J.F.T. & Bhattacharjee, S.S. (1969a) Structures of yeast mannans containing both α- and β-linked D-mannopyranose units. Can. J. Chem. 47, 1499–1505.
Gorin, P.A.J., Spencer, J.F.T. & Eveleigh, D.E. (1969b) Enzymatic degradation of yeast cell-wall mannans and galactomannans to polymeric fragments containing α-(1→6)-linked D-mannopyranose residues. Carbohydr. Res. 11, 387–398.
Gorin, S.E., Sviridov, A.F. & Babjeva, I.P. (1978) Extracellular polysaccharides and the taxonomy of the species of *Lipomyces* (in Russian). Microbiologiya 47, 756–761.
Goto, S. (1963) On a new yeast genus *Naganishia.* J. Ferment. Technol. 41, 459–462.
Goto, S. & Sugiyama, J. (1968) Studies on the Himalayan yeasts and molds (I). A new species of *Debaryomyces* and some asporogenous yeasts. J. Jpn. Bot. 43, 102–108.
Goto, S. & Sugiyama, J. (1970) Studies on Himalayan yeasts and molds (IV). Several asporogenous yeasts, including two new taxa of *Cryptococcus.* Can. J. Bot. 48, 2097–2101.
Goto, S. & Yokotsuka, I. (1962) Classification of the film yeasts maintained in Japan. Bull. Res. Inst. Ferment. Yamanashi Univ. 9, 71–116.
Goto, S., Sugiyama, J. & Iizuka, H. (1969) A taxonomic study of Antarctica yeasts. Mycologia 61, 748–774.
Goto, S., Yamasato, K. & Iizuka, H. (1974) Identification of yeasts isolated from the Pacific Ocean. J. Gen. Appl. Microbiol. 20, 309–316.

Goto, S., Yamakawa, Y. & Yokotsuka, I. (1975) Classification of fragrant odor producing *Cladosporium:* Studies on fragrant odor producing microorganisms: I. J. Agric. Chem. Soc. Japan 49, 377–381.

Gow, N.A.R., Gooday, G.W., Newsam, R.J. & Gull, K. (1980) Ultrastructure of the septum in *Candida albicans.* Curr. Microbiol. 4, 357–359.

Grigoraki, L. & Péju, G. (1921) Sur une nouvelle espèce de levure du genre *Debaryomyces (D. matruchoti).* C.R. Soc. Biol. 85, 459–462.

Grinbergs, J. (1967) Zur Kenntnis einer neuen Hefeart: *Candida tepae* sp. nov. Arch. Mikrobiol. 56, 202–204.

Grinbergs, J. & Yarrow, D. (1970a) Two new *Candida* species: *Candida chilensis* sp. n. and *Candida valdiviana* sp. n. Antonie van Leeuwenhoek 36, 143–148.

Grinbergs, J. & Yarrow, D. (1970b) *Rhodotorula araucariae* sp. n. Antonie van Leeuwenhoek 36, 455–457.

Groenewege, J. (1921a) Meded. Algem. Proefstn. Landbouw Ned. Indië, No. 9.

Groenewege, J. (1921b) Over de oorzaak van rustiness op rubber van *Hevea brasiliensis.* Meded. Algem. Proefstn. Landbouw Ned. Indië No. 11, 5–19.

Grönlund, C. (1892) Eine neue *Torula*-Art und zwei neue *Saccharomyces*-Arten im Neu-Carlsberg Laboratorium untersucht. Z. Gesamte Brauwes. 15, 281.

Groot, G.S.R., Flavell, R.A. & Sanders, J.P.M. (1975) Sequence homology of nuclear and mitochondrial DNAs of different yeasts. Biochim. Biophys. Acta 378, 186–194.

Gropengiesser, C. (1925) Untersuchungen über die Symbiose der Blattiden mit niederen pflanzlichen Organismen. Zentralbl. Bakteriol. Parasitenk., Abt. II, 64, 495–511.

Grosbüsch, J. (1915) Über eine farblose, stark roten Farbstoff erzeugende *Torula.* Zentralbl. Bakteriol. Parasitenk., Abt. II, 42, 625–638.

Grose, E.S. & Marinkelle, C.J. (1968) A new species of *Candida* from Colombian bats. Mycopathol. Mycol. Appl. 36, 225–227.

Grüss, J. (1918) Die Anpassung eines Pilzes *(Anthomyces reukaufii)* an den Blütenbau und den Bienenrüssel. Ber. Dtsch. Bot. Ges. 35, 746–761.

Grüss, J. (1926a) Über einige selten vorkommenden Nektarhefen. Wochenschr. Brau. 43, 57–61.

Grüss, J. (1926b) Genetische und gärungsphysiologische Untersuchungen an Nektarhefen. Jahrb. Wiss. Bot. 66, 109–182.

Grüss, J. (1927) Über hochalpine Nektarhefen. II. *Mikroanthomyces alpinus* nov. gen. et spec. Wochenschr. Brau. 44, 243–246, 253–257.

Guéguen, F. (1904) Les champignons parasites de l'homme et des animaux. Thèse, Paris.

Guého, E. (1979) Deoxyribonucleic acid base composition and taxonomy in the genus *Geotrichum* Link. Antonie van Leeuwenhoek 45, 199–210.

Guerra, P. (1935) Rôle des levures en dermatologie. Thèse, Paris.

Guiart, J. (1910) Saccharomycoses. In: Précis de Parasitologie. Bibliothèque Gilbert et Fournier, pp. 106–109.

Guilliermond, A. (1909) Recherches cytologiques et taxonomiques sur les Endomycétacées. Rev. Gen. Bot. 21, 353–391, 401–419.

Guilliermond, A. (1912) Les Levures. Encyclopédie Scientifique. O. Doin et Fils, Paris.

Guilliermond, A. (1913) Nouvelles observations sur la sexualité des levures. 1. Existence d'une copulation hétérogamique observée dans une espèce nouvelle. 2. Sur la copulation de *Debaryomyces globosus.* 3. Sur les phénomènes de retrogradation de la sexualité constatés dans plusieurs levures. Arch. Protistenk. 28, 52–77.

Guilliermond, A. (1914) Monographie des levures d'Afrique occidentale par la mission chevalière. Ann. Sci. Nat. Bot. Biol. Végétale, Sér. 9, 19, 1–32.

Guilliermond, A. (1918) *Zygosaccharomyces Nadsonii:* Nouvelle espèce de levures à conjugaison hétérogamique. Bull. Soc. Mycol. France 34, 1–16.

Guilliermond, A. (1919) Sur une nouvelle levure à copulation hétérogamique. C.R. Soc. Biol. 82, 466–470.

Guilliermond, A. (1928) Recherches sur quelques Ascomycètes inférieurs isolés de la stigmatomycose des graines de cotonnier: Essai sur la phylogénie des Ascomycètes. Rev. Gen. Bot. 40, 328–342.

Guilliermond, A. & Negroni, P. (1929) Sur la présence d'une copulation hétérogamique dans le *Saccharomyces marxianus.* C.R. Soc. Biol. 101, 564–566.

Guilliermond, A. & Péju, G. (1919) Sur un nouveau champignon présentant des caractères intermédiaires entre les levures et les *Endomyces.* C.R. Soc. Biol. 82, 1343.

Gunge, N. & Tamaru, A. (1978) Genetic analysis of products of protoplast fusion in *Saccharomyces.* Jpn. J. Genetics 53, 41–49.

Gustafson, B.A. (1955) Otitis externa in the dog. A bacteriological and experimental study. Thesis, Stockholm.

Hagler, A.N. & Ahearn, D.G. (1981) A rapid DBB test to detect basidiomycetous affinity of yeasts. Int. J. Syst. Bacteriol. 31, 204–208.

Hajsig, M. (1958) *Torulopsis apicola* nov. spec., new isolates from bees. Antonie van Leeuwenhoek 24, 18–22.

Hammer, B.W. (1919) Studies on formation of gas in sweetened condensed milk. Iowa State Coll. Agr. Exp. St. Research Bull. 54, 211–220.

Hammer, B.W. & Cordes, W.A. (1920) A study of lactose-fermenting yeasts present in 'yeasty' cream. Iowa State Coll. Agric. Exp. Stat. Res. Bull. 61, 1–24.

Hansen, E.C. (1883) Undersøgelser over alkoholgjaersvampenes fysiologi og morfologi. II. Om askosporedannelsen hos slaegten *Saccharomyces.* Medd. Carlsberg Lab. 2, 29–86.

Hansen, E.C. (1886) Les voiles chez le genre *Saccharomyces.* I. C.R. Trav. Lab. Carlsberg 2, 106–136.

Hansen, E.C. (1888a) Recherches sur la physiologie et la morphologie des ferments alcooliques. VII. Action des ferments alcooliques sur les diverses espèces de sucre. C.R. Trav. Lab. Carlsberg 2, 143–167.

Hansen, E.C. (1888b) Undersøgelser over alkoholgjaersvampenes fysiologi og morfologi. VII. Om alkoholgjaersvampenes forhold til sukkerarterne. Medd. Carlsberg Lab. 2, 220–256.

Hansen, E.C. (1889) Ueber die im Schleimflusse lebender Bäume beobachteten Mikroorganismen. Zentralbl. Bakteriol. Parasitenk., Abt. II, 5, 632–640.

Hansen, E.C. (1891) Sur la germination des spores chez les *Saccharomyces.* Ann. Micrographic. 3, 449–474.

Hansen, E.C. (1904) Grundlinien zur Systematik der Saccharomyceten. Zentralbl. Bakteriol. Parasitenk., Abt. II, 12, 529–538.

Hansen, E.C. (1908) Recherches sur la physiologie et la morphologie des ferments alcooliques. XIII. Nouvelles études sur les levures de brasserie à fermentation basse. C.R. Trav. Lab. Carlsberg 7, 179–217.

Harrison, F.C. (1928) A systematic study of some Torulas. Trans. R. Soc. Can., Sect. V, 22, 187–225.

Hart, P.D., Russell, E., Jr. & Remington, J.S. (1969) The compromised host and infection. II. Deep fungal infection. J. Infect. Dis. 120, 169–191.

Hartmann, M. (1903) Eine rassenspaltige *Torula*-Art, welche nur zeitweise Maltose zu vergären vermag *(Torula colliculosa* n. sp.). Wochenschr. Brau. 20, 113–114.

Hasegawa, T. (1958) Studies on the genus *Rhodotorula* (II). The mycological properties of the *Rhodotorula* strains preserved in Japan. J. Ferment. Technol. (Japan) 36, 194–202.

Hasegawa, T. (1959) The lactose assimilating species in the genus *Rhodotorula.* J. Gen. Appl. Microbiol. 5, 30–34.

Hasegawa, T. & Banno, I. (1958) Studies on the genus *Rhodotorula* (III). On the nitrate utilization of *Rhodotorula.* J. Ferment. Technol. (Japan) 36, 403–406.

Hasegawa, T. & Banno, I. (1964) *Rhodotorula infirmo-miniata* (Okunuki) comb. n. Antonie van Leeuwenhoek 30, 65–67.

Hasegawa, T., Banno, I. & Yamauchi, S. (1960) A taxonomic study on the genus *Rhodotorula.* J. Gen. Appl. Microbiol. 6, 196–215.

Hawthorne, D.C. (1963) Directed mutation of the mating type alleles as an explanation of homothallism in yeast. Proc. 11th Int. Congr. Gent. 1, 34–35.

Haynes, W.C., Wickerham, L.J. & Hesseltine, C.W. (1955) Maintenance of cultures of industrially important microorganisms. Appl. Microbiol. 3, 361–368.

Healy, M.E., Dillavon, C.L. & Taylor, G.E. (1977) Diagnostic medium containing inositol, urea, and caffeic acid for selective growth of *Cryptococcus neoformans.* J. Clin. Microbiol. 6, 387–391.

Hedrick, L.R. (1976) *Candida fluviatilis* sp. nov. and other yeasts from aquatic environments. Antonie van Leeuwenhoek 42, 329–332.

Hedrick, L.R. & Burke, G.C. (1951) Two new yeasts from Hawaiian fruit flies. Mycopathol. Mycol. Appl. 6, 92–95.

Hedrick, L.R. & Dupont, P. (1968) Two new yeasts: *Trichosporon aquatile* and *Trichosporon eriense* spp. n. Antonie van Leeuwenhoek 34, 474–482.

Henneberg, W. (1926) Handbuch der Gärungsbakteriologie. II. 2te Aufl., Berlin.

Hennebert, G.L. & Weresub, L.K. (1977) The terms for states and forms of fungi, their names and types. Mycotaxon 6, 207–211.

Henninger, W. & Windisch, S. (1975a) *Pichia lindnerii* sp. n., eine neue Methanol assimilierende Hefe aus Erde. Arch. Microbiol. 105, 47–48.

Henninger, W. & Windisch, S. (1975b) A new yeast of *Sterigmatomyces, S. aphidis* sp. n. Arch. Microbiol. 105, 49–50.

Henninger, W. & Windisch, S. (1976a) *Kluyveromyces blattae* sp. n., eine neue vielsporige Hefe aus *Blatta orientalis.* Arch. Microbiol. 109, 153–156.

Henninger, W. & Windisch, S. (1976b) *Torulopsis spandovensis* sp. n., a new yeast from beer. Arch. Microbiol. 107, 205–206.

Henry, B.S. (1947) The viability of yeast cultures preserved under mineral oil. J. Bacteriol. 54, 264.

Herman, A.I. (1971) Mating responses in the yeast *Hansenula holstii.* Antonie van Leeuwenhoek 37, 275–280.

Herman, A.I., Wickerham, L.J. & Griffin, P. (1966) Allelic determinants of agglutination in the yeast *Hansenula wingei.* Genetics 54, 339.

Herrera, T. & Ulloa, M. (1978) Descripcion de una especie nueva de levadura *Candida queretana,* aislada del tepache de queretaro, Mexico. Bol. Soc. Mex. Microbiol. 12, 13–18.

Herrera, T., Ulloa, M. & Fuentes, I. (1973) Descripcion de una especie nueva de *Hansenula* y una variedada nueva de *Candida tropicalis* aisladas del pozol. Bol. Soc. Mex. Microbiol. 7, 17–26.

Hicks, J.B. & Herskowitz, I. (1976) Interconversion of yeast mating types. I. Direct observations on the action of the homothallism (HO) gene. Genetics 83, 245–258.

Hirosawa, N. & Takada, H. (1969) Salt susceptibility of *Rhodotorula glutinis* var. *salinaria* isolated from salt foam in Japan. Trans. Mycol. Soc. Japan 10, 35–39.

Hjort, A. (1956) Notes on *Saccharomyces rouxii* Boutroux and other yeasts with special regard to their life cycles. C.R. Trav. Lab. Carlsberg, Sér. physiol., 26, 161–181.

Holst, E.C. (1936) *Zygosaccharomyces pini,* a new species of yeast associated with bark beetles in pines. J. Agric. Res. 53, 513–518.

Holzschu, D.L. (1981) Molecular taxonomy and evolutionary relationships among cactophilic yeasts. Thesis, University of California, Davis.

Holzschu, D.L., Presley, H.L., Miranda, M. & Phaff, H.J. (1979a) Identification of *Candida lusitaniae* as an opportunistic yeast in humans. J. Clin. Microbiol. 10, 202–205.

Holzschu, D.L., Chandler, F.W., Ajello, L. & Ahearn, D.G. (1979b) Evaluation of industrial yeasts for pathogenecity. Sabouraudia 17, 71–78.

Holzschu, D.L., Tredick, J. & Phaff, H.J. (1981) Validation of the yeast *Sporidiobolus ruinenii* based on its DNA relatedness to other species of the genus *Sporidiobolus.* Curr. Microbiol. 5, 73–76.

Hopfer, R.L. & Blank, F. (1975) Caffeic acid-containing medium for identification of *Cryptococcus neoformans.* J. Clin. Microbiol. 2, 115–120.

Hopfer, R.L. & Gröschel, D. (1975) Six hour pigmentation test for the identification of *Cryptococcus neoformans.* J. Clin. Microbiol. 2, 96–98.

Hopsu-Havu, V.K., Tunnela, E. & Yarrow, D. (1978) *Candida sonckii* sp. nov. Antonie van Leeuwenhoek 44, 425–438.

Hufschmitt, G., Sartory, A., Sartory, R., & Meyer, J. (1931) Un cas de blastomycose cutanée à foyers multiples. Ann. Dermatol. Syphiligr., Sér. VII, 2, 850–876.

Hunt, D.E. & Carpenter, P.L. (1963) Sterols as components in the mating reaction of *Hansenula wingei.* J. Bacteriol. 86, 845–847.

Hütter, R. & DeMoss, J.A. (1967) Organization of the tryptophan pathway: a phylogenetic study of the fungi. J. Bacteriol. 94, 1896–1907.

Iizuka, H. & Goto, S. (1965) Microbiological studies on petroleum and natural gas. VIII. Determination of red yeasts isolated from oil brines and related materials. J. Gen. Appl. Microbiol. 11, 331–337.

Imšenecky, A. & Kriss, A. (1933) Matériaux pour l'étude des Fungi imperfecti 1. Un nouveau champignon, le *Mastigomyces phillippovi* (nov. gen. nov. sp.) trouvé sur les fruits de *Terminalia chebula.* C.R. Acad. Sci. URSS 7, 111–116 (in Russian with German summary).

Ito, H., Iizuka, H. & Sato, T. (1974) A new radioresistant yeast of *Trichosporon oryzae* nov. sp. isolated from rice. Agric. Biol. Chem. 38, 1597–1620.

Jacob, F. (1969a) *Pichia tannicola,* nouvelle espèce de levure isolée de liqueurs tannantes végétales. Bull. Soc. Mycol. France 85, 111–115.

Jacob, F. (1969b) *Pichia adzetii* et *Pichia abadiae,* nouvelles espèces de levures isolées de liqueurs tannantes végétales. Bull. Soc. Mycol. France 85, 117–127.

Jacob, F. (1970) Deux espèces nouvelles de levures asporogènes isolées de liqueurs tannantes végétales. Ann. Inst. Pasteur 118, 207–213.

Janke, A. (1924) Allgemeine Technische Mikrobiologie. I. Teil: Die Mikroorganismen. Technische Fortschrittberichte Band IV, Verlag Steinkopff, Dresden und Leipzig.

Janke, A. (1928) Über die Formgattung *Kloeckera* Janke. Zentralbl. Bakteriol. Parasitenk., Abt. II, 76, 161.

Janke, A. (1930) *Mycoderma lafarii* n. sp. Arch. Mikrobiol. 1, 176–180.

Janke, A. (1954) *Sporobolomyces roseus* var. *madurae* var. nov. und die Beziehungen zwischen den Genera *Sporobolomyces* und *Bullera.* Zentralbl. Bakteriol. Parasitenk., Orig. 161, 514–520.

Jensen, V. (1967) Taxonomic studies on soil yeasts. 1. The genus *Saccharomyces* (Meyen) Reess. K. Vet. Landboh. Arsskr. 179–194.

Johannsen, E. (1978) Hybridization studies within the genus *Kluyveromyces* van der Walt emend. van der Walt. Thesis, Rhodes University, Grahamstown, South Africa.

Johannsen, E. (1980) Hybridization studies within the genus *Kluyveromyces.* Antonie van Leeuwenhoek 46, 177–189.

Johannsen, E. (1981) Hybridization studies within the genus *Schizosaccharomyces* Lindner. Can. J. Microbiol. 27, 184–191.

Johannsen, E. & Walt, J.P. van der (1978) Interfertility as basis for the delimitation of *Kluyveromyces marxianus.* Arch. Microbiol. 118, 45–48.

Johannsen, E. & Walt, J.P. van der (1980) Hybridization studies within the genus *Schwanniomyces* Klöcker. Can. J. Microbiol. 26, 1199–1203.

Johnson, B.F. (1967) Growth of the fission yeast, *Schizosaccharomyces pombe,* with late, eccentric, lytic fission in an unbalanced medium. J. Bacteriol. 94, 192–195.

Johnson, E.A. & Phaff, H.J. (1978) *Rhodotorula fujisanensis,* a new taxonomic combination. Curr. Microbiol. 1, 223–225.

Johnson, E.A., Conklin, D.E. & Lewis, M.J. (1977) The yeast *Phaffia rhodozyma* as a dietary pigment source for salmonids and crustaceans. J. Fish. Board. Can. 34, 2417–2421.

Johnson, E.A., Villa, T.G., Lewis, M.J. & Phaff, H.J. (1978) A simple method for the isolation of astaxanthin from the basidiomycetous yeast, *Phaffia rhodozyma.* Appl. Environ. Microbiol. 35, 1155–1159.

Johnson, J.L. (1973) Use of nucleic acid homologies in the taxonomy of anaerobic bacteria. Int. J. Syst. Bacteriol. 23, 308–315.

Jones, C.P. & Martin, D.S. (1938) Identification of yeast-like organisms isolated from the vaginal tracts of pregnant and non-pregnant women. Ann. J. Obstet. Gynecol. 35, 98–106.

Jones, E.B.G. & Slooff, W.Ch. (1966) *Candida aquatica* sp. n. isolated from water scums. Antonie van Leeuwenhoek 32, 223–228.

Jong, S.C. & King, D.S. (1976) A nomenclatural note concerning *Sarcinomyces* Lindner and *Sarcinomyces* Oho. Mycotaxon 3, 397–400.

Jong, S.C. & King, D.S. (1977) Identity of *Sterigmatomyces aphidis* and *Trichosporon oryzae.* Mycotaxon 6, 11–16.

Jörgensen, A. (1909) Die Mikroorganismen der Gärungsindustrie. 5te Auflage, P. Parey, Berlin.

Joshi, K.R. & Gavin, J.B. (1974) A simple laboratory method for the rapid identification of *Candida albicans.* Pathology 6, 231–233.

Jülich, W. (1981) Higher taxa of Basidiomycetes. Bibliotheca Mycologica 85, J. Cramer.

Jurzitza, G. (1962) Die Symbiose der Anobiiden und Cerambyciden mit hefeartigen Pilzen. Arch. Mikrobiol. 43, 412–424.

Jurzitza, G. (1970) Über Isolierung, Kultur und Taxonomie einiger Anobiidensymbionten (Insecta, Coleoptera). Arch. Mikrobiol. 72, 203–222.

Jurzitza, G., Kühlwein, C. & Kreger-van Rij, N.J.W. (1960) Zur Systematik einiger Cerambycidensymbionten. Arch. Mikrobiol. 36, 229–243.

Kahanpää, A. (1971) Bronchopulmonary occurrence of fungi in adults. Acta Pathol. Microbiol. Scand. Suppl. 227, 147.

Kahn, S.R. (1976) Ultrastructure of the septal pore apparatus of *Tremella.* J. Gen. Microbiol. 97, 339–342.

Kambayashi, T. (1923) Jpn. J. Dermatol. Urol. 23 (cited from Lodder & Kreger-van Rij, 1952).

Kamienski, T. (1899) Notice préliminaire sur l'espèce de *Metschnikowia (Monospora* Metschn.). Trav. Soc. Imp. Natural. St. Pétersbourg 30, 363–364.

Karamboloff, N. & Krumbholz, G. (1931) Untersuchungen über osmophile Sprosspilze. IV *Zygosaccharomyces gracilis* sp. n. Arch. Mikrobiol. 3, 113–121.

Kato, K., Kurimura, Y., Makiguchi, N. & Asai, Y. (1974) Determination of strongly methanol assimilating yeasts. J. Gen. Appl. Microbiol. 20, 123–127.

Kawakami, N. (1960) Mem. Fac. Eng. Hiroshima Univ. 1, 207–237.

Kawakami, N. & Nehira, T. (1958a) Electron microscopy of fungi. IV. The morphological study of the spore of pellicle group in the genus *Debaryomyces.* Electron Microscopy 7, 33–35.

Kawakami, N. & Nehira, T. (1958b) Electron microscopy of fungi (VI). Sporulation and germination of *Endomycopsis capsularis.* Trans. Mycol. Soc. Japan 8, 3–5.

Kawakami, N., Nehira, T. & Kodama, K. (1961) Electron microscopy of fungi. X. Spore forms of the genus *Pichia.* Mem. Fac. Eng. Hiroshima Univ. 1, 407–414.

Kayser, E. (1892) Contribution à l'étude des levures de vin. Ann. Inst. Pasteur 6, 569–583.

Keddie, F.M. (1967) Tinea versicolor: the electron-microscopic morphology of the genera *Malassezia* and *Pityrosporum.* In: W. Jadassohn & C.G. Shirren (Eds.), XIII Congressus Internationalis Dermatologiae, Vol. 2, Springer Verlag, Berlin.

Keddie, F.M. & Shadomy, S. (1963) Etiological significance of *Pityrosporum orbiculare* in tinea versicolor. Sabouraudia 3, 21–25.

Keilin, D. (1920) On a new Saccharomycete *Monosporella unicuspidata* gen. n. nom., n. sp., parasitic in the body cavity of a dipterous larva *(Dasyhelea obscura* Winnertz). Parasitol. 12, 83–91.

Khan, S.R. (1976) Ultrastructure of the septal pore apparatus of *Tremella.* J. Gen. Microbiol. 97, 339–342.

Khan, S.R. & Kimbrough, J.W. (1980) Septal ultrastructure in some genera of the Tremellaceae. Can. J. Bot. 58, 55–60.

Kiehn, T.E., Edwards, F.F. & Armstrong, D. (1980) The prevalence of yeasts in clinical specimens from cancer patients. Am. J. Clin. Pathol. 73, 518–521.

King, D.S. & Jong, S.C. (1975) *Sarcinosporon:* A new genus to accommodate *Trichosporon inkin* and *Prototheca filamenta.* Mycotaxon 3, 89–94.

King, D.S. & Jong, S.C. (1976a) *Aciculoconidium:* A new hyphomycetous genus to accomodate *Trichosporon aculeatum.* Mycotaxon 3, 401–408.

King, D.S. & Jong, S.C. (1976b) Induction of arthroconidia in *Trichosporon.* Mycopathol. Mycol. Appl. 59, 61–63.

King, D.S. & Jong S.C. (1977) A contribution to the genus *Trichosporon.* Mycotaxon 6, 391–417.

Kirsop, B. (1955) Maintenance of yeasts by freeze-drying. J. Inst. Brew. 61, 466–471.

Kirsop, B. (1974) The stability of biochemical, morphological and brewing properties of yeast cultures maintained by subculturing and freeze-drying. J. Inst. Brew. 80, 585–570.

Kleyn, J.G. (1954) A study of some environmental factors controlling sporulation of yeast using a new sporulation medium. Wallerstein Lab. Commun. 17, 91–104.

Klöcker, A. (1902) Eine neue *Saccharomyces*art *(Saccharomyces saturnus* mihi) mit eigentümlichen Sporen. Zentralbl. Bakteriol. Parasitenk., Abt. II, 8, 129–130.

Klöcker, A. (1909a) Deux nouveaux genres de la famille des Saccharomycètes. C.R. Trav. Lab. Carlsberg 7, 273–278.

Klöcker, A. (1909b) *Endomyces javanensis* nov. sp. C.R. Trav. Lab. Carlsberg 7, 267–272.

Klöcker, A. (1912a) Untersuchungen über einige neue *Pichia*-Arten. Zentralbl. Bakteriol. Parasitenk., Abt. II, 35, 369–375.

Klöcker, A. (1912b) Beschreibungen von 17 *Saccharomyces apiculatus* Formen. Zentralbl. Bakteriol. Parasitenk., Abt. II, 35, 375–388.

Klöcker, A. (1924) Die Gärungsorganismen in der Theorie und Praxis der Alkoholgärungsgewerbe. 3te Aufl., Urban & Schwarzenberg, Berlin–Wien.

Klokke, A.H. & Kamp, H. (1962) Een systematisch onderzoek naar animale mycosen bij vee en mens. Ned. Tijdschr. Geneesk. 106, 1673–1675.

Kluyver, A.J. (1914) Biochemische suikerbepalingen. Thesis, Technische Hoogeschool, Delft.

Kluyver, A.J. & Niel, C.B. van (1925) Über Spiegelbilder erzeugende Hefenarten und die neue Hefengattung *Sporobolomyces.* Zentralbl. Bakteriol. Parasitenk., Abt. II, 63, 1–20.

Kobayashi, T. (1953) Studies on the yeasts and yeast-like fungi which grow well on wood hydrolysate medium. Rep. Wood Saccharification Discussion Committee 2, 1–93.

Kobayashi, Y. (1937) On the genus *Holtermannia* of Tremellaceae. Tokyo Bun. Daigaku Sci. Rep., Sect. B3, 75–81.

Kobayashi, Y. (1939) The genus *Tremella* and its allies from Japan. (Fungorum Ordinis Tremellales Studia Monographica II). Tokyo Bun. Daigaku Sci. Rep. 4, 1–26.

Kobayashi, Y. & Tsubaki, K. (1965) Studies on cultural characters and asexual reproduction of Heterobasidiomycetes, I. Trans. Mycol. Soc. Japan 5, 29–35.

Kocková-Kratochvílová, A. (1976) Taxometric study of the genus *Saccharomyces* (Meyen) Reess, 3rd Part: Small species. Biol. Pracé 22, Slovak Acad. Sci.

Kocková-Kratochvílová, A. & Ondrušová, D. (1971) *Torulopsis*arten aus den Oberflächen höherer Pilze. *Torulopsis kruisii* n. sp. und *Torulopsis schatavii* n. sp. Biologia (Bratislava) 26, 447–485.

Kocková-Kratochvílová, A. & Svobodová-Poláková, Y. (1959) *Nadsonia slovaca* n. sp. isolated in Slovakia. J. Gen. Microbiol. 20, 24–26.

Kocková-Kratochvílová, A., Sláviková, E. & Beránek, J. (1977a) *Torulopsis armenti* n. sp. Z. Allg. Mikrobiol. 17, 429–431.

Kocková-Kratochvílová, A., Sláviková, E., Zemek, J. & Kuniak, L. (1977b) The heterogeneity of the genus *Trichosporon.* Proc. 5th Intern. Spec. Symposium Yeasts, p. 9.

Kocková-Kratochvílová, A., Sláviková, E. & Jensen, V. (1978) Numerical taxonomy of the genus *Debaryomyces* Lodder et Kreger-van Rij. J. Gen. Microbiol. 104, 257–268.

Kocourek, J. & Ballou, C.E. (1969) Method for fingerprinting yeast cell wall mannans. J. Bacteriol. 100, 1175–1181.

Kodama, K. (1969) Electron microscopy of yeasts spore forms of some yeast genera. In: A. Kocková-Kratochvílová (Ed.), Proc. 2nd Symp. Yeasts, Bratislava, 1966, pp. 45–49.

Kodama, K. (1974) Ascosporogenous yeasts isolated from tree exudates in Japan. Ann. Microbiol. Milano 24, 215–231.

Kodama, K. (1975) New species of *Pichia* isolated from tree exudates in Japan. J. Ferment. Technol. 53, 626–630.

Kodama, K. & Kyono, T. (1974a) Ascosporogenous yeasts isolated from tree exudates in Japan. J. Ferment. Technol. 52, 1–9.

Kodama, K. & Kyono, T. (1974b) Ascosporogenous yeasts isolated from tree exudates in Japan (continued). J. Ferment. Technol. 52, 605–613.

Kodama, K., Kyono, T. & Kodama, S. (1955) A study of film-forming yeasts (II). Genus *Pichia* (2). J. Ferment. Technol. 33, 317–324.

Kodama, K., Kyono, T. & Kodama, S. (1962) Taxonomic study on the genus *Pichia* Hansen (2). J. Gen. Appl. Microbiol. 8, 52–55.

Kodama, K., Kyono, T., Iida, K. & Onoyama, N. (1964a) Studies on wild yeasts which thrive in 'sake-moto'. J. Ferment. Technol. 42, 739–745.

Kodama, K., Kyono, T., Naganishi, H. & Takahara, Y. (1964b) *Debaryomyces franciscae* (Capriotti) nov. comb. J. Ferment. Technol. 42, 1–6.

Kodama, K., Kyono, T., Naganishi, H. & Takahara, Y. (1964c) Taxonomic studies on the genus *Debaryomyces.* (IV) *Debaryomyces nilssoni* (Capriotti) nov. comb. J. Ferment. Technol. 42, 655–660.

Kolfschoten, G.A. & Yarrow, D. (1970) *Brettanomyces naardenensis,* a new yeast from soft drinks. Antonie van Leeuwenhoek 36, 458–460.

Komagata, K. & Nakase, T. (1965) New species of the genus *Candida* isolated from frozen foods. J. Gen. Appl. Microbiol. 11, 255–267.

Komagata, K., Nakase, T. & Katsuya, N. (1964) Assimilation of hydrocarbons by yeast. II. Determination of hydrocarbon-assimilating yeasts. J. Gen. Appl. Microbiol. 10, 323–331.

Konokotina, A.G. (1913) Über die neuen Hefepilze mit heterogamer Kopulation – *Nadsonia (Guilliermondia) elongata* und *Debaryomyces tyrocola* (in Russian). Bull. Jard. Imp. Bot. St. Pétersbourg 13, 32–46 (rev. in: Zentralbl. Bakteriol. Parasitenk., Abt. II, 40, 177).

Kostka, G. (1927) Über *Nadsonia Richteri* nov. spec., eine interessante Schleimflusshefe aus Mähren. Verh. Naturforsch. Ver. Brunn, 59 (rev. in: Wochenschr. Brau. 44, 128).

Kozlova, T.M. & Meisel (1975) Possible participation of mitochondria in formation of peroxisomes in yeast organisms. Microbiology 45, 949–950.

Kraepelin, G. & Schulze, U. (1982) *Sterigmatosporidium* gen. n., a new heterothallic basidiomycetous yeast, the perfect state of a new species of *Sterigmatomyces* Fell. Antonie van Leeuwenhoek 48, 471–483.

Krassilnikov, N.A., Babjeva, I.P. & Meavahd, K. (1967) A new genus of soil yeast *Zygolipomyces* nov. gen. Microbiologiya 36, 923–931.

Kreger-van Rij, N.J.W. (1961) Taxonomy of *Cryptococcus neoformans* and its variety *uniguttulatus.* Antonie van Leeuwenhoek 17, 59–64.

Kreger-van Rij, N.J.W. (1964a) *Endomycopsis vini* and *Pichia etchellsii,* spp. n. Antonie van Leeuwenhoek 30, 428–432.

Kreger-van Rij, N.J.W. (1964b) A taxonomic study of the yeast genera *Endomycopsis, Pichia* and *Debaryomyces.* Thesis, Leiden.

Kreger-van Rij, N.J.W. (1965) *Candida ciferrii,* a new yeast species. Mycopathol. Mycol. Appl. 26, 49–52.

Kreger-van Rij, N.J.W. (1966) *Kluyveromyces osmophilus,* a new yeast species. Mycopathol. Mycol. Appl. 29, 137–141.

Kreger-van Rij, N.J.W. (1969a) Taxonomy and systematics of yeasts. In: A.H. Rose & J.S. Harrison (Eds.), The Yeasts. Vol. I, Academic Press, London–New York, pp. 5–78.

Kreger-van Rij, N.J.W. (1969b) A new feature of the ascospores of *Saccharomycodes ludwigii* Hansen. Can. J. Microbiol. 15, 823–825.

Kreger-van Rij, N.J.W. (1969c) Some features of yeast ascospores observed under the electron microscope. A. Kocková-Kratochvílová (Ed.), Proc. 2nd Symp. Yeasts, Bratislava, 1966, pp. 169–176.

Kreger-van Rij, N.J.W. (1970a) The genus *Debaryomyces* Lodder et Kreger-van Rij nom. conserv. In: J. Lodder (Ed.), The Yeasts, a taxonomic study. North-Holland Publ. Co., Amsterdam, pp. 129–156.

Kreger-van Rij, N.J.W. (1970b) The genus *Endomycopsis* Dekker. In: J. Lodder (Ed.), The Yeasts, a taxonomic study. North-Holland Publ. Co., Amsterdam, pp. 166–208.

Kreger-van Rij, N.J.W. (1970c) The genus *Pichia* Hansen. In: J. Lodder (Ed.), The Yeasts, a taxonomic study, North-Holland Publ. Co., Amsterdam, pp. 455–554.

Kreger-van Rij, N.J.W. (1977a) Electron microscopy of sporulation in *Schwanniomyces alluvius*. Antonie van Leeuwenhoek 43, 55–64.

Kreger-van Rij, N.J.W. (1977b) Ultrastructure of *Hanseniaspora* ascospores. Antonie van Leeuwenhoek 43, 225–232.

Kreger-van Rij, N.J.W. (1978a) Ultrastructure of the ascospores of the new yeast genus *Sporopachydermia* Rodrigues de Miranda. Antonie van Leeuwenhoek 44, 451–456.

Kreger-van Rij, N.J.W. (1978b) Electron microscopy of germinating ascospores of *Saccharomyces cerevisiae*. Arch. Microbiol. 117, 73–77.

Kreger-van Rij, N.J.W. (1979) A comparative ultrastructural study of the ascospores of some *Saccharomyces* and *Kluyveromyces* species. Arch. Microbiol. 121, 53–59.

Kreger-van Rij, N.J.W. (1980) Generic differentiation of yeasts. In: F.A. Skinner, S.M. Passmore & R.R. Davenport (Eds.), Biology and Activities of Yeasts. Society for Applied Bacteriology Symposium Series No. 9. Academic Press, London, pp. 29–52.

Kreger-van Rij, N.J.W. & Ahearn, D.G. (1968) Shape and structure of the ascospores of *Hanseniaspora uvarum*. Mycologia 60, 604–612.

Kreger-van Rij, N.J.W. & Veenhuis, M. (1969a) Septal pores in *Endomycopsis platypodis* and *Endomycopsis monospora*. J. Gen. Microbiol. 57, 91–96.

Kreger-van Rij, N.J.W. & Veenhuis, M. (1969b) A study of vegetative reproduction in *Endomycopsis platypodis* by electron microscopy. J. Gen. Microbiol. 58, 341–346.

Kreger-van Rij, N.J.W. & Veenhuis, M. (1970) An electron microscope study of the yeast *Pityrosporum ovale*. Arch. Mikrobiol. 71, 123–131.

Kreger-van Rij, N.J.W. & Veenhuis, M. (1971a) A comparative study of the cell wall structure of basidiomycetous and related yeasts. J. Gen. Microbiol. 68, 87–95.

Kreger-van Rij, N.J.W. & Veenhuis, M. (1971b) Septal pores in *Trichosporon cutaneum*. Sabouraudia 9, 36–38.

Kreger-van Rij, N.J.W. & Veenhuis, M. (1971c) Bipolar budding in yeasts – an electron microscope study. Antonie van Leeuwenhoek 37, 125–136.

Kreger-van Rij, N.J.W. & Veenhuis, M. (1972) Some features of vegetative and sexual reproduction in *Endomyces* species. Can. J. Bot. 50, 1691–1695.

Kreger-van Rij, N.J.W. & Veenhuis, M. (1973a) Electron microscopy of septa in ascomycetous yeasts. Antonie van Leeuwenhoek 39, 481–490.

Kreger-van Rij, N.J.W. & Veenhuis, M. (1973b) Electron microscopy of some special cell contacts in yeasts. J. Bacteriol. 113, 350–356.

Kreger-van Rij, N.J.W. & Veenhuis, M. (1975a) Electron microscopy of ascus formation in the yeast *Debaryomyces hansenii*. J. Gen. Microbiol. 89, 256–264.

Kreger-van Rij, N.J.W. & Veenhuis, M. (1975b) Conjugation in the yeast *Saccharomycopsis capsularis* Schiönning. Arch. Microbiol. 104, 263–269.

Kreger-van Rij, N.J.W. & Veenhuis, M. (1976a) Conjugation in the yeast *Guilliermondella selenospora* Nadson et Krassilnikov. Can. J. Microbiol. 22, 960–966.

Kreger-van Rij, N.J.W. & Veenhuis, M. (1976b) Ultrastructure of the ascospores of some species of the *Torulaspora* group. Antonie van Leeuwenhoek 42, 445–455.

Kroemer, K. & Heinrich, F. (1922) Über eine in überschwefelten Mosten auftretende Hefe der Gattung *Saccharomycodes*. Festschr. Staatl. Lehr. Forschungsanstalt Wein-, Obst- und Gartenbau. Geisenheim, pp. 258–294.

Kroemer, K. & Krumbholz, G. (1931) Untersuchungen über osmophile Sprosspilze. I. Beiträge zur Kenntnis der Gärungsvorgänge und der Gärungserreger der Trockenbeerenauslesen. Arch. Mikrobiol. 2, 352–410.

Krumbholz, G. (1931) Untersuchungen über osmophile Sprosspilze. II. Beiträge zur Kenntnis der Gattung *Zygosaccharomyces*. Arch. Mikrobiol. 2, 411–492.

Krumbholz, G. (1932) Über eine pastoriane Hefe von Weinbeeren, *Saccharomyces elongatus* sp. n. Zentralbl. Bakteriol. Parasitenk., Abt. II, 86, 206–214.

Krumbholz, G. (1933) Ist die Beibehaltung einer Gattung *Torulaspora* berechtigt? Arch. Mikrobiol. 4, 167–169.

Krumbholz, G. & Tauschanoff, W. (1933) *Mycotorula intermedia* n. sp., ein Beitrag zur Kenntnis der Gärungserreger im Wein. Zentralbl. Bakteriol. Parasitenk., Abt. II, 88, 366–373.

Krzemecki, A. (1913) Zentralbl. Bakteriol. Parasitenk., Abt. II, 38, 577.

Kudriavzev, V.I. (1954) Yeast Systematics. U.S.S.R. Academy of Sciences, Moscow (in Russian).

Kudriavzev, V.I. (1960) Die Systematik der Hefen. Akademie Verlag, Berlin.

Kufferath, H. (1920a) Peut-on obtenir du moût de bière alcalin? Ann. Bull. Soc. Roy. Sci. Méd. Nat. Bruxelles 74, 16–46.

Kufferath, H. (1920b) C.R. Soc. Biol. Belge 185 (cited from Custers, 1940).

Kufferath, H. (1925) Études sur les levures du lambic. Isolement, caractères de culture et biologie. Ann. Soc. Zymol. 1, 7–17.

Kufferath, H. & Laer, M.H. van (1921) Études sur les levures de lambic. Leur action chimique sur les milieux de culture. Bull. Soc. Chim. Belge 30, 270–276.

Kumbhojkar, M.S. (1972) *Schizosaccharomyces slooffiae* Kumbhojkar, a new species of osmophilic yeasts from India. Curr. Sci. 41, 151–152.

Kuraishi, H. (1958a) Sci. Rep. Tôhoku Univ. 24, 59.

Kuraishi, H. (1958b) *Pichia megalospora* nov. spec., a new yeast from decomposing mushrooms. Sci. Rep. Tôhoku Univ., IV Ser. Biol. 24, 109–111.

Kurtzman, C.P. (1973) Formation of hyphae and chlamydospores by *Cryptococcus laurentii.* Mycologia 65, 388–395.

Kurtzman, C.P. (1984) Resolution of varietal relationships within the species *Hansenula anomala, Hansenula bimundalis,* and *Pichia nakazawae* through comparisons of DNA relatedness. Mycotaxon, in the press.

Kurtzman, C.P. & Ahearn, D.G. (1976) Sporulation in *Pichia spartinae.* Mycologia 68, 682–685.

Kurtzman, C.P. & Kreger-van Rij, N.J.W. (1976) Ultrastructure of ascospores from *Debaryomyces melissophilus,* a new taxonomic combination. Mycologia 68, 422–425.

Kurtzman, C.P. & Smiley, M.J. (1974) A taxonomic re-evaluation of the round-spored species of *Pichia.* In: H. Klaushofer & U.B. Sleytr (Eds.), Proc. IV. Int. Symp. Yeasts, Vienna, Austria. I. Hochschülerschaft an der Hochschule für Bodenkultur, Vienna, pp. 231–232.

Kurtzman, C.P. & Smiley, M.J. (1976) Heterothallism in *Pichia kudriavzevii* and *Pichia terricola.* Antonie van Leeuwenhoek 42, 355–363.

Kurtzman, C.P. & Smiley, M.J. (1979) Taxonomy of *Pichia carsonii* and its synonyms *P. vini* and *P. vini* var. *melibiosi*: comparison by DNA reassociation. Mycologia 71, 658–662.

Kurtzman, C.P. & Wickerham, L.J. (1972) *Pichia besseyi* sp. n. Antonie van Leeuwenhoek 38, 49–52.

Kurtzman, C.P. & Wickerham, L.J. (1973) *Saccharomycopsis crataegensis,* a new heterothallic yeast. Antonie van Leeuwenhoek 39, 81–87.

Kurtzman, C.P., Smiley, M.J. & Baker, F.L. (1972) Scanning electron microscopy of ascospores of *Schwanniomyces.* J. Bacteriol. 112, 1380–1382.

Kurtzman, C.P., Gentles, J.C., Evans, E.G.V., Slodki, M.E. & Ward, R.M. (1973a) Growth of *Hansenula holstii* on cadavers. Appl. Microbiol. 25, 184–186.

Kurtzman, C.P., Vesonder, R.F. & Smiley, M.J. (1973b) Formation of extracellular C_{14} -C_{18} 2-D-hydroxy fatty acids by species of *Saccharomycopsis.* Appl. Microbiol. 26, 650–652.

Kurtzman, C.P., Baker, F.L. & Smiley, M.J. (1974a) A specimen holder to critical-point dry microorganisms for scanning electron microscopy. Appl. Microbiol. 28, 708–712.

Kurtzman, C.P., Vesonder, R.F. & Smiley, M.J. (1974b) Formation of extracellular 3-D-hydroxypalmitic acid by *Saccharomycopsis malanga.* comb. nov. Mycologia 66, 580–587.

Kurtzman, C.P., Smiley, M.J. & Baker, F.L. (1975) Scanning electron microscopy of ascospores of *Debaryomyces* and *Saccharomyces.* Mycopathol. Mycol. Appl. 55, 29–34.

Kurtzman, C.P., Johnson, C.J. & Smiley, M.J. (1979) Determination of conspecificity of *Candida utilis* and *Hansenula jadinii* through DNA reassociation. Mycologia 71, 844–847.

Kurtzman, C.P., Smiley, M.J., Johnson, C.J., Wickerham, L.J. & Fuson, G.B. (1980a) Two new and closely related heterothallic species, *Pichia amylophila* and *Pichia mississippiensis:* Characterization by hybridization and deoxyribonucleic acid reassociation. Int. J. Syst. Bacteriol. 30, 208–216.

Kurtzman, C.P., Smiley, M.J. & Johnson, C.J. (1980b) Emendation of the genus *Issatchenkia* Kudriavzev and comparison of species by deoxyribonucleic acid reassociation, mating reaction, and ascospore ultrastructure. Int. J. Syst. Bacteriol. 30, 503–513.

Kurtzman, C.P., Smiley, M.J., Johnson, C.J. & Hoffman, M.J. (1984) Deoxyribonucleic acid relatedness among species of the genus *Sterigmatomyces.* Int. J. Syst. Bacteriol., in the press.

Kuttin, E.S. & Müller, J. (1981) *Trichosporon beemeri* sp. nov. Antonie van Leeuwenhoek 47, 255–265.

Kützing, F.T. (1833) Algarum aquae dulcis Germanicarum, Decas III.

Kvasnikov, E.I., Nagornaya, S.S. & Shchelokova, J.F. (1975) Distribution of the yeast *Rhodosporidium diobovatum* in soil and plants. Microbiologiya 44, 753–756 (in Russian; English translation in: Microbiology 44, 679–681).

Kwon-Chung, K.J. (1975) A new genus, *Filobasidiella,* the perfect state of *Cryptococcus neoformans.* Mycologia 67, 1197–1200.

Kwon-Chung, K.J. (1976a) A new species of *Filobasidiella,* the sexual state of *Cryptococcus neoformans* B and C serotypes. Mycologia 68, 942–946.

Kwon-Chung, K.J. (1976b) Morphogenesis of *Filobasidiella neoformans,* the sexual state of *Cryptococcus neoformans.* Mycologia 68, 821–833.

Kwon-Chung, K.J. (1977) Perfect state of *Cryptococcus uniguttulatus.* Int. J. Syst. Bacteriol. 27, 293–299.

Kwon-Chung, K.J. (1978) Heterothallism vs. self-fertile isolates of *Filobasidiella neoformans (Cryptococcus neoformans).* Proc. IV Int. Conf. Mycoses, Pan. Am. Health Org. Sci. Publ. No. 356.

Kwon-Chung, K.J. (1980) Nuclear genotypes of spore chains in *Filobasidiella neoformans (Cryptococcus neoformans).* Mycologia 72, 418–422.

Kwon-Chung, K.J. & Bennett J.E. (1978) Distribution of α and *a* mating types of *Cryptococcus neoformans* among natural and clinical isolates. Am. J. Epidemiol. 108, 337–340.

Kwon-Chung, K.J. & Popkin, T.J. (1976) Ultrastructure of the septal complex in *Filobasidiella neoformans (Cryptococcus neoformans).* J. Bacteriol. 126, 524–528.

Kwon-Chung, K.J., Bennett, J.E. & Theodore, T.S. (1978) *Cryptococcus bacillisporus* sp. nov.: Serotype B-C of *Cryptococcus neoformans.* Int. J. Syst. Bacteriol. 28, 616–620.

Kwon-Chung, K.J., Bennett, J.E. & Rhodes, J.C. (1982a) Taxonomic studies on *Filobasidiella* species and their anamorphs. Antonie van Leeuwenhoek 48, 25–38.

Kwon-Chung, K.J., Polacheck, I. & Bennett, J.E. (1982b) An improved diagnostic medium for separation of *Cryptococcus neoformans* var. *neoformans* (serotypes A and D) and *Cryptococcus neoformans* var. *gattii* (serotypes B and C). J. Clin. Microbiol. 15, 535–537.

Lachance, M.A. (1980) Simple method for determination of deoxyribonucleic acid relatedness by thermal elution in hydroxyapatite microcolumns. Int. J. Syst. Bacteriol. 30, 433–436.

Lachance, M.A. & Phaff, H.J. (1979) Comparative study of molecular size and structure of exo-β-glucanases from *Kluyveromyces* and other yeast genera: evolutionary and taxonomic implications. Int. J. Syst. Bacteriol. 29, 70–78.

Lachance, M.A., Miranda, M., Miller, M.W. & Phaff, H.J. (1976) Dehiscence and active spore release in pathogenic strains of the yeast *Metschnikowia bicuspidata* var. *australis*: possible predatory implication. Can. J. Microbiol. 22, 1756–1761.

Laffer, N.C. (1936) A study of lactose fermenting yeasts. An abstract of a thesis, Urbana, Ill.

Laffin, R.J. & Cutter, V.M. (1959a) Investigations on the life cycle of *Sporidiobolus johnsonii* I. Irradiation and cytological studies. J. Elisha Mitchell Sci. Soc. 75, 89–96.

Laffin, R.J. & Cutter, V.M. (1959b) Investigations on the life cycle of *Sporidiobolus johnsonii* II. Mutants and micromanipulation. J. Elisha Mitchell Sci. Soc. 75, 97–100.

Langeron, M. (1922) Un nouveau *Sporotrichum* malgache, *S. carougeaui* Langeron 1913 et remarques sur les Sporotriches. Bull. Soc. Pathol. Exot. 15, 453–459.

Langeron, M. & Guerra, P. (1935) Bases morphologiques et biologiques de la classification des champignons levuriformes anascospores. Proc. 6th Int. Bot. Congress Amsterdam, 2, 165–167.

Langeron, M. & Guerra, P. (1938) Nouvelles recherches de zymologie médicale. Duex. Pt. Études monographiques sur les Mycotoruloidées. Le genre *Candida* Berkhout 1923. Ann. Parasitol. Hum. Comp. 16, 36–525, 429–476.

Langeron, M. & Guerra, P. (1939a) Remarques sur le *Candida stellatoidea* (Jones et Martin 1938) Ann. Parasitol. Hum. Comp. 17, 257–260.

Langeron, M. & Guerra, P. (1939–1940b) Valeur et nature des variations et dissociations de colonies chez les champignons levuriformes. Ann. Parasitol. Hum. Comp. 17, 447–469.

Langeron, M. & Talice, R.V. (1932) Nouvelles méthodes d'étude et essai de classification des champignons levuriformes. Ann. Parasitol. Hum. Comp. 10, 1–80.

Lapage, S.P., Sneath, P.H.A., Lessel, E.F., Skerman, V.B.D., Seeliger, H.P.R. & Clark, W.A. (1975) International Code of Nomenclature of Bacteria. Published for the International Association of Microbiological Societies by the American Society for Microbiology, Washington, D.C.

LaRue, T.A. & Spencer, J.F.T. (1967) The utilization of imidazoles by yeasts. Can. J. Microbiol. 13, 789–794.

Laurent, E. (1889) Observations sur le champignon du muguet. Bull. Soc. Belge Microscopie 16, 14.

Lavie, P. (1954) Essais de lutte biologique contre l'*Acarapis woodii*, agent de l'acariose de l'abeille. C.R. Acad. Sci. 238, 947–948.

Leask, B.G.S. & Yarrow, D. (1976) *Pichia norvegensis* sp. nov. Sabouraudia 14, 61–63.

Leberle, H. (1909) Beiträge zur Kenntnis der Gattung *Mycoderma*. Kastner u. Callway, Thesis, München.

Lee, J.D. & Komagata, K. (1980) *Pichia cellobiosa, Candida cariosilignicola* and *Candida succiphila*, new species of methanol-assimilating yeasts. Int. J. Syst. Bacteriol. 30, 514–519.

Lee, Ming-Hsia (1974) A new species of yeast: *Rhodotorula sinensis* Lee sp. nov. Acta Microbiol. Sin. 14, 143–156.

Legakis, Ph.A. (1961) A contribution to the study of the yeast flora of apples and apple wine. Thesis, University of Athens (in Greek).

Lembke, A. (1943) Ergebnisse der Theoretischen und Angewandten Mikrobiologie. Bd. I. Systematik der Schimmelpilze. Verlag J. Neumann, Neudamm.

Lempke, P.A. (1969) A reevaluation of homothallism and heterothallism and the species concept in *Sistotrema brinkmanii*. Mycologia 61, 57–76.

Leonard, T.J. & Raper, J.R. (1969) *Schizophyllum commune:* gene controlling induced haploid fruiting. Science 165, 190.

Leupold, U. (1950) Die Vererbung von Homothallic und Heterothallic bei *Schizosaccharomyces pombe*. C.R. Trav. Lab. Carlsberg, Ser. Physiol. 24, 381–480.

Lewin, R.A. & Fogg, G.E. (1978) Typification of microorganisms, a proposal. Taxon 27, 121.

Lewis, Y.S. & Johar, D.S. (1955) Science and Culture 21, 220.

Lindau, G. (1907a) Pflanzliche Parasiten im Sorauer Handbuch der Pflanzenkrankheiten. Bd. II. Parey, Berlin.

Lindau, G. (1907b) Fungi imperfecti: Hyphomycetes. In: L. Rabenhorst (Ed.), Kryptogamen-Flora von Deutschland, Österreich und der Schweiz, 1 Pilze, VIII Abt., Ed. Kummer, Leipzig.

Lindegren, C.C. (1949) The Yeast Cell, its Genetics and Cytology. Educational Publishers Inc., St. Louis.

Lindner, P. (1893) *Schizosaccharomyces pombe* n. sp. neuer Gärungserreger. Wochenschr. Brau. 10, 1298–1300.

Lindner, P. (1894) *Saccharomyces farinosus* und *Saccharomyces Bailii.* Zwei neue Hefenarten aus Danziger Jopenbier. Wochenschr. Brau. 153–156.

Lindner, P. (1895) Mikroskopische Betriebskontrolle in den Gärungsgewerben. P. Parey, Berlin.

Lindner, P. (1898a) Mikroskopische Betriebskontrolle in den Gärungsgewerben. 2te Aufl., P. Parey, Berlin.

Lindner, P. (1898b) *Monilia variabilis*, eine formenreiche und rassenhaltige neue Pilzart. Wochenschr. Brau. 15, 209-213.
Lindner, P. (1901) Mikroskopische Betriebskontrolle in den Gärungsgewerben. 3te Aufl., P. Parey, Berlin.
Lindner, P. (1904) Neue Erfahrungen aus dem letzten Jahre in Bezug der Hefen und Gärung. Jahrb. Vers. Lehranst. Brau. Berlin 7, 441-464.
Lindner, P. (1905a) Die Assimilierbarkeit der Selbstverdauungsprodukte der Bierhefe durch verschiedene Heferassen und Pilze. Mitteilung I. Wochenschr. Brau. 22, 528-530.
Lindner, P. (1905b) Mikroskopische Betriebskontrolle in den Gärungsgewerben. 5te Aufl., P. Parey, Berlin.
Lindner, P. (1907) Wochenschr. Brau. 24, 469.
Lindner, P. & Genoud, E.G. (1913) Zur Charakteristik der *Willia belgica* und einiger Hefen aus belgischem Lambicbier. Wochenschr. Brau. 30, 363-367.
Lloyd, C.J. (1922) Myc. Writ. 7, 1152.
Lochhead, A.G. (1942) *Zygosaccharomyces nectarophilus* n. sp. and *Zygosaccharomyces rugosus* n. sp. Can. J. Res. 20, 89-91.
Lochhead, A.G. & Heron, D.A. (1929) Microbiological studies of honey. Bull. 116-New Series, Dept. Agric. Dom. Canada.
Lodder, J. (1932) Über einige durch das "Centraalbureau voor Schimmelcultures" neuerworbene sporogene Hefearten. Zentralbl. Bakteriol. Parasitenk., Abt. II, 86, 227-253.
Lodder, J. (1934) Die anaskosporogenen Hefen, I. Hälfte. Verh. Kon. Ned. Akad. Wet., Afd. Natuurk., Sect. II, 32, 1-256.
Lodder, J. (1970) General classification of the yeasts. In: J. Lodder (Ed.), The Yeasts, a taxonomic study. North-Holland Publ. Co., Amsterdam, pp. 1-33.
Lodder, J. & Kreger-van Rij, N.J.W. (1952) The Yeasts, a taxonomic study. North-Holland Publ. Co., Amsterdam.
Lodder, J. & Kreger-van Rij, N.J.W. (1978) Proposal (446) for the conservation of the generic name *Debaryomyces* Lodder et Kreger-van Rij against *Debaryomyces* Klöcker. Taxon 27, 306-307.
Lodder, J. & Kreger-van Rij, N.J.W. (1979) *Debaryomyces* or *Debaryozyma*? Int. J. Syst. Bacteriol. 29, 413.
Lodder, J. & Vries, N.F. de (1938) Some notes on *Torulopsis glabrata* (Anderson) nov. comb. Mycopathol. Mycol. Appl. 1, 98-103.
Lodder, J., Slooff, W.Ch. & Kreger-van Rij, N.J.W. (1958) The classification of yeasts. In: A.H. Cook (Ed.), The Chemistry and Biology of Yeasts. Academic Press, New York, pp. 1-62.
Lowy, B. (1971) Tremellales. Flora Neotropica, Monograph No. 6, Hafner Publ. Co., New York.
Ludwig, F. (1886) Ueber Alkoholgärung und Schleimfluss lebender Bäume und deren Urheber. Ber. Dtsch. Bot. Ges. 4, 17-27.
Ludwig, F. (1891) Der Milch- und Rothfluss der Bäume und ihre Urheber. Vorläufige Mitteilung. Centralbl. Bakteriol. Parasitenk. 10, 10-13.
Ludwig, F. (1896) Sur les organismes des écoulements des arbres. Rev. Mycol. Toulouse 18, 45-57, 114-123.
Lund, A. (1954) Studies on the ecology of yeasts. Munksgaard, Copenhagen.
Lurie, H.I. & Shadomy, H.J. (1971) Morphological variations of the hyphae-forming strain of *Cryptococcus neoformans* (Coward strain) in tissues of mice. Sabouraudia 9, 10-14.
Lushbaugh, W.B., Rowton, E.D. & McGhee, R.B. (1976) Redescription of *Coccidiascus legeri* Chatton, 1913 (Nematosporaceae: Hemiascomycetidae), an intracellular, parasitic, yeast-like fungus from the intestinal epithelium of *Drosophila melanogaster*. J. Invertebr. Pathol. 28, 93-107.
Luteraan, P.J. (1954) *Hansenula ellipsoidospora* nov. sp. Rev. Mycol. 19, 85-97.

Mackenzie, D.W.R. & Auret, B.J. (1963) *Rhodotorula nitens* sp. nov. isolated from the atmosphere. J. Gen. Microbiol. 31, 171-177.
Mackinnon, J.E. (1936) Carateres y grado de la virulencia experimental de las Torulopsidaceas de la sub-familia Micotoruleas (Monilias). Ann. Fac. Med. Montevideo 21, 320.
Macy, J.M. & Miller, M.W. (1971) *Endomyces tetrasperma*, sp. n. J. Bacteriol. 105, 637-645.
Magalhães, O. de (1914) Bras. Med. 28 (cited from Lodder & Kreger-van Rij, 1952).
Magalhães, O. de (1918) Nova micose humana. Estudo sobre a morfolojia e biolojia do *Oidium brasiliense*, n. sp., agente etilojico de uma nova molestia do homem. Mem. Inst. Oswaldo Cruz 10, 20-57.
Magalhães, O. de (1932) A classificaçåo do *Oidium brasiliense, Neogeotrichum pulmoneum*, n.g. (O. Magalhães, 1914), emend. O. Magalhães, 1931. Mem. Inst. Oswaldo Cruz 26, 151-167.
Mager, J. & Aschner, M. (1947) Biological studies on capsulated yeasts. J. Bacteriol. 53, 283-295.
Malloch, D., Kane, J. & Lahaie, D.G. (1978) *Filobasidiella arachnophila* sp. nov. Can. J. Bot. 56, 1823-1826.
Manachini, P.L. (1979) DNA sequence similarity, cell wall mannans, and physiological characteristics in some strains of *Candida utilis, Hansenula jadinii* and *Hansenula petersonii*. Antonie van Leeuwenhoek 45, 451-463.
Marcilla, J. & Feduchy, E. (1943) Contribución al estudio de una levadura perteneciente al género *Saccharomycodes*, capaz de fermentar mostos de uva fuertemente sulfitados (mostos azufrados), sin previa desulfitación. Centros de Estudios Generales y de Ampelografía y Viticultura, Cuaderno No. 29.
Marcilla, J., Feduchy, E. & Gomes, J.M. (1945) Contribucion al estudio de levaduras salvajes y cultivadas, de las microfloras española y portuguesa. Trab. Inst. Cajal. Invest. Biol. 3, 161-191.
Marengo, G. (1926) Su di una paronichia dovuta ad una nova specie di *Monilia*: la *Monilia onychophila* Pollacci e Nannizzi. Arch. Biol. 3, 25-36.

Markovetz, A.J. & Kallio, R.E. (1964) Assimilation of alkanes and alkenes by yeasts. J. Bacteriol. 87, 968–969.

Marmur, J. & Doty, P. (1962) Determination of the base composition of DNA from its thermal denaturation temperature. J. Mol. Biol. 5, 109–118.

Martin, D.S., Jones, C.P., Yao, K.F. & Lee, L. (1937) A practical classification of the monilias. J. Bacteriol. 34, 99–123.

Martin, G.W. (1952) Revision of the North Central Tremellales. Univ. Iowa Stud. Nat. Hist. 19, 1–22.

Martin, M.V. & White, F.H. (1981) A microbiological and ultrastructural investigation of germ tube formation by oral strains of *C. tropicalis*. Am. J. Clin. Pathol. 75, 671–676.

Martin, S.M. (1964) Conservation of microorganisms. Ann. Rev. Microbiol. 18, 1–16.

Martini, A. & Phaff, H.J. (1973) The optical determination of DNA-DNA homologies in yeasts. Ann. Microbiol. 23, 59–68.

Martini, A., Phaff, H.J. & Douglass, S.A. (1972) Deoxyribonucleic acid base composition of species in the yeast genus *Kluyveromyces* van der Walt emend. van der Walt. J. Bacteriol. 111, 481–487.

Marzluf, G.A. (1981) Regulation of nitrogen metabolism and gene expression in fungi. Microbiol. Rev. 45, 437–461.

Matile, Ph., Moore, H. & Robinow, C.F. (1969) Yeast cytology. In: A.H. Rose & J.S. Harrison (Eds.), The Yeasts, Vol. 1, Academic Press, London–New York, pp. 219–302.

Matthewson, D.K. & Barnett, J.A. (1974) The effects of different carbon sources and changes in the growth medium on the shape of cells of the yeast *Trigonopsis variabilis*. J. Gen. Microbiol. 83, 427–430.

Mattlet, G. (1926) Mycoses dans l'urundi. Ann. Soc. Belge Méd. Trop. 6, 1–41.

Mayr, E. (1969) Principles of systematic zoology. McGraw-Hill Book Co., New York.

Mazza, S. & Niño, F.L. (1933) 8a Reun. Soc. Arg. Pat. Reg. del Norte 27, 293 (cited from Lodder & Kreger-van Rij, 1952).

Mazza, S., Niño, F.L. & Egües, A. (1930) 5a Reun. Soc. Arg. Pat. Reg. del Norte (Jujay 1929) 284–288.

Mazza, S., Niño, F.L., Quintana, H. & Bernasconi, V. (1931) Bol. Inst. Clin. Quirurg. Univ. Buenos Aires 6, 180–214.

McClary, D.O., Nulty, W.L. & Miller, G.R. (1959) Effect of potassium versus sodium in the sporulation of *Saccharomyces*. J. Bacteriol. 78, 362–368.

McCullough, J. & Herskowitz, I. (1979) Mating pheromones of *Saccharomyces kluyveri*: pheromone interactions between *Saccharomyces kluyveri* and *Saccharomyces cerevisiae*. J. Bacteriol. 138, 146–154.

McCully, E.K. & Robinow, C.F. (1972a) Mitosis in heterobasidiomycetous yeasts. I. *Leucosporidium scottii* (*Candida scottii*). J. Cell Sci. 10, 857–881.

McCully, E.K. & Robinow, C.F. (1972b) Mitosis in heterobasidiomycetous yeasts. II. *Rhodosporidium* sp. (*Rhodotorula glutinis*) and *Aessosporon salmonicolor* (*Sporobolomyces salmonicolor*). J. Cell Sci. 11, 1–31.

McGinnis, M.R., Walker, D.H. & Folds, J.D. (1980) *Hansenula polymorpha* infection in a child with chronic granulomatous disease. Arch. Pathol. Lab. Med. 104, 290–292.

McNabb, R.F.R. (1973) Phragmobasidiomycetidae: Tremellales, Auriculariales, Septobasidiales. In: G.C. Ainsworth, F.K. Sparrow & A.S. Sussman (Eds.), The Fungi, an Advanced Treatise. Vol. 4B, Academic Press, New York, pp. 303–325.

McVeigh, I. & Bracken, E. (1955) The nutrition of *Schizosaccharomyces pombe*. Mycologia 47, 13–25.

Mélard, L. (1910) Sur un organisme isolé d'une bière de fermentation haute. I. Congr. Int. Brasserie, Bruxelles.

Mendonça-Hagler, L.C. & Phaff, H.J. (1975) Deoxyribonucleic acid base composition and deoxyribonucleic acid/deoxyribonucleic acid hybrid formation in psychrophobic and related yeasts. Int. J. Syst. Bacteriol. 25, 222–229.

Menna, M.E. di (1954a) *Debaryomyces marama* n. sp., isolated from the air. J. Gen. Microbiol. 10, 65–67.

Menna, M.E. di (1954b) *Cryptococcus terreus* n. sp. from soil in New Zealand. J. Gen. Microbiol. 11, 195–197.

Menna, M.E. di (1957) The isolation of yeasts from soil. J. Gen. Microbiol. 17, 678–688.

Menna, M.E. di (1958a) Two new species of yeasts from New Zealand. J. Gen. Microbiol. 18, 269–272.

Menna, M.E. di (1958b) *Torulopsis ingeniosa* n. sp., from grass leaves. J. Gen. Microbiol. 19, 581–583.

Menna, M.E. di (1965a) Yeasts in New Zealand soils. N.Z. J. Bot. 3, 194–203.

Menna, M.E. di (1965b) *Schizoblastosporion starkeyi-henricii* Ciferri. Mycopathol. Mycol. Appl. 25, 205–212.

Menna, M.E. di (1966a) Three new yeasts from Antarctic soils: *Candida nivalis, Candida gelida*, and *Candida frigida* spp. n. Antonie van Leeuwenhoek 32, 25–28.

Menna, M.E. di (1966b) Yeasts in Antarctic soils. Antonie van Leeuwenhoek 32, 29–38.

Metschnikoff, E. (1884) Über eine Sprosspilzkrankheit der Daphnien. Beitrag zur Lehre bei dem Kampf der Phagocyten gegen Krankheitserreger. Arch. Pathol. Anat. Physiol. R. Virchow 96, 177–195.

Meyen, J. (1838) Wiegmann Arch. Naturgesch. 4, Bd. 2, 100.

Meyer, M.T. & Phaff, H.J. (1977) Survey for $\alpha(1\rightarrow3)$-glucanase activity among yeasts. J. Bacteriol. 131, 702–706.

Meyer, M.T. & Phaff, H.J. (1980) Purification and properties of $(1\rightarrow3)$-α-glucanases from *Bacillus circulans* WL-12. J. Gen. Microbiol. 118, 197–208.

Meyer, S.A. (1979) DNA relatedness between physiologically similar strains and species of yeasts of medical and industrial importance. In: S. Garratini, S. Paglialunga & N.S. Scrimshaw (Eds.),

Single Cell Protein-Safety for Animal and Human Feeding. Proc. Protein-Calorie Advisory Group of the United Nations System Symp. "Investigations of Single Cell Protein", Milan, Italy. Pergamon Press, New York, pp. 13-19.

Meyer, S.A. & Ahearn, D.G. (1983) Validation of the names of some *Candida* species. In the press.

Meyer, S.A. & Phaff, H.J. (1969) Deoxyribonucleic acid base composition in yeasts. J. Bacteriol. 97, 52-56.

Meyer, S.A. & Phaff, H.J. (1970) Taxonomic significance of the DNA base composition in yeasts. In: D.G. Ahearn (Ed.), Recent Trends in Yeast Research. Spectrum, Vol. I, Georgia State Univ., Atlanta, pp. 1-29.

Meyer, S.A. & Phaff, H.J. (1972) DNA base composition and DNA-DNA homology studies as tools in yeast systematics. In: A. Kocková-Kratochvílová & E. Minarik (Eds.), Yeast Models in Science and Technics. Proc. Spec. Symp. Yeasts, Smolenice. Publishing House of the Slovak Academy of Sciences, Bratislava, pp. 375-387.

Meyer, S.A. & Simione, F.R. (1978) Heterogeneity within the species *Candida sake* and *Candida diddensii*. Abstr. Ann. Meet. Am. Soc. Microbiol., p. 81.

Meyer, S.A. & Simione, F.R. (1983). In the press.

Meyer, S.A., Anderson, K., Brown, R.E., Smith, M.T., Yarrow, D., Mitchell, G. & Ahearn, D.G. (1975) Physiological and DNA characterization of *Candida maltosa*, a hydrocarbon-utilizing yeast. Arch. Microbiol. 104, 225-231.

Meyer, S.A., Brown, R.E. & Smith, M.T. (1977) Species status of *Hanseniaspora guilliermondii* Pijper. Int. J. Syst. Bacteriol. 27, 162-164.

Meyer, S.A., Smith, M.T. & Simione, F.P. (1978) Systematics of *Hanseniaspora* Zikes and *Kloeckera* Janke. Antonie van Leeuwenhoek 44, 79-96.

Migula (1900) System. Bakterien 1, 193 (cited from do Carmo-Sousa, 1970c).

Miller, G.R. & Baharaeen, S. (1979) DNA base composition and photo-reactivation capabilities of six *Hansenula* species. Antonie van Leeuwenhoek 45, 365-368.

Miller, M.W. & Barker, E.R. (1968) *Pichia angophorae*, sp. n. from the exudate of an Australian red gum tree. Antonie van Leeuwenhoek 34, 183-187.

Miller, M.W. & Mrak, E.M. (1953) Yeasts associated with dried-fruit beetles in figs. Appl. Microbiol. 1, 174-178.

Miller, M.W. & Phaff, H.J. (1958) A comparative study of the apiculate yeasts. Mycopathol. Mycol. Appl. 10, 113-141.

Miller, M.W., Phaff, H.J., Miranda, M., Heed, W.B. & Starmer, W.T. (1976a) *Torulopsis sonorensis*, a new species of the genus *Torulopsis*. Int. J. Syst. Bacteriol. 26, 88-91.

Miller, M.W., Yoneyama, M. & Soneda, M. (1976b) *Phaffia* a new yeast genus in the Deuteromycotina (Blastomycetes). Int. J. Syst. Bacteriol. 26, 286-291.

Minami, K., Yamamura, M., Shimizu, S., Ogawa, K. & Sekine, N. (1978) A new methanol-assimilating, high productive, thermophilic yeast. J. Ferment. Technol. 56, 1-7.

Misra, V.C. & Randhawa, H.S. (1976) *Sporobolomyces salmonicolor* var. *fischeri*, a new yeast. Arch. Microbiol. 108, 141-143.

Moens, P.B. (1971) Fine structure of ascospore development in the yeast *Saccharomyces cerevisiae*. Can. J. Microbiol. 17, 507-510.

Mogi, M. (1938) Studies on the yeasts found in "Miso". J. Agric. Chem. Soc. Japan 14, 951-1297.

Mogi, M. (1939a) Studies on the yeasts found in "Miso". J. Agric. Chem. Soc. Japan 15, 921-1221.

Mogi, M. (1939b) Yeasts from a soybean paste "Miso". No. 3. Taxonomy, affinity and characteristic properties of the yeasts isolated from a soybean paste "Miso". J. Agric. Chem. Soc. Japan 15, 1221-1232.

Mogi, M. (1940) J. Agric. Chem. Soc. Japan 16, 7.

Mogi, M. (1942) Studies on the yeasts found in "Miso". J. Agric. Chem. Soc. Japan 18, 543-940, 733-741 (in Japanese).

Möller, A. (1895) Protobasidiomyceten. Vol. 8, Botanische Mitteilungen aus den Tropen. ed. by A.F.W. Schimper, Jena.

Montgomerie, J.Z. & Edwards, J.E., Jr. (1978) Association of infection due to *Candida albicans* with intravenous hyperalimentation. J. Infect. Dis. 137, 197-201.

Montrocher, R. (1967) Quelques nouvelles espèces et variétés du genre *Candida* (levures asporogènes). Rev. Mycol. 32, 69-92, 689-692.

Moore, M. (1935) Head infection caused by a new *Hemispora, H. coremiformis*. Ann. Mo. Bot. Gard. 22, 317-332.

Moore, R.T. (1972) Ustomycota, a new division of higher fungi. Antonie van Leeuwenhoek 38, 567-584.

Moore, R.T. (1978) Taxonomic significance of septal ultrastructure with particular reference to the jelly fungi. Mycologia 70, 1007-1024.

Moore, R.T. (1980) Taxonomic proposals for the classification of marine yeasts and other yeast-like fungi including the smuts. Bot. Mar. 23, 361-373.

Moore, R.T. & Kreger-van Rij, N.J.W. (1972) Ultrastructure of *Filobasidium* Olive. Can. J. Microbiol. 18, 1949-1951.

Moore, R.T. & McAlear, J.H. (1962) Fine structure of mycota. 7. Observations on septa of Ascomycetes and Basidiomycetes. Am. J. Bot. 49, 86-94.

Moraes, R.G. de (1941) Cultura Médica 3, 64 (cited from do Carmo-Sousa, 1970c).

Morais, J.D. Falcâo de & Dália Maia, M.H. (1959) Estudos de microorganismos encontrados em leitos de despéjos de caldas de destilarias de Pernambuco. II. Una nova espécie de *Hansenula: H. polymorpha*. An. Esc. Super. Quim. Univ. Recife 1, 15-20.

Moreau, F. (1949) Le "dangeardien" et les Dangeardiomycètes. Botaniste 34, 315-319.

Moriyon, I. & Ramirez, C. (1974) New species of yeasts isolated from an acid washed brown soil. Proc. 4th Int. Symp. Yeasts Vienna, Part I, 233–234.

Moses, A. & Vianna, G. (1913) Mem. Inst. Oswaldo Cruz 5, 192 (cited from do Carmo-Sousa, 1970c).

Motta, R. (1926) Nuova specie di *Cryptococcus* isolata dalla gola (*Cryptococcus uvae* Pollacci & Nannizzi). Att. R. Accad. Fisiocrit. Siena, IX, 17, 633–637.

Mrak, E.M. & McClung, L.S. (1940) Yeasts occurring on grapes and in grape products in California. J. Bacteriol. 40, 395–407.

Mrak, E.M., Phaff, H.J. & Douglas, H.C. (1942a) A sporulation stock medium for yeasts and other fungi. Science 96, 432.

Mrak, E.M., Phaff, H.J. & Vaughn, R.H. (1942b) Yeasts occurring on dates. J. Bacteriol. 43, 689–700.

Mu, C. & Gao, H. (1979) A new species of yeast *Pichia* Hansen. Act. Microbiol. Sin. 19, 259–264.

Müller, W. (1934) Untersuchungen über die Symbiose von Tieren mit Pilzen und Bakterien. Über die Pilzsymbiose holzfressender Insektenlarven. Arch. Mikrobiol. 5, 84–147.

Mulvany, J.G. (1969) Membrane filter techniques in microbiology. In: J.R. Norris & D.W. Ribbons (Eds.), Methods in Microbiology, Vol. 1, Academic Press, London–New York, pp. 205–253.

Nadakavukaren, M.J. & McCracken, D.A. (1975) *Prototheca filamenta:* a fungus not an alga. Mycopathologia 57, 89–91.

Nadson, G.A. & Konokotina, A.G. (1911) *Guilliermondia*, un nouveau genre de la famille des Saccharomycètes à copulation hétérogamique. Bull. Jard. Imp. St. Pétersbourg 11, 117–143.

Nadson, G.A. & Krassilnikov, N.A. (1928) Un nouveau genre d'Endomycétacées. Compt. Rend. 187, 307–309.

Naganishi, H. (1917) Three new species of yeasts. Bot. Mag. Tokyo 31, 107–115 (in Japanese).

Naganishi, H. (1924) Mycological study of Manchurian kwass. Rep. Cent. Lab. S. Manchurian Railway Co. 10, 425–452 (in Japanese).

Naganishi, H. (1928) J. Zymol. 6 (in Japanese).

Naganishi, H. (1941a) On the yeasts isolated from sediment of spoiled sake. J. Ferment. Technol. 19, 14–42 (in Japanese).

Naganishi, H. (1941b) Studies on the microorganisms in fermenting vinegar mash from rice. Rep. Sci. Assoc. Japan 15, 419–422 (in Japanese).

Nägeli, C. von (1879) Theorie der Gährung, München.

Nakase, T. (1971a) New species of yeasts found in Japan. J. Gen. Appl. Microbiol. 17, 409–419.

Nakase, T. (1971b) Four new yeasts found in Japan. J. Gen. Appl. Microbiol. 17, 469–478.

Nakase, T. (1971c) New species of yeasts resembling *Candida krusei* (Cast.) Berkhout. J. Gen. Appl. Microbiol. 17, 383–398.

Nakase, T. (1972a) A comparative study on *Pichia dispora* and *Pichia saitoi*. J. Gen. Appl. Microbiol. 18, 377–382.

Nakase, T. (1972b) Significance of DNA base composition in the classification of yeasts and yeast-like fungi. In: G. Terui (Ed.), Fermentation Technology Today, Proc. IV Int. Ferment. Symp., Soc. Ferment. Technol., Osaka, Japan, pp. 785–791.

Nakase, T. (1975) Three new asporogenous yeasts found in industrial waste water. Antonie van Leeuwenhoek 41, 201–210.

Nakase, T. & Komagata, K. (1966) New yeasts, *Endomycopsis muscicola* and *Pichia zaruensis*. J. Gen. Appl. Microbiol. 12, 347–352.

Nakase, T. & Komagata, K. (1968a) Taxonomic significance of base composition of yeast DNA. J. Gen. Appl. Microbiol. 14, 345–357.

Nakase, T. & Komagata, K. (1968b) Taxonomic significance of base composition of yeast DNA. Yeast News Letter 17, 4–6.

Nakase, T. & Komagata, K. (1969) DNA base composition of the genus *Hansenula*. J. Gen. Appl. Microbiol. 15, 85–95.

Nakase, T. & Komagata, K. (1970a) Significance of DNA base composition in the classification of the yeast genera *Hanseniaspora* and *Kloeckera*. J. Gen. Appl. Microbiol. 16, 241–250.

Nakase, T. & Komagata, K. (1970b) Significance of DNA base composition in the classification of the yeast genus *Pichia*. J. Gen. Appl. Microbiol. 16, 511–521.

Nakase, T. & Komagata, K. (1971a) Further investigation on the DNA base composition of the genus *Hansenula*. J. Gen. Appl. Microbiol. 17, 77–84.

Nakase, T. & Komagata, K. (1971b) Significance of DNA base composition in the classification of the yeast genera *Cryptococcus* and *Rhodotorula*. J. Gen. Appl. Microbiol. 17, 121–130.

Nakase, T. & Komagata, K. (1971c) Significance of DNA base composition in the classification of the yeast genus *Torulopsis*. J. Gen. Appl. Microbiol. 17, 161–166.

Nakase, T. & Komagata, K. (1971d) Significance of DNA base composition in the classification of the yeast genus *Saccharomyces*. J. Gen. Appl. Microbiol. 17, 227–238.

Nakase, T. & Komagata, K. (1971e) Significance of DNA base composition in the classification of the yeast genus *Debaryomyces*. J. Gen. Appl. Microbiol. 17, 43–50.

Nakase, T. & Komagata, K. (1971f) Significance of DNA base composition in the classification of the yeast genus *Candida*. J. Gen. Appl. Microbiol. 17, 259–279.

Nakase, T. & Komagata, K. (1971g) DNA base composition of some species of yeasts and yeast-like fungi. J. Gen. Appl. Microbiol. 17, 363–369.

Nakase, T., Fukazawa, Y. & Tsuchiya, T. (1972) A comparative study of two forms of *Candida tropicalis* (Cast.) Berkhout. J. Gen. Appl. Microbiol. 18, 349–363.

Nakase, T., Komagata, K. & Fukazawa, Y. (1976) *Candida pseudointermedia* sp. nov., isolated from "Kamaboko", a traditional fish-paste product in Japan. J. Gen. Appl. Microbiol. 22, 177–182.

Nakase, T., Komagata K. & Fukazawa, Y. (1978) *Candida pararugosa*, a new species of asporogenous yeast. J. Gen. Appl. Microbiol. 24, 17–25.

Nakazawa, R. (1909) Zwei Saccharomyceten aus Sakéhefe. Zentralbl. Bakteriol. Parasitenk., Abt. II, 22, 529–540.

Nakazawa, R. (1914) The fission yeasts of canesugar molasses. Taiwan Sotokufu Kenkyujo Hokaku 111, 113–138 (in Japanese).

Nakazawa, R. (1919) Rep. Taiwan Government Lab. 7 (in Japanese).

Nakazawa, R. (1933) Melasse vergärende Hefe (Sammlungs Nr. 396 Forschungsinstitut der Regierung Formosa) *Saccharomyces formosensis* nov. sp. J. Agric. Chem. Soc. 102, 285–291.

Nannizzi, A. (1934) Repertorio sistematico dei miceti dell'uomo e degli animali (Trattato di Micopathologia umana, G. Pollacci, IV) Siena.

Naumov, G.I. (1969a) Comparative genetics of yeasts. I. Complementation of the genes for maltose utilization in maltose-negative species of *Saccharomyces*. Genetika 5, 142–149.

Naumov, G.I. (1969b) Concerning the question of genetic isolation of the species of *Saccharomyces*. Bull. Moscow Univ., Biol. & Soil Sci. 4, 44–46.

Naumov, G.I. (1972a) Revision of the classification of the *Saccharomyces* yeasts on the basis of genetic analyses. 2nd Congr. All-Union Soc. Geneticists and Selectors. I. Gen. Mol. Genet. 34–38.

Naumov, G.I. (1972b) Comparative genetics of yeasts, VII. On identification of mutations blocking the utilization of maltose in natural mutants of *Saccharomyces*. Bull. Univ. Moscow, Biol. & Soil Sci. 3, 34–37.

Naumov, G.I. (1977) Comparative genetics of criteria used in the taxonomy of yeasts. Main Directory of the Microbiological Industry of the Soviet of Ministers U.S.S.R. 1–37.

Naumov, G.I. & Yurkevitsch, V.V. (1970) The variability of the biochemical properties used in the taxonomy of the yeast genus *Saccharomyces*. Progress in Contemporary Biology 70, 315–324.

Nechitsch, A. (1904) Sur les ferments de deux levains de l'Inde, le *Mucor Praini* et le *Dematium Chodati*. Inst. Bot. Univ. Genève 5, 1.

Nel, E.E. & Walt, J.P. van der (1968) *Torulopsis humilis*, sp. n. Mycopathol. Mycol. Appl. 36, 94–96.

Neumann, I. (1972) Biotaxonomische und systematische Untersuchungen an einigen Hefen der Gattung *Saccharomyces*. Beih. Nova Hedwigia 40, 1–79.

Neveu-Lemaire, M. (1912) Recherche des Saccharomycètes et des Blastomycètes dans l'organisme de leur hôte. In: Parasitologie des animaux domestiques. J. Lamarre & Cie., Paris, p. 60.

Neveu-Lemaire, M. (1921) Précis de Parasitologie Humaine. Ed. 5, J. Lamarre, Paris.

Newell, S.Y. & Fell, J.W. (1970) The perfect form of a marine-occurring yeast of the genus *Rhodotorula*. Mycologia 62, 272–281.

Newell, S.Y. & Hunter, I.L. (1970) *Rhodosporidium diobovatum* sp. n., the perfect form of an asporogenous yeast (*Rhodotorula* sp.). J. Bacteriol. 104, 503–508.

Nickerson, W.J. (1943) *Zygosaccharomyces acidifaciens:* A new acetifying yeast. Mycologia 35, 66–78.

Nickerson, W.J. (1944a) Studies on film-forming yeasts. Acid production by *Zygopichia* and *Zygohansenula*. Mycologia 36, 224–233.

Nickerson, W.J. (1944b) Studies in the genus *Zygosaccharomyces* I. Transfer of pellicle-forming yeasts to *Zygopichia*. Farlowia 1, 469–481.

Niehaus, C.J.G. (1932) Untersuchungen über Apiculatushefen. Zentralbl. Bakteriol. Parasitenk., Abt. II, 86, 97–150.

Nieuwdorp, P.J., Bos, P. & Slooff, W.C. (1974) Classification of *Lipomyces*. Antonie van Leeuwenhoek 40, 241–254.

Niño, F.L. (1930) 5a Reun. Soc. Arg. Pat. Reg. del Norte (Jujuy 1929), Buenos Aires, 270.

Niño, F.L. (1933) Prensa Méd. Arg. 20, July 26 and August 2 (cited from do Carmo-Sousa, 1970c).

Niño, F.L. (1934) Blastomycosis humana generalizada por *Cryptococcus* (n. sp.) Buenos Aires (cited from Lodder & Kreger-van Rij, 1952).

Niño, F.L. (1938) Bol. Inst. Clin. Quirurg. Univ. Buenos Aires 14, 591.

Nishimura, T. (1910) Naikokuzei-ishu, Japan No. 45, 47 (in Japanese).

Nishiwaki, Y. (1929) Über eine neue Nachreif-Hefe in dem dunklen Bodensediment des japanischen Saké und über eine neue Hefegattung *Zygosaccharomycodes*. Zentralbl. Bakteriol. Parasitenk., Abt. II, 78, 403–410.

Northam, B.E. & Norris, F.W. (1951) Growth requirements of *Schizosaccharomyces octosporus*, a yeast exacting towards adenine. J. Gen. Microbiol. 5, 502–507.

Novák, E.K. (1964) *Candida soosii* nov. sp., a new yeast related to *Candida requinyii* Szép et Novák 1963. Acta Microbiol. Acad. Sci. Hung. 11, 51–54.

Novák, E.K. & Vitéz, I. (1964) Mycological investigations on clinical materials. II. Description of new yeasts. Zentralbl. Bakteriol. Parasitenk., Abt. I, Orig. 193, 127–133.

Novák, E.K. & Vörös-Felkai, C. (1962) *Rhodotorula slooffii* n. sp. Acta Microbiol. Acad. Sci. Hung. 9, 261–263.

Novák, E.K. & Zsolt, J. (1961) A new system proposed for yeasts. Acta Bot. Acad. Sci. Hung. 7, 93–145.

Nurudeen, T.A. & Ahearn, D.G. (1979) Regulation of melanin production by *Cryptococcus neoformans*. J. Clin. Microbiol. 10, 724–729.

Nyland, G. (1949) Studies on some unusual Heterobasidiomycetes from Washington State. Mycologia 41, 686–701.

Odds, F. (1979) *Candida* and candidosis. University Park Press, Baltimore.

Ogata, K., Nishikawa, H., Ohsugi, M. & Tochikura, T. (1970a) Studies on the production of yeasts. I. A yeast utilizing methanol as a sole carbon source. J. Ferment. Technol. 48, 389-396.

Ogata, K., Nishikawa, H., Ohsugi, M. & Tochikura, T. (1970b) Studies on the production of yeasts. II. The cultural conditions of methanol assimilating yeast, *Kloeckera* sp. No. 2201. J. Ferment. Technol. 48, 470–477.

Ogletree, F.F., Abdelal, A.T. & Ahearn, D.G. (1978) Germ-tube formation by atypical strains of *Candida albicans*. Antonie van Leeuwenhoek 44, 15-24.

Ohara, Y. & Nonomura, H. (1954a) Yeasts occurring in a mash and koji of Tamari-soja. Part 1. *Pichia mogii* nov. sp. J. Agric. Chem. Soc. Japan 28, 122-125.

Ohara, Y. & Nonomura, H. (1954b) A taxonomic study on the yeasts isolated in Japan. Part 1. Genus *Pichia* Hansen. J. Agric. Chem. Soc. Japan 28, 160-165.

Ohara, Y. & Nonomura, H. (1954c) Yeasts occurring in a mash and koji of Tamari-soya. Part 4. On the four strains including *Debaryomyces tamarii* nov. spec. J. Agric. Chem. Soc. Japan 28, 837-840.

Ohara, Y. & Nonomura, H. (1954d) Yeasts occurring in a mash and koji of Tamari-soya. Part 5. On the *Trichosporon cutaneum* and its new variety. J. Agric. Chem. Soc. Japan 28, 886-889.

Ohara, Y., Nonomura, H. & Yunome, H. (1960a) Dynamic aspects of yeast flora during vinous fermentation. Part 5. Bull. Res. Inst. Ferment. Yamanishi Univ. 7, 3–5.

Ohara, Y., Nonomura, H. & Yunome, H. (1960b) New species and varieties of Cryptococcoideae from grape musts. J. Agric. Chem. Soc. Japan 34, 709–711.

Ohara, Y., Nonomura, H. & Yamazaki, T. (1964) *Torulopsis vinacea* sp. nov., a new yeast isolated from grape musts. J. Gen. Appl. Microbiol. 10, 77–78.

Ohara, Y., Nonomura, H. & Yamazaki, T. (1965) *Candida incommunis* sp. nov., a new yeast isolated from grape must. J. Gen. Appl. Microbiol. 11, 273-275.

Oho, O. (1919) Über das sogenannte "Inkin". Kyoto Ingaku Zasshi 16, 7-21.

Oki, T., Kouno, K., Kitai, A. & Ozaki, A. (1972) New yeast capable of assimilating methanol. J. Gen. Appl. Microbiol. 18, 295-305.

Okunuki, K. (1931) Beiträge zur Kenntnis der rosafarbigen Sprosspilze. Jpn. J. Bot. 5, 283-322.

Olive, L.S. (1951) Taxonomic notes on Louisiana fungi. III. Additions to the Tremellales. Mycologia 43, 677–690.

Olive, L.S. (1968) An unusual new heterobasidiomycete with *Tilletia*-like basidia. J. Elisha Mitchell Sci. Soc. 84, 261-266.

Olson, H.C. & Hammer, B.W. (1935) Observations on yeasts causing gas in sweetened condensed milk. Iowa State Coll. J. Sci. 10, 37-43.

Olson, H.C. & Hammer, B.W. (1937) The genera *Sporobolomyces* and *Bullera* from the standpoint of dairy products. Iowa State Coll. J. Sci. 11, 207-213.

Onishi, H. (1957) Studies on osmophilic yeasts. III. Classification of osmophilic soy and miso yeasts. Bull. Agric. Chem. Soc. Japan 21, 151-156.

Onishi, H. (1960) Studies on osmophilic yeasts. IX. Isolation of a new obligate halophilic yeast and some considerations on halophilism. Bull. Agric. Chem. Soc. Japan 24, 226–230.

Onishi, H. (1972) *Candida tsukubaensis* sp. n. Antonie van Leeuwenhoek 38, 365-367.

Onishi, H. & Suzuki, T. (1969) *Torulopsis mannitofaciens* sp. n. isolated from soy-sauce mash. Antonie van Leeuwenhoek 36, 258-260.

Oorschot, C.A.N. van & Hoog, G.S de (1981) Dimorphic behaviour and taxonomy of *Trichosporiella sporotrichoides*. Antonie van Leeuwenhoek 47, 353-366.

Oshima, Y. & Takano, I. (1972) Genetic controlling system for homothallism and a novel method for breeding triploid cells in *Saccharomyces*. Fermentation Technology Today, Proc. 4th Int. Ferment. Symp. pp. 847-852.

Osterwalder, A. (1912) Eine neue Gärungsmonilia: *Monilia vini* n. sp. Zentralbl. Bakteriol. Parasitenk., Abt. II, 33, 257-272.

Osterwalder, A. (1924a) Neue aus Obst- und Traubensäften gewonnene *Saccharomyces*-Arten. Zentralbl. Bakteriol. Parasitenk., Abt. II, 60, 481-528.

Osterwalder, A. (1924b) *Schizosaccharomyces liquefaciens* n. sp., eine gegen freie schweflige Säure widerstandsfähige Gärhefe. Mitt. Geb. Lebensmitt. Hyg. 15, 5-28.

Osterwalder, A. (1934) Von Kaltgärhefen und Kaltgärung. Zentralbl. Bakteriol. Parasitenk., Abt. II, 90, 226-249.

Ota, M. (1923) Cinq levures du genre *Debaryomyces* considérées comme pathogènes. Ann. Parasitol. Hum. Comp. 1, 124-136.

Ota, M. (1924a) Beiträge zur Morphologie, Biologie und Systematik der pathogenen, asporogenen Sprosspilze. Dermatol. Wochenschr. 78, 216-237.

Ota, M. (1924b) Über vier neue pathogene Hefearten von der Gattung *Debaryomyces* (Klöcker). Dermatol. Wochenschr. 78, 284-289, 312-316.

Ota, M. (1924c) Essai de classification des blastomycètes pathogènes. Ann. Parasitol. Hum. Comp. 2, 34-61.

Ota, M. (1925) Remarques complémentaires sur la levure pathogène de Favre (*Myceloblastanon Favrei* n. sp.). Ann. Parasitol. Hum. Comp. 3, 181-184.

Ota, M. (1926a) Champignons parasites de l'homme. Jpn. J. Dermatol. 26, 111-142.

Ota, M. (1926b) Ann. Parasitol. Hum. Comp. 4, 1.

Ota, M. (1928) Champignons parasites de l'homme (Études morphologiques et systématiques). Jpn. J. Dermatol. Urol. 28, 5–6, 101–130.

Otani, Y. (1939) Microbiological studies on the "Nukamiso-pickles". Zentralbl. Bakteriol. Parasitenk., Abt. II, 101, 139-151.

Packer, N.H. & Bersten, A.M. (1977) Effect of Tween 80 on the morphology of *Trigonopsis variabilis*. J. Gen. Microbiol. 101, 233-236.

Palawal, D.K. & Randhawa, H.S. (1978) Evaluation of a simplified *Guizotia abyssinica* seed medium for differentiation of *Cryptococcus neoformans*. J. Clin. Microbiol. 7, 346-348.

Papadakis, C.R. (1922) Sur l'existence d'une copulation hétérogamique dans *Pichia farinosa* Lindner. C.R. Soc. Biol. 86, 447-449.

Pappagianis, D., Remington, J., Collins, M.S. & Hector, R. (1979) Development of resistance to amphotericin B in *Candida lusitaniae* infecting a human. Antimicrob. Agents Chemother. 16, 123-126.

Pedersen, T.A. (1958) *Cryptococcus terricolus* nov. spec. A new yeast isolated from Norwegian soils. C.R. Trav. Lab. Carlsberg 31, 93-103.

Peglion, V. (1897) Sopra un nuovo blastomicete, parassita del frutto del Nocciuela. Atti R. Accad. Naz. Lincei, Ser. 5, 6, 216-278.

Peláez, F.C. & Ramirez Gomez, C. (1956a) El *Sporobolomyces hispanicus*, nueva especie de Sporobolomycetaceae aislado de extracto de corteza de encina. Microbiol. Esp. 9, 37-41.

Peláez Campomanes, F. & Ramirez Gomez, C. (1956b) Una nueva especie de *Endomyces* Reess.: *Endomyces ovetensis* nov. sp. Microbiol. Esp. 9, 191-200.

Pereira Filho (1927) Macroglossia mycotica. J. Trop. Med. Hyg. 30, 8-10.

Persoon, C.H. (1822) Mycologia Eur. 1, 98.

Peynaud, E. & Domercq, S. (1955) Sur les espèces de levures fermentant sélectivement le fructose. Ann. Inst. Pasteur 89, 346-351.

Peynaud, E. & Domercq, S. (1956) Sur les *Brettanomyces* isolés de raisins et de vins. Arch. Mikrobiol. 24, 266-280.

Phaff, H.J. (1956) A proposal for amendment of the diagnosis of the genus *Pichia* Hansen. Antonie van Leeuwenhoek 22, 113-116.

Phaff, H.J. (1970a) The genus *Hanseniaspora* Zikes. In: J. Lodder (Ed.), The Yeasts, a taxonomic study. North-Holland Publ. Co., Amsterdam, pp. 209-225.

Phaff, H.J. (1970b) The genus *Saccharomycopsis* Schiönning. In: J. Lodder (Ed.), The Yeasts, a taxonomic study. North-Holland Publ. Co., Amsterdam, pp. 725-732.

Phaff, H.J. (1970c) The genus *Sporobolomyces* Kluyver et van Niel. In: J. Lodder (Ed.), The Yeasts, a taxonomic study. North-Holland Publ. Co., Amsterdam, pp. 831-862.

Phaff, H.J. (1970d) The genus *Kloeckera* Janke. In: J. Lodder (Ed.), The Yeasts, a taxonomic study. North-Holland Publ. Co., Amsterdam, pp. 1146-1160.

Phaff, H.J. (1970e) The genus *Sporidiobolus* Nyland. In: J. Lodder (Ed.), The Yeasts, a taxonomic study. North-Holland Publ. Co., Amsterdam, pp. 822-830.

Phaff, H.J. (1971) Structure and biosynthesis of the yeast cell envelope. In: A.H. Rose & J.S. Harrison (Eds.), The Yeasts. Vol. 2. Physiology and Biochemistry of Yeasts. Academic Press, London, pp. 135-210.

Phaff, H.J. (1977) Enzymatic yeast cell wall degradation. In: R.E. Feeney & J.R. Whitaker (Eds.), Food proteins – Improvement through chemical and enzymatic modification. Advances in Chemistry Services No. 160. Am. Chem. Soc., Washington, D.C., pp. 244-282.

Phaff, H.J. & Ahearn, D.G. (1970) The genus *Rhodotorula* Harrison. In: J. Lodder (Ed.), The Yeasts, a taxonomic study. North-Holland Publ. Co., Amsterdam, pp. 1187-1223.

Phaff, H.J. & Carmo-Sousa, L. do (1962) Four new species of yeast isolated from insect frass in bark of *Tsuga heterophylla* (Raf.) Sargent. Antonie van Leeuwenhoek 28, 193-207.

Phaff, H.J & Fell, J.W. (1970) The genus *Cryptococcus* Kützing emend. Phaff et Spencer. In: J. Lodder (Ed.), The Yeasts, a taxonomic study. North-Holland Publ. Co., Amsterdam, pp. 1088-1145.

Phaff, H.J. & Holzschu, D.L. (1981) Genome comparison in yeast systematics: Evolutionary divergence among cactus-specific yeasts. In: D. von Wettstein, J. Friis, M. Kielland-Brandt, & A. Stenderup (Eds.). Proc. Benzon Symposium 16, Molecular genetics in yeasts. Munksgaard, Copenhagen, pp. 413-427.

Phaff, H.J. & Knapp, E.P. (1956) The taxonomy of yeasts found in exudates of certain trees and other natural breeding sites of some species of *Drosophila*. Antonie van Leeuwenhoek 22, 117-130.

Phaff, H.J. & Meyer, S.A. (1969) Deoxyribonucleic acid base composition in yeasts. J. Bacteriol. 97, 52-56.

Phaff, H.J. & Miller, M.W. (1961) A specific microflora associated with the fig wasp, *Blastophaga psenes* Linnaeus. J. Insect. Pathol. 3, 233-243.

Phaff, H.J. & Spencer, J.F.T. (1969) Improved parameters in the separation of species in the genera *Rhodotorula* and *Cryptococcus*. Proc. II Int. Symp. Yeasts. Bratislava, 1966, pp. 59-67.

Phaff, H.J. & Yoneyama, M. (1961) *Endomycopsis scolyti*, a new heterothallic species of yeast. Antonie van Leeuwenhoek 27, 196-202.

Phaff, H.J., Mrak, E.M. & Williams, O.B. (1952) Yeasts isolated from shrimp. Mycologia 44, 431-451.

Phaff, H.J., Miller, M.W. & Shifrine, M. (1956) The taxonomy of yeasts isolated from *Drosophila* in the Yosemite region of California. Antonie van Leeuwenhoek 22, 145-161.

Phaff, H.J., Miller, M.W. & Cooke, W.B. (1960) A new species of *Schwanniomyces: Schwanniomyces alluvius*. Antonie van Leeuwenhoek 26, 182-188.

Phaff, H.J., Miller, M.W. & Spencer, J.F.T. (1964) Two new species of *Pichia* isolated from slime fluxes of deciduous trees. Antonie van Leeuwenhoek 30, 132-140.

Phaff, H.J., Miller, M.W., Yoneyama, M. & Soneda, M. (1972) A comparative study of the yeast florae associated with trees from the Japanese islands and on the west coast of North America. In: G. Terui (Ed.), Fermentation Technology Today. Society of Fermentation Technology, Osaka, Japan, pp. 759-774.

Phaff, H.J., Miller, M.W., Miranda, M., Heed, W.B. & Starmer, W.T. (1974) *Cryptococcus cereanus*, a new species of the genus *Cryptococcus*. Int. J. Syst. Bacteriol. 24, 486-490.

Phaff, H.J., Miller, M.W. & Miranda, M. (1976) *Pichia scutulata*, a new species from tree exudates. Int. J. Syst. Bacteriol. 26, 326-331.

Phaff,, H.J., Starmer, W.T., Miranda, M. & Miller, M.W. (1978) *Pichia heedii*, a new species of yeast indigenous to necrotic cacti in the North American Sonoran Desert. Int. J. Syst. Bacteriol. 28, 326-331.

Phaff, H.J., Miller, M.W. & Miranda, M. (1979) *Hansenula alni*, a new heterothallic species of yeast from exudates of alder trees. Int. J. Syst. Bacteriol. 29, 60-63.

Phaff, H.J., Starmer, W.T., Miranda, M. & Miller, M.W. (1980) *Candida mucilagina*, a new species of yeast found in decaying cladodes of *Opuntia inermis* and in necrotic tissue of Cereoid cacti. Int. J. Syst. Bacteriol. 30, 596-900.

Picci, G. & Verona, O. (1963) Su di alcuni fattori che influenzano la morfogenesi di *Trigonopsis variabilis* Sch. Atti. Ist. Bot. Univ. Lab. Cryptogam. Pavia, Ser. V, 3-8.

Pien, F.D., Thompson, R.L., Deye, D. & Roberts, G.D. (1980) *Rhodotorula* septicemia. Mayo Clin. Proc. 55, 258–260.

Pignal, M.-C. (1967) Une nouvelle espèce de levure isolée de larves d'insectes: *Pichia stipitis*. Bull. Mens. Soc. Linn. Lyon 36, 163-168.

Pignal, M.-C. (1970) A new species of yeast isolated from decaying insect-invaded wood. Antonie van Leeuwenhoek 36, 525-529.

Pijper, A. (1928) Een nieuwe *Hanseniaspora*. Verh. Kon. Ned. Akad. Wet. Afd. Natuurk. 37, 868-871.

Pilat, A. (1957) Übersicht der europäischen Auriculariales und Tremellales unter besonderer Berücksichtigung der tschechoslowakischen Arten. Acta Mus. Nat. Pragae 13, B(4), 115-210.

Pitt, J.I. (1974) Resistance of some food spoilage yeasts to preservatives. Food Technol. Austr. 26, 238-241.

Pitt, J.I. & Miller, M.W. (1968) Sporulation in *Candida pulcherrima, Candida reukaufii* and *Chlamydozyma* species: their relationships with *Metschnikowia*. Mycologia 60, 663-685.

Pitt, J.I. & Miller, M.W. (1970a) Speciation in the yeast genus *Metschnikowia*. Antonie van Leeuwenhoek 36, 357-381.

Pitt, J.I. & Miller, M.W. (1970b) The parasexual cycle in yeasts from the genus *Metschnikowia*. Mycologia 62, 462-473.

Polacheck, I. & Kwon-Chung, K.J. (1980) Creatinine metabolism in *Cryptococcus neoformans* and *Cryptococcus bacillisporus*. J. Bacteriol. 142, 15-20.

Pollacci, G. & Nannizzi, A. (1927) Sui caratteri dei generi *Monila* e *Cryptococcus*. Att. R. Accad. Fisiocrit. Siena, Serie X, 2, 35-44.

Pollacci, G. & Nannizzi, A. (1928) I miceti patogeni dell'uomo e degli animali 7, 8.

Pollacci, G. & Nannizzi, A. (1930) I miceti patogeni dell'uomo e degli animali 9, 10.

Pomper, S. & Burkholder, P.R. (1949) Studies on the biochemical genetics of yeast. Proc. Natl. Acad. Sci. U.S.A. 35, 456-464.

Poncet, S. (1973) Incidence de l'introduction d'espèces nouvelles sur les groupes établis dans le genre *Hansenula* (Ascomycètes, Cryptococcacées) par l'application d'une méthode d'analyse factorielle. Ann. Microbiol. (Inst. Pasteur, Paris) 124A, 95-108.

Poncet, S. & Fiol, J.B. (1972) Taxonomy of *Kluyveromyces:* evaluation of DNA base composition. Antonie van Leeuwenhoek 38, 145-152.

Price, C.W. & Phaff, H.J. (1979) *Debaryomyces polymorphus* and *D. pseudopolymorphus*, new taxonomic combinations. Mycologia 71, 444-445.

Price, C.W., Fuson, G.B. & Phaff, H.J. (1978) Genome comparison in yeast systematics: Delimitation of species within the genera *Schwanniomyces, Saccharomyces, Debaryomyces* and *Pichia*. Microbiol. Rev. 42, 161-193.

Pringsheim, E. & Bilewski, H. (1911) Über Rosahefe. Beitr. Biol. Pflanz. 10, 118-133.

Prostoserdov, N.N. & Afrikian, R. (1933) Jerezwein in Armenien. Weinland 5, 389 (cited by Kudriavzev, 1960).

Pulverer, C. & Korth, H. (1971) *Cryptococcus neoformans*: Pigmentbildung aus verschiedenen Polyphenolen. Med. Microbiol. Immunol. 175, 46-51.

Puntoni, V. (1935) Manuale d. Microbiol., Studio, Edit. d. Ist. Univ. Roma, Ed. 2 (cited by Lodder & Kreger-van Rij, 1952).

Queiroz, L.A. de (1973) *Endomycopsis montevidensis* n. sp. Mycopathol. Mycol. Appl. 51, 307-314.

Quinquaud, M. (1868) Nouvelles recherches sur le muguet. Classification du *Syringospora robinii* (dit *Oidium albicans*). Arch. Phys. Norm. Pathol. 1, 290-305.

Quirke, P., Hwang, W. & Validen, C. (1980) Congenital *Torulopsis glabrata* infection in man. Am. J. Clin. Pathol. 73, 137–140.

Rabenhorst, L. (1867) Hedwigia 6, 49 (cited from do Carmo-Sousa, 1970c).

Rambelli, A. (1959) Intorno a qualche aspetto nutrizionale e morfologico di *Trigonopsis variabilis* Schachner. Nuovo G. Bot. Ital. 66, 296-300.

Ramirez Gomes, C. (1953) Estudio sobre nuevas especies de levaduras aisladas de differentes sustratos. Microbiol. Esp. 6, 249–253.

Ramirez, C. (1954) Note sur deux nouvelles espèces de levures isolées de divers milieux. Rev. Mycol. 19, 98-102.

Ramirez Gomez, C. (1957) Contribucion al estudio de la ecologia de las levaduras. I. Estudio de levaduras aisladas de hungos carnosus. Microbiol. Esp. 10, 215-247.

Ramirez Gomez, C. (1959) Nouvelle diagnose d'une levure isolée d'une *Amanita lividopallescens, Geotrichum zambettakesi* n. comb. Rev. Mycol. 24, 434-436.

Ramirez, C. (1974) A compilation of descriptions of new *Candida* species with keys to all species of the genus described up to date. Microbiol. Esp. 27, 15-78.

Ramirez, C. & Boidin, J. (1953a) Trois nouvelles espèces de levures isolées de liqueurs tannantes. Rev. Mycol. 18, 149-156.

Ramirez, C. & Boidin, J. (1953b) Nuevas especies de levaduras aisladas de liquidos curtientes. Microbiol. Esp. 4, 405-414.

Ramirez, C. & Boidin, J. (1954) *Saccharomyces chambardi*, nouvelle espèce de levure isolée de liqueur tannante. Bull. Mens. Soc. Linn. Lyon 23, 151–152.

Ramirez, C. & Boidin, J. (1955) Nueva especie de levadura aislada de liquido curtiente *Saccharomyces chambardi* nov. sp. Microbiol. Esp. 8, 225-228.

Ramirez, C. & Gonzalez, C. (1972) *Candida iberica* sp. n. A new species isolated from Spanish sausages. Can. J. Microbiol. 18, 1778-1780.

Ramirez, C. & Martinez, A.T. (1978) *Torulopsis pampelonensis* sp. nov. A new species of yeast isolated from beech forest soil. Can. J. Microbiol. 24, 433–435.

Ramirez Gomez, C. & Sierra Rico, G. (1956) Una nueva variedad de *Torulopsis holmii* (Jörgensen) y estudio de sus productos metabolícos. Microbiol. Esp. 9, 147–162.

Rankine, B.C. & Fornachon, J.C.M. (1964) *Schizosaccharomyces malidevorans* sp. n., a yeast decomposing L-malic acid. Antonie van Leeuwenhoek 30, 73–75.

Raper, C.A. (1978) Control of development by the incompatibility system in basidiomycetes. In: M.N. Schwalb & P.G. Miles (Eds.), Genetics and Morphogenesis in the Basidiomycetes. Academic Press, London, pp. 3–29.

Raper, C.A., Raper, J.R. & Miller, R.E. (1972) Genetic analysis of the life cycle of *Agaricus bisporus*. Mycologia 64, 1088–1117.

Raper, J.R. (1966) Genetics of sexuality in higher fungi. Ronald Press, New York.

Re, S. (1925) Observations on the cultural and biochemical characters of *Monilia castellanii* (*Cryptococcus castellanii*) and *Monilia macroglossiae* (*Cryptococcus macroglossiae*). J. Trop. Med. Hyg. 28, 317–319.

Recca, J. & Mrak, E.M. (1952) Yeasts occurring in citrus products. Food Technol. 6, 450–454.

Redaelli, P. (1925) I miceti come associazione microbica nelle tubercolosi polmonare cavitaria. Mem. Prem. R. Ist. Lombardo Sci. Lett. Pavia.

Redaelli, P. (1931) Il problema della Torulopsidaceae e dei loro rapporti con l'uomo e con la patologia umana studiato particolarmente in Italia blastomiceti e blastomicosi. Riv. Biol. 13, 171–235.

Redaelli, P. & Ciferri, R. (1947) Un quadriennio di attivita del Centro di Micologia Umana e comparata di Pavia. Mycopathologia 4, 25–47.

Redaelli, P., Ciferri, R. & Giordano, A. (1937) *Debaryomyces neoformans* (Sanfelice) nobis n. comb. pour les espèces du groupe *Saccharomyces hominis – Cryptococcus neoformans – Torula histolytica.* Boll. Sez. Ital. Soc. Int. Microbiol. 9, 24–28.

Redaelli, P., Ciferri, R. & Cavallero, C. (1939) Sul presento *Endomyces albicans* Vuillemin. Mycopathologia 2, 116–121.

Redhead, S.A. & Malloch, D.W. (1977) The Endomycetaceae: new concepts, new taxa. Can. J. Bot. 55, 1701–1711.

Reess, M. (1870) Botanische Untersuchungen über die Alkoholgärungspilze. A. Felix, Leipzig.

Reess, M. (1877) Über den Soorpilz. Sitzungsber. Phys. Med. Soz. Erlangen 9, 190–195.

Reid, D.A. (1970) New or interesting records of British Hymenomycetes, IV. Trans. Brit. Mycol. Soc. 55, 413–441.

Reid, I.D. (1974) Properties of conjugation hormones (erogens) from the basidiomycete, *Tremella mesenterica.* Can. J. Bot. 52, 521–524.

Reid, I.D. & Bartnicki-Garcia, S. (1976) Cell-wall composition and structure of yeast cells and conjugation tubes of *Tremella mesenterica.* J. Gen. Microbiol. 96, 35–50.

Reiersöl, S (1955) Species of *Rhodotorula* isolated from laryngeal swabs. Antonie van Leeuwenhoek 21, 286–289.

Reiersöl, S. (1958) *Torulopsis norvegica* nov. sp. Antonie van Leeuwenhoek 24, 111–112.

Reiersöl, S. & Menna, M. di (1958) A new *Cryptococcus* species. Antonie van Leeuwenhoek 24, 27–30.

Rennerfelt, E. (1937) Undersökningar över svampinfektionen i slipmassa occii des utveckling däri. Sven. Skogsvardsfoerb. Tidskr. 47, 159.

Richter, A.A. von (1912) Über einen osmophilen Organismus den Hefepilz *Zygosaccharomyces mellis acidi* sp. n. Mycol. Zentralbl. 1, 67–76.

Rij, N.J.W. van & Verona, O. (1949) Sopra alcuni lieviti delle olive. Atti Accad. Naz. Lincei 7, 249–253.

Rischin, M. (1921) Über einen Fall von bisher noch nicht beschriebener *Parendomyces*-Erkrankung, die unter dem Bilde des tiefen (Sycosis parasitaria) und oberflächlichen Trichophytie verlief. Arch. Dermatol. Syph. 134, 232–242.

Rivalier, E. & Seydel, S. (1932a) Cultures minces sur lames gélosées, colorées et examinées in situ en préparations définitives, pour l'étude des Cryptogames microscopiques. C.R. Soc. Biol. 110, 181.

Rivalier, E. & Seydel, S. (1932b) Nouveau procédé de culture sur lames gélosées appliqué à l'étude microscopique des champignons des teignes. Ann. Parasitol. Hum. Comp. 10, 444–452.

Roberts, C. (1960) On the occurrence of yeasts in the mountainous soils of Jotunheimen, Norway. C.R. Trav. Lab. Carlsberg 32, 75–87.

Roberts, G.D., Horstmeier, C.D., Land, G.A. & Foxworth, J.H. (1978) Rapid urea broth test for yeasts. J. Clin. Microbiol. 7, 584–588.

Robin, C. (1853) Histoire naturelle des végétaux parasites qui croissent sur l'homme et sur les animaux vivants. Baillière, Paris.

Robinow, C.F. (1961) Mitosis in the yeast *Lipomyces lipofer*. J. Biophys. Biochem. Cytol. 9, 879–892.

Rodrigues de Miranda, L. (1972) *Filobasidium capsuligenum* nov. comb. Antonie van Leeuwenhoek 38, 91–99.

Rodrigues de Miranda, L. (1974) Centraalbureau voor Schimmelcultures Baarn and Delft. Progress Report 1973. Verh. Kon. Ned. Akad. Wetensch., Afd. Natuurk., 2e Reeks 63, 1–20.

Rodrigues de Miranda, L. (1975) Two new species of the genus *Sterigmatomyces*. Antonie van Leeuwenhoek 45, 193–199.

Rodrigues de Miranda, L. (1978) A new genus: *Sporopachydermia*. Antonie van Leeuwenhoek 44, 439–450.

Rodrigues de Miranda, L. (1979) *Clavispora*, a new yeast genus of the Saccharomycetales. Antonie van Leeuwenhoek 45, 479–483.

Rodrigues de Miranda, L. & Batenburg-van der Vegte, W. (1981) *Cryptococcus mollis* Kützing type species of the genus *Cryptococcus*. Antonie van Leeuwenhoek 47, 65–72.

Rodrigues de Miranda, L. & Diem, H.G. (1974) Deux nouvelles espèces de levures isolées de la phyllosphère de l'Orge. Can. J. Bot. 52, 279–282.

Rodrigues de Miranda, L. & Norkrans, B. (1968) *Candida suecica* sp. n. isolated from marine environment. Antonie van Leeuwenhoek 34, 115–118.

Rodrigues de Miranda, L. & Török, T. (1976) *Pichia humboldtii* sp. nov., the perfect state of *Candida ingens*. Antonie van Leeuwenhoek 42, 343–348.

Rodrigues de Miranda, L., Appel, K.R. & Seyfarth, H. (1980) *Pichia sorbitophila* sp. nov. Antonie van Leeuwenhoek 46, 157–159.

Rogers, A.L., Kwon-Chung, K.J. & Flegler, S.L. (1980) A scanning electron microscope comparison of basidial structures in *Filobasidiella neoformans* and *F. bacillispora*. Sabouraudia 18, 85–89.

Rogers, D.P. (1934) The basidium. Stud. Nat. Hist. Univ. Iowa 16, 160–183.

Rogers, D.P. (1971) Patterns of evolution to the homobasidium. In: R.H. Peterson (Ed.), Evolution in the Higher Basidiomycetes. Tennessee Press, Knoxville, pp. 241–257.

Rooney, L. & Moens, P.B. (1973a) Nuclear divisions at meiosis in the ascomycetous yeast *Wickerhamia fluorescens*. Can. J. Microbiol. 19, 1383–1387.

Rooney, L. & Moens, P.B. (1973b) The fine structure of ascospore delimitation in the yeast *Wickerhamia fluorescens.* Can. J. Microbiol. 19, 1389–1392.

Rossi, G. de (1920) I lieviti apiculati nella fermentazione vinaria. Sta. Sper. Agrar. Ital. 53, 233–297.

Rost, K. & Venner, H. (1964) Untersuchungen an Nucleinsäuren. X. Isolierung und Untersuchung von Deoxyribonucleinsäure aus Hefe. Z. Physiol. Chem. 339, 230–237.

Ruinen, J. (1963) The phyllosphere. II. Yeasts from the phyllosphere of tropical foliage. Antonie van Leeuwenhoek 29, 425–438.

Russel, I. & Stewart, G.G. (1979) Sphaeroplast fusion of brewer's yeast strains. J. Inst. Brew. 85, 95–98.

Sabouraud, R. (1904) Maladies du cuir chevelu. II. Les maladies desquamatives. Masson & Cie., Paris.

Saccardo, P.A. (1886) Sylloge Fungorum omnium hucusque cognitorum. Vol. 4, 16.

Saccardo, P.A. (1889) Sylloge Fungorum omnium hucusque cognitorum. Vol. 8, 1042.

Saccardo, P.A. (1895) Sylloge Fungorum omnium hucusque cognitorum. Vol. 11, pars III, 457.

Saccardo, P.A. (1906) Sylloge Fungorum omnium hucusque cognitorum. Vol. 18, pars VII.

Saccardo, P.A. & Sydow, P. (1902) Sylloge Fungorum omnium hucusque cognitorum. Vol. 16, pars V.

Sacchetti, M. (1932a) Contributo alla conoscenza della flora microbica di alcuni formaggi italiano. Arch. Mikrobiol. 3, 650–662.

Sacchetti, M. (1932b) Interno ad alcuni *Zygosaccharomyces*. Rend. R. Accad. Naz. Lincei 15, 762–765.

Sacchetti, M. (1932c) I lieviti della fermentazione del miele italiano. Mem. R. Accad. Sci. Ist. Bologna, Ser. VIII, 9, 2–11.

Sacchetti, M. (1932d) Ricerche sulla fermentazione di un mosto d'uva concentrato. Arch. Mikrobiol. 3, 473–491.

Sacchetti, M. (1933) Contributo alla conoscenza della flora microbica di alcuni formaggi italiana. Arch. Mikrobiol. 4, 427–446.

Sacchetti, M. (1939a) Memoria presentata dall'Acc. Benedetto Prof. V. Peglion nella sessione del 23 Aprile, 1939.

Sacchetti, M. (1939b) Studi su lieviti. Parte I. Descrizione di specie e varietà nuove. Mem. R. Accad. Sci. Ist. Bologna, Serie IX, 6, 3–12.

Saëz, H. (1960) *Debaryomyces coudertii*, nouvelle espèce de levure isolée chez un Manchot royal (*Aptenodytes patagonica* Miller). Bull. Mens. Soc. Linn. Lyon 288–289.

Saëz, H. (1964) *Geotrichum vanriji* nouvelle espèce de champignon arthrospore isolé chez une sarcelle d'été (*Anas querquedula* L.). Bull. Mens. Soc. Linn. Lyon 33, 263–267.

Saëz, H. (1965) *Candida krusei* (Castellani) Berkhout var. *transitoria*, nouvelle variété de levure isolée chez divers mammifères et oiseaux. Bull. Mens. Soc. Linn. Lyon 34, 265–270.

Saëz, H. (1967) *Rhodotorula vuilleminii* n. sp., isolée chez un ibis rouge: *Guara rubra* (L.). Bull. Soc. Mycol. France 83, 953-958.

Saëz, H. (1973) Bull. Soc. Mycol. France 89, 79.

Saggese, V. (1934) Riv. Clin. Pediatr. 9, 104 (cited from van Uden & Buckley, 1970).

Saito, K. (1907) Mikrobiologische Studien über die Zubereitung des Batatenbranntweines auf der Insel Hachijo (Japan). Zentralbl. Bakteriol. Parasitenk., Abt. II, 18, 30-37.

Saito, K. (1909) Preliminary notes on the spore formation of the so-called "Soya-kahmhefe". Bot. Mag. Tokyo 23, 96.

Saito, K. (1910) Notizen über einige koreanische Gärungsorganismen. Zentralbl. Bakteriol. Parasitenk., Abt. II, 26, 369-374.

Saito, K. (1913) Ein neuer *Endomyces* (*Endomyces lindneri*). Z. Gärungsphysiol. 2, 151-153.

Saito, K. (1914) Mikrobiologische Studien über die Bereitung des mandschurischen Branntweins. Rep. Centr. Lab. S. Manchurian Railway Co., 1, 1-60 (rev. in: Bull. Inst. Pasteur, 13, 1, 1915; Zentralbl. Bakteriol. Parasitenk., Abt. II, 55, 323, 1922).

Saito, K. (1922) Untersuchungen über die atmosphärischen Pilzkeime. Mitt. Jpn. J. Bot. 1, 1-54.

Saito, K. (1923) Beschreibung von zwei neuen Hefearten, nebst Bemerkungen über die Sporenbildung bei *Torulaspora Delbrücki*, Lindner. Bot. Mag. Tokyo 37, 63-66.

Saito, K. (1932) Über die in dem in Gärung begriffenen Saft vom Cocoanuss befindlichen Pilze. J. Brew. Sci. 10, 11-15 (in Japanese).

Saito, K. & Ohtani, Y. (1931) On some fermentative yeasts isolated from sap exuded from bark of a tree. Jozogaku Zasshi 9, 6-11 (in Japanese).

Saito, K. & Ota, M. (1932) Studien über die Saccharomyceten in Saké-Moto. I, III. J. Brew. Sci. 10, 787-802 (in Japanese).

Saito, K. & Ota, M. (1934) J. Brew. Sci. 12, 159 (in Japanese).

Sakagami, Y., Isogai, A., Suzuki, A., Tamura, S., Tsuchiya, E. & Fukui, S. (1978a) Isolation of a novel sex hormone, Tremerogen A-10, controlling conjugation tube formation in *Tremella mesenterica* Fries. Agric. Biol. Chem. 42, 1093-1094.

Sakagami, Y., Isogai, A., Suzuki, A., Tamura, S., Tsuchiya, E. & Fukui, S. (1978b) Amino acid sequence of Tremerogen A-10, a peptidal hormone, inducing conjugation tube formation in *Tremella mesenterica* Fr. Agric. Biol. Chem. 42, 1301-1302.

Sakagami, Y., Isogai, A., Suzuki, A., Tamura, S., Kitada, C. & Fujino, M. (1979) Structure of Tremerogen A-10, a peptidal hormone inducing conjugation tube formation in *Tremella mesenterica*. Agric. Biol. Chem. 43, 2643–2645.

Sakagami, Y., Isogai, A., Suzuki, A., Tamura, S., Kitado, C. & Fujino, M. (1980) Structure of Tremerogen A-10, a peptide hormone inducing conjugation tube formation in *Tremella mesenterica*. In: H. Yonehara (Ed.), Peptide Chemistry 1979. Protein Research Foundation (1980).

Salkin, I. & Gordon, M.A. (1977) Polymorphism of *Malassezia furfur*. Can. J. Microbiol. 23, 471-475.

Sand, F.E.M.J. (1973) Recent investigations on the microbiology of fruitjuice concentrates. Proc. Int. Fruit Juice Union, Vienna, pp. 185-216.

Sandhu, R.S. & Randhawa, H.S. (1962) On the reisolation and taxonomic study of *Candida viswanathii*. Mycopathol. Mycol. Appl. 18, 179.

Sanfelice, F. (1894) Contributo alla morphologia e biologia dei blastomiceti che si sviluppano nei succhi di alcuni frutti. Ann. Ist. Ig. R. Univ. Roma 4, 463-469.

Sanfelice, F. (1895a) Sulla azione patogena dei blastomiceta. Ann. Ist. Ig. Univ. Roma 5.

Sanfelice, F. (1895b) Über einen neuen pathogenen Blastomyceten, welcher innerhalb der Gewebe unter Bildung kalkartig aussehender Massen degeneriert. Zentralbl. Bakteriol. Parasitenk., Abt. I, 18, 521-526.

Santa María, J. (1956a) *Hansenula matritensis* nov. spec. An. Inst. Nac. Invest. Agron. 5, 151-162.

Santa María, J. (1956b) *Saccharomyces fragilis* Jörgensen var. *bulgaricus* nov. var. An. Inst. Nac. Invest. Agron. 5, 163-165.

Santa María, J. (1956c) *Pichia membranaefaciens* Hansen var. *sicereum* nov. var. An. Inst. Nac. Invest. Agron. 5, 167-170.

Santa María, J. (1957) Un nuevo género de levaduras: *Citeromyces*. Inst. Nac. Invest. Agron., Seccion Bioquimica, Cuaderno no. 258, 269-276.

Santa María, J. (1958a) Ecología de las levaduras. I. Nuevas especies aisladas de alpechin. Bol. Inst. Nac. Invest. Agron. 38, 301-313.

Santa María, J. (1958b) *Sporobolomyces marcillae* nov. spec., isolated from the air. Arch. Mikrobiol. 32, 29-31.

Santa María, J. (1959a) Oxidación de alcohol etilico a acido acetico por levaduras vivas. I. *Saccharomyces aceti* nov. spec. y *Saccharomyces oxidans* nov. spec., nuevas especies aisladas de vino. An. Inst. Nac. Invest. Agron. 8, 715-735.

Santa María, J. (1959b) Ecología de las levaduras. II. Levaduras del Asucar, de la ciche condensada y de la remolache. An. Inst. Nac. Invest. Agron. 8, 781-795.

Santa María, J. (1959c) *Candida atmosphaerica*, nov. spec., aislada del aire. An. Inst. Nac. Invest. Agron. 8, 799-802.

Santa María, J. (1960) Taxonomy of yeasts. Nature 185, 781-782.

Santa María, J. (1962) *Saccharomyces hienipiensis*, a new melibiosefermenting yeast, unable to assimilate raffinose. J. Gen. Microbiol. 28, 375-378.

Santa María, J. (1963a) New melibiose-utilizing yeasts isolated from "alpechin". Antonie van Leeuwenhoek 29, 329-343.

Santa María, J. (1963b) Sobre la validez del género *"Torulaspora"* (Lindner) en las levaduras ascosporigenas. Bol. Inst. Nac. Invest. Agron. 49, 257-266.

Santa María, J. (1964) *Saccharomyces onubensis*, nov. spec., aislada de mosto de uva, y su relación con *Sacch. capensis* (synom. *Sacch. oviformis*). Bol. Inst. Nac. Invest. Agron. 51, 527-536.

Santa María, J. (1966) Characteristics delimiting the genus *Pichia* Hansen. Antonie van Leeuwenhoek 32, 197-201.

Santa María, J. (1968) *Saccharomyces hispanica* nov. spec. Nueva especie de levadura de "Flor". Bol. Inst. Nac. Invest. Agron. 58, 21-32.

Santa María, J. (1970) *Saccharomyces gaditensis* y *Saccharomyces cordubensis*, dos nuevas especies de levaduras de "flor". Bol. Inst. Nac. Invest. Agron. 62, 57-66.

Santa María, J. (1971) *Candida ergatensis* nov. spec. An. Inst. Nac. Invest. Agron. 1, 85-88.

Santa María, J. (1978) Biotaxonomic studies on yeast. Common. Inst. Nac. Invest. Agron., Serie General 3, 1-61.

Santa María, J. & García Aser, C. (1970) *Pichia castillae*, nov. spec., aislada de insectos. Bol. Inst. Nac. Invest. Agron. 62, 51-55.

Santa María, J. & García Aser, C. (1971) *Debaryomyces yarrowii* nov. spec. An. Inst. Nac. Invest. Agron., Serie General 1, 89-92.

Santa María, J. & García Aser, C. (1977) *Pichia segobiensis*, sp. nov. An. Inst. Nac. Invest. Agron., Serie General 5, 45-50.

Santa María, J. & Sanchez, C. (1970) Significacion taxonomica de las propriedades fisiologicas de las especies incluidas en el genero *Kluyveromyces*. Bol. Inst. Nac. Invest. Agron. 62, 23-36.

Sartory, A. (1922) Champignons parasites de l'homme et des animaux (cited from Lodder & Kreger-van Rij, 1952).

Sartory, A., Sartory, R., Hufschmitt, G. & Meyer, J. (1930) Etude d'un *Cryptococcus* nouveau (*Cryptococcus corallinus*) isolé de lésions rappelant les lésions trichophytiques. C.R. Soc. Biol. 104, 1316-1319.

Sartory, A., Sartory, R. & Meyer, J. (1931) Etude botanique d'une nouvelle levure rose (*Cryptococcus radiatus* n. sp.). C.R. Soc. Biol. 106, 597-600.

Sartory, A., Sartory, R., Weill, J. & Meyer, J. (1932) Un cas de blastomycose invétérée transmissible au cobaye, due à un *Saccharomyces* pathogène (*Saccharomyces jadini* n. sp.). C.R. Acad. Sci. 194, 1688-1690.

Sasakawa, M. (1922) Zur Systematik pathogener und parasitischer Hefen. Zentralbl. Bakteriol. Parasitenk., Abt. I Orig., 88, 269-285.

Sasaki, Y. & Yoshida, T. (1959) A taxonomic reconsideration of lactose fermenting yeast preserving in Japan. Mem. Fac. Agric. Hokkaido Univ. 3, 178-184.

Šašek, V. & Becker, G.E. (1969) Effect of different nitrogen sources on the cellular form of *Trigonopsis variabilis*. J. Bacteriol. 99, 891-892.

Scarr, M.P. & Rose, D. (1966) Study of osmophilic yeasts producing invertase. J. Gen. Microbiol. 45, 9-16.

Schachner, J. (1929) *Trigonopsis variabilis* nov. gen. et spec. Z. Gesamte Brauwes. 52, 137-142.

Schaeffer, A.B. & Fulton, M.D. (1933) A simplified method of staining endospores. Science 77, 194.

Scheda, R. (1966a) Kohlenwasserstoffe zehrende Hefen. Branntweinwirtschaft 106, 373-376.

Scheda, R. (1966b) Merkmalsveränderungen bei Hefen der Gattung *Saccharomyces*. Monatsschr. Brau. 19, 256-258.

Scheda, R. & Bos, P. (1966) Hydrocarbons as substrates for yeasts. Nature 211, 660.

Scheda, R. & Yarrow, D. (1966) The instability of physiological properties used as criteria in the taxonomy of yeasts. Arch. Mikrobiol. 55, 209-225.

Scheda, R. & Yarrow, D. (1968) Variation in the fermentative pattern of some *Saccharomyces* species. Arch. Mikrobiol. 61, 310-316.

Scheffers, W.A. (1966) Stimulation of fermentation in yeasts by acetoin and oxygen. Nature 210, 533-534.

Scheffers, W.A. & Wikén, T.E. (1969) The Custers effect (negative Pasteur effect) as a diagnostic criterion for the genus *Brettanomyces*. Antonie van Leeuwenhoek 35, Suppl.: Yeast Symposium 1969, A31-A32.

Schildkraut, C.L., Marmur, J. & Doty, P. (1962) Determination of the base composition of deoxyribonucleic acid from its buoyant density. J. Mol. Biol. 4, 430-433.

Schimon, O. (1911) Beiträge zur Kenntnis rot gefärbter niederer Pilze. Thesis, Munich.

Schiönning, H. (1903) Nouveau genre de la famille des Saccharomycètes. C.R. Trav. Lab. Carlsberg 6, 101-125.

Schiönning, H. (1908) On *Torula* in English beer manufacture. C.R. Trav. Lab. Carlsberg 7, 138-178.

Schlitzer, R.L. & Ahearn, D.G. (1982) Characterization of atypical *Candida tropicalis* and other uncommon clinical yeast isolates. J. Clin. Microbiol. 15, 511-516.

Schneider, A. (1917) Further note on a parasitic Saccharomycete of the tomato. Phytopathology 7, 52-53.

Schoellhorn, K. (1919) Sur la fermentation de quelques levures des nectars des plantes d'hiver. Bull. Soc. Bot. Genève 11, 154-190.

Schroeter, J. (1886) Krypt. Fl. Schlesien 3, 152 (cited from do Carmo-Sousa, 1970c).

Schweizer, C. (1921) Études sur la fermentation des cerises. Levures du genre *Saccharomyces* isolées de macérations de cerises. Ann. Inst. Pasteur 35, 820-833.

Scott, D.B. & Walt, J.P. van der (1970a) *Hansenula sydowiorum* sp. n. Antonie van Leeuwenhoek 36, 45-48.

Scott, D.B. & Walt, J.P. van der (1970b) Three new yeasts from South African insect sources. Antonie van Leeuwenhoek 36, 389-396.

Scott, D.B. & Walt, J.P. van der (1971a) *Hansenula dryadoides* sp. n., a new species from South African insect sources. Antonie van Leeuwenhoek 37, 171-175.

Scott, D.B. & Walt, J.P. van der (1971b) *Pichia cicatricosa* sp. n., a new auxiliary ambrosia fungus. Antonie van Leeuwenhoek 37, 177-183.

Scrivani, P. (1939) Ricerche sui microorganismi produttori della fioretta nel vino. Arch. Mikrobiol. 10, 446–472.

Seeliger, H.P.R. (1956) Use of a urease test for the screening and identification of cryptococci. J. Bacteriol. 72, 127–131.

Selander, R.K. (1976) Genetic variation in natural populations. In: F.J. Ayala (Ed.), Molecular Evolution. Sinauer Publ., Sunderland, Massachusetts, pp. 21–45.

Sen, K. & Komagata, K. (1979) Distribution of urease of extracellular DNase in yeast species. J. Gen. Appl. Microbiol. 25, 127–137.

Sentheshanmuganathan, S. & Nickerson, W.J. (1962) Nutritional control of cellular form in *Trigonopsis variabilis*. J. Gen. Microbiol. 27, 437–449.

Seymour, F.R., Slodki, M.E., Plattner, R.D. & Stodola, R.M. (1976) Methylation and acetolysis of extracellular D-mannans from yeast. Carbohydr. Res. 48, 225–237.

Shadomy, H.J. (1970) Clamp connections in two strains of *Cryptococcus neoformans*. In: D.G. Ahearn (Ed.), Recent trends in yeast research, Vol. I, Spectrum. Georgia State University, Altanta, pp. 67–72.

Shaw, C.E. & Kapica, L. (1972) Production of diagnostic pigment by phenoloxidase activity of *Cryptococcus neoformans*. Appl. Microbiol. 24, 824–830.

Shaw, F.W. (1926) A new species of *Monilia* pathogenic for man. Science 64, 300.

Shechter, Y. (1973) Electrophoresis and taxonomy of medically important fungi. Bull. Torrey Bot. Club 100, 277–287.

Shehata, E.T.A.M. (1960) Yeasts isolated from sugar-cane and its juice during the production of Aguardente de Cana. Appl. Microbiol. 8, 73–75.

Shehata, E.T.A.M. & Mrak, E.M. (1952) Intestinal yeast floras of successive populations of *Drosophila*. Evolution 6, 325–332.

Shehata, E.T.A.M., Mrak, E.M. & Phaff, H.J. (1955) Yeasts isolated from *Drosophila* and from their suspected feeding places in Southern and Central California. Mycologia 47, 799–811.

Shields, A.B. & Ajello, L. (1966) Medium for selective isolation of *Cryptococcus neoformans*. Science 151, 208.

Shifrine, M. & Phaff, H.J. (1956) The association of yeasts with certain bark beetles. Mycologia 48, 41–55.

Shifrine, M. & Phaff, H.J. (1959) Nutritional requirements of *Saccharomycopsis guttulata* (Robin) Schiönning. Mycologia 51, 318–328.

Shifrine, M., Phaff, H.J. & Demain, A.L. (1954) Determination of carbon assimilation patterns of yeasts by replica plating. J. Bacteriol. 68, 28–35.

Shinoda, T., Kaufman, L. & Padhye, A.A. (1981) Comparative evaluation of the Iatron Serological *Candida* Check Kit and the API 20C Kit for the identification of medically important *Candida*. J. Clin. Microbiol. 13, 513–518.

Shivas, R.G. & Rodrigues de Miranda, L. (1983a) *Cryptococcus phylloplanus* and *Cryptococcus hinnuleus*, two new yeast species. Antonie van Leeuwenhoek 49, 153–158.

Shivas, R.G. & Rodrigues de Miranda, L. (1983b) Two new species of the genus *Sporobolomyces* and a new *Rhodotorula* species from leaf surfaces. Antonie van Leeuwenhoek 49, 159–166.

Siepmann, R. & Höhnk, W. (1962) Über Hefen und einige Pilze (Fungi imp., Hyphales) aus dem Nordatlantik. Veröff. Inst. Meeresforsch. Bremerhaven 8, 79–98.

Sierra, G. (1957) A simple method for the detection of lipolytic activity of micro-organisms and some observations on the influence of the contact between cells and fatty substances. Antonie van Leeuwenhoek 23, 15–22.

Simione, F.R. & Meyer, S.A. (1978) Genetic relatedness between some physiologically similar *Candida* species with high GC contents. Abstr. Ann. Meet. Am. Soc. Microbiol., p. 81.

Sinclair, N.S. & Stokes, S.L. (1965) Obligately psychrophilic yeasts from the polar regions. Can. J. Microbiol. 11, 259–269.

Skinner, C.E. (1947a) The yeast-like fungi: *Candida* and *Brettanomyces*. Bacteriol. Rev. 11, 227–274.

Skinner, C.E. (1947b) Morphology and classification of the yeasts and yeast-like fungi. In: Henrici's Molds, Yeasts, and Actinomycetes. 2nd ed. John Wiley & Sons, Inc., New York, pp. 264–304.

Skinner, C.E. (1950) Generic name for imperfect yeasts, *Cryptococcus* or *Torulopsis*? Am. Midl. Nat. 43, 242–250.

Slodki, M.E. (1966) Hydrolysis products from an extracellular *Tremella* polysaccharide. Can. J. Microbiol. 12, 495–499.

Slodki, M.E. & Wickerham, L.J. (1966) Extracellular polysaccharides and classification of the genus *Lipomyces*. J. Gen. Microbiol. 42, 381–385.

Slodki, M.E., Wickerham, L.J. & Bandoni, R.J. (1966) Extracellular heteropolysaccharides from *Cryptococcus* and *Tremella*: a possible taxonomic relationship. Can. J. Microbiol. 12, 489–494.

Slodki, M.E., Ward, R.M., Boundy, J.A. & Cadmus, M.C. (1972) Extracellular mannans and phosphomannans: Structural and biosynthetic relationships. In: G. Terui (Ed.), Fermentation Technology Today, Proc. IV Int. Ferment. Symp., Soc. Ferment. Technol., Osaka, Japan, pp. 597–601.

Slooff, W.Ch. (1964) Mating types in *Pichia membranaefaciens*. Antonie van Leeuwenhoek 30, 129–131.

Slooff, W.Ch. (1970a) The genus *Lipomyces* Lodder et Kreger-van Rij. In: J. Lodder (Ed.), The Yeasts, a taxonomic study. North-Holland Publ. Co., Amsterdam, pp. 379–402.

Slooff, W.Ch. (1970b) The genus *Schizosaccharomyces* Lindner. In: J. Lodder (Ed.), The Yeasts, a taxonomic study. North-Holland Publ. Co., Amsterdam, pp. 733–755.

Slooff, W.Ch. (1970c) The genus *Pityrosporum* Sabouraud. In: J. Lodder (Ed.), The Yeasts, a taxonomic study. North-Holland Publ. Co., Amsterdam, pp. 1167–1186.

Slooff, W.Ch. (1970d) The genus *Trigonopsis* Schachner. In: J. Lodder (Ed.), The Yeasts, a taxonomic study. North-Holland Publ. Co., Amsterdam, pp. 1353–1357.

Smit, J. (1934) Über die Ursachen des Aufblähens von Belebtschlamm. Arch. Mikrobiol. 5, 550–560.

Smith, M.T. (1973) Centraalbureau voor Schimmelcultures, Baarn and Delft. Progress Report 1972. Verh. Kon. Ned. Akad. Wetensch., Afd. Natuurk. 61, 59–81.

Smith, M.T. (1974) *Hanseniaspora occidentalis* sp. nov. Antonie van Leeuwenhoek 40, 441–444.

Smith, M.T., Walt, J.P. van der & Johannsen, E. (1976) The genus *Stephanoascus* gen. nov. (Ascoideaceae). Antonie van Leeuwenhoek 42, 119–127.

Smith, M.T., Simione, F.P. & Meyer, S.A. (1977) *Kloeckera apis* sp. nov.; the imperfect state of *Hanseniaspora guilliermondii* Pijper. Antonie van Leeuwenhoek 43, 219–223.

Smith, M.T., Batenburg-van der Vegte, W.H. & Scheffers, W.A. (1981) *Eeniella*, a new genus of the Torulopsidales. Int. J. Syst. Bacteriol. 31, 196–203.

Socias Amorós, A., Ramirez Gomez, C. & Genestar Serra, R. (1954a) Una nueva especie de *Endomycopsis: Endomycopsis balearica* nov. sp. Aislada de concentrado de tanino de encina. Microbiol. Esp. 7, 107–110.

Socias Amorós, A., Ramirez Gomez, C. & Peláez Campomanes, F. (1954b) Una nueva especie de *Debaryomyces:* el *Debaryomyces toletanus* nov. sp. Microbiol. Esp. 7, 111–114.

Sonck, C.E. (1974) *Candida savonica* sp. nov. Antonie van Leeuwenhoek 40, 543–545.

Sonck, C.E. & Yarrow, D. (1969) Two new yeast species isolated in Finland. Antonie van Leeuwenhoek 35, 172–177.

Soneda, M. (1959a) Studies on animal-dung inhabiting yeasts. Nagaoa, Mycol. J. Nagao Inst. 6, 1–24.

Soneda, M. (1959b) Nagaoa, Mycol. J. Nagao Inst. 9, 380.

Soneda, M. (1960) On a new yeast genus *Wickerhamia*. Nagaoa, Mycol. J. Nagao Inst. 7, 9–13.

Soneda, M. (1962) An additional paper on animal-dung inhabiting yeasts and on the symbiosis with an amoeba. Trans. Mycol. Soc. Japan 3, 36–42.

Soneda, M. & Uchida, S. (1971) A survey on the yeasts. Bull. Nat. Sci. Mus. Tokyo 14, 367–551, 438–459.

Spencer, J.F.T. & Gorin, P.A.J. (1968) Mannose-containing polysaccharides of the apiculate yeasts *Nadsonia, Hanseniaspora, Kloeckera* and *Saccharomycodes*, and their use as an aid in classification. J. Bacteriol. 96, 180–183.

Spencer, J.F.T. & Gorin, P.A.J. (1969a) Systematics of the genus *Candida* Berkhout: proton magnetic resonance spectra of the mannans and mannose-containing polysaccharides as an aid in classification. Antonie van Leeuwenhoek 35, 33–44.

Spencer, J.F.T. & Gorin, P.A.J. (1969b) Systematics of the genera *Hansenula* and *Pichia*: proton magnetic resonance spectra of their mannans as an aid in classification. Can. J. Microbiol. 15, 375–382.

Spencer, J.F.T., Phaff, H.J. & Gardner, N.R. (1964) *Metschnikowia kamienskii*, sp. n., a yeast associated with brine shrimp. J. Bacteriol. 88, 758–762.

Spencer, J.F.T., Gorin, P.A.J. & Tulloch, A.P. (1970) *Torulopsis bombicola* sp. n. Antonie van Leeuwenhoek 36, 129–133.

Splendore, A. (1912) Bull. Soc. Pathol. Exot. 5, 313 (cited by Lodder & Kreger-van Rij, 1952).

Stadelmann, F. (1975) A new species of the genus *Bullera* Derx. Antonie van Leeuwenhoek 41, 575–582.

Stafleu, F.A., Demoulin, V., Greuter, W., Hiepko, P., Linczevski, I.A., McVaugh, R., Meikle, R.D., Rollins, R.C., Ross, R., Schopf, J.M. & Voss, E.G. (1978) International Code of Botanical Nomenclature. Published for the International Association for Plant Taxonomy. Bohn, Scheltema & Holkema, Utrecht.

Stahl, U. & Esser, K. (1976) Genetics of fruitbody production in higher basidiomycetes I. Monokaryotic fruiting, its correlation with dikaryotic fruiting in *Polyporus ciliatus*. Mol. Gen. Genet. 148, 183–197.

Staib, F. (1962) *Cryptococcus neoformans* and *Guizotia abyssinica* (syn. *G. oleifera* D.C.) (Farbereaktion für *C. neoformans*). Z. Hyg. 148, 466–475.

Staib, F. (1963) Zur Kreatinin-Kreatin Assimilation in der Hefepilzdiagnostik. Zentralbl. Bakteriol. Parasitenk., Abt. I Orig., 191, 429–432.

Stanier, R.Y., Doudoroff, M. & Adelberg, E.A. (1971) General Microbiology. 3rd ed., Macmillan, London.

Starmer, W.T., Phaff, H.J., Miranda, M. & Miller, M.W. (1978a) *Pichia cactophila*, a new species of yeast found in decaying tissue of cacti. Int. J. Syst. Bacteriol. 28, 318–325.

Starmer, W.T., Phaff, H.J., Miranda, M. & Miller, M.W. (1978b) *Pichia amethionina*, a new heterothallic yeast associated with the decaying stems of cereoid cacti. Int. J. Syst. Bacteriol. 28, 433–441.

Starmer, W.T., Phaff, H.J., Miranda, M., Miller, M.W. & Barker, J.S.F. (1979) *Pichia opuntiae*, a new heterothallic species of yeast in the decaying cladodes of *Opuntia inermis* and in the necrotic tissue of cereoid cacti. Int. J. Syst. Bacteriol. 29, 159–167.

Starmer, W.T., Kircher, H.W. & Phaff, H.J. (1980) Evolution and speciation of host plant specific yeasts. Evolution 34, 137–146.

Stautz, W. (1931) Beiträge zur Schleimflussfrage. Phytopathol. Z. 3, 163–229.

Stearn, W.T. (1966) Botanical Latin. London & Edinburgh.

Steiner, J.M. (1924) Etude sur les levures actives des vins valaisins. Thesis, Genève.

Stelling-Dekker, N.M. (1931) Die sporogenen Hefen. Verh. Kon. Ned. Akad. Wetensch., Afd. Natuurk., Sect. II, 28, 1–547.

Stenderup, A. & Bak, A.L. (1968) Deoxyribonucleic acid base composition of some species within the genus *Candida*. J. Gen. Microbiol. 52, 231–236.

Stenderup, A. & Thomsen, J.B. (1964) Identification of *Candida albicans*. Acta Pathol. Microbiol. Scand. 62, 303–304.

Stenderup, A., Meyer, S.A., Bak, A.L. & Christiansen, C. (1972) Taxonomy of *Candida* and *Torulopsis*. In: G. Terui (Ed.), Fermentation Technology Today, Proc. IV Int. Ferment. Symp., Soc. Ferment. Technol., Osaka, Japan, pp. 793–795.

Sternberg, T.H. & Keddie, F.M. (1961) Immunofluorescence studies in Tinea versicolor. Arch. Dermatol. 84, 161–165.

Stoddard, J.L. & Cutler, E.C. (1916) Torula infection in man. Stud. Rock. Inst. Med. Res. 25, 1–98.

Stodola, F.H. & Wickerham, L.J. (1960) Formation of extracellular sphingolipids by microorganisms. II. Structural studies on tetraacetylphytosphingosine from the yeast *Hansenula ciferrii*. J. Biol. Chem. 235, 2584–2585.

Storck, R. (1966) Nucleotide composition of nucleic acids of fungi. II. Deoxyribonucleic acids. J. Bacteriol. 91, 227–230.

Storck, R., Alexopoulos, C.J. & Phaff, H.J. (1969) Nucleotide composition of deoxyribonucleic acid of some species of *Cryptococcus, Rhodotorula* and *Sporobolomyces*. J. Bacteriol. 98, 1069–1072.

Strachan, A.A., Yu, R.J. & Blank, F. (1971) Pigment production of *Cryptococcus neoformans* grown with extracts of *Guizotia abyssinica*. Appl. Microbiol. 22, 478–479.

Streiblová, E. (1963) On the question of the taxonomy of *Endomycopsis javanensis* (Klöcker) Dekker. Folia Microbiol. (Prague) 8, 170–175.

Streiblová, E. (1971) Cell division in yeasts. In: A. Pérez-Miravette & D. Peláez (Eds.), Recent Advances in Microbiology. Proc. 10th Int. Congress Microbiol., Mexico City, pp. 131–140.

Streiblová, E. & Wolf, A. (1972) Cell wall growth during the cell cycle of *Schizosaccharomyces pombe*. Z. Allg. Mikrobiol. 12, 673–684.

Streiblová, E., Beran, K. & Pokorný, V. (1964) Multiple scars, a new type of yeast scar in apiculate yeasts. J. Bacteriol. 88, 1104–1111.

Sugiyama, J. & Goto, S. (1967) Coprophilous fungi from Karakorum I. J. Jpn. Bot. 42, 75–84.

Sugiyama, J. & Goto, S. (1969) Mycoflora in core samples from stratigraphic drillings in middle Japan IV. The yeast genera *Candida* Berkhout, *Trichosporon* Behrend, and *Rhodotorula* Harrison em. Lodder from core samples. J. Fac. Sci. Tokyo Univ., Sect. III, Bot. 10, 97–116.

Sydow, H. (1912) *Guilliermondia*, un nouveau genre de la famille des Saccharomycètes à copulation hétérogamique. Ann. Mycol. 10, 347–348.

Sydow, H. & Sydow, P. (1918) Mykologische Mitteilungen. Ann. Mycol. 16, 240–248.

Sydow, H. & Sydow, P. (1919) Mykologische Mitteilungen. Ann. Mycol. 17, 33–47.

Szep, E. & Novák, E.K. (1963) A new yeast species: *Candida requinyii* n. sp. Acta Bot. Acad. Sci. Hung. 9, 447–453.

Szilvinyi, A. von (1933–1934) *Blastodendrion canis* nov. sp.; ein Beitrag zur Diagnose und Systematik der asporogenen Sprosspilze. Zentralbl. Bakteriol. Parasitenk., Abt. II, 89, 284–299.

Szilvinyi, A. & Kaulich, L. (1948) Zur Kenntnis der Mikroflora österreichischer Kernobstmoste. Mitt. Versuchsanst. Gärungsgewerbe Inst. angew. Mikrobiol. Hochschule Bodenkultur 4, 1–3.

Takahashi, T. & Ikeda, Y. (1959) Bisexual mating reaction in *Saccharomyces chevalieri*. Genetics 44, 375–382.

Takahashi, T. & Yukawa, M. (1915) J. Coll. Agric. Imp. Univ. Tokyo 5, 227 (as cited by Stelling-Dekker, 1931).

Takahashi, T., Saito, H. & Ikeda, Y. (1958) Heterothallic behaviour of a homothallic strain in *Saccharomyces* yeast. Genetics 43, 249–260.

Takano, I. & Oshima, Y. (1967) An allele specific and a complementary determinant controlling homothallism in *Saccharomyces oviformis*. Genetics 57, 875–885.

Takano, I. & Oshima, Y. (1970) Mutational nature of an allele specific conversion of the mating type by the homothallic gene HOa *Saccharomyces*. Genetics 65, 421–427.

Talice, R.V. & MacKinnon, J.E. (1934) 8a Reun. Soc. Arg. Pathol. Reg. Norte (Santiago del Estero 1933), Buenos Aires.

Tanaka, H., Phaff, H.J. & Higgins, L.W. (1966) Enzymatic hydrolysis of yeast cell walls. II. Susceptibilities of isolated cell walls and ascus walls of various yeasts to the actions of bacterial endo-β-glucanases. Symposium on Yeast Protoplasts, Abh. Dtsch. Akad. Wiss. Berlin, Kl. Med. 6, 113–129, 353–357.

Taschdjian, C.L. (1953) A simply prepared identification medium for *Candida albicans*. Mycologia 45, 474–475.

Taylor, N.W. (1965) Purification of sexual agglutination factor from the yeast *Hansenula wingei* by chromatography and gradient sedimentation. Arch. Biochem. Biophys. 111, 181–186.

Taylor, N.W. & Orton, W.L. (1968) Sexual agglutination in yeast. VII. Significance of the 1.7S component from reduced 5-agglutinin. Arch. Biochem. Biophys. 126, 912–921.

Teunisson, D.J., Hall, H.H. & Wickerham, L.J. (1960) *Hansenula angusta*, an excellent species for demonstration of the coexistence of haploid and diploid cells in a homothallic yeast. Mycologia 52, 184–188.

Thaysen, A.C. & Morris, M. (1943) Preparation of a giant strain of *Torulopsis utilis*. Nature 152, 526–528.

Todd, R.L. & Herrmann, W.W. (1936) The life cycle of the organism causing yeast meningitis. J. Bacteriol. 32, 89–97.

Trüper, E. (1928) Über milchzuckervergärende Hefen der Rohmilch. Milchwirtsch. Forsch. 6, 351–402.

Tschalenko, D.K. (1941) Die Ursachen für die Verringerung des Säuregehaltes bei der Gärung von Fruchtweinen. Arb. Zentr. Wiss. Önolog. Labor, Moscow-Leningrad, 1 (as cited by Kudriavzev, 1960).

Tsuchiya, T. & Imai, M. (1968) Serological relationships among various yeast species of *Hansenula, Kloeckera* and *Hanseniaspora*. In: Proc. 10th I.A.M. Symposium on Microbiology, October 1968. Taxonomy of Microorganisms, pp. 147–158.

Tsuchiya, T., Fukazawa, Y., Sano, Y., Shimura, Y. & Murata, T. (1960) Serological classification of the genera *Debaryomyces* and *Schwanniomyces*. Jpn. J. Microbiol. 4, 61–71.

Tsuchiya, T., Kawakita, S., Imai, M. & Miyagawa, K. (1966) Serological classification of the genera *Kloeckera* and *Hanseniaspora*. Jpn. J. Exp. Med. 36, 555–562.

Tsuchiya, T., Fukazawa, Y., Shinoda, T. & Imai, M. (1967) Perfect form of *Candida krusei*. Jpn. J. Exp. Med. 37, 285–290.

Tsuchiya, T., Fukazawa, Y., Taguchi, M., Nakase, T. & Shinoda, T. (1974) Serological aspects of yeast classification. Mycopathol. Mycol. Appl. 53, 77–91.

Tubaki, K. (1958) Studies on the Japanese Hyphomycetes (IV) Miscellaneous group. Bot. Mag. Tokyo 71, 131–137.

Tulasne, L.R. & Tulasne, C. (1847) Mémoire sur les Ustilaginées comparées aux Urédinées. Ann. Sci. Natur. 3, 12–127.

Tulasne, L.R. & Tulasne, C. (1865) Selecta Fungorum Carpologia Vol. 3, pp. 61–62.

Tuynenburg Muys, G. & Willemse, R. (1965) The detection and enumeration of lipolytic microorganisms by means of a modified Eykman-plate method. Antonie van Leeuwenhoek 31, 103–112.

Uden, N. van (1952) Zur Kenntnis von *Torulopsis pintolopesii* sp. nov. Arch. Mikrobiol. 17, 199–208.

Uden, N. van (1962a) On the nomenclature of the genus *Metschnikowia* Kamienski. Rev. Biol. 3, 95–96.

Uden, N. van (1962b) Factors of host-yeast relationship. In: N.E. Gibbons (Ed.) Recent Progress in Microbiology VIII. University of Toronto Press, pp. 635–643.

Uden, N. van & Assis-Lopes, L. (1953a) Zur Kenntnis von *Kloeckera lodderi* sp. nov. Arch. Mikrobiol. 19, 262–266.

Uden, N. van & Assis-Lopes, L. (1953b) Zur Kenntnis von *Candida castellanii* sp. nov. Zugleich ein Beitrag zur Systematik von *Candida parakrusei* und *Candida krusoides.* Arch. Mikrobiol. 18, 356–361.

Uden, N. van & Assis-Lopes, L. (1957) Fermentation of raffinose in the absence of invertase by *Saccharomyces italicus* Castelli var. *melibiosi* nov. var. Port. Acta Biol., Ser. A, 4, 323–327.

Uden, N. van & Buckley, H. (1970) The genus *Candida* Berkhout. In: J. Lodder (Ed.), The Yeasts, a taxonomic study. North-Holland Publ. Co., Amsterdam, pp. 893–1087.

Uden, N. van & Carmo-Sousa, L. do (1957) Yeasts from the bovine caecum. J. Gen. Microbiol. 16, 385–395.

Uden, N. van & Carmo-Sousa, L. do (1959) Further studies on the significance of temperature relations and vitamin deficiency patterns in yeast taxonomy. Port. Acta Biol. 6, 239–256.

Uden, N. van & Carmo-Sousa, L. do (1962) On the intestinal yeast flora of free-living baboons (*Papio cyanocephalus*). Sabouraudia 2, 8–11.

Uden, N. van & Castelo-Branco, R. (1961) *Metschnikowiella zobellii* sp. nov. and *M. krissii*, sp. nov., two yeasts from the Pacific Ocean pathogenic for *Daphnia magna*. J. Gen. Microbiol. 26, 141–148.

Uden, N. van & Farinha, M. (1958) On the significance of temperature relations and vitamin deficiency patterns in the delimitation of yeast species. Port. Acta Biol., Ser. B, 6, 161–178.

Uden, N. van & Kolipinski, M.C. (1962) *Torulopsis haemulonii* nov. spec. a yeast from the Atlantic Ocean. Antonie van Leeuwenhoek 28, 78–80.

Uden, N. van & Vidal-Leiria, M. (1970) The genus *Torulopsis* Berlese. In: J. Lodder (Ed.), The Yeasts, a taxonomic study. North-Holland Publ. Co., Amsterdam, pp. 1235–1308.

Uden, N. van & Windisch, S. (1968) *Candida friedrichii* sp. n., a melibiose-fermenting yeast. Antonie van Leeuwenhoek 34, 270–274.

Uden, N. van & Zobell, C.E. (1962) *Candida marina* nov. spec., *Torulopsis torresii* nov. spec. and *T. maris* nov. spec., three yeasts from the Torres Strait. Antonie van Leeuwenhoek 28, 275–283.

Uden, N. van, Carmo-Sousa, L. do & Farinha, M. (1958) On the intestinal yeast flora of horses, sheep, goats, and swine. J. Gen Microbiol. 19, 435–448.

Ulloa, M. & Herrera, T. (1973) Descripcíon de una variedad de *Kloeckera corticis* aislada del pulque. Bol. Soc. Mex. 7, 27–32.

Unna, P.G. (1896) Two cases of piedra nostra. Lewin's Festschrift, Berlin (abstr. Br. J. Dermatol. Syph. 8, 11, 1896).

Urakami, T. (1975) US Patent 3.929.578, Dec. 30, 1975. Process for cultivating ethanol-assimilating yeasts. Production of yeast from low concentration of ethanol.

Vanbreuseghem, R. (1948) Sur une levure, *Candida truncata*, n. sp. isolée d'une dermatose présentée par le Dr. Eykmans. Arch. Belg. Dermatol. Syphilogr. 4, 307–313.

Vanbreuseghem, R. & Takashio, M. (1970) An atypical strain of *Cryptococcus neoformans* (Sanfelice) Vuillemin 1894. Part II. *Cryptococcus neoformans* var. *gattii* var. nov. Ann. Soc. Belge Méd. Trop. 50, 695–702.

Verdun, P. (1912) *Atelosaccharomyces hominis* (Vuillemin 1901) Verdun. In: P. Verdun, Précis de Parasitologie Humaine, 2me éd., Doin et Fils, Paris, pp. 730–731.

Verona, O. & Ciferri, R. (1938) Revisione dei lieviti asporigeni del genere *Sporobolomyces* Kluyver et van Niel. Atti Ist. Bot. Univ. Pavia, Ser. IV, 10, 241-255.

Verona, O. & Florenzano, G. (1947) Su di una nuova specie di *Brettanomyces* isolata da uva. Rend. Accad. Naz. Lincei, Ser. VIII, 3, 383-387.

Verona, O. & Florenzano, G. (1980) Some taxonomical considerations on *Candida* and allied genera. Mycotaxon 11, 365-368.

Verona, O. & Luchetti, G. (1936) Ricerche microbiologiche su di alcuni vini ed alcune uve delle marche. Boll. Regio Inst. Super. Agrar. Pisa 12, 256-311.

Verona, O. & Luchetti, G. (1941) Descrizione dei lieviti delle uve, dei mosti e dei vini. Garoglio-Tratt. Enol. 2, 288-297.

Verona, O. & Montemartini, A. (1959) Sinossi dei lieviti descritti dal 1952 al 1958. Atti Ist. Bot. Univ. Lab. Critt. Pavia 17, 1-123.

Verona, O. & Picci, G. (1958) A new species of *Trichosporon: Trichosporon hellenicum* n. sp. Ann. Microbiol. Enzimol. 8, 106-108.

Verona, O. & Zardetto de Toledo, O. (1954) Indagini speciologiche sopra alcuni lieviti isolati dai fiori di vite nel territorio di S. Paolo in Brasile. Ann. Fac. Agrar. 15, 1-31.

Verona, O. & Zardetto de Toledo, O. (1959a) Considerazioni interno ad *Endomycopsis fibuliger* (Lindner) Dekker. Mycopathol. Mycol. Appl. 10, 209-215.

Verona, O. & Zardetto de Toledo, O. (1959b) Intorno ad un lievito osmofilo sistematicamente prossimo a *Saccharomyces elegans* Lodder et Kreger-van Rij. Mycopathol. Mycol. Appl. 11, 103-108.

Verrall, A.F. (1943) Fungi associated with certain ambrosia beetles. J. Agric. Res. 66, 135-144.

Vesonder, R.F., Stodola, F.H., Rohwedder, W.K. & Scott, D.B. (1970) 2-D-Hydroxyhexadecanoic acid: A metabolic product of the yeast *Hansenula sydowiorum*. Can. J. Chem. 48, 1985-1986.

Vidal-Leiria, M. (1966a) *Torulopsis mogii* sp. n. Antonie van Leeuwenhoek 32, 312-314.

Vidal-Leiria, M. (1966b) *Torulopsis vanderwaltii* sp. nov. Antonie van Leeuwenhoek 32, 447-449.

Vidal-Leiria, M. (1967) *Candida mogii* sp. n., a halotolerant yeast associated with miso fermentation. Antonie van Leeuwenhoek 33, 341-344.

Vidal-Leiria, M. & Uden, N. van (1963) *Candida silvae* sp. n., a yeast isolated from humans and horses. Antonie van Leeuwenhoek 29, 261-264.

Vidal-Leiria, M., Buckley, H. & Uden, N. van (1979) Distribution of the maximum temperature for growth among yeasts. Mycologia 71, 494-501.

Vishniac, H.S. & Hempfling, W.P. (1979) *Cryptococcus vishniacii* sp. nov., an Antarctic yeast. Int. J. Syst. Bacteriol. 29, 153-158.

Viswanathan, R. & Randhawa, H.S. (1959) *Candida viswanathii* sp. nov. isolated from a case of meningitis. Science and Culture 25, 86.

Vörös-Felkai, G. (1961) Studies on arthrosporogenous yeasts. Acta Microbiol. Hung. 8, 95-102.

Vuillemin, P. (1901) Les blastomycètes pathogènes. Rev. Gen. Sci. 12, 732-751.

Vuillemin, P. (1911) Différence fondamentale entre le genre *Monilia* et les genres *Scopulariopsis, Acrosporium* et *Catenularia*. Bull. Soc. Mycol. France 27, 137-152.

Vuillemin, P. (1931) Les champignons parasites et les mycoses de l'homme. Encyclopédie Mycol. Paris, 2.

Walt, J.P. van der (1956a) *Kluyveromyces* - a new yeast genus of the Endomycetales. Antonie van Leeuwenhoek 22, 265-272.

Walt, J.P. van der (1956b) The yeast *Kluyveromyces africanus* nov. spec. and its phylogenetic significance. Antonie van Leeuwenhoek 22, 321-326.

Walt, J.P. van der (1956c) Kaffircorn malting and brewing studies. II. Studies on the microbiology of Kaffir beer. J. Sci. Food. Agric. 2, 105-113.

Walt, J.P. van der (1956d) *Saccharomyces transvaalensis* nov. spec., a new yeast from soil. Antonie van Leeuwenhoek 22, 190-192.

Walt, J.P. van der (1957) Three new sporogenous yeasts from soil. Antonie van Leeuwenhoek 23, 23-29.

Walt, J.P. van der (1959a) *Pichia robertsii* nov. spec., a new haploid homothallic yeast. Antonie van Leeuwenhoek 25, 337-343.

Walt, J.P. van der (1959b) *Endomycopsis wickerhamii* nov. spec. a new heterothallic yeast. Antonie van Leeuwenhoek 25, 344-348.

Walt, J.P. van der (1959c) *Endomyces reessii* nov. sp. Antonie van Leeuwenhoek 25, 458-464.

Walt, J.P. van der (1961) *Brettanomyces custersianus* nov. spec. Antonie van Leeuwenhoek 27, 332-336.

Walt, J.P. van der (1962) *Schwanniomyces persoonii* nov. spec. Antonie van Leeuwenhoek 28, 81-84.

Walt, J.P. van der (1963) *Fabospora phaffii* sp. n. Antonie van Leeuwenhoek 29, 319-322.

Walt, J.P. van der (1964) *Dekkera*, a new yeast genus of the Saccharomycetaceae. Antonie van Leeuwenhoek 30, 273-280.

Walt, J.P. van der (1965a) *Saccharomyces vafer* and *S. inconspicuus* spp. n. Antonie van Leeuwenhoek 31, 187-192.

Walt, J.P. van der (1965b) *Saccharomyces inusitatus* sp. n. Antonie van Leeuwenhoek 31, 277-281.

Walt, J.P. van der (1965c) The emendation of the genus *Kluyveromyces* van der Walt. Antonie van Leeuwenhoek 31, 341-348.

Walt, J.P. van der (1966a) *Lodderomyces*, a new genus of the Saccharomycetaceae. Antonie van Leeuwenhoek 32, 1-5.

Walt, J.P. van der (1966b) *Pichia acaciae* sp. n. Antonie van Leeuwenhoek 32, 159-161.

Walt, J.P. van der (1967) *Wingea*, a new yeast genus of the Saccharomycetaceae. Antonie van Leeuwenhoek 33, 97-99.

Walt, J.P. van der (1970a) Criteria and methods used in classification. In: J. Lodder (Ed.), The Yeasts, a taxonomic study. North-Holland Publ. Co., Amsterdam, pp. 34–113.

Walt, J.P. van der (1970b) The genus *Dekkera* van der Walt. In: J. Lodder (Ed.), The Yeasts, a taxonomic study. North-Holland Publ. Co., Amsterdam, pp. 157–165.

Walt, J.P. van der (1970c) The genus *Kluyveromyces* (van der Walt) emend. van der Walt. In: J. Lodder (Ed.), The Yeasts, a taxonomic study. North-Holland Publ. Co., Amsterdam, pp. 316–378.

Walt, J.P. van der (1970d) The genus *Lodderomyces* van der Walt. In: J. Lodder (Ed.), The Yeasts, a taxonomic study. North-Holland Publ. Co., Amsterdam, pp. 403–407.

Walt, J.P. van der (1970e) The genus *Saccharomyces* (Meyen) Reess. In: J. Lodder (Ed.), The Yeasts, a taxonomic study. North-Holland Publ. Co., Amsterdam, pp. 555–718.

Walt, J.P. van der (1970f) The genus *Wingea* van der Walt. In: J. Lodder (Ed.), The Yeasts, a taxonomic study. North-Holland Publ. Co., Amsterdam, pp. 772–775.

Walt, J.P. van der (1970g) The genus *Brettanomyces* Kufferath et van Laer. In: J. Lodder (Ed.), The Yeasts, a taxonomic study. North-Holland Publ. Co., Amsterdam, pp. 863–892.

Walt, J.P. van der (1970h) The perfect and imperfect states of *Sporobolomyces salmonicolor*. Antonie van Leeuwenhoek 36, 49–55.

Walt, J.P. van der (1970i) A note on chlamydospore formation in *Trichosporon cutaneum.* Mycopathol. Mycol. Appl. 41, 233–235.

Walt, J.P. van der (1970k) The genus *Syringospora* Quinquaud emend. Mycopathol. Mycol. Appl. 40, 231–243.

Walt, J.P. van der (1971) New combinations in the genera *Brettanomyces, Kluyveromyces, Lodderomyces* and *Wingea*. Bothalia 10, 417–418.

Walt, J.P. van der (1972) The yeast genus *Ambrosiozyma* gen. nov. (Ascomycetes). Mycopathol. Mycol. Appl. 46, 305–316.

Walt, J.P. van der (1973) *Aessosporon dendrophilum* sp. nov., the perfect state of *Bullera dendrophila*. Antonie van Leeuwenhoek 39, 455–460.

Walt, J.P. van der (1978) The genus *Pachytichospora* gen. nov. (Saccharomycetaceae). Bothalia 12, 563–564.

Walt, J.P. van der & Arx, J.A. von (1980) The yeast genus *Yarrowia* gen. nov. Antonie van Leeuwenhoek 46, 517–521.

Walt, J.P. van der & Hopsu-Havu, V.K. (1976) A colour reaction for the differentiation of ascomycetous and hemibasidiomycetous yeasts. Antonie van Leeuwenhoek 42, 157–163.

Walt, J.P. van der & Johannsen, E. (1973) The perfect state of *Torulopsis magnoliae*. Antonie van Leeuwenhoek 39, 635–647.

Walt, J.P. van der & Johannsen, E. (1974a) The dangeardiens of ascomycetous yeasts. Antonie van Leeuwenhoek 40, 185–192.

Walt, J.P. van der & Johannsen, E. (1974b) Ascospores in *Torulopsis dattila* (Kluyver) Lodder. Antonie van Leeuwenhoek 40, 275–279.

Walt, J.P. van der & Johannsen, E. (1975a) The genus *Torulaspora* Lindner. CSIR Res. Rep. No. 325, 1–23.

Walt, J.P. van der & Johannsen, E. (1975b) *Hansenula lynferdii* sp. nov. Antonie van Leeuwenhoek 41, 13–16.

Walt, J.P. van der & Johannsen, E. (1975c) *Pichia philogaea* sp. nov. Antonie van Leeuwenhoek 41, 173–177.

Walt, J.P. van der & Johannsen, E. (1975d) *Trichosporon terrestre* sp. nov. Antonie van Leeuwenhoek 41, 361–365.

Walt, J.P. van der & Johannsen, E. (1978) The genus *Debaryozyma* van der Walt et Johannsen, nom. nov. Persoonia 10, 146–148.

Walt, J.P. van der & Johannsen, E. (1979) A comparison of interfertility and in vitro DNA-DNA reassociation as criteria for speciation in the genus *Kluyveromyces*. Antonie van Leeuwenhoek 45, 281–291.

Walt, J.P. van der & Kerken, A.E. van (1958) The wine yeasts of the Cape. Part I. A taxonomical survey of the yeasts causing turbidity in South African table wines. Antonie van Leeuwenhoek 24, 239–252.

Walt, J.P. van der & Kerken, A.E. van (1959) The wine yeasts of the Cape. Part II. The occurrence of *Brettanomyces intermedius* and *Brettanomyces schanderlii* in South African table wines. Antonie van Leeuwenhoek 25, 145–151.

Walt, J.P. van der & Kerken, A.E. van (1960a) The wine yeasts of the Cape. Part IV. Ascospore formation in the genus *Brettanomyces*. Antonie van Leeuwenhoek 26, 292–296.

Walt, J.P. van der & Kerken, A.E. van (1960b) *Torulopsis domercqii* nov. spec. Antonie van Leeuwenhoek 26, 314–316.

Walt, J.P. van der & Kerken, A.E. van (1961a) The wine yeasts of the Cape. Part V. Studies on the occurrence of *Brettanomyces intermedius* and *Brettanomyces schanderlii*. Antonie van Leeuwenhoek 27, 81–90.

Walt, J.P. van der & Kerken, A.E. van (1961b) Three new anascosporogenous yeasts. Antonie van Leeuwenhoek 27, 206–212.

Walt, J.P. van der & Kerken, A.E. van (1961c) *Candida ingens* nov. spec. Antonie van Leeuwenhoek 27, 284–286.

Walt, J.P. van der & Klift, W.C. van der (1972) *Pichia melissophila* sp. n., a new osmotolerant yeast from apiarian sources. Antonie van Leeuwenhoek 38, 361–364.

Walt, J.P. van der & Liebenberg, N.V.D.W. (1973a) The yeast genus *Wickerhamiella* gen. nov. (Ascomycetes). Antonie van Leeuwenhoek 39, 121–128.

Walt, J.P. van der & Liebenberg, N.V.D.W. (1973b) Agglutinative mating types of *Saccharomyces transvaalensis*. Antonie van Leeuwenhoek 39, 629–633.

Walt, J.P. van der & Nakase, T. (1973) *Candida homilentoma* a new yeast from South African insect sources. Antonie van Leeuwenhoek 39, 449–453.

Walt, J.P. van der & Nel, E.E. (1963) *Saccharomyces vanudenii* nov. spec. Mycopathol. Mycol. Appl. 20, 71–74.

Walt, J.P. van der & Nel, E.E. (1968) *Candida edax* sp. n. Antonie van Leeuwenhoek 34, 106–108.

Walt, J.P. van der & Scott, D.B. (1970) *Bullera dendrophila* sp. n. Antonie van Leeuwenhoek 36, 383–387.

Walt, J.P. van der & Scott, D.B. (1971a) *Pichia ambrosiae* sp. n. Antonie van Leeuwenhoek 37, 15–20.

Walt, J.P. van der & Scott, D.B. (1971b) The yeast genus *Saccharomycopsis* Schiönning. Mycopathol. Mycol. Appl. 43, 279–288.

Walt, J.P. van der & Scott, D.B. (1971c) *Pichia xylopsoci*, a new yeast from South African insect sources. Mycopathol. Mycol. Appl. 44, 321–324.

Walt, J.P. van der & Scott, D.B. (1971d) *Saccharomycopsis synnaedendra*, a new yeast from South African insect sources. Mycopathol. Mycol. Appl. 44, 101–106.

Walt, J.P. van der & Tscheuschner, I.T. (1956a) *Pichia vanriji* n. sp. isolated from soil. J. Gen. Microbiol. 15, 459–461.

Walt, J.P. van der & Tscheuschner, I.T. (1956b) *Saccharomyces delphensis* nov. spec., a new yeast from South African dried figs. Antonie van Leeuwenhoek 22, 162–166.

Walt, J.P. van der & Tscheuschner, I.T. (1956c) *Saccharomyces pretoriensis* n. sp. – from South African soil. J. Gen. Microbiol. 14, 485–486.

Walt, J.P. van der & Tscheuschner, I.T. (1956d) *Saccharomyces capensis* nov. spec. a new yeast from South African grape must. Antonie van Leeuwenhoek 22, 257–260.

Walt, J.P. van der & Tscheuschner, I.T. (1957a) Three new yeasts. Antonie van Leeuwenhoek 23, 184–190.

Walt, J.P. van der & Tscheuschner, I.T. (1957b) *Hanseniaspora vineae* sp. nov. Trans. Br. Mycol. Soc. 40, 211–212.

Walt, J.P. van der, Nel, E.E. & Kerken, A.E. van (1966) *Kluyveromyces wikenii* and *Kluyveromyces cicerisporus* spp. n. Antonie van Leeuwenhoek 32, 393–398.

Walt, J.P. van der, Scott, D.B. & Klift, W.C. van der (1971a) Four new, related *Candida* species from South African insect sources. Antonie van Leeuwenhoek 37, 449–460.

Walt, J.P. van der, Scott, D.B. & Klift, W.C. van der (1971b) Five new *Torulopsis* species from South African insect sources. Antonie van Leeuwenhoek 37, 461–471.

Walt, J.P. van der, Scott, D.B. & Klift, W.C. van der (1972) Six new *Candida* species from South African insect sources. Mycopathol. Mycol. Appl. 47, 221–236.

Walt, J.P. van der, Johannsen, E. & Nakase, T. (1973) *Candida naeodendra*, a new species of the *Candida diddensii* group. Antonie van Leeuwenhoek 39, 491–495.

Walt, J.P. van der, Johannsen, E. & Liebenberg, N.V.D.W. (1974) Cell wall structure, mitosis, and urease activity in *Torulopsis* species of high GC content. Antonie van Leeuwenhoek 40, 417–426.

Walt, J.P. van der, Taylor, M.B. & Liebenberg, N.V.D.W. (1977) Ploidy, ascus formation and recombination in *Torulaspora (Debaryomyces) hansenii*. Antonie van Leeuwenhoek 43, 205–218.

Walt, J.P. van der, Johannsen, E. & Yarrow, D. (1978) *Torulopsis geochares* and *Torulopsis azyma*, two new haploid species of ascomycetous affinity. Antonie van Leeuwenhoek 44, 97–104.

Walt, J.P. van der, Weijman, A.C.M. & Arx, J.A. von (1981) The anamorphic yeast genus *Myxozyma* gen. nov. Sydowia. Ann. Mycol. Ser. II, 34, 191–198.

Walter, J.E. & Coffee, E.G. (1968) Distribution and epidemiologic significance of the serotypes of *Cryptococcus neoformans*. Am. J. Epidemiol. 87, 167–172.

Walters, L.S. (1943) Studies of yeast causing defects in beer II. The distinguishing characters of *Torulopsis cylindrica* nov. sp., a yeast causing a turbidity and unpleasant taste and smell in beer. J. Inst. Brew. 59, 253–256.

Ward, H.M. (1892) The ginger-beer plant, and the organisms composing it: a contribution to the study of fermentation – yeasts and bacteria. Philos. Trans. R. Soc. London 183, 125–197.

Ward, T.J. & Baker, J.L. (1929) *Saccharomyces festinans*. J. Inst. Brew. 35, 466–468.

Warth, A.D. (1977) Mechanism of resistance of *Saccharomyces bailii* to benzoic, sorbic and other weak acids used as food preservatives. J. Appl. Bacteriol. 43, 215–230.

Weber, H. (1979) Substructural studies on sporulation of *Saccharomycopsis lipolytica*. Z. Allg. Mikrobiol. 19, 283–297.

Weber, U. (1922) Beitrag zur Kenntnis der esterbildenden Hefen. Biochem. Z. 129, 208–216.

Wehmer, C. (1906) Durch Pilzenzyme bewirkte Stärkeverzuckerung im Brennereigewerbe. Mycologie der Rumbrennerei und der Arrakbereitung. In: F. Lafar (Ed.), Handbuch der Technischen Mycologie. G. Fischer, Jena, pp. 319–342.

Weidman, F.D. (1925) Exfoliative dermatitis in the Indian rhinoceros (*Rhinoceros unicornis*), with description of a new yeast species, *Pityrosporum pachydermatis*. In: H. Fox (Ed.), Report of the Laboratory and Museum of Comparative Pathology of the Zoological Society of Philadelphia, pp. 36–44.

Weigman, H. & Wolff, A. (1909) Über einige zum "Rübengeschmack" der Butter beitragende Mycelpilze. Zentralbl. Bakteriol. Parasitenk., Abt. II, 22, 657–670.

Weijman, A.C.M. (1977) Carbohydrate composition and taxonomy of the genus *Dipodascus*. Antonie van Leeuwenhoek 43, 323–331.

Weijman, A.C.M. (1979) Carbohydrate composition and taxonomy of *Geotrichum, Trichosporon* and allied genera. Antonie van Leeuwenhoek 45, 119–127.

Weiss, J.J. (1902) J. Med. Res. 7, 208 (cited from Lodder & Kreger-van Rij, 1952).

Went, F.A.F.C. & Prinsen Geerligs, H.C. (1894) Over suiker- en alcoholvorming door organismen in verband met de verwerking der naproducten in rietsuikerfabrieken. Arch. Suikerind. Nederland. Indië III. Mededeel. Proefsta. Java-Suikerind., No. 13.

Wetmur, J.G. & Davidson, N. (1968) Kinetics of renaturation of DNA. J. Mol. Biol. 31, 349–370.

Wettstein, R. von (1885) Untersuchungen über einen neuen pflanzlichen Parasiten des menschlichen Körpers. Sitzungsber. Akad. Wiss. Wien, Abt. I, 19, 33–59.

Whiffen, A.J. (1948) The production, assay and antibiotic activity of Actidione, an antibiotic from *Streptomyces griseus*. J. Bacteriol. 56, 283–291.

Wickerham, L.J. (1946) A critical evaluation of the nitrogen assimilation tests commonly used in the classification of yeasts. J. Bacteriol. 52, 293–301.

Wickerham, L.J. (1951) Taxonomy of yeasts. Techn. Bull. 1029, U.S. Dept. Agr., Washington, D.C.

Wickerham, L.J. (1952) Recent advances in the taxonomy of yeasts. Annu. Rev. Microbiol. 6, 317–332.

Wickerham, L.J. (1956) Influence of agglutination on zygote formation in *Hansenula wingei*, a new species of yeast. C.R. Trav. Lab. Carlsberg, Sér. Physiol. 26, 423–433.

Wickerham, L.J. (1957) Presence of nitrite-assimilating species of *Debaryomyces* in lunch meats. J. Bacteriol. 74, 832–833.

Wickerham, L.J. (1958) Sexual agglutination of heterothallic yeasts in diverse taxonomic areas. Science 128, 1504–1505.

Wickerham, L.J. (1960) *Hansenula holstii*, a new yeast important in the early evolution of the heterothallic species of its genus. Mycologia 52, 171–183.

Wickerham, L.J. (1964a) A preliminary report on a perfect family of exclusively protosexual yeasts. Mycologia 56, 253–266.

Wickerham, L.J. (1964b) A cadaver yeast and related species. Mycologia 56, 398–414.

Wickerham, L.J. (1965a) New heterothallic species of *Hansenula*. I. *Hansenula fabianii*. Mycopathol. Mycol. Appl. 26, 79–86.

Wickerham, L.J. (1965b) New heterothallic species of *Hansenula*. II. *Hansenula bimundalis* and variety *americana*. Mycopathol. Mycol. Appl. 26, 87–103.

Wickerham, L.J. (1965c) Opposite sexes as type specimen for heterothallic haploid yeasts. Taxon 14, 187–188.

Wickerham, L.J. (1966) Validation of the species *Pichia guilliermondii*. J. Bacteriol. 92, 1269.

Wickerham, L.J. (1969a) Hybridization as a basis for speciation in the genus *Hansenula*. Proc. II Int. Symp. Yeasts, 1966, Bratislava, pp. 41–44.

Wickerham, L.J. (1969b) New homothallic taxa of *Hansenula*. Mycopathol. Mycol. Appl. 37, 15–32.

Wickerham, L.J. (1969c) Yeast taxonomy in relation to ecology, genetics and phylogeny. Antonie van Leeuwenhoek 35, Suppl. Yeast Symposium 1969, pp. 31–58.

Wickerham, L.J. (1970a) The genus *Citeromyces* Santa María. In: J. Lodder (Ed.), The Yeasts, a taxonomic study. North-Holland Publ. Co., Amsterdam, pp. 121–127.

Wickerham, L.J. (1970b) The genus *Hansenula* H. et P. Sydow. In: J. Lodder (Ed.), The Yeasts, a taxonomic study. North-Holland Publ. Co., Amsterdam, pp. 226–315.

Wickerham, L.J. (1970c) The genus *Pachysolen* Boidin et Adzet. In: J. Lodder (Ed.), The Yeasts, a taxonomic study. North-Holland Publ. Co., Amsterdam, pp. 448–454.

Wickerham, L.J. & Burton, K.A. (1948) Carbon assimilation tests for the classification of yeasts. J. Bacteriol. 56, 363–371.

Wickerham, L.J. & Burton, K.A. (1952) Occurrence of yeast mating types in nature. J. Bacteriol. 63, 449–451.

Wickerham, L.J. & Burton, K.A. (1954a) A simple technique for obtaining mating types in heterothallic diploid yeasts, with special reference to their uses in the genus *Hansenula*. J. Bacteriol. 67, 303–308.

Wickerham, L.J. & Burton, K.A. (1954b) A clarification of the relationships of *Candida guilliermondii* to other yeasts by a study of their mating types. J. Bacteriol. 68, 594–597.

Wickerham, L.J. & Burton, K.A. (1956) Hybridization studies involving *Saccharomyces lactis* and *Zygosaccharomyces ashbyii*. J. Bacteriol. 71, 290–295.

Wickerham, L.J. & Burton, K.A. (1960) Heterothallism in *Saccharomyces rouxii*. J. Bacteriol. 80, 492–495.

Wickerham, L.J. & Burton, K.A. (1962) Phylogeny and biochemistry of the genus *Hansenula*. Bacteriol. Rev. 26, 382–397.

Wickerham, L.J. & Duprat, E. (1945) A remarkable fission yeast, *Schizosaccharomyces versatilis* nov. sp. J. Bacteriol. 50, 597–607.

Wickerham, L.J. & Flickinger, M.H. (1946) Viability of yeasts preserved two years by the lyophil process. Brew. Dig. 21, 55–59, 65.

Wickerham, L.J. & Kurtzman, C.P. (1971) Two new Saturn-spored species of *Pichia*. Mycologia 63, 1013–1018.

Wickerham, L.J. & Stodola, F.H. (1960) Formation of extracellular sphingolipids by microorganisms. I. Tetraacetylphytosphingosine from *Hansenula ciferrii*. J. Bacteriol. 80, 484–491.

Wickerham, L.J., Flickinger, M.H. & Burton, K.A. (1946) A modification of Henrici's vegetable-juice sporulation medium for yeasts. J. Bacteriol. 52, 611–612.

Wickerham, L.J., Kurtzman, C.P. & Herman, A.I. (1970a) Sexual reproduction in *Candida lipolytica*. Science 167, 1141.

Wickerham, L.J., Kurtzman, C.P. & Herman, A.I. (1970b) Sexuality in *Candida lipolytica*. In: D.G. Ahearn (Ed.), Recent Trends in Yeast Research. Spectrum, Vol. I, Georgia State University, Atlanta, pp. 81-92.

Will, H. (1910) Beiträge zur Kenntnis der Gattung *Mycoderma* nach Untersuchungen von Hans Leberle. Zentralbl. Bakteriol. Parasitenk., Abt. II, 28, 1-37.

Will, H. (1913) *Saccharomyces anamensis*, die Hefe des neueren Amyloverfahrens. Zentralbl. Bakteriol. Parasitenk., Abt. II, 39, 26-53.

Will, H. (1916a) Vergleichende morphologische und physiologische Untersuchungen an vier Kulturen der Gattung *Pseudosaccharomyces* Klöcker (*Saccharomyces apiculatus* Reess). Zentralbl. Bakteriol. Parasitenk., Abt. II, 44, 225-290.

Will, H. (1916b) Beiträge zur Kenntnis der Sprosspilze ohne Sporenbildung, welche in Brauereibetrieben und in deren Umgebung vorkommen. VI. (Schluss)-Mitteilung. Zentralbl. Bakteriol. Parasitenk., Abt. II, 46, 226-281.

Wilson, D.E., Bennett, J.E. & Bailey, J.W. (1968) Serologic grouping of *Cryptococcus neoformans*. Proc. Soc. Exp. Biol. Med. 127, 820-823.

Windisch, S. (1940) Entwicklungsgeschichtliche Untersuchungen an *Torulopsis pulcherrima* (Lindner) Saccardo und *Candida tropicalis* (Castellani) Berkhout. Ein Beitrag zur Systematik der Gärungsmonilien. Arch. Mikrobiol. 11, 368-390.

Windisch, S. (1949) *Sporobolomyces holsaticus* n. sp. Arch. Mikrobiol. 14, 287-290.

Windisch, S. (1952) Zur Biologie und Systematik des Milchschimmels und einiger ähnlicher Formen. II. Milchschimmelähnliche Pilze. Beitr. Biol. Pflanz. 29, 143-170.

Windisch, S. (1953) Zur Biologie und Systematik der *Pseudomonilia albomarginata* Geiger. Zentralbl. Bakteriol. Parasitenk., Abt. II, 107, 84-98.

Windisch, S. (1965) Über zwei neue Pilzarten, *Protendomycopsis domschii* n.g., n. sp. und *Endomyces laibachii* n. sp. Beitr. Biol. Pflanz. 41, 337-358.

Wingard, J.R., Merz, W.G. & Saral, R. (1979) *Candida tropicalis*: a major pathogen in immunocompromised patients. Ann. Intern. Med. 91, 539-543.

Wingard, S.A. (1922) Yeast-spot of Lima beans. Phytopathol. 12, 525-532.

Winge, Ö. & Laustsen, O. (1937) On two types of spore germination and on genetic segregations in *Saccharomyces* demonstrated through single-spore cultures. C.R. Trav. Lab. Carlsberg, Sér. Physiol. 22, 99-116.

Winge, Ö. & Laustsen, O. (1939) On 14 new yeast types, produced by hybridization. C.R. Trav. Lab. Carlsberg, Sér. Physiol. 22, 337-352.

Winge, Ö. & Roberts, C. (1954) Causes of deviations from 2:2 segregations in the tetrads of monohybrid yeasts. C.R. Trav. Lab. Carlsberg, Sér. Physiol. 25, 285-329.

Winter, G. (1884) L. Rabenhorst's Kryptogamen-Flora von Deutschland, Österreich und der Schweiz. Band I. Die Pilze. Abt. I., Verl. Eduard Kummer, Leipzig.

Wolfram, S. & Zach, F. (1933) Über einige durch niedere Pilze verursachte Nagelerkrankungen beim Menschen. Arch. Dermatol. Syph. 169, 95-104.

Wolfram, S. & Zach, F. (1934) Über durch niedere Pilze verursachte Nagelerkrankungen beim Menschen. Arch. Dermatol. Syph. 170, 681-694.

Wynants, J. (1962) Preservation of yeast cultures by lyophilization. J. Inst. Brew. 68, 350-354.

Yabe, K. (1897) On the origin of sake yeast (*Sacch. sake*). Bull. Imp. Univ. Tokyo Coll. Agric. 3, 221-233.

Yamada, K. & Ota, Y. (1963) Studies on the production of lipase by microorganisms. Part IV. Identification of a new strain and determination of its lipase activity. J. Agric. Chem. Soc. Japan 37, 649-652.

Yamada, K., Furukawa, T. & Nakahara, T. (1970) Studies on the utilization of hydrocarbons by microorganisms. Part XVIII. Fermentative production of fumaric acid by *Candida hydrocarbofumarica* n. sp. Agric. Biol. Chem. 34, 670-675.

Yamada, Y. & Kondo, K. (1972) Taxonomic significance of the coenzyme Q system in yeasts and yeast-like fungi. In: K. Iwata (Ed.), Yeasts and Yeast-like Microorganisms in Medical Science. Proc. 2nd Int. Specialized Symp. Yeasts, Tokyo, pp. 63-69.

Yamada, Y. & Kondo, K. (1972a) Taxonomic significance of the coenzyme Q system in yeasts and yeast-like fungi (2). In: G. Terui (Ed.), Fermentation Technology Today. Proc. IVth Int. Ferment. Symp., Soc. Ferment. Technol., Osaka, Japan, pp. 781-784.

Yamada, Y. & Kondo, K. (1973) Coenzyme Q system in the classification of the yeast genera *Rhodotorula* and *Cryptococcus*, and the yeast-like genera *Sporobolomyces* and *Rhodosporidium*. J. Gen. Appl. Microbiol. 19, 59-77.

Yamada, Y., Arimoto, M. & Kondo, K. (1973a) Coenzyme Q system in the classification of the ascosporogenous yeast genus *Schizosaccharomyces* and yeast-like genus *Endomyces*. J. Gen. Appl. Microbiol. 19, 353-358.

Yamada, Y., Kanematsu, Y., Ohashi, M. & Kondo, K. (1973b) On the partly reduced coenzyme Q isolated from *Rhodotorula lactosa* IFO 1058 and its relation to the taxonomic position. Agric. Biol. Chem. 37, 621-628.

Yamada, Y., Okada, T., Ueshima, O. & Kondo, K. (1973c) Coenzyme Q system in the classification of the ascosporogenous yeast genera *Hansenula* and *Pichia*. J. Gen. Appl. Microbiol. 19, 189-208.

Yamada, Y., Arimoto, M. & Kondo, K. (1976a) Coenzyme Q system in the classification of apiculate yeasts in the genera *Nadsonia, Saccharomycodes, Hanseniaspora, Kloeckera* and *Wickerhamia*. J. Gen. Appl. Microbiol. 22, 293-299.

Yamada, Y., Nojiri, M., Matsuyama, M. & Kondo, K. (1976b) Coenzyme Q system in the classification of the ascosporogenous yeast genera *Debaryomyces, Saccharomyces, Kluyveromyces*, and *Endomycopsis*. J. Gen. Appl. Microbiol. 22, 325-339.

Yamada, Y., Arimoto, M. & Kondo, K. (1977) Coenzyme Q system in the classification of some ascosporogenous yeast genera in the families Saccharomycetaceae and Spermophthoraceae. Antonie van Leeuwenhoek 43, 65-71.

Yamada, Y., Takinami-Nakamura, H., Tahara, Y. & Smith, M.T. (1980) The coenzyme Q system in the classification of the ascosporogenous yeast genus *Dekkera* and the asporogenous yeast genus *Brettanomyces*. Antonie van Leeuwenhoek 46, 595-599.

Yamaguchi, (1943) J. Agric. Chem. Soc. Japan 19, 800 (cited from van Uden & Buckley, 1970).

Yamasaki, I. & Fujii, H. (1950) Studies on *Sporobolomyces* red yeast. Part 7. Classification of the genera *Sporobolomyces* and *Bullera*. Bull. Agric. Chem. Soc. Japan 24, 11-15.

Yamazaki, M. & Komagata, K. (1982) Asporogenous yeasts and their supposed ascosporogenous states: an electrophoresis comparison of enzymes. J. Gen. Appl. Microbiol. 28, 119-138.

Yarrow, D. (1965) *Pichia onychis* sp. n. Antonie van Leeuwenhoek 31, 465-467.

Yarrow, D. (1968) *Torulopsis peltata* sp. n. Antonie van Leeuwenhoek 34, 81-84.

Yarrow, D. (1969a) *Candida steatolytica* sp. n. Antonie van Leeuwenhoek 35, 24-28.

Yarrow, D. (1969b) *Selenotila peltata* comb. n. Antonie van Leeuwenhoek 35, 418-420.

Yarrow, D. (1972a) Four new combinations in yeasts. Antonie van Leeuwenhoek 38, 357-360.

Yarrow, D. (1972b) Centraalbureau voor Schimmelcultures, Baarn and Delft. Progress Report 1971. Verh. Kon. Ned. Akad. Wetensch., Afd. Natuurk., Sect. II, 61, 50-69.

Yarrow, D. (1978) *Candida milleri* spec. nov. Int. J. Syst. Bacteriol. 28, 608-610.

Yarrow, D. & Ahearn, D.G. (1971) *Brettanomyces abstinens* sp. n. Antonie van Leeuwenhoek 37, 296-298.

Yarrow, D. & Fell, J.W. (1980) Validation of the name of the yeast *Sporobolomyces holsaticus* Windisch. Mycotaxon 12, 253-254.

Yarrow, D. & Meyer, S.A. (1978) Proposal for amendment of the diagnosis of the genus *Candida* Berkhout nom. cons. Int. J. Syst. Bacteriol. 28, 611-615.

Yarrow, D. & Nakase, T. (1975) DNA base composition of species of the genus *Saccharomyces*. Antonie van Leeuwenhoek 41, 81-88.

Yokote, Y., Sugimoto, M. & Abe, S. (1974) Yeasts utilizing methanol as a sole carbon source. J. Ferment. Technol. 52, 201-209.

Yokotsuka, I. & Goto, S. (1955) Studies on Japanese wine film yeasts. Part 4. On new species of genus *Candida* and *Trichosporon*. J. Agric. Chem. Soc. Japan 29, 132-135.

Yonezawa, W., Aoki, F., Ota, H., Nishi, K. & Matsumoto, K. (1957) J. Vitaminol. 3, 39 (cited from do Carmo-Sousa, 1970c).

Yukawa, M. & Maki, T. (1931) *Schizosaccharomyces japonicus* nov. spec. La. Bul. Sci. Fakultato Terkultura, Kjusu Imp. Univ., Fukuoka, Japan 4, 218-226 (in Japanese with English summary; cited from Wickerham & Duprat, 1945).

Zaake, S. (1960) Über biertrübende wilde Hefen und ihre heutige Bedeutung. Brauerei 14, 842-846.

Zender, J. (1925) Sur la classification des Endomycétacées. Bull. Soc. Bot. Genève 17, 272-302.

Zikes, H. (1911) Zur Nomenclaturfrage der Apiculatushefe. Zentralbl. Bakteriol. Parasitenk., Abt. II, 30, 145-149.

Zimmermann, J. (1938) Sprosspilze im Wein und deren Bestimmung. Zentralbl. Bakteriol. Parasitenk., Abt. II, 98, 36-65.

Zopf, W. (1889) Oxalsäuregährung (an Stelle von Alcoholgährung) bei einem typischen (endosporen) Saccharomyceten (*S. hansenii* n. spec.). Ber. Dtsch. Bot. Ges. 7, 94-97.

Zopf, W. (1890) Die Pilze in morphologischer, physiologischer, biologischer und systematischer Richtung. Trewendt, Breslau.

Zsolt, J. (1957) A new yeast: *Dioszegia hungarica* nov. gen. et sp. Bot. Kozl. 47, 63-66.

Zsolt, J. (1958) *Torulopsis pseudaeria* nov. spec., a new yeast from soil. Antonie van Leeuwenhoek 24, 210-214.

Zundel, G.L. (1953) The Ustilaginales of the World. Contribution No. 176, Dept. Bot., The Pennsylvania State College.

Zwillenberg, L.O. (1966) *Torulopsis anatomiae*, a new yeast resistant against low concentrations of formalin. Antonie van Leeuwenhoek 32, 135-138.

Index of the names of taxa

The names of genera, species and varieties accepted by the authors are indicated with an asterisk. The numbers of the pages referring to diagnoses and standard descriptions are printed in bold type.